ENCYCLOPEDIA OF PHYSICS

EDITED BY

S. FLÜGGE

VOLUME XII

THERMODYNAMICS OF GASES

WITH 253 FIGURES

SPRINGER-VERLAG
BERLIN · GÖTTINGEN · HEIDELBERG
1958

HANDBUCH DER PHYSIK

HERAUSGEGEBEN VON

S. FLÜGGE

BAND XII

THERMODYNAMIK DER GASE

MIT 253 FIGUREN

SPRINGER-VERLAG
BERLIN · GÖTTINGEN · HEIDELBERG
1958

ISBN-13: 978-3-642-45894-1 e-ISBN-13: 978-3-642-45892-7
DOI: 10.1007/978-3-642-45892-7

Softcover reprint of the hardcover 1st edition 1958

Inhaltsverzeichnis.

The Properties of Real Gases.

By

J. S. Rowlinson.

With 40 Figures.

Notation.

The following symbols are used for the principal thermodynamic functions: E (energy), H (heat content), S (entropy), A (Helmholtz free-energy), G (Gibbs free-energy).

Modifications to these symbols have the following significance:

X An extensive property (except for T).

X Value per mole (except for virial coefficients which are conventionally written in italic).

X^c Value at gas-liquid critical point.

X^g Value in saturated vapour (dew-point).

X^l Value in liquid (bubble-point).

X^* Residual value (Sect. 8).

X_a See Sect. 27.

$X_a^\dagger$ See Sect. 27.

X_a^0 See Sect. 27.

No separate symbol is used for the chemical potential, which is required only in Part VI, as the universally used symbol, μ, is that required for the Joule-Thomson coefficient, which occurs throughout this article.

I. Perfect gases and real gases.

1. Introduction. The atmospheric pressure at the surface of the earth is about 10^6 dyne cm^{-2} and is low in the range of pressures now open to experiment. Early studies of the mechanical and thermal behaviour of gases at and near this pressure soon established that all chemically stable gases behave very much alike. At higher pressures this similarity disappears and the relations between pressure, temperature and volume are still the subject of much careful measurement and subtle theory. The simple laws which all gases were found to obey at atmospheric pressure are those associated with the names of Boyle and Charles[1],

$$p V = k_1 m \qquad \text{(constant temperature, Boyle)} \tag{1.1}$$

$$V(t_1) - V(t_2) = k_2 m (t_1 - t_2) \qquad \text{(constant pressure, Charles)} \tag{1.2}$$

where p and V are the pressure and volume of a mass m of gas, where t is the temperature on some convenient but arbitrary temperature scale, such as a mercury thermometer, and where k_1 and k_2 are constants which vary from gas to gas. When atomic and molecular weights came to be known, it was found that k_1 and k_2 were the same for all gases if m was expressed not in grams but in gram-molecular weights (or moles)—an important law which is still often known as Avogadro's hypothesis. If n moles of a substance occupy a volume V, then

[1] For a brief discussion of these laws and their early history, see J. R. Partington [*10*].

the molar volume $\mathrm{V}(=V/n)$ may be used in an equation which combines the laws of BOYLE and CHARLES,

$$\frac{p\mathrm{V}}{R'T'}=1. \tag{1.3}$$

In this equation T' is the temperature on the gas scale, that is the scale whose zero is at that temperature at which the extrapolated pressure of a gas at constant volume becomes zero. The gas-constant R' depends on the size of the degree in this scale and is almost independent of the nature of the gas. If (1.3) were obeyed exactly, it would be quite independent of the gas. It is found experimentally that it does become independent if the limiting value of $p\mathrm{V}$ at zero pressure is used to define the temperature scale, that is, if the ratio of two temperatures T_1' and T_2' is defined by

$$\frac{T_1'}{T_2'}=\lim_{p\to 0}\left[\frac{(p\mathrm{V})_1}{(p\mathrm{V})_2}\right]. \tag{1.4}$$

The temperature scale defined by this equation is called the perfect-gas scale, and a gas which obeys (1.3) at all temperatures and volumes is a perfect gas. This scale is identical with the absolute scale defined by the second law of thermodynamics[1]. The latter may be determined by measuring the changes in the heat content, H, and the Gibbs free-energy, G, for the same isothermal process. The values of ΔH and ΔG are related by the Gibbs-Helmholtz equation which may be solved as an equation for T, the absolute temperature. In the case of a real gas it is found experimentally that the approach of the product $(p\mathrm{V})$ to the perfect-gas value is linear in p at low pressure, as is illustrated, for example, in Fig. 1. That is,

$$\lim_{p\to 0}\left(\mathrm{V}-\frac{\mathrm{R}T'}{p}\right)=B(T'), \tag{1.5}$$

where B is a finite volume which is a function only of the temperature and of the nature of the gas. At low pressures, therefore, the equation of state may be written

$$\mathrm{V}=\frac{\mathrm{R}T'}{p}+B. \tag{1.6}$$

Small changes in absolute temperature and pressure in a system of constant mass and composition produce a change in G given by

$$d\mathrm{G}=-\mathrm{S}\,dT+\mathrm{V}\,dp \tag{1.7}$$

whence

$$\mathrm{G}(p_1,T)-\mathrm{G}(p_2,T)=\int_{p_2}^{p_1}\mathrm{V}\,dp \tag{1.8}$$

$$=\mathrm{R}T'\ln(p_1/p_2)+B(p_1-p_2). \tag{1.9}$$

By differentiation with respect to absolute temperature,

$$\left.\begin{aligned}\mathrm{H}(p_1,T)-\mathrm{H}(p_2,T)=&-\mathrm{R}T^2\ln\left(\frac{p_1}{p_2}\right)\cdot\left(\frac{d}{dT}\right)\left(\frac{T'}{T}\right)\\&-T^2(p_1-p_2)\cdot\left(\frac{d}{dT}\right)\left(\frac{B}{T}\right).\end{aligned}\right\} \tag{1.10}$$

Now it is found for all gases that H, like $p\mathrm{V}$, is a well-behaved function of pressure in the limit of zero pressure. The slope $(\partial\mathrm{H}/\partial p)_T$ has a finite limiting value, as

[1] The simple method of demonstrating this equality which is given here is that used by E. A. GUGGENHEIM: Thermodynamics, 2nd Ed., p. 88–91. Amsterdam: North Holland Publishing Co. 1950.

may be shown by the isothermal Joule-Thomson experiment described in Sect. 4. No measurement of temperature is made in this experiment. Consequently for this experimental result to be consistent with (1.10),

$$\frac{d}{dT}\left(\frac{T'}{T}\right) = 0, \tag{1.11}$$

or T' is proportional to T. The choice of a common unit for the two scales makes the proportionality an equality.

Thus the concept of a perfect gas fulfils three functions. First, it is a convenient but idealised representation of the behaviour of real gases at normal and moderately high pressures. Secondly, it is a statement of the limiting law of behaviour of all gases at sufficiently low pressures. (This statement must occasionally be modified, but in most senses it is true.) Thirdly, it provides a realisable temperature scale which is identical with KELVIN'S absolute scale. Many of the idealised concepts of physics and chemistry perform functions similar to the first of these, but not to the second. For example, the laws of ideal liquid mixtures are a simple and convenient basis for the discussion of the behaviour of real solutions. There are, however, no conditions of temperature or pressure in which the behaviour of a given real solution becomes ideal. To preserve this distinction it is preferable to call a gas which obeys (1.3) a *perfect* gas, and not an *ideal* gas.

2. The choice of independent variables. The state of a one-component one-phase system is completely specified by two of the three variables, pressure, volume and temperature. In practice, the independent variables are chosen to be either pressure and temperature or volume and temperature. The expressions for the thermodynamic functions in terms of these variables are naturally in their most simple form if the energy, E, the Helmholtz free-energy, A, and the heat capacity at constant volume, C_V, are expressed as functions of V and T, and if the heat content, H, the Gibbs free-energy, G, and the heat capacity at constant pressure, C_p, are expressed as functions of p and T. The entropy, S, can be expressed equally readily in either set of variables.

It is formally possible, in classical statistical mechanics[1], to express the Gibbs free-energy in terms of the partition function of an assembly of constant pressure, temperature and number of molecules by an equation analogous to that expressing the Helmholtz free-energy in terms of the partition function of an assembly of constant volume, temperature and number of molecules, namely,

$$f(V, T, N) = e^{-A/kT}, \tag{2.1}$$

$$f(p, T, N) = e^{-G/kT}. \tag{2.2}$$

Nevertheless only the former function (2.1) has been used in practice for calculating the thermodynamic properties in terms of the intermolecular energies, and the term partition function (or phase integral) is restricted in common usage to $f(V, T, N)$.

Practical convenience often makes one set of variables more suitable than the other. In many cases the design of the apparatus determines whether p or V is to be taken as an independent variable. This can be troublesome near the gas-liquid critical point where $(\partial p/\partial V)_T$ and $(\partial^2 p/\partial V^2)_T$ vanish. Here it is preferable to take volume as the variable both when designing an apparatus and when considering the analytical representation of the measurements. The

[1] E.A. GUGGENHEIM: J. Chem. Phys. 7, 103 (1939). — W.B. BROWN: J. Mol. Phys. 1 (1958) (in the press).

shapes of typical isotherms are shown in Fig. 1 as a function of p and T and in Fig. 2 as a function of $(1/V)$ and T. [The function of (pV/RT), plotted in Fig. 1 and 2, is often called the compressibility factor and denoted Z. The name com-

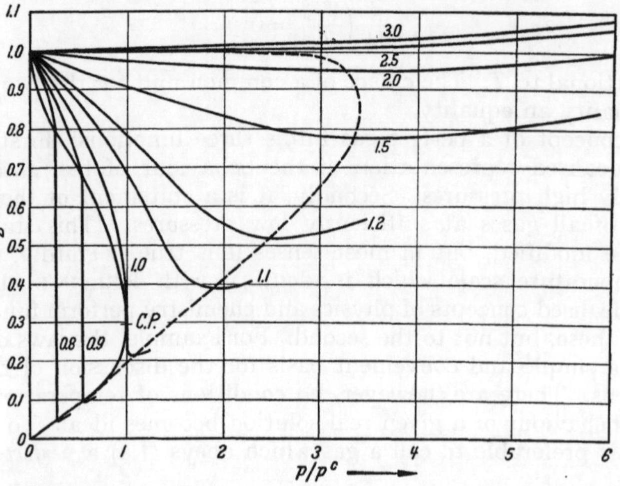

Fig. 1. The compressibility factor, $Z(=pV/RT)$, of the inert gases as a function of pressure for values of T/T^c from 0.3 to 3.0. The tongue-shaped curve on the left of the diagram is the two-phase region. This is crossed by the isotherms below T^c and ends at the critical point (C.P.). The dashed curve is the locus of the Boyle points (see Sect. 10).

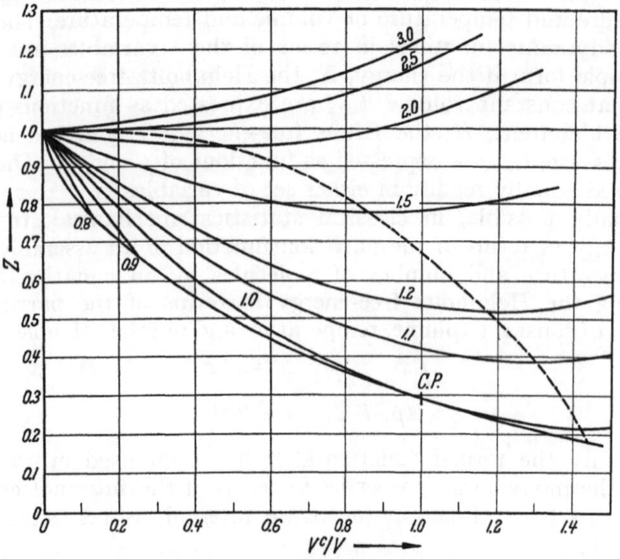

Fig. 2. The compressibility factor, Z, of the inert gases as a function of density for values of T/T^c from 0.8 to 3.0. The two-phase region is that below the lowest curve. The critical isotherm is tangential to this region at the critical point (C.P.). Isotherms at lower temperatures cross this region as rectangular hyperbolae (not shown). The dashed curve is the locus of the Boyle points.

pression factor would be more accurate.] Reduced units have been used in these figures; that is, the ratios p/p^c, T/T^c and V^c/V have been used, where the superscript denotes the value at the critical point. By using these reduced variables the figures can be made to represent qualitatively the behaviour of any fluid.

The curves drawn do, in fact, fit the group of inert gases very closely[1]. It is seen that near the two-phase region the representation of Z as a function of p and T leads to isotherms of very rapidly changing slope and curvature. The representation of Z as a function of $(1/V)$ and T leads to a much smoother family of isotherms whose analytical representation by, say, a polynomial in $(1/V)$ is comparatively easy. The difference between the slopes in the two representations is illustrated by the equations

$$\left(\frac{\partial Z}{\partial (p/p^c)}\right)_T^c = -\infty\,, \tag{2.3}$$

$$\left(\frac{\partial Z}{\partial (V^c/V)}\right)_T^c = -Z^c \approx -0.29\,. \tag{2.4}$$

The choice of the perfect gas as the standard of normal behaviour is unexceptionable at low and moderate pressures (that is, up to a few hundred atmospheres for most gases), but at very high pressures it is less convenient. Above 1000 atmospheres the molar volume and compressibility of a gas are comparable with those of a liquid near its normal boiling-point, and it is now more convenient to consider V as an explicit function of p and T, rather than to consider departures from (1.3). This is shown in Figs. 3 and 4 where it is seen for argon that (pV/RT) differs widely from unity for pressures above 1000 atmospheres, even at temperatures a long way above the critical temperature of 151° K. However V is here only a slowly varying function of p and T and is comparable with the molar volume of liquid and solid argon at zero pressure.

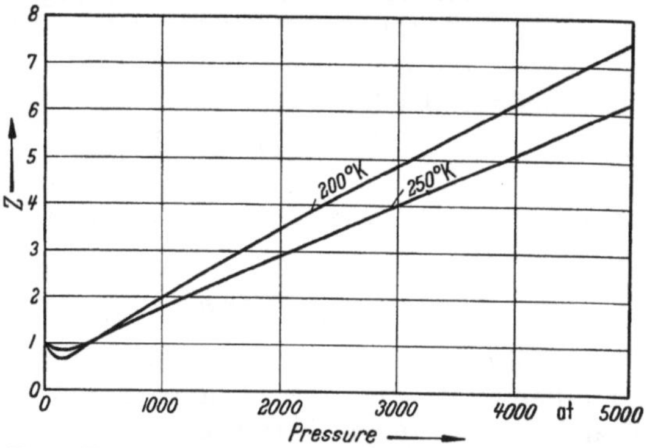

Fig. 3. The compressibility factor, Z, for argon at high pressures (DIN [2]).

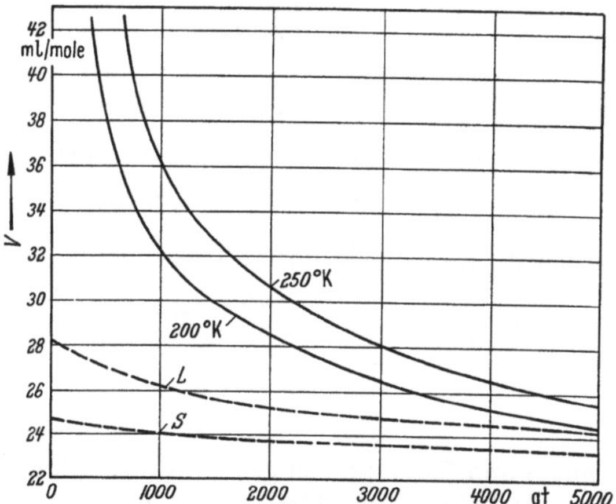

Fig. 4. The molar volume of argon as a function of pressure. The dashed lines are the volumes of the coexisting liquid and solid phases (DIN [2]).

3. The thermodynamic functions of a perfect gas. The pressure, volume and absolute temperature are the three most easily measured properties of a gas. In this section expressions for other thermodynamic properties are derived which contain only these three quantities and their three mutual derivatives. The derivatives are:

$\frac{1}{V}\left(\frac{\partial V}{\partial T}\right)_p$ the coefficient of cubic thermal expansion, which is often denoted α. In the case of a liquid the derivative is usually evaluated along the coexistence curve. This is scarcely distinguishable from the derivative at constant pressure below the normal boiling-point, but is very different at high pressures.

[1] J. S. ROWLINSON [11], see Sect. 24.

$\frac{-1}{V}\left(\frac{\partial V}{\partial p}\right)_T$ the isothermal coefficient of bulk compressibility, which is often denoted β or $\varkappa_T$. This coefficient is sometimes defined for a liquid as $(-1/V_0)\,(\partial V/\partial p)_T$, where V_0 is the volume extrapolated to zero pressure.

$\frac{1}{p}\left(\frac{\partial p}{\partial T}\right)_V$ which is known in German as the *Spannungskoeffizient*, but which has no recognised name in English, although $(\partial p/\partial T)_V$ is often called the thermal pressure coefficient in the case of a liquid.

As a knowledge of any two of p, V and T determines the third in a one-phase system, it follows that

$$dp = \left(\frac{\partial p}{\partial V}\right)_T dV + \left(\frac{\partial p}{\partial T}\right)_V dT, \tag{3.1}$$

whence the three coefficients above are related by

$$\left(\frac{\partial V}{\partial p}\right)_T \left(\frac{\partial T}{\partial V}\right)_p \left(\frac{\partial p}{\partial T}\right)_V = -1 \tag{3.2}$$

The required expressions for other thermodynamic functions may be derived by differentiation of the Helmholtz and Gibbs free-energies with respect to V and T and to p and T respectively.

$$dA = -S\,dT - p\,dV. \tag{3.3}$$

Whence

$$\left(\frac{\partial A}{\partial V}\right)_T = -p, \qquad \left(\frac{\partial A}{\partial T}\right)_V = -S \tag{3.4}$$

$$A(V_1, T) - A(V_2, T) = -\int_{V_2}^{V_1} p\,dV. \tag{3.5}$$

For one mole of a perfect gas this becomes

$$\mathrm{A}(\mathrm{V}_1, T) - \mathrm{A}(\mathrm{V}_2, T) = -\mathrm{R}T \ln (\mathrm{V}_1/\mathrm{V}_2). \tag{3.6}$$

Similarly, from (1.7) and (1.8)

$$\left(\frac{\partial G}{\partial p}\right)_T = V, \qquad \left(\frac{\partial G}{\partial T}\right)_p = -S, \tag{3.7}$$

and for a perfect gas

$$\mathrm{G}(p_1, T) - \mathrm{G}(p_2, T) = \mathrm{R}T \ln\left(\frac{p_1}{p_2}\right). \tag{3.8}$$

Thus A and G of a perfect gas show logarithmic infinities as $(1/V_1)$ and p_1 approach zero. It follows from (1.5) that this is also true for real gases.

From the Gibbs-Helmholtz equation and from (3.4) and (3.7)

$$\left(\frac{\partial E}{\partial V}\right)_T = -p + T\left(\frac{\partial p}{\partial T}\right)_V, \tag{3.9}$$

$$\left(\frac{\partial H}{\partial p}\right)_T = V - T\left(\frac{\partial V}{\partial T}\right)_p. \tag{3.10}$$

These equations, which are sometimes called the *thermodynamic equations of state*, give for isothermal changes in E and H

$$E(V_1, T) - E(V_2, T) = -\int_{V_2}^{V_1}\left[p - T\left(\frac{\partial p}{\partial T}\right)_V\right] dV, \tag{3.11}$$

$$H(p_1, T) - H(p_2, T) = \int_{p_2}^{p_1}\left[V - T\left(\frac{\partial V}{\partial T}\right)_p\right] dp. \tag{3.12}$$

Both integrands vanish for a perfect gas, for which E and H are therefore independent of volume and pressure, and functions only of the temperature. The independence of E of the volume is JOULE's law and is sometimes made the definition of a perfect gas.

From (3.5) and (3.11) and from (1.8) and (3.12)

$$S(V_1, T) - S(V_2, T) = \int_{V_2}^{V_1} \left(\frac{\partial p}{\partial T}\right)_V dV, \tag{3.13}$$

$$S(p_1, T) - S(p_2, T) = -\int_{p_2}^{p_1} \left(\frac{\partial V}{\partial T}\right)_p dp. \tag{3.14}$$

These are integrated forms of MAXWELL's equations

$$\left(\frac{\partial S}{\partial V}\right)_T = \left(\frac{\partial p}{\partial T}\right)_V, \tag{3.15}$$

$$\left(\frac{\partial S}{\partial p}\right)_T = -\left(\frac{\partial V}{\partial T}\right)_p. \tag{3.16}$$

For a perfect gas

$$\mathrm{S}(\mathrm{V}_1, T) - \mathrm{S}(\mathrm{V}_2, T) = \mathrm{R} \ln \left(\frac{\mathrm{V}_1}{\mathrm{V}_2}\right). \tag{3.17}$$

The change with temperature of these functions can be expressed in terms of the molar heat capacities at constant volume and at constant pressure,

$$C_V = \left(\frac{\partial E}{\partial T}\right)_V = T\left(\frac{\partial S}{\partial T}\right)_V = -T\left(\frac{\partial^2 A}{\partial T^2}\right)_V, \tag{3.18}$$

$$C_p = \left(\frac{\partial H}{\partial T}\right)_p = T\left(\frac{\partial S}{\partial T}\right)_p = -T\left(\frac{\partial^2 G}{\partial T^2}\right)_p. \tag{3.19}$$

For a perfect gas C_V and C_p are independent of volume at constant temperature, as are E and H from which they are derived.

The difference between the heat capcities is given by

$$\left(\frac{\partial H}{\partial T}\right)_p = \left(\frac{\partial (E + pV)}{\partial T}\right)_p = \left(\frac{\partial E}{\partial T}\right)_V + \left(\frac{\partial E}{\partial V}\right)_T \left(\frac{\partial V}{\partial T}\right)_p + p\left(\frac{\partial V}{\partial T}\right)_p \tag{3.20}$$

whence, from (3.9),

$$C_p = C_V + T\left(\frac{\partial p}{\partial T}\right)_V \left(\frac{\partial V}{\partial T}\right)_p. \tag{3.21}$$

This equation may be put into two other forms by using (3.2),

$$C_p = C_V - T\left(\frac{\partial V}{\partial T}\right)_p^2 \Big/ \left(\frac{\partial V}{\partial p}\right)_T, \tag{3.22}$$

$$C_p = C_V - T\left(\frac{\partial p}{\partial T}\right)_V^2 \Big/ \left(\frac{\partial p}{\partial V}\right)_T. \tag{3.23}$$

Eq. (3.21) is the most convenient for calculating $(C_p - C_V)$ for a liquid as the isothermal compressibility is not so readily measured as $(\partial p/\partial T)_V$. Eqs. (3.22) and (3.23) are more convenient for gases; the former being used when the independent variables are pressure and temperature, and the latter when they are volume and temperature. For one mole of a perfect gas these equations give

$$\mathrm{C}_p = \mathrm{C}_V + \mathrm{R}. \tag{3.24}$$

By differentiation of (3.9) and (3.10) with respect to T

$$\left(\frac{\partial C_V}{\partial V}\right)_T = T\left(\frac{\partial^2 p}{\partial T^2}\right)_V, \tag{3.25}$$

$$\left(\frac{\partial C_p}{\partial p}\right)_T = -T\left(\frac{\partial^2 V}{\partial T^2}\right)_p. \tag{3.26}$$

The right-hand sides of these equations are both zero for a perfect gas, but for a real gas only that of (3.25) approaches zero at zero density.

Any fluid may be considered as an assembly of negatively charged electrons and positively charged nuclei interacting solely with Coulomb forces. The application[1] of the virial theorem to such an assembly gives the following relation for the mean kinetic energy of the electrons and nuclei

$$K = 3pV - E = 4pV - H, \tag{3.27}$$

whence

$$K(V_1, T) - K(V_2, T) = \int_{V_2}^{V_1} \left[4p + 3V\left(\frac{\partial p}{\partial V}\right)_T - T\left(\frac{\partial p}{\partial T}\right)_V\right] dV. \tag{3.28}$$

The integrand is zero for a perfect gas but in a real gas the increase in K may be considerable and is generally much larger than the decrease in E.

4. The expansion coefficients of a gas. Much useful information about the thermal properties of a gas can be obtained from a study of the several ways of expanding it. Quotients such as $(\Delta y/\Delta x)$ where x is either pressure or volume, may be determined from expansions. It is usually not difficult to reduce the pressure drop sufficiently to obtain the derivative $(\partial y/\partial x)$ with considerable accuracy. The three most important kinds of expansion are known as the simple (or reversible), the Joule, and the Joule-Thomson expansions. In principle, all may be performed adiabatically or isothermally.

α) Simple expansion. Adiabatic coefficient; $\left(\frac{\partial T}{\partial p}\right)_S$,

Isothermal coefficient; $\left(\frac{\partial S}{\partial p}\right)_T$.

If the pressure of gas in a vessel is suddenly reduced by opening and closing a tap then the gas remaining has undergone an adiabatic expansion in which the maximum work is done in expelling the gas that is lost. The temperature change per unit drop of pressure is therefore $(\partial T/\partial p)_S$. This coefficient was first measured[2] indirectly by Desmormes and Clément in 1819 and directly by Lummer and Pringsheim in 1894. If heat q were to be supplied during the expansion to maintain a constant temperature, then the entropy increase would be (q/T). The isothermal coefficient which, in principle, could be so measured is $(\partial S/\partial p)_T$.

β) Joule expansion. Adiabatic coefficient; $\eta = \left(\frac{\partial T}{\partial V}\right)_E$,

Isothermal coefficient; $\left(\frac{\partial E}{\partial V}\right)_T$.

In 1845 Joule[2] attempted to detect the change in temperature of a gas on expanding it from one vessel into a second vessel which was initially evacuated. The gas in the first vessel does work, as in the simple expansion above, but

[1] W. Schottky: Phys. Z. **21**, 232 (1920).
[2] See J. R. Partington [*10*].

the gas which has flowed into the second vessel has a like amount of work done on it. The net work done in the combined system is zero, so that if the process is performed adiabatically it is an expansion at constant energy, E. The coefficient which is measured is $(\partial T/\partial V)_E$, commonly denoted η. If heat were supplied or withdrawn to maintain the temperature constant, then the coefficient $(\partial E/\partial V)_T$ would be measured.

γ) Joule-Thomson expansion. Adiabatic coefficient; $\mu = \left(\frac{\partial T}{\partial p}\right)_H$,

Isothermal coefficient; $\varphi = \left(\frac{\partial H}{\partial p}\right)_T$.

In this expansion[1] the gas is passed continuously through a porous plug or a throttle which maintains a steady pressure drop. In an adiabatic expansion the change in energy, E, must be equal to the work done, that is to the change in $(-pV)$. Thus $H(=E+pV)$ is conserved during the expansion and the coefficient measured is $(\partial T/\partial p)_H$. If, on emerging from the throttle, the gas is heated or cooled at constant pressure to the initial temperature, then the isothermal coefficient is measured, $(\partial H/\partial p)_T$.

The relations between the adiabatic and isothermal coefficients are similar for all three expansions. Equations similar to (3.1) and (3.2) give

$$\left(\frac{\partial S}{\partial p}\right)_T = -\left(\frac{\partial S}{\partial T}\right)_p \left(\frac{\partial T}{\partial p}\right)_S = \frac{-C_p}{T}\left(\frac{\partial T}{\partial p}\right)_S = -\left(\frac{\partial V}{\partial T}\right)_p, \tag{4.1}$$

$$\left(\frac{\partial E}{\partial V}\right)_T = -\left(\frac{\partial E}{\partial T}\right)_V \left(\frac{\partial T}{\partial V}\right)_E = -C_V\left(\frac{\partial T}{\partial V}\right)_E = -p + T\left(\frac{\partial p}{\partial T}\right)_V, \tag{4.2}$$

$$\left(\frac{\partial H}{\partial p}\right)_T = -\left(\frac{\partial H}{\partial T}\right)_p \left(\frac{\partial T}{\partial p}\right)_H = -C_p\left(\frac{\partial T}{\partial p}\right)_H = V - T\left(\frac{\partial V}{\partial T}\right)_p. \tag{4.3}$$

The last parts of this set of equations follow from (3.16), (3.9) and (3.10). Thus, for a perfect gas, the Joule and Joule-Thomson coefficients vanish, while the simple adiabatic coefficient is given by

$$\left(\frac{\partial T}{\partial p}\right)_S = \frac{\mathrm{R}T}{p\mathrm{C}_p}. \tag{4.4}$$

5. The velocity of sound. The adiabatic coefficient of bulk compressibility is defined by

$$\varkappa_S = \frac{-1}{V}\left(\frac{\partial V}{\partial p}\right)_S. \tag{5.1}$$

This coefficient can be measured directly for a liquid or a solid[2] but for a gas it is usually determined from the velocity of sound waves of small amplitude. This velocity, W, is given by[3]

$$W^2 = \frac{\mathrm{V}}{M\varkappa_S}, \tag{5.2}$$

where M is the molecular weight of the gas. This equation cannot be expected to hold in the limit of zero frequency where the compressions are more nearly isothermal than adiabatic. Moreover at high frequencies the thermal properties of a gas no longer have the equilibrium values associated with infinitely slow adiabatic changes and so the velocity again departs from that defined by (5.2). This dispersion at high frequencies is due to the length of time needed for translational energy to be converted to rotational and vibrational energy[3]. However,

---

[1] See footnote 2, p. 8.

[2] D. Tyrer: J. Chem. Soc. Lond. **103**, 1675 (1913); **104**, 2534 (1914). — L. A. K. Staveley and D. N. Parham: Changements de Phase, p. 366. Paris: Société Chim. Phys. 1952.

[3] H. Kneser: This Encyclopedia, Vol. XI.

there is a wide range of frequencies between these lower and upper limits in which W is independent of frequency and is accurately given by (5.2). This value of W may be regarded as an equilibrium property of the gas.

It follows from equations similar to (3.1) and (3.2) that

$$\left(\frac{\partial V}{\partial p}\right)_S = -\frac{(\partial S/\partial p)_V}{(\partial S/\partial V)_p} = -\frac{(\partial S/\partial T)_V\,(\partial T/\partial p)_V}{(\partial S/\partial T)_p\,(\partial T/\partial V)_p} \tag{5.3}$$

and so, from (3.2), (3.18) and (3.19)

$$\left(\frac{\partial V}{\partial p}\right)_S = \frac{C_V}{C_p}\left(\frac{\partial V}{\partial p}\right)_T \tag{5.4}$$

or

$$\frac{\varkappa_S}{\varkappa_T} = \frac{C_V}{C_p} \tag{5.5}$$

and

$$W^2 = \frac{\gamma V^2}{M}\left(\frac{\partial p}{\partial V}\right)_T \tag{5.6}$$

which, for a perfect gas, becomes

$$W^2 = \gamma\frac{RT}{M}. \tag{5.7}$$

where γ is the ratio of C_p to C_V. Thus for a perfect gas W is independent of pressure and approximately proportional to the square root of the absolute temperature.

The difference between $\varkappa_T$ and $\varkappa_S$ may be found from (5.5) and (3.22),

$$\varkappa_T - \varkappa_S = \frac{T}{V C_p}\left(\frac{\partial V}{\partial T}\right)_p^2. \tag{5.8}$$

For a perfect gas this difference is $(V/T C_p)$.

6. Thermodynamic stability. The conditions of mechanical and thermal stability[1] impose certain restrictions on the size and sign of thermodynamic functions. The condition that a phase should be mechanically stable requires that

$$\left(\frac{\partial A}{\partial V}\right)_T < 0, \quad \left(\frac{\partial^2 A}{\partial V^2}\right)_T > 0, \tag{6.1}$$

$$\left(\frac{\partial G}{\partial p}\right)_T > 0, \quad \left(\frac{\partial^2 G}{\partial p^2}\right)_T < 0, \tag{6.2}$$

that is, that p and V must be positive and $(\partial p/\partial V)_T$ negative. At the critical point this derivative is zero. The conditions of stability in this special case are discussed in Sect. 21.

The conditions of thermal stability are

$$\left(\frac{\partial^2 A}{\partial T^2}\right)_V < 0, \quad \left(\frac{\partial^2 G}{\partial T^2}\right)_p < 0, \tag{6.3}$$

that is, that C_p and C_V are both positive. Combining these inequalities with (3.22) gives

$$C_p - C_V \geq 0. \tag{6.4}$$

The two heat capacities are equal at absolute zero and for a fluid, such as water at 4° C, for which the coefficient of thermal expansion is zero. This coefficient is always positive for a gas and so C_p is always greater than C_V.

[1] I. Prigogine and R. Defay: Thermodynamique Chimique, Chap. 15 and 16. Liège: Desoer 1950. (English trans. by D. H. Everett. London: Longmans 1954.)

Eq. (6.4) requires that

$$\left(\frac{\partial S}{\partial T}\right)_p \geq \left(\frac{\partial S}{\partial T}\right)_V > 0, \tag{6.5}$$

and (5.1), (5.4) and (5.8) require that

$$\varkappa_T \geq \varkappa_S > 0. \tag{6.6}$$

II. The thermodynamic properties of real gases.

7. Reference states and standard states. The properties of a gas which are accessible to direct measurement are derivatives, or functions of derivatives, of the free energies. Thus A and G (and, equally, E, H and S) can only be defined and measured with respect to some arbitrary zero. The term *reference state* is used to denote a real or hypothetical state of a system in which a thermodynamic function is chosen to be zero. The term *standard state* is used to describe a state for which it is convenient to tabulate the increase in any function over its value in the reference state.

Two kinds of reference states and standard states are used. They may either be conditions of the system which are realisable in practice or in principle and each defined by a given temperature and volume (or temperature, pressure, and phase), or they may be hypothetical conditions defined by the perfect-gas laws. Both definitions have their advantages. The former provides the most direct link with experiment and offers fewer chances of confusion in its specification. A simple example of such a reference state is that used by Michels and his colleagues [7], 0° C and 1 atmosphere pressure, a temperature and pressure which come within their usual experimental range. However tables which cover solid, liquid and gaseous phases are usually based on a reference state of the perfectly ordered crystal at 0° K and zero pressure. Possible contributions to the entropy from nuclear spin factors or isotopic separation are conventionally omitted, except for hydrogen. The extensive tables for eight gases recently published by Din [2] are referred to this state. This is probably the most convenient state for the presentation of a large body of information covering a wide range of pressure and temperature. However tables which cover wide ranges of temperature but do not give the pressure variation usually have standard states based on the perfect-gas laws. If each function is individually corrected to the perfect-gas state at 1 atmosphere then there is no *real* state of the system in which, for example, G, H and S have simultaneously the values tabulated in the standard state, as the correction is not the same for every function. Such a set of standard states is used for gases in the extensive tables of thermodynamic properties published by Rossini[1] and his colleagues for the American Petroleum Institute[2] and for the National Bureau of Standards[3], and in the tables of Hilsenrath and his colleagues [8]. However the standard states for solids and liquids whose normal boiling points are above 0° C are the real states at 1 atmosphere. The reference state for the heat content in these tables is the hypothetical gaseous state at 0° K. This differs from that of Din by the latent heat of sublimation.

[1] For the definition of standard states, see F. D. Rossini: Chemical Thermodynamics, Chaps. 19 and 23. New York: Wiley 1950. — L. N. Canjar and F. D. Rossini [12]. See also Trans Faraday Soc. **50**, 768 (1954).

[2] Selected Values of Physical and Thermodynamic Properties of Hydrocarbons and Related Compounds. Pittsburgh: Carnegie Press 1950.

[3] Selected Values of Chemical Thermodynamic Properties. Nat. Bur. Stand. Circ. 500, **1952**.

8. Residual functions. The presence of the logarithmic infinity in S, A and G at zero pressure makes the direct tabulation of these functions difficult. It is sometimes convenient to choose as reference states the hypothetical perfect gases at the same temperature and either at the same volume or at the same pressure as the real gas. Thus there is a different reference state for every real state of the system. Thermodynamic functions referred to such states are here called residual functions and denoted by a superscript asterisk. The residual functions $X^*(V, T)$ and $X^*(p, T)$ are therefore defined by[1]

$$X^*(V,T) = \int_\infty^V \left[\left(\frac{\partial X}{\partial V}\right)_T - \left(\frac{\partial X}{\partial V}\right)_T^{\text{perfect}}\right] dV \tag{8.1}$$

$$= \int_0^{1/V} \left[\left(\frac{\partial X}{\partial (1/V)}\right)_T - \left(\frac{\partial X}{\partial (1/V)}\right)_T^{\text{perfect}}\right] d\left(\frac{1}{V}\right), \tag{8.2}$$

$$X^*(p,T) = \int_0^p \left[\left(\frac{\partial X}{\partial p}\right)_T - \left(\frac{\partial X}{\partial p}\right)_T^{\text{perfect}}\right] dp. \tag{8.3}$$

The second expression for $X^*(V, T)$ as an integral over the density is, in practice, more useful than the first. The quantities here called residual functions are sometimes called the internal functions of the gas. The latter name can lead to confusion as the name internal energy is sometimes applied to E, not to E^*, and the name internal pressure to $(\partial E/\partial V)_T$.

The residual pressure defined by (8.1) is

$$p^*(\mathrm{V}, T) = p - \frac{\mathrm{R}T}{\mathrm{V}} = \left(\frac{1}{\mathrm{V}}\right)(p\mathrm{V} - \mathrm{R}T). \tag{8.4}$$

If, however, density is to be used in place of volume, as in (8.2), then it is more useful to introduce a new molar residual, Q^*, defined by an equation analogous to the last part of (8.4),

$$\mathrm{Q}^*(\mathrm{V}, T) = \mathrm{V}(p\mathrm{V} - \mathrm{R}T). \tag{8.5}$$

Q^* is the residual of $(p\mathrm{V}^2)$, less a constant. Thus p^* is the appropriate residual function to use with (8.1) and Q^* the appropriate function with (8.2). By substituting the results of Sect. 3 into (8.1) and (8.2) the following equations for the other molar residual functions are obtained in terms of p^* and Q^*,

$$\mathrm{A}^*(\mathrm{V}, T) = \int_\infty^{\mathrm{V}} \left(\frac{\mathrm{R}T}{\mathrm{V}} - p\right) d\mathrm{V} = -\int_\infty^{\mathrm{V}} p^*\, d\mathrm{V} = \int_0^{1/\mathrm{V}} \mathrm{Q}^*\, d\left(\frac{1}{\mathrm{V}}\right), \tag{8.6}$$

$$\left.\begin{aligned} \mathrm{E}^*(\mathrm{V}, T) &= \int_\infty^{\mathrm{V}} \left[T\left(\frac{\partial p}{\partial T}\right)_{\mathrm{V}} - p\right] d\mathrm{V} \\ &= \int_\infty^{\mathrm{V}} \left[T\left(\frac{\partial p^*}{\partial T}\right)_{\mathrm{V}} - p^*\right] d\mathrm{V} = -\int_0^{1/\mathrm{V}} \left[T\left(\frac{\partial \mathrm{Q}^*}{\partial T}\right) - \mathrm{Q}^*\right] d\left(\frac{1}{\mathrm{V}}\right), \end{aligned}\right\} \tag{8.7}$$

$$\mathrm{S}^*(\mathrm{V}, T) = \int_\infty^{\mathrm{V}} \left[\left(\frac{\partial p}{\partial T}\right)_{\mathrm{V}} - \frac{\mathrm{R}}{\mathrm{V}}\right] d\mathrm{V} = \int_\infty^{\mathrm{V}} \left(\frac{\partial p^*}{\partial T}\right)_{\mathrm{V}} d\mathrm{V} = -\int_0^{1/\mathrm{V}} \left(\frac{\partial \mathrm{Q}^*}{\partial T}\right)_{\mathrm{V}} d\left(\frac{1}{\mathrm{V}}\right), \tag{8.8}$$

---

[1] A. Michels, M. Geldermans and S. R. de Groot: Physica, Haag **12**, 105 (1946).

$$C_V^*(V,T)=\int_\infty^V T\left(\frac{\partial^2 p}{\partial T^2}\right)_V dV=\int_\infty^V T\left(\frac{\partial^2 p^*}{\partial T^2}\right)_V dV=-\int_0^{1/V} T\left(\frac{\partial^2 Q^*}{\partial T^2}\right) d\left(\frac{1}{V}\right), \tag{8.9}$$

$$\left.\begin{aligned} G^*(V,T)&=\int_\infty^V\left[V\left(\frac{\partial p}{\partial V}\right)_T+\frac{RT}{V}\right]dV=\int_\infty^V V\left(\frac{\partial p^*}{\partial V}\right)_T dV\\ &=\int_0^{1/V}\left[\frac{1}{V}\left(\frac{\partial Q^*}{\partial(1/V)}\right)_T+2Q^*\right]d\left(\frac{1}{V}\right),\end{aligned}\right\} \tag{8.10}$$

$$\left.\begin{aligned} H^*(V,T)&=\int_\infty^V\left[T\left(\frac{\partial p}{\partial T}\right)_V+V\left(\frac{\partial p}{\partial V}\right)_T\right]dV=\int_\infty^V\left[T\left(\frac{\partial p^*}{\partial T}\right)_V+V\left(\frac{\partial p^*}{\partial V}\right)_T\right]dV\\ &=\int_0^{1/V}\left[\frac{1}{V}\left(\frac{\partial Q^*}{\partial(1/V)}\right)_T+2Q^*-T\left(\frac{\partial Q^*}{\partial T}\right)_V\right]d\left(\frac{1}{V}\right),\end{aligned}\right\} \tag{8.11}$$

$$C_p^*(V,T)=C_V^*(V,T)-R+T\left(\frac{\partial p}{\partial T}\right)_V^2\Big/\left(\frac{\partial p}{\partial V}\right)_T. \tag{8.12}$$

The limiting values of all these residual functions at infinite volume is zero. This may be seen by using the limiting form of the equation of state, (1.6), which shows that the limiting value of Q^* is finite and equal to $(RT\cdot B)$. These residual functions are inter-related by equations of the usual kind, viz.,

$$E^*(V,T)-A^*(V,T)=H^*(V,T)-G^*(V,T)=T\cdot S^*(V,T), \tag{8.13}$$

$$\left.\begin{aligned} G^*(V,T)-A^*(V,T)&=H^*(V,T)-E^*(V,T)\\ &=V\cdot p^*(V,T)=(1/V)\cdot Q^*(V,T).\end{aligned}\right\} \tag{8.14}$$

A similar set of expressions may be obtained by substituting the results of Sect. 3 into (8.3). However it is again preferable first to define a residual function with a finite limit at zero pressure by an equation analogous to that defining Q^*, or by (8.3)

$$V^*(p,T)=V-\frac{RT}{p}=\left(\frac{1}{p}\right)(pV-RT). \tag{8.15}$$

The other residual functions, defined by (8.3), are expressible in terms of V^*

$$G^*(p,T)=\int_0^p\left(V-\frac{RT}{p}\right)dp=\int_0^p V^*\,dp, \tag{8.16}$$

$$H^*(p,T)=-\int_0^p\left[T\left(\frac{\partial V}{\partial T}\right)_p-V\right]dp=-\int_0^p\left[T\left(\frac{\partial V^*}{\partial T}\right)_p-V^*\right]dp, \tag{8.17}$$

$$S^*(p,T)=-\int_0^p\left[\left(\frac{\partial V}{\partial T}\right)_p-\frac{R}{p}\right]dp=-\int_0^p\left(\frac{\partial V^*}{\partial T}\right)_p dp, \tag{8.18}$$

$$C_p^*(p,T)=-\int_0^p T\left(\frac{\partial^2 V}{\partial T^2}\right)_p dp=-\int_0^p T\left(\frac{\partial^2 V^*}{\partial T^2}\right)_p dp, \tag{8.19}$$

$$A^*(p,T)=-\int_0^p\left[p\left(\frac{\partial V}{\partial p}\right)_T+\frac{RT}{p}\right]dp=-\int_0^p p\left(\frac{\partial V^*}{\partial p}\right)_T dp, \tag{8.20}$$

$$E^*(p,T) = -\int_0^p \left[T\left(\frac{\partial V}{\partial T}\right)_p + p\left(\frac{\partial V}{\partial p}\right)_T\right] dp = -\int_0^p \left[T\left(\frac{\partial V^*}{\partial T}\right)_p + p\left(\frac{\partial V^*}{\partial p}\right)_T\right] dp, \quad (8.21)$$

$$C_V^*(p,T) = C_p^*(p,T) + R + T\left(\frac{\partial V}{\partial T}\right)_p^2 \Big/ \left(\frac{\partial V}{\partial p}\right)_T. \quad (8.22)$$

These functions are inter-related by

$$E^*(p,T) - A^*(p,T) = H^*(p,T) - G^*(p,T) = T \cdot S^*(p,T), \quad (8.23)$$

$$G^*(p,T) - A^*(p,T) = H^*(p,T) - E^*(p,T) = p \cdot V^*(p,T). \quad (8.24)$$

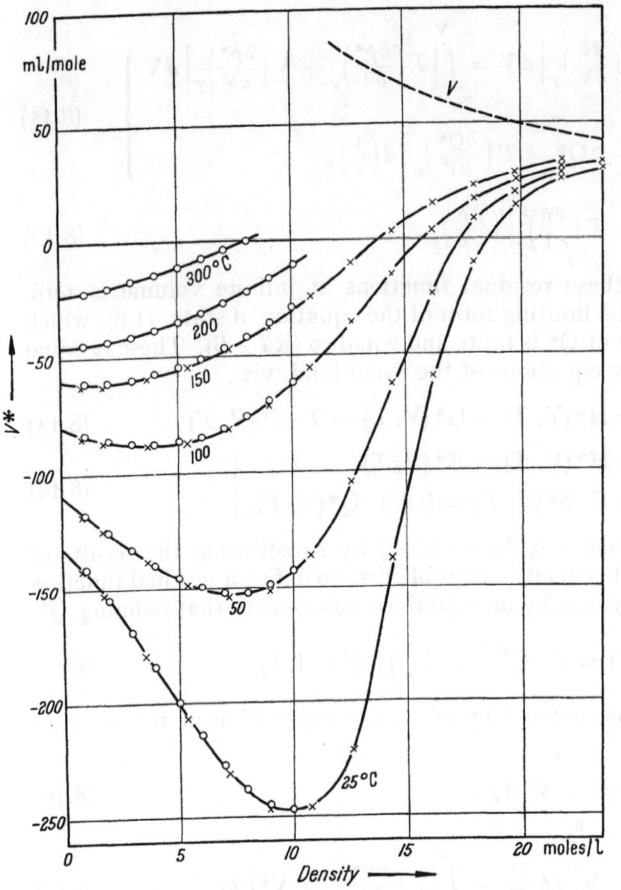

Fig. 5. The residual molar volume of xenon as a function of density. Circles, measurements of Beattie, Barriault and Brierley. Crosses, measurements of Michels, Wassenaar and Louwerse. (See text for references.) The dashed curve shows the total molar volume.

If p', V' and T' define a state of the real gas then the hypothetical perfect gas of the same V' and T' is not identical with that of the same p' and T'. Thus, in general,

$$\left.\begin{array}{l} X^*(V',T') \\ \quad \neq X^*(p',T'). \end{array}\right\} \quad (8.25)$$

If the property X of the perfect gas is independent of density, at constant temperature, as are E, H, C_V and C_p, then $X^*(V',T')$ is equal to $X^*(p',T')$.

The changes in perfect-gas properties, which must be added to the changes in the residual properties to obtain the total changes, are readily calculated analytically. These are usually the greater part of the total changes and so, by using the last parts of the sets of equations above, the errors due to numerical or graphical integration are kept as small as possible. Deming and Shupe[1] have described graphical methods of calculating thermodynamic functions by using two kinds of residual volume which are defined, in the notation of this section, as $(V^* Z)$ and $(-V^*)$, when Z is the compression factor. Din [2] similarly defines his residual volume as $(-V^*)$. The sign convention of (8.15) appears to be more natural.

Fig. 5 to 8 show some of these residual functions for xenon, a gas which has recently been studied by men in two of the laboratories which have been respons-

[1] W. E. Deming and L. E. Shupe: Phys. Rev. **37**, 638 (1931) (nitrogen); **38**, 2245 (1931) (carbon monoxide); **40**, 848 (1932) (hydrogen); **45**, 109 (1934) (calculation of entropy).

ible for many of the accurate modern measurements of compressibility — BEATTIE[1], at the Massachusetts Institute of Technology, and MICHELS[2], at the van der Waals laboratory at Amsterdam. These results cover approximately the temperature range from the critical temperature, (290° K), to twice the critical temperature. They are typical of the behaviour of all gases in this important range and so will be discussed in some detail. In Fig. 5 the residual volume is plotted as a function of the density to show that the agreement between the

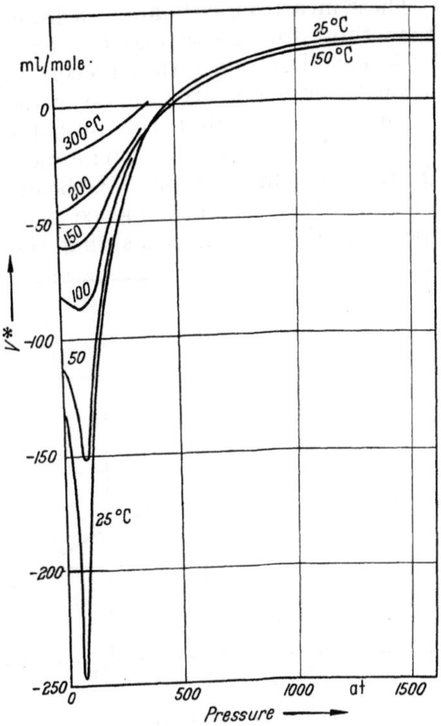

Fig. 6. The residual molar volume of xenon as a function of pressure.

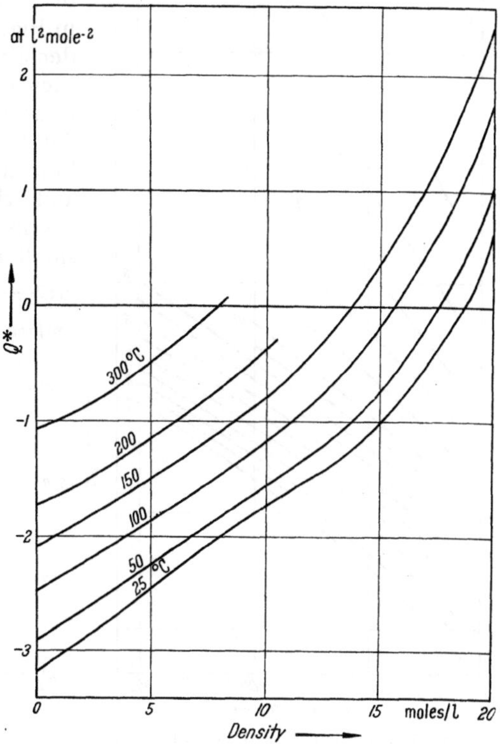

Fig. 7. The residual function, Q*, for xenon as a function of density.

two sets of results is within 1% of V* over most of the common range of density and temperature. At low densities this implies a very much greater accuracy in the actual pressure, volume and temperature measurements. For example, at 50° C and a density of 1 mole/l (that is, $V = 1000$ ml/mole), V* is found by BEATTIE to be -117.4 ml/mole, and by MICHELS to be -118.6 ml/mole. Thus their observed pressures were in the ratio (1118.6/1117.4) or 1.0011—an agreement to about 1 part per 1000. (The difference in V* which is ascribed for convenience to a difference in pressure might, of course, be due to differences in pressure, volume or temperature.) However at a density of 9 mole/l (that is, $V = 111.1$ ml/mole) their values of V* are -149.7 and -150.7 ml/mole respectively. The disagreement in observed pressure has now increased to 4 parts per 1000.

The limiting value of V* at zero density (or zero pressure, in Fig. 6) is seen to be finite at all temperatures. Recent measurements by WHALLEY, LUPIEN

[1] J. A. BEATTIE, R. J. BARRIAULT and J. S. BRIERLEY: J. Chem. Phys. 19, 1219 (1951).
[2] A. MICHELS, T. WASSENAAR and P. LOUWERSE: Physica, Haag 20, 99 (1954).

and Schneider[1] show that the zero pressure limit becomes positive above 500° C. It is from measurements of this kind that it can be confidently asserted that the limiting from of the equation of state at zero pressure is that of (1.5) and (1.6). This behaviour has by now been so often confirmed for so many gases that measurements from which the calculated values of V* do not behave in this way are certainly in error. The residual volume of propane at 104.44° C, as calculated from the early measurements of Sage, Schaafsma and Lacey[2], is plotted in Fig. 9. These measurements are at a lower density than those for xenon and the accuracy of V* is less. It is clear that no certain extrapolation to zero density can be made. All that can be said is that the probable limiting value of V* is about −270 ml/mole and that the sudden rise in V* below 0.4 mole/l is probably a consequence of a small experimental error in measuring the pressure, temperature, the volume of the apparatus, or, in an apparatus which was filled by weight, in the purity of the propane. Similar errors are to be found in some of the early steam tables.

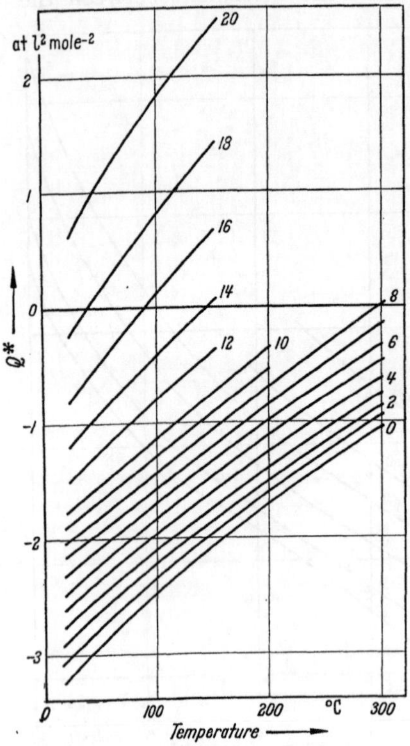

Fig. 8. The residual function, Q*, for xenon as a function of temperature at densities up to 20 moles/l.

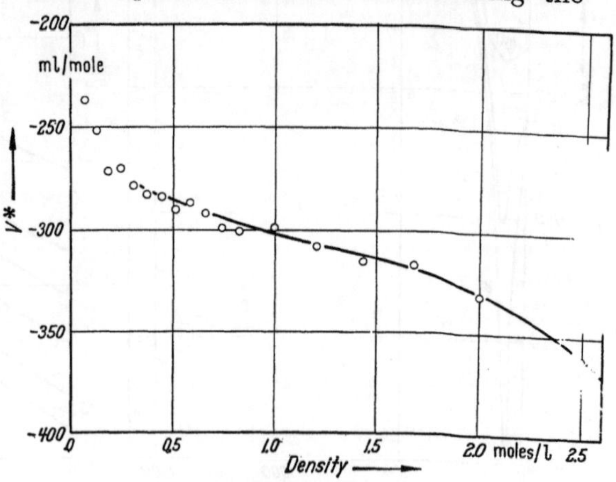

Fig. 9. Inaccurate measurements of the residual molar volume of propane (see text).

A comparison of Fig. 5 and 6 shows that V*, like Z in Fig. 1 and 2, is a more regular function of density than of pressure. Its limiting value at infinite pressure is V, the total molar volume, and Fig. 5 shows that at 23 moles/l, where the pressure is greater than 2000 atmospheres, the residual volume is approaching this upper limit. The rapid variation of V^* with p and T in the neighbourhood of the critical point makes difficult the calculation of thermodynamic functions by means of (8.16) to (8.22). By contrast, the residual Q* (Fig. 7) is a very regular function of density and temperature even at the critical point, and so the calculation by means of (8.6) to (8.12) is much easier. However at very high densities the behaviour of Q* is less convenient than that of V*, as it has no upper

[1] E. Whalley, Y. Lupien and W. G. Schneider: Canad. J. Chem. **33**, 633 (1955).

[2] B. H. Sage, J. G. Schaafsma and W. N. Lacey: Industr. Engng. Chem. **26**, 1218 (1934).

limit. The variation of Q^* with temperature is important in the calculation of E^* and C_V^*. Fig. 8 shows that Q^* is almost a linear function of temperature even at very high densities. Since

$$\left(\frac{\partial^2 Q^*}{\partial T^2}\right)_V = V^2 \left(\frac{\partial^2 p}{\partial T^2}\right)_V \tag{8.26}$$

the curvature of the lines in Fig. 8 is proportional to the curvature of the isochores, that is lines of constant density as a function of temperature. The behaviour of all substances is similar to that shown in Fig. 10 which, for comparison with Fig. 1 and 2, is on a reduced temperature and pressure scale. Careful measurements on some gases have shown that the curvature is negative at low densities, becomes zero at about the critical density and is positive at higher densities. However the determination of this curvature is very difficult and values of C_V^* calculated from (8.9) are not very reliable (see Sect. 23).

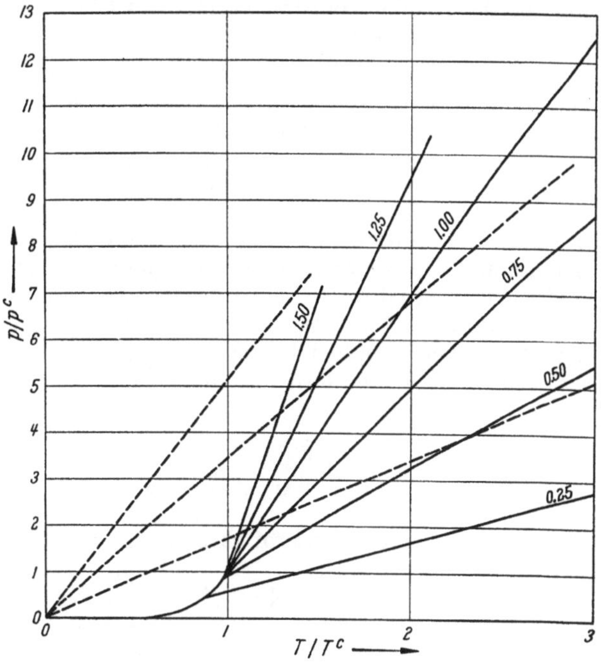

Fig. 10. Isochores of the inert gases at values of V^c/V from 0.25 to 1.50. The three dashed lines are the perfect-gas isochores for reduced densities of 0.50, 1.00 and 1.50.

9. The fugacity. The fugacity is a useful function which has the dimensions of pressure and which may be defined by

$$\left.\begin{aligned} f(p, T) = p \times \\ \times \exp\left[G^*(p, T)/RT\right]. \end{aligned}\right\} \tag{9.1}$$

It was introduced by LEWIS[1] as a function which plays a rôle in an imperfect gas similar to that of pressure in a perfect gas. Re-arranging (9.1) gives

$$G^*(p_1, T) - G^*(p_2, T) = RT \ln\left[\frac{p_2 \cdot f(p_1, T)}{p_1 \cdot f(p_2, T)}\right]. \tag{9.2}$$

Adding this to (3.8) gives for the total Gibbs free-energy

$$G(p_1, T) - G(p_2, T) = RT \ln\left[\frac{f(p_1, T)}{f(p_2, T)}\right]. \tag{9.3}$$

A comparison of this equation with (3.8) shows that the fugacity has taken the part of pressure in the perfect-gas equation. The standard state of 1 atmosphere fugacity is used by ROSSINI[2] for the Gibbs free-energy. Other residual

[1] G. N. LEWIS and M. RANDALL: Thermodynamics, Chap. 17. New York: McGraw-Hill 1923.

[2] See footnotes 1, 2 and 3 on p. 11.

functions can be simply related to the fugacity. Thus,

$$H^*(p, T) = -RT^2 \left(\frac{\partial \ln f}{\partial T}\right)_p \tag{9.4}$$

$$S^*(p, T) = R \ln\left(\frac{f}{p}\right) + RT\left(\frac{\partial \ln f}{\partial T}\right)_p. \tag{9.5}$$

Thus the tabulation of fugacity (or, better, of the ratio f/p which is the activity coefficient) as a function of pressure and temperature is a useful way of summarising the departures from the perfect-gas laws. Such tables are used extensively by chemical engineers.

10. The virial expansions. An open form of the equation of state in which the pressure is expressed in terms of inverse powers of the molar density was first proposed by KAMERLINGH ONNES[1],

$$p = \frac{A'}{V} + \frac{B'}{V^2} + \frac{C'}{V^3} + \cdots \tag{10.1}$$

in which the coefficients A', B' etc. were called the first, second etc. virial coefficients and are functions of the temperature but not of the density. The first coefficient is equal to RT for all gases. He proposed that all the even powers of the density after the first could be omitted. Subsequently, the name virial coefficient has been applied to the coefficients of two similar, and related, power series in which all the terms are included,

$$\frac{pV}{RT} = A + \frac{B}{V} + \frac{C}{V^2} + \cdots, \tag{10.2}$$

$$pV = A'' + B''p + C''p^2 + \cdots. \tag{10.3}$$

The first three coefficients of these series are related by

$$A = \frac{A'}{RT} = \frac{A''}{RT} = 1, \tag{10.4}$$

$$B = \frac{B'}{RT} = B'', \tag{10.5}$$

$$C = \frac{C'}{RT} = C''RT + (B'')^2. \tag{10.6}$$

There are similar but more complicated equations between the higher coefficients[2]. It is modern practice, which is followed here, to restrict the name virial coefficient to the coefficients of (10.2). This series has two advantages, one practical and one theoretical, over the pressure series (10.3). The practical advantage is that the density series is capable of representing a set of experimental results more accurately than the pressure series with a given number of terms. This is clear from Fig. 1 and 2 and from the discussion in Sect. 2. DAVID, HAMANN and PRINCE[3] give a clear example of the danger of using the pressure series to make a poor fit of the experimental results and then using the coefficient C'' to determine the third coefficient of the density series by means of (10.6). In one case this procedure led to a value of C which had the opposite sign to that obtained by fitting the experimental results directly to the density series. The theoretical advantage of the density series is that the second, third, fourth etc.

[1] H. KAMERLINGH ONNES: Comm. Leiden, No. 71, 1901.

[2] L. F. EPSTEIN: J. Chem. Phys. **20**, 1981 (1952); **21**, 762 (1953).

[3] H. G. DAVID, S. D. HAMANN and R. G. H. PRINCE: J. Chem. Phys. **20**, 1973 (1952).

coefficients can be related, by the statistical theory of imperfect gases[1], to the interactions of molecules in pairs, triplets, quadruplets etc. This property is very valuable in the discussion of gas-mixtures (Sect. 28).

The limiting value of the residual volume is the second virial coefficient, B, and the limiting value of Q^* is RTB. The latter residual may be written

$$\frac{Q^*(V,T)}{RT} = B + \frac{C}{V} + \frac{D}{V^2} + \cdots. \tag{10.7}$$

The limits of V^* and Q^* may be used to determine B from experimental results. Another equivalent method is the determination of the limiting slope of (pV) as a function of p or of $(1/V)$.

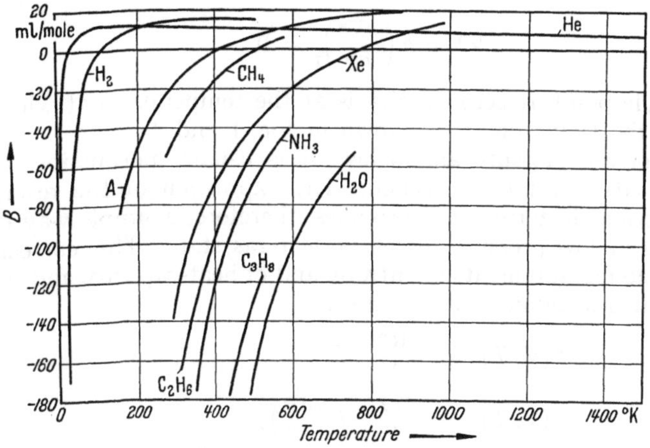

Fig. 11. The second virial coefficients of nine gases.

Two points must be considered in using the virial expansion. The first is the range of density over which such a series is mathematically convergent and the second is the range over which it is possible to determine the coefficients unambiguously from experimental results. The first question is a theoretical one which can only be discussed by the methods of statistical mechanics[1]. The second is a purely practical one whose answer appears to be that only the coefficients B and C can be determined with any precision even from the best measurements. At high temperatures it is usually found that D is positive but at low temperatures even its sign is uncertain. This is clear from Fig. 7. This difficulty in giving unambiguous values to the higher coefficients does not mean that it is not often useful to express (pV) or Q^* as a polynomial in the density with powers up to the sixth or seventh. Such expressions can be very useful for interpolation, but their coefficients are not to be identified with those of (10.2) or (10.7). The higher coefficients of these polynomials usually show a very irregular variation with temperature which means that they are of little use for calculating thermodynamic functions.

Typical graphs showing B as a function of temperature are plotted in Fig. 11. The curves are all very similar. At low temperatures B is negative, with increasing temperature it becomes positive, passes through a very flat maximum and finally decreases very slowly. The maximum has been observed only for helium, hydrogen and neon (not shown), but there is little doubt that all other chemically stable substances would show one if the measurements extended to high enough

[1] See J. E. MAYER: This Volume, Chap. 2.

temperatures. The coefficient C is positive and decreases with increasing temperature. A few negative values have been reported at low temperatures but are probably not reliable.

A point on an isotherm where

$$\left(\frac{\partial(p\mathrm{V})}{\partial(1/\mathrm{V})}\right)_T = 0 \quad \text{and} \quad \left(\frac{\partial(p\mathrm{V})}{\partial p}\right)_T = 0 \tag{10.8}$$

is called a Boyle point as in its vicinity (pV) is a constant, at constant temperature (but not necessarily equal to RT). At a Boyle point

$$\mathrm{Q}^* + \left(\frac{1}{\mathrm{V}}\right)\left(\frac{\partial \mathrm{Q}^*}{\partial(1/\mathrm{V})}\right)_T = 0, \tag{10.9}$$

$$\mathrm{V}^* + p\left(\frac{\partial \mathrm{V}^*}{\partial p}\right)_T = 0. \tag{10.10}$$

Thus the Boyle point at zero density is at the temperature at which B is zero. The locus of the Boyle points is shown in Fig. 1 and 2.

At atmospheric pressure and below, it is usually permissible to ignore all virial coefficients beyond the second. The expression of the residual thermodynamic functions in terms of B provides, therefore, a simple way of correcting measurements at low pressure to the perfect-gas state. This correction is most often applied to measurements of entropy and of heat-capacity and for correcting vapour pressures of liquids to fugacities.

$$\mathrm{A}^*(\mathrm{V}, T) = \frac{\mathrm{R}T}{\mathrm{V}} \cdot B, \tag{10.11}$$

$$\mathrm{E}^*(\mathrm{V}, T) = -\frac{\mathrm{R}T}{\mathrm{V}} \cdot T\left(\frac{dB}{dT}\right), \tag{10.12}$$

$$\mathrm{S}^*(\mathrm{V}, T) = -\frac{\mathrm{R}}{\mathrm{V}}\left[B + T\left(\frac{dB}{dT}\right)\right], \tag{10.13}$$

$$\mathrm{C}_v^*(\mathrm{V}, T) = -\frac{\mathrm{R}}{\mathrm{V}}\left[2T\left(\frac{dB}{dT}\right) + T^2\left(\frac{d^2B}{dT^2}\right)\right], \tag{10.14}$$

$$\mathrm{G}^*(\mathrm{V}, T) = \frac{\mathrm{R}T}{\mathrm{V}} \cdot 2B, \tag{10.15}$$

$$\mathrm{H}^*(\mathrm{V}, T) = \frac{\mathrm{R}T}{\mathrm{V}}\left[B - T\left(\frac{dB}{dT}\right)\right], \tag{10.16}$$

$$\mathrm{C}_p^*(\mathrm{V}, T) = \frac{\mathrm{R}}{\mathrm{V}} \cdot T^2\left(\frac{d^2B}{dT^2}\right), \tag{10.17}$$

$$\mathrm{G}^*(p, T) = p \cdot B, \tag{10.18}$$

$$\mathrm{H}^*(p, T) = p\left[B - T\left(\frac{dB}{dT}\right)\right], \tag{10.19}$$

$$\mathrm{S}^*(p, T) = -p \cdot \left(\frac{dB}{dT}\right). \tag{10.20}$$

$$\mathrm{C}_p^*(p, T) = -p \cdot T\left(\frac{d^2B}{dT^2}\right), \tag{10.21}$$

$$\mathrm{A}^*(p, T) = 0, \tag{10.22}$$

$$\mathrm{E}^*(p, T) = -p \cdot T\left(\frac{dB}{dT}\right), \tag{10.23}$$

$$\mathrm{C}_v^*(p, T) = -p\left[2\left(\frac{dB}{dT}\right) + T\left(\frac{d^2B}{dT^2}\right)\right]. \tag{10.24}$$

The leading term in $A^*(p, T)$ is of the order of p^2. The fugacity is given by

$$f = p \cdot e^{Bp/RT} = p\left(1 + \frac{Bp}{RT}\right) = \frac{RT}{V}\left(1 + \frac{2B}{V}\right). \tag{10.25}$$

This equation is often used for the estimation of fugacities in the form

$$\frac{f}{p} = \frac{p}{(RT/V)} = Z, \tag{10.26}$$

where RT/V is the pressure of a perfect gas at the same density and temperature as the gas under consideration.

The variation of the velocity of sound with pressure may be derived from (5.6) by substituting for $(\partial p/\partial V)_T$ from the virial expansion and for C_V and C_p from (10.21) and (10.24). This gives[1],

$$W = W_0\left(1 + \frac{Sp}{RT}\right), \tag{10.27}$$

$$S = B + \frac{RT}{(C_V)_0}\left(\frac{dB}{dT}\right) + \frac{(RT)^2}{2(C_V)_0 (C_p)_0}\left(\frac{d^2B}{dT^2}\right), \tag{10.28}$$

where the subscript zero indicates the value for the perfect gas. These equations have been used by VAN ITTERBEEK for measuring the second virial coefficients of simple gases at low temperatures, but the method is not a very direct one owing to the derivatives of B which occur in (10.28). The equations have been confirmed within the experimental error of the measurement of these derivatives for the vapours of organic compounds which boil at and above room temperature[2].

11. The Joule-Thompson coefficients. The isothermal and adiabatic Joule-Thomson coefficients, φ and μ, may be expressed in terms of the residual volume by substituting the results of Sect. 8 in (4.3),

$$\varphi = -\mu C_p = \left(\frac{\partial H^*}{\partial p}\right)_T = V^* - T\left(\frac{\partial V^*}{\partial T}\right)_p. \tag{11.1}$$

The limiting molar value of φ at zero pressure is therefore finite and equal to $[B - T(dB/dT)]$. The fact that the limit is finite has been experimentally verified and is used in Sect. 1. However the quantitative comparison of $\varphi_{(p=0)}$ with B is often very difficult. It is particularly so at low temperatures where B is changing rapidly with temperature. The problem is similar to that of verifying (10.28). For example, there are two sets of compressibility measurements on carbon dioxide at 25° C which are apparently in good agreement. However on differentiation, one[3] of these measurements gives a value of $\varphi_{(p=0)}$ of -420 ml/mole and the other[4] a value of -370 ml/mole. The direct measurement[5] of φ gives a limiting value of -390 ml/mole, a value which is confirmed by measurements[6] of the adiabatic coefficient, μ. At high temperatures, where $T(dB/dT)$ is small compared to B, the agreement which may be obtained is usually much better than this. For example, at the maximum of B, the coefficients should be equal. This has been confirmed for helium at 200° K where $B = (11.9 \pm 0.1)$

[1] A. VAN ITTERBEEK and W. H. KEESOM: Comm. Leiden, No. 209a, 1930. See also W. H. KEESOM [4].

[2] J. D. LAMBERT and J. S. ROWLINSON: Proc. Roy. Soc. Lond., Ser. A **204**, 424 (1950).

[3] A. MICHELS and S. R. DE GROOT: Physica, Haag **14**, 218 (1948).

[4] K. E. MACCORMACK and W. G. SCHNEIDER: J. Chem. Phys. **18**, 1269 (1950).

[5] A. CHARNLEY, G. L. ISLES and J. R. TOWNLEY: Proc. Roy. Soc. Lond., Ser. A **218**, 133 (1953).

[6] J. R. ROEBUCK, T. A. MURRELL and E. E. MILLER: J. Amer. Chem. Soc. **64**, 400 (1942).

ml/mole is in perfect agreement with the value of $\varphi_{(p=0)}$ calculated by interpolation of the measurements of μ of ROEBUCK and OSTERBERG[1]. While it is always possible to calculate φ from B, it is not possible to calculate B from φ by integration of

$$B - T\left(\frac{dB}{dT}\right) = \varphi_{(p=0)}. \tag{11.2}$$

The solution of this differential equation, together with the experimental fact that the limit of (B/T) is zero at infinite temperature gives,

$$\frac{B}{T} = \int_0^{1/T} \varphi_{(p=0)}\, d\left(\frac{1}{T}\right). \tag{11.3}$$

The extrapolation of φ to infinite temperature is too uncertain, even for helium for which the measurements extend to 578° K, for this equation to be used for calculating absolute values of B. However the difference of B between two temperatures may be calculated quite accurately.

A point on an isotherm at which φ and μ vanish (C_p is always finite and positive) is called an inversion point and is given by

$$V^* - T\left(\frac{\partial V^*}{\partial T}\right)_p = 0, \tag{11.4}$$

an equation which is similar in form to (10.10), the equation defining a Boyle point. The inversion temperature is that temperature at which the inversion point occurs at zero density and is given by

$$B - T\left(\frac{dB}{dT}\right) = 0. \tag{11.5}$$

It is clear from Fig. 11 that the inversion temperature always lies above the Boyle temperature. Their ratio for most gases is about 1.9. The locus of the inversion points in the p, T plane is a curve concave to the temperature axis, similar to, but outside, the locus of the Boyle points. Within the curve μ is positive, outside the curve it is negative. The position of the curve may be found either from direct measurements of μ or calculated from compressibility measurements. The two methods are compared in Fig. 12 for nitrogen, the gas whose inversion curve has been most widely studied. Here the curve calculated from the measurements of μ by ROEBUCK and OSTERBERG[2] is shown to agree with those calculated from independent sets of compressibility measurements by DEMING and DEMING[3] and by LUNBECK, MICHELS and WOLKERS[4]. The course of the calculated[4] Boyle point curve is also shown. The inversion curve intersects the liquid-vapour boundary from the liquid side. The Boyle curve is almost tangential to the boundary curve near the critical point, as is shown in Fig. 1.

The Joule coefficient, η, is of little importance as, unlike μ, it vanishes at zero pressure. This follows from (4.1) which may be written

$$\left(\frac{\partial E}{\partial V}\right)_T = -\eta\, C_V = -p^* + T\left(\frac{\partial p^*}{\partial T}\right)_V. \tag{11.6}$$

---

[1] J. R. ROEBUCK and H. OSTERBERG: Phys. Rev. **43**, 60 (1933). See W. H. KEESOM [*4*], p. 82–91, for a more detailed comparison for this gas.

[2] J. R. ROEBUCK and H. OSTERBERG: Phys. Rev. **48**, 450 (1935), with corrections J. Chem. Phys. **8**, 627 (1940).

[3] W. E. DEMING and L. S. DEMING: Phys. Rev. **48**, 448 (1935).

[4] R. J. LUNBECK, A. MICHELS and G. J. WOLKERS: Appl. Sci. Res. A **3**, 197 (1952).

The zero-pressure limit is therefore

$$\frac{p^2}{\mathrm{R}}\left(\frac{dB}{dT}\right).$$

The inversion curve for the Joule coefficient, defined by

$$p^* - T\left(\frac{\partial p^*}{\partial T}\right)_V = 0, \tag{11.7}$$

lies outside that for the Joule-Thomson coefficient. Its zero-pressure limit is at the maximum of B, and so above the temperature defined by (11.5). This temperature is probably about 2150° K for nitrogen. The Joule inversion curve

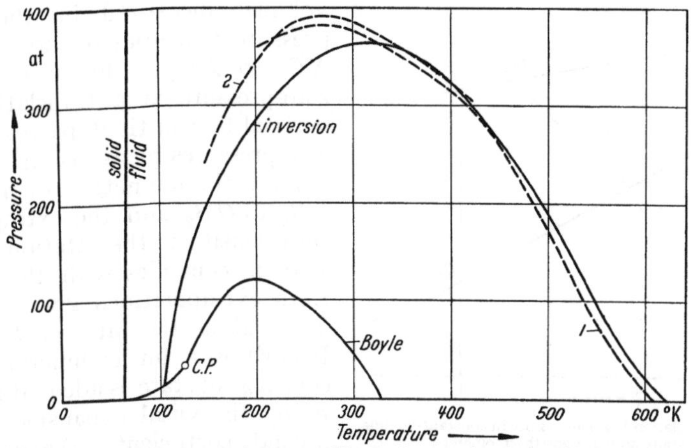

Fig. 12. The Boyle and inversion curves of nitrogen. The Boyle curve is calculated from the isotherms of LUNBECK, MICHELS and WOLKERS. The full-line inversion curve is from the measurements of ROEBUCK and OSTERBERG and the dashed curves are calculated from compressibility measurements by (1) DEMING and DEMING, and (2) LUNBECK, MICHELS and WOLKERS (see text for references). The solid-fluid and liquid-gas phase boundaries are shown.

for this gas[1] appears to pass through a flat maximum at about 500° K and 3700 atm and then to fall slowly reaching 3300 atm at 120° K. The curve probably ends on the solid-liquid boundary, as $(\partial E/\partial V)_T$ is positive along both branches of the liquid-vapour boundary.

12. Gas thermometry. A determination of the ratio of two absolute temperatures on the perfect-gas scale may be made by using (1.4). The assignment of a numerical value to one fixed point is then sufficient to specify the scale completely. GIAUQUE[2] proposed this way of defining the absolute scale of temperature and his proposal has been accepted by the most recent committee[3] of the General Conference on Weights and Measures, who recommend that the fixed point be the triple point of water (not the conventional ice-point which is 0.010° C lower) and that the value asigned be 273.16° K. The absolute temperature of the steam-point thus becomes a matter for experiment but the assigned value of the triple point has been chosen so that the steam-point is, within present experimental error, 373.15° K, thus ensuring that the absolute degree and the Celsius degree are as equal in size as possible.

[1] M. BENEDICT: J. Amer. Chem. Soc. **59**, 2233 (1937).

[2] W. F. GIAUQUE: Nature, Lond. **143**, 623 (1939).

[3] Comité consultatif de Thermométrie auprès du Comité International des Poids et Mesures. Session de 1954. Paris: Gauthiers-Villars 1955. The recommendations will appear in Comptes Rendus des Séances de la Dixième Conférence Générale des Poids et Mesures, to be published (Gauthiers-Villars).

The right-hand side of (1.4) may be evaluated either by taking the ratios of two series of volumes at two temperatures and common pressures and extrapolating the ratios to zero pressure, or else by taking the ratios of two series of pressures at two temperatures and common volumes and extrapolating the ratios to zero density. The final ratio (T_1/T_2) is the same in each case but the experimental arrangements are very different. The apparatus for the first experiment is called a constant-pressure gas-thermometer, and for the second a constant-volume gas-thermometer. The distinction is illustrated in Fig. 13 (after BERTHELOT) which shows two isotherms in the low pressure region where (pV) is a linear function of p. The product (pV) may be calculated for an arbitrary amount of gas and no knowledge of the molecular weight or even of the purity is needed. The volumes measured in the constant-pressure thermometer are in the ratio XC_1 to XC_2. The extrapolation to zero-pressure is a translation of this vertical line until it passes through O. The pressures measured in a constant-volume thermometer are in the ratio OB_1 to OB_2 and the extrapolation to zero density is the rotation of the line until it coincides with the (pV) axis. Both extrapolations give the ratio of T_1 to T_2 as the ratio of OA_1 to OA_2. If (pV) is a linear function of p the equation of state is adequately described by the virial expansion up to the second coefficient. Departures from this equation are not detectable for helium, hydrogen and nitrogen, the three gases most commonly used in gas thermometry, at atmospheric temperature and pressure. The ratio of T_1 to T_2 is related to the ratio of p_1 to p_2 in a constant-volume thermometer by

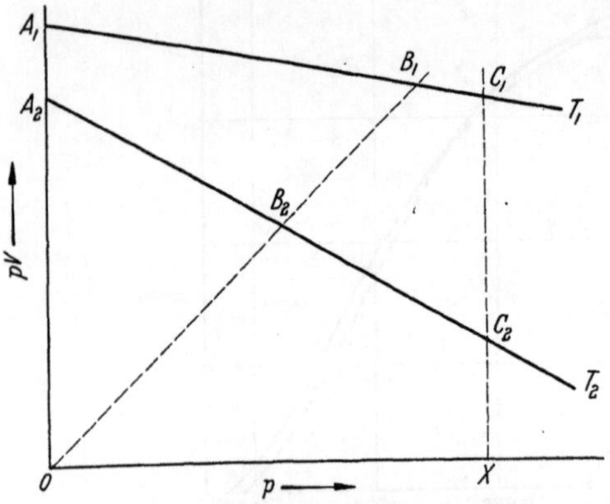

Fig. 13. Extrapolation to zero-pressure in constant-volume and constant-pressure gas-thermometers.

$$\frac{T_1}{T_2} = \left(\frac{p_1}{p_2}\right)\left(1 - \frac{B_1 p_1}{RT_1} + \frac{B_2 p_2}{RT_2}\right) \tag{12.1}$$

The absolute temperatures needed to calculate the correction terms on the right-hand side of this equation need only be approximately correct. The fall of p with decreasing T offsets, to some extent, the increase in the magnitude of (B/T), thus keeping the correction terms reasonably small even at low temperatures. This follows from Fig. 13 where the increasing steepness of the lines AC with decreasing temperature is offset by the slope of OB. In practice, therefore, constant-volume thermometers are more widely used than constant-pressure thermometers. Probably the best is the constant-volume helium thermometer[1]. Owing to its low boiling-point and small virial coefficient helium may be used down to 2° K.

The correction to the constant-pressure thermometer for gas-imperfection may be made directly from a knowledge of the Joule-Thomson coefficient as a function of temperature. If V_1 and V_2 are the two observed volumes at absolute temperatures T_1 and T_2 and constant pressure p, then,

$$\frac{T_1}{T_2} = \frac{V_1 - a}{V_2} \tag{12.2}$$

$$\frac{a}{T_1} = \frac{V_1}{T_1} - \frac{V_2}{T_2}, \tag{12.3}$$

[1] W. H. KEESOM [4], p. 63—82.

where a is a volume small compared to V_1 or V_2. Now (4.3) may be integrated to give an equation analogous to (11.3), and so an expression may be obtained for a in terms of the Joule-Thomson coefficient,

$$a = \int_{1/T_2}^{1/T_1} \varphi(p)\, d\left(\frac{1}{T}\right). \tag{12.4}$$

The integration is carried out at the constant pressure. Only rough values of T_1 and T_2 are needed in this equation.

13. The measurement of molecular weights. The determination of gram-molecular and gram-atomic weights from the densities of gases has been used to supplement gravimetric methods based on chemical reactions in the liquid and solid states. It was the only method available for measuring the atomic weights of the inert gases before the invention of the mass-spectrometer. The method, although simple in principle, is troublesome and tedious in practice and it is only those who have worked many years in this field who have obtained atomic weights correct to a few parts in 10^5. For such work it is clear that corrections must be applied for deviations from the perfect-gas laws at pressures near atmospheric.

The measurement of a molecular weight on the chemical scale (normal $O_2 = 32.0000$) may be made by measuring the density of the gas and that of oxygen at a series of pressures and the extrapolation of the ratios of the densities to zero pressure. This method is therefore usually called the method of limiting densities[1]. The measurement of the ratios of the densities and the extrapolation may be made in several ways. The direct procedure described above has the disadvantage that absolute measurements of density must be made at low pressures. This is difficult, and may be avoided by making only one density measurement on each gas, at atmospheric pressure, and correcting these by means of the second virial coefficients. These may be found from supplementary measurements which need not involve any determination of absolute density.

By using a gas-density microbalance, absolute measurements may be avoided entirely. This is a silica torsion-balance, one arm of which carries a sealed silica bulb and the other a solid silica counterweight. Owing to the buoyancy effect on the bulb, the pressure of the surrounding gas at which the balance is in equilibrium is approximately inversely proportional to its molecular weight. Such an instrument may be used to determine the ratio of the balancing pressure of oxygen to the balancing pressure of the gas whose molecular weight is required. The balancing pressure of oxygen, and so of the other gas, may be changed by adding small riders to the balance arm. In this way a series of ratios of balancing pressures may be obtained which can be extrapolated to zero-pressure to give the exact molecular weight. A particularly careful piece of work of this kind by LAMBERT and PHILLIPS[2] on nitrogen and methane led to values of the atomic weights of

$$\mathrm{N} = 14.0078 \pm 0.0003,$$

$$\mathrm{C} = 12.0112 \pm 0.0005.$$

Results of this precision compare well with what may be achieved with a good mass-spectrometer, and could not, even in principle, be improved without knowing more of the isotopic constitution of the gases. The variation of the amounts

[1] R. WHYTLAW-GRAY: Quart. Rev. Chem. Soc. Lond. **4**, 153 (1950). — F. L. CASADO, D. S. MASSIE and R. WHYTLAW-GRAY: Proc. Roy. Soc. Lond., Ser. A **207**, 483 (1951); **214**, 466 (1952).

[2] B. LAMBERT and C. S. G. PHILLIPS: Phil. Trans. Roy. Soc. Lond., Ser. **A 242**, 415 (1950).

of C^{13} in carbon from natural sources leads to a variation[1] in the chemical atomic weight from 12.0105 to 12.0120. It is doubtful if many more such measurements will be made in view of the skill and time which they need.

The value of the molar gas-constant, R, depends directly on the measurement of the absolute density of oxygen and on the extrapolation of (pV) to zero pressure. BIRGE[2] recommends a normal molar volume (0° C, 1 atmosphere) of (24.4140 ± 0.0006) l/mole for the perfect gas, which, with an absolute temperature of the normal ice-point of 273.15° K, gives

$$R = 0.082057 \text{ l atm mole}^{-1} (°\text{C})^{-1}.$$

14. The liquefaction of gases[3]. The refrigeration and liquefaction of gases is a large subject of which only the least technical part[4] can be discussed here. Two of the methods used in the production of low temperatures are closely related to two of the expansion coefficients discussed in Sects. 4 and 11. The difference between the methods and their relation to the expansion coefficients are best made clear by an entropy-temperature diagram. Fig. 14 shows a typical graph in which isobars are shown by full lines and isenthalps (constant heatcontent) by dashed lines. The isobars cross the two-phase region as straight lines of constant temperature. The isenthalps cross it as curves which are almost continuous with the one-phase isenthalps on the gas-boundary, become continuous with them at the critical point, and are discontinuous on the liquid boundary. Isenthalps for values of H just greater than the critical enter and leave the two-phase region at two pressures on the gas-boundary. The slopes of the isobars and isenthalps are $(\partial S/\partial T)_p$ and $(\partial S/\partial T)_H$. The former is given by (3.19), and the latter by,

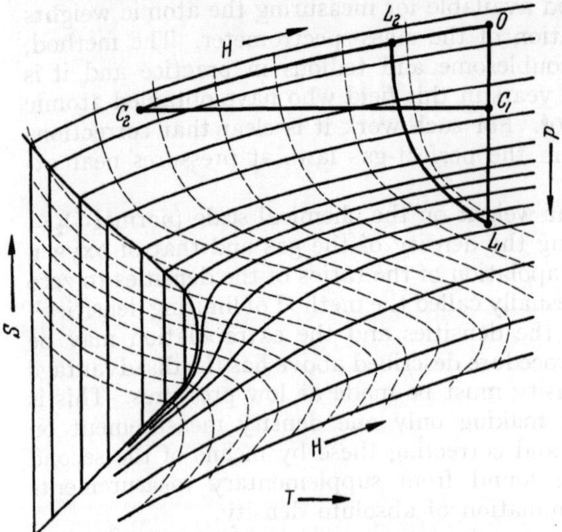

Fig. 14. A typical graph of the variation of entropy with temperature at constant pressure (full lines) and at constant heat-content (dashed lines). The tongue-shaped curve is the boundary of the two-phase region. The cycles OL_1L_2O and OC_1C_2O are those of the simple Linde and Claude schemes for liquefying a gas.

$$\left(\frac{\partial S}{\partial T}\right)_H = \left(\frac{\partial S}{\partial p}\right)_H \left(\frac{\partial p}{\partial T}\right)_H = \left(\frac{\partial S}{\partial p}\right)_H \frac{1}{\mu}. \tag{14.1}$$

From the fundamental differential equation

$$dH = T\,dS + V\,dp, \tag{14.2}$$

$$\left(\frac{\partial S}{\partial p}\right)_H = -\frac{V}{T}, \tag{14.3}$$

[1] See footnote 1, p. 25.
[2] R. T. BIRGE: Rep. Progr. Phys. 8, 90 (1941).
[3] Cf. the article of J. G. DAUNT on the production of low temperatures in Vol. XIV of this Encyclopedia.
[4] M. RUHEMAN, in D. M. NEWITT [9], Chap. 13. — M. DAVIES: Physical Principles of Gas Liquefaction and Low Temperature Rectification. London: Longmans 1949.

whence
$$\left(\frac{\partial S}{\partial T}\right)_H = \frac{-V}{\mu T}. \tag{14.4}$$

The locus of the points where $(\partial S/\partial T)_H$ is infinite is therefore the low-temperature branch of the Joule-Thomson inversion curve (Fig. 12). The slope of the isenthalps at atmospheric pressure is negative (μ positive) at all temperatures below the inversion temperature.

The problem of liquefying a gas is therefore that of taking it from the top right-hand corner of Fig. 14 (high T, high S, atmospheric pressure) to the bottom left-hand corner (low T, low S, atmospheric pressure). The cycle of operations first used by LINDE in 1895 for liquefying air had three stages. First, the gas is

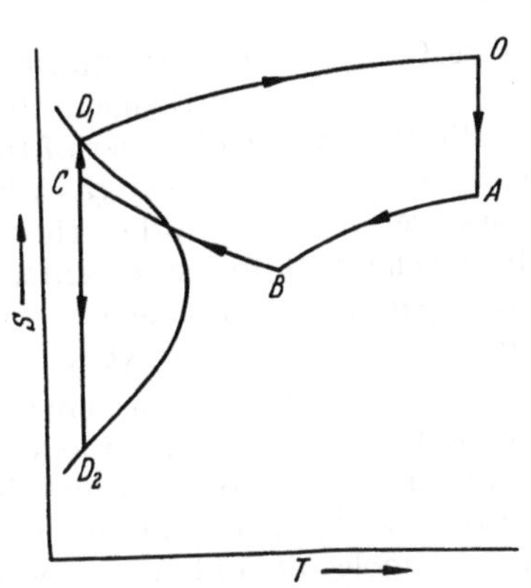

Fig. 15. The steady state of a simple Linde cycle. The proportion of liquid to gas in each cycle is the ratio CD_1 to CD_2.

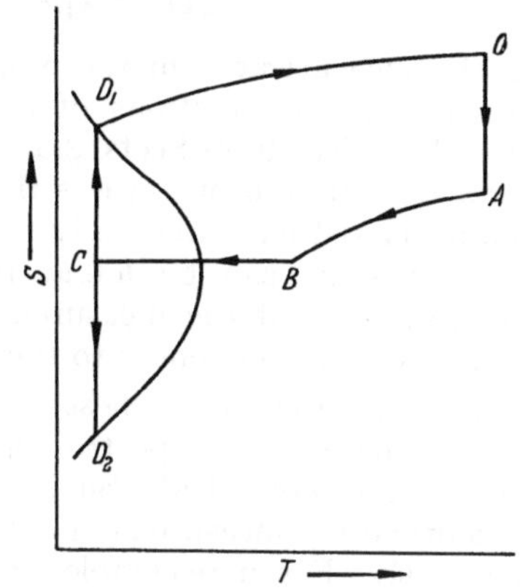

Fig. 16. The steady state of a simple Claude cycle. The proportion of liquid to gas is greater than in a Linde cycle.

compressed isothermally along OL_1, the heat generated being removed at atmospheric temperature. Secondly, it is passed through a valve where, as in a Joule-Thomson experiment, it undergoes an adiabatic expansion at constant heat content, and returns to atmospheric pressure at the point L_2. In the third stage the gas passes through a heat exchanger in which it cools the gas for the next cycle to an initial temperature L_2 and is itself heated to O. This cycle is repeated with the lower starting temperature until liquid is formed which is not re-cycled and the process settles down to a steady state. The steady state cycle is shown in Fig. 15 in which OA represents the initial compression, AB the cooling in the heat exchanger, BC the isenthalpic expansion leading to a mixture of gas (D_1) and liquid (D_2) at atmospheric pressure. The latter is withdrawn from the cycle and the former returns through the heat-exchanger to O. The initial cycle will not produce a cooling unless it lies within the inversion curve, and so hydrogen (inversion temperature 201° K) and helium (50° K) cannot be liquefied without preliminary cooling. KAMERLINGH ONNES[1] first liquefied helium in 1908 by a Linde cycle in which the gas was first cooled by hydrogen boiling under reduced pressure.

The cycle used by CLAUDE in 1902 for liquefying air is based on the simple, or reversible, expansion of Sect. 4. A gas may be cooled by making it do work adiabatically by driving a piston in a well-lagged cylinder. The coefficient $(\partial T/\partial p)_S$ is always positive and so any gas may be cooled and eventually liquefied by a series of such expansions. The course of the initial cycle is shown in Fig. 14

[1] H. KAMERLINGH ONNES: Comm. Leiden No. 108, 1908.

by OC_1, C_1C_2 and C_2O, which represent the isothermal compression, the isentropic expansion, and the passage through the heat exchanger. The cooling produced is seen to be much greater than that in a comparable Linde cycle, but the practical difficulties are greater. KAPITZA found that it was easier to make the expanding gas drive a small turbine than to make it drive a reciprocating engine. The steady state of a simple Claude cycle is shown in Fig. 16 in which the letters have the same significance as in Fig. 15. Both diagrams represent only the most simple forms of the two cycles. In practice, it is found that it is more efficient to divide the gas-stream and to use various sub-cycles whose inter-relations can be quite complicated.

III. Experimental methods.

15. The measurement of the properties of a perfect gas. The calculation of the thermodynamic functions of a perfect gas requires a knowledge only of C_V or C_p. It is clear from Sects. 8 and 11 that the limiting value of either of these heat capacities at zero pressure is the only thermal property of a real gas which must be measured experimentally. If these limits are known, all other thermodynamic functions can be found from purely mechanical measurements. A detailed discussion of the measurement of heat capacities is outside the scope of this chapter, but references to the two principle methods will be given.

For many simple gases these heat capacities, and the perfect-gas entropy, may be calculated from spectroscopic measurements of the normal modes of vibration of the molecules[1]. Such calculations are usually more accurate than direct calorimetric measurements if the vibrational assignment can be made unambiguously. For more complex molecules, and for molecules of low symmetry, calorimetric methods must be used. The experimental arrangements are discussed in several standard works[2].

The simple adiabatic expansion coefficient $(\partial T/\partial p)_S$ is related to C_p by (4.4), but it is only recently that the measurement of this coefficient has been made sufficiently precisely for an accurate value of the heat capacity to be found[3]. Unfortunately the coefficient becomes small as C_p becomes large and so the method is not suitable for complex molecules.

16. Amagat units. The units used in accurate work on the compressibility of gases are either those related directly to c.g.s. units or else the so-called Amagat units.

In the first system the volume is expressed in cubic centimetres or in litres and the pressure in millibars (1 millibar $\equiv$ 1 mb $=$ 1000 dynes cm^{-2}). However for practical purposes the millibar is usually replaced by the mm of mercury at low pressures, and by the standard atmosphere of 760 mm Hg at high pressures. The mm Hg is defined as the pressure exerted by 1 mm of a fluid of density 13.5951 gm cm^{-3} at a standard gravity of 980.665 cm sec^{-2}. This density is as close as present knowledge permits to the density of mercury at 0° C and 760 mm Hg. Thus both the mm Hg and the atmosphere are defined in terms of the fundamental unit, the millibar[4]. From the figures above, 1 atm is 1013.250 mb.

---

[1] E. A. GUGGENHEIM, Vol. III, Part 2. — E. JUSTI [6], p. 45—123. — G. HERZBERG: Infra-Red and Raman Spectra of Polyatomic Molecules. New York: Van Nostrand 1945.

[2] D. M. NEWITT [9], p. 264—275. — J. R. PARTINGTON [10], p. 792—820. — W. H. KEESOM [4], p. 93—101. — K. SCHEEL: Handbuch der Physik, Vol. X, p. 321—349 (1926). — J. K. ROBERTS: Heat and Thermodynamics, p. 162—201 (4th Ed. revised A. R. MILLER). London: Blackie 1951.

[3] G. B. KISTIAKOWSKY and W. W. RICE: J. Chem. Phys. **7**, 281 (1939); **8**, 610, 618 (1940).

[4] F. A. GOULD: Proc. Roy. Soc. Lond., Ser. A **186**, 195 (1946).

The Amagat system uses the same standard atmosphere as its unit of pressure but takes as the unit of volume, the molar volume of the gas at 0° C and unit pressure. Were all gases perfect, this would be a convenient unit (24.4140 l mole^{-1}), but in practice the Amagat unit of volume varies from gas to gas owing to the different departures of different substances from the perfect-gas laws under the standard conditions. Fortunately these departures are usually small enough for the conversion to absolute units to be made from a knowledge of the second virial coefficient only, at 0° C. The procedure is described by MICHELS [7]. Amagat units are used not only for pressure and volume but for density (the reciprocal of the volume) and for derived quantities such as the second virial coefficient, whose dimensions are volume per mole, and for energy, for which the unit is the value of pV at unit pressure and volume and 0° C. Such units have a certain practical convenience but are often difficult to convert to c.g.s. units and lack that close relation to fundamental quantities which is nowadays considered desirable. They cannot be used for highly compressible vapours whose departures from the perfect-gas laws under the standard conditions are large and hard to measure accurately.

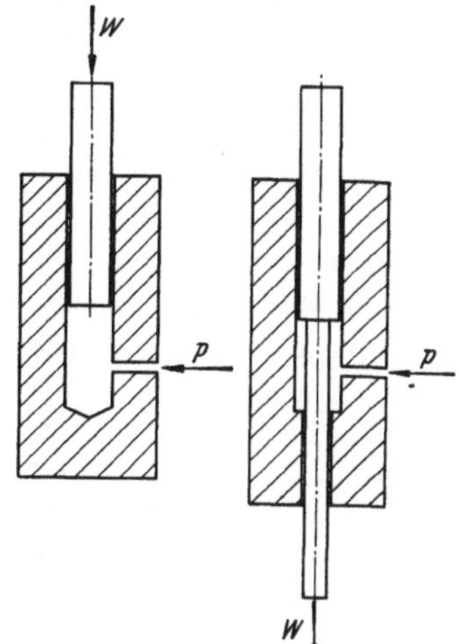

Fig. 17. Simple and differential piston gauges. The pressure to be measured is applied at p and the balancing weights at w.

17. The measurement of high pressures. The definition of the fundamental unit of pressure given above makes a mercury column, at a known temperature and gravity, a possible choice for an instrument to measure high pressures. Such columns were used towards the end of the last century for measuring pressures up to hundreds of atmospheres. It is, however, very difficult to set up a mercury column of a height of several hundred metres in such a way that observations may be made at any point on the column and so that the temperature of the whole may be accurately controlled. Even if the column is split into shorter lengths, connected in series by compressed-gas lines, the temperature control is not easy and the errors of observation are increased by the greater number of mercury lengths to be measured. For most practical purposes, therefore, a piston, or "dead-weight", gauge is now used as a measuring instrument. There are two forms of this gauge, the simple and the differential. They are illustrated in Fig. 17. In the simple gauge an accurately machined piston is maintained in equilibrium with the unknown pressure, p, by the weights, w. The gauge is filled with oil and the clearance of the piston must be large enough for a continuous oil film to be maintained, but not so large that the piston can twist about a horizontal axis under the incipient couple of p and w. Under these conditions, the effective area of the assembly is the mean of that of the piston and that of the cylinder. A gauge of this kind is described by KEYES and BEATTIE [5]. At very high pressures either the diameter of the piston must be very small or the load inconveniently large. In either case the piston will not have the rigidity to resist the couple which p and w exert as soon as any deformation occurs. In these conditions a differential gauge is to be preferred. Here the effective area may be reduced by using coupled pistons of almost equal diameters and the mechanical stability improved by suspending the weights from the lower piston. The difficulties of making such a gauge are less if the two pistons are not on a common shaft, as shown, but are mounted independently in the block and coupled through an external frame. MICHELS[1] has described such a gauge and has discussed the theory of its behaviour at high loads.

The piston gauge can be, and often is, used as an absolute instrument, but it is better to calibrate it against a mercury column as it is difficult to determine accurately the area of the cylinder and to correct for the distortion of the assembly at high pressures. A single test against a short mercury column will give the effective area at zero pressure. At high pressures a pair of gauges may be calibrated against one mercury column by using the method of HOLBORN and SCHULTZE[2]. The two gauges are first calibrated separately against an open mercury column. One gauge is then connected to the top and one to the base of the column and the pressure in the closed system is increased so that the pressure at the top of the column is equal to that at the bottom of the column when it was open to the atmosphere. The lower gauge may now be calibrated at a pressure equal to twice that of the open column. The gauges are interchanged and the operation repeated. This process may be repeated to very high pressures. BETT, HAYES and NEWITT[3] have used this method up to 2500 atmospheres with an accuracy of 1 part in 15000. The limiting factor to the accuracy of this method is our knowledge of the compressibility of mercury.

DADSON[4] has recently described an ingenious method of calibrating piston gauges which is less tedious than that of HOLBORN and SCHULTZE. He made two gauges of different materials, one of steel and one of bronze, whose elastic moduli were in the ratio of 1.44 to 1, and whose POISSON's ratios were equal. The clearances of the pistons were in the ratio of their elastic moduli but otherwise the dimensions of the two gauges were the same. By the principle of similarity, the distortions of the two gauges under high pressures are proportional to the elastic moduli. By balancing one gauge against another when both are connected to a common pressure and noting the change in the difference of the effective areas, both may be calibrated relative to their areas at zero pressure. The latter may be determined by direct measurement or by calibration against a mercury column. The accuracy claimed for this method is 1 part in 10^5 for pressures up to 1500 atmospheres.

18. The measurement of the compressibility of gases at high pressures. The relations between the pressure, volume and temperature of a gas are usually determined either by measuring the pressure as a function of volume at a series of fixed temperatures or by measuring pressure as a function of temperature at a series of fixed volumes. The former, the measurement of $(\partial p/\partial V)_T$ determines an isotherm, and the latter, the measurement of $(\partial p/\partial T)_V$, determines an isochore. Both methods have advantages and disadvantages. The former is preferable if the course of the isotherms changes rapidly with volume and temperature as, for example, near the critical point. It leads more directly to the evaluation of the virial coefficients and so is more convenient for highly compressible vapours. However, the measurement of volume is the observation most liable to error, in work of the highest accuracy, and so the second method is often preferred. It is easier to calibrate the volume of a vessel to a fixed point than to calibrate it throughout its whole length.

---

1 A. MICHELS: Ann. Phys. **72**, 285 (1923); **73**, 577 (1924).

2 L. HOLBORN and H. SCHULTZE: Ann. Phys. **47**, 1089 (1915).

3 K. E. BETT, P. F. HAYES and D. M. NEWITT: Phil. Trans. Roy. Soc. Lond. A **247**, 59 (1954).

4 R. S. DADSON: Nature, Lond. **176**, 188 (1955). — See also [*12*].

The variable volume piezometer used in the early work of the Leiden school has often been described[1] and no further description need be given here. An apparatus which was similar in principle was built at the Massachussetts Institute of Technology and is described by KEYES and BEATTIE [5]. It is shown schematically in Fig. 18, and is used, as follows. The 200 ml vessel A is detached, a weighed amount of gas is introduced and the vessel is sealed at B with a gold disc. It is replaced in the right-hand thermostat and the gold disc is burst by applying pressure to the mercury in the capillary. The pressure is applied by the calibrated injector C which has a capacity of 210 ml. The gas volume in A can be controlled and measured to 0.004 ml by withdrawing or adding mercury. The pressure is measured at D by an oil-filled piston gauge. The constancy of the

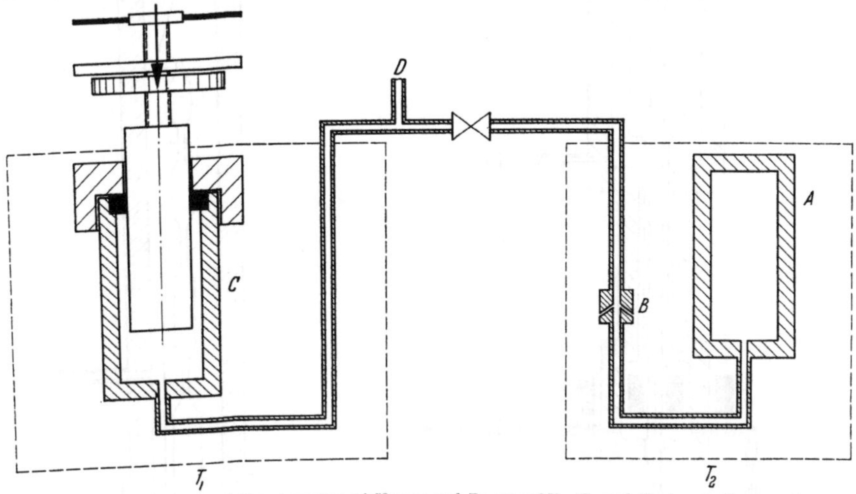

Fig. 18. Schematic diagram of the apparatus of KEYES and BEATTIE [5]. T_1 and T_2 denote the two thermostats.

height of the mercury in this side-arm is maintained by an electrical contact at the oil-mercury interface. A blank run is first made to find the amount of the distortion of the vessel A under pressure. The maximum pressure used is 500 atmospheres. Hysteresis in the deformation of A makes accurate volume measurement impossible beyond this point. The temperature range of this apparatus is 0 to 325° C. The upper limit is fixed by the vapour pressure of mercury which has reached 0.55 atmospheres at this temperature. The correction to be applied for the presence of mercury in the gas phase is not easily calculable, but is discussed below in Sect. 32. This form of apparatus is well suited for measurements in the critical region as the volume of the gas is the variable which is under the observer's direct control. The apparatus is equally well adapted for making measurements along isochores. BEATTIE and his colleagues have studied many gases with this apparatus in the last twenty years[2].

[1] H. KAMERLINGH ONNES and C. BRAAK: Comm. Leiden No. 97a, 1906. — W. H. KEESOM [4], p. 31. — J. D. VAN DER WAALS: Handbuch der Physik, Vol. X, p. 178. 1926.

[2] Krypton, J. Chem. Phys. **20**, 1613 (1952). — Xenon, J. Chem. Phys. **19**, 1219 (1951). — Ethane, J. Chem. Phys. **3**, 93 (1935); J. Amer. Chem. Soc. **61**, 924, 926 (1939). — Propane, J. Chem. Phys. **3**, 96 (1935); J. Amer. Chem. Soc. **59**, 1589 (1937). — n-Butane, J. Amer. Chem. Soc. **61**, 24, 26 (1939). — i-Butane, J. Chem. Phys. **17**, 576 (1949); **18**, 127 (1950). — l-Butene, J. Amer. Chem. Soc. **72**, 1449, 4143 (1950). — n-Pentane, J. Amer. Chem. Soc. **73**, 4431 (1951); **74**, 4778 (1952). — Neopentane, J. Chem. Phys. **19**, 948 (1951); **20**, 1619 (1952). — n-Heptane, J. Amer. Chem. Soc. **59**, 1586, 1587 (1937). — 2,2,4-tri-Methyl Pentane, J. Amer. Chem. Soc. **70**, 3382 (1948). — Methane + n-Butane, J. Chem. Phys. **9**, 871 (1941).

Sage and Lacey have used similar kinds of apparatus in their extensive work on the petroleum hydrocarbons, but in its latest form[1] their piezometer is designed so that the volumes of both gas and liquid phases can be measured as well as the total volume of the fluid. Fig. 19 shows the apparatus. The fluid is contained in the vessel A where it is stirred by the rotary stirrer B which is

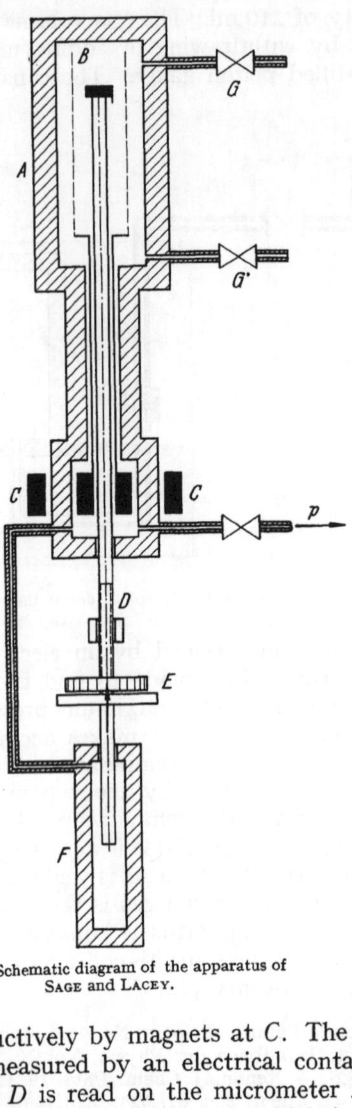

Fig. 19. Schematic diagram of the apparatus of Sage and Lacey.

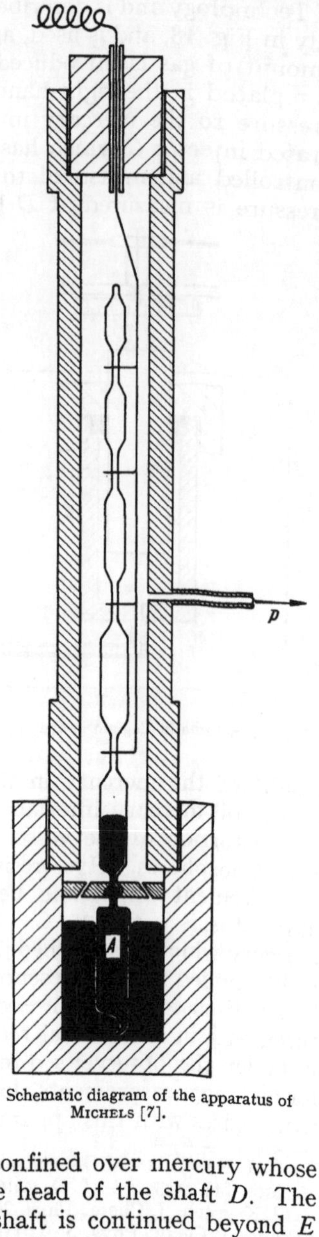

Fig. 20. Schematic diagram of the apparatus of Michels [7].

driven inductively by magnets at C. The fluid is confined over mercury whose height is measured by an electrical contact on the head of the shaft D. The position of D is read on the micrometer E. The shaft is continued beyond E into a compensating cylinder F, which is filled entirely with mercury, so that the movement of the shaft will cause no volume change in the fluid. This cylinder communicates with the main vessel through a capillary. The gas and liquid

[1] B. H. Sage and W. N. Lacey: Trans. Amer. Inst. Min. Metallurg. Engrs. **174**, 102 (1948).

hydrocarbons are admitted to A through the valves G and G'. The relative amount of the two phases is found from a small thermal conductivity gauge fitted to the head of D. The apparatus may be used to 240° C and to 670 atmospheres but its accuracy is probably less than that of the apparatus of KEYES and BEATTIE. SAGE and LACEY and their colleagues have used this apparatus, and its predecessors, for measuring the thermodynamic properties of very many hydrocarbons and their mixtures[1].

The apparatus used by MICHELS at Amsterdam differs from the last two, as the vessel is calibrated only at fixed volumes. Two vessels are used [7], one to cover the range from about 15 to 250 atmospheres and the other from 70 to 3000 atmospheres. The low-pressure assembly, shown in Fig. 20, resembles the apparatus of AMAGAT. The piezometer is made entirely of glass and is mounted in a massive steel cylinder in such a way that its internal and external pressures are always equal. A measured amount of gas is initially confined over mercury in the large bulb A. The pressure is then raised by injecting oil until the mercury makes electrical contact with the first of a series of fine wires sealed through the capillary sections of the piezometer. The volumes between these wires is determined from a previous calibration with mercury[2]. The pressure is measured with a differential piston gauge. The high-pressure apparatus is similar but the lower end of the piezometer is modified to allow it to be filled *in situ* by gas under pressure. In this way measurements may be made to an Amagat density of 600 without either starting with a volume which is inconveniently large or ending with one which is inconveniently small. The mass of gas in the high-pressure apparatus is not measured directly but is found by comparing the results with those obtained from the low-pressure apparatus in the range of densities where the two overlap. The temperature range is usually 0 to 150° C and the precision claimed is 1 part in 2000 at the highest pressures. The results obtained with this equipment by MICHELS and his colleagues in the last twenty years are the most important single contribution to our knowledge of the compressibility of gases. For most of the substances studied, the thermodynamic functions have been calculated from the compressibilities as functions both of density and pressure[3]. This form of apparatus is not suitable for measurements in the critical region.

In 1936 BURNETT[4] proposed an ingenious method which avoided two of the difficulties associated with all the apparatus so far described, namely, the accurate measurement of small volumes and the use of mercury as the confining fluid. His form of apparatus has been used extensively by W G. SCHNEIDER

---

[1] A series of more than 50 papers in Industr. Engng. Chem. of which the first is **26**, 103 (1934).

[2] A. MICHELS and R. O. GIBSON: Ann. Physik **87**, 850 (1928).

[3] Helium, Physica, Haag **8**, 923 (1941). — Argon, Physica, Haag **15**, 627, 689 (1949); Appl. Sci. Res. A **2**, 345 (1951). — Xenon, Physica, Haag **20**, 99 (1954). — Hydrogen and Deuterium, Physica, Haag **8**, 347, 353, 387 (1941). — Nitrogen, Physica, Haag **3**, 585, 597 (1936); **17**, 801 (1951); Appl. Sci. Res. A **3**, 197 (1952). — Oxygen, Physica, Haag **20**, 1209 (1954). — Carbon monoxide, Physica, Haag **18**, 121, 128 (1952); Appl. Sci. Res. A **3**, 253 (1952). — Carbon dioxide, Proc. Roy. Soc. Lond., Ser. A **153**, 201, 214 (1935); **160**, 348, 358, 376 (1937); Appl. Sci. Res. A **1**, 94, 103 (1948); Physica, Haag **14**, 218 (1948). — Methane, Physica, Haag **2**, 1000 (1935); **3**, 569 (1936). — Ethane, Physica, Haag **20**, 17 (1954). — Ethylene, Physica, Haag **3**, 346 (1936); **9**, 967 (1942); **12**, 105 (1946); **13**, 538 (1947); Appl. Sci. Res. A **1**, 55 (1947). — Propylene, Physica, Haag **19**, 287 (1953); Appl. Sci. Res. A **4**, 34 (1953). — Methyl fluoride, Physica, Haag **18**, 114 (1952). — Air, Physica, Haag **18**, 67 (1952); Appl. Sci. Res. A **4**, 52, 381 (1954); A **5**, 121 (1955). — Hydrogen + Nitrogen, Appl. Sci. Res. A **1**, 258, 378 (1949); A **3**, 1 (1951). — Methane + Ethylene, Physica, Haag **6**, 656 (1939). — Ammonia mixtures, Appl. Sci. Res. A **4**, 180 (1954).

[4] E. S. BURNETT: J. Appl. Mech. A **3**, 137 (1936).

and his colleagues in a series of investigations[1] which are notable for the very wide temperature range which they cover. In this apparatus an unknown mass of gas is confined in a thermostatted vessel at a pressure p_1. The gas is then allowed to expand isothermally into a second vessel, initially evacuated, and the pressure measured again, p_2. If the equation of state is written as an expansion in powers of the pressure, (10.3), then,

$$\frac{p_1}{p_2} = r + (r-1)\frac{B'' p_1}{RT} + \left(r - \frac{p_2}{p_1}\right)\frac{C'' p_1^2}{RT} + \cdots \tag{18.1}$$

where r is the ratio of the volumes after and before expansion. By closing a valve between the two vessels, re-evacuating the second, and then repeating the expansion, a second ratio (p_2/p_3) may be measured. These expansions may be repeated as many times as is necessary to obtain a good extrapolation to zero pressure. SCHNEIDER uses an initial pressure of about 80 atmospheres and makes 9 expansions with r equal to about 1.5. The exact value of r and so of B'' and C'' may be determined from a knowledge of (p_i/p_{i+1}) as a function of p_i. The accuracy of B is usually better than 1%. Neither the volumes of the vessels, the mass of the gas nor the exact temperature need be known. However the constancy of the temperature must be maintained as accurately as in more conventional apparatus. SCHNEIDER has used this method to 700° C by transmitting the pressure to the piston gauge through a gas-capillary. By placing only one of the vessels in the high temperature thermostat he has been able to make measurements on helium up to 1200° C. This is probably the highest temperature reached in accurate work on compressed gases. The disadvantages of this method are that it is tedious and can only be used where pV is a slowly changing function of p. An analysis of the results is impossible if this change is too rapid. In practice, this means that the method cannot be used above 200 atmospheres, in the critical region, or near the condensation point of a vapour.

19. The measurement of the compressibility of gases at low pressures. Accurate measurements of the deviations of vapours from the perfect-gas laws at low pressures are needed for the correction of such properties as heat capacities and the velocity of sound to the perfect-gas state and for the correction of vapour pressures to fugacities for the evaluation of the thermodynamic properties of liquids and liquid mixtures. All such corrections can be made if the second virial coefficient is known as a function of the temperature. The pressure range of the measurements used to determined B vary from less than 1 atmosphere to about 20 atmospheres, depending on the size of B and on the vapour pressure of the substance. The apparatus used in these measurements is generally similar to those described above but is built in glass, rather than in steel, and uses a mercury manometer, rather than a piston gauge.

EUCKEN and MEYER[2] used a constant volume apparatus to measure $(\partial p/\partial T)_V$ for organic vapours at pressures close to their condensation point. Known weights of liquid were vapourised into a 1-litre glass bulb and the pressure measured as a function of temperature. Such an apparatus will give the absolute density as well as the second virial coefficient. Its principal drawback is that the absorption of the vapours on the glass wall of the vessel is large and variable, and so gives

[1] Apparatus, Canad. J. Res. B **27**, 339 (1949). — Helium, J. Chem. Phys. **17**, 751 (1949); **18**, 641 (1950). — Neon, Canad. J. Chem. **33**, 589 (1955). — Argon, Canad. J. Chem. **31**, 722 (1953). — Krypton, Trans. Amer. Soc. Mech. Engrs. **76**, 1001 (1954). — Xenon, Canad. J. Chem. **33**, 633 (1955). — Carbon dioxide, J. Chem. Phys. **18**, 1269, 1273 (1950). — Carbon tetrafluoride and Sulphur hexafluoride, J. Chem. Phys. **19**, 845 (1951).

[2] A. EUCKEN and L. MEYER: Z. phys. Chem., Abt. B **5**, 452 (1929).

rise to spurious deviations from the gas laws. A similar apparatus was used by ROPER[1] to measure the second virial coefficient of seven olefins. These vapours were not as close to their condensation points as those studied by EUCKEN and MEYER, and so his results are probably more reliable.

WHYTLAW-GRAY and his colleagues[2] have used an apparatus which might be described as a low-pressure version of that of KEYES and BEATTIE. Mercury is run out from the volumeter and weighed, and the pressure is balanced against a similar volumeter containing nitrogen. This is a gas of small and accurately known departures from the perfect-gas laws. Their results are very precise but are unfortunately restricted to one temperature. S. D. HAMANN, J. F. PEARSE and their colleagues[3] used another differential method, again with nitrogen as the balancing gas, in an apparatus which was similar in principle to that of MICHELS. It is reported to be free from trouble due to absorption as long as the gas is not close to its condensation point.

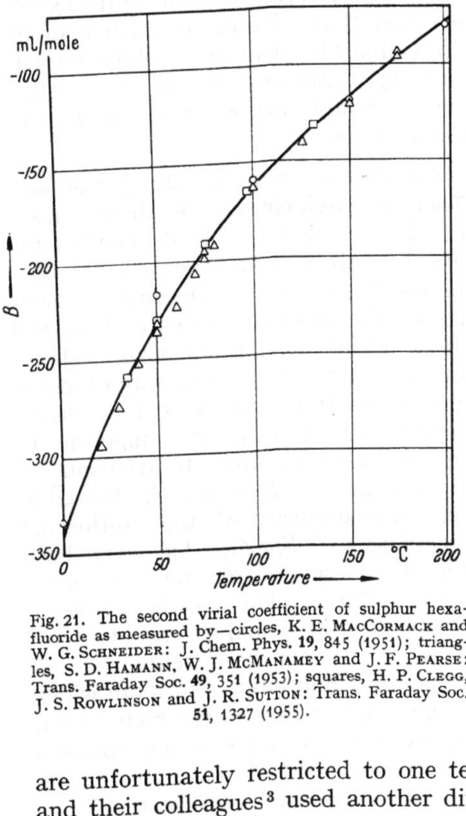

Fig. 21. The second virial coefficient of sulphur hexafluoride as measured by—circles, K. E. MACCORMACK and W. G. SCHNEIDER: J. Chem. Phys. **19**, 845 (1951); triangles, S. D. HAMANN, W. J. MCMANAMEY and J. F. PEARSE: Trans. Faraday Soc. **49**, 351 (1953); squares, H. P. CLEGG, J. S. ROWLINSON and J. R. SUTTON: Trans. Faraday Soc. **51**, 1327 (1955).

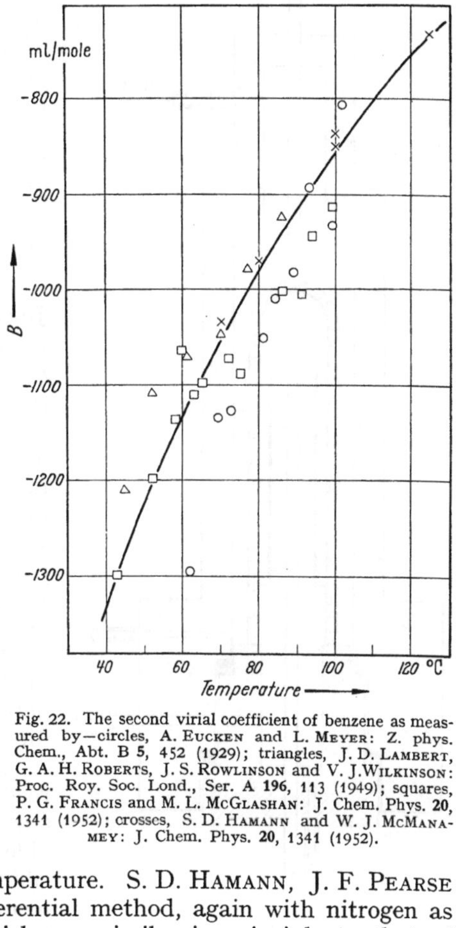

Fig. 22. The second virial coefficient of benzene as measured by—circles, A. EUCKEN and L. MEYER: Z. phys. Chem., Abt. B **5**, 452 (1929); triangles, J. D. LAMBERT, G. A. H. ROBERTS, J. S. ROWLINSON and V. J. WILKINSON: Proc. Roy. Soc. Lond., Ser. A **196**, 113 (1949); squares, P. G. FRANCIS and M. L. MCGLASHAN: J. Chem. Phys. **20**, 1341 (1952); crosses, S. D. HAMANN and W. J. MCMANAMEY: J. Chem. Phys. **20**, 1341 (1952).

J. D. LAMBERT and his colleagues[4] have made very extensive measurements on organic vapours and vapour mixtures in a simple variable-volume apparatus

[1] E. E. ROPER: J. phys. Chem. **44**, 835 (1940); **45**, 321 (1941).

[2] G. A. BOTTOMLEY, D. S. MASSIE and R. WHYTLAW-GRAY: Proc. Roy. Soc. Lond., Ser. A **200**, 201 (1950).

[3] Trans. Faraday Soc. **48**, 101 (1952); **49**, 149, 351, 711 (1953). — Austral. J. Chem. **8**, 149 (1955).

[4] Trans. Faraday Soc. **37**, 421 (1941). — Proc. Roy. Soc. Lond., Ser. A **196**, 113 (1949); **200**, 566 (1950); **210**, 557 (1952); **219**, 490 (1953); **226**, 394 (1954); **232**, 537 (1955).

in which the vapour is confined to a calibrated gas-burette. In some of their measurements it was shown that absorption was not affecting the results by inserting glass-wool into the burette, thus changing its surface-volume ratio.

A recent attempt to use the expansion method of Burnett at low pressures has shown that it has no advantages over conventional methods[1].

The agreement which may be obtained between values of the second virial coefficient determined in different ways is shown in Fig. 21 and 22 for two substances, one of which has a high and one a low vapour pressure in the temperature range shown. It is seen that the relative and absolute agreement between the different methods is much better for sulphur hexafluoride, whose vapour pressure is 23.4 atmospheres at 25° C, than for benzene, whose vapour pressure is 0.12 atmospheres at 25° C.

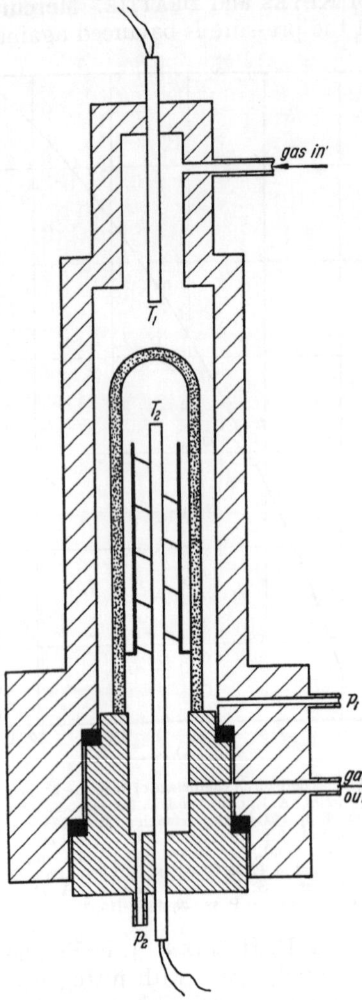

Fig. 23. Schematic diagram of the apparatus of Roebuck.

20. The measurement of the Joule and Joule-Thomson coefficients. The direct measurement of the adiabatic Joule coefficient, η, is very difficult as it vanishes as p^2 at zero pressure (Sect. 11). Few experiments have been described. Those of Keyes and Sears[2] led to little useful information beyond showing that this was not a practical way of measuring deviations from the perfect-gas laws. For example, the temperature change on expanding carbon dioxide from 16 atmospheres to zero pressure was only 0.1° C. Probably no direct measurement of the isothermal Joule coefficient, $(\partial E/\partial V)_T$, has ever been made but it is closely related, by (3.9), to the derivative $(\partial p/\partial T)_V$. This can be measured directly both for gases[3] and for liquids[4].

The Joule-Thomson coefficients are more readily measured and have been more widely studied, although even here the body of reliable results is not as large as for compressibility measurements. Most of the measurements of the adiabatic coefficient, μ, have been made by Roebuck and his colleagues at Wisconsin. Their apparatus is shown schematically in Fig. 23. The pressure drop is maintained by a long thimble of porous porcelain mounted vertically in a steel vessel. The inlet and outlet thermometers, T_1 and T_2, are placed near the top of the plug so as to be insulated by the flowing gas from heat conducted from other parts of the apparatus. The space around the plug is lightly packed

[1] D. Cook: Canad. J. Chem. **35**, 268 (1957).
[2] F. G. Keyes and F.W. Sears: Proc. Nat. Acad. Sci. U.S.A. **11**, 38 (1925).
[3] See Sect. 18.
[4] W. Westwater, H. W. Frantz and J. H. Hildebrand: Phys. Rev. **31**, 135 (1928). — J. H. Hildebrand and J. M. Carter: J. Amer. Chem. Soc. **54**, 3592 (1932).

with asbestos fibre to reduce convection, and the gas inside the plug is made to flow helically around the thermometer to avoid there being pockets of stagnant gas. The pressure drop across the plug is observed at p_1 and p_2. Accurate measurements, although simple in principle, demand much care in the layout and insulation of the apparatus. A simple test to show whether the observed ratio $(\Delta T/\Delta p)$ is the coefficient μ is to vary the flow-rate by using plugs of different porosity. If the apparatus is working properly this ratio is independent of the flow. The apparatus has no moving parts and contains no mercury and so may be used over a wide temperature range. The measurements of J. R. ROEBUCK and his colleagues[1] usually extend from -150 to $+300°$ C and up to 200 atmospheres.

The isothermal coefficient, φ, can be measured by warming the gas, during or after the expansion, back to its initial temperature, and measuring the heat-input. This coefficient has been measured for air[2], nitrogen[3], methane[4], steam[5], carbon dioxide[6], nitrous oxide[6], ethylene[6] and for some binary mixtures[7]. In most of these measurements a valve was used to produce the pressure drop. The valves used by EUCKEN[2,4] and by GUSAK[3] could only be adjusted by dismantling the apparatus but CHARNLEY, ISLES and TOWNLEY[6], describe a valve which is controlled from outside the thermostat, and so allows changes of flow-rate and pressure-drop to be made readily. The advantages of the isothermal method are, first, that it measures directly the departures from the perfect-gas laws and so unlike, μ, it does not need to be supplemented by measurements of C_p, and secondly, that the apparatus is less likely than an adiabatic one to give spurious results due to heat leaks. If the heater is close to the valve the temperature gradients can be kept much lower than is possible in an adiabatic apparatus. The earlier design of COLLINS and KEYES[3] is particularly good in this respect as the expansion was made truly isothermal by using the capillary metal tube which produced the pressure drop as the heater. The disadvantages of the isothermal method are that it can only be used within the inversion curve, where φ is negative, and that the rate of flow of the gas must be accurately measured.

IV. The critical region.

21. The thermodynamics of condensation and the critical point. A small part of that region of the fluid state which is open to experimental observation is formed of two coexisting parts, the liquid phase and the vapour phase. Fig. 12 shows well the relative smallness of this part. Above a certain critical temperature and pressure there is no sub-division of the fluid phase. The change from gas to liquid at temperatures below the critical is a simple first-order transition to which CLAPEYRON's equation applies,

$$\frac{dp}{dT} = \frac{\Delta H}{T\,\Delta V} = \frac{\Delta S}{\Delta V}, \qquad (21.1)$$

---

[1] Helium, Phys. Rev. **43**, 60 (1933); **45**, 332 (1934). — Argon, Phys. Rev. **46**, 785 (1934). — Nitrogen, Phys. Rev. **48**, 450 (1935). — Carbon dioxide, J. Amer. Chem. Soc. **64**, 400 (1942). — Air, Proc. Amer. Acad. Arts Sci. **60**, 537 (1925); **64**, 287 (1930). — Helium + Argon, J. Chem. Phys. **8**, 627 (1940). — Helium + Nitrogen, J. Amer. Chem. Soc. **60**, 341 (1938).

[2] A. EUCKEN, K. CLUSIUS and W. BERGER: Z. techn. Phys. **13**, 267 (1932).

[3] I. M. GUSAK: Phys. Z. Sowjet. **11**, 60 (1937). — S. C. COLLINS and F. G. KEYES: J. phys. Chem. **43**, 5 (1939).

[4] A. EUCKEN and W. BERGER: Z. techn. Phys. **15**, 369 (1934).

[5] S. C. COLLINS and F. G. KEYES: Proc. Amer. Acad. Arts Sci. **72**, 283 (1937).

[6] A. CHARNLEY, G. L. ISLES and J. R. TOWNLEY: Proc. Roy. Soc. Lond., Ser. A **218**, 133 (1953).

[7] A. CHARNLEY, J. S. ROWLINSON, J. R. SUTTON and J. R. TOWNLEY: Proc. Roy. Soc. Lond., Ser. A **230**, 354 (1955).

where the derivative is measured along the coexistence curve and where ΔH, ΔS and ΔV are all finite quantities. The vapour pressure curve ends at the critical point where ΔH, ΔS and ΔV all vanish in such a way that (dp/dT) remains finite.

The behaviour of the thermodynamic functions at pressures and temperatures below, at, and above the critical is shown schematically in Fig. 24a to k. The first three diagrams, a to c, are based on the direct observation of volume as a function of pressure and temperature. The behaviour of the remaining functions is calculated from these observations by the methods of Sect. 8. Fig. 24b shows that $(\partial p/\partial V)_T$ is negative but that it falls to zero during the course of condensation and at the critical point. If p can be represented as an analytical function of V along the critical isotherm, then it is clear that $(\partial^2 p/\partial V^2)_T$ must also vanish at the critical point and that $(\partial^3 p/\partial V^3)_T$ is probably negative. Such a fluid is mechanically stable[1]. If however the third derivative vanishes, the fourth must do likewise, and the fifth, or some higher odd derivative, must be the first non-vanishing derivative. The behaviour of real fluids near this point is not yet established beyond doubt (see next section). Fig. 24c shows that (dV/dT) and (d^2V/dT^2) along the coexistence curves are positive in the liquid and negative in the gas phase. The coefficient $(\partial V/\partial T)_p$ is always positive, for a normal fluid, and becomes infinite at the critical point. The coefficient $(\partial^2 V/\partial T^2)_p$ is positive in the liquid phase, discontinuous at the critical point and negative in the gas phase. [It is equal to (d^2B/dT^2) in the limit of zero density.] It follows from (3.26) that $(\partial C_p/\partial p)_T$ is negative in the liquid and positive in the gas phase, as shown in Fig. 24j and k.

The behaviour of the Gibbs free-energy is shown in Fig. 24d and e. The gas and liquid branches have different slopes and curvatures but meet at the condensation point. At the critical point in Fig. 24d the slopes of the two branches become the same but their curvatures become infinite as

$$\left(\frac{\partial^2 G}{\partial p^2}\right)_T = 1\Big/\left(\frac{\partial p}{\partial V}\right)_T. \qquad (21.2)$$

The curvature in Fig. 24e is

$$\left(\frac{\partial^2 G}{\partial T^2}\right)_p = -\left(\frac{\partial S}{\partial T}\right)_p, \qquad (21.3)$$

which, as shown in Fig. 24i, is also infinite at the critical point.

The heat content and the entropy behave similarly. The following derivatives are all infinite:

$$\left(\frac{\partial H}{\partial p}\right)_T = \varphi = -\mu C_p \quad (4.3); \qquad \left(\frac{\partial H}{\partial T}\right)_p = C_p,$$

$$\left(\frac{\partial S}{\partial p}\right)_T = -\left(\frac{\partial V}{\partial T}\right)_p \quad (3.16); \qquad \left(\frac{\partial S}{\partial T}\right)_p = \frac{C_p}{T}. \quad (3.19)$$

The infinity in C_p is shown in (j) and (k). The isothermal Joule-Thomson coefficient, φ, becomes infinite with C_p, but the adiabatic coefficient, μ, remains finite. The usual expression for this coefficient, Eq. (4.3),

$$\mu = \frac{T(\partial V/\partial T)_p - V}{C_p}$$

[1] See footnote 1, p. 10.

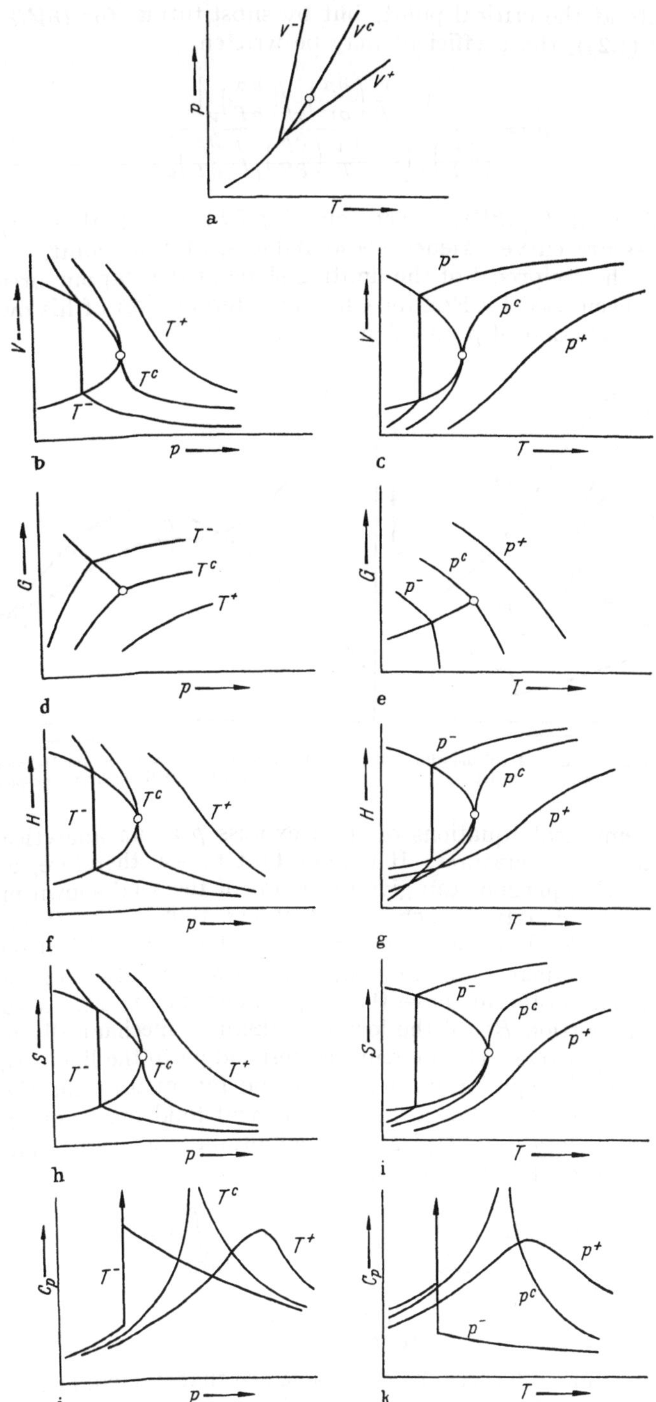

Fig. 24 a–k. Sketches of the changes of thermodynamic functions near the critical point, which is marked with a circle. The symbols T^-, T^c and T^+ indicate temperatures, etc., below, at and above the critical. The heat capacity has an infinity at, and below, T^c.

is indeterminate at the critical point, but by substituting for $(\partial V/\partial T)_p$ by (3.2) and for C_p by (3.23), the coefficient may be written,

$$\mu = \frac{\left[1 + \frac{V}{T}\left(\frac{\partial p}{\partial V}\right)_T \Big/ \left(\frac{\partial p}{\partial T}\right)_V\right]}{\left(\frac{\partial p}{\partial T}\right)_V \left[1 - \frac{C_V}{T}\left(\frac{\partial p}{\partial V}\right)_T \Big/ \left(\frac{\partial p}{\partial T}\right)_V^2\right]}. \tag{21.4}$$

At the critical point $(\partial p/\partial V)_T$ is zero and $(\partial p/\partial T)_V$ is equal to (dp/dT) along the vapour pressure curve. Hence (21.4) reduces, at this point, to the simple result that μ is the reciprocal of the limiting slope of the vapour pressure curve. This result has been used by ROEBUCK for the interpolation of his own measurements of μ as a function of p and T in the critical region.

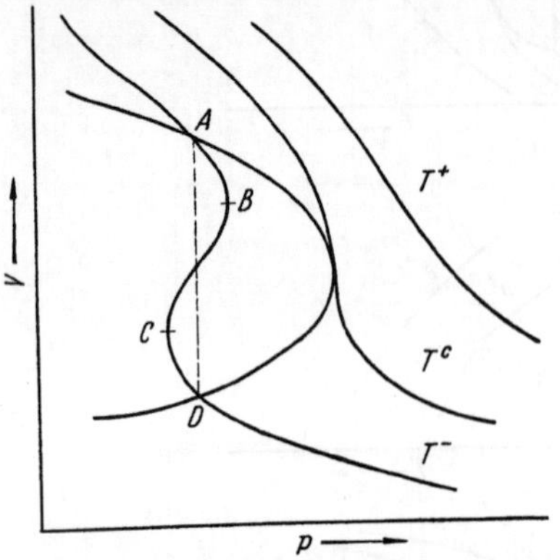

Fig. 25. The continuous variation of volume with pressure through the two-phase region.

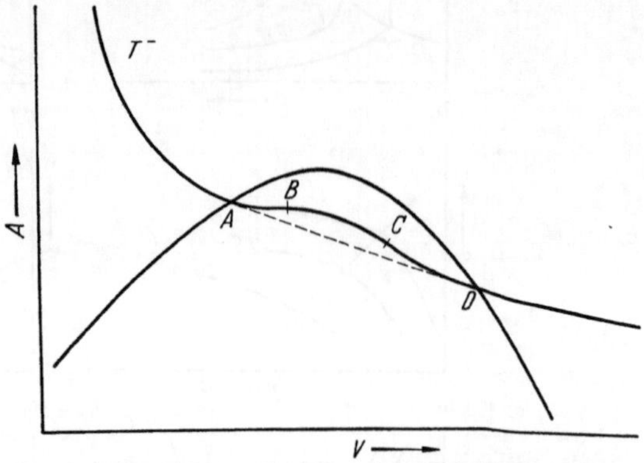

Fig. 26. The continuous variation of the Helmholtz free-energy as a function of volume through the two-phase region.

All simple empirical equations of state express p as an analytical function of V and T at all temperatures. It is important to see, therefore, under what circumstances such equations can give isotherms of the kind shown in Fig. 24b. It is found that for equations such as that of VAN DER WAALS (Sect. 25) there are three solutions for V for a given p and T. Above the critical temperature two of these are imaginary and one real, and below the critical temperature all three are real. The isotherms have the shape shown in Fig. 25. It can be seen at once that the portion BC of the lowest isotherm is mechanically unstable as $(\partial p/\partial V)_T$ is here positive. The pressure corresponding to the line AD, at which the two phases are in equilibrium, may be found by equating the chemical potentials of the two phases. For a single-component fluid these are equal to the molar Gibbs free-energy for two phases in equilibrium at a given p and T. Using superscripts l and g for liquid and gas

$$\mathrm{G}^l = \mathrm{G}^g + \int_{ABCD} \left(\frac{\partial \mathrm{G}}{\partial p}\right)_T dp = \mathrm{G}^g + \int_{ABCD} \mathrm{V}\, dp, \tag{21.5}$$

whence, for equilibrium,

$$\int_{ABCD} \mathrm{V}\, dp = 0. \tag{21.6}$$

This condition requires that AD be constructed so that it divides the continuous curve into two loops of equal area and was discovered by MAXWELL. It is seen, therefore, that the portions of the curve AB and CD are mechanically stable but are unstable with respect to the second phase. A similar division of the continuous curve may be made from a study of the Helmholtz free-energy as a function of V and T, (Fig. 26). Here the mechanically unstable portion BC is the portion where $(\partial^2 \mathrm{A}/\partial \mathrm{V}^2)_T$ is negative. The equilibrium volumes A and D are the points of contact of the common tangent to the continuous isotherm, for on such a tangent,

$$\left(\frac{\partial \mathrm{A}}{\partial \mathrm{V}}\right)_T^A = \left(\frac{\partial \mathrm{A}}{\partial \mathrm{V}}\right)_T^D \tag{21.7}$$

or

$$p^A = p^D, \tag{21.8}$$

and

$$\mathrm{A}^A - \mathrm{A}^D = \mathrm{V}^A \left(\frac{\partial \mathrm{A}}{\partial \mathrm{V}}\right)_T^A - \mathrm{V}^D \left(\frac{\partial \mathrm{A}}{\partial \mathrm{V}}\right)_T^D \tag{21.9}$$

or

$$\mathrm{G}^A = \mathrm{G}^D. \tag{21.10}$$

If this representation of A as a continuous analytical function of V and T in both stable and unstable regions is allowable then the shape of the coexistence curve near the critical point may be calculated by means of a double Taylor expansion about this point,

$$A = A^C + \sum_{n=1} \sum_{m=0}^{n} \frac{1}{n!\,(n-m)!} (V - V^c)^n (T - T^c)^{n-m} \left(\frac{\partial^n A}{\partial V^n \, \partial T^{n-m}}\right)^c. \tag{21.11}$$

A similar expansion may be written for the pressure by differentiating A with respect to volume. This may then be simplified by putting

$$\left(\frac{\partial p}{\partial V}\right)_T^c = 0, \qquad \left(\frac{\partial^2 p}{\partial V^2}\right)_T^c = 0 \tag{21.12}$$

leaving for the first non-vanishing terms,

$$\left.\begin{aligned} p = p^c + (T - T^c)\left(\frac{dp}{dT}\right)^c + (V - V^c)(T - T^c)\left(\frac{\partial^2 p}{\partial V \, \partial T}\right)^c + \\ + \frac{1}{2}(T - T^c)^2 \left(\frac{d^2 p}{dT^2}\right)^c + \frac{1}{6}(V - V^c)^3 \left(\frac{\partial^3 p}{\partial V^3}\right)_T^c. \end{aligned}\right\} \tag{21.13}$$

The other second-order terms vanish and the third-order terms with $(T - T^c)$ as factors have been omitted as small compared to the second-order. Expressing both p^g and p^l by this equation, and equating them at constant temperature gives

$$\left.\begin{aligned} (V^g - V^c)(T - T^c)\left(\frac{\partial^2 p}{\partial V \, \partial T}\right)^c + \frac{1}{6}(V^g - V^c)^3 \left(\frac{\partial^3 p}{\partial V^3}\right)_T^c \\ = (V^l - V^c)(T - T^c)\left(\frac{\partial^2 p}{\partial V \, \partial T}\right)^c + \frac{1}{6}(V^l - V^c)^3 \left(\frac{\partial^3 p}{\partial V^3}\right)_T^c, \end{aligned}\right\} \tag{21.14}$$

whence

$$\left.\begin{aligned} (T - T^c)\left(\frac{\partial^2 p}{\partial V \, \partial T}\right)^c + \\ + \frac{1}{6}\left[(V^g - V^c)^2 + (V^g - V^c)(V^l - V^c) + (V^l - V^c)^2\right]\left(\frac{\partial^3 p}{\partial V^3}\right)_T^c = 0. \end{aligned}\right\} \tag{21.15}$$

Substitute

$$V^c = \alpha (V^g + V^l) \tag{21.16}$$

to give

$$\left.\begin{aligned} (T - T^c)\left(\frac{\partial^2 p}{\partial V \, \partial T}\right)^c + \\ + \frac{1}{6}\left[(V^g - V^l)^2 (1 - 3\alpha + 3\alpha^2) + 3 V^g V^l (1 - 2\alpha)^2\right]\left(\frac{\partial^3 p}{\partial V^3}\right)_T^c = 0. \end{aligned}\right\} \tag{21.17}$$

Taking this to the limit $T = T^c$, it follows that $\alpha = \frac{1}{2}$ to this approximation, or

$$(V^g - V^l)^2 = \frac{-24\,(T - T^c)\,(\partial^2 p/\partial V\,\partial T)^c}{(\partial^3 p/\partial V^3)_T^c}. \tag{21.18}$$

From the observed shape of the isotherms it is clear that both derivatives are negative and so the right-hand-side of this equation is positive if T is below T^c. If $(\partial^3 p/\partial V^3)_T^c$ and $(\partial^4 p/\partial V^4)_T^c$ vanish, then $(\partial^5 p/\partial V^5)_T^c$ is negative and an equation similar to (21.18) can be derived but with $(V^g - V^l)^4$ proportional to $(T^c - T)$. Thus the shape of the coexistence curve near the critical point depends on the order of the first non-vanishing derivative. It must, however, be emphasised that these conclusions all rest on the validity of the expansion (21.11). There is reason to believe that if the Helmholtz free-energy could be calculated exactly by the method of classical statistical mechanics for an infinitely large assembly, it would show the true discontinuities at A and D of Fig. 26 with a straight line joining them[1]. Nevertheless the analysis above is well worth while as all empirical equations of state and all approximate statistical theories require a continuous variation of A with V through the unstable regions. Experimentally[2] it is found that the difference $(V^g - V^l)$ varies as the cube-root of $(T^c - T)$ in the neighbourhood of the critical point, so that neither the square nor the fourth-root is very close to the facts.

22. The measurement of critical constants. The classical view of the critical point is that it is a state of the fluid, described by the three critical constants, T^c, V^c and p^c, at which two following conditions hold:

1. The densities of the two coexisting phases are equal, and

2. the derivatives $(\partial p/\partial V)_T$ and $(\partial^2 p/\partial V^2)_T$ vanish but $(\partial^3 p/\partial V^3)_T$ remains finite and is negative.

It has not been easy to establish experimentally that the two states defined by these conditions are identical. The difficulty arises from the unusual properties of a fluid in the critical region. As the densities of the two phases become more nearly equal the dividing meniscus becomes very hazy and the exact point of its disappearance is hard to determine. At the critical point the heat capacity C_p is infinite, and C_V is very large, so it is hard to reach thermal equilibrium. Even slow changes of pressure and temperature are more nearly adiabatic than isothermal. This leads to hysteresis effects on making slow cyclic changes of pressure and temperature. As $(\partial p/\partial V)_T$ vanishes at the critical point, the compressibility becomes infinite and so the earth's gravitational field is large enough to produce appreciable density gradients in a vessel a few centimetres high. In the critical region there are also large spontaneous fluctuations of the density over macroscopic distances which lead to the scattering of light—the "critical opalescence". All these effects make unusually difficult precise measurements on a system at equilibrium, and good agreement between critical constants measured in different ways is hard to achieve.

The critical temperature has been measured for more substances than the other constants. KOBE and LYNN [6], in a very extensive review of critical properties in 1953, report that it had been measured for 220 substances, excluding mixtures. Most of these measurements were made by observing the temperature at which the meniscus vanished in a system maintained at an overall density

[1] L. VAN HOVE: Physica, Haag **15**, 951 (1949).

[2] H. H. LOWRY and W. R. ERICKSON: J. Amer. Chem. Soc. **49**, 2729 (1927). — E. A. GUGGENHEIM: J. Chem. Phys. **13**, 253 (1945). — M. A. WEINBERGER and W. G. SCHNEIDER: Canad. J. Chem. **30**, 422 (1952).

approximately equal to the critical. If a sealed tube containing liquid and vapour is heated uniformly then one of three things may happen. If the overall density is less than the critical then the meniscus will fall as the liquid evaporates. The pressure will follow the vapour pressure curve from A to B in Fig. 27. When there is no more liquid left the pressure will rise more slowly along the gas-isochore, BC. If the density is greater than the critical the meniscus will rise until the whole tube is full of liquid at D. Further heating produces a very rapid rise in pressure along the liquid-isochore, DE. If the density is close to the critical then the meniscus will rise at first until it is near the centre of the tube where it will remain until the critical point F is reached. Here the meniscus

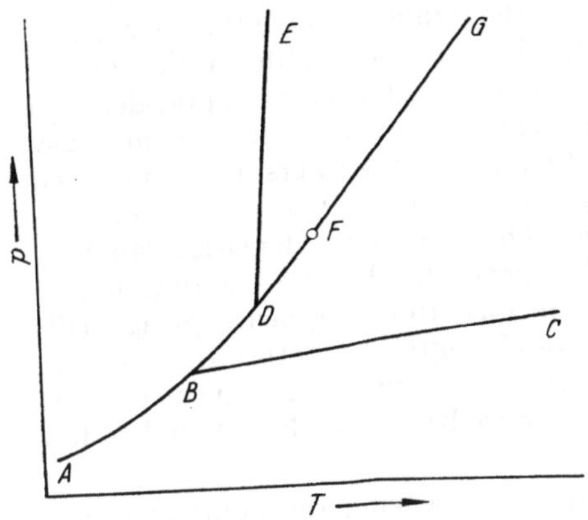

Fig. 27. The pressure in a sealed tube heated beyond the critical temperature.

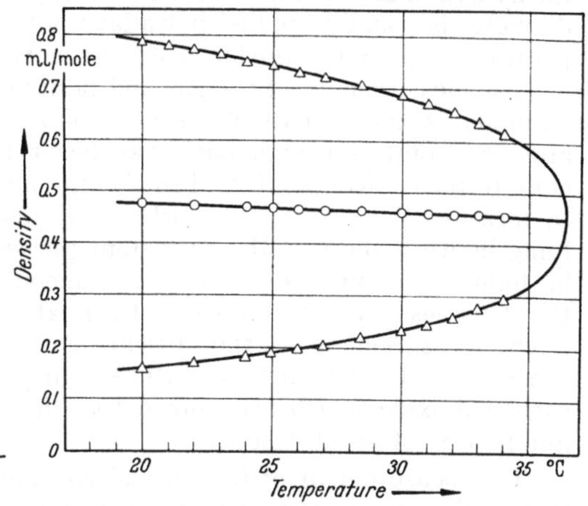

Fig. 28. The orthobaric densities of nitrous oxide [D. Cook: Trans. Faraday Soc. **49**, 716 (1953)]. The triangles are the experimental points and the circles are the means of the orthobaric densities.

vanishes and the pressure will rise uniformly along the critical isochore, FG. As the density difference between the two phases varies only as $(T^c - T)^{\frac{1}{3}}$ no great care is needed to load the tube exactly to the critical density in order to obtain the critical temperature to a few hundredths of a degree. A slight error in loading only causes the meniscus to vanish a little above or a little below the middle of the tube. Conversely, this method cannot be used for measuring the critical density at all accurately.

If the tube is not sealed but is open at its lower end to a reservoir of mercury, then the density may be altered at will and the pressure measured at the same time. This was the apparatus used in the original work of Andrews[1] and later by Young[2]. It has not been significantly altered since. The critical density can be measured in such an apparatus by observing the gas and liquid densities at temperatures from about 3 to 50° C below T^c, and extrapolating to the critical point. The extrapolation is made by using the law of rectilinear diameters discovered by Cailletet and Mathias[3] in 1886. They found that if they drew diameters across a graph (Fig. 28) of the orthobaric densities as a function of temperature then the mid-point of these diameters was very close to a straight

[1] T. Andrews: Proc. Roy. Soc. Lond., Ser. A **18**, 42 (1869).
[2] S. Young: Proc. Roy. Soc. Dublin **12**, 374 (1909/10).
[3] L. Cailletet and E. Mathias: C. R. Acad. Sci., Paris **102**, 1202 (1886); **104**, 1563 (1887).

line. They thought at first that this line was parallel to the temperature axis, but it is now realised that it has a slight slope and can be represented by

$$(\varrho^g + \varrho^l)/2 = a - b\,T \tag{22.1}$$

where ϱ is the density. This line cuts the coexistence curve at the critical point. Therefore

$$(\varrho^g + \varrho^l)/2 = \varrho^c + b\,(T^c - T)\,. \tag{22.2}$$

GUGGENHEIM[1] has shown that the coefficient b is close to $(3\,\varrho^c/4\,T^c)$ for the inert gases, neon, argon, krypton and xenon. The numerical factor is smaller than $\frac{3}{4}$ for hydrogen and helium and a little larger for other gases. The term rectilinear diameter is usually applied nowadays not to the diameters on the density-temperature graph but to the locus of their centres—the straight line defined by (22.2).

The measurement of the gas and liquid densities may be made by finding the volume of a known mass of gas at its bubble-point (the orthobaric liquid-density) and dew-point (the orthobaric gas-density). These measurements are not easy close to the critical point, and so YOUNG, and many later workers, have preferred to measure the relative proportions of gas and liquid at two overall-densities lying between those of the coexistent phases. The orthobaric densities can then be found by solving a pair of simultaneous equations. MAASS[2] has determined the densities directly by means of a float suspended from a quartz spring. Different measurements of the critical density rarely agree to better than 1%, in contrast to the critical temperatures and pressures of simple substances such as carbon dioxide which are known to a few hundredths of a degree and a few hundredths of an atmosphere.

The measurement of the critical constants by observing the point of inflection of the critical isotherm is much more difficult and attempts to make such measurements have led to some notable disagreements. This method has been used mainly by J. A. BEATTIE[3], by O. MAASS, and, more recently, by W. G. SCHNEIDER[4] and his colleagues in a series of papers which are outstanding for the precision and variety of the experimental arrangements. The critical volume is particularly hard to measure from the point of inflection. For example, BEATTIE, POFFENBERGER and HADLOCK[5] found for propane a critical density of 5.13 moles/l from the point of inflection of the critical isotherm, but MEYERS[6] showed that an extrapolation from their orthobaric densities by means of the law of rectilinear diameters gave a critical density of 4.98 moles/l. KOBE and LYNN [*6*] recommend the latter figure.

Because of the slow attainment of equilibrium near the critical point much of the older work is of little value for determining the precise nature of the singularity. Often the fluid sample was not stirred, or, if stirred, it was not kept at a constant temperature and pressure for a long enough time. It was the work of MAASS[2] which first raised serious doubts about the validity of the classical description of the critical point. He found that on heating a fluid through the

---

[1] See footnote 2, p. 42.

[2] O. MAASS: Chem. Rev. **23**, 17 (1938). — S. G. MASON, S. N. NALDRETT and O. MAASS: Canad. J. Res. B **18**, 103 (1940). — S. N. NALDRETT and O. MAASS: Canad. J. Res. B **18**, 118 (1940).

[3] See footnote 2, p. 31.

[4] Changements de Phases. Soc. chim. Phys., Paris 1952, p. 69. — J. Phys. Chem. **54**, 1323 (1950); **55**, 532 (1951). — Canad. J. Chem. **29**, 699 (1951); **30**, 422, 847 (1952). — J. Chem. Phys. **21**, 2080 (1953). — Canad. J. Chem. **32**, 98, 164 (1954).

[5] J. A. BEATTIE, N. POFFENBERGER and C. HADLOCK: J. Chem. Phys. **3**, 96 (1935)

[6] C. H. MEYERS: J. Res. Nat. Bur. Stand. **29**, 157 (1942).

critical point the density gradient persisted after the meniscus had vanished. This gradient could not be destroyed by mechanical stirring but was destroyed by heating the lower end of the tube to a higher temperature than the upper end. Once destroyed, the gradient did not re-appear on cooling the tube uniformly. The temperatures at which the meniscus appeared and disappeared did not always coincide. He found that with ethane and ethylene, in stirred tubes, the top of the coexistence curve was not parabolic, as VAN DER WAALS' theory requires, or even cubic, as is suggested by measurements made some degrees below the critical, but was truncated and had an appreciable flat portion. This portion covered the densities between 0.195 and 0.225 gm/ml for ethane. This flat top coincided with the temperature at which the meniscus vanished but not with the highest temperature at which $(\partial p/\partial V)_T$ vanished (Fig. 29). The region in which this derivative was zero enclosed the coexistence curve and, in ethylene, extended to a temperature 0.7° C higher than its flat top. However it is now thought very doubtful if any of these observations are a description of the true equilibrium behaviour of a real fluid in the absence of a gravitational field. SCHNEIDER and his colleagues have shown by measurements which are more accurate than those of MAASS that true equilibrium can be achieved only by maintaining the temperature constant to within 0.001° C and by continuously stirring the fluid. Under these conditions the coexistence curves of xenon and sulphur hexafluoride have rounded tops. The effect of gravity was eliminated by studying the curve with the tube in both horizontal and vertical positions. The coexistence curve found in the latter position was always flatter than that found in the former position, and lay outside it. The former is undoubtedly the more correct one, for in a tube of negligible height no gravitational density gradient can be established. WHITEWAY and MASON[1] have confirmed these results for ethane and ethylene. By using the radioactive isotope Xe^{127} SCHNEIDER was able to show that the equilibrium density gradients in a vertical tube are no greater than those to be expected from the gravitational field when correct allowance is made for the very great compressibility in this region. He showed also that the meniscus appears and disappears at the same temperature, and that above this temperature $(\partial p/\partial V)_T$ is always negative. The only particular in which he did not verify the classical description was in the precise shape of the rounded top and in the shape of the critical isotherm. HABGOOD and SCHNEIDER[2] found that both $(\partial^3 p/\partial V^3)_T$ and $(\partial^4 p/\partial V^4)_T$ probably vanish at the critical point. If $(\partial^5 p/\partial V^5)_T$ were the first non-vanishing derivative then the coexistence curve would be a quartic (as shown above), a result which is not found in practice. Thus it is quite possible that all the derivatives of p with respect to V vanish exactly at the critical point, so making invalid the Taylor expansion (21.11) and the conclusions based on it.

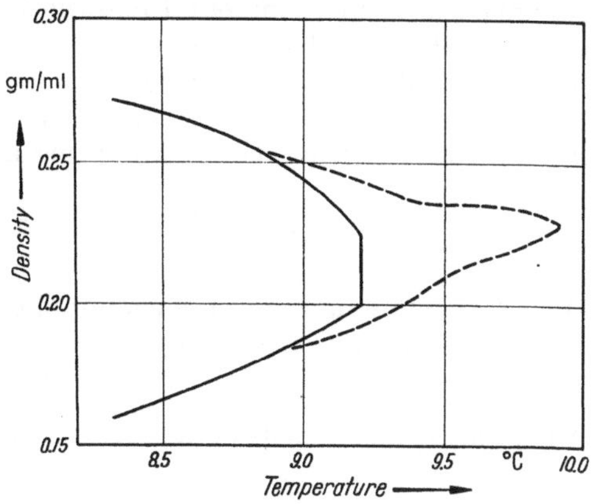

Fig. 29. The orthobaric densities of ethylene, after NALDRETT and MAASS. The dashed curve is the boundary of the region in which $(\partial p/\partial V)_T$ was found to be zero.

[1] S. G. WHITEWAY and S. G. MASON: Canad. J. Chem. **31**, 569 (1953).
[2] H. W. HABGOOD and W. G. SCHNEIDER: Canad. J. Chem. **32**, 98 (1954).

23. Density fluctuations and the scattering of light. Near the critical point large and rapid changes of density occur spontaneously over distances comparable with the wave-length of light. These fluctuations cause such strong scattering of light that the fluid can be opaque at the critical point and can change colour very rapidly near this point, usually between shades of yellow and brown. There are similar fluctuations in the local energy which lead to a strong maximum of C_V at the critical point.

The theory of fluctuations in an equilibrium system can only be developed by using the methods of statistical mechanics, and so lies outside the scope of this chapter. However some of the results of such theories[1] are necessary for a discussion of the experimental observations and are quoted here without proof.

The n-th order fluctuation of a macroscopic property X is defined as $\overline{(x-\bar{x})^n}$, where a bar denotes an average. The first-order fluctuation vanishes, by definition, and it is the second-order one which is usually referred to as the fluctuation, without qualification. The relative fluctuation is defined by

$$\delta_x = \overline{\left(\frac{x-\bar{x}}{\bar{x}}\right)^2} = \frac{\overline{x^2} - 2\overline{x\bar{x}} + \bar{x}^2}{\bar{x}^2} = \frac{\overline{x^2} - 2\left[\overline{(x-\bar{x})}\,\bar{x} + \bar{x}^2\right] + \bar{x}^2}{\bar{x}^2} = \frac{\overline{x^2} - \bar{x}^2}{\bar{x}^2}. \qquad (23.1)$$

The relative fluctuation of the energy of one mole of a pure substance maintained at a temperature T is given by

$$\delta_E = \mathrm{k}T^2\, \mathrm{C}_V/\mathrm{E}^2. \qquad (23.2)$$

If the energy of the substance is distributed amongst ν square terms then

$$\overline{\mathrm{E}} = \nu \mathrm{k}T, \qquad \mathrm{C}_V = \nu \mathrm{k} \qquad (23.3)$$

and so

$$\delta_E = 1/\nu. \qquad (23.4)$$

The number ν will be of the order of Avogadro's number, N, and so in a classical perfect gas, to which (23.3) apply, it is permissible to assign fixed values to T and $\overline{\mathrm{E}}$ at the same time. A fluid of two coexisting phases can absorb large amounts of energy for a small rise in temperature. This property persists at the critical point where the energy is not sharply defined by the volume and temperature. Under these conditions δ_E, and so C_V, are large. Fig. 30 shows the results of Michels and Strijland[2] for carbon dioxide. The calorimetric values of C_V are very large but the maximum has not been located exactly. The values of C_V calculated from the curvature of the isochores (Sect. 8), although also showing a maximum, are not in good agreement with the calorimetric values.

Spatial fluctuations in the density may be expressed by δ_n where n is the fraction of a mole in a sub-volume V_s of an assembly of one mole of total volume V. The average value of n is $(\mathrm{V}_s/\mathrm{V})$. The relative fluctuation is given by

$$\delta_n = \frac{-\mathrm{k}T}{\mathrm{V}_s^2(\partial p/\partial \mathrm{V}_s)_T} = \frac{1}{\bar{n}N}\left[\frac{-\mathrm{R}T}{\mathrm{V}^2(\partial p/\partial \mathrm{V})_T}\right]. \qquad (23.5)$$

For a perfect gas this equation reduces to

$$\delta_n = \frac{1}{\bar{n}N}, \qquad (23.6)$$

[1] R. C. Tolman: Principles of Statistical Mechanics, p. 629–649. Oxford 1938. — J. O. Hirschfelder, C. F. Curtiss and R. B. Bird: Molecular Theory of Gases and Liquids, p. 121–128, 891–898. New York: Wiley 1954. — J. Yvon: Actual. sci. indust. **1937**, No. 542 and 543.

[2] A. Michels and J. Strijland: Physica, Haag **18**, 613 (1952).

thus showing that the relative fluctuation is only of the order of $1/N$ if the sub-volume is macroscopic. The derivation[1] of (23.5) is based on a Taylor expansion of the Helmholtz free-energy of the sub-volume V_s, at constant T and V_s, about $A(\bar{n})$ as a function of $(n-\bar{n})$. All terms have been neglected beyond those of the second order, whose coefficient is $(\partial^2 A/\partial n^2)_T$ or $[-V_s^2(\partial p/\partial V_s)_T/\bar{n}^2 N^2]$. At the critical point the second and third-order terms vanish and (23.5) no longer holds. By using the fourth-order terms it may be shown that,

$$\delta_n^c = \frac{2\sqrt{6}\,\Gamma(\frac{3}{4})}{\Gamma(\frac{1}{4})}\left[\frac{-RT}{V^4(\partial^3 p/\partial V^3)_T}\right]^{\frac{1}{4}}\frac{1}{\bar{n}^{\frac{1}{2}}N^{\frac{1}{4}}}. \tag{23.7}$$

This expression is finite at the critical point as it has been assumed that $(\partial^3 p/\partial V^3)_T$ is the first non-vanishing derivative. These fluctuations are of the order of $N^{\frac{1}{4}}$

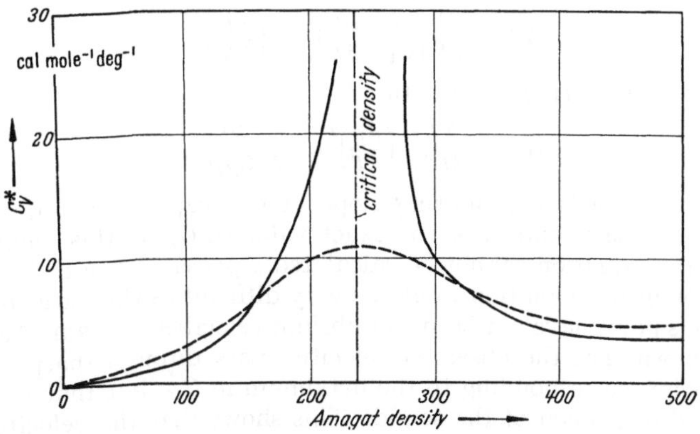

Fig. 30. The residual heat-capacity at constant volume of carbon dioxide at the critical temperature. The full curve epresents the calorimetric measurements of MICHELS and STRIJLAND and the dashed curve the values calculated from the curvature of the isochores.

larger than those of (23.6). Although (23.7) again depends on the validity of a Taylor expansion, there is no doubt of its qualitative correctness. It is this factor of $N^{\frac{1}{4}}$ which is responsible for the very strong opalescence of gas near its critical point.

The theory of the scattering of light (which in this context means electro-magnetic waves of long wavelength) by fluids in which fluctuations occur was developed by ORNSTEIN and ZERNIKE[2]. They found that the intensity of the scattered light should change from the usual fourth-power dependence on frequency found by RAYLEIGH for a dilute gas, to a second-power dependence in the critical region. However their theory predicts that the total scattering, integrated over all angles, becomes infinite at the critical point. This is due to the fact that the correlations of the density fluctuations between one sub-volume and another, which normally fall off very rapidly with their separation, now extend over the whole assembly. One cannot, under these conditions, deal with an infinitely large assembly but must carry integrations only to the boundary of the finite volume. PLACZEK[3] showed that the total scattering then remains finite but becomes a function of the total volume.

[1] See footnote 1, p. 46.

[2] L. S. ORNSTEIN and F. ZERNIKE: Proc. Acad. Sci. Amsterd. **17**, 793 (1914). — Phys. Z. **19**, 134 (1918); **27**, 761 (1926).

[3] G. PLACZEK: Phys. Z. **31**, 1052 (1930).

Precise experimental confirmation of this theory has never been obtained. The recent work of DRICKAMER and his colleagues[1] shows that the scattering from a sample of ethane or ethylene maintained at the critical density follows the fourth-power law when several degrees above the critical temperature and that the dependence on frequency becomes less as the critical temperature is approached. However, very near the critical point the opacity increases but the intensity of the scattered light decreases. This is undoubtedly due to multiple scattering, and it must be concluded that a direct test of the theory is very difficult, if not impossible, at the critical point.

The velocity of sound is considerably affected by the fluctuations which occur in the critical region. The equilibrium velocity, W, is given by (5.6), but this expression becomes indeterminate at the critical point. However by eliminating C_p by (3.23) this expression for W^2 may be transformed to[2],

$$W^2 = \frac{V^2}{M}\left[\frac{T}{C_V}\left(\frac{\partial p}{\partial T}\right)_V^2 - \left(\frac{\partial p}{\partial V}\right)_T\right], \tag{23.8}$$

which reduces at the critical point to

$$W^2 = \frac{T V^2}{M C_V}\left(\frac{dp}{dT}\right)^2 = \frac{T V^2}{M C_V \mu^2}, \tag{23.9}$$

where the derivative is the limiting slope of the vapour pressure curve. Unfortunately, as little is known of the exact value of C_V at this point, the right-hand side of this equation cannot be calculated *a priori*. Even if C_V were known the verification of this equation would be very difficult as the measured velocities of sound at this point have not been equilibrium velocities. HERGET[3] and SCHNEIDER[4] have shown that the observed velocities pass through sharp minima near the critical point, corresponding to the maximum in C_V, but that there is heavy absorption and dispersion of the sound. This shows that the velocity is not the equilibrium velocity, in the sense of Sect. 5, but that some relaxation process is taking place in the fluid with a period comparable with the reciprocal of the frequency of the sound, that is, about 10^{-6} sec. There is little doubt that this relaxation process is the formation and dissolution of large clusters of molecules in the compression and rarefaction parts of the sound wave.

V. Empirical equations of state.

24. The principle of corresponding states. The history of the empirical development of equations to represent the volumetric behaviour of gases is the most disappointing part of this subject. Equation after equation has been proposed, has found some application to a limited range of substances or conditions, and has then been discarded when its limitations became apparent. PARTINGTON [*10*] enumerates eighteen such equations in a list which *excludes* all those that still find some practical application. Before discussing any empirical equation it is useful to examine a principle which has retained its usefulness since it was first enunciated by VAN DER WAALS[5] in 1881.

[1] H. A. CATALDI and H. G. DRICKAMER: J. Chem. Phys. **18**, 650 (1950). — A. L. BABB and H. G. DRICKAMER: J. Chem. Phys. **18**, 655 (1950). — L. G. BLOSSER and H. G. DRICKAMER: J. Chem. Phys. **19**, 1244 (1951).

[2] C. F. CURTISS, C. A. BOYD and H. B. PALMER: J. Chem. Phys. **19**, 801 (1951).

[3] C. M. HERGET: J. Chem. Phys. **8**, 537 (1940).

[4] W. G. SCHNEIDER: J. Chem. Phys. **18**, 1300 (1950). — Canad. J. Chem. **29**, 243 (1951). — H. W. HABGOOD and W. G. SCHNEIDER: Canad. J. Chem. **32**, 164 (1954). — A. G. CHYNOWETH and W. G. SCHNEIDER: J. Chem. Phys. **19**, 1566 (1951); **20**, 1777 (1952). — See also H. TEILSCH and H. TANNEBERGER: Z. Physik **137**, 256 (1954).

[5] See J. R. PARTINGTON [*10*], p. 689 for the history of this principle.

The principle of corresponding states can be stated in the following form. The surfaces representing the volumetric behaviour of substances as a three-dimensional graph in p, V and T can all be reduced to one common surface by multiplying the three sets of coordinates by linear scale factors, that is by three factors for each substance. It follows that any empirical equation of state with three parameters conforms to this principle (unless such numbers as the exponents of V and T are taken as variable parameters). If the equation is to be constrained to approach that of the perfect gas at zero density by fixing the value of the gas-constant, R, then the number of adjustable parameters is two. The best known equation of this kind is that of VAN DER WAALS, and for many years the principle was associated solely with his equation. This was unfortunate, as many substances conform reasonably well to the principle under conditions where they do not obey VAN DER WAALS' equation. The principle has always been widely used by chemical engineers but its rehabilitation with physicists and chemists dates from PITZER's demonstration[1] that for an assembly which obeys the laws of classical statistical mechanics the principle holds exactly for substances whose molecules are all spherical and whose intermolecular potential energies are all of the same functional form and have only two adjustable parameters. GUGGENHEIM[2] showed that the heavier rare gases, argon, krypton and xenon, are a family of substances which, amongst themselves, conform to the principle with great accuracy in all three phases. Other groups, even if composed of chemically related substances, such as the paraffin hydrocarbons, do not conform so well to the principle, either amongst themselves or when compared with the inert gases.

In practice, it is usual to formulate the principle in such a way that the two adjustable scale factors for each substance are chosen from the three critical constants. That is, the p-V-T surfaces are made to coincide in the limit of zero density and, as well as possible, at the critical point. This choice is better than one based on, say, the triple-point constants as substances which conform reasonably well in the fluid states often disagree in the solid. Two of the critical constants, T^c and p^c, can be measured accurately and reasonably simply for a wide range of substances. The scale-factors are therefore usually chosen to be the reciprocals of these constants. The critical volume cannot be accurately measured and its use is rare. However it is often convenient to use a "pseudo-critical volume". This is an estimated volume calculated from p^c and T^c by assuming a value for the critical ratio, $(p^c V^c/\mathrm{R} T^c)$, which, by the principle, should be the same for all substances.

The simplest test that can be made of the principle is the constancy of the critical ratio. The values for 23 substances are shown in Table 1.

Table 1. *The critical ratio* $(p^c V^c/\mathrm{R} T^c)$, *from the constants recommended by* KOBE *and* LYNN [6].

Substance	Ratio	Substance	Ratio	Substance	Ratio
He	0.30	CO_2	0.275	$n-C_5H_{12}$	0.269
Ne	0.307	N_2O	0.272	$n-C_6H_{14}$	0.264
A	0.291	NH_3	0.242	$c-C_6H_{12}$	0.271
Kr	0.291	H_2O	0.230	C_6H_6	0.274
Xe	0.290	CH_4	0.289	C_2H_5OH	0.248
H_2	0.304	C_2H_6	0.285	$(C_2H_5)_2O$	0.261
N_2	0.291	C_3H_8	0.277	SF_6	0.275
CO	0.294	$n-C_4H_{10}$	0.274		

[1] K. S. PITZER: J. Chem. Phys. **7**, 583 (1939).

[2] E. A. GUGGENHEIM: J. Chem. Phys. **13**, 253 (1945).

These ratios may not always be accurate to the last figure shown, owing to errors in V^c, but they reflect clearly the usefulness and limitations of the principle of corresponding states. There is a group of simple substances, A, Kr, Xe, N_2, CO and CH_4, whose intermolecular force fields are spherical, or nearly so, and of similar functional form, for which the ratio is close to 0.291. Very light molecules, He, Ne and H_2, have larger values. This is known to be due to quantum effects which are important with these molecules but negligible with heavier ones. Most substances show a lower value for the ratio, a difference which increases with the complexity or polarity of the molecule. This variation is clearly shown by the series of normal paraffins. A highly symmetric but polyatomic molecule, such as SF_6, also shows a low value, thus showing that its intermolecular force-field cannot be similar to those of the rare gases. These variations are typical of those shown by all the other reduced thermodynamic properties of these substances and indicate the accuracy which can be expected for predictions based directly on the principle. The widely used charts of HOUGEN and WATSON[1] are based on average values of the reduced properties of hydrogen, nitrogen, carbon dioxide, ammonia, methane, propane and n-pentane. They have plotted graphs of the compressibility factor, Z, the residuals H^*, S^* and C_p^*, the isothermal Joule-Thomson coefficient, φ, and similar functions for orthobaric liquid states. They choose reduced pressure and reduced temperature as the independent variables. They estimate the maximum errors of prediction as 15% in Z and 30% in H^* and S^*. GUGGENHEIM[2] has made a correlation of the reduced second and third virial coefficients, that is, B/V^c and $C/(V^c)^2$, of a number of gases and has shown that these conform approximately to the principle but that here again the deviations are greater than the comparatively large experimental errors, particularly for temperatures below the critical.

It has been realised recently that the accuracy of such predictions can be greatly improved by using the fact that the variations from one substance to another are not random. If the heavier inert gases are taken as the standard of normal behaviour then it is found that any reduced thermodynamic property may be used to place other gases in a definite order of increasing departures from this norm. To a good approximation this order is the same whatever property is used to establish it. The critical ratios in Table 1 show the order quite clearly. There are positive deviations for the light molecules, and increasingly negative ones for the increasingly more complex, or more polar, molecules. For the latter class it is thus possible to assign to each substance one parameter which is a measure of its departure from the group of normal substances. This was first proposed by COOK and ROWLINSON[3], who interpreted the parameters in terms of the intermolecular force-fields. In practice, the parameter is best determined from the reduced vapour pressure curve—a property which is widely and accurately measured. Once determined, the parameter may be used to calculate any of the reduced properties of the substance if (1) the property is known for the normal substances, and (2) the departure from the norm, per unit of parameter, is established theoretically or empirically. RIEDEL[4] and, independently, PITZER and his colleagues[5], discovered these correlations empirically. However they use

[1] O. A. HOUGEN and K. M. WATSON: Chemical Process Principles, Part 2, Chap. 12. New York: Wiley 1947.

[2] E. A. GUGGENHEIM: Aust. Rev. Pure Appl. Chem. **3**, 1 (1953).

[3] D. COOK and J. S. ROWLINSON: Proc. Roy. Soc. Lond., Ser. A **219**, 405 (1953). — J. S. ROWLINSON: Trans. Faraday Soc. **50**, 647 (1954).

[4] L. RIEDEL: Chem.-Ing.-Techn. **26**, 83, 259, 679 (1954); **27**, 209 (1955).

[5] K. S. PITZER: J. Amer. Chem. Soc. **77**, 3427 (1955). — K. S. PITZER, D. Z. LIPPMAN, R. F. CURL, C. M. HUGGINS and D. E. PETERSON: J. Amer. Chem. Soc. **77**, 3433 (1955). — K. S. PITZER and R. F. CURL [*12*].

pressure and temperature as variables for the properties of the one-phase system which is inconvenient near the critical point. The theoretical treatment[1] leads to a simple method of calculating Z, as a function of density and temperature, which needs only one table—that for the normal substances. If the parameter, determined from the vapour pressure, which measures the degree of departure from normal is denoted δ^c, then to a good approximation,

$$Z[\vartheta, \varrho] = Z_*[\{\vartheta - 2\delta^c(1-\vartheta)\}, \{\varrho - \delta^c \varrho (1-\vartheta)/\vartheta\}], \quad (24.1)$$

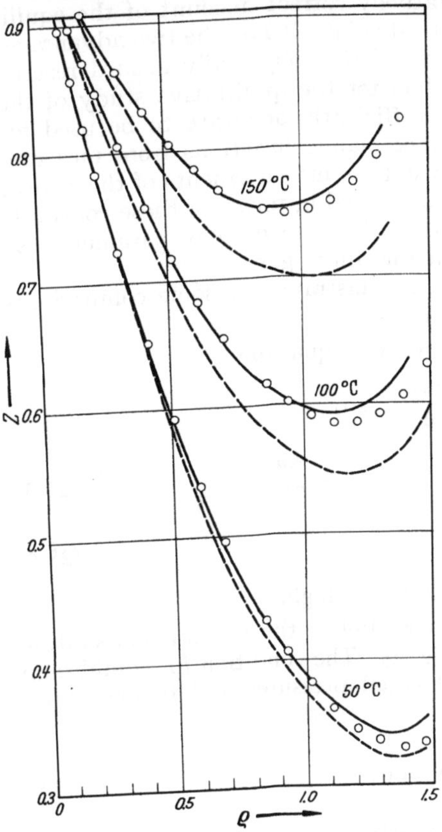

Fig. 31. The compressibility factor, Z, of carbon dioxide as a function of the pseudo-reduced density, ϱ'. The circles are the measurements of MICHELS [7]. The dashed curves are calculated from the properties of the inert gases by means of the principle of corresponding states, and the full curves calculated from these curves by using Eq. (24.1).

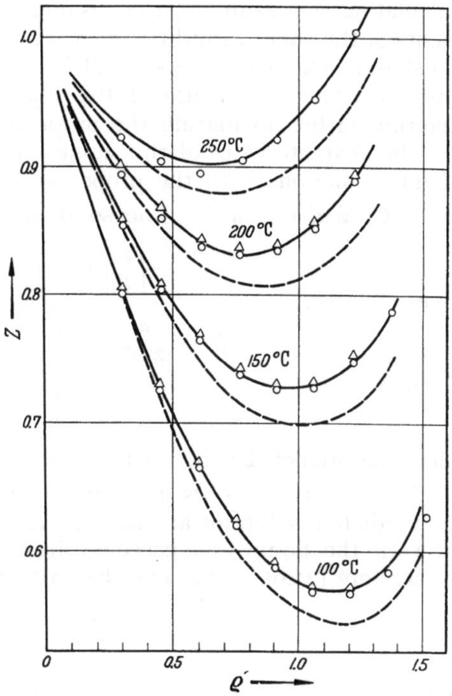

Fig. 32. The compressibility factor of ethane. The circles are the measurements of J. A. BEATTIE, C. HADLOCK and N. POFFENBERGER, J. Chem. Phys. **3**, 93 (1935), and the triangles are the measurements of H. H. REAMER, R. H. OLDS, B. H. SAGE and W. N. LACEY, Industr. Engng. Chem. **36**, 956 (1944). Other conventions are as in Fig. 31.

where ϑ and ϱ are the reduced temperature and pseudo-reduced density, and where the asterisk denotes the normal substance. Examples of the calculation of Z by this method are shown in Fig. 31 and 32. The departure of the reduced vapour-pressure of carbon dioxide from that of the inert gases is twice the departure of ethane. It is seen from these graphs that the departures of Z, at a given reduced temperature and density, are also about twice as large. This method can be used to calculate the compressibility factor with an accuracy of about 1% up to densities of about the critical. It is therefore very much more accurate than the direct use of the principle with no allowance for the individual departures of different gases.

[1] J. S. ROWLINSON [*11*].

25. Van der Waals and similar equations of state. The empirical equation of state of VAN DER WAALS is

$$\left(p + \frac{a}{V^2}\right)(V - b) = RT, \tag{25.1}$$

where a and b are two positive adjustable parameters. The equation is historically important as the first which gave a qualitatively correct account of the equilibrium between liquid and gas and of the critical point. It has the two advantages of being easy to manipulate and of never predicting physically absurd results. These have made it, in the past, a useful tool for the qualitative study of the behaviour of fluids. However, it is rarely sufficiently accurate to be used for calculating, for example, the residual thermodynamic functions from measurements of the compressibility, or, what is worse, from measurements of the critical constants. The parameters a and b may be expressed in terms of these constants and the principal mis-use of this equation has been to employ parameters so determined for calculating the properties of the dilute gas.

The characteristics of this equation may be illustrated by some comparisons of its predictions with the properties of real gases.

α) Critical constants. The solution of the two equations

$$\left(\frac{\partial p}{\partial V}\right)_T^c = 0, \qquad \left(\frac{\partial^2 p}{\partial V^2}\right)_T^c = 0 \tag{25.2}$$

gives

$$p^c = \frac{a}{27b^2}, \qquad V^c = 3b, \qquad RT^c = \frac{8a}{27b}, \tag{25.3}$$

whence

$$\left(\frac{pV}{RT}\right)^c = \frac{3}{8}. \tag{25.4}$$

Thus the predicted value of the critical ratio is too high.

β) Orthobaric volumes and vapour pressures. Below the critical temperature the predicted isotherms are sigmoid, as in Fig. 25. The condition for equilibrium between the liquid and gaseous phases, at equal pressures and temperatures, is therefore (21.6), which may be written,

$$\int_g^l V\left(\frac{\partial p}{\partial V}\right)_T dV = 0. \tag{25.5}$$

Substituting for $(\partial p/\partial V)_T$ from (25.1) and eliminating a and b by (25.3) gives

$$\ln\left[\frac{3V^l - V^c}{3V^g - V^c}\right] + \frac{V^c}{3V^g - V^c} - \frac{V^c}{3V^l - V^c} + \frac{9V^c\, T^c (V^g - V^l)}{4\,T\,V^l\, V^g} = 0. \tag{25.6}$$

The temperature may by eliminated from this equation by substituting from (25.1) into

$$p^l = p^g \tag{25.7}$$

and the equation re-arranged to give

$$\beta^l = \frac{(r+1)\ln r - 2(r-1)}{r^2 - 2r\ln r - 1}, \tag{25.8}$$

$$\beta^g = r\beta^l, \tag{25.9}$$

where

$$\beta^l = \frac{V^c}{3V^l - V^c}, \tag{25.10}$$

$$\beta^g = \frac{V^c}{3V^g - V^c} \tag{25.11}$$

and r is defined by (25.9). The orthobaric volumes may now be found by choosing an arbitary value of the ratio $r(\leq 1)$, calculating V^l and V^g from Eqs. (25.8) to (25.11), and the temperature and vapour pressure by substituting these volumes in (25.1) to give a pair of linear simultaneous equations. The results of these calculations are shown in Fig. 33 and 34. It is seen that qualitatively this equation predicts curves of the correct shape and that the orthobaric densities conform well to the law of rectilinear diameters. The curvature of the density curve near the critical point is given by

$$\left(\frac{V^g - V^l}{V^c}\right)^2 = 16\left(1 - \frac{T}{T^c}\right). \tag{25.12}$$

This equation may be obtained by expanding (25.6) about the critical volume. The derivative $(\partial^3 p/\partial V^3)^c_T$ does not vanish for VAN DER WAALS' equation and so (25.12) can be derived alternatively by substituting for the derivatives in (21.18). The graph of the orthobaric densities of the rare gases[1], which is also shown in Fig. 33, is similar to that predicted by VAN DER WAALS. The most noticeable discrepancy, apart from the intervention of the solid state below $T/T^c = 0.56$, is the cubic dependence on temperature of $(V^g - V^l)$ near the critical point.

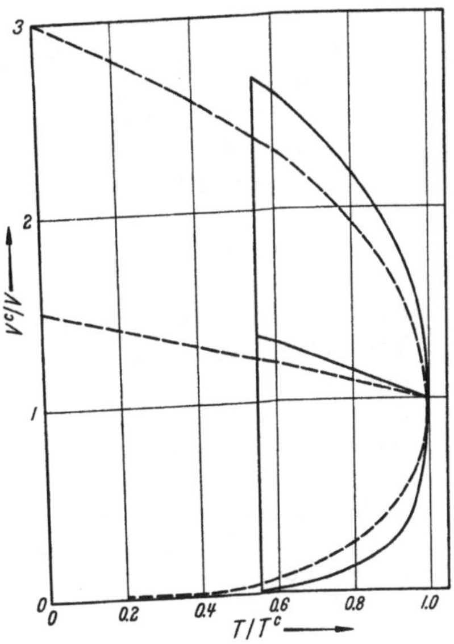

Fig. 33. The reduced orthobaric densities of the inert gases (full-line) and the densities calculated from VAN DER WAALS' equation (dashed line).

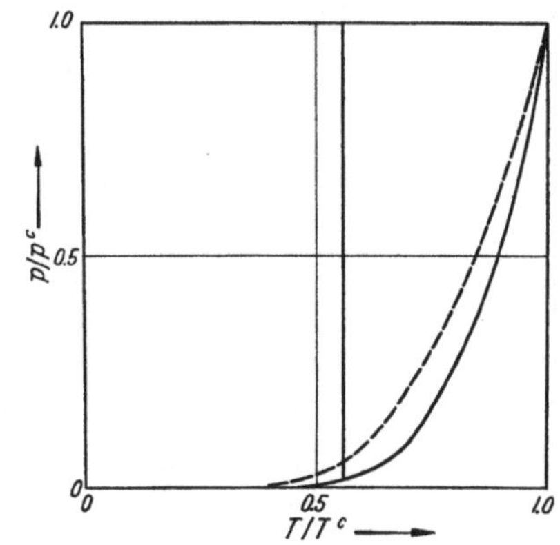

Fig. 34. The reduced vapour-pressure and solid-fluid boundary for the inert gases (full-line) and the vapour pressure calculated from VAN DER WAALS' equation (dashed line).

γ) The residual energy. The residual energy is found by substitution in (8.7) and the residual heat-capacity at constant volume by substitution in (8.9),

$$E^* = -\frac{a}{V} = -\frac{9RT^cV^c}{8V}, \tag{25.13}$$

$$C_V^* = 0. \tag{25.14}$$

Fig. 35 shows the experimental results of MICHELS, GELDERMANS and DE GROOT[2] for ethylene. It is seen that E^* is almost a linear function of density, as required by (25.13), but that the slope is not given by the parameter a if that parameter

---

[1] See footnote 2, p. 49.

[2] A. MICHELS, M. GELDERMANS and S. R. DE GROOT: Physica, Haag **12**, 105 (1946).

is expressed in terms of the critical constants. The vanishing of C_V^*, predicted by (25.14), is not consistent with the results shown in this figure, nor with the behaviour of gases near their critical point shown in Fig. 30.

δ) The virial coefficients. The re-arrangement of (25.1) as a power series in $1/V$ gives,

$$B = b - \frac{a}{RT}, \tag{25.15}$$

$$C = b^2. \tag{25.16}$$

These equations are reasonable representations of the virial coefficients. In fact, (25.15) is of the theoretically correct form for an assembly of rigid spheres with weak attractive forces. Such an equation may be made to fit results such as those of Fig. 11 over quite wide ranges of temperature by suitably choosing a and b. However, it must be emphasised again that the values of a and b which fit the second virial coefficient do not satisfy (25.3).

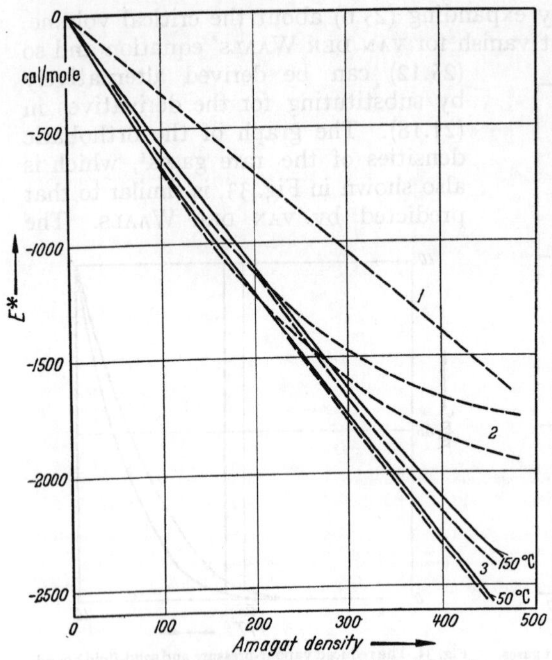

Fig. 35. The residual energy of ethylene (full-lines) from the measurements of A. MICHELS, M. GELDERMANS and S. R. DE GROOT: Physica, Haag **12**, 105 (1946). The dashed curves are values calculated at 50° and 150° C from (1) VAN DER WAALS' equation, (2) the BEATTIE-BRIDGEMAN equation, and (3) the BENEDICT-WEBB-RUBIN equation.

The equations of DIETERICI and of BERTHELOT are similar to that of VAN DER WAALS. The first factor in (25.1) is replaced by $(p\, e^{a'/VRT})$ in the former and by $(p + a''/V^2 T)$ in the latter. DIETERICI's equation is sometimes put forward as a better equation than that of VAN DER WAALS because it gives a better value for the critical ratio,

$$\left(\frac{pV}{RT}\right)^c = \frac{2}{e^2} = 0.2707. \tag{25.17}$$

However the equation is generally unsatisfactory and in certain conditions can lead to physically absurd results in a way that VAN DER WAALS' equation never does. The asymptotic approach of the residual energy to $-\infty$ is an example of this[1].

BERTHELOT's equation is now most frequently used in a modified form at low densities only, where it leads to a second virial coefficient of

$$B = \frac{9}{128} \frac{RT^c}{p^c} \left[1 - 6\left(\frac{T^c}{T}\right)^2\right]. \tag{25.18}$$

This equation represents surprisingly accurately the second virial coefficients of non-polar vapours. Polar vapours show systematic deviations from it[2]. The

[1] J. O. HIRSCHFELDER, C. F. CURTISS and R. B. BIRD: Molecular Theory of Gases and Liquids, p. 257. New York: Wiley 1954.

[2] J. D. LAMBERT, G. A. H. ROBERTS, J. S. ROWLINSON and V. J. WILKINSON: Proc. Roy. Soc. Lond., Ser. A **196**, 113 (1949). — J. S. ROWLINSON: Trans. Faraday Soc. **45**, 974 (1949). — J. D. LAMBERT: Disc. Faraday Soc. **15**, 226 (1953).

equation is accurate enough for the non-polar vapours for it to be used for correcting heat-capacity measurements to the perfect-gas state. Such corrections are sensitive to the exact form of B as they are functions of its first and second temperature derivatives.

26. Modern empirical equations of state. The two modern equations which are most widely used are that of BEATTIE and BRIDGEMAN[1] and that of BENEDICT, WEBB and RUBIN[2]. Other equations find little use nowadays and will not be discussed here. The first of these equations has five adjustable parameters and the second eight. Neither, therefore, can be put into a "corresponding states" form, nor can the parameters be expressed in terms of the critical constants. The parameters can only be determined from actual measurements on a gas and thus, for pure substances, the equations are merely convenient ways of fitting $p-V-T$ results so as to permit easy analytical differentiation and integration.

The Beattie-Bridgeman equation may be written in a form which has some resemblance to VAN DER WAALS' equation

$$\left(p+\frac{A}{V^2}\right)\left(\frac{V}{1+B/V}\right)=RT\left(1-\frac{c}{VT^3}\right), \tag{26.1}$$

where

$$A=A_0\left(1-\frac{a}{V}\right),\qquad B=B_0\left(1-\frac{b}{V}\right). \tag{26.2}$$

(The parameters A and B are not to be confused with the virial coefficients.) The residual function Q^*, (8.5), is given by a quadratic equation in the molar density

$$\frac{Q^*}{RT}=\left[B_0-\frac{A_0}{RT}-\frac{c}{T^3}\right]+\left(\frac{1}{V}\right)\left[-bB_0+\frac{aA_0}{RT}-\frac{cB_0}{T^3}\right]+\left(\frac{1}{V}\right)^2\left[\frac{bcB_0}{T^3}\right]. \tag{26.3}$$

The parameters A_0, a, B_0 and c are always positive. The recommended value of b is often zero, as, for example, for the inert gases, and is never large. It may be positive or negative. Fig. 7 shows that (26.3) is an adequate representation of the change of Q^* with density for xenon at and above room temperature. The parameter b may only be put equal to zero for densities below about 10 moles/l. The three terms in braces in (26.3) are the second, third, and fourth virial coefficients, as may be seen by comparison with (10.7). Thus for those substances for which b is zero, the Beattie-Bridgeman equation is an equation with virial coefficients up to the third only. The temperature dependence of the second coefficient is similar to that of VAN DER WAALS' equation but with the addition of a small term in T^{-3} which allows this equation to be fitted more accurately to experimental results at low temperatures. The temperature dependence of the third coefficient, with b positive or zero, is too rapid. This can lead to serious errors in derived thermodynamic quantities at high densities. Thus, for the residual energy,

$$\left(\frac{\partial E^*}{\partial(1/V)}\right)_T=T\left(\frac{\partial Q^*}{\partial T}\right)_V-Q^* \qquad (8.7)$$

$$=\left[A_0+\frac{3cR}{T^2}\right]+\left(\frac{1}{V}\right)\left[-aA_0+\frac{3cB_0R}{T^2}\right]+\left(\frac{1}{V}\right)^2\left[-\frac{3bcB_0R}{T^2}\right]. \tag{26.4}$$

---

[1] J. A. BEATTIE and O. C. BRIDGEMAN: J. Amer. Chem. Soc. **48**, 1665 (1927). — J. A. BEATTIE and W. H. STOCKMAYER [*1*].

[2] M. BENEDICT, G. R. WEBB and L. RUBIN: J. Chem. Phys. **8**, 334 (1940). — Chem. Engng. Progr. **47**, 419 (1951).

If b is zero, this equation requires that E^* pass through a minimum at quite low densities. The position of the minimum depends but little on temperature and is given by $V=a$ in the limit of infinite temperature. With the constants recommended for ethylene by Beattie and Stockmayer [1], this is a density of 20.1 moles/l, or an Amagat density of 448. At 50° C and 150° C, as shown in Fig. 35, the calculated minima are at Amagat densities of 509 and 473 respectively. The experimental results show no minimum and it is clear that the equation is unsatisfactory at these densities. The change of the residual heat capacity with density is given by

$$\left(\frac{\partial C_V^*}{\partial (1/V)}\right)_T = -T\left(\frac{\partial^2 Q^*}{\partial T^2}\right)_V \quad (8.9),$$

$$= \frac{6cR}{T^3}\left[1+\frac{B_0}{V}\left(1-\frac{b}{V}\right)\right]. \tag{26.5}$$

If b is zero this equation requires the heat capacity to increase linearly with density. This is an improvement on van der Waals' equation which predicts no change of C_V with density, but does not agree with the maximum shown by the experimental results (Fig. 30) and required by the theory of fluctuations. A maximum is predicted for those gases, of which ethylene is one, for which a small positive value of b is recommended. The maximum is approximately at $V=b$, at all temperatures, which is at much too high a density. For ethylene the predicted maximum is at a density of 34.4 moles/l, or an Amagat density of 766, while the experimental maximum is a little below the critical density of 180 Amagat units. The accuracy usually claimed for this equation of state is that it will fit experimental pressures up to the critical density, at temperatures above the critical, with a mean error of about 1%. As the error increases with increasing density, this means that the errors at the critical density should not be more than 2%. This is generally borne out by the experimental results of Beattie and his colleagues[1]. At higher densities the equation quickly becomes unreliable, as shown above, and even at the critical density there may be larger errors than this if the equation is used outside the temperature range covered by the results used to establish the parameters. This is clearly shown if an attempt is made to extrapolate the equation of state of argon down to its critical point. As the recommended value of b is zero the equation becomes one with only second and third virial coefficients, for which,

$$\left(\frac{pV}{RT}\right)^c = \frac{1}{3} \tag{26.6}$$

and

$$B(T^c) = -V^c, \quad C(T^c) = \tfrac{1}{3}(V^c)^2. \tag{26.7}$$

These equations are in error by about 20%. A further warning is necessary against using the parameters of this equation to calculate the second virial coefficient from (26.3). If the parameters are chosen to obtain a good fit of measurements at high densities then the second virial coefficient so calculated may be considerably in error.

The equation of Benedict, Webb and Rubin has three more parameters than that of Beattie and Bridgeman. It has also an exponential term which greatly improves its performance at high densities. It can be written,

$$\left.\begin{aligned}\frac{Q^*}{RT} = \frac{V}{RT}(pV-RT) = \left[B_0 - \frac{A_0}{RT} - \frac{C_0}{RT^3}\right] + \\ + \left(\frac{1}{V}\right)\left[b - \frac{a}{RT}\right] + \left(\frac{1}{V}\right)^4\left[\frac{a\alpha}{RT}\right] + e^{-\gamma/V^2}\left[\frac{c(1+\gamma/V^2)}{VRT^3}\right].\end{aligned}\right\} \tag{26.8}$$

[1] See footnote 2, p. 31.

At low densities this equation resembles closely that of BEATTIE and BRIDGEMAN. The second virial coefficient is the same function of the temperature. However where the Beattie-Bridgeman equation may have a fourth virial coefficient but no higher ones, the Benedict-Webb-Rubin equation never has a fourth but has a fifth and higher coefficients. In practice, the absence of the fourth coefficient is no disadvantage. This equation gives a reasonably satisfactory representation of the fluid in the critical and two-phase region. It can be solved to give good values of the vapour pressure by using (25.5). Fig. 35 shows that it predicts the residual energy with great accuracy up to densities of twice the critical. It predicts a maximum in C_V at a density which satisfies,

$$-\frac{T^4}{6}\left(\frac{\partial^2 Q^*}{\partial T^2}\right)_V = C_0 - \frac{c}{V}\left(1+\frac{\gamma}{V^2}\right)e^{-\gamma/V^2} = 0. \tag{26.9}$$

For ethylene this is a density of 7.0 moles/l or a reduced density of 0.87. In fact, ethylene shows a maximum in C_V at a reduced density of 0.83 at 50° C, followed by a further rise at very high densities. At 150° C the maximum has been replaced by a plateau. The performance of this equation is therefore a great improvement on that of the Beattie-Bridgeman equation at high densities, and so justifies its greater complexity. It is unlikely that it will be superseded by any other empirical equation with greater claims to accuracy.

VI. Gas mixtures.

27. Perfect gas mixtures. The equation of state of a perfect gas of one component may be written

$$p = \frac{nRT}{V}, \tag{27.1}$$

where n is the number of moles and where V is the total volume. By integration of (3.4),

$$A = nRT\ln\left(\frac{1}{V}\right) + \text{constant}, \tag{27.2}$$

where the constant of integration is a function of n and T but not of V. As A is an extensive property, it follows that

$$A = nRT\ln\left(\frac{n}{V}\right) + n g_1(T), \tag{27.3}$$

whence the chemical potential is given by

$$\left(\frac{\partial A}{\partial n}\right)_{V,T} = RT\ln\left(\frac{n}{V}\right) + RT + g_1(T). \tag{27.4}$$

Similarly from

$$V = \frac{nRT}{p} \tag{27.5}$$

and (3.7)

$$G = nRT\ln p + \text{constant} \tag{27.6}$$

$$= nRT\ln\left(\frac{nRT}{V}\right) + n g_2(T), \tag{27.7}$$

and the chemical potential is given by

$$\left(\frac{\partial G}{\partial n}\right)_{p,T} = RT\ln\left(\frac{nRT}{V}\right) + g_2(T). \tag{27.8}$$

By equating the two expressions for the chemical potential,

$$g_2(T) = g_1(T) + \mathrm{R}T - \mathrm{R}T \ln \mathrm{R}T \tag{27.9}$$

and so, by eliminating $g_1(T)$ and omitting the subscript from $g_2(T)$,

$$A = n\,g(T) + n\,\mathrm{R}T\left[\ln\left(\frac{n\mathrm{R}T}{V}\right) - 1\right], \tag{27.10}$$

$$G = n\,g(T) + n\,\mathrm{R}T \ln p. \tag{27.11}$$

The function denoted $g(T)$ is the chemical potential in the standard state of unit pressure at each temperature T.

A perfect-gas mixture is defined as one formed from perfect gases and whose Helmholtz free-energy is given by the sum of the free-energies of its components, when each occupies separately the total volume V at the temperature T. That is

$$A = \sum_\alpha n_\alpha A_\alpha^\dagger + \mathrm{R}T \sum_\alpha n_\alpha\left[\ln\left(\frac{n_\alpha \mathrm{R}T}{V}\right) - 1\right], \tag{27.12}$$

where α is a dummy subscript which is equated successively to the components a, b, c etc., and where A_α is the chemical potential of component α defined by

$$A_\alpha = \left(\frac{\partial A}{\partial n_\alpha}\right)_{V,T,n_b\ldots} = \left(\frac{\partial G}{\partial n_\alpha}\right)_{p,T,n_b\ldots} = G_\alpha. \tag{27.13}$$

The superscript † denotes the pure substance in the standard state of unit pressure. The symbol A_α is used here for chemical potential when the independent variables are V and T, and the symbol G_α when the variables are p and T. By differentiation of (27.12),

$$p = -\left(\frac{\partial A}{\partial V}\right)_{T,n} = \frac{\mathrm{R}T}{V}\sum_\alpha n_\alpha. \tag{27.14}$$

Thus the mixture obeys the same equation of state, (27.1), as the pure substances. The partial pressure of a gas in a mixture is here defined by

$$p_a = \frac{n_a p}{\sum_\alpha n_\alpha} = x_a p, \tag{27.15}$$

where x_a is the mole fraction of species a.

A perfect gas mixture has the following properties:

α) The partial pressure of each gas is the pressure exerted by the same amount of gas when it occupies independently the same volume V. This follows from (27.14) and is Dalton's Law of Partial Pressures.

β) The volume of a mixture is the sum of the volumes of each separate component at the same pressure. This also follows from (27.14) and is Amagat's Law of Additive Volumes.

γ) By differentiation of (27.12) with respect to temperature it follows that the energy and heat capacity of the mixture are linear functions of the number of moles of each component

$$E = \sum_\alpha n_\alpha\left[A_\alpha^\dagger - T\left(\frac{dA_\alpha^\dagger}{dT}\right) - \mathrm{R}T\right], \tag{27.16}$$

$$C_V = \sum_\alpha n_\alpha\left[-T\left(\frac{d^2A_\alpha^\dagger}{dT^2}\right) - \mathrm{R}\right], \tag{27.17}$$

and are the sums of the energies and heat capacities of the pure substances. The entropy is given by

$$S = \sum_\alpha n_\alpha \left[-\left(\frac{dA_\alpha^\dagger}{dT}\right)\right] - \mathrm{R} \sum_\alpha n_\alpha \ln\left(\frac{n_\alpha \mathrm{R} T}{V}\right) \tag{27.18}$$

and is the sum of the entropies of the gases when each occupies separately the total volume V.

δ) The Gibbs free-energy is given by adding pV to (27.12),

$$G = \sum_\alpha n_\alpha A_\alpha^\dagger + \mathrm{R}T \sum_\alpha n_\alpha \ln\left(\frac{n_\alpha \mathrm{R}T}{V}\right) \tag{27.19}$$

$$= \sum_\alpha n_\alpha G_\alpha^\dagger + \mathrm{R}T \sum_\alpha n_\alpha \ln (x_\alpha p), \tag{27.20}$$

whence

$$H = \sum_\alpha n_\alpha \left[G_\alpha^\dagger - T \left(\frac{dG_\alpha^\dagger}{dT}\right)\right], \tag{27.21}$$

$$C_p = \sum_\alpha n_\alpha \left[-T \left(\frac{d^2 G_\alpha^\dagger}{dT^2}\right)\right], \tag{27.22}$$

$$S = \sum_\alpha n_\alpha \left[-\left(\frac{dG_\alpha^\dagger}{dT}\right)\right] - \mathrm{R} \sum_\alpha n_\alpha \ln (x_\alpha p). \tag{27.23}$$

Thus the heat content and heat capacity of the mixture are the sum of the properties of the pure components at the same temperature and pressure, but the free energy and entropy differ from this sum by the terms in $\ln (x_\alpha)$. The change in G on forming a mixture from its components at a constant temperature and pressure is denoted $\Delta_p G$, and is given by

$$\Delta_p G = G(p, T) - \sum_\alpha n_\alpha G_\alpha^0 (p, T), \tag{27.24}$$

where the superscript zero denotes the pure substance but does not imply the state of unit pressure denoted by the superscript †. For a perfect gas

$$G_\alpha^0(p, T) = G_\alpha^\dagger(T) + \mathrm{R}T \ln p, \tag{27.25}$$

whence

$$\Delta_p G = \mathrm{R}T \sum_\alpha n_\alpha \ln x_\alpha \tag{27.26}$$

$$\Delta_p H = 0 \tag{27.27}$$

$$\Delta_p S = -\mathrm{R} \sum_\alpha n_\alpha \ln x_\alpha \tag{27.28}$$

$$\Delta_p V = 0. \tag{27.29}$$

Any mixture, solid, liquid or gas, to which these equations apply over a range of temperature and pressure is an *ideal* mixture, and the entropy change of (27.28) is the *ideal entropy of mixing*. A mixture of perfect gases is the simplest example of an ideal mixture.

28. Real gas mixtures. The object of most of the practical work on compressed gas-mixtures has been to justify the methods, of various degrees of empiricism, which have been proposed for calculating the properties of a mixture from those of its pure components. It is clear, however, that there can be no general solution to this problem. The properties of a pure substance are determined by the forces acting between its molecules. On forming a mixture of two substances, new forces are brought into play which are not present in either of the pure components. It is true that these forces between molecules of different species will generally be close to some "average" of the forces between molecules of the same species, but rarely can they be predicted sufficiently accurately from a knowledge of the latter for the properties of a gas mixture to be calculated from a knowledge of the properties of the pure gases. Even if the forces can be estimated with sufficient accuracy there is no adequately developed theory for satisfactory practical calculations to be made, except for real gases at low densities. The most general simplification that one can hope for, is the calculation of the properties of a ternary, or higher order, mixture from the properties of all the binary mixtures which can be formed from its components. This calculation is possible, in principle, for there are no forces present in a ternary mixture which are not present in the binary mixtures. However classical thermodynamics gives no way of achieving even this modest simplification and so use must be made of statistical thermodynamics and of the virial form of the equation of state which may be derived from it. The successive virial coefficients are related to binary, ternary etc. clusters of molecules[1] and so their dependence on the concentrations in a mixture is known. The second and third virial coefficients of a binary mixture of two gases, a and b, are given by

$$B = x_a^2 B_{aa} + 2x_a x_b B_{ab} + x_b^2 B_{bb} \tag{28.1}$$

$$C = x_a^3 C_{aaa} + 3x_a^2 x_b C_{aab} + 3x_a x_b^2 C_{abb} + x_b^3 C_{bbb}, \tag{28.2}$$

where B_{aa} and C_{aaa} are the coefficients for pure a, B_{bb} and C_{bbb} are the coefficients for pure b, and where the remaining coefficients are characteristic of the mixture and are related to the forces between the molecules in the clusters indicated by the subscripts. Such coefficients are independent of density and composition and functions only of the temperature. These equations can be generalised for the l-th coefficient, J_l, of a mixture of m components,

$$J_2 = \sum_\alpha \sum_\beta x_\alpha x_\beta B_{\alpha\beta} \tag{28.3}$$

$$J_3 = \sum_\alpha \sum_\beta \sum_\gamma x_\alpha x_\beta x_\gamma C_{\alpha\beta\gamma} \quad \text{etc.}, \tag{28.4}$$

where $B_{\alpha\beta} = B_{\beta\alpha}$ etc., and where the dummy subscripts are replaced by all permutations of the m components. It follows that J_l is a polynomial of degree l in the mole fractions and is formed of $[(l+m-1)!/l!\,(m-1)!]$ individual coefficients of the type $B_{\alpha\alpha}$, $B_{\alpha\beta}$ etc. The numerical coefficients in equations such as (28.1) and (28.2) are the numerical coefficients of the polynomial product $\left(\sum_\alpha^m x_\alpha\right)^l$. The equation of state of n moles of a gas mixture $\left(\text{where } n = \sum_\alpha n_\alpha\right)$ in a total volume V is therefore given by

$$\frac{pV}{nRT} = 1 + \sum_{l=2} \frac{J_l n^{l-1}}{V^{-1}}. \tag{28.5}$$

The residual thermodynamic functions may be calculated by substituting from this equation into (8.6) to (8.14). The residual free-energies are given by

$$A^*(V,T) = RT \sum_{l=2} \left(\frac{1}{l-1}\right) \frac{J_l n^l}{V^{l-1}}, \tag{28.6}$$

$$G^*(V,T) = RT \sum_{l=2} \left(\frac{l}{l-1}\right) \frac{J_l n^l}{V^{l-1}}. \tag{28.7}$$

[1] See J. E. MAYER, following chapter.

The total free-energies are obtained by adding these quantities to (27.12) and (27.19). The chemical potential is found by differentiating (28.6) with respect to n_a at constant volume. The differentiation of the virial coefficients gives

$$\left(\frac{\partial}{\partial n_a}\right)_{n_b\ldots} [n^l J_l] = l\, n^{l-1} J_l^a, \tag{28.8}$$

where

$$J_2^a = \sum_\alpha x_\alpha B_{a\alpha} \tag{28.9}$$

$$J_3^a = \sum_\alpha \sum_\beta x_\alpha x_\beta C_{a\alpha\beta} \quad \text{etc.} \tag{28.10}$$

The chemical potential is therefore given by

$$A_a = A_a^\dagger + \mathrm{R}T \ln\left(\frac{n_a \mathrm{R}T}{V}\right) + \mathrm{R}T \sum_{l=2}\left(\frac{l}{l-1}\right)\frac{J_l^a\, n^{l-1}}{V^{l-1}}. \tag{28.11}$$

For the pure substance a, for which J_l^a becomes simply the l-th virial coefficient of pure a, this equation reduces to (G/n_a) where G is given by the sum (27.19) and (28.7). $A_a^\dagger$ is now a standard potential for the pure substance, a function only of T, which is equal to the potential of the hypothetical perfect gas at unit pressure, as in (27.19) and (27.20).

Unfortunately, the chemical potential cannot conveniently be expressed as a function of pressure and temperature. It is only the coefficients of the expansion of pV in powers of the density which are related simply to the composition by Eqs. (28.1) to (28.4). Thus the important differences, Δ_p, of (27.26) to (27.29) cannot be expressed in terms of the virial coefficients in any convenient way. However if the density is low enough for all coefficients beyond the second to be neglected then some useful equations are obtained readily, as the second coefficient is common to both density and pressure series. To this order of approximation,

$$G(p, T) = \sum_\alpha n_\alpha G_\alpha^\dagger + \mathrm{R}T \sum_\alpha n_\alpha \ln(x_\alpha p) + n\, p\, B, \tag{28.12}$$

where B, or J_2, is given by (28.3). The free energy of mixing follows from (27.24)

$$\Delta_p G = \mathrm{R}T \sum_\alpha n_\alpha \ln x_\alpha + p \sum_\alpha n_\alpha (B - B_{\alpha\alpha}). \tag{28.13}$$

Now (28.3) can be re-arranged to give,

$$B = \sum_\alpha x_\alpha^2 B_{\alpha\alpha} + 2 \sum_{\alpha>\beta}\sum x_\alpha x_\beta B_{\alpha\beta} \tag{28.14}$$

$$= \sum_\alpha x_\alpha B_{\alpha\alpha} + 2 \sum_{\alpha>\beta}\sum x_\alpha x_\beta (\delta B)_{\alpha\beta}, \tag{28.15}$$

where

$$(\delta B)_{\alpha\beta} = B_{\alpha\beta} - \tfrac{1}{2} B_{\alpha\alpha} - \tfrac{1}{2} B_{\beta\beta}. \tag{28.16}$$

Substituting (28.15) into (28.13) gives

$$\Delta_p G = \mathrm{R}T \sum_\alpha n_\alpha \ln x_\alpha + 2 p\, n \sum_{\alpha>\beta}\sum x_\alpha x_\beta (\delta B_{\alpha\beta}). \tag{28.17}$$

Thus the excess (or non-ideal) free-energy of mixing vanishes only if all the differences $(\delta B)_{\alpha\beta}$ vanish. By differentiation of (28.17),

$$\Delta_p H = 2 p\, n \sum_{\alpha>\beta}\sum x_\alpha x_\beta \left[(\delta B)_{\alpha\beta} - T\left(\frac{d(\delta B)_{\alpha\beta}}{dT}\right)\right], \tag{28.18}$$

$$\Delta_p S = -\mathrm{R} \sum_\alpha n_\alpha \ln x_\alpha - 2 p\, n \sum_{\alpha>\beta}\sum x_\alpha x_\beta \left[\frac{d(\delta B)_{\alpha\beta}}{dT}\right]; \tag{28.19}$$

$$\Delta_p V = 2 n \sum_{\alpha>\beta}\sum x_\alpha x_\beta (\delta B)_{\alpha\beta}. \tag{28.20}$$

It follows from the last of these equations that AMAGAT's law of additive volumes, although true for a perfect gas-mixture, is not generally true for a real gas-mixture even in the limit of zero pressure[1]. It also follows from this treatment that DALTON's law of partial pressures *is* obeyed by real gases in the limit of zero density[1].

Although there can be no doubt of the correctness of the virial equation at low and moderate densities, the experimental results which can be used for checking these equations are surprisingly meagre. The early work of PENNING[2] and of HOLBORN and OTTO[3] on air suggested that B was a linear function of the oxygen and nitrogen concentrations, that is, that (δB) was zero. This is almost correct for air but is not true in general. VERSCHOYLE[4] verified the quadratic form of B in 1926 for mixtures of hydrogen and nitrogen and showed that (δB) is far from negligible in this system. His values for the coefficients at 0° C are, in AMAGAT units $\times 10^4$,

$$B_{aa} = +6.24, \quad B_{ab} = +5.59, \quad B_{bb} = -4.90, \quad (\delta B)_{ab} = +4.92,$$

where a denotes hydrogen and b denotes nitrogen. This behaviour is typical of mixtures of molecules of very different sizes and energies of interaction, that is, of very different critical volumes and temperatures. Mixtures of helium and nitrogen[5] show an even more extreme behaviour, for here B_{ab} is greater than both B_{aa} and B_{bb}.

The principle of corresponding states has been usefully applied to the calculation of the "mixed" virial coefficient B_{ab}. To do this it is necessary to make some assumption about the strength of the forces acting between molecules a and b which is additional to the principle but which has generally proved sound and useful in practice. When applied to pure substances, the principle relates the coefficient B_{aa} to the coefficient B_{bb} by

$$B_{aa}[T] = (V_a^c/V_b^c) \cdot B_{bb}[T T_b^c/T_a^c]. \tag{28.21}$$

This equation can be extended to B_{ab} by defining the critical constants V_{ab}^c and T_{ab}^c that are appropriate to a hypothetical pure substance whose molecules have an energy of interaction at all separations equal to that between a and b. The "mixed" coefficient is given by

$$B_{ab}[T] = (V_{ab}^c/V_b^c) \cdot B_{bb}[T T_b^c/T_{ab}^c]. \tag{28.22}$$

It is now necessary to relate these hypothetical constants to the constants of pure a and b. The relations which have been found most satisfactory are[6],

$$(V_{ab}^c)^{\frac{1}{3}} = \tfrac{1}{2}(V_a^c)^{\frac{1}{3}} + \tfrac{1}{2}(V_b^c)^{\frac{1}{3}}, \tag{28.23}$$

$$T_{ab}^c = (T_a^c T_b^c)^{\frac{1}{2}}. \tag{28.24}$$

---

[1] J. P. KUENEN: Theorie der Verdampfung und Verflüssigung von Gemischen, p. 99–106. Leipzig: Johann Ambrosius Barth 1906.

[2] F. M. PENNING: Comm. Leiden, No. 166, 1923.

[3] L. HOLBORN and J. OTTO: Z. Physik **10**, 367 (1922); **23**, 77 (1924).

[4] T. T. H. VERSCHOYLE: Proc. Roy. Soc. Lond., Ser. A **111**, 552 (1926).

[5] A. E. EDWARDS and W. E. ROSEVEARE: J. Amer. Chem. Soc. **64**, 2816 (1942). — T. L. COTTRELL and R. A. HAMILTON: Trans. Faraday Soc. **52**, 156 (1956). — T. L. COTTRELL, R. A. HAMILTON and R. P. TAUBINGER: Trans. Faraday Soc. **52**, 1310 (1956). — G. M. KRAMER and J. G. MILLER: J. Phys. Chem. **61**, 785 (1957). — R. C. HARPER and J. G. MILLER: J. Chem. Phys. **27**, 36 (1957).

[6] E. A. GUGGENHEIM and M. L. MCGLASHAN: Proc. Roy. Soc. Lond., Ser. A **206**, 448 (1951). Similar relations between intermolecular force-constants were used earlier by, for example, J. O. HIRSCHFELDER and W. E. ROSEVEARE: J. Phys. Chem. **43**, 15 (1939).

These equations have some basis in the theory of the intermolecular forces of non-polar molecules[1] and are a reasonable approximation for most simple gases. They are, however, best regarded as empirical equations whose justification is their practical success. The values of B_{ab} calculated in this way are often correct to within 2 ml/mole. This calculation always gives positive values of (δB) as the graph of B as a function of temperature is convex upwards (that is, d^2B/dT^2 is negative), except for reduced temperatures above about 40. An approximate extension of this treatment to the higher virial coefficients has been made[2], but this is accurate only for the "mixed" coefficients of molecules of reasonably similar critical constants. The difficulty is that, although the method of statistical thermodynamics gives directly the dependence on composition of C etc., there is no simple relation between the "mixed" coefficients so introduced and those of the pure substances.

The second virial coefficients of mixtures of the vapours of volatile organic substances are given by (28.22) to (28.24) only if both components are non-polar, or only very slightly polar. The equations are correct, for example, for mixtures of chloroform with n-hexane[3], with benzene[4], and with carbon tetrachloride[4]. In these cases, as with mixtures of nitrogen and oxygen, the components of the mixtures have similar enough properties for (δB) to be negligible. However these equations are not correct if one or both of the components is polar. Such mixtures have been studied by LAMBERT[3,5] and his colleagues. They have shown that a mixture of a polar and a non-polar vapour, such as acetonitrile and cyclohexane, has an unusually large positive value of (δB), that is, the absolute value of B_{ab} is much less than that calculated above. This is due to the polar interactions which can occur between molecules a and a, but not between a and b, or between b and b. Conversely if there are strong specific forces between a and b which are not present between a and a or between b and b, then (δB) is negative. Chloroform and diethyl ether form such a mixture in which the specific force is a hydrogen-bond to the oxygen atom of the ether. In the examples quoted, the positive and negative values of (δB) are reflected in the positive and negative deviations from RAOULT's law shown by the liquid mixtures, but this simple correspondence does not hold in all cases.

The experimental methods used in studying gas-mixtures are generally the same as those used for pure substances. However the direct measurement of the volume of mixing at constant pressure has been used occasionally[6] to determine (δB) for a binary mixture by means of (28.20).

Most work on gas mixtures at high pressures has been either on systems of industrial importance, when the composition range studied is often very restricted[7], or else has been undertaken to test some specific point, such as the

---

[1] J. O. HIRSCHFELDER, C. F. CURTISS and R. B. BIRD: Molecular Theory of Gases and Liquids, p. 168, 567, 963. New York: Wiley 1954.

[2] J. S. ROWLINSON, F. H. SUMNER and J. R. SUTTON: Trans. Faraday Soc. **50**, 1 (1954).

[3] J. H. P. FOX and J. D. LAMBERT: Proc. Roy. Soc. Lond., Ser. A **210**, 557 (1951).

[4] P. G. FRANCIS and M. L. McGLASHAN: Trans. Faraday Soc. **51**, 593 (1955).

[5] J. D. LAMBERT, S. J. MURPHY and A. P. SANDAY: Proc. Roy. Soc. Lond., Ser. A **226**, 394 (1954).

[6] A. E. EDWARDS and W. E. ROSEVEARE: J. Amer. Chem. Soc. **64**, 2816 (1942). — J. G. MILLER and R. A. GORSKI: J. Amer. Chem. Soc. **75**, 550 (1953). — A. MICHELS and A. J. H. BOERBOOM: Bull. Soc. chim. Belg. **62**, 119 (1953).

[7] For example, the work on mixtures of hydrogen and nitrogen of R. WIEBE and V. L. GADDY: J. Amer. Chem. Soc. **60**, 2300 (1938), and of A. MICHELS and T. WASSENAAR: Appl. Sci. Res. A **1**, 258 (1948) and the work on air of A. MICHELS, T. WASSENAAR, J. M. LEVELT and W. DE GRAAFF: Appl. Sci. Res. A **4**, 381 (1954).

departures from Dalton's and Amagat's laws[1]. The departures are usually considerable, but there is no large body of experimental work from which any useful generalisations can be made.

29. Chemical equilibrium in real gases. The position of equilibrium in a mixture of imperfect gases is most conveniently discussed in terms of the fugacities of the components. A residual chemical-potential, G_a^*, may be defined by

$$G_a^*(p, T) = \left(\frac{\partial G^*(p, T)}{\partial n_a}\right)_{p, T, n_b \ldots} = \int_0^p \left(\frac{\partial V^*}{\partial n_a}\right)_{p, T, n_b \ldots} dp, \tag{29.1}$$

where the residual partial molar volume is the difference between the partial molar volume in the real mixture and its value in a hypothetical mixture of perfect gases at the same temperature, pressure and composition. The fugacity of component a is then given by an equation analogous to (9.1),

$$f_a(p, T) = p_a \cdot \exp\left[G_a^*(p, T)/RT\right], \tag{29.2}$$

where p_a is the partial pressure. The fugacity approaches the partial pressure in the limit of zero density, and is a convenient representation of the chemical potential. The equilibrium constant K for a reaction between the components of a mixture is given by,

$$\ln K = \sum_\alpha \nu_\alpha \ln (f_\alpha^e), \tag{29.3}$$

where ν_α is the stoichiometric coefficient of component α in the reaction and is conventionally taken as positive for the products and negative for the reactants. The superscript e denotes an equilibrium value. In the limit of zero density

$$\ln K = \left(\sum_\alpha \nu_\alpha\right) \ln p + \sum_\alpha \nu_\alpha \ln (x_\alpha^e), \tag{29.4}$$

where x^e is an equilibrium mole fraction. The last term of this equation is independent of p if $\left(\sum_\alpha \nu_\alpha\right)$ vanishes, but in a real gas mixture, this condition is not sufficient, for the activity coefficient

$$\gamma_\alpha = \frac{f_\alpha}{p_\alpha} \tag{29.5}$$

is itself a function of pressure. The calculation of f_a or γ_a from (29.1) and (29.2) requires a knowledge of the equation of state of the mixture which is rarely achieved. Several approximate ways of calculating the fugacity in a mixture from the properties of the pure components have been proposed[2], but these can never be very satisfactory. The simplest of them is the proposal of Lewis and Randall[3] that

$$f_a(p, T, x_a) = x_a f_a^0(p, T), \tag{29.6}$$

where f_a^0 is the fugacity of the pure substance at the same total pressure. At low pressures (29.6) is correct only if all the differences (δB) are zero[4], and the equation is rarely more than qualitatively correct at high pressures.

---

[1] I. Masson and others: Proc. Roy. Soc. Lond., Ser. A **103**, 524 (1923); A **122**, 253 (1929); A **126**, 268 (1930).

[2] See I. R. Krichevski: J. Amer. Chem. Soc. **59**, 2733 (1937). — J. A. Beattie and W. H. Stockmayer [*1*], p. 262—278. — F. G. Keyes: Commentary on the Scientific Writings of J. W. Gibbs, Vol. 1, p. 337—393 (ed. F. G. Donnan and A. Haas). Yale 1936.

[3] G. N. Lewis and M. Randall: Thermodynamics, p. 225—227. New York: McGraw Hill 1923.

[4] L. J. Gillespie: Phys. Rev. **34**, 352, 1605 (1929).

30. Empirical equations of state of mixtures. The extension to mixtures of the empirical equations discussed in Sects. 25 and 26 is made by representing the parameters of these equations by suitable functions of the composition. Any new parameters, characteristic of the mixture, are then related to the parameters for the pure substances. This last step, in particular, means that such a representation can never be more than a very approximate treatment for some mixtures, and is rarely to be relied upon as a substitute for the direct measurement of their properties.

This two-step substitution is well illustrated by the two pairs of equations used to represent the parameters a and b of van der Waals equation,

$$a = \sum_{\alpha}\sum_{\beta} x_\alpha x_\beta a_{\alpha\beta}, \tag{30.1}$$

$$b = \sum_{\alpha}\sum_{\beta} x_\alpha x_\beta b_{\alpha\beta}, \tag{30.2}$$

$$a_{\alpha\beta} = (a_{\alpha\alpha} a_{\beta\beta})^{\frac{1}{2}}, \tag{30.3}$$

$$(b_{\alpha\beta})^{\frac{1}{3}} = \tfrac{1}{2}(b_{\alpha\alpha})^{\frac{1}{3}} + \tfrac{1}{2}(b_{\beta\beta})^{\frac{1}{3}}. \tag{30.4}$$

These equations[1] are similar in form to those used subsequently for the application of the principle of corresponding states to the virial coefficients B_{ab}, (28.23) and (28.24).

Similar equations have been proposed for the five parameters of the Beattie-Bridgeman equation (26.1) and the eight of the Benedict-Webb-Rubin equation[2]. Calculations based on the former equation have been used to find the equilibrium mole fraction of ammonia in the Haber synthesis[3]. The latter equation is sufficiently accurate to be used for calculating liquid-vapour boundary curves in mixtures near their critical points[4].

It is worth noting that the use of Eqs. (30.1) to (30.4), and similar empirical equations, for calculating second virial coefficients is not entirely consistent with Eqs. (28.22) to (28.24), which are to be preferred.

31. The critical region in binary mixtures[5]. The behaviour of a mixture near the gas-liquid critical point is more complicated than that of a pure substance. There is, as for the pure substance, a condition at which the two coexisting phases of a mixture of fixed total composition have the same pressure, temperature, density and composition. This is called the plait-point of the system and is the point which most closely resembles the critical point of a pure substance. However, it is, in general, neither the maximum temperature nor the maximum pressure at which that mixture can show two phases. The p-T curve of a typical

[1] For full references to the early history of these equations and the reasons which led to the choice of these particular forms, see H. Kamerlingh Onnes and W. H. Keesom: Die Zustandsgleichung. Encyklopädie der mathematischen Wissenschaften, Vol. 5, Part 1, p. 689. 1911.

[2] M. Benedict, G. B. Webb and L. C. Rubin: J. Chem. Phys. **10**, 747 (1942).

[3] L. J. Gillespie and J. A. Beattie: Phys. Rev. **36**, 743, 1008 (1930); **31**, 655 (1931). — J. Amer. Chem. Soc. **52**, 4239 (1930).

[4] See, however, the case of nitrogen-heptane, for a bad failure above 250 atm, G. G. P. Town, Univ. of Wisconsin Tech. Rep. ONR-4, 1953.

[5] This is a subject which has been comparatively neglected for the last fifty years. Its early study, both experimental and theoretical, was made mainly by the Leiden school and was published in a long series of papers reprinted in the Communications. Sixteen of these, starting with No. 59a (1900), have the general title of "Contributions to the knowledge of van der Waals' ψ-surface", (where ψ is the Helmholtz free-energy). The most useful summaries of this work are to be found in a paper by C. M. A. Hartman, Suppl. No. 3, (1901), which is not part of the numbered series, and in Chapt. III and IV of J. P. Kuenen, Theorie der Verdampfung und Verflüssigung von Gemischen. Leipzig: Johann Ambrosius Barth 1906.

mixture is shown in Fig. 36a. The saturated gas and liquid curves meet at the critical or plait-point, C_m. The maximum pressure at which there are two phases is point A and the maximum temperature, point B. The relative positions of these points is a matter for experiment—C_m can lie between A and B or outside them on either branch of the curve. The locus of the plait-points joins the critical points of the pure substances, C_a and C_b, and is the envelope of the p-T loops for mixtures of different total compositions. This follows from the fact that the dew-point branch of one curve cuts the bubble-point branch of another (at point O, Fig. 37) at a pressure and temperature at which the compositions of the two phases are those of the mixtures for which the two p-T curves have been drawn. Thus an infinitely close family of such curves, each of which cuts its neighbours at points infinitesimally close to the plait-point, has as its envelope the plait-point curve.

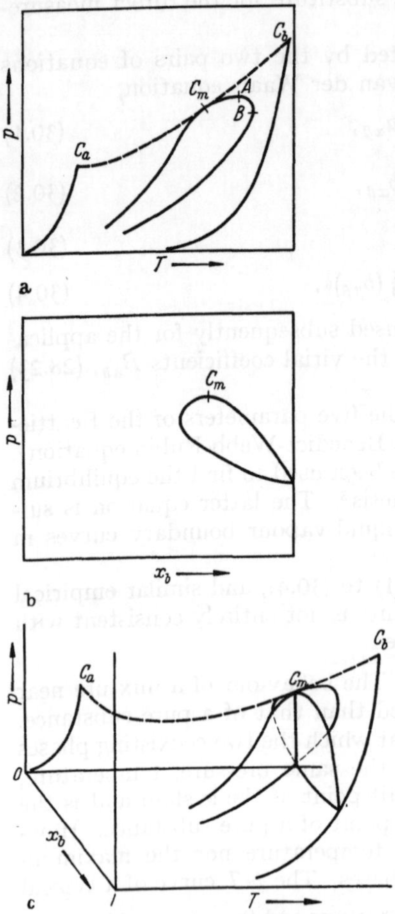

Fig. 36 a—c. The gas-liquid critical point of a binary mixture (see text).

The phases in Fig. 36a are for a mixture of fixed total composition and are not co-existing phases. These are shown in the p-x diagram for constant temperature (Fig. 36b). The plait-point is always the maximum of the p-x loop as the tie-lines joining the co-existing phases are necessarily horizontal. A typical three-dimensional graph, p-T-x, of which Figs. 36a and b are sections, is shown in Fig. 36c.

The volume of a binary system near its critical point is shown in Fig. 38. The tie-lines joining the coexisting phases become, at the plait-point, the tangent to the boundary curve. The stability of the system near this point has been shown to depend only on the ability of the system to withstand spontaneous fluctuations in local composition[1]. The conditions of mechanical and thermal stability are always satisfied. In this, a mixture differs from a single component system.

The most obvious physical difference between a mixture and a pure substance is the occurrence of retrograde condensation and evaporation. These terms were applied by Kuenen in 1892 to regions where condensation or evaporation is caused by changes of pressure or temperature of opposite sign to those needed for a pure substance[2]. Retrograde evaporation may be illustrated by the p-T curve in Fig. 36a. If such a mixture is compressed isothermally at a temperature below C_m then it passes from the gas-phase, through the two-phase region, into the liquid-phase in the normal manner. If, however, the temperature is

---

[1] See, for example, I. Prigogine and R. Defay: Chemical Thermodynamics, p. 239—246. (English trans. D. H. Everett.) London: Longmans 1954.

[2] Qualitative accounts of these phenomena are given by D. L. Katz and F. Kurata: Ind. Engng. Chem. **62**, 817 (1946), and by C. A. Boyd: J. Phys. Chem. **54**, 1347 (1950).

above C_m but below B then the mixture first liquefies in part but on increasing the pressure further the liquid evaporates and the system becomes once more a homogeneous phase. The mixture has passed through two dew-points, but no bubble-point. A second type of retrograde evaporation occurs if a gas is cooled isobarically at a pressure between C_m and A. The example chosen shows only evaporation, but if C_m lies between A and B then the mixture shows retrograde condensation as well. If C_m lies beyond B then the system shows only retrograde condensation. These phenomena are the rule when the vapour pressures of the two pure substances are sufficiently different for there to be an easily observable separation of the dew and bubble-point curves. If the volatility is very different, as in ethane-heptane[1], the retrograde ranges may be as large as 20° C at 10 atm. Generally the ranges are much smaller than this.

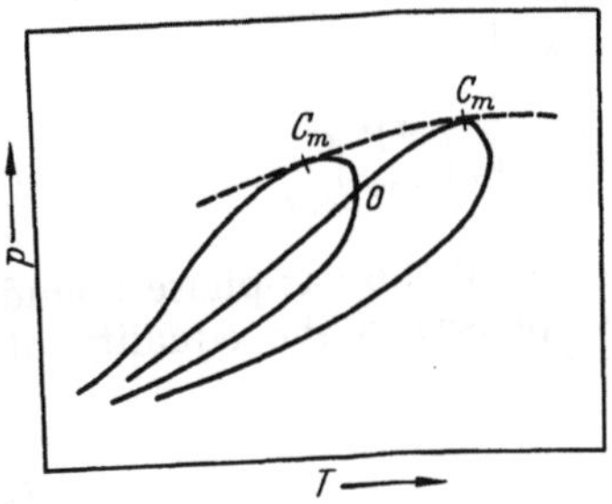

Fig. 37. The intersection of dew and bubble-point curves for mixtures of similar composition.

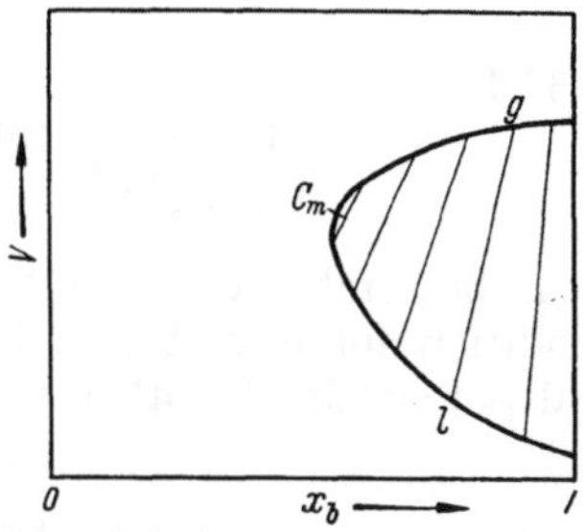

Fig. 38. The orthobaric volumes of a binary mixture near the critical point.

This retrograde behaviour is a necessary consequence of the fact that the p-x loop (Fig. 36b) ends with a rounded top at the plait-point, that is, the slopes of the dew and bubble-point curves, $(dp/dx)^g$ and $(dp/dx)^l$, have opposite signs near the plait-point and both approach zero as the curves meet. This is not so in a liquid mixture below the critical temperatures of both components, when these slopes have the same sign but different magnitudes at the points where the dew and bubble-points coincide on the axes representing the pure substances. Nor is it the case in an azeotropic mixture when, by the Gibbs-Konovalow rule, both slopes are zero, but they approach this limit with the same sign. A Taylor expansion, similar to that used in Sect. 21 and with similar restrictions on its validity, may be used to analyse the shape of the p-x curve near the plait-point.

$$G = G^c + \sum_{n=1} \sum_{m=0}^{n} \frac{1}{m!\,(n-m)!} (p - p^c)^m (x - x^c)^{n-m} \left(\frac{\partial^n G}{\partial p^n\, \partial x^{n-m}}\right)_T^c . \qquad (31.1)$$

At the plait-point, where the mixture is only just stable to diffusion,

$$\left(\frac{\partial^2 G}{\partial x^2}\right)_{p,T}^c = 0, \qquad \left(\frac{\partial^3 G}{\partial x^3}\right)_{p,T}^c = 0, \qquad \left(\frac{\partial^4 G}{\partial x^4}\right)_{p,T}^c > 0, \qquad (31.2)$$

(compare the conditions of stability of a pure substance, Sect. 21). Further

$$\left(\frac{\partial^3 G}{\partial p\, \partial x^2}\right)_T^c = \left(\frac{\partial^2 V}{\partial x^2}\right)_{p,T}^c \neq 0. \qquad (31.3)$$

The last equation follows from the fact that $(\partial^2 G/\partial x^2)_{p,T}$ is positive in the one-phase region and negative in the two-phase region adjoining the plait-point. Hence the change of this derivative with pressure is not zero, and in a normal

[1] W. B. KAY: Ind. Engng. Chem. **30**, 459 (1938).

mixture, such as that shown in Fig. 36, the derivatives in (31.3) are positive. The terms of (31.1) up to $n=4$ are therefore given in

$$\left.\begin{aligned}\left(\frac{\partial^2 G}{\partial x^2}\right)_{p,T} &= (p-p^c)\left(\frac{\partial^2 V}{\partial x^2}\right)^c_{p,T} + \frac{1}{2}(p-p^c)^2\left(\frac{\partial^3 V}{\partial p\,\partial x^2}\right)^c_T + \\ &+ (p-p^c)(x-x^c)\left(\frac{\partial^3 V}{\partial x^3}\right)^c_{p,T} + \frac{1}{2}(x-x^c)^2\left(\frac{\partial^4 G}{\partial x^4}\right)^c_{p,T}.\end{aligned}\right\} \tag{31.4}$$

The terms in $(p-p^c)^2$ and $(p-p^c)(x-x^c)$ can be omitted at once as they must be small compared to the first term. The slope of the gas-phase boundary is given by[1]

$$\left(\frac{dp}{dx}\right)^g = \frac{\left(\frac{\partial^2 G}{\partial x^2}\right)^g_{p,T}}{\left[\frac{V^g - V^l}{x^g - x^l} - \left(\frac{\partial V}{\partial x}\right)^g_{p,T}\right]}, \tag{31.5}$$

From (31.1),

$$\left(\frac{V^g - V^l}{x^g - x^l}\right) - \left(\frac{\partial V}{\partial x}\right)^g_{p,T} = -\frac{1}{2}(x^g - x^l)\left(\frac{\partial^2 V}{\partial x^2}\right)^c_{p,T}, \tag{31.6}$$

which can be combined with (31.4) to give the slope of the gas-phase boundary. The condition for equilibrium between the two phases is the equality of the chemical potentials of both components, that is[1]

$$\left(\frac{\partial G}{\partial x}\right)^g_{p,T} = \left(\frac{\partial G}{\partial x}\right)^l_{p,T} \tag{31.7}$$

and

$$\frac{G^g - G^l}{x^g - x^l} = \left(\frac{\partial G}{\partial x}\right)^g_{p,T}. \tag{31.8}$$

On substitution from (31.4) these give,

$$\frac{1}{6}\left(\frac{\partial^4 G}{\partial x^4}\right)^c_{p,T}\left[(x^g - x^c)^3 - (x^l - x^c)^3\right] + \left(\frac{\partial^2 v}{\partial x^2}\right)^c_{p,T}(p^g - p^c)(x^g - x^l) = 0, \tag{31.9}$$

$$\left(\frac{\partial^4 G}{\partial x^4}\right)^c_{p,T}\left[(x^g - x^c)^2 - (x^l - x^c)^2\right](x^g - x^l) = 0. \tag{31.10}$$

The second of these two equations has two different solutions of which the required one is

$$(x^c - x^g) = -(x^c - x^l), \tag{31.11}$$

and the slopes of the dew and bubble-point curves are given by

$$\left(\frac{dp}{dx}\right)^g = -\left(\frac{dp}{dx}\right)^l = \frac{(x^c - x^g)\,(\partial^4 G/\partial x^4)^c_{p,T}}{3\,(\partial^2 V/\partial x^2)^c_{p,T}}. \tag{31.12}$$

Both derivatives on the right hand side of this equation are positive and so the boundary curve is of the shape shown in Fig. 36, where $(p-p^c)$ is proportional to $(x-x^c)^2$ near the plait-point. If it had been assumed that $(\partial^6 G/\partial x^6)^c_{p,T}$ was the first non-vanishing derivative in (31.2), then $(p-p^c)$ would have been proportional to $(x-x^c)^4$. No experimental test of the exact curvature appears to have been reported. It is seen, therefore, that if G is a differentiatable function of p and x at the critical point, then the occurrence of retrograde condensation is as necessary a consequence of the conditions of thermodynamic stability as is the rounded top of the V-T curve of a pure substance. If the liquid system forms an azeotrope at low temperatures then the azeotropic line may intersect

[1] I. Prigogine and R. Defay: Chemical Thermodynamics, Eq. (18.49). (English trans. D. H. Everett.) London: Longmans 1954.

the plait-point line. Here the p-x curve ends in a cusp and the Taylor expansion of G is no longer valid, as some of the derivatives are discontinuous. Examples of such behaviour are quite common and are cited in the references quoted at the beginning of this section.

If the components in a binary mixture have very different molecular sizes or polarities and if their critical temperatures are widely separated, then the mixture may show a further kind of phase separation at pressures and temperatures above those of the critical points of the pure components. Although the densities of the fluids under these conditions are comparable with those of normal liquids, this new kind of phase separation is best referred to as gas-gas immiscibility. Such a possibility was discussed, in terms of VAN DER WAALS' equation by KAMERLINGH ONNES and KEESOM[1] in 1907 but was not realised

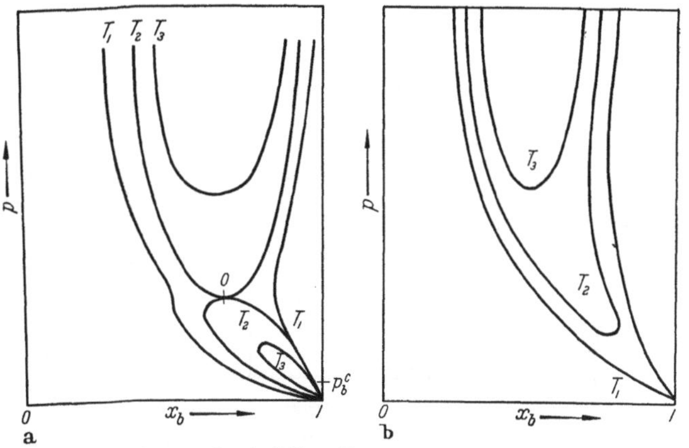

Fig. 39 a and b. Two types of gas-gas immiscibility. The temperatures are in the increasing order, T_1, T_2, T_3.

experimentally until many years later by KRICHEVSKI and TSIKLIS[2,3]. They discovered two kinds of behaviour, illustrated schematically in Figs. 39a and b. In the first[2], the normal p-x loops of a gas-liquid system, shown in the lower half of the diagram fuse with another set of phase-boundary curves, which are the gas-gas immiscibility curves. These show a lower critical solution pressure at each constant temperature, and an upper critical solution temperature at each constant pressure. The left-hand boundary at low temperatures (T_2 and lower) represents the solubility of liquid component b in gaseous a (which is discussed quantitatively in the following section) and the right-hand branch represents the solubility of gaseous a in liquid b. Examples[2] of such systems are ammonia-nitrogen and ammonia-methane. The double contact-point, point O, occurs at 87° C and 1100 atm pressure for the first system. This is 45° C below the critical temperature of pure ammonia but is about 1000 atm higher than its critical pressure. If the temperature at which point O would occur lies above the critical temperature of b then the second, and simpler, type of phase diagram is found, Fig. 39b. This was the type envisaged by KAMERLINGH ONNES and KEESOM, and is shown by mixtures[3] of helium with carbon dioxide, ammonia and propane.

[1] H. KAMERLINGH ONNES and W. H. KEESOM: Comm. Leiden, Suppl. No. 15, 1907.

[2] I. R. KRICHEVSKI and P. E. BOLSHAKOV: Acta physicochim. URSS. **14**, 353 (1941). — I. R. KRICHEVSKI and D. S. TSIKLIS: Acta physicochim. URSS. **18**, 264 (1943). — D. S. TSIKLIS: Dokl. Akad. Nauk SSSR. **86**, 993 (1952).

[3] D. S. TSIKLIS: Dokl. Akad. Nauk SSSR. **86**, 1159 (1952); **101**, 129 (1955).

Gas-gas immiscibility can only occur if the excess Gibbs free-energy of mixing is large and positive. The leading term in the virial expansion for this excess function (28.17) is proportional to (δB) and the mixtures for which immiscibility has been found experimentally are of the type for which this difference is known to be large and positive (Sect. 28). Unfortunately the pressures at which the immiscibility occurs are too high for a quantitative treatment in terms of the virial equation.

32. Solubility in compressed gases. ANDREWS' classical demonstration of the continuity of the liquid and gaseous phases led HANNAY and HOGARTH[1], in 1879, to enquire whether the solvent power of liquids for solids which were involatile, or but slightly volatile persisted, beyond the critical point. They found that it did, and showed that such solutions possessed many of the properties of ordinary liquid solutions, such as the electrical conductivity and characteristic spectra of solutions of electrolytes. Although much of HANNAY'S work at high pressure is now suspect — particularly that on a supposed preparation of diamond published at the same time—these observations have been widely confirmed[2]. It has been shown that such solubility is not confined to the critical region but is common to all compressed gases, above and below their critical temperatures. Some typical examples are the solubility of salts[3] and silica[4] in compressed steam (problems of some technical importance), of solid carbon dioxide in compressed air[5], in compressed helium[6] and in compressed hydrogen[6], of iodine in compressed carbon dioxide[7], and of naphthalene in compressed ethylene[8].

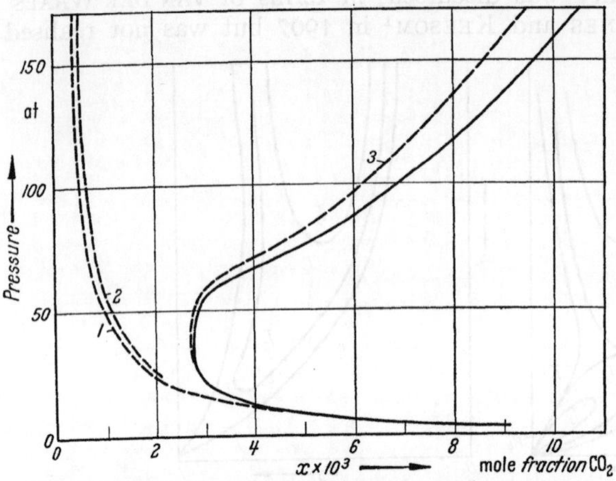

Fig. 40. The composition of solutions of carbon dioxide in compressed air in equilibrium with solid carbon dioxide at −110° C. The full line is the experimental curve of WEBSTER and the three dashed curves are for (1) an ideal gas mixture, (2) an ideal mixture with POYNTING'S correction for the effect of pressure on the activity of the solid, and (3), solubility calculated from Eq. (32.3).

The passage of the solute into the compressed gas can be regarded either as a true solubility or as an enhancement of the normal volatility of the solid by the action of the compressed gas. The distinction is merely verbal. In all reported cases, however, the mixture is believed to be a true molecular solution and not a colloidal dispersion. At low pressures of added gas the enhancement of the vapour pressure is a linear function of the pressure. POYNTING[9] first showed that such

[1] J. B. HANNAY and J. HOGARTH: Proc. Roy. Soc. Lond. **29**, 324 (1879); **30**, 178 (1880). — J. B. HANNAY: Proc. Roy. Soc. Lond. **30**, 484 (1880).
[2] H. S. BOOTH and R. M. BIDWELL: Chem. Rev. **44**, 477 (1949).
[3] A. OLANDER and H. LIANDER: Acta chem. scand. **4**, 1437 (1950). — C. S. COPELAND, T. SILVERMAN and S. W. BENSON: J. Chem. Phys. **21**, 12, 2208 (1953).
[4] C. J. VAN NIEUWENBERG and P. M. VAN ZON: Rec. Trav. chim. Pays-Bas **54**, 129 (1934).
[5] T. J. WEBSTER: Proc. Roy. Soc. Lond., Ser. A **214**, 61 (1952).
[6] A. H. EWALD: Trans. Faraday Soc. **51**, 347 (1955).
[7] H. BRAUNE and F. STRASSMANN: Z. phys. Chem. **143**, 225 (1929).
[8] G. A. M. DIEPEN and F. E. C. SCHEFFER: J. Amer. Chem. Soc. **70**, 4085 (1948).
[9] J. H. POYNTING: Phil. Mag. **12**, 32 (1881).

an increase is to be expected from the purely hydrostatic effect of the applied pressure. The change of free energy of the solid on applying a pressure p is

$$\Delta G = \int_0^p V^s \, dp, \tag{32.1}$$

where V^s is the volume of the solid. If the solid is incompressible its vapour pressure is therefore increased by a factor of exp $[pV^s/RT]$. The term is greater than unity but is never large enough to explain the observed increases, as solids have small molar volumes. This is shown in Fig. 40.

A more satisfactory treatment can be made in terms of the virial equation of state of the gas-mixture[1]. The condition for equilibrium is the equality of the chemical potential, in both phases, of the component present in the pure solid state. The potential of this component in the gas-phase is given by (28.11). Its potential in the solid-phase is given by the value of the potential at zero pressure with the addition of the Poynting term. Equating these two potentials gives the following equation for the ratio of x_a, the observed mole fraction in the gas of the component present in the solid state, to x_a^0, the mole fraction which this component would have were its concentration in the gas-phase the same as that of the pure saturated vapour,

$$\ln\left(\frac{x_a}{x_a^0}\right) = \frac{pV^s}{RT} - \frac{2B_{ab}}{V} - \frac{\frac{3}{2}C_{abb}}{V^2} - \frac{\frac{4}{3}D_{abbb}}{V^3} - \cdots \tag{32.2}$$

$$= \frac{V^s - 2B_{ab}}{V} + \frac{V^s B_{bb} - \frac{3}{2}C_{abb}}{V^2} + \frac{V^s C_{bbb} - \frac{4}{3}D_{abbb}}{V^3} + \cdots \tag{32.3}$$

where b is the added gas and where it has been assumed that $x_a \ll 1$. The successive virial coefficients in (32.2) represent the successive clustering of molecules of species b about the volatalised molecules of species a. The coefficient B_{ab} is negative and an order of magnitude larger than V^s for all systems which have so far been studied. This is to be expected from (28.22). It is the term $(-2B_{ab}/V)$ which is primarily responsible for the large observed solubilities. The next coefficient, C_{abb}, is generally positive[2], and is responsible for the fact that the solubility falls off from a linear proportionality to density at high densities. This is shown in Fig. 40, although the fall-off is there a little exaggerated as the ordinate is pressure and not density. This equation fits excellently the results for simple mixtures and it has been used by EWALD[3] as a direct way of measuring B_{ab}. Solutions of liquids in compressed gas can, in principle, be treated in the same way but it is necessary, in these cases, to correct the chemical potential of the liquids for the amounts of gas dissolved.

These greatly enhanced vapour pressures make it necessary to treat with some reserve the claimed accuracy of some volumetric measurements made on compressed gases at high temperatures and pressures, in which mercury has been used as the confining fluid. For example, measurements made at the Massachusetts Institute of Technology [5] commonly extend to 300° C (and sometimes 325° C) and to 500 atmospheres. At the lower temperature the normal vapour pressure of mercury is 0.325 atm and so the amount which dissolves in the gas at 500 atm total pressure is probably quite large. Preliminary calculations[4], from estimates of the strengths of the intermolecular forces, indicate that the

[1] S. ROBIN and B. VODAR: J. Phys. Radium **13**, 264 (1952). — Disc. Faraday Soc. **15**, 233 (1953). — A. H. EWALD, W. B. JEPSON and J. S. ROWLINSON: Disc. Faraday Soc. **15**, 238 (1953).

[2] J. S. ROWLINSON, F. H. SUMNER and J. R. SUTTON: Trans. Faraday Soc. **50**, 1 (1954).

[3] A. H. EWALD: Trans. Faraday Soc. **49**, 1401 (1953); **51**, 347 (1955).

[4] W. B. JEPSON and J. S. ROWLINSON: J. Chem. Phys. **23**, 1599 (1955).

correction to the observed pressures for the amount of mercury in the gas-phase may be far from negligible in work of the highest accuracy. These calculations have been confirmed recently by measurements[1] of the solubility in propane and *n*-butane up to 30 atm.

Acknowledgements. I am indebted to my colleague Mr. W. B. Brown for his most helpful criticisms of this chapter, and to Mr. C. R. Barber, of the National Physical Laboratory, for some advice on the international recommendations for the temperature scale.

General references.

[1] Beattie, J. A., and W. H. Stockmayer: A Treatise on Physical Chemistry, Vol. **2**, Chap. 2, p. 187–290 (ed. H. S. Taylor and S. Glasstone). New York: Van Nostrand 1951. — An account of the thermodynamic properties of gases, principally in terms of the Beattie-Bridgeman equation.

[2] Din, F.: The thermodynamic functions of gases, Vol. **1**, Ammonia, Carbon monoxide, Carbon dioxide; Vol. **2**, Air, Propane, Acetylene, Ethylene, Argon. London: Butterworth 1956. — A critical compilation of all the experimental results presented in the form of smoothed tables and graphs. It covers all three phases for each pure substance.

[3] Justi, E.: Spezifische Wärme, Enthalpie, Entropie und Dissoziation technischer Gase. Berlin: Springer 1938. — This book covers the calculation of the thermodynamic properties of perfect gases from quantum theory, and the empirical calculation of the properties of real gases. It includes many tables of results.

[4] Keesom, W. H.: Helium, Chap. 2. Amsterdam: Elsevier 1942. This chapter gives experimental details of, for example, the Leiden gas-apparatus, which are relevant to gases other than helium. There is a discussion of the velocity of sound in an imperfect gas and of the use of helium in gas thermometers.

[5] Keyes, F. G.: Proc. Amer. Acad. Arts Sci. **68**, 505–564 (1933). Methods and Procedures used in the Massachusetts Institute of Technology program of investigation of the pressure and volume of water to 460° C. — J. A. Beattie: Proc. Amer. Acad. Arts Sci. **69**, 389–405 (1934). The apparatus and method used for the measurement of the compressibility of several gases in the range 0 to 325° C. — Amongst the apparatus described is a much-copied type of dead-weight gauge.

[6] Kobe, K. A., and R. E. Lynn: Chem. Rev. **52**, 117–236 (1953). The critical properties of elements and compounds. — A review of all work up to 1953 with recommendations of "best" values, and brief descriptions of apparatus.

[7] Michels, A., and C. Michels: Proc. Roy. Soc. A **153**, 201–213 (1935). Isotherms of CO_2 between 0 and 150° C and pressures from 16 to 250 atm. — A. Michels, C. Michels and H. Wouters: Proc. Roy. Soc. Lond., Ser. A **153**, 214–224 (1935). Isotherms of CO_2 between 70 and 3000 atm. — These papers describe the apparatus used at Amsterdam for the last twenty years, and also give an account of the Amagat system of units.

[8] National Bureau of Standards: Tables of Thermal Properties of Gases, Circular 564 (ed. J. Hilsenrath and others). Washington 1955. — These tables cover air, argon, carbon dioxide, carbon monoxide, hydrogen, nitrogen, oxygen and steam. The maximum pressure is only 100 atm but the temperature range is extended to temperatures up to 3000° K by making theoretical, but reliable, extrapolations.

[9] Newitt, D. M.: The design of high pressure plant and the properties of fluids at high pressures. Oxford, England: Clarendon Press 1940. — A full account of experimental methods, and of the thermodynamic and transport properties of compressed gases and liquids.

[10] Partington, J. R.: An advanced treatise on Physical Chemistry, Vol. **1**, The Properties of Gases, Sect. VII A–E, p. 546–848. London: Longmans 1949. — This is principally useful for its references to the early history of the subject.

[11] Rowlinson, J. S.: Trans. Faraday Soc. **51**, 1317–1326 (1955). The reduced equation of state. — A smoothed table of the compressibility factor for the inert gases as a function of reduced temperature and reduced density. The table may also be used for calculating Z for other gases. A similar table with reduced pressure as the second variable is given by K. S. Pitzer, D. Z. Lippman, R. F. Curl, C. M. Huggins and D. E. Peterson: J. Amer. Chem. Soc. **77**, 3433 (1955).

[12] Thermodynamic and Transport Properties of Fluids. London: Inst. Mech. Engng. (in the press). — The proceedings of an international conference held in London in July 1957 by the Institution of Mechanical Engineers and the International Union of Pure and Applied Chemistry.

[1] W. B. Jepson, M. J. Richardson and J. S. Rowlinson: Trans. Faraday Soc. **53**, 1586 (1957).

Theory of Real Gases.

By

JOSEPH E. MAYER.

With 4 Figures.

A. General methods.

1. The fundamental equations. Any one of the relations of GIBBS [1] between the appropriate thermodynamic potential of the system and the corresponding partitio function may be used as a starting equation for deriving the equilibrium properties of a system. The evaluation of the partition function is usually somewhat easier and more direct for the Grand Canonical Ensemble of GIBBS than for either the Macro Canonical or the Micro Canonical. The Grand Canonical Ensemble is appropriate for the description of a system whose thermodynamic state is described by giving the volume, V, the absolute temperature, T, and the values of the chemical potential, μ_a, per molecule for all species, a, of molecules composing the system.

The chemical potential, μ_a, may be defined in terms of the internal energy, E, of the system. The energy is a natural function, $E(V, S, N_a, N_b, \ldots)$, of the volume, V, entropy, S, and numbers, $N_a, N_b, \ldots$, of molecules of species $a, b, \ldots$ respectively. The relation is,

$$dE = T\,dS - P\,dV + \sum_a \mu_a\,dN_a. \tag{1.1}$$

The Gibbs free energy is,

$$G = E + PV - TS. \tag{1.2}$$

From $dG = dE + d(PV) - d(TS)$, using Eq. (1.1) one has,

$$dG = -T\,dS + V\,dP + \sum \mu_a\,dN_a, \tag{1.3}$$

or,

$$\mu_a = \left(\frac{\partial G}{\partial N_a}\right)_{P,T,N_b}, \tag{1.4}$$

and, since G is homogeneous of order unity in the variables, N_a, at fixed P, T,

$$G = \sum_a N_a \mu_a. \tag{1.5}$$

The thermodynamic potential, Θ, which is a natural function of $V, T, \mu_a, \mu_b, \ldots$, is obtained from E as,

$$E - TS - \sum_a \mu_a N_a = \Theta(V, T, \mu_a, \mu_b, \ldots), \tag{1.6}$$

$$d\Theta = -S\,dT - P\,dV - \sum_a N_a\,d\mu_a. \tag{1.7}$$

From Eqs. (1.2), (1.5), and (1.6)

$$\Theta(V, T, \mu_a, \mu_b) = -PV. \tag{1.8}$$

From this and (1.7),

$$\left(\frac{\partial P}{\partial T}\right)_{\mu_a} = \frac{S}{V}, \tag{1.9}$$

$$\left(\frac{\partial P}{\partial \mu_a}\right)_{T,\mu_b} = \frac{N_a}{V} = \varrho_a, \tag{1.10}$$

where the symbol ϱ_a will be used for the number density of molecules of species a. We will use the shorthand notation,

$$\boldsymbol{\mu} = \mu_a, \mu_b, \ldots \tag{1.11}$$

for the set of chemical potentials,

$$\boldsymbol{N} = N_a, N_b, \ldots \tag{1.12}$$

for the number set of molecules, and,

$$\boldsymbol{N}\cdot\boldsymbol{\mu} = \sum_a N_a \mu_a = G. \tag{1.13}$$

The Gibbs equation expresses $-\Theta/\mathrm{k}T$, with k the Boltzmann constant, as the logarithm of a partition function, namely,

$$PV = \mathrm{k}T \log Q_{\mathrm{GC}}(V, T, \boldsymbol{\mu}), \tag{1.14}$$

where Q_{GC} is called the partition function of a Grand Canonical Ensemble. An Ensemble is the collection of a large number of individual systems in the limit that the number approaches infinity. One discusses, then, the average of any property over the members of the ensemble. The Grand Canonical Ensemble is an ensemble of open walled systems, each of the same volume V, immersed in an infinite reservoir of molecules at fixed T and $\boldsymbol{\mu}$. Since T and $\boldsymbol{\mu}$ are specified, both $\boldsymbol{N}$, the number set of molecules in any one system, and E, the energy of any one system containing a given number set $\boldsymbol{N}$, may fluctuate.

The fundamental postulate which we shall use here, and which can be derived by the method of GIBBS from other more intuitive postulates using the laws of mechanics, gives the probability that any member system of the ensemble will contain within its volume, V, exactly the number set $\boldsymbol{N}$ of molecules, and no more, and that these will be in a single quantum state, $\boldsymbol{\nu}$, describing the state of the whole system. If the energy of the system in this quantum state is $E_{\boldsymbol{\nu}}$, the probability is,

$$P(\boldsymbol{N}, \boldsymbol{\nu}) = \mathrm{e}^{[\boldsymbol{N}\cdot\boldsymbol{\mu} - PV - E_{\boldsymbol{\nu}}]/\mathrm{k}T}, \tag{1.15}$$

in which the arbitrary zero's of $\boldsymbol{N}\cdot\boldsymbol{\mu}$ and $E_{\boldsymbol{\nu}}$ are the same. The sum,

$$Q_C(V, T, \boldsymbol{N}) = \sum_{\boldsymbol{\nu}} \mathrm{e}^{-E_{\boldsymbol{\nu}}/\mathrm{k}T}, \tag{1.16}$$

namely the sum over all quantum states, $\boldsymbol{\nu}$, of a system of fixed set $\boldsymbol{N}$ of molecules, in a volume V, of the exponential of the negative energy divided by $\mathrm{k}T$, is known as the partition function of a Macro-Canonical Ensemble, or simply of a Canonical Ensemble. The Canonical Ensemble is one of systems each of volume V, each containing the set $\boldsymbol{N}$ of molecules, all in equilibrium with an infinite reservoir of temperature T.

Since $P(\boldsymbol{N}, \boldsymbol{\nu})$, of Eq. (1.15) is a probability, it follows that the sum over $\boldsymbol{\nu}$ for given $\boldsymbol{N}$,

$$P(\boldsymbol{N}) = \mathrm{e}^{[\boldsymbol{N}\cdot\boldsymbol{\mu} - PV]/\mathrm{k}T} Q_C(V, T, \boldsymbol{N}) \tag{1.17}$$

should be the probability of finding exactly the set $\boldsymbol{N}$ of molecules in a member system of the Grand Canonical Ensemble. The sum of this $P(\boldsymbol{N})$ over all $\boldsymbol{N}$ from zero to infinity must be unity. Using this, and setting,

$$Q_{GC}(V, T, \boldsymbol{\mu}) = \sum_{\boldsymbol{N} \geq 0} e^{(\boldsymbol{N} \cdot \boldsymbol{\mu})/kT} Q_C(V, T, N), \tag{1.18}$$

one obtains Eq. (1.14) for PV.

The single terms for given $\boldsymbol{N}$ under the sum of (1.18) are all positive, and proportional to the probability of finding the set $\boldsymbol{N}$ in a system of given $V, T, \boldsymbol{\mu}$. They rise to a sharp maximum for the set

$$\overline{\boldsymbol{N}} = \overline{N}_a, \overline{N}_b, \ldots \tag{1.19}$$

which is the most probable number set of molecules for such a system. One may write for the logarithm of the sum in (1.18),

$$kT \log Q_{GC} = \overline{\boldsymbol{N}} \cdot \boldsymbol{\mu} + kT \log Q_C(V, T, \overline{\boldsymbol{N}}) + kT \sum_a \log \Delta N_a, \tag{1.20}$$

where the ΔN_a's are certain averages of the fluctuations of the numbers of molecules of species a in a member system of the Grand Canonical Ensemble. The terms $(PV/kT) = \log Q_{GC}$, $(\overline{\boldsymbol{N}} \cdot \boldsymbol{\mu}/kT)$, and $\log Q_C(V, T, \overline{\boldsymbol{N}})$ are all of order $\overline{N} = \sum \overline{N}_a$, the total number of molecules in a member system of maximum probability. We are interested in cases for which this number is of the order 10^{23}, namely a macro system. The fluctuation terms, $\log \Delta N_a$, are at most of order $\log 10^{23} \sim 10^2$, and hence completely negligible. One may then use (1.20) without these terms, and (1.14) to write

$$-[\boldsymbol{N} \cdot \boldsymbol{\mu} - PV] = kT \log Q_C(V, T, \boldsymbol{N}) \tag{1.21}$$

for those systems of the Grand Canonical Ensemble having the most probable number set, $\boldsymbol{N} = \overline{\boldsymbol{N}}$, namely for an equilibrium system. The quantity,

$$A = \boldsymbol{N} \cdot \boldsymbol{\mu} - PV = G - PV = E - TS, \tag{1.22}$$

is the Helmholtz Free Energy,

$$dA = -S\,dT - P\,dV + \sum \mu_a\,d N_a, \tag{1.23}$$

which is the natural thermodynamic potential for a system of fixed V, T, and $\boldsymbol{N}$. With Eq. (1.21), one has,

$$-A(V, T, \boldsymbol{N}) = kT \log Q_C(V, T, \boldsymbol{N}). \tag{1.24}$$

This is the Gibbs equation for the Helmholtz Free Energy in terms of the partition function, $Q_C(V, T, \boldsymbol{N})$, of the corresponding ensemble.

The sum of Eq. (1.16) may be written as an integral. For any macro system the number of allowed quantum states, $\boldsymbol{\nu}$, in an energy range, $\Delta E \approx kT$, for finite T, is always extremely large. Define this number per unit energy range as

$$\Omega(V, E, \boldsymbol{N}) = \lim_{\Delta E \to 0} \left\{ \frac{1}{\Delta E} \sum_{\boldsymbol{\nu}} (E \leq E_{\boldsymbol{\nu}} \leq E + \Delta E) \right\}, \tag{1.25}$$

in which case,

$$Q_C(V, T, \boldsymbol{N}) = \sum_{\boldsymbol{\nu}} e^{-E_{\boldsymbol{\nu}}/kT} = \int^{\infty} e^{-E/kT} \Omega(V, E, \boldsymbol{N})\, dE. \tag{1.26}$$

Analogously to the previous operation we write

$$kT \log Q_C = -\overline{E} + kT \log \Omega(V, \overline{E}, \boldsymbol{N}) + kT \log \Delta E, \tag{1.27}$$

with $\bar{E}$ the most probable (equilibrium) energy of the member systems of the Canonical Ensemble. Again neglect the term $\log \Delta E$ to write, from (1.24),

$$-A+E = TS(V,E,\boldsymbol{N}) = \mathrm{k}T\log\Omega(V,E,\boldsymbol{N}), \tag{1.28}$$

for the entropy in terms of Ω, the partition function of the Micro-Canonical Ensemble of systems of fixed V, E and $\boldsymbol{N}$.

Of the three equations, (1.14), (1.24), and (1.28), any one of which may be used to compute the thermodynamic behavior of a macro system from a knowledge of the mechanical properties of the molecules composing it, the first for PV is usually the one most easy to manipulate, in spite of its apparent complexity compared to the other two.

The various thermodynamic relationships obtained by taking derivatives all follow simply from these equations. Using (1.14) we have,

$$V\left(\frac{\partial P}{\partial \mu_a}\right)_{V,T,\mu_b} = \mathrm{k}T\left[\frac{\partial \log Q_{\mathrm{CG}}(V,T,\boldsymbol{\mu})}{\partial \mu_a}\right]_{V,T,\mu_b,\ldots}, \tag{1.29}$$

$$V\left(\frac{\partial P}{\partial T}\right)_{V,\mu} = \frac{PV}{T} + \mathrm{k}T\left[\frac{\partial \log Q_{\mathrm{CG}}(V,T,\boldsymbol{\mu})}{\partial T}\right]_{V,\boldsymbol{\mu}}. \tag{1.30}$$

In these expressions use (1.18) with (1.16). We have, using (1.13) for $\boldsymbol{N}\cdot\boldsymbol{\mu}$, that,

$$\mathrm{k}T\left(\frac{\partial \log Q_{\mathrm{GC}}}{\partial \mu_a}\right)_{V,T,\mu_b} = \frac{1}{Q_{\mathrm{GC}}}\sum_{\boldsymbol{N}\geq 0} N_a\, e^{\boldsymbol{N}\cdot\boldsymbol{\mu}/\mathrm{k}T}\, Q_{\mathrm{C}}(V,T,\boldsymbol{N}),$$

and with (1.14),

$$\mathrm{k}T\left(\frac{\partial \log Q_{\mathrm{GC}}}{\partial \mu_a}\right)_{V,T,\mu_b} = \sum_{\boldsymbol{N}\geq 0} N_a\, e^{[\boldsymbol{N}\cdot\boldsymbol{\mu}-PV]/\mathrm{k}T}\, Q_{\mathrm{C}}(V,T,\boldsymbol{N}),$$

which, from (1.17) is,

$$\mathrm{k}T\left(\frac{\partial \log Q_{\mathrm{GC}}}{\partial \mu_a}\right)_{V,T,\mu_b} = \sum_{\boldsymbol{N}\geq 0} N_a\, P(\boldsymbol{N}), \tag{1.31}$$

with $P(\boldsymbol{N})$ the probability of finding exactly the set $\boldsymbol{N}$ of molecules in the ensemble. From the definition of an average this is,

$$\sum_{\boldsymbol{N}\geq 0} N_a\, P(N) = \overline{N}_a, \tag{1.32}$$

where $\overline{N}_a$ is the average number of molecules of species a in the members of the ensemble. From this and (1.29),

$$\left(\frac{\partial P}{\partial \mu_a}\right)_{V,T,\mu_b} = \frac{\overline{N}_a}{V} = \varrho_a, \tag{1.33}$$

in agreement with the thermodynamic relationshiop (1.10).

Operating similarly with the temperature derivative, one has from (1.18)

$$\mathrm{k}T\left(\frac{\partial \log Q_{\mathrm{GC}}}{\partial T}\right)_{V,T,\boldsymbol{\mu}}$$
$$= \frac{1}{T Q_{\mathrm{GC}}}\sum_{\boldsymbol{N}\geq 0}\left[-\boldsymbol{N}\cdot\boldsymbol{\mu} + \mathrm{k}T^2\left(\frac{\partial \log Q_{\mathrm{C}}(V,T,\boldsymbol{N})}{\partial T}\right)_{V,\boldsymbol{N}}\right] e^{\boldsymbol{N}\cdot\boldsymbol{\mu}/\mathrm{k}T}\, Q_{\mathrm{C}}(V,T,\boldsymbol{N}),$$

or with (1.14) and (1.17),

$$\left.\begin{aligned} \mathrm{k}T\left(\frac{\partial \log Q_{GC}}{\partial T}\right)_{V,T,\mu} &= \frac{1}{T}\sum_{\boldsymbol{N}\geq 0}\left[-\boldsymbol{N}\cdot\boldsymbol{\mu} + \mathrm{k}T^2\left(\frac{\partial \log Q_C}{\partial T}\right)_{V,\boldsymbol{N}}\right] P(\boldsymbol{N}) \\ &= \frac{1}{T}\left[-G + \sum_{\boldsymbol{N}\geq 0} \mathrm{k}T^2\left(\frac{\partial \log Q_C}{\partial T}\right)_{V,\boldsymbol{N}} P(\boldsymbol{N})\right]. \end{aligned}\right\} \tag{1.34}$$

Now from (1.16),

$$\mathrm{k}T^2\left(\frac{\partial \log Q_C}{\partial T}\right)_V = \frac{1}{Q_C}\sum_{\boldsymbol{\nu}} E_{\boldsymbol{\nu}}\, \mathrm{e}^{-E_{\boldsymbol{\nu}}/\mathrm{k}T} = \overline{E}(V, T, \boldsymbol{N}), \tag{1.35}$$

since $\mathrm{e}^{-E_{\boldsymbol{\nu}}/\mathrm{k}T}/Q_C$ is the probability that, if the system of volume V has the set $\boldsymbol{N}$ of molecules, it will be in the quantum state $\boldsymbol{\nu}$ of energy $E_{\boldsymbol{\nu}}$, so that $\overline{E}$ is the average energy of a system of given $V, T, \boldsymbol{N}$. The right hand side of (1.34) is then,

$$\mathrm{k}T\left(\frac{\partial \log Q_{GC}}{\partial T}\right)_{V,T,\mu} = \frac{1}{T}\left[\overline{E}(V, T, \overline{\boldsymbol{N}}) - G\right], \tag{1.36}$$

and from (1.34) and (1.30)

$$V\left(\frac{\partial P}{\partial T}\right)_{V,\mu} = \frac{1}{T}\left[PV + \overline{E} - G\right] = S, \tag{1.37}$$

in agreement with (1.9).

These manipulations merely serve to demonstrate the self consistency of the statistical mechanical equations with the usual thermodynamic relations. A less trivial relation, is that since

$$\begin{aligned} S &= V\left(\frac{\partial P}{\partial T}\right)_{V,\mu} \\ &= \frac{PV}{T} - \frac{\mathrm{k}}{Q_{GC}}\sum_{\boldsymbol{N}\geq 0}\sum_{\boldsymbol{\nu}}\left(\frac{\boldsymbol{N}\cdot\boldsymbol{\mu} - E_{\boldsymbol{\nu}}}{\mathrm{k}T}\right)\mathrm{e}^{(\boldsymbol{N}\cdot\boldsymbol{\mu} - E_{\boldsymbol{\nu}})/\mathrm{k}T} \end{aligned}$$

and with (1.14) to eliminate Q_{GC}, and (1.15) for the probability, $P(\boldsymbol{N}, \boldsymbol{\nu})$, that a member of the ensemble will contain exactly the set $\boldsymbol{N}$ of molecules, and be in the quantum state $\boldsymbol{\nu}$, one has

$$\left.\begin{aligned} S &= -\mathrm{k}\sum_{\boldsymbol{N}\geq 0}\sum_{\boldsymbol{\nu}\geq 0}\left[\frac{\boldsymbol{N}\cdot\boldsymbol{\mu} - PV - E_{\boldsymbol{\nu}}}{kT}\right] P(\boldsymbol{N}, \boldsymbol{\nu}) \\ &= -\mathrm{k}\sum_{\boldsymbol{N}\geq 0}\sum_{\boldsymbol{\nu}\geq 0} P(\boldsymbol{N}, \boldsymbol{\nu}) \log P(\boldsymbol{N}, \boldsymbol{\nu}). \end{aligned}\right\} \tag{1.38}$$

This equation is sometimes used as the fundamental definition of entropy.

Another useful equation gives interpretive significance to the heat capacity as a fluctuation. Use (1.35) for the energy to obtain,

$$\left.\begin{aligned} C_V &= \left(\frac{\partial E}{\partial T}\right)_{V,N} = \frac{1}{\mathrm{k}T^2}\left[\frac{1}{Q_C}\sum_{\boldsymbol{\nu}} E_{\boldsymbol{\nu}}^2\, \mathrm{e}^{-E_{\boldsymbol{\nu}}/\mathrm{k}T} - \frac{1}{Q_C^2}\left(\sum_{\boldsymbol{\nu}} E_{\boldsymbol{\nu}}\, \mathrm{e}^{-E_{\boldsymbol{\nu}}/\mathrm{k}T}\right)^2\right] \\ &= \frac{1}{\mathrm{k}T^2}\left(\overline{E^2} - \overline{E}^2\right), \end{aligned}\right\} \tag{1.39}$$

showing that $\mathrm{k}T^2 C_V$ is the average of the square of the energy minus the square of the average energy of the system.

We may similarly use the method by which (1.31) was obtained, taking the second derivative of $\log Q_{GC}$ with respect to μ in a one component system, to find a relationship between the compressibility and a fluctuation in the numbers of molecules. From (1.33),

$$\left(\frac{\partial^2 P}{\partial \mu^2}\right)_T = \left(\frac{\partial \varrho}{\partial \mu}\right)_T = \left(\frac{\partial \varrho}{\partial P}\right)_T \left(\frac{\partial P}{\partial \mu}\right)_T = \varrho \left(\frac{\partial \varrho}{\partial P}\right)_T = \varrho^2 \varkappa, \tag{1.40}$$

where $\varkappa$ is the isothermal compressibility. Following the sequence before (1.31)

$$\left.\begin{aligned} V\,\mathrm{k}T\left(\frac{\partial^2 P}{\partial \mu^2}\right)_T &= (\mathrm{k}T)^2\left(\frac{\partial^2 \log Q_{GC}}{\partial \mu^2}\right)_{V,T} \\ &= \frac{1}{Q_{GC}}\sum_{N\geq 0} N^2\, e^{N\mu/\mathrm{k}T}\, Q_C(N) - \left[\frac{1}{Q_{GC}}\sum_{N\geq 0} N\, e^{N\mu/\mathrm{k}T}\, Q_C(N)\right]^2 \\ &= \sum_{N\geq 0} N^2 P(N) - \left[\sum_{N\geq 0} N\,P(N)\right]^2 \\ &= \overline{N^2} - \overline{N}^2. \end{aligned}\right\} \tag{1.41}$$

From these two equations we have

$$\varkappa = \frac{1}{V\,\mathrm{k}T\,\varrho^2}\left[\overline{N^2} - \overline{N}^2\right], \tag{1.42}$$

showing that the isothermal compressibility is $(V\,\mathrm{k}T\,\varrho^2)^{-1}$ times the average of the square of the number of molecules minus the square of the average number.

Both C_V and $\varkappa$ must necessarily be positive, since the average of the square can never be less than the square of the average. At a phase transition for which $\varkappa$ becomes infinite the fluctuations, $\overline{N^2} - \overline{N}^2$, which normally are of order $\overline{N}$, approach infinity more rapidly as the system approaches infinite size.

We have shown how, by using the fact that the logarithm of a sum of positive terms can be approximated by the logarithm of the largest term, one may proceed from the equation for the Grand Canonical Partition Function to that for the Canonical, and then to the Micro Canonical. The reverse steps are equally easy. Take as fundamental postulate that, for system of fixed V, E, and $\boldsymbol{N}$, the entropy is,

$$S(V, E, \boldsymbol{N}) = \mathrm{k} \log \Omega(V, E, \boldsymbol{N}), \tag{1.43}$$

with Ω the total number of quantum states. The condition for equilibrium is that S shall be a maxijmum.

Consider, now, a super-system, or ensemble, of energy ME, volume MV, containing the number set $M\overline{\boldsymbol{N}}$ of molecules, partitioned into M equal volumes, each of volume V. Ask the total number, $\Omega(m)$, of quantum states of this super-system, when m of the member systems contain a fixed set $\boldsymbol{N}$ of molecules in a single definite quantum state $\boldsymbol{\nu}$ of energy $E_{\boldsymbol{\nu}}$. This number, $\Omega(m)$, will be the number of ways one can select the m member systems out of the total of M, namely $M!/(M-m)!\,m!$, times the number of quantum states of a system of volume $(M-m)\,V$, of energy $ME - mE_{\boldsymbol{\nu}}$, and of number set $M\overline{\boldsymbol{N}} - m\boldsymbol{N}$, subject only to the negligible restriction of the exclusion of the single quantum state $\boldsymbol{\nu}$. The entropy, $S(m)$, of the total Ensemble, when m systems are in $\boldsymbol{\nu}$ is k times the logarithm of this number or,

$$S(m) = \mathrm{k}\sum_{n=1}^{m} \log \frac{M+1-n}{n} + S(MV - mV, ME - mE_{\boldsymbol{\nu}}, M\overline{\boldsymbol{N}} - m\boldsymbol{N}). \tag{1.44$_1$}$$

This is to be a maximum at the equilibrium value of m,

$$m = M\,P(\boldsymbol{N}, \boldsymbol{\nu}). \tag{1.44$_2$}$$

We set the derivative of the right hand expression equal to zero, remembering that by letting $M \to \infty$ the changes, V, E and $\boldsymbol{N}$, for unit change of m, are infinitesimal compared to the total volume, energy, and number set. The thermodynamic expressions derivable immediately from (1.1),

$$\left(\frac{\partial S}{\partial E}\right)_{V,\boldsymbol{N}} = \frac{1}{T}, \quad \left(\frac{\partial S}{\partial V}\right)_{E,\boldsymbol{N}} = \frac{P}{T}, \quad \left(\frac{\partial S}{\partial N_a}\right)_{V,E,N_b} = -\frac{\mu_a}{T}, \tag{1.45}$$

are used, to find, since $m/M = P(\boldsymbol{N}, \boldsymbol{\nu}) \ll 1$,

$$\left.\begin{aligned} \mathrm{k}\log\frac{m}{M-m+1} &= \mathrm{k}\log\frac{m}{M} = \mathrm{k}\log P(\boldsymbol{N},\boldsymbol{\nu}) \\ &= \frac{PV}{T} - \frac{E_{\nu}}{T} + \sum_a \frac{\mu_a N_a}{T}, \end{aligned}\right\} \tag{1.46}$$

which is the same as Eq. (1.15).

Had all member systems of the ensemble been restricted to have the same number set $\boldsymbol{N} = \overline{\boldsymbol{N}}$, one would find, with

$$\left.\begin{aligned} \sum_a \mu_a N_a - PV &= A, \\ P(\boldsymbol{\nu}) &= \mathrm{e}^{A/\mathrm{k}T}\,\mathrm{e}^{-E_{\nu}/\mathrm{k}T}, \end{aligned}\right\} \tag{1.47}$$

for the probability of the single quantum state $\boldsymbol{\nu}$ in a system of fixed V, T, and number set $\boldsymbol{N}$. The condition that the sum over $\boldsymbol{\nu}$ of $P(\boldsymbol{\nu})$ must be unity leads to Eq. (1.24) for $A(V, T, \boldsymbol{N})$ with (1.16) for the partition function of the Canonical Ensemble.

The three ensemble equations,

$$S(V,E,\boldsymbol{N}) = \mathrm{k}\log\Omega(V,E,\boldsymbol{N}), \tag{1.48$_1$}$$

for the Micro Canonical Ensemble,

$$A(V,T,\boldsymbol{N}) = -\mathrm{k}T\log Q_C(V,T,\boldsymbol{N}), \tag{1.48$_2$}$$

for the Canonical Ensemble, and

$$VP(T,\boldsymbol{\mu}) = \mathrm{k}T\log Q_{GC}(V,T,\boldsymbol{\mu}), \tag{1.48$_3$}$$

are thus three equally basic equivalent formulations of the fundamental equation of Statistical Mechanics. The Ensemble Partition Functions are,

$$\Omega(V,E,\boldsymbol{N}) = \lim_{\Delta E\to 0}\left[\sum_{\nu}(E \le E_{\nu} \le E + \Delta E)\right], \tag{1.49$_1$}$$

$$\left.\begin{aligned} Q_C(V,T,\boldsymbol{N}) &= \sum_{\nu}\mathrm{e}^{-E_{\nu}/\mathrm{k}T} \\ &= \int \mathrm{e}^{-E/\mathrm{k}T}\,\Omega(V,E,\boldsymbol{N})\,dE, \end{aligned}\right\} \tag{1.49$_2$}$$

$$Q_{GC}(V,T,\boldsymbol{\mu}) = \sum_{\boldsymbol{N}}\mathrm{e}^{\boldsymbol{N}\cdot\boldsymbol{\mu}/\mathrm{k}T}\,Q_C(V,T,\boldsymbol{N}). \tag{1.49$_3$}$$

2. The classical and Boltzmann limits. The original equations of GIBBS for the partition functions were, of course, not cast in the form of summations over quantum states of the system, but as integrals over the coordinate and momentum space. For many purposes the classical limiting equations are more convenient. These are valid only if two essentially independent conditions are fulfilled. The first condition is that the density of molecules in their own phase space must not be too high. The second condition requires that the spacing of the quantum states in energy along any single coordinate momentum pair be small compared to $\mathrm{k}T$.

The micro-state of a quantum mechanical system is completely defined by the specification of the quantum number, $\boldsymbol{\nu}$, describing the state function, $\Psi_{\boldsymbol{\nu}}$, of the system, where $\Psi_{\boldsymbol{\nu}}$ may, for instance be taken as a function of the complete coordinate set of the system. The quantum number, $\boldsymbol{\nu}$, defining the state, will have the same dimensionality as the coordinate set. For the purposes of description of an equilibrium system we may restrict ourselves to the functions $\Psi_{\boldsymbol{\nu}}$ which are the allowed stationary state solutions of the time independent Schroedinger equation:

$$\boldsymbol{H} \cdot \Psi_{\boldsymbol{\nu}} = E_{\boldsymbol{\nu}} \Psi_{\boldsymbol{\nu}}, \tag{2.1}$$

with $\boldsymbol{H}$ the Hamiltonian operator.

The micro-state in classical mechanics is described by the specification of the momenta and coordinates. The momentum coordinate space of the whole system, consisting of the (numbered) momenta, $\boldsymbol{p}$, and coordinates, $\boldsymbol{q}$, of all the N (numbered) molecules, is called the γ-space (gas-space). The state of the system is described by a single point in this space, whose path in time is completely determined by the classical equations of motion, which are,

$$\dot{p}_\alpha = -\frac{\partial H}{\partial q_\alpha}, \quad \dot{q}_\alpha = \frac{\partial H}{\partial p_\alpha}, \tag{2.2}$$

for each single conjugated momentum coordinate pair, p_α, q_α. One may alternately use the μ-space (molecule space) which is the phase space of the momenta, $\boldsymbol{p}_a$, and coordinates, $\boldsymbol{q}_a$, of a single molecule of species a, and describe the state of the gas consisting of this single species by N_a points representing the states of the single molecules.

The dimensionality of the μ-space, $\boldsymbol{p}_a \boldsymbol{q}_a$, is $2f_a$ where f_a, the number of degrees of freedom of molecules of species a, is the number of coordinates required to completely specify the position and configuration of a single molecule of this species. Since the classical description is completely inadequate if the electronic configurations are excited, we can restrict ourself to cases where only the positions of the atoms composing the molecule are specified, so that

$$f_a = 3n_a, \tag{2.3}$$

where n_a is the number of atoms composing a single molecule of a. The dimensionality of the γ-space, $\boldsymbol{p}\,\boldsymbol{q}$, for fixed number set $\boldsymbol{N}$, is

$$F = \sum_a f_a N_a = \boldsymbol{f} \cdot \boldsymbol{N} = \sum_a 3 n_a N_a. \tag{2.4}$$

Each stationary solution, $\Psi_{\boldsymbol{m}}$, of the time independent Schroedinger equation for the complete system corresponds to a limited region of γ-space of volume $\Delta\boldsymbol{p}\Delta\boldsymbol{q} = h^F$, in which the uncertainty, $\Delta p_\alpha \Delta q_\alpha$ for each single conjugated momentum coordinate pair, is of area h. This is seen in more detail in Sect. 26. These regions, or cells, are contiguous in the γ-space for adjacent quantum numbers, and fill up the complete γ-space. The region of $\boldsymbol{p}, \boldsymbol{q}$ contained in one such cell is one of approximately constant value of the classical Hamiltonian, namely one for which,

$$H(\boldsymbol{p}, \boldsymbol{q}) \approx E_{\boldsymbol{m}}, \tag{2.5}$$

with $E_{\boldsymbol{m}}$ the energy of the state $\Psi_{\boldsymbol{m}}$. Provided consecutive solutions, $\boldsymbol{m}, \boldsymbol{m}+1$, along every one of the F different dimensions of the quantum number $\boldsymbol{m}$ have energy differences small compared to $\mathbf{k}T$,

$$E_{\boldsymbol{m}+1} - E_{\boldsymbol{m}} \ll \mathbf{k}T, \tag{2.6}$$

one may replace the sum of any not too rapidly varying function of $E/\mathbf{k}T$ over the stationary states $\boldsymbol{m}$ by an integral,

$$\sum_{\boldsymbol{m}} \approx \frac{1}{h^F} \iint \cdots \int d\boldsymbol{p}\, d\boldsymbol{q}. \tag{2.7}$$

Conversely, if regions $\Delta\boldsymbol{p}\,\Delta\boldsymbol{q}$ with every pair component $\Delta p_\alpha \Delta q_\alpha = h$ exist in the γ-space such that the variation of the classical Hamiltonian within the cross section $\Delta p_\alpha \Delta q_\alpha$ for fixed values of the other momenta and coordinates is small compared to $\mathbf{k}T$,

$$\delta H\,[\text{in } \Delta p_\alpha \Delta q_\alpha = h,\ 1 \leq \alpha \leq F] \ll \mathbf{k}T, \tag{2.8}$$

condition (2.6) will be satisfied, and (2.7) may be used.

However not every solution of the time dependent Schroedinger equation, $\Psi_{\boldsymbol{m}}$, is an allowed state, Ψ_ν, of the system. The allowed states, Ψ_ν, must be symmetric or antisymmetric in the exchanges of coordinates of identical molecules.

The molecular species fall into two classes[1], those whose allowed states are symmetric, and which form Bose-Einstein systems (or are said to obey "Bose-Einstein Statistics"), and those for which the allowed states are antisymmetric, and form Fermi-Dirac systems (are said to obey "Fermi-Dirac Statistics"). Molecules composed of an even number of Fermions, that is those for which the sum of the numbers of neutrons plus protons plus electrons are even, form Bose-Einstein systems. Those composed of an odd number of Fermions form Fermi-Dirac systems. Since the number of protons and electrons are equal in a neutral molecule, it follows that neutral molecules are BOSE-EINSTEIN if their number of neutrons is even, and FERMI-DIRAC if the neutron number is odd.

For cells, $\boldsymbol{m}$, in the γ-space, in which no two identical molecules have the same momenta and coordinate (within h^f) the corresponding $\Psi_{\boldsymbol{m}}$ lacks any symmetry in the exchange of the coordinates of identical molecules. These are the cells of γ-space for which the corresponding μ-space representation has no two molecules in the same μ-space cell. For such a cell in γ-space there exist altogether

$$\boldsymbol{N}! = N_a!\,N_b!\ldots \tag{2.9}$$

cells of identical energy, including the one in question, which differ only by permutations of the coordinates and momenta together, of identical molecules. In the μ-space these correspond merely to the permutations of the numbers on the points representing the molecules. (Since the same μ-space is not, in general, useful for the representation of different molecular species, the μ-space representation is only pertinent to the case of a single component species.)

The $\boldsymbol{N}!$ different solutions, $\Psi_{\boldsymbol{m}}, \Psi_{\boldsymbol{m}'}, \ldots$ of the Schroedinger equation corresponding to these cells differ only in the permutations of the coordinates of identical molecules, and all have the same energies, $E_{\boldsymbol{m}} = E_{\boldsymbol{m}'} = \cdots$. Only one linear combination of these $N!$ solutions is an allowed state, Ψ_ν, of the system. The sum of all of them is totally symmetric in exchanges, and is the allowed state for Bose-Einstein systems, the sum with alternating signs for the odd and even permutations is totally antisymmetric, and is the allowed state for Fermi-Dirac systems. For those regions of γ-space for which the individual identical molecules

[1] P. EHRENFEST and J. R. OPPENHEIMER: Phys. Rev. **37**, 333 (1931).

are not close together (compared to h^F) one may replace the sum over the allowed states $\boldsymbol{\nu}$ by

$$\sum_{\boldsymbol{\nu}} \approx \frac{1}{\boldsymbol{N}!\, h^F} \iint \cdots \int d\boldsymbol{p}\, d\boldsymbol{q}. \tag{2.10}$$

The cells for which two or more identical molecules have the same coordinates and momenta (within h^F) correspond to configurations in μ-space with two or more molecules in the same μ-cell. These γ-cells correspond to solutions which are symmetric in permutations of the coordinates of the molecules in question. No non-zero totally antisymmetric linear combination can be formed from this and the other solutions $\Psi_{\boldsymbol{m}'}$, belonging to the cells differing in permutations of coordinates and momenta of the other identical molecules. The corresponding region should be excluded from the integration of Eq. (2.10) for Fermi-Dirac systems. For Bose-Einstein systems the symmetric linear combination can be formed from the sum. The total number of solutions which differ only in permutations of the other identical moleculs is now *less* than $N!$ so that the integral over $d\boldsymbol{p}\,d\boldsymbol{q}$ should have a higher weight in this region than that given by (2.10). The approximation of Eq. (2.10) is, thus, always invalid in the region of phase space along the diagonals where two or more identical particles are close together in their own μ-space.

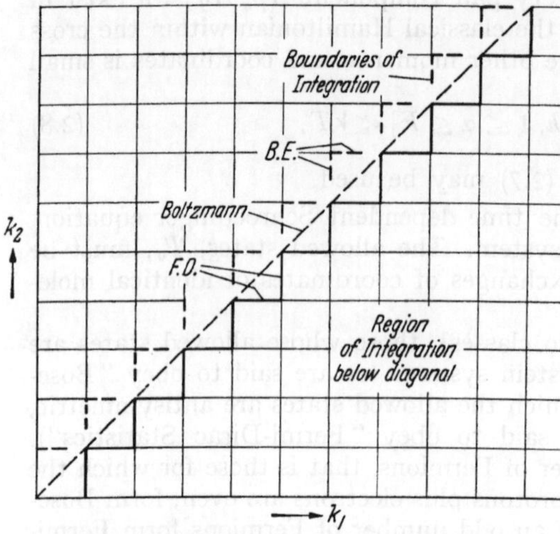

Fig. 1. The "cells" for two identical particles moving independently in one dimension.

The situation may be described graphically, Fig. 1, for the case of two identical mass points of mass m, constrained to move in field free space along a single coordinate x between $x=0$ and $x=l$. The stationary solutions are

$$\Psi_{\boldsymbol{m}} = \frac{4}{l} \sin \frac{\pi k_1 x}{l} \sin \frac{\pi k_2 x}{l},$$

$$\boldsymbol{m} \equiv k_1, k_2,$$

with the positive integer k_1, k_2 related to the magnitude of momenta:

$$|p_1| = \frac{h}{2l} k_1, \quad |p_2| = \frac{h}{2l} k_2.$$

The prescription of Eq. (2.07) is to write

$$\sum_{\boldsymbol{m}} = \sum_{k_1} \sum_{k_2} \approx \int_0^\infty \int_0^\infty dk_1\, dk_2 = \frac{4}{h^2} \iint_0^l dx_1\, dx_2 \iint_0^\infty d|p_1|\, d|p_2|$$

$$= \frac{1}{h^2} \iint_0^l dx_1\, dx_2 \iint_{-\infty}^{+\infty} dp_1\, dp_2.$$

This is valid if the summation is over any function which varies sufficiently slowly with consecutive $k_1 k_2$ values, in which case the sum over the function can be replaced by the integral over the function. The cell (Fig. 1) with $k_1 = k$, $k_2 = k'$, $k \neq k'$, and that with $k_1 = k'$, $k_2 = k$ differ only in the exchange of the identical particles. They represent the same physical situation, and there is only *one* allowed solution corresponding to these two solutions. One divides the integral by $2 = 2!$, and, corresponding to the prescription (2.10) writes

$$\sum_{\nu} = \int_0^{\infty} d k_1 \int_0^{k_1} d k_2 = \frac{1}{2} \iint_0^{\infty} d k_1 \, d k_2 = \frac{1}{2 h^2} \iint_0^{l} d x_1 \, d x_2 \iint_{-\infty}^{+\infty} d p_1 \, d p_2,$$

which is equivalent to the integral below the diagonal of Fig. 1.

The states $k_1 = k_2$, however, are unique. For Fermi-Dirac systems they should be excluded, and the integration only extended to the edges of the squares around these points on the diagonal. For Bose-Einstein systems the complete squares should be included and the integration extended beyond the diagonal to include them. For distances far from the origin compared to the cell size the correction is immaterial.

The case that one sums over all allowed stationary states of the Schroedinger equation, and then corrects for the requirement of symmetry in the eigenfunction by division with $N!$, *either* using the quantum mechanical sum, *or* using the classical integration over phase-space, is called "Boltzmann Statistics". It should be emphasized that no real system obeys "Boltzmann Statistics". This is a mathematical limit which is approached very closely in most cases by both Bose-Einstein and Fermi-Dirac systems.

To summarize this section; one may often approximate the summations over the quantum mechanical states by integrals over phase space. The Micro Canonical partition function, $\Omega(V, E, \boldsymbol{N})$ of Eq. (1.25) is then,

$$\left.\begin{aligned} \Omega(V, E, \boldsymbol{N}) &= \lim_{\Delta E \to 0} \left[\frac{1}{\Delta E} \sum_{\nu} (E \leq E_{\nu} \leq E + \Delta E) \right] \\ &\cong \frac{1}{\boldsymbol{N}! \, h^F} \lim_{\Delta E \to 0} \left[\frac{1}{\Delta E} \iint_{E \leq H(\boldsymbol{p}, \boldsymbol{q}) \leq E + \Delta E} \cdots \int d\boldsymbol{p} \, d\boldsymbol{q} \right]. \end{aligned}\right\} \tag{2.11}$$

The Canonical partition function, $Q_C(V, T, \boldsymbol{N})$ becomes

$$Q_C(V, T, \boldsymbol{N}) = \sum_{\nu} e^{-E_{\nu}/kT} = \frac{1}{\boldsymbol{N}! \, h^F} \iint \cdots \int e^{-H(\boldsymbol{p}, \boldsymbol{q})/kT} d\boldsymbol{p} \, d\boldsymbol{q}. \tag{2.12}$$

This procedure is numerically valid provided two conditions are satisfied. These are

a) The integration over $d\boldsymbol{p} d\boldsymbol{q}$ must be sufficiently negligible in the region where two or more identical molecules have the same momenta and coordinates in units of h, which in practice requires that the representation of the system in μ-space have a low density of molecules of all species, a, per unit volume h^{f_a} in this space.

b) In all regions which contribute appreciably to the integral there must be areas $\Delta p_{\alpha} \Delta q_{\alpha} = h$ for every single conjugate momentum coordinate pair, within which $H(\boldsymbol{p}, \boldsymbol{q})$ varies negligibly compared to kT.

For real gases condition (a) is almost always satisfied for the coordinates of the center of mass of the molecule. The density in the momentum component of the μ-space decreases with increasing mass and temperature. It is probable that only for the two isotopes of Helium, He^3 which is FERMI-DIRAC, and He^4

which is BOSE-EINSTEIN, do significant deviations from the Boltzmann computation occur at temperatures and pressures for which the systems exist as gases. The deviations at the density of the *liquid* are certainly significant, and probably account for the abnormal behavior of liquid Helium II for the He^4 isotope. Another case frequently treated as a gas is that of electrons in metals, which form a Fermi-Dirac system, and for which, due to the low mass and high density, the Boltzmann approximation is completely invalid.

On the other hand condition (b) is seldom satisfied for gases other than the monatomic noble gases. The internal coordinates of vibration only become "classical" at very high temperatures. As will be discussed in the next section the different coordinate momentum pairs can be handled separately, and the motion of the centers of mass are usually treated by the classical equations.

In Sect. 24 a more formal, and less intuitive, development of the classical equations from the quantum mechanical summation is given.

Whereas we have started with the quantum mechanical formulation, the original expressions of GIBBS were derived from considerations of classical mechanics. Gibbs used the Liouville Theorem which states that the motions of all isolated systems whose micro-state points in γ-space at a time t occupy some volume element $W = \Delta\boldsymbol{p}\Delta\boldsymbol{q}$ will be such that at a later time, t', they will be in another portion of γ-space, but occupy the same numerical volume, W. From this it follows that a uniform distribution in γ-space of the members of the micro-canonical ensemble, which are isolated systems of constant $V, E, \boldsymbol{N}$, will be stationary in time. From this one postulates that a randomly chosen system about which only the quantities $V, E, \boldsymbol{N}$ are known will have a probability that its micro-state lies in a volume $W = \Delta\boldsymbol{p}\Delta\boldsymbol{q}$ proportional only to W, and independent of $\boldsymbol{p}, \boldsymbol{q}$ provided only $\boldsymbol{p}$ and $\boldsymbol{q}$ are such that $H(\boldsymbol{p}, \boldsymbol{q}) = E$.

It is then shown that a quantity S proportional to the logarithm of the γ-space volume between E and $E + \Delta E$ has the properties of the thermodynamic entropy. This, then establishes Eq. (1.28) with Ω defined by (2.11) except for the factors $1/\boldsymbol{N}!\,h^F$. From this one can use the arguments of Sect. 1 to obtain the equations for $-A$ and PV in terms of the other two partition functions, Eqs. (1.48_2), and (1.48_3).

In order to obtain expressions for the entropy that are extensive (proportional to the size of the system) it was necessary to introduce the factor $1/\boldsymbol{N}!$ in Eq. (2.11) for Ω. Gibbs justified this by pointing out that points in γ-space differing only in permutations of identical molecules correspond to the same physical situation if the molecules were truly identical.

The numerical values of the partition functions depend, then, on the units used for the momentum coordinate space. This appears as an arbitrary zero of the entropy in the final equations for the thermodynamic properties. With the introduction of Planck's constant h into physics one had a universal constant of just the dimensions, energy time, of the product $p_\alpha q_\alpha$ of any conjugated coordinate momentum pair [see Eq. (2.2)]. SACKUR and TETRODE took the natural step of making the partition functions dimensionless by using the factor h^{-F} in Eq. (2.11), that is essentially of measuring $p_\alpha q_\alpha$ in units h. This led to a zero of entropy consistent with the "Third Law" of thermodynamics.

However, the entropy at absolute zero would then become negative infinity, since the classical formulation permits zero phase space to be available as $T \to 0$. Correspondingly the heat capacity must remain finite, and indeed still large as $T \to 0$ in the classical equation. One can see that the heat capacity at constant volume must always be equal to or larger than $\frac{1}{2}k$ per degree of freedom, as follows. Use Cartesian coordinates of the centers of mass of the atoms. The potential is then a very complicated function of the distances, but the kinetic energy is simply the sum,

$$\mathrm{KE} = \sum_{i=1}^{F} \frac{1}{2m_i} p_i^2, \tag{2.13}$$

over the Cartesian momenta, p_x, p_y, p_z, for each atom. The Hamiltonian is the KE plus the potential, which latter depends only on the coordinates. The partition function, (2.12), becomes a product of F terms due to the kinetic energy of each degree of freedom,

$$\frac{1}{h}\int_{-\infty}^{+\infty} e^{-p_i^2/2m_i \mathrm{k}T}\, dp = \sqrt{\frac{2\pi m_i \mathrm{k}T}{h^2}} \tag{2.14}$$

times one term due to the potential energy,

$$Q_\tau(V,T,N) = \iint\ldots\int e^{-U/\mathrm{k}T}\, dx_1 \ldots dx_F, \tag{2.15}$$

namely,

$$Q_C(V,T,\boldsymbol{N}) = \frac{1}{\boldsymbol{N}!}\, Q_\tau(V,T,\boldsymbol{N}) \prod_{l=1}^{F} \sqrt{\frac{2\pi m_i \mathrm{k}T}{h^2}}. \tag{2.16}$$

The energy, Eq. (1.35), is

$$E(V,T,\boldsymbol{N}) = \mathrm{k}T^2 \left(\frac{\partial \log Q_C}{\partial T}\right)_V = \frac{1}{2} F\,\mathrm{k}T + \mathrm{k}T^2 \left(\frac{\partial \log Q_\tau}{\partial T}\right)_V. \tag{2.17}$$

The first term, $\frac{1}{2}F\,\mathrm{k}T$, due to the kinetic energy, is just $\frac{1}{2}\mathrm{k}T$ per degree of freedom and contributes $\frac{1}{2}F\mathrm{k}$ to the heat capacity. The second term can be treated by the method used to obtain (1.39) and contributes

$$C_\tau = \frac{1}{\mathrm{k}T^2}\left[\overline{U^2} - \overline{U}^2\right] \tag{2.18}$$

to the heat capacity, a quantity which is always positive.

At the present stage of development of Physics one would prefer to base the development of statistical mechanics on the laws of quantum mechanics. However the method closely parallels that of GIBBS. The statement equivalent to the deduction from the Liouville Theorem of the equal probability of equal volumes in phase space, is the equality of probabilities of all allowed quantum states of the systems of a Micro-Canonical Ensemble. From these one proceeds as did GIBBS, but with the partition functions defined by sums over quantum states. The original equations involving integrations over phase space then become asymptotically valid under the conditions (a) and (b) above.

3. Separation of degrees of freedom, perfect gases. If the complete set p, q of momenta and coordinates of the system can be broken into non-overlapping subsets,

$$\left.\begin{aligned} \boldsymbol{p} &= \boldsymbol{p}_\alpha, \boldsymbol{p}_\beta, \ldots, \boldsymbol{p}_\gamma, \\ \boldsymbol{q} &= \boldsymbol{q}_\alpha, \boldsymbol{q}_\beta, \ldots, \boldsymbol{q}_\gamma, \end{aligned}\right\} \tag{3.1}$$

such that the Hamiltonian, $H(\boldsymbol{p},\boldsymbol{q})$, can be written as a sum,

$$H(\boldsymbol{p},\boldsymbol{q}) = \sum_\alpha H_\alpha(\boldsymbol{p}_\alpha, \boldsymbol{q}_\alpha), \tag{3.2}$$

the system is said to be separable. In this case the solutions of the Schroedinger equation are products of functions,

$$\Psi_{\boldsymbol{m}} = \prod_\alpha \psi_{\boldsymbol{m}_\alpha}(\boldsymbol{q}_\alpha); \quad \boldsymbol{m} = \boldsymbol{m}_\alpha, \boldsymbol{m}_\beta, \ldots, \boldsymbol{m}_\gamma, \tag{3.3}$$

and the energies, $E_{\boldsymbol{m}}$, are sums,

$$E_{\boldsymbol{m}} = \sum_{\alpha} E_{\boldsymbol{m}_\alpha}. \tag{3.4}$$

The sum over all Schroedinger equation solutions, $\boldsymbol{m}$, of $e^{-E_{\boldsymbol{m}}/kT}$ is then a product,

$$\left.\begin{aligned} \sum_{\boldsymbol{m}} e^{-E_{\boldsymbol{m}}/kT} &= \sum_{\boldsymbol{m}_\alpha} \sum_{\boldsymbol{m}_\beta} \cdots \sum_{\boldsymbol{m}_\gamma} e^{-\sum_\alpha E_{\boldsymbol{m}_\alpha}/kT} \\ &= \prod_\alpha \Big(\sum_{\boldsymbol{m}_\alpha} e^{-E_{\boldsymbol{m}_\alpha}/kT} \Big). \end{aligned}\right\} \tag{3.5}$$

If one uses the "Boltzmann Statistics" approximation,

$$\sum_{\nu} = \frac{1}{\boldsymbol{N}!} \sum_{\boldsymbol{m}} \tag{3.6}$$

one may write

$$\left.\begin{aligned} Q_C(V, T, \boldsymbol{N}) &= \sum_{\nu} e^{-E_\nu/kT} \\ &= \frac{1}{\boldsymbol{N}!} \prod_\alpha \Big(\sum_{\boldsymbol{m}_\alpha} e^{-E_{\boldsymbol{m}_\alpha}/kT} \Big). \end{aligned}\right\} \tag{3.7}$$

The thermodynamic functions are given by logarithms of the partition functions. One therefore obtains additive contributions to the thermodynamic potentials from the different separable degrees of freedom.

For a single molecule consisting of n atoms in field free space of volume V one may start with a coordinate system of the n different three-vectors, $\boldsymbol{r}_j$ of the individual atoms,

$$\boldsymbol{r}_j = x_j, y_j, z_j; \quad 1 \leq j \leq n. \tag{3.8}$$

If the mass of atom j is m_j, and,

$$m = \sum_{j=1}^{n} m_j, \tag{3.9}$$

is the total mass of the molecule, the three dimensional vector coordinate,

$$\boldsymbol{r} = \frac{1}{m} \sum_{j=1}^{n} m_j \boldsymbol{r}_j \tag{3.10}$$

of the center of mass of the molecule is a separable coordinate. The Hamiltonian of the molecule, in the absence of an external potential, becomes,

$$H = \frac{1}{2m} \boldsymbol{p} \cdot \boldsymbol{p} + H(\boldsymbol{p}_i, \boldsymbol{q}_i), \tag{3.11}$$

where $\boldsymbol{p}$ is the three component vector momentum conjugate to $\boldsymbol{r}$,

$$\boldsymbol{p} = m \dot{\boldsymbol{r}}, \tag{3.12}$$

and $\boldsymbol{p}_i, \boldsymbol{q}_i$ are the $3(n-1)$ dimensional internal momenta and internal coordinates of the molecule.

If, now, for a system consisting of the set $\boldsymbol{N}$ of molecules one assumes the mutual potential energies of the molecules to be negligible in the pertinent region of integration in the phase space, and arbitrarily set equal to zero, the total Hamiltonian of the system becomes separable in the coordinate-momenta of the single molecules,

$$H_{\boldsymbol{N}} = \sum_a \sum_{j_a=1}^{N_a} \left[\frac{1}{2m_a} \boldsymbol{p}_{ja} \cdot \boldsymbol{p}_{ja} + H_{ai}(\boldsymbol{p}_{ija}, \boldsymbol{q}_{ija}) \right], \tag{3.13}$$

where, in general, the form of the "internal" Hamiltonians, H_{ai}, will be different for the different species, a, of molecules, but identical for all members, j_a, of one species.

The Hamiltonians H_{ai} will usually not satisfy the condition b of the previous section of leading to quantum states with energies closely spaced compared to kT along every quantum number. Actually the internal coordinates are, at least approximately, separable, (Sect. 6) and the terms due to the rotational coordinates and momenta are usually sufficiently classical, but the vibrational terms are seldom so. In any case, one defines an internal partition function, $Q_i(T)$, for each species of molecule,

$$Q_i(T) = \sum_n \mathrm{e}^{-\varepsilon_n/\mathrm{k}T}, \tag{3.14}$$

where the sum runs over all quantum states, n, of the internal degrees of freedom of the molecule, and ε_n is the (internal) energy of the state n. (The symbol ε is used for energy per molecule, here and later, whereas E is used for the energy of the complete macro-system.)

The Hamiltonian of the translational degrees of freedom certainly obeys the condition that the energies of adjacent quantum states are closely spaced, since a region $\Delta p_x \Delta x = h$ with $\Delta x = L_x$, the macroscopic distance along the x-axis between the two containing walls, has a very small Δp_x. Following the prescription of replacing summation over states by integration, we define a translational partition function, Q_{tr}, for each species of molecule, which now depends on V, as

$$\left.\begin{aligned} Q_{\mathrm{tr}}(V,T) &= \frac{1}{h^3}\iiint_V d\boldsymbol{r} \iiint_{-\infty}^{\infty} \mathrm{e}^{-(1/2m\mathrm{k}T)\boldsymbol{p}\cdot\boldsymbol{p}}\, d\boldsymbol{p} \\ &= \frac{V}{h^3}\int_0^{\infty} 4\pi p^2 \mathrm{e}^{-(1/2m\mathrm{k}T)p^2}\, dp \\ &= V\left(\frac{2\pi m \mathrm{k}T}{h^2}\right)^{\frac{3}{2}} = V\lambda^{-3}, \end{aligned}\right\} \tag{3.14'}$$

where λ, of dimension length is,

$$\lambda = \sqrt{\frac{h^2}{2\pi m \mathrm{k}T}}, \tag{3.15}$$

and sometimes called the DE BROGLIE wavelength at T.

If now the separable Hamiltonian, $H_{\boldsymbol{N}}$, of Eq. (3.13) is used in Eq. (1.16) for the partition function, Q_C, of the Canonical Ensemble, and the Boltzmann limit, that one divides by $\boldsymbol{N}!$ and sums over all states $\boldsymbol{m}$ of Schroedinger solutions, is used, one finds

$$Q_C(V,T,\boldsymbol{N}) = \prod_a \frac{[V\lambda_a^{-3} Q_{ia}(T)]^{N_a}}{N_a!}. \tag{3.16}$$

With this in Eq. (1.18) for the partition function of the Grand Canonical Ensemble, inverting the summation and product,

$$\left.\begin{aligned} Q_{GC}(V,T,\boldsymbol{\mu}) &= \prod_a \left[\sum_{N_a \geq 0} \frac{[\mathrm{e}^{\mu_a/\mathrm{k}T} V\lambda_a^{-3} Q_{ia}]^{N_a}}{N_a!}\right] \\ \log Q_{GC}(V,T,\boldsymbol{\mu}) &= \sum_a V\lambda_{ia}^{-3} \mathrm{e}^{\mu_a/\mathrm{k}T} Q_{ia}(T). \end{aligned}\right\} \tag{3.17}$$

The Gibbs equation, (1.14), for the pressure

$$PV = \mathrm{k}T \log Q_{\mathrm{GC}}$$

is seen to lead to a total pressure, P, made up additively by a contribution, P_a, from each constituent,

$$P = \sum_a P_a, \tag{3.18}$$

with the partial pressure of constituent a given by,

$$P_a = \mathrm{k}T\, e^{\mu_a/\mathrm{k}T}\, \lambda_a^{-3}\, Q_{ia}(T). \tag{3.19}$$

The relation for every molecular species,

$$\mu_a = \mathrm{k}T\left[\log \frac{P_a \lambda_a^3}{\mathrm{k}T} - \log Q_{ia}(T)\right], \tag{3.20}$$

leads to the expression for G, the Gibbs Free Energy,

$$G = N\mu = N\mathrm{k}T\left[\log \frac{P\lambda^3}{\mathrm{k}T} - \log Q_i\right], \tag{3.21}$$

for every one component perfect gas. In a mixture of perfect gases the free energy is the sum of that of the component species at their own partial pressures.

The thermodynamic relation,

$$(\partial G/\partial P)_{T,N} = V, \tag{3.22}$$

leads to the perfect gas law,

$$V = \frac{N\mathrm{k}T}{P}. \tag{3.23}$$

The Helmholtz free energy, $A = G - PV$, is found to be

$$A = N\mathrm{k}T\left[\log \frac{\lambda^3 N}{\mathrm{e}V} - \log Q_i\right], \tag{3.24}$$

where $-PV = -N\mathrm{k}T = N\mathrm{k}T\log(1/\mathrm{e})$ has been used. This can also be obtained directly from (1.24) giving A as $-\mathrm{k}T\log Q_{\mathrm{C}}$, using (3.16) for Q_{C}, and the Stirling approximation,

$$\log N! = N(\log N - 1) = N\log\frac{N}{\mathrm{e}}. \tag{3.25}$$

The entropy $S = -(\partial A/\partial T)_V = -(\partial G/\partial T)_P$ is,

$$S = N\mathrm{k}\left[\log\left(\frac{\mathrm{e}^{\frac{5}{2}}}{\lambda^3}\frac{V}{N}\right) - \frac{d}{dT}\,T\log Q_i(T)\right], \tag{3.26}$$

and the energy, $E = A + TS$, is,

$$E = N\mathrm{k}T\left[\frac{3}{2} + T\frac{d\log Q_i}{dT}\right]. \tag{3.27}$$

The Heat Content, or enthalpy, $H = E + PV$, is

$$H = N\mathrm{k}T\left[\frac{5}{2} + T\frac{d\log Q_i}{dT}\right]. \tag{3.28}$$

The total number, $\Omega(V, E_{\mathrm{tr}}, N)$, of quantum states of translation of total energy E_{tr}, for a gas of identical molecules in volume V, is, from (2.11),

$$\Omega(V, E, N) = \frac{V^N}{N!\,h^{3N}} \underset{\Sigma p_i^2 = 2mE_{\mathrm{tr}}}{\iint\cdots\int} \prod_{i=1}^{3N} dp_i, \tag{3.29}$$

in which the integral is the surface area, $A_{3N}(R)$, of a $3N$ dimensional sphere of radius $R=\sqrt{2mE_{\text{tr}}}$. The asymptotic expression, for large N, $\log A_{3N}(R)=\frac{3N}{2}\log\frac{2\pi e R^2}{3N}$, may be used in this to find, from (1.28),

$$S_{\text{tr}} = k\log\Omega_{\text{tr}} = k N \log\left[\frac{V}{N}\,e^{\frac{5}{2}}\left(\frac{2\pi m}{h^2}\,\frac{2}{3}\,\frac{E_{\text{tr}}}{N}\right)^{\frac{3}{2}}\right]. \tag{3.30}$$

If, in this, one uses $E_{\text{tr}}=\frac{3}{2}N\,\mathsf{k}T$, from (3.27) for the translational contribution to the energy one arrives at the contribution to the translational entropy expressed in Eq. (3.26).

One general result should be pointed out here. If the gas consists of several species a, of the same, or closely similar masses, $m_a=m$ for all a, and the mechanical properties of the molecules are sufficiently identical that the internal partition functions are the same, $Q_{ia}=Q_i$ for all a, then the expressions (3.27) and (3.28) for E and H may be regarded as those for $N=\sum N_a$ identical molecules. However since $N_a=x_a N$, with x_a the number fraction of species a,

$$\sum_a x_a = 1, \tag{3.31}$$

the entropy, Eq. (3.26), will contain a positive additive term,

$$S_{\text{mix}} = N\,\mathsf{k}\sum_a\left[-\,x_a\log x_a\right] \tag{3.32}$$

in addition to the entropy of N identical molecules of this mass, m, and mechanical properties. This was regarded as a paradox by GIBBS. It is less paradoxical in the light of more modern understanding that no conceivable physical adiabatic process permits two unlike species, a and b, to approach absolute identity.

However, such a situation actually often exists in the case that one of the elements composing a single molecular species consists of several isotopes. Since it will later be shown (Sect. 11) that this entropy of mixing term, at least at high temperatures, retains the same value in all systems containing the same number of atoms of the element in question, it can often be, and usually is, discarded in tabulations.

The summary of this section is that the thermodynamic potentials, E, H, A, and G, of a given number of moles of a perfect gas may always be written as a sum of two terms, one due to translation, and one due to the internal degrees of freedom. The terms due to translation in S, A, and G, depend on P, T, [or (V/N), T] and differ for different molecular species only in the mass, m. For E and H the translational term depends only on T. The term due to the internal degrees of freedom depends only on T and the internal energy states, ε_n, of the molecule.

The equations for one mol of gas are given below. We use[1],

$$h = (6.62517 \pm 0.00023)\times 10^{-27}\,\text{erg sec},$$
$$\mathsf{k} = (1.38044 \pm 0.00007)\times 10^{-16}\,\text{erg deg}^{-1},$$
$$N_0 = (6.02486 \pm 0.00016)\times 10^{23}\,(\text{gm mole})^{-1}.$$

Derived constants:

$$\mathrm{R} = \mathsf{k}N_0 = 8.31696\times 10^{7}\,\text{erg mole}^{-1}\,\text{deg}^{-1},$$
$$= 1.98709\,\text{cal}_{15}\,\text{mole}^{-1}\,\text{deg}^{-1},$$

[1] The fundamental constants are those of E. R. COHEN, J. W. M. DUMOND, TH. W. LAYTON and J. S. ROLLET: Rev. Mod. Phys. 27, 363 (1955).

$$\mathrm{R}\log x = (19.1505\times 10^{7})\log_{10} x \,\text{erg mole}^{-1}\text{deg}^{-1},$$
$$= 4.57544\log_{10} x \,\text{cal}_{15}\,\text{mole}^{-1}\text{deg}^{-1},$$
$$1\text{ atm} = 1.013246\times 10^{6}\,\text{dyne cm}^{-1} \approx (10^{3}\,\text{gm/cm}^{3}),$$
$$m = \text{mass per molecule},$$
$$M = \text{molecular weight (Physical Scale)},$$
$$V = \text{volume per mole}.$$

$$\lambda = \sqrt{\frac{h^2}{2\pi m \mathrm{k} T}} = \frac{17.46}{\sqrt{MT}}\times 10^{-8}\text{ cm}.$$

$$E_{\mathrm{tr}} = \tfrac{3}{2}\mathrm{R}T, \tag{3.33}$$

$$H_{\mathrm{tr}} = \tfrac{5}{2}\mathrm{R}T, \tag{3.34}$$

$$\left.\begin{aligned} A_{\mathrm{tr}} &= \mathrm{R}T\log\left[\lambda^3\frac{N_0}{V}\frac{1}{\mathrm{e}}\right]\\ &= \mathrm{R}T\left[-\tfrac{3}{2}\log M - \tfrac{3}{2}\log T - \log V + K_{AV}\right]\\ &= \mathrm{R}T\left[-\tfrac{3}{2}\log M - \tfrac{5}{2}\log T + \log P + K_{AP}\right],\end{aligned}\right\} \tag{3.35}$$

$$\left.\begin{aligned} G_{\mathrm{tr}} &= \mathrm{R}T\log\left[\lambda^3\frac{P}{kT}\right]\\ &= \mathrm{R}T\left[-\tfrac{3}{2}\log M - \tfrac{3}{2}\log T - \log V + K_{GV}\right]\\ &= \mathrm{R}T\left[-\tfrac{3}{2}\log M - \tfrac{5}{2}\log T + \log P + K_{GP}\right],\end{aligned}\right\} \tag{3.36}$$

$$\left.\begin{aligned} S_{\mathrm{tr}} &= \mathrm{R}\log\left[\frac{1}{\lambda^3}\frac{V}{N_0}e^{\frac{5}{2}}\right]\\ &= \mathrm{R}\left[\tfrac{3}{2}\log M + \tfrac{3}{2}\log T + \log V + K_{SV}\right]\\ &= \mathrm{R}\left[\tfrac{3}{2}\log M + \tfrac{5}{2}\log T - \log P + K_{SP}\right].\end{aligned}\right\} \tag{3.37}$$

Table 1. *Constants for computation of translational functions*[1].

Constant	Units	Value	R × Constant joule/mol.deg	R × Constant cal/mol.deg
K_{AV}	V cm³	+ 7.07324	+ 58.8279	+ 14.0552
	V liters	+ 0.16548	+ 1.3763	+ 0.3288
K_{AP}	P atm	+ 2.66556	+ 22.1694	+ 5.2967
	P mm	− 3.96775	− 32.9996	− 7.8843
	P dynes/cm²	− 11.16312	− 92,8432	− 22.1821
K_{GV}	V cm³	+ 8.07324	+ 67.1448	+ 16.0423
	V liters	+ 1.16548	+ 9.6933	+ 2.3159
K_{GP}	P atm	+ 3.66556	+ 30.4863	+ 7.2838
	P mm	− 2.96775	− 24.6827	− 5.8972
	P dynes/cm²	− 10.16312	− 84.5263	− 20.1950
K_{SV}	V cm³	− 5.57324	− 46.3524	− 11.0745
	V liters	+ 1.33452	+ 11.0991	+ 2.6518
K_{SP}	P atm	− 1.16556	− 9.6939	− 2.3161
	P mm	+ 5.46775	+ 45.4751	+ 10.8649
	P dynes/cm²	+ 12.66312	+ 105.3187	+ 25.1628

[1] These are for use with the Physical Scale of Atomic Weights.

The internal contributions are given in terms of the internal partition function per molecule,

$$Q_i = \sum_n e^{-\varepsilon_n/kT}, \tag{3.38}$$

where the sum runs over all internal energy states, n, of energy ε_n.

$$\left.\begin{aligned} E_i = H_i &= R\,T^2 \frac{d \log Q_i}{dT} \\ &= \frac{1}{Q_i} N_0 \sum_n \varepsilon_n e^{-\varepsilon_n/kT}, \end{aligned}\right\} \tag{3.39}$$

$$A_i = G_i = -R\,T \log Q_i, \tag{3.40}$$

$$\left.\begin{aligned} S_i &= R \frac{d}{dT} T \log Q \\ &= k\,N_0 \sum_n \frac{e^{-\varepsilon_n/kT}}{Q_i} \left[\frac{\varepsilon_n}{kT} - \log Q_i\right], \end{aligned}\right\} \tag{3.41}$$

$$\left.\begin{aligned} C_{vi} = C_{pi} &= R \frac{d}{dT} T^2 \frac{d \log Q_i}{dT} \\ &= R\left[2T \frac{d \log Q_i}{dT} + T^2 \frac{d^2 \log Q_i}{dT^2}\right] \\ &= R \sum_{n>m}\sum \left[\frac{\varepsilon_n - \varepsilon_m}{kT}\right]^2 e^{-(\varepsilon_n + \varepsilon_m)/kT}. \end{aligned}\right\} \tag{3.42}$$

It is customary to choose the energy, ε_0, of the internal quantum state of lowest energy to be zero. In this case it is seen from (3.37) that in the limit $T \to 0$ the internal energy, E_i, is zero, as is the translational energy, Eq. (3.33). This choice of the zero of energy sets the energy of the gas $E(T)$ to be zero at $T = 0$.

The further treatment of the contribution of the internal energy states to the thermodynamic properties is given in Sect. 6 to 12 inclusive.

4. Imperfect gases. The neglect of the mutual potential energy of the molecules invariably leads to the equations for a perfect gas. In order to investigate the properties of an imperfect gas, or a condensed phase, one must carry out the summations over the quantum states of the whole system, or the integral over coordinates and momentum, without neglect of this mutual interaction potential. The Hamiltonian is now no longer separable in the coordinates of the molecules, and one is forced to find some approximate method to reduce the problem to integrations over a more manageable number than 10^{23} coordinates.

Separability of the internal coordinates of the molecules is usually assumed at the outset, that is one assumes that one adds to the sum of the Hamiltonians of the individual molecules composing the system a potential energy depending only on the center of mass coordinates of the molecules. Summation over the quantum states of the internal degrees of freedom, for all molecules, then gives, as in the perfect gas case, a factor Q_{ia} for each molecule of species a, in the partition function of the system. These factors, in turn, lead to the additive terms, E_i, H_i, A_i, G_i, and S_i listed in Eqs. (3.38) to (3.42), all of which are functions of the temperature, T, only, and independent of the state of aggregation of the molecules. The problem of finding the dependence of the thermodynamic properties of the system on the density of molecules is then essentially the same for every system as it is for the monatomic gases, except in so far as the actual mutual potential energy is different for different kinds of molecules.

The potential energy, $U_N(\boldsymbol{r}_1, \boldsymbol{r}_2, \ldots, \boldsymbol{r}_N)$, for the complete system of the set $\boldsymbol{N}$ of molecules is usually assumed to be a sum of terms, each of which depends only on the distance,

$$\boldsymbol{r}_{\alpha\beta} = |\boldsymbol{r}_\alpha - \boldsymbol{r}_\beta|, \tag{4.1}$$

between a pair, α and β, of the numbered molecules,

$$U_{\boldsymbol{N}}(\boldsymbol{r}_1, \boldsymbol{r}_2, \ldots, \boldsymbol{r}_N) = \sum_{N \geq \alpha > \beta \geq 1} \sum u_{\alpha\beta}(r_{\alpha\beta}). \tag{4.2}$$

The individual pair potentials, $u_{\alpha\beta}(r_{\alpha\beta})$, will depend on the species a, b, of the molecules α and β, but will, of course, be all identical in their functional dependence on $r_{\alpha\beta}$ for a given species pair.

The pair potential is often assumed to have the analytical form,

$$u(r) = u_0\left[\left(\frac{r_0}{r}\right)^{12} - 2\left(\frac{r_0}{r}\right)^6\right], \tag{4.3}$$

known as the LENNARD-JONES 6—12 potential, which has a minimum value,

$$u(r = r_0) = -u_0, \tag{4.3$_1$}$$

at $r = r_0$, both u_0 and r_0 depending on the molecular species. This equation may also be written as

$$u(r) = 4u_0\left[\left(\frac{r^*}{r}\right)^{12} - \left(\frac{r^*}{r}\right)^6\right], \tag{4.3$_2$}$$

$$r^* = \frac{r_0}{2^{\frac{1}{6}}}, \tag{4.3$_3$}$$

where r^* is the distance at which $u(r) = 0$. The form (4.3$_2$) has the characteristic that, in reasonably close approximation, $(r^*)^3 = v_0$, where v_0 is the volume per molecule at zero pressure in a close packed lattice at $T = 0$. This arises simply because the volume per molecule in a close packed lattice is $(r')^3/\sqrt{2}$ where r' is the distance between nearest neighbors, which, at zero pressure will be nearly that of the distance of minimum potential energy, r_0. Small deviations due to the attraction of more distant than neighbor interactions will tend to decrease the volume of the crystal at $T = 0$, whereas the quantum mechanical zero point energy tends to increase it. The heat of sublimation of the crystal, extrapolated to $T = 0$ will be approximately $6u_0$ per molecule, again greater than this due to the influence of distant neighbor interactions, and reduced by the quantum mechanical zero point energy.

If the Hamiltonian, $H_{\boldsymbol{N}}$, of the set $\boldsymbol{N}$ of molecules, is assumed to be separable in the internal and translational degrees of freedom, that is if the mutual potential energy of the molecules is written as a function, $U_{\boldsymbol{N}}(\boldsymbol{r}_1, \ldots, \boldsymbol{r}_N)$, of the coordinates, $\boldsymbol{r}_\alpha$, of the centers of mass only, one writes

$$H_{\boldsymbol{N}} = \sum_{\alpha=1}^{N} \frac{1}{2m_\alpha} \boldsymbol{p}_\alpha \cdot \boldsymbol{p}_\alpha + U_{\boldsymbol{N}} + \sum_{\alpha=1}^{N} H_{i\alpha}(\boldsymbol{p}_{i\alpha}, \boldsymbol{q}_{i\alpha}). \tag{4.4}$$

In this $\boldsymbol{r}_\alpha$ and $\boldsymbol{p}_\alpha$, are the three dimensional Cartesian vector coordinates of the center of mass of the molecules α, and the conjugate vector momentum, respectively. The $\boldsymbol{q}_{i\alpha}$ and $\boldsymbol{p}_{i\alpha}$ are the internal coordinates and conjugate momenta of molecule α. The Hamiltonian, $H_{i\alpha}$, of the internal coordinates and momenta of molecule α will depend, in its functional form, on the species a, to which molecule α belongs. The Hamiltonian, H_N, is separable into $N+1$ parts, N internal Hamiltonians, $H_{i\alpha}$, of the individual molecules, and one translational term,

$$H_{N\,\mathrm{tr}} = \sum_{\alpha=1}^{N} \frac{1}{2m_\alpha} \boldsymbol{p}_\alpha \cdot \boldsymbol{p}_\alpha + U_N(\boldsymbol{r}_1, \ldots, \boldsymbol{r}_\alpha, \ldots \boldsymbol{r}_N), \tag{4.5}$$

depending on the centers of mass of all of the molecules.

The partition function, $Q_C(V, T, \boldsymbol{N})$, is then a product of $N+1$ terms,

$$Q_C(V, T, \boldsymbol{N}) = Q_{\mathrm{tr}}(V, T, \boldsymbol{N}) \prod_a [Q_{ia}(T)]^{N_a}, \tag{4.6}$$

where the translational term, Q_{tr}, is, for almost all real conditions, given adequately by the classical and Boltzmann approximation, as,

$$Q_{\mathrm{tr}} = \frac{1}{\boldsymbol{N}!} \iint_V \cdots \int d\boldsymbol{r}_1 \ldots d\boldsymbol{r}_N \, e^{-\frac{U_N}{kT}} \prod_{\alpha=1}^{N} \left[\frac{1}{h^3} \iiint_{-\infty}^{+\infty} e^{-\frac{\boldsymbol{p}_\alpha \cdot \boldsymbol{p}_\alpha}{2m_\alpha kT}} d\boldsymbol{p}_\alpha\right]. \tag{4.7}$$

The integration over the momenta can be performed readily, and combined with the factor h^{-3}, as in (3.14). There remains the integral,

$$Q_\tau(V, T, \boldsymbol{N}) = \iint_V \cdots \int e^{-U_N(\boldsymbol{r}_1, \ldots \boldsymbol{r}_N)/kT} d\boldsymbol{r}_1 \ldots d\boldsymbol{r}_N. \tag{4.8}$$

One has,

$$Q_{\mathrm{tr}}(V, T, \boldsymbol{N}) = \frac{1}{\boldsymbol{N}!} Q_\tau(V, T, \boldsymbol{N}) \prod_a \lambda_a^{-3N_a}, \tag{4.9}$$

where λ_a is given by (3.15), $\lambda_a = h/\sqrt{2\pi m_a kT}$. The Canonical partition function, Q_C, is then

$$Q_C(V, T, N) = \frac{1}{\boldsymbol{N}!} Q_\tau(V, T, \boldsymbol{N}) \prod_a (Q_{ia}(T)\, \lambda_a^{-3})^{N_a}. \tag{4.10}$$

The expression, (1.18), for the Grand Canonical partition function now becomes

$$Q_{GC}(V, T, \boldsymbol{\mu}) = \sum_{\boldsymbol{N} \geq 0} \frac{1}{\boldsymbol{N}!} Q_\tau(V, T, \boldsymbol{N}) \prod_a \left[Q_{ia} \lambda_a^{-3} e^{\frac{\mu_a}{kT}}\right]^{N_a}. \tag{4.11}$$

The quantity,

$$z_a = \frac{Q_{ia}}{\lambda_a^3} e^{\frac{\mu_a}{kT}}, \tag{4.12}$$

is an activity, namely it has the dimensions of a reciprocal volume, and at fixed temperature its concentration dependence is given by a constant times $e^{\mu/kT}$. In the limit $z_a \to 0$ for all species, a, of molecules, the series of positive terms,

$$e^{PV/kT} = Q_{GC}(V, T, \boldsymbol{\mu}) = \sum_{\boldsymbol{N} \geq 0} \frac{\boldsymbol{z}^{\boldsymbol{N}}}{\boldsymbol{N}!} Q_\tau(V, T, \boldsymbol{N}), \tag{4.13}$$

$$\boldsymbol{z}^{\boldsymbol{N}} \equiv \prod_a z_a^{N_a}, \tag{4.14}$$

$$\boldsymbol{N}! \equiv \prod_a N!, \tag{4.15}$$

has large numerical contributions only from terms for which $N = \sum N_a$ is small. For these terms the regions of integration of (4.8) in which molecules are close together are of negligible weight compared to the region in which all molecular pair distances, $r_{\alpha\beta}$, are great, and for which $U_N \to 0$. That is, for these terms the perfect gas approximation, $U_N=0$, is justified, $Q_\tau = V^N$, and one obtains the perfect gas equation, (3.18) and (3.19) where, from (4.12) for z_a,

$$\frac{P_a}{\mathrm{k}T} = \frac{Q_{ia}\,e^{\mu_a/\mathrm{k}T}}{\lambda_a^3} = z_a, \tag{4.16}$$

and the total pressure, P, is the sum of the partial pressures, $P = \sum_a P_a$. Since, from (1.10) that $(\partial P/\partial \mu_a)_{T,\mu_b} = N_a/V = \varrho_a$, and from (4.12) for z_a, we have, from (4.16) that

$$\left(z_a \frac{\partial}{\partial z_a}\right)_{T,z_b}\left(\frac{P}{\mathrm{k}T}\right) = \varrho_a = z_a, \quad \text{(perfect Gas)}, \tag{4.17$_1$}$$

one finds that the activity, z_a, defined by (4.12) is such that in the limit that all densities, ϱ_a, approach zero, and the system approaches a perfect gas in properties, the activity z_a becomes equal to the number density, ϱ_a, of mulecules of species a.

If $P/\mathrm{k}T$ is regarded as a function of T, $\boldsymbol{z}$ we have for the temperature derivative the equation,

$$\left(\frac{\partial}{\partial T}\right)_{\boldsymbol{z}}\left(\frac{P}{\mathrm{k}T}\right) = \frac{1}{V\mathrm{k}T^2}(E - E_0), \tag{4.17$_2$}$$

with E the energy of the system, and E_0 the energy of the same composition in the perfect gas limit. This may be derived as follows. From (1.9),

$$\begin{aligned}\mathrm{k}T^2\left(\frac{\partial}{\partial T}\right)_{\mu}\left(\frac{P}{\mathrm{k}T}\right) &= \frac{1}{V}[TS - PV]\\ &= \mathrm{k}T^2\left(\frac{\partial}{\partial T}\right)_{\boldsymbol{z}}\left(\frac{P}{\mathrm{k}T}\right) + \sum_a\left[\left(\frac{\partial}{\partial \log z_a}\right)_{T,z_b}\left(\frac{P}{\mathrm{k}T}\right)\right]\mathrm{k}T^2\left(\frac{\partial \log z_a}{\partial T}\right)_{\mu}.\end{aligned}$$

Rearrange this, using (4.17$_1$), to write

$$\begin{aligned}\mathrm{k}T^2\left(\frac{\partial}{\partial T}\right)_{\boldsymbol{z}}\left(\frac{P}{\mathrm{k}T}\right) = \frac{1}{V}[TS - PV] + \sum_a \varrho_a \mathrm{k}T \log z_a -\\ - \sum_a \varrho_a T\left(\frac{\partial}{\partial T}\right)_{\mu_a}(\mathrm{k}T\log z_a).\end{aligned}$$

With $\mathrm{k}T\log z_a = \mu_a - \mu_{0a}$, $\varrho_a\mu_a = \overline{N}_a\mu_a/V$, and $\overline{N}_a\mu_a = G$, $TS - PV + G = E$, we have

$$\mathrm{k}T^2\left(\frac{\partial}{\partial T}\right)_{\boldsymbol{z}}\left(\frac{P}{\mathrm{k}T}\right) = \frac{E}{V} - \sum_a \varrho_a\left(\mu_{0a} - T\frac{d\mu_{0a}}{dT}\right).$$

The quantity μ_{0a} is the chemical potential μ_a of species a in the perfect gas limit, and at number density $\varrho_a = 1$. The temperature derivative, at constant density of the perfect gas is $PV - TS$ per molecule, and this subtracted from μ_a is the energy per molecule. The sum over a of the product of ϱ_a times the energy per molecule of species a in its perfect gas limit is E_0/V.

The summarized results of this section may be stated as follows. In order to express the equation of state of a system it is convenient to use as variables the absolute temperature, T, and the activities, z_a, of the species, a, of molecules present. The activity z_a may be defined in terms of the chemical potential, μ_a, as,

$$z_a = e^{(\mu_a - \mu_{0a})/\mathrm{k}T}, \tag{4.18}$$

with μ_{0a} so chosen that in the limit of the perfect gas, at zero density of all chemical species,

$$\lim_{\varrho\to 0}\left[\frac{z_a}{\varrho_a}\right] = 1. \tag{4.19}$$

The pressure, P, as a function of T and the set,

$$\boldsymbol{z} = \boldsymbol{z}_a, \ldots, \boldsymbol{z}_b, \ldots \tag{4.20}$$

of activities can be expressed by the equation,

$$e^{PV/kT} = \sum_{\boldsymbol{N} \geq 0} \frac{\boldsymbol{z}^{\boldsymbol{N}}}{\boldsymbol{N}!} \int\int_V \cdots \int F_{\boldsymbol{N}}^{(0)}(\boldsymbol{r}_1, \ldots, \boldsymbol{r}_{\boldsymbol{N}})\, d\boldsymbol{r}_1 \ldots d\boldsymbol{r}_N, \tag{4.21}$$

with,

$$F_{\boldsymbol{N}}^{(0)} = e^{-U_{\boldsymbol{N}}(\boldsymbol{r}_1, \ldots, \boldsymbol{r}_N)/kT}. \tag{4.22}$$

In this $U_{\boldsymbol{N}}$ is the mutual potential energy of the set $\boldsymbol{N} = N_a, \ldots, N_b, \ldots,$ of molecules. The notation $\boldsymbol{z}^{\boldsymbol{N}} = \prod_a z_a^{N_a}$ and $\boldsymbol{N}! = \prod_a N_a!$ is used. The zero of potential is such that $U_{\boldsymbol{N}} \to 0$ for all pair distances, $r_{\alpha\beta} \to 0$. The zero of energy is thus chosen as that of the perfect gas at absolute zero. The function $F_{\boldsymbol{N}}^{(0)}$ is proportional to the probability density that the set $\boldsymbol{N}$ of molecules, if they alone are in the volume V, are at $\boldsymbol{r}_1, \ldots, \boldsymbol{r}_N$. It is normalized to approach unity in value if all pair distances approach infinity.

5. Discussion of the imperfect gas equation. Eq. (4.21) is valid for an open system, that is a system of infinite extent in which the temperature and activities, z_a are specified. The volume, V, over which the integration of (4.21) is extended may be chosen to have any shape or size, but is not to be thought of as being limited by physical container walls of the system. The potential energy, $U_{\boldsymbol{N}}(\boldsymbol{r}_1, \ldots, \boldsymbol{r}_N)$, should now contain a term due to the averaged effect of the molecules outside of this mathematically defined volume. This term will depend on the average density, that is on $\boldsymbol{z}$, as well as the temperature. Differently stated the partition function, $Q_C(V, T, \boldsymbol{N})$ of the Canonical Ensemble strictly depends on the nature of the surface container, since this, in turn, influences the potential energy of the molecules in the neighborhood of the surface of V. The correct Q_C to insert in the expression (1.18) for the Grand Canonical Partition Function, $Q_{GC}(V, T, \boldsymbol{\mu})$, is that in which the walls are composed of molecules at the temperature T and activity set $\boldsymbol{z}$.

This surface term would be fantastic to evaluate. Since, however, it can only contribute to the integrand for molecules that are within molecular distances of the surface it can happily be neglected by choosing V sufficiently large. However, this automatically requires that the sum of (4.21) will not converge for any reasonable value of N. The individual terms for each set $\boldsymbol{N}$ in the sum of (4.21) are proportional to the probability of finding the set $\boldsymbol{N}$ in V. These terms rise to a maximum in value at the set $\boldsymbol{N}$ which is the most probable number set of molecules to be found in V. If V is chosen as large enough that the molecules at the surface are a negligible fraction, the $\overline{N}$ must be enormous, indeed usually with a total number of the order of 10^{20} or greater. Thus the sum of (4.21) is far from being convergent in the simple numerical sense of being computable by evaluation of a reasonable number of the first terms. It is this characteristic that requires the manipulations given in Sects. 13 and 15 in order to obtain equations that are subject to numerical evaluation.

However, the sum of (4.21) for any finite V, and for any real system is certainly convergent in the strict mathematical sense of eventual convergence. This is most easily seen from (4.22). All real potentials, $U_{\boldsymbol{N}}$, become positive if all the molecules are crowded too close together. The physical situation would not be altered if one assumed that $U_{\boldsymbol{N}}$ became infinity, and hence $e^{-U_{\boldsymbol{N}}/kT}$ identically zero, if any pair distance, $r_{\alpha\beta}$ becomes smaller than some r_0. There will then be a maximum total $N = \Sigma N_a$, namely $N = N_m = \sqrt{2}\, V/r_0^3$ such that *all* configurations, $\boldsymbol{r}_1, \ldots, \boldsymbol{r}_N$ for $N > N_m$ have an identically zero integrand, and all terms for $N > N_m$ will be zero. The sum will then have a finite limit, $N = N_m$, linearly proportional to V, the volume over which the integration is extended.

However, the absolute convergence of the sum in (4.21) for finite V presents us with an apparent dilemma. Any function which can be expressed as a power series in $z_1, z_2, \ldots, z_a, \ldots$, which is absolutely convergent in the range $0 \leq z_a \leq \infty$, for all a, is necessarily analytic in this region. The function, and all its derivatives will be continuous. The experimental function, $P(T, z)$, and hence $e^{PV/kT}$, is known to have experimentally observable discontinuities in the first derivatives at those values of the set, $\boldsymbol{z}$, of activity, for which phase transitions, such as condensation, occur. This is clear from (4.17), since the first derivative of P/kT with respect to $\log z_a$ is the number density, $\varrho_a = N_a/V$, which increases discontinuously in going from the gas to the liquid or crystalline phase. The dilemma is readily removed[1] by considering (4.21) to be valid, with the omission of the $\boldsymbol{z}$ dependent surface terms in U_N, only in the limit $V \to \infty$. In the limit $V \to \infty$ there is no necessity that the series (4.21) be absolutely convergent, and singularities in the function $e^{PV/kT}$ can, and do, occur.

However, the description of the series (4.21) for macroscopic V, as leading to an expression for $P(T, z)$ which is analytic, although correct in the strict mathematical sense, is apt to be misleading in numerical terms. There is no difficulty in constructing, heuristically, values for the quantity $Q_\tau(V, T, \boldsymbol{N})$, Eq. (4.8), that, even for volumes as small as (1 mm³) lead to abrupt changes in the slope of P versus z_a, so abrupt that they would appear to be discontinuous within the limits of error of any experimental measurement.

It is probably instructive to demonstrate this for the one component system. The quantity, $Q_\tau(V, T, N)$, may be written as

$$\frac{1}{N!} Q_\tau(V, T, N) = e^{-A_\tau/kT}, \tag{5.1}$$

where $A_\tau(V, T, N)$, apart from an additive constant dependent on T and proportional to N, is the Helmholtz free energy. This quantity can be written as,

$$A_\tau = N a_\tau(T, v), \qquad v = V/N. \tag{5.2$_1$}$$

The thermodynamic relation between a_τ and $\log z$ is given by:

$$a_\tau = kT \log[z(v)] - v P(v). \tag{5.2$_2$}$$

The function $a_\tau(v)$ at fixed T is always monotonically decreasing with v, and always with a non-negative second derivative, since

$$(\partial a_\tau/\partial v)_T = -P, \tag{5.3$_1$}$$

$$(\partial^2 a_\tau/\partial v^2)_T = -(\partial P/\partial v)_T \geq 0. \tag{5.3$_2$}$$

The sum (4.21) can be expressed as,

$$e^{PV/kT} = \sum_{N \geq 0} e^{N[\log z - (a_\tau/kT)]}. \tag{5.4}$$

The (positive) terms, T_N, of the sum rise to a maximum value when the derivative with respect to N is zero or when,

$$\frac{d \log T_N}{dN} = \log z - \frac{a_\tau}{kT} - \frac{N}{kT}\left(\frac{d a_\tau}{dN}\right)_{V,T} = 0. \tag{5.5$_1$}$$

---
[1] L. VAN HOVE: Physica, Haag **15**, 951 (1949).

Since, from (5.2), $v = V/N$, this is when

$$\log z - \frac{a_\tau}{\mathsf{k}T} + \frac{v}{\mathsf{k}T}\left(\frac{\partial a_\tau}{\partial v}\right)_T = 0, \tag{5.5$_2$}$$

or from (5.3$_1$) when (5.2$_2$) is satisfied.

The terms of the sum in (5.4) have a_τ dependent on N through the relationship (5.2$_1$) that $v = V/N$, and the negative volume derivative, Eq. (5.3$_1$), of a_τ is the pressure $P(T, v)$ which varies from term to term and is not, generally, the pressure, $P(T, z)$, appearing on the left. It is only for the largest term of the sum that (5.5$_2$) and (5.2$_2$) are satisfied. Using the principle that the logarithm of the sum in (5.4) is adequately given by the logarithm of the largest term one again arrives at (5.2$_2$) for $P(T, z)$, namely

$$\frac{V}{\mathsf{k}T}P(T, z) = N\log z - \frac{a_\tau}{\mathsf{k}T},$$

showing that the pressure computed from the (5.4) is the same as the pressure computed from (5.3$_1$) for the term with the most probable number of molecules.

Now turn to the second derivative, $d^2\log T_N/dN^2$, which from (5.5$_1$) and $v = V/N$ is found to be,

$$\frac{d^2\log T_N}{dN^2} = \frac{V}{N^2\mathsf{k}T}\left(\frac{\partial}{\partial v}\right)_T\left[a_\tau - v\left(\frac{\partial a_\tau}{\partial v}\right)_T\right] = -\frac{v^2}{N\mathsf{k}T}\left(\frac{\partial^2 a_\tau}{\partial v^2}\right)_T, \tag{5.6}$$

which, from (5.3$_2$), is never positive. The extremum we found by setting $d\log T_N/dN = 0$ is indeed never a minimum! However, there may exist a range of volumes per molecule, say between the volume per molecule in the liquid, v_l, and that for the gas, v_g, for which $(\partial P/\partial v)_T = -(\partial^2 a_\tau/\partial v^2)_T = 0$. There exists, then, for one unique value of z, namely $z = z_c$, a range of v values $v_l \leq v \leq v_g$ for which $a_\tau = a_\tau(v_l) - P_c(v - v_l)$ and Eq. (5.5$_2$)

$$\frac{d\log T_N}{dN} = \log z_c - \frac{1}{\mathsf{k}T}\left[a_\tau(v_l) + P_c v_l\right] = 0$$

is satisfied independently of v.

For this unique $z = z_c$, then, there is no longer *one* largest term in the sum of (5.4) but a range of N values,

$$V/v_g = N_g \leq N \leq N_l = V/v_l,$$

for which the terms are of equal magnitude. Putting in a realistic case of a gas condensing at about room temperature at one atmosphere, and as small a V as 0.5 mm³, one may have,

$$N_g \approx 10^{16}, \qquad N_l \approx 10^{19}.$$

Now examine the relative magnitudes of these terms at $z = z_c(1 + \varepsilon)$. The ratio of the magnitudes of the terms at $N = N_l$ and at $N = N_g$, will be

$$\frac{T_{N_l}}{T_{N_g}} = (1 + \varepsilon)^{N_l - N_g} \approx e^{10^{19}\varepsilon}.$$

If ε is as small as $+10^{-9}$, which would be considered rather extreme experimental accuracy, the ratio of probability that one would observe the value $N = N_l$ to the probability of $N = N_g$ will shift to $e^{10^{10}}$, a sufficiently large value to be regarded as a practical certainty. For $\varepsilon = -10^{-9}$ the shift in the opposite direction is equal. It is equally easy to demonstrate that, for this behavior of $Q_\tau/N!$, the shift of $(\partial(P/\mathrm{k}T)/\partial \log z)_T$ from the gas density, $\varrho_g = 1/v_g$, to the liquid density $\varrho_l = 1/v_l$, is complete within this experimentally negligible range of the variable z.

One thus sees that although the mathematical language, "in the limit $V \to \infty$", is convenient to apply to Eq. (4.21), since it permits us to discuss true singularities in the function $P(T, z)$, it is not to be assumed that computations made for finite experimental cases would lead to results discordant with experimental experience.

Several conclusions can be drawn from Eq. (4.21) without recourse to the procedure by which it is put into more computable form. One of these is the "law of corresponding states". We shall assume that the mutual potential energy, U_N, for all single species of molecule, say species a, has one universal functional dependence of the type,

$$U_{N_a}(\boldsymbol{r}_1, \dots \boldsymbol{r}_{N_a}) = u_{0a}\, \mathscr{U}_{N_a}^{(a)}\left[\frac{\boldsymbol{r}^1}{r_{0a}}, \dots, \frac{\boldsymbol{r}_{N_a}}{r_{0a}}\right],$$

where the function $\mathscr{U}_N^{(a)}$ is a universal function which is the same for all species a of the class of molecules which obey the law. Consider the function $Q_{\tau a}(V, T, N)$ for a single component system composed of molecules of species a, and introduce a dimensionless measure of temperature,

$$t = \mathrm{k}T/u_{0a} \tag{5.7}$$

and a dimensionless measure of volume,

$$\mathscr{V} = V/r_{0a}^3 = V/v_{0a}, \tag{5.8$_1$}$$

$$v_{0a} = r_{0a}^3, \tag{5.8$_2$}$$

n the integral of (4.8) for Q_τ. Using the dimensionless coordinate,

$$\boldsymbol{R}_i = \frac{\boldsymbol{r}_i}{r_{0a}}. \tag{5.9}$$

The integral becomes,

$$\left.\begin{aligned} Q_{\tau a}(V, T, N) &= v_{0a}^N \iint_{\mathscr{V}} \dots \int e^{-t^{-1}\mathscr{U}_N^{(a)}(\boldsymbol{R}_1, \dots \boldsymbol{R}_N)}\, d\boldsymbol{R}_1 \dots d\boldsymbol{R}_N \\ &= v_{0a}^N\, \mathscr{Q}_\tau(\mathscr{V}, t, N), \end{aligned}\right\} \tag{5.10}$$

where $\mathscr{Q}(\mathscr{V}, t, N)$ is a universal function, independent of the species of molecule, of the reduced volume, $\mathscr{V}$, and reduced temperature, t. The Grand Partition Function, (4.13) is now,

$$e^{PV/\mathrm{k}T} = \sum_{N \ge 0} \frac{(z_a v_{0a})^N}{N!}\, \mathscr{Q}(\mathscr{V}, t, N). \tag{5.11}$$

With the reduced activity,

$$\zeta = z_a v_{0a}, \tag{5.12}$$

and reduced pressure,

$$p = \frac{P v_{0a}}{u_{0a}} \tag{5.13}$$

Eq. (5.11) becomes,

$$e^{p\mathscr{V}/t} = \sum_{N \ge 0} \frac{\zeta^N}{N!}\, \mathscr{Q}(\mathscr{V}, t, N), \tag{5.14}$$

independent of the molecular species.

It follows that all species of molecules for which the mutual potential energies, U_N, can be superimposed by adjusting the scales of length and energy, the equation of state expressed in terms of the reduced dimensionless temperature, t, pressure, p, volume, $\mathscr{V}$, or activity, ζ, is a universal function, $p(t, \zeta)$, or $p(t, \mathscr{V}/N)$, independent of the species.

The reduced equation of state will depend, of course, on the analytical form of $\mathscr{U}_N^{(a)}$. Without knowing this form no significance in terms of the potential can be assigned to u_{0a} or v_{0a}. One may, however, choose them in terms of the observed equation of state. One frequently chooses u_{0a} in such a way that at the critical temperature t is unity, and v_{0a} as the critical volume per molecule,

$$u'_{0a} = \mathrm{k} T_{ca}, \tag{5.15}$$

$$v'_{0a} = v_{ca}. \tag{5.16}$$

Alternatively, if a universal potential function is assumed, one might choose u_{0a} and r_{0a} so as to have simple significance in terms of this potential. One of the most frequent choices, which appears to be good for a large class of molecules is the sum of pair potential, Eq. (4.2), with the individual pair potential given by the Lennard-Jones 6-12 potential, (4.3$_2$). This choice is generally assumed to be good for spherical or nearly spherical non-polar molecules. Diatomic and many polyatomic molecules are, in this sense, sufficiently "nearly spherical". One then chooses u_{0a} to be the magnitude of the minimum of the pair potential, and $r_{0a} = r^*$, the value of r for which the potential is zero in the repulsive side. For this choice v_{0a} is approximately the volume per molecule of the crystal at 0° K, if the crystal form is close packed. Empirically one finds for this choice,

$$3.16\, v_{0a} \approx v_{ca}, \tag{5.17}$$

$$1.26\, u_{0a} \approx \mathrm{k} T_{ca}, \tag{5.18}$$

as approximate relationships.

Eq. (4.21) of the previous section has been derived with a minimum of amplification of the assumptions, or discussion of them, in order to attain a concise and brief statement. With appropriate modifications of interpretation this equation is far more generally applicable than the derivation given in the previous section would appear to indicate. We shall now give some amplification of this point.

The purely formal manipulation of Eq. (4.21) is carried out in Sects. 13 and 15 of this article and leads finally to an expression for the pressure P as a power series in density, ϱ. The coefficient of ϱ^ν known as the ν'th virial coefficient, is there expressed as an integral over the mutual coordinates of ν molecules. This formal procedure requires, for its validity, only very general properties of the functions, $F_N^{(0)}$, occurring as the integrands of (4.21). The most essential property that is used is the following. If the coordinates, $\boldsymbol{r}_1, \ldots, \boldsymbol{r}_{N+M}$ of the function $F_{\boldsymbol{N}+\boldsymbol{M}}$ take such values that the molecules fall naturally into two subsets, $\boldsymbol{N}$, and $\boldsymbol{M}$, such that all pair distances, r_{nm}, become very large if molecule n is a member of $\boldsymbol{N}$, and m a member of $\boldsymbol{M}$, then $F_{\boldsymbol{N}+\boldsymbol{M}}$ approaches in value the product $F_{\boldsymbol{N}} F_{\boldsymbol{M}}$. This may be written as,

$$\lim_{\substack{r_{nm}\to\infty \\ n\subset \boldsymbol{N},\, m\subset \boldsymbol{M}}} \text{all}\; [F_{\boldsymbol{N}+\boldsymbol{M}}^{(0)}(\boldsymbol{r}_1, \ldots, \boldsymbol{r}_{N+M})] = F_{\boldsymbol{N}}^{(0)}(\boldsymbol{r}_1, \ldots, \boldsymbol{r}_N)\, F_{\boldsymbol{M}}^{(0)}(\boldsymbol{r}_{N+1}, \ldots, \boldsymbol{r}_{N+M}). \tag{5.19}$$

In terms of the potential energy, $U_{\boldsymbol{N}+\boldsymbol{M}}$ this means that

$$\lim_{\substack{r_{nm}\to\infty\\ n\subset\boldsymbol{N},\, m\subset\boldsymbol{M}}} \text{all } [U_{\boldsymbol{N}+\boldsymbol{M}}(\boldsymbol{r}_1,\ldots,\boldsymbol{r}_{N+M})] = U_{\boldsymbol{N}}(\boldsymbol{r}_1,\ldots,\boldsymbol{r}_N) + U_{\boldsymbol{M}}(\boldsymbol{r}_{N+1},\ldots,\boldsymbol{r}_{N+M}), \tag{5.20}$$

which is obviously satisfied if the forces between the molecules are short range. Thus the formal results of Sects. 13 and 15 are valid for a very wide variety of possible potentials, and are by no means limited to the Lennard-Jones 6-12 potential given by (4.3), nor to the assumption that the total potential, $U_{\boldsymbol{N}}$, is given as a sum of pair potentials as in Eq. (4.2). In fact these general developments do not even make use of the property that the potential, $U_{\boldsymbol{N}}$, is independent of temperature.

Of course any numerical evaluation of the virial coefficients or other quantities obtained in this formal manipulation depends on an exact assumption about the form of the potential, and for this purpose the Lennard-Jones 6-12 potential between pairs, and a sum of pair potentials is frequently assumed.

In Sects. 22 to 27 a formal treatment of the exact quantum mechanical problem is presented. In Sect. 24 a function $F_N^{(0)}(\boldsymbol{r}_1,\ldots,\boldsymbol{r}_N)$ is defined as the diagonal element of the density matrix for the equilibrium ensemble, integrated over the internal coordinates of the molecules. This quantity $F_N^{(0)}(\boldsymbol{r}_1,\ldots,\boldsymbol{r}_N)$ is proportional to the probability density of finding the set $\boldsymbol{N}$ of molecules at the coordinates $\boldsymbol{r}_1,\ldots,\boldsymbol{r}_N$, for a system containing exactly the set $\boldsymbol{N}$ of molecules in some volume V which includes all the coordinate values. The proportionality constant is such that $F_{\boldsymbol{N}}^{(0)}\to 1$ if all pair distances, $r_{\alpha\beta}\to\infty$. In the classical limit the function $F_{\boldsymbol{N}}^{(0)}$ is given by Eq. (4.22) as $e^{-U_{\boldsymbol{N}}/kT}$.

Even when the classical limit is not valid the function $F_{\boldsymbol{N}}^{(0)}$ will still have the property of Eq. (5.19), as well as the other characteristics necessary for the development given in Sects. 13 and 15. The equation for the Grand Canonical Partition Function, and hence for the pressure is exactly (4.21), even when (4.22) is invalid, and hence (4.21) and the formal results of Sects. 13 and 15 are completely general. However, it is to be emphasized that numerical evaluation of the virial coefficients, or other experimentally observable quantities depend on a knowledge of the functions $F_{\boldsymbol{N}}^{(0)}$, at least for small sets $\boldsymbol{N}$. This is not simply solved in the non-classical case.

B. The internal partition function.

6. Introduction and method. The internal partition function for a single isolated molecule,

$$Q_i = \sum_n e^{-\varepsilon_n/kT} \tag{6.1}$$

is a sum over all internal quantum states, n, of $e^{-\varepsilon_n/kT}$, in which ε_n is the energy of the state n. The zero of energy is conventionally chosen to be that of the lowest energy level of the molecule, in which case the thermodynamic zero of energy is that of the perfect gas at 0° K. The word "state" is used to designate a single allowed independent quantum state of the molecule corresponding to an allowed single independent solution of the time independent Schroedinger equation with the separated Hamiltonian, $H_i(\boldsymbol{q}_i)$, of the internal coordinates of the molecule. One "level" of given energy may be degenerate, and include several states. For instance any level of total angular momentum j, $\left[P^2 = j(j+1)\frac{h^2}{4\pi^2}\right]$, will consist of $2j+1$ states of the same energy, corresponding to the $2j+1$ possible values of m_j, $-j\leq m_j\leq +j$, where the projection, P_z, of the angular momentum on a unique axis, say the z-axis, is $P_z = m_j \frac{h}{2\pi}$. The sum in (6.1) will then include all $2j+1$ different states, m_j, for given j.

For monatomic molecules, that is for free atoms, the different energy levels are due to electronic excitation, and are, as a rule widely spaced in energy compared to $\mathbf{k}T$. Except for the hydrogen atom, the values of the energies, ε_n, do not obey any simple analytical expression. The summation of (6.1) must usually be done numerically, inserting the observed energies of the spectroscopic terms, and remembering to count each term of given j value $2j+1$ times. Fortunately the wide spacing of the terms usually means that only very few, and very often only the first level, need be counted.

For instance the noble gases, which are the only atoms that form stable monatomic gases at ordinary densities and temperatures, have a single $^1S_0$ lowest electronic level, for which $J=0$, and which is chosen as having zero energy. This term contributes unity to the sum. The next excited level in Helium has an energy $\frac{\varepsilon}{hc} = 1.6 \times 10^5$ cm^{-1}, $\varepsilon/\mathbf{k} = 2.3 \times 10^5$ °K, so that the exponential $e^{-\varepsilon/\mathbf{k}T}$ is appreciable only at about 10^5 °K. Even for Xenon the first excitation energy, which is the lowest of any of the noble gases, is $\frac{\varepsilon}{hc} = 6.7 \times 10^4$ cm^{-1}, $\varepsilon/\mathbf{k} = 9.6 \times 10^4$ °K. For all ordinary temperatures one may safely set $Q_i = 1$, $\log Q_i = 0$, for the noble gases. (See Sect. 8 for a discussion of nuclear spin contributions.)

Among the relatively volatile alkalis, caesium has the lowest excitation potential. The lowest level is $^2S_{1/2}$ with $J = 1/2$, $2J+1 = 2$. The levels $^2P_{1/2}$, $^2P_{3/2}$ with $J = 1/2$ and $J = 3/2$ respectively have energies 11178.3 cm^{-1} and 11732.3 cm^{-1} or with $\varepsilon/\mathbf{k} = 1.60 \times 10^4$ °K and 1.69×10^4 °K. The next level has $\varepsilon/\mathbf{k} = 2.1 \times 10^4$ °K. The partition function,

$$Q_i(\mathrm{Cs}) = 2 + 2e^{-(1.60 \times 10^4/T)} + 4e^{-(1.69 \times 10^4/T)} + \cdots$$

differs from 2 appreciably only above several thousand degrees Kelvin.

One example of an atomic gas for which the partition function is not trivial is that of fluorine. The molecule F_2 dissociates appreciably at 2000° K so that the atoms constitute an appreciable fraction of the gas at this or higher temperatures. The lowest level of the atom is $^2P_{3/2}$, $J = 3/2$, $2J+1 = 4$, but, the first excited level, $^2P_{1/2}$, $J = 1/2$, $2J+1 = 2$, has $\frac{\varepsilon}{hc} = 407.0$ cm^{-1} $\varepsilon/\mathbf{k} = 587°$ K. The next excited level can be neglected. We have

$$Q_i(F) = 4 + 2e^{-587/T}.$$

For diatomic and polyatomic molecules there are many low energy excited levels due to rotation and vibration. For a given electronic state of the molecule the energies of these levels are given, with satisfactory accuracy in most cases, by relatively simple analytic expressions in terms of the quantum numbers, and within this approximation, the energies are additive in terms due to the different quantum numbers. This arises, of course, due to the fact that the internal Hamiltonian can be separated in reasonable approximation. As discussed in Sect. 3 when the Hamiltonian is separable into additive terms, each dependent on only a small subset of the coordinates and momenta not occuring in other terms, the wave function solution is a product, and the quantum mechanical energies are sums. In this case the partition function is a product, Eq. (3.5), and its logarithm a sum. One thus obtains additive contributions to the thermodynamic potentials due to the different degrees of freedom.

For diatomic molecules there are six degrees of freedom of motion of the point nuclei, of which three can be separated as the coordinates of the center of mass. The remaining three internal coordinates can be chosen as two polar angles, and one distance, r, between the two nuclei. The additive part of the Hamiltonian dependent on the angles ϑ and φ contains r implicitly as the moment of inertia, I, which is $[m_1 m_2/(m_1+m_2)]\, r^2$. As an approximation one inserts for r the value, r_0, at which the potential is a minimum. The remaining terms in the Hamiltonian depend only on r. For small displacements the potential above the minimum value at r_0 is assumed to be quadratic in $(r-r_0)$, so that the solution for this coordinate, $r-r_0$, is that of a simple one dimensional harmonic oscillator. The molecule is thus idealized as an independent two dimensional rigid rotator plus a harmonic oscillator. One obtains two additive contributions to the thermodynamic potentials, one due to rotation containing only one parameter, the moment of inertia I_0 at r_0, (see however Sect. 7 on symmetry number), and one due to vibration containing, as parameter, the frequency of vibration, ν (Sect. 10).

Instead of attempting to actually solve the original quantum mechanical problem, which would involve obtaining the potential function for the nuclei by solving the many electron problem at fixed nuclear distance, r, one now takes recourse to spectroscopic observations which give values of I_0 and ν. The spectroscopic information not only gives the parameters needed, but usually also gives information justifying the approximation, namely the fact the observed energy levels do actually agree reasonably well with the simple equations used.

Indeed the spectroscopic data on most diatomic molecules give, in addition to I and ν, the empirical parameters for the corrections to be applied to the simple formula normally used for the energies of the various quantum mechanical levels. These are corrections which arise due to the anharmonicity of the potential, due to the stretching of the molecule at high angular momenta, and due to the effect of vibration on the moment of inertia. By a perturbation method [3] the effect of these corrections on the partition function and on the thermodynamic potentials can be evaluated. The additional terms are actually found, in most cases, to be small.

The treatment of polyatomic molecules is similar. In a single electronic state a molecule containing n atoms has $3n$ degrees of freedom, of which $3n-3$ are internal. If the equilibrium configuration is linear one uses the two polar angles ϑ, φ, and there remain $3n-5$ coordinates, upon which the internal potential energy depends. If the equilibrium configuration is non-linear one requires three Eulerian angles, ϑ, φ, ψ, and there are left $3n-6$ "vibrational" degrees of freedom. By assuming the moments of inertia to be those of the equilibrium configuration the rotational term of the Hamiltonian is separated. For linear molecules one obtains an additive rotational contribution to the thermodynamic potentials which is the same as for diatomic molecules, and depends on one parameter only, the one moment of inertia. For the general non-linear molecule with all three moments unequal there is no closed solution for the energies of the rotational levels in terms of the quantum numbers. However the spacing of the levels, in all real cases, is now so close, due to the relatively large moments (as it actually also is for most diatomic molecules), that one may use a classical integration which can be performed (Sect. 9). One obtains an additive term to the thermodynamic potentials containing the three moments of inertia as parameters.

The internal potential energy depends on the remaining $3n-5$ or $3n-6$ internal coordinates. Again one assumes the potential to be quadratic in displacements from equilibrium. A choice of "normal" coordinates can now always be found as a linear combination of the Cartesian displacement coordinates of the individual atoms such that the total Hamiltonian is separable in the normal coordinates, that is such that the kinetic energy is the sum of the squares of the conjugated momenta, and the potential energy a sum of coefficients times the squares of the normal coordinates. The quantum energies are then additive, and one obtains an additive oscillator contribution to the thermodynamic potentials from each of the $3n-5$ or $3n-6$ vibrational degrees of freedom, each depending on the characteristic frequency of that degree of freedom. If now all frequencies are known experimentally they can be used in the equations without attempting to actually solve for the normal coordinates.

In practice it is relatively simple, for most diatomic molecules, to obtain thermodynamic functions which are probably superior in accuracy to those which can be obtained by direct calorimetric measurements[1]. This is much less often the case with polyatomic molecules. Various complications enter. In the first place spectroscopic values for the moments of inertia are frequently not available. Recourse must be made to the less precise information obtained from electron diffraction measurements, and the moments computed from the known atomic masses and the derived geometry. Or even the geometry of the molecule may be assumed from chemical rules and the empirical "bond-distances" of the types of chemical bonding occurring. Presumably the correction terms due to centrifugal stretching of the molecule at high rotational quantum numbers are more important than in diatomic molecules, especially if the distortion involves only the "bending" of chemical bonds. Complete spectroscopic information on all frequencies of the molecule are often not available, since for molecules having any degree of symmetry many of the transitions are "forbidden" both in the infra-red spectra and the Raman. In these cases it is necessary to construct some model of "force constants" which are fitted to the observed frequencies and used to compute the unknown ones. Last, but not least, in many molecules some of the frequencies may be very far from harmonic, indeed so much so that their description as vibrational frequencies may be more than misleading. This is particularly true of molecules in which "libration" around chemical bonds can occur. A simple example is that of the ethane molecule, H_3C-CH_3, in which the two methyl groups can rotate in opposite directions. Finally one might mention, at this point, that in symmetric molecules many of the frequencies may be degenerate, that is several of the different $3n-5$ or $3n-6$ degrees of freedom may have the same identical values of the frequency. It is necessary, then, to know which of the observed frequencies correspond to these degenerate modes which must be counted two or more times. In order to solve this problem it usually is necessary to actually investigate the normal coordinate analysis, which then also gives information as to which modes of vibration appear in the infra-red and which appear in the Raman spectra. This complete analysis of the spectral data is normally necessary before the observed frequencies can be used in the equations with any assurance of validity.

We have discussed molecules without reference to possible excited electronic levels. Fortunately the vast majority of chemically stable molecules have a single lowest electronic state, with no excited level at low enough energy to play any role in the partition function at temperatures for which the molecule is not almost entirely dissociated. If, however, an excited electronic state must be considered, it is, in principle at least, necessary to compute the vibrational and rotational contributions to Q_i for the lowest and for the excited electronic states separately, and to add them. That is, the complete Hamiltonian is not

[1] E. B. WILSON jr.: Chem. Rev. **27**, 17 (1940).

separable in the coordinates of the electrons and of the nuclei. The "potential energy" of the nuclei in fixed configuration are determined by the total electronic energy, potential plus kinetic, in a given electronic state[1]. These potentials are generally different for different states. The moments of inertia and the vibrational frequencies are then different in two different electronic states.

7. The symmetry number. There is one significant correction which must be made to the prescription for computing the internal contributions to the thermodynamic potentials which was implied in Sect. 6. The rotational factor in the internal partition function is not identical for a symmetric molecule and for a non-symmetric molecule, even with the same moments of inertia. The usually adequate prescription is to divide the partition function computed for the non-symmetric molecule by the symmetry number of the molecule, γ.

For a rigid symmetrical molecule there will be, in general, several different values of the internal coordinates which correspond to no other change of configuration than the permutation of some identical nuclei. The number of these values is called the symmetry number γ.

For instance, consider the case of a symmetric diatomic molecule, such as O_2. The two nuclei a, and b, have Cartesian coordinates x_a, y_a, z_a, and x_b, y_b, z_b. The coordinates in which one describes the positions of the nuclei for the solution of the quantum states are:

$$X = \tfrac{1}{2}(x_a + x_b),$$

$$Y = \tfrac{1}{2}(y_a + y_b),$$

$$Z = \tfrac{1}{2}(z_a + z_b),$$

$$\vartheta = \arccos \frac{z_a - z_b}{\sqrt{(x_a - x_b)^2 + (y_a - y_b)^2 + (z_a - z_b)^2}},$$

$$\varphi = \arcsin \frac{y_a - y_b}{\sqrt{(x_a - x_b)^2 + (y_a - y_b)^2}},$$

$$q = r - r_0 = \sqrt{(x_a - x_b)^2 + (y_a - y_b)^2 + (z_a - z_b)^2} - r_0.$$

The interchange of a with b does not alter X, Y, Z, or q. It changes the sign under the arc cos and arc sin of ϑ and φ, so that ϑ becomes $\vartheta - \pi$ and φ becomes $-\varphi$. The total eigenfunction, Ψ, of the molecule must be either symmetric or antisymmetric in the exchange a with b, and in the case of the $O^{16}O^{16}$ molecule must be symmetric. The electronic part of the function will have a definite parity, positive or negative, and for the ground state of a symmetric molecule is always positive, as it is for O_2. This means that the exchange a with b, which, since the electronic coordinates are measured from the body fixed coordinate system determined by r_a and r_b, corresponds to an inversion of the electronic coordinates through the position of the center of mass, $r = \frac{1}{2}(r_a + r_b)$, leaves the function unchanged. (Negative parity means that the function changes sign, is antisymmetric, in this inversion.) The coordinates X, Y, Z, and q are unchanged by the exchange, but ϑ goes to $\vartheta - \pi$ and φ to $-\varphi$. The rotational functions have the parity $(-1)^j$, that is those with $j = 0, 2, 4, \ldots$ are unchanged by this transposition, whereas those with odd $j = 1, 3, \ldots$, go into their negative. Thus only the rotational states of even j in the lowest electronic level of O_2 have eigenfunctions of the correct allowed symmetry, and only these occur as allowed states of the molecule.

[1] M. BORN and J. R. OPPENHEIMER: Ann. d. Phys. **84**, 457 (1927).

The summation over the rotational levels then runs over even j values only in the partition function of $O^{16}O^{16}$, and not over all of them, as it does in the case of a non symmetrical molecule, such as CO. The separation of the rotational levels is small, except in the one case of the very light molecule H_2. The energy difference between the lowest $j=0$ state and the first excited $j=1$ level in a molecule with the moment of inertia of O_2 is 2.9 cm^{-1} or $\varepsilon/\mathbf{k}=4°$ K. The effect of counting only the even j levels in the rotational partition function is to divide the sum over all levels by 2, the symmetry number of the symmetric diatomic molecule.

In general for any symmetric diatomic molecule whatever symmetry of the eigenfunctions is required in exchange of the nuclei, and whatever symmetry the electronic and nuclear states have in this exchange, only the even *or* only the odd rotational levels are allowed for each quantum state of the other degrees of freedom. Hence always at high temperatures the partition function is smaller by a factor one-half than that for a non symmetric molecule of the same moment of inertia and other mechanical properties. The requirement of high temperatures in this statement is satisfied for all molecules, except H_2, above or near to their boiling points.

In H_2 the rotational levels are widely spaced, due to its low moment of inertia. The energy of the $j=1$ level is 118.7 cm^{-1} or 169.9° K, so that appreciably below room temperature the sums over the even j values and the sum over the odd j values differ considerably.

For polyatomic molecules of non unit symmetry number, the situation is similar. The detailed discussion in terms of the symmetry of the eigenfunction is relatively complicated[1], but the result is simple enough. If there are γ different configurations of the internal coordinates (usually only the rotational coordinates, see later this section) which differ only in corresponding to different permutations of identical nuclei, then only the fraction $1/\gamma$ of the solutions of the Schroedinger equation will be permitted as allowed states of the molecule. Since the rotational levels are always closely spaced compared to $\mathbf{k}T$ at all temperatures for which these molecules exist as gases at appreciable density, the partition function is smaller by the factor $1/\gamma$ than that for a molecule posessing no symmetry.

In the high temperature limit, then, the effect of symmetry is to add a temperature independent negative term $-\mathrm{R}\log\gamma$ to the entropy per mol of the molecule, and has no effect on the energy.

The introduction of the symmetry number, in the general case, is seen to be necessary by consideration of the classical limit[2] in which the partition function becomes an integral over the coordinates and momenta divided by $\boldsymbol{N}!$ (Sect. 2). Consider a gas composed of N molecules of one species, each of which contains n atoms, n_1 of type $1, \ldots, n_a$ atoms of type a, and we use the symbol $\boldsymbol{N}=N_1, \ldots, N_a, \ldots$ for the set of numbers of atoms. We may now use the Cartesian coordinates, $x_\alpha, y_\alpha, z_\alpha$, of the atoms, $1 \leq \alpha \leq \sum N_a = nN$, and their conjugated momenta $p_{x\alpha}, p_{y\alpha}, p_{z\alpha}$, for the description of the system. The partition function, $Q_C(V, T, N)$ for the system is now given, in the classical limit, as

$$Q_C(V, T, \boldsymbol{N}) = \frac{1}{h^{3nN}\boldsymbol{N}!}\iint\cdots\int \mathrm{e}^{-H/\mathbf{k}T}\prod_{\alpha=1}^{nN} d x_\alpha\, d y_\alpha\, d z_\alpha\, d p_{x\alpha}\, d p_{y\alpha}\, d p_{z\alpha}. \tag{7.1}$$

Now, of course, the Hamiltonian is absurdly complicated in these coordinates. It takes large negative values only in those regions of the configuration space

---

[1] J. E. MAYER, S. BRUNAUER and M. G. MAYER, jr.: Amer. Chem. Soc. **55**, 37 (1933).
[2] P. EHRENFEST and TRKAL: Proc. Sect. Sci. Amsterdam **23**, 169 (1920).

of the atoms for which they form the N molecules each with n_a atoms of type a in some configuration close to the equilibrium configuration of the molecule. We adjust the arbitrary zero of energy to be the potential energy of such a position with N molecules, each in its equilibrium configuration, and transform to molecular coordinates, the coordinates $\boldsymbol{r}_1, \ldots, \boldsymbol{r}_j, \ldots, \boldsymbol{r}_N$, of the centers of mass of the molecules, and some internal coordinates, $\boldsymbol{q}_1, \ldots, \boldsymbol{q}_j, \ldots, \boldsymbol{q}_N$, for each molecule.

However, the original integral of (7.1) runs over all configurations of the atoms. In our molecular coordinate system we integrate only over configurations in which the same atoms are associated in one molecule. We must multiply this integral by the factor,

$$\frac{\boldsymbol{N}!}{N!\left(\prod_a n_a!\right)^N}$$

which is the number of ways that the set $\boldsymbol{N}$ of atoms can be assigned to N molecules. We than have,

$$Q_c(V,T,N) = \frac{1}{h^{3N} N!}\left[\frac{1}{h^{3(n-1)}\prod_a n_a!}\right]^N \iint\cdots\int e^{-H/kT}\prod_{j=1}^{N} d\boldsymbol{r}_j\, d\boldsymbol{p}_j\, d\boldsymbol{q}_j\, d\boldsymbol{p}_{q_j}, \tag{7.2}$$

for the partition function of the system. Now assume the Hamiltonian to be separable in the internal degrees of freedom of each molecule separately. One finds as in Eq. (4.6)

$$Q_c(V,T,N) = Q_{\text{tr}}(V,T,N)\, Q_i^N \tag{7.3}$$

where,

$$Q_i = \frac{1}{h^{3(n-1)}\boldsymbol{n}!}\iint\cdots\int e^{-H_i/kT} d\boldsymbol{q}\, d\boldsymbol{p}, \tag{7.4}$$

$$\boldsymbol{n}! = \prod_a n_a!. \tag{7.5}$$

The formula to divide the integral over the coordinate and momenta by the product of the factorials of the numbers of identical molecules (and by h to the number of degrees of freedom) holds even for the internal partition function. However, it was tacitly assumed, in deriving the weighting factor, that the integration in (7.4) included all possible configurations of the atoms within the molecule. In practice one integrates only over small displacements from some equilibrium position. There are actually $\boldsymbol{n}!$ different but physically identical equilibrium configurations, not all of which may be covered by the practical choice of the coordinate system. (More strictly expressed the coordinate system is usually such that all possible configurations would actually be covered, but the potential, which is generally taken to be quadratic in the vibrational displacement coordinates may be artificially assigned a fictitiously large value at some positions for which it actually descends again to the zero value.) If only γ different positions are covered correctly by the integral in (7.4) one must multiply its value by $\boldsymbol{n}!/\gamma$, finally obtaining

$$Q_i = \frac{1}{h^{3(n-1)}\gamma}\iint\cdots\int e^{-H_i/kT} d\boldsymbol{q}\, d\boldsymbol{p}. \tag{7.6}$$

Thus for symmetric diatomic molecules the rotational coordinates, $\vartheta\varphi$, do correctly cover the $2! = 2$ positions differing only in permutations of the two identical atoms, and Eqs. (7.4) and (7.6) are identical, since $\gamma = 2$. However the methane molecule, CH_4 with $\boldsymbol{n}! = 1!\,4! = 24$, has 24 different possible permutation of the four identical hydrogens. Only $\gamma = 12$ of these are covered by the rotational coordinates. The factor $2 = n!/\gamma$ is the number of (optically active) isomers that a molecule of the same topological symmetry would have if all four hydrogens were replaced by different atoms.

At first sight it might appear that the symmetry number use might introduce discontinuous transitions in the thermodynamic properties of molecules with almost the same geometrical properties. This is not the case. Consider the pyramidal molecule NH_3 for which $n! = 6$ and $\gamma = 3$. A planar symmetric molecule AB_3 would have $\gamma = 6$. Now one of the internal normal vibrational coordinates of NH_3 is the amplitude of the motion in which the nitrogen moves normal to the plane of the three hydrogens, the hydrogen atoms spreading somewhat as the nitrogen approaches the plane. This coordinate has two equal minima corresponding to the $n!/\gamma = 2$ "isomers". The potential energy diagram is sketched in Fig. 2. As long as the potential hump, at the position of the nitrogen in the plane of the hydrogens, is sufficiently high, it is adequate to represent the potential by a parabola around one minimum, integrate only over this minimum, and divide the total internal partition function by 3. One might alternately integrate over both minima, in which case one has chosen a coordinate representation which includes $\gamma = 6$ permutation positions, and divide the partition function by 6. The result will be the same. If the molecule is nearly planar the hump in the middle will be low, and the approximation of representing the potential by a single parabola at only one minimum becomes poor. It is then natural to attempt to integrate with the true potential, using $\gamma = 6$. In the limit that the molecule is actually planar the hump has disappeared, and one parabola with the minimum at the planar position, and $\gamma = 6$, is the obvious approximation.

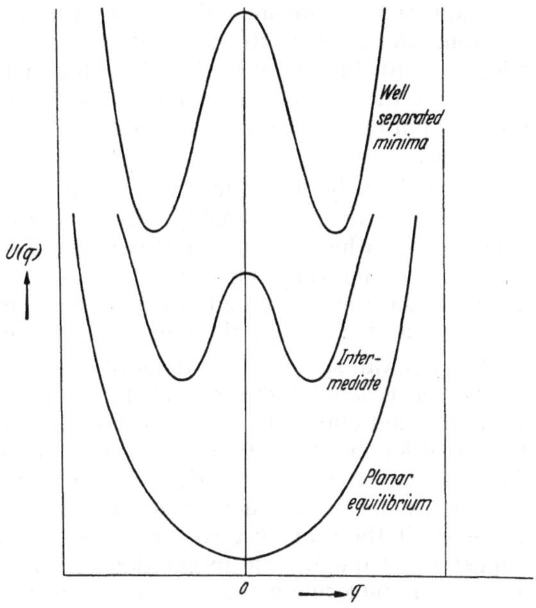

Fig. 2. Schematic potentials for the "Umbrella" frequencies of AB_3.

The quantum mechanical picture is similar. If one "forgets" about the two minima and solves the equations assuming only one parabolic potential function one obtains equally spaced vibrational levels. There will be another corresponding set of vibrational levels in the other minimum. Since the separating potential is not infinite these two sets of levels interact, leading to doublets for each level, the separation of the two members of the doublet increasing as the levels approach the energy of the hump. The separation of the doublet members in the lowest vibrational state has been measured in NH_3 by microwave absorption, and is known to be $0.8\,\text{cm}^{-1}$. If one computes the contribution to the factor in the partition function due to this degree of freedom forgetting the existence of the second minimum, and the doubling of the levels, one must divide the final partition function by $\gamma = 3$ since only 3 of the 6 permutations were covered in the rotational coordinates. If, however, one counts the doublet states separately, one obtains essentially a two fold greater contribution from this mode of vibration, but divides the final partition function by 6, since all $n! = 6$ permutations have been included in the summation. Many similar situations exist in other cases.

8. Nuclear spin contribution. The summation over states in the internal partition function of Eq. (6.1) includes states differing in the orientations of the nuclear spins if any atoms of the molecule have non-zero nuclear spin. The effect of the inclusion of these states is normally to multiply the partition function by a temperature independent factor, $2s_a + 1$, for each atom of type a having nuclear spin s_a, in the molecule. This factor of the partition function becomes an additive temperature independent term, in the entropy. Since the number of atoms involved in the two sides of any chemical reaction is constant, the additive contribution, $\mathrm{R}\log(2s_a + 1)$, for each atom of type a, in each molecule, cancels in the ΔS of the reaction. It is therefore customary to neglect this nuclear spin entropy term in calculations and tabulations of the thermodynamic functions.

At sufficiently low temperatures the term will no longer be temperature independent, and, presumably, always becomes zero at $T = 0°$ K. At temperatures for which the material is gaseous, except in the case of H_2, the term does have its limiting value, and the neglect is numerically justified, as indeed it also is for the condensed phase of most molecules, except some containing hydrogen, and except at the lowest experimentally available temperatures.

The nuclear spin of all atoms of even atomic number, Z, and even mass number, A, is zero. Atoms of odd A have half integer, non zero spins, s_a. Atoms of even A and odd Z (odd-odd nuclei) have integer spin, usually non zero. The $2s_a + 1$ different orientations of the nucleus of spin s_a have energy differences, $\Delta\varepsilon/\mathbf{k}$, of the order of 10^{-3} °K, due to the interaction of the nuclear magnetic moments and nuclear quadrupole moments with electronic magnetic and field moments, if these are present, and even smaller differences due to the direct magnetic and quadrupole interaction of the different nuclei in the same molecule. If the molecule contains n_a atoms of type a there are, then, $\prod_a (2s_a + 1)^{n_a}$ different nuclear spin orientations of negligibly different energy, and if the molecule has symmetry number, γ, equal to unity, each of these nuclear states occurs with every electronic, rotational, vibrational state of the molecule. For temperatures above 10^{-3} °K the effect of these nuclear states is to multiply the Q_i computed without their inclusion by the factor $\prod_a (2s_a + 1)^{n_a}$. The $\log Q_i$ has then an added term $\sum n_a \log(2s_a + 1)$ for each molecule containing n_a atoms of type a. The nuclear spin does not affect the energy, but appears in the molar entropy, S, as an additive $\mathrm{R}\sum n_a \log(2s_a + 1)$. Since it cancels in the entropy changes of all chemical reactions it is convenient to neglect it completely, especially since not all nuclear spins are known.

For molecules of non unit symmetry number, γ, the situation is different, although again the neglect is usually numerically justifiable. The $\prod_a (2s_a + 1)^{n_a}$ different nuclear spin states will, in general, have different symmetry character with respect to permutations of the identical nuclei of type a. Each such state can combine only with a rotational level of a given symmetry character, in order that the final eigenfunction should have the allowed symmetry of being totally antisymmetric, or totally symmetric, with respect to permutations of identical nuclei. Of the rotational states only the fraction $1/\gamma$ will have the required symmetry to combine with a given nuclear spin state. Each nuclear spin configuration will then occur, and will occur with the fraction $1/\gamma$ of the rotational states. If the energy differences between the rotational levels are small compared to $\mathrm{k}T$, so that the use of the symmetry number γ as a factor, $1/\gamma$, in the internal partition function is justified, then the total partition function computed for a non symmetrical molecule without nuclear spin is to be multiplied by the factor $(1/\gamma) \prod_a (2s_a+1)^{n_a}$. The nuclear spin entropy, equal to $\mathrm{R} \sum n_a \log(2s_a+1)$ per mol, appears as before.

However, the energy difference between differing configurations of mutual nuclear spin is in this case no longer determined by the very small magnetic and quadrupole interactions of the Hamiltonian, but by the symmetry requirements, and thus is of the magnitude of the energy differences between the differing rotational states of the molecule, which may be considerable. In practice, as discussed in Sect. 7, the use of the symmetry number as a factor, $1/\gamma$, in the internal partition function, is numerically justified for gases other than H_2 at temperatures for which the gas has appreciable densities. Thus in the entropy computation for gases other than H_2 the nuclear spin entropy has its high temperature limiting value.

For monatomic crystals, and for liquids other than He^3, as well as for crystals and liquids of non-symmetrical molecules, the nuclear spin entropy persists down to temperatures below those for which heat capacity measurements have yet been made. As a result the "third law entropies", obtained by integration of $\int_0^T C_p \, d \log T$ using the observed C_p values, will also not include the nuclear spin entropy, and will be consistent with gas entropies computed by neglect of this term. This is not necessarily the case with symmetric molecules, particularly those containing H-atoms, which may have low moments of inertia and relatively large energy differences of the rotational levels. In these molecules the condensed phase at the lowest temperatures for which heat capacities have been measured may well exist almost entirely in the lowest allowed rotational level, and hence in one specified nuclear spin orientation, so that the "third law entropy" may include this term, and not be consistent with gas entropies computed by its omission.

The detailed discussion of the symmetry of the eigenfunctions for polyatomic molecules of high γ is complicated. The example of a diatomic molecule is simple. We may pick H_2, since in this case, due to the low moment of intertia, the details become significant in the computation of the thermodynamic functions. The proton has spin, $s=\frac{1}{2}$, and can exist in two spin states, say the state α corresponding to a $+\frac{1}{2}$ projection on some unique axis, or the state β corresponding to the projection $-\frac{1}{2}$ on this axis. Each of the two nuclei, 1 and 2, may exist in either of these states, making $4=(2s+1)^2$ spin states of the molecule. The three states,

$$\begin{gathered} \alpha(1)\,\alpha(2), \\ \frac{1}{\sqrt{2}}\left[\alpha(1)\,\beta(2)+\beta(1)\,\alpha(2)\right], \\ \beta(1)\,\beta(2), \end{gathered}$$

are symmetric in permutations of the nuclei 1 with 2. These three states all correspond to total nuclear spin vector unity, with the three projections along the unique axis of $+1$, 0, -1, respectively. The fourth state,

$$\frac{1}{\sqrt{2}}\left[\alpha(1)\,\beta(2)-\beta(1)\,\alpha(2)\right],$$

is antisymmetric in exchange of the nuclei, has total spin zero, and zero projection on the unique axis.

Since the lowest electronic state has even parity (see Sect. 7 on the symmetry number) the symmetric spin triplet combines only with the odd antisymmetric rotational levels, whereas the antisymmetric spin singelt combines only with the even rotational levels, in order to give an antisymmetric total eigenfunction. The molecules with parallel nuclear spin, that is those in the spin triplet, are called ortho-hydrogen, those with the opposed spin in the antisymmetric spin singlet are called para-hydrogen. Ortho-hydrogen has rotational quantum number, j, odd only, and para-hydrogen has j even only. The complete internal partition function of the equilibrium hydrogen mixture has, as factor, the sum over even j level plus three times the sum over the odd j levels. At high temperatures the sums over j even and over j odd are equal, and both equal to half the sum over both, which latter would give the normal rotational partition function factor for a non-symmetric molecules of the same moment of inertia. The total internal partition function is then $\frac{1}{2}+\frac{3}{2}=2=(2s+1)^2/\gamma$ times the partition function of the non-symmetric molecules.

At low temperatures the appreciable energy, $\varepsilon_1/\mathrm{k}=169.9°$ K, of the $j=1$ level makes the sum over odd j values much smaller than that of the even sum, which starts with the term $j=0$ of energy zero. At the boiling point, 20° K, the Boltzmann factor $e^{-\varepsilon_1/\mathrm{k}T}=e^{-8.5}$ of the $j=1$ level has become almost negligible, and the total internal partition function at equilibrium is practically unity. Wheras the equilibrium gas at high temperatures is three-fourths ortho-hydrogen and one fourth para-hydrogen, at low temperatures the equilibrium mixture is nearly one hundred percent para.

The situation in hydrogen is complicated by the fact that the ortho-para transition is not induced by normal collisions between the molecules. An inhomogeneous magnetic field, acting differently on the two nuclear magnetic moments, is effective in producing transitions. Collisions with magnetically active molecules or surfaces catalyse the transitions. In the absence of such catalysts hydrogen gas cooled from room temperature to the boiling point will retain the high temperature composition of three fourths ortho-hydrogen, and this non equilibrium distribution will persist for long periods.

The prescription for treating the thermodynamic properties of this nonequilibrium mixture is simply that of computing as though the two gases, ortho-hydrogen and para-hydrogen, were two different chemical species existing at fixed mol fractions, each having its own separate internal partition functions. Assume the two internal partition functions at T are Q_{0i} and Q_{pi} respectively, and the mol fractions are x_0 and $x_p=1-x_0$. If the total pressure is P, so that the partial pressures are $x_0 P$ and $(1-x_0)P$ the translational contributions to the Gibbs free energy per mol, G_{tr}, is the sum of the contributions of the two gases, which from Eq. (3.36) is

$$G_{\mathrm{tr}}=G_{\mathrm{tr}}(T,P)+\mathrm{R}T\left[x_0\log x_0+(1-x_0)\log(1-x_0)\right]. \tag{8.1}$$

To this must be added the internal contributions, which from Eq. (3.39) are

$$G_{\mathrm{int}}=-\mathrm{R}T\left[x_0\log Q_{0i}+(1-x)\log Q_{pi}\right]. \tag{8.2}$$

The total Gibbs free energy per mol is

$$G=G_{\mathrm{tr}}+G_i=G_{\mathrm{tr}}(T,P)+\mathrm{R}T\left[x_0\log\frac{x_0}{Q_{0i}}+(1-x_0)\log\frac{1-x_0}{Q_{pi}}\right]. \tag{8.3}$$

At equilibrium, x_0, the mol fraction of ortho-hydrogen, will take such a value, x_{0e}, that G is a minimum, $\partial G/\partial x_0 = 0$,

$$\log \frac{x_{0e}}{Q_{0i}} - \log \frac{1 - x_{0e}}{Q_{pi}} = 0, \tag{8.4$_1$}$$

$$x_{0e} = \frac{Q_{0i}}{Q_{0i} + Q_{pi}}. \tag{8.4$_2$}$$

Using this equilibrium value of the mol fraction Eq. (8.3) becomes,

$$G = G_{\mathrm{tr}}(T, P) - RT \log (Q_{0i} + Q_{pi}), \tag{8.5}$$

which is the expression given by Eq. (3.39) for the gas with internal partition function,

$$Q_i = Q_{0i} + Q_{pi}. \tag{8.6}$$

9. Rotational partition functions. *α) Diatomic molecules.* The normal coordinates of a diatomic molecule consisting of atoms of mass m_1 and m_2, $m = m_1 + m_2$, are

$$\left.\begin{aligned}
x &= \frac{1}{m}(m_1 x_1 + m_2 x_2), \\
y &= \frac{1}{m}(m_1 y_1 + m_2 y^2), \\
z &= \frac{1}{m}(m_1 z_1 + m_2 z_2), \\
\vartheta &= \arccos\left[\frac{z_1 - z_2}{\sqrt{(x_1 - x_2)^2 + (y_1 - y_2)^2 + (z_1 - z_2)^2}}\right], \\
\varphi &= \arcsin\left[\frac{y_1 - y_2}{\sqrt{(x_1 - x_2)^2 + (y_1 - y_2)^2}}\right], \\
\xi &= \sqrt{(x_1 - x_2)^2 + (y_1 - y_2)^2 + (z_1 - z_2)^2} - r_0
\end{aligned}\right\} \tag{9.1}$$

in which x_1, y_1, z_1, and x_2, y_2, z_2, are the Cartesian coordinates of the atoms 1 and 2 respectively. The potential energy, u, in these coordinates, depends only on the one coordinate, ξ, and r_0 is chosen as the distance of minimum potential, so that u has a minimum value at $\xi = 0$,

$$u = u(\xi), \quad \left(\frac{du}{d\xi}\right)_{\xi=0} = 0. \tag{9.2}$$

The conjugated momenta are,

$$\left.\begin{aligned}
p_x &= m\dot{x}, \\
p_y &= m\dot{y}, \\
p_z &= m\dot{z}, \\
p_\vartheta &= I(\xi)\,\dot{\vartheta}, \\
p_\varphi &= I(\xi)\sin^2\vartheta\,\dot{\varphi}, \\
p_\xi &= \mu\dot{\xi},
\end{aligned}\right\} \tag{9.3}$$

where μ is the reduced mass,

$$\mu = \frac{m_1 m_2}{m}, \tag{9.4}$$

and $I(\xi)$ is the moment of inertia,

$$I(\xi) = \mu[r_0 + \xi]^2. \tag{9.5}$$

The Hamiltonian, in these coordinates, is now

$$H = \frac{1}{2m}\left(p_x^2 + p_y^2 + p_z^2\right) + \frac{1}{2\mu} p_\varrho^2 + u(\xi) + \frac{1}{2I(\xi)}\left(p_\vartheta^2 + \frac{1}{\sin^2\vartheta} p_\varphi^2\right). \tag{9.6}$$

The separation of the center of mass coordinates is rigorous. The approximation,

$$I(\xi) = I(\xi = 0) = \mu r_0^2 = I_0, \tag{9.7}$$

separates the internal part into a sum, namely if

$$H_{\text{rot}} = \frac{1}{2I_0}\left(p_\vartheta^2 + \frac{1}{\sin^2\vartheta} p_\varphi^2\right), \tag{9.8$_1$}$$

$$H_{\text{vib}} = \frac{1}{2\mu} p_\xi^2 + u(\xi), \tag{9.8$_2$}$$

then,

$$H = H_{\text{trans}} + H_{\text{vib}}(\xi, p_\xi) + H_{\text{rot}}(\vartheta, \varphi, p_\vartheta, p_\varphi). \tag{9.9}$$

The time independent Schroedinger equation for the rotational Hamiltonian,

$$H_{\text{rot}}\,\Psi(\vartheta, \varphi) = \varepsilon_{\text{rot}}\,\Psi(\vartheta, \varphi) \tag{9.10}$$

has solutions,

$$\Psi(\vartheta, \varphi) = P_{j,m}(\cos\vartheta)\,e^{im\varphi}, \tag{9.11}$$

with $P_{j,m}$ the j, m spherical harmonic,

$$j = 0, 1, 2, \ldots, \qquad -j \leq m \leq +j. \tag{9.12}$$

The energies depend only on j, and are,

$$\varepsilon_{\text{rot}}(j, m) = \frac{h^2}{8\pi^2 I_0} j(j+1). \tag{9.13}$$

The rotational factor in the partition function is, for this two dimensional rotator,

$$Q_{\text{rot}}^{(2)} = \sum_{j=0}^{\infty} \sum_{m=-j}^{m=+j} e^{-\sigma j(j+1)} = \sum_{j=0}^{j=\infty} (2j+1)\,e^{-\sigma j(j+1)}, \tag{9.14}$$

$$\sigma = \frac{h^2}{8\pi^2 I_0 kT}. \tag{9.15}$$

For large values of σ one must numerically evaluate,

$$Q_{\text{rot}}^{(2)} = 1 + 3e^{-2\sigma} + 5e^{-6\sigma} + 7e^{-12\sigma} + \cdots \tag{9.16}$$

which converges reasonably rapidly for $\sigma \gtrsim 1$. For small values of σ the Euler-MacLaurin summation formula,

$$\sum_{n=0}^{\infty} f(n) = \int_0^{\infty} f(x)\,dx + \frac{1}{2} f(0) - \frac{1}{12}\left(\frac{df}{dx}\right)_{x=0} + \frac{1}{720}\left(\frac{d^3 f}{dx^3}\right)_{x=0} - \frac{1}{30240}\left(\frac{d^5 f}{dx^5}\right)_{x=0}$$

with the substitution of $j(j+1) = z$, $(2j+1)\,dj = dz$ for the integral, yields,

$$Q_{\text{rot}}^{(2)} = \frac{1}{\sigma}\left(1 + \frac{\sigma}{3} + \frac{\sigma^2}{15} + \frac{4\sigma^3}{315} + \cdots\right) \tag{9.17}$$

which is numerically valid to about 1% for σ unity or less.

For diatomic molecules not containing hydrogen the values of σ are usually below 0.03 at the boiling point of the gas, so that even the approximation $Q_{\mathrm{rot}} = \frac{1}{\sigma}$ is good to one percent. The quantity

$$B_0 = \frac{h}{8\pi^2 I c} = \sigma \frac{\mathrm{k} T}{h c}, \tag{9.18$_1$}$$

of dimensions l^{-1} is the quantity usually listed in spectroscopic tables. This has the value 1.989 cm^{-1} for N_2, which is probably the highest of any non-hydrogenic diatomic molecule, so that,

$$\frac{B_0 h c}{\mathrm{k}} = \frac{h^2}{8\pi^2 I \mathrm{k}} = \sigma T \tag{9.18$_2$}$$

has the value $\sigma T = 2.847°$ K for N_2. At the boiling point of N_2, $T = 77.3°$ K, the value of σ is 0.037. For other molecules not containing hydrogen the values of σ are lower at their boiling points. For instance σT is 2.059° K for O_2, 0.3477° K for Cl_2 and 0.05340° K for I_2. For HF, the value of σT is 29.409° K, and at its boiling point of 292.4° K, $\sigma = 0{,}1$; the correction terms in (9.17) are not negligible, and for HCl $\sigma T = 14.946°$ K and $\sigma = 0.079$ at the boiling point, $T = 189.4°$ K. Only for H_2 with $\sigma T = 84.971°$ K is the use of the exponential sum required at temperatures for which the gas exists at appreciable pressures.

The rotational partition function obtained by symmation over even j values only,

$$Q^{(2)}_{\mathrm{rot.even}} = 1 + 5\mathrm{e}^{-6\sigma} + 9\mathrm{e}^{-20\sigma} + 13\mathrm{e}^{-42\sigma} + \cdots, \tag{9.19$_1$}$$

is to be used for para-hydrogen, whereas that summed over odd j values,

$$Q^{(2)}_{\mathrm{rot.odd}} = 3\mathrm{e}^{-2\sigma} + 7\mathrm{e}^{-12\sigma} + 11\mathrm{e}^{-30\sigma} + 15\mathrm{e}^{-56\sigma} + \cdots \tag{9.19$_2$}$$

applies to ortho-hydrogen, and must be multiplied by three for the three nuclear spin orientations (Sect. 8). The application of the Euler-MacLaurin summation formula to the sum over the even j-values leads only to the approximation that it is equal to one-half the sum over all j-values. This approximation is poor for σ values nearly as large as unity. The apparent convergence of the series at these high values of σ is due, in this case, to the peculiar grouping of terms in powers of σ.

The approximation $Q^{(2)}_{\mathrm{rot}} = 1/\sigma$, (9.17), obtained by substitution of the integral alone for the sum is exactly the same as that obtained by the use of the classical Hamiltonian and integration over coordinate and momentum. The classical expression with (9.8$_1$) for H_{rot} of the two degrees of freedom is

$$Q^{(2)}_{\mathrm{rot}} = \frac{1}{h^2} \int_0^{\pi} d\vartheta \iint_{-\infty}^{+\infty} d p_\vartheta \, d p_\varphi \int_0^{2\pi} d\varphi \, \mathrm{e}^{-\frac{1}{2 I_0 \mathrm{k} T}\left(p_\vartheta^2 + \frac{1}{\sin^2\vartheta} p_\varphi^2\right)}.$$

Integration over $d\varphi$ gives the factor 2π, the integration over $d p_\varphi$ the factor $\sqrt{2\pi I_0 \mathrm{k} T}\,\sin\vartheta$, and that over $d p_\vartheta$ gives $\sqrt{2\pi I_0 \mathrm{k} T}$. Finally $\int_0^{\pi} \sin\vartheta \, d\vartheta = 2$, so that

$$Q^{(2)}_{\mathrm{rot}} = 8\pi^2 I_0 \mathrm{k} T / h^2 = 1/\sigma.$$

The additive contribution, Eq. (3.38),

$$E^{(2)}_{\mathrm{rot}} = \mathrm{R} T^2 \, d \log Q^{(2)}_{\mathrm{rot}} / d T,$$

to the energy per mol due to rotation is seen, from (9.17) for $Q^{(2)}_{\mathrm{rot}}$, and (9.15) for σ, to be,

$$E^{(2)}_{\mathrm{rot}} = \mathrm{R} T \left[1 - \frac{\sigma}{3} - \frac{\sigma^2}{45} - \frac{8\sigma^3}{945} - \cdots\right]. \tag{9.20}$$

At high temperatures, $\sigma \to 0$, the energy is just the classical kinetic energy of two degrees of freedom per molecule (Sect. 2).

β) Polyatomic molecules. The rotational partition function for a linear polyatomic molecule is, of course, just that for a diatomic molecule of the same moment of inertia, I_0,

$$I_0 = \sum_i m_i r_{0i}^2 \tag{9.21}$$

in which m_i is the mass of the i-th atom, and r_{0i} the distance along the linear axis at equilibrium of the i-th atom from the position of center of mass of the molecule.

For a non-linear polyatomic molecule there are, in general, three principle moments of inertia. If the positions of the atoms composing the molecule are measured in Cartesian coordinates from the center of mass of the molecule, so that,

$$\sum_i m_i x_i = \sum_i m_i y_i = \sum_i m_i z_i = 0 \tag{9.22}$$

the moments about the Cartesian axes passing through the center of mass are,

$$\left.\begin{aligned} I_{xx} &= \sum_i m_i (y_i^2 + z_i^2), \quad I_{yy} = \sum_i m_i (z_i^2 + x_i^2), \\ I_{zz} &= \sum_i m_i (x_i^2 + y_i^2), \end{aligned}\right\} \tag{9.23}$$

and the products of inertia are,

$$I_{yz} = \sum_i m_i y_i z_i, \quad I_{zx} = \sum_i m_i z_i x_i, \quad I_{xy} = \sum_i m_i x_i y_i. \tag{9.24}$$

The three principle moments, A, B, C, of the molecule are the three roots for η of the three equations,

$$\left.\begin{aligned} \alpha(I_{xx} - \eta) - \beta I_{xy} - \gamma I_{xz} &= 0, \\ -\alpha I_{yx} + \beta(I_{yy} - \eta) - \gamma I_{yz} &= 0, \\ -\alpha I_{zx} - \beta I_{zy} + \gamma(I_{zz} - \eta) &= 0, \end{aligned}\right\} \tag{9.25}$$

with $\alpha^2 + \beta^2 + \gamma^2 = 1$. The corresponding α, β, γ are the direction cosines of the three principle axes.

The coordinates for the polyatomic molecule are then chosen as the three coordinates of the center of mass, the three Eulerian angles, ϑ, φ, ψ, describing the orientation of the rigid body of principle moments A, B, C, in space, and $3n-6$ coordinates upon which the potential energy will depend. The moments A, B, and C will depend explicitly on the values of these last vibrational coordinates. As in the case of the diatomic molecule one uses the approximation that A, B, and C are constants, given by the values at the position of the minimum of potential. The Hamiltonian is then separable, and the rotational part[1] is

$$\left.\begin{aligned} H_{\text{rot}}^{(3)} = {} & \frac{\sin^2\psi}{2A}\left\{p_\vartheta - \frac{\cos\psi}{\sin\vartheta\sin\psi}\,(p_\varphi - \cos\vartheta\, p_\psi)\right\}^2 \\ & + \frac{\cos^2\psi}{2B}\left\{p_\vartheta + \frac{\sin\psi}{\sin\vartheta\sin\psi}\,(p_\varphi - \cos\vartheta\, p_\psi)\right\}^2 \\ & + \frac{1}{2C}\,p_\psi^2, \end{aligned}\right\} \tag{9.26}$$

for the three-dimensional rigid rotator.

[1] H. B. G. CASIMIR: Diss. Leiden 1931.

The solution of the Schroedinger equation for the general case that $A \neq B \neq C$ does not lead to an analytic closed form for the energies in terms of the quantum numbers. However, since moments of inertia of polyatomic molecules are large, and their boiling points relatively high, one can use the classical approximation,

$$Q_{\text{rot}}^{(3)} = \frac{1}{h^3 \gamma} \int_0^{\pi} d\vartheta \iint_0^{2\pi} d\psi\, d\varphi \iiint_{-\infty}^{+\infty} dp_\psi\, dp_\varphi\, dp_\vartheta\, e^{-H_{\text{rot}}/kT}, \tag{9.27}$$

in which γ is the symmetry number of the molecule.

The transformation of the Hamiltonian, H_{rot}, of (9.26) into the form,

$$\left.\begin{aligned} \frac{1}{kT} H_{\text{rot}} = {} & \frac{1}{2kT}\left(\frac{\sin^2\psi}{A} + \frac{\cos^2\psi}{B}\right) \times \\ & \times \left[p_\vartheta + \left(\frac{1}{B} - \frac{1}{A}\right) \frac{\sin\psi\cos\psi}{\sin\vartheta\left(\frac{\sin^2\psi}{A} + \frac{\cos^2\psi}{B}\right)} (p_\varphi - \cos\vartheta\, p_\psi)\right]^2 + \\ & + \frac{1}{2kT} \frac{1}{AB\sin^2\vartheta} \frac{1}{\frac{\sin^2\psi}{A} + \frac{\cos^2\psi}{B}} (p_\varphi - \cos\vartheta\, p_\psi)^2 + \\ & + \frac{1}{2kT} \frac{1}{C} p_\psi^2, \end{aligned}\right\} \tag{9.28}$$

makes direct integration in the order dp_ϑ first, followed by dp_φ and dp_ψ quite simple if it is remembered that

$$\int_{-\infty}^{+\infty} e^{-a(x+b)^2} dx = \int_{-\infty}^{+\infty} e^{-ay^2} dy = \sqrt{\frac{\pi}{a}}.$$

Integration over p_ϑ leads to the factor,

$$\sqrt{2\pi kT}\left[\frac{\sin^2\psi}{A} + \frac{\cos^2\psi}{B}\right]^{-\frac{1}{2}}, \tag{a}$$

and over p_φ to the factor

$$\sqrt{2\pi kT}\sqrt{AB}\left[\frac{\sin^2\psi}{A} + \frac{\cos^2\psi}{B}\right]^{+\frac{1}{2}} \sin\vartheta. \tag{b}$$

The terms in ψ cancel in the two factors.

Integration over p_ψ gives the factor,

$$\sqrt{2\pi kT}\sqrt{C}. \tag{c}$$

Finally the two angles φ and ψ give factors 2π each and $\int_0^{\pi} \sin\vartheta\, d\vartheta = 2$, or together the three angular coordinates give the factor

$$8\pi^2. \tag{d}$$

The product of the factors (a), (b), (c), (d) divided by γh^3 can be written as,

$$Q_{\text{rot}}^{(3)} = \frac{\sqrt{\pi}}{\gamma}\left(\frac{8\pi^2 A kT}{h^2}\right)^{\frac{1}{2}}\left(\frac{8\pi^2 B kT}{h^2}\right)^{\frac{1}{2}}\left(\frac{8\pi^2 C kT}{h^2}\right)^{\frac{1}{2}}. \tag{9.29}$$

The energy per mol due to rotation, $RT^2\, d\log Q_{\text{rot}}^{(3)}/dT$, Eq. (3.38), is

$$E_{\text{rot}}^{(3)} = \tfrac{3}{2} RT, \tag{9.30}$$

corresponding to the kinetic energy of three classical degrees of freedom per molecule.

10. Vibrational partition function. The classical Hamiltonian of an oscillator is

$$H_{\text{osc}} = \frac{1}{2\mu} p_\xi^2 + \frac{a}{2} \xi^2, \tag{10.1}$$

in which ξ is a coordinate of dimensions length, p_ξ its conjugated momentum, $p_\xi = \mu \dot{\xi}$. The quantity μ is a mass, and a a force constant. The choice of coordinates of Eq. (9.1) for a diatomic molecule leads to the additive contribution,

$$H_{\text{vib}} = \frac{1}{2\mu} p_\xi^2 + u(\xi), \tag{10.2}$$

from the coordinate ξ, with μ the reduced mass, $\mu = m_1 m_2/(m_1 + m_2)$. The coordinate ξ is a measure of the distance apart of the two atoms, but with the zero so chosen that the distance, r_0, when $\xi = 0$ is that of the minimum of potential, Eq. (9.2). If $u(\xi)$ is expanded as a power series in ξ the linear term is missing

$$u(\xi) = u_0 + \frac{a}{2} \xi^2 + \cdots, \tag{10.3}$$

and if powers higher than quadratic are omitted one has the oscillator Hamiltonian, (10.1).

The classical frequency of vibration of the oscillator given by (10.1) is,

$$\nu = \frac{1}{2\pi} \sqrt{\frac{a}{\mu}}, \tag{10.4}$$

and the solution of the Schroedinger equation gives allowed non-degenerate stationary state energies at

$$\varepsilon(n) = u_0 + (n + \tfrac{1}{2})\, h\nu \tag{10.5}$$

with n integral, $n \geqq 0$. It is customary in quantum mechanics to measure the energy from the minimum of the potential, u_0, that is to set $u_0 = 0$ in (10.5). For purposes of computation of the thermodynamic quantities it is more convenient to choose the zero of energy as that of the lowest state of the molecule, or $u_0 = -\frac{1}{2} h\nu$, so that (10.5) becomes,

$$\varepsilon(n) = n\, h\nu. \tag{10.5'}$$

This choice of the zero of energy is such that the perfect gas at absolute zero of temperature is assigned zero energy. However molecules containing different isotopic species of the same chemical element have, to very high degree of approximation, the same functional dependence of $u(\xi)$ on ξ, (and the same value of r_0). The value of u_0 is thus the same for the various isotopically different molecules if the separated atoms are chosen as the arbitrary zero of energy. The frequencies, and hence $\frac{1}{2} h\nu$, depend on the reduced masses, through (10.4). The energies of the different isotopic molecules at absolute zero are hence different, if measured from some more absolute zero, such as that of the isolated atoms at infinite separation, but in a known way. This fact is important in using the partition functions to compute isotopic equilibria, and the somewhat less arbitrary choice of $u_0 = 0$, is more convenient for the discussion of isotopic equilibria (Sect. 11). We shall nevertheless adopt the convention of (10.5') for the zero of energy, assigning each perfect gas zero energy at $T = 0°$ K.

With the approximation of assuming the harmonic forces of (10.3) the vibrational partition function is,

$$Q_{\text{vib}} = \sum_{n \geqq 0} e^{-n h\nu/kT} = \frac{1}{1 - e^{-n h\nu/kT}}, \tag{10.6}$$

which goes to unity at $T \to 0°$ K. For high temperatures, using $e^{-x} = 1 - x + \frac{1}{2!} x^2 - \cdots$, one has

$$\left.\begin{aligned} Q_{\text{vib}} &= \frac{1}{\frac{h\nu}{kT} - \frac{1}{2}\left(\frac{h\nu}{kT}\right)^2 + \frac{1}{6}\left(\frac{h\nu}{kT}\right)^3 - \frac{1}{24}\left(\frac{h\nu}{kT}\right)^4 + \cdots} \\ &= \frac{kT}{h\nu}\left[1 + \frac{1}{2}\frac{h\nu}{kT} + \frac{1}{12}\left(\frac{h\nu}{kT}\right)^2 - \frac{1}{720}\left(\frac{h\nu}{kT}\right)^4 + \cdots\right]. \end{aligned}\right\} \tag{10.7}$$

The classical Q_{vib} is obtained as,

$$Q_{\text{vib}}^{(0)} \approx \frac{1}{h} \iint_{-\infty}^{+\infty} e^{-H_{\text{vib}}/kT}\, dp_\xi\, d\xi, \tag{10.8}$$

and with (10.1) for $H_{\text{vib}} = H_{\text{osc}}$ leads immediately to,

$$Q_{\text{vib}}^{(0)} = \frac{1}{h}\sqrt{2\pi\mu kT}\sqrt{\frac{2\pi kT}{a}} = \frac{kT}{h\nu} \tag{10.9}$$

from (10.4) for ν. In using (10.1) for the classical Hamiltonian, however, we have used the minimum of potential for the zero of energy, and in order to convert to the zero $u_0 + \frac{1}{2}h\nu$ we must multiply by $e^{+h\nu/kT} \cong 1 + \frac{1}{2}(h\nu/kT) + \cdots$. The classical expression to be compared with (10.7) is then,

$$Q_{\text{vib}} = Q_{\text{vib}}^{(0)} e^{h\nu/2kT} = \frac{kT}{h\nu}\left[1 + \frac{1}{2}\left(\frac{h\nu}{kT}\right) + \frac{1}{8}\left(\frac{h\nu}{kT}\right)^2 + \cdots\right],$$

which differs from (10.7) only in quadratic and higher terms.

The contribution to the internal energy, per mol, due to vibration, is, from Eq. (3.38) that $E_{\text{vib}} = RT^2\, d\log Q_{\text{vib}}/dT$, and using form (10.7) for high temperatures,

$$E_{\text{vib}} = -\frac{1}{2}N_0 h\nu + RT\left[1 + \frac{1}{12}\left(\frac{h\nu}{kT}\right)^2 - \frac{1}{720}\left(\frac{h\nu}{kT}\right)^4 + \cdots\right]. \tag{10.10}$$

At high temperatures the energy above the zero of potential, $u_0 = 0$, is then

$$E_{\text{vib}} - \frac{1}{2}N_0 h\nu = RT\left[1 + \frac{1}{12}\left(\frac{h\nu}{kT}\right)^2 - \cdots\right], \tag{10.11}$$

in agreement with general considerations (Sect. 27) that the deviations from the classical value, RT, should be quadratic in h/kT.

Polyatomic molecules, as discussed in Sect. 6 have $3n-6$ or $3n-5$ coordinates, upon which the potential depends, the number being dependent on whether the three angles for a non-linear configuration, or the two for a linear molecule are used. If the zero's of the coordinates are chosen as the values at the potential minimum a development of the potential in powers of the coordinates will contain no linear terms. If, then, cubic and higher order terms are neglected a linear transformation of the coordinates always exists such that in the new coordinates, ξ_i, $1 \leq i \leq 3n-6$ or $3n-5$, the Hamiltonian is separable, with each additive term of the form (10.1), or by scaling with $\zeta_i = \sqrt{\mu_i}\,\xi_i$ the equivalent,

$$H_{\text{osc}} = p_{\zeta i}^2 + 2\pi^2 \nu_i^2 \zeta_i^2. \tag{10.12}$$

Each such degree of freedom then contributes a factor Q_{vib} dependent on its characteristic frequency, ν_i, and the additive term $\log Q_{\text{vib}}\left(\frac{h\nu_i}{kT}\right)$ to the logarithms of the partition function.

However, whereas, the approximation that the potential is quadratic in deviations of the coordinate from the position of the minimum is usually adequate for diatomic molecules, there are cases where it is utterly inadequate in polyatomic molecules.

11. Isotope effects. For atoms heavier than hydrogen the presence of different isotopic species has little direct effect on the properties of the molecule. The internal partition function may, however, be considerably altered by the change in the symmetry number. This change is just such as to preserve equal, or essentially equal, isotopic composition in different molecular species at equilibrium. The thermodynamic properties of the gas are altered mainly by the addition of a temperature independent entropy of mixing per molecule equal to

$$s_{\text{mix}} = -\,\mathrm{k} \sum_a n_a \sum_\alpha x_{a\alpha} \log x_{a\alpha}, \tag{11.1}$$

where the sum runs over all elements a having n_a atoms in the molecule and over the isotopes α of element a with $x_{a\alpha}$ the mol fraction of isotope α in element a. Since, in all chemical reactions the total number of atoms of each element is equal in reactants and products, this term cancels in the total ΔS of the reaction. The isotopic entropy of mixing can therefore, and frequently is, neglected in computing or tabulating thermodynamic values.

Actually, as we shall prove in this section, the isotopic composition does remain identical in all phases at equilibrium in the classical limit, and the thermodynamic properties of the system are then altered only by the additional entropy of mixing term. However, when quantum effects are appreciable this is no longer true, and the equilibrium constants of chemical isotopic exchange reactions differ from their classical limit. We shall show how this difference can be simply computed.

Consider first the monatomic perfect gas of element a having mol fraction $x_{a\alpha}$ of isotope α. From Eq. (3.20) with P_α, the partial pressure of isotope α given as $x_{a\alpha} P$,

$$\mu_{a\alpha} = \mathrm{k}T\left[\log \frac{P\lambda_{a\alpha}^3 x_{a\alpha}}{\mathrm{k}T} - \log Q_i(T)\right]. \tag{11.2}$$

The internal partition function, Q_i, will be the same for all isotopes. The de Broglie wavelengths, $\lambda_{a\alpha} = h/\sqrt{2\pi m_{a\alpha}\mathrm{k}T}$, differ slightly in the difference of mass, $m_{\alpha a}$. Define a geometric mean mass, m_a, by

$$m_a = \prod_\alpha m_{a\alpha}^{x_{a\alpha}}, \tag{11.3$_1$}$$

$$\log m_a = \sum_\alpha x_{a\alpha} \log m_{a\alpha} \tag{11.3$_2$}$$

and a de Broglie wave length $\lambda_a = h/\sqrt{2\pi m_a \mathrm{k}T}$. The free energy per molecule of this gas will be,

$$\left.\begin{aligned} \sum_\alpha x_{a\alpha}\mu_{a\alpha} &= \mathrm{k}T\left[\log \frac{P\lambda_a^3}{\mathrm{k}T} - \log Q_i\right] + \mathrm{k}T\sum_\alpha x_{a\alpha}\log x_{a\alpha} \\ &= \mu_a + \mathrm{k}T\sum_a x_{a\alpha}\log x_{a\alpha}, \end{aligned}\right\} \tag{11.3$_3$}$$

with μ_a the free energy per molecule, or the chemical potential of element a that would be computed if we were unaware of the existence of the different isotopes. In terms of μ_a and the $x_{a\alpha}$'s the chemical potentials $\mu_{a\alpha}$ of the isotopic species are

$$\mu_{a\alpha} = \mathrm{k}T \log x_a + 3 \log \frac{\lambda_{a\alpha}}{\lambda_a} + \mu_a, \tag{11.3_4}$$

$$\lambda_{a\alpha}^{-3} \, e^{\frac{\mu_{a\alpha}}{\mathrm{k}T}} = x_{a\alpha} \lambda_a^{-3} e^{\frac{\mu_a}{\mathrm{k}T}}. \tag{11.3_5}$$

Now turn to the expression for the Canonical partition function, $Q_C(V, T, \boldsymbol{M})$, for a system of $\boldsymbol{M} = M_{a\alpha}, M_{a\beta}, \ldots, M_{b\alpha}, M_{b\beta}, \ldots$ atoms of different elements $a, b, \ldots$, and different isotopes, $a\alpha, a\beta, \ldots, b\alpha, b\beta, \ldots$ of each element. Use the classical limit, Eqs. (2.15) and (2.16) that,

$$Q_C(V, T, \boldsymbol{M}) = \frac{1}{\boldsymbol{M}!} \Big[\prod_\alpha \prod_a \lambda_{a\alpha}^{-3M_{a\alpha}} \Big] Q_\tau(V, T, \boldsymbol{M}, \tag{11.4_1}$$

with

$$Q_\tau(V, T, \boldsymbol{M}) = \iint_V \cdots \int e^{-U_{\boldsymbol{M}}\{\boldsymbol{M}\}/\mathrm{k}T} d\boldsymbol{r}_1 \cdots d\boldsymbol{r}_M, \tag{11.4_2}$$

and the integration made over the Cartesian vectors of the atoms. The potential, $U_{\boldsymbol{M}}$, will depend only on the set,

$$\boldsymbol{N} = N_a, N_b, \cdots, \tag{11.4_3}$$

$$N_a = \sum_\alpha M_{a\alpha}, \tag{11.4_4}$$

of numbers, N_a, of atoms of the element a, and not on the isotopic composition. Now use this in the computation of the Grand Canonical partition function,

$$\left.\begin{aligned} e^{PV/\mathrm{k}T} &= Q_{GC}(V, T, \boldsymbol{\mu}) = \sum_{\boldsymbol{M}} e^{\boldsymbol{M}\cdot\boldsymbol{\mu}/\mathrm{k}T} Q_C(V, T, \boldsymbol{M}) \\ &= \sum_{\boldsymbol{M}} \frac{1}{\boldsymbol{M}!} \Big[\prod_a \prod_\alpha (\lambda_{a\alpha}^{-3} e^{\mu_{a\alpha}/\mathrm{k}T})^{M_{a\alpha}} \Big] Q_\tau(V, T, \boldsymbol{N}). \end{aligned}\right\} \tag{11.4_5}$$

If this system is in chemical equilibrium with a gas of monatomic atoms of element a with composition $x_{a\alpha}$ and total μ_a value (11.3₃), the individual chemical potentials, $\mu_{a\alpha}$, of the isotopic species $a\alpha$ will have the values of Eq. (11.3₅). Since Q_τ depends on $\boldsymbol{N}$ only, we can sum over all $\boldsymbol{M}$ values consistent with a given $\boldsymbol{N}$ first, Eq. (11.4₄), and with (11.3₅) immediately find,

$$e^{PV/\mathrm{k}T} = \sum_{\boldsymbol{N}} \frac{1}{\boldsymbol{N}!} \Big[\prod_a (\lambda_a^{-3} e^{\mu_a/\mathrm{k}T})^{N_a} \Big] Q_\tau(V, T, \boldsymbol{N}). \tag{11.4_6}$$

Thus, in the classical limit, we can afford to forget about the isotopes, using the overall fictitious chemical potentials, μ_a, of the elements. The chemical potentials of molecules in the system having n_a atoms of element a will be $\sum_a n_a \mu_a$ plus the term $-\mathrm{k}T s_{\mathrm{mix}}$ of (11.1).

One may now show that the isotopic composition of the ensemble described by (11.4₆) is actually the same as that of the monatomic gas with which it is in equilibrium. To show this we use the thermodynamic relation $[\partial(PV)/\partial\mu_a]_{T,\mu_b\ldots} = \overline{N}_\alpha$, Eq. (1.33) and $[\partial(PV)/\partial\mu_{a\alpha}]_{T,\mu_{a\beta},\ldots,\mu_b} = \overline{M}_{a\alpha}$. It is only necessary to prove that $(\partial\mu_a/\partial\mu_{a\alpha})_{T,\mu_a\ldots} = x_a$. This follows from (11.3₃).

In any chemical reaction the equilibrium constants for the various possible isotopic species involving the same molecule, will, in the classical limit, be such that the relative fractions of the isotopes in the different molecular species are equal. Small deviations due to quantum effects exist. An isotopic exchange reaction such as

$$C^{14}O + C^{12}O_2 \rightleftharpoons C^{12}O + C^{14}O_2$$

will have unit value for the equilibrium constant in the classical limit. However one of the type

$$CO^{18} + CO^{16}O^{16} \rightleftharpoons CO^{16} + CO^{16}O^{18}$$

will have equilibrium constant 2, since if x_{18} is to be the same for CO and CO_2 the fraction of CO^{18} molecules in CO must be x_{18} but the fraction of $CO^{16}O^{18}$ in CO_2 must be $2x_{18}$. The classical value of the equilibrium constant is the ratio of the products of the symmetry numbers, γ, for the reactants divided by that of the reaction products. (Two different "optical isomers" such as can exist in $CX_\alpha X_\beta X_\gamma X_\delta$ are considered as two different molecules.) In general the equilibrium constant, K, of a chemical reaction is $K = e^{-\Delta\mu_0/kT}$, with $\Delta\mu_0$ the sum of the chemical potentials of the product molecules, multiplied by their number, minus the same sum for the reactants, all chemical potentials computed at the standard pressure for which K is used. From (3.20) for μ, we have, for a reaction in which the total numbers of molecules does not change,

$$K = \frac{\prod_i \lambda_i'^{-3} Q_i'}{\prod_i \lambda_i^3 Q_i}, \tag{11.5$_1$}$$

where the product in the numerator runs over the molecules of the reaction products, and that in the denominator over those in the reactants. We seek the value of

$$K'_{ex} = \frac{\prod_i \gamma_i}{\prod_i \gamma_i'} K = \frac{\prod_i \gamma_i'^{-1} \lambda_i'^{-3} Q_i'}{\prod_i \gamma_i^{-1} \lambda_i^{-3} Q_i}, \tag{11.5$_2$}$$

for an isotopic exchange reaction, a number which will be unity in the classical limit.

Since the total numbers of atoms of each isotopic species is the same in reactants and products we can multiply both numerator and denominator by $\lambda_{a\alpha}^3$ for each atom, and include with each factor i above, the product $\prod_a \prod_\alpha (\lambda_{a\alpha}^3)^{n_{a\alpha i}}$ of the atoms in molecule i. In an isotopic exchange reaction one can pair the molecules in reactants and products such that for every i the same chemical species occurs in reactant and product, differing only in isotopic composition. Thus if,

$$f_i = \frac{\prod_a \prod_\alpha (\lambda_{a\alpha}^3)^{n'_{a\alpha i}} \lambda_i'^{-3} \gamma_i'^{-1} Q_i'}{\prod_a \prod_\alpha (\lambda_{a\alpha}^3)^{n_{a\alpha i}} \lambda_i^{-3} \gamma_i^{-1} Q_i}, \tag{11.5$_3$}$$

for each molecular species i involved in the reaction, where the numerator has the isotopic composition of the product molecule, and the denominator that of the reactant, then

$$K'_{ex} = \prod_i f_i. \tag{11.5$_4$}$$

Now we know that in the classical limit numerator and denominator of (11.5$_3$) are identical. We need then only to write the ratio of each Q_i to its classical expression to evaluate f_i. Except for H_2, HD, D_2, etc., the rotational factors will be classical and hence may be neglected. Only the vibrational degrees of freedom need be considered. For these, however, we must be cautious and use the quantum mechanical contribution, Q_{vib}, when the energy is measured from the zero of potential energy, since these energies are identical in the two molecules, and not the energies of the lowest vibrational state of energy $\frac{1}{2}h\nu$ above this. Using, then, Eq. (10.9) for the classical Q_{vib}, and $e^{-\frac{1}{2}h\nu/kT}$ times (10.6) for the quantum mechanical, we find the ratio,

$$R(\nu) = \frac{(h\nu'/kT)\,e^{-\frac{1}{2}h\nu'/kT}\,(1-e^{-h\nu'/kT})^{-1}}{(h\nu/kT)\,e^{-\frac{1}{2}h\nu/kT}\,(1-e^{-h\nu/kT})^{-1}}, \tag{11.5$_5$}$$

from each vibrational frequency. Let $\nu'=\nu+\Delta\nu$, and if $\Delta\nu/kT$ is small, one finds,

$$R(\nu) = 1 - \left[\frac{1}{2} - \frac{kT}{h\nu} + \frac{1}{e^{h\nu/kT}-1}\right]\frac{h\Delta\nu}{kT}. \tag{11.5$_6$}$$

The effective K'_{ex} of (11.5$_4$) differs from unity by,

$$K'_{ex} = 1 - \sum_j \left[\frac{1}{2} - \frac{kT}{h\nu_j} + \frac{1}{e^{h\nu_j/kT}-1}\right]\frac{h\Delta\nu_j}{kT} \tag{11.5$_7$}$$

where the sum runs over all frequencies j of the molecules of the reactants, and $\Delta\nu_j$ is the frequency of the same mode of vibration in the product molecule minus that in the reactant.

At low temperatures the sum is $\frac{1}{2}\frac{h}{kT}\sum_j \Delta\nu_j$, namely the change in energy due to the difference in zero point energies of the molecules only. At high temperatures, $(e^x-1)^{-1} = x^{-1} - \frac{1}{2} + \frac{1}{12}x$, one has

$$K'_{ex} = 1 - \sum_j \frac{1}{12}\frac{h\nu_j}{kT}\frac{h\Delta\nu_j}{kT}. \tag{11.5$_8$}$$

12. Summary of internal partition functions. The internal partition function

$$Q_i = \sum_n e^{-\varepsilon_n/kT} \tag{12.1}$$

for a molecule, is the sum over all internal quantum states n, of energy ε_n. In so far as different degrees of freedom are separable, it is the product of contributions from the separable degrees of freedom, and its logarithm is a sum of the logarithms of these.

For temperatures at which molecules exist as gases, with the single exception of hydrogen, there is a nuclear spin contribution of a simple factor, $(2s_\alpha+1)$, for every atom, α, of the molecule, having nuclear spin s_α. This contribution leads to a nuclear spin entropy per mole,

$$S_{\text{nucle. spin}} = R\sum_\alpha n_\alpha \log(2s_\alpha+1) \tag{12.2}$$

which is normally neglected.

For normal temperature, and again excepting molecules containing H, D and T, it is approximately adequate to neglect differences due to different isotopes. An entropy term due to mixing of isotopes,

$$S_{\text{mix}} = -R\sum_\alpha x_{a\alpha}\log x_{a\alpha}, \tag{12.3}$$

per formula weight of element a, with $x_{a\alpha}$ the mol fraction of isotope α in the atoms of element a should then be added to the total entropy. This, again is normally ignored.

For monatomic atoms the sum (12.1) must be summed numerically for the electronic energy levels, counting $2J+1$ states for every level of total angular momentum quantum number J. Usually excited states contribute negligibly.

For molecules the contributions due to rotation, and to each normal mode of vibration can be factored approximately as independent factors. If, as is very rarely the case, more than one electronic state of the molecule contributes appreciably, the rotational and vibrational contributions for each state must be computed independently, since moments of inertia and frequencies depend on the electronic state.

If the classical equations for rotation given below are used, and they are valid in practice for all molecules except hydrogen, and if the molecule is symmetric, a factor $1/\gamma$ must be used in Q_i. This leads to a negative additive entropy per mole,

$$S_{\text{sym}} = -\,\mathrm{R} \log \gamma, \tag{12.4}$$

with γ, the symmetry number, equal to the total number of permutations of identical atoms covered by the rotational coordinates. Thus $\gamma = 2$ for diatomic elements, O_2, N_2, etc.

For linear molecules the factor, $Q_{\text{rot}}^{(2)}$ due to the two rotational degrees of freedom depends on the moment of inertia, I, in the equilibrium position. The quantity,

$$B = \frac{h}{8\pi^2 I c}$$

of dimension reciprocal length (wave number) is listed in spectroscopic tables for most diatomic molecules. The factor $Q_{\text{rot}}^{(2)}$ depends on

$$\left.\begin{aligned} \sigma &= \frac{B h c}{\mathrm{k} T} = \frac{h^2}{8\pi^2 I \mathrm{k} T} = 1.4390 \frac{B\,(\text{in cm}^{-1})}{T} \\ &= 35.6 \times 10^{-4} \frac{1}{I T}, \end{aligned}\right\} \tag{12.5$_1$}$$

namely,

$$\left.\begin{aligned} Q_{\text{rot}}^{(2)} &= 1 + 3\,\mathrm{e}^{-2\sigma} + 5\,\mathrm{e}^{-6\sigma} + 7\,\mathrm{e}^{-12\sigma} + \cdots \\ &= \frac{1}{\sigma}\left(1 + \frac{\sigma}{3} + \frac{\sigma^2}{15} + \frac{4\sigma^3}{315} + \cdots\right) 0.1\,\%,\ \sigma \leq 0.7. \end{aligned}\right\} \tag{12.5$_2$}$$

The additive contributions to the thermodynamic functions due to rotation for linear molecules are, per mole of gas,

$$\left.\begin{aligned} G_{\text{rot}}^{(2)} &= A_{\text{rot}}^{(2)} = -\,\mathrm{R} T \log Q_{\text{rot}}^{(2)} \\ &= \mathrm{R} T \left[\log \sigma - \frac{\sigma}{3} - \frac{\sigma^2}{90} - \frac{8\sigma^3}{2835} - \cdots\right], \end{aligned}\right\} \tag{12.5$_3$}$$

$$\left.\begin{aligned} H_{\text{rot}}^{(2)} &= E_{\text{rot}}^{(2)} = \mathrm{R} T \frac{d \log Q_{\text{rot}}^{(2)}}{d \log T} \\ &= \mathrm{R} T \left[1 - \frac{\sigma}{3} - \frac{\sigma^2}{45} - \frac{8\sigma^3}{945} - \cdots\right], \end{aligned}\right\} \tag{12.5$_4$}$$

$$\left.\begin{aligned} S_{\text{rot}}^{(2)} &= \mathrm{R} \frac{d}{dT}\left[T \log Q_{\text{rot}}^{(2)}\right] \\ &= \mathrm{R}\left[1 - \log \sigma - \frac{\sigma^2}{90} - \frac{16\sigma^3}{2835} - \cdots\right], \end{aligned}\right\} \tag{12.5$_5$}$$

$$\left.\begin{aligned} C_{(2)}^{\text{rot}} &= \mathrm{R} \frac{d}{dT}\, T^2 \frac{d}{dT} \log Q_{\text{rot}}^{(2)} \\ &= \mathrm{R}\left[1 + \frac{\sigma^2}{45} + \frac{16\sigma^3}{945} + \cdots\right]. \end{aligned}\right\} \tag{12.5$_6$}$$

For non-linear polyatomic molecules the three degrees of freedom of rotation contribute a factor, $Q_{\text{rot}}^{(3)}$, to Q_i, dependent on the values, A, B, C, of the three principle moments of inertia. Using

$$\sigma_1 = \frac{h^2}{8\pi^2 A\,\mathrm{k}T}, \quad \sigma_2 = \frac{h^2}{8\pi^2 B\,\mathrm{k}T}, \quad \sigma_3 = \frac{h^2}{8\pi^2 C\,\mathrm{k}T}, \tag{12.6$_1$}$$

the equation,

$$Q_{\text{rot}}^{(3)} = \sqrt{\frac{\pi}{\sigma_1\sigma_2\sigma_3}}, \tag{12.6$_2$}$$

gives the classical limit of $Q_{\text{rot}}^{(3)}$ which is adequate for all temperatures for which such molecules exist as gases. The additive contribution to the thermodynamic functions are, per mole of gas,

$$G_{\text{rot}}^{(3)} = A_{\text{rot}}^{(3)} = \frac{1}{2}\,\mathrm{R}T\log\frac{\sigma_1\sigma_2\sigma_3}{\pi}, \tag{12.6$_3$}$$

$$H_{\text{rot}}^{(3)} = E_{\text{rot}}^{(3)} = \tfrac{3}{2}\mathrm{R}T, \tag{12.6$_4$}$$

$$S_{\text{rot}}^{(3)} = \mathrm{R}\left[\frac{3}{2} - \frac{1}{2}\log\frac{\sigma_1\sigma_2\sigma_3}{\pi}\right], \tag{12.6$_5$}$$

$$C_{\text{rot}}^{(3)} = \tfrac{3}{2}\mathrm{R}. \tag{12.6$_6$}$$

A linear molecule containing n atoms will have $3n-5$ degrees of freedom depending on the potential, and if the molecule is non-linear, there will be $3n-6$ of them. In so far as these can be regarded as a system of uncoupled linear harmonic oscillators each such degree of freedom contributes a factor Q_{vib} to Q_i, a factor which depends on the frequency ν of that mode of oscillation. We have,

$$Q_{\text{vib}} = \frac{1}{1-\mathrm{e}^{-h\nu/\mathrm{k}T}}. \tag{12.7$_1$}$$

The added contributions to the thermodynamic functions per mole, and per degree of freedom of frequency ν are

$$G_{\text{vib}} = A_{\text{vib}} = \mathrm{R}T\log\left(1-\mathrm{e}^{-h\nu/\mathrm{k}T}\right), \tag{12.7$_2$}$$

$$H_{\text{vib}} = E_{\text{vib}} = \mathrm{R}T\,\frac{h\nu/\mathrm{k}T}{\mathrm{e}^{h\nu/\mathrm{k}T}-1}, \tag{12.7$_3$}$$

$$S_{\text{vib}} = \mathrm{R}\left[\frac{h\nu/\mathrm{k}T}{e^{h\nu/\mathrm{k}T}-1} - \log\left(1-\mathrm{e}^{-h\nu/\mathrm{k}T}\right)\right], \tag{12.7$_4$}$$

$$C_{\text{vib}} = \mathrm{R}\,\frac{(h\nu/\mathrm{k}T)^2\,\mathrm{e}^{h\nu/\mathrm{k}T}}{(\mathrm{e}^{h\nu/\mathrm{k}T}-1)^2}. \tag{12.7$_5$}$$

C. Classical imperfect gases.

13. *P* as a power series in *z*. The starting equation for the treatment of systems in which the interaction between molecules is taken into account, and which therefore means any system not obeying the Perfect Gas equation of state, $PV = N\mathrm{k}T$, can be conveniently chosen as Eq. (4.21),

$$\left.\begin{aligned} \mathrm{e}^{PV/\mathrm{k}T} &= \sum_{\boldsymbol{N}\geq 0}\frac{\boldsymbol{z}^{\boldsymbol{N}}}{\boldsymbol{N}!}\iint_V\cdots\int F_{\boldsymbol{N}}^{(0)}(\boldsymbol{r}_1,\ldots,\boldsymbol{r}_N)\,d\boldsymbol{r}_1\ldots d\boldsymbol{r}_N,\\ \boldsymbol{z}^{\boldsymbol{N}} &= z_a^{N_a}z_b^{N_b}\ldots,\qquad \boldsymbol{N}! = N_a!\,N_b!\ldots, \end{aligned}\right\} \tag{13.1}$$

with z_a the activity of species a, of molecule defined in terms of the chemical potential, μ_a, as

$$z_a = e^{(\mu_a - \mu_{0a})/kT}, \tag{13.2}$$

normalized by the choice of μ_{0a} to become equal to the number density, ϱ_a, of molecules of species a in the perfect gas limit of all densities approaching zero,

$$\lim_{p_a, p_b, \ldots \to 0} \left[\frac{z_a}{\varrho_a} \right] = 1. \tag{13.3}$$

The quantity,

$$e^{-PV/kT} \boldsymbol{z}^{\boldsymbol{N}} F_{\boldsymbol{N}}^{(0)} (\boldsymbol{r}_1, \ldots, \boldsymbol{r}_N)\, d\boldsymbol{r}_1 \ldots d\boldsymbol{r}_N = P_{\boldsymbol{N}}^{(0)} (\boldsymbol{r}_1, \ldots, \boldsymbol{r}_N)\, d\boldsymbol{r}_1 \ldots d\boldsymbol{r}_N, \tag{13.4}$$

is the probability, that in a member of the Grand Canonical Ensemble of volume V, specified by T and $\boldsymbol{z}$, there will be exactly the set $\boldsymbol{N}$ of molecules, and no more, and that these be ar $\boldsymbol{r}_1, \ldots, \boldsymbol{r}_N$ in the volume element $d\boldsymbol{r}_1 \ldots d\boldsymbol{r}_N$.

As discussed in Sect. 5, Eq. (13.1) with $F_{\boldsymbol{N}}^{(0)}$ defined by (13.4) is general. For a classical monatomic gas $F_{\boldsymbol{N}}^{(0)}$ is given by

$$F_{\boldsymbol{N}}^{(0)} = e^{-U_{\boldsymbol{N}}/kT} \tag{13.5}$$

with $U_{\boldsymbol{N}}$ the mutual interaction potential of the molecules, with the zero of energy chosen as that when all pair distances approach infinity.

The development of Eq. (13.1) into a tractable form for numerical computation for gases depends on the use of the fact that

$$\lim_{\substack{\text{all pair distances} \\ r_{\alpha\beta} \to \infty}} [F_{\boldsymbol{N}}^{(0)}] = 1. \tag{13.6}$$

We develop $F_{\boldsymbol{N}}^{(0)}$ as unity plus a sum of terms, each of which is a product of functions that depend on the coordinates of only a subset of the molecules, and which are zero when the molecules of the subset are far distant from each other. The sum of the integrals over the leading term (unity) leads to the perfect gas equation, and the integrals over the correction terms finally give the pressure, P, as a power series in the fugacities.

The simplest way to see the structure of the correction terms is to consider the case of one component molecular species, and in which U_N is a sum of pair interactions,

$$U_N = \sum_{N \geq \alpha > \beta \geq 1} \sum u(r_{\alpha\beta}), \tag{13.7}$$

with

$$r_{\alpha\beta} = |\boldsymbol{r}_\alpha - \boldsymbol{r}_\beta|. \tag{13.7$_1$}$$

For this case,

$$\left.\begin{aligned} F_N^{(0)} &= e^{-U_N/kT} = \prod_{N \geq \alpha > \beta \geq 1} \prod e^{-u(r_{\alpha\beta})/kT} \\ &= \prod_{N \geq \alpha > \beta \geq 1} \prod [1 + f_{\alpha\beta}], \end{aligned}\right\} \tag{13.8}$$

where,

$$f_{\alpha\beta} = f^{(0)}(r_{\alpha\beta}) = e^{-u(r_{\alpha\beta})/kT} - 1, \tag{13.9}$$

is a function which approaches zero as $r_{\alpha\beta} \to 0$. The product of (13.8) is expanded as a sum,

$$
\begin{aligned}
F_N^{(0)} = 1 &+ \sum_{N \geq \alpha > \beta \geq 1}\sum f_{\alpha\beta} + \sum_{N \geq \alpha > \beta > \gamma \geq 1}\sum\sum \times \\
&\times [f_{\gamma\alpha} f_{\alpha\beta} + f_{\alpha\beta} f_{\beta\gamma} + f_{\beta\gamma} f_{\gamma\alpha} + f_{\alpha\beta} f_{\beta\gamma} f_{\gamma\alpha}] + \sum_{N \geq \alpha > \beta > \gamma > \delta \geq 1}\sum\sum\sum \times \\
&\times [f_{\alpha\beta} f_{\gamma\delta} + f_{\alpha\gamma} f_{\beta\delta} + f_{\delta\alpha} f_{\beta\gamma}] + \\
&+ [f_{\alpha\beta} f_{\gamma\delta} f_{\alpha\gamma} + f_{\alpha\beta} f_{\gamma\delta} f_{\alpha\delta} + \cdots (16 \text{ terms in all}) + \\
&+ f_{\alpha\beta} f_{\gamma\delta} f_{\alpha\gamma} f_{\alpha\delta} + \cdots (15 \text{ terms in all}) + f_{\alpha\beta} f_{\gamma\delta} f_{\alpha\gamma} f_{\alpha\delta} f_{\beta\gamma} + \cdots \times \\
&\times (6 \text{ terms in all}) + f_{\alpha\beta} f_{\alpha\gamma} f_{\alpha\delta} f_{\beta\gamma} f_{\beta\delta} f_{\gamma\delta}] + \sum \cdots.
\end{aligned}
\tag{13.10}
$$

The first term is unity, the second sum contains all $\frac{1}{2}N(N-1)$ pair terms, $f_{\alpha\beta}$, and the sum over the $N(N-1)(N-2)/3!$ triples each contains four terms. The

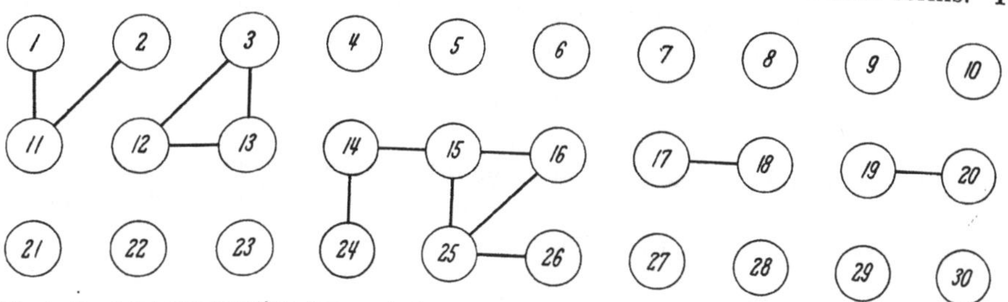

Fig. 3. Graphical representation of one product contributing to the term, $g_3(1, 2, 11)\, g_3(3, 12, 13)\, g_1(4) \ldots g_1(10)\, g_6(14, 15, 16, 24, 25, 26)\, g_2(17, 18)\, g_2(19, 20)\, g_1(21)\, g_1(22)\, g_1(23)\, g_1(27) \ldots g_1(30)$, in $F_{30}^{(0)}(1, \ldots, 30)$.

sum over the $\binom{N}{4}$ quadruples contains two kinds of terms. The first three written out explicitly, cluster the quadruplet into two independent pairs, whereas the other 38 terms all approach zero in value unless *all* of the quadruplet are clustered close together in the coordinate space.

The general terms may be grouped together according to the number, N_n, of clusters containing n molecules each. They may be most easily visualized by reference to a figure (Fig. 3). In this figure each numbered molecule is represented by a numbered circle, and the function $f_{\alpha\beta}$ by a line connecting circles α and β. There is a one to one correspondence between each product occurring in the sum of (13.10) and each possible figure, drawing either no line or one line between each pair. The general figure may be specified as having N_1 specified molecules with no line connecting them with any other, N_2 specified pairs connected to each other but to no other molecule, ..., N_n clusters each containing n specified molecules in which a path of lines connects any pair of molecules in the cluster. Sum all the products that correspond to such a specification. This sum can be written as a product of cluster functions, $g_n^{(0)}$, of the coordinates of n specified molecules

$$
g_n^{(0)}(\boldsymbol{r}_1, \ldots, \boldsymbol{r}_n) = \sum_{\substack{\text{sum over all} \\ \text{connected products}}} \prod f^{(0)}(r_{\alpha\beta}), \tag{13.11}
$$

in which the sum includes all possible connected graphs of the n molecules. By connected we mean that every pair of molecules are connected directly, or indirectly by a sequence of lines in the graph.

The cluster integrals,

$$b'_n(V,T) = \frac{1}{Vn!}\iint_V \cdots \int g_n^{(0)}(\boldsymbol{r}_1,\ldots,\boldsymbol{r}_n)\,d\boldsymbol{r}_1\ldots d\boldsymbol{r}_n, \tag{13.12$_1$}$$

$$b'_1(V,T) \equiv 1, \tag{13.12$_2$}$$

will have a limit,

$$b_n^{(0)} = b_n^{(0)}(T) = \lim_{V\to\infty}[b'_n(V,T)], \tag{13.13}$$

for all finite n values, if the mutual pair potential, $u(r)$, goes to zero more rapidly than r^{-3}. This follows, since, if one molecule, say the n-th, is held at the origin, the integrand, g_n, of (13.12) approached zero if any of the other coordinates, r, approaches infinity. The integral over all but the n-th will be $n!\,b_n^{(0)}$ independently of V for large V. The integration over the coordinates of the last n-th molecule will give $Vn!\,b_n^{(0)}$ in the limit that V becomes sufficiently large.

The sum of all products corresponding to having N_1 cluster of one specified molecule each, N_2 clusters containing each 2 specified molecules, ..., N_n clusters each of n specified molecules, $\sum_n n N_n = N$, contributes a term,

$$\prod_n (V n!\, b_n^{(0)})^{N_n}, \tag{a}$$

to the integral over $F_N^{(0)}$. The number of such terms for a specified set of numbers N_n, $\sum n N_n = N$, is the number of ways N numbered objects can be clustered in N_n clusters of size n each, which is,

$$\frac{N!}{\prod_n N_n!\,(n!)^{N_n}}. \tag{b}$$

Use the product of $(a)\times(b)\times(z^N/N!)$ in (13.1) to find

$$\left.\begin{aligned} e^{PV/kT} &= \sum_{N\geq 0} z^N \sum_{N_n(\sum n N_n = N)} \prod_n \frac{(V b_n^{(0)})^{N_n}}{N_n!}\\ &= \prod_n \sum_{N_n} \frac{(V b_n^{(0)} z^n)^{N_n}}{N_n!}\\ &= e^{V\sum_{n\geq 0} b_n^{(0)} z^n}, \end{aligned}\right\} \tag{13.14}$$

or,

$$P = kT\sum_{n\geq 1} b_n^{(0)} z^n \tag{13.15}$$

in which we have taken the limit, (13.13), that $V\to\infty$.

The expression for finite volume,

$$P(z,T,V) = kT\sum_{n\geq 1} b'_n(T,V)\, z^n, \tag{13.15$_1$}$$

is rigorous and exact in all cases, the limiting process,

$$\left.\begin{aligned} \lim_{V\to\infty}[P(z,T,V)] &= kT \lim_{V\to\infty}\Big[\sum_{n\geq 1} b'_n(V,T)\, z^n\Big]\\ &= kT\sum_{n\geq 1} b_n^{(0)}(T)\, z^n, \end{aligned}\right\} \tag{13.15$_2$}$$

is also certainly correct as long as the sum of $b_n^{(0)} z^n$ converges. The limit of convergence of the series is at the activity of condensation, as is discussed more fully in Sect. 16.

The general use of the cluster functions does not at all depend on the assumption of Eq. (13.7) that the potential may be written as a sum of pair functions, nor on the treatment of a one component system.

It becomes desirable, before discussing the more general case to attain a greater compactness of writing coordinates, at the expense of using a somewhat formalized notation. We have used $\boldsymbol{N}$ to represent a set of numbers of molecules, N_1 of kind $1, \ldots, N_a$ of kind $a, \ldots$. Use the symbol $\{\boldsymbol{N}\}$ to represent the coordinates of these molecules,

$$\{\boldsymbol{N}\} = \boldsymbol{r}_1, \ldots, \boldsymbol{r}_N, \tag{13.16$_1$}$$

and $d\{\boldsymbol{N}\}$ for the volume element,

$$d\{\boldsymbol{N}\} = d\boldsymbol{r}_1 \ldots d\boldsymbol{r}_N. \tag{13.16$_2$}$$

Let $\{\boldsymbol{n}\}_{\boldsymbol{N}}$ stand for the coordinates of $\boldsymbol{n} = n_1, n_2, \ldots, n_a, \ldots,$ molecules which are a specified subset of $\boldsymbol{N}$. For given numerical value of the numbers, n_a, of $\boldsymbol{n}$, there are

$$\binom{\boldsymbol{N}}{\boldsymbol{n}} = \prod_a \frac{N_a!}{(N_a - n_a)!\, n_a!} \tag{13.17}$$

different subsets $\{\boldsymbol{n}\}_{\boldsymbol{N}}$, for given $\boldsymbol{n}$.

Use the symbol $\{k\{\boldsymbol{n}_i\}_{\boldsymbol{N}}\}_u$ to designate a partition of the set $\boldsymbol{N}$ into k different subsets, $\{\boldsymbol{n}_i\}_{\boldsymbol{N}}$, $1 \leq i \leq k$, the different subsets being unconnected. By unconnected we mean that no two different subsets have any member in common, but the sum of all these subsets constitutes the whole set $\boldsymbol{N}$,

$$\sum_{l=1}^{k} \{\boldsymbol{n}_i\}_{\boldsymbol{N}} \equiv \{\boldsymbol{N}\}. \tag{13.18}$$

Such a partition may contain $N_{\boldsymbol{n}}$ different subsets for a given $\boldsymbol{n} = n_1, \ldots, n_a$, with $\sum_{\boldsymbol{n}} \boldsymbol{n} N_{\boldsymbol{n}} = \boldsymbol{N}$. There are,

$$\frac{\boldsymbol{N}!}{\prod_{\boldsymbol{n}} N_{\boldsymbol{n}}!\,(\boldsymbol{n}!)^{N_{\boldsymbol{n}}}} = \frac{\prod_a N_a!}{\prod_{\boldsymbol{n}} N_{\boldsymbol{n}}! \left[\prod_a n_a!\right]^{N_{\boldsymbol{n}}}} \quad \text{different partitions } \{k\{\boldsymbol{n}_i\}_{\boldsymbol{N}}\}_u \text{ for given } N_{\boldsymbol{n}} \text{ values.} \tag{13.19}$$

Using this notation one writes,

$$F_{\boldsymbol{N}}^{(0)}\{\boldsymbol{N}\} = \sum_{\{k\{\boldsymbol{n}_i\}_{\boldsymbol{N}}\}_u} \prod_{i=1}^{k} g_{\boldsymbol{n}}^{(0)}\{\boldsymbol{n}_i\}_{\boldsymbol{N}}, \tag{13.20}$$

where the sum runs over all partitions $\{k\{\boldsymbol{n}_i\}_{\boldsymbol{N}}\}_u$ from the single identical partition of $k=1$, $\boldsymbol{n}_1 = \boldsymbol{N}$, to the single partition with $k = N = \sum_a N_a$, and all $\boldsymbol{n}_i$ consisting of a single molecule. For a single component system the first terms are

$$\left.\begin{aligned}
F_1^{(0)}(\boldsymbol{r}_1) &= 1 = g_1^{(0)}(\boldsymbol{r}_1),\\
F_2^{(0)}(\boldsymbol{r}_1, \boldsymbol{r}_2) &= g_2^{(0)}(\boldsymbol{r}_1, \boldsymbol{r}_2) + g_1^{(0)}(\boldsymbol{r}_1)\, g_1^{(0)}(\boldsymbol{r}_2),\\
F_3^{(0)}(\boldsymbol{r}_1, \boldsymbol{r}_2, \boldsymbol{r}_3) &= g_3^{(0)}(\boldsymbol{r}_1, \boldsymbol{r}_2, \boldsymbol{r}_3) + g_2^{(0)}(\boldsymbol{r}_1)\, g_1^{(0)}(\boldsymbol{r}_2, \boldsymbol{r}_3) + \\
&\quad + g_1^{(0)}(\boldsymbol{r}_2)\, g_2^{(0)}(\boldsymbol{r}_3, \boldsymbol{r}_1) + g_1^{(0)}(\boldsymbol{r}_3)\, g_2^{(0)}(\boldsymbol{r}_1, \boldsymbol{r}_2) + \\
&\quad + g_1^{(0)}(\boldsymbol{r}_1)\, g_1^{(0)}(\boldsymbol{r}_2)\, g_1^{(0)}(\boldsymbol{r}_3),
\end{aligned}\right\} \tag{13.21}$$

so that

$$\left.\begin{aligned} g_1^{(0)}(\boldsymbol{r}_1) &= F_1^{(0)}(\boldsymbol{r}_1),\\ g_2^{(0)}(\boldsymbol{r}_1,\boldsymbol{r}_2) &= F_2^{(0)}(\boldsymbol{r}_1,\boldsymbol{r}_2) - F_1^{(0)}(\boldsymbol{r}_1)\,F_1^{(0)}(\boldsymbol{r}_2),\\ g_3^{(0)}(\boldsymbol{r}_1,\boldsymbol{r}_2,\boldsymbol{r}_3) &= F_3^{(0)}(\boldsymbol{r}_1,\boldsymbol{r}_2,\boldsymbol{r}_3) - F_1^{(0)}(\boldsymbol{r}_1)\,F_2^{(0)}(\boldsymbol{r}_2,\boldsymbol{r}_3) -\\ &\quad - F_1^{(0)}(\boldsymbol{r}_2)\,F_2^{(0)}(\boldsymbol{r}_3,\boldsymbol{r}_4) - F_1^{(0)}(\boldsymbol{r}_3)\,F_2^{(0)}(\boldsymbol{r}_1,\boldsymbol{r}_2) +\\ &\quad + 2F_1^{(0)}(\boldsymbol{r}_1)\,F_1^{(0)}(\boldsymbol{r}_2)\,F_1^{(0)}(\boldsymbol{r}_3).\end{aligned}\right\} \qquad (13.22)$$

The case that there are several kinds of molecules is included in (13.21) and (13.22) by remembering that the analytical form of the $F_{\boldsymbol{N}}^{(0)}$'s and $g_{\boldsymbol{n}}^{(0)}$'s depend on which species the numbered molecules belong to, the functions no longer being symmetrical in permutations of the coordinates if the molecules belong to different species.

The general expression,

$$g_{\boldsymbol{n}}^{(0)}\{\boldsymbol{n}\} = \sum_{\{l\{\nu_i\}\boldsymbol{n}\}_u} (-)^{l-1}(l-1)!\prod_{j=1}^{k} F_{\nu_j}^{(0)}\{\boldsymbol{\nu}_j\}_{\boldsymbol{n}}, \qquad (13.23)$$

may be inferred from (13.22), and checked by noting that if used in (13.20) the result is an identity. Namely, we write

$$F_{\boldsymbol{N}}^{(0)}\{\boldsymbol{N}\} = \sum_{\{k\{\boldsymbol{n}_i\}\boldsymbol{N}\}_u}\prod_{i=1}^{k}\Bigg[\sum_{\{l_i\{\nu_i\}\boldsymbol{n}_i\}_u}(-)^{l_i-1}(l_i-1)!\prod_{j=1}^{l_i}F_{\nu_j}^{(0)}\{\boldsymbol{\nu}_i\}_{\boldsymbol{n}_i}\Bigg]. \qquad (13.24_1)$$

In this use $l=\sum_{i=1}^{k} l_i$ and number the subsets $\{\boldsymbol{\nu}_j\}_{\boldsymbol{n}_i}$ according to j from $j=1$ to $j=l$. The subset $\{\boldsymbol{\nu}_j\}_{\boldsymbol{n}_i}$ is a subset of $\boldsymbol{N}$, $\{\boldsymbol{\nu}_j\}_{\boldsymbol{N}}$ so that the equation is of the form,

$$F_{\boldsymbol{N}}^{(0)}\{\boldsymbol{N}\} = \sum_{\{l\{\nu_j\}\boldsymbol{N}\}_u} C\{l\{\boldsymbol{\nu}_j\}_{\boldsymbol{N}}\}_u \prod_{j=1}^{l} F_{\nu_j}\{\boldsymbol{\nu}_j\}_{\boldsymbol{N}}, \qquad (13.24_2)$$

where $C\{l\{\boldsymbol{\nu}_j\}_{\boldsymbol{N}}\}_u$ is a numerical coefficient. This coefficient is,

$$\left.\begin{aligned} C\{l\{\boldsymbol{\nu}_j\}_{\boldsymbol{N}}\}_u &= \sum_{\{k\{l_i\}l\}_u}\prod_{i=1}^{k}(-)^{l_i-1}(l_i-1)!\\ &= \sum_{n_\lambda(\sum\lambda n_\lambda=l)}\frac{l!\,(-)^l}{\prod_\lambda n_\lambda!\,(-\lambda)^{n_\lambda}},\end{aligned}\right\} \qquad (13.24_3)$$

where the last expression follows from (13.19), and n_λ is the number of i values for which $l_i=\lambda$. The coefficient, C, depends only on l and is

$$C(l) = l!\,(-)^l x \quad \text{coeff. of } x^l \text{ in } \left\{ e^{-\sum_{\lambda\ge 1}\frac{x^\lambda}{\lambda}} = e^{\log(1-x)} = 1-x\right\}, \qquad (13.24_4)$$

and hence zero except for $l=1$, $C(1)=1$. Eq. (13.24_2) is thus an identity, proving the validity of (13.23).

The decomposition of the functions $F_{\boldsymbol{N}}^{(0)}\{\boldsymbol{N}\}$ into a sum of products of functions of subsets $g_{\boldsymbol{n}}^{(0)}\{\boldsymbol{n}\}_{\boldsymbol{N}}$, by (13.20), with the inverse Eq. (13.23) for the functions $g_{\boldsymbol{N}}^{(0)}\{\boldsymbol{n}\}$, is general for any set of functions, $F_{\boldsymbol{N}}^{(0)}$, defined for all sets $\boldsymbol{N}$. However one property of the functions $F_{\boldsymbol{N}}^{(0)}\{\boldsymbol{N}\}$ is necessary to make this representation useful, and this is a general property of physically significant probability density functions. This property is that the probability density functions for the sum of two infinitely distant sets of molecules is the product of the probability densities for the two independent sets,

$$\lim_{R_{\boldsymbol{N},\boldsymbol{M}}\to\infty} [F^{(0)}_{\boldsymbol{N}+\boldsymbol{M}}\{\{\boldsymbol{N}\}+\{\boldsymbol{M}\}\}] = F^{(0)}_{\boldsymbol{N}}\{\boldsymbol{N}\}\, F^{(0)}_{\boldsymbol{M}}\{\boldsymbol{M}\}. \tag{13.25}$$

Insert this in (13.23) for $g_{\boldsymbol{n}+\boldsymbol{m}}\{\boldsymbol{n}+\boldsymbol{m}\}$ when $R_{\boldsymbol{n},\boldsymbol{m}}\to\infty$. The only products of F functions which remain are those of subsets $\{\boldsymbol{\nu}_i\}_{\boldsymbol{n}}$, and of subsets $\{\boldsymbol{\mu}_j\}_{\boldsymbol{m}}$, but none $F_{\boldsymbol{\nu}+\boldsymbol{\mu}}$ of any sum of subsets from $\boldsymbol{n}$ and from $\boldsymbol{m}$. That is, we can write,

$$\left.\begin{aligned}&\lim_{R_{\boldsymbol{n},\boldsymbol{m}}\to\infty} [g^{(0)}_{\boldsymbol{n}+\boldsymbol{m}}\{\{\boldsymbol{n}\}+\{\boldsymbol{m}\}\}]\\ &= \sum_{\{k_1\{\boldsymbol{\nu}_i\}\boldsymbol{n}\}_u}\ \sum_{\{k_2\{\boldsymbol{\mu}_j\}\boldsymbol{m}\}_u} C(k_1,k_2)\prod_{i=1}^{k_1}\ \prod_{j=k_1+1}^{k_1+k_2} F^{(0)}_{\boldsymbol{\nu}_i}\{\boldsymbol{\nu}_i\}_{\boldsymbol{n}}\, F^{(0)}_{\boldsymbol{\mu}_j}\{\boldsymbol{\mu}_j\}_{\boldsymbol{m}},\end{aligned}\right\} \tag{13.26$_1$}$$

with C a numerical coefficient, symmetric in k_1 and k_2. Without loss of generality we can assume that $k_1 \geq k_2$. The coefficient C is then the sum over all l values from k_1 to $k=k_1+k_2$ inclusive, of $(-)^{l-1}(l-1)!$ times the number of ways that $k-l$ of the first k_1 subsets, i, can be paired with an equal number of the second k_2 subsets, j. This number is

$$C = \sum_{l=k_1}^{k_1+k_2} \frac{k_1!\,k_2!\,(-)^{l-1}(l-1)!}{(l-k_1)!\,(k_1+k_2-l)!\,(l-k_2)!}.$$

Use $l=k_1+\lambda$ to write this as,

$$\left.\begin{aligned}C &= (-)^{k_1-1}k_1!\sum_{\lambda=0}^{k_2}\frac{k_2!\,(-)^{\lambda}}{(k_2-\lambda)!\,\lambda!}\,\frac{(k_1+\lambda-1)!}{(k_1-k_2+\lambda)!}\\ &= (-)^{k_1-1}k_1!\left[\left(\frac{d}{dx}\right)^{k_2-1} x^{k_1-1}(1-x)^{k_2}\right]_{x=1} = 0.\end{aligned}\right\} \tag{13.26$_2$}$$

We therefore find that if the functions $F^{(0)}_N$ obey condition (13.25) we have that,

$$\lim_{R_{\boldsymbol{n},\boldsymbol{m}}\to\infty} [g^{(0)}_{\boldsymbol{n}+\boldsymbol{m}}\{\{\boldsymbol{n}\}+\{\boldsymbol{m}\}\}] = 0, \tag{13.27}$$

namely the functions $g^{(0)}_{\boldsymbol{n}}$ approach zero in value unless the coordinates have such values that the molecules are close together in space. We refer to the functions $g^{(0)}_{\boldsymbol{n}}\{\boldsymbol{n}\}$ as cluster functions.

As in the one component case, Eqs. (13.12) define the integrals,

$$b'_{\boldsymbol{n}}(V,T) = \frac{1}{V\boldsymbol{n}!}\iint_V \cdots \int g^{(0)}_{\boldsymbol{n}}\{\boldsymbol{n}\}\, d\{\boldsymbol{n}\}. \tag{13.28$_1$}$$

Since $g'^{(0)}_1 \equiv 1$ for all species of molecules we have,

$$b'_1 \equiv 1, \qquad \text{all molecular species.} \tag{13.28$_2$}$$

The integrals $b_{\boldsymbol{n}}$ for $n \geq 2$ will, in general, depend on the species of molecules occurring in the set $\boldsymbol{n}$, that is on the separate numbers $n_1, n_2, \ldots, n_a, \ldots$ of $\boldsymbol{n} = n_1, n_2, \ldots, n_a, \ldots$.

Provided the approach to zero of Eq. (13.27) is sufficiently rapid, namely more rapidly than R^{-3}, the limit

$$b^{(0)}_{\boldsymbol{n}} = \lim_{V\to\infty} [b'_n(V,T)], \tag{13.28$_3$}$$

will exist. (This, however, will not be true for the case of charged, that is ionic molecules.)

The treatment of the general multicomponent system is now almost identical to that of the one-component system. With Eq. (13.29) for $F_{\boldsymbol{N}}^{(0)}\{\boldsymbol{N}\}$ the integral of the single terms of (13.1) becomes a product

$$\prod_{\boldsymbol{n}} [V \boldsymbol{n}\, b'_{\boldsymbol{n}}(V, T)]^{N_{\boldsymbol{n}}}$$

with $\sum \boldsymbol{n} N_{\boldsymbol{n}} = \boldsymbol{N}$. The number of such terms for a specified set of the numbers $N_{\boldsymbol{n}}$ is given by (13.19). The product divided by $\boldsymbol{N}!$ and multiplied by $\boldsymbol{z}^{\boldsymbol{N}}$ can be written as

$$\prod_{\boldsymbol{n}} \frac{[V b'_n(V, T)\, \boldsymbol{z}^{\boldsymbol{n}}]^{N_{\boldsymbol{n}}}}{N_{\boldsymbol{n}}!}$$

which must be summed over all values of $N_{\boldsymbol{n}}$, $0 \leq N_{\boldsymbol{n}} \leq \infty$, to give $e^{PV/kT}$ in Eq. (13.1). The sum leads to the expression,

$$P(\boldsymbol{z}, T, V) = kT \sum_{\boldsymbol{n}} b'_{\boldsymbol{n}}(V, T)\, \boldsymbol{z}^{\boldsymbol{n}}, \tag{13.29$_1$}$$

and

$$\left.\begin{aligned}\lim_{V\to\infty} [P(\boldsymbol{z}, T, V)] &= kT \lim_{V\to\infty} \Big[\sum_{\boldsymbol{n}} b'_n(V, T)\, \boldsymbol{z}^{\boldsymbol{n}}\Big] \\ &= kT \sum_{\boldsymbol{n}} b_{\boldsymbol{n}}^{(0)}\, \boldsymbol{z}^{\boldsymbol{n}}.\end{aligned}\right\} \tag{13.29$_2$}$$

Since the first terms, $b_1^{(0)}$, for all species are unity, the limit for low activities is

$$P = kT \sum_a z_a + O(z^2), \tag{13.30}$$

which is the perfect gas limit.

Another, and apparently completely independent treatment of the original equation, (13.1), leads to (13.29$_2$) in which the cluster integrals, $b'_n(V)$, have a different definition. This definition is not, as yet, as numerically useful for the evaluation of the lower coefficients as that of (13.28$_1$), but is of considerable theoretical interest. The method is due to YANG and LEE[1].

We have seen in Sect. 5, that, without altering the physical situation in any real manner, namely by assuming that the pair potential terms $u(r_{\alpha\beta})$ become infinite for $r_{\alpha\beta} < r_0$, we can assign an N_m such that for all terms $N > N_m$ the terms in (13.1) are identically zero. We then have that the right hand side of (13.1) is a polynomial of order N_m, which can be written, for the one-component case, as

$$e^{PV/kT} = \prod_{i=1}^{N_m} \left(1 - \frac{z}{z_i}\right), \tag{13.31$_1$}$$

since the term at $z = 0$ is unity. In this the z_i's are the N_m roots of the polynomial. Since the coefficients of the polynomial are all positive none of these roots lie on the positive real axis of z.

Now write, from (13.31$_1$),

$$PV = kT \sum_{i=1}^{N_m} \log\left(1 - \frac{z}{z_i}\right). \tag{13.31$_2$}$$

If z_1 is the first root, namely $|z_i| \geq |z_1|$ for all $2 \leq i \leq N_m$, one can develop the $\log[1 - (z/z_i)]$ as long as,

$$|z| < |z_1|, \tag{13.32}$$

to obtain

$$PV = kT \sum_{i=1}^{N_m} \sum_{n=1}^{\infty} -\frac{1}{n}\left(\frac{z}{z_i}\right)^n. \tag{13.33$_1$}$$

[1] C. N. YANG and T. D. LEE: Phys. Rev. 87, 404, 410 (1952).

The order of the two summations can be interchanged, namely one writes,

$$P = \mathrm{k}T \sum_{n=1} \left[-\frac{1}{Vn} \sum_{i=1}^{N_m(v)} \left(\frac{1}{z_i(V)} \right)^n \right] z^n, \tag{13.33$_2$}$$

which is Eq. (13.29$_2$) with

$$b_n'(V) = \frac{1}{nV} \sum_{i=1}^{N_m(V)} - \left(\frac{1}{z_i(V)} \right)^n, \tag{13.34}$$

and, as before, $b_n^{(0)}$ the limit of $b_n'(V)$.

The N_m roots, z_i, will not be on the positive real axis, for finite V, but will occur in conjugate pairs at symmetric angles, φ_i and $-\varphi_i$. The sums of (13.34) will all be real. Under certain conditions of U these will crowd the real axis as $V \to \infty$, and the limit $[b_n^{(0)}]^{-1/n}$ $(n \to \infty) = z_0$ which determines the value $z = z_0$ at which (13.29$_2$) has a singularity on the positive real axis of z will be the value of these z_is which approach the real axis.

14. Interpretation of the power series $\Sigma b_n^{(0)} z^n$. The general equation (13.29$_2$), or, more simply, the one-component special case, (13.15), can be given a very simple interpretation.

Consider any specified set, $\{N\}$, of a large number, N, of three-dimensional coordinates in a volume V. In such a set there may be an apparent randomness of the coordinate values, except that there are, say, a larger than random number of pairs, and perhaps triples and higher clusters of coordinates, which are close together. In general, however, it will be impossible to define precisely and uniquely the number of pairs, since one would have to define arbitrarily the distance, r, within which two molecules must be, in order to be called a pair. For a purely random distribution of N_1 coordinates in V, the expected random number of pairs within the distance r of each other would be,

$$N_2^{(\mathrm{ran})}(r) = \frac{1}{2} N_1 (N_1 - 1) \left(\frac{4\pi}{3} r^3 \right) \frac{1}{V}. \tag{14.1}$$

If now $N_2^{(\mathrm{obs})}(r)$ is the observed number of pair distances within r or less, then the number,

$$N_2'(r) = N_2^{(\mathrm{obs})}(r) - N_2^{(\mathrm{ran})}(r), \tag{14.2$_1$}$$

might be expected to have a limit,

$$N_2 = \lim_{r \to \infty} [N_2'(r)], \tag{14.2$_2$}$$

independent of r, for large values of r, and might be defined as the number of clusters of two molecules, meaning by this term the *excess* number of pairs above the expected random number.

This number of pair clusters, N_2, may be negative, if the coordinate values are such as to exclude near neighbors.

In a similar manner the number of triple clusters, N_3, could be defined as the observed number of triples, minus the random distribution of independent single coordinates, and minus the expected correction for the three independent sets of one of the coordinates randomly distributed times the excess of pairs for the other pair distance.

The Eqs. (13.22) for g_2 and g_3 are seen to be proportional to just these "probability densities" for pair and triple clusters respectively, at the fixed coordinate values, and the integrals, $b_n^{(0)}$, are therefore related to the probabilities of finding clusters at any distance. The term,

$$\prod_n \frac{(V b_n z^n)^{N_n}}{N_n!},$$

which occurs in $z^N/N!$ times the integral over F_N, then corresponds to the contribution to the exponential of $PV/\mathrm{k}T$ of the term of N_1 clusters of single particles, N_2 pair cluster, ..., and N_n clusters of n-mers. That is, in Eq. (13.1),

$$e^{PV/\mathrm{k}T} = \prod_n \sum_{N_n \geq 0} \frac{(V b_n^{(0)} z^n)^{N_n}}{N_n!} \tag{14.3}$$

the term for given N_n is proportional to the "probability" of having N_n of these excess clusters of n molecules each. The average number, $\overline{N}_n$, of clusters of n is then,

$$\left.\begin{aligned} \overline{N}_n &= \sum_{N_n \geq 0} N_n \frac{(V b_n^{(0)} z^n)^{N_n}}{N_n!} \left[e^{-PV/\mathrm{k}T} \prod_{m \neq n} \sum_{N_m} \frac{(V b_m^{(0)} z^m)^{N_m}}{N_m!} \right] \\ &= \frac{V b_n^{(0)}}{\mathrm{k}T} \frac{\partial P}{\partial b_n^{(0)}} = V b_n^{(0)} z^n, \end{aligned}\right\} \tag{14.4}$$

where the last expression follows from Eq. (13.15).

Using (14.4) in (13.15) we see that,

$$PV = \mathrm{k}T \sum_{n \geq 1} \overline{N}_n, \tag{14.5}$$

namely $PV = \overline{N}_c \mathrm{k}T$ where $\overline{N}_c = \sum_n \overline{N}_n$ is the average total number of clusters of all sizes. Thus the pressure can formally be considered to be given by the perfect gas expression in terms of the total numbers of the independent clusters.

It is important to notice that what we have termed "probability densities" of certain cluster configurations are actually the differences between the true probability densities of these configurations and a computed probability density assuming independence of the positions of clusters of smaller size. These so-called cluster "probability densities" may then be negative, and the single average number, $\overline{N}_n$, of clusters of n-mers can take negative values. Thus, in the case of repulsive potentials, the distribution in which molecules never approach closely, is represented by a number, N_1, of independently positioned molecules, with $N_1 > N$ the total number of molecules, plus a *negative* number, N_2, N_3, etc., of pair and higher clusters, meaning only that the numbers observed in close proximity to each other is *less* than that expected from a random distribution. Actually, at temperatures below the Boyle point, the pair cluster number, $\overline{N}_2$ is positive, and at temperatures below the critical point the higher cluster numbers, $\overline{N}_n$, $n \geq 2$, are usually positive.

In the case that the potential, U, is a sum of pair interactions, Eq. (13.7) there is no very long range of activity values for which the number of pairs, $\overline{N}_2$, takes values comparable with $\overline{N}_1$, and for which also the higher cluster numbers ,$\overline{N}_3$, $\overline{N}_4$, etc., remain negligible. In the case, however, that the potential is of the nature of that between, say, hydrogen atoms, in which there is a saturation effect of a definite valence chemical bond, pair clusters will form at densities for which the number of triples may be truly negligible. Indeed, in this case, at reasonable temperatures, the equilibrium configuration may include almost complete pairing at densities for which the random number of singles within molecular distances would be negligible. In such a case the distinction between "pair clusters" as the excess above random of the number of atoms close together in pairs, and the actual number of molecules, is numerically insignificant. It is common practice in such a case to treat the system as one of chemical equilibrium between the two chemical species, H and H_2, and, if necessary to include the mutual potentials between molecules and between molecules and atoms. However, it is clear, that in such an approximation the mutual potentials between atoms must be neglected, since when two atoms approach closely our arbitrary definition labels them as molecules.

The interpretation of $V b_n^{(0)} z^n$ as the average number, $\overline{N}_n$, of clusters of n-mers seems to be particularly valuable in making clear the behavior at the activity of condensation for which the series $\sum b_n^{(0)} z^n$ becomes singular.

At low temperatures one may expect $b_n^{(0)}$ for large n values to approach in value the expression (see also Sect. 16),

$$b_n^{(0)} \approx \frac{b_0^{n-1}}{\sigma^{n^{\frac{2}{3}}}}, \quad \text{(large } n\text{)}, \tag{14.6}$$

in which the "surface energy" term σ, arises from the fact that a number of molecules proportional to $n^{\frac{2}{3}}$ cannot have as many f-bonds at the distance for which f is a maximum as those in the "interior" of the cluster. This term, σ, will be expected to have a numerical value greater than unity. The sum $\sum b_n z^n$ will be singular at $z = z_0 = 1/b_0$, but if σ is greater than unity the sum will approach a finite value as z approaches z_0 from $z < z_0$.

The average number, $\overline{N}_n$, of clusters of size n divided by V, which will be the number density, ϱ_n, of clusters of this size, will be, as $z \to z_0$,

$$\frac{\overline{N}_n}{V} = \varrho_n (z = z_0) = \frac{1}{b_0 \sigma^{n^{\frac{2}{3}}}}, \quad \left(z_0 = \frac{1}{b_0}\right) \tag{14.7}$$

which, for $\sigma > 1$ and n of the order of $n \approx 1000$, can be quite negligible. Indeed this number can well be so small that no clusters of n would be expected in any reasonable macroscopic volume.

Now increase z to $z_0(1+\varepsilon)$. The limit for any finite ε and infinite n is, of course,

$$\lim_{n\to\infty} [\overline{N}_n] = \frac{V}{b_0} \lim_{n\to\infty} \left[\frac{(1+\varepsilon)^n}{\sigma^{n^{\frac{2}{3}}}}\right] = \infty \tag{14.8}$$

expressing the singularity of the series expression. However, for any finite V and for sufficiently large n the functions $b_n'(V)$ become dependent on V, and in general, since at low temperatures the integrand is positive for most parts of the configuration space, $d b_n'/d V > 0$. Hence for sufficiently large n the cluster integral $b_n'(V)$ in a finite volume will be smaller than $b_n^{(0)}$ given by (14.6). Now one may assume the volume dependence of $b_n'(V)$ to become significant only at volumes of the order of that occupied by the molecules in the condensed phase, say at $n \approx 10^{24}$ for $V = 100\ \text{cm}^3$, so that (14.6) may be assumed to be correct up to this value of n. With $V/b_0 \approx 10^{24}$ we find $\overline{N}_{10^{24}}$ of the order of unity when

$$\log(1+\varepsilon_0) = \varepsilon_0 = 10^{-8} \log\sigma - 10^{-24} \log 24, \tag{14.9}$$

a number of the order of 10^{-8}.

Thus for z values between z_0 and $z_0(1+\varepsilon_0)$ the pressure,

$$P = \frac{kT}{V} \sum_{n \geq 1} \overline{N}_n \tag{14.10}$$

will stay constant, while $\overline{N}_{10^{24}}$ climbs towards unit value. The density,

$$\varrho = \sum_{n \geq 0} n\, b_n^{(0)} z^n = \sum n \frac{\overline{N}_x}{V}, \tag{14.11}$$

however, climbs during this change of z towards the density of the large single cluster. For higher z values the behavior of the functions depends completely on the volume dependence of $b_n'(V)$ and the cluster concept completely loses its significance.

It is, however, of some interest to examine the equilibrium behavior of the clusters of intermediate size, say $n \approx 1000$. The equilibrium number, $\overline{N}_{1000}$, of these at $z = z_0(1+\varepsilon)$ will be, if $V/b_0 = 10^{24}$,

$$\overline{N}_{1000} = 10^{24} \frac{(1+\varepsilon)^{1000}}{\sigma^{100}} \tag{14.12}$$

and at $\varepsilon = \varepsilon_0 = 10^{-8}$ if $\sigma \approx e$, from (14.9), $\overline{N}_n = \mathrm{e}^{-45}$. The intermediate clusters do not grow during condensation. It is only when ε takes such values that

$$\log(1+\varepsilon) = \frac{\sigma}{10} \frac{1}{1000} \log 10^{24}$$

or $\varepsilon \sim 1/4$, that, with the numbers chosen, there would be, at equilibrium, one cluster of a thousand molecules in a volume of 100 cm³.

In the absence of surfaces or nucleating particles the large clusters can only grow dynamically by the collision of single molecules, or the relatively abundant small clusters of two or three, with clusters of intermediate size. The intermediate clusters, due to the surface tension term, are unstable except at activities considerably in excess of that of condensation. For this reason supersaturation is possible. The equation of state of the supersaturated gas is given by the series of (14.10) or (13.15) which converges asymptotically beyond the activity, z_0, of condensation. However, this asymptotic series does *not* represent the equilibrium function $P(z)/\mathrm{k}T$, which ascends steeply with increasing z beyond $z = z_0$.

15. Conversion to power series in density. Although the natural thermodynamic variables for the pressure P, are T, and the chemical potential, $\boldsymbol{\mu}$, or the convenient activity set, $\boldsymbol{z}$, in place of $\boldsymbol{\mu}$, [see Sect. 1, Eqs. (1.9) and (1.10)] the experimental description of a system is more often given in terms of T and volume or the concentrations. For units of concentration we find it most convenient to use the number densities, $\varrho_a = N_a/V$.

Eq. (13.15), for the one component system,

$$P = \mathrm{k}T \sum_{n \geq 1} b_n^{(0)} z^n, \tag{15.1}$$

can be used to find, from (4.17),

$$\left.\begin{aligned} z\left(\frac{\partial}{\partial z}\right)_T \frac{P}{\mathrm{k}T} = \varrho &= \sum_{n\geq 1} n\, b_n^{(0)} z^n, \\ &= z + 2b_2^{(0)} z^2 + 3b_3^{(0)} z^3 + \cdots. \end{aligned}\right\} \tag{15.2}$$

In this, set

$$z = \varrho + a_1 \varrho^2 + a_2 \varrho^2 + \cdots, \tag{15.3}$$

and equate coefficients to find,

$$z = \varrho - 2b_2^{(0)} \varrho^2 - [3b_3^{(0)} - 8(b_2^{(0)})^2]\, \varrho^3 - \cdots. \tag{15.4}$$

Use this in (15.1) to write,

$$P = \mathbf{k} T \varrho [1 - b_2^{(0)} \varrho - 2 [b_3^{(0)} - 2 (b_2^{(0)})^2] \varrho^2 - \cdots], \tag{15.5}$$

up to terms of order ϱ^3.

Now turn to the simple case of the sum of pair potentials, Eq. (13.11) for g_3 with (13.12) for b_3. One has

$$g_3(\boldsymbol{r}_1, \boldsymbol{r}_2, \boldsymbol{r}_3) = f_{31} f_{12} + f_{12} f_{23} + f_{23} f_{31} + f_{12} f_{23} f_{31}, \tag{15.6}$$

which must be integrated to give $6 V b_3^{(0)}$. The integral,

$$\begin{aligned} &\frac{1}{6V} \iiint\limits_V f_{31} f_{12} \, d\boldsymbol{r}_1 \, d\boldsymbol{r}_2 \, d\boldsymbol{r}_3 \\ &= \frac{1}{6V} \int\limits_V d\boldsymbol{r}_1 \int\limits_0^\infty 4\pi r_{13}^2 f(r_{13}) \, dr_{13} \int\limits_0^\infty 4\pi r_{12}^2 f(r_{12}) \, dr_{12} \\ &= \frac{1}{6} [2 b_2^{(0)}]^2 = \frac{2}{3} [b_2^{(0)}]^2, \end{aligned}$$

since,

$$\left.\begin{aligned} b_2 = \frac{1}{2V} \iint\limits_V f_{12} \, d\boldsymbol{r}_1 \, d\boldsymbol{r}_2 &= \frac{1}{2V} \int\limits_V d\boldsymbol{r}_1 \int\limits_0^\infty 4\pi r_{12}^2 f(r_{12}) \, dr_{12} \\ &= \frac{1}{2} \int\limits_0^\infty 4\pi r^2 f(r) \, dr. \end{aligned}\right\} \tag{15.7}$$

The first three terms in (15.6) contribute together $2(b_2^{(0)})^2$ to b_3, so that,

$$b_3 = 2 [b_2^{(0)}]^2 + \tfrac{1}{3} \beta_2, \tag{15.8}$$

where,

$$\beta_2 = \frac{1}{2! V} \iiint\limits_V f_{12} f_{23} f_{31} \, d\boldsymbol{r}_1 \, d\boldsymbol{r}_2 \, d\boldsymbol{r}_3. \tag{15.9}$$

If we now define,

$$\beta_1 = \frac{1}{V} \iint\limits_V f_{12} \, d\boldsymbol{r}_1 \, d\boldsymbol{r}_2 = 2 b_2, \tag{15.10}$$

and use these in (15.5) one may write,

$$P = \mathbf{k} T \varrho [1 - 2\beta_1 \varrho - \tfrac{2}{3} \beta_2 \varrho^2 - \cdots]. \tag{15.11$_1$}$$

If the algebra is carried out to higher powers of ϱ a similarly simple result is obtained, and can be shown to be general. The higher b_n's always contain additive contributions which can be computed as products of b_ν's for $\nu < n$. The remaining additive term may be defined as β_{n-1}/n. If this is done, the equation,

$$P = \mathbf{k} T \varrho \left[1 - \sum_{\nu \geq 1} \frac{\nu}{\nu + 1} \beta_\nu \varrho^\nu \right], \tag{15.11$_2$}$$

is obtained for P as a power series in ϱ.

To see this for b_4 turn to the expansion of $\prod (1 + f_{\alpha\beta})$ given in (13.10). The first three terms, $f_{\alpha\beta} f_{\gamma\delta}$ etc., in the quadruple sum correspond to two pair clusters, and do not enter b_4. In b_4, the 16 terms containing products of three f's, lead to integrals proportional to β_1^3. Of the 15 terms containing products of four f's, 12 have integrals proportional to $\beta_2 \beta_1$, namely those of the form

$f_{\alpha\beta} f_{\beta\gamma} f_{\gamma\delta} f_{\delta\beta}$ (see Fig. 4). The remaining terms are to be grouped in β_3. In terms of the graphs representing the connected products composing b_4 those in which two or more independent paths of lines exist between all 6 pairs, $\alpha\beta$, that is paths which do not intersect at any common vertex, compose the terms of β_3.

For the case of the sum of pair potentials, we define β_ν as

$$\beta_\nu = \lim_{V\to\infty}\left[\frac{1}{V\nu!}\int\!\!\int_V\!\!\cdots\!\int \sum \prod f_{\alpha\beta}\, d\boldsymbol{r}_1 \cdots d\boldsymbol{r}_{\nu+1}\right] \tag{15.12}$$

sum over all more than singly connected products between $\nu + 1$ vertices.

With this definition, Eq. (15.11_2) is valid.

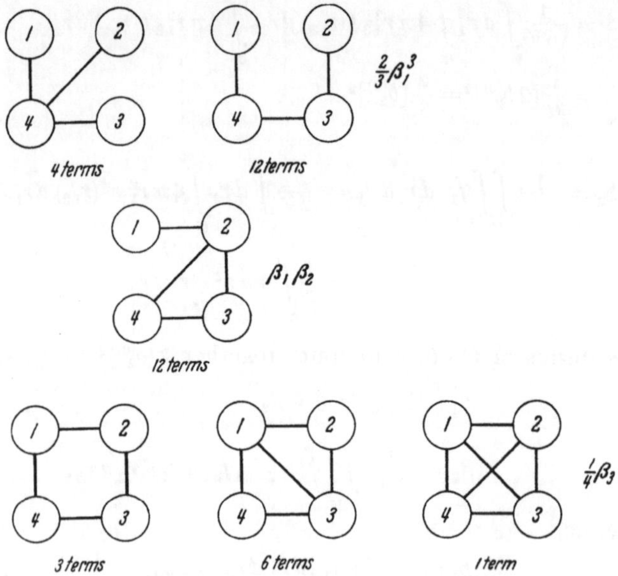

Fig. 4. Products occurring in the integrand of $b_4^{(0)}$.

The proof that (15.11_2) with (15.12) for β_ν, is generally correct, even for the case of the sum of pair potentials and a single component system, is comparatively complicated.

For this one component case the steps are as follows. First, show that one can write, for all n, that,

$$b_n^{(0)} = \frac{1}{n^2} \sum_{m_\nu\left[\sum_\nu \nu m_\nu = n-1\right]} \prod_\nu \frac{(n\beta_\nu)^{m_\nu}}{m_\nu!}. \tag{15.13$_1$}$$

With this as a general expression, prove Eq. (15.11_2).

The proof that (15.13_1) is general, with the β_ν defined by (15.12) is given in several places in the literature[1]. The proofs are relatively involved. The combinatorial counting of the number of ways graphs can be constructed for which the corresponding products integrate to give $V\prod_\nu (\nu!\,\beta_\nu)^{m_\nu}$ is by no means simple.

[1] M. BORN and K. FUCHS: Proc. Roy. Soc. Lond., Ser. A **166**, 391 (1938).

For the purposes of this article we prefer to use (15.13₁) for which the inversion,

$$\beta_\nu = \sum_{m_\mu \geq 0 \left[\sum_\mu (\mu-1) m_\mu = \nu\right]} (-)^{\sum m_\mu + 1} \frac{[\nu - 1 + \sum m_\mu]!}{\nu!} \prod_{\mu \geq 2} \frac{[\mu\, b_\mu^{(0)}]^{m_\mu}}{m_\mu!}, \qquad (15.13_2)$$

holds, and regard these equations as the definition of the quantities β_ν. One may then check, by counting the types of graphs occurring in the representation of the functions g_n, for low n-values, that (15.12) is indeed correct for $\nu = 1, 2, 3$, and 4, or higher.

The first few terms are

$$\left.\begin{aligned} b_2 &= \tfrac{1}{2}\beta_1, \\ b_3 &= \tfrac{1}{3}\beta_2 + \tfrac{1}{2}\beta_1^2, \\ b_4 &= \tfrac{1}{4}\beta_3 + \beta_1\beta_2 + \tfrac{2}{3}\beta_1^3, \\ b_5 &= \tfrac{1}{5}\beta_4 + \beta_1\beta_3 + \tfrac{1}{2}\beta_2^2 + \tfrac{5}{2}\beta_1^2\beta_2 + \tfrac{25}{24}\beta_1^4, \end{aligned}\right\} \qquad (15.13_3)$$

$$\left.\begin{aligned} \beta_1 &= 2b_2, \\ \beta_2 &= 3b_3 - 6b_2^2, \\ \beta_3 &= 4b_4 - 24 b_2 b_3 + \tfrac{80}{3} b_2^3, \\ \beta_4 &= 5b_5 - 40 b_2 b_4 - \tfrac{45}{2} b_3^2 + 180 b_2^2 b_3 - 140 b_2^4. \end{aligned}\right\} \qquad (15.13_4)$$

Using, then, (15.13₁) as a definition of β_ν, we shall now undertake to demonstrate the general validity of Eq. (15.11₂).

The expression, (15.13), for $b_n^{(0)}$ is equivalent to

$$\left.\begin{aligned} b_n^{(0)} &= \frac{1}{n^2} \quad \text{coeff. of } y^{n-1} \text{ in } \left[e^{n\sum_\nu \beta_\nu y^\nu}\right] \\ &= \frac{1}{n^2}\,\frac{1}{2\pi i} \oint \frac{e^{n\sum \beta_\nu y^\nu}}{y^n}\, dy, \end{aligned}\right\} \qquad (15.14)$$

where the integration is to be taken in a small circle of the complex plane around the origin. Eq. (15.14) is also valid for $b_1^{(0)} \equiv 1$. Define the functions,

$$H_\varkappa(z) = \sum_{n \geq 1} n^\varkappa\, b_n^{(0)} z^n, \qquad (15.15_1)$$

$$H_{\varkappa+1} = z \frac{d}{dz} H_\varkappa. \qquad (15.15_2)$$

From (15.14) one has that,

$$H_2(z) = \frac{1}{2\pi i} \sum_{n \geq 1} \oint \left[\frac{z}{y}\, e^{\sum \beta_\nu y^\nu}\right]^n dy. \qquad (15.16_1)$$

By changing the order of summation and integration, and using $\sum_{n \geq 1} x^n = x/(1-x)$, one has,

$$H_2(z) = \frac{1}{2\pi i} \oint \frac{z\, e^{\sum \beta_\nu y^\nu}}{y - z\, e^{\sum \beta_\nu y^\nu}}\, dy. \qquad (15.16_2)$$

This step is valid provided the path of integration is chosen so that,

$$\left|\frac{z}{y}\, e^{\sum \beta_\nu y^\nu}\right| < 1, \qquad (15.16_3)$$

and this can always be done for sufficiently small z-values.

The path of integration of (15.16$_3$) encloses the singularity in the integrand at y_0, for which the equation,

$$z = y_0\, e^{-\Sigma \beta_\nu y_0^\nu}, \tag{15.17}$$

is satisfied, and no other singularity. The residue at this singularity is found by inserting the derivative of the denominator, namely $1 - \sum \nu \beta_\nu y_0^\nu$ from (15.17), for the denominator, so that the residue is,

$$H_2(z) = \sum_{n \geq 1} n^2 b_n^{(0)} z^n = \frac{y_0}{1 - \sum\limits_{\nu \geq 1} \nu \beta_\nu y_0^\nu}. \tag{15.18}$$

The quantity y_0 is determined by the implicit Eq. (15.17), and from that is seen to obey the differential equation,

$$\frac{d \log z}{d y_0} = \frac{1 - \sum \nu \beta_\nu y_0^\nu}{y_0},$$

so that, from (15.18) and (15.15$_2$),

$$y_0 + c = \int H_2(z)\, d \log z = H_1(z) = \sum n\, b_n^{(0)} z^n = \varrho,$$

from (15.2). Apart from a possible additive constant y_0 is the number density ϱ. From (15.17) it is seen that

$$\lim_{y_0 \to 0} \left[\frac{z}{y_0}\right] = 1, \tag{15.19}$$

which fixes the constant, c, as zero. Since,

$$\frac{P}{\mathrm{k}T} = H_0(z) = \int_0^1 H_1(z)\, d \log z = \int_0^\varrho y_0 \frac{d \log z}{d y_0} d y_0$$

$$= \int_0^\varrho \left[1 - \sum \nu \beta_\nu y_0^\nu\right] d y_0 = \varrho \left[\left[1 - \sum \frac{\nu}{\nu + 1} \beta_\nu \varrho^\nu\right]\right],$$

we have established Eq. (15.11$_2$) for sufficiently small values of z.

The equations may be summarized here. As explicit expressions in terms of the number density ϱ we have,

$$z = \varrho\, e^{-\sum\limits_\nu \beta_\nu \varrho^\nu}, \tag{15.20}$$

$$\left.\begin{aligned} \mu &= \mathrm{k}T \log z + \mu_0 \\ &= \mathrm{k}T \left[\log \varrho - \sum_{\nu \geq 1} \beta_\nu \varrho^\nu\right] + \mu_0, \end{aligned}\right\} \tag{15.21}$$

$$P = \mathrm{k}T \varrho \left[1 - \sum_{\nu \geq 1} \frac{\nu}{\nu + 1} \beta_\nu \varrho^\nu\right], \tag{15.22}$$

and for the compressibility, $\varkappa$,

$$\varkappa = \frac{1}{\varrho}\left(\frac{\partial \varrho}{\partial P}\right)_T = \frac{1}{\mathrm{k}T \varrho} \frac{1}{1 - \sum\limits_{\nu \geq 1} \nu \beta_\nu \varrho^\nu}. \tag{15.23}$$

The equations are valid at small z and ϱ values, since, from (15.19) it is seen that at small enough z values the circle of integration of (15.16$_2$) can be chosen as close to the origin as desired and still satisfy (15.16$_3$). From (15.17) it is seen that $z = 0$ when $y_0 = 0$, and that the positive real axis of z corresponds to the positive real axis of y_0, since the coefficients, β_ν, of the sum $\sum \beta_\nu y_0^\nu$ are all real. The function $H_2(z)$ is seen from (15.18) to be regular as long as $\sum \nu \beta_\nu y_0^\nu$ is both regular and less than unity. The first singularity along the positive real axis of z in $H_2(z)$ and hence in any of the functions $H_\varkappa(z)$ occurs at the lower value of

$y_0=\varrho_1$ or $y_0=\varrho_2$ on the positive real axis for which

$$\sum_{\nu\geq 1}\nu\beta_\nu\varrho_1^\nu = \text{singular}, \tag{A}$$

$$\sum_{\nu\geq 1}\nu\beta_\nu\varrho_2^\nu = 1. \tag{B}$$

As we have previously discussed in Sect. 5, the pressure P will be singular (on the positive real axis of z) at the activity value, z_0, of a phase change, if the computation is made for P in the limit of infinite volume. This we have done by using the limit, $b_n^{(0)}$, Eq. (13.28_3). Since $P/\mathsf{k}T=H_0(z)$, the conditions either A or B, whichever occurs at lower density, will determine the first phase change, namely that of condensation. Normally, at low temperatures, condition A, namely the singularity in the series, occurs at the lower density, and determines condensation. The point will be discussed in greater detail in Sect. 16.

The rigorous and detailed derivation of the equations which are equivalent to (15.20) to (15.23) for the multicomponent system is very formidable. A plausibility argument, which is so compelling as to leave little doubt as to their correctness will be substituted here.

Turn to the integrand of (15.12) which is a sum of multiply connected products of functions $f_{\alpha\beta}=f(r_{\alpha\beta})$ of the distance $r_{\alpha\beta}$ between molecules α and β for $\nu+1\geq\alpha>\beta\geq 1$. If the molecules belong to varying species, a, b, etc., the analytic form of the function $f_{\alpha\beta}$ will depend on the species of α and of β. The integral of (15.12) will depend on the set of numbers of molecules of the different species. We renumber $\nu+1=\boldsymbol{m}$, and, as before write $\boldsymbol{m}=m_\alpha, m_b, \ldots$, and define,

$$B_{\boldsymbol{m}}=\lim_{V\to\infty}\left[\frac{1}{V\boldsymbol{m}!}\iint\limits_V\cdots\int\sum\prod f_{\alpha\beta}\,d\boldsymbol{r}_1,\ldots,d\boldsymbol{r}_{\boldsymbol{m}}\right]. \tag{15.24}$$

sum over all more than singly connected products between the set $\boldsymbol{m}$ of m molecules.

As before we can write $b_{\boldsymbol{n}}$ as a sum of products of $B_{\boldsymbol{m}}$'s.

Now consider the case that the mutual potentials between all the different molecular species happen to be identical, so that $b_{\boldsymbol{n}}$ and $B_{\boldsymbol{m}}$ depend only on the numbers $n=\sum n_a$ $m=\sum m_a$ respectively, except for the factorials $\boldsymbol{n}!=n_a!n_b!\ldots$ and $\boldsymbol{m}!=m_a!\,m_b!\,\ldots$, in the denominators of the defining equations, (13.28_1) and (15.24), so that

$$b_{\boldsymbol{n}}^{(0)}=\frac{n!}{\boldsymbol{n}!}\,b_n^{(0)}, \tag{15.25$_1$}$$

$$B_{\boldsymbol{m}}=\frac{m!}{\boldsymbol{m}!}\,B_m=\frac{m!}{\boldsymbol{m}!}\,\frac{1}{m}\,\beta_{-1}. \tag{15.25$_2$}$$

We define quantities x_a for all species, and an overall single activity $z=\sum z_a$, by

$$\sum_a x_a=1, \tag{15.26$_1$}$$

$$z_a=x_a z. \tag{15.26$_2$}$$

The equation for the pressure is now from (13.29_2), (15.25_1) and (15.26),

$$\left.\begin{aligned} P&=\mathsf{k}T\sum_{\boldsymbol{n}\geq 1} b_{\boldsymbol{n}}^{(0)}\boldsymbol{z}^{\boldsymbol{n}}\\ &=\mathsf{k}T\sum_{n\geq 1} b_n^{(0)}z^n\left[\sum_{\substack{n_a\\(\sum n_a=n)}}\frac{n!}{\prod n_a!}\,x_a^{\,n_a}\right]\\ &=\mathsf{k}T\sum_{n\geq 1} b_n^{(0)}z^n\Big(\sum_a x_a\Big)^n\\ &=\mathsf{k}T\sum_{n\geq 1} b_n^{(0)}z^n, \end{aligned}\right\} \tag{15.27}$$

as for a one component system of activity z. The number density ϱ_a, of species a is

$$\left.\begin{aligned} \varrho_a &= z_a \left(\frac{\partial}{\partial z_a}\right)_{T, z_b} \left(\frac{P}{\mathrm{k}T}\right) = x_a z \left(\frac{\partial}{\partial z}\right)_T \left(\frac{P}{\mathrm{k}T}\right) \\ &= x_a \sum_{n \geq 1} n\, b_n^{(0)} z^n, \end{aligned}\right\} \tag{15.28$_1$}$$

or since $\sum x_a = 1$,

$$\varrho_a = x_a \varrho \tag{15.28$_2$}$$

$$\varrho = \sum_a \varrho_a = z \left(\frac{\partial}{\partial z}\right)_T \left(\frac{P}{\mathrm{k}T}\right) = \sum_{n \geq 1} n\, b_n^{(0)} z^n. \tag{15.28$_3$}$$

Since Eq. (15.27) is that for a single component system, with the total density, ϱ, defined in the same way by (15.28$_3$) we can immediately write the equation (15.22) for P in terms of $\beta_\nu \varrho^\nu$. However, from (15.28$_2$) and (15.25$_2$) we can always replace $\beta_\nu \varrho^\nu$ with $B_{\boldsymbol{m}} \boldsymbol{\rho}^{\boldsymbol{m}}$ by using,

$$\beta_{m-1} \varrho^m = \sum_{\substack{m_a \\ (\Sigma m_a = m)}} m\, B_{\boldsymbol{m}} \boldsymbol{\rho}^{\boldsymbol{m}}. \tag{15.29}$$

We thus have,

$$P = \mathrm{k}T \left[\varrho - \sum_{\boldsymbol{m}} (m-1)\, B_{\boldsymbol{m}} \boldsymbol{\rho}^{\boldsymbol{m}}\right]. \tag{15.30}$$

The equation is then proved for the case of equality of the mutual potentials between different species. If the mutual potentials were not equal, so that (15.25) were not valid, then we could still, in principle, transform (13.29$_2$) from a power series in $\boldsymbol{z} = z_a, z_b, \ldots$, to one in $\boldsymbol{\rho} = \varrho_a, \varrho_b, \ldots$, with ϱ_a defined as $z_a [\partial (P/\mathrm{k}T)/\partial z_a]_{T, Z_b}$. The coefficients must involve only the $b_{\boldsymbol{n}}^{(0)}$'s. Since these, in turn, are expressible in terms of the $B_{\boldsymbol{m}}$'s, we can write the power series in terms of these quantities. The final results must, however, reduce to Eq. (15.30) for the case of equality of the mutual potentials.

This, then proves that, in the general case, the coefficients of $\boldsymbol{\rho}_{\boldsymbol{m}}$ in any of the expressions must have the dependence on $\boldsymbol{m}$ given in the above equations multiplied by some linear combination of the integrals $B_{\boldsymbol{m}}, B_{\boldsymbol{m}'}, B_{\boldsymbol{m}''}$, etc., with $\boldsymbol{m}, \boldsymbol{m}', \boldsymbol{m}''$ all being sets of a total of m molecules. The requirement that if the density ϱ_a of any species a of molecules is zero the potential characteristic of that species cannot enter the expression for the thermodynamic function makes the identification of $B_{\boldsymbol{m}} \boldsymbol{\rho}^{\boldsymbol{m}}$ unique for small values of $\boldsymbol{m}$. Any other assignment than $B_{\boldsymbol{m}} \boldsymbol{\rho}^{\boldsymbol{m}}$ in general would seem to be sufficiently ridiculous to be discarded without further proof.

In any case the detailed analysis for the general case has been given by FUCHS[1], and the Eq. (15.30) is that derived by him.

To obtain the other thermodynamic properties we use (15.30) noting that $\varrho_a = N_a/V$, $(\partial \varrho_a/\partial N_a)_V = 1/V$ and $(\partial \varrho_a/\partial V)_{\boldsymbol{N}} = -\varrho_a V^{-1}$, to write

$$\left.\begin{aligned} V(dP)_T &= -\mathrm{k}T \left[\varrho - \sum_{\boldsymbol{m}} m(m-1)\, B_{\boldsymbol{m}} \boldsymbol{\rho}^{\boldsymbol{m}}\right] dV + \\ &\quad + \sum_a \frac{\mathrm{k}T}{\varrho_a} \left[\varrho_a - \sum_{\boldsymbol{m}} m_a (m-1)\, B_{\boldsymbol{m}} \boldsymbol{\rho}^{\boldsymbol{m}}\right] dN_a. \end{aligned}\right\} \tag{15.31}$$

[1] K. FUCHS: Proc. Roy. Soc. Lond., Ser. A **179**, 408 (1942).

Setting $(dP)_T = 0$, $dN_b = 0$, $b \neq a$, we have, for the partial molecular volume of species a,

$$\left.\begin{aligned} v_a &= \left(\frac{\partial V}{\partial N_a}\right)_{P,T,N_b,\ldots} \\ &= \frac{1}{\varrho}\,\frac{\varrho_a - \sum\limits_m m_a(m-1)\,B_m\,\rho^m}{\varrho_a - x_a\sum\limits_m m(m-1)\,B_m\,\rho^m}\,. \end{aligned}\right\} \tag{15.32}$$

In the case of a single species, or if all potentials are identical so that (15.25_2) holds, this reduces to $v_a = \varrho^{-1}$. In any case $\sum x_a v_a = \varrho^{-1}$.

Since $(dG)_T = V(dP)_T$ we integrate Eq. (15.31) at constant $\boldsymbol{N}$ from $\varrho = 1$ to find the Gibbs free energy, G. At $\varrho = 1$, $\varrho_a = x_a$ the system is a perfect gas with $G_0 = \sum\limits_a N_a(\mu_{0a} + \mathrm{k}T\log x_a)$. Since $(dV)_{\boldsymbol{N}} = -(N/\varrho^2)\,d\varrho$ we have

$$G = \sum_a N_a\,\mu_{0a} + N\,\mathrm{k}T\left[\sum_a x_a\log x_a + \log\varrho - \frac{1}{\varrho}\sum_m m\,B_m\,\rho^m\right]. \tag{15.33}$$

The chemical potential, μ_a, of the single species a, are

$$\mu_a = \left(\frac{\partial G}{\partial N_a}\right)_{P,T,N_b,\ldots} = \left(\frac{\partial G}{\partial N_a}\right)_{V,T,N_b,\ldots} + \left(\frac{\partial G}{\partial V}\right)_{T,\boldsymbol{N}}\left(\frac{\partial V}{\partial N_a}\right)_{P,T,N_b,\ldots}. \tag{15.34}$$

The derivatives are more easily performed by rewriting (15.33) as,

$$\left.\begin{aligned} G &= \sum_a N_a(\mu_{0a} + \mathrm{k}T\log N_a) - N\,\mathrm{k}T\log V \\ &\quad - \mathrm{k}T\sum_m m\,B_m\,\boldsymbol{N}^m\,V^{-(m-1)}\,. \end{aligned}\right\} \tag{15.35}$$

With this in (15.34) one has

$$\begin{aligned} \mu_a &= \mu_{0a} + \mathrm{k}T\log\varrho_a + \mathrm{k}T\left[1 - \frac{1}{\varrho_a}\sum_m m_a\,m\,B_m\,\rho^m\right] - \\ &\quad - \mathrm{k}T\left[\varrho - \sum_m m(m-1)\,B_m\,\rho^m\right] v_a. \end{aligned}$$

Use v_a from (15.32) in this to obtain,

$$\mu_a = \mu_{0a} + \mathrm{k}T\left[\log\varrho_a - \frac{1}{\varrho_a}\sum_m m_a\,B_m\,\rho^m\right], \tag{15.36}$$

or

$$\log z_a = \log\varrho - \frac{1}{\varrho_a}\sum_m m_a\,B_m\,\rho^m\,. \tag{15.37}$$

The compressibility is, directly from (15.23)

$$\varkappa = \frac{1}{\mathrm{k}T}\,\frac{1}{\varrho - \sum\limits_m m(m-1)\,B_m\,\rho^m}\,. \tag{15.38}$$

Eqs. (15.30) to (15.38) then serve to give the thermodynamic properties of a gaseous mixture in terms of the integrals B_m.

16. Condensation and critical behavior. The virial development,

$$\frac{P}{\mathrm{k}T} = \varrho\left[1 - \sum_{\nu\geq 1}\frac{\nu}{\nu+1}\,\beta_\nu\,\varrho^\nu\right], \tag{16.1}$$

for the single component system is valid for all densities less than both ϱ_1 and ϱ_2. The two densities $\varrho_1(T)$ and $\varrho_2(T)$ are defined as those for which the function

$\vartheta(\varrho, T)$ represented by the series $\Sigma \nu \beta_\nu(T) \varrho^\nu$ is

$$\vartheta(\varrho_1, T) = \text{singular}, \quad \text{with } \varrho_1 \text{ positive real}, \tag{A}$$

$$\vartheta(\varrho_2, T) = 1, \quad \text{with } \varrho_2 \text{ positive real}, \tag{B}$$

respectively. At low temperatures condition (A) is satisfied at densities $\varrho_1(T)$ for which the value $\vartheta_s(T)$, approached by ϑ as $\varrho \to \varrho_1$ from-low values on the positive real axis, is less than unity. This may be deduced, either from an estimation of the values of β_ν, or, *a posteriori*, from a knowledge of the experimental behavior of saturated vapors.

The integrand, (15.12) of β_ν, contains all multiply connected products $\Pi f_{\alpha\beta}$, of the pair functions, $f_{\alpha\beta} = e^{-u_{\alpha\beta}/kT} - 1$, of $\nu+1$ molecules. These functions have a maximum value at the pair distance, $r_{\alpha\beta} = r_0$ for which $u_{\alpha\beta} = u(r_{\alpha\beta})$ has its minimum value of $-u_0$. For low temperatures this maximum is sharp, and large compared to unity. The main contribution to the integral will come from that part of the configuration space for which as many pair distances as possible have the value $r_{\alpha\beta} = r_0$. Since, for large ν-values, most (although not all) of the products with many pair functions, $f_{\alpha\beta}$, missing will be multiply connected, and hence included in the integrand of β_ν, the integrand will be closely approximated, for large ν-values and low temperatures, by the product $\Pi(f_{\alpha\beta}+1)$ for *all* pairs, namely $e^{-U_{\nu+1}/kT}$, provided this integration is limited to a volume in which a condensed phase of $\nu+1$ molecules would be stable, and not to the total volume, which might be much larger. Since one may find, for large ν values, $(\nu+1)!$ configurations, namely those of a close packed lattice, for which each molecule, except a number proportional to $\nu^{\frac{2}{3}}$ on the surface, are surrounded by 12 others at the distance r_0, one will have, neglecting the interaction of next to neighbors, a maximum value of the integrand of about $e^{+(6\nu u_0 - 3\nu^{\frac{2}{3}} u_0)/kT}$.

Writing,

$$b_0 = e^{6u_0/kT} \Delta v, \tag{16.2}$$

with Δv some three dimensional amplitude of vibration, and,

$$\sigma \approx e^{3u_0/kT} \gg 1, \tag{16.3}$$

a surface term, one may approximate β_ν as,

$$\beta_\nu \approx (\nu+1) \frac{b_0^\nu}{\sigma^{\nu^{\frac{2}{3}}}}, \tag{16.4}$$

where $\nu+1 = (\nu+1)!/\nu!$ comes from the two factorials.

This term, β_ν, divided by $\nu+1$ [see Eq. (15.13)] with $m_{n-1}=1$, $m_\nu=0$ for $\nu<n-1$, will be the leading term of $b_{\nu+1}$ at very low temperatures. This same approximation, then, was used in the discussion of Sect. 14.

The radius of convergence, ϱ_1, of $\Sigma \nu \beta_\nu \varrho^\nu$, condition A, will be, from (16.4), given by,

$$\varrho_1 = \frac{1}{\Delta v} e^{-6u_0/kT} = \frac{1}{b_0}. \tag{16.5}$$

This, in turn, is seen to be the usual approximate expression for the number density, ϱ_1, of a perfect gas at low densities in equilibrium with a close packed array of molecules, provided, as here, only nearest neighbor interaction is included in the estimation.

For low values of ν Eq. (16.4) will be a poor approximation, but since a much smaller number than 6 "bonds" per molecule can be formed we will always have,

$$\beta_\nu \ll b_0^\nu, \quad \text{small } \nu \text{ values}, \tag{16.6}$$

and, at least at very low temperatures, β_ν will be less for small ν than the expression given in (16.4). Hence $\vartheta_s(T)$ willl be expected not to exceed in value the sum,

$$\sum_{\nu \geq 1} \nu(\nu+1)\,\sigma^{-\nu^{\frac{2}{3}}} = S(\sigma) > \vartheta_s. \tag{16.7}$$

For large values of σ the sum $S(\sigma)$ takes very low values.

Whatever scepticism one may have about the above estimate, it is certainly known *a posteriori*, from the observed behavior of saturated vapors, that $\vartheta_s(T)$ *is* small at low temperatures. At low temperatures, and hence very low densities, saturated vapors are nearly Perfect Gases. The compressibility, $\varkappa$ of a Perfect Gas is $\varkappa = 1/\mathbf{k}T\varrho$, so that, from (15.23), $\varkappa = (1/\mathbf{k}T\varrho)(1-\vartheta_s)^{-1}$, one finds ϑ_s to be negligible at low temperatures.

Furthermore, both from the examination of the behavior of the integrand of β_ν, and from the experimental behavior of saturated vapors, one can say that as T increases the density, ϱ_1, at saturation increases rapidly, and the value of $\vartheta_s(T)$ also increases. However, for all temperatures less than the critical, $T < T_c$, the compressibility of the gas approaches a finite value as the density of saturation, ϱ_1, is approached, and hence from (15.13),

$$\vartheta_s(T) < 1, \qquad T < T_c, \tag{16.8}$$

namely condition (A) determines the point of condensation below the critical point.

Now no thermodynamic argument, nor any mathematical deduction from the behavior of β_ν seems to exclude the possibility that as T increases the singularity on the real axis of the function $\vartheta(\varrho, T)$ should branch into the complex plane at some $T = T_0$, and that the branch at $\varrho = \varrho_0$, $T = T_0$ should occur when $\vartheta_s(T_0)$ remains below unity. Any such temperature would be a critical temperature, since directly above T_0 the function $\vartheta(\varrho, T)$ would remain regular on the positive real axis and less than unity for $\varrho > \varrho_0$, so that the pressure would be regular. However this type of critical point would be one for which the compressibility would be finite, and this is certainly *not* observed in any known experimental case.

The experimental behavior at the critical point, $T = T_c$, $\varrho = \varrho_c$ is that the compressibility, $\varkappa$, is infinite, so that, from (15.23),

$$\vartheta_s(T_c) = 1, \tag{16.9}$$

namely *both* conditions (A) and (B) are satisfied, $\varrho_1 = \varrho_2 = \varrho_c$ at $T = T_c$.

There remains then, the problem of the behavior above T_c. At $T = T_c$, $\varrho = \varrho_c$, the sum,

$$\varrho\left(\frac{\partial}{\partial\varrho}\right)_T \vartheta(\varrho, T) = \lim_{\substack{\varrho\to\varrho_c \\ T\to T_c}} \left[\sum_{\nu\geq 1} \nu^2 \beta_\nu \varrho^\nu\right] \tag{16.10}$$

cannot be negative, since otherwise, in view of (16.9), the value of $\vartheta(\varrho, T)$ would exceed unity for $\varrho < \varrho_c$, and a prior singularity in P for $\varrho_2 < \varrho_c$ from condition (B) would exist. If, however, the sum (16.10) is greater than zero at T_c, then condition (B) will be satisfied for some finite temperature range, ΔT, above T_c at $\varrho = \varrho_2(T)$, and the pressure function, $P(\varrho, T)$ would have a singularity with infinite compressibility, $\varkappa$, at this density, $\varrho_2(T)$. The limit of this temperature range, $T_c + \Delta T$, would be that for which condition (B), $\vartheta(\varrho_2, T + \Delta T) = 1$, and the condition that $\sum \nu^2 \beta_\nu \varrho_2^\nu = 0$ were simultaneously valid, so that above $T_c + \Delta T$ the function $\vartheta(\varrho, T)$ would never reach unity.

If, however, the function $P(\varrho, T)$ is to remain regular above T_c we must require that the singularity of condition (A) has a branch point at T_c, so that above T_c the singularity does not exist on the positive real axis of ϱ.

Thus if the function $P(\varrho, T)$ is to be regular above the critical temperature, T_c, for all values of ϱ in the neighborhood of the critical density, ϱ_c, (a singularity at a very much higher density values, $\varrho_l \gg \varrho_c$, due to solidifcation to a lattice may be expected) one must require *three* simultaneous conditions on the function $\vartheta(\varrho, T)$:

The singularity of $\vartheta(\varrho, T)$ branches at ϱ_c, T_c, (a)

$$\vartheta(\varrho_c, T_c) = 1, \tag{b}$$

$$\varrho\left(\frac{\partial\vartheta}{\partial\varrho}\right)_{T=T_c,\,\varrho=\varrho_c} = 0. \tag{c}$$

It would seem almost incredible that, for all the wide variety of intra-molecular potentials existing in nature, the function $\vartheta(\varrho, T)$ would always obey simultaneously conditions (a), (b), and (c). For instance, turn to our estimation, (16.5), of ϱ_1. The Δv occurring in this expression can be written as $\Delta v = \Delta r^3$ in which Δr is some averaged distance around the minimum of $u(r_{\alpha\beta})$ at r_0 for which one can replace $u(r_{\alpha\beta})$ by $-u_0$. On the other hand, β_1 will be of the order $\beta_1 \approx 4\pi r_0^2 \times \Delta r\, e^{u_0/kT}$. The term $\beta_1 \varrho_1$ will depend on the ratio $r_0/\Delta r$ as well as on u_0/kT. Although the estimate of (16.5) for ϱ_1 can hardly be valid at critical temperatures, it would indeed seem strange that the conditions determining the branch point, which must be determined solely by β_ν for $\nu \to \infty$, should depend in the same way on r_0, Δr, and u_0 as the values of β_ν for small values of ν, which contribute to conditions (b) and (c).

One may counter that this argument is invalid, as follows. Fundamentally we are investigating the properties of the branch point in the function $\varphi(z, T) = P/kT$, represented by the power series in z of (13.15). The function φ is known (experimentally) to approach equal finite values from above and below on the positive real axis of z, but to have a discontinuity in its derivative, ϱ/z. By the transformation to a function of ϱ, with coefficients β_ν defined in terms of (15.13) we deduce that either all three conditions (a), (b), (c) hold at the branch point of $\vartheta(\varrho, T)$ or that there is a finite range ΔT below the branch point for which $\varrho(\partial\varphi/\partial\varrho)_T$ is zero. Since

$$\varrho\left(\frac{\partial\varphi}{\partial\varrho}\right)_T = \frac{\varrho^2}{[\partial^2\varphi/(\partial\log z)^2]_T}, \tag{16.11}$$

we therefore conclude that $[\partial^2\varphi/(\partial\log z)^2]_T$ is necessarily infinite at the singularity for a finite range of T, instead of only reaching infinite values exactly at the branch point. Now since the coefficients β_ν are only formally defined in terms of the $b_n^{(0)}$'s, our "proof" would be equally valid for any function $\varphi(z, T)$ having a parameter T, and is, therefore, ridiculous.

This objection seems to us to be wrong. Although, in the interest of avoiding a complicated proof which is otherwise unnecessary, we have chosen to formally define β_ν by (15.13), it is still true, that, for a system of pair potentials, the β_ν are computable in the general case by the integration of a relatively simply definable integrand, Eq. (15.12). Even if the potential contains explicit triple terms,

$$U_N\{N\} = \sum_{N\geq\alpha>\beta\geq 1}\sum u_2(r_{\alpha\beta}) + \sum_{N\geq\alpha>\beta>\gamma\geq 1}\sum\sum u_3(\boldsymbol{r}_\alpha, \boldsymbol{r}_\beta, \boldsymbol{r}_\gamma) + \cdots$$

an equation similar to (15.12) may be used with u_3 appearing only in the integrand for all β_ν with $\nu \geq 2$, and not in β_1. The coefficients, β_ν, of the power series in ϱ

seem to have a simple significance which transcends their purely formal definition by (15.13). If they do have such a fundamental meaning in terms of the potentials, it seems to us extraordinarily unlikely that all three conditions, (a), (b), and (c), are simultaneously valid for *all* naturally occurring potentials.

On the other hand, the experimental behavior above the critical temperature does *not* seem to be consistent with the existence of a density $\varrho_2(T)$, for a finite temperature range, for which the compressibility, $\varkappa$, is infinite, and $(\partial P/\partial \varrho)_T$ equal to zero.

One is therefore led to re-examine the mathematical rigor of the complete development. Eq. (13.29$_1$), for the one component system,

$$P(z, T, V) = \mathsf{k}T \sum_{n\geq 1} b'_n(V, T)\, z^n, \tag{16.12}$$

is certainly valid for any system of finite volume, V. As was discussed in Sect. 5, the function $P(z, T, V)$ is necessarily analytic and regular, in the strict mathematical sense, if V is finite. It is only the limiting function,

$$\left.\begin{aligned} P(z, T) &= \lim_{V\to\infty} [P(z, T, V)] \\ &= \lim_{V\to\infty} \Big[\sum_{n\geq 1} b'_n(V, T)\, z^n\Big], \end{aligned}\right\} \tag{16.13}$$

which shows a singularity. As long as $P(z, T)$ is regular on the positive real axis the change in order of the summation and the limiting process,

$$\left.\begin{aligned} \lim_{V\to\infty}\Big[\sum_{n\geq 1} b'_n(V, T)\, z^n\Big] &= \sum_{n\geq 1}\Big[\lim_{V\to\infty} b'_n(V, T)\Big] z^n \\ &= \sum_{n\geq 1} b_n^{(0)} z^n = P'(z, T), \end{aligned}\right\} \tag{16.14}$$

is certainly valid. It is, however, *not* true that the singularity on z of the expression on the left of (16.14), namely $P(z, T)$, and that of the function on the right, $P'(z, T)$, represented by the sum $\sum b_n^{(0)} z^n$, need necessarily be identical.

However, at low temperatures $b_n^{(0)}$ is positive, and has a positive limit,

$$\lim_{n\to\infty} [b_n^{(0)}(T)]^{1/n} = b_0(T) > 0. \tag{16.15}$$

Also, for low T values $b'_n(V)$ is bounded,

$$-[b_0(T)]^n < b'_n(V, T) \leq b_n^{(0)}(T). \tag{16.16}$$

In this case the singularities of $P(z, T)$ and of the function $P'(z, T)$ *will* be the same. The detailed case is discussed by IKEDA[1].

This is also borne out *a posteriori* by the fact that ϱ_1 of (16.5) is the usual approximate expression for the density of the saturated vapor at low temperatures.

Since we can deduce that the expression $\sum b_n^{(0)} z^n$ is correct up to the first singularity on the positive real axis of z at low temperatures, one is inclined to assume it generally. Indeed there is no reason to believe that the expression $P(z, T) = P'(z, T) = \sum b_n^{(0)} z^n$ is incorrect up to, and including the singularity, below T_c. However, exactly when condition (B), that $\vartheta_s(T) = 1$, becomes valid, the character of $b'_n(T, V)$, and of the volume dependence of these integrals, changes drastically. At this temperature, and above, it may well be that the singularities in $P(z, T)$ and $P'(z, T)$ are no longer identical.

[1] K. IKEDA: Prog. Theor. Phys. **16**, 341 (1956).

The pronounced change in the volume dependence of $b'_n(T, V)$ at $T = T_c$ arises as follows. Below T_c the leading term in $b_n^{(0)}$ for large n is one with a single β_ν factor of (15.13) with ν of the order of n. Since the integrand of β_ν contains multiply connected products, $\Pi f_{\alpha\beta}$, the main contribution to the integral comes from that portion of configuration space in which as many molecules as possible are surrounded by others. Up to some $n \leq V/v_0$, with v_0 some appropriate volume per molecule, one will expect $b'_n(T, V) \approx b_n^{(0)}$. Above T_c, however, the graphs, of the simple form $f(r_{12})\, f(r_{23})\, f(r_{34}) \ldots f(r_{n-1,n})$ contribute appreciably to $b_n^{(0)}$. For such products the integral over the molecules 2 to $n-1$ remains large even if the distance between the two end molecules, r_{1n}, is of the order of $n^{\frac{1}{2}} v_0^{\frac{1}{3}}$. Presumably, then, $b'_n(T, V)$ may depend appreciably on V unless $n \lesssim (V/v_0)^{\frac{2}{3}}$, namely at a much smaller n-value for given V than below T_c.

If the identification of $P(z, T)$ with $P'(z, T)$ is not made, one may still formally define the coefficients $\beta_\nu(T, V)$, for any $b'_n(T, V)$ by (15.13), but the integral of (15.12), will no longer be correct for β_ν at high values of ν. The integral (15.12) for β_ν and the definition (15.13) are only simultaneously valid for $b_n^{(0)}$. One would, it is true, still be able to define a function, $\vartheta(\varrho, T)$, as the limit of $\sum \nu \beta_\nu(T, V)\, \varrho^\nu$ for infinite V, and conditions (A) and (B) would still determine singularities of $P(z, T)$. The coefficients $\beta_\nu(T, V)$ would indeed, still be volume independent, and given by (15.12) for low ν values, or even to very large values for macroscopic V values. However, above T_c, the β_ν for $\nu \gtrsim (V/v_0)^{\frac{2}{3}}$ would have a very fancy volume dependence, not at all that to be expected by extending the integration of the integrand of (15.12) over some finite volume, V; but one dictated by (15.13_2) and the peculiar volume dependence of $b_n^{(0)}$. In view of this rather complicated relationship of $\beta_\nu(T, V)$ on V above T_c, it would be rash to conclude anything about the possibility of interchanging the order of summation and going to the limit $V \to \infty$ at the singularity.

However, it remains true that if the function $P(z, T)$ is completely analytic for $T > T_c$ in the neighborhood of the critical activity, z_c, then the interchange of summation and limit, $P(z, T) = P'(z, T)$ is permitted, and we would be required to conclude that conditions (a), (b), and (c) were simultaneously satisfied by $\vartheta(\varrho, T)$ defined in terms of the simple volume independent β_ν's of Eq. (15.12). Since we are inclined to reject this, we must conclude that an irregularity does exist in the $P(z, T)$ function above T_c, presumably one in which some of the higher derivatives, $(\partial^\nu P/\partial \varrho^\nu)_T$ undergo discontinuous changes.

Such a lack of analytical behavior would be impossible to verify or disprove experimentally. It appears to be true that no simple empirical expression at temperatures just above T_c represents the pressure-density function for both ranges of density, above the critical, and below the critical. This, however, need not indicate that the function is non-analytic, since, in any case it is clear that a singularity must exist rather close to the real axis in this temperature range, and the function may very well be too complicated to be represented simply so close to a singularity.

17. Summary of the series developments. The series developments discussed in the previous sections lead directly to equations for the quantity P/kT as a series in the activity z, or for a multicomponent system the activity set $\mathbf{z} = z_a, z_b, \ldots$, or, alternately, as a series in ϱ or $\boldsymbol{\varrho}$ respectively, with ϱ the number density, $\varrho = \overline{N}/V$. The coefficients are simply definable integrals of functions of the intermolecular pair potential. In this section we will summarize the results and discuss the evaluation of the integrals for some simple cases.

The activity z is defined in terms of the chemical potential, μ, as

$$z = e^{(\mu-\mu_0)/kT}, \qquad \lim_{\varrho\to 0}\left[\frac{z}{\varrho}\right] = 1, \tag{17.1}$$

with μ_0 the chemical potential of the perfect gas at number density, $\varrho = 1$. The quantity P/kT as a function of $\boldsymbol{z}$, T, has derivatives [Eqs. (4.17_1) and (4.17_2)],

$$\left(\frac{\partial}{\partial \log z_a}\right)_{T, z_b \ldots}\left(\frac{P}{kT}\right) = \varrho_a, \tag{17.2}$$

$$\left(\frac{\partial}{\partial \log T}\right)_{\boldsymbol{z}}\left(\frac{P}{kT}\right) = \frac{1}{VkT}(E - E_0), \tag{17.3}$$

with E the energy and E_0 the energy of the same composition in the perfect gas limit. The series developments in $\boldsymbol{z}$ have cluster integrals $b_{\boldsymbol{n}}^{(0)}$ [see Eq. (17.12_1)] as coefficients, with $b_1^{(0)} \equiv 1$ for all species. The equation for P/kT is, [Eq. (13.29_2)],

$$\left.\begin{aligned} \frac{P}{kT} &= \sum_{\boldsymbol{n}\geq 1} b_{\boldsymbol{n}}^{(0)} \boldsymbol{z}^{\boldsymbol{n}} \\ &= \sum_a z_a + \sum_{a\geq b}\sum b_{ab}^{(0)} z_a z_b + \sum_{a\geq t\geq c}\sum\sum b_{abc}^{(0)} z_a z_b z_c + \cdots, \end{aligned}\right\} \tag{17.4}$$

where the subscripts refer to the different molecular species. The normalization used here for $b_{ab} = b_{ba}$ is such that this integral appears only once in the summations, as do the six identical integrals b_{abc} for $a > b > c$. The equation for the number density is

$$\left.\begin{aligned} \varrho_a &= \sum_{\boldsymbol{n}\geq 1} n_a b_{\boldsymbol{n}}^{(0)} \boldsymbol{z}^{\boldsymbol{n}} \\ &= z_a + 2 b_{aa} z_a^2 + \sum_{b\neq a} b_{ab} z_a z_c + 3 b_{aaa} z_a^3 + \\ &\quad + \sum_{b\neq a} 2 b_{aab} z_a^2 z_b + \sum_{b>c,\, b\neq a,\, c\neq a}\sum b_{abc} z_a z_b z_c + \cdots. \end{aligned}\right\} \tag{17.5}$$

The mutual potential energy per unit volume from (17.3), is,

$$\frac{1}{V}[E - E_0] = kT \sum_{\boldsymbol{n}\geq 2} \frac{d b_{\boldsymbol{n}}^{(0)}}{d \log T} \boldsymbol{z}^{\boldsymbol{n}}. \tag{17.6}$$

In terms of the number density, the series have the irreducible integrals, $B_{\boldsymbol{m}}$, as coefficients. The usual nomenclature for a one-component system uses a normalization that is inconvenient in the multicomponent system, the irreducible integral β_ν for a single component is [Eq. (15.29)]

$$\beta_{m-1} = m B_m. \tag{17.7}$$

The integrals, $B_{\boldsymbol{m}}$, are defined only for $m \geq 2$. The quantity $P/\varrho kT$, which is unity for the perfect gas is [Eqs. (15.22) and (15.30)], $m = \sum_a m_a$,

$$\frac{P}{\varrho kT} = 1 - \sum_{\nu\geq 1} \frac{\nu}{\nu+1} \beta_\nu \varrho^\nu, \quad \text{(one component)}, \tag{17.8$_1$}$$

$$\left.\begin{aligned} \frac{P}{\varrho kT} &= 1 - \frac{1}{\varrho} \sum_{\boldsymbol{m}\geq 2} m B_{\boldsymbol{m}} \boldsymbol{\rho}^{\boldsymbol{m}} \\ &= 1 - 2\varrho \sum_{a\geq b}\sum B_{ab} x_a x_b - 3\varrho^2 \sum_{a\geq b\geq c}\sum\sum B_{abc} x_a x_b x_c + \cdots, \end{aligned}\right\} \tag{17.8$_2$}$$

with x_a, x_b, etc. the mol fraction of component a and b respectively. The chemical potential, μ, is [Eqs. (15.17) and (17.1)],

$$\mu = \mu_0 + \mathrm{k}T \log \varrho - \mathrm{k}T \sum_{\nu \geq 1} \beta_\nu \varrho^\nu, \tag{17.9$_1$}$$

for the one-component system, and [Eq. (15.36)],

$$\left.\begin{aligned} \mu_a &= \mu_{0a} + \mathrm{k}T \log \varrho_a - \frac{\mathrm{k}T}{\varrho_a} \sum_{\boldsymbol{m} \geq 2} m_a B_{\boldsymbol{m}} \boldsymbol{\rho}^{\boldsymbol{m}} \\ &= \mu_{0a} + \mathrm{k}T \log \varrho + \mathrm{k}T \log x_a - \mathrm{k}T \varrho [2 B_{aa} x_a + \sum_{b \neq a} B_{ab} x_b] - \\ &\quad - \mathrm{k}T \varrho^2 [3 B_{aaa} x_a^2 - \sum_{b \neq a} B_{aab} x_a x_b - \sum_{b>c, b \neq a, c \neq a} \sum B_{abc} x_b x_c] - \cdots, \end{aligned}\right\} \tag{17.9$_2$}$$

for multicomponent systems.

The compressibility is [Eq. (15.23)],

$$\varkappa = \frac{1}{\varrho} \left(\frac{\partial \varrho}{\partial P}\right)_{T, \boldsymbol{x}} = \frac{1}{\mathrm{k}T \varrho} \frac{1}{1 - \sum_{\nu \geq 1} \nu \beta_\nu \varrho^\nu}, \tag{17.10$_1$}$$

for the one-component system, and [Eq. (15.38)]

$$\varkappa = \frac{1}{\mathrm{k}T \varrho} \frac{1}{1 - \sum_{\boldsymbol{m}} m(m-1) B_{\boldsymbol{m}} \boldsymbol{x}^{\boldsymbol{m}} \varrho^{m-1}}, \tag{17.10$_2$}$$

for the multicomponent case.

The partial molecular volume of species a in the multicomponent system is [Eq. (15.32)],

$$\left(\frac{\partial V}{\partial N_a}\right)_{P, T, N_b \ldots} = \frac{1}{\varrho} \frac{1 - x_a^{-1} \sum_{\boldsymbol{m}} m_a (m-1) B_{\boldsymbol{m}} \boldsymbol{x}^{\boldsymbol{m}} \varrho^{m-1}}{1 - m \sum_{\boldsymbol{m}} m(m-1) B_{\boldsymbol{m}} \boldsymbol{x}^{\boldsymbol{m}} \varrho^{m-1}}. \tag{17.11}$$

The cluster integrals b_n are related to the irreducible integrals [Eq. (15.13$_1$)], by,

$$\left.\begin{aligned} b_1 &= 1, \\ b_2 &= \tfrac{1}{2} \beta_2, \\ b_3 &= \tfrac{1}{3} \beta_3 + \tfrac{1}{2} \beta_1^2, \end{aligned}\right\} \tag{17.12$_1$}$$

in the one-component case, and,

$$\left.\begin{aligned} &b_{aa} = B_{aa}, \quad b_{ab} = B_{ab}, \\ &b_{aaa} = B_{aaa} + 2 B_{aa}^2, \\ &b_{aab} = b_{aba} = b_{baa} = B_{aab} + 2 B_{ab} B_{aa} + \tfrac{1}{2} B_{ab}^2, \\ &b_{abc} = b_{acb} = \cdots = B_{abc} + B_{ca} B_{ab} + B_{ab} B_{bc} + B_{bc} B_{ca}, \end{aligned}\right\} \tag{17.12$_2$}$$

for the multicomponent case.

The irreducible integrals are, for $m = 2$,

$$B_{aa} = \tfrac{1}{2} \beta_1 = \tfrac{1}{2} \int_0^\infty 4\pi r^2 [\mathrm{e}^{-u_{aa}(r)/\mathrm{k}T} - 1] \, dr, \tag{17.13$_1$}$$

$$B_{ab} = B_{ba} = \int_0^\infty 4\pi r^2 [\mathrm{e}^{-u_{ab}(r)/\mathrm{k}T} - 1] \, dr, \tag{17.13$_2$}$$

with u_{aa} and u_{ab} the pair interaction potentials of two molecules of species a and of one molecule of species a with one of species b respectively. The integrals for three molecules are,

$$\left.\begin{aligned} B_{aaa} &= \tfrac{1}{3}\beta_2 = \tfrac{1}{6}\iint\limits_{-\infty}^{+\infty} \left[e^{-u_{aa}(r_{01})/kT}-1\right]\left[e^{-u_{aa}(r_{02})/kT}-1\right]\times \\ &\qquad\times\left[e^{-u_{aa}(r_{12})/kT}-1\right] d\boldsymbol{r}_1\, d\boldsymbol{r}_2, \\ B_{aab} &= B_{aba} = \cdots = \tfrac{1}{2}\iint\limits_{-\infty}^{+\infty} \left[e^{-u_{aa}(r_{01})/kT}-1\right]\left[e^{-u_{ab}(r_{02})/kT}-1\right]\times \\ &\qquad\times\left[e^{-u_{ab}(r_{12})/kT}-1\right] d\boldsymbol{r}_1\, d\boldsymbol{r}_2, \\ B_{abc} &= B_{acb} = \cdots = \iint\limits_{-\infty}^{+\infty} \left[e^{-u_{ab}(r_{01})/kT}-1\right]\left[e^{-u_{ac}(r_{02})/kT}-1\right]\times \\ &\qquad\times\left[e^{-u_{bc}(r_{12})/kT}-1\right] d\boldsymbol{r}_1\, d\boldsymbol{r}_2 \end{aligned}\right\} \quad (17.14_1)$$

with $r_{01}=|\boldsymbol{r}_1|$, $r_{02}=|\boldsymbol{r}_2|$, and $r_{12}=|\boldsymbol{r}_1-\boldsymbol{r}_2|$. The expressions of (17.14) are given for the case that the total potential for three molecules is assumed to be the sum of the three pair potentials. If this is not assumed the component potential, $u_3(r_1, r_2, r_3)$, which must approach zero if any of the three molecules becomes distant from the other two, is defined by

$$U_3(\boldsymbol{r}_1, \boldsymbol{r}_2, \boldsymbol{r}_3) = u(r_{12}) + u(r_{23}) + u(r_{31}) + u_3(\boldsymbol{r}_1, \boldsymbol{r}_2, \boldsymbol{r}_3). \qquad (17.14_2)$$

The term

$$\left[e^{-\frac{u_3(0,\boldsymbol{r}_1\boldsymbol{r}_2)}{kT}}-1\right] e^{-\frac{u(r_{01})+u(r_{02})+u(r_{12})}{kT}}, \qquad (17.14_3)$$

must then be added to the integrand in each of the expressions of (17.14_1).

The normalization used here divides each integral by the product of the factorials of the number of identical molecules. If the integrals were all normalized by division with the factorial of the total number of molecules then each possible term $B_{ab}+B_{ba}$, or $B_{abc}+B_{acb}+B_{bca}+B_{cab}+B_{cba}$ could be used in the summations with the same answer. The distinction requires some caution in using tabulated values from the literature.

The irreducible integrals for pairs, β_1 or B_{aa}, B_{ab} have been integrated and tabulated for a variety of assumed potentials. For many potentials also the integrals for triples, β_2, for the one component case have been performed. The best compilation is in the book by HIRSCHFELDER, CURTISS and BIRD [*4*].

A few simple examples will indicate the nature of the values. For the "hard sphere" potential, $u(r)=\infty$, $r<2r_0$, $u(r)=0$, $r\geq 2r_0$, the value is

$$\frac{1}{2}\beta_1 = -\frac{16\pi}{3}r_0^3 = -4v_0' \quad \text{(hard sphere potential)} \qquad (17.15_1)$$

independent of T, since the integrand of (17.13_1) is then -1 for $r<r_0$, and zero for $r\geq r_0$. The volume, v_0', of (17.15) is here the volume, $v_0'=\frac{4\pi}{3}r_0^3$, of the spheres, whose closest distance of approach is $2r_0$. The pressure equation then starts as,

$$\frac{P}{\varrho kT} = 1 + 4v_0'\varrho + \cdots, \quad \text{(hard spheres)}. \qquad (17.15_2)$$

If the mutual interaction potential $u(r)$ is assumed to have a negative region, $u(r)=-u_0$, from $2r_0\leq r\leq 2r_0+\Delta r$, outside of the hard sphere core, and to be zero for $r>2r_0+\Delta r$, there is an additional region between $2r_0$ and $2r_0+\Delta r$

where the integrand is $e^{u_0/kT}-1$, and,

$$\left.\begin{aligned}\frac{1}{2}\beta_1 &= 4v_0'\left\{\left[\left(1+\frac{\Delta r}{2r_0}\right)^3-1\right][e^{u_0/kT}-1]-1\right\},\\ &\begin{bmatrix} u=\infty, & r<2r_0,\\ u=-u_0, & 2r_0\leq r\leq 2r_0+\Delta r,\\ u=0, & 2r_0+\Delta r<r.\end{bmatrix}\end{aligned}\right\}\tag{17.15$_3$}$$

The Lennard-Jones 6—12 potential,

$$u(r)=4u_0\left[\left(\frac{r^*}{r}\right)^{12}-\left(\frac{r^*}{r}\right)^{6}\right],\tag{17.16}$$

is usually assumed to be the standard pair potential for nearly spherical non polar chemically saturated molecules. The integral is,

$$\left.\begin{aligned}\frac{1}{2}\beta_1 &= \frac{2\pi}{3}v_0\int_0^\infty\left[e^{-\frac{4}{\tau}[x^{-12}-x^{-6}]}-1\right]3x^2\,dx,\\ v_0 &= r^{*3},\qquad \tau=kT/u_0.\end{aligned}\right\}\tag{17.17}$$

The volume v_0 is approximately the volume of a close packed lattice of these molecules at 0° K. A partial integration,

$$\int_0^\infty[e^{-\varphi(x)}-1]\,3x^2\,dx=\int_0^\infty x^3\frac{d\varphi}{dx}e^{-\varphi(x)}\,dx,$$

brings this into the form,

$$\frac{1}{2}\beta_1=\frac{2\pi}{3}v_0\frac{24}{\tau}\int_0^\infty x^2(x^{-6}-2x^{-12})\,e^{-\frac{4}{\tau}(x^{-12}-x^{-6})}\,dx.$$

Expansion of the term $e^{-\frac{4}{\tau}x^{-6}}$ and transformation of $x^{-12}=z$ brings the series into terms that can be integrated analytically. The result is

$$\left.\begin{aligned}\frac{1}{2}\beta_1 &= \frac{2\pi}{3}v_0\tau^{-\frac{1}{4}}\sqrt{2}\sum_{\nu\geq 0}\frac{2^\nu}{4\nu!}\Gamma\left(\frac{2\nu-1}{4}\right)\tau^{-\nu/2}\\ &= \frac{2\pi}{3}v_0\tau^{-\frac{1}{4}}[-1.733+2.564\,\tau^{-\frac{1}{2}}+8.66\,\tau^{-1}+4.27\,\tau^{-\frac{3}{2}}+2.17\,\tau^{-2}+\cdots].\end{aligned}\right\}\tag{17.18}$$

The convergence is rapid only for rather large values of $\tau=kT/u_0$. At low τ values $\frac{1}{2}\beta_1$ is large and positive, decreasing from $-28\frac{2\pi}{3}v_0$ at $\tau=0.3$ to $1.68\frac{2\pi}{3}v_0$ at $\tau=1.26$, about the critical temperature, and passing zero at $\tau=3.42$, the Boyle Point. For higher τ values $\frac{1}{2}\beta_1$ decreases to a minimum, $-0.53\frac{2\pi}{3}v_0$ at $\tau=30$ and then very slowly rises toward zero. The qualitative behavior at very low τ values is not unlike that of (17.15$_3$), but with Δr approximately proportional to $\sqrt{\tau}$. At high τ values the "soft" r^{-12} core of the Lennard-Jones potential, in contrast to the "hard core" impenetrable sphere used in arriving at (17.15$_3$), prevents $\frac{1}{2}\beta_1$ from approaching values as low as $\frac{2\pi}{3}v_0$, which would be the limit of a hard core potential extended to the first zero of the potential (17.16) at $r=r^*$. This soft core also accounts for the gradual rise toward zero at very high τ values.

Originally LENNARD-JONES compared a variety of two-term inverse power law potentials,

$$u(r) = u_0 \left(\frac{m}{n}\right)^{m/(m-n)} \left(\frac{n}{m-n}\right) \left[\left(\frac{r^*}{r}\right)^m - \left(\frac{r^*}{r}\right)^n\right], \tag{17.19}$$

with observed virial coefficients. [The potential of (17.19) is zero at $r = r^*$, and has a minimum, $-u_0$, at $r = r^* \left(\frac{m}{n}\right)^{1/m-n}$.] The experimental evidence indicated that the attractive power, n, was nearly $n=6$, a value which is required as a limit at large enough r by the theory of the London dispersion forces. Values of m between about 10 and 16 are almost equally satisfactory, and the predilection for $m=12$ is undoubtedly due largely to the fact that Eq. (17.19) is particularly simple for 6—12.

A considerably more elaborate, four parameter function, the Buckingham-Corner Potential, has been used for the computation of $\frac{1}{2}\beta_1$, and values are tabulated in HIRSCHFELDER, CURTISS and BIRD [4]. This potential has an exponential repulsive term, and both r^{-6} and r^{-8} attractive terms. In order to avoid the infinities at $r \to 0$ of the latter they are multiplied by an exponential of $-4\left(\frac{r_0}{r} - 1\right)^3$ for $r < r_0$, the minimum of the potential. The four parameters can be chosen as u_0, the negative of the potential at the minimum, r_0, the value of r at the minimum, the ratio, β, of the inverse eighth to the inverse sixth power term at the minimum, and α, the steepness of the potential. The analytical form is then,

$$\left.\begin{aligned} u(r) &= b\,\mathrm{e}^{-\alpha\frac{r}{r_0}} - \left(\frac{c}{r^6} + \frac{c'}{r^8}\right)\mathrm{e}^{-4\left(\frac{r_0}{r}-1\right)^3}, \quad & r \leq r_0, \\ u(r) &= b\,\mathrm{e}^{-\alpha\frac{r}{r_0}} - \left(\frac{c}{r^6} + \frac{c'}{r^8}\right), & r \geq r_0, \end{aligned}\right\} \tag{17.20$_1$}$$

with

$$\left.\begin{aligned} b &= \left[(1+\beta)\frac{c}{r_0^6} - u_0\right]\mathrm{e}^{\alpha}, \\ c &= \frac{u_0 \alpha r_0^6}{\alpha(1-\beta) - 6 - 8\beta}, \\ c' &= \frac{\beta}{r_0^2}\,c. \end{aligned}\right\} \tag{17.20$_2$}$$

The interaction energy of two dipoles, $\boldsymbol{p}_a$ and $\boldsymbol{p}_b$, directed so as to make angles ϑ_a and ϑ_b with the vector $\boldsymbol{r}_{ab} = |\boldsymbol{r}_a - \boldsymbol{r}_b|$ connecting their centers, is

$$\Delta u(r_{ab}, \vartheta_a, \varphi_a, \vartheta_b, \varphi_b) = -\frac{1}{r_{ab}^3}\left[3 p_a p_b \cos\vartheta_a \cos\vartheta_b - \boldsymbol{p}_a \cdot \boldsymbol{p}_b\right]. \tag{17.21$_1$}$$

One can write,

$$\boldsymbol{p}_a \cdot \boldsymbol{p}_b = p_a p_b \left[\cos\vartheta_a \cos\vartheta_b + \sin\vartheta_a \sin\vartheta_b \cos(\varphi_a - \varphi_b)\right], \tag{17.21$_2$}$$

so that $\Delta u(r_{ab})$ may also be written as

$$\Delta u = -\frac{p_a p_b}{r^3}\left[2\cos\vartheta_a \cos\vartheta_b - \sin\vartheta_a \sin\vartheta_b \cos(\varphi_a - \varphi_b)\right]. \tag{17.21$_3$}$$

If this potential, $\Delta u(r)$, is added to a LENNARD-JONES one obtains what is known as the Stockmayer Potential, which now depends on the angles $\vartheta_a \varphi_a$ and $\vartheta_b \varphi_b$ of the two molecules. This is, presumably, a fairly realistic potential for two polar molecules whose shape is not too far from spherical. The integration over the angles must, of course, be normalized by division with 4π for each molecule, in order to obtain correctly normalized values of the irreducible integrals, B_{ab}, B_{abc}, etc.

If $\Delta u(r)$ of (17.21$_3$) is added to the $u(r)$ of (17.16) and the contribution to $e^{-(u+\Delta u)/kT}$ due to integration over the angles at large r values is computed by expanding the exponential, the linear term in $\Delta u(r)$ contributes zero, since the average potential $\Delta u(r)$ is equally positive and negative. The quadratic term contributes a net added attraction proportional to the inverse sixth power of r, that is the net averaged attraction of two dipoles at large distances is proportional to r^{-6}. However, this extra term corresponds to a temperature dependent potential. Computed values of $\frac{1}{2}\beta_1$ for this potential are also listed in HIRSCHFELDER, CURTISS and BIRD [4].

Third virial coefficients, $-\frac{2}{3}\beta_2$, for these potentials are also tabulated in HIRSCHFELDER, CURTISS and BIRD. In general β_2 is positive only at lower temperatures than β_1, crossing zero at a reduced temperature $\tau \cong 0.9$, and remaining of value approximately $-0.5\left(\frac{2\pi}{3}v_0\right)^2$ for a considerable range of τ above unity. The values for the 6—12 potential are, in units of $\left(\frac{2\pi}{3}v_0\right)^2$, -0.43 at $\tau=1$, -0.59 at $\tau=1.25$, and then very gradual increase to 0.43 at $\tau=2$ and 0.35 at $\tau=3$.

Computations of the fourth virial coefficient, $-\frac{3}{4}\beta_3$, have not been carried out for any realistic potential. The values,

$$\frac{1}{2}\beta_1 = -4v_0' = \frac{2\pi}{3}v_0,$$

$$\tfrac{2}{3}\beta_2 = -\tfrac{5}{8}(4v_0')^2,$$

$$\tfrac{3}{4}\beta_3 = -0.2869(4v_0')^3,$$

$$\tfrac{4}{5}\beta_4 = -(0.115 \pm 0.005)(4v_0')^4,$$

for hard spheres of radius r_0, $v_0' = \frac{4\pi}{3}r_0^3$, having collision diameter $2r_0$, and with $v_0=(2r_0)^3$, have been computed.

18. Distribution functions. *α) Concepts.* The thermodynamic relations say nothing about the configurational structure of the system. The spatial arrangement of molecules in the perfect gas is completely random, the probability that a given molecule will be found at a specified position $\boldsymbol{r}$ within the volume element $d\boldsymbol{r}$ is independent of the positions of all other molecules. The more physical statement is that the probability that the volume element $d\boldsymbol{r}$ at $\boldsymbol{r}$ shall be occupied by a molecule is $\varrho\, d\boldsymbol{r}$, and independent of whether other volume elements $d\boldsymbol{r}_1, d\boldsymbol{r}_2, \ldots$ at $\boldsymbol{r}_1, \boldsymbol{r}_2, \ldots$ etc., are occupied or not.

Actually, of course, this statement is true only of the model, without potentials of interaction, which leads to the equations of the perfect gas. The real gas, even at the densities for which it behaves as a perfect gas, has potentials of interaction. However to behave like a perfect gas the density must be sufficiently low that the range of these interactions, the distance $r_{\max}$ at which $u(r_{\max})/kT$ becomes negligible compared to unity, is small compared to $\varrho^{-\frac{1}{3}}$, $r_{\max} \ll \varrho^{-\frac{1}{3}}$. The statement of the first paragraph about perfect gases is then true only when the distances from $\boldsymbol{r}$ to $\boldsymbol{r}_1, \boldsymbol{r}_2, \ldots$ etc., are larger than $r_{\max}$, which may still be true even if these distances are relatively small compared to $\varrho^{-\frac{1}{3}}$, the average distance between molecules.

The structure of a crystalline lattice is already partially described by specifying the probability density, $\varrho_1(\boldsymbol{r})$, that a single molecule be found at $\boldsymbol{r}$, provided the coordinate axes are chosen to have fixed angles and fixed origin with respect to the lattice. For a fluid $\varrho_1(\boldsymbol{r})$ is constant, and equal to the average number density, ϱ, as in the perfect gas. Nevertheless it has sense to inquire about the "structure" of a real fluid, liquid or dense gas.

The pertinent logically defined functions which appear to be most appropriate for the description of the structure of a system are the hierarchy of reduced probability densities for differing numbers of molecules, or the distribution functions, obtained from these by normalization with the average number density. We define

$$\varrho_{\boldsymbol{n}}(\boldsymbol{r}_1, \ldots, \boldsymbol{r}_n) = \varrho_{\boldsymbol{n}}\{\boldsymbol{n}\} = \boldsymbol{\rho}^{\boldsymbol{n}} F_{\boldsymbol{n}}\{\boldsymbol{n}\}, \tag{18.1}$$

by the statement that $\varrho_{\boldsymbol{n}}\{\boldsymbol{n}\}\, d\{\boldsymbol{n}\}$ is the probability that the set of coordinate positions $\{\boldsymbol{n}\}$ are simultaneously occupied by a set $\boldsymbol{n}$ of molecules, within the volume element $d\{\boldsymbol{n}\}$, independently of the positions of other molecules in the system. The set $\{\boldsymbol{n}\}$ of coordinate positions specifies the type of molecules, species a at the positions $\{n_a\}$, species b at $\{n_b\}$, ... etc. It does not, however, regard the molecules as numbered explicitly, that is the probability $\varrho_{\boldsymbol{n}}\{\boldsymbol{n}\}\, d\{\boldsymbol{n}\}$ sums the $\boldsymbol{n}!$ probabilities that any of the numbered n_a molecules of species a in the system shall occupy any of the positions in $\{n_a\}\, d\{n_a\}$, independently of order in the numbered n_a positions, for all a.

For any single thermodynamic phase of a fluid the probability density $\varrho_{\boldsymbol{n}+\boldsymbol{m}}\{\{\boldsymbol{n}\}+\{\boldsymbol{m}\}\}$, becomes the independent product of the two probability densities $\varrho_{\boldsymbol{n}}\{\boldsymbol{n}\}\,\varrho_{\boldsymbol{m}}\{\boldsymbol{m}\}$ if the distance $r_{\boldsymbol{n}\boldsymbol{m}}$ between *all* the molecules of the set $\boldsymbol{n}$ and *all* of the set $\boldsymbol{m}$ becomes large,

$$\lim_{r_{\boldsymbol{n},\boldsymbol{m}}\to\infty} \left[\varrho_{\boldsymbol{n}+\boldsymbol{m}}\{\{\boldsymbol{n}\}+\{\boldsymbol{m}\}\}\right] = \varrho_{\boldsymbol{n}}\{\boldsymbol{n}\}\,\varrho_{\boldsymbol{m}}\{\boldsymbol{m}\}, \tag{18.2$_1$}$$

$$\lim_{r_{\boldsymbol{n}\boldsymbol{m}}\to\infty} \left[F_{\boldsymbol{n}+\boldsymbol{m}}\{\{\boldsymbol{n}\}+\{\boldsymbol{m}\}\}\right] = F_{\boldsymbol{n}}\{\boldsymbol{n}\}\,F_{\boldsymbol{m}}\{\boldsymbol{m}\}. \tag{18.2$_2$}$$

Since, for a fluid, $\varrho_1(\boldsymbol{r}) = \varrho$, the average number density in a one component system, and $\varrho_{1a}(\boldsymbol{r}) = \varrho_a$ the number density of species a, in general, and $\varrho_{\boldsymbol{n}}\{\boldsymbol{n}\} = \boldsymbol{\rho}^{\boldsymbol{n}}$ from (18.2$_1$) if *all* distances $r_{\alpha\beta}\to\infty$, we have from (18.1), that,

$$\lim_{\text{all } r_{\alpha\beta}\to\infty} \left[F_{\boldsymbol{n}}\{\boldsymbol{n}\}\right] = 1. \tag{18.2$_3$}$$

This leads to the normalization for finite $\boldsymbol{n}$,

$$\lim_{V\to\infty}\left[\frac{1}{V^n}\iint_V\cdots\int F_{\boldsymbol{n}}\{\boldsymbol{n}\}\, d\{\boldsymbol{n}\}\right] = 1, \tag{18.2$_4$}$$

provided the approach to unity in (18.2$_3$) is sufficiently rapid. It is seen that the functions $F_{\boldsymbol{n}}$ have the qualitative properties discussed in Sect. 13 for $F_{\boldsymbol{N}}^{(0)}$, and for a fluid, $F_1 \equiv 1$. Indeed, as we shall immediately show, the functions $F_{\boldsymbol{N}}^{(0)}$ are the limit, at zero densities, of the functions $F_{\boldsymbol{n}}$.

The classical equation, (4.21), of the Grand Canonical Ensemble, gives the probability density, $P_{\boldsymbol{N}}(V, T, \boldsymbol{z}, \{\boldsymbol{N}\})$, that in an open system of infinite extent, with specified temperature and activity set $\boldsymbol{z}$, there will be exactly the set $\boldsymbol{N}$ of molecules (no more and no less) in a volume V, and that these will be at the coordinate position $\{\boldsymbol{N}\}$, as,

$$P_{\boldsymbol{N}}(V, T, \boldsymbol{z}, \{\boldsymbol{N}\}) = \mathrm{e}^{-PV/kT}\, \boldsymbol{z}^{\boldsymbol{N}} F_{\boldsymbol{N}}^{(0)}\{\boldsymbol{N}\}. \tag{18.3$_1$}$$

In this, if the system is classical,

$$F_N^{(0)}\{N\} = e^{-U_N\{N\}/kT}. \tag{18.3$_2$}$$

In (18.3$_1$), as in the definition of ϱ_n, the probability density is given for the sum of all $N!$ permutations of the numbered molecules among the set $\{N\}$ of coordinates. This expression must be divided by $N!$ to give the probability density, $P'_N\{N\} = (1/N!)\, P_N\{N\}$, that the numbered molecule 1 be at the first numbered coordinate position, molecule 2 at the second, etc. If there are exactly the set N of molecules in V we can select from the numbered set $\{N\}$ a subset $\{n\}_N$ in $N!/(N-n)!\,n!$ different ways, which may occupy the positions $\{n\}$. This numbered set $\{n\}_N$ can occupy the position $\{n\}$ in $n!$ different ways. The additive contribution to $\varrho_n\{n\}$ due to having exactly the set N of molecules in V is thus $[1/(N-n)!]\, P_N\{N\}$, integrated over all positions of the set $N-n$. Sum this over all number sets $N \geqq n$. Then renumber, to find with (17.3$_1$),

$$\left.\begin{aligned} \varrho_n\{n\} = \boldsymbol{\rho}^n F_n(z, T, \{n\}) &= \sum_{N\geqq 0} \frac{1}{N!} \iint_V \cdots \int P_{n+N}\{\{n\} + \{N\}\}\, d\{N\} \\ &= e^{-PV/kT} z^n \sum_{N\geqq 0} \frac{z^N}{N!} \iint_V \cdots \int F^{(0)}_{n+N}\{\{n\} + \{N\}\}\, d\{N\}. \end{aligned}\right\} \tag{18.4$_1$}$$

In the limit all $z_a \to 0$, $(z/\boldsymbol{\rho}) = 1$, and $e^{-PV/kT} = 1$, we see that,

$$\lim_{\text{all } \varrho_a = z_a \to 0} [F_n(z, T, \{n\})] = F_n^{(0)}(T, \{n\}), \tag{18.4$_2$}$$

since only the single term, $N \equiv 0$, remains in the sum.

The integral of ϱ_n over V is seen to be

$$\left.\begin{aligned} \iint_V \cdots \int \varrho_n\{n\}\, d\{n\} = \boldsymbol{\rho}^n \iint_V \cdots \int F_n(z, T, \{n\})\, d\{n\} \\ = \sum_N \frac{N!}{(N-n)!} \iint_V \cdots \int \frac{1}{N!} P_N\{N\}\, d\{N\} \\ = \sum_N \frac{N!}{(N-n)!} P(N) = \left\langle \frac{N!}{(N-n)!} \right\rangle_{Av}, \end{aligned}\right\} \tag{18.5}$$

since $\iint \cdots \int \frac{1}{N!} P_N\{N\}\, d\{N\} = P(N)$ the probability of set N in V. In a single thermodynamic phase, for which fluctuations are finite,

$$\left\langle \frac{N!}{(N-n)!} \right\rangle_{Av} = \langle N^n \rangle_{Av} \left[1 + O\left(\frac{1}{\overline{N}}\right)\right] = \prod_a \overline{N}_a^{n_a} \left[1 + O\left(\frac{1}{\overline{N}_a}\right)\right].$$

Since $\boldsymbol{\rho}^n V^n = \prod_a \overline{N}_a^{n_a}$ we indeed find the normalization of F_n given by (18.2$_4$) in the limit $V \to \infty$, $\overline{N}_a \to \infty$, for finite sets n.

β) The cluster function for two molecules. In particular, for a single component system, consider the function,

$$g_2(\boldsymbol{r}_\alpha, \boldsymbol{r}_\beta) = F_2(\boldsymbol{r}_\alpha, \boldsymbol{r}_\beta) - F_1(\boldsymbol{r}_\alpha)\, F_1(\boldsymbol{r}_\beta) = F_2(r_{\alpha\beta}) - 1. \tag{18.6}$$

For a fluid g_2 will depend only on the distance, $r_{\alpha\beta}$, between molecule α and β. This function is the first significant member of a set of functions, $g_1 = F_1 = 1$, $g_2, g_3, \ldots$ etc., defined analogously to the $g_n^{(0)}$ of Eqs. (13.21) and (13.22). Since

$F_1 \equiv 1 \int F_1 \, d\boldsymbol{r} = V$, and $\varrho V = \overline{N}$, we have, from (18.5),

$$\left.\begin{aligned} b_2 &= \frac{1}{2!\,V} \iint g_2(\boldsymbol{r}_\alpha, \boldsymbol{r}_\beta)\, d\boldsymbol{r}_\alpha\, d\boldsymbol{r}_\beta \\ &= \frac{1}{2V\varrho^2}\left[\langle N(N-1)\rangle_{Av} - \overline{N}^2\right] \\ &= \frac{1}{2V\varrho^2}\left[\overline{N^2} - \overline{N}^2\right] - \frac{1}{2\varrho}. \end{aligned}\right\} \tag{18.7$_1$}$$

We found in Sect. 1, Eq. (1.42), that $\varkappa = (V\mathrm{k}T\,\varrho^2)^{-1}[\overline{N^2} - \overline{N}^2]$. One therefore has, for the isothermal compressibility, $\varkappa$,

$$\varkappa = \frac{1}{\varrho\,\mathrm{k}T}(1 + 2b_2\varrho). \tag{18.7$_2$}$$

For the perfect gas g_2 differs from zero only at distances $r < r_{\max} \ll \varrho^{-\frac{1}{3}}$, so that $\varrho b_2 \ll 1$. The compressibility of the perfect gas is $(\varrho\,\mathrm{k}T)^{-1}$.

γ) Open versus closed systems. The reduced density probability functions $\varrho_{\boldsymbol{n}}$, and the corresponding distribution functions $F_{\boldsymbol{n}}$, defined in this section are for an open system, corresponding to the Grand Canonical Ensemble, which is an Ensemble of systems, each of definite volume V, but of which each is at equilibrium, having open walls, with an infinite reservoir of fixed activity set $\boldsymbol{z}$ and temperature T. One may, of course, similarly define reduced probability densities for a closed system, corresponding to a Macro Canonical Ensemble of fixed V and number set $\boldsymbol{N}$, in equilibrium with an infinite thermostat of fixed temperature T. As is to be expected the two definitions lead to almost identical functions for large systems at the corresponding conditions. There is, however, a significant difference in the two definitions, which becomes important in certain manipulations[1]. Since this difference has occasionally led to confusion in the literature we shall discuss it here in some detail. The essential feature of the difference, however, arises from the simple fact that in the closed system of fixed number, N, there is no distinction between $\overline{N^2}$ and $\overline{N}^2$, $\overline{N^2} - \overline{N}^2 = N^2 - N^2 = 0$, and the integral b_2 has a fixed value, $2\varrho b_2 = -1$.

For the closed system we can write the expression for $\varrho_{\boldsymbol{n}}$ from that given in terms of P_N in (18.4$_1$) simply by omission of the sum over $\boldsymbol{N}$, setting $P_{\boldsymbol{n}} \equiv 0$ for all $\boldsymbol{N}$ except one, the number set fixed for the system, or for one component,

$$\varrho_n\{n\} = \frac{1}{(N-n)!} \iint_V \cdots \int P\{N\}\, d\{N-n\}. \tag{18.8}$$

Since the probability density is so defined that,

$$\frac{1}{N!} \iint_V \cdots \int P\{N\}\, d\{N\} \equiv 1, \tag{18.8$_1$}$$

we have,

$$\iint_V \cdots \int \varrho_n\{n\} = \frac{N!}{(N-n)!}, \tag{18.8$_2$}$$

with N a fixed number. Proceeding through the steps that led to (18.7$_1$) one finds $2\varrho b_2 = -1$.

---

[1] J. Yvon: Fluctuations en Densité. Paris: Hermann & Cie. 1937.

Now in each case one has

$$\varrho_2(r_{\alpha\beta}) \to \varrho^2 \quad \text{if} \quad r_{\alpha\beta} \gg r_{\max}. \tag{18.8$_3$}$$

However, in the case of the closed system the equation is not an identity, but

$$\varrho_2(r_{\alpha\beta}) - \varrho^2 = O\left(\frac{1}{N}\right), \qquad r_{\alpha\beta} \gg r_{\max}. \tag{18.8$_4$}$$

The function $g_2(r)$, defined by (18.6) thus approaches zero only to order N^{-1} for large values of r, and the integral,

$$b_2 = \tfrac{1}{2}\int 4\pi r^2 g_2(r)\,dr,$$

has an added contribution of order $V/N = \varrho^{-1}$, not present in the case of the open system. Naturally, similar terms occur in the case of higher n values, and Eqs. (18.2$_1$) and (18.2$_2$) have additional terms of order N^{-1} dependent on the coordinate sets $\{\boldsymbol{m}\}$ and $\{\boldsymbol{n}\}$ in closed systems.

The origin of the discrepancy is clear. The quantity $\varrho_2(\boldsymbol{r}_\alpha, \boldsymbol{r}_\beta)[\varrho_1(\boldsymbol{r}_\alpha)]^{-1} d\boldsymbol{r}_\beta$ is the probability that, if a molecule is known to be at $\boldsymbol{r}_\alpha$, there is a molecule at $\boldsymbol{r}_\beta$ in $d\boldsymbol{r}_\beta$. Now, in a closed system the fact that one molecule is known to be at $\boldsymbol{r}_\alpha$ leaves only $N-1$ other molecules, so that the average density at large distances from $\boldsymbol{r}_\alpha$ is no longer exactly $(N/V) = \varrho$. It will also, in general, not be $(N-1)/V$, since the molecule at $\boldsymbol{r}_\alpha$ excludes other molecules from its immediate neighborhood, in addition to attracting some into its first coordination sphere. The effective volume for the $N-1$ other molecules is no longer V but has some undetermined correction. Similarly $\varrho_{n+1}\{\{n\} + \boldsymbol{r}_\beta\}[\varrho_n\{n\}]^{-1} d\boldsymbol{r}_\beta$ is the probability that a molecule be at $\boldsymbol{r}_\beta$ in $d\boldsymbol{r}_\beta$ when n are known to be at the positions $\{n\}$. It is clear that at large distances, $\boldsymbol{r}_\beta$, far from any of the coordinates $\{n\}$, the correction will depend on the coordinate set $\{n\}$, since the effective volume left free for the other $N-n$ molecules depends on $\{n\}$.

This effect does not exist in the open system under the normal situation of a single thermodynamic phase. The system is immersed in an infinite bath, and molecules are free to move through the walls into and out of the system. The average density at $\boldsymbol{r}_\beta$ in the cases discussed above is independent of whether or not one or many molecules are known to be at fixed positions far from $\boldsymbol{r}_\beta$, since the normal flow through the open walls adjust the average density at $\boldsymbol{r}_\beta$ to normal values, $\varrho = \overline{N}/V$.

δ) Potentials of average force. The distribution functions, F_n, may be written, in analogy with $F_n^{(0)}$, as,

$$F_{\boldsymbol{n}}(\boldsymbol{z}, T, \{\boldsymbol{n}\}) = e^{-W_{\boldsymbol{n}}\{\boldsymbol{n}\}/kT}, \tag{18.9}$$

which defines the functions $W_{\boldsymbol{n}}\{\boldsymbol{n}\}$ of dimension energy. These functions depend on the temperature and activity set, $\boldsymbol{z}$, or density set $\boldsymbol{\rho}$ in the system. In view of (18.4$_2$) that $F_{\boldsymbol{n}} \to F_{\boldsymbol{n}}^{(0)}$ for zero density, and Eq. (18.3$_2$) for $F_{\boldsymbol{n}}^{(0)}$, we have that,

$$\lim_{\text{all } \varrho_a = z_a \to 0} [W_{\boldsymbol{n}}(\boldsymbol{z}, T, \{\boldsymbol{n}\})] = U_{\boldsymbol{n}}\{\boldsymbol{n}\}, \tag{18.10}$$

for a classical system. We shall now show that the functions $W_{\boldsymbol{n}}$ have the simple interpretation in a classical system of being potentials of average force.

We mean by this the following. Define

$$\bar{f}_{x\alpha}(\boldsymbol{z}, T, \{\boldsymbol{n}\}) = -\frac{\partial}{\partial x_\alpha} W_{\boldsymbol{n}}(\boldsymbol{z}, T, \{\boldsymbol{n}\}), \tag{18.11$_1$}$$

where x_α is any coordinate of the α-th molecule, which is one of the set $\boldsymbol{n}$. The quantity $\bar{f}_{x\alpha}$ is the average force acting on the α-th molecule along the coordinate x_α when the set $\boldsymbol{n}$ of molecules is at the position $\{\boldsymbol{n}\}$.

This average force will be due to two parts, a direct constant force,

$$f_{x\alpha}^{(0)}\{\boldsymbol{n}\} = -\frac{\partial}{\partial x_\alpha} U_{\boldsymbol{n}}\{\boldsymbol{n}\}, \tag{18.11$_2$}$$

due to the set $\boldsymbol{n}$ of molecules fixed at $\{\boldsymbol{n}\}$, plus the average value of a fluctuating force exerted by the other molecules not in the set $\boldsymbol{n}$. If the molecule α is isolated in the set $\boldsymbol{n}$, that is if the coordinates of the other $n-1$ molecules are all very distant from $\boldsymbol{r}_\alpha$, then this average force will be zero, since the others not of the set $\boldsymbol{n}$ will surround $\boldsymbol{r}_\alpha$ isotropically in a fluid. However if some or all of the $\boldsymbol{n}-1$ other molecules are close to $\boldsymbol{r}_\alpha$ they will influence the average positions of the molecules of the rest of the gas, and they will no longer surround $\boldsymbol{r}_\alpha$ isotropically. This average fluctuating force will be proportional to the number density, ϱ, and will vanish at $\varrho \to 0$. This leads to Eq. (18.10).

The demonstration of (18.11$_1$) is simple. Turn to Eq. (18.4$_1$) for $F_{\boldsymbol{n}}$, with (18.3$_2$) for $F_{\boldsymbol{n}}^{(0)}$. Differentiate both sides with respect to x_α which is one of the coordinates of $\{\boldsymbol{n}\}$. From (18.9) for $F_{\boldsymbol{n}}\{\boldsymbol{n}\}$ one obtains the factor $-\frac{1}{\mathbf{k}T}\frac{\partial}{\partial x_\alpha}W_{\boldsymbol{n}}$, which from (18.11$_1$) is $(1/\mathbf{k}T)\,\bar{f}_{x\alpha}\{\boldsymbol{n}\}$, multiplying $\varrho_{\boldsymbol{n}}$ on the left hand side. On the right hand side of (18.4$_1$) the differentiation operates only on $U_{\boldsymbol{n}+\boldsymbol{N}}$, and can be performed under the summation and integration. The original integrand is then multiplied by

$$-\frac{1}{\mathbf{k}T}\frac{\partial}{\partial x_\alpha} U_{\boldsymbol{n}+\boldsymbol{N}}\{\{\boldsymbol{n}\}+\{\boldsymbol{N}\}\} = \frac{1}{\mathbf{k}T} f_{x\alpha}^{(0)}\{\{\boldsymbol{n}\}+\{\boldsymbol{N}\}\}. \tag{18.12}$$

Cancel the terms $\mathbf{k}T$, write the original integrand as $P_{\boldsymbol{n}+\boldsymbol{N}}$. One finds,

$$\bar{f}_{x\alpha}\{\boldsymbol{n}\} = \frac{\sum\limits_{\boldsymbol{N}\geq 0}\frac{1}{\boldsymbol{N}!}\iint\limits_V\cdots\int f_{\alpha x}^{(0)}\{\{\boldsymbol{n}\}+\{\boldsymbol{N}\}\}\,P_{\boldsymbol{n}+\boldsymbol{N}}\{\{\boldsymbol{n}\}+\{\boldsymbol{N}\}\}\,d\{\boldsymbol{N}\}}{\varrho_{\boldsymbol{n}}\{\boldsymbol{n}\}}. \tag{18.13}$$

This is, however, just the average force acting along x_α: namely the total force $f_{x\alpha}^{(0)}$ multiplied by the probability $P_{\boldsymbol{n}+\boldsymbol{N}}/\varrho_{\boldsymbol{n}}$ that if molecules are at $\{\boldsymbol{n}\}$ there are the set $\boldsymbol{N}$ of other molecules at $\{\boldsymbol{N}\}$, integrated over all positions and summed over $\boldsymbol{N}$.

ε) The Kirkwood closure. The total potential of average force for the set $\boldsymbol{n}$ of molecules, $W_{\boldsymbol{n}}$, can always be decomposed into a sum of terms due to potentials depending on single coordinates, $w_1(\boldsymbol{r}_\alpha)$, terms depending on pairs alone, $w_2(\boldsymbol{r}_\alpha, \boldsymbol{r}_\beta)$, terms due to triples, $w_3(\boldsymbol{r}_\alpha, \boldsymbol{r}_\beta, \boldsymbol{r}_\gamma)$, ... etc. For the fluid we always have $w_1 \equiv 0$. The pair terms will, in general, depend on the species a and b of α and β, as will the triple terms, but be of the same analytic form for different molecules of the same species set. The equation can be written as,

$$W_{\boldsymbol{n}}\{\boldsymbol{n}\} = \sum_{\{\boldsymbol{\nu}\}\boldsymbol{n}} w_\nu\{\boldsymbol{\nu}\}_{\boldsymbol{n}}, \tag{18.14$_1$}$$

where the sum runs over all subsets $\{\boldsymbol{\nu}\}_{\boldsymbol{n}}$ of the set $\{\boldsymbol{n}\}$ of coordinates, from $\nu=1$, (for which $w_1\equiv 0$) to $\{\boldsymbol{\nu}\}_{\boldsymbol{n}} \equiv \{\boldsymbol{n}\}$, the improper subset of all. The inversion of (18.14$_1$) is,

$$w_{\boldsymbol{n}}\{\boldsymbol{n}\} = \sum_{\{\boldsymbol{\nu}\}\boldsymbol{n}} (-)^{n-\nu}\, W_\nu\{\boldsymbol{\nu}\}_{\boldsymbol{n}}. \tag{18.14$_2$}$$

Since the overall potentials, W_{n+m} become

$$\lim_{r_{nm}\to\infty}\left[W_{n+m}\{\{n\}+\{m\}\}\right] = W_n\{n\} + W_m\{m\}, \tag{18.14$_3$}$$

the component potentials, $w_\nu\{\nu\}$ necessarily approach zero in value if the set $\{\nu\}$ breaks into two distant subsets,

$$\lim_{r_{\nu\mu}\to\infty}\left[w_{\nu+\mu}\{\{\nu\}+\{\mu\}\}\right] = 0. \tag{18.14$_4$}$$

Since it is customary to assume that for saturated molecules the similar decomposition of U_n, Eq. (4.2) contains pair terms only,

$$u_\nu \equiv 0, \qquad \nu \neq 2, \tag{18.14$_5$}$$

and since $w_\nu \to u_\nu$ at low densities, it is natural to assume the same for w_ν, namely that w_ν can be neglected except for $\nu = 2$.

Were this assumption correct the hierarchy of functions F_n would be completely determined by the functions for pairs, F_{ab}, of all different pair specie combinations, or for the one-component system by the one-pair distribution function, F_2, namely,

$$F_n\{n\} = \prod_{n\geq\alpha>\beta\geq 1}\prod F_2(r_{\alpha\beta}). \tag{18.15}$$

The closure of (18.15) was first suggested by KIRKWOOD[1], and called by him superposition. It is undoubtedly poor for very large n-values, but may be quite good for $n=3$ in particular[2].

For a one-component system the pair distribution function, $F_2(r)$, is related to the experimentally measurable scattering of X-rays or neutrons, of wavelength λ, if λ is comparable with the significant distances $\varrho^{-\frac{1}{3}}$ in the system.

The three-dimensional Fourier transform, $\gamma_2(t)$ of the cluster function, $g_2 = F_2 - 1$, defined as

$$\gamma_2(t) = \int_0^\infty 4\pi r^2 g_2(r)\frac{\sin 2\pi t r}{2\pi t r}\,dr, \tag{18.16$_1$}$$

$$t\gamma_2(t) = 2\int_0^\infty r g_2(r)\sin 2\pi t r\,dr, \tag{18.16$_2$}$$

$$g_2(r) = \int_0^\infty 4\pi t^2\gamma_2(t)\frac{\sin 2\pi t r}{2\pi t r}\,dt, \tag{18.16$_3$}$$

$$r g_2(r) = 2\int_0^\infty t\gamma_2(t)\sin 2\pi t r\,dt, \tag{18.16$_4$}$$

is related to the scattering intensity, I at an angle ϑ related to t by,

$$2\pi t = s = \frac{4\pi}{\lambda}\sin\frac{\vartheta}{2}, \tag{18.17$_1$}$$

$$I(s) \approx \varrho + \varrho\gamma_2\left(\frac{s}{2\pi}\right). \tag{18.17$_2$}$$

Actually for X-rays the scattering intensity $I(s)$ must first be corrected by the form factors for the isolated atoms (or molecules), $F(s)$, so that if $F(s)$ is the

[1] J. G. KIRKWOOD: J. Chem. Phys. 3, 300 (1935).

[2] See, however, the dissenting opinion of B. R. A. NIJBOER and L. VAN HOVE: Phys. Rev. 85, 777 (1952).

intensity of scattering at s from the incident beam due to a single atom

$$I_{\exp}(s) = c F^2(s)\left[\varrho + \varrho\,\gamma_2\left(\frac{s}{2\pi}\right)\right], \tag{18.17$_3$}$$

with c a proportionality constant proportional to the volume of the scattering sample.

Thus $g_2(r)$ and $F_2(r) = g_2(r) + 1$ may be determined experimentally. No method for experimentally determining the distribution functions for three or more molecules has been found.

Since only $F_2(r)$ can at present be measured there is no direct experimental method of evaluating the validity of (18.15), even for F_3. Theoretical investigations succeed only in showing that it cannot be exact, even if, or perhaps particularly if, the corresponding equation is exact, for $F_3^{(0)}$, that is, if U_3 is an exact sum of pair potentials. The magnitude of the deviation is difficult to evaluate or interpret. There exists one heuristic justification. The integral equation method, discussed in section 20, requires a closure method to express F_3 in terms of F_2. The only method used, so far, by KIRKWOOD and co-workers, has been that or (18.15) and the pair distribution functions for liquids, and for hard sphere gases at high densities obtained by this method appear to be quite excellent.

ζ) Equations involving F_2. In the case of a potential, U_N, that is strictly a sum of pair terms, the function F_2 (with ϱ) determines both the average potential energy of the system and the pressure. We have already seen [Eq. (18.7$_2$)] that it also determines the compressibility.

The relation of F_2 or ϱ_2 to the potential energy is almost self evident. Since the total potential function is a sum of pair terms the total potential energy will be the sum over all pairs of molecules of the integral of the pair interaction potential times the probability density that the two molecules of the pair will have the distance r. The total number of pairs is $\frac{1}{2}N(N-1)$. The correctly normalized probability density is seen from (18.5$_1$) to be $\frac{1}{2}\varrho_2 = \frac{1}{2}\varrho^2 F_2$. The pair interaction potential $u_2(r)$ is short range so that the integration over the first $d\boldsymbol{r}$ will give a term independent of volume, and that over the second will give V as a factor. Since the kinetic energy for the classical system is a function only of T, and the potential is zero in the perfect gas, this is the difference $E - E_0$ of the energy of the system, E, and that, E_0 of the system of the same composition and temperature in the perfect gas limit. One finds

$$\frac{E - E_0}{\overline{N}} = \frac{1}{2}\varrho \int_0^\infty 4\pi r^2 u(r) F_2(r)\, dr, \tag{18.18$_1$}$$

or for more components,

$$\frac{E - E_0}{\overline{N}} = \frac{1}{2}\varrho \sum_a \sum_b x_a x_b \int_0^\infty 4\pi r^2 u_{ab}(r) F_{ab}(r)\, dr, \tag{18.18$_2$}$$

where x_a is the mol fraction, $x_a = \varrho_a/\varrho$, of species a, and u_{ab}, F_{ab} are the pair potentials and pair distribution functions of molecules of type a and b respectively.

These equations can also be derived readily from Eq. (4.17$_2$) by differentiation of the expression (4.21) with respect to T at constant z, using (18.4$_1$), for ϱ_2.

The expression for the pressure is more interesting. Write (4.21) with (4.22) as

$$e^{PV/kT} = \sum_{\boldsymbol{N}} \frac{\boldsymbol{z}^{\boldsymbol{N}}}{\boldsymbol{N}!} \int\!\!\int_V \!\cdots\! \int e^{-U_{\boldsymbol{N}}\{\boldsymbol{N}\}/kT}\, d\{\boldsymbol{N}\}. \tag{18.19$_1$}$$

This equation expresses P as a function of T and z, and P is known to be independent of V. We make a transformation to dimensionless coordinates, $\boldsymbol{R}_\alpha = \boldsymbol{r}_\alpha / V^{\frac{1}{3}}$, so that the limits of integration are in a dimensionless box of unit volume. Since,

$$d\{\boldsymbol{N}\} = \prod_{\alpha=1}^{N} d\boldsymbol{r}_\alpha = V^N \prod_{\alpha=1}^{N} d\boldsymbol{R}_\alpha, \tag{18.19$_2$}$$

we have,

$$e^{PV/kT} = \sum_{\boldsymbol{N}} \frac{\boldsymbol{z}^{\boldsymbol{N}} V^N}{\boldsymbol{N}!} \int\int_1 \cdots \int e^{-U_{\boldsymbol{N}}(V^{\frac{1}{3}}\boldsymbol{R})} \, d\boldsymbol{R}^{(N)}. \tag{18.19$_3$}$$

Now differentiate both sides with respect to V, and multiply by $V e^{-PV/kT}$. On the left one gets PV/kT, since $(\partial P/\partial V)_{z,T} = 0$. On the right the logarithmic differentiation of V^N gives a factor N in each term. Since the integral of each term multiplied by $e^{-PV/kT}$ is $P_{\boldsymbol{N}}$, the probability of the set $\boldsymbol{N}$, the sum of all these terms is $\bar{N}$. The other occurrence of V is in U_N, which depends on the $\boldsymbol{r}_\alpha = V_{\frac{1}{3}} \boldsymbol{R}_\alpha$. The differentiation with respect to log V multiplies the integrand by

$$\left.\begin{aligned} &-\frac{1}{kT}\,\frac{1}{3}\sum_\alpha R_\alpha \cdot \nabla_\alpha U_{\boldsymbol{N}}(V^{\frac{1}{3}}\boldsymbol{R}_\alpha) \\ &= -\frac{1}{3\,kT}\sum_\alpha \boldsymbol{r}_\alpha \cdot \nabla_\alpha U_{\boldsymbol{N}}(\boldsymbol{r}_\alpha). \end{aligned}\right\} \tag{18.19$_4$}$$

Again multiplication by $e^{-PV/kT}$ converts the remainder of the term into $P_{\boldsymbol{N}}\{\boldsymbol{N}\}$, the probability density of the set $\boldsymbol{N}$ at the position $\{\boldsymbol{N}\}$. Integration and summation means that we evaluate the average of (18.19$_4$). We have

$$PV = \bar{N} kT - \tfrac{1}{3}\Big\langle \sum_\alpha \boldsymbol{r}_\alpha \cdot \nabla_\alpha U_{\boldsymbol{N}}\{\boldsymbol{N}\} \Big\rangle_{Av}. \tag{18.19$_5$}$$

The first term, $\bar{N}kT$, is the perfect gas expression, due to the momentum transfer to the walls, the second term is the famous virial of Clausius.

Now use the expression for $U_{\boldsymbol{N}}$ as a sum of pair terms, $u_{\alpha\beta}(r_{\alpha\beta})$. The operator $\sum \boldsymbol{r}_\alpha \cdot \nabla_\alpha$ acts on each of the $\frac{1}{2}N(N-1)$ terms twice, once with $\boldsymbol{r}_\alpha \cdot \nabla_\alpha$ and once with $\boldsymbol{r}_\beta \cdot \nabla_\beta$. Since $r_{\alpha\beta} = |\boldsymbol{r}_{\alpha\beta}| = |\boldsymbol{r}_\alpha - \boldsymbol{r}_\beta|$ we have,

$$x_\alpha \frac{\partial}{\partial x_\alpha} + x_\beta \frac{\partial}{\partial x_\beta} = x_\alpha - x_\beta \frac{\partial}{\partial x_{\alpha\beta}} = \frac{(x_\alpha - x_\beta)^2}{r_{\alpha\beta}} \frac{d}{dr_{\alpha\beta}}$$

and summing the three components,

$$\frac{1}{3}\Big\langle \sum_\alpha \boldsymbol{r}_\alpha \cdot \nabla_\alpha U_{\boldsymbol{N}} \Big\rangle_{Av} = \frac{1}{3} \sum\sum_{N \geq \alpha > \beta \geq 1} \Big\langle r_{\alpha\beta} \frac{d u_{\alpha\beta}(r_{\alpha\beta})}{d r_{\alpha\beta}} \Big\rangle_{Av}. \tag{18.19$_6$}$$

The sum of averages of the pair positions leads to,

$$P = \varrho\, kT - \frac{\varrho^2}{6} \sum_a \sum_b x_a x_b \int_0^\infty 4\pi r^2 \left(r \frac{d u_{ab}(r)}{dr} \right) F_{ab}(r)\, dr. \tag{18.20}$$

η) Summary. The "structure" of a dense gas may be described by a hierarchy of probability density functions $\varrho_{\boldsymbol{n}}$ giving the probability density that position sets $\{\boldsymbol{n}\}$ in the gas are occupied. For any fluid the functions for single coordinates are trivial constants, the average densities ϱ_a of the species a. The pair functions, $\varrho_{aa}, \varrho_{ab}, \varrho_{bb}$, etc., are functions only of the distance between the two molecules. The functions for three or more molecules are not expreimentally available, nor easily computable, but are often assumed to be proportional to the products of the pair functions properly normalized,

$$\varrho_{abc}(\boldsymbol{r}_a, \boldsymbol{r}_b, \boldsymbol{r}_c) = \frac{\varrho_{ab}(r_{ab})\,\varrho_{bc}(r_{bc})\,\varrho_{ca}(r_{ca})}{\varrho_a\,\varrho_b\,\varrho_c}. \tag{18.21}$$

For many purposes it is easier to use the normalized distribution functions, $F_{\boldsymbol{n}}$,

$$\varrho_{\boldsymbol{n}}\{\boldsymbol{n}\} = \boldsymbol{\rho}^{\boldsymbol{n}} F_{\boldsymbol{n}}\{\boldsymbol{n}\}, \tag{18.22}$$

which approach unity in value when all distances are large. The function,

$$W_{\boldsymbol{n}}\{\boldsymbol{n}\} = -\,\mathrm{k}T \log F_{\boldsymbol{n}}\{\boldsymbol{n}\}, \tag{18.23}$$

is a potential of the average force between the molecules of the set $\boldsymbol{n}$ in the fixed position $\{\boldsymbol{n}\}$, for a classical system.

The pair distribution function, F_2, for a one component system is related to the average potential energy per molecule, by

$$\frac{E - E_0}{\overline{N}} = \frac{1}{2}\,\varrho \int\limits_0^\infty 4\pi r^2 u(r)\, F_2(r)\, dr, \tag{18.24}$$

to the pressure by,

$$P = \varrho\,\mathrm{k}T - \frac{\varrho^2}{6}\int\limits_0^\infty 4\pi r^2 \left(r\,\frac{du(r)}{dr}\right) F_2(r)\, dr, \tag{18.25}$$

and to the isothermal compressibility by,

$$\varkappa = \frac{1}{\varrho}\left(\frac{\partial\varrho}{\partial P}\right)_r = \frac{1}{\varrho\,\mathrm{k}T}\left[1 + \varrho\int\limits_0^\infty 4\pi r^2\,[F_2(r) - 1]\, dr\right]. \tag{18.26}$$

19. Transformation to differences of activity. The set of Eqs. (18.4_1),

$$\left.\begin{aligned} e^{PV/\mathrm{k}T}\left(\frac{\boldsymbol{\rho}}{\boldsymbol{z}}\right)^{\boldsymbol{n}} F_{\boldsymbol{n}}(\boldsymbol{z}, T, \{\boldsymbol{n}\}) &= G_{\boldsymbol{n}}(\boldsymbol{z}, T, \{\boldsymbol{n}\}) \\ &= \sum_{\boldsymbol{N}\geq 0}\frac{\boldsymbol{z}^{\boldsymbol{N}}}{\boldsymbol{N}!}\iint\limits_V\cdots\int F^{(0)}_{\boldsymbol{n}+\boldsymbol{N}}\{\{\boldsymbol{n}\} + \{\boldsymbol{N}\}\}\, d\{\boldsymbol{N}\}, \end{aligned}\right\} \tag{19.1}$$

can be solved readily for $F^{(0)}_{\boldsymbol{n}}$ in terms of the $G_{\boldsymbol{n}}$'s. We have seen [Eq. (18.4_2)] that $F_{\boldsymbol{n}}(\boldsymbol{z}\to 0)$ approaches $F^{(0)}_{\boldsymbol{n}}$ in value. Since $\frac{\boldsymbol{\rho}}{\boldsymbol{z}}\to 1$ as all $\varrho_a\to 0$, and $e^{PV/\mathrm{k}T}$ approaches unity in value as all $\varrho_a\to 0$, $\overline{N}\to 0$, the function $F^{(0)}_{\boldsymbol{N}}$ can be written as

$$F^{(0)}_{\boldsymbol{N}} = G_{\boldsymbol{N}}(\boldsymbol{z}\to 0, T, \{\boldsymbol{N}\}). \tag{19.2}$$

The structure of the set (19.1) is similar to that of a Taylor expansion, and the inverse[1],

$$\left.\begin{aligned} F^{(0)}_{\boldsymbol{n}}\{\boldsymbol{n}\} &= G_{\boldsymbol{n}}(\boldsymbol{z}\to 0, T, \{\boldsymbol{n}\}) \\ &= \sum_{\boldsymbol{N}\geq 0}\frac{(-\boldsymbol{z})^{\boldsymbol{N}}}{\boldsymbol{N}!}\iint\limits_V\cdots\int G_{\boldsymbol{n}+\boldsymbol{N}}(\boldsymbol{z}, T, \{\{\boldsymbol{n}\} + \{\boldsymbol{N}\}\}\, d\{\boldsymbol{N}\}, \end{aligned}\right\} \tag{19.3}$$

can readily be proved correct by inserting (19.1) for the integrands of (19.3). One then has a double summation over sets, say $\boldsymbol{N}_1$ and $\boldsymbol{N}_2$ of $F^{(0)}_{\boldsymbol{n}+\boldsymbol{N}_1+\boldsymbol{N}_2}$ to be integrated over $d\{\boldsymbol{N}\}_1 d\{\boldsymbol{N}_2\}$. Use the symbol $\boldsymbol{N} = \boldsymbol{N}_1 + \boldsymbol{N}_2$ and integrate over

[1] J. E. Mayer: J. Chem. Phys. **10**, 629 (1942).

$d\{\boldsymbol{N}\}$, summing over $\boldsymbol{N}$ and $\boldsymbol{N}_1$, the latter sum running from zero to $\boldsymbol{N}$. The sum over $\boldsymbol{N}_1$ is simply

$$\sum_{\boldsymbol{N}_1 \geq 0}^{\boldsymbol{N}} \frac{(-\boldsymbol{z})^{\boldsymbol{N}_1} (\boldsymbol{z})^{\boldsymbol{N}-\boldsymbol{N}_1}}{\boldsymbol{N}_1!\,(\boldsymbol{N}-\boldsymbol{N}_1)!} = \prod_a \left[\frac{z_a^{N_a}}{N_a!} \sum_{N_{1a}=0}^{N_{ja}=N_a} \frac{(-)^{N_{1a}} N_a!}{(N_a - N_{1a})!\,N_{1a}!}\right] = \prod_a \delta(N_a),$$

namely zero unless $\boldsymbol{N}=0$, and then unity. Since the series are absolutely convergent for all finite V the identity is established.

One may now, of course, use (19.3) for any activity set $\boldsymbol{z}$ to eliminate $F^{(0)}_{\boldsymbol{n}+\boldsymbol{N}}$ from (19.1) witten for a different activity set, say $\boldsymbol{z}+\boldsymbol{y}$, $=z_a+y_a$, $z_b+y_b, \ldots$. Again one obtains a double sum over $\boldsymbol{N}_1$ and $\boldsymbol{N}_2$, and by setting $\boldsymbol{N}_1+\boldsymbol{N}_2=\boldsymbol{N}$ may sum over $\boldsymbol{N}$ and $0 \leq \boldsymbol{N}_1 \leq \boldsymbol{N}$. The latter sum involves powers, $(\boldsymbol{z}+\boldsymbol{y})^{\boldsymbol{N}-\boldsymbol{N}_1}(-z)^{\boldsymbol{N}_1}$ divided by the factorials $\boldsymbol{N}_1!\,(\boldsymbol{N}-\boldsymbol{N}_1)!$ The summation is $\boldsymbol{y}^{\boldsymbol{N}}/\boldsymbol{N}!$. The result,

$$G_{\boldsymbol{n}}(\boldsymbol{z}+\boldsymbol{y}, T, \{\boldsymbol{n}\} = \sum_{\boldsymbol{N}\geq 0} \frac{\boldsymbol{y}^{\boldsymbol{N}}}{\boldsymbol{N}!} \iint \cdots \int G_{\boldsymbol{n}+\boldsymbol{N}}(\boldsymbol{z}, T, \{\{\boldsymbol{n}\}+\{\boldsymbol{N}\}\})\, d\{\boldsymbol{N}\}, \qquad (19.4)$$

then includes all previous equations as special cases, namely if $\boldsymbol{z}\equiv 0$ it is (19.1) and if $\boldsymbol{y}=-\boldsymbol{z}$ it is (19.3). The equations permit one to compute distribution unctions at any activity from those at any other.

The special case $\boldsymbol{n}\equiv 0$, $G_0=\mathrm{e}^{PV/kT}$ gives,

$$\mathrm{e}^{[P(\boldsymbol{y}+\boldsymbol{z})-P(\boldsymbol{z})]\frac{V}{kT}} = \sum \frac{1}{\boldsymbol{N}!}\left[\boldsymbol{y}\frac{\boldsymbol{\rho}(\boldsymbol{z})}{\boldsymbol{z}}\right]^{\boldsymbol{N}} \iint_V \cdots \int F_{\boldsymbol{n}}(\boldsymbol{z}, T, \{\boldsymbol{N}\})\, d\{\boldsymbol{N}\}, \qquad (19.5_1)$$

which, of course is identical to (4.21) when $\boldsymbol{z}=0$, $[\varrho_a/z_a]=1$, and $P(\boldsymbol{z})=0$.

Writing

$$\eta_a = y_a \frac{\varrho_a(\boldsymbol{z})}{z_a}, \qquad (19.5_2)$$

$$\boldsymbol{\eta} = \eta_a, \eta_b, \ldots, \qquad (19.5_3)$$

as before, the equation is,

$$\mathrm{e}^{[P(\boldsymbol{y}+\boldsymbol{z})-P(\boldsymbol{z})]\frac{V}{kT}} = \sum_{\boldsymbol{N}\geq 0} \frac{\boldsymbol{\eta}^{\boldsymbol{N}}}{\boldsymbol{N}!} \iint_V \cdots \int F_{\boldsymbol{N}}(\boldsymbol{z}, T, \{\boldsymbol{N}\})\, d\{\boldsymbol{N}\}, \qquad (19.5_4)$$

which is identical in form to (4.21) with $P(\boldsymbol{y}+\boldsymbol{z})-P(\boldsymbol{z})$ replacing P, and $\boldsymbol{\eta}$ replacing $\boldsymbol{z}$. The functions $F_{\boldsymbol{N}}$ have all the properties of $F^{(0)}_{\boldsymbol{N}}$ used in deriving the series developments of the previous sections.

One defines cluster functions, $g_{\boldsymbol{n}}(\boldsymbol{z}, T, \{\boldsymbol{n}\})$, by the same Eqs. (13.20) to (13.23) as those defining $g^{(0)}_{\boldsymbol{n}}$, using $F_{\boldsymbol{n}}$ to replace $F^{(0)}_{\boldsymbol{n}}$. The cluster integrals, $b_{\boldsymbol{n}}(z)$, of $(1/V\boldsymbol{n}!)$ times the integration of $g_{\boldsymbol{n}}$ over all coordinates, $d\{\boldsymbol{n}\}$, in V, are defined, as were the $b^{(0)}_{\boldsymbol{n}}$-s previously by (13.12$_1$) and (13.13).

The expression, (13.29$_2$), previously derived for the pressure, now becomes

$$P(\boldsymbol{z}+\boldsymbol{y}) = P(\boldsymbol{z}) + kT \sum_{\boldsymbol{n}\geq 1} b_{\boldsymbol{n}}(\boldsymbol{z})\, \boldsymbol{\eta}^{\boldsymbol{n}}. \qquad (19.6)$$

This may be handled in various ways. The expression is valid for any phase, including a condensed phase, providing the series converges, or more strictly provided the function represented by the series on the right has no singularity on the real axis between $\boldsymbol{\eta}=0$, $\boldsymbol{y}+\boldsymbol{z}=\boldsymbol{z}$, and $\boldsymbol{y}$, $\eta_a=y_a\varrho_a(\boldsymbol{z})/z_a$. This will, then be true if there is no phase change between $\boldsymbol{y}+\boldsymbol{z}$ and $\boldsymbol{z}$. For a gas we can assume this to be the case.

If, in (19.6) one sets $\boldsymbol{y} = -\boldsymbol{z}$, so that $\eta_a = (-z_a \varrho_a(\boldsymbol{z})/z_a) = -\varrho_a(\boldsymbol{z})$, and $P(\boldsymbol{y}+\boldsymbol{z}) = 0$, one has,

$$P(\boldsymbol{\rho}) = \mathrm{k}T \sum_{\boldsymbol{n}\geq 1} (-)^{\boldsymbol{n}-1} b_{\boldsymbol{n}}(\boldsymbol{z})\, \boldsymbol{\rho}^{\boldsymbol{n}}. \tag{19.7}$$

As before, one may also introduce the irreducible integrals, $B_{\boldsymbol{m}}$, evaluated at $\boldsymbol{\rho}$ or $\boldsymbol{z}$. With these, the quantity $(1/\mathrm{k}T)[P(\boldsymbol{y}+\boldsymbol{z}) - P(\boldsymbol{z})]$ is expressed as a power series in the quantities,

$$\zeta_a = \left(\frac{\partial}{\partial \log \eta_a}\right)_{T,\boldsymbol{z},\eta_b,\ldots} \left[\frac{P(\boldsymbol{y}+\boldsymbol{z}) - P(\boldsymbol{z})}{\mathrm{k}T}\right], \tag{19.8$_1$}$$

$$\boldsymbol{\zeta} = \zeta_a, \zeta_b, \ldots. \tag{19.8$_2$}$$

From (19.5$_2$) $(\partial/\partial \log \eta_a)_{T,\boldsymbol{z},\eta_b}\ldots = (\partial/\partial \log y_a)_{T,\boldsymbol{z},y_b,\ldots}$ and,

$$\zeta_a = y\left(\frac{\partial}{\partial y_a}\right)_{T,\boldsymbol{z},y_b} \left[\frac{P(\boldsymbol{y}+\boldsymbol{z}) - P(\boldsymbol{z})}{\mathrm{k}T}\right] = \frac{y_a}{z_a + y_a}\, \varrho_a(\boldsymbol{y}+\boldsymbol{z}). \tag{19.8$_3$}$$

The expression, (15.30), is then,

$$P(\boldsymbol{z}+\boldsymbol{y}) = P(\boldsymbol{z}) + \mathrm{k}T\left[\zeta - \sum_{\boldsymbol{m}\geq 2} (m-1)\, B_{\boldsymbol{m}}(\boldsymbol{z})\, \zeta^{\boldsymbol{m}}\right]. \tag{19.9}$$

Again set $\boldsymbol{y} = -\boldsymbol{z}$ so that $P(\boldsymbol{z}+\boldsymbol{y}) = 0$. The ratio $\varrho_a(\boldsymbol{y}+\boldsymbol{z})/(z_a + y_a)$ is now unity as all activities go to zero. One has,

$$P(\boldsymbol{z}) = \sum_a z_a + \sum_{\boldsymbol{m}\geq 2} (-)^m (m-1)\, B_{\boldsymbol{m}}(\boldsymbol{z})\, \boldsymbol{z}^{\boldsymbol{m}}. \tag{19.10}$$

We see that in terms of the integrals evaluated at $\boldsymbol{z}$ the pressure is determined by a series in $-\varrho_a$ with the cluster integrals, $b_{\boldsymbol{n}}(\boldsymbol{\rho})$, as coefficients, and as a series in $-z_a$ with $B_{\boldsymbol{m}}(\boldsymbol{z})$ as coefficients. Of course, Eqs. (19.7) and (19.10) are not true series developments, in that the coefficients depend on $\boldsymbol{z}$ or $\boldsymbol{\rho}$.

Eq. (19.6) may readily be manipulated to obtain relations for the cluster integrals, $b_{\boldsymbol{n}}(\boldsymbol{z})$, in terms of thermodynamic derivatives or in terms of the cluster integrals $b_{\boldsymbol{n}}^{(0)}$, evaluated at $\boldsymbol{z} \equiv 0$. Use the operator,

$$\left.\begin{aligned} \boldsymbol{O}_{\boldsymbol{n}} &= \frac{1}{\mathrm{k}T}\left[\frac{\boldsymbol{\rho}(\boldsymbol{z})}{\boldsymbol{z}}\right]^{\boldsymbol{n}} \frac{1}{\boldsymbol{n}!}\left(\frac{\partial}{\partial \boldsymbol{\eta}}\right)^{\boldsymbol{n}}_{T,\boldsymbol{z}} \\ &= \frac{1}{\mathrm{k}T} \prod_a \left[\frac{1}{n_a!}\left(\frac{\varrho_a(\boldsymbol{z})}{z_a}\,\frac{\partial}{\partial \eta_a}\right)^{n_a}_{T,\boldsymbol{z},\eta_b\cdots},\right.\end{aligned}\right\} \tag{19.11$_1$}$$

which is

$$\boldsymbol{O}_{\boldsymbol{n}} = \frac{1}{\mathrm{k}T}\,\frac{1}{\boldsymbol{n}!}\left(\frac{\partial}{\partial \boldsymbol{y}}\right)^{\boldsymbol{n}} = \frac{1}{\mathrm{k}T} \prod_a \left[\frac{1}{n_a!}\left(\frac{\partial}{\partial y_a}\right)^{n_a}_{T,\boldsymbol{z},y_b,\cdots}\right]. \tag{19.11$_2$}$$

Apply $\boldsymbol{O}_{\boldsymbol{n}}$ to (19.6) to find,

$$\boldsymbol{O}_{\boldsymbol{n}} P(\boldsymbol{y}+\boldsymbol{z}) = \left[\frac{\boldsymbol{\rho}(\boldsymbol{z})}{\boldsymbol{z}}\right]^{\boldsymbol{n}} \sum_{\boldsymbol{m}\geq 0} \frac{(\boldsymbol{n}+\boldsymbol{m})!}{\boldsymbol{n}!}\, b_{\boldsymbol{n}+\boldsymbol{m}}(\boldsymbol{z})\, \boldsymbol{\eta}^{\boldsymbol{m}}. \tag{19.12$_1$}$$

In the limit $\boldsymbol{y} \to 0$, $\boldsymbol{\eta} \to 0$ this is,

$$[\boldsymbol{O}_{\boldsymbol{n}} P(\boldsymbol{y}+\boldsymbol{z})]_{\boldsymbol{y}\to 0} = \frac{1}{\boldsymbol{n}!}\left(\frac{\partial}{\partial \boldsymbol{z}}\right)_T \left(\frac{P(\boldsymbol{z})}{kT}\right) = b_{\boldsymbol{n}}(\boldsymbol{z})\left(\frac{\boldsymbol{\rho}}{\boldsymbol{z}}\right)^{\boldsymbol{n}}, \tag{19.12$_2$}$$

or with (19.12$_1$),

$$\left.\begin{aligned} &b_{\boldsymbol{n}}(\boldsymbol{y}+\boldsymbol{z})\left[\frac{\boldsymbol{\rho}(\boldsymbol{y}+\boldsymbol{z})}{\boldsymbol{y}+\boldsymbol{z}}\right]^{\boldsymbol{n}} \\ &= \sum_{\boldsymbol{m}\geq 0} \frac{(\boldsymbol{n}+\boldsymbol{m})!}{\boldsymbol{n}!}\, b_{\boldsymbol{n}+\boldsymbol{m}}(\boldsymbol{z})\left[\boldsymbol{y}\,\frac{\boldsymbol{\rho}(\boldsymbol{z})}{\boldsymbol{z}}\right]^{\boldsymbol{n}}.\end{aligned}\right\} \tag{19.13}$$

In this set $\boldsymbol{z}\equiv 0$, and $\varrho_a/z_a=1$. Replace $\boldsymbol{y}$ by $\boldsymbol{z}$ to find,

$$b_n(\boldsymbol{z})\,\boldsymbol{\rho}^{\boldsymbol{n}}=\sum_{\boldsymbol{m}\geq 0}\frac{(\boldsymbol{n}+\boldsymbol{m})!}{\boldsymbol{n}!}\,b^{(0)}_{\boldsymbol{n}+\boldsymbol{m}}\,\boldsymbol{z}^{\boldsymbol{n}+\boldsymbol{m}}\,, \tag{19.14}$$

as an expression for any cluster integral at $\boldsymbol{z}$ in terms of those at $\boldsymbol{z}\equiv 0$.

Since $b_1\equiv 1$ for all components a, Eq. (19.13) with $\boldsymbol{n}=a$ becomes,

$$\varrho_a(\boldsymbol{y}+\boldsymbol{z})=\sum_{\boldsymbol{m}}\left[\varrho_a(\boldsymbol{z})\,b_{\boldsymbol{m}+a}(\boldsymbol{z})+b_{\boldsymbol{m}}(\boldsymbol{z})\right]\left[\boldsymbol{y}\,\frac{\boldsymbol{\rho}(\boldsymbol{z})}{\boldsymbol{z}}\right]^{\boldsymbol{m}} \tag{19.15}$$

where we mean by $\boldsymbol{m}+a$ the set with m_a+1 members of a, m_b of b, ..., etc. Take derivatives of (19.15) with respect to y_a or y_b, in the limit $\boldsymbol{y}\to 0$, and one finds,

$$\left(\frac{\partial\varrho_a}{\partial\log z_a}\right)_{T,z_b,\ldots}=\varrho_a+2b_{aa}\,\varrho_a^2, \tag{19.16}$$

$$\left(\frac{\partial\varrho_a}{\partial\log z_b}\right)_{T,z_a,z_c,\ldots}=\left(\frac{\partial\varrho_b}{\partial\log z_a}\right)_{T,z_b,z_c,\ldots}=b_{ab}\,\varrho_a\,\varrho_b\,. \tag{19.17}$$

These give us relations for the integrals, b_{ab}, in terms of thermodynamic quantities.

In principle, given $(\partial\varrho_a/\partial\log z_b)_{T,z_a,z_c,\ldots}$ we can determine the more usually measured thermodynamic functions, such as isothermal compressibility at constant composition, and partial molal volumes. The general set of equations,

$$\left.\begin{aligned}x_a(d\log\varrho_a)_T&=(d\,x_a)_T+x_a(d\log\varrho)_T\\&=[x_a+2b_{aa}x_a^2\varrho]\,(d\log z_a)_T+\sum_{b\neq a}b_{ab}\,x_a\,x_b\,\varrho\,(d\log z_b)_T\,,\end{aligned}\right\} \tag{19.18}$$

of which there are as many as there are components, can be solved for $d\,x_a=0$, using,

$$\left(\frac{\partial\log z_a}{\partial\log\varrho}\right)_{T,\boldsymbol{x}}=\left[\frac{\partial\log z_a}{\partial\left(\dfrac{P}{\mathrm{k}T}\right)}\right]_{T,\boldsymbol{x}}\left[\frac{\partial\left(\dfrac{P}{\mathrm{k}T}\right)}{\partial\log\varrho}\right]_{T,\boldsymbol{x}}=v_a\,\frac{1}{\varkappa\,\mathrm{k}T}\,, \tag{19.19}$$

with $v_a=(\partial V/\partial N_a)_{T,P,N_b,\ldots}$, the partial molecular volume of species a, and $\varkappa$ the isothermal compressibility at constant composition.

The set becomes, for each a,

$$x_a\varkappa\,\mathrm{k}T=(x_a+2b_{aa}x_a^2\varrho)\,v_a+\sum_{b\neq a}(b_{ab}\,x_a\,x_b\,\varrho)\,v_b\,, \tag{19.20}$$

which can be solved for both $\varkappa\,\mathrm{k}T$ and v_a with the use of,

$$\varrho\sum x_a\,v_a=1\,. \tag{19.21}$$

The solution for a two component system for $\varkappa$ is

$$\varrho\,\varkappa\,\mathrm{k}T=\frac{1+2\varrho\,[b_{aa}\,x_a+b_{bb}\,x_b]+x_a\,x_b\,\varrho^2\,[4b_{aa}\,b_{bb}-b_{ab}^2]}{1+2x_a\,x_b\,\varrho\,[b_{aa}-b_{ab}+b_{bb}]}\,, \tag{19.22}$$

which reduces to (18.7_2) when $b_{ab}=2b_{aa}=2b_{bb}=2b_2$. The partial molecular volumes are,

$$\varrho\,v_a=\frac{1+2x_b\,\varrho\,[b_{bb}-\tfrac{1}{2}\,b_{ab}]}{1+2x_a\,x_b\,\varrho\,[b_{aa}-b_{ab}+b_{bb}]}\,, \tag{19.23$_1$}$$

$$\varrho\,v_b=\frac{1+2x_a\,\varrho\,[b_{aa}-\tfrac{1}{2}\,b_{ab}]}{1+2x_a\,x_b\,\varrho\,[b_{aa}-b_{ab}+b_{bb}]}\,. \tag{19.23$_2$}$$

20. The use of integral equations. In principle, were the pair distribution function, $F_2(r)$, known at all densities, the equation of state would be determined, since from F_2 the pressure can be computed. The function $F_2(r)$ can be shown to obey various integral equations, and can be computed, although with considerable difficulty, from any of these.

The simplest of these integral equations can be written from inspection, and the relations (18.9) and (18.11$_1$), between F_2 and the potential of average force,

$$F_2(r_{12}) = e^{-w(r_{12})/kT} \tag{20.1}$$

$$-\frac{\partial w}{\partial r} = \bar{f}_r \tag{20.2}$$

with $\bar{f}_r$ the average force along the line connecting two molecules. This force is composed additively of the direct force,

$$f_r^{(0)} = -\frac{du}{dr} \tag{20.3}$$

acting between the two molecules at a distance r apart, plus a term due to other molecules. The second term will be the integral over all positions $\boldsymbol{r}_3$ of the component,

$$f_r^{(0)}(r_{13})\,\frac{\boldsymbol{r}_{12}\cdot\boldsymbol{r}_{13}}{r_{12}r_{13}} = -\frac{\boldsymbol{r}_{12}\cdot\boldsymbol{r}_{13}}{r_{12}r_{13}}\,\frac{du(r_{13})}{dr_{13}}, \tag{20.4}$$

acting along r_{12} due to a third molecule at $\boldsymbol{r}_3$, times the conditional probability density,

$$\frac{\varrho_3(\boldsymbol{r}_1,\boldsymbol{r}_2,\boldsymbol{r}_3)}{\varrho_2(\boldsymbol{r}_1,\boldsymbol{r}_2,\boldsymbol{r}_3)}, \tag{20.5}$$

that a third molecule is at $\boldsymbol{r}_3$ if the two are at $\boldsymbol{r}_1$ and $\boldsymbol{r}_2$.

We therefore have,

$$\frac{dw}{dr} = \frac{du}{dr} + \int \frac{\varrho_3(\boldsymbol{r}_1,\boldsymbol{r}_2,\boldsymbol{r}_3)}{\varrho_2(\boldsymbol{r}_1,\boldsymbol{r}_2)}\,\frac{(\boldsymbol{r}_{12}\cdot\boldsymbol{r}_{13})}{r_{12}r_{13}}\,\frac{du(r_{13})}{dr_{13}}\,d\boldsymbol{r}_3. \tag{20.6}$$

The probability density, (20.5), is

$$\left.\begin{aligned}\frac{\varrho_3}{\varrho_2} &= \varrho\,\frac{F_3(\boldsymbol{r}_1,\boldsymbol{r}_2,\boldsymbol{r}_3)}{F_2(\boldsymbol{r}_1,\boldsymbol{r}_2)}\\ &= \varrho\,e^{-\frac{1}{kT}[W_3(\boldsymbol{r}_1,\boldsymbol{r}_2,\boldsymbol{r}_3)-w(r_{12})]},\end{aligned}\right\} \tag{20.7}$$

where W_3 is the total potential of average force of the three molecules at $\boldsymbol{r}_1$, $\boldsymbol{r}_2$, and $\boldsymbol{r}_3$. If now one assumes the Kirkwood closure,

$$W_3(\boldsymbol{r}_1,\boldsymbol{r}_2,\boldsymbol{r}_3) = w(r_{12}) + w(r_{23}) + w(r_{31}), \tag{20.8}$$

the equation takes the form,

$$\frac{dw}{dr} = \frac{du}{dr} + \varrho\int e^{-\frac{1}{kT}[w(r_{13})+w(r_{23})]}\left(\frac{\boldsymbol{r}_{12}\cdot\boldsymbol{r}_{13}}{r_{12}r_{13}}\right)\frac{du(r_{13})}{dr_{13}}\,d\boldsymbol{r}_3. \tag{20.9}$$

This is an integral differential equation for the unknown function $w(r)$. The equation was first developed by YVON[1], and later by BORN and GREEN[2].

---

[1] J. YVON: La Théorie Statistique des Fluides et L'équation d'Etat. Paris: Hermann & Cie. 1935.

[2] M. BORN and H. S. GREEN: Proc. Roy. Soc. Lond., Ser. A **188**, 10 (1946), and H. S. GREEN: Molecular Theory of Fluids. Amsterdam: North Holland Publishing Co. 1952.

An alternative, but similar equation, (20,26$_3$), has been developed by KIRKWOOD, using the concept of a single molecule whose interactions with all the others is $\lambda u(r)$ instead of $u(r)$. In the limit $\lambda \to 0$ this molecule is completely free, but at $\lambda = 1$ it becomes one of the molecules of the system.

We will show here that a variety of equations can be derived, all of which are closely related. The derivation serves the additional purpose of showing how, by a perturbation method, the effect of an alteration in the pair potential can be computed.

Consider the set of functions, G_n^* and G_n, related by an equation such as (19.4),

$$G_n\{n\} = \sum_{N \geq 0} \frac{y^N}{N!} \iint_V \cdots \int G_{n+N}^*\{\{n\} + \{N\}\}\, d\{N\}, \qquad (20.10_1)$$

$$G_n^*\{n\} = \sum_{N \geq 0} \frac{(-y)^N}{N!} \iint_V \cdots \int G_{n+N}\{\{n\} + \{N\}\}\, d\{N\}. \qquad (20.10_2)$$

Now consider some operator, $\boldsymbol{O}$, for which

$$\boldsymbol{O}\, G_N = G_N \sum_{\{m\}_N \geq 0} \psi_m\{m\}_N, \qquad (20.11_1)$$

$$\boldsymbol{O}\, G_N^* = G_N^* \sum_{\{m\}_N \geq 0} \psi_m^*\{m\}_N. \qquad (20.11_2)$$

For instance, if the potential energy function U_N is $U_N^{(0)} + \lambda \Delta U_N$, and ΔU_N is a sum of component potentials $\Delta u_1(\boldsymbol{r}_\alpha)$, $\Delta u_2(\boldsymbol{r}_\alpha, \boldsymbol{r}_\beta), \ldots$, then for the operator,

$$\boldsymbol{O} = -\frac{1}{kT}\frac{\partial}{\partial \lambda}, \qquad (20.12_1)$$

and if the functions G_N^* were

$$G_N^* = G_N(z \to 0, T, \{N\}) = e^{-\frac{1}{kT} U_N\{N\}}, \qquad (20.12_2)$$

we would have

$$\psi_m^*\{m\} = \Delta u_m\{m\}. \qquad (20.12_3)$$

The corresponding $\psi_m\{m\}$ would be,

$$\left.\begin{aligned} \psi_0 = \frac{\partial}{\partial \lambda}\left(\frac{PV}{kT}\right), \quad \psi_a(\boldsymbol{r}_\alpha) = \frac{\partial}{\partial \lambda}[\log \varrho_a + w_1(\boldsymbol{r}_\alpha)], \\ \psi_m\{m\} = \frac{\partial}{\partial \lambda} w_m\{m\}, \end{aligned}\right\} \qquad (20.12_4)$$

where the w_m are the component potentials of average force. However, for the moment we wish to leave the meaning of the functions, ψ_m^* and ψ_m, to be assigned later.

Operate on both sides of (20.10$_1$) by the operator $\boldsymbol{O}$. On the left one obtains (20.11$_1$) with $N = n$. On the right one multiplies the integrand by the sum

$$\sum_{\{\nu\}_n \geq 0} \sum_{\{m\}_N \geq 0} \psi_{\nu+m}^*\{\{\nu\}_n + \{m\}_N\}. \qquad (20.13_1)$$

Consider one term for fixed subset $\{\nu\}_n$, and fixed value of the set m. There are

$$\frac{N!}{(N-m)!\,m!} \qquad (20.13_2)$$

different subsets $\{m\}_N$ of given m in $\{N\}$. Multiply by this number, and write $M = N - m$. Now integrate over $d\{M\}$, and sum over all M values. From

(20.10_1) one obtains,

$$\frac{y^n}{m!}\iint\limits_V\cdots\int G_{n+m}\{\{n\}+\{m\}\}\,\psi^*_{\nu+m}\{\{\nu\}_n+\{m\}\}\,d\{m\}, \tag{20.13$_3$}$$

which must be summed over all $\{\nu\}_n$ and m. Divide both sides of the equation by G_n. On the left one has

$$\Psi_n\{n\} = \sum_{\{k\}n\geq 0} \psi_k\{k\}_n, \tag{20.13$_4$}$$

and this can be solved for ψ_k by,

$$\psi_k\{k\} = \sum_{\{n\}k\geq 0} (-)^{k-n}\,\Psi_n\{n\}_k. \tag{20.13$_5$}$$

For given ψ_k the function $\psi^*_{\{\nu\}k+m}$ in (20.13_3) occurs multiplied by

$$(-)^{k-\nu-\varkappa}\,\frac{G_{\nu+\varkappa+m}\{\{\nu\}_k+\{\varkappa\}_{k-\nu}+\{m\}\}}{G_{\nu+\varkappa}\{\{\nu\}_k+\{\varkappa\}_{k-\nu}\}}, \tag{20.13$_6$}$$

in the term $(-)^{k-n}\Psi_n$ of (20.13_5), with $\{n\}_k=\{\nu\}_k+\{\varkappa\}_{k-\nu}$. Sum (20.13_6) over all $\{\varkappa\}_{k-\nu}$, and define,

$$K_{k,\nu,m} = \sum_{\{\varkappa\}_{k-\nu}\geq 0} (-)^{k-\nu-\varkappa}\,\frac{G_{\nu+\varkappa+m}\{\{\nu\}_k+\{\varkappa\}_{k-\nu}+\{m\}\}}{G_{\nu+\varkappa}\{\{\nu\}_k+\{\varkappa\}_{k-\nu}\}}. \tag{20.14}$$

With this kernel we have,

$$\left.\begin{aligned}\psi_k\{k\} = \sum_{m\geq 0}\frac{y^m}{m!}\sum_{\{\nu\}_k\geq 0}\iint\cdots\int K_{k,\nu,m}(\{k-\nu\}_k,\{\nu\}_k,\{m\})\times\\ \times\,\psi^*_{\nu+m}\{\{\nu\}_k+\{m\}\}\,d\{m\}.\end{aligned}\right\} \tag{20.15}$$

At the cost of making the expressions for the kernels, K, more complicated, one can simplify the appearance of (20.15). For two identical sets $n\equiv m$, $n_a=m_a$, $n_b=m_b$, ..., define

$$\delta(\{n\},\{m\}) = \sum P_m \prod_{\alpha=1}^{n}\delta(r_\alpha-r_\beta), \tag{20.16}$$

as the sum of all $n!=m!$ permutations of products of the n delta functions connecting the identical molecules of the two different sets in pairs. Use (19.1) for G_n and define,

$$\left.\begin{aligned}&K_{n,m}(\{n\},\{m\})\\ &=\sum_{\{\nu\}n}\sum_{\{\nu\}m}\frac{\left[y\,\frac{\rho(z)}{z}\right]^{m-\nu}}{m!}\,\delta(\{\nu\}_n,\{\nu\}_m)\times\\ &\quad\times\sum_{\{\varkappa\}_{n-\nu}}(-)^{n-\nu-\varkappa}\,\frac{F_{\varkappa+m}\{\{\varkappa\}_{n-\nu}+\{m\}\}}{F_{\varkappa+\nu}\{\{\varkappa\}_{n-\nu}+\{\nu\}_m\}}.\end{aligned}\right\} \tag{20.17}$$

Eq. (20.15) is now

$$\psi_n\{n\} = \sum_{m\geq 0}\iint\cdots\int K_{n,m}(\{n\},\{m\})\,\psi^*_m\{m\}\,d\{m\}. \tag{20.18$_1$}$$

The corresponding operations, can of course, be made with the Eq. (20.10_2), reciprocal to (20.1), obtaining

$$\psi^*_n\{n\} = \sum_{m\geq 0}\iint\cdots\int K_{n,m}{}^*(\{n\},\{m\})\,\psi_m\{m\}\,d\{m\}, \tag{20.18$_2$}$$

where the kernels $K^*_{n,m}$ are defined analogously to (20.17) but with $-y_a$ replacing y_a, $\rho_m(z+y)$ and $z+y$ replacing $\rho(z)$ and z, and the distribution functions, F_n, those of the activity set $z+y$.

Setting the solution (20.18$_1$) for $\psi_n\{n\}$ in (20.18$_2$), or *vice versa*, and remembering that any *one* of the sets of functions, ψ^*_n *or* ψ_n, (but not both) can be chosen completely arbitrarily, one has,

$$\left.\begin{aligned}&\sum_{k\geq 0}\iint\cdots\int K^*_{n,k}K_{k,m}\,d\{k\}\\ &\quad=\sum_{k\geq 0}\iint\cdots\int K_{n,k}K^*_{k,m}\,d\{k\}=\delta(n-m)\frac{1}{n!}\,\delta(\{n\},\{m\}).\end{aligned}\right\}\quad(20.19)$$

We thus see that the general method does not depend on the particular operation O, involved in producing the functions ψ^*_n and ψ_n, but that a reciprocal relation between the kernels K and K^* is involved in the fundamental equations. However the actual practical use made of the equations has always been by another procedure than to examine (20.19). Instead one chooses an operator O such that a known relation between ψ_n and F_n as well as between ψ^*_n and F^*_n is involved. One of the methods leads to an expression equivalent to (20.9) and the other to the equation used by KIRKWOOD.

Before discussing these methods let us return to the kernels of (20.14) and list some of the special values. If $m=0$ the ratio, $G_{\nu+\varkappa+m}/G_{\nu+\varkappa}$, is unity. The sum of $(-)^{k-\nu-\varkappa}$ over all $(k-\nu)!/\varkappa!(k-\nu-\varkappa)!$ subsets $\{\varkappa\}_{k-\nu}$ is zero unless $k-\nu\equiv 0$, or $\nu\equiv k$. Hence $K_{k,\nu,0}$ is zero except for $\nu=k$, and is then unity. We have

$$K_{0,0,0}=1\,,\qquad(20.20_1)$$

$$K_{k,k,0}=1\,,\qquad(20.20_2)$$

$$K_{k,k-\varkappa,0}=0,\quad \varkappa>0\,,\qquad(20.20_3)$$

$$K_{0,0,m}=\frac{G_m}{G_0}=\left(\frac{\rho}{z}\right)^m F_m(z)\,.\qquad(20.20_4)$$

In the limit $y_a\to 0$ for all a, the kernels $K_{n,m}$ of (20.17) contain only the one non zero term $\{\nu\}_m\equiv\{m\}$, which is $\frac{1}{m!}\,\delta(\{\nu\}_n,\{m\})\,K_{n,\nu,0}$, and hence zero unless $\{\nu\}_n\equiv\{n\}$, from (20.20$_3$). Hence from (20.20$_2$),

$$K_{n,m}(y\to 0)=\frac{1}{m!}\,\delta(\{n\},\{m\})\,\delta(n-m)\,.\qquad(20.21)$$

Thus at $y\to 0$ Eq. (20.19) involves only the single term $k\equiv m\equiv n$ and since

$$\iint\cdots\int\delta(\{n\},\{k\})\,\delta(\{k\},\{m\})\,d\{k\}=m!\,\delta(\{n\},\{m\}),$$

it is trivially satisfied.

The usual use of the integral equations is such that one attempts to eliminate all but one non-trivial kernel not listed in (20.20), namely,

$$K_{2,1,1}=K_{ab,a,c}=\frac{\varrho_c}{z_c}\left[\frac{F_{abc}}{F_{ab}}-\frac{F_{ac}}{F_a}\right].\qquad(20.22_1)$$

For a fluid $F_a\equiv 1$. If one then makes the approximation of the Kirkwood closure, $F_{abc}=F_{ab}F_{bc}F_{ca}$, the expression is

$$K_{ab,a,c}\approx\frac{\varrho_c}{z_c}\,e^{-\frac{1}{kT}w_{ac}(r_{ac})}\left[e^{-\frac{1}{kT}w_{ab}(r_{ab})}-1\right].\qquad(20.22_2)$$

The Yvon equation is obtained from our formalism as follows. Choose the G^* in (20.10_1) to be taken at zero activity, $G^*_{\boldsymbol{N}} = F^{(0)}_{\boldsymbol{N}} = \mathrm{e}^{-U_{\boldsymbol{N}}/\mathrm{k}T}$, and G_n at activity $\boldsymbol{y}$. Let the system consist of two components, a and b, with $y_a = z$ but y_b vanishingly small, and let the potential energy, $U_{\boldsymbol{N}}$, be a sum of pair potentials all of which are identical, $u_{aa} = u_{ab} = u_{bb} = u$. Let the operator 0 be $-\mathrm{k}T$ times the sum $\sum_\beta (\partial/\partial x_\beta)$ for all molecules of species b. We have, from (20.11_2)

$$\psi^*_{ab} = \frac{\partial u(|\boldsymbol{r}_\alpha - \boldsymbol{r}_\beta|)}{\partial x_\beta} \tag{20.23}$$

and *only* these terms are non-zero. Now write the equation for,

$$\psi_{ab} = -\mathrm{k}T\, G^{-1}_{ab} \frac{\partial}{\partial x_\beta} G_{ab}(|\boldsymbol{r}_\alpha - \boldsymbol{r}_\beta|) = \frac{\partial w_{ab}(|\boldsymbol{r}_\alpha - \boldsymbol{r}_\beta|)}{\partial x_\beta}, \tag{20.24}$$

which, from (20.15) involves $\boldsymbol{m} = 0$, $\boldsymbol{m} = a$, $\boldsymbol{m} = b$ and $\boldsymbol{m} = 2 = a + b$. The terms $\boldsymbol{m} = b$ and $\boldsymbol{m} = a + b$ drop out, since $y_b = 0$. The two remaining terms, with (20.22_2) are, when $\boldsymbol{r}_\alpha - \boldsymbol{r}_\beta$ is chosen to lie on the x-axis,

$$\frac{\partial w(r_{12})}{d r_{12}} = \frac{d u(r_{12})}{d r_{12}} + \varrho \int \mathrm{e}^{-\frac{1}{\mathrm{k}T} w(r_{13})} \left[\mathrm{e}^{-\frac{1}{\mathrm{k}T} w(r_{23})} - 1\right] \left[\frac{\boldsymbol{r}_{12} \cdot \boldsymbol{r}_{13}}{r_{12} r_{13}}\right] \frac{d u(r_{13})}{d r_{13}}\, d\boldsymbol{r}_3, \tag{20.25}$$

since $z\varrho/z = \varrho$. The difference,

$$\varrho \int \mathrm{e}^{-\frac{1}{\mathrm{k}T} w(r_{13})} \left(\frac{\boldsymbol{r}_{12} \cdot \boldsymbol{r}_{13}}{r_{12} r_{13}}\right) \frac{d u(r_{13})}{d r_{13}}\, d\boldsymbol{r}_3 = 0$$

between this and (20.9) is zero since $(\boldsymbol{r}_{12} \cdot \boldsymbol{r}_{13})$ averages to zero when integrated over the angles.

The Kirkwood equation may be obtained as follows. Again let G^*_N refer to zero activity, and G_N to a two-component system with $y_a = z$ and $y_b = 0$. Let

$$U_{\boldsymbol{N}}\{\boldsymbol{N}\} = \sum_{N_a \geq \alpha > 1} \sum_{\gamma \geq} u(r_{\alpha\gamma}) + \lambda \sum_{\beta = N_a + 1}^{\beta = N_a - N_b = N_a} \sum_{\alpha = 1} u(r_{\alpha\beta}) + \lambda^2 \sum_{N \geq \delta > \beta \geq N_a + 1} \sum u(r_{\beta\delta}). \tag{20.26$_1$}$$

At $\lambda = 0$, then, the species b has molecules that interact with no others, and which will move independently of the others. At $\lambda = 1$ the species b become identical to species a in all but name.

The operator $\boldsymbol{0}$ is taken as $-\mathrm{k}T \frac{\partial}{\partial \lambda}$. We have,

$$\psi^*_{ab} = u(|\boldsymbol{r}_\alpha - \boldsymbol{r}_\beta|), \tag{20.26$_2$}$$

and although ψ^*_{bb} is non-zero at $\lambda \neq 0$, we will not require it. We again compute ψ_{ab} which is $d w_{ab}/d\lambda$, and again note that since $y_b \to 0$ we need only two terms of (20.15), namely $\boldsymbol{m} = 0$, $\boldsymbol{m} = a$. We have, with (20.22_2),

$$\frac{d w_{ab}(r_{12})}{d\lambda} = \frac{d u(r_{12})}{d\lambda} + \varrho \int \mathrm{e}^{-\frac{1}{\mathrm{k}T} w_{ab}(r_{13})} \left[\mathrm{e}^{-\frac{1}{\mathrm{k}T} w_{aa}(r_{23})} - 1\right] u(r_{13})\, d\boldsymbol{r}_3. \tag{20.26$_3$}$$

The boundary conditions are,

$$w_{ab}(\lambda = 0) = 0, \qquad w_{ab}(\lambda = 1) = w_{aa}. \tag{20.26$_4$}$$

In neither case, (20.25) nor (20.26_3), are the equations exact, since the closure, assuming $W_3(r_1, r_2, r_3) = w(r_{12}) + w(r_{23}) + w(r_{31})$, was needed. It is therefore not true that the two methods, even if solved exactly, need lead to identical expressions for $w(r)$, since the approximation of the closure may lead to different neg lects in the two cases and probably does. The solutions for $w(r)$ obtained from

(20.26_3) have been evaluated by KIRKWOOD and coworkers for realistic potentials at liquid densities and temperatures, and for the hard sphere potential. The results for the liquids resemble closely those obtained from X-ray scattering data, and those for the hard sphere potential agree well with those determined by digital computor methods.

In both cases the same method can be employed with the reciprocal equation, (20.10_2), that is by setting the distribution functions at z under the integral, with the known functions at zero activity on the left. In this case the kernels are known combinations of the pair potentials. The unknown functions, w, occur only linearly in the equations. The logic of the closure is now different than before.

It is, of course, true that for real gases the potential energy function, U_N, is not strictly a sum of pair potentials. It is convenient, however, for the sake of a simple language, to assume that we compute for a model in which this is always strictly true. In this case, for the two operations outlined, leading to (20.25) and (20.26_3), the functions ψ_m^* at zero activity were strictly zero for $m > 2$. The computation was then made for ψ_{ab} in both cases, and the terms that occurred on the right were strictly limited to the one ψ_{ab}^* plus the single integral, involving F_3. The only approximation is then in setting this equal to the product of F_2's.

If, however, the reciprocal equation is used, the situation is different. The integral corresponding to that occurring in (20.25) or (20.26_3) now involves $e^{-u(r)/kT}$, and the integrand is strictly correct. However the sum of (20.15) involves terms of higher m values, since ψ_m which now occurs under the integrals is not zero for $m > 2$. Logically, if the function W_3 is always the sum of the three pair functions, then ψ_3 must always be zero, so that the omission of higher terms in the sum is now equivalent to assuming the Kirkwood closure. Now there is no reason, however, for assuming that the omission leads to the *same* neglect as in (20.25) or in (20.26_3). However, there is also no *a priori* reason to assume that the error introduced is larger, or even as large.

With this closure the equations are obtained from (20.25) or (20.26_2) by interchanging ϱ with $-z$, and u with w. One has,

$$\frac{du(r_{12})}{dr_{12}} = \frac{dw(r_{12})}{dr_{12}} - z\int e^{-\frac{u(r_{13})}{kT}}\left[e^{-\frac{u(r_{23})}{kT}} - 1\right]\left[\frac{\boldsymbol{r}_{12}\cdot\boldsymbol{r}_{13}}{r_{12}r_{13}}\right]\frac{dw(r_{13})}{dr_{13}}\,d\boldsymbol{r}_3, \tag{20.27}$$

and,

$$u(r_{12}) = \frac{dw(r_{12})}{d\lambda} - z\int e^{-\frac{\lambda u(r_{13})}{kT}}\left[e^{-\frac{u(r_{23})}{kT}} - 1\right]\frac{dw(r_{13})}{d\lambda}\,d\boldsymbol{r}_3, \tag{20.28}$$

which are both linear in the unknown functions.

21. Gas crystal transition. The normal phase diagram, at temperatures below the critical, and above the triple point, shows a phase transition at some activity, z_0, to the liquid phase, and at a higher activity, z_l, to a crystalline lattice. Above the critical temperature, T_c, the phase transition to the liquid at $z_0(T)$ disappears. One may well ask whether the other persists to high temperatures, or disappears at some critical temperature, T_c. In other words does the gas phase, at high temperatures, extend to infinite pressure, or is it bounded by a sharp transition to the lattice at all temperatures?

In so far as an experimental answer exists it points toward the latter answer, namely that the sharp phase transition to the lattice persists to high temperatures, and presumably, to all temperatures for which the molecules retain their identity.

The experimental evidence is limited to the single material, helium, for which both critical temperature and pressure are so low that the lattice transition can be found at temperatures exceeding the critical by many fold. Not only does the transition to the crystalline state exist at high temperatures, but there is no indication that it is diminishing in importance, that is $\Delta V/V$ and $\Delta S/S$ do not decrease as T increases.

Nevertheless there persists the question whether a system, in which there were only repulsive forces between the molecules, could even exhibit a phase change, or whether the system would show only one thermodynamic phase, the gaseous, to infinite pressure. No definite rigorous answer to the question is known at the present time. However, this author at least, is convinced that a phase change is possible in such a case. Probably the most convincing argument is the experimental evidence, quoted above, for helium, but some theoretical remarks may be added to the experimental information.

Consider the hypothetical case of molecules for which the mutual pair potential becomes infinity at a distance $2r_0$. The distribution function for pairs, $F_2(r)$, will always be zero for $r < 2r_0$, and the integral

$$b_2 = \tfrac{1}{2} \int_0^\infty 4\pi r^2 [F_2(r) - 1]\, dr, \tag{21.1}$$

will always contain a contribution $-\frac{2\pi}{3}(2r_0)^3$ due to the integration between $0 \leq r \leq 2r_0$. Now assume that $F_2(r)$ never exceeds unity,

$$F_2(r) \leq 1, \quad \text{all } r. \tag{21.2}$$

In this case we would have,

$$b_2 \leq -\frac{2\pi}{3}(2r_0)^3. \tag{21.3}$$

The isothermal compressibility, $\varkappa$, is

$$\varkappa \mathbf{k} T \varrho = 1 + 2 b_2 \varrho, \tag{21.4}$$

and cannot be negative. Such a material, at infinite pressure, will compress to a close-packed lattice with distance $2r_0$ between neighbors, having a minimum volume per molecule of $(2r_0)^3/\sqrt{2}$ or a maximum density, $\varrho_l = \sqrt{2}(2r_0)^{-3}$. For this density, with $\varkappa = 0$, we have

$$2b_2\varrho = -1, \qquad b_2 = -2^{-\frac{3}{2}}(2r_0)^3, \tag{21.5}$$

clearly more positive than the limit of (21.3). It follows that, for a considerable range of density less than ϱ_l, the pair distribution function, $F_2(r)$, must show values exceeding unity for some $r > 2r_0$. The potential of average force, $w(r)$ for pairs,

$$w(r) = -\mathbf{k} T \log F_2(r), \tag{21.6}$$

must then take negative values, and at the higher densities there is an attractive average force between pairs at some distance, presumably just above $2r_0$.

We have seen in Sect. 19 that the equations for activity $z + y$ can be based on the distribution functions at z, and that the form of these is the same as those for the activity at $\eta = y[\varrho(z)/z]$ in terms of $F_n^{(0)}$. Now even if $F_2^{(0)} \leq 1$ for all r, that is if the pair potential is purely repulsive, $u(r) \geq 0$, it must be that at high densities the corresponding $w(r)$ becomes negative for some r-values. Since we know that a phase transition does exist when $u(r)$ has a sufficiently large negative region, it does not seem to be unreasonable that a phase transition may exist for the purely repulsive potential. In other words, the effect of a high pressure is to create an effective attractive average force, which may cause a phase transition.

Since the experimental evidence shows that helium far above the critical temperature, with an extremely weak region of $u(r)<0$, does have a transition, one is inclined to accept the existence of such a transition for all soft repulsive potentials, and hence for all molecules up to any temperature for which the concept of identity of the molecules with short range forces is valid.

The question of whether hypothetical hard spheres would show a transition is somewhat different, and an entirely different kind of argument seems to speak against it. The argument of the past few paragraphs, of course, applies equally to the hard sphere case, and makes it permissive that a transition might occur.

The special argument against a transition with finite ΔV for the hard spheres is as follows. A close packed lattice of N hard spheres, each of radius r_0, has volume $V_0 = N(2r_0)^3/\sqrt{2}$. Increase the volume by lengthening one edge by an amount $\Delta l = n\left[\frac{2}{\sqrt{3}} - 1\right](2r_0)$ and there will be n possible planes along which the configuration of the lattice can change by a random amount. For an infinite system this volume change, $\Delta V/V_0 \sim (n/N^{\frac{1}{3}})$ may approach zero for finite n. At volumes percentually infinitessimally greater than V_0 there can be no true infinite range order in the lattice, and the average distance, $(N^{\frac{1}{3}}/n)$, between dislocations would appear to decrease linearly with $\Delta V/V_0$. Since we are accustomed to associate the transition to a crystalline phase with the sharp appearance of long range order, one would be inclined to assume that the transition would occur at volumes infinitessimally greater than V_0.

One might therefore be inclined to postulate that whereas transitions of finite $\Delta V/V$ occur in going from gas to lattice for "soft" repulsive potentials, the effective volume change of the transition, $\Delta V/V$, shrinks to zero as the repulsive potential approaches the hard sphere limit.

However, several objections to the above argument may be made. One of these is that equilibrium dislocations in real crystals presumably exist, and that truly infinite range order is not the distinguishing characteristic of the crystalline phase. Another, and more annoying objection, because its significance is hard to assess, is that the close packed lattice of true hard spheres has no simply definable long range order, since a random arrangement of every second plane perpendicular to the hexagonal axis occurs.

In any case there exists good evidence from computations that a transition at a considerably larger volume than V_0 does occur in a hard sphere system. This is indicated by the results of the integral equation solutions, and also from computations made using digital computors on hard sphere models.

D. Quantum gases.

22. General considerations. The treatment of gases used so far explicitly considered the quantum mechanical sum in the internal partition function, Sects. 6 to 12, although for the rotational contribution it was pointed out that the classical approximation usually suffices. The assumption of separability of the internal and translational degrees of freedom was made, namely that the Hamiltonian of the system could be written as a sum of terms, one for each molecule involving only the internal degrees of freedom of that molecule, and another involving only the translational degrees of freedom of all molecules, namely the coordinates and momenta of the centers of mass of the molecules. This latter term was handled purely classically.

The separability of the Hamiltonian, strictly interpreted, means that the mutual potential of interaction of two molecules is a function of the distance, r, of their centers of mass, and depends in no way on their relative angles of orientation, nor on such other internal coordinates as the vibrational amplitudes. Actually, at least by an approximate method, this assumption could be removed, and its removal has been already implied in the discussion of the angular dependent Stockmayer potential, in Sect. 17. The essential requirement of the previous treatment is that the function $e^{-U_N/kT}$, regarded as a function of the coordinates of centers of mass, $\{N\} = \boldsymbol{r}_1, \boldsymbol{r}_2, \ldots, \boldsymbol{r}_N$, should be averaged, for each value of $\{N\}$, over all the internal states of the molecules, appropriately weighted with the energies $e^{-\varepsilon_n/kT}$. If, as in the case of two dipole molecules, the potential depends purely on the relative angular coordinates, and is assumed not to depend on the vibrational coordinates, and if the classical equation for rotation is assumed, since the internal rotational kinetic energy is indeed separable, one can legitimately average $e^{-U_N/kT}$ over the angles, and express the resulting function as one depending only on the coordinates, $\{N\}$, of the centers of mass. In this case, however, the averaged energy, U_N, will not be independent of the temperature.

As we shall see in Sect. 24, a formal reduction to the problem of integration over the coordinates, $\{N\}$, of the centers of mass, of a function $F_N^{(0)}\{N\}$, can also be made for any quantum mechanical problem. In the most general case, however, the actual evaluation, from first principles, of the function $F_N^{(0)}$ would be most formidable. Fortunately, however, the cases that actually arise are not too terrifying in this respect. Most molecules, other than those of H_2, and of the two isotopes of helium, actually satisfy the classical conditions quite well. It is essentially only for these gases of extremely light molecules (as well as for gases consisting of electrons and protons) that quantum mechanical considerations become of singificance in handling the center of mass coordinates. For helium the internal electronic degrees of freedom can truly be neglected at all temperatures for which the quantum effects of the center of mass motion are of any importance. The molecule H_2 is as close to spherical symmetry as any of the many truly classical diatomic molecules such as N_2, O_2, F_2, ..., etc., which appear to follow the same law of corresponding states as the noble gases A, Kr, etc., so that one is inclined to believe that the separation of degrees of freedom is really justified. Actually the specific volumes of ortho and para liquid hydrogen are somewhat different, as well as their heats of vaporization, so that the assumption is by no means perfect. Thus in practice, except for certain effects which will be discussed later, and which are not too difficult to take into account, one tends to simplify the treatment of quantum gases to that of gases without internal degrees of freedom.

As discussed in Sect. 2 there are two distinguishable, and to a considerable extent separable, assumptions necessary in going to the classical limit from the quantum mechanical equations. These are:

(a) that the density of identical molecules in their μ-space must be low compared to unity, if the volume $d\boldsymbol{p}_\alpha\, d\boldsymbol{r}_\alpha$ is measured in units h^3, and

(b) that for every degree of freedom there must necessarily be regions $\Delta p_i \Delta q_i =$ in which the energy changes little compared to kT.

For a perfect monatomic gas, defined as one in which the mutual potentials of interaction, U_N, are identically zero, condition (b) is easily satisfied at all macroscopic volumes. The coordinates may then be chosen as the cartesian coordinates of the centers of mass for each molecule, and $\Delta q_i = \Delta x$ is the length,

l, of the vessel. Choosing a very small vessel, say $l=10^{-2}$ cm, the neighboring quantum states differ by $|\Delta p|=h/2l\approx 3\times 10^{-25}$ gm cm sec^{-1}. The energy difference, $\Delta\varepsilon$, is $\Delta\varepsilon=\frac{1}{m}p\,\Delta p=\sqrt{\frac{2\varepsilon}{m}}\,\Delta p$. For states of energy $\mathbf{k}T$, $\varepsilon=\mathbf{k}T$ we have,

$$\frac{\Delta\varepsilon}{\mathbf{k}T}=\sqrt{\frac{2}{m\mathbf{k}T}}\,\Delta p=\sqrt{\pi}\sqrt{\frac{h^2}{2\pi m\mathbf{k}T}}\,\frac{1}{l}=\sqrt{\pi}\,\frac{\lambda}{l}\,.$$

Choosing the lowest mass, that of an electron, and $T=1^\circ$ K, $\Delta\varepsilon/\mathbf{k}T\sim 1.5\times 10^{-3}$, and for the mass of a proton $\Delta\varepsilon/\mathbf{k}T\sim 4\times 10^{-5}$. The condition of continuity of energy states in the phase space is amply justified. This is, however, no longer true for the imperfect gas of very light molecules at such low temperatures. The coordinate along a line connecting a molecule with two diagonally opposite neighbors is of the order of the mean free path, and the corresponding $\Delta\varepsilon/\mathbf{k}T$ ncreases inversely with this distance, which, when the gas becomes measurably mperfect is by factors of 10^4 or more smaller than the assumed l chosen above.

However, even for perfect gases the density in μ-space, may, at least hypothetically, be taken as high as one chooses, and deviations would be appreciable from the Boltzmann equations. These deviations are in different directions for systems obeying Bose-Einstein "statistics", that is having state functions symmetric in pair permutations of the identical molecules, than in those for Fermi-Dirac systems possessing only antisymmetric state functions.

The treatment of Bose-Einstein and Fermi-Dirac perfect gases can be made in an extremely simple manner, not involving more than a simple counting of states. The treatment of interacting gases is more complicated. We choose here to give the very simple treatments for the two perfect gases in Sect. 23. A more elaborate method, suitable for the examination of interacting molecules is discussed in Sect. 24 and applied to the perfect gases in Sect. 25. In Sect. 26 the use of this method for Boltzmann gases with interaction is discussed, and in Sect. 27 the combination with the two cases of Fermi and Einstein systems is elaborated.

23. Simple Bose-Einstein and Fermi-Dirac perfect gases. *α) General relations.* The fundamental equation for the Grand Canonical Ensemble, (1.14) with (1.16) and (1.18)

$$e^{PV/\mathbf{k}T}=\sum_{N\geq 0}e^{N\mu/\mathbf{k}T}\sum_{\mathbf{v}}e^{-E_{\mathbf{v}}/\mathbf{k}T}\tag{23.1}$$

is used for a one-component system. The sum N runs over all numbers of molecules, and the sum over $\mathbf{v}$ is over all quantum states of energy $E_{\mathbf{v}}$ of a system of N identical molecules. The Hamiltonian of the system is separable as the sum of Hamiltonians of the individual molecules, if the gas is perfect, so that the quantum state, of the system, consists in the set of numbers,

$$\mathbf{v}=n_1,n_2,\ldots,n_\nu\ldots,\tag{23.2}$$

$$\sum_\nu n_\nu=N,\tag{23.3}$$

with n_ν giving the number of molecules of the system which are in the quantum state ν of the molecules. The energy of the single molecule in the state ν is ε_ν, and of the system is then,

$$E_{\mathbf{v}}=\sum_\nu n_\nu\,\varepsilon_\nu.\tag{23.4}$$

We have, from (23.1),

$$\left.\begin{aligned} e^{PV/kT} &= \sum_N e^{N\mu/kT} \sum_{n_1,\ldots,n_\nu,\ldots,\sum_\nu n_\nu=N} \cdots \sum \cdots \; e^{-\sum_\nu n_\nu \frac{\varepsilon_\nu}{kT}} \\ &= \sum_{n_1}\sum_{n_2}\cdots\sum_{n_\nu}\cdots\prod_\nu e^{-n_\nu\frac{\varepsilon_\nu-\mu}{kT}} \\ &= \prod_\nu\left[\sum_{n_\nu} e^{n_\nu\frac{\mu-\varepsilon_\nu}{kT}}\right], \end{aligned}\right\} \tag{23.5$_1$}$$

or

$$PV = kT\sum_\nu \log\sum_{n_\nu} e^{n_\nu\frac{(\mu-\varepsilon_\nu)}{kT}}, \tag{23.5$_2$}$$

Two cases arise. If the system has symmetric states any number of molecules, n_ν, may occupy the state ν, that is the summation under the logarithm is from $0 \leq n_\nu \leq \infty$. The other case is that of the Fermi-Dirac system, for which never more than one molecule can occupy one state ν. In this case the summation has only $n_\nu = 0, 1$. For the first case, since

$$\sum_{n=0}^{\infty} e^{nx} = \frac{1}{1-e^{nx}},$$

we find $-\log[1-e^{(\mu-\varepsilon_\nu)/kT}]$, and for the second case $+\log[1+e^{(\mu-\varepsilon_\nu)/kT}]$. One may write,

$$PV = kT\sum_\nu \mp\log[1\mp e^{(\mu-\varepsilon_\nu)/kT}], \qquad \begin{matrix}-\text{B.E.}\\ +\text{F.D.},\end{matrix} \tag{23.6}$$

for both the Bose-Einstein and Fermi-Dirac systems. The thermodynamic relation, (1.33),

$$\left(\frac{\partial P}{\partial\mu}\right)_T = \frac{\overline{N}}{V} = \varrho, \tag{23.7$_1$}$$

with (23.6) gives

$$\overline{N} = \sum_\nu \frac{1}{e^{(\varepsilon_\nu-\mu)/kT}\mp 1}, \qquad \begin{matrix}-\text{B.E.}\\ +\text{F.D.}\end{matrix} \tag{23.7$_2$}$$

The sum runs over all quantum states, ν, of the molecules, internal and translational. The average number, $\overline{N}_\nu$, of molecules in the state ν is,

$$\overline{N}_\nu = \frac{1}{e^{(\varepsilon_\nu-\mu)/kT}\mp 1} \qquad \begin{matrix}-\text{B.E.}\\ +\text{F.D.}\end{matrix} \tag{23.7$_3$}$$

For the Fermi-Dirac systems, with the plus sign, this number can never exceed unity. If the number of quantum states is N_0 up to some energy $\mu_0 > kT$, then if the average number of molecules in the system is to be $\overline{N} = N_0$, the states of energy lower than μ_0 will be nearly filled, and those above largely empty. The value of μ will be close to this μ_0, and identical to it at $T = 0$. For the Bose-Einstein system, with the minus sign, on the other hand, the opposite is true, the numbers $\overline{N}_\nu$ decrease with increasing energy from the energetically lowest state of energy $\varepsilon_\nu = 0$ *more* rapidly than the exponential $e^{-\varepsilon_\nu/kT}$ of the Boltzmann equation. If the sum, for all states ν other than the lowest, with $\varepsilon_0 = 0$,

$$\sum_{\nu\geq 1}\frac{1}{e^{\varepsilon_\nu/kT}-1} = \overline{N}_\varepsilon(T), \tag{23.7$_4$}$$

is finite, then at μ values infinitessimally greater than zero the single lowest state of zero energy will accomodate all the molecules of the system greater than $\overline{N}_\varepsilon(T)$, namely there will be $\overline{N}_0 = \overline{N} - \overline{N}_\varepsilon(T)$ molecules in this lowest state, for any $\overline{N} > \overline{N}_\varepsilon(T)$, and the density can increase without limit at essentially constant $\mu = 0$. This is known as the Bose-Einstein condensation.

In order to put numbers into the equations one must know the energy levels, ε_ν, of the molecules. If there are no internal degrees of freedom these will be purely translational states having $V h^{-3} 4\pi p^2\, dp$ different states between the magnitudes p and $p + dp$ of momentum. The energies of these states are at $\varepsilon = \frac{1}{2m} p^2$, $p = \sqrt{2m\varepsilon}$, $dp = \frac{1}{2}\sqrt{2m}\,\varepsilon^{-\frac{1}{2}}\, d\varepsilon$, or,

$$N(\varepsilon)\, d\varepsilon = 2\pi V h^{-3} (2m)^{\frac{3}{2}} \varepsilon^{\frac{1}{2}}\, d\varepsilon, \tag{23.7$_5$}$$

would be the number of quantum states ν of energies between ε and $\varepsilon + d\varepsilon$.

Actually all Fermi particles must have half integer spins, and hence non-zero angular momenta; there will be, then, $(2s+1)$ internal states, differing only in spin orientation, and having the same energy, for every translational state. Thus for Fermi-Dirac molecules there are necessarily at least two internal states. For Bose-Einstein systems s must be integer, and is most usually zero. If, then, there are no other internal degrees of freedom one writes (23.7$_2$) as

$$\varrho = \frac{\overline{N}}{V} = 2\pi (2s+1) \left(\frac{2m}{h^2}\right)^{\frac{3}{2}} \int_0^\infty \frac{\sqrt{\varepsilon}\, d\varepsilon}{e^{\varepsilon - \mu/kT} \mp 1} \qquad \begin{matrix} -\text{B.E.} \\ +\text{F.D.} \end{matrix} \tag{23.7$_6$}$$

Introduce,

$$\lambda = \frac{h}{\sqrt{2\pi m k T}}, \tag{23.8}$$

$$z = \frac{1}{\lambda^3} e^{\mu/kT}, \tag{23.9}$$

which is an activity. With $\varepsilon/kT = x$, the equation becomes

$$\left.\begin{aligned} \varrho &= \frac{2}{\sqrt{\pi}} (2s+1) \frac{1}{\lambda^3} \int_0^\infty \frac{\sqrt{x}\, z\lambda^3 e^{-x}}{1 \mp z\lambda^3 e^{-x}}\, dx \\ &= \frac{2}{\sqrt{\pi}} (2s+1) \frac{1}{\lambda^3} \sum_{n\geq 1} \pm (\pm z\lambda^3)^n \int_0^\infty \sqrt{x}\, e^{-nx}\, dx \\ &= (2s+1) \frac{1}{\lambda^3} \sum_{n\geq 1} \pm \frac{(\pm z\lambda^3)^n}{n^{\frac{3}{2}}}, \qquad \begin{matrix} +\text{B.E.} \\ -\text{F.D.} \end{matrix} \end{aligned}\right\} \tag{23.10}$$

since

$$\int_0^\infty \sqrt{y}\, e^{-y}\, dy = \tfrac{1}{2}\sqrt{\pi}.$$

At activity $z \to 0$ the density approaches zero, but

$$\lim_{\varrho \to 0} \left[\frac{z}{\varrho}\right] = 2s + 1. \tag{23.11}$$

We have previously always chosen our scale of activity so that this limit is unity, which it is in (23.11) for spinless molecules of $s = 0$. It is much more convenient in the case of these equations to retain the normalization (23.11) rather than $z = \varrho$ as both approach zero. For given z, T, then, the density is always proportional to $2s+1$, the density of each single internal state m of fixed spin projection is,

$$\varrho_m = \frac{1}{\lambda^3} \sum_{n \geq 1} (\pm)^{n-1} \frac{(z\lambda^3)^n}{n^{\frac{3}{2}}}, \qquad \begin{matrix} +\text{B.E.} \\ -\text{F.D.} \end{matrix} \tag{23.12}$$

and,

$$\lim_{z \to 0} \left[\frac{z}{\varrho_m}\right] = 1. \tag{23.13}$$

The equations have a formal identity to the imperfect gas equations of Sects. 13 and 14 [Eq. (14.11) for ϱ], namely,

$$\varrho_m = \sum_{n \geq 1} n\, b_n^{(0)} z^n, \tag{23.14}$$

with

$$\left.\begin{aligned} b_1^{(0)} &\equiv 1, \\ b_n^{(0)} &= (\pm)^{n-1} \frac{\lambda^{3(n-1)}}{n^{\frac{5}{2}}} \qquad \begin{matrix} +\text{B.E.} \\ -\text{F.D.} \end{matrix} \end{aligned}\right\} \tag{23.15}$$

The formal identity is shown to be very real in Sect. 25.

For spin $s=0$, one internal state only, the equations are identical to those of an ordinary imperfect gas. For two or more internal states, $s \geq \frac{1}{2}$, each internal quantum state acts as an imperfect gas, with cluster integrals $b_n^{(0)}$ of (23.15), but the different internal states do not interact with each other. The gas can be treated by the formalism of Sects. 13 to 21, as consisting of independent species a and b, with all cluster integrals $b_{\boldsymbol{n}}$ for which the set $\boldsymbol{n}$ contains more than one kind of molecule identically zero, $b_{\boldsymbol{n}} = 0$, unless $\boldsymbol{n} = n_a$ only.

We can immediately write the pressure equation

$$\left.\begin{aligned} P &= (2s+1)\, \mathrm{k}T \sum_{n \geq 1} b_n^{(0)} z^n \\ &= (2s+1)\, \mathrm{k}T z \sum_{n \geq 1} (\pm)^{n-1} \frac{(z\lambda^3)^{n-1}}{n^{\frac{5}{2}}} \end{aligned}\right\} \qquad \begin{matrix} +\text{B.E.} \\ -\text{F.D.} \end{matrix} \tag{23.16}$$

with the partial pressure, P_m, of each internal spin state given without the factor $(2s+1)$. The total density $\varrho = \sum_{m=-s}^{m=+s} \varrho_m = (2s+1)\,\varrho$ is

$$\varrho = (2s+1)\, z \sum_{n \geq 1} (\pm)^{n-1} \frac{(z\lambda^3)^{n-1}}{n^{\frac{3}{2}}} \qquad \begin{matrix} +\text{B.E.} \\ -\text{F.D.} \end{matrix} \tag{23.17}$$

In the limit $z \to 0$ the gas behaves classically, and the kinetic energy per molecule is $\frac{3}{2}\mathrm{k}T$. Eq. (4.17$_2$) for the difference of this energy and that at z, gives the energy per molecule as

$$\varepsilon = \frac{\mathrm{k}T}{\varrho} \left(\frac{\partial}{\partial \log T}\right)_z \frac{P}{\mathrm{k}T} + \frac{3}{2}\mathrm{k}T. \tag{23.18$_1$}$$

Since $(\partial \log \lambda / \partial \log T)_z = -\frac{1}{2}$ one finds, from (23.16),

$$\varepsilon = \frac{3}{2}\mathrm{k}T \left[1 - (2s+1) \frac{z}{\varrho} \sum_{n \geq 1} (\pm)^{n-1} (n-1) \frac{(z\lambda^3)^{n-1}}{n^{\frac{5}{2}}}\right]. \tag{23.18$_2$}$$

Using this with (23.16) and (23.17) one has

$$\varepsilon = \frac{3}{2} \frac{P}{\varrho}, \tag{23.18$_3$}$$

an equation which is general for the kinetic energy of translation of all perfect gases.

The smaller cluster integrals can be evaluated numerically from the equations of (15.13$_4$). They are,

$$\left.\begin{aligned}\beta_1 &= \pm 0.707107\,\lambda^3,\\ \beta_2 &= +(4.950\times 10^{-3})\,\lambda^6,\\ \beta_3 &= \pm(0.148\times 10^{-3})\,\lambda^9,\\ \beta_4 &= +(0.477\times 10^{-6})\,\lambda^{12}\end{aligned}\right\}\quad \begin{matrix}+\text{B.E.}\\ \mp\text{F.D.}\end{matrix} \tag{23.18$_4$}$$

Using (17.8$_1$) one has

$$\left.\begin{aligned}P &= \frac{2}{3}\,\varrho\,\varepsilon = \mathrm{k}T\sum_{m=-s}^{m=+s}\Big(\varrho_m - \sum_{\nu\geq 1}\frac{\nu}{\nu+1}\,\beta_\nu\,\varrho_m^{\nu+1}\Big)\\ &= \mathrm{k}T\,\varrho\Big[1 \mp 0.35355\,\frac{\varrho\,\lambda^3}{2s+1} - (3.30\times 10^{-3})\Big(\frac{\varrho\,\lambda^3}{2s+1}\Big)^2 \mp\\ &\quad \mp(0.111\times 10^{-3})\Big(\frac{\varrho\,\lambda^3}{2s+1}\Big)^3 - (0.382\times 10^{-6})\Big(\frac{\varrho\,\lambda^3}{2s+1}\Big)^4 \mp \cdots\Big]\end{aligned}\right\}\quad \begin{matrix}-\text{B.E.}\\ \pm\text{F.D.}\end{matrix} \tag{23.18$_5$}$$

Eq. (17.9$_1$) may be used for μ, with $\mu_0 = \mathrm{k}T\log\lambda^3$ and replacing ϱ in that equation by $\varrho_m = \varrho/(2s+1)$. One finds

$$\left.\begin{aligned}\mu &= \mathrm{k}T\log\Big(\frac{\varrho\,\lambda^3}{2s+1}\Big) + \mathrm{k}T\Big[\mp 0.707107\Big(\frac{\varrho\,\lambda^3}{2s+1}\Big) -\\ &\quad -(4.950\times 10^{-3})\Big(\frac{\varrho\,\lambda^3}{2s+1}\Big)^2 \mp (0.148\times 10^{-3})\Big(\frac{\varrho\,\lambda^3}{2s+1}\Big)^3 -\\ &\quad -(0.477\times 10^{-6})\Big(\frac{\varrho\,\lambda^3}{2s+1}\Big)^4 \mp \cdots\Big].\end{aligned}\right\}\quad \begin{matrix}-\text{B.E.}\\ \pm\text{F.D.}\end{matrix} \tag{23.19}$$

β) Bose-Einstein condensation. For the Bose-Einstein perfect gas the series $\Sigma b_n^{(0)} z^n$ of (23.16) with the positive coefficients, $b_n^{(0)} = n^{-\frac{5}{2}}\lambda^{3(n-1)}$, has a singularity on the positive real axis at

$$z_c = \lambda^{-3} = \Big(\frac{2\pi m\,\mathrm{k}T}{h^2}\Big)^{\frac{3}{2}}. \tag{23.20}$$

The sums,

$$\sum_{n\geq 1}\frac{1}{n^{\frac{5}{2}}} = 1.341, \tag{23.21$_1$}$$

$$\sum_{n\geq 1}\frac{1}{n^{\frac{3}{2}}} = 2.612, \tag{23.21$_2$}$$

are finite, and the pressure approaches the definite positive value

$$\left.\begin{aligned}P_c &= 1.341\,(2s+1)\,\mathrm{k}T\,\lambda^{-3}\\ &= 1.341\,(2s+1)\Big(\frac{2\pi m}{h^2}\Big)^{\frac{3}{2}}(\mathrm{k}T)^{\frac{5}{2}},\end{aligned}\right\} \tag{23.21$_3$}$$

while ϱ becomes

$$\left.\begin{aligned}\varrho_c &= 2.612\,(2s+1)\,\lambda^{-3}\\ &= 2.612\,(2s+1)\Big(\frac{2\pi m\,\mathrm{k}T}{h^2}\Big)^{\frac{3}{2}}.\end{aligned}\right\} \tag{23.21$_4$}$$

The number $P/\varrho\,\mathrm{k}T$ at this singularity is independent of T or any other parameter of the system,

$$\frac{P_c}{\varrho_c\,\mathrm{k}T} = \frac{1.341}{2.612} = 0.5134. \tag{23.21$_5$}$$

The second derivative of $\frac{1}{\mathrm{k}T}P$ with respect to $\log z$, which is proportional to the isothermal compressibility, $\varkappa$,

$$\frac{1}{\mathrm{k}T}\left(\frac{\partial^2 P}{\partial \log z^2}\right)_T = \left(\frac{\partial \varrho}{\partial \log z}\right)_T = \mathrm{k}T\,\varrho\left(\frac{\partial \varrho}{\partial P}\right)_T = \mathrm{k}T\,\varkappa,$$

is then

$$\mathrm{k}T\,\varkappa = \frac{1}{\lambda^3}\sum_{n\geq 1}\frac{1}{n^{\frac{1}{2}}}(z\,\lambda^3)^n, \tag{23.21$_6$}$$

which approaches infinite values as the singularity is approached. The condensation is one in which the compressibility becomes gradually infinite, rather than abruptly, as in normal condensation of a classical gas.

Numerically,

$$\lambda = \frac{17.46}{\sqrt{MT}}\times 10^{-8}\,\mathrm{cm}, \tag{23.21$_7$}$$

with M the molecular weight. The relation (23.21$_4$) gives for $s=0$, $N_0=$ AVOGADRO's number,

$$\frac{N_0}{\varrho_c} = V_c = \frac{1227}{(MT)^{\frac{3}{2}}}\,\mathrm{cm}^3\,\mathrm{mole}^{-1} \tag{23.22$_1$}$$

for the volume per mole at which the condensation occurs at T, or

$$T_c = \frac{114.7}{V^{\frac{2}{3}}M} \tag{23.22$_2$}$$

for the temperature in terms of the volume per mole in cm³. For molecular weight, $M=4$, and the volume per mole of liquid helium, $V=27.6\,\mathrm{cm}^3$, one finds $T_c=3.14^\circ$ K.

For densities less than those of condensation Eqs. (23.18) and (23.19) converge quite rapidly. From (23.21$_4$) one can compute the critical density at any temperature, and set

$$\frac{\varrho\,\lambda^3}{2s+1} = 2.612\frac{\varrho}{\varrho_c}. \tag{23.23}$$

We find

$$\frac{P}{\varrho\,\mathrm{k}T} = \frac{2}{3}\frac{\varepsilon}{\mathrm{k}T} = 1 - 0.4618\frac{\varrho}{\varrho_c} - 0.0225\left(\frac{\varrho}{\varrho_c}\right)^2 - 0.00196\left(\frac{\varrho}{\varrho_c}\right)^3 \ldots. \tag{23.24}$$

γ) Fermi-Dirac gases. The series $\sum b_n^{(0)} z^n$ with alternating signs for the Fermi-Dirac gas has no singularity on the positive real axis. The various equations of this section can be used for numerical computation at low densities, and are completely adequate for gases other than electrons at any density for which they actually exist as gases. Indeed the effects due to the potential of interaction, which is here neglected, is always more important than the effects discussed here.

However an electron gas, particularly at the high densities of electrons in metals[1], would be completely degenerate, in that the quantum states would be filled to energies far higher than $\mathrm{k}T$ at any ordinary temperatures. Eq. (23.7$_6$) at $T=0$ has a denominator in the integrand which is unity, for the Fermi-Dirac case, when $\varepsilon<\mu$, but infinity for $\varepsilon>\mu$. The integral is then $\int_0^\mu \sqrt{\varepsilon}\,d\varepsilon = \frac{2}{3}\mu^{\frac{3}{2}}$.

---

[1] A. SOMMERFELD: Z. Physik **47**, 1 (1928) also A. SOMMERFELD and H. BETHE: Handbuch der Physik, Bd. XXIV, Teil 2. 1934.

One finds for the chemical potential, μ_0, at $T=0$,

$$\varrho = \frac{4\pi}{3}(2s+1)\left(\frac{2m\mu_0}{h^2}\right)^{\frac{3}{2}}, \tag{23.25$_1$}$$

$$\mu_0 = \frac{h^2}{2m}\left(\frac{3}{4\pi}\frac{\varrho}{2s+1}\right)^{\frac{2}{3}}. \tag{23.25$_2$}$$

This energy, the energy of the top filled level, is called the Fermi energy. At $T=0^\circ$ K all states of lower energy are completely filled, and all states of higher energy are empty.

The average energy per molecule is obtained by multiplying the integrand of (23.7$_6$) with ε, integrating, and dividing by ϱ. One finds

$$\varepsilon(T=0) = \tfrac{3}{5}\mu_0. \tag{23.26}$$

Since, in general $\frac{P}{\varrho} = \frac{3}{2}\varepsilon$, Eq. (23.18$_3$), we have,

$$\varepsilon(T=0) + \frac{P(T=0)}{\varrho} = \mu_0, \tag{23.27}$$

which is correct, since the entropy at absolute zero is zero.

The usual method of mathematical treatment for the Fermi gas at low temperatures, or high densities, starts from the solution at $T=0$ and evaluates the integral of (23.7$_6$) and similar integrals by series development in temperature. The integrand is some smooth function, $f(\varepsilon)$, of the energy ε, multiplied by a factor $[e^{(\varepsilon-\mu)/kT}+1]^{-1}$, which is close to a step function at low temperatures, namely unity for $\varepsilon<\mu$, and zero for $\varepsilon>\mu$. Write the integral as

$$I = \int_0^\infty f(\varepsilon)\, g(\varepsilon)\, d\varepsilon, \tag{23.28$_1$}$$

$$g(\varepsilon) = \frac{1}{e^{\frac{\varepsilon-\mu}{kT}}+1}, \tag{23.28$_2$}$$

with

$$\frac{dg}{d\varepsilon} = -\frac{1}{kT}\,\frac{1}{e^{\frac{\varepsilon-\mu}{kT}}+1}\,\frac{1}{e^{-\frac{\varepsilon-\mu}{kT}}+1}. \tag{23.28$_3$}$$

The derivative, $dg/d\varepsilon$, is a symmetric function of $\varepsilon-\mu$, with a sharp minimum of value $-1/4kT$ at $\varepsilon-\mu=0$. By a partial integration of (23.28$_1$) one has

$$I = -\int_0^\infty \frac{dg}{d\varepsilon} F(\varepsilon)\, d\varepsilon, \tag{23.28$_4$}$$

with

$$F(\varepsilon) = \int_0^\varepsilon f(\varepsilon')\, d\varepsilon'. \tag{23.28$_5$}$$

Develop $F(\varepsilon)$ about $\varepsilon=\mu$, so that with

$$F_0(\mu) = \int_0^\mu f(\varepsilon)\, d\varepsilon, \tag{23.28$_6$}$$

$$F(\varepsilon) = F_0(\mu) + \sum_{n\geq 0}^{\infty} \frac{(\varepsilon-\mu)^{n+1}}{(n+1)!}\left[\frac{d^n f(\varepsilon)}{d\varepsilon^n}\right]_{\varepsilon=\mu}, \tag{23.28$_7$}$$

in (23.28$_4$) one has

$$I = -F_0(\mu)\int_0^\infty \frac{dg}{d\varepsilon}\,d\varepsilon - \sum_{n\geq 0}^{\infty}\left(\frac{d^n f}{d\varepsilon^n}\right)_{\varepsilon=\mu}\int_0^\infty \frac{(\varepsilon-\mu)^{n+1}}{(n+1)!}\,\frac{dg}{d\varepsilon}\,d\varepsilon. \qquad (23.28_8)$$

The first integral is $g(\infty)-g(0)\approx -1$ since $g(\infty)=0$ and $g(0)\approx 1$. We extend the lower limit of the integrals in the sum to $-\infty$, since $dg/d\varepsilon$ is so close to zero for $\varepsilon<0$. The function $dg/d\varepsilon$ is even in $(\varepsilon-\mu)$ so that, in the summation, only odd n terms are non zero, $n+1=2m$ with $m\geq 1$. Change the variable of integration to $x=(\varepsilon-\mu)/\mathsf{k}T$, so that the integrands are,

$$\frac{1}{2m!}(\mathsf{k}T)^{2m}x^{2m}\,\mathsf{k}T\,\frac{dg}{d\varepsilon} = -(\mathsf{k}T)^{2m}\frac{1}{(2m)!}\,\frac{x^{2m}}{e^x(1+e^{-x})^2}.$$

Expand $e^{-x}(1+e^{-x})^{-2} = -\sum\nu(-)^\nu e^{-\nu x}$. The integrations over x are then $-2(\mathsf{k}T)^{2m}J_m$, with

$$-J_m = \frac{1}{(2m)!}\int_0^\infty \frac{x^{2m}\,dx}{e^x(1+e^{-x})^2} = -\sum_{\nu\geq 1}\nu(-)^\nu\frac{1}{(2m)!}\int_0^\infty x^{2m}e^{-\nu x}\,dx = -\sum_{\nu\geq 1}\frac{(-)^\nu}{\nu^{2m}},$$

$$J_1 = \frac{\pi^2}{12},\qquad J_2 = \frac{7\pi^4}{720}. \qquad (23.28_9)$$

One has

$$I = \int_0^\mu f(\varepsilon)\,d\varepsilon + \frac{\pi^2}{6}(\mathsf{k}T)^2\left(\frac{df}{d\varepsilon}\right)_{\varepsilon=\mu} + \frac{7\pi^4}{720}(\mathsf{k}T)^4\left(\frac{d^3f}{d\varepsilon^3}\right)_{\varepsilon=\mu} + \cdots. \qquad (23.29)$$

The application of this to (23.7$_6$) with $f(\varepsilon)=\sqrt{\varepsilon}$ leads to

$$\varrho = \frac{4\pi}{3}(2s+1)\left(\frac{2m\mu}{h^2}\right)^{\frac{3}{2}}\left[1+\frac{\pi^2}{8}\left(\frac{\mathsf{k}T}{\mu}\right)^2 + \frac{7\pi^4}{640}\left(\frac{\mathsf{k}T}{\mu}\right)^4+\cdots\right]$$

or with (23.25$_2$) for μ_0,

$$\left(\frac{\mu}{\mu_0}\right)^{\frac{3}{2}}\left[1+\frac{\pi^2}{8}\left(\frac{\mathsf{k}T}{\mu}\right)^2+\frac{7\pi^4}{640}\left(\frac{\mathsf{k}T}{\mu}\right)^4+\cdots\right]=1,$$

which can be inverted to obtain

$$\mu = \mu_0\left[1-\frac{\pi^2}{12}\left(\frac{\mathsf{k}T}{\mu_0}\right)^2-\frac{\pi^4}{80}\left(\frac{\mathsf{k}T}{\mu_0}\right)^4-\cdots\right] \qquad (23.30)$$

for the chemical potential μ in terms of μ_0, the chemical potential at $T=0°$ K given by (23.25$_2$) in terms of the number density. The energy per molecule, ε, is found to be

$$\varepsilon = \frac{3}{5}\mu_0\left[1+\frac{5\pi^2}{12}\left(\frac{\mathsf{k}T}{\mu_0}\right)^2-\frac{\pi^4}{16}\left(\frac{\mathsf{k}T}{\mu_0}\right)^4+\cdots\right], \qquad (23.31)$$

and the pressure is, as always for a perfect gas,

$$P = \tfrac{2}{3}\varepsilon\varrho. \qquad (23.32)$$

The heat capacity at constant volume is, from (23.31), seen to be

$$c_v = \left(\frac{\partial\varepsilon}{\partial T}\right)_\varrho = \mathsf{k}\left[\frac{\pi^2}{2}\,\frac{\mathsf{k}T}{\mu_0}-\frac{3\pi^4}{20}\left(\frac{\mathsf{k}T}{\mu_0}\right)^3+\cdots\right], \qquad (23.33)$$

and to approach zero linearly at $T\to 0$. The entropy $\int_0 c_v\,d\log T$, is, per molecule,

$$s = \mathsf{k}\left[\frac{\pi^2}{2}\,\frac{\mathsf{k}T}{\mu_0}-\frac{\pi^4}{20}\left(\frac{\mathsf{k}T}{\mu_0}\right)^3+\cdots\right]. \qquad (23.34)$$

δ) Summary. Perfect gases, defined as those lacking any potential energy of interaction between molecules, show deviations from the perfect gas equation $P = \varrho kT$, at sufficiently low temperatures. The deviations are appreciable only if $\varrho\lambda^3$ is not entirely negligible compared to unity, with λ the de Broglie wave length, $\lambda = h/\sqrt{2\pi m kT}$. The quantity $\varrho\lambda^3$ is unity only at volumes per mole, V_0,

$$V_0 = \frac{3.2}{(MT)^{\frac{3}{2}}} \text{ liters mole}^{-1}$$

so that for real gases the effect is generally negligible, and of interest only for its theoretical implications.

The deviations from perfect gas behavior can be put into the formalism of the treatment by series development used for classical gases, with the difference that the coefficients, $b_n^{(0)}$, in the activity series for P/kT can be evaluated exactly for all finite n.

The coefficients for Bose-Einstein systems and for Fermi-Dirac systems differ only in that the former have all positive sign, whereas the latter systems have alternating signs.

Bose-Einstein systems show a condensation at $\varrho\lambda^3 = 2.612$ (for single internal states) a condensation in which the excess molecules for higher densities are all in the single translational quantum state of zero energy, and which, for the hypothetical perfect gas, would persist with constant pressure to infinite density. Energy, entropy, and pressure of the Bose-Einstein gases all become zero at $T = 0°$ K.

Fermi-Dirac systems have finite non-zero energy and pressure at $T = 0°$ K, although the entropy, of course, becomes zero. At low temperatures the energy increases quadratically, and the heat capacity and entropy linearly, with T.

24. Density matrix formulation. *α) General statement.* The most general method of Quantum Statistical Mechanics uses the concept of the density matrix, a formulation which has the great advantage of bringing out most closely the analogy with the classical methods [2], [5]. The partition function,

$$Q_c = \sum_{\nu} e^{-E_\nu/kT}, \tag{24.1}$$

of the Canonical Ensemble, is proportional to the trace of the diagonal representation of the density matrix, namely the sum over the diagonal elements, $\boldsymbol{\nu}, \boldsymbol{\nu}$, of the representation in the quantum numbers $\boldsymbol{\nu}$ of the stationary states of fixed energy $E_{\boldsymbol{\nu}}$. The equilibrium density matrix is proportional to the matrix

$$\boldsymbol{\rho}_{\boldsymbol{N}} = A\, e^{-\boldsymbol{H}_{\boldsymbol{N}}/kT}, \tag{24.2}$$

with $\boldsymbol{H}$ the Hamiltonian operator of the system. The trace, $A Q_c$, of the matrix is invariant under the representation of the matrix. The diagonal elements, $\varrho_{\boldsymbol{N}}(\{\boldsymbol{N}\}, \{\boldsymbol{N}\})$, of the coordinate representation are proportional to the probability density of finding the system at the coordinate $\{\boldsymbol{N}\}$. These elements, in the classical limit go over to

$$\lim_{h\to 0} [\varrho_{\boldsymbol{N}}(\{\boldsymbol{N}\}, \{\boldsymbol{N}\}) = A \left[\prod_a \lambda_a^{-f_a N_a}\right] e^{-U_{\boldsymbol{N}}/kT}. \tag{24.3}$$

The trace is the integral of this over the coordinates, and in the classical limit gives the classical equations. We shall develop these equations from the formulation presented in Sect. 1.

β) The density matrix. The state of a quantum mechanical system is most fully described by a continuous single-valued normalized function, Φ, of time and of the coordinates, or of the momenta of the system. Choose the coordinates as variables, and represent them as before by $\{\boldsymbol{N}\}$ for a set $\boldsymbol{N}$ of molecules, but include in $\{\boldsymbol{N}\}$ all $f_\alpha - 3$ internal coordinates, $q_{\alpha i}$, for the molecules, α, $1 \leq \alpha \leq N$, as well as the 3 coordinates $\boldsymbol{r}_\alpha$ of the center of mass. For an ensemble consisting of systems $i = 1, 2, \ldots$, each containing the same set $\boldsymbol{N}$ of molecules, the density matrix, $\varrho_{\boldsymbol{N}}(\{\boldsymbol{N}\}', \{\boldsymbol{N}\}'')$, in the coordinate representation, depends on two sets $\{\boldsymbol{N}\}'\,\{\boldsymbol{N}\}''$ of coordinates, and is defined as

$$\varrho_{\boldsymbol{N}}(\{\boldsymbol{N}\}', \{\boldsymbol{N}\}'') = \sum_i \Phi_i^* \{\boldsymbol{N}\}' \, \Phi_i \{N\}'', \tag{24.4}$$

with Φ_i^* the conjugate complex of Φ_i. The functions will generally be time dependent, and it is by no means necessary that the states Φ_i, Φ_j of different members be orthogonal.

The state, Φ_i, of any member could always be represented as a linear combination,

$$\Phi_i \{\boldsymbol{N}\} = \sum_{\boldsymbol{\nu}} a_{i\boldsymbol{\nu}} \Psi_{\boldsymbol{\nu}} \{\boldsymbol{N}\}, \tag{24.5}$$

of the orthonormal set, $\Psi_{\boldsymbol{\nu}}$, of stationary solutions of the time independent Schrödinger equation,

$$\boldsymbol{H}_{\boldsymbol{N}} \Psi_{\boldsymbol{\nu}} \{\boldsymbol{N}\} = E_{\boldsymbol{\nu}} \Psi_{\boldsymbol{\nu}} \{\boldsymbol{N}\}, \tag{24.6$_1$}$$

$$\int\int \cdots \int \Psi_{\boldsymbol{\nu}}^* \{\boldsymbol{N}\} \Psi_{\boldsymbol{\mu}} \{\boldsymbol{N}\} \, d\{\boldsymbol{N}\} = \delta(\boldsymbol{\nu} - \boldsymbol{\mu}), \tag{24.6$_2$}$$

with $\boldsymbol{H}_{\boldsymbol{N}}$ the Hamiltonian operator of the systems. The normalization of Φ_i requires that,

$$\sum_{\boldsymbol{\nu}} a_{i\boldsymbol{\nu}}^* a_{i\boldsymbol{\nu}} = 1. \tag{24.7}$$

Use of (24.5) in (24.4) enables us to write,

$$\varrho_{\boldsymbol{N}}(\{\boldsymbol{N}\}', \{N\}'') = \sum_{\boldsymbol{\nu}} \sum_{\boldsymbol{\mu}} P_{\boldsymbol{\nu\mu}} \Psi_{\boldsymbol{\nu}}^* \{\boldsymbol{N}\}' \Psi_{\boldsymbol{\mu}} \{\boldsymbol{N}\}'', \tag{24.8$_1$}$$

$$P_{\boldsymbol{\nu},\boldsymbol{\mu}} = \sum_i a_{i\boldsymbol{\nu}}^* a_{i\boldsymbol{\mu}}. \tag{24.8$_2$}$$

In general there would be off-diagonal elements $P_{\boldsymbol{\nu\mu}}$ for $\boldsymbol{\nu} \neq \boldsymbol{\mu}$, and the elements $P_{\boldsymbol{\nu},\boldsymbol{\mu}}$ and $P_{\boldsymbol{\nu\nu}}$ may be time dependent.

For the equilibrium system at temperature T we assume that the off-diagonal elements are zero, and that the diagonal elements, $P_{\boldsymbol{\nu\nu}}$, are proportional to $e^{-E_{\boldsymbol{\nu}}/kT}$. This would be the case, for instance, for an ensemble in which every member i is in some single stationary state $\boldsymbol{\nu}_i$, and for which the fraction of members of the ensemble in the state $\boldsymbol{\nu}$ is proportional to $e^{-E_{\boldsymbol{\nu}}/kT}$. One writes

$$\varrho_{\boldsymbol{N}}(\{\boldsymbol{N}\}', \{N\}'') = A \sum_{\boldsymbol{\nu}} \Psi_{\boldsymbol{\nu}}^* \{\boldsymbol{N}\}' e^{-E_{\boldsymbol{\nu}}/kT} \Psi_{\boldsymbol{\nu}} \{\boldsymbol{N}\}'' \tag{24.9}$$

for the density matrix, in coordinate representation, of the equilibrium ensemble.

The question of normalization of $\boldsymbol{\rho}_{\boldsymbol{N}}$ is arbitrary. The $\boldsymbol{\rho}_{\boldsymbol{N}}$ of (24.4) is so defined that its trace would be the number of member systems of the ensemble. The normalization of (24.9) is, in view of (24.6$_2$)

$$\left.\begin{aligned} \operatorname{Tr}(\boldsymbol{\rho}_{\boldsymbol{N}}) &= \int\int \cdots \int \varrho_{\boldsymbol{N}}(\{\boldsymbol{N}\}, \{\boldsymbol{N}\}) \, d\{\boldsymbol{N}\} \\ &= A \sum_{\boldsymbol{\nu}} e^{-E_{\boldsymbol{\nu}}/kT} = A \, Q_c(V, T, \boldsymbol{N}), \end{aligned}\right\} \tag{24.10}$$

with Q_c the partition function of the canonical ensemble of the set $\boldsymbol{N}$ of molecules in V at T, defined in Eq. (1.16). The matrix defined as that of (24.9) with $A = Q_c^{-1}$ would have trace unity, and is the usual normalized density matrix. We shall later choose another normalization.

The probability density that a system in state $\boldsymbol{\nu}$ would be observed at the coordinate configuration $\{\boldsymbol{N}\}$ is $\Psi_\nu^*\{\boldsymbol{N}\}\,\Psi_\nu\{\boldsymbol{N}\}$, and since $\frac{1}{Q_c}\,\mathrm{e}^{-E_\nu/kT}$ is the probability that a member system be in state $\boldsymbol{\nu}$, it follows from (24.9) that

$$\frac{1}{A\,Q_c}\,\varrho_{\boldsymbol{N}}(\{\boldsymbol{N}\},\{\boldsymbol{N}\})\,d\{\boldsymbol{N}\}$$

is the probability that a member system of the ensemble be found at the coordinate value $\{\boldsymbol{N}\}$ in $d\{\boldsymbol{N}\}$.

The operator, $\mathrm{e}^{-\boldsymbol{H}_{\boldsymbol{N}}/kT}$, is defined as

$$\mathrm{e}^{-\boldsymbol{H}_{\boldsymbol{N}}/kT} = \sum_{n\geq 0}\frac{1}{n!}\,\frac{(-)^n}{(kT)^n}\,(\boldsymbol{H}_{\boldsymbol{N}})^n, \tag{24.11}$$

in which $(\boldsymbol{H}_{\boldsymbol{N}})^n$ means n-fold repetition of the operation $\boldsymbol{H}_{\boldsymbol{N}}$. In view of (24.6_1) the operator $\mathrm{e}^{-\boldsymbol{H}_{\boldsymbol{N}}/kT}$ operating on Ψ_ν gives the sum (24.11) with the number E_ν replacing $\boldsymbol{H}_{\boldsymbol{N}}$, multiplied by Ψ_ν. The sum is $\mathrm{e}^{-E_\nu/kT}$. We have

$$\mathrm{e}^{-\boldsymbol{H}_{\boldsymbol{N}}/kT}\,\Psi_\nu\{\boldsymbol{N}\} = \mathrm{e}^{-E_\nu/kT}\,\Psi_\nu\{\boldsymbol{N}\}, \tag{24.12_1}$$

so that (24.9) can also be written as

$$\varrho_{\boldsymbol{N}}(\{\boldsymbol{N}\}',\{\boldsymbol{N}\}'') = A\sum_{\nu}\Psi_\nu^*\{\boldsymbol{N}\}'\,\mathrm{e}^{-\boldsymbol{H}_{\boldsymbol{N}}/kT}\,\Psi_\nu\{\boldsymbol{N}\}''. \tag{24.12_2}$$

The advantage of the formulation (24.12_2) is that the value of $\varrho_{\boldsymbol{N}}$ is now invariant under the choice of functions Ψ_ν, and that any complete orthonormal set, $\Phi_k\{\boldsymbol{N}\}$, of functions may be used to replace Ψ_ν in (24.12_2), namely that,

$$\varrho_{\boldsymbol{N}}(\{\boldsymbol{N}\}',\{\boldsymbol{N}\}'') = A\sum_{k}\Phi_k^*\{\boldsymbol{N}\}'\,\mathrm{e}^{-\boldsymbol{H}_{\boldsymbol{N}}/kT}\,\Phi_k\{\boldsymbol{N}\}'', \tag{24.13}$$

with Φ_k any complete orthonormal set of functions of $\{\boldsymbol{N}\}$ in V, obeying the same boundary conditions, and having the same symmetry as the allowed solutions, $\Psi_\nu\{\boldsymbol{N}\}$, of the time independent Schrödinger equation (24.6_1).

This invariance, which is basic to matrix algebra, may be proved readily. The functions Φ_k may always be written as a linear combination,

$$\Phi_k = \sum_{\nu} a_{k\nu}\Psi_\nu, \tag{24.14_1}$$

with

$$a_{k,\nu} = \int \Psi_\nu^*\{\boldsymbol{N}\}\,\Phi_k\{\boldsymbol{N}\}\,d\{\boldsymbol{N}\}, \tag{24.14_2}$$

as can be seen by multiplying both sides of (24.14_1) by Ψ_ν^* and integrating, using the orthonormality relation (24.6_2). Similarly if

$$\Psi_\nu = \sum_{k} a_{\nu k}^\dagger\,\Phi_k \tag{24.14_3}$$

the corresponding orthonormality condition for the Φ_k's gives an equation for $a_{\nu k}^\dagger$ which is just the conjugate complex of (24.14_2), or

$$a_{\nu k}^\dagger = a_{k\nu}^*. \tag{24.14_4}$$

Use (24.14$_3$) for both Ψ_ν^* and Ψ_μ in (24.6$_2$) or (24.14$_1$) for both $\Phi_{\boldsymbol{k}}^*$ and $\Phi_{\boldsymbol{n}}$ in a similar orthonormality equation for the Φ's, to prove that

$$\sum_\nu a_{\boldsymbol{k}\nu} a_{\nu\boldsymbol{n}}^\dagger = \delta(\boldsymbol{k}-\boldsymbol{n}), \tag{24.14$_5$}$$

$$\sum_{\boldsymbol{k}} a_{\nu\boldsymbol{k}}^\dagger a_{\boldsymbol{k}\mu} = \delta(\nu-\mu), \tag{24.14$_5'$}$$

namely that the matrix $\boldsymbol{A}$ of elements $a_{\boldsymbol{k}\nu}$, and its conjugate transposed matrix $\boldsymbol{A}^\dagger$, of elements $a_{\nu\boldsymbol{k}}^\dagger = a_{\boldsymbol{k}\nu}^*$, are reciprocal,

$$\boldsymbol{A}^\dagger \boldsymbol{A} = \boldsymbol{A}\boldsymbol{A}^\dagger = \boldsymbol{1}, \tag{24.14$_6$}$$

with $\boldsymbol{1}$ the unit matrix. The transformation matrix $\boldsymbol{A}$ is said to be unitary.

Use (24.14$_1$) for $\Phi_{\boldsymbol{k}}^*$ and $\Phi_{\boldsymbol{k}}$ in (24.13) to find

$$\left.\begin{aligned}
&\sum_{\boldsymbol{k}}\sum_\nu a_{\nu\boldsymbol{k}}^\dagger \Psi_\nu^*\{\boldsymbol{N}\}' \,\mathrm{e}^{-\boldsymbol{H}_{\boldsymbol{N}}/kT} \sum_\mu a_{\boldsymbol{k}\mu} \Psi_\mu\{\boldsymbol{N}\}'' \\
&\quad = \sum_\nu\sum_\mu \Big(\sum_{\boldsymbol{k}} a_{\nu\boldsymbol{k}}^\dagger a_{\boldsymbol{k}\mu}\Big) \Psi_\nu^*\{\boldsymbol{N}\}' \,\mathrm{e}^{-\boldsymbol{H}_{\boldsymbol{N}}/kT} \Psi_\mu\{\boldsymbol{N}\}'' \\
&\quad = \sum_\nu \Psi_\nu^*\{\boldsymbol{N}\}' \,\mathrm{e}^{-\boldsymbol{H}_{\boldsymbol{N}}/kT} \Psi_\mu\{\boldsymbol{N}\}''
\end{aligned}\right\} \tag{24.14$_7$}$$

from (24.14$_5$), where the transference of the *numbers* $a_{\boldsymbol{k},\mu}$ to the front of the operator is completely legitimate. This proves (24.13) to be identical to (24.12$_2$).

γ) Plane wave representation. A particularly simple set of functions, $\Phi_{\boldsymbol{k}}$, for monatomic molecules whose coordinate set $\{\boldsymbol{N}\} = \boldsymbol{r}_1, \boldsymbol{r}_2, \ldots, \boldsymbol{r}_N$, are the cartesian vector set, is the set of plane wave functions,

$$\Phi_{\boldsymbol{k}} = \frac{1}{L^{3N/2}}\,\mathrm{e}^{(2\pi i/L)\sum_\alpha \boldsymbol{k}_\alpha\cdot\boldsymbol{r}_\alpha}, \tag{24.15$_1$}$$

which are orthonormal in a cubic box of side L, $V = L^3$. These (or real linear combinations of them) are the stationary solutions of the Schrödinger equation for the perfect gas with $U_{\boldsymbol{N}} \equiv 0$. The functions $\Phi_{\boldsymbol{k}}$ are states of fixed momentum vector, $\boldsymbol{p}_\alpha$,

$$\boldsymbol{p}_\alpha = \frac{h}{L}\boldsymbol{k}_\alpha, \tag{24.15$_2$}$$

$$d\boldsymbol{k}_\alpha = \frac{V}{h^3}\,d\boldsymbol{p}_\alpha. \tag{24.15$_3$}$$

Since, for large L, the states lie very close in p space, the summation over $\boldsymbol{k}$ in (24.13) can be replaced by integration over $\prod_\alpha d\boldsymbol{k}_\alpha$. Replacing $\boldsymbol{k}_\alpha$ by the more meaningful momenta, $\boldsymbol{p}_\alpha$ one may write (24.13) for a one component system of monatomic molecules as

$$\left.\begin{aligned}
&\varrho_{\boldsymbol{N}}^{(u)}(\{\boldsymbol{N}\}', \{\boldsymbol{N}\}'') \\
&\quad = \frac{A}{h^{3N}} \iint_{-\infty}^{+\infty}\!\!\cdots\!\int \mathrm{e}^{-2\pi i/h}\sum_\alpha \boldsymbol{p}_\alpha\cdot\boldsymbol{r}_\alpha' \,\mathrm{e}^{-\boldsymbol{H}_{\boldsymbol{N}}/kT}\,\mathrm{e}^{-2\pi i/h\sum_\alpha \boldsymbol{p}_\alpha\cdot\boldsymbol{r}_\alpha''} \prod_\alpha d\boldsymbol{p}_\alpha.
\end{aligned}\right\} \tag{24.15$_4$}$$

As discussed in the next paragraphs, we have not, in (24.15$_4$) satisfied the symmetry conditions on the allowed states for Bose-Einstein or Fermi-Dirac systems. For this reason a superscript (u) for unsymmetrized has been placed on the $\varrho_{\boldsymbol{N}}$ on the left.

δ) Symmetrization of ϱ_N. The elements $\varrho_N(\{\boldsymbol{N}\}', \{\boldsymbol{N}\}'')$ of $\boldsymbol{\rho}_N$, are seen from (24.9) to be Hermitian,

$$\varrho_N(\{\boldsymbol{N}\}', \{\boldsymbol{N}\}'') = \varrho_N^*(\{\boldsymbol{N}\}'', \{\boldsymbol{N}\}'), \tag{24.16}$$

and the diagonal elements, $\varrho_N(\{\boldsymbol{N}\}, \{\boldsymbol{N}\})$, are, of course, positive real. If only allowed functions, $\Psi_\nu\{\boldsymbol{N}\}$, which are either symmetric or antisymmetric in exchange of pairs of identical molecules, are used in (24.9), as should be the case, then $\varrho_N(\{\boldsymbol{N}\}', \{\boldsymbol{N}\}'')$ will be properly symmetric or antisymmetric in exchange of pairs of identical molecules in the set $\{\boldsymbol{N}\}'$, *or* in a similar pair exchange in $\{\boldsymbol{N}\}''$. A simultaneous exchange of corresponding pairs in both $\{\boldsymbol{N}\}'$ *and* $\{\boldsymbol{N}\}''$ will multiply the function by $(+1)^2 = (-1)^2$ in both the case of symmetric or antisymmetric functions, so that the elements are always symmetric in permutations of identical molecules in both $\{\boldsymbol{N}\}'$ and $\{\boldsymbol{N}\}''$ together.

In general, any function, $F\{N\}$ of coordinates $N = \boldsymbol{r}_1, \boldsymbol{r}_2, \ldots, \boldsymbol{r}_N$, can be uniquely decomposed into additive components which have definite symmetry character with respect to permutations of sets of the coordinates, that is symmetric or antisymmetric in permutations within certain subsets. Only one such component will be totally symmetric and one totally antisymmetric. The sum of all $N!$ permutations of the totally symmetric component will be $N!$ times itself, but this sum for all other components is zero. The sum of all $N!$ permutations, p, of the totally antisymmetric component, multiplied by $(-)^{P_p}$, with P_p the number of pairs permuted, is $N!$ times itself, and again the sum is zero for all other components. We may thus select the symmetric component $F^{(S)}\{N\}$, or the antisymmetric component $F^{(A)}\{N\}$, by

$$F^{(S)}\{N\} = \frac{1}{N!}\sum_{p=1}^{N!} F\{N_p\}, \tag{24.17$_1$}$$

$$F^{(A)}\{N\} = \frac{1}{N!}\sum_{p=1}^{N!} (-)^{P_p} F\{N_p\}, \tag{24.17$_2$}$$

where the sum runs over all permutations, $1 \leq p \leq N!$ for which $\{N_p\}$ are the permuted coordinates,

$$\{N_p\} = \boldsymbol{P}_p\{N\}, \tag{24.17$_3$}$$

with $\boldsymbol{P}_p$ the permutation operator of permutation p. The number P_p in the exponent of $(-)$ in (24.17$_2$) is the number of pairs permuted in the permutation p, and is uniquely even or odd.

We may thus ignore the symmetry requirements on the set Ψ_ν or $\boldsymbol{\Phi}_k$ of (24.12$_2$) or (24.13) and select the correctly symmetrized component of $\varrho_N(\{N\}', \{N\})$ by use of (24.17$_1$) or (24.17$_2$), summing over all independent permutations $\{N_{p'}\}'$ and $\{N_{p''}\}''$ of ϱ_N, and dividing by $(N!)^2$. Actually the double summation is totally unnecessary. Every pair permutation, $\{N_{p'}\}', \{N_{p''}\}''$, can be attained by first permuting both sets $\{N\}'$ and $\{N\}''$ together by one of the permutations, p' (or p''), an operation that leaves ϱ_N unaltered, and then permuting only $\{N\}''$ (or $\{N\}'$) by the permutation which results in $\{N_{p''}\}''$. It is therefore only necessary to sum over all permutations of one of the sets $\{N\}'$ or $\{N\}''$.

In the general case of a set $\boldsymbol{N}$ of molecules there are $\boldsymbol{N}!$ permutations of identical molecules, some species may require symmetric, and other species the antisymmetric functions. Symbolically we write,

$$\varrho_{\boldsymbol{N}}(\{\boldsymbol{N}\}', \{\boldsymbol{N}\}'') = \frac{1}{\boldsymbol{N}!}\sum_{p=1}^{\boldsymbol{N}!}\left[\prod_{\alpha} (\pm)^{P_{ap}}\right] \varrho_{\boldsymbol{N}}^{(u)}(\{\boldsymbol{N}_p\}', \{\boldsymbol{N}\}''), \tag{24.17$_4$}$$

with P_{ap} the numbers of pairs of a permuted in p.

The permuted coordinates, $\{N_p\}'$, for the one identical permutation, $p=1$, $P_{ap}=0$, is of course,

$$\{N_1\}' \equiv \{N\}', \tag{24.17$_5$}$$

so that the first term of $\varrho_{\boldsymbol{N}}$ is just $\varrho_{\boldsymbol{N}}^{(u)}$ divided by $\boldsymbol{N}!$, or,

$$\left.\begin{aligned}\varrho_{\boldsymbol{N}}(\{\boldsymbol{N}\}',\{\boldsymbol{N}\}'') &= \frac{1}{\boldsymbol{N}!}\,\varrho_{\boldsymbol{N}}^{(u)}(\{\boldsymbol{N}\}',\{\boldsymbol{N}\}'') + \\ &+ \frac{1}{\boldsymbol{N}!}\sum_{p=2}^{\boldsymbol{N}!}\Big[\prod_{\alpha}(\pm)^{P_{ap}}\Big]\,\varrho_{\boldsymbol{N}}^{(u)}(\{\boldsymbol{N}_p\}',\{\boldsymbol{N}\}'').\end{aligned}\right\} \tag{24.17$_6$}$$

In the classical limit the function $\varrho_{\boldsymbol{N}}^{(u)}(\{\boldsymbol{N}_p\}',\{\boldsymbol{N}\}'')$, $p\geqq 2$, is zero, or at least of zero weight. The sum for $p\geqq 2$ is then zero. The properly symmetrized function $\varrho_{\boldsymbol{N}}$, is then

$$\lim_{h\to 0}\,[\varrho_{\boldsymbol{N}}(\{\boldsymbol{N}\}'\,\{\boldsymbol{N}\}'')] = \frac{1}{\boldsymbol{N}!}\,\varrho_{\boldsymbol{N}}^{(u)}(\{N\}',\{N\}''), \tag{24.17$_7$}$$

which, of course, corresponds to the factor $1/\boldsymbol{N}!$ introduced in the phase space integration, Eq. (2.10), in Sect. 2, and arises from the fact that approximately only $1/\boldsymbol{N}!$ of the solutions, Ψ_ν, of (24.6$_1$) has the required symmetry.

ε) Reduction to $\{\boldsymbol{N}\}=\boldsymbol{r}_1,\ldots,\boldsymbol{r}_N$. The elements $\varrho_{\boldsymbol{N}}(\{\boldsymbol{N}\}',\{\boldsymbol{N}\}'')$, of the density matrix were defined with $\{\boldsymbol{N}\}'$ and $\{\boldsymbol{N}\}''$ including all coordinates, $\boldsymbol{r}_\alpha$, $\boldsymbol{q}_{i\alpha}$ of the molecules $1\leqq\alpha\leqq N$, with $\boldsymbol{r}_\alpha$ the three dimensional vector of the center of mass, and $\boldsymbol{q}_{i\alpha}$ the internal coordinates of molecule α. We can always integrate over the part diagonal in the $\boldsymbol{q}_{i\alpha}$'s, namely

$$\left.\begin{aligned}&\varrho_{\boldsymbol{N}}^{(\mathrm{tr})}(\boldsymbol{r}_1',\ldots,\boldsymbol{r}_N';\boldsymbol{r}_1'',\ldots,\boldsymbol{r}_N'')\\ &\quad=\iint\cdots\int\varrho_{\boldsymbol{N}}(\{N\}',\{N\}'')\prod_{\alpha}\delta(\boldsymbol{q}_{i\alpha}'-q_{i\alpha}'')\,d\boldsymbol{q}_{i\alpha}'\,dq_{i\alpha}'',\end{aligned}\right\} \tag{24.18$_1$}$$

and so obtain a function $\varrho_{\boldsymbol{N}}^{(\mathrm{tr})}$, depending only on the translational center of mass coordinates. Now, always approximately, and in certain cases exactly, this integration serves only to bring a factor of the internal partition function $Q_{i\alpha}$ for each molecule, or, for all molecules,

$$\prod_{\alpha=1}^{N}Q_{i\alpha}=\prod_{a}Q_{ia}^{N_a}, \tag{24.18$_2$}$$

multiplied by the independent factor in $\varrho_{\boldsymbol{N}}(\{\boldsymbol{N}\}'\,\{\boldsymbol{N}\}'')$ which depends on the center of mass coordinates, $\boldsymbol{r}_1,\ldots,\boldsymbol{r}_{\boldsymbol{N}}$, alone and which is that characteristic of a monatomic gas. This is exactly true if the potential energy $U_{\boldsymbol{N}}\{\boldsymbol{N}\}$ is separable in the center of mass coordinates,

$$U_{\boldsymbol{N}}\{\boldsymbol{N}\}=U_{\mathrm{tr}}(\boldsymbol{r}_1,\ldots\boldsymbol{r}_N)+\sum_{\alpha}u_\alpha(\boldsymbol{q}_{i\alpha}), \tag{24.19$_1$}$$

since in this case the solutions $\Psi_\nu\{\boldsymbol{N}\}$ of (24.6$_1$) are products,

$$\Psi_\nu\{\boldsymbol{N}\}=\chi_{\boldsymbol{k}}(\boldsymbol{r}_1,\ldots\boldsymbol{r}_N)\prod_{\alpha=1}^{N}\psi_{\nu\alpha}(\boldsymbol{q}_{i\alpha}), \tag{24.19$_2$}$$

and the energies are sums,

$$E_\nu=E_{\boldsymbol{k}}+\sum_{\alpha=1}^{N}\varepsilon_{\nu\alpha}. \tag{24.19$_3$}$$

The integration of (24.18_1) for one α over $d\boldsymbol{q}'_{i\alpha}\, d\boldsymbol{q}''_{i\alpha}$ then contributes, in $\varrho_{\boldsymbol{N}}^{(u)}$, the factor

$$\left.\begin{aligned} \int \cdots \int\int \sum_{\nu\alpha} \mathrm{e}^{-\frac{\varepsilon_{\nu\alpha}}{kT}} \psi^*_{\nu\alpha}(\boldsymbol{q}'_{i\alpha})\, \psi_{\nu\alpha}(\boldsymbol{q}''_{i\alpha})\, \delta(\boldsymbol{q}'_{i\alpha} - \boldsymbol{q}''_{i\alpha})\, d\boldsymbol{q}'_{i\alpha}\, d\boldsymbol{q}''_{i\alpha} \\ = \sum_{\nu\alpha} \mathrm{e}^{-\frac{\varepsilon_{\nu\alpha}}{kT}} = Q_{ia}. \end{aligned}\right\} \tag{24.19$_4$}$$

with Q_{ia} the internal partition function of a molecule of the species a to which α belongs, Eq. (3.14).

Even if the potential is not strictly separable, the function $\Psi_\nu\{\boldsymbol{N}\}$ in a coordinate position $\boldsymbol{r}_1, \ldots, \boldsymbol{r}_N$ for which *all* pair distances $r_{\alpha\beta}$ are large will asymptotically approach (24.19_2) with $\chi_{\boldsymbol{k}}$ a plane wave function, and the energies will be those of (24.19_3). The function $\varrho_{\boldsymbol{N}}^{(\mathrm{tr})}$ will thus approach the values,

$$\lim_{\text{all } r_{\alpha\beta}\to 0} \varrho_{\boldsymbol{N}}^{(\mathrm{tr})} = \Big[\prod_a Q_{ia}^{N_a}\Big]\, \varrho_{\boldsymbol{N}} \text{ (monatomic gas)}, \tag{24.19$_5$}$$

in which limit, of course, the $\varrho_{\boldsymbol{N}}$ (monatomic gas) is purely that of a perfect gas.

However, this contribution of the factor Q_{ia} by integration over the internal coordinates, $\boldsymbol{q}_{i\alpha}$, of every molecule α of species a, is only true for the unpermuted density matrix, $\varrho_{\boldsymbol{N}}^{(u)}$. In any one of the terms $\varrho_{\boldsymbol{N}}^{(u)}(\{\boldsymbol{N}_p\}', \{\boldsymbol{N}\}'')$, $p \geqq 2$, of (24.17_6) the integrals occurring in (24.19_4) involve integration over $\psi^*_{\nu p\alpha}\, \psi_{\nu\alpha}$. If the internal quantum state, $\psi_{p\nu\alpha}$, of the atom p_α into which α has been permuted, is different than the state $\psi_{\nu\alpha}$, then, because of the orthogonality condition, the contribution of the integral is zero. As a result only permutations of identical molecules of the same internal quantum states should be made in the symmetrization, (24.17_6), of the density matrix $\varrho_{\boldsymbol{N}}^{(\mathrm{tr})}$ of the center of mass coordinates.

Actually the complete discussion of this case would be highly artificialr Essentially all molecules having any complexity of internal states are rather heavy, or the temperature is relatively high, or both. In either case the density in the μ-space of the centers of mass is low, and the contribution of the terms for $p \geqq 2$ from the sum in (24.17_6) is negligible. It suffices to take the single term of the identical permutation, $p = 1$, which is correctly multiplied by the product of the internal partition functions for all molecules. The only cases for which attention to the permutations $p \geqq 2$ is required are the cases of very light molecules, He or H_2, and at low temperatures, and even for these examples the effects are small for the gases. For He^4 there is a single internal state, $Q_i = 1$. For He^3 there are two internal states, the two projections of the nuclear spin of value $\frac{1}{2}$. For H_2 at sufficiently low temperatures for the symmetrization to become significant only the single lowest state of zero rotation and zero resultant nuclear spin is present at equilibrium. If ortho hydrogen is present in non-equilibrium quantities it must anyway be handled as a separate molecular species of higher energy, and fixed mol fraction, and the formalism of a two component system in which permutations of identical molecules only are allowed, is then used. The cases of He^3, or electrons, in which $s = \frac{1}{2}$, $Q_i = 2s + 1 = 2$ does cause some awkwardness which, however, is not serious. One may proceed either of two ways. One rather simple method is just to treat the system, for purposes of antisymmetrization, as a two component system, each spin component treated as a separate molecule. An alternative method is to note that the factor $Q_i = 2$ for each molecule in $\varrho_{\boldsymbol{N}}^{(u)}$ is reduced by exactly the factor $Q_i^{P_p}$ with P_p the numbers of pairs permuted in the permutations if *all* $N!$ permutations are carried out.

Henceforth we shall always assume that the integration of (24.18_1) has been performed, and restrict our attention to functions $\varrho_{\boldsymbol{N}}(\{\boldsymbol{N}\}'\,\{\boldsymbol{N}\}'')$ in which the coordinates $\{\boldsymbol{N}\}'$ and $\{\boldsymbol{N}\}''$ are those of the centers of mass only. The plane wave choice, (24.15_4), for the orthonormal set, seems then to be particularly convenient, although there are extreme cases where this choice is poor.

We shall now fix the arbitrary normalization of the functions, and in such a way that the diagonal elements $\varrho_{\boldsymbol{N}}(\{\boldsymbol{N}\},\{\boldsymbol{N}\})$ become the functions $\frac{1}{\boldsymbol{N}!}F_{\boldsymbol{N}}^{(0)}\{\boldsymbol{N}\}$, used in the imperfect gas equations. We introduce the now familiar de Broglie wave length,

$$\lambda_a = \sqrt{\frac{h^2}{2\pi m_a \mathrm{k} T}}, \tag{24.20$_1$}$$

define

$$\{\boldsymbol{N}\} = \boldsymbol{r}_1, \boldsymbol{r}_2, \ldots, \boldsymbol{r}_N, \tag{24.20$_2$}$$

and the unsymmetrized density matrix, reduced to the center of mass coordinates, now normalized as

$$\varrho_{\boldsymbol{N}}^{(u)}(\{\boldsymbol{N}\}',\{\boldsymbol{N}\}'') = \Big[\prod_a \lambda_a^{3N_a}\Big] \sum_{\boldsymbol{k}} \Phi_{\boldsymbol{k}}^*\{\boldsymbol{N}\}'\, \mathrm{e}^{-\boldsymbol{H}_{\boldsymbol{N}}/\mathrm{k}T}\, \Phi_{\boldsymbol{k}}\{\boldsymbol{N}\}'', \tag{24.20$_3$}$$

with $\Phi_{\boldsymbol{k}}$ any orthonormal set of functions of $\{\boldsymbol{N}\}$ in V, having the correct boundary conditions, but not selected for symmetry. The Hamiltonian operator $\boldsymbol{H}_{\boldsymbol{N}}$, is

$$\boldsymbol{H}_{\boldsymbol{N}} = \sum_\alpha \frac{h^2}{8\pi^2 m_a} \nabla_\alpha \cdot \nabla_\alpha + U_{\boldsymbol{N}}\{\boldsymbol{N}\}, \tag{24.20$_4$}$$

$$\nabla_\alpha \cdot \nabla_\alpha = \frac{\partial^2}{\partial x_\alpha^2} + \frac{\partial^2}{\partial y_\alpha^2} + \frac{\partial^2}{\partial z_\alpha^2}. \tag{24.20$_5$}$$

If plane-wave functions are used for $\Phi_{\boldsymbol{k}}$, we have, from (24.20_3) and $(24.15_4$

$$\left.\begin{aligned} \varrho_{\boldsymbol{N}}^{(u)}(\{\boldsymbol{N}\}'\,\{\boldsymbol{N}\}'') = \Big[\prod_a \Big(\frac{1}{2\pi m_a \mathrm{k}T}\Big)^{3N_a/2}\Big] \iint_{-\infty}^{+\infty}\!\!\cdots\!\int \mathrm{e}^{-(2\pi i/h)\{\boldsymbol{N}\}'\cdot \boldsymbol{p}^{(\boldsymbol{N})}} \times \\ \times\, \mathrm{e}^{-\boldsymbol{H}_{\boldsymbol{N}}/\mathrm{k}T}\, \mathrm{e}^{(2\pi i/h)\{\boldsymbol{N}\}''\cdot \boldsymbol{p}^{(\boldsymbol{N})}}\, d\boldsymbol{p}^{(\boldsymbol{N})}, \end{aligned}\right\} \tag{24.20$_6$}$$

$$\boldsymbol{p}^{(\boldsymbol{N})} = \boldsymbol{p}_1, \boldsymbol{p}_2, \ldots, \boldsymbol{p}_\alpha, \ldots \boldsymbol{p}_N, \tag{24.20$_7$}$$

$$\{\boldsymbol{N}\}\cdot \boldsymbol{p}^{(\boldsymbol{N})} = \sum_\alpha \boldsymbol{p}_\alpha \cdot \boldsymbol{r}_\alpha. \tag{24.20$_8$}$$

The density matrix of proper symmetry, $\boldsymbol{\rho}_{\boldsymbol{N}}$, is obtained from $\varrho_{\boldsymbol{N}}^{(u)}$ by $(24.17_4$ or (24.17_6) noting the requirements for the $Q_{ia} \neq 1$. The function $F_{\boldsymbol{N}}^{(0)}\{\boldsymbol{N}\}$ i defined as $\boldsymbol{N}!$ times the diagonal element of $\varrho_{\boldsymbol{N}}$.

$$F_{\boldsymbol{N}}^{(0)}\{\boldsymbol{N}\} = \boldsymbol{N}!\,\varrho_{\boldsymbol{N}}(\{\boldsymbol{N}\},\{\boldsymbol{N}\}). \tag{24.21}$$

The partition function $Q_c(V, T, \boldsymbol{N})$ for a Canonical Ensemble of systems, each having the set $\boldsymbol{N}$ of molecules in volume V at temperature T, is then, with this normalization,

$$Q_c(V, T, \boldsymbol{N}) = \Big[\prod_a \Big(\frac{Q_{ia}}{\lambda_a^3}\Big)^{N_a}\Big] \iint_V\!\cdots\!\int \frac{1}{\boldsymbol{N}!} F_{\boldsymbol{N}}^{(0)}\{\boldsymbol{N}\}\, d\{\boldsymbol{N}\} \tag{24.22}$$

with Q_{ia} the internal partition function of molecules of species a, Eq. (3.14).

The Grand Canonical Ensemble partition function is now, from (1.18) and (1.14),

$$\left.\begin{aligned} e^{PV/kT} &= Q_{GC}(V,T,\boldsymbol{\mu}) \\ &= \sum_{\boldsymbol{N}\geq 0} e^{\boldsymbol{N}\cdot\boldsymbol{\mu}/kT} Q_C(V,T,\boldsymbol{N}) \\ &= \sum_{\boldsymbol{N}\geq 0} \frac{\boldsymbol{z}^{\boldsymbol{N}}}{\boldsymbol{N}!} \int\!\!\int_V \cdots \int F_{\boldsymbol{N}}^{(0)}\{\boldsymbol{N}\}\, d\{\boldsymbol{N}\} \end{aligned}\right\} \tag{24.23}$$

with

$$\boldsymbol{z} = z_a, z_b, \ldots \tag{24.24}$$

and z_a the activity of species a, see Eq. (4.16),

$$z_a = \frac{Q_{ia}}{\lambda_a^3} e^{\mu_a/kT}, \tag{24.25}$$

which approaches the number density, ϱ_a, of species a in the perfect classical gas limit of zero density.

ζ) Other representations of $\boldsymbol{\rho}_N$. Before proceeding to show how the density matrix, $\varrho_N^{(u)}$, of (24.20$_6$) may actually be computed and used, we should like to make some remarks about other representations of $\boldsymbol{\rho}_N$.

We have chosen to emphasize the coordinate representation of the matrix $\boldsymbol{\rho}_N$, with elements $\varrho_N(\boldsymbol{r}'_1, \ldots, \boldsymbol{r}'_N; \boldsymbol{r}''_1, \ldots, \boldsymbol{r}''_N)$ since this leads most readily to comparison with the classical case. The matrix is Hermitian, that is each element ϱ_N is the conjugate complex of the element with the two sets $\{\boldsymbol{N}\}'$ and $\{\boldsymbol{N}\}''$ exchanged. Any such matrix can be brought into another representation by a unitary transformation matrix, $\boldsymbol{A}$, through the operation $\boldsymbol{A}\,\boldsymbol{\rho}\,\boldsymbol{A}^\dagger$. At least two such representations other than the coordinate representation are useful, and one peculiar mixed representation is particularly valuable.

A unitary transformation matrix $\boldsymbol{A}$ is one obeying Eq. (24.14$_6$), that is, with elements obeying Eqs. (24.14$_4$), (24.14$_5$) and (24.14$_{5'}$). Since the indices $\{\boldsymbol{N}\}'$ $\{\boldsymbol{N}\}''$ of the coordinate representation of $\boldsymbol{\rho}_N$ are continuous variables, the unitary matrix elements of a transformation from the coordinates to some other representation must have these continuous variables $\{\boldsymbol{N}\}$ as one index. The other index may be continuous or not. The operation of summation over the discontinuous indices in (24.14$_5$) is replaced by integration over continuous indices, in this case over $d\{\boldsymbol{N}\}$, the $3N$ fold coordinate space of the N molecules. The unitary transformation matrix, $\boldsymbol{A}$, that transforms $\boldsymbol{\rho}$ to the representation $\boldsymbol{\nu}, \boldsymbol{\mu}$, with $\boldsymbol{\nu}$ the quantum states of the system in which the Hamiltonian matrix $\boldsymbol{H}$ is diagonal, is just the matrix of elements $a(\boldsymbol{\nu}, \{\boldsymbol{N}\})$ which are the eigenfunctions $\Psi_\nu\{\boldsymbol{N}\}$. The conjugate transpose $A^\dagger$ has elements $a^\dagger(\{\boldsymbol{N}\}, \boldsymbol{\nu}) = a^*(\boldsymbol{\nu}, \{\boldsymbol{N}\}) = \Psi_\nu^*\{\boldsymbol{N}\}$. This matrix is unitary since

$$\int a(\boldsymbol{\nu}, \{\boldsymbol{N}\})\, a^\dagger(\{\boldsymbol{N}\}, \boldsymbol{\nu})\, d\{\boldsymbol{N}\} = \int \Psi_\nu\{\boldsymbol{N}\} \Psi_\mu^*\{\boldsymbol{N}\}\, d\{\boldsymbol{N}\} = \delta(\boldsymbol{\nu}\cdot\boldsymbol{\mu}), \tag{24.26}$$

$$\sum_{\nu=0}^{\infty} a^\dagger(\{\boldsymbol{N}\}', \{\boldsymbol{\nu}\})\, a(\{\boldsymbol{\nu}\}, \{\boldsymbol{N}\}'') = \sum_{\nu\geq 0} \Psi_\nu^*\{\boldsymbol{N}\}' \Psi_\nu\{\boldsymbol{N}\}'' = \delta(\{\boldsymbol{N}\}' - \{\boldsymbol{N}\}''). \tag{24.26'}$$

Using (24.9) for $\varrho_N(\{\boldsymbol{N}\}' \{\boldsymbol{N}\}'')$, the element $\boldsymbol{\nu}, \boldsymbol{\mu}$ is, using (24.26), just

$$\varrho_N(\boldsymbol{\nu}, \boldsymbol{\mu}) = A\, \delta(\boldsymbol{\nu} - \boldsymbol{\mu})\, e^{-E_\nu/kT}, \tag{24.27}$$

namely zero unless $\boldsymbol{\nu} \equiv \boldsymbol{\mu}$, and then $e^{-E_\nu/kT}$. The only non-zero elements are the diagonal elements, the equilibrium density matrix is diagonal in the same representation as the energy.

The unitary transformation to momentum space $\mathbf{p}^{(N)} = \boldsymbol{p}_1, \ldots, \boldsymbol{p}_N$, is the matrix whose elements are the plane wave functions, $\Phi_{\boldsymbol{k}}$ of (24.15_1) with (24.15_2) and (24.15_3). The elements,

$$\left.\begin{aligned} \varrho_{\boldsymbol{N}}(\boldsymbol{p}'^{(\boldsymbol{N})}\boldsymbol{p}''^{(\boldsymbol{N})}) = \frac{1}{h^{3N}} \iint \cdots \int \mathrm{e}^{(2\pi i/h)\,\boldsymbol{p}'^{(\boldsymbol{N})}\cdot\{\boldsymbol{N}\}'}\, \varrho_{\boldsymbol{N}}(\{\boldsymbol{N}\}'\{\boldsymbol{N}\}'') \times \\ \times\, \mathrm{e}^{-(2\pi i/h)\,\boldsymbol{p}''^{(\boldsymbol{N})}\cdot\{\boldsymbol{N}\}''}\, d\{\boldsymbol{N}\}'\, d\{\boldsymbol{N}\}'', \end{aligned}\right\} \tag{24.28}$$

are those of the momentum space representation of the matrix, and the diagonal elements $\varrho_{\boldsymbol{N}}(\boldsymbol{p}^{(\boldsymbol{N})}, \boldsymbol{p}^{(\boldsymbol{N})})$ are proportional to the probability density of the momentum values $\boldsymbol{p}^{(\boldsymbol{N})}$. In the classical limit these elements are proportional to $\mathrm{e}^{-(\boldsymbol{p}^{(\boldsymbol{N})}\cdot\boldsymbol{p}^{(\boldsymbol{N})})/2mkT}$.

The trace of the matrix is invariant under the representation so that the sum or integral of the diagonal elements may be used with any representation to obtain the partition function.

A particularly useful representation is a mixed one, first introduced by WIGNER[1], which gives the quantum mechanical analogue of the probability density in coordinate and momentum space. The elements of this representation are functions of $\{\boldsymbol{N}\}$ and $\boldsymbol{p}^{(\boldsymbol{N})}$. Use

$$\left.\begin{aligned} \{\boldsymbol{N}\} &= \tfrac{1}{2}(\{\boldsymbol{N}\}' + \{\boldsymbol{N}\}'') \\ &= \tfrac{1}{2}(\boldsymbol{r}_1' + \boldsymbol{r}_1''), \tfrac{1}{2}(\boldsymbol{r}_2' + \boldsymbol{r}_2''), \ldots, \tfrac{1}{2}(\boldsymbol{r}_N' + \boldsymbol{r}_N''), \end{aligned}\right\} \tag{24.29$_1$}$$

$$\{\boldsymbol{N}\}''' = \{\boldsymbol{N}\}' - \{\boldsymbol{N}\}'' = \boldsymbol{r}_1' - \boldsymbol{r}_1'', \ldots, \tag{24.29$_2$}$$

and form

$$\left.\begin{aligned} \sigma_{\boldsymbol{N}}(\{\boldsymbol{N}\}, \boldsymbol{p}^{(\boldsymbol{N})}) = \frac{1}{h^{3N}} \iint \cdots \int \varrho_{\boldsymbol{N}}(\{\{\boldsymbol{N}\} + \tfrac{1}{2}\{\boldsymbol{N}\}'''\}, \{\{\boldsymbol{N}\} - \tfrac{1}{2}\{\boldsymbol{N}\}'''\}) \times \\ \times\, \mathrm{e}^{(2\pi i/h)\,\boldsymbol{p}^{(\boldsymbol{N})}\cdot(\boldsymbol{N})'''}\, d\{\boldsymbol{N}\}'''. \end{aligned}\right\} \tag{24.29$_3$}$$

This function has the property that

$$\iint_{-\infty}^{+\infty} \cdots \int \sigma_{\boldsymbol{N}}(\{\boldsymbol{N}\}, \boldsymbol{p}^{(\boldsymbol{N})})\, d\boldsymbol{p}^{(\boldsymbol{N})} = \varrho_{\boldsymbol{N}}(\{\boldsymbol{N}\}, \{\boldsymbol{N}\}), \tag{24.30$_1$}$$

$$\iint_V \cdots \int \sigma_{\boldsymbol{N}}(\{\boldsymbol{N}\}, \boldsymbol{p}^{(\boldsymbol{N})})\, d\{\boldsymbol{N}\} = \varrho_{\boldsymbol{N}}(\boldsymbol{p}^{(\boldsymbol{N})}, \boldsymbol{p}^{(\boldsymbol{N})}). \tag{24.30$_2$}$$

These relations follow from the fact that

$$\frac{1}{h^{3N}} \iint_{-\infty}^{+\infty} \cdots \int \mathrm{e}^{(2\pi i/h)\,\boldsymbol{p}^{(\boldsymbol{N})}\cdot\{\boldsymbol{N}\}}\, d\boldsymbol{p}^{(\boldsymbol{N})} = \delta\{\boldsymbol{N}\}, \tag{24.30$_3$}$$

$$\frac{1}{h^{3N}} \iint_V \cdots \int \mathrm{e}^{(2\pi i/h)\,\boldsymbol{p}^{(\boldsymbol{N})}\cdot\{\boldsymbol{N}\}}\, d\{\boldsymbol{N}\} = \delta(\boldsymbol{p}^{(\boldsymbol{N})}), \tag{24.30$_4$}$$

where the second is strictly valid only in the limit $V \to \infty$. Use (24.29_3) for $\sigma_{\boldsymbol{N}}$ in the integral of (24.30_1). Integrate first over $d\boldsymbol{p}^{(\boldsymbol{N})}$, using (24.30_3) which multiplies $\varrho_{\boldsymbol{N}}(\{\{N\} + \frac{1}{2}\{N\}'''\}, \{\{N\} - \frac{1}{2}\{N\}'''\})$ by $\delta\{N\}'''$, so that integration over $d\{N\}'''$ gives (24.30_1). Similarly use (24.29_3) for $\sigma_{\boldsymbol{N}}$ in the integral of (24.30_1). Change back to $\{\boldsymbol{N}\}'$ and $\{\boldsymbol{N}\}''$ from (24.29_1) and (24.29_2) noting that $d\{\boldsymbol{N}\}'''\, d\{\boldsymbol{N}\} = d\{\boldsymbol{N}\}'\, d\{\boldsymbol{N}\}''$. One has, then (24.28) with $\boldsymbol{p}'^{(\boldsymbol{N})} = \boldsymbol{p}''^{(\boldsymbol{N})} = \boldsymbol{p}^{(\boldsymbol{N})}$.

[1] E. WIGNER: Phys. Rev. **40**, 749 (1932). — MOYAL: Proc. Cambridge Phil. Soc. **45**, 99 (1949). — IRVING and ZWANZIG: J. Chem. Phys. **19**, 1173 (1951).

From (24.30_1) or (24.30_2) we have

$$\left.\begin{aligned}\iint\cdots\int \sigma_{\boldsymbol{N}}(\{\boldsymbol{N}\},\boldsymbol{p}^{(\boldsymbol{N})})\, d\{\boldsymbol{N}\}\, d\boldsymbol{p}^{(\boldsymbol{N})} &= \iint\cdots\int \varrho_{\boldsymbol{N}}(\{\boldsymbol{N}\},\{N\})\, d\{\boldsymbol{N}\}\\ &= \iint\cdots\int \varrho_{\boldsymbol{N}}(\boldsymbol{p}^{(\boldsymbol{N})},\boldsymbol{p}^{(\boldsymbol{N})})\, d\boldsymbol{p}^{(\boldsymbol{N})} = \operatorname{tr}\boldsymbol{\rho}_{\boldsymbol{N}}.\end{aligned}\right\} \tag{24.31}$$

Since $\varrho_{\boldsymbol{N}}(\{\boldsymbol{N}\},\{\boldsymbol{N}\})/\operatorname{tr}\boldsymbol{\rho}_{\boldsymbol{N}}$ is a probability density, the average of any function, $f\{\boldsymbol{N}\}$, of the coordinates is

$$\left.\begin{aligned}\langle f\{\boldsymbol{N}\}_{\mathrm{Av}}\rangle &= \frac{1}{\operatorname{tr}\boldsymbol{\rho}_{\boldsymbol{N}}}\iint\cdots\int \varrho_{\boldsymbol{N}}(\{\boldsymbol{N}\},\{\boldsymbol{N}\})\, f\{\boldsymbol{N}\}\, d\{\boldsymbol{N}\}\\ &= \frac{1}{\operatorname{tr}\boldsymbol{\rho}_{\boldsymbol{N}}}\iint\cdots\int \sigma_{\boldsymbol{N}}(\{\boldsymbol{N}\},\boldsymbol{p}^{(\boldsymbol{N})})\, f\{\boldsymbol{N}\}\, d\{\boldsymbol{N}\}\, d\boldsymbol{p}^{(\boldsymbol{N})},\end{aligned}\right\} \tag{24.32$_1$}$$

and similarly,

$$\langle f(\boldsymbol{p}^{(\boldsymbol{N})})\rangle_{\mathrm{Av}} = \frac{1}{\operatorname{tr}\boldsymbol{\rho}_{\boldsymbol{N}}}\iint\cdots\int \sigma_{\boldsymbol{N}}(\{\boldsymbol{N}\},\boldsymbol{p}^{(\boldsymbol{N})})\, f(\boldsymbol{p}^{(\boldsymbol{N})})\, d\{\boldsymbol{N}\}\, d\boldsymbol{p}^{(\boldsymbol{N})}. \tag{24.32$_2$}$$

The average value of any function of coordinate and momentum involving a product of coordinate and its conjugate momentum, $x_\alpha p_{x\alpha}$, is now less trivial, since always

$$\langle p_i q_i - q_i p_i\rangle_{\mathrm{Av}} = \frac{h}{2\pi i}$$

and the correct average depends on the order of writing, namely either $p_i q_i$ or $q_i p_i$. Since multiplication of such a product by $\sigma_{\boldsymbol{N}}$ and integration over $d\boldsymbol{p}^{(\boldsymbol{N})} d\{\boldsymbol{N}\}$ would lead to identical values whatever the order of writing, it is clear that the obvious generalization of (24.32_1) and (2.32_1) to any function $f(\{\boldsymbol{N}\},\boldsymbol{p}\{\boldsymbol{N}\})$, of coordinates and momenta requires some elaboration. Actually it can be shown that,

$$\langle f(\{\boldsymbol{N}\},\boldsymbol{p}^{(\boldsymbol{N})})\rangle_{\mathrm{Av}} = \frac{1}{\operatorname{tr}\boldsymbol{\rho}_{\boldsymbol{N}}}\iint\cdots\int \sigma_{\boldsymbol{N}}(\{\boldsymbol{N}\},\boldsymbol{p}^{(\boldsymbol{N})})\, f(\{\boldsymbol{N}\},\boldsymbol{p}^{(\boldsymbol{N})})\, d\{\boldsymbol{N}\}\, d\boldsymbol{p}^{(\boldsymbol{N})} \tag{24.32$_3$}$$

is the correct classical analogue of the average value, namely for $q_i p_i$ or $p_i q_i$ it gives,

$$\tfrac{1}{2}\langle q_i p_i + p_i q_i\rangle_{\mathrm{Av}}.$$

The Eqs. (24.32_1) to (24.32_3) are exactly analogous to the useful equations of a function proportional to probability density in coordinate-momentum space.

On the other hand it is fundamental to the quantum mechanics that the uncertainty principle forbids a specific answer to any question involving the exact simultaneous specification of position and conjugated momentum. It is therefore logically excluded to refer to $\sigma_{\boldsymbol{N}}(\{\boldsymbol{N}\},\boldsymbol{p}^{(\boldsymbol{N})})$ as a probability density. This also is demonstrated by the fact that the function defined by (24.29_3) can, and does in some instances, take negative values. However the function is always real, since it can immediately be seen that it is equal to its own conjugate complex. In the classical limit it does go to the classical probability density proportional to $e^{-H/kT}$.

25. Density matrices for perfect gases. The unsymmetrized density matrix element $\varrho_N^{(u)}(\{N\}'\{N\}'')$ (24.20_6), for a one-component monatomic perfect gas, is constructed from the simple Hamiltonian,

$$\frac{1}{kT}\boldsymbol{H}_N = \frac{1}{2mkT}\sum_{i=1}^{3N}\left(\frac{h}{2\pi i}\frac{\partial}{\partial x_i''}\right)^2, \tag{25.1}$$

and is

$$\left.\begin{aligned}&\varrho_N^{(u)}(\{N\}',\{N\}'')\\&\quad=\left(\frac{1}{2\pi m\,\mathrm{k}T}\right)^{3N/2}\iint\cdots\int \mathrm{e}^{-\frac{2\pi i}{h}\{N\}\cdot\boldsymbol{p}^{(N)}}\,\mathrm{e}^{-\boldsymbol{H}_N/\mathrm{k}T}\,\mathrm{e}^{\frac{2\pi i}{h}\{N\}''\cdot\boldsymbol{p}^{(N)}}\,d\boldsymbol{p}^{(N)}.\end{aligned}\right\}\tag{25.2}$$

The Hamiltonian (25.1) is a sum of commuting operators, so that the exponential is

$$\mathrm{e}^{-\boldsymbol{H}_N/\mathrm{k}T}=\prod_{i=1}^{3N}\mathrm{e}^{-\boldsymbol{H}_i/\mathrm{k}T},\tag{25.3$_1$}$$

$$\frac{1}{\mathrm{k}T}\boldsymbol{H}_i=\frac{1}{2m\,\mathrm{k}T}\left(\frac{h}{2\pi i}\,\frac{\partial}{\partial x_i''}\right)^2.\tag{25.3$_2$}$$

The plane wave $\mathrm{e}^{(2\pi i/h)\{N\}''\cdot\boldsymbol{p}^{(N)}}$ is a product of plane waves,

$$\mathrm{e}^{(2\pi i/h)\{N\}''\cdot\boldsymbol{p}^{(N)}}=\prod_{i=1}^{3N}\mathrm{e}^{(2\pi i/h)\,p_i x_i''}.\tag{25.3$_3$}$$

The function $\varrho_N^{(u)}$ is then a product of functions for each i, $1\leq i\leq 3N$, of x_i' and x_i'',

$$\varrho_N^{(u)}(\{N\}',\{N\}'')=\prod_{i=1}^{3N}f(x_i',x_i''),\tag{25.3$_4$}$$

$$f(x_i',x_i'')=\left(\frac{1}{2\pi m\,\mathrm{k}T}\right)^{\frac12}\int\limits_{-\infty}^{+\infty}\mathrm{e}^{-(2\pi i/h)\,p\,x_i'}\,\mathrm{e}^{-\boldsymbol{H}_i/\mathrm{k}T}\,\mathrm{e}^{(2\pi i/h)\,p\,x_i''}\,dp.\tag{25.3$_5$}$$

By definition,

$$\mathrm{e}^{-\boldsymbol{H}_i/\mathrm{k}T}=\sum_{n\geq 0}\frac{(-1)^n}{n!}\left(\frac{\boldsymbol{H}_i}{\mathrm{k}T}\right)^n=\sum_{n\geq 0}\frac{(-1)^n}{n!}\left(\frac{1}{2m\,\mathrm{k}T}\right)^n\left(\frac{h}{2\pi i}\,\frac{\partial}{\partial x_i''}\right)^{2n}\tag{25.3$_6$}$$

and since

$$\left(\frac{h}{2\pi i}\,\frac{\partial}{\partial x_i''}\right)^2\cdot\mathrm{e}^{(2\pi i/h)\,p\,x_i''}=p^2\,\mathrm{e}^{(2\pi i/h)\,p\,x_i''},\tag{25.3$_7$}$$

we find

$$f(x_i',x_i'')=\left(\frac{1}{2\pi m\,\mathrm{k}T}\right)^{\frac12}\int\limits_{-\infty}^{+\infty}\mathrm{e}^{(2\pi i/h)\,p\,(x_i''-x_i')}\,\mathrm{e}^{-(p^2/2m\,\mathrm{k}T)}\,dp.\tag{25.3$_8$}$$

The integration can be performed by completing the square in the exponent, and is simply,

$$f(x_i',x_i'')=\mathrm{e}^{-(\pi/\lambda^2)\,(x_i'-x_i'')^2}\tag{25.3$_9$}$$

with $\lambda=h/\sqrt{2\pi m\,\mathrm{k}T}$. The sum $(x_i'-x_i'')^2$ for the three Cartesian coordinates of one molecule is $|\boldsymbol{r}_\alpha'-\boldsymbol{r}_\alpha''|^2$. We have from (25.3$_4$),

$$\varrho_N^{(u)}(\{N\}',\{N\}'')=\prod_{\alpha=1}^{N}\mathrm{e}^{-(\pi/\lambda^2)\,|\boldsymbol{r}_\alpha'-\boldsymbol{r}_\alpha''|^2}.\tag{25.4}$$

The classical limit is obtained by letting $h=0$, which is $\lambda\to 0$, in which case (25.4) becomes unity for $\{N\}'\equiv\{N\}''$, and zero otherwise,

$$\left.\begin{aligned}&\lim_{h\to 0}\varrho_N^{(u)}(\{N\},\{N\})=1\\&\lim_{h\to 0}\varrho_N^{(u)}(\{N\}',\{N\}'')=0,\qquad \{N\}'\not\equiv\{N\}''.\end{aligned}\right\}\tag{25.5}$$

We wish now, for the case of a single internal state, $Q_i = 1$, to symmetrize or antisymmetrize $\varrho_N^{(u)}$ by the use of (24.17₄), and find $N!$ times the diagonal element $\varrho_N(\{N\}, \{N\})$ of the symmetrized matrix, which is the $F_N^{(0)}\{N\}$ of Eq. (24.21) namely,

$$\left.\begin{aligned} F_N^{(0)}\{N\} &= \sum_{p=1}^{N!} (\pm)^{P_p} \left\{ e^{-(\pi/\lambda^2)\sum_\alpha |\boldsymbol{r}'_{\alpha p} - \boldsymbol{r}''_\alpha|^2} \right\}_{\boldsymbol{r}'_\alpha = \boldsymbol{r}''_\alpha = \boldsymbol{r}_\alpha} \\ &= \sum_{p=1}^{N!} (\pm)^{P_p} e^{-(\pi/\lambda^2)\sum_\alpha |\boldsymbol{r}_{\alpha p} - \boldsymbol{r}_\alpha|^2}, \end{aligned}\right\} \begin{matrix} +\text{B.E.} \\ -\text{F.D.} \end{matrix} \tag{25.6}$$

where $\boldsymbol{r}_{\alpha p}$ is the coordinate of the molecule αp to which α is permuted in the permutation p.

Any permutation p can be analyzed as a product of cyclic permutations. A cyclic permutation of a subset $\{n\}_N$ is one in which the n molecules, in some order, are permuted so that the first replaces the second, the second replaces the third, ..., the $(n-1)$-th replaces the n-th and n-th replaces the first. A general permutation is then one for which the set $\{N\}$ is partitioned into k non overlapping subsets $\{n_i\}_N$, $1 \le i \le k$, $\sum\limits_1 \{n_i\}_N \equiv \{N\}$, and in each subset i one of the $(n_i - 1)!$ possible cyclic permutations has occurred. We may have, for some values of i, that $n_i = 1$, meaning *no* permutation of the single molecule of that subset.

Define a function, $g_n^{(0)}\{n\}$ of the coordinates of n molecules, as the sum of the $(n-1)!$ different possible cyclic permutations of these n molecules. Namely if

$$C_n\{n\} = e^{-(\pi/\lambda^2)[r_{12}^2 + r_{23}^2 + \cdots + r_{n-1,n}^2 + r_{n1}^2]}, \tag{25.7₁}$$

then summing over all $(n-1)!$ permutations of the molecules other than the first,

$$g_n^{(0)}\{n\} = \sum_{p=1}^{(n-1)!} C_n(\boldsymbol{r}_1, \boldsymbol{r}_{2p}, \boldsymbol{r}_{3p}, \ldots, \boldsymbol{r}_{np}), \tag{25.7₂}$$

$$g_1 = 1. \tag{25.7₃}$$

We may then write,

$$F_N^{(0)}\{N\} = \sum_{\{k\{n_i\}_N\}_u} \prod_{i=1}^{k} [(\pm)^{n-1} g_n\{n_i\}_N], \quad \begin{matrix} +\text{B.E.} \\ -\text{F.D.} \end{matrix} \tag{25.8}$$

using the notation of Sect. 13 that $\{k\{n_i\}_N\}_u$ is a partition of $\{N\}$ in k unconnected subsets, $\{n_i\}_N$.

The functions $g_n^{(0)}\{n\}$ are positive. They are zero, for finite n values, if any pair distance $r_{\alpha\beta}$ of members of $\{n\}$ goes to infinity. Their maximum value is $(n-1)!$ at the position for which all molecules are at one place. In the classical limit for which $\lambda \to 0$ they remain non-zero, but only in the region of zero weight where all molecules have identical coordinates, and of course for a real gas this position is excluded by high positive repulsive potentials.

The significant result of this analysis is that the probability density function $F_N^{(0)}\{N\}$ is not uniform in space for even a perfect Bose-Einstein or Fermi-Dirac gas[1]. There is a deviation from unity wherever the molecules approach closely. For Bose-Einstein systems the cluster functions are all positive, the molecules have a net statistical attraction in ordinary coordinate space. For the Fermi-Dirac system the cluster functions alternate in sign, and the net effect is that of a repulsion.

In the Fermi-Dirac case, with a single internal state, this repulsion is absolute, the value of $F_N^{(0)}\{N\}$ is zero if the coordinates of two molecules α and β are identical,

[1] B. KAHN and G. E. UHLENBECK: Physica, Haag 5, 399 (1939).

$\boldsymbol{r}_\alpha = \boldsymbol{r}_\beta$. This may be deduced *a priori* from the fact that the eigenfunctions are all antisymmetric in exchange of $\boldsymbol{r}_\alpha$ and $\boldsymbol{r}_\beta$, and since, when these two coordinates are identical the exchange involves no change in coordinates, the functions are all equal to their own negatives, and hence zero. The diagonal of the density matrix, $F_N^{(0)}\{N\}$, is the sum of the squares of the eigenfunctions, weighted by their probability factors, and must be zero when $\boldsymbol{r}_\alpha = \boldsymbol{r}_\beta$ for any pair. However, it is interesting to see that this also follows from (25.8). The terms of (25.8) can then all be paired in exactly cancelling terms. Consider any partition $\{k\{n_i\}_N\}_u$ of (25.8) in which the molecules α and β occur in the same subset i. Each cycle function, $C_{n+2}\{\{n_i\}+\boldsymbol{r}_\alpha+\boldsymbol{r}_\beta\}$, of this subset is, if $\boldsymbol{r}_\alpha=\boldsymbol{r}_\beta$, exactly equal in value, for any value of the coordinates of other molecules, to a product $C_{\nu+1}\{\{\nu\}_{n_i}+\boldsymbol{r}_\alpha\}\times C_{n-\nu+1}\{\{n_i\}-\{\nu\}_{n_i}+\boldsymbol{r}_\beta\}$ of two cycle functions occurring in the partition $\{k+1\{n_i\}_N\}_u$ which differs from the first only in having $\nu+\boldsymbol{r}_\alpha$ and $n-\nu+\boldsymbol{r}_\beta$ replace $n+\alpha+\beta$. The coefficients of these cycle functions differ in sign, and the total sum is zero.

To proceed, one inserts (25.8) in Eq. (24.23) for the Grand Canonical Ensemble, and defines, in the same manner as before for the imperfect gas,

$$b_n^{(0)} = \lim_{V\to\infty}\left[\frac{1}{n!\,V}\int\!\!\int_V\!\cdots\!\int g_n^{(0)}\{n\}\,d\{n\}\right], \tag{25.9}$$

to obtain

$$P = \mathbf{k}T\sum_{n\geq 1} b_n^{(0)} z^n. \tag{25.10}$$

As we shall immediately demonstrate the cluster functions have the value $b_n^{(0)} = \lambda^{-3(n-1)} n^{-\frac{5}{2}}$ of Eq. (23.15), and the other equations of Sect. 23 follow.

The integration of (25.9) can be performed readily for any finite n. The method is applicable to any cycle function, even if the individual functions of the different pair distances differ. Consider the integral

$$I_n(r_{10}) = \int\!\!\int\cdots\int f_1(r_{12})\,f_2(r_{23})\cdots f_n(r_{n0})\,d\boldsymbol{r}_2\ldots d\boldsymbol{r}_n, \tag{25.11$_1$}$$

in which $r_{ij} = |\boldsymbol{r}_i - \boldsymbol{r}_j|$. Form the Fourier transforms, for each function $f_\nu(r)$,

$$g_\nu(t) = \int f(r)\,e^{2\pi i\boldsymbol{t}\cdot\boldsymbol{r}}\,d\boldsymbol{r}. \tag{25.11$_2$}$$

The reciprocal transform is

$$f_\nu(r) = \int g_\nu(t)\,e^{2\pi i\boldsymbol{t}\cdot\boldsymbol{r}}\,d\boldsymbol{t}. \tag{25.11$_3$}$$

Use (25.11$_3$) for $f_\nu(r) = f(|\boldsymbol{r}_\nu - \boldsymbol{r}_{\nu+1}|)$ in (25.11$_1$) with a different $\boldsymbol{t}$, namely $\boldsymbol{t}_\nu$, for each transform. The integration is now over $d\boldsymbol{r}_2\ldots d\boldsymbol{r}_n\,d\boldsymbol{t}_1\ldots d\boldsymbol{t}_n$. The integrand is $g_1(t_1)\,g_2(t_2)\ldots g_n(t_n)$ times an exponential of

$$\begin{aligned}&2\pi i\left[\boldsymbol{t}_1\cdot(\boldsymbol{r}_1-\boldsymbol{r}_2)+\boldsymbol{t}_2\cdot(\boldsymbol{r}_2-\boldsymbol{r}_3)+\cdots+\boldsymbol{t}_n\cdot(\boldsymbol{r}_n-\boldsymbol{r}_0)\right]\\&\quad=2\pi i\left[\boldsymbol{r}_1\cdot\boldsymbol{t}_1+\boldsymbol{r}_2(\boldsymbol{t}_2-\boldsymbol{t}_1)+\cdots+\boldsymbol{r}_n(\boldsymbol{t}_n-\boldsymbol{t}_{n-1})-\boldsymbol{r}_0\,\boldsymbol{t}_n\right].\end{aligned}$$

Since

$$\int e^{2\pi i\boldsymbol{t}\cdot\boldsymbol{r}}\,d\boldsymbol{r} = \delta(\boldsymbol{t}),$$

the integration over $d\boldsymbol{r}_2\ldots d\boldsymbol{r}_n$ gives a factor $\delta(\boldsymbol{t}_2-\boldsymbol{t}_1)(\boldsymbol{t}_3-\boldsymbol{t}_2)\ldots\delta(\boldsymbol{t}_n-\boldsymbol{t}_{n-1})$, and integration over $d\boldsymbol{t}_2\ldots d\boldsymbol{t}_n$ merely sets all $\boldsymbol{t}$'s equal to $\boldsymbol{t}_1=\boldsymbol{t}$. One has

$$I_n(r_{10}) = \int\left[\prod_{\nu=1}^{n} g_\nu(t)\right] e^{2\pi i\boldsymbol{t}\cdot(\boldsymbol{r}_1-\boldsymbol{r}_0)}\,d\boldsymbol{t}, \tag{25.11$_4$}$$

namely the Fourier transform of $I(r_{10})$ is the product of the Fourier transforms of the n functions in the chain, $f_1(r_{12})\ldots f_n(r_{n0})$. The integrals we seek are those of the cycle functions, C_n, which are of the form of $I_n(r_{10}=0)$, with all f's identical. Since there are $(n-1)!$ cycle functions in each $g_n^{(0)}$, and the integral $b_n^{(0)}$ is $1/n!$ times that over $g_n^{(0)}$ we have

$$b_n^{(0)} = (\pm)^{n-1} \frac{1}{n} \int g^n(t)\, dt, \tag{25.11$_5$}$$

with

$$g(t) = \int e^{-(\pi/\lambda^2) r^2} e^{2\pi i \boldsymbol{t}\cdot\boldsymbol{r}}\, d\boldsymbol{r}. \tag{25.11$_6$}$$

In (25.11$_6$) write $\boldsymbol{t}\cdot\boldsymbol{r} = \text{tr}\cos\vartheta$, and $d\boldsymbol{r} = 2\pi r^2 \sin\vartheta\, d\vartheta\, dr$. With $\cos\vartheta = x$, $\sin\vartheta\, d\vartheta = -dx$ the integration over x extends from 1 to -1. Since

$$\int_{-1}^{+1} e^{iax}\, dx = 2\frac{\sin a}{a},$$

we can rewrite (25.11$_6$) as

$$g(t) = \int_0^\infty 4\pi r^2 e^{-(\pi/\lambda^2) r^2} \frac{\sin 2\pi t r}{2\pi t r}\, dr. \tag{25.11$_7$}$$

The integration is standard, and one finds,

$$g(t) = \lambda^3 e^{-\pi\lambda^2 t^2}. \tag{25.11$_8$}$$

Using this in (25.11$_5$) with $d\boldsymbol{t} = 4\pi t^2\, dt$,

$$b_n^{(0)} = (\pm)^{n-1} \frac{\lambda^{3n}}{n} \int_0^\infty 4\pi t^2 e^{-n\pi\lambda^2 t^2}\, dt = \frac{(\pm 1)^{n-1}}{n^{\frac{5}{2}}} \lambda^{3(n-1)}, \tag{25.11$_9$}$$

in agreement with (23.15).

To summarize this section, we find first that the classical limit, $h\to 0$, of the function $F_N^{(0)}\{N\}$ defined in (24.21) as $N!$ times the diagonal element of the coordinate representation of the density matrix is indeed the same as the corresponding classical function, $e^{-U_N/kT}$, for a perfect gas, namely unity, since $U_N \equiv 0$. We also find that symmetrization of the matrix leads to the same equations for the Bose-Einstein and Fermi-Dirac perfect gases derived in Sect. 23. In so doing, however, we see that the distribution in coordinate space of the molecules is not uniform, but there is a net attraction in the Bose-Einstein case, and a net repulsion in the Fermi-Dirac case. The coefficients, $b_n^{(0)}$, of the development of P/kT in a power series in activity z, arise exactly as in the classical case, as integrals of the cluster functions for n molecules.

26. Density matrix with a potential. *α) The general equation.* In the last section we found that the classical limit, $h\to 0$, of the function $F_N^{(0)}\{N\}$, which is $N!$ times the diagonal element of the density matrix in coordinate representation, did indeed go over into the classical analogue, namely unity, in the case of a perfect gas. In this section we shall discuss the unsymmetrized matrix with a mutual potential of interaction between molecules. This function,

$$\left.\begin{aligned} &\varrho_N^{(u)}(\{\boldsymbol{N}\}', \{\boldsymbol{N}\}'') \\ &= \left[\prod_a \left(\frac{1}{2\pi m_a kT}\right)^{3N_a/2}\right] \iint\cdots\int e^{-(2\pi i/h)\{\boldsymbol{N}\}'\boldsymbol{p}^{(\boldsymbol{N})}} e^{-(\boldsymbol{H}_N/kT)} e^{(2\pi i/h)\{\boldsymbol{N}\}''\cdot\boldsymbol{p}^{(\boldsymbol{N})}}\, d\boldsymbol{p}^{(\boldsymbol{N})}, \end{aligned}\right\} \tag{26.1}$$

replaces $e^{-U_N/kT}$ in the classical equations.

The Hamiltonian operator divided by kT is

$$\frac{1}{kT}\boldsymbol{H}_N = \sum_\alpha \frac{1}{2m_\alpha kT}\left(\frac{h}{2\pi i}\nabla_\alpha \cdot \frac{h}{2\pi i}\nabla_\alpha\right) + \frac{1}{kT} U_N\{\boldsymbol{N}\}, \tag{26.2}$$

and the operator $e^{-\boldsymbol{H}_N/\mathrm{k}T}$ is a sum of powers of this. The equations look much simpler if a change of scale of coordinates and momenta is made. This in turn, is simpler if all molecules are identical, that is at least have identical masses although this is not necessary. We use then the case of one component, and introduce dimensionless coordinates,

$$\boldsymbol{R} = \frac{2\sqrt{\pi}\,\boldsymbol{r}_1}{\lambda}, \quad \frac{2\sqrt{\pi}\,\boldsymbol{r}_2}{\lambda}, \ldots, \quad \frac{2\sqrt{\pi}\,\boldsymbol{r}_N}{\lambda}, \tag{26.3$_1$}$$

with $\lambda = h/\sqrt{2\pi m \mathrm{k}T}$, the usual de Broglie wave length. The dimensionless momenta are chosen as

$$\boldsymbol{P} = \frac{\boldsymbol{p}_1}{\sqrt{2m\mathrm{k}T}}, \quad \frac{\boldsymbol{p}_2}{\sqrt{2m\mathrm{k}T}}, \ldots, \quad \frac{\boldsymbol{p}_N}{\sqrt{2m\mathrm{k}T}}, \tag{26.3$_2$}$$

so that

$$\boldsymbol{P}\cdot\boldsymbol{R} = \frac{2\pi}{h}\{N\}\cdot\boldsymbol{p}^{(N)}, \tag{26.3$_3$}$$

$$d\boldsymbol{P} = \left(\frac{1}{2m\mathrm{k}T}\right)^{3N/2} d\boldsymbol{p}^{(N)}. \tag{26.3$_4$}$$

With

$$U^*(R) = \frac{1}{\mathrm{k}T}\,U_N\{N\}, \tag{26.3$_5$}$$

the Hamiltonian operator divided by $\mathrm{k}T$ is

$$\boldsymbol{H}^* = \frac{1}{\mathrm{k}T}\,\boldsymbol{H}_N = \frac{\nabla}{i}\cdot\frac{\nabla}{i} + U^*(R), \tag{26.3$_6$}$$

and the density matrix becomes

$$\varrho_N^{(u)}(\boldsymbol{R}', \boldsymbol{R}'') = (\pi)^{-3N/2}\int\!\!\int\cdots\int e^{-i\boldsymbol{R}'\cdot\boldsymbol{P}}\, e^{-\boldsymbol{H}^*}\, e^{i\boldsymbol{R}''\cdot\boldsymbol{P}}\, d\boldsymbol{P}. \tag{26.3$_7$}$$

The expansion,

$$e^{-\boldsymbol{H}^*} = \sum_{n\geq 0}\frac{(-)^n}{n!}(\boldsymbol{H}^*)^n, \tag{26.4$_1$}$$

is used in (26.3$_7$), where $\boldsymbol{H}^*$ operates on $\varphi = e^{i\boldsymbol{R}''\cdot\boldsymbol{P}}$. The first operation, $\boldsymbol{H}^*\varphi$ gives $\boldsymbol{H}^*\varphi = \varphi(\boldsymbol{P}\cdot\boldsymbol{P} + U^*)$. The quantity

$$H^* = \frac{1}{\mathrm{k}T}\,H = \boldsymbol{P}\cdot\boldsymbol{P} + U^*(R''), \tag{26.4$_2$}$$

is the classical Hamiltonian divided by $\mathrm{k}T$ at momentum $\boldsymbol{P}$ and coordinate $\boldsymbol{R}''$, that is in the dimensional coordinates,

$$H = \frac{1}{2m}\sum_\alpha \boldsymbol{p}_\alpha\cdot\boldsymbol{p}_\alpha + U(\boldsymbol{r}_1, \ldots, \boldsymbol{r}_N). \tag{26.4$_3$}$$

A second operation of $\boldsymbol{H}^*$ on φH^* gives,

$$(\boldsymbol{H}^*)^2\varphi = \boldsymbol{H}^*\varphi H^* = \varphi\left(H^* + 2\boldsymbol{P}\cdot\frac{\nabla}{i} - \nabla\cdot\nabla\right)H^*, \tag{26.4$_4$}$$

and for n operations,

$$\left.\begin{aligned}(\boldsymbol{H}^*)^n\varphi &= \varphi\left(H^* + 2P\cdot\frac{\nabla}{i} - \nabla\cdot\nabla\right)^{n-1} H^*\\ &= \varphi\left(H^* + 2P\cdot\frac{\nabla}{i} - \nabla\cdot\nabla\right)^{n},\end{aligned}\right\} \tag{26.4$_5$}$$

since the addition of the two operators $2P\cdot\frac{\nabla}{i}-\nabla\cdot\nabla$ to the last factor add zero. One may thus formally write

$$\varrho_N^{(u)}(\boldsymbol{R}',\boldsymbol{R}'')=\pi^{-3N/2}\int\int\cdots\int e^{i\boldsymbol{P}\cdot(\boldsymbol{R}''-\boldsymbol{R}')}\,e^{-\left(H^*+2P\cdot\frac{\nabla}{i}-\nabla\cdot\nabla\right)''}\,dP, \qquad (26.4_6)$$

where the $''$ on the operator in the exponent means that it is to be evaluated at $\boldsymbol{P}$ and $\boldsymbol{R}''$.

β) Classical limit. The classical limit is obtained by neglecting the operators ∇ in this, since $h\to 0$. Translated back to the dimensional coordinates these add terms of the type

$$\left(\frac{h^2}{8\pi^2 m\,\mathrm{k}T}\nabla\cdot\nabla\right)^{\nu}\left(\frac{h}{4\pi m\,\mathrm{k}T}\boldsymbol{p}^{(N)}\cdot\nabla\right)^{\mu}\frac{1}{\mathrm{k}T}U_N(\boldsymbol{r}_1'',\ldots,\boldsymbol{r}_N''), \qquad (26.5_1)$$

which are zero when $h\to 0$. The equation

$$\left.\begin{aligned}\lim_{h\to 0}\left[\varrho_N^{(u)}(\boldsymbol{R}',\boldsymbol{R}'')\right]&=\pi^{-3N/2}\int\int\cdots\int e^{i\boldsymbol{P}\cdot(\boldsymbol{R}''-\boldsymbol{R}')}\,e^{-[\boldsymbol{P}\cdot\boldsymbol{P}+U^*]}\,d\boldsymbol{P}\\&=e^{-U^*-\frac{1}{4}(\boldsymbol{R}''-\boldsymbol{R}')\cdot(\boldsymbol{R}''-\boldsymbol{R}')}\\&=\left[e^{-(1/\mathrm{k}T)\,U_N\{N\}''-(\pi/\lambda^2)\sum_{\alpha}(|\boldsymbol{r}_\alpha''-\boldsymbol{r}_\alpha'|)^2}\right]\end{aligned}\right\} \qquad (26.5_2)$$

is then obtained for the classical limit. Since $\lambda\to 0$ one has

$$\lim_{h\to 0}\left[\varrho_N^{(u)}(\{N\},\{N\})\right]=\lim_{h\to 0}\left[F_N^{(0)}\{N\}\right]=e^{-(1/\mathrm{k}T)\,U_N\{N\}}, \qquad (26.5_3)$$

and the off-diagonal elements of $\varrho_N^{(u)}$ are all zero. The permuted elements in the symmetrization will also be zero, since the positive repulsive potential $U_N\{N\}$ requires zero values of $\varrho_N^{(u)}(\{N\},\{N\})$ when two molecules have identical coordinate values, and the diagonal elements of a permutation of α with β will have the factor $e^{-(2\pi/\lambda^2)\,r_{\alpha\beta}^2}$ which is zero for $r_{\alpha\beta}>0$, $\lambda=0$.

γ) High temperature expansion. For small values of h the exponent of the operator occurring in (26.4_2) can be expanded[1] in powers of h, and the resultant function written as the exponential[2] of a function of $\boldsymbol{P}$ and $\boldsymbol{R}''$. That is one may write (26.4_6) as

$$\varrho_N^{(u)}(\boldsymbol{R}',\boldsymbol{R}'')=\pi^{-3N/2}\int\int\cdots\int e^{i\cdot\boldsymbol{P}(\boldsymbol{R}''-\boldsymbol{R}')-H^*-\varphi^*(\boldsymbol{R}'',\boldsymbol{P})}\,d\boldsymbol{P}, \qquad (26.6_1)$$

with $\varphi^*(\boldsymbol{R},\boldsymbol{P})$ a function, namely,

$$\left.\begin{aligned}\varphi^*(\boldsymbol{R},\boldsymbol{P})&=-\boldsymbol{P}\cdot\frac{\nabla}{i}U^*+\frac{1}{2}\nabla\cdot\nabla U^*-\\&\quad-\frac{1}{3}\nabla U^*\cdot\nabla U^*-\frac{2}{3}(\boldsymbol{P}\cdot\nabla)^2U^*+\cdots.\end{aligned}\right\} \qquad (26.6_2)$$

Integration over $d\boldsymbol{P}$ gives the diagonal element,

$$\varrho_N^{(u)}(\{N\},\{N\})=e^{-(1/\mathrm{k}T)\,[U_N\{N\}+\Delta U_N\{N\}+\cdots}, \qquad (26.6_3)$$

with $\Delta U_N\{N\}$ a power series in λ of which the first term is

$$\Delta U_N\{N\}=\frac{\lambda^2}{24\pi}\nabla\cdot\nabla U_N-\frac{\lambda^2}{48\pi\,\mathrm{k}T}\nabla U_N\cdot\nabla U_N+\cdots, \qquad (26.6_4)$$

and $\lambda=h/\sqrt{2\pi m\mathrm{k}T}$. The equation is also correct for a multicomponent system if, for $\lambda\nabla$ we substitute $\sum_\alpha\lambda_\alpha\nabla_\alpha$, that is, multiply each derivative with respect to a coordinate of an atom or molecule α with the λ_α having the mass, m_α, of α.

[1] E. WIGNER: Phys. Rev. **40**, 749 (1932).
[2] J. E. MAYER and W. BAND: J. Chem. Phys. **15**, 141 (1947).

The algebra necessary to obtain (26.6$_4$) is relatively long. It is simpler to obtain the form

$$\varrho_N^{(u)}(\boldsymbol{R},\boldsymbol{R}) = \pi^{-3N/2}\int\int\cdots\int e^{-H^*}\left[1-\varphi^*(\boldsymbol{R},\boldsymbol{P}) - \tfrac{1}{2}(\boldsymbol{P}\cdot\boldsymbol{\nabla} U)^2+\cdots\right]d\boldsymbol{P}, \quad (26.7_1)$$

and from this,

$$\varrho_N^{(u)}(\boldsymbol{R},\boldsymbol{R}) = e^{-(1/kT)\,U_N\{N\}}\left[1-\frac{1}{kT}\Delta U_N\{N\}+\cdots\right]. \quad (26.7_2)$$

Since the factors $[1-\varphi^*-\frac{1}{2}(\boldsymbol{P}\cdot\nabla U)^2]$ and $\left[1-\frac{1}{kT}\Delta U_N\right]$ are the first terms in the expansion of the exponentials $e^{-\varphi^*}$ or $e^{-\Delta U/kT}$, it follows that if the functions can be written with a convergent expansion in the exponent the terms to λ^2 must be those given in (26.6$_1$) to (26.6$_4$).

For the diagonal element, $\boldsymbol{R}''=\boldsymbol{R}'=\boldsymbol{R}$, the integrand of (26.4$_6$) for $\varrho_N^{(u)}(\boldsymbol{R},\boldsymbol{R})$ is the sum of products,

$$S=\sum_{n\geq 0}\frac{(-)^n}{n!}\left(H^*+2\boldsymbol{P}\cdot\frac{\boldsymbol{\nabla}}{i}-\boldsymbol{\nabla}\cdot\boldsymbol{\nabla}\right)^n. \quad (26.7_3)$$

The operations on H^* yield

$$2\boldsymbol{P}\cdot\frac{\boldsymbol{\nabla}}{i}H^* = 2\left[\left(\boldsymbol{P}\cdot\frac{\boldsymbol{\nabla}}{i}U^*\right)+H^*\boldsymbol{P}\cdot\frac{\boldsymbol{\nabla}}{i}\right], \quad (26.7_4)$$

and

$$-\boldsymbol{\nabla}\cdot\boldsymbol{\nabla}H^* = -(\boldsymbol{\nabla}\cdot\boldsymbol{\nabla}U^*)-(\boldsymbol{\nabla}\cdot U^*)\cdot\boldsymbol{\nabla}-H^*(\boldsymbol{\nabla}\cdot\boldsymbol{\nabla}). \quad (26.7_5)$$

We are interested in terms where the operation of (26.7$_4$) has acted once and twice, but that of (26.7$_5$) only once. These give us all terms up to h^2, since the first contains h, and the second h^2 from (26.5$_1$). The sum S then contains one term,

$$\sum_{n\geq 0}\frac{(-)^n}{n!}(H^*)^n = e^{-H^*} \quad \text{(a)}$$

which is independent of h. There is one term from (26.7$_4$) linear in h, namely

$$2\sum_{n\geq 2}\frac{(-)^n}{n!}(H^*)^{n-2}C_1(n)\,(\boldsymbol{P}\cdot\boldsymbol{\nabla}U^*), \quad \text{(b)}$$

and two terms quadratic in h, namely

$$-4\sum_{n\geq 3}\frac{(-)^n}{n!}(H^*)^{n-3}C_2(n)\,[(\boldsymbol{P}\cdot\boldsymbol{\nabla})^2U^*], \quad \text{(c)}$$

and,

$$-4\sum_{n\geq 4}\frac{(-)^n}{n!}(H^*)^{n-4}C_3(n)\,[\boldsymbol{P}\cdot\boldsymbol{\nabla}U^*]^2. \quad \text{(d)}$$

From (26.7$_5$) there are two terms quadratic in h,

$$-\sum_{n\geq 2}\frac{(-)^n}{n!}(H^*)^{n-2}C_4(n)\,(\boldsymbol{\nabla}\cdot\boldsymbol{\nabla}U^*), \quad \text{(e)}$$

and,

$$-\sum_{n\geq 3}\frac{(-)^n}{n!}(H^*)^{n-3}C_5(n)\,(\boldsymbol{\nabla}\cdot U^*)\cdot(\boldsymbol{\nabla}U^*). \quad \text{(f)}$$

We wish to show that

$$C_1 = \tfrac{1}{2} n(n-1), \tag{a'}$$

$$C_2 = \tfrac{1}{6} n(n-1)(n-2), \tag{b'}$$

$$C_3 = \tfrac{1}{8} n(n-1)(n-2)(n-3), \tag{c'}$$

$$C_4 = \tfrac{1}{2} n(n-1), \tag{d'}$$

$$C_5 = \tfrac{1}{3} n(n-1)(n-2). \tag{e'}$$

If this is accomplished we will have demonstrated the validity of (26.7_1), with (26.6_2) for φ^*.

In a single product term

$$T_n = \left(H^* + 2\boldsymbol{P}\cdot\frac{\nabla}{i} - \nabla\cdot\nabla\right)^n, \tag{26.7_6}$$

occurring in (26.7_3) there will be one term linear in $\boldsymbol{P}\cdot\frac{\nabla}{i}$, namely

$$2\sum_{\nu\geq 0}^{n-1} (H^*)^{n-1-\nu}\left(\boldsymbol{P}\cdot\frac{\nabla}{i}\right)(H^*)^{\nu}.$$

This term leads to the n-th term in (b). Since $\left(\boldsymbol{P}\cdot\frac{\nabla}{i}\right)(H^*)^{\nu} = \nu(H^*)^{\nu-1}\left(\boldsymbol{P}\cdot\frac{\nabla}{i}U^*\right)$ the coefficient $C_1(n)$ is

$$C_1(n) = \sum_{\nu\geq 0}^{n-1}\nu = \tfrac{1}{2} n(n-1).$$

The terms that lead to (c) and (d) are

$$-4\sum_{\nu=0}^{n-2}\sum_{\mu=0}^{\nu}(H^*)^{n-2-\nu}\left(\boldsymbol{P}\cdot\frac{\nabla}{i}\right)(H^*)^{\nu-\mu}\left(\boldsymbol{P}\cdot\frac{\nabla}{i}\right)(H^*)^{\mu}$$

so that

$$C_2(n) = \sum_{\nu=1}^{n-2}\sum_{\mu=1}^{\nu}\mu = \sum_{\nu=1}^{\nu=n-2}\tfrac{1}{2}\nu(\nu+1) = \tfrac{1}{6} n(n-1)(n-2)$$

and

$$C_3 = \sum_{\nu=2}^{n-2}\nu\sum_{\mu=1}^{\nu}\mu = \sum_{\nu=2}^{\nu-n-2}\tfrac{1}{2}(\nu+1)\nu(\nu-1) = \tfrac{1}{8} n(n-1)(n-2)(n-3).$$

The coefficient C_4 is to be computed exactly like C_2, $C_4 = C_2$. The coefficient C_5 is

$$C_5 = \sum_{\nu\geq 2}^{\nu=n-1}\nu(\nu-1) = \tfrac{1}{3} n(n-1)(n-2).$$

To obtain (26.7_2) from (26.7_1) one must integrate,

$$\left.\begin{aligned}\varrho_N^{(u)}(\boldsymbol{R},\boldsymbol{R}) = \mathrm{e}^{-u^*}\pi^{-3N/2}\int\int\cdots\int \mathrm{e}^{-\boldsymbol{P}\cdot\boldsymbol{P}}\Big[1 + \boldsymbol{P}\cdot\frac{\nabla}{i}U^* - \frac{1}{2}\nabla\cdot\nabla U^* + \\ + \frac{1}{3}(\nabla U^*\cdot\nabla U^*) + \frac{2}{3}(\boldsymbol{P}\cdot\nabla)^2 U^* - \frac{1}{2}(\boldsymbol{P}\cdot\nabla U^*)^2 + \cdots\Big]\,d\boldsymbol{P}.\end{aligned}\right\}\tag{26.7_7}$$

The integral of the terms independent of $\boldsymbol{P}$ is $\pi^{3N/2}$, cancelling the factor in front of the integral. The integral over any single component, P_i of $\boldsymbol{P}$ gives zero, since $\mathrm{e}^{-\boldsymbol{P}\cdot\boldsymbol{P}}$ is even. The terms in $(\boldsymbol{P}\cdot\nabla)^2 U$ which are products of one component $P_i^2\left(\frac{\partial^2}{\partial x_i}U^*\right)$ contribute $\frac{1}{2}\pi^{3N/2}$ each, as do those of $(\boldsymbol{P}\cdot\nabla U^*)^2$. The result is

$$\varrho_N^{(u)}(\boldsymbol{R},\boldsymbol{R}) = \mathrm{e}^{-U^*}\left[1 - \tfrac{1}{6}\nabla\cdot\nabla U^* + \tfrac{1}{12}(\nabla U^*)\cdot(\nabla U^*)\right]. \tag{26.7_8}$$

Using

$$\mathrm{k}T(\nabla U^*(R)) = \frac{\lambda}{2\sqrt{\pi}} \nabla U_N(\boldsymbol{r}_1, \ldots, \boldsymbol{r}_N), \tag{26.7$_9$}$$

from the coordinate transformation (26.3$_1$), one does indeed obtain (26.6$_4$) for $\Delta U_N\{N\}$.

δ) Pair potentials. At relatively high temperatures, then, for which λ is small, the effect of quantum mechanics is to alter the effective potential energy to be used in the classical equations, by introducing a temperature dependent term ΔU_N as an addition to the classical potential. It is interesting to examine this term in the case that the total potential is a sum of pair terms,

$$U_N\{N\} = \sum_{N\geq\alpha>\beta\geq 1}\sum u(r_{\alpha\beta}), \tag{26.8}$$

each depending on the distance between molecule pairs. The algebra of examining (26.6$_4$) for this case is not complicated but will not be detailed here. Terms in $\Delta U_N\{N\}$ depending on triples of molecules occur. One can write

$$\Delta U_N\{N\} = \sum_{N\geq\alpha>\beta\geq 1}\sum \vartheta(r_{\alpha\beta}) + \sum_{N\geq\alpha>\beta\geq\gamma\geq 1}\sum\sum \vartheta_3(r_{\alpha\beta}, r_{\beta\gamma}, r_{\gamma\alpha}). \tag{26.9}$$

One finds, for identical molecules,

$$\vartheta(r) = \frac{\lambda^2}{12\pi}\frac{d^2u}{dr^2} + \frac{\lambda^2}{6\pi}\frac{1}{r}\frac{du}{dr} - \frac{\lambda^2}{24\pi\mathrm{k}T}\left(\frac{du}{dr}\right)^2. \tag{26.10$_1$}$$

$$\vartheta_3 = \frac{\lambda^2}{4\pi}\sum_{\alpha=1}^{3}\frac{du(r_{\alpha\beta})}{dr_{\alpha\beta}}\frac{du(r_{\alpha\gamma})}{dr_{\alpha\gamma}}\frac{r_{\alpha\beta}^2 + r_{\alpha\gamma}^2 - r_{\gamma\beta}^2}{r_{\alpha\beta}r_{\alpha\gamma}}. \tag{26.10$_2$}$$

For the Lennard-Jones, 6—12 potential,

$$u(r) = 4u_0\left[\left(\frac{r_0}{r}\right)^{12} - \left(\frac{r_0}{r}\right)^6\right], \tag{26.11$_1$}$$

one has for $u(r)+\vartheta(r)$, the new effective pair potential, a negative term proportional to r^{-26} from the last term, $(du/dr)^2$. This, of course, arises only from the lack of convergence of the development in powers of λ^2 of the series with such an inverse power series potential, but means that computations using the series development must employ a cut-off at short distances of approach. The general nature of the new effective pair potential is to be more positive in the neighborhood of the minimum of $u(r)$, but to increase at small r-values less rapidly than the classical potential. The potential is "smoothed" compared to the classical $u(r)$.

ε) Summary. The diagonal element of the coordinate representation of the density matrix, normalized as in (24.20$_3$) does become equal, in the classical limit to $\mathrm{e}^{-U/\mathrm{k}T}$. For relatively high temperatures one may obtain an approximation to this diagonal element by replacing U of the classical equation with $U+\Delta U$, with ΔU given by (26.6$_4$) up to terms in $\lambda^2 = h^2/2\pi m\mathrm{k}T$. The development is valid as long as λ is small compared to distances within which the potential changes appreciably compared to $\mathrm{k}T$.

27. Symmetrization of the density matrix with a potential. The combined effect of symmetrization and that of a potential of interaction is difficult to carry out with any rigor. However a rather qualitative treatment is possible. It is of greatest interest in the treatment of He^4 liquid, since the transition is presumably connected with the Bose-Einstein condensation which, for a perfect gas of molecular weight 4, having the experimental density of the liquid, should occur at a slightly higher temperature than the observed λ-point transition. Several authors have therefore undertaken to examine the Bose-Einstein transition in an imperfect gas[1].

Using the Eq. (26.6_1) for $\varrho_N^{(u)}(\boldsymbol{R}', \boldsymbol{R}'')$ and neglecting the function $\varphi^*(\boldsymbol{R}'', \boldsymbol{P})$, the only difference between this and the perfect gas equation derived in Sect. 25 is in the multiplicative factor e^{-u^*} not involving $\boldsymbol{P}$. The integration over $d\boldsymbol{P}$ then yields the perfect gas unsymmetrized density matrix, (25.4) multiplied by e^{-u^*},

$$\varrho_N^{(u)}(\{N\}', \{N\}'') = e^{-(\pi/\lambda^2)|\boldsymbol{r}'_\alpha - \boldsymbol{r}''_\alpha|^2} e^{-(1/kT) U_N\{N\}''}. \tag{27.1}$$

It is of course clear that in all real cases if λ is large enough to make the off-diagonal terms significant in symmetrization, the neglect of $\varphi^*(\boldsymbol{R}'', \boldsymbol{P})$ in (26.6_1) is completely unjustified. Indeed this is clear from (27.1) since the function represented there is not symmetric in exchange of $\{N\}'$ with $\{N\}''$ as it must be. Nevertheless there are many good reasons for believing that the qualitative form of (27.1) is adequate for computing the diagonal elements of the symmetrized matrix.

One may of course define

$$F_N^{(u)}\{N\} = \varrho_N^{(u)}(\{N\}, \{N\}), \tag{27.2$_1$}$$

a function which will certainly have the qualitative properties of the functions $F_N^{(0)}\{N\}$ of the classical case: namely this function is positive real, symmetric in permutations, will be zero if any pair distance is zero provided the potentials become infinite in this case, will go to unity for all pair distances sufficiently large, and will become the product, $F_{N-M}^{(u)} F_M^{(u)}$ if the two subsets $\{N-M\}$ and $\{M\}$ are sufficiently separated. One may then define,

$$\{N\} = \tfrac{1}{2}\{\{N\}' + \{N\}\}'', \tag{27.2$_2$}$$

$$\{N\}''' = \{\{N\}' - \{N\}''\}, \tag{27.2$_3$}$$

and a function G_N by

$$\varrho_N^{(u)}(\{N\}', \{N\}'') = \sqrt{F_N^{(u)}\{N\}' F_N^{(u)}\{N\}''}\, G_N(\{N\}, \{N\}'''). \tag{27.2$_4$}$$

In view of (27.2_1) and (27.2_4) we have so defined G_N that,

$$G_N(\{N\}, \{N\}''' \equiv 0) = 1. \tag{27.2$_5$}$$

If only forces derived from a potential were assumed in the original Hamiltonian, as we have always done, that is if magnetic forces are neglected, the original wave functions, Ψ_ν, can always be chosen to be real, and ϱ_N is then real. In this case it is symmetric in exchange of $\{N\}'$ with $\{N\}''$ or when $\{N\}''' \to -\{N\}'''$, so that G_N is even in $\{N\}'''$. The density matrix, ϱ_N, is certainly a maximum on the diagonal in all important configurations, and decreases toward zero for large distances $|\boldsymbol{r}'_\alpha - \boldsymbol{r}''_\alpha|$ of any molecule α. Since G_N is independent of $\{N\}$ at its maximum, one may assume it not to depend greatly upon $\{N\}$, and approximate it by a product of even functions of $|\boldsymbol{r}'_\alpha - \boldsymbol{r}''_\alpha|$ for all α. A reasonable analytic form of these functions is a Gaussian.

One therefore writes,

$$\varrho_N^{(u)}(\{N\}', \{N\}'') = \sqrt{F_N^{(u)}\{N\}' F_N^{(u)}\{N\}''} \prod_{\alpha=1}^{N} e^{-(\pi/\lambda_e^2)|\boldsymbol{r}'_\alpha - \boldsymbol{r}''_\alpha|^2}, \tag{27.2$_6$}$$

[1] S. T. BUTLER and M. H. FRIEDMAN: Phys. Rev. 98, 287, 294 (1955). — G. V. CHESTER: Phys. Rev. 99, 1062 (1955).

as an approximate form for the off-diagonal elements of $\varrho_N^{(u)}$ in terms of the diagonals, (27.2$_1$). One expects λ_e to be not greatly different than $\lambda = h/\sqrt{2\pi m \mathrm{k} T}$, and FEYNMAN[1] gives arguments why one should believe it to be somewhat smaller, $\lambda_e < \lambda$.

The symmetrization of Eq. (24.17$_4$) now affects only the Gaussian in (27.2$_6$), since $F_N^{(u)}$ is symmetric in permutations. The procedure of Sect. 25 may be used, and with functions $g_n^{(s)}\{n\}$ defined by (25.7$_1$), (25.7$_2$), and (25.7$_3$) with λ_e replacing λ one has for $F_N^{(0)}$

$$F_N^{(0)}\{N\} = F_N^{(u)}\{N\} \sum_{\{k\{n_i\}Nu\}} \prod_{i=1}^{k} g_n^{(s)}\{n_i\}_N. \tag{27.3}$$

This function, in turn, can be decomposed into new cluster functions, $g_n^{(0)}\{n\}$, by the method of Sect. 13, Eq. (13.23), and cluster integrals, $b_n^{(0)}$ defined as $(1/n!\,V)$ times the integrals of $g_n^{(0)}\{n\}\,d\{n\}$ in V, Eq. (13.28$_1$). The equation

$$P = \mathrm{k}T \sum_{n \geq 1} b_n^{(0)} z^n \tag{27.4}$$

is of course obtained.

We may however, now examine the complete cluster functions, $g_n^{(0)}$, and from this examination see certain relatively simple additive contributions to $b_n^{(0)}$. For instance suppose the function $F_n^{(u)}$ were analyzed into cluster functions, $g_n^{(u)}\{n\}$. One then has

$$F_N^{(0)}\{N\} = \sum_{\{k\{n_i\}N\}_u} \sum_{\{l\{m_j\}N\}_u} \prod_{i=1}^{k} \prod_{j=1}^{l} g_n^{(s)}\{n_i\}_N \, g_m^{(u)}\{m_j\}_N. \tag{27.5}$$

The cluster function $g_m^{(u)}\{m_j\}_N$ will always occur as an additive term in $g_m^{(0)}$, namely from that term of the sum of products of (27.5) in which the partition $\{_k\{n_i\}_N\}_u$ includes the members of $\{m_j\}_N$ in single clusters, $g_1^{(s)} = 1$, only. Similarly $g_n^{(s)}$ will occur additively in $g_n^{(0)}$. There will also be many other terms of products $g_{m+\nu}^{(u)}\{\{m\}+\{\nu\}\}\, g_{n+\nu}^{(s)}\{\{n\}+\{\nu\}\}$, and worse, in $g_{m+\nu+n}^{(0)}$. A formal analysis of this is not extremely difficult, but unrewarding, since the integrations cannot be performed. However, the selection of two additive contributions to $b_n^{(0)}$ gives equations susceptible to interpretation, and probably represents the exact equation not too distantly.

For each of the $n-1$ cyclic functions in $g_n^{(s)}$ add all contributions for which either $g_1^{(u)} = 1$ for all members of $\boldsymbol{r}_\alpha < \{n\}$, or pair functions $g_2^{(u)}(r_{\alpha\beta})$ for pairs connected in the cycle function. The additive contribution, $b_n^{(0)}$ is $(n-1)!$ times the integral of a cycle in which the functions

$$[1 + g_2^{(u)}(r_{\alpha\beta})]\, \mathrm{e}^{-(\pi/\lambda_e^2) r_{\alpha\beta}^2} = F_2^{(u)}(r_{\alpha\beta})\, \mathrm{e}^{-(\pi/\lambda_e^2) r_{\alpha\beta}^2}$$

occur in the cycle. These terms can be computed if $F_2^{(u)}$ is known, or estimated. Following the method of Sect. 25, Eq. (25.11$_7$) we compute the Fourier transform,

$$g(t) = \frac{2}{t} \int_0^\infty r\, F_2^{(u)}(r)\, \mathrm{e}^{-(\pi/\lambda_e^2) r^2} \sin 2\pi t r\, dr, \tag{27.6}$$

so that

$$b_n^{(s)} = \frac{1}{n} \int_0^\infty 4\pi t^2 [g(t)]^n\, dt. \tag{27.7}$$

[1] R. P. FEYNMAN: Phys. Rev. **91**, 1291, 1301 (1953); **94**, 262 (1954).

The other additive contribution we choose to consider is simply $b_n^{(u)}$ due to the integration of $g_n^{(u)}\{n\}$. We would then have

$$P = \mathrm{k}T\,z\left[1 + \sum_{n\geq 2} b_n^{(u)} z^{n-1} + \sum_{n\geq 2} b_n^{(s)} z^{n-1}\right]. \tag{27.8}$$

In this approximation, then, the deviations from perfect classical gas behavior, $P=\mathrm{k}Tz$, are additively due to two sums, one originating from the unsymmetrized density matrix, $F_N^{(u)}$, and the other from the symmetrization, modified from the perfect gas equation by the $F_2^{(u)}$ in (27.6). Condensation will occur at the first singularity of the positive real axis in either of the two sums.

There is little doubt that for the real gas, He^4, the condensation to the liquid is to be ascribed essentially to the singularity in the sum over $b_n^{(u)}$, since the symmetrization plays little role at the activities for which the gas actually condenses. However, hypothetical cases in which $F_N^{(u)}$ had only a repulsive hard core would show Bose-Einstein condensation with the added effect of the hard core repulsion. The case has been treated by BUTLER and FRIEDMAN.

It should be remarked that far more elegant and rigorous treatments based on Eq. (27.3) are not only possible, but not too difficult. In particular, using this in the Grand Canonical partition function one can obtain equations formally applicable to the liquid in which the distribution functions $F_n^{(u)}(z, T, \{n\})$ of a hypothetical Boltzmann liquid occur combined with the cyclic permutation functions.

Bibliography.

[1] GIBBS, WILLARD: Elementary Principles in Statistical Mechanics. New Haven: Yale University Press 1902.
[2] TOLMAN, RICHARD C.: The Principles of Statistical Mechanics. Oxford: University Press 1938.
[3] MAYER, J. E., and M. G. MAYER: Statistical Mechanics. New York: John Wiley & Sons 1940.
[4] HIRSCHFELDER, J. O., C. F. CURTISS and R. B. BIRD: Molecular Theory of Gases and Liquids. New York: John Wiley & Sons, London: Chapman and Hall 1954.
[5] NEUMANN, J. v.: Mathematische Grundlagen der Quantenmechanik. Berlin: Springer 1932. In English translated by R. T. BEYER, Princeton University Press 1955.

Principles of the Kinetic Theory of Gases.

By

HAROLD GRAD.

With 6 Figures.

Introduction.

For the purposes of this article, the subject of the kinetic theory of gases is considered to be coextensive with the theory of the Boltzmann equation. We consider only the original equation of MAXWELL and of BOLTZMANN for classical point molecules and short range forces, putting aside the equally interesting but distinct questions which arise from the inclusion of internal degrees of freedom, quantum interactions, inverse square forces, and imperfect gases. The special case of a KNUDSEN gas of freely streaming particles is only touched on, mainly for purposes of comparison.

In Chap. I, a critical account is presented of the relation between the two descriptions of a dilute gas: (1) via a molecular distribution function and the Boltzmann equation, and (2) via particle dynamics of an n-body system where n is large.

Solution of the Boltzmann equation has proceeded along two lines. One is the investigation of the so-called *normal* solutions which exhibit classical fluid-like behavior. This theory has been developed by HILBERT, ENSKOG, and CHAPMAN and a critical account together with some new features is presented in Chap. IV. The problem of obtaining general solutions of the Boltzmann equation has been looked at from the mathematical viewpoint of existence theory by CARLEMAN, WILD, and MORGENSTERN —a survey of the present status is given in Chap. III—and by more practical expansion techniques following MAXWELL by GRAD, WANG CHANG and UHLENBECK, MOTT-SMITH, and GROSS, JACKSON and ZIERING, described in Chap. V. A distinct line, iteration and expansion in the neighborhood of free streaming, associated with the names JAFFÉ, (HEINEMAN) ROSE, KELLER, and WANG CHANG and UHLENBECK is only touched on because only initial results have been obtained by this difficult method.

A strong attempt is made to exhibit the interrelation and basic unity of these seemingly disparate techniques of kinetic theory.

I. The place of the Boltzmann equation in kinetic theory.

Two basic questions, (1) the degree of validity of the Boltzmann equation and (2) its relation to a gas which is governed by the laws of particle dynamics, are discussed in this chapter. The principal tool is a limiting process in which the ratio of the mean-free-path to macroscopic dimensions is held fixed while the gas becomes more and more rarefied. In the limit, we find that although one aspect of intermolecular forces, namely, imperfection of the gas, disappears, the effect of collisions on the temporal evolution of the gas remains finite and is given by the Boltzmann equation. It is confirmed that the Boltzmann equation becomes more nearly valid as the density decreases and that it is valid for arbitrarily large deviations from equilibrium within a mean-free-path and arbitrarily rapid fluctuations compared with the mean collision time; the limiting length is the diameter of a molecule and the limiting time is the mean duration of a collision.

A very significant point is that the Boltzmann equation emerges from particle dynamics without the intrusion of anything stochastic or random; (we distinguish the introduction of an *a priori* probability, which refers merely to repeated trials and is completely deterministic, from such acausal concepts as Markov processes, statistical independence, etc.). Something like randomness is found to be a property of strictly deterministic systems when the number of degrees of freedom is large. We can say that a reversible system will sometimes approximate the behavior of an irreversible system to an arbitrary degree of accuracy.

Many of the paradoxes and subtleties which are rife in this subject are reduced if not removed by the interpretation of the molecular distribution function as an expectation rather than as a physical number density of particles. In particular there is no compelling reason to smooth or coarse-grain the one-particle distribution either in time or in space. The Boltzmann H-function is a well-behaved function of time and very likely approaches a constant value (its minimum) for unbounded time even for a finite system of n particles.

While the philosophical subtleties are faced as resolutely as possible, no attempt has been made to supply complete mathematical details in this article[1].

a) Elementary properties of LIOUVILLE's equation.

1. LIOUVILLE's equation. The position of a particle will be denoted by $\boldsymbol{x}$ or x^i, $i=1, 2, 3$, and its velocity by $\boldsymbol{\xi}$ or ξ^i. In a system of n particles we distinguish the particles by $\boldsymbol{x}_1 \ldots \boldsymbol{x}_n$ or by x_n^i if necessary, and the same for the velocities, $\boldsymbol{\xi}_r$ or ξ_r^i. The $6n$-dimensional phase space with representative point $(\boldsymbol{x}_1 \ldots \boldsymbol{x}_n, \boldsymbol{\xi}_1 \ldots \boldsymbol{\xi}_n) \equiv Z$ will be denoted by Γ, and the 6-dimensional phase space with representative point $(\boldsymbol{x}, \boldsymbol{\xi}) \equiv z$ will be denoted[2] by γ or by γ_r if it refers specifically to the particle r with coordinates $(\boldsymbol{x}_r, \boldsymbol{\xi}_r) \equiv z_r$; clearly $Z=(z_1 \ldots z_n)$. We note the abbreviations $\int \ldots d\boldsymbol{x}$, $\int \ldots d\boldsymbol{\xi}$, $\int \ldots dz_1$, $\int \ldots dZ$ which stand for 3-fold, 3-fold, 6-fold, and $6n$-fold integrations respectively.

The motion of the system of n particles is governed by equations

$$m_r \frac{d^2 \boldsymbol{x}_r}{dt^2} = \boldsymbol{X}_r(Z). \tag{1.1}$$

For simplicity we assume that the particles are identical, in particular $m_r = m$, and that there are no external forces except for the ones implicit in the fact that the gas is contained in a box. We take

$$\boldsymbol{X}_r(Z) = \sum_{s=1}^{n} \boldsymbol{X}_{rs}(z_r, z_s), \tag{1.2}$$

where

$$\boldsymbol{X}_{rs}(z_r, z_s) = \boldsymbol{X}_{12}(z_r, z_s) \tag{1.3}$$

[1] Many of the proofs have not yet been published. The formal derivation, including the limiting process, was described in a course on Statistical Mechanics at New York University in 1950—1951; the proof of the initial chaos theorem was presented at a colloquium at New York University in February, 1952; a major portion of the entire theory, including the subsequent chaos theorem, was presented at an adress before the American Physical Society in State College, Pennsylvania in July, 1953. The theorem on the Knudsen gas in a perfectly reflecting box was stated in an address before the American Physical Society in Mexico City, August 1955.

[2] This notation differs from the more common one of μ for 6-space and γ for $6n$-space.

and[1]

$$\left.\begin{aligned} \boldsymbol{X}_{12}(z_1, z_2) &= \frac{\boldsymbol{r}}{r}\,\varphi'(r) \\ \boldsymbol{r} &= \boldsymbol{x}_2 - \boldsymbol{x}_1 . \end{aligned}\right\} \tag{1.4}$$

In other words, the forces are derivable from the two-particle potential φ.

We denote by $F_n(Z)$ a probability density defined on Γ; we have $F_n \geq 0$ and $\int F_n\, dZ = 1$. It is implicitly assumed that there is some experimental procedure which, when repeated, gives a frequency count of microscopic states Z which corresponds to the distribution function $F_n(Z)$; the significance of this introduction of probability into the problem will be discussed in Sect. 9. We assume that F_n is symmetric with respect to the n particles; i.e., $F_n(z_1 \ldots z_n)$ is a symmetric function of the n groups of variables.

A phase function is a function, $\psi(Z)$, on the space Γ. Any observable quantity can be represented as a phase function; its value is certainly determined if we know the positions and velocities of all particles. The expected value of the observable ψ is

$$\overline{\psi}(t) = \int \psi(Z)\, F_n(Z, t)\, dZ . \tag{1.5}$$

The distribution function, F_n, allows us to compute not only the expected value of ψ but its whole probability distribution. For, if we define $\psi_a(Z)$ to be equal to ψ when $\psi \leq a$ and zero otherwise, $\overline{\psi}_a$ considered as a function of the parameter a is exactly the *cumulative* distribution function of the values taken by ψ.

The equations of motion (1.1), written in the form

$$\left.\begin{aligned} \frac{d\boldsymbol{x}_r}{dt} &= \boldsymbol{\xi}_r , \\ m\frac{d\boldsymbol{\xi}_r}{dt} &= \boldsymbol{X}_r , \end{aligned}\right\} \tag{1.6}$$

can be interpreted geometrically as defining a flow in the phase space Γ; the succession of states, Z, is represented by a curve, $Z(t)$. In general[2], one such curve passes through each point Z. Given this motion of the phase space, it is intuitively clear that the evolution of the probability density in time, $F_n(Z, t)$, is determined once it is known initially. This can be made precise. If we observe the system to be in the state $Z(0)$ initially, we shall certainly find it in the state $Z(t)$ at time t. Moreover, if we know that Z is somewhere within the domain D_0 at $t = 0$, we are certain to find $Z(t)$ in D_t (the image of D_0 under the flow) at the time t. From this remark we are led to postulate the conservation of probability, namely,

$$\frac{d}{dt}\int\limits_{D_t} F_n(Z, t)\, dZ = 0 . \tag{1.7}$$

The total probability assigned to any domain D which is carried by the flow in the phase space is constant in time. The differential form of the conservation law, (1.7), is

$$\frac{\partial F_n}{\partial t} + \sum_{r=1}^{n} \frac{\partial}{\partial z_r}\cdot\left(\frac{dz_r}{dt}\, F_n\right) = 0 . \tag{1.8}$$

---

[1] We use ordinary type, r, to refer to the absolute magnitude of a vector, $\boldsymbol{r}$.

[2] Positions of static equilibrium are exceptions.

Here dz_r/dt is the flow velocity, and the dot $(\cdot)$ is used to indicate a summation over the six components of z_r and dz_r/dt. Rewriting (1.8) in terms of $\boldsymbol{x}_r$, and $\boldsymbol{\xi}_r$, we obtain *Liouville's equation*,

$$\frac{\partial F_n}{\partial t} + \sum_{r=1}^{n} \boldsymbol{\xi}_r \cdot \frac{\partial F_n}{\partial \boldsymbol{x}_r} + \frac{1}{m} \sum_{r=1}^{n} \boldsymbol{X}_r \cdot \frac{\partial F_n}{\partial \boldsymbol{\xi}_r} = 0; \tag{1.9}$$

here the dots represent sums over the three components of the various vectors. This partial differential equation implies that the function F_n is constant on every curve $Z(t)$ which represents a motion of the system. This is a property possessed by the probability density F_n, for any Hamiltonian system. For, as is well known, the phase space flow is incompressible for a Hamiltonian system, and this implies that the "density" F_n is constant in time following any "fluid" element.

Not only is F_n constant on a path, $Z(t)$, but so is any function of F_n. For example this is true of the function $h(F_n) = F_n \log F_n$. From the incompressibility of the flow we conclude that $\int h\, dZ$, integrated over all of Γ, is independent of the time. We denote the value of this integral by H_Γ (assuming that it exists),

$$H_\Gamma(t) \equiv \int F_n(Z, t) \log F_n(Z, t)\, dZ \tag{1.10}$$

and state the theorem

$$H_\Gamma = \text{const}. \tag{1.11}$$

The significance of this theorem will be discussed at length.

2. Reduced and truncated distribution functions. The *reduced* or *marginal* distribution functions are defined as follows:

$$F_1(z_1) = \int F_n(z_1 \ldots z_n)\, dz_2 \ldots dz_n, \tag{2.1}$$

$$F_r(z_1 \ldots z_r) = \int F_n(z_1 \ldots z_n)\, dz_{r+1} \ldots dz_n. \tag{2.2}$$

By the assumed symmetry of the function F_n, we conclude that F_r is symmetric in its r arguments and, furthermore, that the functional form of F_r is independent of any particular choice of r molecules. All the functions F_r have the normalization $\int F_r(z_1 \ldots z_r)\, dz_1 \ldots dz_r = 1$. The significance of F_1 is that $F_1(z_1)\, dz_1$ is the expectation of finding molecule 1 in the state dz_1 about z_1. It is easy to show[1] that $n F_1(z)\, dz$ is the expected number of molecules with coordinates within dz of the point z. Somewhat more loosely, $n F_1(z)$ can be termed the density of molecules in the phase space γ. Similarly $n m F_1\, dz$ is the expected value of the mass within dz, or we may refer to

$$f(z) \equiv n\, m\, F_1(z) \tag{2.3}$$

as the *mass density* in γ.

It will be convenient to introduce certain *truncated* distribution functions. These are defined in terms of a parameter σ, dimensionally a length, which can be thought of as having the order of magnitude of a molecular diameter. The exact choice of a value for σ will be made later. Consider $\boldsymbol{x}_1$, the location of molecule 1, to be fixed. The subdomain D_r of the space γ_r is defined by the inequality

$$|\boldsymbol{x}_r - \boldsymbol{x}_1| \geq \sigma \qquad (\text{all } \xi_r). \tag{2.4}$$

D_r contains all states of molecule r which are farther removed than σ from molecule 1. The $6(n-1)$-dimensional domain D is defined to be the product $D_2 \times$

[1] Define the phase function $\varphi_D(Z)$ as the number of coordinates $\boldsymbol{x}_1 \ldots \boldsymbol{x}_n$ which lie in a given domain D of γ. An elementary computation yields $\overline{\varphi}_D = \int_D n F_1\, dz_1$.

$D_3 \ldots \times D_n$; i.e., D contains all states of molecules $2 \ldots n$ in which *no molecule is closer than* σ to molecule 1. The truncated distribution F_1^σ is defined by

$$F_1^\sigma(z_1) \equiv \int_D F_n(z_1 \ldots z_n)\, dz_2 \ldots dz_n. \tag{2.5}$$

This function is the expectation of finding no molecule within a distance σ of molecule 1, granted that molecule 1 has the state z_1. An alternative expression for F_1^σ is

$$\left.\begin{aligned} F_1^\sigma(z_1) = F_1(z_1) - (n-1)\int_{\bar{D}_2} F_2(z_1, z_2)\, dz_2 + \\ + \tfrac{1}{2}(n-1)(n-2) \int_{\bar{D}_2 \times \bar{D}_3} F_3(z_1, z_2, z_3)\, dz_2\, dz_3 - \cdots, \end{aligned}\right\} \tag{2.6}$$

where $\bar{D}_r$ is the complementary domain to D_r; it includes those z_r which satisfy $|\boldsymbol{x}_r - \boldsymbol{x}_1| < \sigma$. From (2.6) it is clear that F_1^σ approximates F_1 if $n\sigma^3$, the occupied volume, is small compared to the volume of the physical box containing the gas.

Just as before we can interpret $n F_1^\sigma(z)\, dz$ as the expected number of molecules in dz which have no close neighbors (as measured by the distance σ) and

$$f^\sigma(z) \equiv n\, m\, F_1^\sigma(z) \tag{2.7}$$

as the mass density in γ of those molecules with no close neighbors.

One further truncated distribution that will arise is

$$\left.\begin{aligned} F_2^\sigma(z_1, z_2) \equiv \int_{D'} F_n(z_1 \ldots z_n)\, dz_3 \ldots dz_n, \\ D' = D_3 \times D_4 \times \cdots D_n. \end{aligned}\right\} \tag{2.8}$$

It should be noted that F_2^σ is not symmetric in its arguments z_1 and z_2 since the domain of integration involves $\boldsymbol{x}_1$ but not $\boldsymbol{x}_2$.

3. Reduced equations. Equations giving the time evolution of any of the distribution functions defined in Sect. 2 can be obtained by integrating out the appropriate variables in LIOUVILLE's equation. In what follows we operate purely formally and assume that all the integrals and derivatives which are written down exist and that the integrands are sufficiently well-behaved to allow the interchange of integration and differentiation with respect to non-integrated variables. Boundary terms will arise from integration by parts and from integration of derivatives of various functions. All the distribution functions are assumed to approach zero rapidly for large ξ and to be identically zero for large x (i.e., outside the physical container). The boundary terms will therefore vanish.

First we integrate through LIOUVILLE's equation, (1.9), with respect to all variables except z_1.

$$\int \left\{ \frac{\partial F_n}{\partial t} + \sum_{r=1}^{n} \frac{\partial}{\partial \boldsymbol{x}_r} \cdot (\boldsymbol{\xi}_r F_n) + \frac{1}{m} \sum_{r=1}^{n} \frac{\partial}{\partial \boldsymbol{\xi}_r} \cdot (\boldsymbol{X}_r F_n) \right\} dz_2 \ldots dz_n = 0.$$

In the time derivative term we interchange the order of differentiation and integration to obtain $\partial F_1/\partial t$. In the remainder, all terms in the summation except for $r = 1$ drop out as boundary terms. The $\partial/\partial \boldsymbol{x}_1$ and $\partial/\partial \boldsymbol{\xi}_1$ can be taken outside the integral sign and we are left with

$$\frac{\partial F_1}{\partial t} + \boldsymbol{\xi}_1 \cdot \frac{\partial F_1}{\partial \boldsymbol{x}_1} + \frac{1}{m} \frac{\partial}{\partial \boldsymbol{\xi}_1} \cdot \int \boldsymbol{X}_1 F_n\, dz_2 \ldots dz_n = 0.$$

Expanding $\boldsymbol{X}_1 = \sum_s \boldsymbol{X}_{1s}$ and using the symmetry of F_n yields

$$\frac{\partial F_1}{\partial t} + \boldsymbol{\xi}_1 \cdot \frac{\partial F_1}{\partial \boldsymbol{x}_1} + \frac{n-1}{m} \frac{\partial}{\partial \boldsymbol{\xi}_1} \cdot \int \boldsymbol{X}_{12} F_2(z_1, z_2)\, dz_2 = 0. \tag{3.1}$$

In exactly the same way we obtain an equation for $\partial F_r/\partial t$ by integrating out $z_{r+1} \ldots z_n$. From

$$\int \left\{ \frac{\partial F_n}{\partial t} + \sum_{s=1}^{n} \frac{\partial}{\partial \boldsymbol{x}_s} \cdot (\boldsymbol{\xi}_s F_n) + \frac{1}{m} \sum_{s=1}^{n} \frac{\partial}{\partial \boldsymbol{\xi}_s} \cdot (\boldsymbol{X}_s F_n) \right\} dz_{r+1} \ldots dz_n = 0$$

we obtain

$$\frac{\partial F_r}{\partial t} + \sum_{s=1}^{r} \boldsymbol{\xi}_s \cdot \frac{\partial F_r}{\partial \boldsymbol{x}_s} + \frac{1}{m} \sum_{s=1}^{r} \frac{\partial}{\partial \boldsymbol{\xi}_s} \cdot \int \boldsymbol{X}_s F_n\, dz_{r+1} \ldots dz_n = 0$$

and then

$$\left.\begin{aligned} &\frac{\partial F_r}{\partial t} + \sum_{s=1}^{r} \boldsymbol{\xi}_s \cdot \frac{\partial F_r}{\partial \boldsymbol{x}_s} + \frac{1}{m} \sum_{s,k=1}^{r} \boldsymbol{X}_{sk} \cdot \frac{\partial F_r}{\partial \boldsymbol{\xi}_s} + \\ &\quad + \frac{n-r}{m} \sum_{s=1}^{r} \frac{\partial}{\partial \boldsymbol{\xi}_s} \cdot \int \boldsymbol{X}_{12}(z_s, z_{r+1})\, F_{r+1}(z_1 \ldots z_{r+1})\, dz_{r+1} = 0. \end{aligned}\right\} \tag{3.2}$$

Several remarks are in order. First of all, the equation for F_r is not a differential equation in the same sense that LIOUVILLE's equation is. Not only does F_r appear but also F_{r+1}. This means that the equation for F_r does not by itself serve to determine the evolution of $F_r(t)$ from initial data. The same is true of the system of r equations for $F_1 \ldots F_r$. The reason is intuitively clear. $F_n(t)$ is uniquely determined by giving $F_n(0)$ and using LIOUVILLE's equation. Initial specification of F_r does not determine $F_n(0)$. We could expect $F_r(0)$ to determine $F_r(t)$ only if the problem somehow separated into an r-particle problem whose motion was unaffected by the others. Specifically the equation for F_r, (3.2), is an r-particle LIOUVILLE's equation together with additional terms giving the interaction of the remaining particles in terms of a typical $(r+1)$-*st* particle. These comments reveal the crucial distinction between Eq. (3.1) and the Boltzmann equation. The latter does serve to determine $F_1(t)$ in terms of $F_1(0)$ (see Chap. III).

The computation of an equation for the truncated distribution, F_1^σ, is slightly more complicated because the variable $\boldsymbol{x}_1$ appears in the limits of integration and this does not allow a simple interchange of integration and differentiation. For the computation we need the following elementary three-dimensional identities

$$\int_{|\boldsymbol{y}-\boldsymbol{x}|>\sigma} \frac{\partial}{\partial \boldsymbol{y}} \cdot \boldsymbol{A}(\boldsymbol{x}, \boldsymbol{y})\, d\boldsymbol{y} = \int_{|\boldsymbol{y}-\boldsymbol{x}|>\sigma} \operatorname{div} \boldsymbol{A}\, d\boldsymbol{y} = -\oint_{|\boldsymbol{y}-\boldsymbol{x}|=\sigma} \boldsymbol{A} \cdot d\boldsymbol{S}, \tag{3.3}$$

$$\frac{\partial}{\partial \boldsymbol{x}} \cdot \int_{|\boldsymbol{y}-\boldsymbol{x}|>\sigma} \boldsymbol{A}(\boldsymbol{x}, \boldsymbol{y})\, d\boldsymbol{y} = \int_{|\boldsymbol{y}-\boldsymbol{x}|>\sigma} \frac{\partial}{\partial \boldsymbol{x}} \cdot \boldsymbol{A}(\boldsymbol{x}, \boldsymbol{y})\, d\boldsymbol{y} - \oint_{|\boldsymbol{y}-\boldsymbol{x}|=\sigma} \boldsymbol{A} \cdot d\boldsymbol{S}. \tag{3.4}$$

The first is GAUSS' formula, and the second is easily verified. Note that the outer normal to the domain $|\boldsymbol{y}-\boldsymbol{x}|<\sigma$ has been taken to be positive. Now, integrating out LIOUVILLE's equation[1] over the domain D defined in Sect. 2

$$\int_D \left\{ \frac{\partial F_n}{\partial t} + \sum_{r=1}^{n} \frac{\partial}{\partial \boldsymbol{x}_r} \cdot (\boldsymbol{\xi}_r F_n) + \frac{1}{m} \sum_{r=1}^{n} \frac{\partial}{\partial \boldsymbol{\xi}_r} \cdot (\boldsymbol{X}_r F_n) \right\} = 0,$$

[1] Differentials are omitted where the domain of integration is shown explicitly.

we obtain

$$\frac{\partial F_1^\sigma}{\partial t} + \sum_{r=1}^{n} \int_D \frac{\partial}{\partial \boldsymbol{x}_r} \cdot (\boldsymbol{\xi}_r F_n) + \frac{1}{m} \frac{\partial}{\partial \boldsymbol{\xi}_1} \cdot \int_D \boldsymbol{X}_1 F_n = 0.$$

We denote by S_r the sphere $|\boldsymbol{x}_r - \boldsymbol{x}_1| = \sigma$ and evaluate

$$\int_D \frac{\partial}{\partial \boldsymbol{x}_1} \cdot (\boldsymbol{\xi}_1 F_n) = \frac{\partial}{\partial \boldsymbol{x}_1} \cdot (\boldsymbol{\xi}_1 F_1^\sigma) + \sum_{r=2}^{n} \oint_{S_r} F_2^\sigma(z_1, z_r)\, \boldsymbol{\xi}_1 \cdot d\boldsymbol{S}\, d\boldsymbol{\xi}_2$$

and

$$\sum_{r=2}^{n} \int_D \frac{\partial}{\partial \boldsymbol{x}_r} \cdot (\boldsymbol{\xi}_r F_n) = -\sum_{r=2}^{n} \oint_{S_r} F_2^\sigma(z_1, z_r)\, \boldsymbol{\xi}_r \cdot d\boldsymbol{S}\, d\boldsymbol{\xi}_2$$

using the formulas (3.3) and (3.4) and the definitions (2.5) and (2.8). Also,

$$\int_D \boldsymbol{X}_1 F_n = \sum_{r=2}^{n} \int_D \boldsymbol{X}_{12}(z_1, z_r) F_n = (n-1) \int_{D_2} \boldsymbol{X}_{12}(z_1, z_2)\, F_2^\sigma(z_1, z_2)\, dz_2.$$

Combining these results,

$$\frac{\partial F_1^\sigma}{\partial t} + \boldsymbol{\xi}_1 \cdot \frac{\partial F_1^\sigma}{\partial \boldsymbol{x}_1} + (n-1) \oint_{S_2} F_2^\sigma (\boldsymbol{\xi}_1 - \boldsymbol{\xi}_2) \cdot d\boldsymbol{S}\, d\boldsymbol{\xi}_2 + \frac{n-1}{m} \frac{\partial}{\partial \boldsymbol{\xi}_1} \cdot \int_{D_2} \boldsymbol{X}_{12} F_2^\sigma\, dz_2 = 0. \tag{3.5}$$

In the first integral $\boldsymbol{x}_2$ is integrated over the sphere $|\boldsymbol{x}_2 - \boldsymbol{x}_1| = \sigma$ while in the second integral it is integrated over the domain $|\boldsymbol{x}_2 - \boldsymbol{x}_1| > \sigma$; $\boldsymbol{\xi}_2$ is in both cases integrated over all values.

Eq. (3.5) becomes transparent in the special case

$$\boldsymbol{X}_{12} = 0 \quad \text{for} \quad |\boldsymbol{x}_2 - \boldsymbol{x}_1| > \sigma, \tag{3.6}$$

i.e., with a finite cut-off for the intermolecular force. It takes the form

$$\frac{\partial F_1^\sigma}{\partial t} + \boldsymbol{\xi}_1 \cdot \frac{\partial F_1^\sigma}{\partial \boldsymbol{x}_1} + (n-1) \oint_{S_2} F_2^\sigma (\boldsymbol{\xi}_1 - \boldsymbol{\xi}_2) \cdot d\boldsymbol{S}\, d\boldsymbol{\xi}_2 = 0. \tag{3.7}$$

This equation states that the number of free particles (those not near any other particles) changes only in virtue of the loss of a pair of particles as they enter each other's sphere of influence or as the gain of a pair of particles when they separate. This is in contrast to the equation for F_1 which indicates a smooth change in the number of particles in a given state under the continuous influence of forces. This property of the equation for F_1^σ, that its rate of change is governed by completed collisions rather than by continuously varying molecular states, is held in common with the Boltzmann equation. If there is no finite cut-off, the additional term in (3.5) can be interpreted as a correction term which includes the effect of grazing encounters. Of course, (3.7) is not a determined equation for F_1^σ because it also contains F_2^σ.

The formal similarity between (3.7) and the Boltzmann equation can be made even closer by a simple change of variables. As a parameter domain for the integration over the sphere S_2 we introduce that diametral plane of the sphere which is perpendicular to the direction of the relative velocity,

$$\boldsymbol{V} \equiv \boldsymbol{\xi}_2 - \boldsymbol{\xi}_1. \tag{3.8}$$

Introduce polar coordinates (r, ε) in this plane, $0 < r < \sigma$, $0 < \varepsilon < 2\pi$, and write

$$d\omega = r\,dr\,d\varepsilon \tag{3.9}$$

for the area element; ε is measured from an arbitrary axis[1]. The transformation is taken to be a projection of S_2 onto the disc ω. Since ω is covered twice, it is necessary to distinguish the two hemispheres

$$\left.\begin{aligned} S_2^+ &: \quad \boldsymbol{V}\cdot d\boldsymbol{S} > 0, \\ S_2^- &: \quad \boldsymbol{V}\cdot d\boldsymbol{S} < 0. \end{aligned}\right\} \tag{3.10}$$

S_2^+ refers to receding particles, and S_2^-, to approaching particles. The point on S_2^+ which projects onto (r, ε) we call $\boldsymbol{x}_2^+(r, \varepsilon)$, and the point on S_2^- we call $\boldsymbol{x}_2^-(r, \varepsilon)$. We have[2]

$$\left.\begin{aligned} (\xi_1 - \xi_2)\cdot d\boldsymbol{S} &= -V\,d\omega \quad \text{on} \quad S_2^+, \\ (\xi_1 - \xi_2)\cdot d\boldsymbol{S} &= V\,d\omega \quad \text{on} \quad S_2^-. \end{aligned}\right\} \tag{3.11}$$

In these variables, (3.7) takes the form

$$\frac{\partial F_1^\sigma}{\partial t} + \xi_1 \cdot \frac{\partial F_1^\sigma}{\partial \boldsymbol{x}_1} = (n-1)\int \left(F_2^\sigma(z_1, z_2^+) - F_2^\sigma(z_1, z_2^-)\right) V\,d\omega\,d\xi_2. \tag{3.12}$$

The argument z_2^+ is $(\xi_2, \boldsymbol{x}_2^+)$, and z_2^- is $(\xi_2, \boldsymbol{x}_2^-)$. The surface integral in Eq. (3.5) can be transformed in exactly the same way.

b) Unique features of the Boltzmann equation.

4. The level of description. Perhaps the most distinguishing feature of the Boltzmann equation is that its argument is the one-particle distribution, F_1. This is to be compared with LIOUVILLE's equation on the one hand which describes the behavior of F_n, and the equations of fluid dynamics on the other which involve a local thermodynamic description. In the amount of detail it gives about a state, the one-particle description is midway between the thermodynamic and the n-particle description. The question arises whether we are necessarily throwing away information by adopting a less detailed description or whether, in some sense, the information which is ignored may sometimes be irrelevant. The answer is that in certain limiting cases a lower level of description may be as complete as what seems to be a more detailed description.

The simplest example of this insensitivity to a detailed description of a state is in equilibrium thermodynamics. We known that all equilibrium properties of a simple substance are determined by just two parameters—energy and volume. This statement can be made with mathematical precision but involves a limit in which the size o the system becomes infinite [65], [25].

Another example which is more intuitive is afforded by a *Knudsen gas*. This is a gas which is sufficiently rarefied so that intermolecular forces can be completely ignored. If, purely formally, we drop the intermolecular force term in (3.1), we obtain a one-particle LIOUVILLE's equation for F_1. This implies that the state $F_1(t)$ is uniquely determined by $F_1(0)$.

There is a significant difference between the above two examples. In the first example the thermodynamic description is in a certain sense complete in that the specification of energy and volume even serves to determine the distributions F_r for r small compared to n (although the actual computation of F_r provides formid-

[1] See Fig. 6, p. 236.
[2] See footnote 1, p. 207.

able difficulties). However, in the second example F_1 does not serve to determine F_r or even F_2, but the problem separates so that if one chooses to be satisfied with a certain incomplete description of a state, this description is found to be self-contained.

Just as the Knudsen gas is obtained by a limiting process, we find that a *Boltzmann gas* is obtained by a different limiting process. In this limit (described in detail in Sect. 7), it is found that intermolecular forces retain a finite effect on the evolution of the system in time (as the collision term in the Boltzmann equation) but become negligible in terms of any intermolecular potential energy contribution to the equation of state or non-kinetic terms in the stresses or heat flow. The Boltzmann gas is therefore a perfect gas but is different from the Knudsen gas.

The way in which a description in terms of F_1 becomes self-contained is as follows. To a given initial $F_1(0)$ corresponds a whole class of compatible n-particle distributions, say $\{F_n(0)\}$. To each $F_n(0)$ corresponds a unique solution of LIOUVILLE's equation $F_n(t)$. From each $F_n(t)$ we obtain an $F_1(t)$ by integration. The result is a class, $\{F_1(t)\}$, which reduces to the single given function $F_1(0)$ at $t=0$. As a consequence of the limit process, the class of functions $\{F_1(t)\}$ converges (in some sense) upon a *single* function $F_1(t)$ which is a solution of BOLTZMANN's equation.

The way in which F_1 becomes uncoupled from $F_2 \ldots F_n$ in this limit is more subtle than in the two examples given above. As in the thermodynamic case, F_1 does not determine all the distributions $F_2 \ldots F_n$, nor do they evolve independently. What happens is that, although the highest order distributions $F_n, F_{n-1}, \ldots$ retain their individuality, at the other end F_2, F_3, and successively higher distributions become dependent on F_1 as $n \to \infty$. However, this dependence of F_2 on F_1 is rather singular; this will be elaborated upon in Sect. 11.

5. Irreversibility. The second distinguishing feature of the Boltzmann equation is the precise way in which irreversibility is manifested. BOLTZMANN's H-function is defined as

$$H_\gamma = \int F_1 \log F_1 \, dz. \tag{5.1}$$

The subscript γ distinguishes this from H_Γ, defined in (1.10) in terms of the n-particle distribution. The values of H_Γ and H_γ are related by the inequality[1]

$$H_\Gamma \geqq n H_\gamma; \tag{5.2}$$

the equal sign holds if and only if F_n separates into a product,

$$F_n(Z) = F_1(z_1) F_1(z_2) \ldots F_1(z_n). \tag{5.3}$$

For a solution of LIOUVILLE's equation, we recall that H_Γ is a constant in time No simple statement can be made about the behavior of H_γ as computed from a solution of LIOUVILLE's equation except that is not generally constant. On the other hand, for a solution of BOLTZMANN's equation, it is easily shown that $H_\gamma(t)$ is a monotone decreasing function of the time (see Chap. II). Turning to the limiting process mentioned in the previous section, insofar as $F_1(t)$ computed from a solution of LIOUVILLE's equation approximates a solution of BOLTZMANN's equation, we know that $H_\gamma(t)$ for a solution of LIOUVILLE's equation is, at least to some approximation, monotone decreasing. To be specific, suppose we choose an initial $F_1(0)$ and an initial $F_n(0)$ which is a product as in (5.3). Initially we have $H_\Gamma = n H_\gamma$. For all time we have $H_\Gamma =$ constant and $H_\Gamma \geqq n H_\gamma$. We

[1] The proof is elementary; similar inequalities are found in [*60*].

have proved a weaker statement than the one made above (but by strictly elementary means); namely, that H_γ can only decrease from its initial value. The more general result will be shown later.

The dual nature of solutions of LIOUVILLE's equation, one reversible (exemplified by $H_\Gamma =$ constant) and the other irreversible (H_γ approximately monotone decreasing) will be elaborated upon in Sect. 12. Actually, the irreversible description is more accurate, and the reversible one turns out to be a mathematical fine point. Unfortunately the latter lies much closer to the surface!

6. Departure from equilibrium. It is a simple matter to verify that the basic length scale in the Boltzmann equation is the mean-free-path and the basic time scale is the molecular collision time. This constrasts sharply with almost all other equations of mathematical physics which exhibit the irreversibility of fluid behavior over distances which must be large compared to the mean-free-path and times which must be large compared to the mean collision time. Another way of stating this difference is that the usual thermodynamics of irreversible processes, for example, is concerned with small (linear) deviations from equilibrium while the Boltzmann equation permits large (nonlinear) deviations. We must be careful to distinguish the nonlinearity of the equations of fluid dynamics from the linear irreversibility mechanism (e.g., heat flow proportional to temperature gradient).

The principal use of the Boltzmann equation is when there are significant changes in gas flow properties over a mean-free-path or over a collision time. There is no reason to doubt its validity in shock waves of arbitrary strength or in sound waves of arbitrary frequency; it is, indeed, the only way of studying such problems. From this viewpoint we must consider the use of the Boltzmann equation to compute the values of transport coefficients as a secondary application.

c) Outline of a derivation of the Boltzmann equation.

7. The Boltzmann gas and the Knudsen gas. For simplicity we first consider elastic sphere molecules and then generalize to arbitrary intermolecular potentials. We are given a fixed domain as a container for the gas and a fixed function $F_1(\boldsymbol{\xi}, \boldsymbol{x})$ defined for $\boldsymbol{x}$ in the given domain. The number of particles, n, is allowed to approach infinity. The mass of each particle is allowed to approach zero in such a way that

$$m\,n = \text{const}. \tag{7.1}$$

The total mass of the system is constant. Moreover, the distribution function $f = m n F_1$ is constant in this limit as are all the conventional macroscopic quantities which are derived from f (e.g., temperature, pressure, fluid flow velocity, etc.). Finally the spherical molecules are decreased in size in this limit according to the rule

$$n\,\sigma^2 = \text{const}; \tag{7.2}$$

σ is the molecular diameter. This relation implies that the mean-free-path ($\sim 1/n\sigma^2$) is held constant in this limit. The same is true of the collision time since velocities are not scaled at all ($F_1(\boldsymbol{\xi})$ is fixed). Finally we observe that

$$n\,\sigma^3 \to 0. \tag{7.3}$$

This, however, is the volume occupied by the molecules themselves. We conclude that the gas is becoming more rarefied and that the perfect gas law, $pV = n\,\mathrm{k}\,T$, is obeyed in the limit. We see that we have lost one of the aspects of intermolecular forces, namely, internal potential energy, but we have retained the collision process as a finite effect. Although at any instant there are relatively fewer and fewer particles undergoing collision as $n \to \infty$, these particles determine the evolution of F_1. For example, the stress tensor is given by the momentum transfer across a surface [Eq. (17.5)] and has no contribution from intermolecular forces. Still, its behavior is intimately connected with collisions via the mechanism of viscosity. The first effect has the order $n\sigma^3$, and the second has the order $1/n\sigma^2$.

If we examine the Boltzmann equation for the mass density distribution in the case of elastic spheres, we find a factor σ^2/m multiplying the collision term. This, however, is a constant, so the Boltzmann equation itself is invariant under this limit. It is justifiable to consider the limiting gas to be a *continuum* since the graininess has been made arbitrarily small, and this limiting continuum which satifies the Boltzmann equation *exactly* is termed a Boltzmann gas.

In case the above-described limit is thought to be nonphysical in that molecular masses and dimensions are tampered with, it can be replaced by an exactly equivalent but physically realizable limit. Take σ and m to be constant. As the total number of particles, n, is increased, the container is scaled in linear dimension proportional to $\sqrt{n}$, and F_1 is scaled similarly as a function of x; i.e., $n^{\frac{3}{2}} F_1(\boldsymbol{\xi}, \boldsymbol{x}/\sqrt{n})$ is held fixed (the factor $n^{\frac{3}{2}}$ is required for normalization). In this limit, the mean-free-path scales in proportion to the size of the container, and the gas becomes more rarefied. What we are doing is scaling all macroscopic flow gradients as well as the container size in proportion to the mean-free-path. This limit is not so formally elegant as the previous one since all the physically interesting distributions, F_1, nF_1, and f approach zero. Because of this, one might be tempted to drop the collision term in the Boltzmann equation since it is quadratic in f. This would leave one with the equation for a Knudsen gas. In a certain sense this approximation is proper since the collision time becomes infinitely long in the limit. The question at stake is an inversion of limits. If one observes the gas for only a fixed time interval, the behavior will become increasingly Knudsen-like, but if one observes the gas for infinite time for each value of n (or for a time which increases directly as the mean collision time), one must consider it to be a Boltzmann gas.

In terms of the original scaling, the Knudsen gas limit can be described tentatively as

$$n\sigma^2 \to 0 \tag{7.4}$$

instead of (7.2). This allows one, at least formally, to drop the collision term in the Boltzmann equation. If this limit is converted into more physical terms keeping m and σ fixed but increasing the size of the container, we find a situation which is at first glance not much different from the previous one. It would seem that whether or not collisions can be ignored depends on how long one observes the system. The correct conclusion, surprisingly, depends on the boundary conditions which are imposed at the container walls as well as on the length of observation. If the boundary conditions are stochastic (e.g., diffuse reflection) and a molecule forgets its past after an encounter with the boundary, then the gas is truly Knudsen in the limit (7.4); one should solve the Knudsen gas equations together with the boundary conditions, and the result is uniformly valid for all time. However, with specular reflection at the boundary, gas collisions will always dominate given enough time. The fault in the original analysis given above (i.e., that it did not perceive this distinction) was in assuming that, because a term in an equation is small, dropping it produces a small error in the solution to the equation. This point is quite general. In deriving an equation (e.g., the Boltzmann equation), one cannot deal with the equation but must consider the solutions as well. In summary, the limit (7.4) defines a Knudsen gas if there is a stochastic boundary condition; otherwise it is either a Knudsen gas or a Boltzmann gas depending on the time of observation.

In the case of an intermolecular potential, $\varphi(r)$, where r is the intermolecular distance, we introduce the same basic limit, (7.1) and (7.2), but σ is now a scaling parameter in the potential instead of a strict molecular diameter. Specifically we assume that

$$\varphi(r, \sigma) = \sigma^2 \psi\left(\frac{r}{\sigma}\right), \tag{7.5}$$

where ψ is a fixed function of its argument. The factor σ^2 is chosen so that the various two body orbits scale geometrically. In particular, it is easy to verify that the collision cross-section (see Sect. 15) remains a fixed function of its arguments and is merely multiplied by a factor σ^2 just as for elastic sphere molecules. This guarantees that the Boltzmann equation is invariant during the approach to the limit. Again it is easy to verify that the mean-free-path and the mean collision time are fixed but the imperfect gas corrections drop out.

For comparison we note that the limit

$$\left.\begin{aligned} m\,n &= \text{const}, \\ n\,\sigma^3 &= \text{const} \end{aligned}\right\} \tag{7.6}$$

in which the mean-free-path approaches zero and the equation of state remains imperfect leads to the non-dissipative Euler equations of fluid dynamics [*51*][1]. In this limit not only do the distributions $F_2, F_3, \ldots$ become dependent on F_1, but they, as well as F_1, become dependent on the much less detailed fluid dynamical description of a state. A different procedure which yields the Euler equations only for a perfect gas is to take the Boltzmann gas limit and then let $n\sigma^2 \to 0$ (see Sect. 26). In each case, the dissipative Navier-Stokes equations can be obtained as a first order correction to the Euler equations, i.e., by keeping first order terms in the expansion parameter. It is easy to see that the Navier-Stokes equations can be obtained as a zero order limit only if the intermolecular forces become long range (i.e., soft) in the limit[2].

8. Formal derivation. First consider the case of a finite cut-off, (3.6); the exact equation which F_1^σ satisfies is (3.12). To obtain the Boltzmann equation we take four steps which we describe as follows:

(1) truncation,
(2) binary collisions,
(3) molecular chaos,
(4) slow variation of F_1.

Proceeding operationally, first we drop the index σ from F_1^σ and F_2^σ in (3.12). Next we replace the arguments in $F_2(z_1, z_2^+, t)$ by the values they would need to have at the start of a *binary collision* (say $\bar{z}_1, \bar{z}_2, \bar{t}$) in order to end up with the given values z_1, z_2^+ at the time t (the start of the collision is defined by $|\boldsymbol{x}_2 - \boldsymbol{x}_1| = \sigma$,

[1] This analysis is very illuminating, but some gaps still remain. The easier problem of obtaining the primitive conservation equations with full stress tensor and heat flow (which is not a determined system) is treated formally in [*34*] and [*25*] and with somewhat more care in [*54*].

[2] H. GRAD: Address before the American Physical Society, Mexico City, August, 1955.

and the molecules are approaching). Step (3) involves replacing $F_2(z_1, z_2^-)$ by $F_1(z_1)\,F_1(z_2^-)$ and $F_2(\bar z_1, \bar z_2)$ by $F_1(\bar z_1)\,F_1(\bar z_2)$. Finally, in the four functions $F_1(z_1, t)$, $F_1(z_2^-, t)$, $F_1(\bar z_1, \bar t)$, $F_1(\bar z_2, \bar t)$, we replace $\boldsymbol{x}_2^-$, $\bar{\boldsymbol{x}}_1$, $\bar{\boldsymbol{x}}_2$ by $\boldsymbol{x}_1$ and $\bar t$ by t. Replacing $n-1$ by n in (3.12), we obtain

$$\frac{\partial F_1}{\partial t} + \boldsymbol{\xi}_1 \cdot \frac{\partial F_1}{\partial \boldsymbol{x}_1} = n \int \{F_1(\bar{\boldsymbol{\xi}}_1)\, F_1(\bar{\boldsymbol{\xi}}_2) - F_1(\boldsymbol{\xi}_1)\, F_1(\boldsymbol{\xi}_2)\}\, V\, d\omega\, d\boldsymbol{\xi}_2; \tag{8.1}$$

$\boldsymbol{x}_1$ and t are the same in all arguments. This is the Boltzmann equation. In terms of the mass density distribution, the equation is

$$\frac{\partial f}{\partial t} + \boldsymbol{\xi}_1 \cdot \frac{\partial f}{\partial \boldsymbol{x}_1} = \frac{1}{m} \int \{f(\bar{\boldsymbol{\xi}}_1)\, f(\bar{\boldsymbol{\xi}}_2) - f(\boldsymbol{\xi}_1)\, f(\boldsymbol{\xi}_2)\}\, V\, d\omega\, d\boldsymbol{\xi}_2. \tag{8.2}$$

The change of scale $d\omega = \sigma^2\, d\Omega$ shows that the collision integrals in (8.1) and (8.2) are invariant under the limit described in Sect. 7. The integration for $d\Omega$ is over the unit disc, and the factors $n\sigma^2$ and σ^2/m which appear outside the integral are fixed.

We now present a heuristic justification for this transition from (3.12) to (8.1); the mathematical justification is described in the remainder of this chapter. We recall (Sect. 3) that when $n\sigma^3$ is small, F_1^σ approximates F_1 and F_2^σ approximates F_2. This justifies step (1). This statement describes the number of particles actually in collision at any instant as small compared to n. We extend this by claiming that the number of ternary or higher order collisions is small compared to the number of binary collisions. This statement can be made more precise. Consider Eq. (3.2) specialized to the case $r=2$. If $|\boldsymbol{x}_2 - \boldsymbol{x}_1| < \sigma$, i.e., if molecules 1 and 2 are in each other's field of force, the integral (i.e., ternary collision) term in (3.2) is small (by a factor σ) compared to the binary collision term involving the force $\boldsymbol{X}_{12}(z_1, z_2)$ (although in the long run the two terms would exert a comparable influence since a particle spends only a small part of its lifetime near any other). We conclude that for a small time interval on the order of the duration of a collision (order of σ) the two particle *Liouville equation* is a good approximation to the exact equation for F_2. The solution of this equation is $F_2 = \text{const}$ on a two particle trajectory. This justifies step (2), the replacement of $F_2(z_1, z_2^+, t)$ by $F_2(\bar z_1, \bar z_2, \bar t)$. It is crucial that the binary collision assumption is applied to $F_2(z_1, z_2^+, t)$ before the chaos assumption. Step (3) is the intuitively reasonable assumption that two particles which are about to collide are statistically independent whereas particles which have just collided may very well be correlated. If this distinction is ignored and we set $F_2(z_1, z_2) = F_1(z_1)\, F_1(z_2)$ wherever it appears in Eq. (3.12), we find that collisions have disappeared and F_1 satisfies the equation for a Knudsen gas.

Finally we turn to step (4). The justification of this step follows from the observation that the various $\boldsymbol{x}$-coordinates differ from $\boldsymbol{x}_1$ and $\bar t$ differs from t by amounts on the order of σ. On the other hand, the space (and presumably time) dependence of F_1 is fixed as $\sigma \to 0$.

The inclusion of intermolecular potentials which extend to infinity (but which die down rapidly enough) can be accomplished by a simple expedient. The cutoff parameter which defines the truncated distribution, F_1^σ, and the scaling parameter in $\varphi(r, \sigma)$ are, in principle, distinct. Let us call the former σ' and the latter σ and assume that both are small. Now consider Eq. (3.5) for $F_1^{\sigma'}$; it contains the grazing collision term which was previously dropped. As in the previous argument, we drop the distinction between truncated and nontruncated distributions. However, we must keep the parameter σ' in the integration limits, S_2 and D_2. We end up with the same collision integral as in (8.1) but with $d\omega$ integrated over a disc of diameter σ' and, in addition, a correction term. We now allow σ' to approach zero but at a slower rate than σ so that $\sigma'/\sigma \to \infty$ while $n(\sigma')^3 \to 0$ and $n\sigma^2 = \text{const}$. If we introduce the change of scale which was previously described, $d\omega = \sigma^2 d\Omega$, we obtain the same collision integral as before except that $d\Omega$ is integrated over the whole plane instead of over the unit disc. In other words, although the cutoff parameter becomes macroscopically small, it is large when compared to molecular dimensions. As to the error term,

$$\frac{n}{m}\frac{\partial}{\partial \boldsymbol{\xi}_1}\cdot\int_{D_2} \mathbf{X}_{12} F_2\, dz_2,$$

from the scaling we conclude that the major contribution to this integral (ignoring the restriction to D_2) comes from $|\boldsymbol{x}_2 - \boldsymbol{x}_1|$ on the order of σ. Since $\sigma' \gg \sigma$ and the sphere σ' is excluded from the integration, we conclude that the error term is small, provided that the integral converges. This is guaranteed if the intermolecular force decays faster than an inverse cube of the radius.

There is one important mathematical distinction between a Boltzmann equation with a cutoff and one without; the collision term of the latter is not absolutely convergent. The positive and negative terms do not converge individually but only when combined. The total collision cross-section, $\int d\omega$, is infinite.

In summary, the transition from the exact equation for F_1^σ in an n-particle system to the Boltzmann equation was made by applying certain intuitively appealing assumptions. It does no violence to one's intuition about a deterministic dynamical system that F_1^σ may approximate to F_1 or that binary encounters should dominate in a rarefied gas (even though the mathematical justification happens to be exceedingly difficult—cf. Sect. 10). The molecular chaos asumption is usually handled more gingerly since, as a matter of principle, it should not be necessary to make random assomptions about a dynamical system. That this is a legitimate *approximation* rather than an *ad hoc* assumption will be shown later in this chapter. Apart from this, what is disguised by the above formal demonstration is the complex problem of estimating errors uniformly for entire classes of functions F_1 and F_2 which arise from a sequence of Liouville equations with more and more independent variables.

d) Critique.

9. Two viewpoints. There are two distinct ways of introducing statistical mechanics, i.e., a distribution function, into a dynamics problem. It is necessary to distinguish them very carefully since the formal analysis may be very similar in the two cases but the interpretation is not. The distinction is a difference in the definition of the distribution function. The one which was adopted in Sect. 1 is that F_n is a probability density and refers to the possibility of repeated experiments. In a mathematical theory properties of the function F_n are almost entirely a matter of postulation since they depend in a very complicated way on the mechanism for repeating experiments. For this to be a satisfactory representation of nature, it must be shown that macroscopic conclusions depend very little on the specific choice of an initial function F_n. From F_n we obtain the one-particle distribution, F_1, by integration.

The second possibility is as follows. A point Z (i.e., a state) can be represented in γ-space as n points, $(z_1 \ldots z_n)$. If n is large, we can suppose that this "gas" of points approximates a smooth distribution which, despite the previous use of this symbol, we again call $n F_1(z)$. The "density" F_1 is defined only for a volume in γ which is "not too small and not too large". More precisely, a smooth distribution $F_1(z)$ can be defined only as a limit when $n \to \infty$[1]. This F_1 is entirely distinct from the previous one. It is an actual number density not a probability density. It refers to a single system rather than to the combined behavior of a statistical assemblage. However, as a smooth function it really refers only to a limiting system with infinitely many particles. To illustrate this, consider the integral $\int_D n F_1(z)\, dz$ over a domain, D, which is so small that the value of the integral is less than one. With the first interpretation of F_1 as a probability, this means that the expected number of molecules within D is a small quantity; i.e., if one repeats the experiment many times, sometimes a molecule will be found in D and sometimes not, and this integral gives the fraction of the number of experiments in which a molecule will be found there. With the second definition of F_1, no conclusion whatsoever is drawn; F_1 is meaningful only for domains which are larger than D. This is a defect of the second definition. Mathematically, the number density is much harder to work with because a complicated limiting process has to be treated before one even succeeds in defining F_1 properly. Of course, similar difficulties may arise with the probability density, for example when one tries to show that macroscopic results are independent of the choice of initial F_n, but there is much to be said in favor of postponing difficulties!

In addition to the question of mathematical convenience there are real differences. In principle the two methods apply to two different experimental realizations, namely, repeated experiments performed on a finite system of n particles vs. observation of a single system of many (strictly speaking, infinitely many) particles. It seems likely that both methods will lead to the Boltzmann equation in an appropriate limit. But even here, although the equation is the same the argument, F_1, is different (we could compare this with LAPLACE's equation which is satisfied by entirely different physical quantities such as temperature and electrostatic potential). It is possible that conditions for the validity of BOLTZMANN's equation differ in the two cases; e.g., whether or not nonlinear deviations from equilibrium are permitted might have different answers. The nature of the convergence to BOLTZMANN's equation as $n \to \infty$ is certainly different in the two cases. It seems likely that, as a number density, the F_1 in the Boltzmann equation represents a smoothed out average of the F_1's which converge to it, whereas this smoothing is probably not required to insure convergence of the probability density (see Sect. 12). Also certain subtleties related to the Poincaré recurrence theorem which are present when F_1 is a number density are absent when it is a probability density. The Poincaré theorem states that if a dynamical system is observed to be in a given state, then it must eventually return arbitrarily close to this same state. Thus there can be no naïve approach to equilibrium as $t \to \infty$ for the number density F_1; one must either eliminate some exceptional time intervals or perform some kind of smoothing. It will be seen in Sect. 12 that, at least in certain simple examples, the probability density F_1 can approach equilibrium unambiguously and without exception. How does one reconcile this difference in behavior? This is a mathematical question, not a physical one. If the Poincaré recurrence time is sufficiently large, the two predictions may well be physically indistinguishable. The answer is that the probability density, F_1, does not predict the outcome of any observation but only its expectation. Instead of repeating experiments consecutively in time as

[1] It does not help to introduce the concept of cumulative distribution function or to consider nF_1 to be a sum of δ-functions. Neither the Boltzmann equation nor the quantity $\int F_1 \log F_1$ can be given any precise meaning unless F_1 exists as a conventional function.

would actually be done in the laboratory, we suppose that all possible repetitions (i.e., the elements of the sample space) are started simultaneously and are observed simultaneously. At a later time any particular sample which happens to be in an unusual state corresponding to the possibilities of the Poincaré theorem will be in a minority; more precisely, it will be of measure zero or have zero probability. If the sample space is infinite, specifically a continuum, then one cannot be surprised if at *no* subsequent time are there enough exceptional samples to disturb the smooth approach to equilibrium of the probability distribution. The more complicated description in terms of the number density F_1 gives a more detailed picture of the actual physical possibilities. But one is not normally interested in *all* the possibilities; it is sufficient to be cognizant of the overwhelming majority of possibilities. Introduction of the probability density accomplishes this; presumably smoothing of the number density accomplishes the same purpose but more clumsily. An example is given in Sect. 12 in which smoothing of the number density (by letting $n \to \infty$) is exactly equivalent to introducing a probability.

It should be mentioned that probability is usually introduced into the problem even with the second definition of F_1 as a number density in order to eliminate certain awkward and unusual occurrences which are "unlikely". This is done by stating theorems about "almost all" points in Γ (and sometimes about almost all times) and clearly implies that a probability has been placed on Γ. The two methods are still distinct, however, because of the different definitions of observable quantities (in one case the value of a phase function and in the other, its expectation) and also because, while "unlikely" in the second case refers to the small measure of a $6n$-dimensional set, in the first case it refers to some sort of measure in a function space, i.e., to an unlikely choice of initial F_n. The two concepts are related but only in a subtle and complex way.

In principle both procedures should be carried to completion since the results will be different and carry different interpretations; we concentrate on the first method because of the simpler formal manipulations, but we shall compare the two whenever necessary.

Perhaps a word should be said about the distinction between dealing with probability in the fully deterministic sense of observing the evolution of solutions of LIOUVILLE's equation and the concept of "randomness". A stochastic process is uniquely determined by giving the flow in phase space (e.g., by specifying a Hamiltonian) and by specifying an initial probability $F_n(Z, 0)$. By this we mean that not only is the probability distribution of the values of any phase function $\psi(Z)$ determined for all time, but also all joint probabilities of observing $\psi(Z, t_1) < a_1$, $\psi(Z, t_2) < a_2$, etc. are fully determined. Well developed mathematical techniques exist only for certain special types of stochastic processes, e.g., Markov processes. These special stochastic assumptions, which usually imply some degree of forgetfulness of the past behavior of the system, are almost inevitably incompatible with the original, dynamically defined, stochastic process, but the hope is that they may be verified to a certain degree of approximation or in some limit.

10. "Almost all" considerations. Any statement that solutions of LIOUVILLE's equation converge to a solution of the Boltzmann equation is rather subtle, and the meaning of "converge" will be analyzed in the next few sections.

The first point to be made is that, unless one is careful, the solutions of LIOUVILLE's equation will converge much more strongly than is desired i.e. to the equilibrium, Maxwellian solution. This is, of course, a solution of BOLTZMANN's equation but an uninteresting one. For example, we know that if F_n is microcanonical, i.e., a singular distribution which is constant on an energy surface[1], then, as n becomes large, F_1 will approach the Maxwellian distribution which corresponds to the given energy surface. This result can be greatly extended. For example, if F_n is merely *bounded* on this energy surface, $a < F_n < b$, but is otherwise arbitrarily distributed, the same conclusion follows; F_1 becomes Maxwellian for large n. As a matter of fact, the bounds can increase as any power of n, for example,

$$0 \leq F_n < b\, n^5,$$

and we still conclude that F_1 can be nothing but Maxwellian. It is easily seen that a non-Maxwellian F_1 can make its appearance only if F_n increases *exponentially* with n. More precisely, if we adopt a bound of the type $F_n < K^n$, we find that the F_1's which are thereby admitted can be described by a condition roughly like $H_\gamma < \log K$. In other words, the deviation of the given F_1 from equilibrium, which is measured by the value of H_γ (see Sect. 18), determines how rapidly F_n must be allowed to grow as a function of n.

Another way of looking at this is in terms of the alternative definition of F_1 given in Sect. 9, i.e., as a particle density rather than as an expectation. In this case F_1 is defined in some rough sense at each point of Γ. We know that, not only is the mean value of F_1 on an energy surface approximately Maxwellian for large n, but also that F_1 is approximately Maxwellian for most points on the energy surface. As a matter of fact, the relative measure of the set of points on which F_1 differs appreciably from a Maxwellian becomes small on the order of K^{-n}. In order to have an expectation F_1 which is appreciably non-Maxwellian, we must concentrate the probability density on such a small set; since F_n must integrate to unity, it must grow at a rate which is at least K^n.

These remarks bring out some of the mathematical difficulties of the problem. It is quite clear that we cannot expect *every* choice of initial F_n to give an $F_1(t)$ which converges to a solution of BOLTZMANN'S equation. We must somehow eliminate exceptional and poorly behaved functions $F_n(t)$. However, in allowing F_n to be wild enough to allow a non-Maxwellian F_1 at $t=0$, we have opened the gates to solutions of LIOUVILLE'S equation which can be very ill behaved indeed. For example, no uniform pointwise bounds independent of n exist on the distributions F_r (and in particular F_1). The inequality (5.2) does allow one to obtain bounds in measure. However, this difficulty is compounded by the fact that the evolution of F_1 is determined by the values taken by F_2 on a set whose measure approaches zero as $n \to \infty$ (see Sect. 11).

This difficulty also appears with F_1 as a number density. Exceptional behavior, e.g. non-convergence to the Boltzmann equation, would be represented by a set of points of small measure. But the whole set of points in which we are interested is already extremely small, of order K^{-n}. We must show that the exceptional set is still smaller.

11. Molecular chaos. The chaos property is intuitively a sort of random affair. Also intuitively, one feels that collisions exert a randomizing influence. Thus one might expect the collision process to produce chaos even if it is not there initially. This mechanism obviously requires a number of collision times for it to take effect and even a large number if chaos is to be satisfied with some degree of precision. Such an argument can never lead to a Boltzmann equation

[1] The density is constant not with respect to the element of surface area but with respect to the measure induced by the volume in a thin energy shell.

in which the collision time is the relevant time scale. However, it is exactly this type of argument which leads to *linear* dissipative processes. If we wish to justify the validity of the Boltzmann equation in enough generality to describe the evolution of an arbitrary initial function $F_1(z)$ with a resolution comparable to the collision time, then we must know *a priori* that the initial F_n is chaotic. Granted this, one would still have to verify that LIOUVILLE's equation does not destroy the chaos property. We therefore separate the problem into two parts. The *initial chaos* problem is one of probability theory alone. It is necessary to show that choosing an initial F_n "at random" is very likely to yield one which is chaotic. This problem has nothing to do with particle dynamics or LIOUVILLE's equation. The *subsequent chaos* problem is one in the theory of partial differential equations. Is a certain property propagated in time when it happens to be present initially? Actually, this is an asymptotic problem since it is easy to see that it cannot be true unless n is large. The answer to these two questions demonstrates that an isolated dynamical system unperturbed by any external influences can and in most cases will behave irreversibly provided that n is large.

Consider first the question of initial chaos. For this purpose, it is sufficient to take the simplest definition of chaos, viz.,

$$F_2(z_1, z_2) = F_1(z_1)\, F_1(z_2)\,. \tag{11.1}$$

We consider the function $F_1(z)$ to be fixed. In the class of all functions $F_2(z_1, z_2)$ which reduce to the given function F_1 on integration, it is clear that in no reasonable sense do most of them satisfy (11.1). However, consider all functions $F_n(z_1 \ldots z_n)$ which are consistent with F_1. We claim that the class of functions $\{F_2^{(n)}\}$ which is obtained by integration from the class $\{F_n\}$ converges to the single function $F_1(z_1)\, F_1(z_2)$ as $n \to \infty$. This statement can be made precise and the theorem proved[1]. The crucial point is that the functions F_n are assumed to be symmetric in their arguments $z_1 \ldots z_n$. In general, a given value of F_n is taken at $n!$ different points in Γ. The effect of this is to concentrate the probability density in those regions of Γ where (11.1) tends to be satisfied. Of course, the result cannot be proved for *all* sequences F_n, $n \to \infty$, but it is proved to hold for almost all sequences, "almost all" being defined in an appropriate way.

The simplest way to describe the situation is in terms of number densities F_1 and F_2 rather than probabilities. Roughly speaking there is a function F_2 as well as a function F_1 assigned to each point $Z = (z_1 \ldots z_n)$ in Γ-space. The set on which F_1 is not approximately Maxwellian has a relative measure which is exponentially small in n, say K^{-n}; the value of K depends on the degree of precision by which we define "approximately" Maxwellian. The measure of the subset in which $F_2(z_1, z_2) \neq F_1(z_1)\, F_1(z_2)$ to some degree of precision has the order K_1^{-n} where $K_1 > K$. This set is exponentially smaller than the first set; the relative measure is $(K/K_1)^n$. This statement can be made even stronger in terms of probability densities. To any given F_1 there corresponds a minimum value of the growth constant K, say K_0, which will allow a sequence $\{F_n\}$ to reproduce this F_1. For the sequence to converge to an F_2 which is not a product of F_1's, K must be larger than K_0, say K_1. Now among all sequences $\{F_n\}$ bounded by any K^n, $K \geq K_1$, (these are already a small subset of all sequences which are compatible with the given F_1) only a small subset (this measure is in a function space) will yield an F_2 which is not a product.

The more general result, that $F_r(z_1 \ldots z_r)$ converges to $F_1(z_1) \ldots F_1(z_r)$ for any fixed r, is proved in the same way.

[1] Not yet published.

Now we turn to the subsequent chaos problem. It is evident that the chaos property (11.1) is almost immediately destroyed by LIOUVILLE's equation. To see this, take an initially chaotic F_n and allow LIOUVILLE's equation to operate on it for a short time. If we now choose z_1 and z_2 so that, tracing the two particles backward in time, they have just collided, then (11.1) will in general not be satisfied. However, in order to derive the Boltzmann equation we do not need the full property (11.1); we need it only for particles which are about to collide. We term this weaker condition *one-sided* chaos and (11.1) *two-sided*. While there is no reason to expect the one-sided chaos property to be instantly destroyed, it is not at all clear intuitively how it is affected by the complex motion of the system. What must be shown is that it is not destroyed for an *arbitrarily long time* (i.e., for arbitrarily many collision times) if n is sufficiently large.

The basic idea is that $F_2(z_1, z_2)$ and $F_1(z_1)\,F_1(z_2)$ behave similarly in time if $\boldsymbol{x}_1$ and $\boldsymbol{x}_2$ are not adjacent. Specifically, in terms of

$$\left.\begin{aligned}\delta_{1,2} &= F_2(z_1, z_2) - F_1(z_1)\,F_1(z_2),\\ \delta_{1,23} &= F_3(z_1, z_2, z_3) - F_1(z_1)\,F_2(z_2, z_3),\\ &\cdots\cdots\end{aligned}\right\}\qquad(11.2)$$

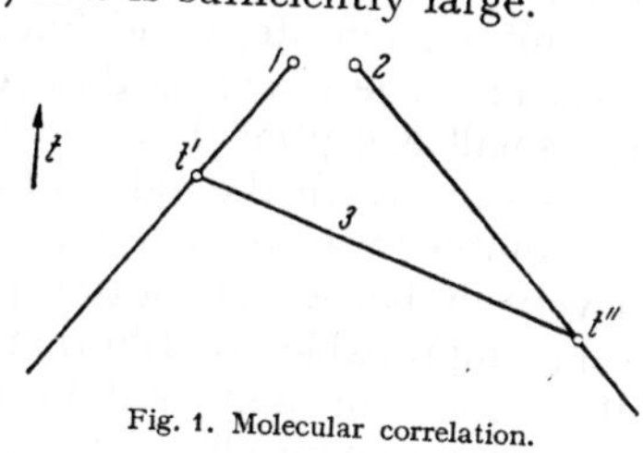

Fig. 1. Molecular correlation.

we have

$$\left.\begin{aligned}\frac{\partial\delta_{1,2}}{\partial t} &+ \boldsymbol{\xi}_1\cdot\frac{\partial\delta_{1,2}}{\partial\boldsymbol{x}_1} + \boldsymbol{\xi}_2\cdot\frac{\partial\delta_{1,2}}{\partial\boldsymbol{x}_2} + \frac{1}{m}\boldsymbol{X}_{12}\cdot\frac{\partial F_2}{\partial\boldsymbol{\xi}_1} + \frac{1}{m}\boldsymbol{X}_{21}\cdot\frac{\partial F_2}{\partial\boldsymbol{\xi}_2},\\ &+\frac{n}{m}\frac{\partial}{\partial\boldsymbol{\xi}_1}\cdot\int\boldsymbol{X}_{12}(z_1, z_3)\,\delta_{13,2}\,dz_3,\\ &+\frac{n}{m}\frac{\partial}{\partial\boldsymbol{\xi}_2}\cdot\int\boldsymbol{X}_{12}(z_2, z_3)\,\delta_{1,23}\,dz_3,\end{aligned}\right\}\qquad(11.3)$$

(for simplicity we have identified n, $n-1$, and $n-2$). If $\boldsymbol{x}_1$ and $\boldsymbol{x}_2$ are separated by more than the distance σ, then the rate of change of $\delta_{1,2}$ ($\dot{\delta}_{1,2}$) following the unperturbed rectilinear paths of z_1 and z_2 is given by an integral which involves $\delta_{13,2}$ and $\delta_{1,23}$. The same type of result obtains for the time derivative of $\delta_{1,23}$ etc. We can compute $\dot{\delta}_{1,2}$ in terms of $\delta_{1,23}$; $\dot{\delta}_{1,23}$ in terms of $\delta_{1,234}$; etc. and eventually try to relate $\delta_{1,2}$ at the time t to the initial values of a certain number of δ's. But these initial values become small (F_r is chaotic) when $n\to\infty$, so $\delta_{1,2}$ is small at the finite time t. This analysis ignores the extra terms which appear when two particles collide. The latter situation is most easily seen in terms of a diagram, Fig. 1. Particles 1 and 2 do not intersect when traced back in time[1]. In evaluating $\dot{\delta}_{13,2}$ at the time t', we find that, for some values of $\boldsymbol{\xi}_3$, the backward path intersects particle 2 at a time t''. This part of the domain of integration dz_3 can give rise to a large contribution. On the other hand, these large contributions are found only in sets of small measure when σ is small. A strong indication can be given of a proof of the theorem that the δ's remain small if they are initially small[2], but there are still gaps because of the difficulty discussed in Sect. 10, that F_r can be extremely large on certain sets of small measure.

From this analysis we can see that, even if F_1 converges very nicely to a solution of BOLTZMANN's equation, the convergence of F_2 is, in a certain sense, very singular. The convergence of F_2 has to be described on two length scales, macroscopic and microscopic. Suppose it is proved that $F_2(z_1, z_2)$ converges to $F_1(z_1)\,F_1(z_2)$

[1] With a specular reflection boundary condition, they may intersect, but this can only occur on a set of small measure.

[2] Unpublished.

for almost all values of z_1 and z_2 (i.e., macroscopically). This is not a strong enough result to imply the subsequent chaos theorem! We need to verify the chaos relation for only those sets of values z_1, z_2 which are about to collide, but this is a set of small measure which approaches zero with σ. However, it is possible to specify the exceptional set on which $F_2(z_1, z_2)$ does *not* converge to $F_1(z_1)\, F_1(z_2)$ rather precisely and verify that this set can in fact be ignored.

This exceptional set, on which chaos is not satisfied, has its own special interest. It consists, roughly speaking, of pairs of points z_1, z_2 which have recently collided. It is the existence of this exceptional set which provides the connection between the reversibility of LIOUVILLE's equation and the irreversibility of BOLTZMANN's equation. Suppose we suddenly "turn off" LIOUVILLE's equation, reverse all velocities, and proceed again with the solution of LIOUVILLE's equation. We retrace our steps and find, for example, that H_γ is increasing instead of decreasing. The reason is that, even though chaos is satisfied almost everywhere, the small exceptional set on which it is not satisfied has now become exactly the set on which the value of F_2 determines the course of F_1.

Another way of describing the situation is that LIOUVILLE's equation takes a very probable distribution (two-sided) and almost instantly converts it into a very improbable distribution (one-sided). The proof which was referred to above indicates only that LIOUVILLE's equation does not destroy forward chaos for a macroscopically long time. It is likely (cf. Sect. 12) that after a while the collision process will act so as to preserve forward chaos *indefinitely*, but this would seem to be very difficult to prove.

A word should be said about boundary conditions. We saw that in connection with the subsequent chaos theorem it was necessary to trace back a number of particle paths. If a boundary intercedes before reaching $t=0$, the argument must be modified. With a specular boundary we merely continue tracing back the reflected particle paths until they reach $t=0$. With a stochastic boundary it is necessary to supplement the usual form of boundary condition which describes only the one-particle distribution of emitted particles with the statement that particles emitted from different points of the boundary and at different times are statistically independent[1]. This is a necessary supplement to the initial chaos condition.

12. Coarse graining. It has long been understood that the transition from microscopic to macroscopic variables requires some kind of smoothing or coarse-graining, but the exact amount is subject to differences in opinion. One can smooth with respect to space variables (usually phase space) or with respect to time, or if one is cautious, with respect to both. In this context, the distinction between the two interpretations of F_n of Sect. 9 is crucial. First let us consider the case where F_n is a probability. All expectations, being expressed as integrals, $\overline{\psi} = \int \psi(Z)\, F_n(Z)\, dZ$, are already smoothed with respect to Z. The distribution F_1, by its very definition as an integral of F_n, is automatically somewhat smoothed although this is a very mild type of smoothing since the integration is not extended over the full dimensionality of Γ. We wish to indicate, by means of an example, that the amount of smoothing which is implicit in the very definition of F_1 is probably sufficient to insure that it is a well behaved function for any value of n. The question of whether a *sequence* of functions F_1 converges smoothly to a solution of the Boltzmann equation as $n \to \infty$ is not referred to by this example.

[1] It is sufficient to assume that the correlation decays to zero in a distance (or time) comparable to σ.

Consider a rectangular box in which there is a pure Knudsen gas, i.e. a gas of particles which are moving independently and are reflected specularly by the walls of the box. The motion of an individual particle is well known. For example, for almost all initial conditions a particle will come arbitrarily close to every point in the box and will in the long run even visit different regions for times exactly in proportion to their volumes. Introduce a probability density F_n. As mentioned earlier (see Sect. 4), the problem separates, and F_1 satisfies a one-particle Liouville equation. Insofar as we are interested in the behavior of F_1, the number of particles, n, is irrelevant. About $F_1(\boldsymbol{\xi}, \boldsymbol{x}, 0)$ we make only the assumption that it is Lebesgue integrable in $(\boldsymbol{\xi}, \boldsymbol{x})$; what we have eliminated is the possibility that F_1 is singular, e.g., a δ-function. Since $F_1(\boldsymbol{\xi}, \boldsymbol{x}, t)$ is constant along a trajectory (a trajectory is a sequence of straight lines joined where they intersect the boundary), at any time t, F_1 takes on the same numerical values that it was given initially; clearly $F_1(\boldsymbol{\xi}, \boldsymbol{x}, t)$ approaches no limit as $t \to \infty$. On the other hand, the density in physical space,

$$F_x(\boldsymbol{x}, t) = \int F_1(\boldsymbol{\xi}, \boldsymbol{x}, t)\, d\boldsymbol{\xi}, \tag{12.1}$$

can without difficulty be shown to converge to a limit as $t \to \infty$. Furthermore, this limit is a constant independent of $\boldsymbol{x}$, say $\overline{F}$[1]. A stronger result can be proved. The function

$$H_x(t) = \int F_x \log F_x\, d\boldsymbol{x} \tag{12.2}$$

can be shown to approach a limit which is exactly $\overline{F} \log \overline{F}$ multiplied by the volume of the container[2]. This limit is the smallest value taken by H_x. We recall that $H_\gamma = \int F_1 \log F_1\, dz$ is a constant in time. The physical conclusion is that, no matter what is the initial density in physical space, it will eventually approach a uniform distribution in space. What we have found is that the amount of smoothing inherent in the definition of F_x, viz., integration over velocity, has made F_x converge although F_1 does not. One conclusion we cannot draw is that H_x approaches its limit monotonely; this is not true in general.

The descent from the phase space $\gamma = (\boldsymbol{\xi}, \boldsymbol{x})$ to the physical space $(\boldsymbol{x})$ can be compared to the descent from Γ to γ. Collisions with the boundary (which is the convergence mechanism in the example) might be compared to the combination of boundary and intermolecular collisions (for simplicity we can think of elastic spheres). Although a proof seems to be quite difficult, it should not be surprising to find that, although F_n cannot converge, F_1 may converge to a limit as $t \to \infty$. This would be true for every value of n. Although the fact of an approach to a limit has not been proved, the limiting function that would be approached is easily computed and shown to approximate a Maxwellian distribution for large n. Continuing the analogy with the Knudsen gas example, although H_Γ is a constant in time, we should expect H_γ to approach its minimum value as $t \to \infty$, but not necessarily monotonely. Non-monotonicity is a definite possibility if the initial F_n is not chaotic. However consider the case of initial chaos. Using the analysis of Sect. 11, one would expect H_γ to decrease approximately monotonely for a long macroscopic time if n is large. Since the chaos property involves the smoothed low order distributions F_1 and F_2, it could be expected to approach its equilibrium state (which is rigorously chaotic[3]) as $t \to \infty$. Furthermore, the mechanism

[1] Convergence is in measure (or in probability); i.e. the measure of the set on which (probability that) $|F_x - \overline{F}| > \varepsilon$ becomes arbitrarily small. Convergence is pointwise if F_1 is continuous.

[2] Not yet published.

[3] Ref. [36], p. 59.

of this ultimate approach to equilibrium is collisions (by the analogy with the Knudsen gas), and this time constant is macroscopically finite. The two mechanisms overlap, and we conclude that the chaos property should hold for *all time*. Summarizing, we can expect to find that H_γ is an approximately monotone decreasing function which settles down to its minimum value and never again departs from it. We do not expect any complications such as are associated with the Poincaré recurrence theorem, e.g., eventual returns to large values of H_γ. Moreover, we may expect F_1 to converge to a solution of the Boltzmann equation *uniformly in time*, $0<t<\infty$. The difficulties associated with Poincaré recurrences are eliminated by the minimal smoothing associated with the reduction of a non-singular probability density F_n to the single particle distribution F_1.

There is one feature of the problem with intermolecular collisions which is not present in the Knudsen gas problem and which might necessitate additional smoothing. This is the fact (cf. Sect. 11) that the behavior of F_1 is determined by the values of F_2 on a set of very small measure. Although it is not at all difficult to show that F_2 is a well behaved function in the large, it is exceedingly difficult to eliminate the mathematical possibility that the extremely small regions in which F_2 may behave violently might concentrate on exactly those small regions which determine the behavior of F_1. However, at worst this requires smoothing over small distances comparable to σ, and this does not affect any of the other conclusions. A more serious smoothing question arises in the justification of step (4) of Sect. 8.

In order to bring the relationship between H_Γ and H_γ into stronger focus, the concept of coarse-graining must be made more precise. The mathematical term is *weak convergence*. The definition can be presented in two equivalent forms. Consider a sequence of functions $f_n(x)$ (the dimensionality of the space, x, is immaterial). The sequence is said to converge weakly to the limit $f(x)$ if

$$\int_D f_n(x)\, dx \to \int_D f(x)\, dx \tag{12.3}$$

for every measurable set D or if

$$\int \psi(x) f_n(x)\, dx \to \int \psi(x) f(x)\, dx \tag{12.4}$$

for every integrable function $\psi(x)$. The first criterion is closer to the intuitive concept of coarse-graining while the second relates directly to the expected value of an observable. As an example, the function $\sin(nx)$ converges weakly to zero.

It is important to realize with regard to H_γ that if a sequence of functions $F_1^{(n)}$ converges *weakly* to a limit F_1, the sequence of values $H_\gamma^{(n)}$ does not necessarily converge to the value of H_γ for the limit function. For the same reason, if $F_n(t)$ converges weakly to a limit as $t\to\infty$, this does not imply anything about the convergence of H_Γ.

Consider again the example of the Knudsen gas in a box. It is clear that, although F_1 does not convergence pointwise, it does converge weakly to the constant $\bar{F}$[1]. Because of the weak convergence, we do not expect (and do not find) that H_γ (which is constant) converges to the same limit as does H_x. Similarly, for an actual gas, we expect F_n to converge weakly to its equilibrium value whereas F_1 may converge strongly. This is reflected in the fact that H_Γ is a constant while H_γ approaches its minimum value.

---

[1] Weak convergence is a general property of strongly mixing ergodic flows (cf. ref. [*31*]). However, it should be noted that mixing is not the reason for weak convergence in this example which happens to be completely non-mixing.

Finally, we turn to the alternative definition of F_1 as an actual density of particles rather than a probability. Again we try to clarify the situation by reference to the Knudsen gas in a box. For a finite number of particles in the box we can take F_1 to be a sum of δ-functions, say $F_1^{(n)}$. Choose a fixed function $F_1(\boldsymbol{\xi}, \boldsymbol{x})$ (non-singular), and choose the sequence of initial distributions $F_1^{(n)}$ so that it approximates F_1 as $n \to \infty$. This is, of course, weak convergence of $F_1^{(n)}$ to F_1. It can be shown that, if $F_1^{(n)}$ converges weakly to F_1 at $t=0$ then, for any given t, $F_1^{(n)}(t)$ converges weakly to $F_1(t)$. Moreover, given $F_1(0)$, we can find a large value of t such that $F_x(t)$ (the density in physical space) approximates a constant as closely as we please. If we now pick n so large that $F_1^{(n)}(0)$ closely approximates $F_1(0)$, then $F_x^{(n)}(t)$ will approximate a uniform distribution of points in physical space. In other words, the discrete distribution $F_1^{(n)}$ mimics the approach to equilibrium of F_1. But if we now let t grow, keeping n fixed, we know that eventually almost anything can happen to the distribution $F_x^{(n)}(t)$. This is an elementary example of the significance of interchanging the limits $t \to \infty$ and $n \to \infty$. If we let $n \to \infty$ first, then we get an approach to equilibrium as $t \to \infty$. But if we let $t \to \infty$ first, there can be no approach to equilibrium. More precisely, for a fixed large n there are two time scales. At first there is observed an approach to equilibrium, but, after a much longer time, this trend can be observed to reverse.

In this example of a Knudsen gas, F_1 as a probability and F_1 as the limit of a sequence of number densities $F_1^{(n)}$ both satisfy the same partial differential equation.

We would expect to find the same difference in behavior of the two types of F_1 in an actual gas. If F_1 is a singular particle density, the most we can expect is convergence to a solution of the Boltzmann equation in some fixed (but arbitrary) time interval $0 < t < T$. But even more, by the very definition of the singular distribution $F_1^{(n)}$, we can only expect weak convergence to a solution of the Boltzmann equation. Unfortunately it does not seem possible to formulate a Boltzmann equation or to even define $H_\gamma^{(n)}$ directly for a singular distribution $F_1^{(n)}$.

13. Time reversal. There is one experimental situation where the *a priori* probability distribution F_n is known. In this case repeated experiments are defined as repeated observations of a given isolated dynamical system. If the energy is the only time-independent integral of the motion, F_n must be the microcanonical distribution on an energy surface [*24*]. This follows from the ergodic theorem; the sample space of successive observations (not necessarily at uniform intervals) is uniformly spread over the whole energy surface[1]. F_n is by definition time-independent. Nevertheless, the number density distribution F_1 does vary in time and permits the study of nonequilibrium properties. One of the basic postulates of the theory of irreversible thermodynamics is that of ONSAGER-CASIMIR, which states that a prepared nonequilibrium state will react in the same way as the "same" nonequilibrium state if it is observed to arise spontaneously by the fluctuations of an isolated dynamical system [*13*]. This is trivially true if "same" refers to the complete microscopic state. If it refers to some macroscopic description (e.g. an approximate specification of the number density F_1), then the claim is that the sample space of externally prepared states which are compatible with the given macroscopic state is the same as the sample space of recurrences of the given state in a system governed by LIOUVILLE'S equation. This postulate seems to contradict the statement made in Sect. 11

[1] See footnote 1, p. 221.

that LIOUVILLE's equation instantly converts a likely (two-sided) distribution into an unlikely (one-sided) distribution. The initial chaos theorem of Sect. 11 provides the way out of this dilemma and, moreover, affords a proof of the Onsager-Casimir postulate in this case. We remarked in Sect. 11 that, except for a set of small measure, all points on an energy surface correspond to an equilibrium value of F_1. A still smaller set of points violates the two-sided chaos property. Furthermore, at any given (large) deviation from equilibrium, one-sided chaos is very infrequent compared to two-sided chaos. This verifies the Onsager-Casimir postulate. For, in the case under consideration, the probability distribution induced on the sample space of successive observations of the given nonequilibrium state is given by the relative measure on the energy surface; most recurrences of the given state will be found coupled with two-sided chaos

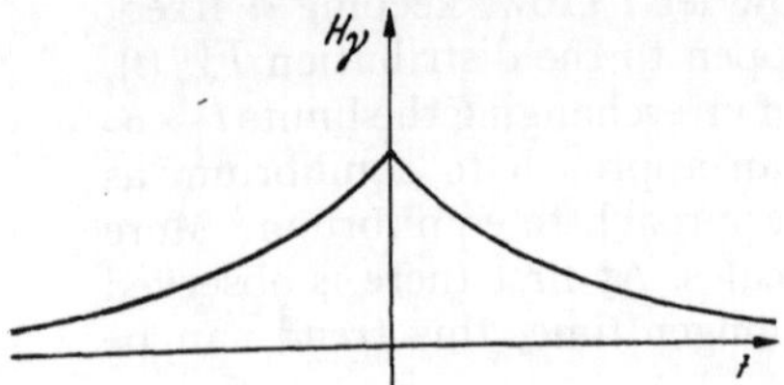

Fig. 2. Variation of H with time.

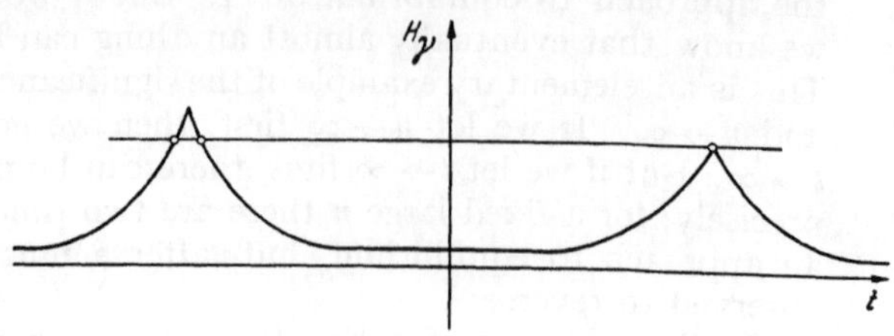

Fig. 3. Variation of H with time.

just as would be found with an externally prepared state. We see that LIOUVILLE's equation has the remarkable property that it seems to take a very "unlikely" path in leading from a "likely" nonequilibrium state back to equilibrium. Of course, two-sided chaos is restored as equilibrium is approached.

The description of this situation by the EHRENFESTS [15] is intuitively appealing. Let us suppose that H_γ is somehow defined at each point of Γ (e.g. by taking F_1 to be a step function after dividing the physical domain into boxes). If an initial state is two-sidedly chaotic, H_γ will decrease in either the positive or the negative time direction (see Fig. 2). If we blindly apply the Onsager-Casimir postulate, we find the same symmetric figure every time there is a large spontaneous departure from equilibrium. The probability of a departure, H_γ, from the equilibrium value, H_γ^0, can be estimated by something like $\exp n\,(H_\gamma^0 - H_\gamma)$. In other words, if H_γ is an unlikely value, a value slightly larger is much more unlikely. Consider the sample space of times at which a certain large value of H_γ is observed. For some of these occurences, H_γ is on a rising curve, for some on a descending curve, and for some, the value H_γ is just about at the peak (see Fig. 3). In each of the first two cases, a much more unlikely event, namely a larger value of H_γ exists either slightly before or slightly after the given value is observed. We conclude that the sample space consists almost exclusively of observations of H_γ which are just about at the peak. Thus, although a state of one-sided chaos is very unlikely compared to other states with the same in stantaneous value of H_γ, it is not at all unlikely compared to the state of higher H_γ from which it originated.

As a final description of this situation we remark that *there is no time's arrow*. If we prepare a state we will observe that H_γ decreases; the same will be true if we follow the prepared state backwards in time. Or, if we observe an isolated system until it achieves a certain preassigned, unlikely state, we will observe that H_γ immediately begins to decrease; or if on observing this state, we reverse all velocities, we will again observe that H_γ decreases. No matter which direction in time we look, H_γ can only decrease. We can obtain an increase of H_γ when we reverse the time only if we have specifically selected the given state not on its own merits but as the successor to another unlikely state. This analysis is applicable to an idealized mathematical model of a gas in a box. The answer could conceivably by different in an unbounded universe with gravitation instead of a box.

14. Survey of other derivations. It is necessary to distinguish several levels on which the Boltzmann equation can be derived. There are the original heuristic physical arguments of MAXWELL and BOLTZMANN and their more sophisticated counterpart today, heuristic formal mathematical arguments. Then there are the attempts at instilling some degree of mathematical rigor into the subject. These take several forms depending on the level of mathematical rigor and on the level of physical generality. Finally, there are completely stochastic treatments which bypass the connection with deterministic particle dynamics and attempt to relate the Boltzmann equation to a more general stochastic particle model.

The derivation given in Sect. 8, stripped of its justification, is probably the most direct formal derivation starting from LIOUVILLE's equation. Its simplicity arises from the fact that the exact Eq. (3.12) which is satisfied by the truncated distribution F_1^σ can be formally transformed into the Boltzmann equation almost at sight. This derivation is similar to the one of BORN and GREEN[1] except that the large number of error terms which in the latter must be individually considered and dropped (some of which are even dropped without being aware of their existence) are here lumped together in the difference between F_1 and F_1^σ. Even as an intuitive physical derivation, this would seem to be superior to the conventional one[2]. The function $n F_1^\sigma$ has a simple interpretation as the density of particles which are not undergoing collisions at a given instant. Eq. (3.7) for F_1^σ has an immediate intuitive meaning when one realizes that F_1^σ changes only as particles pass through the sphere $|\boldsymbol{x}_2-\boldsymbol{x}_1|=\sigma$, and the transition to (3.12) and thence to the Boltzmann equation is a triviality.

A central point in all of the critical analyses is the treatment of molecular chaos; although there are other points of comparable if not greater mathematical difficulty, none are as important philosophically since it is with this step that irreversibility makes an obvious appearance. Perhaps a word should be said about the emphasis on mathematical rigor. Since the historic article of the EHRENFESTS [*15*], the intuitive picture has been quite clear although with the passage of time it has been modified in detail and in precision. The only point that remains to be verified is a mathematical one (ignoring for the moment the question of generalizations of the Boltzmann equation): does the irreversible Boltzmann equation follow from reversible particle dynamics alone or must one insert extraneous random influences?

It is surprising how little of the basic content of the Ehrenfest article has been superseded in a half century. The need for coarse-graining of the number density, nF_1, and the concept of an unlikely state being observed almost inevitably at its turning point were clearly stated. However, this exposition is inevitably entangled in difficulties centering around Poincaré recurrences and ergodicity because of the choice of F_1 as a number density rather than the simpler probability density. The ergodic problem is of course much clearer today, and

[1] The revised version in [*5*] is more direct than the original derivation in [*4*].
[2] For example, see [*11*].

in particular it is evident that the question at hand, namely, a universal monotone decay to equilibrium (whether it holds for $t \to \infty$ or only for not too large a time) has nothing at all to do with ergodicity. The justifiable necessity for coarse graining F_1 when it is taken as a singular number density and the concomitant ambiguity in defining H has possibly led to an overemphasis on the necessity for meddling with the equations of motion in order to observe irreversibility.

Before proceeding, a few general remarks are in order. First let us describe a certain awkward and *ad hoc* procedure which will turn out to be helpful. Instead of posing initial values for LIOUVILLE's equation and then watching it evolve, we interrupt it periodically and perturb the system. Specifically, at each interruption we observe F_1 but do not alter it; however, we replace $F_2(z_1, z_2)$ by $F_1(z_1) F_1(z_2)$. In other words we reintroduce chaos periodically. The period of interruption is taken to be large compared to the duration of a representation collision but small compared to the mean collision time. Between successive interruptions, ternary and higher order collisions can be neglected compared to binary collisions, so F_2 satisfies a two particle Liouville equation. It is an elementary matter to verify that the chaos property is not disturbed by the two particle Liouville equation over times as short as the interruption time. This is an artificial procedure to overcome the problem of subsequent chaos (Sect. 11). Incidentally, too frequent interruptions would yield an approximation to the equations of fluid dynamics instead of the Boltzmann equation.

In Sect. 11 it was stated that the chaos property propagates for a finite time when $n \to \infty$. We wish to refine this statement by estimating the error in the chaos property in terms of n (or $\sigma \sim 1/\sqrt{n}$). First of all, for two approaching particles on the order of σ apart, the correlation between their velocities is at least on the order of σ. This represents the likelihood that there was a recent encounter by a single third particle with each of the given particles (it may very well be that the numerical value of this correlation is a very small number multiplied by σ, but this does not alter the order of the term as $\sigma \to 0$). For two particles which are on the order of a mean-free-path apart and aimed at each other, this correlation has the order of σ^2. These estimates arise directly from the formulas of Sect. 11; in particular, see Fig. 1. For particles which are separated by several mean-free-paths, it is reasonable to expect the correlation to die down exponentially in the separation distance. Since the mean-free-path is the macroscopic reference length, this does not alter the estimate $O(\sigma^2)$ for the correlation between particles at any fixed separation as $n \to \infty$. Specifically, even if chaos is assumed to be identically satisfied at $t = 0$, at any fixed later time we can expect to find a correlation $O(\sigma^2)$ in general and $O(\sigma)$ for adjacent particles.

Probably the first serious mathematical attempt at a derivation of the Boltzmann equation is KIRKWOOD's [*40*]. The starting point is LIOUVILLE's equation, and the basic tools are time-smoothing and the connection with Brownian motion. Time-smoothing and the resulting restriction to linear departures from equilibrium are probably essential for the general theory of fluids which is the ultimate goal of this series of papers, and no attempt was made to remove this stringent restriction in the case of gases. The smoothing serves the same purpose as the introduction of the truncated distribution, namely, to provide an equation in which only complete collisions appear; the possibility of periodic orbits was explicitly considered. The molecular chaos question was correctly posed but was left largely unanswered. In general one can say that many of the crucial questions were brought into focus by these pioneering papers.

An entirely different approach to the problem was initiated by BOGOLUBOV [3]. Formally the problem is attacked by a method which parallels the virial expansion in powers of the density for the state of equilibrium. It is postulated that the higher distributions, F_r, $r > 1$, are dependent on F_1 only (this is similar to the treatment of normal solutions of the Boltzmann equation, cf. Chap. IV). The crucial point is the adoption of a certain "boundary condition" that a group of r particles which is sufficiently dispersed satisfies $F_r(z_1 \dots z_r) = F_1(z_1) \dots F_1(z_r)$; i.e., the individual particles are statistically independent. The result is a generalization of the Boltzmann equation which presumably applies to real gases. The collision term is a power series in the density; the first term is essentially the Boltzmann collision term (except that the $\boldsymbol{x}$ coordinate is not the same in all arguments), and the higher terms involve successively three particle collisions, four particle collisions, etc. The crucial argument is that any higher order distribution, F_r, of r particles which may be in contact, when traced back in time ultimately reaches a state of dispersion for the particles which allows F_r to be expressed in terms of F_1. It is this independence condition which allows one to conclude that F_r is actually a functional of F_1 and to deduce an equation for the time evolution of F_1 alone, independent of the initial values of F_r for $r > 1$.

Although this procedure is formally quite ingenious, it is not described sufficiently precisely to enable one to estimate its value or its possible validity. Although a limit is taken, $n \to \infty$, not much use is made of it. The size of the domain is increased with n so as to keep the number density fixed. Although it is not explicitly stated, the presumption is that in any fixed domain, F_1 is not varied during the limit; this requires F_1 to be continuously redefined in the newly acquired volume. In this limit the molecular size, mean-free-path, and macroscopic scale of variation in $\boldsymbol{x}$ all remain fixed, and only the size of the domain increases. This alone does not seem to be a radical enough limiting process to produce a functional dependence of F_r on F_1. The use that BOGOLUBOV makes of the independence boundary condition is very similar to the "interruption" process described earlier in this section (and is scarcely better justified). In order to carry his expansion to order r in the density, $\nu = n/V$, the independence condition must be correct to order ν^{r-1} at a slightly earlier time when the particles are dispersed. However, even if the independence condition is exactly satisfied initially, we have just seen that collisions will destroy it, at least to the order $(\sigma/L)^2$ (L is the mean-free-path which was constant in the previous limit), but this is $O(\nu^2)$. At best we can hope that F_2 and F_3 are functionals of F_1. For the higher order distributions this cannot be so. Whether or not this restriction to two terms of the Bogolubov expansion makes it valid, depends on a more precise analysis of the subsequent chaos (or subsequent independence) problem than has yet been made with this limiting process.

Even to first order in ν this problem is not the same as the one formulated in Sect. 7. The expansion in ν is an expansion in the neighborhood of mean-free-path *large* compared to macroscopic variations. This is a very unusual limiting case which excludes all the problems to which one usually applies the Boltzmann equation! However, as a consequence of holding macroscopic gradients fixed while the mean-free-path grows, BOGOLUBOV gets a generalization of the Boltzmann equation in which the $\boldsymbol{x}$ arguments of f, f_1, $\bar{f}$, and $\bar{f}_1$ are not the same. It is possible that what BOGOLUBOV has in mind is a limit in which finite macroscopic gradients are superposed on gradients which scale as the mean-free-path (as in Sect. 7); this would yield an equation which is a true generalization of the conventional Boltzmann equation. However, a proof is still lacking and even as a formal equation it would seem to be very difficult to extract practical results from this generalization of the Boltzmann equation. Incidentally, the fact that the same method of expansion (the virial) is actually convergent in equilibrium is of no significance with respect to the present nonequilibrium problem because in the equilibrium case it is known beforehand that F_r is not only a functional of F_1 alone, it is a point function of the thermodynamic state.

Another attempt along somewhat similar lines is the one by M. S. GREEN [26]. The limiting process is the same as BOGOLUBOV'S, followed again by a power series expansion in density, but only the spatially homogeneous case is treated. This is a strong limitation, but it eliminates the necessity of redefining F_1 as the domain is increased. For exactly the same reasons given above, the postulated dependence of F_r on F_1 for $r > 3$ is suspect. It is clear from GREEN'S formulas that the density expansion is similar to an expansion in powers of the time. Consequently, one would expect the solution up to ternary collisions to be a good approximation for only a small number of collisions times. However, strong use is made of the limiting behavior of these terms for large time. This seems to be a question of inversion of limits which requires extreme care. There is an excellent and detailed treatment of ternary collisions, and it is quite possible that to this order the modification of the Boltzmann equation is formally sound. The analysis of the independence condition (i.e. subsequent chaos) is more careful than BOGOLUBOV'S, but no specific error estimates are given. In particular, the difficulties connected with an F_n which must grow exponentially with n (cf. Sect. 10) are not faced.

Returning to the limit of Sect. 7, if it is true that the evolution of F_1 in time is self-determined up to the first order in σ, then ternary collisions can be included by the following interruption scheme. In addition to the exact equation for F_1 and F_2 we use the isolated 3 particle Liouville equation for F_3. The solution is interrupted at intervals which are large compared to σ but small compared to unity and the initial conditions

$$F_2(z_1, z_2) = F_1(z_1)\, F_1(z_2),$$
$$F_3(z_1, z_2, z_3) = F_1(z_1)\, F_1(z_2)\, F_1(z_3)$$

are reinserted each time. Insofar as this procedure is legitimate, it would provide the generalization of GREEN'S ternary collision equation to the spatially inhomogeneous case. It is not directly comparable to BOGULUBOV'S procedure.

A potentially valuable idea proposed by H. L. FRISCH [18] is to insert a parameter λ in the equations for the distributions F_r in such a way that it "partially shields" a group of r particles from the $r+1$'st. A power series in λ gives formulas which are roughly similar to an expansion in powers of the time. In order to get away from the neighborhood of $t=0$, it would seem to be necessary to insert a chaos or independence argument.

In quite a different direction are the methods of VAN HOVE [66] in quantum mechanics and of BROUT and PRIGOGINE [6], [7] in classical mechanics. In [6] BROUT derives a "Master" equation rather than a Boltzmann equation. This is an equation which replaces LIOUVILLE'S equation but strongly resembles the Boltzmann equation. It can be described only in the spatially homogeneous case. In Γ space there is strict determinism; each point $Z(0)$ has a unique successor $Z(t)$. This does not hold in a lower dimensional space, e.g. in the space of velocities, $\Xi = (\xi_1 \ldots \xi_n)$. A point $\Xi(0)$ has many images $Z(0)$ and the subsequent $\Xi(t)$ is not unique. BROUT shows, by a formal argument, that in the limit $n \to \infty$

and $\nu = n/V \to 0$, the "motion" of a point Ξ satisfies a stochastic Markov process which states that the probability that in a time dt $(\xi_1, \xi_2, \xi_3 \ldots \xi_n)$ becomes $(\xi_1', \xi_2', \xi_3 \ldots \xi_n)$, where (ξ_1', ξ_2') represents a possible outcome of a complete binary collision between ξ_1 and ξ_2, is given by the conventional collision cross-section as found in the Boltzmann equation. This argument is very illuminating but ignores the mathematical difficulties mentioned in Sect. 10. Disturbing possibilities involving, for example, higher order collisions, are dismissed if they are found only on sets of small measure. However, if F_1 is not the equilibrium distribution, only sets of small measure are of interest, and, in the absence of more elaborate error estimates, the Master equation can be considered to have been verified only in equilibrium.

The Master equation has its own intrinsic interest as a postulated stochastic equation. From it KAC [37] has been able to derive the spatially homogeneous Boltzmann equation (chiefly by verifying subsequent chaos for this stochastic equation) on the assumption of initial chaos. This establishes an intriguing relation between two interesting stochastic processes, but the relation to LIOUVILLE's equation has still to be clarified.

In [7] BROUT and PRIGOGINE introduce a small coupling parameter, λ, into the intermolecular potential which is assumed to be soft so there are only grazing collisions. By Fourier analysis and expansion in powers of λ, they are able to obtain a Fokker-Planck equation for F_1 in the spatially homogeneous case. This is relevant to the theory of the Boltzmann equation since it can be formally verified that with grazing collisions alone, the collision term of the Boltzmann equation takes the form of a second order differential operator as in a Fokker-Planck equation. These results are obtained for large times defined by $\lambda^2 t =$ const; interpreting λ as a measure of the molecular diameter, we see that this is consistent with the limiting process of Sect. 7. The basic idea is that higher order correlations (given by higher order Fourier coefficients) remain small if they are initially small. This should be compared with the subsequent chaos property.

This is a very ingenious and powerful tool but, again, the limiting process is treated somewhat cavalierly. The annoying exponential growth of F_n would here be reflected in the possibility of very large Fourier coefficients of high order.

II. Elementary properties of the Boltzmann equation[1].

15. Properties of binary collisions. General properties of a binary collision, independent of the specific intermolecular forces, are governed by the conservation of momentum and of energy. These can be written

$$\xi + \xi_1 = \xi' + \xi_1', \tag{15.1}$$

$$\xi^2 + \xi_1^2 = \xi'^2 + \xi_1'^2 \tag{15.2}$$

for two molecules of equal mass; (ξ, ξ_1) can represent the velocities before collision and (ξ', ξ_1') the velocities after, but it is clear that from these equations alone there is complete symmetry between (ξ, ξ_1) and (ξ', ξ_1'). If (ξ, ξ_1) is considered to be given with (ξ', ξ_1') to be determined, we have a system of four Equations [(15.1) is vector and (15.2) is scalar] for six unknowns. We can expect a two parameter family of solutions. The two parameters are most easily introduced as the unit vector, $\boldsymbol{\alpha}$, in the direction of the change in velocity of ξ,

$$\xi' - \xi = \boldsymbol{\alpha} A. \tag{15.3}$$

[1] For a more complete treatment see the article by L. WALDMANN in this volume.

We can eliminate A and obtain an explicit parametric solution of (15.1), (15.2) in terms of the parameter $\boldsymbol{\alpha}$. From (15.1) we get

$$\xi_1' - \xi_1 = -\boldsymbol{\alpha} A \tag{15.4}$$

and from (15.2),

$$\xi^2 + \xi_1^2 = (\xi + \boldsymbol{\alpha} A)^2 + (\xi_1 - \boldsymbol{\alpha} A)^2;$$

provided that $A \neq 0$, we find

$$A = \boldsymbol{\alpha} \cdot \boldsymbol{V}, \tag{15.5}$$

where

$$\boldsymbol{V} \equiv \xi_1 - \xi. \tag{15.6}$$

Substituting in (15.3) and (15.4), we obtain

$$\left.\begin{aligned} \xi' &= \xi + \boldsymbol{\alpha}(\boldsymbol{\alpha} \cdot \boldsymbol{V}), \\ \xi_1' &= \xi_1 - \boldsymbol{\alpha}(\boldsymbol{\alpha} \cdot \boldsymbol{V}), \end{aligned}\right\} \tag{15.7}$$

which is the desired solution of (15.1), (15.2). This parametrization of the solutions of (15.1), (15.2) is not one-to-one. The great circle $\boldsymbol{\alpha} \cdot \boldsymbol{V} = 0$ corresponds to a single solution (the identity). All other solutions are covered twice, once by each hemisphere, $\boldsymbol{\alpha} \cdot \boldsymbol{V} > 0$ and $\boldsymbol{\alpha} \cdot \boldsymbol{V} < 0$. Both α and $-\alpha$ correspond to the same solution. However, if we insist on the definition of α as having *the direction* $\xi' - \xi$, we must have

$$\boldsymbol{\alpha} \cdot (\xi' - \xi) > 0. \tag{15.8}$$

This removes the ambiguity and makes the parametrization one-to-one (except for $\boldsymbol{\alpha} \cdot \boldsymbol{V} = 0$). The relation between the parametrization in terms of α (the *apse-direction*) and in terms of ω [the impact parameter plane—see Eq. (3.9)] will be determined later.

We remark that (15.7) represents a linear transformation of the six-space (ξ, ξ_1) into itself. The inverse transformation is easily found. We have

$$\boldsymbol{V}' \equiv \xi_1' - \xi' = \xi_1 - \xi - 2\boldsymbol{\alpha}(\boldsymbol{\alpha} \cdot \boldsymbol{V}) = \boldsymbol{V} - 2\boldsymbol{\alpha}(\boldsymbol{\alpha} \cdot \boldsymbol{V}),$$

from which

$$\boldsymbol{\alpha} \cdot \boldsymbol{V}' = -\boldsymbol{\alpha} \cdot \boldsymbol{V}. \tag{15.9}$$

Inserting this in (15.7) we find

$$\left.\begin{aligned} \xi &= \xi' + \boldsymbol{\alpha}(\boldsymbol{\alpha} \cdot \boldsymbol{V}'), \\ \xi_1 &= \xi_1' - \boldsymbol{\alpha}(\boldsymbol{\alpha} \cdot \boldsymbol{V}'), \\ \boldsymbol{V}' &= \xi_1' - \xi'. \end{aligned}\right\} \tag{15.10}$$

In other words, with $\boldsymbol{\alpha}$ fixed (or reversed in sign), the linear transformation between (ξ, ξ_1) and (ξ', ξ_1') is its own inverse. For the parametrization to be unique as provided by (15.8) requires a change in sign of $\boldsymbol{\alpha}$ in (15.10) as compared to (15.7). An immediate conclusion is that for α fixed (or reversed) the Jacobian of the transformation is unity,

$$\frac{\partial(\xi', \xi_1')}{\partial(\xi, \xi_1)} = 1. \tag{15.11}$$

We recall that there enters in the Boltzmann equation the molecular pair state $(\bar{\xi}, \bar{\xi}_1)$ which, after a collision, emerges at the given state (ξ, ξ_1); $(\bar{\xi}, \bar{\xi}_1)$ can now be identified with the state (ξ', ξ_1') which *results from* a collision by (ξ, ξ_1)

provided, however, that α is reversed. The relation between these transformations is shown in Fig. 4.

There is a very simple (and useful) geometric interpretation of the transformation (15.7). Consider (ξ, ξ_1) to be a pair of points in three-space rather than a single point in six-space. It follows immediately from (15.7) that the points ξ, ξ_1, ξ', ξ_1' are in the simple geometrical relation shown in Fig. 5. Given ξ and ξ_1, we construct a sphere with $\xi_1 - \xi$ as diameter. The locus of pairs (ξ', ξ_1') which are possible outcomes of a collision between ξ and ξ_1 is given by all pairs of antipodal points on this sphere.

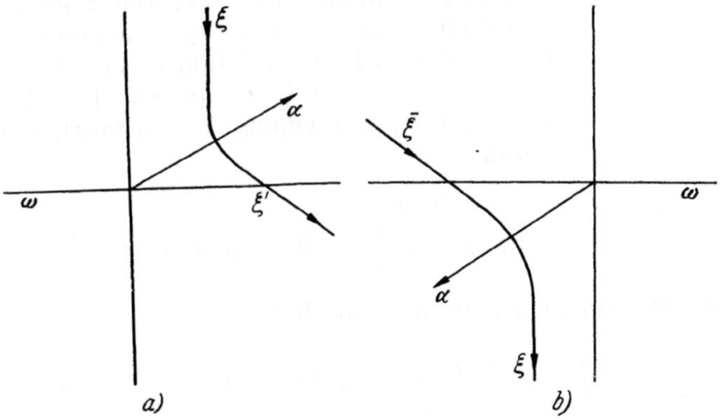

Fig. 4 a and b. Inverse collisions.

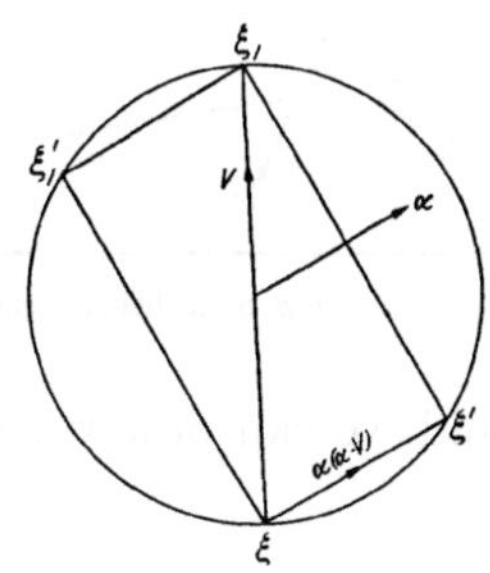

Fig. 5. Transformation of velocities.

It is convenient to restate Eqs. (15.1) and (15.2) and their geometrical interpretation in the following way. Each of the five functions ψ_r, $r = 0 \ldots 4$, defined by

$$\left.\begin{aligned} \psi_0 &= 1 \\ \psi_i &= \xi_i \\ \psi_4 &= \xi^2 \end{aligned}\right\} i = 1, 2, 3 \tag{15.12}$$

satisfies the property

$$\psi(\xi) + \psi(\xi_1) = \psi(\xi') + \psi(\xi_1'). \tag{15.13}$$

Geometrically the sum of the values of ψ at antipodal points is constant on every sphere. Such a function ψ is called a *summational invariant.* The converse is also true. Any summational invariant is a linear combination of the five functions ψ_r[1]. This fact will be used repeatedly.

The Boltzmann equation (8.1), is not actually an equation until the relation between the parameter $\boldsymbol{\alpha}$ which is found in the argument of $\bar{f}$ and $\bar{f}_1$[2] and the integration variable $d\omega$ is made definite. To do this, we introduce spherical coordinates with origin at molecule ξ and take $\boldsymbol{V}$ to be the polar axis. Specifically, we define the colatitude, ϑ, by

$$\boldsymbol{\alpha} \cdot \boldsymbol{V} = \boldsymbol{V} \cos\vartheta \tag{15.14}$$

and introduce an arbitrary origin for the longitude, ε. We have

$$d\omega = r\, dr\, d\varepsilon, \tag{15.15}$$

[1] Ref. [22], Appendix 1 and ref. [39], paragraph 27.

[2] We shall use the notation $\bar{f}$ for $f(\bar{\xi})$, $\bar{f}_1$ for $f(\bar{\xi}_1)$, f' for $f(\xi')$, and f_1' for $f(\xi_1')$.

and we recall that the *impact parameter*, r, is the coordinate of the point at which ω is intersected by the asymptote of the trajectory $\boldsymbol{\xi}_1$, see Fig. 6 (the trajectory is drawn for an attractive force).

It must be borne in mind that the coordinate axes are attached to a *representative* molecule $\boldsymbol{\xi}$ at the position $\boldsymbol{x}$ (cf. the derivation of the collision term in Sect. 3), and they do not move with any particular molecule $\boldsymbol{\xi}$ through an encounter.

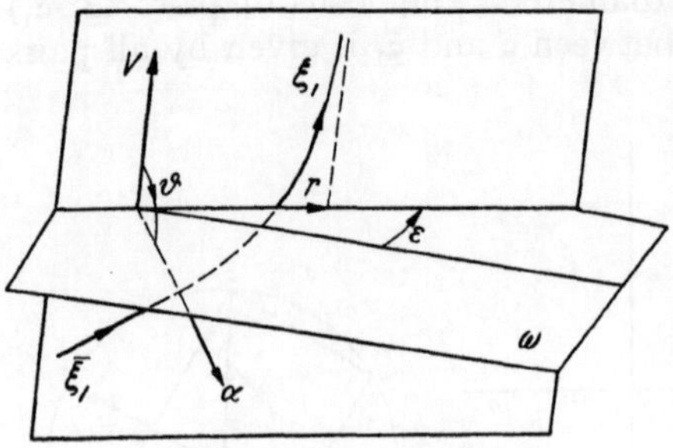

Fig. 6. Collision parameters.

The variables ε, ϑ (and also V) enter explicitly into the argument of f. Solution of the ordinary differential equations of a binary encounter will give the angle of deflection (which is $\pi - 2\vartheta$) as a function of V and of r. If this relation is monotone, it can be inverted to obtain $r(\vartheta, V)$. Replacing r by ϑ (keeping V constant), we obtain

$$\left.\begin{aligned} d\omega &= r\,dr\,d\varepsilon \\ &= r(\vartheta, V)\left|\frac{\partial}{\partial\vartheta} r(\vartheta, V)\right| d\vartheta\, d\varepsilon, \end{aligned}\right\} \tag{15.16}$$

and we can now write the Boltzmann equation in the form

$$\frac{\partial f}{\partial t} + \boldsymbol{\xi}\cdot\frac{\partial f}{\partial \boldsymbol{x}} = \frac{1}{m}\int B(\vartheta, V)\,(f' f_1' - f f_1)\, d\vartheta\, d\varepsilon\, d\boldsymbol{\xi}_1 \tag{15.17}$$

where

$$B(\vartheta, V) = V r\left|\frac{\partial r}{\partial\vartheta}\right|. \tag{15.18}$$

We recall that f is the mass density distribution. The variables of integration are now explicitly shown. In this formula we have chosen to identify $\boldsymbol{\xi}$ with $\boldsymbol{\xi}'$ in accordance with conventional usage. The function $B(\vartheta, V)$, which depends on the intermolecular potential, is closely related to the differential collision cross-section.

If the intermolecular force is purely repulsive, then ϑ is monotone in r, and the transformation (15.16) is legitimate. If the force changes sign, then the mapping of the plane ω into the sphere $\boldsymbol{\alpha}$ is not one-to-one. One possibility is to keep the original variable r instead of introducing ϑ, in which case the argument of f' is a complicated function of the integration variables. Or it is possible to define the Jacobian $|r\,\partial r/\partial\vartheta|$ as a multivalued function of ϑ and regain formula (15.17) by defining $B(\vartheta, V)$ to be the sum of the several values that may be taken by $|r\,\partial r/\partial\vartheta|$ at a given ϑ. This sum is, of course, what would be observed experimentally since one could not distinguish between two molecules emerging at the same angle ϑ even though their trajectories (i.e., impact parameters, r) might be different.

It is easily verified for rigid elastic spheres that $\boldsymbol{\alpha}$ is the unit vector along the line of centers at the instant of impact and, if σ is the molecular diameter [*46*], [*11*],

$$B(\vartheta, V) = \sigma^2 V \sin\vartheta\cos\vartheta. \tag{15.19}$$

For molecules with a force of repulsion which is a power law,

$$X = \frac{K}{r^s}, \qquad s > 3, \tag{15.20}$$

it can be shown that

$$B(\vartheta, V) = k\, m\, V^{\varkappa}\, b\,\frac{db}{d\vartheta}, \tag{15.21}$$

where k and $\varkappa$ are molecular constants related to K and s by

$$\left.\begin{aligned} k &= \frac{1}{m}\left(\frac{2K}{m}\right)^{2/s-1}, \\ \varkappa &= \frac{s-5}{s-1}, \end{aligned}\right\} \tag{15.22}$$

and the function $b(\vartheta)$ is defined implicitly by

$$\left.\begin{aligned} \vartheta(b) &= \int\limits_0^{\beta_0} \frac{d\beta}{\left[1-\beta^2-\frac{2}{s-1}\left(\frac{\beta}{b}\right)^{s-1}\right]^{\frac{1}{2}}}, \\ 1-\beta_0^2 &- \frac{2}{s-1}\left(\frac{\beta_0}{b}\right)^{s-1} = 0. \end{aligned}\right\} \tag{15.23}$$

In this case the Boltzmann equation takes the form,

$$\frac{\partial f}{\partial t} + \boldsymbol{\xi}\cdot\frac{\partial f}{\partial \boldsymbol{x}} = k\int V^{\varkappa}(f' f_1' - f f_1)\, b\, db\, d\varepsilon\, d\boldsymbol{\xi}_1. \tag{15.24}$$

The advantage of this special force law is the separation of $B(\vartheta, V)$ into a product of a power of V and a function of ϑ alone. In particular, the special case $\varkappa=0$, i.e. $s=5$, the so-called *Maxwellian molecule*, has the feature that B is a function of ϑ alone.

It should be noted that, except for molecules with a finite cutoff, $\int d\omega = \int r\,dr\,d\varepsilon$ is infinite; this implies that $\int B(\vartheta, V)\, d\vartheta$ is infinite. This is merely the statement that the total collision cross-section is infinite. One consequence of this fact is that the collision term in the Boltzmann equation is only conditionally convergent; the positive and negative parts each diverge.

16. Symmetry relations. It is convenient to introduce the bilinear form

$$J(f, g) \equiv \frac{1}{2m}\int V(f' g_1' + f_1' g' - f g_1 - f_1 g)\, d\omega\, d\boldsymbol{\xi}_1. \tag{16.1}$$

In terms of this, the collision term of the Boltzmann equation is $J(f, f)$. Given a function $\varphi(\boldsymbol{\xi})$, we define

$$J_\varphi(f, g) \equiv \int \varphi(\boldsymbol{\xi})\, J(f, g)\, d\boldsymbol{\xi} = \frac{1}{2m}\int \varphi\, V(f' g_1' + f_1' g' - f g_1 - f_1 g)\, d\omega\, d\boldsymbol{\xi}\, d\boldsymbol{\xi}_1. \tag{16.2}$$

From (15.7), we observe that an interchange of $\boldsymbol{\xi}$ with $\boldsymbol{\xi}_1$ produces an interchange of $\boldsymbol{\xi}'$ with $\boldsymbol{\xi}_1'$ (V changes sign). Also we recall that the transformation (15.7) is its own inverse; thus replacement of $(\boldsymbol{\xi}, \boldsymbol{\xi}_1)$ by $(\boldsymbol{\xi}', \boldsymbol{\xi}_1')$ transforms $(\boldsymbol{\xi}', \boldsymbol{\xi}_1')$ into $(\boldsymbol{\xi}, \boldsymbol{\xi}_1)$. From this we see that the quantity

$$(f' g_1' + f_1' g' - f g_1 - f_1 g)$$

is unaltered by an interchange of $\boldsymbol{\xi}$ with $\boldsymbol{\xi}_1$ and changes sign by an interchange of $(\boldsymbol{\xi}, \boldsymbol{\xi}_1)$ with $(\boldsymbol{\xi}', \boldsymbol{\xi}_1')$. In each case the value of V is unaltered (from Fig. 5, we see that $V' = V$). Finally, recalling that the Jacobian of the transformation between $(\boldsymbol{\xi}, \boldsymbol{\xi}_1)$ and $(\boldsymbol{\xi}', \boldsymbol{\xi}_1')$ is unity, we arrive at the following alternative forms for J_φ,

$$\left.\begin{aligned} J_\varphi &= \frac{1}{2m}\int \varphi(\boldsymbol{\xi})\, V(f' g_1' + f_1' g' - f g_1 - f_1 g)\, d\omega\, d\boldsymbol{\xi}\, d\boldsymbol{\xi}_1 \\ &= \frac{1}{2m}\int \varphi(\boldsymbol{\xi}_1)\, V(f' g_1' + f_1' g' - f g_1 - f_1 g)\, d\omega\, d\boldsymbol{\xi}\, d\boldsymbol{\xi}_1 \\ &= -\frac{1}{2m}\int \varphi(\boldsymbol{\xi}')\, V(f' g_1' + f_1' g' - f g_1 - f_1 g)\, d\omega\, d\boldsymbol{\xi}\, d\boldsymbol{\xi}_1 \\ &= -\frac{1}{2m}\int \varphi(\boldsymbol{\xi}_1')\, V(f' g_1' + f_1' g' - f g_1 - f_1 g)\, d\omega\, d\boldsymbol{\xi}\, d\boldsymbol{\xi}_1. \end{aligned}\right\} \tag{16.3}$$

To obtain the second line, we substitute $(\boldsymbol{\xi}, \boldsymbol{\xi}') \leftrightarrow (\boldsymbol{\xi}_1, \boldsymbol{\xi}_1')$; to obtain the third line, we substitute $(\boldsymbol{\xi}, \boldsymbol{\xi}_1) \leftrightarrow (\boldsymbol{\xi}', \boldsymbol{\xi}_1')$; to obtain the final line, we perform both substitutions. From these, we obtain

$$J_\varphi = \frac{1}{2m} \int \frac{1}{2} (\varphi + \varphi_1) V(f' g_1' + f_1' g' - f g_1 - f_1 g)\, d\omega\, d\boldsymbol{\xi}\, d\boldsymbol{\xi}_1, \tag{16.4}$$

$$= \frac{1}{2m} \int \frac{1}{4} (\varphi + \varphi_1 - \varphi' - \varphi_1') V(f' g_1' + f_1' g' - f g_1 - f_1 g)\, d\omega\, d\boldsymbol{\xi}\, d\boldsymbol{\xi}_1. \tag{16.5}$$

Reversing the process, we also verify

$$J_\varphi = \frac{1}{2m} \int \frac{1}{2} (\varphi' + \varphi_1' - \varphi - \varphi_1) V(f g_1 + f_1 g)\, d\omega\, d\boldsymbol{\xi}\, d\boldsymbol{\xi}_1, \tag{16.6}$$

$$= \frac{1}{2m} \int (\varphi' - \varphi) V(f g_1 + f_1 g)\, d\omega\, d\boldsymbol{\xi}\, d\boldsymbol{\xi}_1. \tag{16.7}$$

In particular, from (16.5) we observe that J_φ vanishes if φ is a summational invariant (15.13).

17. The conservation equations. The mass density and macroscopic fluid flow velocity are defined by

$$\varrho(\boldsymbol{x}, t) \equiv \int f(\boldsymbol{\xi}, \boldsymbol{x}, t)\, d\boldsymbol{\xi}, \tag{17.1}$$

$$\boldsymbol{u}(\boldsymbol{x}, t) \equiv \frac{1}{\varrho} \int \boldsymbol{\xi} f(\boldsymbol{\xi}, \boldsymbol{x}, t)\, d\boldsymbol{\xi}. \tag{17.2}$$

Higher moments of f are most conveniently defined in terms of the velocity relative to the mean,

$$\boldsymbol{c} \equiv \boldsymbol{\xi} - \boldsymbol{u}. \tag{17.3}$$

We note that

$$\int \boldsymbol{c} f\, d\boldsymbol{\xi} = 0. \tag{17.4}$$

Up to fourth moments we have

$$P_{ij} \equiv \int c_i c_j f\, d\boldsymbol{\xi}, \tag{17.5}$$

$$S_{ijk} \equiv \int c_i c_j c_k f\, d\boldsymbol{\xi}, \tag{17.6}$$

$$Q_{ijkl} \equiv \int c_i c_j c_k c_l f\, d\boldsymbol{\xi}. \tag{17.7}$$

In particular, P_{ij} is the *stress* tensor and the contracted third moment

$$q_i = \tfrac{1}{2} S_{irr} = \int \tfrac{1}{2} c_i c^2 f\, d\boldsymbol{\xi} \tag{17.8}$$

is the *heat-flow* vector. Actually P_{ij} represents the mean rate of flow per unit area in the j-direction of the i-component of momentum, and q_j represents the mean rate of flow of kinetic energy in the j-direction; both flow rates are measured in a coordinate system moving at the fluid velocity $\boldsymbol{u}$. Consistent with the fact that we are treating a perfect gas, there are no contributions to stresses or energy flow from intermolecular forces.

It is also interesting to compute the momentum and energy flow through an element of surface which is at rest, i.e., to compute moments of f with respect to $\boldsymbol{\xi}$ rather than $\boldsymbol{c}$. We have

$$\int \xi_i \xi_j f\, d\boldsymbol{\xi} = \int (c_i + u_i)(c_j + u_j) f\, d\boldsymbol{\xi} = P_{ij} + \varrho u_i u_j \tag{17.9}$$

and

$$\int \tfrac{1}{2} \xi_i \xi^2 f\, d\boldsymbol{\xi} = \int \tfrac{1}{2} (c_i + u_i)(c^2 + 2\boldsymbol{c} \cdot \boldsymbol{u} + u^2) f\, d\boldsymbol{\xi} = q_i + u_j P_{ji} + \varrho u_i (e + \tfrac{1}{2} u^2) \tag{17.10}$$

where

$$e = \frac{1}{\varrho} \int \frac{1}{2} c^2 f \, d\xi \tag{17.11}$$

is the *internal energy* per unit mass (i.e., energy measured in a frame moving with the fluid). In (17.9) we see that the momentum flow across a surface is the sum of the macroscopically carried momentum flow, $\varrho u_i u_j$, and the momentum flow measured in a frame moving with the fluid, P_{ij}. In (17.10) we see that the total energy flow through a surface is what is macroscopically carried by the fluid, $\varrho u_i (e + \frac{1}{2} u^2)$, plus the work done across the surface, $u_j P_{ji}$ plus the heat flow, q_i.

We recall that each of the functions ψ_r of (15.12) is a summational invariant. If we multiply both sides of the Boltzmann equation by any one of these functions and integrate over ξ, the contribution from the collision term vanishes. For example, taking $\psi_0 = 1$

$$\int \left\{ \frac{\partial f}{\partial t} + \xi_r \frac{\partial f}{\partial x_r} \right\} d\xi = 0,$$

and interchanging the integration and differentiation, we get

$$\frac{\partial \varrho}{\partial t} + \frac{\partial}{\partial x_r} (\varrho u_r) = 0. \tag{17.12}$$

This is the macroscopic equation of conservation of mass. Similarly, taking $\psi_i = \xi_i$ and $\psi_4 = \frac{1}{2} \xi^2$, making use of (17.9) and (17.10), we obtain

$$\frac{\partial}{\partial t} (\varrho u_i) + \frac{\partial}{\partial x_r} (\varrho u_i u_r + P_{ir}) = 0 \tag{17.13}$$

and

$$\frac{\partial}{\partial t} \left[\varrho \left(e + \frac{1}{2} u^2 \right) \right] + \frac{\partial}{\partial x_r} \left[\varrho u_r \left(e + \frac{1}{2} u^2 \right) + u_s P_{sr} + q_r \right] = 0; \tag{17.14}$$

these are the equations of conservation of momentum and energy respectively. Eqs. (17.12) to (17.14) are quite general macroscopically but are restricted here to the case of a perfect gas in virtue of the definitions (17.5), (17.8), and (17.11).

For later use we interpolate the definition of scalar pressure,

$$p = \tfrac{1}{3} P_{rr}, \tag{17.15}$$

of the reduced stress tensor or stress deviator,

$$p_{ij} \equiv P_{ij} - p \, \delta_{ij} \tag{17.16}$$

which has the property

$$p_{rr} = 0, \tag{17.17}$$

and of the temperature

$$\mathrm{R} T \equiv \frac{p}{\varrho} = \frac{2}{3} e. \tag{17.18}$$

18. The H-theorem. We define[1]

$$H(\boldsymbol{x}, t) \equiv \int f \log f \, d\xi, \tag{18.1}$$

$$\overline{H}(t) \equiv \int H \, d\boldsymbol{x}, \tag{18.2}$$

$$H_i(\boldsymbol{x}, t) \equiv \int \xi_i f \log f \, d\xi. \tag{18.3}$$

[1] This is an exception to the convention (footnote 1, p. 207) that the magnitude of the vector H_i is denoted by H.

H and H_i are local fluid properties while $\overline{H}$ is a property of the whole system[1]. H_i represents the rate of flow of $f \log f$ through a surface. In order to find an equation for the rate of change of H we multiply through BOLTZMANN's equation by $(\log f + 1)$ and integrate with respect to $\boldsymbol{\xi}$,

$$\int (1+\log f)\left\{\frac{\partial f}{\partial t}+\xi_r\frac{\partial f}{\partial x_r}\right\}d\boldsymbol{\xi}=\int (1+\log f)\,J(f,f)\,d\boldsymbol{\xi}.$$

On the left side we note that

$$(1+\log f)\frac{\partial f}{\partial t}=\frac{\partial}{\partial t}(f\log f)$$

$$(1+\log f)\,\xi_r\frac{\partial f}{\partial x_r}=\frac{\partial}{\partial x_r}(\xi_r f\log f),$$

and, interchanging integration and differentiation, we obtain

$$\frac{\partial H}{\partial t}+\frac{\partial H_r}{\partial x_r}.$$

On the right side we can drop the 1 since it is a summational invariant and, from (16.5) obtain

$$\int \log f\, J(f,f)\,d\boldsymbol{\xi}=-\frac{1}{8m}\int V(f'f_1'-ff_1)\log\left(\frac{f'f_1'}{ff_1}\right)d\omega\,d\boldsymbol{\xi}\,d\boldsymbol{\xi}_1\equiv -G.$$

Clearly G is non-negative and is positive unless

$$f'f_1'=ff_1 \tag{18.4}$$

holds identically. Combining the two results, we have

$$\frac{\partial H}{\partial t}+\frac{\partial H_r}{\partial x_r}=-G. \tag{18.5}$$

Integrating this over a domain D with boundary Σ yields

$$\frac{d\overline{H}}{dt}+\oint_{\Sigma}\boldsymbol{H}\cdot d\boldsymbol{S}=-\int_D G\,dV. \tag{18.6}$$

If the boundary condition on Σ is such that there is no flow of $\boldsymbol{H}$ (e.g., specular reflection), then we conclude that $\overline{H}$ is a decreasing function of the time unless (18.4) is satisfied for all $\boldsymbol{x}$ and all $\boldsymbol{\xi}$. Eq. (18.4) implies that $\log f$ is a summational invariant. From this we conclude that

$$\log f = a+\boldsymbol{b}\cdot\boldsymbol{\xi}+c\,\xi^2 \tag{18.7}$$

where the parameters a, $\boldsymbol{b}$, and c can be functions of $\boldsymbol{x}$ and t. This implies that f is *locally Maxwellian*; the parameters a, $\boldsymbol{b}$, c can easily be related to the density, velocity, and temperature [see Eq. (18.8)].

It can be shown that the only locally Maxwellian solutions of Boltzmann's equation, i.e., locally Maxwellian solutions of

$$\frac{\partial f}{\partial t}+\xi_r\frac{\partial f}{\partial x_r}=0,$$

are an *absolute Maxwellian* (i.e., one which is space and time independent), or a *rigid body rotation* (i.e., one in which the temperature is constant but in which

[1] $\overline{H}$ is essentially the same as H_γ of Chap. I, except that it is defined with respect to the mass density f instead of the probability density F_1.

the pressure and velocity vary in an obvious way—this can occur only if permitted by the boundary conditions), or certain other rather peculiar solutions which cannot occur in a bounded domain[1]. We conclude that with appropriate boundary conditions $\overline{H}$ decreases monotonely until equilibrium is reached.

From the definitions (17.1), (17.2), (17.11), and (17.18), it is easy to verify that a Maxwellian (or locally Maxwellian) distribution takes the form

$$f^{(0)} = \frac{\varrho}{(2\pi \mathsf{R}\, T)^{\frac{3}{2}}} \exp\left\{-\frac{(\boldsymbol{\xi}-\boldsymbol{u})^2}{2\mathsf{R}\, T}\right\}. \tag{18.8}$$

A simple computation shows that, for a locally Maxwellian $f^{(0)}$, $H^{(0)}$ is given by

$$H^{(0)} = \int f^{(0)} \log f^{(0)}\, d\boldsymbol{\xi} = -\frac{1}{\mathsf{R}}\, \varrho\, S; \tag{18.9}$$

here S is the thermodynamic entropy per unit mass and R is the gas constant per unit mass. A similar computation yields the result

$$H_r^{(0)} = u_r H^{(0)}; \tag{18.10}$$

i.e., $H_r^{(0)}$ is merely the rate at which the stream carries $H^{(0)}$. A better approximation to H_r can be obtained by replacing $\log f$ by $\log f^{(0)}$ but leaving f in (18.3),

$$H_r^{(1)} = \int \xi_i f \log f^{(0)}\, d\boldsymbol{\xi} = H_r^{(0)} - \frac{q_r}{\mathsf{R}\, T}. \tag{18.11}$$

In addition to the convection of S there is the classical heat flow term. In general nonequilibrium states, there is no simple connection between H and S and between H_r and the convection of S and heat flow; cf. Eqs. (28.38), (28.39).

III. Existence theory.

19. Survey. From the physical interpretation of the Boltzmann equation, it is easy to guess the general type of initial value and boundary value problem which we might expect to be well posed. Of course such conjectures must be proved mathematically in order to know that the Boltzmann equation faithfully reflects the physical problem. It is possible to draw the same heuristic conclusions directly from the equation. We write it in the form

$$\frac{\partial f}{\partial t} + \boldsymbol{\xi} \cdot \frac{\partial f}{\partial \boldsymbol{x}} = J, \tag{19.1}$$

where

$$J \equiv \frac{1}{m} \int V(f' f_1' - f f_1)\, d\omega\, d\boldsymbol{\xi}_1. \tag{19.2}$$

The variables $(\boldsymbol{x}, t)$ occur only as parameters in the integral part, whereas $\boldsymbol{\xi}$ occurs only as a parameter in the differential part. In other words, the collision term, J, operates on the function $f(\boldsymbol{\xi})$ at the same time that it is "flowing" in $\boldsymbol{x}$ and t according to the differential terms. This suggests that appropriate initial and boundary values would be governed by the differential part alone and would be the same as for the equation of a Knudsen gas,

$$\frac{\partial f}{\partial t} + \boldsymbol{\xi} \cdot \frac{\partial f}{\partial \boldsymbol{x}} = 0. \tag{19.3}$$

Specifically one would expect to give $f(\boldsymbol{\xi}, \boldsymbol{x})$ initially and give a boundary condition which determines the reflected "half" of the distribution in terms of the

[1] Ref. [22], Appendix 2.

incident "half". The simplest example is specular reflection for which the boundary condition is

$$f(\boldsymbol{\xi} - 2\boldsymbol{n}\,(\boldsymbol{n}\cdot\boldsymbol{\xi}), \boldsymbol{x}_0, t) = f(\boldsymbol{\xi}, \boldsymbol{x}_0, t), \qquad \boldsymbol{\xi}\cdot\boldsymbol{n} > 0. \tag{19.4}$$

Here $\boldsymbol{x}_0$ is a point on the boundary and $\boldsymbol{n}$ is the outer normal at the point $\boldsymbol{x}_0$; the condition $\boldsymbol{\xi}\cdot\boldsymbol{n} > 0$ specifies incident molecules. One frequently considers *stochastic* boundary conditions; by this we mean a boundary condition which is appropriate to a microscopic situation in which the emergent velocity of a molecule is not uniquely determined by its incident velocity. Such boundary conditions will, in general, contain some parameters which refer to the state of the boundary and others which refer to its composition. The state of the boundary is usually expressed thermodynamically, i.e., in terms of its temperature, but this restrictive description could be dropped if necessary[1]. The boundary substance is usually defined in terms of a small number of *accomodation coefficients*[2], or, more abstractly, it can be described by a stochastic kernel

$$f(\boldsymbol{\xi}, \boldsymbol{x}_0, t) = \int K(\boldsymbol{\xi}, \boldsymbol{\eta})\, f(\boldsymbol{\eta}, \boldsymbol{x}_0, t)\, d\boldsymbol{\eta}, \qquad \boldsymbol{\eta}\cdot\boldsymbol{n} > 0, \qquad \boldsymbol{\xi}\cdot\boldsymbol{n} < 0. \tag{19.5}$$

In order to include the specular boundary condition, (19.4), as a special case, K must be allowed to be singular. A common boundary condition is to assume that a certain fraction, α, of the incident molecules are specularly reflected and the remainder, $1-\alpha$, are emitted as a Maxwellian distribution at the temperature of the wall. This corresponds to the kernel

$$K(\boldsymbol{\xi}, \boldsymbol{\eta}) = \alpha\, \delta(\boldsymbol{n}\cdot\boldsymbol{\xi} + \boldsymbol{n}\cdot\boldsymbol{\eta}) + (1-\alpha)\, \frac{\boldsymbol{n}\cdot\boldsymbol{\eta}}{2\pi\, \mathrm{R}^2 T_0^2} \exp\left(-\frac{\xi^2}{2\mathrm{R}T_0}\right). \tag{19.6}$$

This kernel contains a single parameter, α, which describes the nature of the wall and a single parameter, T_0, which reflects its state. The factor multiplying the exponential has been adjusted so that there is no net accumulation of particles at the surface (one can imagine porous walls or transient conditions where this might be violated).

A generalization of (19.6) to include two wall parameters (i.e. two accomodation coefficients) is

$$K(\boldsymbol{\xi}, \boldsymbol{\eta}) = \alpha\, \delta(\boldsymbol{n}\cdot\boldsymbol{\xi} + \boldsymbol{n}\cdot\boldsymbol{\eta}) + (1-\alpha)\, \frac{\boldsymbol{n}\cdot\boldsymbol{\eta}}{2\pi\, \mathrm{R}^2 T_0^2} \exp\left(-\frac{(\boldsymbol{\xi}-\boldsymbol{U})^2}{2\mathrm{R}T_0}\right). \tag{19.7}$$

It has been tacitly assumed up to now that the wall is stationary. The generalization of (19.4) and (19.6) to moving walls is obvious; for example, if the wall moves only tangentially, (19.7) is the correct generalization of (19.6) with U identified as the wall velocity. Still considering only tangential motion of the wall, U can be taken as a second parameter representing the wall material; e.g., in this formula, U might represent some mean value of the wall velocity and the gas velocity adjacent to the wall thereby indicating partial accommodation with respect to momentum.

The stochastic boundary condition (19.5) is sometimes written in terms of $\boldsymbol{\xi} f$ rather than f since $\boldsymbol{\xi} f$ represents the flow rate across an element of surface; the two are, of course, equivalent provided that proper identification of the kernels is made.

What we would like to be able to do for the Boltzmann equation is show the existence of a unique solution in an arbitrary domain, given the distribution

[1] From recent experiments [*47*] it would appear that the state of an adsorbed film might have to be specified as well as the state of the boundary material proper.

[2] For a more complete treatment see Mechanics of Rarefied Gases by S. A. SCHAAF, Vol. IX of this Encyclopedia.

function initially together with some boundary condition of the type just discussed. Furthermore, one would like to demonstrate the approach to equilibrium as $t \to \infty$ in the case of boundary conditions which are compatible with this possibility (e.g., it could not hold for moving walls in general). It should be possible to verify the constancy of the total mass and energy as well as the decrease of H when these are allowed by the boundary conditions. The question of the degree of mathematical generality is double-edged. If very few restrictions are imposed on the initial distribution, $f(\xi, \boldsymbol{x}, 0)$, in general it will not be possible to prove very much about the solution, $f(\xi, \boldsymbol{x}, t)$. For example, one would not expect to be able to demonstrate that the energy of the system is constant if the second moment of f, $\int \xi^2 f(\xi, \boldsymbol{x}, 0)\, d\xi\, d\boldsymbol{x}$, were not assumed to be finite initially. On the other hand, there may be certain features inherent in the Boltzmann equation which would provide the solution, $f(\xi, \boldsymbol{x}, t)$, with properties which are not enjoyed by $f(\xi, \boldsymbol{x}, 0)$. For example, in the special case treated by CARLEMAN (see below) it is shown that f is strictly positive, $f > 0$, at any positive time, $t > 0$, even though f may have had stretches where it was identically zero initially. This property is essential in order to deduce that the time derivative of H exists, cf. Eq. (18.5). It is even possible that the Boltzmann equation has smoothing properties similar to the heat equation, but nothing like this has been proved. For the latter we know that, no matter how wild the temperature distribution may be initially, it becomes analytic an instant later. In another direction, we would expect to be able to estimate the approach to equilibrium in terms of exponential decay times. More precisely (cf. Sect. 22), there should be a decay on the scale of the collision time for certain features of f followed by an ultimate decay to equilibrium at an entirely different time scale. Finally, one should be able to prove all of these properties under only slight restrictions as to the nature of the intermolecular forces.

At the present time the program outlined above is largely unaccomplished[1]. Nevertheless, enough special cases can be proved to illustrate the likelihood that such a general theory may eventually develop even though each example contains only a few of the difficulties which are combined in the general problem[2].

In order to treat a problem which is spatially inhomogeneous, i.e., one in which f depends on $\boldsymbol{x}$, we shall have to be content with a solution *in the small*, i.e., for only a finite time interval, $0 < t < T$. The reason for this restriction is the quadratic dependence of the collision term on f. The solution of the ordinary differential equation $df/dt = f^2$ blows up after a finite time. One cannot hope to get a solution of BOLTZMANN'S equation *in the large*, i.e., for all time, without making strong use of the cancellation implicit in the fact that J is a difference of two positive terms which approach one another as $t \to \infty$; ($J \to 0$ if f becomes Maxwellian). This is more subtle than it might appear at first glance. For, under many physically important circumstances (see Sect. 26), we can expect f to become approximately *locally* Maxwellian long before homogeneity in $\boldsymbol{x}$ is attained. The subsequent behavior should be similar to that of the solution of some form of hydrodynamical equations; for these, growth estimates are known to be exceedingly difficult. This difficulty evaporates in a spatially homogeneous problem for which the fluid dynamics is trivial.

[1] *Note added in proof:* The existence of a solution in the large (infinite time) for space-dependent problems of the elastic-sphere Boltzmann equation has been published (posthumously) by CARLEMAN (T. CARLEMAN: Problèmes Mathématiques dans la théorie cinétique des gaz. Publ. Sci. Inst. Mittag-Leffler **2**. Uppsala: Almquist & Wiksells 1957). The methods are quite intricate, and unfortunately in the editing some of the arguments have been left incomplete.

[2] For a contrary opinion, see [*50*].

The first problem treated (Sect. 20) is the spatially inhomogeneous case in the small. The initial f is assumed to be uniformly bounded by a Maxwellian which is independent of $\boldsymbol{x}$. This is general enough to include essentially all fluid dynamic (i.e., thermodynamic) initial states. For simplicity of presentation we take an unbounded domain and a *pseudo-Maxwellian* molecule; these restrictions are unnecessary, and the proof can be extended to cover the case of an arbitrary force law with a cut-off together with a bounded domain and stochastic boundary condition provided that the domain is not too large[1]. The pseudo-Maxwellian molecule is an artificial (i.e., nonexistent) molecule which combines the best mathematical features of the Maxwellian inverse-fifth-power molecule and the elastic sphere; it has a finite total cross-section and the function B [Eq. (15.17)] is independent of V. It is quite unlikely that there exists an intermolecular potential which combines these features. Nevertheless, this is very useful as a mathematical expedient (the physical artificiality of this model does not seem to have been previously noticed).

It is interesting to note that the elementary methods of Sect. 20 are sufficient to prove existence *in the large* for the linearized Boltzmann equation which applies to small perturbations about an equilibrium distribution. This linear equation has been treated mainly by expansion methods (discussed in Chap. V) which offer much more precise information in special cases but which are not amenable to general analysis of existence theorems. We shall not consider the existence theory of the linearized equation.

In Sect. 21 we turn to the solution of the Boltzmann equation in the large but restrict the problem to the spatially homogeneous case and again take pseudo-Maxwellian molecules. The first existence theorem in this general category was proved by CARLEMAN [*9*] who treated a distribution which was isotropic as well as homogeneous (f is a function of the scalar magnitude of the velocity and of time) for the problem of elastic spheres. The homogeneous case with a pseudo-Maxwellian molecule was considered by WILD [*70*], and the techniques were greatly simplified by MORGENSTERN [*49*]. The existence proof in Sect. 21 is essentially that of MORGENSTERN. CARLEMAN'S analysis is too elaborate to reproduce here. The principal feature of his paper is the deduction of some of the more subtle properties of the solutions in addition to their existence. For example, it is shown that $H=\int f\log f$ and $\partial H/\partial t$ exist and that H decreases monotonely and f becomes Maxwellian as $t\to\infty$. Moreover, although f itself does not vary monotonely, an upper bound to f valid for all time is found in terms of the initial f alone.

The two examples in Sects. 20 and 21 have been chosen for their mathematical simplicity. This simplicity is directly attributable to the choice of a pseudo-Maxwellian molecule. The extension of the homogeneous problem (Sect. 21) to arbitrary molecules can probably be done with some effort using methods similar to those of CARLEMAN; the special case of elastic spheres and isotropic f is already done by CARLEMAN. On the other hand, the extension of the solution of the inhomogeneous problem (Sect. 20) to infinite time presents great difficulties even for pseudo-Maxwellian molecules. Keeping the small-time restriction, the extension to more general molecules and bounded domains is only moderately difficult.

All the existence methods mentioned are iterative and are based on a reduction of the Boltzmann equation to a pure integral equation by integrating out the differential terms. The first procedure is directly suggested by Eq. (19.1),

---

[1] Unpublished.

which we solve in terms of $J(\xi, \boldsymbol{x}, t)$ as a given inhomogeneous term in the differential equation,

$$f(\xi, \boldsymbol{x}, t) = f(\xi, \boldsymbol{x} - \xi t, 0) + \int_0^t J(\xi, \boldsymbol{x} - \xi(t-s), s)\, ds. \tag{19.8}$$

With J as defined in (19.2), this is an integral equation for f. It is assumed that $f(\xi, \boldsymbol{x} - \xi t, 0)$ is given. Although this is the simplest procedure, it is hard to incorporate in it the cancellation which exists in J as a difference between two positive terms. Even for analysis in the small, the factor V which is found in J makes things awkward unless the cancellation is taken into account to some extent. An important exception is the case of pseudo-Maxwellian molecules where a change in variables makes V disappear.

A slightly more subtle procedure is to split J into two terms and transpose the negative one,

$$\frac{\partial f}{\partial t} + \xi \cdot \frac{\partial f}{\partial \boldsymbol{x}} + \nu f = L; \tag{19.9}$$

here

$$\nu \equiv \frac{1}{m} \int V f_1\, d\omega\, d\xi_1 \tag{19.10}$$

and

$$L \equiv \frac{1}{m} \int V f' f_1'\, d\omega\, d\xi_1. \tag{19.11}$$

In order to write the equation in this form, it is first necessary to cut off the intermolecular force. To get an integral equation, we merely solve (19.9) considering ν and L to be given functions of ξ, $\boldsymbol{x}$, and t. We obtain[1]

$$\left.\begin{aligned} f(\xi, \boldsymbol{x}, t) &= f(\xi, \boldsymbol{x} - \xi(t-t_0), t_0) \exp\left\{-\int_{t_0}^t \nu(\xi, \boldsymbol{x} - \xi(t-s), s)\, ds\right\} + \\ &\quad + \int_{t_0}^t L(\xi, \boldsymbol{x} - \xi(t-\tau), \tau) \exp\left\{-\int_\tau^t \nu(\xi, \boldsymbol{x} - (t-s), s)\, ds\right\} d\tau. \end{aligned}\right\} \tag{19.12}$$

The value t_0 has been used instead of the initial time $t=0$ in order to include in this formula the case of a bounded domain. Given $(\xi, \boldsymbol{x}, t)$, if we trace back a particle with this speed, we find that it either intersects a boundary at a time $t_0 > 0$ or it extends to the initial manifold in which case we set $t_0 = 0$. The physical interpretation of (19.12) is clear. The number of particles at $(\boldsymbol{x}, t)$ having velocity ξ is made up of those which started out heading for $\boldsymbol{x}$ at the previous time s from the point $\boldsymbol{x} - \xi(t-s)$ and were not lost in the process. The exponential factor measures the probability of survival. In addition to the integration over s, there is a finite contribution from those particles which started out at $t=0$ or else from the wall at some time t_0 depending on ξ and $\boldsymbol{x}$. It is clear that we must supplement (19.12) with a boundary condition. Mathematically the simplest boundary condition would be one in which the emergent half of the distribution is specified *a priori* independent of what is the influx of gas particles. With this type of (physically artificial) boundary condition, the function $f(\xi, \boldsymbol{x} - \xi(t-t_0), t_0)$ in (19.12) is known explicitly just as for a pure initial value problem. With a more practical boundary condition such as (19.5), for example, the function $f(\xi, \boldsymbol{x} - \xi(t-t_0), t_0)$ must be found as part of the solution.

It is worth remarking that with a stochastic boundary condition even the Knudsen gas presents a nontrivial problem. The solution of the differential

[1] This was observed by Enskog [*17*].

equation locally is
$$f(\boldsymbol{\xi}, \boldsymbol{x}, t) = f(\boldsymbol{\xi}, \boldsymbol{x} - \boldsymbol{\xi}(t - t_0), t_0), \tag{19.13}$$

where the path is traced back until either $t_0 = 0$ or $\boldsymbol{x} - \boldsymbol{\xi}(t - t_0) = \boldsymbol{x}_0$ intersects the boundary. Since the distribution of molecules leaving the wall depends on the incident stream, (19.13) merely relates f at one point in terms of f at another. This can be written as an integral equation, but a rather awkward one, using the boundary condition (19.5). The existence of solutions in the large is easily established because the problem is linear. However, the approach to equilibrium (or to a stationary state) which is governed by the boundary conditions is more subtle [45]. In the special case of a time-independent problem subject to the boundary condition (19.6) the problem simplifies considerably. For simplicity take $\alpha = 0$. The determination of f can be reduced to the determination of the boundary function

$$Q(S) = \int\limits_{\boldsymbol{\xi}\cdot\boldsymbol{n}>0} (\boldsymbol{\xi}\cdot\boldsymbol{n})\, f\, d\boldsymbol{\xi} \tag{19.14}$$

by means of an integral equation of the form

$$Q(S) = P(S) + \int L(S, S')\, Q(S')\, dS' \tag{19.15}$$

where P represents the flow through any entrances and L is a kernel depending on the geometry [14].

Another useful interpretation of (19.12) is as follows. Introduce the function

$$\mu(\tau) \equiv \nu(\boldsymbol{\xi}, \boldsymbol{x} - \boldsymbol{\xi}(t-\tau), \tau) \exp\left\{-\int\limits_{\tau}^{t} \nu(\boldsymbol{\xi}, \boldsymbol{x} - \boldsymbol{\xi}(t-s), s)\, ds\right\}. \tag{19.16}$$

We verify that

$$\int\limits_{t_0}^{t} \mu(\tau)\, d\tau = 1 - \exp\left\{-\int\limits_{t_0}^{t} \nu(\boldsymbol{\xi}, \boldsymbol{x} - \boldsymbol{\xi}(t-s), s)\, ds\right\}. \tag{19.17}$$

Eq. (19.12) states that $f(\boldsymbol{\xi}, \boldsymbol{x}, t)$ is a mean value with combined weight 1 of two functions, a distributed mean of the function

$$g(\tau) \equiv \frac{L(\boldsymbol{\xi}, \boldsymbol{x} - \boldsymbol{\xi}(t-\tau), \tau)}{\nu(\boldsymbol{\xi}, \boldsymbol{x} - \boldsymbol{\xi}(t-\tau), \tau)} \tag{19.18}$$

taken with weight $\mu(\tau)\, d\tau$, $t_0 < \tau < t$, and of the function $f(\boldsymbol{\xi}, \boldsymbol{x} - \boldsymbol{\xi}(t - t_0), t_0)$ taken with the remaining weight.

The formula (19.12) has been written in such form that it applies also to time-independent steady-flow solutions of BOLTZMANN's equation. We merely drop the third (time) argument in all functions and set $t = 0$ wherever else it occurs. The parameter t_0 has the same significance as before; it is the time interval required to reach the boundary from $\boldsymbol{x}$ at a velocity $-\boldsymbol{\xi}$. The quantities s and τ are merely integration variables.

The precise integral equation used by Carleman takes several forms depending on the estimates to be made. The quantity ν is further subdivided into a sum, $\nu_1 + \nu_2$, where ν_1 is an explicit function of $\boldsymbol{\xi}$ and ν_2 can be dropped to get upper estimates on f, or into a difference $\nu_1' - \nu_2'$, with ν_1' also explicit and ν_2' dropped to get lower estimates on f. Of course the space dependence is not present.

It is possible to describe a stepwise approximation to solutions of the Boltzmann equation which has particular interest even though it has not been used in any actual problem. Divide the time into intervals, $0 = t_0 < t_1 < t_2 < \dots$. First take $f(\boldsymbol{\xi}, \boldsymbol{x}) = f_0$ and solve the space-independent Boltzmann equation ($\boldsymbol{x}$ is a parameter) until $t = t_1$; call this $f_1^*(\boldsymbol{\xi}, \boldsymbol{x})$. Then write $f_1(\boldsymbol{\xi}, \boldsymbol{x}) = f_1^*(\boldsymbol{\xi}, \boldsymbol{x} -$

$\xi(t_1 - t_0))$. Continue, alternating the solution of the spatially homogeneous Boltzmann equation and the free flow (Knudsen gas) equation. One would expect convergence to a solution of the Boltzmann equation on refining the time subdivision. As an approximation, instead of refining the time, let us take the mean collision time as the common interval $t_n - t_{n-1}$, and, instead of solving the space-independent Boltzmann equation for just this time interval, solve it for an infinite time at each step. In other words, at each $\boldsymbol{x}$ we replace f by the equivalent locally Maxwellian distribution and then continue with the free flow for another step. This will be a rough approximation to the correct solution insofar as collisions can be neglected entirely in one collision time (free-flow step) and at the same time can be expected to give a complete approach to equilibrium in one collision time (local-Maxwellian step). The interesting feature of this procedure is that it is very similar to the mean-free-path analysis of elementary kinetic theory.

20. Space dependent solutions in the small[1]. We introduce the abbreviations

$$[f] = f' f_1' - f f_1, \tag{20.1}$$

$$\{f\} = f' f_1' + f f_1. \tag{20.2}$$

For a pseudo-Maxwellian molecule, we can write

$$J = \int B(\vartheta)\,[f]\,d\vartheta\,d\varepsilon\,d\xi_1, \tag{20.3}$$

and

$$\int B(\vartheta)\,d\vartheta\,d\varepsilon = \beta, \tag{20.4}$$

where β is a finite constant. We assume that f is initially uniformly bounded by a space-independent Maxwellian

$$f(\xi, \boldsymbol{x}, 0) < f^{(0)}(\xi). \tag{20.5}$$

From (19.8) we construct the following iteration process,

$$\left.\begin{aligned} f_{n+1}(\xi, \boldsymbol{x}, t) &= f_0(\xi, \boldsymbol{x} - \xi t) + \int_0^t J_n(\xi, \boldsymbol{x} - \xi(t-s), s)\,ds, \\ J_n(\xi, \boldsymbol{x}, t) &= \int B(\vartheta)\,[f_n]\,d\vartheta\,d\varepsilon\,d\xi_1. \end{aligned}\right\} \tag{20.6}$$

To start the iteration, we compute J_0 using

$$f_0 = f(\xi, \boldsymbol{x}, 0). \tag{20.7}$$

The proof of the existence of a solution can be made to depend on the following estimate. If, at a given $\boldsymbol{x}$ and t, f satisfies the inequality

$$f < 2 f^{(0)}, \tag{20.8}$$

where $f^{(0)}$ is the Maxwellian bound (20.5), then

$$|J| < 8\beta R f^{(0)} \tag{20.9}$$

where R is the known constant

$$R = \int f^{(0)}\,d\xi. \tag{20.10}$$

The proof is immediate; we have

$$\begin{aligned} |J| &< \int B\{f\}\,d\vartheta\,d\varepsilon\,d\xi_1 \\ &< 4\int B\{f^{(0)}\}\,d\vartheta\,d\varepsilon\,d\xi_1 \\ &= 8\int B f^{(0)} f_1^{(0)}\,d\vartheta\,d\varepsilon\,d\xi_1 \\ &= 8\beta R f^{(0)}. \end{aligned}$$

[1] Not previously published.

Now we restrict t to the interval

$$0 < t < T = 1/8\beta R. \tag{20.11}$$

In this time interval, we can show that all iterates satisfy the bound

$$f_n < 2f^{(0)}. \tag{20.12}$$

We use induction. First of all, $f_0 < f^{(0)} < 2f^{(0)}$ satisfies the inequality. Next, assuming that f_n satisfies this inequality, from (20.6) we obtain

$$\begin{aligned} f_{n+1} &< f_0(\boldsymbol{\xi}, \boldsymbol{x} - \boldsymbol{\xi} t) + \int_0^t |J_n|\, ds \\ &< f^{(0)} + \int_0^t 8\beta R f^{(0)}\, ds \\ &= f^{(0)} + 8\beta R t f^{(0)} \\ &< 2f^{(0)}; \end{aligned}$$

this completes the induction.

It is convenient to introduce as the norm of a function $g(\boldsymbol{\xi}, \boldsymbol{x}, t)$ the maximum value with respect to all $\boldsymbol{\xi}$, $\boldsymbol{x}$, and t $(0 < t < T)$ of $|g|/f^{(0)}$,

$$\|g\| = \max_{\boldsymbol{\xi}, \mathbf{x}, t} \left(\frac{|g|}{f^{(0)}}\right). \tag{20.13}$$

From (20.6), by subtraction,

$$f_{n+1} - f_n = \int_0^t (J_n - J_{n-1})\, ds, \tag{20.14}$$

$$J_n - J_{n-1} = \int B([f_n] - [f_{n-1}])\, d\vartheta\, d\varepsilon\, d\boldsymbol{\xi}_1. \tag{20.15}$$

We have

$$\begin{aligned} [f_n] - [f_{n-1}] &= f'_n f'_{1,n} - f_n f_{1,n} - f'_{n-1} f'_{1,n-1} + f_{n-1} f_{1,n-1} \\ &= f'_n (f'_{1,n} - f'_{1,n-1}) + f'_{1,n-1}(f'_n - f'_{n-1}) - f_n (f_{1,n} - f_{1,n-1}) - f_{1,n-1}(f_n - f_{n-1}), \end{aligned}$$

from which

$$\begin{aligned} |[f_n] - [f_{n-1}]| < f'_n |f'_{1,n} - f'_{1,n-1}| + f'_{1,n-1} |f'_n - f'_{n-1}| + \\ + f_n |f_{1,n} - f_{1,n-1}| + f_{1,n-1} |f_n - f_{n-1}|. \end{aligned}$$

Dividing both sides by $f^{(0)} f_1^{(0)} = f^{(0)\prime} f_1^{(0)\prime}$ and using (20.12) as well as (20.13), we get

$$\begin{aligned} \frac{|[f_n] - [f_{n-1}]|}{f^{(0)} f_1^{(0)}} &< 2\|f'_{1,n} - f'_{1,n-1}\| + 2\|f'_n - f'_{n-1}\| + 2\|f_{1,n} - f_{1,n-1}\| + 2\|f_n - f_{n-1}\| \\ &= 8\|f_n - f_{n-1}\|. \end{aligned}$$

Now, returning to (20.15), we have

$$\left.\begin{aligned} |J_n - J_{n-1}| &< \int B |[f_n] - [f_{n-1}]|\, d\vartheta\, d\varepsilon\, d\boldsymbol{\xi}_1 \\ &< 8\|f_n - f_{n-1}\| \int B f^{(0)} f_1^{(0)}\, d\vartheta\, d\varepsilon\, d\boldsymbol{\xi}_1 \\ &= 8\beta R \|f_n - f_{n-1}\| f^{(0)}, \end{aligned}\right\} \tag{20.16}$$

or

$$\|J_n - J_{n-1}\| < 8\beta R \|f_n - f_{n-1}\|, \tag{20.17}$$

while from (20.14),

$$|f_{n+1} - f_n| < \int_0^t |J_n - J_{n-1}|\, ds$$

or

$$\|f_{n+1} - f_n\| < t \|J_n - J_{n-1}\|. \tag{20.18}$$

Finally,

$$\|f_{n+1}-f_n\| < 8\beta R t \|f_n - f_{n-1}\|,$$
$$\|J_{n+1}-J_n\| < 8\beta R t \|J_n - J_{n-1}\|,$$

and, for any choice of the number α, $0<\alpha<1$, we have

$$\left.\begin{aligned} \|f_{n+1}-f_n\| &< \alpha \|f_n - f_{n-1}\|, \\ \|J_{n+1}-J_n\| &< \alpha \|J_n - J_{n-1}\|, \\ 0<t&<\alpha T. \end{aligned}\right\} \tag{20.19}$$

By comparison with the geometric series $1/(1-\alpha)$, we conclude that there exist functions $f(\boldsymbol{\xi}, \boldsymbol{x}, t)$ and $J(\boldsymbol{\xi}, \boldsymbol{x}, t)$ with the property that

$$\left.\begin{aligned} \|f-f_n\| &\to 0, \\ \|J-J_n\| &\to 0. \end{aligned}\right\} \tag{20.20}$$

It is now an easy matter to conclude from (20.6) (by subtracting f from both sides of the first equation and J from the second equation, then taking norms) that f and J satisfy

$$\left.\begin{aligned} f(\boldsymbol{\xi}, \boldsymbol{x}, t) &= f_0(\boldsymbol{\xi}, \boldsymbol{x}-\boldsymbol{\xi} t) + \int_0^t J(\boldsymbol{\xi}, \boldsymbol{x}-\boldsymbol{\xi}(t-s), s)\, ds, \\ J(\boldsymbol{\xi}, \boldsymbol{x}, t) &= \int B(\vartheta)\, [f]\, d\vartheta\, d\varepsilon\, d\boldsymbol{\xi}_1. \end{aligned}\right\} \tag{20.21}$$

If $f_0(\boldsymbol{\xi}, \boldsymbol{x})$ is continuous in its variables, it is easy to verify that $f_1(\boldsymbol{\xi}, \boldsymbol{x}, t)$ and all the functions f_n and J_n in turn are continuous in $(\boldsymbol{\xi}, \boldsymbol{x})$ and in t as well. By the uniform convergence, f and J are continuous. From (20.21) we conclude that the directional derivative $\partial f/\partial t + \boldsymbol{\xi}\cdot\partial f/\partial \boldsymbol{x}$ exists (to show that $\partial/\partial t$ and $\partial/\partial \boldsymbol{x}$ exist separately requires additional estimates of derivatives similar to those made above) and f satisfies the original Boltzmann equation together with the assigned initial values. Similar conclusions can be drawn if f_0 is only assumed to be Lebesgue measurable; the solution $f(\boldsymbol{\xi}, \boldsymbol{x}, t)$ will be Lebesgue measurable, and the directional derivative $\partial f/\partial t + \boldsymbol{\xi}\cdot\partial f/\partial \boldsymbol{x}$ will exist.

It is very easy to show that this solution is unique among all functions $f(\boldsymbol{\xi}, \boldsymbol{x}, t)$ which are bounded by some Maxwellian function.

By using the integral equation (19.12) in this problem instead of (19.8), one can obtain lower estimates on f as well as upper estimates. This makes it possible to estimate integrals of the form $\int f \log g\, d\boldsymbol{\xi}$ and thereby show the existence of H and of $\partial H/\partial t$ as well as the monotone decrease of the former.

It is clear that the elementary nature of this proof depends crucially on the exponential bounds and the pseudo-Maxwellian assumption which, taken together, allow J to be easily estimated.

21. Spatially homogeneous solutions in the large [*49*]. This time we shall want to iterate (19.12) rather than (19.8). More precisely, we take the equations

$$f(\boldsymbol{\xi}, t) = f(\boldsymbol{\xi}, 0)\, e^{-\nu t} + \int_0^t L(\boldsymbol{\xi}, \tau)\, e^{-\nu(t-\tau)}\, d\tau, \tag{21.1}$$

where

$$L = \int B(\vartheta)\, f' f_1'\, d\vartheta\, d\varepsilon\, d\boldsymbol{\xi}_1 \tag{21.2}$$

$$\nu = \int B f_1\, d\vartheta\, d\varepsilon\, d\boldsymbol{\xi}_1 = \beta\varrho, \tag{21.3}$$

$$\varrho = \int f\, d\boldsymbol{\xi}, \tag{21.4}$$

$$\beta = \int B\, d\vartheta\, d\varepsilon. \tag{21.5}$$

For a pseudo-Maxwellian molecule, β is a given constant. In writing (21.1), we have already made use of a certain fact that should really be proved to be a property of the solution; namely, that the mass density, ϱ, is a constant in time. What we shall do is solve Eq. (21.1) assuming that the constant ν is given without connecting ν with the quantity $\varrho(t)=\int f(\xi,t)\,d\xi$ which has to be computed. Then we shall find that, if the constant ν is chosen so that $\nu=\beta\varrho$ initially, the solution has the property $\varrho=$ const and $\nu=\beta\varrho$ for all time. This will establish that (21.1) is actually the Boltzmann equation for this solution.

We assume that $f(\xi,0)$ is given and is integrable,

$$\int f(\xi,0)\,d\xi=\varrho(0), \tag{21.6}$$

and take the constant ν to be

$$\nu=\beta\,\varrho(0). \tag{21.7}$$

An iteration is defined by

$$f_{n+1}=f(\xi,0)\,e^{-\nu t}+\int_0^t L_n\,e^{-\nu(t-\tau)}\,d\tau \tag{21.8}$$

$$L_n=\int B f_n' f_{1,n}'\,d\vartheta\,d\varepsilon\,d\xi_1. \tag{21.9}$$

To start the iteration we take $f_0=0$ and define

$$\varrho_n(t)=\int f_n(\xi,t)\,d\xi. \tag{21.10}$$

We wish to show that the iteration is monotone, i.e.,

$$f_n(\xi,t)<f_{n+1}(\xi,t) \tag{21.11}$$

but with uniformly bounded $\varrho_n(t)$,

$$\varrho_n(t)<\varrho(0). \tag{21.12}$$

This will establish the convergence of f_n to a limit. The monotonicity is trivial. Since $f_0=0$ and $f_1=f(\xi,0)\,e^{-\nu t}$, we have $f_1>f_0$. Using induction, $f_n>f_{n-1}$ implies $L_n>L_{n-1}$ by (21.9) which implies $f_{n+1}>f_n$ by (21.8), thus completing the proof of monotonicity. To establish the uniform bound, we integrate (21.8)

$$\varrho_{n+1}=\varrho(0)\,e^{-\nu t}+\int_0^t e^{-\nu(t-\tau)}\left\{\int L_n\,d\xi\right\}d\tau.$$

But

$$\begin{aligned}\int L_n\,d\xi&=\int B f_n' f_{1,n}'\,d\vartheta\,d\varepsilon\,d\xi\,d\xi_1\\&=\int B f_n f_{1,n}\,d\vartheta\,d\varepsilon\,d\xi\,d\xi_1\\&=\beta\,\varrho_n^2,\end{aligned}$$

from which

$$\varrho_{n+1}=\varrho(0)\,e^{-\nu t}+\int_0^t \beta\,\varrho_n^2(\tau)\,e^{-\nu(t-\tau)}\,d\tau. \tag{21.13}$$

Again we use induction. We have $\varrho_0(t)=0<\varrho(0)$. Assuming that $\varrho_n(t)<\varrho(0)$, we find, from (21.13),

$$\varrho_{n+1}<\varrho(0)\,e^{-\nu t}+\beta\,\varrho^2(0)\int_0^t e^{-\nu(t-\tau)}\,d\tau=\varrho(0).$$

From the Lebesgue theorem on monotone convergence [*63*], we now conclude that $f_n(\xi,t)$ converges to a function $f(\xi,t)$ which also satisfies the bound

$$\varrho(t)=\int f(\xi,t)\,d\xi\le\varrho(0). \tag{21.14}$$

From the convergence of $f_n(\xi)$ to $f(\xi)$ follows the convergence of $f_n(\xi')\,f_n(\xi_1')$ to $f(\xi')\,f(\xi_1')$ and, using FUBINI's theorem [*57*], the convergence of L_n to $\int B f' f_1'\times$

$d\vartheta\, d\varepsilon\, d\xi_1$. We conclude that f and L satisfy (21.1) and (21.2). It remains to be shown that $\varrho(t) = \text{const}$. Integrating (21.1), we have

$$\varrho(t) = \varrho(0)\, e^{-\nu t} + \int_0^t \beta\, \varrho^2(\tau)\, e^{-\nu(t-\tau)}\, d\tau. \tag{21.15}$$

For the moment restrict t to the interval $0 < t < T$ where T is chosen so $e^{-\beta T} = \frac{1}{2}$. If $\bar{\varrho}$ is the greatest lower bound of $\varrho(t)$ for $0 < t < T$, (21.15) implies

$$\begin{aligned}\varrho(t) &\geq \varrho(0)\, e^{-\nu t} + \int_0^t \beta\, \bar{\varrho}^2\, e^{-\nu(t-\tau)}\, d\tau \\ &= \varrho(0)\, e^{-\nu t} + \frac{\bar{\varrho}^2}{\varrho(0)}\,(1 - e^{-\nu t}) \\ &\geq \frac{1}{2}\left\{\varrho(0) + \frac{\bar{\varrho}^2}{\varrho(0)}\right\}\end{aligned}$$

since the right hand side is a monotone decreasing function of t. However, this implies that

$$\bar{\varrho} \geq \frac{1}{2}\left\{\varrho(0) + \frac{\bar{\varrho}^2}{\varrho(0)}\right\}$$

from which it follows that $\bar{\varrho} = \varrho(0)$ and, together with (21.14), we have $\varrho(t) = \varrho(0)$ in the interval $0 < t < T$. Repetition of this argument for $T < t < 2T$, etc. shows that $\varrho(t)$ is constant for all time. In particular, $\nu = \beta\varrho$ for all time. From (21.1) we conclude that $f(\xi, t)$ is differentiable in t and therefore satisfies the original Boltzmann equation.

Within the framework of this method, it can be easily shown that the solution obtained is unique only if it is compared with all possible solutions for which $\int f(\xi, t)\, d\xi$ is known to be a constant. In other words, it has not been shown that an initially homogeneous solution cannot become inhomogeneous [*49*]. This is a consequence of taking ν to be a constant in (21.1) rather than equal to $\beta\varrho$ whatever ϱ might be. However, from the results of Sect. 20 we know that the solution is unique even among spatially inhomogeneous solutions (but only if they are bounded by a Maxwellian).

This existence theorem is very general insofar as restrictions on the initial function are concerned. One should verify that more well-behaved initial functions retain their properties. It is a simple matter to show that the energy $\int \frac{1}{2}\xi^2 f(\xi, t)\, d\xi$ exists and is constant, provided that this integral exists initially. It is not clear that assuming only that $H = \int f \log f\, d\xi$ exists initially (rather than that f is bounded by some well-behaved function) is enough to guarantee its existence and that of dH/dt for $t > 0$. From Sect. 20 we know that if f is bounded by a Maxwellian, this is enough to guarantee the existence of H but only for a short time. The proof of such properties for all time is probably feasible using Carleman's methods.

It is interesting to note that, if f is expanded in Hermite polynomials of ξ, the variation of each Hermite coefficient in time is explicit and exponential for the actual Maxwellian molecule (cf. Chap. V); a convergence proof along these lines could probably be obtained without great difficulty.

IV. Normal solutions of the Boltzmann equation.

22. Introduction and survey. It has been mentioned that the mean-free-path is the basic length scale and the collision time is the basic time scale for the Boltzmann equation. On the other hand, in problems of conventional gas dynamics, both of these quantities are very small compared to macroscopic sizes.

There is a more striking difference. For the Boltzmann equation a state is defined by giving a distribution function, $f(\boldsymbol{\xi}, \boldsymbol{x})$; for a set of fluid equations a state is defined in terms of the first few moments of f, the density, $\varrho(\boldsymbol{x})$, the velocity, $\boldsymbol{u}(\boldsymbol{x})$, and the energy, $e(\boldsymbol{x})$, or temperature, $T(\boldsymbol{x})$. This reduction in the amount of detail which is required to specify a state fluid dynamically suggests that a limiting process is involved. We recall (Chap. I) that there exist several limiting cases which result in a reduction in amount of detail from the maximum which is present in LIOUVILLE'S equation. One of these yields the Boltzmann equation and another one yields fluid dynamical equations. It should be possible to find a direct mathematical transition from the Boltzmann equation to a set of fluid equations for the special case of a perfect gas. This transition can only be found by letting the collision time approach zero in an appropriate way. We would like to find that the very large manifold of solutions $f(\boldsymbol{\xi}, \boldsymbol{x}, t)$ of the Boltzmann equation (one for each initial value $f(\boldsymbol{\xi}, \boldsymbol{x})$) condenses down to a much smaller manifold which associates a single $f(\boldsymbol{\xi}, \boldsymbol{x}, t)$ to each set of initial values $\varrho(\boldsymbol{x})$, $\boldsymbol{u}(\boldsymbol{x})$, $T(\boldsymbol{x})$. Or, since any time t can be called an initial time, we must expect that in the limit a single distribution function $f(\boldsymbol{\xi}, \boldsymbol{x})$ will be assigned to every fluid dynamical state $\varrho(\boldsymbol{x})$, $\boldsymbol{u}(\boldsymbol{x})$, $T(\boldsymbol{x})$. Another way of putting this is that in the limit one should find that the time derivative of f is determined not by f itself (which is what the Boltzmann equation states), but it is determined by $(\varrho, \boldsymbol{u}, T)$. The next step would be to eliminate f entirely and deduce that the time derivative of $(\varrho, \boldsymbol{u}, T)$ is determined by the instantaneous value of $(\varrho, \boldsymbol{u}, T)$; this is precisely a fluid dynamical description. One cannot expect that $f(\boldsymbol{\xi}, \boldsymbol{x})$ will be determined by the numerical values of ϱ, $\boldsymbol{u}$, and T at the same point $\boldsymbol{x}$; in general $f(\boldsymbol{\xi})$ will not be a point function of the five variables $\varrho, \boldsymbol{u}, T$ but $f(\boldsymbol{\xi}, \boldsymbol{x})$ will be a *functional* of the five functions $\varrho(\boldsymbol{x})$, $\boldsymbol{u}(\boldsymbol{x})$, $T(\boldsymbol{x})$. Actually, if one is satisfied with the simplest set of fluid equations, viz. the nondissipative Euler equations, then $f(\boldsymbol{\xi})$ does become a point function of ϱ, $\boldsymbol{u}$, and T in the limit. The complications arise in trying to obtain more realistic fluid equations.

The Hilbert expansion (Sect. 23) is a formal expansion of solutions of the Boltzmann equation in powers of a certain parameter [*30*]. Ordinarily one would expect to be able to approximate to *all* the solutions of an equation (possibly some much better than others) by such an expansion. The amazing feature of this expansion, implied by the Hilbert uniqueness theorem, is that any solution which is expandable in the form assumed is uniquely determined by the *fluid dynamical state alone*. In other words, this expansion picks out exactly those *normal solutions* of the Boltzmann equation which are appropriate to a fluid dynamical description. By a formal extension of HILBERT'S analysis [*20*], the stresses and heat flow can be expressed directly as successively higher order gradients of ϱ, $\boldsymbol{u}$, and T (Sect. 24). Inserting these expressions into the conservation equations, (17.12) to (17.14) accomplishes the elimination of f which was mentioned above and yields a sequence of self-consistent fluid equations. This task of eliminating the intermediary, f, once it is known to depend only on the fluid variables and their gradients, can be performed more easily by the method of ENSKOG [*16*] (Sect. 25). The two formal procedures are shown to be identical in Sect. 25. The essential difference is that ENSKOG *postulates*, the existence of a functional dependence of $f(\boldsymbol{\xi}, \boldsymbol{x})$ on $\varrho(\boldsymbol{x})$, $\boldsymbol{u}(\boldsymbol{x})$, and $T(\boldsymbol{x})$ which is expandable in a parameter, whereas HILBERT *deduces* the dependence of f on ϱ, $\boldsymbol{u}$, and T by merely assuming that f is expandable in this parameter.

There are a number of mathematical subtleties hidden within these formal manipulations. Although the results of Sect. 24 (which are identical to ENSKOG'S but differently derived) are formally equivalent to the ones obtained by HILBERT

(Sect. 23), they are quite distinct mathematically. For example, the number of boundary conditions which may be imposed in HILBERT'S method does not usually increase with the order of approximation; although it is not inevitable (cf. Sect. 33), it is likely that an increasing number of boundary conditions is required as the order is increased in the Chapman-Enskog sequence of equations. This extraordinary difference is created by what seem to be only slightly different procedures for truncating an infinite series into a sequence of finite stages.

In order to allow the collision time to become small, it is convenient to introduce the small parameter ε into the Boltzmann equation as follows:

$$\frac{\partial f}{\partial t} + \boldsymbol{\xi} \cdot \frac{\partial f}{\partial \boldsymbol{x}} = \frac{1}{\varepsilon} J. \tag{22.1}$$

We are interested in the properties of solutions of this equation for small ε. If the left hand side of the equation remains finite for small ε (it reflects macroscopic variation of the fluid variables), we conclude that $J \to 0$ or f becomes locally Maxwellian. Insofar as f is locally Maxwellian, it is completely and explicitly determined by the fluid state $\varrho(\boldsymbol{x})$, $\boldsymbol{u}(\boldsymbol{x})$, $T(\boldsymbol{x})$. However, it is exactly the small deviation of f from Maxwellian which, after division by ε in (22.1) determines the finite fluid flow properties of the gas. This small deviation from local Maxwellian must therefore be found.

The essential feature of this phenomenon of the reduction to fluid dynamics is that there exist two different time scales. The first, the collision time, represents the order of magnitude of time required for an initially non-Maxwellian flow to settle down. After this the fluid equations take over. Again there is a decay time but a much longer one. It is easy to estimate that the macroscopic decay time, conventionally deemed to result from viscosity and heat conductivity, is on the order of the square of a representative macroscopic time divided by the collision time. In other words, as $\varepsilon \to 0$, we simultaneously decrease the collision time and increase the macroscopic decay time. The first period could be described as one in which the quantity H decays to the thermodynamic entropy, S. The second period is governed by the decay of S to its maximum value. This separation of the two time scales is only possible when ε is sufficiently small.

In Sect. 26 a mathematically heuristic asymptotic evaluation of the solution of BOLTZMANN'S equation for small ε is presented. The most significant feature is the connection between the smooth fluid behavior over most of the domain and several types of boundary layers. There are boundary layers near actual physical boundaries, initial "boundary" layers near $t=0$, and a third type found in the midst of the fluid (shock waves). Near one of these boundary layers the behavior of the solution is roughly exponential in $1/\varepsilon$. Elsewhere it is approximately a power series. The formal evaluation of these power series is most directly done by following HILBERT or ENSKOG (Sects. 24 and 25). However, what this asymptotic analysis makes clear is the transition from general solutions of the Boltzmann equation to normal solutions as $\varepsilon \to 0$. It is possible that a rigorous evaluation of the range of validity of normal solutions could eventually emerge from this type of analysis.

It is important to note that, when ε is not small, entirely different procedures such as the ones discussed in Chaps. III and V are necessary in order to solve the Boltzmann equation. If ε is not small, the two time scales merge, and the various boundary layers cannot be separated from the flow as a whole.

23. The Hilbert expansion [*30*]. For purposes of obtaining a formal asymptotic expansion, we insert the small parameter ε in the Boltzmann equation,

$$\dot{f} = \frac{1}{\varepsilon} J(f, f), \tag{23.1}$$

with the abbreviation

$$\dot{f} \equiv \frac{\partial f}{\partial t} + \xi_r \frac{\partial f}{\partial x_r}. \tag{23.2}$$

We shall make much use of the quadratic character of $J(f, f)$. In particular [see Eq. (16.1)],

$$J(f_1 + f_2, g_1 + g_2) = J(f_1, g_1) + J(f_1, g_2) + J(f_2, g_1) + J(f_2, g_2). \tag{23.3}$$

Proceeding formally, we adopt the expansion

$$f = f^{(0)} + \varepsilon f^{(1)} + \varepsilon^2 f^{(2)} + \cdots, \tag{23.4}$$

substitute it into (23.1), and equate coefficients of powers of ε. We obtain

$$J(f^{(0)}, f^{(0)}) = 0, \tag{23.5}$$

$$2 J(f^{(0)}, f^{(1)}) = \dot{f}^{(0)}, \tag{23.6}$$

$$\sum_{m=0}^{n} J(f^{(m)}, f^{(n-m)}) = \dot{f}^{(n-1)}. \tag{23.7}$$

We rewrite (23.7) in the form

$$2 J(f^{(0)}, f^{(n)}) = \dot{f}^{(n-1)} - \sum_{m=1}^{n-1} J(f^{(m)}, f^{(n-m)}) \tag{23.8}$$

in order to separate the term in $f^{(n)}$ from the ones involving lower orders. Eq. (23.5) states that $f^{(0)}$ is locally Maxwellian. We define the five parameters in it by

$$\left.\begin{aligned} \varrho^{(0)} &= \int f^{(0)}\, d\xi, \\ \varrho^{(0)} \boldsymbol{u}^{(0)} &= \int \xi f^{(0)}\, d\xi, \\ 3\varrho^{(0)} \mathrm{R} T^{(0)} = 2\varrho^{(0)} e^{(0)} &= \int (\xi - \boldsymbol{u}^{(0)})^2 f^{(0)}\, d\xi. \end{aligned}\right\} \tag{23.9}$$

The superscripts are used because there is no compelling reason to identify the five parameters which occur in $f^{(0)}$ with the moments of f; this possibility is left open.

In order to find $f^{(1)}$ from (23.6) one must solve an integral equation. From (23.8), we see that exactly the same integral equation (with different inhomogeneous terms) must be solved to obtain $f^{(n)}$, assuming that $f^{(0)} \ldots f^{(n-1)}$ have already been found. It can be shown that this is a Fredholm equation with a symmetric kernel[1] when expressed in terms of the variable $f^{(1)}/f^{(0)}$ or, in general, in terms of

$$g^{(n)} = f^{(n)}/f^{(0)}. \tag{23.10}$$

The situation is complicated by the fact that *zero* is a five-fold degenerate eigenvalue. The homogeneous equation

$$J(f^{(0)}, f^{(0)} g) = 0 \tag{23.11}$$

has the five linearly independent solutions

$$g = \sum_{r=0}^{4} \gamma_r \psi_r, \tag{23.12}$$

[1] Ref. [*30*] for elastic spheres and [*11*] for the general case.

where the ψ_r are defined as in (15.12). To prove this, consider

$$J_g(f^{(0)}, f^{(0)} g) \equiv \int g\, J\, d\xi = 0.$$

Using (16.5) and recalling that

$$f^{(0)} f_1^{(0)} = f^{(0)\prime} f_1^{(0)\prime},$$

we find that J_g can be put into the form

$$\int V f^{(0)} f_1^{(0)} (g + g_1 - g' - g_1')^2\, d\omega\, d\xi\, d\xi_1 = 0.$$

We conclude that (23.11) is satisfied only if g is a summational invariant. The coefficients γ_r can be functions of $\boldsymbol{x}$ and t. From the theory of the Fredholm equation, we know that a solution of the inhomogeneous equation

$$J(f^{(0)}, f^{(0)} g) = h \tag{23.13}$$

exists only if h satisfies the orthogonality conditions

$$\int \psi_r\, h\, d\xi = 0. \tag{23.14}$$

If these conditions are satisfied, there is a five parameter family of solutions,

$$g = \bar{g} + \sum \gamma_r \psi_r. \tag{23.15}$$

Finally, a particular solution, say $\bar{g}$, can be made unique by imposing on it the conditions

$$\int \psi_r \bar{g} f^{(0)}\, d\xi = 0. \tag{23.16}$$

Summarizing, we have found that for each $f^{(n)}$ (including $f^{(0)}$), there is a five parameter family of solutions. However, to guarantee the existence of a solution for $f^{(1)}$, five conditions must be put on $\dot{f}^{(0)}$, and, in turn, to solve for $f^{(n)}$, five conditions will have to be imposed on $\dot{f}^{(n-1)}$. These conditions will take the form of partial differential equations in $\boldsymbol{x}$ and t for each set of five parameters.

The case of $f^{(0)}$ is distinct from the others since the five parameters enter differently. The compatibility condition on $f^{(1)}$ is

$$\int \psi_r\, \dot{f}^{(0)}\, d\xi = 0. \tag{23.17}$$

But these are exactly the five conservation equations, (17.12) to (17.14), for the special case of a locally Maxwellian f. The fact that $f^{(0)}$ is locally Maxwellian implies that

$$\left.\begin{aligned} P_{ij} &= p\,\delta_{ij}, \\ q_i &= 0. \end{aligned}\right\} \tag{23.18}$$

Putting this into the conservation equations yields the *Euler equations* of fluid dynamics (i.e., without dissipation and with the entropy constant on a particle path provided it does not cross a shock). These are the equations satisfied by $\varrho^{(0)}$, $\boldsymbol{u}^{(0)}$, and $T^{(0)}$.

We continue by induction, assuming that $f^{(0)} \ldots f^{(n-1)}$ have already been uniquely determined. We have

$$f^{(n)} = \bar{f}^{(n)} + \sum \gamma_r^n \psi_r f^{(0)} \tag{23.19}$$

where $\bar{f}^{(n)}$ is a solution of (23.8) which is uniquely determined in terms of $f^{(0)} \ldots f^{(n-1)}$ by the conditions

$$\int \psi_r \bar{f}_n\, d\xi = 0. \tag{23.20}$$

The integral equation for $f^{(n+1)}$ has a solution only if

$$\int \psi_r \left\{ \dot{f}^{(n)} - \sum_{m=1}^{n} J(f^{(n+1)}, f^{(n+1-m)}) \right\} d\xi = 0 .$$

The terms in J drop out by (16.5) since the ψ_r are summational invariants. We are left with

$$\int \psi_r \dot{f}^{(n)} d\xi = 0$$

or

$$\int \psi_r \bar{\dot{f}}^{(n)} d\xi + \sum_s \int (\dot{\gamma}_s^n \psi_r \psi_s f^{(0)} + \gamma_s^n \psi_r \psi_s \dot{f}^{(0)}) d\xi = 0 . \tag{23.21}$$

Introducing the notation,

$$\left.\begin{aligned} a_{rs} &= \int \psi_r \psi_s f^{(0)} d\xi , \\ b_{rs}^k &= \int \xi_k \psi_r \psi_s f^{(0)} d\xi , \\ c_{rs} &= \int \psi_r \psi_s \dot{f}^{(0)} d\xi = \frac{\partial a_{rs}}{\partial t} + \frac{\partial b_{rs}^k}{\partial x_k} , \\ d_r^n &= \int \psi_r \bar{\dot{f}}^{(n)} d\xi = \frac{\partial}{\partial x_k} \int \psi_r \xi_k \bar{f}^{(n)} d\xi = \frac{\partial e_{rk}^n}{\partial x_k} , \\ & r, s = 0, 1, 2, 3, 4 \qquad k = 1, 2, 3 \end{aligned}\right\} \tag{23.22}$$

we find that (23.21) can be written as

$$a_{rs} \frac{\partial \gamma_s^n}{\partial t} + b_{rs}^k \frac{\partial \gamma_s^n}{\partial x_k} + c_{rs} \gamma_s^n + d_r^n = 0 \tag{23.23}$$

or, using the alternative forms in (23.22),

$$\frac{\partial}{\partial t} (a_{rs} \gamma_s^n) + \frac{\partial}{\partial x_k} (b_{rs}^k \gamma_s^n + e_{rk}^n) = 0 . \tag{23.24}$$

This is a set of five differential equations, $r = 0, 1, 2, 3, 4$, for γ_r^n. It should be remarked that the d_r^n (or e_{rk}^n) are known in terms of $f^{(0)} \ldots f^{(n-1)}$ only. The determinant of a_{rs} can be easily computed and is not zero unless $f^{(0)}$ is identically zero. We conclude that $\gamma_r^n(\boldsymbol{x}, t)$ [and therefore $f^{(n)}(\xi, \boldsymbol{x}, t)$] is uniquely determined by the solution of the system (23.23) provided that initial values $\gamma_r^n(\boldsymbol{x}, 0)$ are given.

Now let us introduce the notation

$$\varrho_r^n = \int \psi_r f^{(n)} d\xi \tag{23.25}$$

as well as

$$\varrho_r = \int \psi_r f \, d\xi . \tag{23.26}$$

We have

$$\varrho_r = \sum_{n=0}^{\infty} \varepsilon^n \varrho_r^n \tag{23.27}$$

and also

$$\left.\begin{aligned} \varrho_0 &= \varrho , \\ \varrho_i &= \varrho \, u_i , \qquad i = 1, 2, 3, \\ \varrho_4 &= 2\varrho \left(e + \tfrac{1}{2} u^2\right) . \end{aligned}\right\} \tag{23.28}$$

From (23.19) and (23.20), we find that

$$\varrho_r^n = a_{rs} \gamma_r^n ; \tag{23.29}$$

since the determinant of a_{rs} does not vanish, we can use ϱ_r^n or γ_r^n interchangeably [Eq. (23.24) gives the time derivative of ϱ_r^n explicitly]. In particular we conclude that a unique solution exists to the formal Hilbert expansion up to the n'th step provided that we assign initial values to *all* the quantities $\varrho_r^0(\boldsymbol{x}), \varrho_r^1(\boldsymbol{x}) \ldots \varrho_r^n(\boldsymbol{x})$. This is, at first glance, in contrast to a fluid dynamical formulation in which only five quantities, namely, $\varrho_r(\boldsymbol{x})$, can be given initial values. Nonetheless, these two formulations are compatible. What we have shown is that the functions $f^{(n)}(\boldsymbol{\xi}, \boldsymbol{x}, t)$ are uniquely determined by the initial values $\varrho_r^n(\boldsymbol{x}, 0)$. Alternatively, we have found that the power series in ε,

$$f(\boldsymbol{\xi}, \boldsymbol{x}, t; \varepsilon) = \sum_{n=0}^{\infty} \varepsilon^n f^{(n)}(\boldsymbol{\xi}, \boldsymbol{x}, t), \tag{23.30}$$

is uniquely determined by the values of

$$\varrho_r(\boldsymbol{x}, 0; \varepsilon) = \sum_{n=0}^{\infty} \varepsilon^n \varrho_r^n(\boldsymbol{x}, 0). \tag{23.31}$$

What we wish to show is that for a *fixed* value of ε, $f(\boldsymbol{\xi}, \boldsymbol{x}, t; \varepsilon)$ is uniquely determined by the initial values $\varrho_r(\boldsymbol{x}, 0; \varepsilon)$. We assume that the series (23.30) and (23.31) converge or otherwise represent a one-parameter family of solutions of the Boltzmann equation. Now consider a Boltzmann equation in which the parameter μ appears instead of ε. If we follow the procedure just described, we obtain a unique solution to the equation

$$\dot{F} = \frac{1}{\mu} J(F, F) \tag{23.32}$$

when we impose the initial conditions

$$\left.\begin{aligned} \int F^{(0)} \psi_r \, d\boldsymbol{\xi} &= \varrho_r(\boldsymbol{x}, 0; \varepsilon), \\ \int F^{(n)} \psi_r \, d\boldsymbol{\xi} &= 0, \qquad n > 0. \end{aligned}\right\} \tag{23.33}$$

This solution is a function of both parameters, $F(\boldsymbol{\xi}, \boldsymbol{x}, t; \varepsilon, \mu)$. Now consider the function $F(\boldsymbol{\xi}, \boldsymbol{x}, t; \varepsilon, \varepsilon)$ for any *fixed* value of ε. As the solution of Eq. (23.32) with initial values (23.33) it was constructed with a knowledge of only $\varrho_r(\boldsymbol{x}, 0; \varepsilon)$ for the given value of ε (μ was the expansion parameter). On the other hand, expanding this function with respect to ε we see that it has the same initial values $\varrho_r^n(\boldsymbol{x}, 0)$ as the original function $f(\boldsymbol{\xi}, \boldsymbol{x}, t; \varepsilon)$ and is therefore identical to it. This proves Hilbert's uniqueness theorem. Specifically, we see that the values of $\varrho_r(\boldsymbol{x}, 0)$ can be distributed more or less arbitrarily among the initial values $\varrho_r^n(\boldsymbol{x}, 0)$.

It is evident that this uniqueness theorem cannot hold for the Boltzmann equation in all generality, for we have seen in Chap. III that, in general, the distribution function itself must be specified initially. What seems to be true is that the class of solutions of the Boltzmann equation which can be represented by *power series* in ε is very restricted, and in this class there is a one-to-one correspondence between solutions $f(\boldsymbol{\xi}, \boldsymbol{x}, t)$ and initial values $\varrho_r(\boldsymbol{x}, 0)$. We shall refer to this as the *Hilbert class* of solutions or the class of *normal solutions*. We see the far-reaching consequences of the apparently innocuous assumption that f could be expanded in a power series. Now, if the solution $f(\boldsymbol{\xi}, \boldsymbol{x}, t)$ is determined by $\varrho_r(\boldsymbol{x}, 0)$, then certainly $f(\boldsymbol{\xi}, \boldsymbol{x}, 0)$ is determined by $\varrho_r(\boldsymbol{x}, 0)$. Clearly the time $t = 0$ cannot be special. Therefore, in the Hilbert class there is a unique correspondence between a distribution function $f(\boldsymbol{\xi}, \boldsymbol{x})$ and its five moments $\varrho_r(\boldsymbol{x})$. More precisely, there is a family of functions $f_H(\boldsymbol{\xi}, \boldsymbol{x})$, one to each choice of the set of functions $\varrho_r(\boldsymbol{x})$, with the property that if an f_H is taken as an initial value for the Boltzmann equation, the resulting solution will remain from then on in the class f_H.

If $f(\boldsymbol{\xi}, \boldsymbol{x}, t)$ is determined by $\varrho_r(\boldsymbol{x}, 0)$, then so is $\varrho_r(\boldsymbol{x}, t) = \int \psi_r f(\boldsymbol{\xi}, \boldsymbol{x}, t)\, d\boldsymbol{\xi}$. We should like to be able to determine $\varrho_r(\boldsymbol{x}, t)$ directly from $\varrho_r(\boldsymbol{x}, 0)$ without use of the intermediary, $f(\boldsymbol{\xi}, \boldsymbol{x}, t)$—this would then be fluid dynamics. But if $f(\boldsymbol{\xi}, \boldsymbol{x})$ is determined as a functional of $\varrho_r(\boldsymbol{x})$, it must follow that the higher moments $p_{ij}(\boldsymbol{x})$ and $q_i(\boldsymbol{x})$ (see Sect. 17) are also determined as functionals of $\varrho_r(\boldsymbol{x})$. If they can be evaluated, then the conservation equations become a determined system and provide the variation of $\varrho_r(\boldsymbol{x})$ in time directly, and the Boltzmann equation has been eliminated. This will be done in the next section and again in Sect. 25.

The question of how many boundary conditions can be taken by a normal solution is much more complex than the question of initial conditions. The parameters in $f^{(0)}$ satisfy the Euler equations for which the answer is more or less known but is rather complicated depending on whether the flow is rotational or irrotational or whether it is subsonic or supersonic or transonic, etc. The equations for γ_r^n are linear and have the same characteristics (i.e. sound speeds) as the Euler equations for $\varrho^{(0)}$, $\boldsymbol{u}^{(0)}$, and $T^{(0)}$. They can therefore take the same number and type of boundary conditions. Thus one can impose the same number of boundary conditions on each of $f^{(0)}, f^{(1)}, \ldots$. If these sets of boundary conditions are additive, e.g. if they are linear in the ϱ_r^n, then, just as in the case of initial values, they amount to a *single* set of boundary conditions on ϱ_r itself. The duplication of boundary values applied to $f^{(0)}, f^{(1)}, \ldots$ in this case amounts to allowing the boundary condition to depend on ε. On the other hand, the situation is clouded by the fact that the equations for γ_r^n are inhomogeneous. Also, there is an added flexibility in most steady state problems because the solution of the Euler equations is not uniquely defined by the specification of boundary conditions alone. These matters will be returned to in the next section and again in Sect. 33.

24. Continuation of the Hilbert theory [*20*]. In the previous section, we were led to the conclusion that, for solutions which are expandable in power series, it should be possible to express the stress tensor, p_{ij}, and the heat flow vector, q_i, in terms of ϱ_r, i.e., in terms of ϱ, $\boldsymbol{u}$, and T.

First we recall that there is no loss of generality if we apply the entire initial values, ϱ_r, on ϱ_r^0 alone,

$$\left.\begin{aligned} \varrho_r^0(\boldsymbol{x}, 0) &= \varrho_r(\boldsymbol{x}), \\ \varrho_r^n(\boldsymbol{x}, 0) &= 0, \quad n > 0. \end{aligned}\right\} \tag{24.1}$$

This is permissible because of the uniqueness theorem. It is important to note that, despite the fact that $\varrho_r^0 = \varrho_r$ initially (which means that the parameters in the locally Maxwellian $f^{(0)}$ have been identified with those of the initial f), this will not remain true as time progresses. The differential equations for ϱ_r^n are inhomogeneous and $\varrho_r^n(\boldsymbol{x}, t)$ will deviate from zero as time progresses; consequently, ϱ_r^0 will no longer equal ϱ_r.

In order to find $f^{(1)}$ we must evaluate $\dot{f}^{(0)}$. Since $f^{(0)}$ is a function of $\boldsymbol{x}$ and t only in its dependence on ϱ_r^0, we have

$$\left.\begin{aligned} \dot{f}^{(0)} &= \frac{\partial f^{(0)}}{\partial t} + \xi_j \frac{\partial f^{(0)}}{\partial x_j} \\ &= \frac{\partial f^{(0)}}{\partial \varrho_r^0}\left(\frac{\partial \varrho_r^0}{\partial t} + \xi_j \frac{\partial \varrho_r^0}{\partial x_j}\right). \end{aligned}\right\} \tag{24.2}$$

The distribution $f(\xi, x)$ is a functional of $\varrho_r(x)$ which does not involve the time. Therefore, in solving for $f^{(1)}(\xi, x)$ (for which we need $\dot f^{(0)}$), we are free to set $t=0$. This allows us to replace ϱ_r^0 in (24.2) by ϱ_r and $\partial\varrho_r^0/\partial x_j$ by $\partial\varrho_r/\partial x_j$; it does *not* permit the replacement of the *time derivative* $\partial\varrho_r^0/\partial t$ by $\partial\varrho_r/\partial t$. But we know that ϱ_r^0 satisfies the Euler equations. From them we can evaluate $\partial\varrho_r^0/\partial t$ in terms of space derivatives of ϱ_r^0, and the latter are the same as for ϱ_r at $t=0$. In this way we evaluate $\dot f^{(0)}$ explicitly at $t=0$ in terms of only space derivatives of ϱ_r. A simple calculation yields the result,

$$\dot f^{(0)}|_{t=0} = f^{(0)}\left\{\frac{1}{2T}\frac{\partial T}{\partial x_i}c_i\left(\frac{c^2}{RT}-5\right)+\frac{1}{RT}\frac{\partial u_i}{\partial x_j}\left(c_i c_j-\frac{1}{3}c^2\delta_{ij}\right)\right\}. \tag{24.3}$$

Since x and t are merely parameters in the integral equation, and it is a linear equation, we see at once that the solution, $f^{(1)}$, is a linear homogeneous function of the parameters $\frac{1}{T}\frac{\partial T}{\partial x_i}$ and $\frac{1}{T}\frac{\partial u_i}{\partial x_j}$. Consideration of parity of the polynomials in c which multiply these two coefficients shows that $p_{ij}^{(1)}=\int c_i c_j f^{(1)}\,d\xi$ will be a linear homogeneous function of $\partial u_i/\partial x_j$ alone while $q_i^{(1)}=\int\frac{1}{2}c_i c^2 f^{(1)}\,d\xi$ will be proportional to $\partial T/\partial x_i$. These are the Navier-Stokes relations. The computation of $f^{(1)}$ as a function of ξ, i.e., the evaluation of the factors which multiply the coefficients $\frac{1}{T}\frac{\partial T}{\partial x_i}$ and $\frac{1}{T}\frac{\partial u_i}{\partial x_j}$, and from these the computation of the coefficients of viscosity and of heat conduction, is a formidable task and is considered in detail by L. WALDMANN in a companion article in this volume[1].

Now let us turn to the general case. It is crucial that we do not set $t=0$ too soon. Suppose that $f^{(0)}\ldots f^{(n-1)}$ have been evaluated in terms of $\varrho_r^{(0)}\ldots\varrho_r^{(n-1)}$ and their space derivatives. The inhomogeneous term in the integral equation for $f^{(n)}$, (23.8), involves all of these and time derivatives as well (arising from $\dot f^{(n-1)}$). The time derivatives are replaced by space derivatives by use of the differential equations for the γ_r^n (which are equivalent to ϱ_r^n). After solving for $f^{(n)}$, we can set $t=0$ which leaves us with only space derivatives of $\varrho_r^0=\varrho_r$ ($\varrho_r^n=0$ initially for $n>0$). But, if we are to continue to find $f^{(n+1)}$, then we must not set $t=0$ yet but must keep the original expression in terms of all the ϱ_r^n since further time derivatives will occur.

The final result is that $f^{(n)}(\xi, x)$ as a functional of $\varrho_r(x)$ is a point function of up to n-th order space derivatives of $\varrho_r(x)$. Since the γ_r^n all vanish at $t=0$, we have

$$\left.\begin{aligned} p_{ij}^n &= \int c_i c_j \bar f^{(n)}\,d\xi,\\ q_i^n &= \int \tfrac{1}{2} c_i c^2 \bar f^{(n)}\,d\xi. \end{aligned}\right\} \tag{24.4}$$

25. The Chapman-Enskog method. The standard method of evaluating stresses and heat flow from the Boltzmann equation is that of CHAPMAN and ENSKOG [*16*], [*10*], [*8*]. What we shall discuss here is the abstract formal structure of this technique (following ENSKOG) in particular in its relation to the Hilbert expansion.

Motivated by the Hilbert uniqueness theorem and the observation that f depends not only on ϱ_r but on its derivatives we postulate the following form of f,

$$f = f(\xi, \varrho, \nabla\varrho, \varepsilon) \tag{25.1}$$

where

$$\varrho = \int \psi f\,d\xi \tag{25.2}$$

[1] See also [*29*], [*42*], and [*12*] for alternative methods.

is assumed to hold identically. We are using the abbreviated notation ϱ for ϱ_r, ψ for ψ_r, and $\nabla\varrho$ for $(\nabla\varrho, \nabla^2\varrho, \ldots)$. We also postulate that $\partial\varrho/\partial t$ takes the form

$$\frac{\partial \varrho}{\partial t} = \Phi(\varrho, \nabla\varrho, \varepsilon). \tag{25.3}$$

This is an abstract form of the conservation equations. Space and time do not enter explicitly in any of the arguments, but only implicitly through ϱ and $\nabla\varrho$. Finally we assume that the dependence on ε is as a power series; no claim is made as to analyticity (i.e. convergence) or even asymptotic representation.

From (25.1),

$$f = \sum_{n=0}^{\infty} \varepsilon^n f^{(n)}(\xi, \varrho, \nabla\varrho) \tag{25.4}$$

and from (25.3),

$$\frac{\partial \varrho}{\partial t} = \sum_{n=0}^{\infty} \varepsilon^n \Phi^{(n)}(\varrho, \nabla\varrho). \tag{25.5}$$

Differentiating (25.1),

$$\frac{\partial f}{\partial t} = \frac{\partial f}{\partial \varrho}\frac{\partial \varrho}{\partial t} + \frac{\partial f}{\partial \nabla\varrho}\frac{\partial \nabla\varrho}{\partial t} \tag{25.6}$$

and inserting (25.4) and (25.5),

$$\frac{\partial f}{\partial t} = \sum_{r,s=0}^{\infty} \varepsilon^{r+s}\left(\frac{\partial f^{(r)}}{\partial \varrho}\Phi^{(s)} + \frac{\partial f^{(r)}}{\partial \nabla\varrho}\nabla\Phi^{(s)}\right). \tag{25.7}$$

Inserting this into the Boltzmann equation and equating coefficients of ε^n yields

$$\sum_{r+s=n-1}\left(\frac{\partial f^{(r)}}{\partial \varrho}\Phi^{(s)} + \frac{\partial f^{(r)}}{\partial \nabla\varrho}\nabla\Phi^{(s)}\right) + \xi_j\frac{\partial f^{(n-1)}}{\partial x_j} = \sum_{r+s=n} J(f^{(r)}, f^{(s)}). \tag{25.8}$$

This is a sequence of integral equations for $f^{(0)}, f^{(1)}, \ldots$ which looks similar to (23.7) in the Hilbert expansion. In fact, they are identical provided that all the time derivatives in (23.7) are eliminated by use of the differential equations (23.23) as described in Sect. 24. To confirm this directly is an intricate task because of the different formalism. It is easier to do this by verifying that the solution just obtained is of HILBERT's type and then identifying the two solutions by application of the uniqueness theorem. We have a certain amount of freedom just as in the Hilbert expansion, and we are free to set

$$\left.\begin{aligned} &\int \psi f^{(0)}\, d\xi = \varrho, \\ &\int \psi f^{(n)}\, d\xi = 0, \quad n > 0. \end{aligned}\right\} \tag{25.9}$$

These are not initial conditions; they are constraints which are applied for all time. From this point on the procedure is unique. Using (25.9) and recalling that ϱ and $\nabla\varrho$ are independent variables in the various differentiations of $f^{(n)}$, i.e.

$$\left.\begin{aligned} &\int \psi_r \frac{\partial f^{(0)}}{\partial \varrho_s}\, d\xi = \frac{\partial \varrho_r}{\partial \varrho_s} = \delta_{rs}, \\ &\int \psi_r \frac{\partial f^{(n)}}{\partial \varrho_s}\, d\xi = 0, \\ &\int \psi_r \frac{\partial f^{(n)}}{\partial \nabla\varrho}\, d\xi = 0, \end{aligned}\right\} \tag{25.10}$$

we obtain the following compatibility conditions on the integral equations (25.8),

$$\Phi_r^{(n-1)} + \frac{\partial}{\partial x_j}\int \psi_r \xi_j f^{(n-1)}\, d\xi = 0. \tag{25.11}$$

The conditions (25.9) guarantee that the solution of the integral equation for $f^{(n)}$ is unique. The undetermined "coefficients" in the expansions (25.4) and (25.5) have been determined, the $f^{(n)}$ by solving integral equations (25.8) and the $\Phi^{(n)}$ as in (25.11). The result is a solution of HILBERT's type, i.e. expandable in power series in ε, and the initial values are the same, (25.9), as for the series computed in Sect. 24 by elimination of time derivatives of γ_r^n. The results are therefore identical.

The Enskog procedure is sometimes presented as a completely *ad hoc* recipe for juggling terms in an expansion $f = f^{(0)} + f^{(1)} + \cdots$ in order to get results which have the desired form [*11*]. Although the expansion in powers of ε just given is not any more justified on mathematical grounds at the present time, it is somewhat more concise.

We have described two distinct procedures for obtaining normal solutions of the Boltzmann equation. They are formally equivalent but in actuality exceedingly different. The original procedure described in Sect. 23 (HILBERT), was to specify initial values for ϱ_r^0 (for simplicity equal to ϱ_r) and for each set ϱ_r^n, $n = 1, 2, \ldots$. The variation in time is given by the Euler equations for ϱ_r^0 and the linear equations (23.24) for ϱ_r^n, $n > 0$. The second procedure (CHAPMAN-ENSKOG) of which two versions were presented, one in Sect. 24 and the second just preceeding, was to find the successive approximations to p_{ij} and q_i and insert them into the conservation equations which then yield the variation in time. These two procedures can be easily compared. They differ in the way in which the infinite expansion $\sum_0^\infty f^{(n)}$ is truncated at a finite n. The sum of the first n compatibility equations (23.24) together with the Euler equations takes the form

$$\frac{\partial}{\partial t}\left(\sum_{m=0}^{n} \varrho_r^m\right) + \frac{\partial}{\partial x_k}\left(\int \xi_k \psi_r f^{(0)}\, d\xi + \sum_{m=1}^{n} (b_{rs}^k \gamma_s^m + e_{rk}^m)\right) = 0. \tag{25.12}$$

At $t=0$, with $\gamma_s^n = 0$ and with e_{rk}^m explicitly written out, this becomes

$$\frac{\partial}{\partial t}\left(\sum_{m=0}^{n} \varrho_r^m\right) + \frac{\partial}{\partial x_k}\left(\int \xi_k \psi_r f^{(0)}\, d\xi + \sum_{m=1}^{n} \int \xi_k \psi_r \bar{f}^{(m)}\, d\xi\right) = 0. \tag{25.13}$$

From (23.20) and at $t=0$,

$$\left.\begin{aligned} &\int \xi_k \psi_0 \bar{f}^{(n)}\, d\xi = 0, \\ &\int \xi_k \psi_j \bar{f}^{(n)}\, d\xi = p_{jk}^n, \qquad j = 1, 2, 3, \\ &\int \xi_k \psi_4 \bar{f}^{(n)}\, d\xi = 2\,(u_j p_{jk}^n + q_k^n). \end{aligned}\right\} \tag{25.14}$$

We see that the sum in the second term of (25.13) separates out the successive approximations to the stresses and heat flow (the interpretation is not so simple for $t > 0$). Thus the Chapman-Enskog procedure is to take Eq. (25.12) (in which one is permitted to replace ϱ^0 by ϱ at $t=0$ by the Hilbert method) and replace ϱ^0 by ϱ *for all t* obtaining Eq. (25.13). The two procedures agree at $t=0$ and, presumably, for some small time t which may become larger as $n \to \infty$. The extent of this difference in treatment is indicated by the question of boundary conditions. We saw in Sect. 23 that, although the situation is complicated, in many cases we can expect the number of boundary conditions to remain fixed in the Hilbert scheme as $n \to \infty$ (this is borne out by an explicit example of WANG CHANG and UHLENBECK; see Sect. 33). Although the question of the correct number of boundary conditions for the sequence of fluid equations in the Chapman-Enskog method is again very complex, one should not be surprised at the possibility of imposing an increasing number of boundary conditions as $n\to\infty$. The precise meaning of normal solutions and of the various interpretations is not entirely clear at this time, but a partial clarification will be attempted in Sect. 26.

26. Asymptotic behavior[1]. It is convenient to take the integral form (19.12) but with the parameter ε inserted. We write the Boltzmann equation in the form

$$\frac{\partial f}{\partial t}+\boldsymbol{\xi}\cdot\frac{\partial f}{\partial \boldsymbol{x}}+\frac{1}{\varepsilon}\nu f=\frac{1}{\varepsilon}L \tag{26.1}$$

[cf. Eq. (19.9)] and then obtain

$$\left.\begin{aligned} f(\boldsymbol{\xi},\boldsymbol{x},t)&=f(\boldsymbol{\xi},\boldsymbol{x}-\boldsymbol{\xi}t,0)\exp\left\{-\frac{1}{\varepsilon}\int_0^t\nu(\boldsymbol{\xi},\boldsymbol{x}-\boldsymbol{\xi}(t-s),s)\,ds\right\}+\\ &+\int_0^t g(\boldsymbol{\xi},\boldsymbol{x}-\boldsymbol{\xi}(t-\tau),\tau)\,\mu(\boldsymbol{\xi},\boldsymbol{x}-\boldsymbol{\xi}(t-\tau),\tau)\,d\tau\end{aligned}\right\} \tag{26.2}$$

where we have introduced the functions μ and g of (19.16) and (19.18),

$$\left.\begin{aligned}\mu(\boldsymbol{\xi},\boldsymbol{x}-\boldsymbol{\xi}(t-\tau),\tau)&=\frac{1}{\varepsilon}\nu(\boldsymbol{\xi},\boldsymbol{x}-\boldsymbol{\xi}(t-\tau),\tau)\times\\ &\times\exp\left\{-\frac{1}{\varepsilon}\int_\tau^t\nu(\boldsymbol{\xi},\boldsymbol{x}-\boldsymbol{\xi}(t-s),s)\,ds\right\},\end{aligned}\right\} \tag{26.3}$$

$$g(\boldsymbol{\xi},\boldsymbol{x},t)=\frac{L(\boldsymbol{\xi},\boldsymbol{x},t)}{\nu(\boldsymbol{\xi},\boldsymbol{x},t)}. \tag{26.4}$$

The value of this particular form is that it expresses f as a mean value of the initial f and of the function g with combined weight equal to one. For small ε the weight is concentrated mostly on g, and on g it is concentrated near $\tau=t$. In the limit, we have

$$f(\boldsymbol{\xi},\boldsymbol{x},t)=g(\boldsymbol{\xi},\boldsymbol{x},t). \tag{26.5}$$

This can be written $\nu f=L$ which is the same as $J=0$; in the limit f is locally Maxwellian. This is a more complicated way of observing what has already been found by more direct means. However, the integral form (26.2) is particularly well suited for asymptotic approximation when ε is small. A standard technique is to integrate by parts, differentiating the function whose mean value is desired and integrating the peaked weight function. This gives an expansion in powers of ε as well as additional exponential terms. Using an abbreviated notation, we obtain

$$\left.\begin{aligned} f&=f_0\,e^{-\frac{1}{\varepsilon}\bar\nu}+g\,e^{-\frac{1}{\varepsilon}\bar\nu(\tau)}\Big|_0^t-\\ &-\frac{\varepsilon}{\nu}\frac{dg}{d\tau}e^{-\frac{1}{\varepsilon}\bar\nu(\tau)}\Big|_0^t+\frac{\varepsilon}{\nu}\frac{d}{d\tau}\left(\frac{\varepsilon}{\nu}\frac{dg}{d\tau}\right)e^{-\frac{1}{\varepsilon}\bar\nu(\tau)}\Big|_0^t+\cdots.\end{aligned}\right\} \tag{26.6}$$

We have written f_0 for $f(\boldsymbol{\xi},\boldsymbol{x}-\boldsymbol{\xi}t,0)$, g for $g(\boldsymbol{\xi},\boldsymbol{x}-\boldsymbol{\xi}(t-\tau),\tau)$, ν for $\nu(\boldsymbol{\xi},\boldsymbol{x}-\boldsymbol{\xi}(t-\tau),\tau)$, $\bar\nu(\tau)$ for $\int_\tau^t\nu(\boldsymbol{\xi},\boldsymbol{x}-\boldsymbol{\xi}(t-s),s)\,ds$, $\bar\nu$ for $\bar\nu(0)$, and $\frac{d}{d\tau}$ for $\frac{\partial}{\partial\tau}+\boldsymbol{\xi}\cdot\frac{\partial}{\partial\boldsymbol{x}}$. At first let us drop the exponential terms and keep terms up to first order in ε only,

$$f=g-\frac{\varepsilon}{\nu}\dot g \tag{26.7}$$

[1] H. GRAD, previously unpublished; see also ref. [17].

(we recall that $\dot{g} = \partial g/\partial t + \boldsymbol{\xi} \cdot \partial g/\partial \boldsymbol{x}$). Next we write

$$f = f^{(0)} + \varepsilon f^{(1)} \tag{26.8}$$

where $f^{(0)}$ is locally Maxwellian, according to (26.5). To be consistent up to the order retained, we replace g by $g^{(0)} = f^{(0)}$ in the term multiplied by ε in (26.7), but the zero order g in (26.7) is expanded to two terms. As one might expect, we get

$$2J(f^{(0)}, f^{(1)}) = \dot{f}^{(0)}, \tag{26.9}$$

which is exactly Eq. (23.6). It is quite clear that, as long as we retain only powers of ε, we shall obtain the same results which were obtained more elegantly in Sects. 23 to 25.

The advantage of this procedure over the expansion methods already described is that it makes it possible to study the validity of the power series expansion and also indicates the possibility of improvement. If the error term is kept after integrating by parts a certain number of times, one can in principle estimate the accuracy of the expansion up to this order. The difficulty is that in the absence of very complete existence theorems one can only estimate the error by making *ad hoc* assumptions about the properties that solutions are likely to have. However, even qualitative information is valuable for the insight it gives in this difficult subject. Consider a definite problem in which we are given fixed initial values and boundary values for $f(\boldsymbol{\xi}, \boldsymbol{x})$ independent of ε. Also suppose that in the neighborhood of a given point $(\boldsymbol{x}, t)$, the solution $f(\boldsymbol{\xi}, \boldsymbol{x}, t; \varepsilon)$ converges as $\varepsilon \to 0$. In this neighborhood, f will become locally Maxwellian, and the fluid dynamical Euler equations become a good approximation. If, as $\varepsilon \to 0$, we move closer to $t = 0$, keeping t on the order of ε, this conclusion will be invalidated because of the exponential term $f_0\, e^{-\bar{\nu}/\varepsilon}$. In other words, there is an initial boundary layer of thickness ε (on the order of the collision time). In exactly the same way, if we had taken the more complicated integral form (19.12) which includes boundary conditions, we would have obtained a spatial boundary layer of thickness ε (on the order of the mean-free-path).

We made the tacit assumption that $f(\boldsymbol{\xi}, \boldsymbol{x}, t; \varepsilon)$ converges at a fixed interior point $(\boldsymbol{x}, t)$ which is not on a boundary (for verbal simplicity we consider $t = 0$ as part of the boundary of the space-time manifold). This is too difficult to prove. On the other hand, it is easy to find cases where it can be disproved. By assuming convergence we obtain the Euler equations. It is well known that solutions of these equations do not always exist in the large because of the appearance of shocks. It is conjectured (this is, in fact, the mainstay of compressible gas dynamics) that the correct procedure is to piece together solutions of the Euler equations at judiciously chosen shock discontinuities. "Judicious" here refers to the present state of the art; one suspects that the appearance of shocks is uniquely defined. The behavior of the shock as well as the flow on each side is determined by the Euler equations combined with the shock jump conditions which are merely the expression of the conservation of mass, momentum, and energy across the shock. We conclude that there exist internal "boundary" layers in addition to the ones previously considered. Their location is determined by the Euler equations alone (making the conventional fluid dynamical conjecture), i.e., without reference to the Boltzmann equation or to ε. Furthermore, studies of shock structure show that the thickness of a shock layer is on the order of ε (the mean-free-path) just as for the other boundary layers.

Our modified conjecture is as follows. Given a fixed $(\boldsymbol{x}, t)$ which is not on a boundary (either of $\boldsymbol{x}$ or of t) and not on a shock (this is, in principle, determinable by solution of the Euler equations), then we may expect $f(\boldsymbol{\xi}, \boldsymbol{x}, t; \varepsilon)$

to converge in the neighborhood of $(\boldsymbol{x}, t)$, and the limiting behavior in this neighborhood will be given by a solution of the Euler equations. There is one more difficulty that must be mentioned. Solution of the Euler equations requires initial values and boundary values. The initial values $f(\boldsymbol{\xi}, \boldsymbol{x})$ surely determine initial values for $\varrho(\boldsymbol{x})$, $\boldsymbol{u}(\boldsymbol{x})$, $T(\boldsymbol{x})$. However, a boundary layer in which the Euler equations are not satisfied intervenes between $t=0$ and the domain where the Euler equations take over. It must be verified that the "asymptotic initial values" at $t \sim \varepsilon$ approach the actual initial values at $t=0$. This is almost certainly true for the problem just discussed, namely, the Euler equations considered as a limiting representation of the Boltzmann equation. Assuming this, we can make a conjecture in the large. As $\varepsilon \to 0$ the solution of BOLTZMANN's equation converges to the generalized solution (with shocks) of the Euler equations with corresponding initial and boundary values. The situation becomes more complex when we turn to the question of a complete normal solution expansion as an asymptotic representation of an actual solution. In order to use the actual initial values of ϱ_r in a normal solution expansion up to ε^n, we must verify that these initial values are unaltered up to the same order by the intervening boundary layer. Even if the normal solution is a legitimate local asymptotic approximation to a true solution, it is possible to lose track of the initial values on passing through the boundary layer. There are two paths to follow. First, there will be no initial boundary layer if the initial $f(\boldsymbol{\xi}, \boldsymbol{x})$ is a normal f, i.e., one in the Hilbert class f_H. We can say that the theory applies in these special cases. Adherence to this restriction would considerably reduce the importance of normal solutions. The alternative is to investigate the initial layer and supplement the asymptotic differential equations with asymptotic initial values computed somehow from the actual initial values. This is yet to be done in any sort of generality[1].

The question of matching boundary conditions for normal solutions is more complex than that of initial conditions. First of all, it is not known in advance how many boundary conditions can be fitted to a given degree of approximation (cf. Sect. 25). Second, the two methods of expanding normal solutions (HILBERT and CHAPMAN-ENSKOG) may differ in this respect. In general we can expect to lose some of the ability that general solutions of the Boltzmann equation have in matching boundary conditions. In this case it becomes necessary to compute asymptotic boundary values from the actual boundary values, just as for initial values. An explicit example of this loss of boundary conditions due to WANG CHANG and UHLENBECK is discussed in Sect. 33.

A general picture of the internal structure of the initial layer might be obtained as follows. We return to (26.6) and this time retain some of the exponential terms in addition to terms of the first order in ε. We have

$$f = g - \frac{\varepsilon}{\nu} \dot{g} (1 - e^{-\bar{\nu}/\varepsilon}) + (f_0 - g_0)\, e^{-\bar{\nu}/\varepsilon} \tag{26.10}$$

where

$$g_0 = g(\boldsymbol{\xi}, \boldsymbol{x} - \boldsymbol{\xi} t, 0). \tag{26.11}$$

This can be put in the form,

$$J = J_0\, e^{-\bar{\nu}/\varepsilon} + \varepsilon \dot{g} (1 - e^{-\bar{\nu}/\varepsilon}) \tag{26.12}$$

where

$$J_0 = J(f_0, f_0). \tag{26.13}$$

[1] A special initial value problem for which the explicit solution of the Boltzmann equation is known has been found by TRUESDELL [64], and in this case one can state explicitly what is the correct modification of the initial values.

Instead of (26.8), we write

$$f = f^{(0)} + f^{(1)}$$

and assume that $f^{(1)}$ is small compared to $f^{(0)}$ leaving open in what way it is small. On the right side of (26.12) we replace $\dot{g}$ by $\dot{g}^{(0)} = \dot{f}^{(0)}$ and $\bar{\nu}$ by $\bar{\nu}^{(0)}$; we obtain

$$2J(f^{(0)}, f^{(1)}) = J(f_0, f_0) \exp\left(-\frac{1}{\varepsilon}\bar{\nu}^{(0)}\right) + \varepsilon \dot{f}^{(0)}\left(1 - \exp\left(-\frac{1}{\varepsilon}\bar{\nu}^{(0)}\right)\right). \tag{26.14}$$

It is now clear why the term $\varepsilon \dot{g} e^{-\bar{\nu}/\varepsilon}$ was retained even though it might be considered higher order than either ε or $e^{-\frac{1}{\varepsilon}\bar{\nu}}$ separately; if we let $t \to 0$, the solution can be made to take the correct initial values, f_0. Since the integral equation (26.14) is linear, the solution will be a sum of terms depending on $\dot{f}^{(0)}$ (i.e., Navier-Stokes terms) and on the initial values, $J(f_0, f_0)$. The dependence on the exponential, however, is very complicated. One reason is that $\bar{\nu}^{(0)}$ is a function of $\boldsymbol{\xi}$, which is the argument of the integral equation. This dependence on $\boldsymbol{\xi}$ cannot be eliminated by taking a Maxwellian molecule (in which case ν is independent of $\boldsymbol{\xi}$), because $\boldsymbol{\xi}$ also enters in the argument $\boldsymbol{x} - \boldsymbol{\xi} t$. The solution will therefore not be of simple exponential type; nevertheless, it can be expected to behave in about the same way, qualitatively. For a solution of (26.14) to exist, the right side must satisfy the compatibility conditions, (23.14). These will be conditions on the five parameters in $f^{(0)}$, namely, $\varrho_r^{(0)}(\boldsymbol{x}, t)$, which conditions will approximate the Euler equations if the exponentials are small. Unfortunately, we are interested in these equations only when the exponentials are not small.

Qualitatively, one can expect the solution of (26.14) [after the solution of the compatibility conditions for $\varrho_r^{(0)}(\boldsymbol{x}, t)$] to describe the state of affairs in the boundary layer and to give a reasonably accurate transition from the initial value f_0 to the locally Maxwellian $f^{(0)}$ which satisfies the Euler equations. This has not been done. It is quite possible that a direct attack on (26.2) following some of the methods of Chaps. III or V might be simpler. However, from this analysis, it is easy to estimate the size of the boundary layer which turns out to be on the order of $\varepsilon \log(1/\varepsilon)$ instead of ε. We reach this estimate by setting equal the two "small" coefficients ε and $e^{-\bar{\nu}/\varepsilon}$, noting that $\bar{\nu} \sim t\nu$. The distinction between the two thickness estimates, ε and $\varepsilon \log(1/\varepsilon)$, is evident. If we are satisfied with the locally Maxwellian, $f^{(0)}$, as a good approximation, the boundary layer effect becomes negligible after a distance ε. But, if a more refined approximation, $f^{(0)} + \varepsilon f^{(1)}$, is desired, the boundary effect becomes negligible compared to $\varepsilon f^{(1)}$ only at the larger distance $\varepsilon \log(1/\varepsilon)$. Alternatively, the Euler equations are a satisfactory approximation for $t \gg \varepsilon$, whereas the Navier-Stokes refinement is not warranted until $t \gg \varepsilon \log(1/\varepsilon)$.

It is now clear that the so-called *normal* solutions of the Boltzmann equation are probably justifiable in an asymptotic sense only if one is not too close to a boundary layer or a shock layer. Their use in the *slip-flow* regime[1], i.e., near a boundary, is permissible only if one is not interested in the behavior within a layer which may be many [i.e., $\log(1/\varepsilon)$] mean-free-paths wide, and if one can verify that the asymptotic "boundary" conditions which exist beyond such a layer agree with the boundary conditions at the actual wall to the proper order. Using the methods of Chap. V, it may be possible to relate the actual boundary condition at the wall with the boundary values seen by the normal solution somewhat away from the wall. In certain special cases, e.g. near a plane boundary and for small values of the heat flow and stresses, the boundary conditions are

[1] See article by S. A. Schaaf in Vol. IX of this Encyclopedia.

contained implicitly in the already known exact solutions of these problems (cf. Sect. 33).

The possibility of analyzing a shock layer by such methods is a different question. Analytically, the mean value of a function g must be computed [Eq. (26.2)] when both the function g and the weight function steepen simultaneously. Integration by parts is worthless. However, in this case it is easy to see by a simple change of the scale of length that the parameter ε can be made to disappear from the problem entirely[1]. After removing ε, a new parameter measuring the strength of the shock can be inserted, and it enters in almost the same way as did ε. For example, we can take the density ratio across the shock ($\varepsilon = \varrho_1/\varrho_0 - 1$) as the parameter in the expansion (23.4). The Hilbert expansion becomes an expansion in powers of the shock strength. In this example, no initial values or boundary values (except implicitly at infinity) appear in the integral equation (26.2). There is no *a priori* reason to suspect that a power series expansion restricts the totality of solutions of the Boltzmann equation to a special class or even to suspect that the series does not converge. This question is open. On the other hand, it is perfectly possible that there are two *different* asymptotic expansions, one on each side of the shock, and the connection between them may not be determinable by a normal solution analysis but only by a complete analysis of the transition zone. It is frequently found that relating asymptotic expansions which are valid in different regions is a difficult problem.

V. Other methods of solution.

27. Introduction. The theory of normal solutions of the Boltzmann equation was given in Chap. IV. The question of general solutions was looked at from the abstract point of view of existence theory in Chap. III. Here we wish to consider the question of general solutions from a more practical standpoint. There are two more or less distinct lines of attack. One is to obtain new phenomenological equations which generalize the usual fluid dynamical equations in cases where it is no longer permissible to rely on normal solutions. The object is to show the transition from the Boltzmann equation in which a state is given by $f(\boldsymbol{\xi}, \boldsymbol{x})$ to the conventional fluid description in which a state is given by $\varrho(\boldsymbol{x})$, $\boldsymbol{u}(\boldsymbol{x})$, and $T(\boldsymbol{x})$ in sufficient generality to cover a broad class of problems. This might be described as interpolation between the two extreme specifications of a state. This approach was initiated by MAXWELL [*46*] and continued by GRAD [*22*][2] (Sect. 28). The second line is to attempt to solve the Boltzmann equation for the distribution $f(\boldsymbol{\xi}, \boldsymbol{x}, t)$ itself in specific problems. This is of course the ultimate goal. A number of attempts have been made to obtain results in the neighborhood of free flow [*35*], [*58*], [*38*]. More recently, attempts have been made to cover the whole range of values of the mean-free-path. This line has been pursued mainly by WANG CHANG and UHLENBECK [*68*], [*69*] and by MOTT-SMITH [*53*].

The principal tool up to the present time has been some form of expansion with somewhat less attention paid to iterations[3]. With regard to expansion techniques, the distinction between the two lines of attack mentioned above

[1] The parameter ε can be removed by a change of the length scale in any time-independent problem. However, it reappears in the location of the boundary and the problem is essentially the same as in the original variables. The shock problem is unique in that there are no finite boundaries so ε can be legitimately removed.

[2] The intervening period shows an almost exclusive preoccupation with normal solutions.

[3] Ref. [*58*] and [*32*]; also see [*22*] and Sect. 28 for a combination of iteration and expansion.

reduces to the question of how many terms can be carried conveniently; problems with special symmetries or other simplifications allow more complete expansions. There are several expansion techniques which have been used. Two which are in principle similar but in practice are not are expansion of f in the neighborhood of a locally Maxwellian $f^{(0)}$ with parameters ϱ, $\boldsymbol{u}$, T depending on $\boldsymbol{x}$ and t (Sect. 28) or in the neighborhood of an absolute Maxwellian $f^{(0)}$ with constant parameters (Sect. 30). By the choice of a Maxwellian, local or absolute, as a starting point, one has almost no alternative to the use of a Hermite polynomial expansion[1]. The expansion is in the variable $\boldsymbol{\xi}$, and the coefficients are functions of $\boldsymbol{x}$ and t. These coefficients can be considered to be *macroscopic state variables*. This is the interpolation between a fluid state and a molecular state which was described above. If only a few additional Hermite coefficients are taken, this is most naturally interpreted as a macroscopic generalization of fluid dynamics. If many terms in the expansion are taken, it is more appropriate to consider this as a microscopic approximation to f itself, but the line of demarcation is to some extent arbitrary.

Whether expansion about a local or absolute Maxwellian is preferable must be decided individually for each problem. The absolute Maxwellian has the advantage of yielding simpler equations. The local Maxwellian is clearly preferable in cases where the solution happens to be approximately locally Maxwellian but with definitely non-constant $\varrho(\boldsymbol{x})$, $\boldsymbol{u}(\boldsymbol{x})$, $T(\boldsymbol{x})$. The latter situation occurs when the two time scales, collision time vs. viscous decay time, or the two length scales, mean-free-path vs. flow dimensions (see Sect. 22) differ by an order of magnitude or more. The absolute Maxwellian expansion is probably preferable when these parameters have the same order of magnitude. The two expansions are essentially identical for the linearized problem of small perturbations about equilibrium. In the linearized problem a substantially complete solution can be obtained if there are no physical boundaries or if the boundaries are parallel planes (Sect. 33).

Another expansion technique is to split the distribution function $f(\xi_1, \xi_2, \xi_3)$ into two halves, say for $\xi_1 > 0$ and $\xi_1 < 0$, and to expand each half separately over the half space [*28*]. The disadvantage is an evident increase in complexity. The advantage is an improved rate of convergence near a boundary where the two halves of the distribution function (incident vs. emergent molecules) may look entirely different. This type of expansion is touched on in Sect. 31.

With each of the above-mentioned expansions, any reasonable distribution function can be approximated arbitrarily closely. However, in virtue of the expected rapid increase in labor with an increase in the order of approximation, it becomes quite important to look for special features in any given problem before deciding what is the best type of expansion. It should be mentioned that, even though one can approximate arbitrarily closely to a given f by a sequence of approximating functions f_n, there is no guarantee that the solution of the equations satisfied by the f_n will in any way approximate a solution of the Boltzmann equation. All of these methods (at least in their application to the nonlinear Boltzmann equation) are purely formal, and there is as yet no reason to suspect any *qualitative* differences in convergence.

[1] Ref. [*22*]. This Hermite expansion is substantially identical to the subsequently used expansions in Sonine polynomials, Refs. [*68*] and [*53*]. Previously Sonine polynomials had been used, not in the Boltzmann equation itself, but to solve the Fredholm equation which arises in the theory of normal solutions, Ref. [*8*]. The distinction between the two expansions is merely one of a choice of coordinates; a Hermite expansion in rectangular coordinates becomes a product of spherical harmonics and Sonine polynomials when expressed in spherical coordinates.

Another type of expansion which could prove useful in special cases (although it has not been carried very far) is to take as weight function an anisotropic Maxwellian $f^{(0)}$; i.e. include the anisotropy in the exponent of $f^{(0)}$ as well as in the polynomials which follow [44]. This might be valuable in treating strongly anisotropic problems, e.g., a strong shock wave. Carrying this idea further, one could put the whole approximation sequence in the exponent, i.e., expand $\log f$. This has the advantage (not possessed by the other methods) that the approximating functions are all positive but the disadvantage of greatly complicating the derivation of the equations satisfied by the approximating functions.

The general question of boundary conditions for any expansion method is considered in Sect. 29.

Because of the difficulty in obtaining useful results from the Boltzmann equation, various Boltzmann-like equations have at times been proposed. This question is touched on in Sect. 32.

A brief discussion of the relative merits of some of these methods, comparing the solution of special problems concludes the chapter (Sect. 33).

28. Expansion about a locally Maxwellian distribution [22]. We recall that the Maxwellian distribution is given by

$$f^{(0)} = \frac{\varrho}{(2\pi \mathrm{R} T)^{\frac{3}{2}}} e^{-c^2/2\mathrm{R}T} \tag{28.1}$$

where $\boldsymbol{c} = \boldsymbol{\xi} - \boldsymbol{u}$ is the velocity with respect to the mean. As a weight function in an expansion, it is permissible to select the five parameters in $f^{(0)}$ arbitrarily. We choose them to be the local values of the respective moments of the actual distribution function, f. If we expand an arbitrary function f with this $f^{(0)}$ as weight function, five of the expansion parameters will be fixed because of the above identification. The function $f(\boldsymbol{\xi}, \boldsymbol{x}, t)$ is replaced by an infinite set $\varrho(\boldsymbol{x}, t)$, $\boldsymbol{u}(\boldsymbol{x}, t)$, $T(\boldsymbol{x}, t)$ and $a^{(n)}(\boldsymbol{x}, t)$ (to be defined). The expansion is to be substituted for f in the Boltzmann equation and corresponding coefficients equated. The result will be an infinite sequence of differential equations for ϱ, $\boldsymbol{u}$, T and the $a^{(n)}$. There is a single time derivative in each equation. We list the equations in sequence according to the time derivatives, first the conservation equations, $\partial\varrho/\partial t$, $\partial\boldsymbol{u}/\partial t$, $\partial T/\partial t$ and then the equations with $\partial a^{(n)}/\partial t$ in order. These differential equations are first order and quasilinear (i.e. linear in derivatives) and contain terms quadratic in $a^{(n)}$ arising from the collision term. In general no finite set of equations is uncoupled from the rest of the system. To obtain a determined system up to a certain order $a^{(n)}$, we retain the equations up to $\partial a^{(n)}/\partial t$ and in them set $a^{(m)} = 0$, $m > n$, wherever such terms occur. This procedure is formal but of a general type which can be expected to be convergent. Before expanding, it is convenient to introduce the dimensionless velocity

$$\boldsymbol{v} = \frac{\boldsymbol{c}}{\sqrt{\mathrm{R}T}} \tag{28.2}$$

and the dimensionless distribution function

$$g = \frac{(\mathrm{R}T)^{\frac{3}{2}}}{\varrho} f. \tag{28.3}$$

The definitions of ϱ, $\boldsymbol{u}$, and T imply the following conditions on g

$$\left.\begin{aligned} \int g\, d\boldsymbol{v} &= 1, \\ \int \boldsymbol{v}\, g\, d\boldsymbol{v} &= 0, \\ \int v^2 g\, d\boldsymbol{v} &= 3. \end{aligned}\right\} \tag{28.4}$$

The Maxwellian, $g^{(0)}$, is given by

$$g^{(0)} = \frac{1}{(2\pi)^{\frac{3}{2}}} e^{-v^2/2}. \tag{28.5}$$

We expand g in Hermite polynomials,

$$g(\boldsymbol{v}, \boldsymbol{x}, t) = g^{(0)}(\boldsymbol{v}) \sum_{n=0}^{\infty} \frac{1}{n!} a_i^{(n)}(\boldsymbol{x}, t) \mathscr{H}_i^{(n)}(\boldsymbol{v}) \tag{28.6}$$

or, in less abbreviated form,

$$g = g^{(0)} \left(a^{(0)} \mathscr{H}^{(0)} + a_i^{(1)} \mathscr{H}_i^{(1)} + \frac{1}{2!} a_{ij}^{(2)} \mathscr{H}_{ij}^{(2)} + \cdots \right). \tag{28.7}$$

For f we have

$$f = f^{(0)} \sum_{n=0}^{\infty} \frac{1}{n!} a_i^{(n)} \mathscr{H}_i^{(n)}. \tag{28.8}$$

A few words should be said about this notation for Hermite polynomials [*21*]. $\mathscr{H}_i^{(n)}$ is a tensor with n subscripts, $i = (i_1 \ldots i_n)$, as well as a polynomial of n-th degree. The coefficient $a_i^{(n)}$ is also a tensor of order n, and summation over the entire index set i is implied in (28.6). As might be expected, the use of a Cartesian tensor notation (invariant under rotation) is of great help in deriving and interpreting general formulas. The inversion formula for the coefficients $a_i^{(n)}$ is

$$a_i^{(n)} = \int g \mathscr{H}_i^{(n)} d\boldsymbol{v} = \frac{1}{\varrho} \int f \mathscr{H}_i^{(n)} d\boldsymbol{\xi}. \tag{28.9}$$

The first few polynomials are,

$$\left.\begin{aligned}
\mathscr{H}^{(0)} &= 1, \\
\mathscr{H}_i^{(1)} &= v_i, \\
\mathscr{H}_{ij}^{(2)} &= v_i v_j - \delta_{ij}, \\
\mathscr{H}_{ijk}^{(3)} &= v_i v_j v_k - (v_i \delta_{jk} + v_j \delta_{ik} + v_k \delta_{ij}), \\
\mathscr{H}_{ijkl}^{(4)} &= v_i v_j v_k v_l - (v_i v_j \delta_{kl} + v_i v_k \delta_{jl} + v_i v_l \delta_{jk} + \\
&\quad + v_j v_k \delta_{il} + v_j v_l \delta_{ik} + v_k v_l \delta_{ij}) + (\delta_{ij} \delta_{kl} + \delta_{ik} \delta_{jl} + \delta_{il} \delta_{jk}).
\end{aligned}\right\} \tag{28.10}$$

Each component of the tensor $\mathscr{H}_i^{(n)}$ is a product of three ordinary (one variable) Hermite polynomials with total degree equal to n. For a given n, all the polynomials $\mathscr{H}_i^{(n)}$ are independent and orthogonal except in the trivial case of permutation of the subscripts. Specifically, each component of $\mathscr{H}_i^{(n)}$ is orthogonal to each component of $\mathscr{H}_j^{(m)}$ unless $n = m$ and in addition the set of subscripts $i_1 \ldots i_n$ is a permutation of $j_1 \ldots j_n$. As indicated by (28.10), no numerical coefficients other than ± 1 occur in the explicit form of $\mathscr{H}^{(n)}$. Every $a^{(n)}$ is a dimensionless polynomial moment of f by (28.9); the first few, expressed in terms of the elementary moments defined in Sect. 17 are,

$$\left.\begin{aligned}
a^{(0)} &= 1, \\
a_i^{(1)} &= 0, \\
a_{ij}^{(2)} &= p_{ij}/p, \\
a_{ijk}^{(3)} &= S_{ijk}/p\sqrt{RT}, \\
a_{ijkl}^{(4)} &= Q_{ijkl}/p\,RT - \frac{1}{p}(p_{ij}\delta_{kl} + p_{ik}\delta_{jl} + p_{il}\delta_{jk} + \\
&\quad + p_{jk}\delta_{il} + p_{jl}\delta_{ik} + p_{kl}\delta_{ij}) - (\delta_{ij}\delta_{kl} + \delta_{ik}\delta_{jl} + \delta_{il}\delta_{jk}).
\end{aligned}\right\} \tag{28.11}$$

These are combinations of moments which vanish (except for $a^{(0)}$) when f is Maxwellian. By contraction of $a^{(3)}$, we get

$$a_i^{(3)} = a_{irr}^{(3)} = 2q_i/p\sqrt{RT}. \tag{28.12}$$

We see that the first nontrivial Hermite coefficients are dimensionless forms of the stress tensor, p_{ij}, and the heat flow vector, q_i. The coefficients $a^{(2)}$, $a^{(3)}$, and $a^{(4)}$ should be considered as being ultimately equal in physical significance although $a_{ij}^{(2)}$ and $a_i^{(3)}$ have the psychological advantage of familiarity.

We insert the expansion (28.8) into the Boltzmann equation and equate coefficients of $H_i^{(n)}$. The result is an infinite sequence of differential equations in the coefficients $a_i^{(n)}$ [1]. As a sample we write the equation for $a^{(2)}$ [dRT/dt is defined in (28.36)].

$$\left.\begin{aligned}
&\frac{\partial a_{ij}^{(2)}}{\partial t} + u_r\frac{\partial a_{ij}^{(2)}}{\partial x_r} + a_{ir}^{(2)}\frac{\partial u_j}{\partial x_r} + a_{rj}^{(2)}\frac{\partial u_i}{\partial x_r} + (a_{ij}^{(2)} + \delta_{ij})\frac{1}{RT}\frac{dRT}{dt} + \\
&+\sqrt{RT}\frac{\partial a_{ijr}^{(3)}}{\partial x_r} + \frac{\sqrt{RT}}{\varrho}a_{ijr}^{(3)}\frac{\partial\varrho}{\partial x_r} + \frac{3}{2RT}a_{ijr}^{(3)}\frac{\partial RT}{\partial x_r} + \\
&+\frac{\partial u_i}{\partial x_j} + \frac{\partial u_j}{\partial x_i} = J_{ij}^{(2)}.
\end{aligned}\right\} \tag{28.13}$$

The equation in which $\partial a_i^{(n)}/\partial t$ occurs contains a contribution from the collision term which is

$$\left.\begin{aligned}
&J_i^{(n)} = \frac{\varrho}{2}\sum_{r,s=0}^{\infty}\beta_{ijk}^{(nrs)}a_j^{(r)}a_k^{(s)},\\
&i=(i_1\ldots i_n),\qquad j=(j_1\ldots j_r),\qquad k=(k_1\ldots k_s).
\end{aligned}\right\} \tag{28.14}$$

The coefficients $\beta^{(nrs)}$ are given by

$$\beta_{ijk}^{(nrs)} = \frac{1}{r!\,s!}\int\frac{1}{m}B(\vartheta,V)\,g^{(0)}(\boldsymbol{v})\,g^{(0)}(\boldsymbol{v}_1)\,\mathscr{H}_j^{(r)}(\boldsymbol{v})\,\mathscr{H}_k^{(s)}(\boldsymbol{v}_1)\,[\mathscr{H}_i^{(n)}]\,d\vartheta\,d\varepsilon\,d\boldsymbol{v}\,d\boldsymbol{v}_1 \tag{28.15}$$

or by the equivalent expression

$$\left.\begin{aligned}
\beta_{ijk}^{(nrs)} = \frac{1}{r!\,s!}\int\frac{1}{m}&B(\vartheta,V)\,g^{(0)}(\boldsymbol{v})\,g^{(0)}(\boldsymbol{v}_1)\,\mathscr{H}_i^{(n)}(\boldsymbol{v})\times\\
&\times[\mathscr{H}_j^{(r)}(\boldsymbol{v})\,\mathscr{H}_k^{(s)}(\boldsymbol{v}_1)]\,d\vartheta\,d\varepsilon\,d\boldsymbol{v}\,d\boldsymbol{v}_1;
\end{aligned}\right\} \tag{28.16}$$

we recall the notation

$$[\varphi] = \varphi' + \varphi_1' - \varphi - \varphi_1. \tag{28.17}$$

These expressions are obtained by a simple application of the symmetry relations of Sect. 16. In general, $J^{(n)}$ is an infinite quadratic sum of Hermite coefficients. The coefficients β are functions of the temperature because $B(\vartheta,V) = B(\vartheta,|\boldsymbol{v}_1-\boldsymbol{v}|\sqrt{RT})$. The case of Maxwellian molecules where B is a function of ϑ alone is particularly simple. First of all, the coefficients β become pure numbers (i.e., dependent only on molecular constants but not on the state of the gas). Further, the velocity integrations in (28.15) can be performed explicitly; the integrands are polynomials multiplied by exponentials. Most important, the sum in (28.14) is *finite* in the case of Maxwellian molecules [22]. Since $[\mathscr{H}_i^{(n)}]$ in (28.15) is a polynomial in $\boldsymbol{v}$ and $\boldsymbol{v}_1$ of combined degree n, the integral vanishes when $r+s>n$ (by orthogonality). On the other hand, from (28.16), since $[\mathscr{H}_j^{(r)}(\boldsymbol{v})\,\mathscr{H}_k^{(s)}(\boldsymbol{v}_1)]$ is

---

[1] Reference [22], equation (4.20).

of combined degree $r+s$, the integral vanishes when $n > r+s$. In other words, $\beta^{(nrs)}$ vanishes except for $n=r+s$, and $J^{(n)}$ is a finite quadratic sum of Hermite coefficients of order not exceeding n.

From (28.13), we see that the differential equations do not separate; $a^{(3)}$ occurs in the equation for $\partial a^{(2)}/\partial t$, and, in general, $a^{(n+1)}$ is found in the equation for $\partial a^{(n)}/\partial t$. All Hermite coefficients occur in each $J^{(n)}$ except in the case of Maxwellian molecules. In order to get a determined system of equations we must decide to approximate f by a *finite* sum of Hermite polynomials, say up to $a^{(n)}$. In the system of equations up to $\partial a^{(n)}/\partial t$ we set all higher order Hermite coefficients equal to zero. On the left (differential) side, only the final equation for $a^{(n)}$ itself is altered; in particular, the conservation equations are retained complete with stress tensor and heat flow vector. On the right (collision) side, we must cut off the sum (28.14) at n except in the case of Maxwellian molecules where the exact expression for $J^{(n)}$ already has the proper form. In [22] it is shown for $J^{(2)}$ and $J^{(3)}$ and in [53] for a number of other cases that very little error is made even in the case of non-Maxwellian molecules by restricting the sum (28.14) to $r+s=n$ (rather than $r+s\leq n$).

As an example, suppose we keep up to third order terms. A state is defined by specifying $a_{ij}^{(2)}(\boldsymbol{x})$ and $a_{ijk}^{(3)}(\boldsymbol{x})$ as well as $\varrho(\boldsymbol{x})$, $u_i(\boldsymbol{x})$, and $T(\boldsymbol{x})$. The variation in time of these state variables is governed by a set of *twenty* differential equations[1] (the number is considerably reduced in problems with symmetry). Or, we can decide to keep only part of the third order terms, namely, $a_i^{(3)}$; in this case, we prefer to describe the *thirteen* state variables in terms of the familiar dimensional moments p_{ij} and q_i in addition to ϱ, u_i, and T. To this approximation, f takes the form

$$f=f^{(0)}\left\{1+\frac{p_{ij}}{2p\,\mathrm{R}T}\,c_i\,c_j-\frac{q_i}{p\,\mathrm{R}T}\,c_i\left(1-\frac{c^2}{5\,\mathrm{R}T}\right)\right\}, \tag{28.18}$$

and the equations for p_{ij} and q_i are

$$\left.\begin{aligned}&\frac{\partial p_{ij}}{\partial t}+\frac{\partial}{\partial x_r}(u_r p_{ij})+\frac{2}{5}\left(\frac{\partial q_i}{\partial x_j}+\frac{\partial q_j}{\partial x_i}-\frac{2}{3}\delta_{ij}\frac{\partial q_r}{\partial x_r}\right)+\\&+p_{ir}\frac{\partial u_j}{\partial x_r}+p_{jr}\frac{\partial u_i}{\partial x_r}-\frac{2}{3}\delta_{ij}\,p_{rs}\frac{\partial u_r}{\partial x_s}+\\&+p\left(\frac{\partial u_i}{\partial x_j}+\frac{\partial u_j}{\partial x_i}-\frac{2}{3}\delta_{ij}\frac{\partial u_r}{\partial x_r}\right)+\beta\,\varrho\,p_{ij}=0,\end{aligned}\right\} \tag{28.19}$$

$$\left.\begin{aligned}&\frac{\partial q_i}{\partial t}+\frac{\partial}{\partial x_r}(u_r\,q_i)+\frac{7}{5}q_r\frac{\partial u_i}{\partial x_r}+\frac{2}{5}q_r\frac{\partial u_r}{\partial x_i}+\frac{2}{5}q_i\frac{\partial u_r}{\partial x_r}+\\&+\mathrm{R}T\frac{\partial p_{ir}}{\partial x_r}+\frac{7}{2}p_{ir}\frac{\partial\,\mathrm{R}T}{\partial x_r}-\frac{p_{ir}}{\varrho}\frac{\partial P_{rs}}{\partial x_s}+\\&+\frac{5}{2}p\frac{\partial\,\mathrm{R}T}{\partial x_i}+\frac{2}{3}\beta\,\varrho\,q_i=0.\end{aligned}\right\} \tag{28.20}$$

It turns out that only the single collision parameter, β, enters[2]; it is given by

$$\beta(\mathrm{R}T)=\frac{2}{5}\sqrt{2\pi}\int\limits_0^\infty x^6\,\mathrm{e}^{-\frac{1}{2}x^2}\left\{\int\frac{1}{m}\,B(\vartheta,x\sqrt{2\,\mathrm{R}T})\sin^2\vartheta\cos^2\vartheta\,d\vartheta\right\}dx. \tag{28.21}$$

---

[1] Ref. [22], Eq. (5.12).

[2] This is a consequence of the restriction $r+s=n$ in the sum (28.14). This is exact for Maxwellian molecules and a good approximation in general; for the general case, see Ref. [22].

Note that p_{ij} and q_i are on a par with ϱ, u_i, and T; they are not explicitly given in terms of the fluid variables or any of their derivatives. The effect of stresses and heat flow on the distribution function is clearly seen in (28.18). The stresses make f anisotropic, and heat flow makes it skew. Furthermore, these are the smallest number of parameters that permit this degree of flexibility.

The parameter β in (28.19) and (28.20) has a number of interpretations. In a purely time dependent problem with no space variation we have

$$\left.\begin{aligned} \frac{\partial p_{ij}}{\partial t} + \beta\varrho p_{ij} &= 0, \\ \frac{\partial q_i}{\partial t} + \frac{2}{3}\beta\varrho q_i &= 0. \end{aligned}\right\} \tag{28.22}$$

This exhibits the Maxwell relaxation effect[1]. If there are no macroscopic gradients, the stresses and heat flow decay exponentially. The time $1/\beta\varrho$ is on the order of the mean collision time. Next, we apply the method of obtaining normal solutions but to the moment equations instead of to the Boltzmann equation itself. The expansions of f and of $\partial f/\partial t$ can be easily translated into expansions for the various moments of f. Solution for $f^{(1)}$ in terms of $f^{(0)}$[cf. Eq. (23.6)] is equivalent to writing $p_{ij}^{(1)}$ and $q_i^{(1)}$ on the right side of (28.19) and (28.20) while replacing all quantities on the left side by their locally Maxwellian values. We get

$$p\left(\frac{\partial u_i}{\partial x_j} + \frac{\partial u_j}{\partial x_i} - \frac{2}{3}\delta_{ij}\frac{\partial u_r}{\partial x_r}\right) + \beta\varrho p_{ij}^{(1)} = 0, \tag{28.23}$$

$$\frac{5}{2}p\frac{\partial(\mathrm{R}T)}{\partial x_i} + \frac{2}{3}\beta\varrho q_i^{(1)} = 0. \tag{28.24}$$

These are the Navier-Stokes relations from which we identify the viscosity and heat conductivity coefficients as

$$\mu(\mathrm{R}T) = \frac{p}{\beta\varrho} = \frac{\mathrm{R}T}{\beta(\mathrm{R}T)}, \tag{28.25}$$

$$\lambda(\mathrm{R}T) = \frac{15}{4}\frac{p\mathrm{R}}{\beta\varrho} = \frac{15}{4}\mathrm{R}\mu = \frac{5}{2}C_V\mu. \tag{28.26}$$

We see that at one and the same time β acts like a time constant, a viscosity coefficient, and a heat conductivity coefficient. In actuality it is none of these precisely. As a *parameter* in the differential equations (28.19) and (28.20) it has a very complex effect on the solutions of these equations and only on occasion will there exist a simple interpretation.

The values of μ and λ given here are what CHAPMAN and COWLING [*11*] call the first approximation to their true values. They are exact for Maxwellian molecules and very good approximations for any force law. The term "first approximation" refers to a sequence of approximations made in solving the Fredholm integral equation for $f^{(1)}$ and does not refer to further approximations, $f^{(n)}$, in the normal solution expansion. Apparently, by taking moments of the Boltzmann equation, we have succeeded in solving the integral equations for $f^{(1)}$ exactly for Maxwellian molecules and to a good approximation in general[2]. The reason is that in the special case of Maxwellian molecules the eigenfunctions of the linear Fredholm equation which occurs in the Hilbert theory are the Hermite polynomials[3]. Consequently, expansion of $f^{(1)}$ in Hermite polynomials yields the

[1] Ref. [*46*], Vol. II, p. 703.

[2] This procedure is similar to the original method of CHAPMAN, Ref. [*10*], in contrast to that of ENSKOG, Ref. [*16*].

[3] This was first noted by WANG CHANG and UHLENBECK [*68*], although strong use of this fact had been made previously by BURNETT [*8*], and others.

explicit solution. This fact can be easily deduced from the statement that only terms with $r+s=n$ occur in (28.14); see Sect. 30. Furthermore, since the off-diagonal terms of $\beta^{(nrs)}$ are small for all types of molecule, their neglect [which gives the viscosity and heat conductivity values in (28.25) and (28.26)] gives an approximate solution of the integral equation. The exact solution for μ and λ in these terms involves inversion of the infinite matrix $\beta^{(nrs)}$. Since it is almost diagonal, various simple iterations can provide quick convergence[1].

The higher order normal solutions are also very easily interpreted in terms of the moment equations for the case of Maxwellian molecules. On the right side we have $p_{ij}^{(2)}$, $p_{ij}^{(3)}$ ... and $q_i^{(2)}$, $q_i^{(3)}$ ... in turn (and similarly for higher order moments). On the left side, we insert the results of the previous order calculation of p_{ij} and q_i in the space derivative terms but follow a slightly more complicated procedure for the terms $\partial p_{ij}/\partial t$ and $\partial q_i/\partial t$. Insofar as the space derivative terms are concerned, this is precisely an iteration process, solving for the right side in terms of the left side. It is easy to verify that all the terms required for the computation of $p_{ij}^{(2)}$ and $q_i^{(2)}$ (i.e. for the Burnett equations) are contained within the equations of the thirteen moment approximation[2]. Nevertheless, the thirteen moment system [which consists of (28.19), (28.20) and the conservation equations, (17.12) to (17.14)] is not directly comparable to any stage of approximation of the Chapman-Enskog-Hilbert theory. It is quite likely that the thirteen moment approximation is more accurate than $f^{(1)}$ or $f^{(2)}$ (i.e., than the Navier-Stokes or the Burnett equations). The determination of $f^{(3)}$ (in the moment formalism) would require taking some information from higher moment equations. On the other hand, to *no* degree of approximation does the normal solution analysis lose the primitive property that the future is determined solely by the present state of ϱ, u, and T, whereas the thirteen moment formulation has exceeded this low level of approximation.

The relation between the Navier-Stokes equations and the thirteen moment equations can be made more precise than by the formal procedure of applying the normal solution expansion to the moment equations. Let us consider a flow problem in which the reference time $1/\beta\varrho$ is short compared to the time required for any appreciable macroscopic changes. Write (28.19) and (28.20) in the form

$$\frac{\partial p_{ij}}{\partial t} + A_{ij} + \beta\varrho p_{ij} = 0, \tag{28.27}$$

$$\frac{\partial q_i}{\partial t} + B_i + \frac{2}{3}\beta\varrho q_i = 0. \tag{28.28}$$

A_{ij} and B_i stand for all the remaining terms which, by assumption, are almost constant in a time comparable to $1/\beta\varrho$. Over such a short period of time, these equations can be considered as ordinary differential equations with inhomogeneous constant terms A_{ij} and B_i. The solution consists of a transient plus a steady state; the latter is

$$\left.\begin{aligned} p_{ij} &= -A_{ij}/\beta\varrho, \\ q_i &= -\tfrac{3}{2}B_i/\beta\varrho. \end{aligned}\right\} \tag{28.29}$$

In other words, no matter what values p_{ij} and q_i have initially, within a short time they approach the equilibrium values (28.29). Still treating (28.27) and (28.28) as ordinary differential equations, if A_{ij} and B_i are slowly varying functions of the time, then, except for a small error, we obtain the solution of the differential equation as the *quasi-equilibrium* (28.29). In many cases (but not in

[1] See [*11*]. For alternative methods, see [*29*], [*42*] and [*12*].
[2] Ref. [*11*], Note G.

all—this requires numerical estimates in any given example) the dominant parts of A_{ij} and B_i are exactly the velocity gradient and temperature gradient terms which yield the Navier-Stokes relations. This is a refinement of the arguments of Sect. 26 which indicated an exponential decay to a solution of the nonviscous Euler equations. Of course, if $1/\beta\varrho$ is not small compared to a representative time, then the thirteen moment equations do not approximate the Navier-Stokes equations and should give better results. One would expect the thirteen moment equations to demonstrate, at least to some extent, the effect of an overlap between the two time scales (see Sect. 26), viz., the collision time and the macroscopic decay time.

It is interesting to note that heat flow in a gas at rest will give rise to stresses in exactly the same pattern that would result from a velocity field [cf. (28.19)].

The Hermite expansion, in all probability, gives rise to *hyperbolic* systems of differential equations, i.e., equations in which there is a finite speed of propagation (the thirteen moment case is discussed in Sect. 29). On the other hand, the Navier-Stokes equations as well as all other equations arising from normal solutions (with the exception of the Euler equations) are *parabolic* and exhibit infinite speed of propagation of signals. For mathematical simplicity and for simple qualitative description of the properties of a set of equations, a hyperbolic system is frequently preferable. On the other hand, the solution of a hyperbolic system can be interrupted by the appearance of shocks (see Sect. 33 for the thirteen moment case). Thus for some purposes it would seem to be preferable to have a suitable parabolic system. This can be achieved by an interpolation process which mimics the interpolation of the Navier-Stokes equations between the five moment Euler equations and the thirteen moment equations. The parallel procedure would be to obtain an approximation for the fourth moments (and the remaining third moments not contained in q_i) from the higher moment equations. For example, one can replace the left side of the higher moment equations by the thirteen moment f, (28.18). It would not do to use $f^{(0)}$ on the left side since, to this approximation, all Hermite coefficients of higher order than $a_{ij}^{(2)}$ and $a_i^{(3)}$ would be zero[1]. The moments which must be computed for insertion into the thirteen moment equations are those components of S_{ijk} which are distinct from $q_i = \frac{1}{2} S_{irr}$ (these are seven in number) and the contracted fourth order tensor Q_{ijrr} (six components). It is convenient to compute the following combinations of moments (the reason is explained below):

$$s_{ijk} = S_{ijk} - \tfrac{2}{5}(q_i\,\delta_{jk} + q_j\,\delta_{ik} + q_k\,\delta_{ij}), \tag{28.30}$$

$$q = Q_{rrss} - 15\,p\,\mathrm{R}\,T, \tag{28.31}$$

$$q_{ij} = Q_{ijrr} - 5\,p\,\mathrm{R}\,T\,\delta_{ij} - 7\,\mathrm{R}\,T\,p_{ij} \tag{28.32}$$

(q is distinguished from the heat flow, q_i, by the subscript). The results are[2]

$$\left.\begin{aligned}\frac{3}{2}\beta\varrho\, s_{ijk}/\mathrm{R}T = {} & \frac{2}{5}\left(\frac{\partial p_{ir}}{\partial x_r}\delta_{jk} + \frac{\partial p_{jr}}{\partial x_r}\delta_{ki} + \frac{\partial p_{kr}}{\partial x_r}\delta_{ij}\right) - \\ & - \left(\frac{\partial p_{ij}}{\partial x_k} + \frac{\partial p_{ik}}{\partial x_j} + \frac{\partial p_{jk}}{\partial x_i}\right),\end{aligned}\right\} \tag{28.33}$$

$$\frac{2}{3}\beta\varrho\, q/\mathrm{R}T = -8\left(p_{rs}\frac{\partial u_r}{\partial x_s} + \frac{\partial q_r}{\partial x_r}\right) - 4 q_r\left(2\frac{d u_r}{dt} + 7\frac{\partial \mathrm{R}T}{\partial x_r}\right), \tag{28.34}$$

[1] Ref. [22], Eq. (4.20).
[2] The result for s_{ijk} is given in Ref. [22].

$$\left.\begin{aligned}\beta'\varrho q_{ij}/\mathrm{R}T = {} & \frac{8}{3}p_{rs}\frac{\partial u_r}{\partial x_s}\delta_{ij} - 2p_{ir}\left(\frac{\partial u_j}{\partial x_r}+\frac{\partial u_r}{\partial x_j}\right) - 2p_{jr}\left(\frac{\partial u_i}{\partial x_r}+\frac{\partial u_r}{\partial x_i}\right) - \\ & - p_{ij}\left(2\frac{\partial u_r}{\partial x_r} + 7\frac{d\mathrm{R}T}{dt}\Big/\mathrm{R}T\right) - \\ & - \frac{14}{5\mathrm{R}T}\left(q_i\frac{du_j}{dt} + q_j\frac{du_i}{dt} - \frac{2}{3}q_r\frac{du_r}{dt}\delta_{ij}\right) - \\ & - \frac{14}{5\mathrm{R}^2T^2}\left(\frac{\partial}{\partial x_i}(q_j\mathrm{R}^2T^2) + \frac{\partial}{\partial x_j}(q_i\mathrm{R}^2T^2) - \frac{2}{3}\frac{\partial}{\partial x_r}(q_r\mathrm{R}^2T^2)\,\delta_{ij}\right).\end{aligned}\right\}\quad (28.35)$$

The Lagrangian time derivatives, du_r/dt and $d\mathrm{R}T/dt$, which appear in these formulas are defined by

$$\frac{d}{dt} = \frac{\partial}{\partial t} + u_r\frac{\partial}{\partial x_r} \quad (28.36)$$

and may be eliminated by use of the conservation equations if desired. The new collision integral β' is defined by

$$\left.\begin{aligned}\beta'(\mathrm{R}T) = {} & \frac{\sqrt{2\pi}}{420}\int\limits_0^\infty x^6(3x^4 - 42x^2 + 301)\,\mathrm{e}^{-\frac{1}{2}x^2}\times \\ & \times\left\{\int\frac{1}{m}B\left(\vartheta, x\sqrt{2\mathrm{R}T}\right)\sin^2\vartheta\cos^2\vartheta\,d\vartheta\right\}dx.\end{aligned}\right\}\quad (28.37)$$

It should be noted that $\frac{3}{2}\beta\varrho$, $\frac{2}{3}\beta\varrho$, and $\beta'\varrho$ are the time constants for the decay of s_{ijk}, q, and q_{ij} respectively, just as $\beta\varrho$ and $\frac{2}{3}\beta\varrho$ are the time constants for p_{ij} and q_i [cf. Eq. (28.22)]. For a Maxwellian molecule, we find $\beta' = \frac{7}{6}\beta$. The moment combinations (28.30) to (28.32) are unique in that they exhibit a simple exponential decay in the case of a Maxwellian molecule. More generally, p_{ij} and q_{ij} will be coupled as will be q_i and s_{ijk} with the equivalent "layers" of the fifth order moments, etc., but the coupling is quite small even for non-Maxwellian molecules.

The formulas (28.33) to (28.35) give extra terms in Eqs. (28.19) and (28.20) which we shall not write down. The formulas are not overwhelming in actual problems of interest. However, if one looks for the simplest modification of the thirteen moment equations, one should take $s_{ijk}=0$, $q_{ij}=0$, and only insert q from (28.34).

The question of iterations is considered in some generality in [*32*]. However, although a *single* iteration, such as the one which yields the Navier-Stokes equations or the modification (28.33) to (28.35) of the thirteen moment equations, is likely to significantly increase the accuracy, further iteration would not seem to be justified, and one should rather include higher order moments. However, this is a matter of taste rather than mathematics at the present time.

We conclude this section with a word about entropy and the H-function to the thirteen moment order of approximation. We have, approximately,

$$H^{(13)} = -\frac{1}{\mathrm{R}}\varrho S + \frac{p_{rs}p_{rs}}{2p\mathrm{R}T} + \frac{2}{5}\frac{q_r q_r}{p\mathrm{R}^2T^2} \quad (28.38)$$

and

$$H_i^{(13)} = u_i H^{(13)} - \frac{q_i}{\mathrm{R}T} + \frac{4}{5}\frac{p_{ir}q_r}{p\mathrm{R}T}; \quad (28.39)$$

cf. Eqs. (7.13) and (7.14) of Ref. [*22*]. In general, neither S nor H nor even the approximation (28.38) to H has any monotonicity properties for solutions of the thirteen moment equations. However, the Eqs. (28.38) and (28.39) are significant in themselves and indicate, to a certain extent, what is the difference between S and H. In particular, (28.39) states that there is a flow of H into a domain if there are present both stresses and heat flow.

29. Boundary conditions. The problem of supplying boundary conditions to match a set of equations must be considered in two parts: how many conditions should be imposed and how are they to be obtained? From the boundary conditions for the Boltzmann equation itself, Sect. 19, it is easy to get an infinite sequence of boundary conditions for any type of expansion. However, the number of boundary conditions to be selected from this sequence is determined by what is appropriate for the system of equations under consideration and is not immediately evident from the order of the approximation or from the number of equations.

Consider the thirteen moment system. The variables are

$$\varrho,\ u_i,\ p,\ p_{ij},\ q_i \tag{29.1}$$

and are thirteen in number. In a one-dimensional flow problem this reduces to five variables

$$\varrho,\ u_1,\ p,\ p_{11},\ q_1, \tag{29.2}$$

and in two dimensions there are nine variables

$$\varrho,\ u_1,\ u_2,\ p,\ p_{11},\ p_{12},\ p_{22},\ q_1,\ q_2. \tag{29.3}$$

We conclude immediately that the requisite numbers of *initial* conditions to go with problems (29.1) to (29.3) are thirteen, five, and nine respectively. To count boundary conditions, one must first compute the characteristics, (roughly speaking, the sound speeds), of the partial differential equations. In the one-dimensional case these turn out to be (approximately, for small stresses and heat flow) [22]

$$\left.\begin{aligned} V &= u\,, \\ V &= u \pm a_1 \sqrt{\mathrm{R}T}\,, \\ V &= u \pm a_2 \sqrt{\mathrm{R}T} \end{aligned}\right\} \tag{29.4}$$

where

$$\left.\begin{aligned} a_1^2 &= \tfrac{1}{15}\left(13 + \sqrt{94}\right) \sim 4.54\,, \\ a_2^2 &= \tfrac{1}{15}\left(13 - \sqrt{94}\right) \sim 0.661\,. \end{aligned}\right\} \tag{29.5}$$

For the Euler equations we would get $a^2 = \frac{5}{3}$. In the present example there are two finite sound speeds and a third which is zero. Taking into account plus and minus signs there are five characteristics. Now, for simplicity, take one boundary condition to be $u=0$ at a fixed wall. The boundary surface will be intersected by two backward characteristics and two forward characteristics; the conclusion is that one must impose a total of two boundary conditions or one more in addition to $u=0$. Intuitively this might relate to the heat flow through the boundary. It is important to note that the zero speed characteristic does not contribute a boundary condition.

Next we turn to the two-dimensional case. Here we find the same speeds a_1 and a_2 as before and, in addition,

$$a_3^2 = \tfrac{7}{5}. \tag{29.6}$$

This time the zero sound speed occurs as a triple root; the total is nine since a_1, a_2, and a_3 are each counted twice. If we again take a fixed wall, say perpendicular to the 1-direction, and take $u_1=0$ as one of the boundary conditions, we discover that *two* additional boundary conditions must be imposed. They could relate to the normal component of heat flow and the tangential stress.

For the Navier-Stokes equations, the correct number of boundary conditions is the same as for the thirteen moment equations (at least for one and two dimensions; the three-dimensional thirteen moment characteristics have not been computed). What the answer is for the higher order normal solution approximations is a matter of conjecture.

It is clear from these examples that mere counting of equations or of order of differentiation is not enough. Also, in principle, physical intuition is helpless in this matter. The physical situation does not change depending on how accurately one approximates to the distribution function, or depending on whether one assumes f to be analytic in a parameter ε or not, or whether one chooses to iterate equations rather than to increase their number; but all of these may have an effect on the relevant number of boundary conditions.

Our procedure will be to describe the construction of an infinite sequence of boundary conditions without trying to coordinate orders of approximation in the boundary condition expansion and in the Boltzmann equation expansion. Given a set of equations, one can select the appropriate number of boundary conditions. It should be mentioned that the differential equations can impose restrictions on *what* should be specified at a boundary as well as how much; this kind of information might help in the selection of boundary conditions, but it usually requires elaborate computation for any appreciable number of equations.

We recall the boundary condition

$$f(\xi) = \int_{\eta\cdot\boldsymbol{n}>0} K(\xi,\eta)\, f(\eta)\, d\eta\,, \qquad \xi\cdot\boldsymbol{n}<0 \tag{29.7}$$

[see Eq. (19.5)]. For the moment suppose that K is non-singular; i.e., there is no specular reflection. We introduce the notation

$$\left.\begin{aligned} f^+(\xi) &= \begin{cases} f(\xi), & \xi\cdot\boldsymbol{n}>0\\ 0, & \xi\cdot\boldsymbol{n}<0, \end{cases}\\ f^-(\xi) &= \begin{cases} 0, & \xi\cdot\boldsymbol{n}>0\\ f(\xi), & \xi\cdot\boldsymbol{n}<0. \end{cases} \end{aligned}\right\} \tag{29.8}$$

Clearly,

$$f^+(\xi) + f^-(\xi) = f(\xi). \tag{29.9}$$

In (29.7), we can write $f^+(\eta)$ on the right and $f^-(\xi)$ on the left. Eq. (29.7) maps the space of functions f^+ into a space of functions f^-. Alternatively, in the space of all functions f, (29.7) singles out a linear subspace of functions $f=f^+ + f^-$ which are possible boundary functions. One way of specifying this subspace is to list a set of functions φ_r (e.g. polynomials) which are orthogonal to every member of the subspace,

$$\int \varphi_r(\xi)\, f(\xi)\, d\xi = 0. \tag{29.10}$$

For many purposes it is even preferable to consider (29.10) as the basic form in which to state boundary conditions rather than in terms of a kernel as in (29.7).

Let us now try to find a set of functions φ_r which are appropriate to any given kernel K. Take any function $\varphi = \varphi^+ + \varphi^-$ as in (29.8). We have

$$\int f(\xi)\, \varphi(\xi)\, d\xi = \int f^+ \varphi^+\, d\xi + \int f^- \varphi^-\, d\xi. \tag{29.11}$$

From (29.7),

$$\left.\begin{aligned}\int f^- \varphi^- d\xi &= \int \varphi^-(\xi)\, K(\xi,\eta)\, f^+(\eta)\, d\eta\, d\xi\\ &= \int \psi^+(\eta)\, f^+(\eta)\, d\eta\end{aligned}\right\} \tag{29.12}$$

where

$$\psi^+(\eta) \equiv \int \varphi^-(\xi)\, K(\xi,\eta)\, d\xi. \tag{29.13}$$

Consequently,

$$\int f(\xi)\, \varphi(\xi)\, d\xi = \int f^+ (\varphi^+ + \psi^+)\, d\xi. \tag{29.14}$$

We are looking for functions, φ, for which $\int f\varphi\, d\xi = 0$ when f^+ is arbitrary. Consequently we take $\varphi^+ + \psi^+ = 0$. The class of functions φ_r satisfying (29.10) can now be specified. Choose φ_r^- arbitrarily; choose φ_r^+ as

$$\varphi^+(\eta) = -\int \varphi^-(\xi)\, K(\xi,\eta)\, d\xi. \tag{29.15}$$

The sum, $\varphi_r^+ + \varphi_r^- = \varphi_r$, is what is desired. The set of functions φ_r^- can be taken as a complete set of polynomials in the half space $\xi\cdot\boldsymbol{n} < 0$ or any other convenient functions. If we wish the boundary condition for a fixed wall to be satisfied,

$$\int (\xi\cdot\boldsymbol{n})\, f(\xi)\, d\xi = 0, \tag{29.16}$$

then $\varphi_1 = \xi\cdot\boldsymbol{n}$ should be included in the list φ_r; the construction (29.15) yields $\varphi = \xi\cdot\boldsymbol{n}$ if we start from $\varphi^- = \xi\cdot\boldsymbol{n}$.

The boundary condition (29.10) is equally applicable to the case in which K is singular. For example, if we take the K of (19.6), and $\psi_r^\pm$ represents the two sets of half-range orthogonal polynomials with weight function $A\,\mathrm{e}^{-B\xi^2}$ in the two half spaces $\xi\cdot\boldsymbol{n} \gtrless 0$, then we can take $\varphi_1 = \xi\cdot\boldsymbol{n}$ and

$$\varphi_r = \psi_r^- - \alpha\,\psi_r^+, \quad r > 1. \tag{29.17}$$

For $\alpha = 1$, we can take any set of polynomials which is odd in $\xi\cdot\boldsymbol{n}$; one such set is given by (29.17).

It is clear that there is a tremendous amount of flexibility in the choice of the functions φ_r. For example, with specular reflection, any set of odd functions will do. Furthermore, one need not include polynomials of all orders in the construction of a complete set of ψ_r^- in (29.17); for example, it is sufficient (but possibly not wise) to choose a set of odd polynomials ψ_r^-, in which case φ_r is continuous at the origin. Of course, in any given case one should try to select functions φ_r which allow a good fit, are computationally manageable, and are simply related to the functions which are used to expand f.

It was mentioned in Sect. 19 that the kernel K will have a certain number of parameters, *accomodation coefficients*, which define the nature of the boundary as distinguished from its thermodynamic state. Independently, there is a certain definite number of boundary conditions which is appropriate to a given set of equations. If these two numbers are equal, then it is possible to bypass the whole question of the construction of a kernel or of a set of functions φ_r and construct the boundary conditions directly in terms of the definition of the accomodation coefficients.

We consider the case of two accomodation coefficients, α_M for momentum and α_E for energy. The coefficient α_M is defined to be the fraction of the incident flow of tangential momentum which is lost to the wall. Any value $0 \leq \alpha_M \leq 1$ is consistent with the equilibrium state since both the incident and emergent tangential momenta are zero. The exact analogue for energy flow is inappropriate since at equilibrium the incident and emergent energy flows are equal but not zero. Instead, we define a coefficient α_E as the ratio of the difference (in absolute value) between the actual incident and emergent energy flow to the difference between the incident energy flow and the emerging flow that would exist if it emerged in equilibrium at the temperature of the wall (i.e., with perfect accomodation). We define M^+ and M^- to be the incident and emergent tangential momentum flow and E^+ and E^- to be the incident and emergent energy flow, signed consistently so that $M^+ + M^-$ and $E^+ + E^-$ are the net flows. The definitions of α_M and α_E can be written

$$\alpha_M M^+ = M^+ + M^-, \tag{29.18}$$

$$\alpha_E (E^+ + E_0) = E^+ + E^-. \tag{29.19}$$

E_0 is the energy flow that would emerge in equilibrium at the temperature of the wall. $E^+ + E^-$ is the net energy flow; it is the heat flow into the wall only if the tangential gas velocity is zero. The boundary condition (19.6) amounts to taking

$$\alpha_E = \alpha_M = 1 - \alpha. \tag{29.20}$$

For simplicity take the normal to the wall in the 1-direction, with the gas to the left. The mass flow into the wall is

$$m = \int \xi_1 f^+ \, d\xi = - \int \xi_1 f^- \, d\xi$$

and the quantities $M^\pm$ and $E^\pm$ are given by

$$\left.\begin{aligned} M_j^\pm &= \int \xi_1 \xi_j f^\pm \, d\xi \\ &= \int \xi_1 c_j f^\pm \, d\xi + u_j \int \xi_1 f^\pm \, d\xi \\ &= p_{1j}^\pm \pm m u_j , \end{aligned}\right\} \tag{29.21}$$

$$\left.\begin{aligned} E_j^\pm &= \int \tfrac{1}{2} \xi_1 \xi^2 f^\pm \, d\xi \\ &= \int \tfrac{1}{2} \xi_1 c^2 f^\pm \, d\xi + \int \xi_1 (\mathbf{c} \cdot \mathbf{u}) f^\pm \, d\xi + \tfrac{1}{2} u^2 \int \xi_1 f^\pm \, d\xi \\ &= q_1^\pm + u_r p_{1r}^\pm \pm \tfrac{1}{2} m u^2 . \end{aligned}\right\} \tag{29.22}$$

The "half" moments are defined as indicated, and the velocity u_j is in the (2, 3) direction. To the thirteen moment approximation, m, $p_{ij}^\pm$, and $q_i^\pm$ are evaluated in Ref. [*22*],

$$m = \frac{p + \frac{1}{2} p_{11}}{(2\pi \mathrm{R} T)^{\frac{1}{2}}}, \tag{29.23}$$

$$p_{1j}^\pm = \frac{1}{2} p_{1j} \pm \frac{q_j}{5 (2\pi \mathrm{R} T)^{\frac{1}{2}}}, \tag{29.24}$$

$$q_1^\pm = \frac{1}{2} q_1 \pm \left(2p + \frac{3}{2} p_{11}\right) \left(\frac{\mathrm{R} T}{2\pi}\right)^{\frac{1}{2}} \tag{29.25}$$

and it is easily shown that

$$E_0 = -2 m \mathrm{R} T_0 \tag{29.26}$$

where T_0 is the boundary temperature. Insertion of these formulas into (29.18) and (29.19) yields

$$\frac{p_{1j}}{p} = \sqrt{\frac{2}{\pi}} \frac{\alpha_M}{2 - \alpha_M} \left\{ \left(1 + \frac{1}{2} \frac{p_{11}}{p}\right) \frac{u_j}{\sqrt{\mathrm{R} T}} + \frac{q_j}{5 p \sqrt{\mathrm{R} T}} \right\}, \tag{29.27}$$

$$\frac{q_1 + p_{1r} u_r}{p \sqrt{\mathrm{R} T}} = \sqrt{\frac{8}{\pi}} \frac{\alpha_E}{2 - \alpha_E} \left\{ \left(1 + \frac{1}{2} \frac{p_{11}}{p}\right) \left(1 - \frac{T_0}{T} + \frac{1}{2} \frac{u^2}{\mathrm{R} T}\right) + \frac{1}{4} \frac{p_{11}}{p} + \frac{q_r u_r}{10 p \mathrm{R} T} \right\}. \tag{29.28}$$

These boundary conditions are qualitatively similar to MAXWELL's[1]; in a rough, intuitive way a tangential stress is associated with a slip velocity (but also with a tangential heat flow), and the normal component of heat flow is associated with a temperature jump (and with many other things). The term in p_{1r} can be eliminated from the second boundary condition,

$$\left.\begin{aligned} -\frac{q_1}{p\sqrt{RT}} &= \sqrt{\frac{8}{\pi}}\,\frac{\alpha_E}{2-\alpha_E}\left\{\left(1+\frac{1}{2}\,\frac{p_{11}}{p}\right)\left(1-\frac{T_0}{T}\right)+\frac{1}{4}\,\frac{p_{11}}{p}\right\}+\\ &+\sqrt{\frac{8}{\pi}}\,\frac{\alpha_E-\alpha_M}{(2-\alpha_E)\,(2-\alpha_M)}\left\{\left(1+\frac{1}{2}\,\frac{p_{11}}{p}\right)\frac{u^2}{RT}+\frac{q_r\,u_r}{5pRT}\right\}.\end{aligned}\right\} \tag{29.29}$$

A word should be said about the significance of the moments p_{ij} and q_j as observables since the velocity of the gas is not the same as the velocity of the wall. The stresses actually measured at the wall are given by $\int \xi_i\,\xi_j\,f = P_{ij} + \varrho u_i u_j$, but this reduces to P_{ij} for the only relevant case, $i=1$. However, the energy flow through the wall is not q_1 but $\int \frac{1}{2}\xi_1\xi^2 f = q_1 + u_r p_{1r}$. If one were to measure the heat flow through the boundary, this distinction would have to be made.

The stress boundary condition (29.27), does not exist in a one-dimensional problem, has a single component in two dimensions and has two components in three dimensions. The heat flow boundary condition (29.28) or (29.29) is in all cases a single scalar condition. The total number of boundary conditions (including $u_1=0$) is correct for the thirteen moment equations and for the Navier-Stokes equations. If a greater number of boundary conditions is desired, it is necessary to construct a kernel, $K(\boldsymbol{\xi}, \boldsymbol{\eta})$, with α_M and α_E built in. This might be done as indicated in Sect. 19.

30. Expansion about absolute equilibrium. It is obviously a more elementary procedure to expand f about an $f^{(0)}$ which is constant rather than a function of $\boldsymbol{x}$ and t. This leads to great simplifications in certain respects; but it is extremely awkward in others. We first describe the general procedure and then specialize to the case of the linearized Boltzmann equation [*68*], [*53*]. Let us denote by $\varrho^{(0)}$, $\boldsymbol{u}^{(0)}$ (usually zero), and $T^{(0)}$ the constant parameters of $f^{(0)}$. They may or may not be related to mean values or other representative values of these quantities in the domain. Consider Hermite polynomials in the dimensionless velocity

$$\boldsymbol{v} = \frac{\boldsymbol{\xi}}{\sqrt{RT^{(0)}}} \tag{30.1}$$

for the dimensionless distribution function

$$g = \frac{(RT^{(0)})^{\frac{3}{2}}}{\varrho^{(0)}}\,f. \tag{30.2}$$

For example, the second Hermite coefficient is

$$\begin{aligned} A_{ij}^{(2)} &= \int (v_i v_j - \delta_{ij})\,g\,d\boldsymbol{v}\\ &= \frac{1}{\varrho^{(0)}}\int\left(\frac{\xi_i\,\xi_j}{RT^{(0)}}-\delta_{ij}\right)f\,d\boldsymbol{\xi}\\ &= \frac{1}{p^{(0)}}\,(P_{ij}+\varrho\,u_i\,u_j-\varrho\,RT^{(0)}\,\delta_{ij}),\end{aligned}$$

where the moments P_{ij}, ϱ, u_i have their usual meaning. We know from Sect. 28 that the collision integrals are most easily represented in terms of the *local* Hermite coefficients, $a^{(n)}$. It is quite unnatural to express the collision integrals in

[1] Ref. [*46*], Vol. II, p. 703.

terms of the *absolute* Hermite coefficients $A^{(n)}$. For example, the stress equation contains the quantity

$$\frac{p_{ij}}{p^{(0)}} = A^{(2)}_{ij} - \frac{1}{3} A^{(2)}_{rr}\,\delta_{ij} - \frac{A^{(1)}_i A^{(1)}_j}{A^{(0)}} + \frac{1}{3}\,\frac{A^{(1)2}}{A^{(0)}}\,\delta_{ij}.$$

However, the differential equations are considerably simpler, and this might compensate. Using the identity [21]

$$\xi_j \mathscr{H}^{(n)}_{i_1\ldots i_n} = \mathscr{H}^{(n+1)}_{i_1\ldots i_n j} + \delta_{i_1 j}\mathscr{H}^{(n-1)}_{i_2\ldots i_n} + \delta_{i_2 j}\mathscr{H}^{(n-1)}_{i_1 i_3\ldots i_n} + \cdots + \delta_{i_n j}\mathscr{H}^{(n-1)}_{i_1\ldots i_{n-1}} \tag{30.3}$$

and the definitions

$$\int \mathscr{H}^{(n)}_{i_1\ldots i_n} f\, d\boldsymbol{\xi} = \varrho^{(0)} A^{(n)}_{i_1\ldots i_n}, \tag{30.4}$$

$$\int \mathscr{H}^{(n)}_{i_1\ldots i_n} J(f, f)\, d\boldsymbol{\xi} = \varrho^{(0)} \tilde{J}^{(n)}_{i_1\ldots i_n}, \tag{30.5}$$

we obtain

$$\frac{\partial}{\partial t} A^{(n)}_{i_1\ldots i_n} + \frac{\partial}{\partial x_j} A^{(n+1)}_{i_1\ldots i_n j} + \frac{\partial}{\partial x_{i_1}} A^{(n-1)}_{i_2\ldots i_n} + \cdots + \frac{\partial}{\partial x_{i_n}} A^{(n-1)}_{i_1\ldots i_{n-1}} = \tilde{J}^{(n)}_{i_1\ldots i_n}. \tag{30.6}$$

It is possible that this expansion might be useful in cases where the mean free path is comparable to flow dimensions. However, the most important application of this method is to the *linearized* Boltzmann equation for which not only are the differential equations simple, as in (30.6), but also the collision integrals are simple.

In this case we write

$$f = f^{(0)}(1 + \varepsilon h), \tag{30.7}$$

substitute this into the Boltzmann equation, and drop terms in ε^2. This is not the beginning of a normal solution expansion because there is no $1/\varepsilon$ in front of the collision term; thus there are first order contributions on both sides of the equation. We obtain

$$\left.\begin{aligned} \frac{\partial h}{\partial t} + \xi_r \frac{\partial h}{\partial x_r} &= \frac{1}{f^{(0)}} J(f^{(0)}, f^{(0)} h) \\ &= \frac{1}{m}\int B\, f^{(0)}(\boldsymbol{\xi}_1)\,[h(\boldsymbol{\xi}') + h(\boldsymbol{\xi}_1') - h(\boldsymbol{\xi}) - h(\boldsymbol{\xi}_1)]\, d\vartheta\, d\varepsilon\, d\boldsymbol{\xi}_1 \\ &\equiv L(h). \end{aligned}\right\} \tag{30.8}$$

With the exception of a factor $f^{(0)}$ which has been removed, the integral operator L is the same one that occurs in the Hilbert theory. This is now a *linear* integro-differential equation, and expansion methods can take on more than just a formal meaning.

It was mentioned in Sect. 28 that the eigenfunctions of the operator L are simply contractions of Hermite polynomials when B is a function of ϑ alone. We shall now verify this fact after first introducing the dimensionless notation (30.1). We have

$$L(h) = \varrho^{(0)} \int \beta(\vartheta)\, g^{(0)}(\boldsymbol{v}_1)\,[h]\, d\vartheta\, d\varepsilon\, d\boldsymbol{v}_1 \tag{30.9}$$

where

$$\beta(\vartheta) = \frac{1}{m} B(\vartheta), \tag{30.10}$$

$$g^{(0)}(\boldsymbol{v}) = \frac{1}{(2\pi)^{\frac{3}{2}}}\, e^{-\frac{1}{2}v^2}, \tag{30.11}$$

and the notation $[h] = h' + h_1' - h - h_1$ is used. To obtain the solutions of

$$L(h) = \lambda h \tag{30.12}$$

we multiply by $H^{(n)}(\boldsymbol{v})\, g^{(0)}(\boldsymbol{v})$ and integrate,

$$\lambda a^{(n)} = \varrho^{(0)} \int \beta\, g^{(0)}(\boldsymbol{v})\, g^{(0)}(\boldsymbol{v}_1)\, \mathscr{H}^{(n)}(\boldsymbol{v})\, [h]\, d\vartheta\, d\varepsilon\, d\boldsymbol{v}\, d\boldsymbol{v}_1, \tag{30.13}$$

$$= \varrho^{(0)} \int \beta\, g^{(0)}(\boldsymbol{v})\, g^{(0)}(\boldsymbol{v}_1)\, h(\boldsymbol{v})\, [\mathscr{H}^{(n)}]\, d\vartheta\, d\varepsilon\, d\boldsymbol{v}\, d\boldsymbol{v}_1. \tag{30.14}$$

We have set

$$a^{(n)} = \int g^{(0)} \mathscr{H}^{(n)} h\, d\boldsymbol{v}. \tag{30.15}$$

It is worth remarking that the reason why the analysis of Sect. 28 can be exactly duplicated despite the fact that $a^{(n)}$ is here an absolute Hermite coefficient while previously it was a local Hermite coefficient is that the equation has been linearized. In particular, $\frac{1}{\varrho}\int \boldsymbol{\xi}\, f\, d\boldsymbol{\xi} = \boldsymbol{u}$ is of order ε (there is no loss of generality in taking $\boldsymbol{u}^{(0)} = 0$). Consequently the leading term (with respect to ε) in any polynomial is the same when expressed in terms of $\boldsymbol{\xi} = \boldsymbol{c} + \boldsymbol{u}$ and when expressed in terms of $\boldsymbol{c}$. This eliminates the complication mentioned at the beginning of this section.

Next, substitute the expansion

$$h = \sum_{m=0}^{\infty} \frac{1}{m!}\, a_i^{(m)}\, \mathscr{H}_i^{(m)} \tag{30.16}$$

into (30.13) and (30.14). Exactly as in Sect. 28, we observe that (30.13) vanishes for terms $m < n$ in the expansion while (30.14) vanishes for terms $m > n$. We are left with

$$\lambda a_i^{(n)} = \varrho^{(0)} \beta_{ij}^{(n)}\, a_j^{(n)} \tag{30.17}$$

where

$$\beta_{ij}^{(n)} = \int \beta(\vartheta)\, g^{(0)}(\boldsymbol{v})\, g^{(0)}(\boldsymbol{v}_1)\, \mathscr{H}_i^{(n)}(\boldsymbol{v})\, [\mathscr{H}_j^{(n)}]\, d\vartheta\, d\varepsilon\, d\boldsymbol{v}\, d\boldsymbol{v}_1. \tag{30.18}$$

The indices i and j stand for $(i_1 \ldots i_n)$ and $(j_1 \ldots j_n)$, and a summation over j is understood in (30.17). This formula is a linearized special case of (28.14), but the derivation has been repeated in order to have a self-contained presentation in the linear case.

The eigenvalue problem for the operator L has been split into a finite dimensional problem for each value of n; we must find the eigentensors a_i of (30.17) given the "matrix" β_{ij}. The solution can be given explicitly. Take an arbitrary symmetric tensor of order $n - 2m$, $a^{(n-2m)}$, which vanishes when contracted with respect to two of its indices. Construct a tensor of order n by multiplying $a^{(n-2m)}$ by m Kronecker δ's and symmetrize it by summing over all permutations of the indices. The result is a solution of (30.17). This result follows from the fact that two tensors of order n which are constructed from base tensors $a^{(n-2m)}$ and $a^{(n-2m')}$ are orthogonal if $m \neq m'$. Furthermore, the eigenvalue $\lambda_m^{(n)}$ is a function of m alone; it depends only on the order of the base tensor $a^{(n-2m)}$ from which the construction was made and not on the specific tensor $a^{(n-2m)}$. If n is even, there is one more solution, namely, the symmetric sum of a product of $\frac{1}{2}n$ Kronecker δ's. If n is odd, there is an extra solution constructed by taking any vector $a^{(1)}$ and building it up with δ's. It is now clear how to rewrite the expansion of h, (30.16), as a sum of eigenfunctions of L. This is done by a process of "peeling off layers" of the Hermite coefficients. For example, for a second order tensor there are two layers, $p\,\delta_{ij}$ and p_{ij} defined by

$$P_{ij} = p\,\delta_{ij} + p_{ij}, \quad p = \tfrac{1}{3} P_{rr}. \tag{30.19}$$

For a third order tensor there are also two layers,

$$S_{ijk} = \tfrac{1}{5}(S_i\,\delta_{jk} + S_j\,\delta_{ik} + S_k\,\delta_{ij}) + s_{ijk} \tag{30.20}$$

where

$$S_i = S_{irr}. \tag{30.21}$$

For a fourth order tensor, there are three layers,

$$\left.\begin{aligned}Q_{ijkl} = {} & \tfrac{1}{15} Q(\delta_{ij}\delta_{kl} + \delta_{ik}\delta_{jl} + \delta_{il}\delta_{jk}) + \\ & + \tfrac{1}{7}(Q_{ij}\delta_{kl} + Q_{ik}\delta_{jl} + Q_{il}\delta_{jk} + Q_{jk}\delta_{il} + \\ & + Q_{jl}\delta_{ik} + Q_{kl}\delta_{ij}) + q_{ijkl}\end{aligned}\right\} \tag{30.22}$$

where

$$\left.\begin{aligned}Q &= Q_{rrss}, \\ Q_{ij} &= Q_{ijrr} - \tfrac{1}{3} Q\,\delta_{ij}.\end{aligned}\right\} \tag{30.23}$$

Each of the layers is constructed from a subtensor with zero trace (e.g., p_{ij}, s_{ijk}, Q_{ij}, q_{ijkl}) or else from a scalar (p or Q) or a vector (S_i). Any of the Hermite combinations

$$\left.\begin{gathered}\delta_{ij}\mathscr{H}^{(2)}_{ij}, \\ p_{ij}\mathscr{H}^{(2)}_{ij}, \\ (S_i\delta_{jk} + S_j\delta_{ik} + S_k\delta_{ij})\mathscr{H}^{(3)}_{ijk}, \\ s_{ijk}\mathscr{H}^{(3)}_{ijk}, \\ (\delta_{ij}\delta_{kl} + \cdots)\mathscr{H}^{(4)}_{ijkl}, \\ (q_{ij}\delta_{kl} + \cdots)\mathscr{H}^{(4)}_{ijkl}, \\ q_{ijkl}\mathscr{H}^{(4)}_{ijkl}\end{gathered}\right\} \tag{30.24}$$

are eigenfunctions of the operator L. There is a single eigenvalue for each layer. Note that the quantities q and q_{ij} introduced in (28.31) and (28.32) are just the differences between Q and Q_{ij} and their respective thirteen moment approximations.

The eigenfunctions are more commonly given in terms of Sonine polynomials. These are basically the same as Hermite polynomials but expressed in spherical coordinates instead of rectangular[1]. The symmetries inherent in the invariant Cartesian tensor notation are frequently advantageous. There is no need to explicitly list the special linear combinations (30.24) of the basic Hermite polynomials; all computations can be carried out in the invariant notation, and the linear combinations in (30.24) arise naturally.

Since the Hermite (or Sonine) polynomials are at one and the same time eigenfunctions of the operator L and orthogonal with respect to the unperturbed $f^{(0)}$, they are evidently the logical expansion functions in any problem with Maxwellian molecules. Some examples of the use of this technique are given in Sect. 33. Convergence is (empirically) fairly rapid. However, the question of fitting boundary values is not particularly suited for a Hermite expansion, and the general remarks of the last section cannot be greatly simplified in the linear case or by the use of an absolute Maxwellian as weight function. In the case of non-Maxwellian molecules, the Hermite expansion does not diagonalize the operator L. However, the resulting infinite matrix is symmetric [this is seen by comparison of the alternative expressions (30.13) and (30.14)]. This symmetry can be interpreted as the expression of the Onsager reciprocity law in terms of the "thermodynamic" coordinates $a^{(n)}$ (cf. [*43*]). Specifically, the (linearized)

---

[1] In the notation of Ref. [*68*], the eigenvalues are λ_{rl} and the eigenfunctions are ψ_{rl}. To compare with our notation it is only necessary to note that l is the order of the base tensor and r is the number of supplementary Kronecker δ's; the total tensor order (also the degree of the polynomial) is $l+2r$.

entropy is given by

$$
\begin{aligned}
H &= \int f \log f \, d\xi \\
&= \int f \left\{ \log f^{(0)} + \log \left(1 + \sum_{n=1}^{\infty} \frac{1}{n!} a_i^{(n)} \mathcal{H}_i^{(n)} \right) \right\} d\xi \\
&\sim \int f \left\{ \log f^{(0)} + \sum_{n=1}^{\infty} a_i^{(n)} \mathcal{H}_i^{(n)} \right\} d\xi \\
&= H^{(0)} + \varrho \sum_{n=1}^{\infty} \frac{1}{n!} a_i^{(n)} a_i^{(n)} .
\end{aligned}
$$

In other words, the choice of Hermite coordinates always diagonalizes the entropy but diagonalizes the entropy production only in the case of Maxwellian molecules.

31. Half-range expansion [*28*]. The Hermite expansion has been seen to be particularly appropriate in generalizing fluid type equations (near a locally Maxwellian state) and for the linearized Boltzmann equation. Boundary conditions can best be taken care of by an entirely different type of expansion. It is abundantly clear from the way in which boundary conditions are given that the two halves of the distribution, $f^+(\xi)$ and $f^-(\xi)$, may be entirely different, and f may even be discontinuous at $\xi \cdot \boldsymbol{n} = 0$ near a boundary. The best way to approximate f near a boundary is to approximate f^+ and f^- separately. For example, we can introduce weight functions (possibly even different ones) and expand each of f^+ and f^- in half range polynomials. This, however, will wreak havoc with both the differential equations and the collision integrals. Once the weight functions are chosen, a function f can be described in terms of two sequences of coefficients, $a_+^{(n)}$ and $a_-^{(n)}$ corresponding to two sequences of functions $\varphi_+^{(n)}(\xi)$ defined for $\xi \cdot \boldsymbol{n} > 0$ and $\varphi_-^{(n)}(\xi)$ defined for $\xi \cdot \boldsymbol{n} < 0$. Equations for the coefficients $a_\pm^{(n)}$ can be obtained by integrating through the Boltzmann equation after multiplication by $\varphi_\pm^{(n)}$. In principle the method is similar to the previous ones; in practice it is not.

This is clearly an example of a method which must be fitted to the problem. For simple geometries it is perfectly feasible and can be more quickly convergent than full range expansions near a boundary. In problems in which the boundaries are not parallel planes (or possibly cylinders or spheres), or in which the state of the gas changes from point to point at the boundary, the difficulties would seem to be too great.

A possible application of half range expansions would be to investigate the boundary layer for a distance of a few mean free paths from a wall. If this could be done in some generality, it might be possible to connect the state in terms of full range coefficients $a^{(n)}$ outside a boundary layer with the exact state at the wall. This concept (which has a more crucial importance with regard to normal solutions, cf. Sect. 26), could greatly extend the accuracy of the full range theories.

32. Boltzmann-like equations. The solution of problems with the Boltzmann equation is sufficiently formidable to make it worthwhile to look for alternatives and modifications. It is evidence of the inherent simplicity of the Boltzmann equation itself that not very many suggestions of this type have been made. We shall discuss two suggested modifications of the Boltzmann equation, one which has been made from the point of view of achieving practical solutions to problems, and another from the point of view of the proof of existence.

In the case of existence, D. MORGENSTERN [50] has constructed a modification of the Boltzmann equation which allows the nonlinearity to be handled; (this is the chief difficulty in proving existence in the large, cf. Chap. III). The idea is to introduce a mollifier or smoothing kernel. The arguments of the functions appearing in the usual collision term differ in the velocity variable only; $\boldsymbol{x}$ is the same for all. MORGENSTERN keeps the variable $\boldsymbol{x}$ in $f(\boldsymbol{\xi}, \boldsymbol{x})$ and $f(\boldsymbol{\xi}', \boldsymbol{x})$ but replaces it by $\boldsymbol{y}$ in $f(\boldsymbol{\xi}_1, \boldsymbol{y})$ and $f(\boldsymbol{\xi}_1', \boldsymbol{y})$. The quantity $(f' f_1' - f f_1)$ is then multiplied by a kernel $k(\boldsymbol{x}, \boldsymbol{y})$ and is integrated with respect to y as well as the other conventional variables. With the aid of some very fine analysis, MORGENSTERN is able to prove the existence of a solution in the large for several types of boundary conditions.

Two claims are made which do not seem warranted. The first is that it is doubtful that there exist solutions to the classical equation in general. There is not much evidence in favor of this opinion other than the difficulty of the problem. The second is that the mollified equation is as consistent with the physical picture as is the Boltzmann equation. In a certain abstract sense this is correct. However, the particular form of mollifier which is adopted does not exhibit this. Insofar as molecules $\boldsymbol{\xi}$ and $\boldsymbol{\xi}_1$ are at different locations, the two states of a molecule, $\boldsymbol{\xi}$ and $\boldsymbol{\xi}'$ or $\boldsymbol{\xi}_1$ and $\boldsymbol{\xi}_1'$, are just as far removed. Possibly this can be taken care of without difficulty. A more serious objection, from the point of view of the derivation of the Boltzmann equation, is that this mollifier includes a term on the order of σ, the molecular diameter, and does not represent the final limit of a Boltzmann gas (cf. Sect. 7). If it is kept, many other terms of a similar order should also be kept. But even more important, it is not clear that any Boltzmann equation at all exists (in the sense that the one particle distribution completely defines a state) unless the limit is approached. This is only an argument for the *desirability* of dealing with the classical Boltzmann equation; a general existence theory must eventually be attained in order to ascertain its physical significance.

From the practical point of view, an equation which has been suggested as a replacement for the Boltzmann equation is the following [2]

$$\frac{\partial f}{\partial t} + \xi_r \frac{\partial f}{\partial x_r} = \nu (f^{(0)} - f). \tag{32.1}$$

Only the collision term has been altered; $f^{(0)}$ is the locally Maxwellian function corresponding to f, and ν is a parameter which is allowed to be a function of ϱ, $\boldsymbol{u}$, T and possibly of $\boldsymbol{\xi}$. The basic idea of this modification of the collision term is at first glance very appealing; this form exhibits explicitly the tendency for an approach to local equilibrium and eventually to absolute equilibrium. With ν independent of $\boldsymbol{\xi}$, this equation exhibits all the elementary qualitative features of the Boltzmann equation. We have

$$\int \psi_r (f^{(0)} - f)\, d\boldsymbol{\xi} = 0 \tag{32.2}$$

if ψ_r is a summational invariant. The conservation equations follow immediately Also, since

$$\int f \log f^{(0)}\, d\boldsymbol{\xi} = \int f^{(0)} \log f^{(0)}\, d\boldsymbol{\xi},$$

(this is verified by explicit substitution of $f^{(0)}$), we have

$$\int \log f (f^{(0)} - f)\, d\boldsymbol{\xi} = \int \log (f/f^{(0)})\, (f^{(0)} - f)\, d\boldsymbol{\xi} \leq 0. \tag{32.3}$$

The H theorem is therefore valid for (32.1). There does not seem to be any point in considering the case when ν is a function of $\boldsymbol{\xi}$ since neither the conservation equations nor the H theorem are then valid. Also, the chief simplification due to (32.1), that of easy approximation to the collision term by expansion of f, is lost.

Possibly the best justification of (32.1) (from the Boltzmann equation rather than *ab initio*) is as follows. We recall [Eq. (28.14)] that

$$J_i^{(n)} = \tfrac{1}{2}\varrho \sum_{r,s=0}^{\infty} \beta_{ijk}^{(nrs)} a_j^{(r)} a_k^{(s)}, \tag{32.4}$$

from which we conclude that the collision term $J(f, f)$ has the expansion

$$J(f,f) = \frac{1}{2}\varrho f^{(0)} \sum_{n=1}^{\infty} \sum_{r,s=0}^{\infty} \frac{1}{n!} \beta_{ijk}^{(nrs)} \mathscr{H}_i^{(n)} a_j^{(r)} a_k^{(s)}. \tag{32.5}$$

For Maxwellian molecules only the terms $r+s=n$ enter; for the linearized equation only the terms $r=0$, $s=n$ or $r=n$, $s=0$ enter. Making these approximations,

$$J(f,f) = \varrho f^{(0)} \sum_{n=1}^{\infty} \frac{1}{n!} \beta_{ik}^{(n0n)} \mathscr{H}_i^{(n)} a_k^{(n)}. \tag{32.6}$$

The final approximation,

$$\varrho \beta_{ik}^{(n0n)} = \nu\, \delta_{ik} \tag{32.7}$$

yields the desired result, (32.1); (the symbol δ_{ik} is an abbreviation for a product of n Kronecker δ's, each with a single i and a single k). Roughly speaking, (32.1) results from the assumption that $\beta^{(nrs)}$ is virtually diagonal. There is no coupling between different Hermite modes and all relaxation times are equal.

The Eq. (32.1) has been used successfully in a number of specific problems [*2*], [*27*], [*28*]. However, it would seem that solution of the actual Boltzmann equation in any specific problem is not very much more difficult. In cases where the requisite collision integrals have already been evaluated, there is hardly any difference at all in labor. The most significant application of this equation at the present time would seem to be in problems which require half-range expansions for which the collision integrals are not known. For a quick survey of possibilities (32.1) is valuable, but with moderate additonal effort the results can be made quantitatively correct by using the actual Boltzmann equation.

The situation would be radically different if strong analytical use could be made of the specific form of the collision term in (32.1). This has not yet been done. So far as existence theory goes, the Eq. (32.1) seems to be as intractable as the Boltzmann equation itself; even if ν is taken to be a constant, $f^{(0)}$ is a very strongly nonlinear term.

33. Solution of special problems. In this section we shall survey the solutions of a number of representative problems in order to cast some light on the relative merits of the various techniques for solving the Boltzmann equation. A much more complete analysis of these and other problems including the relation to experiment is given in Vol. IX of this Encyclopedia[1].

First we list the problems to be considered. There are the dispersion of sound, the shock layer, heat flow between parallel plates, plane Couette flow (shear flow), the Stokes problem of flow around a sphere, and the Rayleigh problem of an impulsively started plate. These problems can be classified in several ways. There are problems without boundary and those with boundary; there are linear problems and nonlinear ones; there are simple geometries and more complex ones; finally we consider various methods of solution.

[1] S. A. SCHAAF: Mechanics of Rarefied Gases.

Not every problem has been solved by every method, and we shall emphasize those problems which provide interesting comparisons. The Knudsen or free flow solution and the Navier-Stokes solution hold a unique position as universal standards of comparison. They represent the two extremes which the Boltzmann equation is intended to bridge, and they are simple enough mathematically to have been solved in each of the examples to be considered. The problems we are considering are so primitive from the fluid dynamical viewpoint that in most of these problems the solution of the Euler equations is trivial.

The mathematical difficulty of a problem is governed primarily by the number of independent variables, the complexity of the geometry, and the fact of linearity or nonlinearity. The Stokes and Rayleigh problems are the most difficult because they involve two independent variables (i.e. partial differential equations) and nontrivial boundaries. No attempt has been made on the nonlinear problems in these cases (we exclude from consideration purely fluid dynamical treatments which have no bearing on the solution of the Boltzmann equation).

The heat flow, shear flow, and shock problems are one-dimensional (involving ordinary differential equations) and therefore allow the treatment of the nonlinear versions but only in the more primitive approximations to the Boltzmann equation. In the linear versions, the heat flow and shear flow problems are simple enough to allow an essentially complete solution to be obtained. The shock problem does not exist in a linearized version.

We can summarize the present state of the subject as follows. Linear two-dimensional and nonlinear one-dimensional problems can be solved to some low order of approximation to the Boltzmann equation. One-dimensional linear problems can be solved to a high order of approximation in the simplest geometrical configurations, i.e. parallel planes or possibly concentric cylinders.

The problem of sound dispersion is special because, although it is two-dimensional (x and t), the variables can be separated and, at least at the outset, the problem can be treated without boundaries. This problem has been solved to a high degree of approximation for Maxwellian molecules by Wang Chang and Uhlenbeck [68]. The method is the Hermite or Sonine polynomial expansion described in Sect. 30. The results are computed in three stages. The "zeroth approximation" is the Euler equations. The "first approximation" is the thirteen moment equations. The "second approximation" includes s_{ijk}, q, and q_{ij}. The "third approximation" (which is only named but not computed) adds the outer shell of fourth moments, q_{ijkl}, as well as the first two layers of fifth moments (vector and third order tensor). This somewhat arbitrary subdivision into successive approximations is chosen because it insures that an expansion of the various dispersion coefficients in powers of the frequency is unaltered up to some definite order by all further approximations. For some reason, the first approximation is called "Navier-Stokes" and the second "Burnett". They are evidently much more elaborate approximations than their appelations indicate and, in principle, this expansion in orthogonal functions is entirely distinct from the normal solution method. In the second approximation the variables are eight in number, ϱ, u_1, p, p_{11}, S_1, s_{111}, q, and q_{11}. The dispersion relation is a polynomial which is eighth degree in frequency and sixth degree in wave number. There are eight time-like modes and six space-like. This means that eight initial values and, in all likelihood, three boundary values can be imposed. These results should turn out to be very accurate if they are coupled with appropriate boundary conditions. At wavelengths which are comparable to the mean-free-path, it can be verified[1] that the mode which is the analytical continuation of the ordinary

[1] H. Grad, address before the American Physical Society, Pittsburgh, March, 1951.

undamped soundspeed at large wavelength is damped as highly as the other modes. Consequently, all three modes will be observed simultaneously in any high frequency experiment. Their relative amplitudes can be found analytically only by solving a boundary value problem. Also, the speed of convergence of the Wang Chang-Uhlenbeck approximations cannot be accurately assessed without solving the relevant boundary value problems. This has not yet been done; consequently comparison with experimental is not yet definitive.

A problem which has apparently not been posed before is the "Knudsen" problem for sound dispersion. Within one mean-free-path or so of an oscillating boundary, a free molecular flow solution can probably be computed with no great difficulty. The damping will quite likely turn out to be linear (the damping mechanism is the shift in phase of particles which hit the wall at different times). This solution should be compared to the initial slope of the exponentially damped solutions near the boundary.

The simplest shock problem is stated as follows. We look for a one-dimensiona steady flow in which a transition occurs from one constant state at $x=-\infty$ to another constant state at $x=+\infty$. The two states will be related by the conventional jump conditions. This solution confirms the traditional discontinuous approximation only when the transition layer is macroscopically thin. Early Navier-Stokes treatments using constant coefficients of heat conduction and viscosity were made by RANKINE [*55*], TAYLOR [*61*], RAYLEIGH [*56*], and BECKER [*1*]. BECKER noted in a special case that the shock thickness became vanishingly small for strong shocks (smaller than the mean-free-path). In the same special case, THOMAS [*62*] observed that taking $\mu \sim T^{\frac{1}{2}}$ (elastic sphere model) made the limiting thickness finite, and MORDUCHOW and LIBBY [*48*] observed that with a variation $\mu \sim T^s$, the limiting thickness is zero for $s<\frac{1}{2}$ but it is infinite for $s>\frac{1}{2}$. The explanation for this anomalous behavior is as follows [*41*], [*23*]. The mean-free-path is a function of temperature as well as density except in the case of elastic spheres, $s=\frac{1}{2}$, where it depends on density alone. The density ratio across a shock is at most four; the temperature ratio approaches infinity for strong shocks. Except in the case of elastic spheres, the mean-free-path varies enormously as one crosses a strong shock, and whether the front or back sees a larger mean-free-path depends on whether $s<\frac{1}{2}$ or $s>\frac{1}{2}$. The anomalous results follow from the convention of using the mean-free-path in front of the shock as a reference length. If the reference point is chosen appropriately within the shock, then the thickness remains bounded for all strengths and for any temperature dependence of μ and λ. Moreover, if $x/L=y$ is introduced as a new distance variable where L is the *variable* mean-free-path (this is a nonuniform stretching), the shock profile becomes almost universal and varies by only a few percent for an arbitrary choice of $\mu(T)$ [*23*].

Solutions of this problem have also been obtained using the Burnett equations and the next approximation in the Enskog expansion [*41*], [*67*], [*72*], the thirteen moment equations [*23*], and by a special approximation method of MOTT-SMITH [*52*]. Mott-Smith approximates the distribution function by a sum of two locally Maxwellian distributions with different parameters ϱ, $\boldsymbol{u}$, and T. Differential equations for these parameters are obtained by taking appropriate moments of the Boltzmann equation after inserting this approximation for f. For weak shocks where the other theories are known to be correct, the Mott-Smith result is poor. However, it is intended as a theory applicable to strong shocks and, if the sum of two Maxwellians does approximate the distribution function of a strong shock (the idea is that it is a superposition of two streams one from $+\infty$ and the other from $-\infty$), it could turn out to be quite accurate for strong shocks.

It is quite difficult to compare results because there are so many parameters; the results depend on the choice of molecular force law and on the choice of observable (the temperature profile will look entirely different from the density profile) as well as on the strength. With a certain definition of shock thickness, the thickness, δ, of the velocity profile for a Maxwellian gas is found to be as follows[1]: for the Navier-Stokes equations

$$\delta = 1 - 0.497\,\alpha^2 + \cdots, \tag{33.1}$$

for the Burnett equations

$$\delta = 1 + 0.032\,\alpha^2 + \cdots, \tag{33.2}$$

for the next order Chapman-Enskog approximation

$$\delta = 1 + 0.094\,\alpha^2 + \cdots, \tag{33.3}$$

and for the thirteen moment equations

$$\delta = 1 + 0.712\,\alpha^2 + \cdots. \tag{33.4}$$

The parameter α is a measure of the shock strength; α ranges from zero for weak shocks to one for infinitely strong shocks. The Chapman-Enskog expansion shows a steady increase in thickness with successive approximations, but the thirteen moment thickness is considerably larger. The Mott-Smith zero-strength thickness is about 1.25 in these units. The Wang-Chang results go up to order α^2 (the expansion is even in α) for the case of Maxwellian molecules. The thirteen moment results (as well as the Navier-Stokes results) have been computed up to order α^4 for arbitrary molecular forces [*23*]. In the Navier-Stokes case, this expansion up to terms in α^4 is found to be a very good approximation for all strengths including infinite strength ($\alpha = 1$).

The fact that the Navier-Stokes equations are parabolic and the thirteen moment equations are hyperbolic is brought out in a striking way. The speed of a weak shock is comparable to the ordinary gas soundspeed, $a^2 = \frac{5}{3}$ in the notation of Sect. 29. With increasing strength the speed increases, and at a certain strength ($\alpha = 0.518$) the shock speed reaches the value of the fastest characteristic, $a^2 = 4.54$, of the thirteen moment equations [Eq. (29.5)]. For greater shock strengths, a true discontinuity shock develops in the thirteen moment theory; heuristically, no smooth solution can propagate at a speed faster than that of small disturbances. The thirteen moment equations are ultraconservative; they refuse to predict results which may be inaccurate. On the other hand, the Navier-Stokes equations sometimes predict smooth solutions for strengths greater than infinity (with a transition from positive to negative density)! The Burnett solution has only been evaluated as a power series in the shock strength, and its behavior for strong shocks is not known. The Mott-Smith thickness decreases with increasing shock strength and just about crosses the thirteen moment value where the thirteen moment solution breaks down.

Better results in the shock problem could be obtained by the use of the higher order approximation (28.33) to (28.35) or by expanding about an anisotropic $f^{(0)}$, or by iteration using (19.12).

[1] Ref. [*67*], [*41*] and [*23*]. The Wang-Chang expansion is found to converge much more rapidly when expressed in terms of this expansion parameter $\alpha\,(=\frac{1}{3}\varepsilon$ in the notation of [*23*]) used by Kohler and Grad.

The problem of linear one-dimensional heat flow has been solved to a high degree of approximation by WANG CHANG and UHLENBECK [69] and also by MOTT-SMITH [53] using the polynomial expansion of Sect. 30. Because the problem is linear, the precise form of the solution can be explicitly written. The distribution function is a sum of a term which is linear in x and an infinite sum of exponentials. The linear term is the conventional heat flow term. The exponents must be found by solving an infinite determinant. The numerical coefficients are obtained by inversion of an infinite matrix which depends on the boundary condition applied on f. What is done in the papers quoted is to compute the solution to a certain order by dropping all polynomials above that order. The meaning of this solution is intuitively clear. There is a finite temperature jump at the wall with magnitude depending on the mean-free-path, and there is an exponential boundary layer on the order of the width of the mean-free-path as discussed in Sect. 26.

In addition, WANG CHANG and UHLENBECK compute the solution using the Hilbert expansion (in this special problem it happens to be identical to the Chapman-Enskog solution of the problem). It turns out that the first order solution, i.e.,

$$q_i = -\lambda \frac{\partial T}{\partial x_i}, \tag{33.5}$$

is the exact normal solution to all orders. Moreover, this solution is identical to the term linear in x of the complete solution. We can say that the exact solution to the linearized heat flow problem is the sum of the Hilbert-Enskog-Chapman solution (which converges trivially in this problem) and an exponential boundary layer. If the separation of the plates is not significantly larger than the mean-free-path, then the boundary layer cannot be distinguished from the flow as a whole.

The question of boundary conditions is particularly interesting. As a general rule, we expect the number of boundary conditions in the Hilbert expansion to be independent of the order of approximation and to be the same as for the Euler equations (Sect. 24). The Chapman-Enskog expansion is expected to require an increasing number of boundary conditions in general. Neither expectation is fulfilled in this example. In the given geometry, the Euler equations do not allow the specification of anything more than zero velocity at the two ends. However, with velocity *identically zero*, the problem is degenerate and no amount of boundary restriction can determine the solution which allows the temperature to be an arbitrary function of x. The higher order compatibility conditions in the Hilbert expansion cut down this arbitrariness to a single parameter in each order; this could be the heat flow or the value of the temperature gradient. The degree of flexibility is thus the same as for the Navier-Stokes equations. In the Chapman-Enskog expansion, the result is also degenerate, presumably because the velocity and pressure are identically constant, and no higher derivatives enter into approximations beyond the Navier-Stokes.

The nonlinear problem, for large values of the heat flow, is easily solved for the Navier-Stokes equations and for the thirteen moment equations; in this example the two are identical. It is not clear how difficult it would be to carry out high order approximations using the methods of Sects. 28 or 30.

The problem of plane Couette flow is very similar to the heat flow problem. By an expansion in Sonine polynomials, MOTT-SMITH obtains a general form for the solution [53] which is again the sum of a linear term and an infinite sum of exponentials. The linear term is the conventional solution with a constant velocity gradient and heat flow linear in x, and the exponential terms represent a boundary layer when the plates are separated by a number of mean-free-paths. MOTT-SMITH gives a very simple approximate formula to take into account the boundary layer by adding to the width of the apparatus a multiple of the mean-free-path which is almost independent of the parameters. This could be considered to be the "asymptotic" boundary condition referred to in Sect. 26 for normal solutions.

This problem has also been solved by GROSS, JACKSON, and ZIERING [28] using half-range polynomials and the modified Boltzmann equation which was noted in Sect. 32. The same type of exponential boundary layer is found. It would be interesting to compare the convergence of this technique and the effort required with that of the full-range method just described for the actual Boltzmann equation.

An intriguing result has been found by TRUESDELL [64]. It is an exact solution of the Boltzmann equation for a certain type of shear flow. He observes that by inserting into the infinite set of exact nonlinear moment equations a linear velocity profile (e.g., $u = k y$, $v = w = 0$) and assuming that all other quantities are space-independent, there results a set of ordinary differential equations in time alone. These are linear with constant coefficients and can be solved for each higher moment in turn in the case of Maxwellian molecules ($a^{(n+1)}$ enters into the equation for $a^{(n)}$ only as a space derivative). The exponential growth of temperature and pressure is a consequence of the absence of heat flow to compensate the viscous dissipation. As might be expected, there is an initial boundary layer on the order of the collision time followed by a macroscopic exponential growth at a time scale governed by the viscous dissipation. It is interesting to note that the initial decay is not monotone; there are complex exponents. The normal solution expansion converges to the "dominant" solution, i.e., with the initial decay removed. However, the normal solution takes on the actual initial values only up to first order. The thirteen moment equations are exact for this example.

In this example, the parameter in the normal solution expansion is the collision time compared to the viscous growth exponent. Linearization of this problem corresponds to making the viscous growth time large. Consequently, linearization coincides with taking only the first term in the normal solution expansion; thus the two processes cannot be separated in this example as they can in the time-independent shear flow.

The nonlinear Couette flow, i.e., with large velocity gradients, can be solved by use of the Navier-Stokes equations [33] and the thirteen moment equations[1]. In this case the results are different but are very easily related. Introduce

$$\alpha = \frac{p_{xy}}{p} \tag{33.6}$$

and

$$\beta = \frac{\mu}{p}\frac{du}{dy}. \tag{33.7}$$

In the Navier-Stokes case, $\alpha + \beta = 0$, while in the thirteen moment case, there is a polynomial relation between α and β which reduces to $\alpha + \beta = 0$ for small α. The profile is determined implicitly by the equations

$$\gamma p y = \pm \int \frac{\mu(\tau)\, d\tau}{\sqrt{\tau_0 - \tau}}, \tag{33.8}$$

$$u^2 = \delta^2 (\tau_0 - \tau) \tag{33.9}$$

[1] See footnote 1, p. 287.

where $\tau = \mathsf{R}T$, μ is the viscosity coefficient, τ_0 is the maximum value attained by τ, and γ and δ are certain algebraic functions of α and β. For the Navier-Stokes equations they reduce to $\gamma^2 = \frac{8}{15}\alpha^2$ and $\delta^2 = 30$. The stresses are given by

$$p_{xx} : p_{yy} : p_{zz} : p_{xy} = -8\beta : 6\beta : 2\beta : 5. \tag{33.10}$$

It is hard to foresee how much difficulty is involved in extending such nonlinear results to higher order approximations.

The problem of cylindrical Couette flow has no simple explicit solution in the nonlinear case, and practical results can only be obtained numerically or by expansion in terms of various parameters [*59*].

Next we turn to the problem of slow flow around a sphere. The Navier-Stokes equations can be solved explicitly in this problem with a slip boundary condition [*44*] which for simplicity we take in the form

$$p_{ij} n_j = \beta u_i; \tag{33.11}$$

the tangential stress is proportional to the slip velocity. The drag on the sphere turns out to be

$$F = 6\pi\mu U a \frac{2\mu + \beta a}{3\mu + \beta a} \tag{33.12}$$

where U is the flow speed at infinity and a is the radius of the sphere. The slip coefficient β is proportional to the density. The mean-free-path has the order μ/β. For large mean-free-path compared to a, $F \to 4\pi\mu U a$, while for small mean-free-path, $F \to 6\pi\mu U a$ (STOKES' formula). The first result is evidently nonsense, because there can be no drag on a sphere in a vacuum. This is in direct contrast to the Couette flow problem just mentioned in which the Navier-Stokes equations combined with a slip boundary condition give excellent results over the whole range of mean-free-path; in particular the stress reduces to zero as the mean-free-path becomes large. It is easy to see that the accurate result in the case of the Couette flow is accidental. In any geometry in which a constant flow or a rigid body rotation is not permitted by the boundary conditions, the stresses

$$p_{ij} = -\mu \left(\frac{\partial u_i}{\partial x_j} + \frac{\partial u_j}{\partial x_i} - \frac{2}{3} \delta_{ij} \frac{\partial u_r}{\partial x_r} \right) \tag{33.13}$$

cannot vanish. The value of the density (i.e., mean-free-path) is irrelevant. About the only cases in which the Navier-Stokes equations are not immediately ruled out for low densities are geometries with parallel plane or concentric cylinder boundaries, or else no boundaries at all. However, almost all the solvable geometries fall into this very special category!

The linearized problem of the flow around a sphere has been solved for the thirteen moment equations and the boundary conditions (29.27) and (29.28). The solution is explicit and the drag is a complicated algebraic expression in the mean-free-path [*19*]. It exhibits the correct drag, viz., zero, for infinite mean-free-path, and it yields the correct Stokes limit for small mean-free-path, but it is not very accurate in between. However, it is the only solution to date which covers the whole range of mean-free-path.

Finally, we note the very interesting treatment of the impulsively started plate by YANG and LEES [*71*]. They use the linearized thirteen moment equations with slip boundary conditions and solve by the use of the Laplace transform. Since the equations are hyperbolic, a disturbance can only move away from the plate at a finite speed. A short time after the impulse, there is only a thin layer of gas near the plate which has started to move; this layer moves as a whole at an intermediate speed which is determined by the slip boundary condition. As time goes on, the wave front moves out and decays in magnitude, and the profile approaches the Rayleigh result

$$\frac{1}{\sqrt{t}} e^{-\frac{x^2}{t}} \tag{33.14}$$

asymptotically. This problem has the unique feature of a transition from free flow for small times to classical fluid dynamics for large times, and it avoids the initial infinity of the Navier-Stokes solution.

It is clear from this brief summary that the problems that can be solved by these techniques are far from exhausted. However, the relatively complete results which have been found in the linear problems of sound dispersion, heat flow, and shear flow are unlikely to be easily extensible to many other problems. Even in these problems, it is only the form of the solution which is obtained in complete generality; there is considerable computation required for each additional step in the approximation.

References.

[1] Becker, R.: Z. Physik **8**, 321 (1922).
[2] Bhatnager, P. L., E. P. Gross and M. Krook: Phys. Rev. **94**, 511 (1954).
[3] Bogolubov, N.: J. Phys. USSR. **10**, 265 (1946).
[4] Born, M., and H. S. Green: Proc. Roy. Soc. Lond., Ser. A **188**, 10 (1946).
[5] Born, M., and H. S. Green: A General Kinetic Theory of Liquids. Cambridge 1949.
[6] Brout, R.: Physica, Haag **22**, 509 (1956).
[7] Brout, R., and I. Prigogine: Physica, Haag **22**, 621 (1956).
[8] Burnett, D.: Proc. Lond. Math. Soc. **40**, 382 (1935).
[9] Carleman, T.: Acta math., Stockh. **60**, 91 (1933).
[10] Chapman, S.: Phil. Trans. Roy. Soc. Lond. A **217**, 115 (1917).
[11] Chapman, S., and T. G. Cowling: The Mathematical Theory of Non-Uniform Gases, Cambridge University Press 1952.
[12] Cotter, J. R.: Proc. Roy. Irish Acad. A **55**, 1 (1952).
[13] deGroot, S. R.: Thermodynamics of Irreversible Processes. New York: Interscience 1951.
[14] deMarcus, W. C.: The Problem of Knudsen Flow. U.S. Atomic Energy Commission, Report K-1302, ORGDP.
[15] Ehrenfest, P., u. T. Ehrenfest: Encykl. math. Wiss. **4**, 2 (1911).
[16] Enskog, D.: Kinetische Theorie der Vorgänge in mäßig verdünnten Gasen. Uppsala 1917.
[17] Enskog, D.: Ark. Mat., Astronom. Phys., Ser. A **21**, *1* (1928).
[18] Frisch, H. L.: J. Chem. Phys. **22**, 1713 (1954).
[19] Goldberg, R.: Slow Flow of a Rarefied Gas Past a Spherical Obstacle. Thesis, New York University, 1954.
[20] Grad, H.: Kinetic Theory and Statistical Mechanics, mimeographed notes. Institute of Mathematical Sciences, New York University, 1949.
[21] Grad, H.: Comm. Pure Appl. Math. **2**, 325 (1949).
[22] Grad, H.: Comm. Pure Appl. Math. **2**, 331 (1949).
[23] Grad, H.: Comm. Pure Appl. Math. **5**, 257 (1952).
[24] Grad, H.: Comm. Pure Appl. Math. **5**, 455 (1952).
[25] Grad, H.: J. Phys. Chem. **56**, 1039 (1952).
[26] Green, M. S.: J. Chem. Phys. **25**, 836 (1956).
[27] Gross, E. P.,and M. Krook: Phys. Rev. **102**, 593 (1956).
[28] Gross, E. P., E. A. Jackson and S. Ziering: Ann. of Physics **1**, 141 (1957).
[29] Hecke, E.: Math. Z. **12**, 274 (1922).
[30] Hilbert, D.: Grundzüge einer allgemeinen Theorie der linearen Integralgleichungen. Wien: J. B. Teubner 1924.
[31] Hopf, E.: Ergodentheorie. New York: Chelsea Publ. Co. 1948.
[32] Ikenberry, E., and C. Truesdell: J. Rat. Mech. Analysis **5**, 1 (1956).
[33] Illingworth, C. R.: Proc. Cambridge Phil. Soc. **46**, 469 (1950).
[34] Irving, J. H., and J. G. Kirkwood: J. Chem. Phys. **18**, 817 (1950).
[35] Jaffe, G.: Ann. der Phys. **6**, 195 (1930).

[36] JEANS, J. H.: The Dynamical Theory of Gases. Cambridge 1916.
[37] KAC, M.: Proceedings of Third Berkeley Symposium on Mathematical Statistics and Probability. U. of Cal. Press 1956.
[38] KELLER, J. B.: Comm. Pure Appl. Math. **1**, 275 (1948).
[39] KENNARD, E. H.: Kinetic Theory of Gases. New York: McGraw-Hill 1938.
[40] KIRKWOOD, J. G.: J. Chem. Phys. **15**, 72 (1947).
[41] KOHLER, M.: Théorie moléculaire de l'onde de choc dans les gaz monoatomiques. Laboratoire d'etudes balistiques de Saint Louis, 1946.
[42] KOHLER, M.: Z. Physik **124**, 772 (1947).
[43] KOHLER, M.: Z. Physik **27**, 215 (1950).
[44] LAMB, H.: Hydrodynamics. New York: Dover Publ. 1945.
[45] LEBOWITZ, J. L., and H. L. FRISCH: Phys. Rev. **107**, 917 (1957).
[46] MAXWELL, J. C.: Scientific Papers. New York: Dover Publ. n. d.
[47] MERLIC, E., and F. C. HURLBUT: The Determination of slip flow coefficients for various Histories. Inst. of Engineering Res., U. of Calif., 1956.
[48] MORDUCHOW, M., and P. A. LIBBY: J. Aer. Sci. **16** (1949).
[49] MORGENSTERN, D.: Proc. Nat. Acad. Sci. **40**, 719 (1954).
[50] MORGENSTERN, D.: J. Rat. Mech. Analysis **4**, 533 (1955).
[51] MORREY, C. B.: Comm. Pure a. Appl. Math. **8**, 279 (1955).
[52] MOTT-SMITH, H. M.: Phys. Rev. **82**, 885 (1951).
[53] MOTT-SMITH, H. M.: A New Approach in the Kinetic Theory of Gases. Lincoln Laboratory, M.I.T. 1954.
[54] NOLL, W.: J. Rat. Mech. Analysis **4**, 627 (1955).
[55] RANKINE, W. J. M.: Trans. Roy. Soc. **160** (1870).
[56] RAYLEIGH, J.: Proc. Roy. Soc. Lond., Ser. A **84** (1910).
[57] ROGOSINSKI, W. W.: Volume and Integral. New York: Interscience 1952.
[58] ROSE, M. (HEINEMAN): Comm. Pure Appl. Math. **1**, 259 (1948).
[59] ROSE, M. (HEINEMAN): Phys. Rev. **91**, 469 (1953).
[60] SHANNON, C. E., and W. WEAVER: The Mathematical Theory of Communication. Univ. of Illinois Press 1949.
[61] TAYLOR, G. I.: Proc. Roy. Soc. Lond., Ser. A **84** (1910).
[62] THOMAS, L. H.: J. Chem. Phys. **12**, 449 (1944).
[63] TITCHMARSH, E. C.: The Theory of Functions. Oxford University Press 1939.
[64] TRUESDELL, C.: J. Rat. Mech. Analysis **5**, 55 (1956).
[65] VAN HOVE, L.: Physica, Haag **15**, 951 (1949).
[66] VAN HOVE, L.: Physica, Haag **21**, 517 (1955).
[67] WANG CHANG, C. S.: On the Theory of the Thickness of Weak Shock Waves. Dept. of Engr. Research, U. of Mich., 1948.
[68] WANG CHANG, C. S., and G. E. UHLENBECK: On the Propagation of Sound in Monotonic Gases. Engr. Research Inst., U. of Mich., 1952.
[69] WANG CHANG, C. S. and G. E. UHLENBECK: The Heat Transport Between two Parallel Plates as Functions of the Knudsen Number. Engr. Research Inst., U. of Mich., 1953.
[70] WILD, E.: Proc. Cambridge Phil. Soc. **47**, 602 (1951).
[71] YANG, H. T., and L. LEES: RAYLEIGH's problem at low mach number according to the kinetic theory of gases. Guggenheim Aeron. Lab., Cal. Inst. of Tech. 1955.
[72] ZOLLER, K.: Z. Physik **130**, 1 (1951).

Transporterscheinungen in Gasen von mittlerem Druck.

Von

Ludwig Waldmann.

Mit 56 Figuren.

A. Phänomenologie der Transportvorgänge.

Im Teil A werden, nach einer qualitativen Übersicht über die Erscheinungen in Gasen, die phänomenologischen Ansätze für Reibung, Wärmeleitung und Diffusion besprochen, welche später nach der kinetischen Theorie eine genauere Begründung erfahren.

1. Einteilung der Erscheinungen in Gasen. Begrenzung des Themas. Historisches. Die kinetische Gastheorie hat in der Entwicklung der neueren Physik eine wesentliche Rolle gespielt. Bevor es durch die Deutung der Brownschen Bewegung, der Röntgenstrahlinterferenzen und der Radioaktivität einen augenfälligen Beweis für die atomare Struktur der Materie gab, hatte die Atomvorstellung ihre große Kraft bei der Erklärung der Eigenschaften von Gasen bewährt. Der enge Zusammenhang zeigt sich sogleich, wenn man versucht, eine Einteilung der Erscheinungen in Gasen vorzunehmen: dies kann nur geschehen, indem man von Anfang an Begriffe der Atomistik benutzt.

Das Gas befinde sich in einem Gefäß von der linearen Abmessung L. Die Anzahl der Moleküle pro Volumeinheit (Konzentration) werde mit n bezeichnet. Die Moleküle seien als starre Kugeln vom Durchmesser d idealisiert[1]. Die Erscheinungen in dem Gas sind also durch die drei Längen gekennzeichnet

$$\left.\begin{array}{ll} \text{Gefäßabmessung} & L\,, \\ \text{mittlerer Molekülabstand} & a = n^{-\frac{1}{3}}\,, \\ \text{Moleküldurchmesser} & d\,. \end{array}\right\} \tag{1.1}$$

Anstelle des mittleren Molekülabstandes kann man auch die freie Weglänge

$$l = \frac{1}{n\,\pi\,d^2} = \frac{a^3}{\pi\,d^2} \tag{1.2}$$

benutzen [Clausius (1858); erste zahlenmäßige Berechnung aus der Viscosität durch Maxwell 1860 [38], Bd. 1, S. 391].

Je nach dem Verhältnis dieser Längen zueinander hat man verschiedenes Verhalten des Gases zu erwarten; die Betrachtung der möglichen Grenzfälle gibt die Einteilung der Erscheinungen. Selbstverständlich ist dabei stets $L \gg d$ anzunehmen und wegen der Undurchdringlichkeit der Moleküle muß $a \gtrsim d$, d.h. auch $l \gtrsim d$, sein. Gemeinsam gilt in allen im folgenden genannten Fällen im

[1] Bei allgemeiner Wechselwirkung hat man unter d etwa den kleinsten Abstand zu verstehen, bis zu dem sich zwei Moleküle bei der betreffenden Temperatur im Mittel nähern können. Da es hier nur auf Größenordnungen ankommt, genügt diese qualitative Festlegung.

thermischen Gleichgewicht für die Translationsgeschwindigkeit der Moleküle das universelle Verteilungsgesetz von MAXWELL (1860; [38], Bd. 1, S. 377), welches erst dann außer Kraft tritt, wenn die Translationsbewegung quantenmechanisch beschrieben werden muß.

Wir betrachten zunächst

Fall I: $a \gg d$. Äquivalent mit dieser Beziehung ist nach (1.2) $l \gg d$. Dies bedeutet, daß die Moleküle sich meistens geradlinig bewegen und, wenn die Wechselwirkung der Moleküle untereinander überhaupt eine Rolle spielt, höchstens Zweierstöße in Betracht kommen. Es handelt sich um den eigentlichen Gaszustand und um das Erscheinungsgebiet, welches durch die Transportgleichung von MAXWELL (1866; [38], Bd. 2, S. 26ff.) und die Fundamentalgleichung von BOLTZMANN (1872) beherrscht wird.

Je nach der Gefäßabmessung sind weiter die Unterfälle zu unterscheiden

A. $L \gg l$ oder $n L d^2 \gg 1$

B. $l \gg L$.

Die *Bedingung A* besagt, daß ein Molekül beim Durchqueren des Gefäßes viele andere Moleküle trifft. Es handelt sich um ein *Gas von mittlerem Druck*, oder, wie man auch sagt, um den Gültigkeitsbereich der hydrodynamischen Näherung. Die molekularen Zweierstöße sind wesentlich. Es können generelle lineare phänomenologische Gesetze für Reibung, Wärmeleitung und Diffusion aufgestellt werden. Es ist, zumindest für einatomige Gase, gelungen durch Auswertung der Boltzmann-Gleichung den Zusammenhang zwischen der molekularen Wechselwirkung und den phänomenologischen Konstanten lückenlos herzustellen [ENSKOG (Dissertation 1917 [11]; 1921); CHAPMAN (1916, 1917); vorbereitet durch LORENTZ (1905 und [36]) und HILBERT [23]]. Auch die durch die Quantenmechanik bedingten Modifikationen konnten berücksichtigt werden [UEHLING und UHLENBECK (1933)]. Typisch für diesen Druckbereich ist, daß der Wärmestrom zwischen zwei planparallelen Platten verschiedener Temperatur vom Druck unabhängig und dem Plattenabstand umgekehrt proportional ist; entsprechendes gilt für die Reibung.

Die *Bedingung B* besagt, daß fast nur Stöße der Moleküle gegen die Gefäßwand vorkommen und den Zustand des Gases bestimmen; auf die molekularen Zweierstöße kommt es nicht an. Es handelt sich um ein *hochverdünntes Gas*, welches die Erscheinungen der Molekularströmung zeigt (KNUDSEN 1909, 1911). Hierfür können keine generellen phänomenologischen Gesetze aufgestellt werden; man hat jede experimentelle Anordnung für sich nach gaskinetischen Methoden zu behandeln. Die Wechselwirkung mit der Wand ist dabei entscheidend. Typisch für diesen Druckbereich ist, daß der Wärmestrom zwischen zwei parallelen Platten vom Plattenabstand unabhängig und dem Druck direkt proportional ist; entsprechendes gilt wieder für die Reibung.

Zu bemerken ist, daß die Fälle A und B nicht vollständig getrennt sind. Im Fall A sind zwar im Innern des Gases die molekularen Zweierstöße praktisch allein maßgebend, aber in der Randzone von der Dicke l bestimmen die Wandstöße den lokalen Zustand wesentlich mit. Dies führt zu den Oberflächenerscheinungen der mäßig verdünnten Gase: (Reibungs-)Gleitung (KUNDT und WARBURG 1875a), thermische Gleitung (MAXWELL 1879, [38], Bd. 2, S. 681), Diffusionsgleitung (KRAMERS und KISTEMAKER 1943), Temperatursprung (v. SMOLUCHOWSKI 1898). Bei gleichbleibender Gefäßabmessung und abnehmender Konzentration breitet sich die Randzone immer mehr aus. So geht Fall A kontinuierlich in Fall B über.

Für die Theorie besonders schwierig ist das Zwischengebiet $l \approx L$. Hierhin gehören auch die bei Überschallströmungen auftretenden Stoßwellen; als „Gefäßabmessung" ist in diesem Fall die Dicke der Stoßwelle zu denken, welche bei starken Stößen nicht viel größer als l ist (R. BECKER 1922).

Es bleibt noch zu betrachten

Fall II: $a \approx d$. Äquivalent damit ist nach (1.2) $l \approx d$. Dies bedeutet, daß ein Molekül fast ständig mit anderen in Wechselwirkung steht. Es handelt sich um ein *stark komprimiertes Gas* oder um kondensierte Materie. Die Boltzmann-Gleichung ist nicht mehr ausreichend, an ihre Stelle treten Erweiterungen, die von ENSKOG (1922) für Gase aus starren Kugeln und später von BORN und GREEN (1946, 1947) sowie von KIRKWOOD (1946) und IRVING und KIRKWOOD (1950) für den allgemeinen Fall aufgestellt wurden. Die Unterscheidung der obigen Fälle A und B gibt es natürlich hier nicht mehr; für komprimierte Gase und Flüssigkeiten gilt die Hydrodynamik, aber die theoretische Aufgabe, den Zusammenhang zwischen der molekularen Wechselwirkung und den phänomenologischen Konstanten herzustellen, wird sehr verwickelt und ist nicht so vollständig lösbar wie für die verdünnten Gase.

Dieser Artikel ist auf die Vorgänge in Gasen von mittleren Drucken beschränkt. Die Transportphänomene im Knudsen-Gebiet werden nicht behandelt (s. dazu HERZFELD [*21*], KENNARD [*31*], DUSHMAN [*10*] und PATTERSON [*41*]). Von komprimierten Gasen ist nur in Ziff. 5 noch kurz die Rede; es sei diesbezüglich auf die Bücher von CHAPMAN und COWLING [*4*], besonders Kap. 16, sowie HIRSCHFELDER, CURTISS und BIRD [*25*], besonders Kap. 9, verwiesen.

2. Die phänomenologischen Ansätze bei einfachen Gasen. Zur makroskopischen Beschreibung eines einfachen Gases dienen die Dichte ϱ, die Gasgeschwindigkeit $\boldsymbol{v} = (v_1, v_2, v_3) = (v_\mu)$ [1], die spezifische Energie u, der Drucktensor $p_{\mu\nu}$ und der Wärmestrom q_μ. Alle diese Größen sind Funktionen des Ortes $\boldsymbol{r} = (x_\mu)$ und der Zeit t.

Unter der spezifischen Energie u sei die Energie der Masseneinheit in dem lokalen, mit der Gasgeschwindigkeit bewegten Bezugssystem verstanden. u ist also von der Größenordnung des Quadrats der Schallgeschwindigkeit. Anstelle der Energie u, deren physikalische Bedeutung keiner weiteren Erläuterung bedarf, kann auch die lokale kinetische Temperatur $T(t, \boldsymbol{r})$ benutzt werden. T wird erhalten durch Auflösen der Beziehung

$$u = u_{\text{Gleichgewicht}}(T) \tag{2.1}$$

bei gegebenem u. Mit der Funktion $u_{\text{Gleichgewicht}}(T)$ ist der durch die Thermodynamik gegebene, für ideale Gase gültige Zusammenhang von Energie und Temperatur gemeint. Die kinetische Temperatur T ist also identisch mit der thermodynamischen Temperatur, welche sich in einem Massenelement einstellt, wenn man dieses ohne Änderung seiner Energie isoliert und das thermische Gleichgewicht abwartet.

Wie die kinetische Temperatur durch eine realisierbare Anordnung, z.B. mit dem Thermoelement, genau gemessen werden kann, ist eine Frage für sich und bedarf unter Umständen sorgfältiger Überlegung. Wegen Literatur über Temperaturmessung in schnell strömenden Gasen sei z.B. auf LIENEWEG [*35*], S. 253 verwiesen.

Nun sollen die Erhaltungssätze für Masse, Impuls, Energie und Drehimpuls formuliert und die hydrodynamische Näherung (s. Ziff. 1) für Reibung und

[1] Der Index $\mu (= 1, 2, 3)$ bezeichnet in diesem Artikel stets die drei Raumrichtungen eines cartesischen Koordinatensystems; bei Bedarf steht an Stelle von μ auch ν oder λ.

Wärmeleitung angegeben werden. Dabei können wir uns kurz fassen, da wir nochmals darauf zurückkommen, in Ziff. 31 auf die Erhaltungssätze, in Ziff. 43 ff. auf die hydrodynamische Näherung.

α) Die Erhaltung der Masse kommt in der *Kontinuitätsgleichung*

$$\frac{\partial \varrho}{\partial t} = -\frac{\partial \varrho v_\mu}{\partial x_\mu} \tag{2.2}$$

zum Ausdruck[1]. Mit dem Operator der substantiellen Differentiation

$$\frac{d}{dt} = \frac{\partial}{\partial t} + v_\mu \frac{\partial}{\partial x_\mu} \tag{2.3}$$

hat man auch

$$\frac{d\varrho}{dt} = -\varrho \frac{\partial v_\mu}{\partial x_\mu} \tag{2.4}$$

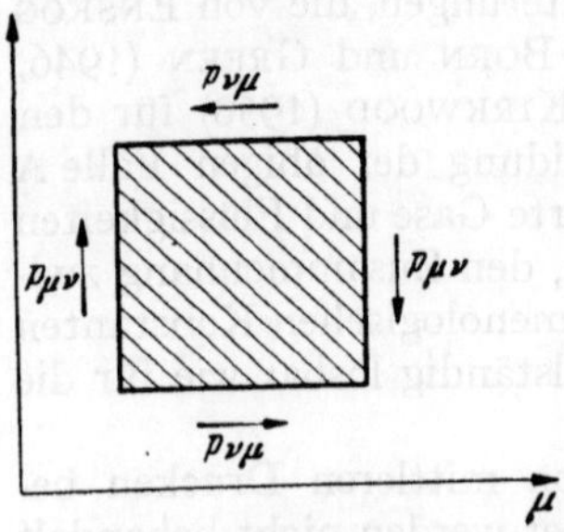

Fig. 1. Zur Definition des Drucktensors.

β) Zur Formulierung der *Bewegungsgleichung* werde zunächst die Druckdefinition erläutert. Aus der Gasmasse sei ein Würfel $(dx)^3$ herausgeschnitten gedacht. Um die Bewegung nicht zu stören, müssen dann an den Würfelflächen Ersatzkräfte angreifen. Es sei die Würfelfläche $(dx)^2$, deren in das Würfelinnere weisende Normale die negative x_μ-Richtung hat, ins Auge gefaßt. Die an ihr angreifende Ersatzkraft habe in der x_ν-Richtung die Komponente $-p_{\mu\nu}(dx)^2$. Zur Veranschaulichung dient Fig. 1; die Pfeile bedeuten die Richtung der Kraftkomponente bei positiver Tensorkomponente. Es gilt

$$p_{\mu\nu}(t, \boldsymbol{r}) = p_{\nu\mu}(t, \boldsymbol{r}); \tag{2.5}$$

der Drucktensor ist symmetrisch. Denn das Trägheitsmoment des Würfels ist $\propto (dx)^5$, das Drehmoment der Flächenkräfte $\propto (p_{\mu\nu} - p_{\nu\mu})(dx)^3$; damit die Winkelbeschleunigung endlich bleibt für $dx \to 0$, muß also (2.5) gelten. Voraussetzung ist dabei, daß die Moleküle keinen Eigendrehimpuls besitzen, s. Abschnitt δ dieser Ziffer. — Wegen der Druckmessung an strömenden Gasen, Pitot- und Prandtl-Staurohr, s. z.B. E. SCHMIDT [*48*], S. 257. — Die Bewegungsgleichung eines Gases, das sich in einem äußeren Kraftfeld befindet, $\boldsymbol{b}$ = Kraft pro Masseneinheit (Beschleunigung), lautet nun

$$\frac{\partial \varrho v_\mu}{\partial t} = -\frac{\partial}{\partial x_\nu}(\varrho v_\mu v_\nu) - \frac{\partial p_{\mu\nu}}{\partial x_\nu} + \varrho b_\mu \tag{2.6}$$

oder nach (2.2) und (2.3) auch

$$\varrho \frac{dv_\mu}{dt} = -\frac{\partial p_{\mu\nu}}{\partial x_\nu} + \varrho b_\mu . \tag{2.7}$$

Zum *Reibungsansatz* der hydrodynamischen Näherung gelangt man durch die Forderung, daß der Unterschied zwischen dem im thermischen Gleichgewicht bei der kinetischen Temperatur T gültigen Drucktensor und dem tatsächlichen Drucktensor dem Geschwindigkeitsgefälle proportional sein soll. Im thermischen Gleichgewicht gilt (ideale Zustandsgleichung)

$$p_{\mu\nu,\,\text{Gleichgewicht}} = n\,\mathbf{k}T\,\delta_{\mu\nu} . \tag{2.8}$$

[1] Über einen doppelten Index μ oder ν oder λ ist stets stillschweigend zu summieren. Es ist also z.B.

$$v_\mu v_\mu = v_1^2 + v_2^2 + v_3^2 = \boldsymbol{v}^2; \quad \frac{\partial v_\mu}{\partial x_\mu} = \frac{\partial v_1}{\partial x_1} + \frac{\partial v_2}{\partial x_2} + \frac{\partial v_3}{\partial x_3} = \operatorname{div} \boldsymbol{v}; \quad p_{\mu\nu} v_\nu = p_{\mu 1} v_1 + p_{\mu 2} v_2 + p_{\mu 3} v_3 .$$

Dabei ist n die Konzentration der Moleküle, k die Boltzmann-Konstante und $\delta_{\mu\nu}$ ($=1$ für $\mu=\nu$; $=0$ für $\mu\neq\nu$) der Einheitstensor. (In der makrophysikalischen Darstellung sollte man eigentlich von der Mol-Konzentration und der molaren Gaskonstanten sprechen; um Bezeichnungen zu sparen, ziehen wir es vor, die molekularen Größen zu verwenden.) Mit der Abkürzung

$$p = n\,\mathsf{k}T \tag{2.9}$$

— p sei als „Gleichgewichtsdruck" bezeichnet — lautet also der Reibungsansatz

$$p_{\mu\nu} = p\,\delta_{\mu\nu} - \eta'\,\frac{\partial v_\lambda}{\partial x_\lambda}\,\delta_{\mu\nu} - \eta\left(\frac{\partial v_\mu}{\partial x_\nu} + \frac{\partial v_\nu}{\partial x_\mu}\right). \tag{2.10}$$

η und η' sind die im allgemeinen von der Konzentration und der Temperatur abhängigen Viscositätskonstanten; ihre physikalische Einheit ist das „Poise" $=$ 1 g/cm sec.

Zur Erläuterung von (2.10) sei zunächst bemerkt, daß dies der allgemeinste, im Geschwindigkeitsgefälle lineare und mit (2.5) verträgliche Ansatz ist. Ein „Wirbelreibungsterm" $\partial v_\mu/\partial x_\nu - \partial v_\nu/\partial x_\mu$ kann bei Molekülen ohne Eigendrehimpuls nicht vorkommen. Man sieht das auch durch Betrachtung der Strömung mit verschwindender Deformation

$$\frac{\partial v_\mu}{\partial x_\nu} + \frac{\partial v_\nu}{\partial x_\mu} = 0.$$

Dann gilt auch $\partial v_\lambda/\partial x_\lambda = 0$ und die Reibung verschwindet. Das muß aber so sein, denn aus dem Verschwinden der Deformation schließt man, daß

$$v_\mu = -\,\omega_{\mu\nu}\,x_\nu \quad \text{mit} \quad \omega_{\mu\nu} = -\,\omega_{\nu\mu} = \text{const},$$

das Gas also wie ein starrer Körper, somit reibungsfrei, rotiert. Anläßlich einer Diskussion über „Wirbelreibung" hat FUES (1941) diese Dinge für einatomige Gase noch einmal erläutert.

Ferner ist eine Bemerkung zu der zweiten Viscositätskonstanten η' zu machen. Es sei eine allseitige Expansion des Gases

$$v_\mu = \omega\,x_\mu, \qquad \omega = \text{const} \tag{2.10a}$$

betrachtet. Der Druck beträgt in diesem Fall

$$p_{\mu\nu} = \left[p - \left(\eta' + \tfrac{2}{3}\eta\right) 3\omega\right]\delta_{\mu\nu}.$$

Den Koeffizienten

$$\eta_V = \eta' + \tfrac{2}{3}\eta$$

bezeichnet man als Volumviscosität. Diese verschwindet bei den einatomigen Gasen, was man sofort aus der gaskinetischen Definition für Druck und Temperatur in diesem Fall erkennt [s. bei Gl. (31.23)]. Mit Benutzung von η_V statt η' lautet (2.10)

$$p_{\mu\nu} = p\,\delta_{\mu\nu} - \eta_V\,\frac{\partial v_\lambda}{\partial x_\lambda}\,\delta_{\mu\nu} - \eta\left(\frac{\partial v_\mu}{\partial x_\nu} + \frac{\partial v_\nu}{\partial x_\mu} - \frac{2}{3}\,\frac{\partial v_\lambda}{\partial x_\lambda}\,\delta_{\mu\nu}\right). \tag{2.11}$$

Für die Spur des Drucktensors gilt

$$\frac{1}{3}\,p_{\mu\mu} = p - \eta_V\,\frac{\partial v_\lambda}{\partial x_\lambda}. \tag{2.12}$$

Nur bei einatomigen Gasen ist sie also mit dem Gleichgewichtsdruck p identisch.

Schließlich sei noch erwähnt, daß in allen praktischen Fällen die Reibungsterme in (2.10) sehr klein sind im Vergleich zu dem statischen Druck p. Das haben W. und G. MEISSNER (1939) und neuerdings SLAWSKI, MICHELS und JACOBS (1954) überlegt. MICHELS, JACOBS und SLAWSKI (1954) haben experimentell nachgewiesen, daß die Expansion eines Gases bei schnell bewegtem Kolben (fast Schallgeschwindigkeit) bis 1000 atm isentropisch ist.

Setzt man nun (2.11) in die Bewegungsgleichung (2.7) ein und nimmt dabei η und η_V als von x unabhängig an, so erhält man die Stokes-Navierschen Gleichungen für das Gas in der Form

$$\varrho \frac{dv_\mu}{dt} = -\frac{\partial p}{\partial x_\mu} + \eta \frac{\partial^2 v_\mu}{\partial x_\nu \partial x_\nu} + \left(\eta_V + \frac{1}{3}\eta\right) \frac{\partial^2 v_\nu}{\partial x_\mu \partial x_\nu} + \varrho b_\mu. \tag{2.13}$$

Man nennt

$$\mu = \frac{\eta}{\varrho} \tag{2.14}$$

die kinematische Zähigkeit (Dimension cm²/sec). Die charakteristische Zeit zum Abklingen einer Gasströmung in einem Gefäß der linearen Abmessung L beträgt

$$t_0 = \frac{L^2}{\mu}. \tag{2.15}$$

γ) *Die Energiegleichung* besagt, daß die Änderung pro Zeiteinheit der im Einheitsvolumen befindlichen Energie $\varrho(u+\frac{1}{2}v^2)$ durch die Divergenz des Energiestroms $\varrho(u+\frac{1}{2}v^2)\,v_\mu + q_\mu$, durch die Leistung des Drucks an der Volumeinheit $-\frac{\partial}{\partial x_\mu}(p_{\mu\nu}v_\nu)$ und durch diejenige des äußeren Kraftfeldes $\varrho b_\mu v_\mu$ hervorgerufen wird. Es ist also

$$\frac{\partial}{\partial t}\left[\varrho\left(u+\frac{1}{2}v^2\right)\right] = -\frac{\partial}{\partial x_\mu}\left[\varrho\left(u+\frac{1}{2}v^2\right)v_\mu + q_\mu\right] - \frac{\partial}{\partial x_\mu}(p_{\mu\nu}v_\nu) + \varrho b_\mu v_\mu \tag{2.16}$$

oder nach (2.3) und (2.7) auch

$$\varrho \frac{du}{dt} = -\frac{\partial q_\mu}{\partial x_\mu} - p_{\mu\nu}\frac{\partial v_\mu}{\partial x_\nu}. \tag{2.17}$$

Daraus erhält man die Gleichung für die Temperaturänderung mittels (2.1). Danach ist nämlich

$$du = c_v\, dT, \quad c_v = \text{spezifische Wärme bei konstanten Volumen}. \tag{2.18}$$

Also lautet (2.17) auch

$$\varrho c_v \frac{dT}{dt} = -\frac{\partial q_\mu}{\partial x_\mu} - p_{\mu\nu}\frac{\partial v_\mu}{\partial x_\nu}. \tag{2.19}$$

Mit der spezifischen Enthalpie

$$\left.\begin{aligned} &i = u + \frac{kT}{m}, \qquad di = c_p\, dT,\\ &c_p = \text{spezifische Wärme bei konstantem Druck} \end{aligned}\right\} \tag{2.20}$$

ergibt sich die äquivalente Beziehung

$$\varrho c_p \frac{dT}{dt} - \frac{dp}{dt} = -\frac{\partial q_\mu}{\partial x_\mu} - (p_{\mu\nu} - p\,\delta_{\mu\nu})\frac{\partial v_\mu}{\partial x_\nu}. \tag{2.21}$$

Der *Wärmestromansatz* der hydrodynamischen Näherung lautet

$$q_\mu = -\lambda \frac{\partial T}{\partial x_\mu}. \tag{2.22}$$

λ ist der im allgemeinen von der Konzentration und der Temperatur abhängige Wärmeleitkoeffizient. Er wird meistens in cal/grad cm sec angegeben. Zu dem Ansatz ist zu bemerken, daß er der einzig mögliche ist, wenn gefordert wird, daß der Wärmestrom linear vom Temperaturgefälle abhängt. Das Geschwindigkeitsgefälle kann in einem linearen Ansatz nicht vorkommen; daß auch das Konzentrationsgefälle nicht eingeht, wird in der kinetischen Theorie begründet (s. Ziff. 46). Setzt man (2.11) in (2.21) ein und vernachlässigt die Ortsabhängigkeit von p und λ sowie die Reibung [zu letzterem s. auch MADELUNG (1943)], so erhält man

$$\varrho\, c_p \frac{dT}{dt} - \frac{\partial p}{\partial t} = \lambda \frac{\partial^2 T}{\partial x_\mu \partial x_\mu}. \tag{2.23}$$

Man nennt

$$K = \frac{\lambda}{\varrho\, c_p} \tag{2.24}$$

die Temperaturleitfähigkeit (Dimension cm^2/sec). Die charakteristische Zeit zum Abklingen von Temperaturunterschieden eines Gases unter konstantem Druck in einem Gefäß der linearen Abmessung L beträgt

$$t_0 = \frac{L^2}{K}. \tag{2.25}$$

Aus dem Vorangehenden ist ersichtlich, daß in einem wärmeleitenden Gas, das in einem ruhenden Gefäß eingeschlossen ist, bei Abwesenheit eines äußeren Kraftfelds im stationären Zustand kein Druckgefälle und keine Strömung vorhanden ist. Dies ist typisch für den Gültigkeitsbereich der hydrodynamischen Näherung. Bei stark verdünnten Gasen gilt das nicht; hier sind Temperaturunterschiede im allgemeinen mit Strömungen verbunden (thermomolekulare Strömung und Druckdifferenz; Radiometereffekt).

Schließlich sei noch angemerkt, daß nach (2.15) und (2.25) die kinematische Zähigkeit μ und die Temperaturleitfähigkeit dimensionsgleich sind, daß also die reine Zahl

$$\frac{\mu}{K} = \frac{\eta\, c_p}{\lambda} \quad \text{(Prandtl-Zahl)} \tag{2.26}$$

existiert. Sie ist bei Gasen von der Größenordnung 1; ihre Berechnung ist ein wichtiges gaskinetisches Problem (s. Ziff. 50 und 52).

δ) Der Drehimpulssatz wird am deutlichsten, wenn man die bei (2.5) gemachte Annahme fallen läßt, daß die Moleküle keinen Eigendrehimpuls besitzen. Es sei $J_{\mu\nu}(t, \boldsymbol{r})$ der mittlere Eigendrehimpuls pro Molekül, im folgenden kurz Spin genannt, $j_{\mu\nu\lambda}$ die zugehörige Stromdichte. $J_{\mu\nu}$ ist ein schiefsymmetrischer Tensor, der mit einem axialen Vektor in der bekannten Weise $J_{12} = J_3$, cyclisch zusammenhängt. Ferner ist bei Spin-Molekülen zu unterscheiden zwischen der mittleren Molekulargeschwindigkeit v_μ und dem mittleren kanonischen Impuls pro Masseneinheit p_μ/m. Die Kontinuitäts-, Impuls- und Spinänderungs-Gleichungen haben ohne äußere Felder die Gestalt

$$\frac{dn}{dt} + n \frac{\partial v_\lambda}{\partial x_\lambda} = 0, \tag{2.27}$$

$$n \frac{dp_\mu}{dt} + \frac{\partial p_{\mu\lambda}}{\partial x_\lambda} = 0, \tag{2.28}$$

$$n \frac{dJ_{\mu\nu}}{dt} + \frac{\partial j_{\mu\nu\lambda}}{\partial x_\lambda} = D_{\mu\nu}, \tag{2.29}$$

wo die Drehmomentdichte durch

$$D_{\mu\nu} = p_{\mu\nu} - p_{\nu\mu} + n(p_\mu v_\nu - p_\nu v_\mu) \tag{2.30}$$

gegeben ist.

Diese Beziehungen, die man beispielsweise mittels des Diracschen Wellenfelds für den Spin $\frac{1}{2}$ begründen kann, garantieren die Erhaltung des Gesamtdrehimpulses. Um das zu sehen, werde mittels (2.28) zunächst die Änderungsgleichung des Bahndrehimpulses pro Molekül

$$l_{\mu\nu} = x_\mu p_\nu - x_\nu p_\mu$$

mit der zugehörigen Stromdichte

$$l_{\mu\nu\lambda} = x_\mu p_{\nu\lambda} - x_\nu p_{\mu\lambda}$$

aufgestellt. Man findet ohne weiteres aus (2.28)

$$n \frac{d l_{\mu\nu}}{dt} + \frac{\partial l_{\mu\nu\lambda}}{\partial x_\lambda} = - D_{\mu\nu}. \tag{2.31}$$

Addition zu (2.29) liefert für den Gesamtdrehimpuls $m_{\mu\nu} = l_{\mu\nu} + J_{\mu\nu}$ den Erhaltungssatz

$$n \frac{d m_{\mu\nu}}{dt} + \frac{\partial (l_{\mu\nu\lambda} + j_{\mu\nu\lambda})}{\partial x_\lambda} = 0. \tag{2.32}$$

Zugleich wird nochmals die Voraussetzung von (2.5) deutlich. Für spinlose Moleküle muß nämlich bereits (2.31) den vollständigen Drehimpulssatz darstellen. Daher muß hier $D_{\mu\nu} = 0$ sein und, weil jetzt überdies $p_\mu \propto v_\mu$, ist das mit der Symmetrie (2.5) des Drucktensors gleichbedeutend.

Es folgen noch einige Andeutungen über die Verallgemeinerung der phänomenologischen Ansätze auf Spinflüssigkeiten. Wir nehmen näherungsweise an, daß p_μ zu v_μ proportional sei. Den Drucktensor zerlegen wir in seinen symmetrischen und antisymmetrischen Anteil:

$$p_{\mu\nu} = p_{\mu\nu}^{(s)} + p_{\mu\nu}^{(a)}. \tag{2.33}$$

Der Reibungsansatz (2.11) bezieht sich jetzt auf den symmetrischen Anteil. Für den schiefsymmetrischen Anteil wird gesetzt

$$p_{\mu\nu}^{(a)} = - \frac{1}{2} \frac{n}{\tau} \left[J_{\mu\nu} - \frac{1}{2} \Theta \left(\frac{\partial v_\nu}{\partial x_\mu} - \frac{\partial v_\mu}{\partial x_\nu} \right) \right]. \tag{2.34}$$

Hier hat τ, eine Transportkonstante, die Bedeutung der Relaxationszeit des Spins. Sie ist bestimmt durch die Wechselwirkung der Spins mit der Bahnbewegung der Moleküle. Die Konstante Θ, von der Dimension eines molekularen Trägheitsmoments, ist thermodynamischer Natur. Sie bestimmt die gyromechanische Einstellung der Spins im thermischen Gleichgewicht bei gleichmäßiger Rotation der ganzen Substanz (Wirkung der Coriolis-Kraft). Durch den g-Faktor des Moleküls ist mit dieser Einstellung der Spins ein magnetisches Moment verknüpft (Magnetisierung durch Rotation, Barnett-Effekt).

Außer dem Eigendrehimpuls $J_{\mu\nu}$ oder statt seiner können mit den Molekülen andere Größen verknüpft sein; z.B. wird man einen symmetrischen Tensor zweiter Stufe (Quadrupolmoment) einzuführen haben um die Strömungsdoppelbrechung zu beschreiben. Die Aufstellung der entsprechenden Änderungsgleichungen und phänomenologischen Ansätze sei übergangen.

3. Die phänomenologischen Ansätze bei binären Gasgemischen. Beim binären Gemisch hat man es mit zwei Dichten ϱ_1, ϱ_2 und zwei (mittleren Molekular-) Geschwindigkeiten $v_{\mu 1}, v_{\mu 2}$ zu tun. Statt der Dichten kann man auch die

Konzentrationen n_1, n_2 der Moleküle zugrunde legen; sie hängen über die Molekülmassen m_1, m_2 miteinander zusammen:

$$\varrho_i = n_i m_i, \quad i = 1, 2. \tag{3.1}$$

Auf das Gemisch als Ganzes beziehen sich die Gesamtdichte und -konzentration

$$\varrho = \varrho_1 + \varrho_2, \quad n = n_1 + n_2, \tag{3.2}$$

sowie die mittlere Massengeschwindigkeit

$$v_\mu = \frac{1}{\varrho}(\varrho_1 v_{\mu 1} + \varrho_2 v_{\mu 2}) \tag{3.3}$$

und die mittlere Teilchengeschwindigkeit

$$w_\mu = \frac{1}{n}(n_1 v_{\mu 1} + n_2 v_{\mu 2}). \tag{3.4}$$

Die Geschwindigkeit jeder Komponente in dem einen oder anderen Bezugssystem kann man Massen-Diffusionsgeschwindigkeit

$$V_{\mu i} = v_{\mu i} - v_\mu \tag{3.5}$$

oder Teilchen-Diffusionsgeschwindigkeit

$$W_{\mu i} = v_{\mu i} - w_\mu \tag{3.6}$$

nennen. Ähnlich ist zu unterscheiden zwischen dem Wärmestrom q_μ und $q_\mu^{(w)}$ in den beiden Bezugssystemen.

Welches der genannten Bezugssysteme man anwendet, ist gleichgültig und nur eine Frage der Bequemlichkeit. Wie in Ziff. 32δ näher ausgeführt wird, schreibt sich die Bewegungsgleichung am einfachsten an Hand der mittleren Massengeschwindigkeit, während die Diffusions- und Wärmeleitungsvorgänge zweckmäßig von dem mittleren Teilchenruhsystem aus betrachtet werden.

Von einem äußeren Kraftfeld werde abgesehen. Die *Bewegungsgleichung* lautet dann ebenso wie die entsprechende für ein einfaches Gas (2.7) oder (2.13).

Die Diffusion bewirkt eine Änderung der Zusammensetzung des Gemisches. Es ist deshalb angebracht, die *Diffusionsgleichung* als Änderungsgleichung für den einen der Molenbrüche

$$\gamma_1 = \frac{n_1}{n} = \gamma; \quad \gamma_2 = \frac{n_2}{n} = 1 - \gamma \tag{3.7}$$

zu schreiben. Aus der Erhaltung der Teilchenzahlen beider Komponenten folgt

$$\left(\frac{d\gamma}{dt}\right)_w = -\frac{1}{n}\frac{\partial}{\partial x_\mu} n_1 W_{\mu 1} \tag{3.8}$$

mit

$$\left(\frac{d}{dt}\right)_w = \frac{\partial}{\partial t} + w_\mu \frac{\partial}{\partial x_\mu}. \tag{3.9}$$

Wegen der Herleitung sei auf Gl. (32.33) verwiesen. Für den Teilchendiffusionsstrom wird, wenn man Linearität in den Gradienten der in Betracht kommenden Größen verlangt, ein Ansatz gelten müssen der Art

$$n_1 W_{\mu 1} = -n D\left(\frac{\partial \gamma}{\partial x_\mu} + k_T \frac{\partial T}{T\,\partial x_\mu}\right). \tag{3.10}$$

Dieser Ansatz geht auf ENSKOG (1911, 1917) [*11*] und CHAPMAN (1917) zurück, die bei der Ausarbeitung der kinetischen Theorie binärer Gasgemische zu ihm gelangten [s. (56.1) und (59.6); nach (54.10) ist im kräftefreien Fall $d_{\mu 1} = \partial\gamma/\partial x_\mu$].

Der Diffusionsstrom besteht also aus zwei Anteilen, dem Konzentrations- und dem Thermodiffusionsstrom, entsprechend den beiden Summanden der Klammer. Statt des Thermodiffusionsverhältnisses k_T benutzt man häufig auch

$$\alpha = \frac{k_T}{\gamma(1-\gamma)}, \tag{3.11}$$

den Thermodiffusionsfaktor, womit (3.10) die Form annimmt

$$n_1 W_{\mu 1} = -n D\left(\frac{\partial \gamma}{\partial x_\mu} + \alpha \gamma (1-\gamma) \frac{\partial T}{T \partial x_\mu}\right). \tag{3.12}$$

Die Thermodiffusion wurde auf Grund der theoretischen Voraussage experimentell gesucht und gefunden von CHAPMAN und DOOTSON (1917) an CO_2/H_2-Gemisch.

Durch Einsetzen von (3.10) in (3.8) unter Vernachlässigung der Ortsabhängigkeit von n und D sowie der Thermodiffusion erhält man als Diffusionsgleichung

$$\left(\frac{d\gamma}{dt}\right)_w \approx D \frac{\partial^2 \gamma}{\partial x_\mu \partial x_\mu}. \tag{3.13}$$

Die charakteristische Einstellzeit des Konzentrationsgleichgewichts in einem Gefäß der Abmessung L beträgt also

$$t_0 = \frac{L^2}{D}. \tag{3.14}$$

Lösungen der Diffusionsgleichung für viele Versuchsanordnungen findet man in den Büchern von JOST [*29*] und CRANK [*7*].

Die *Temperaturgleichung* hat nach (32.37) mit ϱc_p statt $\frac{5}{2} n \mathrm{k}$ bei konstantem Druck und ohne Reibung sowie ohne äußeres Kraftfeld die Form

$$\varrho c_p \left(\frac{dT}{dt}\right)_w \approx -\frac{\partial q_\mu^{(w)}}{\partial x_\mu}. \tag{3.15}$$

Darin bedeutet $q_\mu^{(w)}$ den Wärmestrom, gemessen im Bezugssystem der mittleren Teilchengeschwindigkeit. Ein linearer Ansatz für $q_\mu^{(w)}$ wird ebenfalls die Gradienten des Molenbruches und der Temperatur enthalten; anstelle des ersteren kann man auch ebensogut einen der Diffusionsströme, etwa $n_1 W_{\mu 1}$ nehmen. So hätte man nochmals zwei phänomenologische Koeffizienten zu erwarten. Tatsächlich zeigt aber die kinetische Theorie, daß in dem Ansatz für den Wärmestrom wiederum der Thermodiffusionsfaktor auftritt und nur eine neue Konstante, der Wärmeleitkoeffizient, erscheint [s. die zu (57.1) führenden Überlegungen]. Es ist nach (59.9) beim binären Gemisch

$$q_\mu^{(w)} = -\lambda \frac{\partial T}{\partial x_\mu} + \alpha \mathrm{k} T \cdot n_1 W_{\mu 1}. \tag{3.16}$$

Auch diese Beziehung geht auf ENSKOG (1917) [*11*] und CHAPMAN (1917) zurück. Der erste Anteil stellt den (Temperatur-) Wärmestrom dar. Der Koeffizient λ kann mittels der stationären Wärmeleitung in ruhendem Gas gemessen werden, denn der zweite Anteil, der Diffusionswärmestrom, verschwindet bei ruhendem Gas im stationären Zustand. Wenn aber die Diffusion im Gang ist, so fließt nach (3.16) ein Wärmestrom, auch wenn zunächst gar kein Temperaturgefälle vorhanden ist. Dieser Wärmestrom ist die Ursache des Diffusionsthermoeffekts, welcher darin besteht, daß während der Diffusion von Gasen ineinander spontan Temperaturdifferenzen sich ausbilden, die — bei ruhendem Gas — wieder abklingen in dem Maß, wie die Diffusion abklingt. Dies wurde erstmals von DUFOUR (1873) beobachtet, aber in der Folge nicht weiter untersucht, bis CLUSIUS und WALDMANN (1942) den Effekt wiederentdeckten und WALDMANN (1943 bis 1947) auch experimentell nachwies, daß in (3.12) und (3.16) derselbe Koeffizient α vorkommt. Die Vorzeichen sind dabei so, daß dasjenige Gas, welches infolge Thermodiffusion die warme Region bevorzugt, sich auf Grund des Diffusionthermoeffekts erwärmt. Näheres über die Messungen findet sich in Ziff. 20 und 21.

Nun werde (3.16) in (3.15) eingesetzt und die Ortsabhängigkeit von λ und $\alpha \mathrm{k} T$ vernachlässigt. Das gibt die Temperaturänderungsgleichung

$$\varrho\, c_p \left(\frac{dT}{dt}\right)_w \approx \lambda \frac{\partial^2 T}{\partial x_\mu \partial x_\mu} - \alpha\, \mathrm{k} T \frac{\partial n_1 W_{\mu 1}}{\partial x_\mu} \tag{3.17}$$

oder nach (3.8), mit der Temperaturleitfähigkeit $K = \lambda/\varrho c_p$, auch

$$\left(\frac{dT}{dt}\right)_w \approx K \frac{\partial^2 T}{\partial x_\mu \partial x_\mu} + \frac{\alpha p}{\varrho\, c_p} \left(\frac{d\gamma}{dt}\right)_w . \tag{3.18}$$

Diese Gleichung ist als Ausgangspunkt bei der Diskussion des Temperatureffekts in diffundierenden Gasen geeignet.

Es sei noch bemerkt, daß nach (2.15) und (3.14) auch die kinematische Zähigkeit und der Diffusionskoeffizient dimensionsgleich sind, daß also bei Gemischen zu (2.26) weiter die reine Zahl

$$\frac{D}{\mu} = \frac{\varrho D}{\eta} \tag{3.19}$$

hinzukommt. Auch sie ist bei Gasen von der Größenordnung 1.

4. Dimensionsbetrachtungen über die Transportkoeffizienten. Massenabhängigkeit. Es sollen hier, ohne Benutzung der detaillierten kinetischen Theorie, einstweilen durch einfache Dimensionsbetrachtungen die Abhängigkeit des Reibungs-, Wärmeleit- und Diffusionskoeffizienten von der Temperatur und der Molekülmasse ermittelt werden für zwei Molekülmodelle, das der starren Kugeln und das der Potenzkraft-Zentren. Dazu hat man sich zunächst geeignete Einheiten für Länge, Masse und Zeit zu verschaffen, dann zu sehen, welche dimensionslosen Zahlen außerdem zur Verfügung stehen, und schließlich die Transportgrößen, d.h. Reibungsdruck, Wärme- und Diffusionsstrom in diesen Einheiten unter Berücksichtigung der dimensionslosen Zahlen auszudrücken. Es ist zweckmäßig, auf mäßig verdünnte Gase nicht gleich am Anfang zu spezialisieren, sondern zunächst Gase beliebiger Konzentration ins Auge zu fassen. Solche Dimensionsbetrachtungen wurden zuerst von RAYLEIGH (1900) angestellt.

α) *Gase beliebiger Konzentration.* 1. Starre Kugeln. Das Gas bestehe aus starren Kugeln vom (Wirkungs-)Querschnitt $\sigma \approx d^2$ (d = Kugeldurchmesser), der Masse m und der mittleren Geschwindigkeit c. Bis auf Zahlenfaktoren gilt

$$m\, c^2 = \mathrm{k} T. \tag{4.1}$$

Es können gewählt werden als

$$\text{Einheit der} \begin{Bmatrix} \text{Länge} & d \\ \text{Masse} & m \\ \text{Zeit} & d/c. \end{Bmatrix} \tag{4.2}$$

Außerdem muß die Konzentration n des Gases und im Fall des binären Gemischs der Molenbruch γ bekannt sein, womit dann alle zur Verfügung stehenden Daten aufgeführt sind. Aus der Konzentration läßt sich eine weitere Länge, $n^{-\frac{1}{3}}$, ableiten, deren Verhältnis zu der Einheit d die reine Zahl

$$d\, n^{\frac{1}{3}} \tag{4.3a}$$

bildet. Im Fall des Gemischs kämen noch γ sowie m_1/m_2 und d_1/d_2 hinzu. Zu bemerken ist, daß die bisherige Argumentation die klassische Mechanik voraussetzt. Nach der Quantenmechanik gibt es eine weitere dimensionslose Zahl, nämlich das Verhältnis von de Broglie-Wellenlänge zu Atomdurchmesser. Die Wellenlänge eines Teilchens der Geschwindigkeit c beträgt $h/m c \approx h/\sqrt{m k T}$, $h =$ Plancksches Wirkungsquantum. Also hat man die reine Zahl

$$\Lambda = \frac{h}{d\sqrt{m\,\mathrm{k}T}}. \tag{4.3b}$$

Man sieht daraus, daß die Welleneigenschaften besonders im Grenzfall leichter Atome He, H_2 und tiefer Temperatur ($\lesssim 100°$ K) sich bemerkbar machen werden (Genaueres s. in Teil G). Tatsächlich hat in allen sonstigen Fällen Λ einen so kleinen Wert, daß man in guter Näherung überhaupt $h=0$ setzen, also klassisch rechnen darf. Der quantenmechanische Parameter Λ wird im folgenden nur bei Bedarf wieder erwähnt.

Jetzt sollen Impulsstrom (= Druck), Wärmestrom und Teilchenstrom in den Einheiten (4.2) ausgedrückt werden. Druck ist Impuls pro Flächen- und Zeiteinheit oder Energie pro Volumeneinheit, Wärmestrom ist Energie pro Flächen- und Zeiteinheit, Teilchenstrom ist Anzahl pro Flächen- und Zeiteinheit. Demnach können gewählt werden als

$$\text{Einheit von} \begin{cases} \text{Druck} & m\,c^2\,d^{-3} \\ \text{Wärmestrom} & m\,c^3\,d^{-3} \\ \text{Teilchenstrom} & c\,d^{-3}. \end{cases} \tag{4.4}$$

In den phänomenologischen Ansätzen kommen ferner die Gasgeschwindigkeit und die Temperatur vor. Wir nehmen als

$$\text{Einheit der} \begin{cases} \text{Geschwindigkeit} & c \\ \text{Temperatur} & m\,c^2\,\mathrm{k}^{-1}. \end{cases} \tag{4.5}$$

Das Reibungsgesetz $p_{xy} = -\eta\,(\partial v_x/\partial y + \partial v_y/\partial x)$ schreibt sich nun in diesen Einheiten wie folgt

$$p_{xy}/m\,c^2\,d^{-3} = -\eta^* \left(\frac{\partial v_x}{\partial y} + \frac{\partial v_y}{\partial x}\right) \Big/ \frac{c}{d},$$

wo η^* eine von der Zahl (4.3a) abhängige dimensionslose Größe ist. Somit ist durch

$$\eta = \eta^*(d\,n^{\frac{1}{3}})\,\frac{m\,c}{d^2} \tag{4.6}$$

die Abhängigkeit der Viscosität eines reinen Gases von d, m, c und n gegeben. Über die Funktion η^* kann zunächst keine weitere Aussage gemacht werden.

Das Wärmeleitungsgesetz $q_x = -\lambda \frac{\partial T}{\partial x}$ lautet in unseren Einheiten

$$q_x/m\,c^3\,d^{-3} = -\lambda^* \frac{\partial T}{\partial x} \Big/ \frac{m\,c^2\,\mathrm{k}^{-1}}{d}.$$

Also gilt für den Wärmeleitkoeffizienten

$$\lambda = \lambda^*(d\,n^{\frac{1}{3}})\,\frac{\mathrm{k}\,c}{d^2}, \tag{4.7}$$

wo wieder λ^* eine dimensionslose Größe ist.

Für den Diffusionsstrom $j_x = -n D\,\frac{\partial \gamma}{\partial x}$ gilt in unseren Einheiten

$$j_x/c\,d^{-3} = -D^* \frac{\partial \gamma}{\partial x} \Big/ \frac{1}{d}.$$

Also hat man für den Selbst-Diffusionskoeffizienten

$$D = D^*(d n^{\frac{1}{3}}) \frac{c}{n d^2}. \tag{4.8}$$

D^* ist wieder dimensionslos. — Beim binären Gemisch werden η^*, λ^* und D^* auch von γ, m_1/m_2 und d_1/d_2 abhängen.

2. Potenzkraft-Zentren. Ähnliche Betrachtungen kann man machen, wenn sich die Moleküle mit der Kraft

$$K = \frac{\varkappa}{r^\nu}, \qquad \varkappa > 0$$

abstoßen. Eine Längeneinheit steht hier nicht unmittelbar zur Verfügung. Man kann sie aber gewinnen, indem man den Abstand r_0 aufsucht, bei dem die zu K gehörige potentielle Energie Φ gleich ist der kinetischen Energie zweier Atome:

$$\Phi_0 = \frac{\varkappa}{(\nu-1)\, r_0^{\nu-1}} = m c^2.$$

r_0 ist also der kleinste Abstand, bis zu dem sich die Atome bei zentralem Stoß nähern. Wir wählen sodann als Einheit der Länge

$$\left(\frac{\varkappa}{m c^2}\right)^{\frac{1}{\nu-1}} \propto \left(\frac{\varkappa}{\mathrm{k} T}\right)^{\frac{1}{\nu-1}} \tag{4.9}$$

und haben statt (4.3a) die reine Zahl

$$\left(\frac{\varkappa}{\mathrm{k} T}\right)^{\frac{1}{\nu-1}} n^{\frac{1}{3}}. \tag{4.10}$$

Diese Längeneinheit ist anstelle von d in (4.6) bis (4.8) einzusetzen.

β) Mäßig verdünnte Gase. Speziell werde nun der Fall der verdünnten Gase betrachtet. Wir lassen $n \to 0$ gehen, wobei gleichzeitig die Gefäßabmessung zunehmen soll, derart, daß sie groß gegen die freie Weglänge bleibt. Die Zahlenfaktoren η^*, λ^*, D^* in (4.6) bis (4.8) streben dabei konstanten Grenzwerten zu (s. unten) und man hat für ein Gas aus starren Kugeln mit der freien Weglänge (Zahlenfaktoren weggelassen)

$$l = \frac{1}{n d^2} \tag{4.11}$$

und der mittleren Stoßzeit

$$\tau = \frac{l}{c} = \frac{1}{n d^2 c} \tag{4.12}$$

die folgenden Zusammenhänge

$$\left.\begin{aligned} \eta &\propto \frac{m c}{d^2} = n m c l \propto p \tau \qquad && (p = \text{Druck}), \\ \lambda &\propto \frac{\mathrm{k} c}{d^2} = n \mathrm{k} c l \propto c_v \eta \qquad && (c_v = \text{spezifische Wärme}), \\ D &\propto \frac{c}{n d^2} = c l \propto \frac{\eta}{\varrho} \qquad && (\varrho = \text{Dichte}). \end{aligned}\right\} \tag{4.13}$$

Die Proportionalitätsfaktoren sind reine Zahlen, die unabhängig sind von den Molekül- und Gasdaten.

Die freie Weglänge ist hier gewissermaßen von selbst aufgetaucht. In den älteren gastheoretischen Arbeiten, die nicht von der Boltzmann-Gleichung ausgingen, stand gewöhnlich der Begriff der freien Weglänge im Mittelpunkt, und es wurden viele Versuche unternommen, ihn zu präzisieren und zu verfeinern. Es sei diesbezüglich auf das Buch von JEANS [*28*] oder den Artikel von JÄGER [*26*] verwiesen.

Die Beziehungen (4.13) gelten auch für das Potenzkraftmodell, wofern man nur anstelle der die starren Kugeln kennzeichnenden Länge d die Länge aus (4.9) setzt. Somit hat man hier

$$\left.\begin{aligned} \eta &= \eta^* \varkappa^{-\frac{2}{\nu-1}} m^{\frac{1}{2}} (\mathrm{k}T)^{\frac{\nu+3}{2(\nu-1)}} \\ \lambda &= \lambda^* \mathrm{k} \varkappa^{-\frac{2}{\nu-1}} m^{-\frac{1}{2}} (\mathrm{k}T)^{\frac{\nu+3}{2(\nu-1)}} \\ D &= D^* \varkappa^{-\frac{2}{\nu-1}} m^{-\frac{1}{2}} n^{-1} (\mathrm{k}T)^{\frac{\nu+3}{2(\nu-1)}} = D^* \varkappa^{-\frac{2}{\nu-1}} p^{-1} m^{-\frac{1}{2}} (\mathrm{k}T)^{\frac{3\nu+1}{2(\nu-1)}}. \end{aligned}\right\} \quad (4.14)$$

Für $\nu \to \infty$ gehen die Formeln (4.14) in (4.13) über. Zahlenwerte für η^* und D^* sind in (Ziff. 51, Tabelle 5) und (Ziff. 71, Tabelle 10) angegeben.

Tabelle 1a. *Viscosität von Wasserstoff und Deuterium in Abhängigkeit vom Druck bei 25° C.* Nach MICHELS, SCHIPPER und RINTOUL (1953).

p (atm)	1	150	1700
$\sqrt{2}\,\eta_{H_2}$ (μ-Poise)	126,1	129,5	215,1
η_{D_2} (μ-Poise)	126,1	129,5	217,1

Es bleibt die obige Behauptung zu begründen, daß die Zahlenfaktoren in (4.6), (4.7), (4.8) und (4.14) für verdünnte Gase konstante, von der Konzentration unabhängige Grenzwerte annehmen. Das liegt daran, daß bei den verdünnten Gasen nur die Zweierstöße eine Rolle spielen, wie es in der Boltzmann-Gleichung zum Ausdruck kommt. Wenn man für die Verteilungsfunktion des Gases den Ansatz macht

$$f = F + F\,\Phi,$$

wo F eine Maxwell-Verteilung ist und also $F\Phi$ die Abweichungen vom lokalen thermischen Gleichgewicht beschreibt, so gilt für Φ nach der Boltzmann-Gleichung in der hydrodynamischen Näherung

$$\frac{\partial F}{\partial t} + \cdots \approx -\iint F F_1 (\Phi + \Phi_1 - \Phi' - \Phi_1')\, \sigma g \, d e' \, d\mathbf{c}_1.$$

(Näheres s. Ziff. 45.) Daraus folgt, daß $F\Phi$ von der Konzentration unabhängig ist. Aus $F\Phi$ berechnen sich aber der Reibungsdruck (als ein Teil des gesamten Impulsstroms), der Wärme- und der Teilchenstrom. Demnach sind auch diese Ströme alle von der Konzentration unabhängig. Das gleiche gilt dann aber für die genannten Zahlenfaktoren.

γ) Massenabhängigkeit der Transportkoeffizienten. Nach (4.6) müssen sich die Zähigkeiten zweier Gase mit den Molekulargewichten $M_{1,2}$, aber derselben zwischenmolekularen Kraft bei gleicher Temperatur und gleichem, beliebig hohen Druck genau wie $\sqrt{M_1/M_2}$ verhalten; für die Wärmeleit- und Diffusionskoeffizienten sollte nach (4.7) und (4.8) das Verhältnis $\sqrt{M_2/M_1}$ sein, wieder unabhängig vom Druck. Man kann das prüfen durch Vergleich der Werte für Wasserstoff und Deuterium, für die $\sqrt{M_2/M_1} = 1{,}414$ ist.

Die *Viscosität* von D_2 wurde erstmals durch TORREY (1935) sowie von CLEAVE und MAASS (1935) gemessen, bei 1 at, 25° C mit dem Ergebnis $\eta_{D_2}/\eta_{H_2} = 1{,}41 \pm 0{,}03$. Bis zu 2000 at haben MICHELS, SCHIPPER und RINTOUL (1953) gemessen und bei 25° C die in Tabelle 1a angegebenen Werte gefunden.

Gl. (4.6) ist dadurch hinsichtlich der Massenabhängigkeit bestätigt und zugleich ist gezeigt, daß D_2 und H_2 praktisch dieselbe zwischenmolekulare Kraft haben. — Bei tiefen Temperaturen gilt aber die einfache Massenabhängigkeit (4.6) nicht mehr, da dann der quantenmechanische Parameter Λ aus (4.3b) merklich von Null verschieden ist und dadurch das klassische Ähnlichkeitsgesetz außer Kraft tritt. Die Viscosität von H_2 und D_2 ist durch VAN ITTERBEEK und CLASS (1938) sowie durch VAN ITTERBEEK und VAN PAEMEL (1940) gemessen worden und wiederum, mit fast denselben Ergebnissen, von E. W. BECKER und MISENTA (1955). Deren Werte sind in Tabelle 1b aufgeführt. Man sieht, daß das Verhältnis η_{D_2}/η_{H_2} bei den niedrigen Temperaturen deutlich vom klassischen Wert 1,41 abweicht. Entsprechendes gilt nach RIETVELD und VAN ITTERBEEK (1957) für das Verhältnis η_{HD}/η_{H_2}.

Tabelle 1b. *Viscosität von Wasserstoff und Deuterium in Abhängigkeit von der Temperatur.* Nach E. W. BECKER und MISENTA (1955).

T(°K)	15	20	64	90	273 [1]
η_{H_2} (μ-Poise)	7,9	10,80	31,4	41,1	83,6
η_{D_2} (μ-Poise)	9,6	13,45	42,4	56,6	117,2
η_{D_2}/η_{H_2}	1,21	1,24	1,35	1,38	1,40

Die *Wärmeleitfähigkeit* von D_2 bei Zimmertemperatur ist zuerst von CLEAVE und MAASS (1935), sodann von ARCHER (1936) und NOTHDURFT (1937) gemessen worden. UBBINK (1948) hat die Messungen (nach der Plattenmethode, s. Ziff. 10) auf tiefe Temperaturen ausgedehnt. Tabelle 1c zeigt einige Ergebnisse. Zunächst

Tabelle 1c. *Wärmeleitkoeffizient von Wasserstoff und Deuterium in Abhängigkeit von der Temperatur.* Nach UBBINK (1948).

T (°K)	15	20	65	90	273
λ_{H_2} (μ-cal/° cm sec)	28,7	38,6	104	136	417
λ_{D_2} (μ-cal/° cm sec)	15,5	22,7	106	125	302
$\lambda_{H_2}/\lambda_{D_2}$	1,85	1,70	0,98	1,09	1,38
$\frac{\lambda_{H_2}}{\lambda_{D_2}} \cdot \frac{\eta_{D_2}}{\eta_{H_2}}$	2,24	2,10	1,32	1,50	1,93

sieht man, daß bei 273° K das Verhältnis $\lambda_{H_2}/\lambda_{D_2}$ den klassischen Wert $\sqrt{2}$ für einatomige Gase fast erreicht. Das ist in der Tat nach der Euckenschen Beziehung (s. Ziff. 98) bei genügender Anregung der Rotation zu erwarten. Bei den beiden tiefsten Temperaturen wird der klassische Wert erheblich überschritten; die Welleneigenschaften der H_2- und D_2-Moleküle machen sich bemerkbar. Im übrigen passen aber die bei 15, 20 und 273° K gefundenen Werte von $\lambda_{H_2}/\lambda_{D_2}$ gut zu den zugehörigen Meßwerten von η_{D_2}/η_{H_2} aus Tabelle 1a. Bei 15 und 20° K können nämlich leichter und schwerer Wasserstoff als einatomiges Gas behandelt werden und demnach gilt [vgl. (50.16)] die Beziehung $\lambda/\eta c_v = 2{,}5$. Dabei ist $c_v = \frac{3}{2}\mathsf{R}/M$ die spezifische Wärme, R die molare Gaskonstante, M die Molmasse. Sonach hat man bei niedrigen Temperaturen zu erwarten

$$\frac{\lambda_{H_2}}{\lambda_{D_2}} \cdot \frac{\eta_{D_2}}{\eta_{H_2}} = \frac{M_{D_2}}{M_{H_2}} = 2. \tag{4.15}$$

Das wird aber, nach dem über das Verhältnis $\lambda_{H_2}/\lambda_{D_2}$ Gesagten, auch bei hohen Temperaturen gelten. In der Tat ist die Beziehung am Anfang und am Ende des Temperaturintervalls einigermaßen erfüllt (s. letzte Zeile von Tabelle 1c). In dem Zwischengebiet um 90° K, wo die Molwärme von $\frac{3}{2}\mathsf{R}$ auf $\frac{5}{2}\mathsf{R}$ ansteigt, sind die Verhältnisse nicht so einfach. Wir kommen darauf in Ziff. 98 zurück. Es sei hier nur noch darauf hingewiesen, daß bei 65° K die Wärmeleitkoeffizienten von H_2 und D_2 praktisch übereinstimmen. Das rührt davon her, daß beim D_2 der Anstieg der Rotationswärme bei niedrigerer Temperatur einsetzt als beim H_2; infolgedessen trägt bei 65° K die Rotationswärme beim D_2 mehr zur

[1] Die Werte zu 273° K sind der Zusammenstellung bei UBBINK (1948) entnommen; dort weitere Quellenangaben.

Wärmeleitfähigkeit bei als beim H_2, wodurch der Einfluß der Masse auf den Translationsanteil der Wärmeleitfähigkeit kompensiert wird.

Der *Selbstdiffusionskoeffizient* $D_{H_2/H_2} = 1{,}44\ \mathrm{cm^2/sec}$ bei 760 Torr, 20° C wurde von HARTECK und SCHMIDT (1933) gemessen mittels Ortho- und Parawasserstoff. Entsprechende Messungen an Deuterium sind nicht bekannt. Jedoch ist $D_{D_2/H_2} = 1{,}30$ bzw. $1{,}21\ \mathrm{cm^2/sec}$ bei 760 Torr, 20° C gemessen von HEATH, IBBS und WILD (1941) bzw. von WALDMANN (1944), im Mittel also $1{,}25_5\ \mathrm{cm^2/sec}$. Demnach wäre gemittelt $D_{D_2/H_2}:D_{H_2/H_2} = 1/1{,}15$. In (4.8) ist diesmal in beiden Fällen die reduzierte Masse einzusetzen, d.h. $\frac{1}{2} m_{H_2}$ für H_2/H_2 und $\frac{2}{3} m_{H_2}$ für D_2/H_2. Danach ist das theoretische Verhältnis der beiden Diffusionskoeffizienten $= \sqrt{\frac{3}{4}} = 1/1{,}15$, in Übereinstimmung mit dem experimentellen Wert. —

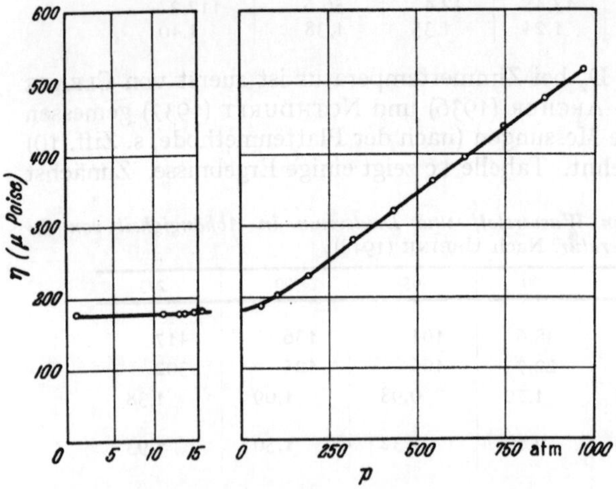

Fig. 2. Viscosität η von N_2 bei 25° C in Abhängigkeit vom Druck. Nach MICHELS und GIBSON (1931).

Die in (4.14) zum Ausdruck kommende Temperaturabhängigkeit der Koeffizienten wird später diskutiert (η und λ in Ziff. 51, D in Ziff. 71); von der Druckabhängigkeit handelt die folgende Ziffer.

Über den Thermodiffusionsfaktor können, da dieser eine reine Zahl ist, ohne detaillierte Kinetik keine Aussagen gemacht werden außer der, daß er druckunabhängig sein muß bei nicht zu hohem Druck.

5. Druckabhängigkeit von η, λ, pD und α. Druckunabhängigkeit bei mäßigen Drucken. Daß die *Viscosität* der Gase bei mäßigen Drucken vom Druck unabhängig ist, erkannte MAXWELL zuerst theoretisch 1860 ([*38*], Bd. 1, S. 391) und wies es dann 1866 ([*38*], Bd. 2, S. 1), gleichzeitig mit O. E. MEYER (1865), mittels der schwingenden Scheibe (s. Ziff. 8) qualitativ experimentell nach. Genauere Versuche von KUNDT und WARBURG (1875b) ergaben, daß die Viscosität der Luft bei 15° C von 20 bis 760 Torr sich um weniger als 1% ändert. Umfangreiche Messungen mit der Strömungsmethode (s. Ziff. 6) machten MICHELS und GIBSON (1931) an Stickstoff. Danach erhöht sich dessen Viscosität bei 25° C bis 15 atm nur um 1%; bei höheren Drucken wächst sie linear an (s. Fig. 2). In derselben Arbeit wird die interessante Beobachtung gemacht, daß bei N_2 für $p \lessapprox 300$ atm $\left(\frac{\partial \eta}{\partial T}\right)_p > 0$ gilt, wie es für Gase typisch ist, und daß für $p \gtrapprox 300$ atm $\left(\frac{\partial \eta}{\partial T}\right)_p < 0$ gilt, wie es für Flüssigkeiten typisch ist. Ferner wurde gefunden, daß $\mu = \eta/\varrho$ bei fester Temperatur in Abhängigkeit vom Druck ein Minimum durchläuft, wie es von ENSKOG (1922) vorausgesagt war. — Weitere Literatur s. bei HERZFELD ([*22*], S. 163) und bei HIRSCHFELDER et al. ([*25*], S. 613). Die Viscosität von Argon bis 2000 atm wurde kürzlich von MICHELS, BOTZEN und SCHUURMAN (1954) gemessen.

Auch die Druckunabhängigkeit des *Wärmeleitkoeffizienten* bei mäßigen Drucken ist experimentell verifiziert. Nach GREGORY und ARCHER (1926) und nach MILVERTON (1934) ändert sich λ für Luft von 4° C zwischen 200 und 700 Torr um weniger als 1‰. Das wurde für dipollose Gase im wesentlichen bestätigt von LAMBERT, STAINES und WOODS (1950) (s. Fig. 3). Die starke Druckabhängigkeit beim Methylalkohol hängt damit zusammen, daß dieser ein Dipolgas ist, welches zur Doppelmolekülbildung neigt, so daß die einfachen Annahmen der Theorie einatomiger Gase nicht zutreffen. Man kann diese Druckabhängigkeit nach SCHÄFER und FOZ GAZULLA (1942) quantitativ deuten auf Grund des örtlich verschiedenen Grades der Doppelmolekülbildung, der durch die Temperaturabhängigkeit des chemischen Gleichgewichts bedingt ist. Denn die Folge davon ist ein (Diffusions-) Transport von Reaktionsenergie, welcher wegen der Druckabhängigkeit der Doppelmolekülbildung ebenfalls vom Druck abhängt. Dies ist ganz analog zu der von NERNST (1904) gefundenen erhöhten Wärmeleitfähigkeit bei dissoziierenden Gasen (vgl. EUCKEN [*13*], S. 325). Weitere Literatur über die Wärmeleitfähigkeit bei hohen Drucken s. bei E. U. FRANK [*16*]. λ_{N_2} und λ_A bis 2000 atm wurden von MICHELS und BOTZEN (1953) und von MICHELS, BOTZEN, FRIEDMAN und SENGERS (1956) gemessen.

Tabelle 2. *Selbstdiffusions-Koeffizient von CO_2 bei 23° C und verschiedenen Drucken.* Nach TIMMERHAUS und DRICKAMER (1951).

p atm	D cm²/sec	pD atm cm²/sec
1,243	0,0847	0,105
9,33	0,0104	0,098
18,59	0,0049	0,092

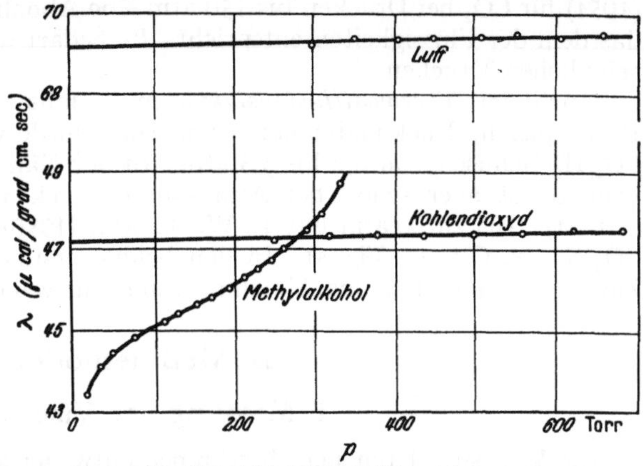

Fig. 3. Wärmeleitkoeffizient λ bei 66° C in Abhängigkeit vom Druck. Nach LAMBERT, STAINES und WOODS (1950).

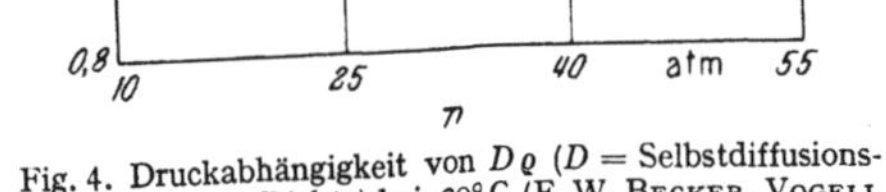

Fig. 4. Druckabhängigkeit von $D\varrho$ (D = Selbstdiffusionskoeffizient, ϱ = Dichte) bei 20° C (E. W. BECKER, VOGELL und ZIGAN, 1953).

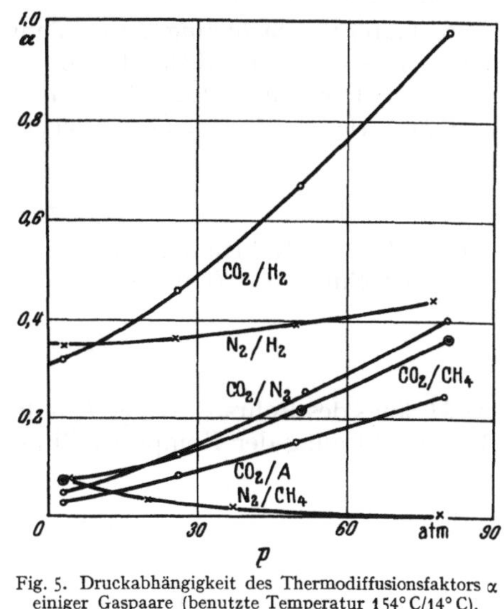

Fig. 5. Druckabhängigkeit des Thermodiffusionsfaktors α einiger Gaspaare (benutzte Temperatur 154° C/14° C). (E.W. BECKER 1950).

Die Druckabhängigkeit eines *Selbstdiffusions-Koeffizienten* und zwar an Kohlendioxyd ist zuerst von TIMMERHAUS und DRICKAMER (1951) mittels des radioaktiven ^{14}C untersucht worden. Es ergab sich (s. Tabelle 2) eine annähernde Konstanz des Produkts pD, wie es nach (4.14) sein soll, bis etwa 20 atm. Bis zu 52 atm haben E. W. BECKER, VOGELL und ZIGAN (1953) die Selbstdiffusion von CO_2 und N_2 gemessen mittels angereichertem ^{13}C und ^{15}N (s. Fig. 4). Sie finden bei CO_2 eine Zunahme von $D\varrho$ mit wachsendem Druck und schließen daraus, da dies der Theorie dichter Gase von ENSKOG (1922) widerspricht, auf eine Nicht-Addivität der Molekularkräfte. Jedoch geben ROBB und DRICKAMER (1951) für CO_2 bei Drucken bis 150 atm eine Abnahme von $D\varrho$ an, ein Verhalten, das dem der Flüssigkeiten entspricht. Es bedarf wohl weiterer Versuche bei den sehr hohen Drucken.

Auch der *Thermodiffusionsfaktor* muß, falls nur Zweierstöße im Gas eine Rolle spielen, druckunabhängig sein. Dies wurde von GREW (1945) bestätigt an CO_2/H_2-Gemisch von 0,2 bis 3,5 atm. In der Nähe des Siedepunkts einer Komponente ist aber schon bei Atmosphärendruck α merklich druckabhängig (s. Ziff. 21). Bis zu 90 atm hat E. W. BECKER (1950) α gemessen, u. a. für N_2/H_2, CO_2/N_2, N_2/CH_4 (s. Fig. 5). In den beiden ersten Fällen beobachtet man eine Zunahme, im letzten eine Abnahme von α mit wachsendem Druck.

B. Meßmethoden.

I. Messung der Viscosität.

Die Viscosität kann man bestimmen entweder aus der Geschwindigkeit einer laminaren, durch ein bekanntes Druckgefälle hervorgerufenen Strömung oder aus der Kraft, welche eine Strömung bekannter Geschwindigkeit tangential auf eine Wand ausübt. Es sollen nur die wichtigsten Prinzipien angedeutet und einige neuere Arbeiten besprochen werden; wegen weiterer Einzelheiten und Literatur sei auf HERZFELD [22] verwiesen.

6. Laminare Strömung durch Capillaren. Für die stationäre Geschwindigkeit v eines Gases im Abstand r von der Rohrachse gilt bei kleinem Druckgefälle — so daß die Kompressibilität des Gases und die kinetische Energie vernachlässigt werden dürfen — die aus (2.13) folgende *Poisseuillesche Formel*

$$v = -\frac{1}{4\eta}\frac{dp}{dz}\left(r_0^2 - r^2\right).$$

r_0 bedeutet den Rohrradius; das Gas soll in der z-Richtung strömen. Die durch den Querschnitt pro Zeiteinheit strömende Masse beträgt demnach

$$\dot{M} = -\frac{\pi}{8} r_0^4 \frac{\varrho}{\eta}\frac{dp}{dz}.$$

Da $\dot{M}$ längs des Rohrs — Länge L — konstant ist, kann man dafür, in genäherter Berücksichtigung der Kompressibilität, auch schreiben

$$\dot{M} = \frac{\pi}{8} r_0^4 \cdot \frac{1}{L}\int_{p_2}^{p_1} \frac{\varrho}{\eta}\, dp.$$

Darin bedeuten $p_{1,2}$ den Druck an der Ein- und Austrittsstelle des Rohrs. Korrekturen sind weiter nötig wegen der kinetischen Energie (HAGENBACH) und der Gleitung (s. HERZFELD [22], S. 164). Die Methode kann zur Relativmessung oder — nach genauer Kalibrierung der Capillaren — auch zur Absolutbestimmung der Viscosität benutzt werden. Von der Trautzschen Schule wurde hauptsächlich diese Methode in zahlreichen exakten Arbeiten angewandt (grundlegend: TRAUTZ und WEIZEL 1925). Auch Präzisionsmessungen der Viscosität von Luft im Zusammenhang mit der Millikanschen Bestimmung der Elementarladung wurden damit gemacht (RAPP 1913, BOND 1937). BOND fand $\eta_{\text{Luft}, 23^\circ \text{C}} = 183{,}47\ \mu$ Poise.

Zur Feststellung kleiner Änderungen des Strömungswiderstandes einer Capillare infolge einer Änderung der Viscosität ist die *Brückenmethode* von ENGELHARDT und SACK (1932) besonders geeignet. Sie beruht auf demselben Prinzip wie die Wheatstone-Brücke. Anstelle der vier elektrischen Widerstände treten vier Strömungswiderstände, anstelle des Brückengalvanometers ein empfindliches Manometer. Die Autoren haben damit nachgewiesen, daß sich die Viscosität von Sauerstoff im Magnetfeld ändert, was für die Wärmeleitfähigkeit schon vorher von SENFTLEBEN (1930) entdeckt worden war. Neuerdings haben E. W. BECKER und STEHL (1952a) nach dieser Methode den sehr geringen Unterschied der Viscosität von Normal- und Parawasserstoff gemessen. Fig. 6 zeigt ihre Anordnung. $K_{1,\ldots,4}$ sind die vier Capillaren, D ist ein elektrostatisches Differentialmanometer mit $2 \cdot 10^{-6}$ Torr/Skt. Empfindlichkeit (BECKER und STEHL 1952b); die Umwandler U — weite Rohre mit einem elektrisch heizbaren axialen Pt-Draht — dienen dazu, den von links in die Brücke einströmenden Parawasserstoff wahlweise in Normalwasserstoff zu verwandeln. Die Capillaren K_2 und K_4 befinden sich auf Zimmertemperatur, K_1 und K_3 haben die Temperaturen des Kältebades (15 bis 90° K). Die Messung geschieht in der Weise, daß man zunächst bei ungeheizten Umwandlern Parawasserstoff einströmen läßt und die Brücke mittels einer Regelcapillare abgleicht. Heizt man sodann etwa den vor K_1 liegenden Umwandler, so werden jetzt K_1 und K_2 von Normalwasserstoff (75% Ortho-H_2, 25% Para-H_2) durchströmt und da bei Zimmertemperatur, also in K_2 und K_4, die Viscositäten der Wasserstoffmodifikationen praktisch gleich sind, zeigt sich der nun zwischen K_1 und K_3 bei tiefer Temperatur herrschende Viscositätsunterschied an dem Manometer (Größenordnung 1% bei 15° K). Bei Heizung des anderen Umwandlers muß sich der Manometerausschlag umkehren. Daß die Viscositäten bei Zimmertemperatur gleich sind, haben die Autoren eigens mit einem direkt vor K_2 geschalteten Umwandler bewiesen.

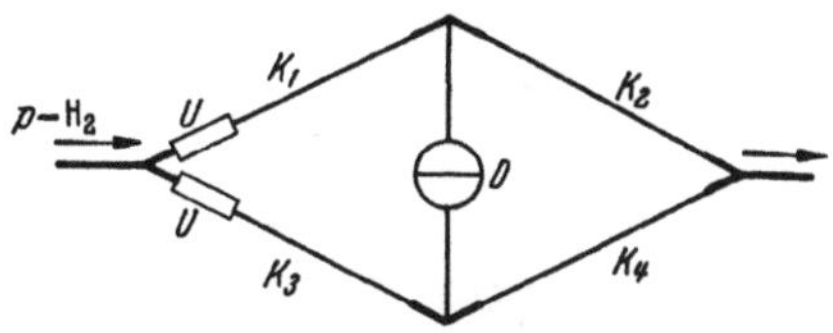

Fig. 6. Schema der Brückenanordnung von E. W. BECKER und O. STEHL (1952a) zur Messung des Viscositätsunterschieds von Normal- und Parawasserstoff.

7. Methode der stationären Ablenkung. Diese Methode geht auf MALLOCK (1888) und COUETTE (1890) zurück, die damit die Viscosität von Flüssigkeiten bestimmten; GILCHRIST (1913) und TIMIRIAZEFF (1913) wandten sie zuerst auf Gase an. Fig. 7 zeigt den Apparat, den KELLSTRÖM (1937) zur Präzisionsbestimmung der Viscosität von Luft benutzte. Er besteht aus einem inneren Zylinder *a*, der an einem Torsionsfaden *e* drehbar aufgehängt ist, und einem äußeren Zylinder *b*, der durch ein Zahnrad und die Schnecke *f* in gleichmäßige Rotation versetzt wird. Infolge der Reibung des zwischen den Zylindern befindlichen Gases wird der innere Zylinder aus der Ruhelage herausgedreht, was an dem Spiegel *d* abzulesen ist. Die Zylinder *cc* sind feststehende Schutzzylinder und bezwecken die Elimination der Störeffekte an den Enden.

Zur Berechnung der Couetteschen Strömung dient wieder (2.13). Im stationären Zustand und bei Vernachlässigung quadratischer Glieder gilt

$$\eta \Delta v_\mu = \frac{\partial p}{\partial x_\mu}.$$

Für das zylindersymmetrische Strömungsfeld macht man den Ansatz

$$v_x = -v(r)\frac{y}{r}, \quad v_y = v(r)\frac{x}{r}, \quad p = p(r)$$

und findet dann, daß $dp/dr = 0$ sein muß und für die Tangentialgeschwindigkeit v die Gleichung gilt

$$\frac{d^2v}{dr^2} + \frac{1}{r}\frac{dv}{dr} - \frac{v}{r^2} = 0.$$

Die Lösung mit den Randbedingungen $v = 0$ für $r = a$ (Radius des inneren Zylinders), $v = \omega b$ für $r = b$ (Radius des äußeren Zylinders; ω die Winkelgeschwindigkeit der Rotation) lautet

$$v = \frac{\omega b^2}{r}\,\frac{r^2 - a^2}{b^2 - a^2}.$$

Der Reibungsdruck am inneren Zylinder für $x = r$, $y = 0$ hat den Wert

$$p_{xy} = -\frac{2b^2}{b^2 - a^2}\,\omega\eta,$$

woraus sich das auf den inneren Zylinder (Länge L) wirkende Drehmoment zu

$$\frac{4\pi a^2 b^2 L}{b^2 - a^2}\,\omega\eta$$

ergibt. Es ist aus dem Drehwinkel bekannt, wenn man die Direktionskraft des Aufhängefadens bestimmt hat (aus Schwingungsdauer und Trägheitsmoment). Für die Genauigkeit der Messung ist die Exaktheit der Zylinder und Lager maßgebend; KELLSTRÖM gibt die Längen a ($\approx$ 2,8 cm), b ($\approx$ 3,2 cm), L ($\approx$ 10 cm) auf 1 bis 5 μ an. Die Umdrehungsdauer des Zylinders war etwa 1 min und auf $1:10^4$ konstant. — KELLSTRÖM fand $\eta_{\text{Luft, }23^\circ\text{C}} = 183{,}49$ μ Poise in Übereinstimmung mit dem nach der Capillarenmethode gewonnenen Wert von BOND (1937) (s. Ziff. 6). MILLIKAN hatte 1916 die Viscosität 182,26 bei 23° C benützt. Durch die Neubestimmung der Luftviscosität verschwand die Diskrepanz zwischen den nach der Öltröpfchen- und der Kristallgitter-Methode (BÄCKLIN) erhaltenen Werten der Elementarladung, die sich nun praktisch übereinstimmend zu $e = 4{,}804 \cdot 10^{-10}$ el. stat. Ldg. Einh. ergibt[1].

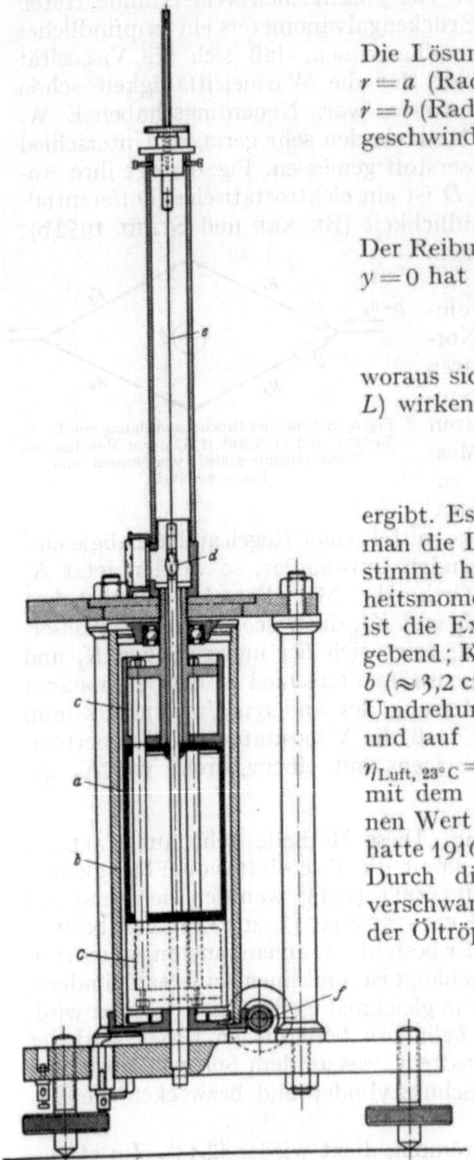

Fig. 7. Zylinderapparat zur Präzisionsmessung der Viscosität (aus KELLSTRÖM, 1937).

8. Methode der gedämpften Schwingung. Diese Methode wurde bereits von MAXWELL (1866, [38], Bd. 2, S. 1) benutzt bei Versuchen, die von ihm theoretisch vorhergesagte Druckunabhängigkeit der Viscosität von Gasen zu bestätigen.

Seit dem Bericht von HERZFELD [22] wurde die durch die Viscosität eines Gasgemischs bewirkte Dämpfung schwingender Scheiben zur Gasanalyse anläßlich von Thermodiffusionsversuchen bei tiefen Temperaturen viel benutzt von

[1] Vgl. auch den Artikel von E. R. COHEN und J. W. M. DUMOND in Bd. XXXV dieses Handbuches.

VAN ITTERBEEK u. Mitarb. (VAN ITTERBEEK und VAN PAEMEL 1938, 1940; dieselben und VAN LIERDE 1947). Wir besprechen die etwas abgewandelte Methode an Hand des von E. W. BECKER, MISENTA und SCHMEISSNER (1954) benutzten Apparats zur Bestimmung der Viscosität von gasförmigem ^{3}He und ^{4}He zwischen 1,3 und 4,2° K. Fig. 8 zeigt das Meßprinzip. Eine an Spannbändern aufgehängte, sehr dünnwandige Aluminiumglocke C führt Drehschwingungen zwischen dem Kupfermantel A und dem Kupferkern B aus. Die Glocke besitzt einen halben Deckel D, der zusammen mit der direkt darüber befindlichen isolierten Platte E einen Drehkondensator bildet. Dieser Kondensator liegt in einer Hochfrequenzbrücke. Über den Verstärker V werden die mechanischen Schwingungen durch den Lichtschreiber S registriert. Der Ringspalt zwischen A und B war 0,9 mm weit; die Länge von C betrug 36 mm. Die Auswertung geschah unter der Annahme, daß das logarithmische Dekrement der Schwingung (Dauer der Periode 1,38 sec) einfach proportional ist zur Viscosität des Gases. Zur Eichung wurde die bekannte Viscosität $\eta = 83{,}68\,\mu$ Poise vom ^{4}He bei 77,3° K (KAMERLINGH ONNES und WEBER 1913) herangezogen. Der Fülldruck für die Versuche bei 4° K war 0,5 Torr; die Druckunabhängigkeit des logarithmischen Dekrements Λ bei diesem Druck wurde nachgeprüft. Um ein Bild von der Meßgenauigkeit zu geben, seien einige Λ- und η-Werte für ^{4}He notiert:

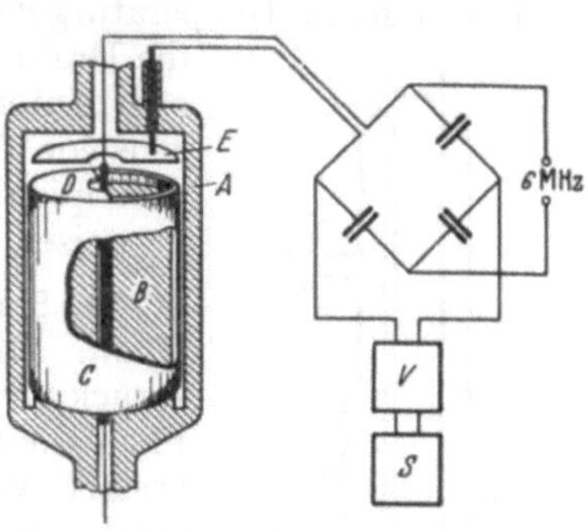

Fig. 8. Apparatur zur Viscositätsmessung an Helium bei tiefer Temperatur (E.W. BECKER, MISENTA und SCHMEISSNER, 1954).

	77,3° K	4,20° K	1,29° K	„Vakuum"
$\Lambda \cdot 10^4$	684	88,2	27,8	1
η (μ-Poise)	83,68	10,78	3,40	—

Es wurde angenommen, daß die Dämpfung im „Vakuum" von Gasresten, nicht von Reibung der Aufhängung herrührt. Ein dadurch möglicher Fehler wäre bei den kleinsten η-Werten höchstens 3%.

Zur Absolutmessung ist die Methode der schwingenden Scheibe nicht geeignet, weil die hydrodynamische Theorie — obwohl von MAWOOD (1938) ausgearbeitet — kompliziert ist; das dürfte wegen der Endeffekte ebenso auch für den eben besprochenen schwingenden Zylinder gelten.

9. Methode der fallenden Kugel. Bei dem auf HOEPPLER (1933) zurückgehenden Viscosimeter wird die Zähigkeit aus der Zeit t des unbeschleunigten Falles einer Kugel in einem geneigten, kalibrierten Glasrohr entnommen. Die Kugel soll den gleichen thermischen Ausdehnungskoeffizienten haben wie das Rohr um den Temperatureinfluß auszuschalten. Dieser Apparat wird wegen seiner Einfachheit viel zur Bestimmung der Viscosität von Flüssigkeiten benutzt, fand aber auch schon bei Gasen Anwendung, so z. B. in der umfangreichen Arbeit von WOBSER und FR. MÜLLER (1941). Es werden dort theoretische Betrachtungen angestellt, die zur Begründung der Beziehung $\eta = a t$ führen, wo a eine vom Neigungswinkel abhängige Apparatekonstante ist. Die Absolutberechnung von a ist wegen der verwickelten Verhältnisse nicht möglich. Gleichwohl ist der Apparat, da keine Korrekturen nötig sind, zu präzisen Relativmessungen sehr geeignet. Die Autoren fanden durch Eichung an Hand der genau bekannten Viscosität des Wassers für $\eta_{\text{Luft},23^\circ\text{C}}$ einen Wert, der praktisch mit den am Schluß von Ziff. 7 zitierten Angaben übereinstimmt. In der gleichen Arbeit werden eine Reihe weiterer Gas-Viscositäten und deren Temperaturabhängigkeit von 20 bis 100° C gemessen und die Zuverlässigkeit der Methode betont.

II. Messung der Wärmeleitfähigkeit.

Es gibt hierfür zwei Möglichkeiten: Messung des stationären Wärmestroms bei bekanntem Temperaturgefälle, wobei je nach der Geometrie der Anordnung die Draht- und die Plattenmethode zu unterscheiden sind, oder Untersuchung des quasi- oder nichtstationären Abklingens von Temperaturunterschieden. Wir besprechen die Meßprinzipien wieder an Hand einiger neuerer Arbeiten und verweisen im übrigen auf den Bericht von Herzfeld [*22*].

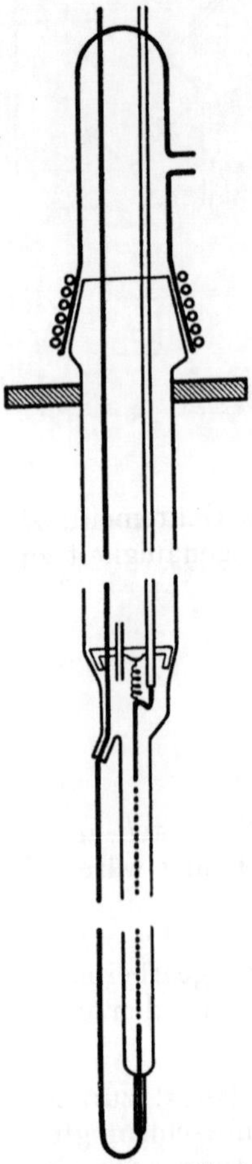

Fig. 9. Zelle zur Absolutmessung der Wärmeleitfähigkeit (aus Gardiner und Schäfer, 1956).

10. Stationäre Methoden. *α) Die Methode des koaxialen Drahts.* Diese Anordnung geht auf Schleiermacher (1888) zurück, wurde von Goldschmidt (1911; Anwendung auf Flüssigkeiten) abgewandelt, von Eucken (1911 a und b) verbessert, von S. Weber (1917, 1927) theoretisch und experimentell weiter untersucht und von Kannuluik und Martin (1933, 1934) nochmals variiert. Die Hauptaufgabe der Theorie war dabei die Fehlerquellen so genau zu erfassen, daß die Methode für Absolutmessungen brauchbar ist.

Die Methode beruht auf der Messung des Wärmeverlustes eines elektrisch geheizten, in der Achse eines gasgefüllten Rohres eingespannten Drahtes, wobei der Draht zugleich als Widerstandsthermometer dient. Fig. 9 zeigt eine gegenwärtige Ausführung (Gardiner und Schäfer 1956). Die eigentliche Meßzelle — unterer Teil der Figur — besteht aus einem 6,334 mm weiten Glasrohr mit einem axialen Platindraht von 31 μ Durchmesser, 12,60 cm Länge und etwa 24 Ω Widerstand, der durch eine Wolframfeder gespannt ist. Oben an der Meßzelle ist ein weiteres Glasrohr angesetzt, so daß die Zelle in einen Dampf-Thermostaten gesteckt werden kann. Durch dieses Rohr werden auch das Gas und die Drähte zugeführt. Die Zuführung der Drähte geschieht so, daß keine Thermokräfte auftreten und sich in der Wheatstone-Brücke die Widerstände der Zuführungen bei jeder Temperatur fast kompensieren. Der Widerstand des Pt-Drahts ist das Maß für dessen Temperatur. Mit einem Präzisions-Ampèremeter wird außerdem der Gesamtstrom gemessen, so daß man die im Platin-Draht entwickelte Stromwärme kennt. Die Temperaturdifferenz zwischen Draht und Wand war 3 bis 5°, der Druck 350 bis 450 Torr. Die Thermostat-Temperatur mußte jeweils etwa $\frac{1}{2}$ Std auf 0,01° C konstant sein, um eine Meßgenauigkeit in λ von 1% zu haben.

Es seien r_0 und T_0 Radius und Temperatur des Rohres, r, L, T und $W = W_0[1+\alpha(T-T_0)]$ Radius, Länge, Temperatur und Widerstand des Drahtes, J der Strom. Für sehr langes Rohr ist T längs des Drahts fast konstant und es gilt idealisiert

$$T - T_0 = \frac{\dot{Q}}{2\pi L\lambda} \ln\frac{r_0}{r} \quad \text{mit} \quad \dot{Q} = J^2 W. \tag{10.1}$$

Dabei sind aber folgende Fehlerquellen bzw. Korrekturen zu beachten: a) Konvektion, b) Temperatursprung, c) Wärmeleitung längs des Drahts, d) Strahlung.

a) wird nach EUCKEN vermieden durch Wahl enger, langgestreckter, vertikaler Rohre und nicht zu hohen Druck; die Wärmeleitung durch Konvektion ist proportional der Temperaturdifferenz und dem Quadrat des Drucks (ULSAMER 1936).

b) ist vernachlässigbar bei nicht zu niedrigem Druck oder bei Vermeidung großer Temperaturgradienten durch Wahl dicker Drähte. Der Temperatursprung ist dem Temperaturgradienten an der betreffenden Grenzfläche proportional

$$\delta T = -\gamma \frac{dT}{dr},$$

wobei der Temperatursprungkoeffizient γ (v. SMOLUCHOWSKI 1898, vgl. KENNARD [31], S. 311) die Größenordnung der freien Weglänge hat, also dem Druck umgekehrt proportional ist. Berücksichtigung des Temperatursprungs ergibt anstelle von (10.1)

$$T - T_0 = \frac{\dot{Q}}{2\pi L \lambda}\left[\ln\frac{r_0}{r} + \gamma\left(\frac{1}{r} + \frac{1}{r_0}\right)\right], \tag{10.2}$$

woraus das Gesagte folgt.

c) wurde von WEBER (1917) und noch eingehender von KANNULUIK und MARTIN (1933) diskutiert. Sei z die Koordinate längs des Drahts, λ_D dessen Wärmeleitkoeffizient. Dann gilt bei Annahme radialer Wärmeleitung im Gas die Energiebilanz

$$\pi r^2 \lambda_D \frac{d^2 T}{dz^2} - \frac{2\pi\lambda}{\ln\frac{r_0}{r}}(T - T_0) + J^2 \frac{W_0}{L}[1 + \alpha(T - T_0)] = 0. \tag{10.3}$$

Diese Gleichung ist leicht zu lösen und somit kann der Einfluß der Wärmeleitung im Draht rechnerisch berücksichtigt werden.

d) Die Strahlungsverluste sind bei kleinen Temperaturdifferenzen geringfügig.

In den älteren Arbeiten hat man die Endkorrektur, d.i. c) im vorigen Absatz, experimentell zu eliminieren versucht. Das geschieht nach SCHLEIERMACHER (1888) dadurch, daß man mittels zweier sehr dünner, seitlich zugeführter Potentialdrähte den Widerstand und damit die Temperatur im Mittelteil des axialen Heizdrahts bestimmt. Nach GOLDSCHMIDT (1911) schaltet man zwei verschieden lange, sonst gleichartige Meßzellen in der Wheatstone-Brücke gegeneinander, so daß sich die Endeffekte beider gegeneinander wegheben.

Um die Temperatursprung-Korrektur zu vermeiden, benutzten KANNULUIK und MARTIN (1934) sowie KANNULUIK und CARMAN (1952) einen dicken Meßdraht. Die letzteren Autoren hatten ein 0,7169 cm weites Platin-Iridium-Rohr mit axialem Platindraht von 0,1465 cm Durchmesser, 11,631 cm Länge und 0,00687701 Ω bei 0° C, etwa 3° Übertemperatur, keine Druckabhängigkeit von 350 bis 760 Torr. Zur Widerstandsmessung diente ein Wolffscher thermokraftfreier Fünfdekaden-Kompensationsapparat mit kleinen Widerständen. Gemessen wurde die Wärmeleitfähigkeit der Edelgase von −180 bis 300° C. Die Wärmeleitung längs des Drahts bedingte bei den Messungen an Argon eine etwa 50%ige Korrektur.

In der Arbeit von GARDINER und SCHÄFER (1956) wird die Endkorrektur nochmals ausführlich dargestellt. Bei diesen Messungen war die Wärmeableitung längs des dünnen Drahts nur etwa 3%, der Strahlungsverlust etwa 0,5% der Wärmeableitung im Gas.

Ebenfalls Absolutmessungen mit der Drahtmethode hat Schottky jr. (1952) ausgeführt. E. U. Franck (1951) dagegen bestimmt in einer sehr reichhaltigen Arbeit die Wärmeleitfähigkeit einer Reihe von Gasen bei Temperaturen von 100 bis 700° K durch Relativmessung in bezug auf Argon, das als einatomiges Gas die Beziehung $\lambda = 2{,}5\, c_v \eta$ (s. Ziff. 50) gut erfüllt und dessen Zähigkeit genau bekannt ist.

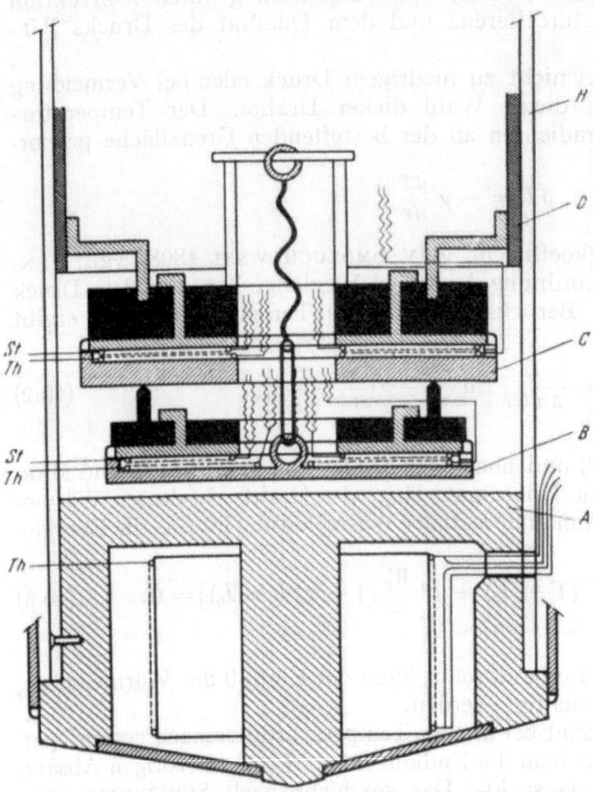

Fig. 10. Plattenapparat zur Bestimmung des Wärmeleitkoeffizienten von Gasen, besonders bei tiefen Temperaturen. Aus Ubbink und de Haas (1943a).

β) Die Methode der horizontalen Platten. Diese Methode ist apparativ nicht so einfach wie die Methode des koaxialen Drahts, hat aber den großen Vorteil, daß sich Konvektion mit Sicherheit vermeiden läßt. Wenn die obere Platte wärmer ist als die untere, dann ist die Schichtung stabil und es besteht keine Gefahr, daß sich das Gas in Bewegung setzt. Dazu kommt, daß der Temperatursprung an den Platten bei gleicher Temperaturdifferenz kleiner ist als am Draht und sich durch Variation des Plattenabstands exakt eliminieren läßt. Deshalb ist nach Ubbink und de Haas (1943a) die Plattenmethode bei Tieftemperaturmessungen anzuwenden, wo wegen hoher Gasdichte und niedriger Viscosität bei dem Rohr mit koaxialem Draht starke Konvektion eintritt. Ferner wurde die Plattenmethode neuerdings von Michels et al. (1956) zur Messung der Wärmeleitfähigkeit von Argon bis zu 2500 atm benutzt. Die Drahtmethode wäre auch hier nicht mehr zuverlässig. Der bei diesen Versuchen benutzte Plattenapparat ist in der Arbeit von Michels und Botzen (1952) beschrieben.

Die Plattenmethode ist erstmals von Hercus und Laby (1919) sowie weiter von Laby (1934) und Hercus und Sutherland (1934) angewandt worden. Fig. 10 zeigt eine neuere Ausführung, die sich aber im Prinzip von den früheren Anordnungen nicht viel unterscheidet. A ist die ungeheizte Platte, welche die Temperatur des umgebenden Wärmebads hat. B ist die obere, geheizte Platte (Heizelement St). Über B und thermisch dagegen möglichst gut isoliert befindet sich die ebenfalls geheizte Schutzplatte C (Heizelement St). Deren Heizung wird so geregelt, daß C dieselbe Temperatur hat wie B, so daß von B aus alle Wärme nach unten fließt. Zur Temperaturmessung dienen die drei Thermoelemente Th. Die Platten, etwa 4 cm Durchmesser, bestehen aus Kupfer, die schwarz gezeichneten Teile aus Ebonit. Zur Elimination des Temperatursprungs kann der Abstand AB durch Verschieben des Rohrs D in dem äußeren Hüllrohr

H verändert werden. Die wirksame Fläche der Platten *B* wird nach HERCUS und LABY (1919) durch Kapazitätsmessung an einem Modell ermittelt. Wegen weiterer theoretischer und apparativer Einzelheiten sei auf die Originalarbeit von UBBINK und DE HAAS verwiesen.

Die Messungen von UBBINK und DE HAAS (1943b) an Helium bei tiefen Temperaturen beseitigten eine Diskrepanz mit der Theorie, welche in diesem Fall bestanden hatte (s. Ziff. 52).

11. Abkühlungsmethoden. Wie schon eingangs angedeutet gibt es grundsätzlich zwei Möglichkeiten. Man kann entweder das exponentielle Abklingen der Temperatur eines erwärmten Körpers (Quecksilberthermometer) beobachten, dessen Wärmekapazität groß ist gegen diejenige des Gases, von welchem er umgeben ist. Die Wärmeleitung im Gas ist in diesem Fall quasistationär, die Abklingzeit ist ein Maß für den Wärmeleitkoeffizienten. Diese Methode benutzten CURIE und LEPAPE (1931) für Relativmessungen. Oder man beobachtet das Abklingen von Temperaturdifferenzen im Gas selbst, indem man das von isothermen Wänden eingeschlossene Gas momentan, z.B. durch eine kleine adiabatische Expansion, etwas abkühlt und den Wiederanstieg der Temperatur registriert. Die Abklingzeit ist in diesem Fall aber ein Maß für die Temperaturleitfähigkeit [vgl. (2.24)], also für das Verhältnis λ/c_p. Derartige Messungen scheinen nicht gemacht zu sein.

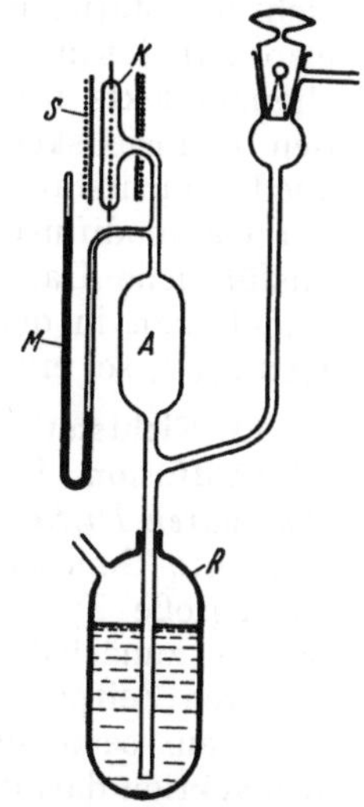

Fig. 11. Wärmeleitzelle zur Mikroanalyse von Gasgemischen. Nach GREW (1947).

12. Gasanalyse. Kurz sei noch auf die vielfache Verwendung der Wärmeleitzellen zur Gasanalyse hingewiesen. Da der Wärmeleitkoeffizient eines binären Gemisches bei fester Temperatur vom Molenbruch — allerdings im allgemeinen nicht linear — abhängt, kann man nach Eichung mit bekannten Gemischen umgekehrt aus der Wärmeleitfähigkeit auf die Zusammensetzung schließen (vgl. dazu DAYNES [9]).

Temperaturänderungen der Wand würden bei einer einzigen Wärmeleitzelle die Messung fälschen. Deshalb haben (SHAKESPEAR und) DAYNES (1920) zwei völlig gleichartige Zellen, je 1,5 cm lang, 0,6 cm im Durchmesser und mit einer Platin-Spirale von je 10 Ω versehen, in einem Kupferblock vereinigt und derart in die Wheatstone-Brücke gelegt, daß Temperaturänderungen des wassergekühlten Kupferblocks den Abgleich der Brücke nicht beeinflussen. Das zu analysierende Gas wird in die eine Zelle gefüllt, die andere bleibt völlig abgeschlossen. Diese als Katharometer bezeichnete Anordnung wurde viel bei Thermodiffusionsversuchen benutzt, z.B. von IBBS (1921, 1925) und NETTLEY (1954) (vgl. Fig. 19).

Wärmeleitzellen für Mikroanalysen sind u. a. von SACHSSE und BRATZLER (1935) und MELVILLE und BOLLAND (1937) angegeben worden. Fig. 11 zeigt eine Ausführung, die GREW (1947) bei der Untersuchung der Thermodiffusion von Edelgasen benutzte. Die eigentliche Zelle besteht aus einer Capillare *K* von 2,5 cm Länge und 2 mm Weite mit einer axialen Wolframwendel (20 Ω). Durch Heraufpressen des Quecksilbers aus dem Reservoir *R* wird das zu analysierende Gas aus dem Gefäß *A* in die Capillare *K* hinein verdichtet auf den am Manometer *M* abzulesenden Meßdruck (50 Torr). Es wird wieder das Katharometer-Prinzip angewandt: Die Drahtspule *S*, welche gleichen Widerstand und Temperaturkoeffizienten wie die Wendel besitzt und die Temperatur der Capillarenwand hat, bildet ebenfalls einen Zweig der Wheatstone-Brücke zwecks

Kompensation der Temperaturschwankungen des Thermostaten. Wegen weiterer experimenteller Einzelheiten s. auch EBBINGHAUS [*10a*].

Die Genauigkeit der Wärmeleitfähigkeitsmethode zur Analyse von Isotopengemischen geben GROTH und SUSSNER (1944) auf 0,01% des Atomgewichts an. Äußerste Reinheit der Gasproben ($1:10^6$) ist dabei vorauszusetzen.

III. Messung des Diffusionskoeffizienten.

Der Diffusionskoeffizient ist maßgebend für die Einstellzeit des Gleichgewichts in einem inhomogenen ruhenden Gemisch (nichtstationäre Diffusion) sowie für die Eindringtiefe eines Gases in den Strom eines anderen entgegen der Strömungsrichtung (stationäre Diffusion). Jede der beiden damit angedeuteten Meßmethoden erfordert Gasanalysen. Da ferner die Diffusion in Gasen auf Grund des Diffusionsthermoeffekts fast immer von Temperaturdifferenzen begleitet ist, kann man den Diffusionskoeffizienten auch durch deren Messung ermitteln, im Prinzip wieder nichtstationär oder stationär. Schließlich wurde auch die Einstellzeit der Thermodiffusion zur Messung des Diffusionskoeffizienten herangezogen. Eine ausführliche Darstellung des experimentellen Materials über die Gasdiffusion findet man in der Monographie von JOST [*29*]; auch ein Bericht von WALDMANN [*52*] sei erwähnt.

13. Nichtstationäre Diffusion (Methode von LOSCHMIDT, 1870). Quasistationäre Diffusion. Bei der ältesten und am meisten angewandten Methode, *der nichtstationären Diffusion,* füllt man die zwei verschiedenen Gase (oder verschiedene bekannte Gemische zweier Gase) in zwei vertikale, übereinander angeordnete, gleichgroße Zylinderrohre. Die Diffusion beginnt, wenn die Verbindung der beiden, zunächst getrennten Gasräume hergestellt wird. In einem geeigneten späteren Zeitpunkt wird die Verbindung unterbrochen und das Gas in jedem Rohr auf seine mittlere Zusammensetzung untersucht. Zur Vermeidung von Konvektion darf der Rohrdurchmesser nicht zu groß gewählt werden und ist das schwerere Gas in das untere Rohr zu füllen.

Zur Auswertung benötigt man eine von LOSCHMIDT aufgestellte Formel. Es sei zu Beginn der Diffusion ($t=0$) der Molenbruch des Gases 1 im unteren Rohr ($z<0$) $\gamma=\gamma_u(0)$, im oberen Rohr ($z>0$) $\gamma=\gamma_o(0)$. Die Länge jedes Rohres betrage $L/2$. Dann lautet die Lösung der Diffusionsgleichung (3.13)

$$\left.\begin{aligned}\gamma &= \tfrac{1}{2}\left[\gamma_u(0)+\gamma_o(0)\right] - \\ &\quad -\left[\gamma_u(0)-\gamma_o(0)\right]\cdot\frac{2}{\pi}\sum_{\nu=0}^{\infty}\frac{1}{2\nu+1}\,e^{-\left(\frac{2\nu+1}{L}\pi\right)^2 Dt}\sin\frac{2\nu+1}{L}\pi z\end{aligned}\right\}\tag{13.1}$$

(s. z. B. FRANK-V. MISES [*17*], S. 578). Daraus entnimmt man die für jedes Rohr gemittelten Molenbrüche zur Zeit t, die gasanalytisch zu bestimmen sind:

$$\left.\begin{aligned}\bar{\gamma}_u-\bar{\gamma}_o &= \left[\gamma_u(0)-\gamma_o(0)\right]\cdot\frac{8}{\pi^2}\sum_{\nu=0}^{\infty}\frac{1}{(2\nu+1)^2}\,e^{-\left(\frac{2\nu+1}{L}\pi\right)^2 Dt} \\ &= \left[\gamma_u(0)-\gamma_o(0)\right]\cdot\frac{8}{\pi^2}\left(e^{-\pi^2\frac{Dt}{L^2}}+\frac{1}{9}\,e^{-9\pi^2\frac{Dt}{L^2}}+\cdots\right).\end{aligned}\right\}\tag{13.2}$$

Diese Reihe konvergiert sehr schnell und man kann die Diffusionszeit ohne weiteres so wählen, daß schon das zweite Glied in der letzten Klammer zu vernachlässigen ist.

Beim Loschmidtschen Apparat wurden die Rohre durch einen Hahn großer Bohrung getrennt und verbunden. Um den Hahn zu vermeiden benutzte v. WOGAU (1907) bei der Untersuchung der Diffusion von Flüssigkeiten Rohre mit scheibenförmigen, geschliffenen Flanschen an den offenen Enden; durch Drehung des einen Flansches gegen den anderen wird die Verbindung der Rohre hergestellt. Fig. 12 zeigt die von BOARDMAN und WILD (1937) benutzte Anordnung zur Messung der Diffusion von Gaspaaren gleichen Molekulargewichts. $R_{1,2}$ sind die beiden Kupferrohre, $F_{1,2}$ die Flansche, bei S liegt der Drehpunkt des oberen Flansches, C sind Leitungen zum Füllen und Evakuieren. Die Rohre hatten 1,3 cm Innendurchmesser und waren zusammen 90 cm lang. Einen ähnlichen Apparat benutzten BRAUNE und ZEHLE (1941) zur Messung der Selbstdiffusion von HCl und HBr unter Zuhilfenahme der deuterierten Verbindungen. Die Diffusionszeiten betragen bei Gasen mittleren Molekulargewichts und der genannten Apparatlänge etwa 1 Std bei 0° C, 1 atm. Aus der bekannten Diffusionszeit und der gemessenen Vermischung kann dann nach (13.2) der Diffusionskoeffizient gefunden werden.

Zur Messung der — bei Gasen ziemlich geringen — Abhängigkeit des Diffusionskoeffizienten vom Mischungsverhältnis hat man die Diffusion von Gemischen ineinander zu untersuchen (LONIUS 1909) (s. Ziff. 73).

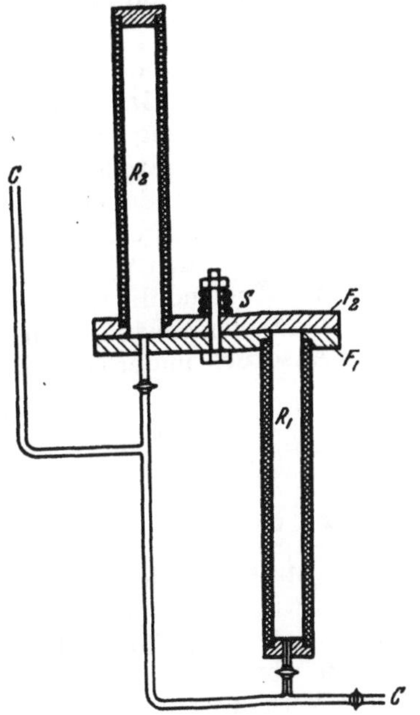

Fig. 12. Diffusionsapparat von BOARDMAN und WILD (1937).

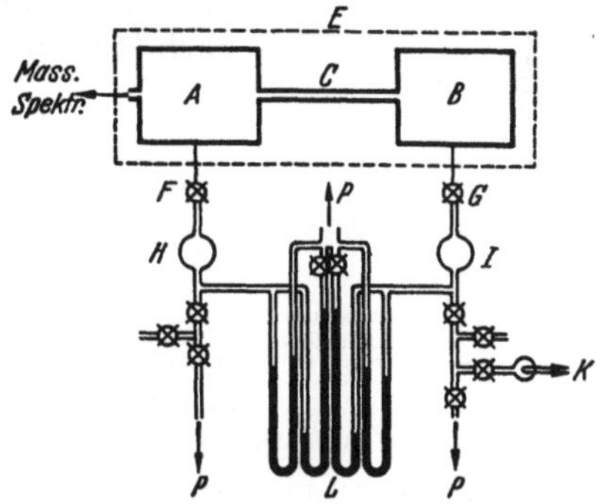

Fig. 13. Diffusionsapparat ohne bewegliche Teile nach NEY und ARMISTEAD (1947). Figur nach der Arbeit von WINN (1950).

Eine modifizierte Methode, welche auf der *quasistationären Diffusion* beruht (vgl. auch Ziff. 17), haben NEY und ARMISTEAD (1947) sowie, nach deren Vorbild, WINN (1950) angewandt. Erstere untersuchten Uranhexafluorid, letzterer die Temperaturabhängigkeit der Selbstdiffusion von A, Ne usw. (s. Ziff. 72). In beiden Fällen waren bewegliche Teile im Diffusionsapparat unerwünscht. Dieser besteht deshalb (s. Fig. 13) aus zwei gleichgroßen Gefäßen A und B, die fest durch ein ziemlich langes und dünnes Rohr C verbunden sind. Innerhalb des Ofens E befinden sich keine Hähne. Bei WINN waren die Gefäße je 357 cm³ groß, das Rohr war 8 cm lang und 0,3 cm im Durchmesser. Die Gefäße werden über P evakuiert (dazu Hitzdrahtmanometer bei K) und durch gleichzeitiges Öffnen der Ventile F, G aus den Vorratsbehältern H und I, welche die beiden Gase unter gleichem Druck (Manometer L) enthalten, über gleichdimensionierte Capillaren gefüllt. Dann werden F, G wieder geschlossen und sofort das Einsetzen der Diffusion an dem angeschlossenen Massenspektrometer — es handelt sich um Isotopenverbindungen — verfolgt. Nach den Formeln von Ziff. 17 kann aus dem alsbald einsetzenden, zeitlich exponentiellen Abfall des Unterschieds der Molenbrüche in beiden Gefäßen der Diffusionskoeffizient ermittelt werden.

Ebenfalls die quasistationäre Diffusion benutzten KL. SCHÄFER, CORTE und MOESTA (1951). Diese Autoren untersuchten die Abhängigkeit der Fremddiffusion vom Mischungsverhältnis der Gase und von der Temperatur (s. Ziff. 73); sie hatten einen Apparat mit Quecksilberverschluß, erprobten aber auch einen mechanischen Verschluß, der sich bei tiefen Temperaturen gut bewegen ließ.

14. Stationäre Diffusion. G. HERTZ (1923) hat die stationäre Diffusion in strömenden Gasen zur Isotopentrennung benutzt; HARTECK und SCHMIDT (1933) haben diese Methode angewandt um den Diffusionskoeffizienten von Para- gegen Normalwasserstoff zu messen.

Zur Erläuterung des Prinzips betrachten wir die eindimensionale, stationäre Diffusion von Gas 1 gegen Gas 2. Bei $x = -\infty$ soll reines Gas 2 mit der Geschwindigkeit w in der $+x$-Richtung einströmen. Der Molenbruch γ_0 des Gases 1 an der Stelle $x = 0$ sei künstlich vorgegeben. Wenn Druck und Temperatur überall konstant sind, so hat die mittlere Teilchengeschwindigkeit des Gemischs überall den Wert w. Die Bedingung, daß im stationären Zustand die Geschwindigkeit v_1 von Gas 1 für $x < 0$ verschwindet, findet nach (3.6) und (3.10) ihren Ausdruck in der Gleichung

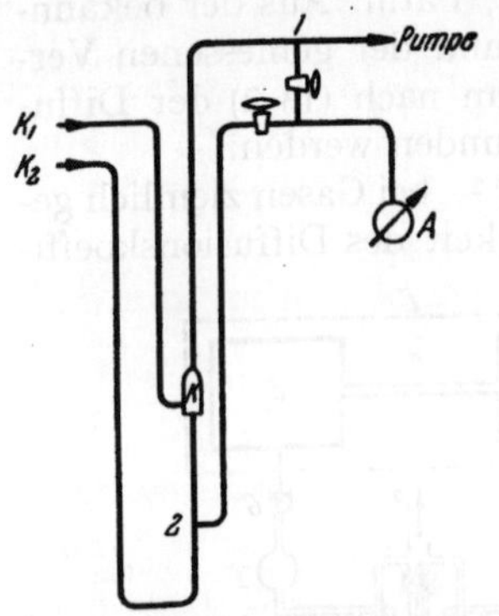

Fig. 14. Apparat von HARTECK und SCHMIDT (1933) zur stationären Messung von Diffusionskoeffizienten.

$$v_1 = w + W_1 = w - \frac{1}{\gamma} D \frac{\partial \gamma}{\partial x} = 0$$

oder

$$\gamma = \gamma_0 e^{\frac{w x}{D}} \quad (\text{für } x < 0). \tag{14.1}$$

Dies gibt die stromaufwärts stattfindende Verarmung an Komponente 1 an. Dabei wurde die bei Gasen gut erfüllte Annahme gemacht, daß der Diffusionskoeffizient nicht vom Mischungsverhältnis abhängt. Bei bekannter Strömungsgeschwindigkeit kann man also D entnehmen aus dem Verhältnis γ/γ_0 der Molenbrüche des Gases 1 an zwei verschiedenen Stellen in dem Gasstrom.

Fig. 14 zeigt das Schema der Apparatur, mit welcher HARTECK und SCHMIDT (1933) die Diffusion von p-H_2 (Gas 1) gegen n-H_2 bei tiefen Temperaturen untersuchten. Durch die Capillaren $K_{1,2}$ treten p-H_2 bzw. n-H_2 in das Kölbchen K ein und strömen von da zur Pumpe. In dem Kölbchen und in der zur Pumpe führenden Capillare stellt sich nach Maßgabe der zuströmenden Menge ein bestimmter Molenbruch γ_0 des p-H_2 ein. γ_0 kann an der Stelle 1 durch die angeschlossene Analysenapparatur A, im wesentlichen eine Wärmeleitzelle, gemessen werden. Von K aus diffundiert der p-H_2 nun aber auch stromaufwärts in die Capillare K_2. Der an der Stelle 2 herrschende Molenbruch kann wieder mittels A festgestellt werden. Die Capillaren mit dem Kölbchen tauchten in ein Dewar-Gefäß mit der Kühlflüssigkeit; keine beweglichen Teile befinden sich auf tiefer Temperatur.

Ein gewisser Nachteil des Verfahrens ist es, daß man den Apparat mit einem Gemisch von bekanntem Diffusionskoeffizienten eichen muß. Die Formel (14.1) trifft ja für die Strömung in Capillaren, für die ein Poiseuillesches Geschwindigkeitsprofil gilt, nicht genau zu, auch ist die Entfernung x bei einer Glasapparatur nicht exakt definiert. (Bemerkt sei, daß eine weitere vereinfachende, oben gemachte Annahme, die der Druckkonstanz, harmlos ist, da das Verhältnis w/D nach der Kontinuitätsgleichung vom Druck unabhängig ist.) Zur Eichung benutzten die Autoren die Diffusion von p-H_2 gegen n-H_2 bei Zimmertemperatur, nachdem sie den zugehörigen Diffusionskoeffizienten vorher nach der Loschmidtschen Methode gemessen hatten.

Stationäre Diffusionsmessungen an radioaktivem Tracer-Gas, welches an einem Punkt in einen laminaren Gasstrom eingeleitet wird, haben WESTENBERG und WALKER (1957) vorgenommen.

15. Methode des nichtstationären Diffusionsthermoeffekts. Für die Diffusion und Wärmeleitung im ruhenden, binären Gemisch ($w=0$) gilt nach (3.13) und (3.18)

$$\frac{\partial\gamma}{\partial t} = D\,\Delta\gamma, \tag{15.1}$$

$$\left.\begin{aligned}\frac{\partial T}{\partial t} &= K\,\Delta T + \\ &+ \delta T\,\frac{\partial\gamma}{\partial t},\end{aligned}\right\} \tag{15.2}$$

wo zur Abkürzung

$$\delta T = \alpha\,\frac{\mathsf{R}}{C_{p,\,\mathrm{mol}}}\,T \tag{15.3}$$

gesetzt ist (R = molare Gaskonstante, $C_{p,\,\mathrm{mol}}$ = Molwärme) und Δ den Laplace-Operator bedeutet. Die Thermodiffusion spielt für das Jetzige keine Rolle und ist weggelassen. Es soll gezeigt werden, daß die während der Diffusion vorübergehend auftretenden Temperaturunterschiede asymptotisch nach Maßgabe des Diffusionskoeffizienten abklingen und dadurch dessen Bestimmung ermöglichen (WALDMANN 1947b).

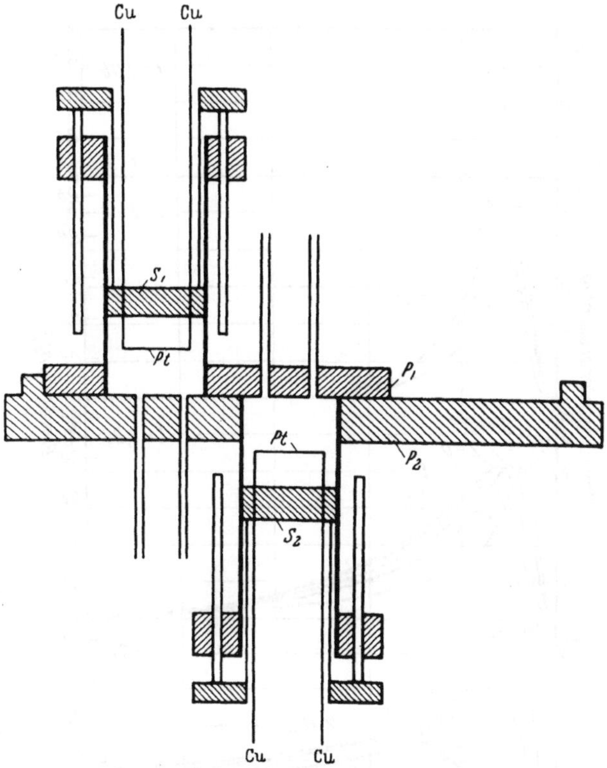

Fig. 15. Apparat zur Messung des nichtstationären Diffusionsthermoeffektes (WALDMANN 1947b).

Wir nehmen die Lösung der Diffusionsgleichung (15.1) als bekannt an:

$$\gamma = \gamma(D\,t,\boldsymbol{r}).$$

Sie soll die Anfangs- und die Randbedingung (Verschwinden des Diffusionsstroms an der Oberfläche) erfüllen. Dann kann man sogleich die Lösung der Temperaturgleichung (15.2) angeben, welche als Anfangsbedingung $T = \mathrm{const} = T_0$ hat und als Randbedingung das Verschwinden des Wärmestroms (adiabatische Wand). Sie lautet

$$T = T_0 + \delta T\cdot\frac{D}{D-K}\left[\gamma(D\,t,\boldsymbol{r}) - \gamma(K\,t,\boldsymbol{r})\right]. \tag{15.4}$$

Man verifiziert dies durch Einsetzen. Nun nähert sich der Molenbruch nach (15.1) zeitlich exponentiell seinem Endwert; ist z.B. das Gasgemisch in einem Rohr der Gesamtlänge $2l$ eingeschlossen, so gilt nach (13.1)

$$\left[\gamma(D\,t,\boldsymbol{r}) - \gamma(\infty,\boldsymbol{r})\right]_{t\to\infty} \propto \mathrm{e}^{-\frac{\pi^2 D t}{4l^2}}. \tag{15.5}$$

Nach (15.4) ist demnach auch

$$[T - T_0]_{t\to\infty} \propto e^{-\frac{\pi^2 D t}{4 l^2}}. \qquad (15.6)$$

Dabei ist allerdings $K \gg D$ angenommen [damit das letzte Glied in (15.4) weggelassen werden kann], während tatsächlich K und D etwa gleich groß sind. Trotzdem ist (15.6) auf das Experiment anwendbar, weil man es ja in Wirklichkeit nicht mit adiabatischen, sondern mit isothermen Wänden zu tun hat und die dadurch bedingte radiale Wärmeableitung — vorausgesetzt daß der Diffusionszylinder nicht zu flach ist — sich ähnlich auswirkt wie die Annahme $K \gg D$.

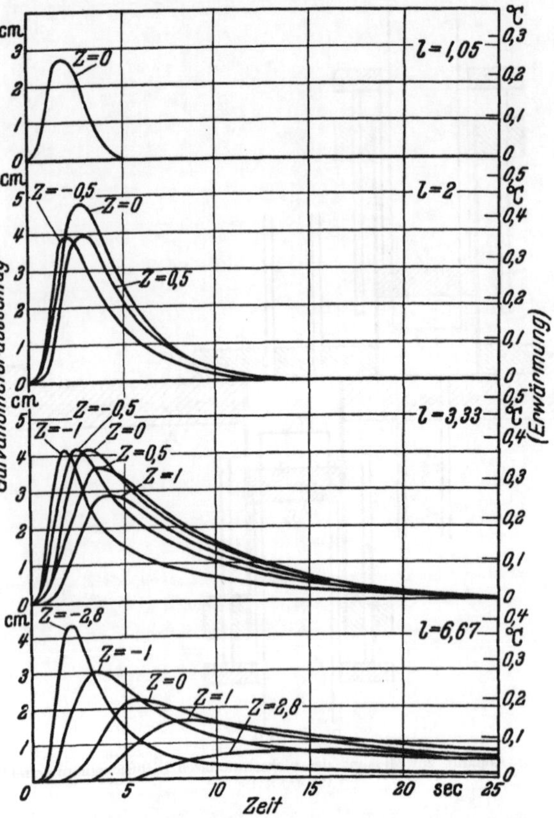

Fig. 16. Diffusionsthermoeffekt, beobachtet bei der Diffusion von N_2/80% N_2, 20% H_2 in der oberen Kammer. l Höhe jeder Kammer in cm; z Höhe des Meßdrahts über Kammermitte in cm, Beginn der Diffusion als Zeitnullpunkt (Waldmann 1947b).

Den Apparat zeigt Fig. 15. Die beiden Messingrohre — 4 cm Innendurchmesser —, welche die Diffusionskammern bilden, tragen an den offenen Enden zwei Platten $P_{1,2}$; durch Verschieben von P_1 wird der Apparat in die Füllstellung oder die Diffusionsstellung gebracht. Mittels der Stempel $S_{1,2}$ kann die Höhe der Kammern verstellt werden. Zur Temperaturmessung dienen die ebenfalls verschiebbaren, 15 μ dicken Platindrähte Pt, die wahlweise in einer Wheatstone-Brücke liegen. Fig. 16 zeigt einige der mit diesem Apparat gewonnenen Registrierkurven. Die logarithmische Umzeichnung dieser Kurve ist in Fig. 17 dargestellt und zeigt die Gültigkeit des exponentiellen Abfalls (15.6). Aus der Neigung der asymptotischen Geraden in Fig. 17 und den bekannten Kammerhöhen kann man sodann den Diffusionskoeffizienten entnehmen. Wegen der Kürze der Meßdauer spielt Konvektion bei dem Versuch keine Rolle.

16. Methode des stationären Diffusionsthermoeffekts. Auch bei der Diffusion von stationären Gasströmen ineinander treten Temperaturunterschiede auf, die man zur Bestimmung des Diffusionskoeffizienten benutzen kann (Waldmann 1950b). Der benutzte Apparat war dem in Fig. 23 (Ziff. 21) dargestellten ähnlich. Nur waren die Rohre länger — je 24 cm —, ihr gemeinsamer Schlitz konnte durch einen Schieber, welcher mit einer Längenskala versehen war, beliebig abgedeckt werden und die Meßdrähte waren nicht längs der Achse, sondern quer zu ihr fest eingespannt. So war es möglich, verschiedene Entfernungen z der Meßdrähte von der Schieberkante einzustellen.

Wieder erhält man einen exponentiellen Abfall der Temperatur, diesmal in Abhängigkeit von der Koordinate z. Dies folgt aus den Gleichungen

$$w \frac{\partial \gamma}{\partial z} = D \Delta \gamma; \quad w \frac{\partial T}{\partial z} = K \Delta T + \delta T \, w \frac{\partial \gamma}{\partial z}, \tag{16.1}$$

welche für mit der Geschwindigkeit $w(x, y)$ in der z-Richtung strömendes Gemisch an Stelle von (15.1) und (15.2) gelten. Der asymptotische Temperaturverlauf stromabwärts ist

$$[T - T_0]_{z-\infty} \propto e^{-z/L}, \tag{16.2}$$

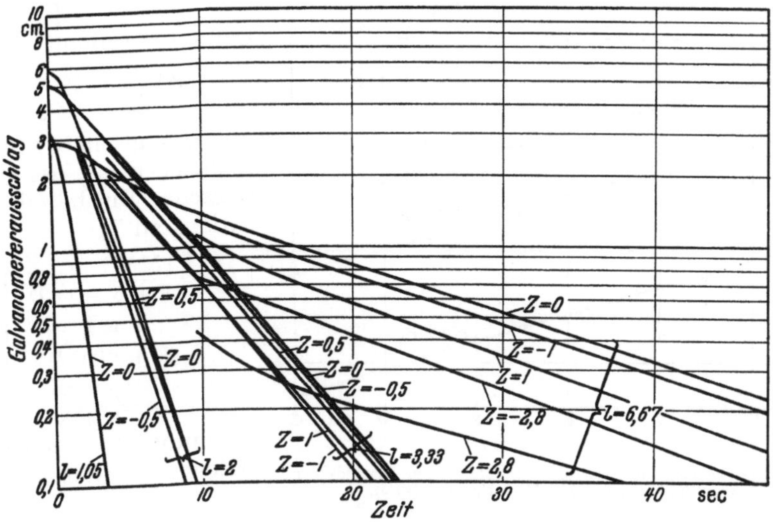

Fig. 17. Logarithmische Umzeichnung der Kurven aus Fig. 16; Zeitpunkt des Maximalausschlags als Zeitnullpunkt.

wobei für die Abfallslänge L bei nicht zu kleiner Geschwindigkeit $\overline{w}$ (Mittelwert über den Rohrquerschnitt)

$$L \propto \frac{\overline{w}}{D} \tag{16.3}$$

gilt. Durch Bestimmung der Abfallslänge L findet man bei bekannter Geschwindigkeit $\overline{w}$ den Diffusionskoeffizienten, wenn man sich durch Eichung mittels eines Gemischs von bekanntem D-Wert den Proportionalitätsfaktor in (16.3) verschafft hat.

17. Bestimmung des Diffusionskoeffizienten aus der Einstellzeit des stationären Zustands der Thermodiffusion. Die in einem ruhenden Gasgemisch durch Thermodiffusion auftretenden Unterschiede der Zusammensetzung sind im allgemeinen recht klein. Wenn man aus der Einstellzeit des stationären Zustands den Diffusionskoeffizienten ermittelt, so erhält man dessen Wert bei sehr genau definierter Zusammensetzung des Gases. Diese Methode ist daher besonders geeignet, um die Abhängigkeit des Diffusionskoeffizienten vom Molenbruch zu studieren. Sie wurde von NETTLEY (1954) entwickelt, nachdem IBBS schon früher die exponentielle Einstellung des stationären Zustands der Thermodiffusion beobachtet hatte.

Um das Prinzip der Methode zu erläutern, seien zwei Gasbehälter mit den Volumen V, V' betrachtet, die durch ein Rohr von gleichmäßigem Durchmesser miteinander verbunden sind. Molekülzahlen und Molenbrüche in den Behältern seien N, γ bzw. N', γ'; L und Q seien Länge und Querschnitt des Verbindungsrohres. Von der Thermodiffusion werde zunächst abgesehen und wir fragen, wie sich die Molenbrüche γ, γ' zeitlich einander durch Diffusion angleichen, wenn vorausgesetzt werden darf, daß das Volumen des Rohrs klein ist im Vergleich zu dem der Behälter. Dann herrscht in dem Rohr stets der Zustand der quasistationären Diffusion (vgl. CLARK JONES und FURRY [5], S. 164). Der von V nach V' gerichtete Teilchendiffusionsstrom beträgt $QnD(\gamma-\gamma')/L$ und für die Teilchenzahlen in jedem Behälter gelten die Gleichungen

$$N\frac{d\gamma}{dt} = \frac{dN\gamma}{dt} = -\frac{QnD}{L}(\gamma-\gamma') = -\frac{dN'\gamma'}{dt} = -N'\frac{d\gamma'}{dt}.$$

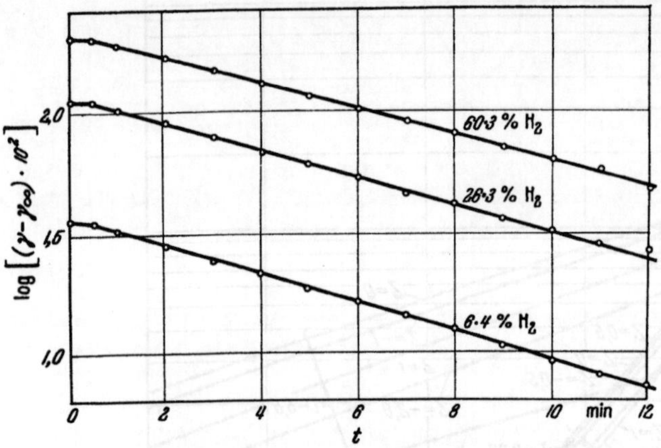

Fig. 18. Die exponentielle Einstellung des stationären Zustands der Thermodiffusion von N_2/H_2. Nach NETTLEY (1954).

Daraus entnimmt man

$$\frac{d(\gamma-\gamma')}{dt} = -\frac{\gamma-\gamma'}{t_0}. \quad (17.1)$$

Also erfolgt ein exponentielles Angleichen mit der charakteristischen Einstellzeit („Relaxationszeit")

$$\left.\begin{aligned} t_0 &= \frac{NN'}{N+N'}\frac{L}{QnD} \\ &= \frac{VV'L}{(V+V')QD}. \end{aligned}\right\} \quad (17.2)$$

Mit $\gamma_\infty = \frac{N\gamma + N'\gamma'}{N+N'}$, dem sich asymptotisch einstellenden Molenbruch, hat man auch

$$\gamma - \gamma_\infty \propto \gamma' - \gamma_\infty \propto e^{-\frac{t}{t_0}}. \quad (17.3)$$

Fast dasselbe gilt für die Einstellung der Thermodiffusion; der Unterschied besteht nur darin, daß hier γ_∞ und γ'_∞ zu unterscheiden sind, entsprechend der im stationären Zustand vorhandenen Entmischung.

Für die Versuche benutzte NETTLEY den Apparat von IBBS (s. Fig. 19); es war lediglich das untere, kalte Volumen etwas größer gewählt ($V = 82{,}5$ cm³, $V' = 7{,}6$ cm³) und das Verbindungsrohr so eng, daß sich eine genügend große Einstellzeit ergab (Rohrlänge 5,0 cm, lichte Weite 0,315 cm). Die Temperaturen waren $T = 288°$ K und $T' = 373°$ K. Das Verbindungsrohr muß einheitliche Temperatur haben, damit der Diffusionskoeffizient einen definierten Wert besitzt. Die Einstellung des stationären Zustands wurde durch laufende Ablesungen an der Wärmeleitzelle (Katharometer) verfolgt. Fig. 18 zeigt ein Beispiel. NETTLEY konnte mit dieser Methode sehr genau die geringe Abnahme des Diffusionskoeffizienten von N_2/H_2 mit zunehmendem H_2-Gehalt bestätigen (s. Ziff. 73).

IV. Messung des Thermodiffusionsfaktors.

Der Thermodiffusionfaktor α (oder das Thermodiffusionverhältnis k_T) ist das quantitative Maß für zweierlei Erscheinungen in binären Gasgemischen: für die durch ein Temperaturgefälle bewirkte partielle Entmischung der Komponenten und für die durch einen Diffusionsstrom hervorgerufenen Temperaturunterschiede. Dementsprechend stehen auch zwei verschiedene Meßmethoden für α zur Verfügung. Der Vergleich der nach den beiden Methoden erhaltenen Werte miteinander ist ein wichtiger Test für die kinetische Gastheorie und die Thermodynamik der irreversiblen Prozesse.

Einen sehr vollständigen Überblick über die Thermodiffusion gibt die Monographie von GREW und IBBS [18]. Auf den störenden Einfluß der Thermodiffusion bei physikalisch-chemischen Messungen haben EMMETT und SHULTZ (1932) hingewiesen; neuerdings hat WHALLEY (1951) darüber gearbeitet.

18. Die Zweigefäßmethode. Ein binäres Gemisch sei in zwei durch ein Rohr verbundene Gefäße eingeschlossen, deren jedes sich in einem anderen Temperaturbad befindet. Das warme Gefäß sei über dem kalten angebracht, so daß keine thermische Konvektion im Schwerefeld eintreten kann. Die Vermeidung der Konvektion ist wichtig. Nach GREW (1945) hat bei den Versuchen von SCHMAHL und SCHEWE (1940) an CO_2/H_2-Gemisch die Konvektion eine Rolle gespielt und eine starke Druckabhängigkeit des Thermodiffusionsfaktors vorgetäuscht. Ohne Konvektion verschwindet im stationären Zustand der Diffusionsstrom jeder Komponente und man hat nach (3.10) und (3.12)

$$\left.\begin{aligned} \frac{\partial\gamma}{\partial x_\mu} &= -k_T\frac{\partial T}{T\,\partial x_\mu} = -\alpha\gamma(1-\gamma)\frac{\partial T}{T\,\partial x_\mu} \\ \text{oder}\quad \frac{d\gamma}{dT} &= -\frac{k_T}{T} = -\frac{\alpha\gamma(1-\gamma)}{T}. \end{aligned}\right\}\qquad(18.1)$$

Dies besagt, daß die Zusammensetzung an jeder Raumstelle eine Funktion der dort herrschenden Temperatur ist. Da k_T oder α im allgemeinen von T und γ abhängt, kann man (18.1) nicht exakt integrieren und hat zur Messung möglichst kleine Temperaturdifferenzen anzuwenden oder auf solche zu extrapolieren. Da die erzielte Entmischung gewöhnlich klein ist, macht die γ-Abhängigkeit von k_T oder α praktisch keine Schwierigkeit.

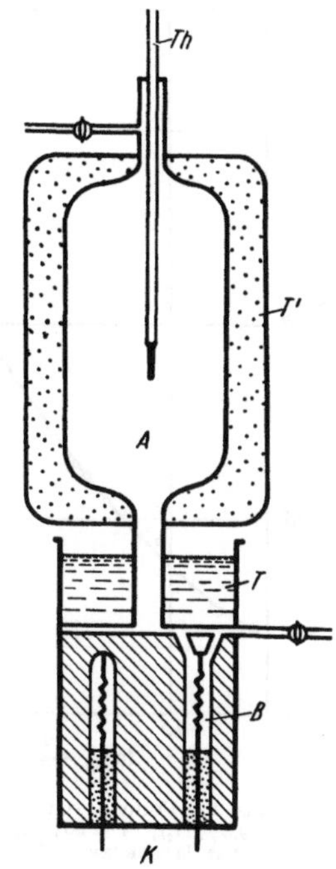

Fig. 19. Apparat von IBBS (1925) zur Messung der Thermodiffusion.

Ein Beispiel einer Meßapparatur, die auf IBBS (1925) zurückgeht, zeigt Fig. 19. Die beiden Gefäße sind verschieden groß gewählt. In dem oberen warmen Gefäß A von etwa 12 cm³ (10 bis 300° C) ändert sich daher die Gaszusammensetzung nur relativ wenig; das untere kalte Gefäß B wird von einem 1,5 cm³ großen Hohlraum in einem wassergekühlten Kupferblock K (10° C) gebildet und enthält eine etwa 1 cm lange Platin-Spirale (etwa 10 Ω). In dem Gefäß B findet hauptsächlich die Änderung des Mischungsverhältnisses statt; der Pt-Draht dient zu deren Feststellung nach dem Prinzip der Wärmeleitzelle. Die zweite, völlig abgeschlossene Zelle, links in dem Kupferblock, enthält eine gleichartige Pt-Spirale (Katharometer, s. Ziff. 12).

Außer der Wärmeleitzelle wurden bei Thermodiffusionsversuchen zur Gasanalyse auch Viscositätsmessungen mittels schwingender Scheibe (VAN ITTERBEEK, VAN PAEMEL und VAN LIERDE 1947), die Interferometermethode (G. und O. BLÜH 1934), Massenspektrometer (NIER 1939, 1940, E. W. BECKER und BEYRICH 1952) und — bei der Untersuchung von Radon — die Radioaktivität (HARRISON 1937, 1942) benutzt.

Wenn man k_T als Konstante behandelt, so folgt aus (18.1) für die zwischen den Temperaturen T und T' sich einstellende Entmischung

$$\gamma - \gamma' = k_T \log\frac{T'}{T}.\qquad(18.2)$$

Die damit behauptete Proportionalität von Entmischung und $\log T'/T$ gilt, wie Fig. 20 zeigt, nur bei genügend hohen Temperaturen T' (verglichen mit der Siedetemperatur der höher siedenden Komponente), während bei niedrigeren Temperaturen T' deutlich eine Änderung von k_T — im allgemeinen eine Abnahme des absoluten Betrages mit abnehmender Temperatur — stattfindet. Den wahren, zur Temperatur T gehörigen k_T- oder α-Wert erhält man aus der Tangente in dem Punkt $T' = T$ an eine solche Kurve, wie sie in Fig. 20 dargestellt ist; bei starker Krümmung der Kurve — starker Temperaturabhängigkeit von k_T — wird das Verfahren ungenau, ebenso dann, wenn wegen der Kleinheit der Entmischung überhaupt nur bei großem Temperaturunterschied $T' - T$ gemessen werden kann, wie z.B. bei Isotopen. Die Unsicherheit würde wegfallen, wenn man die Temperaturabhängigkeit von k_T oder α kennte. Bei Isotopen sind die Verhältnisse genügend einfach, daß man entsprechende Ansätze machen kann. Einmal ist hier α praktisch vom Mischungsverhältnis unabhängig und nach Messungen sowohl wie nach dem Lennard-Jonesschen Modell gilt für die Temperaturabhängigkeit in brauchbarer Näherung

$$\alpha = \alpha_\infty \cdot \left(1 - \frac{T_0}{T}\right) \tag{18.3}$$

mit geeigneten Konstanten α_∞, T_0 (= Temperatur, bei welcher α sein Vorzeichen umkehrt). Brown (1940) hat vorgeschlagen, diese Temperaturabhängigkeit bei der Auswertung der Messungen anzunehmen. Schreibt man dann nach (18.1) für die Entmischung

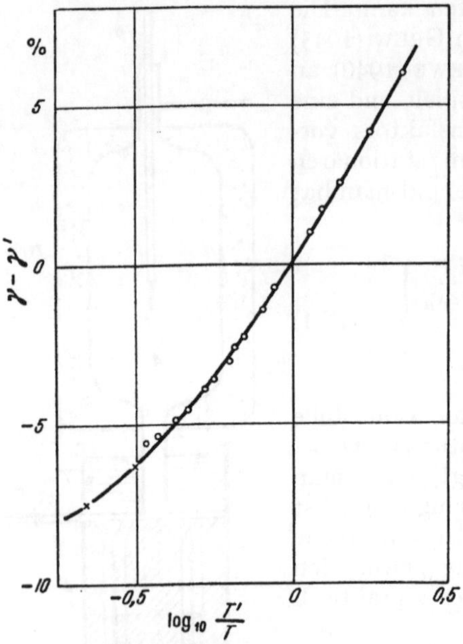

Fig. 20. Prüfung von (18.2) an N_2/H_2-Gemisch (39,8% H_2); $T = 293°$ K, $T' = 12$ bis 600° K. × van Itterbeek *et al.* (1947); ○ Grew (1949).

$$\gamma - \gamma' = \gamma(1-\gamma)\int_T^{T'} \frac{\alpha(T'')}{T''}\,dT'' = \gamma(1-\gamma)\bar{\alpha}\log\frac{T'}{T}$$

so gilt, wie man leicht nachrechnet, für den dadurch definierten Mittelwert $\bar{\alpha}$ die Zuordnung

$$\bar{\alpha} = \alpha(\bar{T}) \quad \text{wo} \quad \bar{T} = \frac{T\,T'}{T' - T}\log\frac{T'}{T}. \tag{18.4}$$

Damit hat man ein zusammenhängendes Wertepaar $\bar{\alpha}$ und $\bar{T}$ aus einer einzigen Entmischungsmessung. Etwas andere Vorschläge für die Zuordnung haben Srivastava und Madan (1953) gemacht (s. auch Holleran 1953).

Bei der Bestimmung von α mittels des Diffusionsthermoeffekts tritt eine Unsicherheit in der zuzuordnenden Temperatur von vornherein nicht auf (s. Ziff. 20 und 21).

19. Vervielfachung der Thermodiffusions-Anreicherung. Als erster wies Chapman (1919) auf die Möglichkeit hin, die Thermodiffusion zur Isotopentrennung zu benutzen. Praktische Bedeutung erlangte dies aber erst, als Clusius und Dickel (1938) auf den entscheidenden Gedanken kamen, den durch (18.1) gegebenen Elementareffekt durch das Gegenstromverfahren zu vervielfachen. Der dazu dienende Apparat — das Thermodiffusionstrennrohr — ist von bestrickender Einfachheit, da das für die Thermodiffusion nötige Temperaturgefälle — erzeugt durch einen Heizdraht in der Achse eines vertikal aufgestellten, außen gekühlten Rohres — zugleich den Gegenstrom hervorruft. Mit diesem Apparat gelang sehr rasch zum ersten Male die fast vollständige Trennung der Chlorisotope in wägbarer Menge; es wurden je etwa 0,5 n-Liter HCl mit 99,6% ^{35}Cl bzw. 99,4% ^{37}Cl gewonnen (CLUSIUS und DICKEL 1939). Viele weitere Reindarstellungen von Isotopen folgten. Eine der neuesten Arbeiten (CLUSIUS, HUBER, HÜRZELER und SCHUMACHER 1956) behandelt die Reindarstellung des seltenen Isotops ^{21}Ne aus dem natürlichen polynären Isotopengemisch mittels des Kunstgriffs der Hilfsgase, der zuerst an Kr—X-Gemisch von SCHUMACHER und CLUSIUS (1953) angewandt wurde. Im Fall des Neons dienten die deuterierten Methane als Hilfsgase. Auch das Auftreten von Isotopen-Austauschreaktionen im Trennrohr ist studiert worden, insbesondere durch T. I. TAYLOR u. Mitarb. (1947, 1948). In den Fällen genügend rascher Einstellung des Austausch-Gleichgewichts kann man bei zwei Isotopen den aus Diffusion einschließlich Thermodiffusion und Austauschreaktion bestehenden, verwickelten Vorgang in dem wirklichen, polynären Molekülgas reduzieren auf einen einfachen Diffusionsprozeß in einem gedachten, binären „Atomgas“ (s. dazu die bei Gl. (74.8) zitierten Arbeiten).

Der Rückschluß von der mit dem Trennrohr erzielten Entmischung auf den Thermodiffusionsfaktor ist mit Schwierigkeiten verbunden. Die Vorgänge im Trennrohr sind nämlich nach der quantitativen Theorie, die auf WALDMANN (1939), VAN DER GRINTEN (1939), FURRY, JONES und ONSAGER (1939) und DEBYE (1939) zurückgeht (Zusammenfassungen durch FLEISCHMANN und JENSEN [*15*]; CLARK JONES und FURRY [*5*]), mitbestimmt durch die Zähigkeit und den Diffusionskoeffizienten sowie deren Temperaturabhängigkeit. All diese Größen müßten bekannt sein. Ferner treten parasitäre Vermischungen auf, die experimentell mühsam zu unterdrücken und theoretisch schwer erfaßbar sind. Gleichwohl wurde öfters versucht, Trennrohrmessungen in bezug auf α auszuwerten. So erhielten CLUSIUS und DICKEL (1942) für ^{86}Kr/^{84}Kr zunächst den Wert $\alpha = 0{,}0034$ bei 450° K und korrigierten ihn dann durch Schätzung auf 0,0047; neuerdings haben CORBETT und WATSON (1956) auf diese Weise für ^{40}A/^{36}A den Wert $\alpha = 0{,}013$ bei 553° K gefunden, während STIER (1942) nach der Zweigefäßmethode für fast dieselbe Mitteltemperatur $\alpha = 0{,}022$ maß. Daraus erhellt die Schwierigkeit der Auswertung. Bei sehr kleinem Thermodiffusionsfaktor wird man diese Nachteile allerdings in Kauf nehmen, um überhaupt messen zu können. Das einfachste ist dann, das Trennrohr mit einem Gemisch von bekanntem α zu eichen, wie es E. W. BECKER, HENKES und SEIFERT (1955) taten, um den Thermodiffusionsfaktor von Ortho- gegen Parawasserstoff zu bestimmen, der nach ihren Vergleichsmessungen mit He/H_2-Gemisch bei 90° K, 700 Torr den sehr kleinen Wert $\alpha = 1{,}1 \cdot 10^{-3}$ hat.

Eine genau bekannte Vervielfachung der Trennung (18.1) und damit eine Erhöhung der Analysengenauigkeit wird dadurch erreicht, daß man nach Einstellung des stationären Zustands der Thermodiffusion das eine der beiden gleichgroßen Gefäße evakuiert, das restliche angereicherte Gemisch wieder beide Gefäße erfüllen läßt und neuerdings den stationären Zustand der Thermodiffusion abwartet, usw. Die damit verknüpfte Druckverminderung setzt schließlich dem Verfahren ein Ende, einmal im Hinblick auf die Analyse und ferner deshalb, weil man ja im Bereich kleiner freier Weglängen bleiben muß, um nicht schließlich einen ganz anderen Effekt zu messen. Nach diesem Verfahren (mit bis zu 12maliger Verdünnung) haben u. a. NIER (1939, 1940) und STIER (1942) die α-Werte bestimmt für $^{13}CH_4$/$^{12}CH_4$ sowie ^{22}Ne/^{20}Ne und ^{40}A/^{36}A.

Man kann daran denken, das Trennrohr für Meßzwecke nutzbar zu machen dadurch, daß man es in einzelne wohldefinierte Trennglieder zerlegt, welche durch genügend große Strömungswiderstände verbunden sind und den diskreten Böden einer Rektifikationskolonne vergleichbar sind. Ein solches Verfahren ist von CLUSIUS und HUBER (1955a) ausgearbeitet worden. Die von den Autoren als Trennschaukel bezeichnete Anordnung (s. Fig. 21) besteht aus einer Anzahl vertikaler Rohre, die oben erwärmt, unten gekühlt sind. Das obere Ende des ersten Rohrs ist mit dem unteren Ende des zweiten Rohrs durch eine Capillare verbunden, usw.; schließlich steht das obere Ende des letzten (z-ten) Rohrs mit dem unteren Ende des ersten Rohres in Verbindung durch eine Capillare, welche eine zylindrische Erweiterung Zy mit einem Kolben K hat. Periodisches Hin- und Herbewegen des Kolbens erzwingt an den durch Capillaren verbundenen Rohrenden die gleiche Zusammensetzung des Gases. Da im stationären Zustand in jedem Rohr die Entmischung (18.1) sich einstellt, hat man zwischen dem unteren Ende des ersten und dem oberen Ende des letzten Rohres den z-fachen Trenneffekt. Nimmt man α in (18.1) als konstant an, so ist

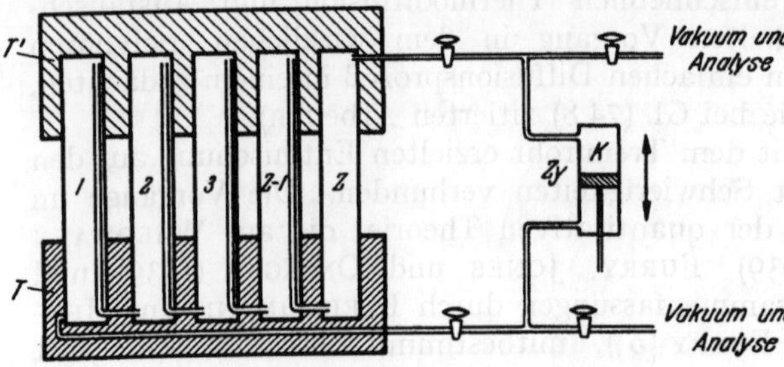

Fig. 21. Prinzip der Trennschaukel nach CLUSIUS und HUBER (1955a).

$$d \log \frac{\gamma}{1-\gamma} = -\alpha\, d \log T.$$

Man kann integrieren und hat für den Trennfaktor eines einzigen Rohres

$$\frac{\gamma/(1-\gamma)}{\gamma'/(1-\gamma')} = \left(\frac{T'}{T}\right)^{\alpha},$$

wo γ den Molenbruch am unteren, γ' den am oberen Ende des ersten Rohrs bedeutet. Für den vervielfachten Trennfaktor der ganzen Anordnung gilt

$$\frac{\gamma/(1-\gamma)}{\gamma^{(z)}/(1-\gamma^{(z)})} = \left(\frac{T'}{T}\right)^{z\alpha},$$

wo $\gamma^{(z)}$ den Molenbruch am oberen Ende des z-ten Rohrs bezeichnet. Natürlich ist die exponentielle Formel nur dann konsequent, wenn α tatsächlich temperaturunabhängig ist. Man hat, da dies im allgemeinen nicht zutrifft, für Messungen wieder kleine relative Temperaturdifferenzen zu benutzen und schreibt dann besser nach (18.1) direkt

$$\gamma - \gamma^{(z)} = z\,\alpha\,\gamma\,(1-\gamma)\,\frac{T'-T}{T}.$$

CLUSIUS und HUBER haben die Wirkungsweise dieser Anordnung an CO_2/H_2-Gemisch untersucht und den α-Wert bei 220, 294 und 387° K in Abhängigkeit vom Mischungsverhältnis bestimmt mit einem vermutlichen Fehler von nicht mehr als 2 bis 3%. Eine theoretische Untersuchung der Fehlerquellen hat VAN DER WAERDEN (1957) gemacht.

20. Bestimmung des Thermodiffusionsfaktors aus dem nichtstationären Diffusionsthermoeffekt. Die experimentellen Grundlagen des Diffusionsthermoeffekts sind in Ziff. 15 und 16 dargelegt, bei der Besprechung der Messung von Diffusionskoeffizienten. In erster Linie wird man den Effekt aber zur Bestimmung von Thermodiffusionsfaktoren heranziehen. Auch hierzu gibt es wieder zwei Möglichkeiten: nichtstationäre Beobachtung am ruhenden Gas (WALDMANN 1947b), stationäre Beobachtung am strömenden Gas (s. nächste Ziffer).

Für eine Absolutbestimmung von α ist es nötig, auf Grund der Gln. (15.1) und (15.2) eine exakte Aussage über die Diffusionstemperatur zu machen. Die direkte Integration dieser Gleichungen für eine praktisch ausführbare Anordnung, etwa die der Fig. 15 (Ziff. 15), wäre schwierig. Man kommt viel einfacher zum Ziel durch Einführung des Temperaturzeitintegrals

$$\Phi(\boldsymbol{r}) = \int_0^\infty [T(t, \boldsymbol{r}) - T_0]\, dt, \tag{20.1}$$

wo T_0 die konstante Temperatur der Gefäßwände bezeichnet. Für diese Raumfunktion Φ gilt nämlich nach (15.2) und (15.3)

$$\Delta\Phi = \frac{\alpha p}{\lambda}[\gamma(0, \boldsymbol{r}) - \gamma(\infty, \boldsymbol{r})], \tag{20.2}$$

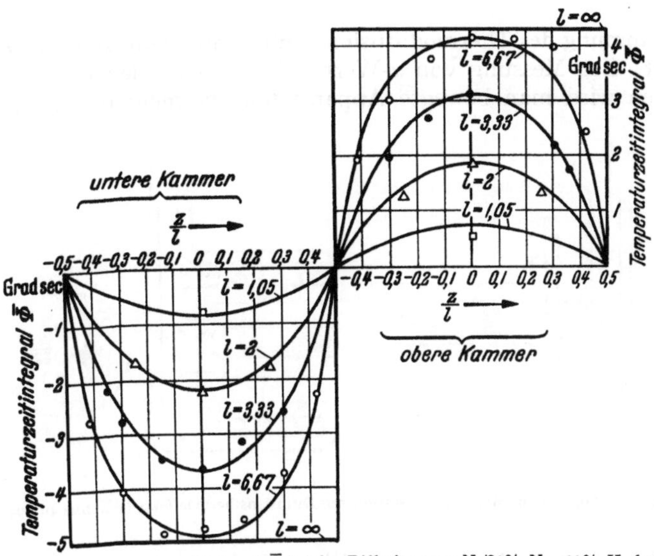

Fig. 22. Meßpunkte für das Temperaturzeitintegral $\bar{\Phi}$ bei der Diffusion von N_2/80% N_2, 20% H_2 bei 20° C, 760 Torr Kurven berechnet nach (20.2) (WALDMANN 1947b).

also eine Poissonsche Differentialgleichung mit bekannten Quellen; die Quellenfunktion rechts in (20.2) ist sogar besonders einfach, weil gebietsweise konstant. Im eindimensionalen Fall — Gasgemisch zwischen zwei planparallelen isothermen Platten im Abstand $2l$ eingeschlossen — ergibt sich aus (20.2) sofort

$$\Phi = \frac{\alpha p}{\lambda}(\gamma_\infty - \gamma_0) \cdot \frac{1}{2}|x|\,(l - |x|), \tag{20.3}$$

wo x den Abstand von der Mittelebene der Platten bedeutet. Das Temperaturintegral verschwindet also exakt in der Mittelebene $x = 0$ und natürlich an den Wänden $x = \pm l$; in der oberen und unteren „Kammer" zeigt es je einen parabelförmigen Verlauf, von entgegengesetztem Vorzeichen in den beiden Kammern. Auch für die zylindrische Anordnung aus Fig. 15 kann das Temperaturzeitintegral leicht genau berechnet werden (WALDMANN 1947a), womit dann die Methode zur Absolutmessung von α gegeben ist. Ein Beispiel für eine solche Messung und die zugehörigen berechneten Kurven zeigt Fig. 22 ($\bar{\Phi}$ = Mittelwert über den Zylinderdurchmesser; z = Höhe des Meßdrahts über der Mittelebene der oberen bzw. unteren Kammer). Übrigens kann man das Temperaturzeitintegral aus den Registrierkurven von Fig. 16 (Ziff. 15) ohne weiteres entnehmen, trotz der Verzerrung der Anzeige durch Galvanometerträgheit und -dämpfung; wie man an Hand der Bewegungsgleichung des Galvanometers überlegt, ist diese Verzerrung nämlich ohne Einfluß auf das Zeitintegral des Stroms, der durch das Galvanometer fließt.

Ein Vorteil der Diffusionswärme-Methode besteht darin, daß man α-Werte erhält, die zu einer ganz bestimmten Temperatur und Zusammensetzung (letzteres bei Diffusion wenig verschiedener Gemische ineinander) gehören. Ein weiterer Vorteil besteht darin, daß die Meßgenauigkeit im Gegensatz zu den Thermodiffusionsmessungen praktisch unabhängig ist vom Mischungsverhältnis; α läßt sich für $\gamma \to 0$ oder 1 ebenso gut bestimmen wie für mittlere γ. Ein Nachteil ist, daß man zur Auswertung den Wärmeleitkoeffizienten kennen muß, wie (20.3) zeigt.

21. Bestimmung des Thermodiffusionsfaktors aus dem stationären Diffusionsthermoeffekt. Zur Messung von α-Werten bei hohen oder tiefen Temperaturen oder Drucken wird man bewegte Apparateteile vermeiden wollen und deshalb

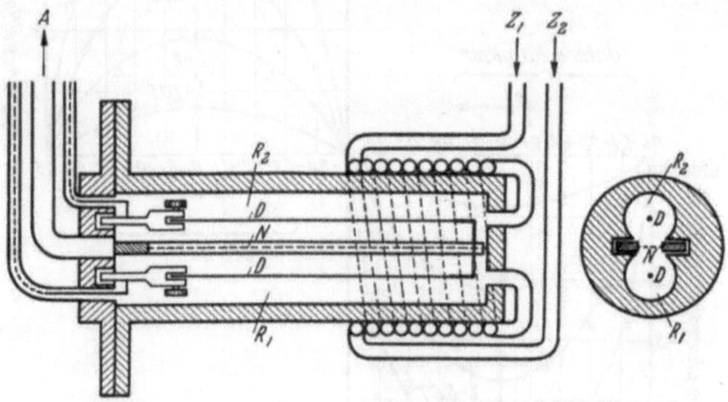

Fig. 23. Apparat zur Messung des stationären Diffusionsthermoeffekts (WALDMANN 1949).

die stationäre Methode bevorzugen. Eine hierfür geeignete Anordnung ist in Fig. 23 dargestellt. Die beiden Gase — das leichtere oben — treten durch die Zuführungen $Z_{1,2}$ in die Rohre $R_{1,2}$, je 0,75 cm Durchmesser und 4 cm lang, ein und durch Rohr A vermischt wieder aus. $R_{1,2}$ stehen durch einen Schlitz, der mit einem Netz N bedeckt ist, auf der ganzen Länge in Verbindung. Die Temperaturmeßdrähte D, je ein 8 μ-Golddraht von 15 Ω bei 20° C, sind in der Achse jedes Rohrs gespannt.

Man kann die Theorie — durch Einführung des Temperaturlinienintegrals, erstreckt in der Längsrichtung des Rohrs — ebenso exakt durchführen wie im nichtstationären Fall; die Rechnung ist bei Annahme Poiseuilleschen Geschwindigkeitsprofils ganz elementar (WALDMANN 1947c). Aber wegen des nötigen langen Meßdrahts ist die Temperaturempfindlichkeit dabei gering. Deshalb war für Untersuchungen bei tiefen Temperaturen und niedrigem Druck (WALDMANN 1949) die Methode des Temperaturlinienintegrals nicht benutzt worden; es wurde vielmehr die Strömungsgeschwindigkeit so gewählt, daß der Galvanometerausschlag maximal wird. Um dann noch auf den α-Wert schließen zu können, muß man mit Gemischen von bekanntem α eichen.

In der Arbeit von 1949 wurde die Druckabhängigkeit der Diffusionstemperaturen theoretisch auf Grund der Thermodynamik der irreversiblen Prozesse und auch experimentell untersucht. Es gibt zwei Gründe für eine Druckabhängigkeit. Einmal werden bei genügend hohem Druck allein auf Grund der Abweichung des Gases von der idealen Zustandsgleichung bei der Diffusion vorübergehende Temperaturänderungen eintreten. Dies findet in der gegenüber (15.2) erweiterten Temperaturgleichung

$$\frac{dT}{dt} = K\,\Delta T + \delta T\,\frac{d\gamma}{dt} + \delta T_B \cdot 2\,D\,(\operatorname{grad}\gamma)^2 \tag{21.1}$$

seinen Ausdruck. Darin hat die Temperatur δT_B die Bedeutung

$$\left.\begin{aligned} \delta T_B &= \frac{p}{C_{p,\mathrm{mol}}}\left(b - T\,\frac{db}{dT}\right) \\ \text{mit}\qquad b &= B_{11} - 2B_{12} + B_{22}. \end{aligned}\right\} \tag{21.2}$$

Die B_{ik} sind die molaren zweiten Virialkoeffizienten des Gemischs, $C_{p,\mathrm{mol}}$ die isobare Molwärme. δT_B ist also druckproportional. Zweitens ist aber α und somit auch der Koeffizient δT in (21.1) druckabhängig, und zwar macht sich dies ungefähr bei denselben Drucken bemerkbar, bei welchen das Realitätsglied δT_B eine Rolle zu spielen beginnt. Das war in der Arbeit von WALDMANN (1949) übersehen, wurde aber von HAASE (1950) und von E. W. BECKER (1950) ergänzt.

Es gilt für N_2/H_2-Gemisch bei 273°K, 1 atm schätzungsweise $\delta T/\delta T_B \approx -0{,}01$. In diesem Fall sind die Diffusionstemperaturen also praktisch vom Druck unabhängig. Bei CO_2/A dagegen schätzt man bei 200° K, 1 atm den Wert $\delta T/\delta T_B \approx -1{,}6$. Hier ist die Diffusionstemperatur in der Tat stark druckabhängig, wie der Versuch zeigte. Ein Beispiel einer solchen Messung ist in Fig. 24 dargestellt. Man sieht, daß in der Nähe der Kondensationstemperatur einer Komponente beim Druck von 1 atm tatsächlich die Diffusionstemperaturen stark vom Druck abhängen und zwar sowohl im Sinne des δT_B-Glieds aus (21.1) als auch im Sinne einer Druckabhängigkeit von α (oder δT). Das δT_B-Glied hat nämlich in beiden Rohren gleiches Vorzeichen und fast gleichen Betrag; es wird wegen der Abhängigkeit von C_p im unteren Rohr eher absolut etwas kleiner sein als im oberen. Die Fig. 24 zeigt aber, daß der Druckeinfluß unten bedeutend größer ist als oben und überdies von entgegengesetztem Vorzeichen. Daraus folgt, daß sich der Thermodiffusionsfaktor für CO_2/A von 100 bis 700 Torr erheblich ändert; für 194° K schätzt man aus der Figur $\frac{1}{\alpha_0}\frac{d\alpha}{dp} \approx 1\ \mathrm{atm}^{-1}$ ab, wo $\alpha_0 = 0{,}026$ den auf den Druck $p = 0$ extrapolierten Thermodiffusionsfaktor bedeutet. Das paßt größenordnungsmäßig zu der von BECKER und SCHULZEFF (1948) und von BECKER (1950) in ähnlichen Fällen gefundenen Druckabhängigkeit des α.

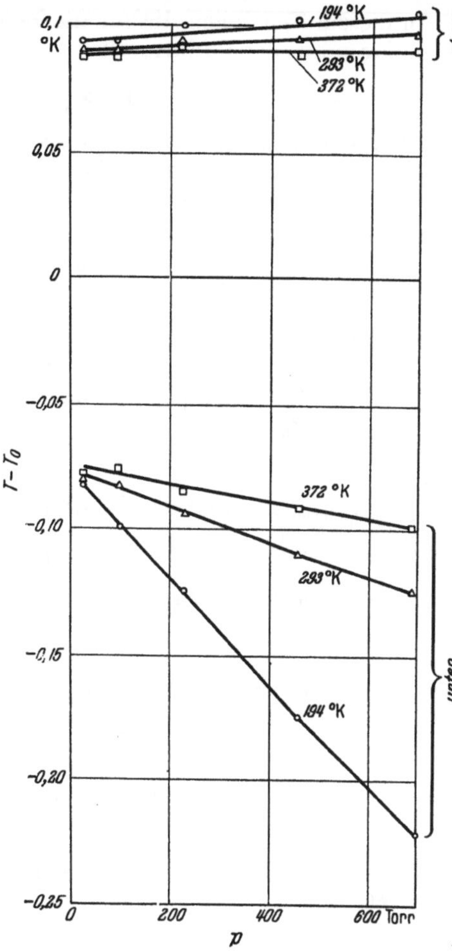

Fig. 24. Stationäre Temperaturdifferenz bei Diffusion von CO_2 (unteres Rohr) gegen A (oberes Rohr) in Abhängigkeit vom Druck. Strömungsgeschwindigkeit so, daß Temperaturdifferenz maximal (WALDMANN 1949).

Es ist also bei der Messung des Thermodiffusionsfaktors auf dessen Druckabhängigkeit selbst bei Atmosphärendruck in manchen Fällen zu achten.

C. Grundlagen der kinetischen Theorie einatomiger Gase.

In diesem Teil sollen die für die Behandlung der Transportvorgänge nötigen Grundlagen aus der Punktmechanik und der Statistik zusammengestellt werden.

I. Punktmechanik.

22. Die Gleichungen des elastischen Stoßes. Liouvillescher Satz. Zum bequemen späteren Gebrauch behandeln wir die beiden Fälle gleicher und verschiedener Masse gesondert.

α) *Atome gleicher Masse.* Die Geschwindigkeiten der beiden Atome vor dem Stoß sollen $\boldsymbol{c}$ und $\boldsymbol{c}_1$, nach dem Stoß $\boldsymbol{c}'$ und $\boldsymbol{c}_1'$ sein. Der Stoß sei elastisch. Die Erhaltungssätze lauten

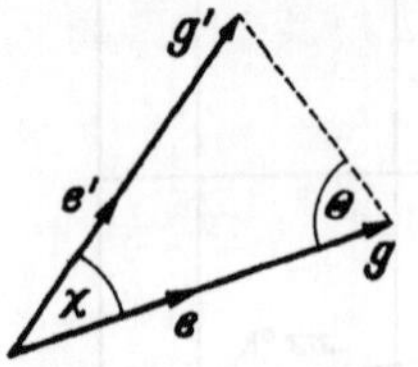

Fig. 25. Die Einheitsvektoren $\boldsymbol{e}$, $\boldsymbol{e}'$, welche wahlweise den Stoß festlegen.

$$\text{Impuls:} \quad \boldsymbol{c} + \boldsymbol{c}_1 = \boldsymbol{c}' + \boldsymbol{c}_1' \tag{22.1}$$

$$\text{Energie:} \quad c^2 + c_1^2 = c'^2 + c_1'^2. \tag{22.2}$$

Nun werde die Schwerpunktsgeschwindigkeit vor oder nach dem Stoß

$$\boldsymbol{c}_S = \tfrac{1}{2}(\boldsymbol{c} + \boldsymbol{c}_1) = \boldsymbol{c}_S' \tag{22.3}$$

und die Relativgeschwindigkeit vor und nach dem Stoß

$$\boldsymbol{g} = \boldsymbol{c} - \boldsymbol{c}_1; \quad \boldsymbol{g}' = \boldsymbol{c}' - \boldsymbol{c}_1'. \tag{22.4}$$

eingeführt. Die Umkehrung lautet

$$\boldsymbol{c}_{(1)} = \boldsymbol{c}_S \,{}_{(}\pm{}_{)}\, \tfrac{1}{2}\boldsymbol{g}; \quad \boldsymbol{c}_{(1)}' = \boldsymbol{c}_S \,{}_{(}\pm{}_{)}\, \tfrac{1}{2}\boldsymbol{g}'. \tag{22.5}$$

Dem Impulssatz ist durch $\boldsymbol{c}_S = \boldsymbol{c}_S'$ bereits genügt. Für die kinetische Energie gilt

$$\frac{m}{2}(c^2 + c_1^2) = m c_S^2 + \frac{m}{4} g^2 \tag{22.6}$$

und Entsprechendes nach dem Stoß. Der Energiesatz verlangt also

$$g^2 = g'^2. \tag{22.7}$$

Nun seien die Einheitsvektoren $\boldsymbol{e}$ und $\boldsymbol{e}'$ in Richtung der Relativgeschwindigkeiten eingeführt:

$$\boldsymbol{g} = g\,\boldsymbol{e}; \quad \boldsymbol{g}' = g'\,\boldsymbol{e}'. \tag{22.8}$$

Nach (22.5) und (22.7) kann man die Geschwindigkeiten nach dem Stoß als Funktion derjenigen vor dem Stoß angeben, wenn man die Richtung $\boldsymbol{e}'$ kennt:

$$\left.\begin{aligned} \boldsymbol{c}' &= \boldsymbol{c}_S + \tfrac{1}{2} g\,\boldsymbol{e}' = \tfrac{1}{2}(\boldsymbol{c} + \boldsymbol{c}_1 + |\boldsymbol{c} - \boldsymbol{c}_1|\,\boldsymbol{e}') \\ \boldsymbol{c}_1' &= \boldsymbol{c}_S - \tfrac{1}{2} g\,\boldsymbol{e}' = \tfrac{1}{2}(\boldsymbol{c} + \boldsymbol{c}_1 - |\boldsymbol{c} - \boldsymbol{c}_1|\,\boldsymbol{e}'). \end{aligned}\right\} \tag{22.9}$$

Der Einheitsvektor $\boldsymbol{e}'$ bezeichnet also die Freiheit, welche die Erhaltungssätze für die Geschwindigkeiten nach dem Stoß noch zulassen, wenn diejenigen vor dem Stoß gegeben sind. Ebensogut kann man den Stoß auch durch den Einheitsvektor $\boldsymbol{e}$ festlegen. Es gilt dann entsprechend

$$\left.\begin{aligned} \boldsymbol{c} &= \boldsymbol{c}_S + \tfrac{1}{2} g'\,\boldsymbol{e} = \tfrac{1}{2}(\boldsymbol{c}' + \boldsymbol{c}_1' + |\boldsymbol{c}' - \boldsymbol{c}_1'|\,\boldsymbol{e}) \\ \boldsymbol{c}_1 &= \boldsymbol{c}_S - \tfrac{1}{2} g'\,\boldsymbol{e} = \tfrac{1}{2}(\boldsymbol{c}' + \boldsymbol{c}_1' - |\boldsymbol{c}' - \boldsymbol{c}_1'|\,\boldsymbol{e}). \end{aligned}\right\} \tag{22.10}$$

Die Beziehungen (22.9) oder (22.10) seien als *Stoßgleichungen* bezeichnet. Es sei bemerkt, daß sich (22.9) nicht nach $\boldsymbol{c}$, $\boldsymbol{c}_1$ auflösen läßt; als unabhängige Variable in (22.9) sind eben eigentlich $\boldsymbol{c}_S$, g und $\boldsymbol{e}'$ aufzufassen. Entsprechendes gilt für (22.10). Wichtig ist ferner die Tatsache, daß (22.9) und (22.10) richtig bleiben, wenn man die gestrichenen und die ungestrichenen Größen miteinander vertauscht, und auch wenn man $\boldsymbol{c}$, $\boldsymbol{c}'$ mit $\boldsymbol{c}_1$, $\boldsymbol{c}_1'$ vertauscht und zugleich $\boldsymbol{e}$, $\boldsymbol{e}'$ durch $-\boldsymbol{e}$, $-\boldsymbol{e}'$ ersetzt.

β) Atome verschiedener Masse. Wir bezeichnen die Geschwindigkeiten der beiden Stoßpartner mit den Massen m_i und m_j wieder bzw. mit $\boldsymbol{c}$ und $\boldsymbol{c}_1$ vor dem Stoß, mit $\boldsymbol{c}'$ und $\boldsymbol{c}_1'$ nach dem Stoß. Die Erhaltungssätze lauten

$$\text{Impuls:} \qquad m_i\,\boldsymbol{c} + m_j\,\boldsymbol{c}_1 = m_i\,\boldsymbol{c}' + m_j\,\boldsymbol{c}_1' \tag{22.11}$$

$$\text{Energie:} \qquad \frac{m_i}{2}\,c^2 + \frac{m_j}{2}\,c_1^2 = \frac{m_i}{2}\,c'^2 + \frac{m_j}{2}\,c_1'^2. \tag{22.12}$$

Die Schwerpunktsgeschwindigkeit vor und nach dem Stoß ist

$$\boldsymbol{c}_S = \frac{m_i}{m_i+m_j}\,\boldsymbol{c} + \frac{m_j}{m_i+m_j}\,\boldsymbol{c}_1 = \boldsymbol{c}_S'. \tag{22.13}$$

Die Definition (22.4) der Relativgeschwindigkeiten ist beizubehalten. Als Umkehrung hat man

$$\boldsymbol{c} = \boldsymbol{c}_S + \frac{m_j}{m_i+m_j}\,\boldsymbol{g}, \qquad \boldsymbol{c}_1 = \boldsymbol{c}_S - \frac{m_i}{m_i+m_j}\,\boldsymbol{g} \tag{22.14}$$

und

$$\boldsymbol{c}' = \boldsymbol{c}_S + \frac{m_j}{m_i+m_j}\,\boldsymbol{g}', \qquad \boldsymbol{c}_1' = \boldsymbol{c}_S - \frac{m_i}{m_i+m_j}\,\boldsymbol{g}'. \tag{22.15}$$

Für die kinetische Energie, einmal durch die Geschwindigkeiten im Laboratoriumssystem ausgedrückt und zum anderen durch Schwerpunkts- und Relativgeschwindigkeit, gilt:

$$\frac{m_i}{2}\,c^2 + \frac{m_j}{2}\,c_1^2 = \frac{1}{2}\,(m_i+m_j)\,c_S^2 + \frac{1}{2}\,m_{ij}\,g^2. \tag{22.16}$$

Dabei ist $m_{ij} = \frac{m_i m_j}{m_i+m_j}$ die reduzierte Masse. Entsprechendes gilt für die Geschwindigkeiten nach dem Stoß. Wegen $\boldsymbol{c}_S = \boldsymbol{c}_S'$ bleiben die beiden Glieder auf der rechten Seite, das sind die kinetische Energie der Schwerpunkts- und der Relativbewegung, beim elastischen Stoß jedes für sich erhalten; man hat also wieder $g = g'$.

Mit Benutzung des Einheitsvektors $\boldsymbol{e}'$ aus (22.8) erhält man nun die *Stoßgleichungen*

$$\left.\begin{aligned} \boldsymbol{c}' &= \boldsymbol{c}_S + \frac{m_j}{m_i+m_j}\,g\,\boldsymbol{e}' = \frac{1}{m_i+m_j}\,(m_i\,\boldsymbol{c} + m_j\,\boldsymbol{c}_1 + m_j\,|\boldsymbol{c}-\boldsymbol{c}_1|\,\boldsymbol{e}') \\ \boldsymbol{c}_1' &= \boldsymbol{c}_S - \frac{m_i}{m_i+m_j}\,g\,\boldsymbol{e}' = \frac{1}{m_i+m_j}\,(m_i\,\boldsymbol{c} + m_j\,\boldsymbol{c}_1 - m_i\,|\boldsymbol{c}-\boldsymbol{c}_1|\,\boldsymbol{e}'). \end{aligned}\right\} \tag{22.17}$$

Wieder bleiben diese Gleichungen richtig, wenn man die gestrichenen und die ungestrichenen Größen vertauscht oder wenn man $\boldsymbol{c}$, $\boldsymbol{c}'$ mit $\boldsymbol{c}_1$, $\boldsymbol{c}_1'$ sowie m_i mit m_j vertauscht und zugleich $\boldsymbol{e}'$ durch $-\boldsymbol{e}'$ ersetzt.

γ) Der Liouvillesche Satz beim Stoß. Für später wird ein wichtiger Zusammenhang zwischen den Spielräumen der Geschwindigkeiten vor dem Stoß und denen nach dem Stoß benötigt. Um ihn aufzustellen, sei die folgende lineare Abbildung der Größen $\boldsymbol{c}$, $\boldsymbol{c}_1$, $\boldsymbol{g}'$ auf die Größen $\boldsymbol{c}'$, $\boldsymbol{c}_1'$, $\boldsymbol{g}$ betrachtet:

$$\left.\begin{aligned} \boldsymbol{c}' &= \frac{1}{m_i+m_j}\,(m_i\,\boldsymbol{c} + m_j\,\boldsymbol{c}_1 + m_j\,\boldsymbol{g}') \\ \boldsymbol{c}_1' &= \frac{1}{m_i+m_j}\,(m_i\,\boldsymbol{c} + m_j\,\boldsymbol{c}_1 - m_i\,\boldsymbol{g}') \\ \boldsymbol{g} &= \qquad\qquad \boldsymbol{c} - \quad \boldsymbol{c}_1. \end{aligned}\right\} \tag{22.18}$$

Weil sich die Umkehrung dieser Transformation durch Vertauschen der gestrichenen und ungestrichenen Größen ergibt und weil die Transformation linear ist, muß für die Funktionaldeterminante gelten

$$\frac{\partial(\boldsymbol{c}'\boldsymbol{c}_1'\boldsymbol{g})}{\partial(\boldsymbol{c}\boldsymbol{c}_1\boldsymbol{g}')} = \frac{\partial(\boldsymbol{c}\boldsymbol{c}_1\boldsymbol{g}')}{\partial(\boldsymbol{c}'\boldsymbol{c}_1'\boldsymbol{g})}.$$

Für jede beliebige Transformation gilt aber identisch

$$\frac{\partial(\boldsymbol{c}'\boldsymbol{c}_1'\boldsymbol{g})}{\partial(\boldsymbol{c}\boldsymbol{c}_1\boldsymbol{g}')} \cdot \frac{\partial(\boldsymbol{c}\boldsymbol{c}_1\boldsymbol{g}')}{\partial(\boldsymbol{c}'\boldsymbol{c}_1'\boldsymbol{g})} = 1.$$

Also ist

$$\left(\frac{\partial(\boldsymbol{c}'\boldsymbol{c}_1'\boldsymbol{g})}{\partial(\boldsymbol{c}\boldsymbol{c}_1\boldsymbol{g}')}\right)^2 = 1$$

und man hat

$$d\boldsymbol{c}\,d\boldsymbol{c}_1\,d\boldsymbol{g}' = d\boldsymbol{c}'\,d\boldsymbol{c}_1'\,d\boldsymbol{g}. \tag{22.19}$$

Darin bedeuten $d\boldsymbol{c} = dc_1\,dc_2\,dc_3$, usw. die Volumelemente der jeweiligen dreidimensionalen Geschwindigkeitsräume. Mit der Bezeichnung

$$d\boldsymbol{g} = g^2\,dg\,d\boldsymbol{e}, \qquad d\boldsymbol{g}' = g'^2\,dg'\,d\boldsymbol{e}', \tag{22.20}$$

wo die Differentiale $d\boldsymbol{e}, d\boldsymbol{e}'$ sinngemäß die Flächenelemente auf den Einheitskugeln $\boldsymbol{e}^2 = \boldsymbol{e}'^2 = 1$ bedeuten, lautet (22.19) auch

$$d\boldsymbol{c}\,d\boldsymbol{c}_1\,g'\,d\boldsymbol{e}' = d\boldsymbol{c}'\,d\boldsymbol{c}_1'\,g\,d\boldsymbol{e} \cdot \frac{dg^2}{dg'^2}. \tag{22.21}$$

In dieser Form ist die Beziehung (22.19) auch brauchbar, wenn man nun für den Betrag g', wie es der Energiesatz verlangt, keinen Spielraum zuläßt, sondern annimmt, daß g' eine Funktion von g ist (oder umgekehrt[1]). Außerdem sei betont, daß die bisherigen Betrachtungen von (22.18) ab allgemein auch für den unelastischen Stoß gelten.

Jetzt werde auf den elastischen Stoß, $g = g'$, spezialisiert. Hier lautet (22.21) einfach

$$d\boldsymbol{c}\,d\boldsymbol{c}_1\,d\boldsymbol{e}' = d\boldsymbol{c}'\,d\boldsymbol{c}_1'\,d\boldsymbol{e}, \tag{22.22}$$

was als Liouvillescher Satz für den elastischen Stoß bezeichnet sei. Die Werte der Massen gehen hier nicht ein.

23. Der differentielle Streuquerschnitt. Bei molekularen Zusammenstößen sind die Bahnen der Teilchen nach der Quantenmechanik prinzipiell, nach der klassischen Mechanik praktisch nicht genau bekannt. Erst nach erfolgtem Stoß kann man den Stoßvektor $\boldsymbol{e}'$ angeben. Streuexperimente und Quantenmechanik liefern für die verschiedenen Richtungen aber eine Häufigkeitsstatistik, die man durch den differentiellen Streuquerschnitt beschreibt. Für die Zwecke der Kinetik einatomiger Gase ist die Kenntnis des Streuquerschnitts notwendig und hinreichend.

Wegen ausführlicher Darstellungen von Streuexperimenten an Molekularstrahlen siehe bei KENNARD [*31*], HERZFELD [*22*], ESTERMANN [*12*] und RAMSEY [*43*]; über die Berücksichtigung der Geschwindigkeitsverteilung bei Streuexperimenten haben HIBY und PAHL (1951) gearbeitet. Mit der Herstellung intensiver Molekularstrahlen haben sich NÖLLER, OETJEN und JAECKEL (1949) sowie E. W. BECKER und BIER (1954) — Methode der Lavaldüse — befaßt. Die Streuung neutraler Partikel mit Energien von etwa 1000 eV haben AMDUR und PEARLMAN (1941), AMDUR und HARKNESS (1954) — beide Male He/He — sowie

[1] Macht man diese Einschränkung gleich zu Beginn, so hat man es anstelle von (22.18) mit einer nicht-linearen Transformation $\boldsymbol{c}, \boldsymbol{c}_1, \boldsymbol{e}' \to \boldsymbol{c}', \boldsymbol{c}_1', \boldsymbol{e}$ zu tun und der Beweis von (22.21) wird umständlicher.

AMDUR und MASON (1954) — A/A — untersucht und mit gaskinetischen Daten verglichen. Wegen der Quantenmechanik des Stoßes sei auf die Monographie von MOTT und MASSEY [39] verwiesen, sowie auf Ziff. 82.

Wir geben im folgenden einige Erläuterungen zur Definition des Streuquerschnitts. Es seien zwei unbegrenzte, sich durchdringende Atomstrahlen betrachtet mit den Konzentrationen n_i, n_j und den Partikelmassen m_i, m_j. Wir greifen ein Volumen V, das Streuvolumen, heraus und fassen die von ihm ausgehenden Streustrahlen ins Auge. Mehrfachstreuung bleibe unberücksichtigt. Zur Vereinfachung betrachtet man das zum Stoß (ij) gehörige äquivalente Einkörperproblem. Danach hat man es mit fiktiven Teilchen der Masse $m_{ij} = m_i m_j/(m_i + m_j)$ zu tun, n_i pro Volumeneinheit, welche sich gegen unendlich schwere, ruhende Teilchen, n_j pro Volumeneinheit, mit der (Relativ-) Geschwindigkeit $\boldsymbol{g}$ vor dem Stoß bewegen. Bei diesem äquivalenten Streuproblem (s. Fig. 26) ist im großen Abstand r vom Volumen V die Konzentration n'_{ij} des Streustrahls proportional r^{-2}, n_i, n_j und V. Wir setzen

$$n'_{ij} = \frac{|a_{ij}|^2}{r^2} n_i n_j V .$$

Der Proportionalitätsfaktor $|a_{ij}|^2$ hat die Dimension einer Fläche; die Schreibweise soll darauf hinweisen, daß er sich quantenmechanisch aus der Streuamplitude ergibt. Die Anzahl $d\dot{N}'_{ij}$ der fiktiven Teilchen m_{ij}, welche aus V kommend in der Zeiteinheit durch das Kugelflächenelement $r^2\, d\boldsymbol{e}'$ (s. Fig. 26) treten, beträgt

$$d\dot{N}'_{ij} = |a_{ij}|^2 g'\, d\boldsymbol{e}'\, n_i n_j V .$$

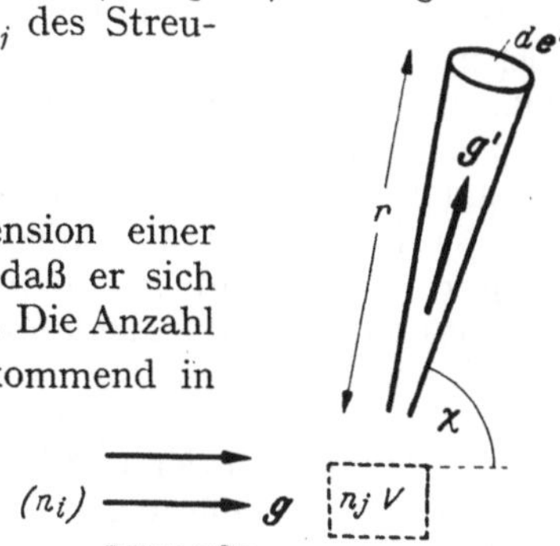

Fig. 26. Zur Definition des Streuquerschnitts mittels des äquivalenten Einkörperproblems.

Nun wird weiter definitionsgemäß

$$d\dot{N}'_{ij} = \sigma_{ij}\, g\, d\boldsymbol{e}'\, n_i n_j V \qquad (23.1)$$

gesetzt. Der differentielle Streuquerschnitt σ_{ij} hängt also gemäß

$$\sigma_{ij} = \frac{g'}{g} |a_{ij}|^2 \qquad (23.2)$$

mit dem Absolutquadrat der Streuamplitude zusammen. Bei elastischer Streuung ist überdies $g' = g$. Aus Gründen der Drehinvarianz kann bei kugelsymmetrischen Teilchen σ_{ij} nur vom Betrag der Relativgeschwindigkeit und dem Ablenkwinkel abhängen:

$$\sigma_{ij} = \sigma_{ij}(\chi, g), \quad \text{wo} \quad \cos\chi = \boldsymbol{e}\boldsymbol{e}'. \qquad (23.3)$$

Mit dem Azimutwinkel ε von $\boldsymbol{g}'$ um $\boldsymbol{g}$, d.h. dem Winkel der Ebene $(\boldsymbol{g}, \boldsymbol{g}')$ gegen eine feste Ebene durch $\boldsymbol{g}$, hat man auch

$$d\boldsymbol{e}' = \sin\chi\, d\chi\, d\varepsilon, \quad \text{wobei} \quad 0 \leq \chi \leq \pi. \qquad (23.4)$$

Es sei noch darauf hingewiesen, daß die durch (23.1) gegebene Anzahl von Streuprozessen pro Zeiteinheit auch für das Laboratoriumssystem gilt. Es ändert sich zwar beim Übergang zum Schwerpunktssystem der Bewegungszustand des Streubezirks, nicht aber dessen Größe V, auf die es wegen der räumlichen Homogenität allein ankommt.

Gl. (23.1) wird in Ziff. 28 und 29 als Ausgangspunkt bei der Aufstellung der Boltzmann-Gleichung dienen. Der Streuquerschnitt σ wird dabei als bekannt vorausgesetzt werden. Die Atommechanik liefert gerade dieses σ, wenn man durch Übergang zum Schwerpunktssystem das Zweikörper- auf das Einkörperproblem reduziert. Dagegen muß man, um σ aus Streuexperimenten an gekreuzten Atomstrahlen zu ermitteln, erst an Hand von (22.9) bzw. (22.17) den Zusammenhang von dc', dc_1' mit de entnehmen. Wir gehen darauf nicht ein, da die zugehörigen Formeln hier nicht weiter benötigt werden. Einen einfachen Spezialfall betrachtet Kennard [31], S. 122. Da diese Umrechnung sehr häufig in der Kernphysik zu machen ist, hat man Tabellen dafür angelegt[1].

24. Allgemeines über Molekularkräfte. Durch die Erhaltungssätze und den Streuquerschnitt ist der Molekularstoß ausreichend beschrieben. Der Streuquerschnitt ist aber selbst wieder eine abgeleitete Größe, die sich auf die molekulare Wechselwirkung zurückführen läßt, eine Aufgabe der Mechanik, die in Ziff. 26, 27 und 82 besprochen wird. Hier sollen einige qualitative Bemerkungen über Molekularkräfte gemacht werden.

Man beschränkt sich in der Gastheorie meist auf die Betrachtung einatomiger Gase. Dementsprechend sei hier nur die Rede von der Wechselwirkung zwischen elektrisch neutralen, kugelsymmetrischen Teilchen, die sich nicht chemisch verbinden (z.B. Edelgasatome). Ganz allgemein läßt sich sagen, daß sich solche unpolaren Atome bei großem Abstand anziehen (Kondensation), bei kleinem Abstand abstoßen (Inkompressibilität der Flüssigkeit).

Die Theorie der Molekularkräfte ist weitläufig und — als Teil der Atomtheorie — nur auf quantenmechanischer Grundlage möglich (Heitler und F. London 1927; Hellman [20], Pauling [42], Hartmann [19]). Es ergibt sich dabei, daß es zweckmäßig ist die Gesamtkraft in zwei Anteile, Valenzkraft und Dispersionskraft, zu zerlegen. Die Valenzkraft ist maßgebend bei Abständen der beiden Atome von der Größenordnung ihrer Durchmesser, die Dispersionskraft ist maßgebend bei größeren Abständen; beide Kräfte sind universell.

Die Valenzkraft kommt durch Überdeckung der Elektronenhüllen zustande und kann, je nach Art und Zustand der beiden Atome, eine Anziehung oder Abstoßung sein. Im Fall der Valenzanziehung (Beispiel: zwei Wasserstoffatome mit entgegengesetzten Elektronenspins) kann es zur chemischen Bindung kommen. Bei zwei Atomen mit abgeschlossenen Elektronenhüllen, etwa zwei Edelgasatomen, also in dem für die Gastheorie typischen Fall, gibt es aber nur Valenzabstoßung. Erwähnt sei noch, daß allen Valenzkräften gemeinsam ein exponentieller Abfall $e^{-r/a}$ ist bei großem Abstand r der Atome (a = Durchmesser der Elektronenhülle). Dies ergibt sich aus dem erwähnten Zustandekommen durch Überdeckung der Elektronenhüllen und dem bekannten exponentiellen Abfall der quantenmechanischen Aufenthaltswahrscheinlichkeit der Elektronen des einzelnen Atoms.

Das Potential der Dispersionskraft ist generell proportional r^{-6}, der Proportionalitätsfaktor läßt sich nach F. London (1930) durch die Polarisierbarkeit der Atome oder durch die Koeffizienten der Dispersionsformel darstellen (s. auch Eucken [13], § 99). Ein einfaches Modell hierfür geben nach F. London zwei Oszillatoren ab, jeder bestehend aus einer positiven Ladungswolke mit einem darin harmonisch oszillierenden Elektron (J. J. Thomsonsches Atommodell). Zur Vereinfachung sei angenommen, daß die Oszillation nur in Richtung der Verbindungslinie beider Atommittelpunkte stattfinde. Letztere seien bei x_1, x_2 gelegen, die Koordinaten der Elektronen seien $x_1+\xi_1$, $x_2+\xi_2$, m sei die Elektronenmasse, $\varkappa$ die Kraftkonstante jedes Oszillators. Dann hat man die Bewegungsgleichungen

$$m\ddot{\xi}_1 = -\varkappa\xi_1 + K_{12}; \qquad m\ddot{\xi}_2 = -\varkappa\xi_2 + K_{21}.$$

[1] J. B. Marion u. A. S. Ginzbarg: Tables for the Transformation of Angular Distribution Data from the Laboratory System to the Centre-of-Mass System. Houston: Shell Development Company 1955 (unveröff.). Siehe auch Hanson, Taschek u. Williams: Rev. Mod. Phys. **21**, 635 (1949).

Dabei bedeutet K_{12} die auf das Elektron „1" vom Atom „2" ausgeübte Kraft, bestehend aus einer Coulombschen Anziehung durch die positive Ladung e des Atoms und einer Abstoßung durch dessen Elektron:

$$K_{12} = \frac{e^2}{(x_2 - x_1 - \xi_1)^2} - \frac{e^2}{(x_2 + \xi_2 - x_1 - \xi_1)^2}.$$

Fig. 27. Oszillatormodell zum r^{-6}-Gesetz.

Bei großem Abstand $r = |x_2 - x_1|$ gilt also

$$K_{12} \approx 2\frac{e^2}{r^3}\xi_2, \quad \text{analog} \quad K_{21} \approx 2\frac{e^2}{r^3}\xi_1,$$

und man hat

$$m\ddot{\xi}_1 = -\varkappa\xi_1 + 2\frac{e^2}{r^3}\xi_2; \quad m\ddot{\xi}_2 = -\varkappa\xi_2 + 2\frac{e^2}{r^3}\xi_1.$$

Das sind zwei linear gekoppelte Oszillatoren. Für deren Normalschwingungsfrequenzen liefert der Ansatz $\xi_{1,2} \sim e^{i\omega t}$

$$\omega = \sqrt{\frac{\varkappa}{m}\left(1 \pm \frac{2e^2}{\varkappa r^3}\right)}.$$

Mit

$$\omega_0 = \sqrt{\frac{\varkappa}{m}}$$

hat man bei großem r also die zwei Frequenzen

$$\omega_{\mathrm{I}} \approx \omega_0\left(1 + \frac{e^2}{\varkappa r^3} - \frac{1}{2}\frac{e^4}{\varkappa^2 r^6}\right), \quad \omega_{\mathrm{II}} \approx \omega_0\left(1 - \frac{e^2}{\varkappa r^3} - \frac{1}{2}\frac{e^4}{\varkappa^2 r^6}\right).$$

Dasselbe erhält man nach der Quantenmechanik, da sich die Methode der Normalschwingungen unverändert in diese überträgt. Für die weitere Diskussion, nämlich die Frage nach der Wechselwirkungsenergie, muß man die klassische Vorstellung nun natürlich beiseite lassen. Dem Grundzustand der beiden isolierten Atome ($r = \infty$!) ist die Nullpunktsenergie

$$E_0(\infty) = 2 \cdot \tfrac{1}{2}\hbar\omega_0$$

zuzuschreiben; die Nullpunktsenergie des Grundzustandes der beiden gekoppelten Atome, welche zwei entkoppelten Oszillatoren mit Frequenzen ω_{I} und ω_{II} äquivalent sind, beträgt

$$E_0(r) = \tfrac{1}{2}\hbar\omega_{\mathrm{I}} + \tfrac{1}{2}\hbar\omega_{\mathrm{II}}.$$

Als potentielle Wechselwirkungsenergie ergibt sich somit

$$\Phi(r) = E_0(r) - E_0(\infty) = -\frac{1}{2}\frac{\hbar\omega_0 e^4}{\varkappa^2 r^6}.$$

Man bezeichnet sie nach ihrer Herkunft auch als induzierte Dipol-Dipol-Wechselwirkung. Soviel zur Begründung des Londonschen r^{-6}-Gesetzes.

25. Molekülmodelle der Gastheorie. In der Gastheorie benutzt man verschiedene Molekülmodelle, die je nach dem Grad der Verfeinerung durch geeignete Wahl der Parameter eine mehr oder weniger gute Anpassung an die wirklichen Molekularkräfte erlauben. Die gebräuchlichen Modelle seien kurz zusammengestellt (s. auch Fig. 28).

α) Starre Kugeln vom Durchmesser d. Der Ansatz für die potentielle Energie Φ lautet:

$$\left.\begin{array}{ll}\Phi=\infty & \text{für } r<d, \\ \Phi=0 & \text{für } r>d.\end{array}\right\} \tag{25.1}$$

Dieses Modell gibt nur die Valenzabstoßung bei kleinem Abstand wieder. Da hiernach die Atome einen genau definierten Durchmesser haben, kann man für ein solches Gas eindeutig eine freie Weglänge definieren. Bei den folgenden Modellen entfällt diese Möglichkeit. — Die Kugeln sind ohne Trägheitsmoment zu denken.

Dieses Modell ist das älteste, welches in der Gastheorie Verwendung fand (CLAUSIUS 1858) (s. Ziff. 51).

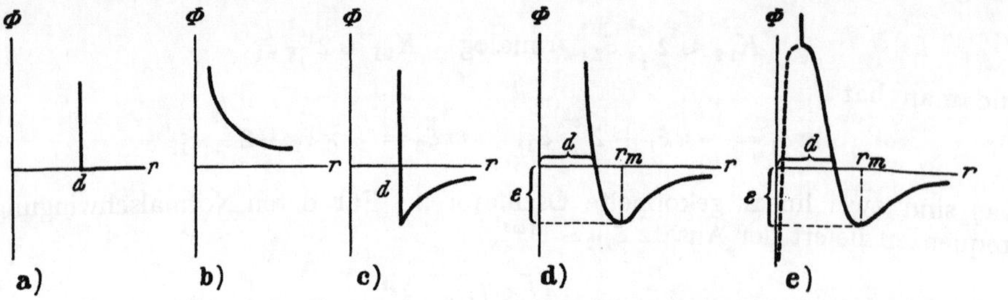

Fig. 28a—e. Potentielle Energie Φ in Abhängigkeit vom Abstand r. a starre Kugeln, b Potenzkraft-Zentren, c Sutherland-Modell, d Lennard-Jones-Modell, e Exponential-Potential.

β) Potenzgesetz. Der Ansatz für die Abstoßungskraft K lautet

$$K=-\frac{\partial\Phi}{\partial r}=\frac{\varkappa}{r^{\nu}} \quad \text{bzw.} \quad \Phi=\frac{\varkappa}{\nu-1}\,r^{-\nu+1}. \tag{25.2}$$

Dieses Modell gibt ebenfalls nur die Abstoßung wieder, enthält aber bereits zwei Parameter und erlaubt zwischen harten (z.B. $\nu\approx 14$ aus der Zähigkeit um 20° C von Helium) und weichen (z.B. $\nu\approx 5$ aus der Zähigkeit von Chlorgas um 20° C) Molekülen zu unterscheiden. Der spezielle Fall $\nu=5$ ist für die Gastheorie aus rechnerischen Gründen von besonderer Bedeutung (Maxwellsches Gas, s. Ziff. 27 und 38ff.)

Das Modell der Potenzkraft-Zentren wurde vor allem von ENSKOG und CHAPMAN bei der Anwendung ihrer grundlegenden Methoden benutzt (s. Ziff. 51).

γ) Abstoßungs- und Anziehungskraft sind enthalten in dem allgemeinen Ansatz von Lennard-Jones (1924)

$$K=\frac{\varkappa}{r^{\nu}}-\frac{\varkappa'}{r^{\nu'}}. \tag{25.3}$$

Dieser Ansatz bietet den Vorteil, daß sich der zweite Virialkoeffizient der Zustandsgleichung durch eine konvergente Reihe darstellen läßt. Es seien zwei Spezialfälle von (25.3) hervorgehoben:

a) Modell von Sutherland (1893). Mit $\nu=\infty$ erhält man

$$\left.\begin{array}{ll}\Phi=\infty & \text{für } r<d, \\ \Phi=-\dfrac{\varkappa'}{\nu'-1}\,r^{-\nu'+1} & \text{für } r>d.\end{array}\right\} \tag{25.4}$$

Bei diesem Modell ist die Abstoßung durch starre Kugeln idealisiert, dazu kommt aber eine Anziehung bei größerem Abstand; es enthält also bereits die beiden charakteristischen Anteile der Molekularkräfte und war dementsprechend erfolgreich in der kinetischen Theorie. Nach dem über die Dispersionskraft Gesagten wäre eigentlich $\nu=6$ zu setzen. Dann hat das Modell zwei Parameter.

b) Das spezielle (6; 12)-Potential von Lennard-Jones

$$\Phi = 4\varepsilon\left[\left(\frac{d}{r}\right)^{12} - \left(\frac{d}{r}\right)^{6}\right]. \tag{25.5}$$

Hier ist über den Exponenten der Anziehung bei großem Abstand im Sinne der Theorie der Dispersionskraft verfügt; der Exponent der Abstoßung ist für die Rechnung bequem gewählt. Die Bedeutung der Parameter ε und d geht aus Fig. 28d hervor. Das Minimum des Potentials liegt bei $r_m=\sqrt[6]{2}\,d$. Mit diesem Potential sind in neuerer Zeit besonders viele numerische Rechnungen durchgeführt worden, die auf quantitativen Vergleich mit dem Experiment abzielen, vor allem von HIRSCHFELDER, BIRD und SPOTZ (1948, 1949) (s. Ziff. 51, 71 und 75). Werte von ε und d für eine Reihe von Gasen sind in Ziff. 52 (Tabelle 7) angegeben. Schon früher haben HASSÉ und COOK (1929) das spezielle Lennard-Jones-Potential $\Phi=4\varepsilon\,[(d/r)^8-(d/r)^4]$ numerisch ausgewertet, was besonders wichtig war für die Aufklärung der Thermodiffusion durch CLARK JONES (1941) (s. Ziff. 75).

Streuexperimente an schnellen He-Atomen (0,3 bis 1 keV), bei denen die Form des Abstoßungspotentials wesentlich ist, werden durch den Ansatz (25.5) nach AMDUR und PEARLMAN (1941) nicht gut wiedergegeben. Diese Autoren erzielen eine brauchbare Darstellung mit einer Abstoßung $\Phi \propto e^{-a\sqrt{r}}$.

δ) Exponential-Potential. Für das Wechselwirkungspotential zweier Heliumatome hat SLATER (1928) angesetzt

$$\Phi = b\,e^{-r/a} - \frac{c}{r^6} \tag{25.6}$$

und die Konstanten quantenmechanisch absolut berechnet [s. auch SLATER und KIRKWOOD (1931)]. Dasselbe taten BLEICK und MAYER (1934) für die Neonatome. Auch in der Theorie der kubischen Ionenkristalle fand dieser Ansatz Verwendung (BORN und MAYER 1932); für die Anwendung in der Theorie der Zustandsgleichung wurde er von BUCKINGHAM (1938) noch um ein weiteres Anziehungsglied $\propto r^{-8}$ erweitert entsprechend einer induzierten Dipol-Quadrupol-Wechselwirkung.

Der Exponentialansatz (25.6) beschreibt die Valenzabstoßung im Sinne der Atommechanik und ist insofern besser als die vorhergehenden Ansätze, hat aber, da der Dispersionsanteil jetzt bei kleinem Abstand nicht kompensiert wird, den Nachteil, daß $\Phi\to-\infty$ geht für $r\to 0$ (gestrichelter Teil in Fig. 28e). Dies ist aber bei nicht zu hoher kinetischer Energie der Teilchen harmlos, denn es kommt in diesem Fall auf den Potentialverlauf bei sehr kleinem Abstand wegen des vorgelagerten Potentialbergs nicht an. Gleichwohl wurde von BUCKINGHAM und CORNER (1947) für die Berechnung des zweiten Virialkoeffizienten der Ansatz erweitert, indem der Dispersionsanteil durch einen Faktor für $r\to 0$ zum Verschwinden gebracht wurde.

Die Verwertung des Ansatzes (25.6) für die Transporterscheinungen in Gasen ist MASON (1954) zu verdanken (s. Ziff. 75). MASON schreibt das (6; Exp)-Potential in der Form

$$\Phi = \frac{\varepsilon}{1-\frac{6}{s}}\left[\frac{6}{s}\,e^{s\left(1-\frac{r}{r_m}\right)} - \left(\frac{r_m}{r}\right)^6\right]. \tag{25.7}$$

Die Bedeutung von ε und r_m ist aus Fig. 28e abzulesen. Der Parameter[1] s, der bei den Anwendungen die Werte 12, 13, 14, 15 erhält, ist ein Maß für die Steilheit des Abstoßungspotentials, wie aus der Masonschen Aufstellung

s	12	13	14	15	LENNARD-JONES (6; 12)
d/r_m	0,8761	0,8832	0,8891	0,8942	0,8909

(25.8)

hervorgeht. Auch MASON benutzt bei seinen Rechnungen den gestrichelten Teil von Fig. 28e nicht; er ersetzt ihn durch $\Phi = \infty$, wie es in der Zeichnung angedeutet ist.

Speziell für H_2-Moleküle haben neuerdings EVETT und MARGENAU (1953) quantenmechanisch das Wechselwirkungspotential berechnet; der Wert $s = 12$ paßt hierzu am besten [s. GREW, JOHNSON und NEAL (1954)].

Bei dem späteren Vergleich der theoretischen Transportkoeffizienten, besonders des Thermodiffusionsfaktors (siehe Ziff. 52, 75 und 77), mit den Messungen wird sich zeigen, daß dieser Vergleich um so besser auszufallen scheint, je genauer das Wechselwirkungspotential den Ergebnissen der Quantenmechanik gerecht wird.

26. Zusammenhang von Streuquerschnitt und Stoßparameter nach der klassischen Mechanik. Nach den Vorstellungen der klassischen Mechanik besitzt ein Massenpunkt eine exakt definierte Bahnkurve. Demnach kann man den Stoß, statt durch den Ablenkwinkel χ aus Fig. 25, auch durch den Stoßparameter b charakterisieren. Dies ist der Abstand der Bahnasymptoten der beiden Teilchen im Schwerpunktssystem vor und nach dem Stoß (Fig. 29a). Es gibt einen eindeutigen Zusammenhang:

$$\chi = \chi_{ij}(b, g). \tag{26.1}$$

Wenn diese Funktion bekannt ist, hat man ebenfalls alles Wissenswerte über den Stoß; in der klassischen Theorie ist es daher möglich, sie anstelle von σ_{ij} zu benutzen. Um den Zusammenhang von beidem zu finden, betrachten wir wie in Ziff. 23 das äquivalente Einkörperproblem. Auf das Flächenelement $b\,db\,d\varepsilon$, welches an einem der $n_j V$ ruhenden, unendlich schweren Atome seitlich, senkrecht zu $\boldsymbol{g}$, befestigt sei, würden pro Zeiteinheit $n_i g\, b\, db\, d\varepsilon$ fiktive Atome der Masse m_{ij} auftreffen, wenn keine Ablenkung stattfände. Insgesamt finden also in dem Streuvolumen V

$$d\dot{N}_{ij} = g\, b\, db\, d\varepsilon\, n_i n_j V \tag{26.2}$$

Stöße pro Zeiteinheit statt mit dem Stoßparameter innerhalb db und dem Azimut innerhalb $d\varepsilon$. Dabei wird die Relativgeschwindigkeit $\boldsymbol{g}$ in den Raumwinkel

$$d\boldsymbol{e}' = \sin\chi_{ij}\, d\chi_{ij}\, d\varepsilon = \sin\chi_{ij} \left|\frac{\partial \chi_{ij}}{\partial b}\right| db\, d\varepsilon$$

abgelenkt. Wenn auch die Umkehrung $b = b_{ij}(\chi, g)$ von (26.1) eine eindeutige Funktion ist — bei reiner Abstoßung ist dies der Fall — so gibt (26.2) sogleich auch die Anzahl $d\dot{N}'_{ij}$ der in den Winkel $d\boldsymbol{e}'$ gestreuten Teilchen:

$$d\dot{N}'_{ij} = g\, b\, db\, d\varepsilon\, n_i n_j V = \frac{g}{\sin\chi} \left| b_{ij} \frac{\partial b_{ij}}{\partial \chi}\right| d\boldsymbol{e}'\, n_i n_j V. \tag{26.3}$$

[1] Bei BUCKINGHAM und CORNER sowie bei MASON heißt er α; um eine mit dem Thermodiffusionsfaktor gleichlautende Bezeichnung zu vermeiden, wurde hier der Buchstabe s gewählt.

Der Vergleich mit (23.1) zeigt, daß

$$\sigma_{ij}(\chi, g) = \frac{1}{\sin\chi}\left|b_{ij}\frac{b_{ij}}{\partial\chi}\right|. \tag{26.4}$$

Dies ist der gesuchte Zusammenhang von Streuquerschnitt und Stoßparameter, welche in der klassischen Theorie wahlweise benutzt werden können. — Ist b_{ij} eine mehrdeutige Funktion von χ, so hat man rechts in (26.4) über alle Werte von b_{ij}, die zum selben χ führen, zu summieren. Als Beispiel sei das Modell der starren Kugeln vom Durchmesser d_i, d_j betrachtet. Nach Fig. 29b ist

$$\left.\begin{aligned} \chi &= 2 \arccos \frac{2b}{d_i + d_j} \quad &&\text{für} \quad b < \frac{d_i + d_j}{2}, \\ \chi &= 0 &&\text{für} \quad b > \frac{d_i + d_j}{2}, \end{aligned}\right\} \tag{26.5}$$

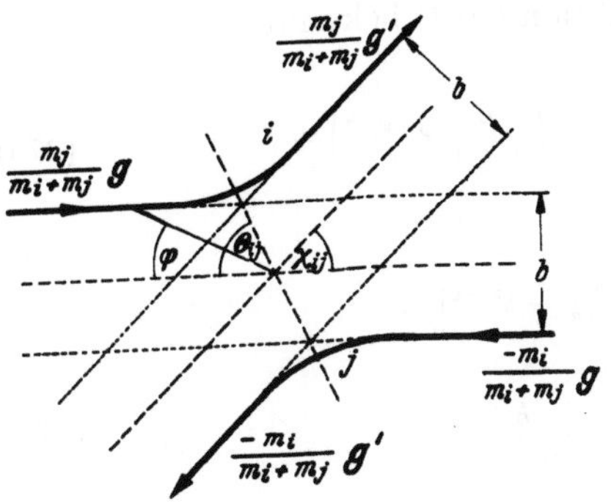

Fig. 29a. Klassische Bahnen im Schwerpunktssystem; einfachster Fall: reine Abstoßungskräfte.

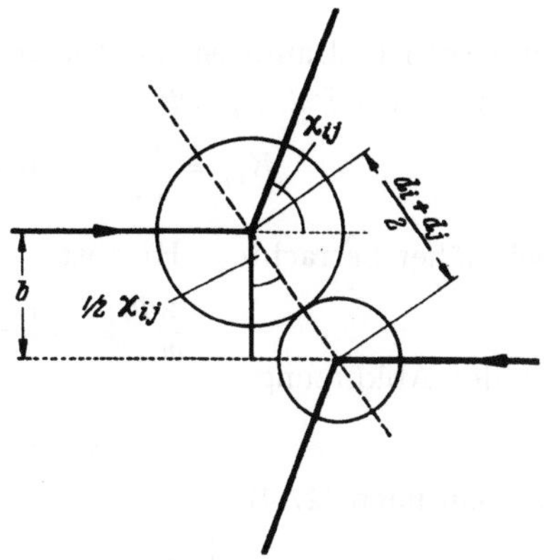

Fig. 29b. Zwei starre Kugeln beim Stoß; Schwerpunktssystem.

also nach (26.4)

$$\sigma_{ij,\,\text{starr}} = \frac{1}{4}\left(\frac{d_i + d_j}{2}\right)^2. \tag{26.6}$$

Der differentielle Streuquerschnitt ist in diesem Fall eine Konstante.

27. Integration der klassischen Bewegungsgleichung. Im allgemeinen Fall einer beliebigen potentiellen Energie $\Phi_{ij}(\boldsymbol{r})$, $\boldsymbol{r}$ = Radiusvektor von einem Atom zum anderen, hat man im Schwerpunktssystem die Bewegungsgleichung

$$m_{ij}\frac{d^2\boldsymbol{r}}{dt^2} = -\frac{\partial\Phi_{ij}}{\partial\boldsymbol{r}} \tag{27.1}$$

mit der reduzierten Masse

$$m_{ij} = \frac{m_i m_j}{m_i + m_j}. \tag{27.2}$$

Für zwei gleichartige Atome der Masse m ist

$$m_{\text{red}} = \tfrac{1}{2}m. \tag{27.3}$$

Mittels Flächen- und Energiesatz kann (27.1) durch Quadratur gelöst werden. Zwischen dem Polarwinkel φ (vgl. Fig. 29a) und der Größe

$$s = \frac{b}{r}$$

erhält man, wenn $\Phi_{ij}(\infty)=0$ normiert wird, den Zusammenhang

$$\frac{d\varphi}{ds}=\frac{1}{\sqrt{1-s^2-\frac{2\Phi_{ij}}{m_{ij}g^2}}}.$$

Der Ablenkwinkel χ bzw. der Winkel Θ (Fig. 29a) beträgt also

$$\Theta_{ij}(b,g)=\frac{1}{2}(\pi-\chi_{ij})=\int\limits_0^{s_{ij}^*}\frac{ds}{\sqrt{1-s^2-\frac{2\Phi_{ij}}{m_{ij}g^2}}},\tag{27.4}$$

wo $s_{ij}^*=\frac{b}{r_{\min}}=s_{ij,\,\max}(b,g)$ durch

$$1-s_{ij}^{*2}-\frac{2\Phi_{ij}(b/s_{ij}^*)}{m_{ij}g^2}=0\tag{27.5}$$

festgelegt ist. Damit ist im Prinzip die Funktion (26.1) bekannt.

Es sei das *Potenzgesetz*

$$K_{ij}=\frac{\varkappa_{ij}}{r^{\nu_{ij}}},\qquad \text{d.h.}\qquad \Phi_{ij}=\left(\frac{\varkappa}{\nu-1}\,r^{-\nu+1}\right)_{ij}\tag{27.6}$$

noch näher betrachtet. Hier ist

$$\frac{2\Phi_{ij}}{m_{ij}g^2}=\left[\frac{2}{\nu-1}\,\frac{\varkappa}{m\,g^2\,b^{\nu-1}}\,s^{\nu-1}\right]_i.$$

Mit der Abkürzung

$$s_{0,ij}=\left[\left(\frac{m\,g^2}{\varkappa}\right)^{\frac{1}{\nu-1}}b\right]_{ij}\tag{27.7}$$

hat man nach (27.4)

$$\Theta_{ij}=\left[\int\limits_0^{s^*}\frac{ds}{\sqrt{1-s^2-\frac{2}{\nu-1}\left(\frac{s}{s_0}\right)^{\nu-1}}}\right]_{ij}=\Theta_{ij}(s_{0,ij}),\tag{27.8}$$

wobei nach (27.5)

$$1-s_{ij}^{*2}-\left[\frac{2}{\nu-1}\left(\frac{s^*}{s_0}\right)^{\nu-1}\right]_{ij}=0.\tag{27.9}$$

Der Streuwinkel hängt hier also nicht von b und g einzeln, sondern nur von deren Kombination $s_{0,ij}$ ab.

Für das in (26.2) vorkommende Integrationselement gilt im Fall des Potenzgesetzes nach (27.7)

$$g\,b\,db=\left[\left(\frac{\varkappa}{m}\right)^{\frac{2}{\nu-1}}g^{\frac{\nu-5}{\nu-1}}\,s_0\,ds_0\right]_{ij}.\tag{27.10}$$

Ist speziell $\nu=5$ — *Maxwellsche Moleküle* —, so hat man [s. auch (23.1), (23.4) und (26.3)]

$$g\,b\,db=\sigma_{ij}\,g\sin\chi\,d\chi=\left[\left(\frac{\varkappa}{m}\right)^{\frac{1}{2}}s_0\,ds_0\right]_{ij}.\tag{27.11}$$

Hier kommt die Geschwindigkeit g in dem Integrationselement also nicht vor. Vielmehr bedeutet (27.11) mit (27.8), daß in

$$\sigma_{ij}\,g\,d\mathbf{e}'=\sigma_{ij}\,g\sin\chi\,d\chi\,d\varepsilon=-\Gamma_{ij}\sin\Theta\,d\Theta\,d\varepsilon\tag{27.12}$$

der Faktor

$$\Gamma_{ij}(\Theta)=4\cos\Theta\,\sigma_{ij}\,g=\left[\left(\frac{\varkappa}{m}\right)^{\frac{1}{2}}\frac{s_0\,ds_0}{\sin\Theta\,d\Theta}\right]_{ij}\tag{27.13}$$

nur von Θ, nicht von g abhängt. Darauf beruhen die mathematischen Vereinfachungen, die dieses spezielle Modell für die Gastheorie mit sich bringt (s. Ziff. 38f.)

II. Statistische Grundlagen der kinetischen Gastheorie.

Die von BOLTZMANN (1872) aufgestellte Integrodifferentialgleichung für die Verteilungsfunktion der molekularen Geschwindigkeiten bildet den Ausgangspunkt der exakten Gastheorie. Die folgende anschauliche Herleitung (Ziff. 28 und 29) unterscheidet sich von der klassischen Herleitung nur insofern, als sie den Begriff des Streuquerschnitts statt den des Stoßparameters benutzt und dadurch auch zur Quantenmechanik paßt, soweit es sich um verdünnte, nichtentartete Gase handelt. [Der quantenmechanische Gesichtspunkt der Gasentartung wurde von UEHLING und UHLENBECK (1933) in die kinetische Theorie eingeführt.]

Aus der allgemeinen statistischen Mechanik, welche im Prinzip jede Materie, auch Flüssigkeiten und feste Körper, umfaßt, haben BORN und H. S. GREEN (1946, 1947) sowie KIRKWOOD (1946) und IRVING und KIRKWOOD (1950) die Boltzmann-Gleichung als Grenzgesetz für den verdünnten Gaszustand abgeleitet. Ferner haben BOGOLJUBOFF (1946) und M. S. GREEN (1956) das Problem behandelt. Auf diese Arbeiten soll hier nicht eingegangen werden; es sei diesbezüglich auf den Bericht von DE BOER [*1*] und auf das Buch von HIRSCHFELDER *et al.* [*25*] verwiesen. Mit der anschaulichen Herleitung haben aber diese allgemeinen Methoden doch gemeinsam, daß auch sie zur Boltzmann-Gleichung führen nur mittels einer zusätzlichen, nicht weiter begründeten Annahme, der Superpositionsannahme der Verteilungsfunktion, welche der Boltzmannschen Unordnungsannahme äquivalent ist. Ebenfalls beiseite gelassen seien die Modifikationen, welche die Relativitätstheorie an der Boltzmann-Gleichung bewirkt; einiges darüber findet sich bei MARROT (1946).

Im Anschluß an die Besprechung der Boltzmann-Gleichung werden sodann die Erhaltungssätze (Ziff. 31 und 32) sowie der Entropiesatz und das thermische Gleichgewicht (Ziff. 33 und 34) diskutiert.

28. Die Boltzmann-Gleichung für reine einatomige Gase. Die kinetische Theorie beschreibt den Gaszustand mittels der Verteilungsfunktion $f(t, \boldsymbol{r}, \boldsymbol{c})$. Es ist $f(t, \boldsymbol{r}, \boldsymbol{c})\, d\boldsymbol{r}\, d\boldsymbol{c}$ die Anzahl der Atome[1], welche zur Zeit t im Raumelement $d\boldsymbol{r}$ bei $\boldsymbol{r}$ und im Geschwindigkeitselement $d\boldsymbol{c}$ bei $\boldsymbol{c}$ im statistischen Mittel — man denke sich in bekannter Weise eine große Anzahl von Ausfertigungen desselben mechanischen Systems — vorhanden sind (s. auch KOGA 1955). Zur Beschreibung weiterer Eigenschaften der mittleren Verteilung, die bei höherer Konzentration wesentlich werden, wie z.B. der Korrelation der Örter zweier Atome infolge der Wechselwirkung, wären erweiterte Verteilungsfunktionen $f(t, \boldsymbol{r}_1, \boldsymbol{c}_1, \boldsymbol{r}_2, \boldsymbol{c}_2)$ usw. nötig (vgl. BORN und GREEN 1946). Bei verdünnten Gasen spielen sie keine Rolle. Und generell werden auch die statistischen Schwankungen ignoriert, denen die Verteilungsfunktion unterliegt. Fragen wie z.B. die nach der mittleren Schwankung des Wärmestroms bleiben durch die Boltzmann-Gleichung unbeantwortet (vgl. zu diesem Problem ORNSTEIN und MILATZ 1939).

Es sei zunächst angenommen, daß keine atomaren Stöße stattfinden, daß sich die Gasatome der Masse m also nur unter dem Einfluß der äußeren Kraft $\boldsymbol{K} = m\boldsymbol{b}$, z.B. der Schwerkraft oder einer elektrischen Kraft, bewegen. Ort und Geschwindigkeit eines Atoms zur Zeit t seien $\boldsymbol{r}$ und $\boldsymbol{c}$, zur Zeit $\bar{t}$ entsprechend $\bar{\boldsymbol{r}}$ und $\bar{\boldsymbol{c}}$. Dann muß sein

$$f(\bar{t}, \bar{\boldsymbol{r}}, \bar{\boldsymbol{c}})\, d\bar{\boldsymbol{r}}\, d\bar{\boldsymbol{c}} = f(t, \boldsymbol{r}, \boldsymbol{c})\, d\boldsymbol{r}\, d\boldsymbol{c}.$$

---

[1] Es ist abgekürzt $d\boldsymbol{r} = dx\, dy\, dz = dx_1\, dx_2\, dx_3$; $d\boldsymbol{c} = dc_1\, dc_2\, dc_3$. Indices $\mu, \nu, \ldots = 1, 2, 3$ bezeichnen die drei Raumrichtungen, s. a. Anm. 1 von S. 297 und 298.

Nach dem Liouvilleschen Satz der Mechanik ist aber $d\bar{\boldsymbol{r}}\, d\bar{\boldsymbol{c}} = d\boldsymbol{r}\, d\boldsymbol{c}$. Mit $\bar{t} = t + dt$, $\bar{\boldsymbol{r}} = \boldsymbol{r} + \boldsymbol{c}\, dt$, $\bar{\boldsymbol{c}} = \boldsymbol{c} + \boldsymbol{b}\, dt$ erhält man also

$$f(t + dt, \boldsymbol{r} + \boldsymbol{c}\, dt, \boldsymbol{c} + \boldsymbol{b}\, dt) = f(t, \boldsymbol{r}, \boldsymbol{c})$$

oder

$$\mathscr{D}(f) \equiv \frac{\partial f}{\partial t} + c_\mu \frac{\partial f}{\partial x_\mu} + b_\mu \frac{\partial f}{\partial c_\mu} = 0. \tag{28.1}$$

Durch diese partielle Differentialgleichung erster Ordnung ist die zeitliche Änderung der Verteilungsfunktion eines hochverdünnten Gases gegeben. f kann demnach berechnet werden, wenn der Anfangszustand und die Randbedingungen an den Gefäßwänden bekannt sind.

Gl. (28.1), auf Gemische erweitert (s. Ziff. 29), bildet auch den Ausgangspunkt für die Diskussion hochverdünnter ionisierter Gase; nur ist $\boldsymbol{b}$ in diesem Fall nicht von vornherein gegeben, da das elektromagnetische Kraftfeld selbst wieder durch den Zustand des Gasgemischs mitbestimmt ist (WATSON 1956; BRUECKNER und WATSON 1956). Auf dieses Gebiet der Plasmatheorie und Magnetohydrodynamik kann in diesem Artikel nicht eingegangen werden (s. CHAPMAN und COWLING [4], Kap. 18; SPITZER [50]; COWLING [6]).

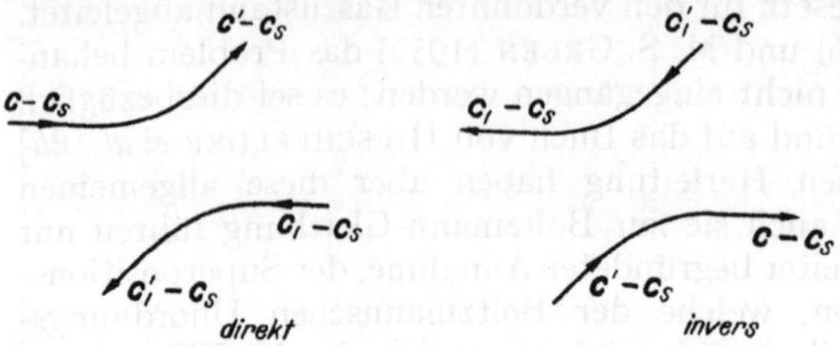

Fig. 30. Direkter und inverser Stoß; klassische Teilchenbahnen im Schwerpunktssystem.

Die atomaren Stöße bewirken zusätzliche Änderungen der Verteilungsfunktion. Wir betrachten zuerst die Zusammenstöße der beiden Teilchenströme mit Geschwindigkeiten $\boldsymbol{c}$ in $d\boldsymbol{c}$ und $\boldsymbol{c}_1$ in $d\boldsymbol{c}_1$, welche im Volumen $d\boldsymbol{r}$ mit Stoßvektor in $d\boldsymbol{e}'$ stattfinden. Für deren Anzahl pro Zeiteinheit ergibt sich nach (23.1), worin jetzt $n_i = f\, d\boldsymbol{c}$, $n_j = f_1\, d\boldsymbol{c}_1$, $V = d\boldsymbol{r}$ zu setzen ist, der Ausdruck

$$\sigma g\, d\boldsymbol{e}'\, f f_1\, d\boldsymbol{c}\, d\boldsymbol{c}_1\, d\boldsymbol{r}.$$

Dabei ist zur Abkürzung $f(t, \boldsymbol{r}, \boldsymbol{c}_1) = f_1$ geschrieben. Durch Integration über alle Streurichtungen $\boldsymbol{e}'$ und alle Geschwindigkeiten $\boldsymbol{c}_1$ des Stoßpartners erhalten wir somit für die Anzahl der Teilchen in $d\boldsymbol{r}$ mit Geschwindigkeit in $d\boldsymbol{c}$, welche pro Zeiteinheit durch irgendwelche Stöße ihre Geschwindigkeit ändern,

$$\left(\frac{\delta f}{\delta t}\right)_- d\boldsymbol{r}\, d\boldsymbol{c} = d\boldsymbol{r}\, d\boldsymbol{c} \iint f f_1\, \sigma g\, d\boldsymbol{e}'\, d\boldsymbol{c}_1. \tag{28.2}$$

Durch andere Streuprozesse erhalten gleichzeitig Atome eine Geschwindigkeit in $d\boldsymbol{c}$. Dazu sei bemerkt, daß die Grundgleichungen der Punktmechanik invariant sind gegen die Inversion, d.h. gleichzeitige Vorzeichenumkehr der Zeit- und Ortskoordinaten. Dies ist aus den klassischen Bewegungsgleichungen des Zweierstoßes

$$m\ddot{\boldsymbol{r}} = -\frac{\partial \Phi(|\boldsymbol{r} - \boldsymbol{r}_1|)}{\partial \boldsymbol{r}}, \quad m\ddot{\boldsymbol{r}}_1 = -\frac{\partial \Phi(|\boldsymbol{r} - \boldsymbol{r}_1|)}{\partial \boldsymbol{r}_1} \tag{28.3}$$

ersichtlich; durch die Transformation $t, \boldsymbol{r}, \boldsymbol{r}_1 \to -t, -\boldsymbol{r}, -\boldsymbol{r}_1$ erhält man wieder eine mögliche Bahn. Entsprechendes gilt in der Quantenmechanik. Mit der Wellenfunktion $\psi(t, \boldsymbol{r})$ ist zugleich $\psi^*(-t, -\boldsymbol{r})$ eine Lösung der Schrödinger-Gleichung und Entsprechendes gilt für die Erwartungswerte jedes quantenmechanischen Operators. Der zum direkten Stoß $(\boldsymbol{c}, \boldsymbol{c}_1) \to (\boldsymbol{c}', \boldsymbol{c}_1')$ mit dem Richtungsvektor $\boldsymbol{e}'$ gehörige Streuquerschnitt gilt also auch für den inversen Stoß $(\boldsymbol{c}', \boldsymbol{c}_1') \to (\boldsymbol{c}, \boldsymbol{c}_1)$ mit dem Richtungsvektor $\boldsymbol{e}$. Fig. 30 veranschaulicht dies. Wesentlich ist dabei, daß es sich um einfache Massenpunkte mit kugelsymmetrischem

Kraftfeld handelt ohne weitere Kennzeichen wie z. B. Eigendrehimpuls (Spin), verknüpft mit einem magnetischen Dipolmoment. Bei Spinteilchen gibt es den inversen Stoß nicht. Wegen einer ausführlichen Diskussion sei auf TOLMAN [*51*], S. 119, WATANABE [*54*] und auf Teil H dieses Artikels verwiesen.

Aus (23.1) entnimmt man für die Anzahl pro Zeiteinheit der inversen Stöße, bei welchen die Geschwindigkeiten vor dem Stoß in $d\boldsymbol{c}'$ und $d\boldsymbol{c}_1'$ und der Richtungsvektor $\boldsymbol{e}$ in $d\boldsymbol{e}$ liegen,

$$\sigma g\, d\boldsymbol{e}\, f' f_1'\, d\boldsymbol{c}'\, d\boldsymbol{c}_1'\, d\boldsymbol{r}.$$

Zur Abkürzung ist $f(t, \boldsymbol{r}, \boldsymbol{c}') = f'$ und $f(t, \boldsymbol{r}, \boldsymbol{c}_1') = f_1'$ geschrieben. Mittels des Liouvilleschen Satzes (22.22) findet man daraus als Gesamtzahl der Teilchen in $d\boldsymbol{r}$, welche pro Zeiteinheit durch Stoß eine Geschwindigkeit in $d\boldsymbol{c}$ erhalten,

$$\left(\frac{\delta f}{\delta t}\right)_+ d\boldsymbol{r}\, d\boldsymbol{c} = d\boldsymbol{r}\, d\boldsymbol{c} \iint f' f_1' \sigma g\, d\boldsymbol{e}'\, d\boldsymbol{c}_1. \tag{28.4}$$

Die Zusammenfassung von (28.1), (28.2) und (28.4) ergibt nun die *Boltzmann-Gleichung*

$$\frac{\partial f}{\partial t} + c_\mu \frac{\partial f}{\partial x_\mu} + b_\mu \frac{\partial f}{\partial c_\mu} = \iint (f' f_1' - f f_1)\, \sigma g\, d\boldsymbol{e}'\, d\boldsymbol{c}_1. \tag{28.5}$$

Die Argumente $\boldsymbol{c}'$, $\boldsymbol{c}_1'$ in f', f_1', sowie in σ, vgl. (23.3). sind gemäß (22.9) durch $\boldsymbol{c}$, $\boldsymbol{c}_1$, $\boldsymbol{e}'$ ausgedrückt zu denken.

Bei Benutzung der klassischen Mechanik ist es nach (26.3) und (26.4) möglich, den Stoß statt durch die Streurichtung $\boldsymbol{e}'$ durch den Stoßparameter b und das Azimut ε zu kennzeichnen. Es ist im einzelnen

$$\int \cdots \sigma g\, d\boldsymbol{e}' = \int_0^\pi \int_0^{2\pi} \cdots \sigma g \sin\chi\, d\chi\, d\varepsilon = \int_0^\infty \int_0^{2\pi} \cdots g\, b\, db\, d\varepsilon. \tag{28.6}$$

Dementsprechend kann (28.5) umgeformt werden.

Die statistische Beschreibung des Stoßprozesses durch den Streuquerschnitt ist dem Wesen der Boltzmann-Gleichung besonders gut angepaßt. R. v. MISES ([*38a*], S. 220) schließt sogar umgekehrt, daß der statistische Charakter der Boltzmann-Gleichung nur mit statistischen Elementargesetzen der Mechanik logisch verträglich sei. Auf einen Punkt ist nochmals hinzuweisen. Stillschweigend wurde bei der obigen Herleitung die an die Voraussetzung des mäßig verdünnten Gases geknüpfte Boltzmannsche Unordnungsannahme gemacht, welche behauptet, daß die beim Stoß nach Maßgabe des Streuquerschnitts sich ausbildende lokale Ordnung der Molekulargeschwindigkeiten sich vor dem nächsten Stoß wieder völlig verwischt hat (keine „Schattenwirkung" der Atome). Bei der eingangs erwähnten Herleitung der Boltzmann-Gleichung aus der allgemeinen statistischen Mechanik steckt die Unordnungsannahme in der Superpositionsannahme für die Verteilungsfunktion. Letztere besagt, daß die am Anfang dieser Ziffer erwähnte Paar-Verteilungsfunktion angesetzt werden darf als Produkt der Einzel-Verteilungsfunktionen zweier Teilchen:

$$f(t, \boldsymbol{r}_1, \boldsymbol{c}_1, \boldsymbol{r}_2, \boldsymbol{c}_2) = f(t, \boldsymbol{r}_1, \boldsymbol{c}_1) \cdot f(t, \boldsymbol{r}_2, \boldsymbol{c}_2).$$

Die Superpositionsannahme konnte ebensowenig wie die Unordnungsannahme bisher näher begründet werden, so daß der Ursprung der Boltzmann-Gleichung letzten Endes in ein gewisses Dunkel gehüllt ist (s. dazu KÜMMEL 1955, 1956). Diese Frage hat ein praktisches Interesse im Hinblick auf Vorgänge, wie die Stoßwellen in Überschallströmungen, wo sich Dichte, Geschwindigkeit und Temperatur des Gases auf einer Strecke von der Größenordnung der freien Weglänge merklich ändern können. Gleichwohl erscheint es physikalisch plausibel, daß die Boltzmann-Gleichung auch diese Vorgänge richtig wiedergibt.

29. Die Boltzmann-Gleichungen für einatomige Gasgemische. Die Komponenten des Gemischs seien durch den Index i (oder j) unterschieden. Jede Komponente hat ihre eigene Verteilungsfunktion $f_i(t, \boldsymbol{r}, \boldsymbol{c})$. Ohne Stöße hat man analog zu (28.1)

$$\mathcal{D}_i(f_i) \equiv \frac{\partial f_i}{\partial t} + c_\mu \frac{\partial f_i}{\partial x_\mu} + b_{\mu i} \frac{\partial f_i}{\partial c_\mu} = 0. \tag{29.1}$$

Bei mäßig verdünnten Gasen sind die Zweierstöße zu berücksichtigen. Es können die Atome einer Sorte i nicht nur mit ihresgleichen, sondern auch mit Atomen jeder anderen Sorte zusammenstoßen. All diese Möglichkeiten tragen gleichzeitig zur Änderung der Verteilungsfunktion f_i bei, all diese Änderungen sind zu addieren. Mit den in (22.11) eingeführten Bezeichnungen $\boldsymbol{c}'$, $\boldsymbol{c}_1'$ für die Geschwindigkeit nach einem Stoß (ij) und der Bezeichnung σ_{ij} für den dabei maßgebenden Streuquerschnitt hat man als Verallgemeinerung von (28.5), das Boltzmann-Gleichungssystem

$$\mathcal{D}_i[f_i(\boldsymbol{c})] = \sum_j \iint [f_i(\boldsymbol{c}') f_j(\boldsymbol{c}_1') - f_i(\boldsymbol{c}) f_j(\boldsymbol{c}_1)] \sigma_{ij} g \, d\boldsymbol{e}' \, d\boldsymbol{c}_1. \tag{29.2}$$

Dabei sind im Argument der Verteilungsfunktionen t und $\boldsymbol{r}$ der Kürze halber weggelassen. Die Geschwindigkeiten nach dem Stoß sind nach (22.17) durch die Geschwindigkeiten vor dem Stoß und den Richtungsvektor $\boldsymbol{e}'$ ausgedrückt zu denken[1]. Bei klassischer Mechanik kann wieder gemäß (28.6), mit σ_{ij} anstelle von σ, umgeschrieben werden.

30. Eine andere Schreibweise der Boltzmann-Gleichung. Für die spätere Diskussion der linearisierten Boltzmann-Gleichung ist es zweckmäßig, noch eine andere Schreibweise der Boltzmann-Gleichung anzugeben, die deren Struktur besonders deutlich macht.

Gesucht ist der Ausdruck für die Änderung pro Zeiteinheit der Verteilungsfunktion durch die atomaren Zweierstöße. Beim Zweierstoß gehen die Geschwindigkeiten $\boldsymbol{c}$, $\boldsymbol{c}_1$ der beiden Atome in irgendwelche Geschwindigkeiten $\boldsymbol{c}'$, $\boldsymbol{c}_1'$ über. Die Häufigkeit pro Volumeinheit eines solchen Stoßes (mit entsprechenden Spielräumen für die Geschwindigkeiten) muß dem Produkt $f f_1 \, d\boldsymbol{c} \, d\boldsymbol{c}_1 \, d\boldsymbol{c}' \, d\boldsymbol{c}_1'$ proportional sein. Über den Proportionalitätsfaktor machen wir zunächst weiter keine Aussage als die, daß er eine Funktion aller vier Geschwindigkeiten ist. Er sei mit $W(\boldsymbol{c}, \boldsymbol{c}_1 | \boldsymbol{c}', \boldsymbol{c}_1')$ bezeichnet. Dieselbe Funktion W, genommen für die entsprechenden Variablen, nämlich $W(\boldsymbol{c}', \boldsymbol{c}_1' | \boldsymbol{c}, \boldsymbol{c}_1)$ gibt dann die Wahrscheinlichkeit dafür, daß nach dem Stoß ein Atom die Geschwindigkeit $\boldsymbol{c}$ besitzt (und der Partner die Geschwindigkeit $\boldsymbol{c}_1$). Mit Benutzung dieser Funktion W lautet also die Boltzmann-Gleichung für reines Gas:

$$\frac{\partial f}{\partial t} + c_\mu \frac{\partial f}{\partial x_\mu} + b_\mu \frac{\partial f}{\partial c_\mu} = \iiint [W(\boldsymbol{c}', \boldsymbol{c}_1' | \boldsymbol{c}, \boldsymbol{c}_1) f' f_1' - W(\boldsymbol{c}, \boldsymbol{c}_1 | \boldsymbol{c}', \boldsymbol{c}_1') f f_1] \, d\boldsymbol{c}_1 \, d\boldsymbol{c}' \, d\boldsymbol{c}_1'. \tag{30.1}$$

Die Stoßwahrscheinlichkeit W kann natürlich nicht beliebig angenommen werden. Sie muß ja unter anderem auch die Erhaltungssätze von Impuls und Energie zum Ausdruck bringen. Zuvor sind aber einige noch allgemeinere Eigenschaften von W zu konstatieren:

1. Als *Wahrscheinlichkeit* kann W nicht negativ sein

$$W \geqq 0. \tag{30.2}$$

2. Bei zwei *gleichartigen unterscheidbaren* (klassischen) Teilchen kann die Stoßwahrscheinlichkeit nicht von der Numerierung abhängen; es muß also gelten

$$W(\boldsymbol{c}, \boldsymbol{c}_1 | \boldsymbol{c}', \boldsymbol{c}_1') = W(\boldsymbol{c}_1, \boldsymbol{c} | \boldsymbol{c}_1', \boldsymbol{c}'). \tag{30.3a}$$

Bei gleichartigen, nicht-unterscheidbaren (quantenmechanischen) Teilchen hat man sogar die Symmetrie innerhalb jedes der beiden Variablenpaare vor und nach dem Stoß:

$$W(\boldsymbol{c}, \boldsymbol{c}_1 | \boldsymbol{c}', \boldsymbol{c}_1') = W(\boldsymbol{c}_1, \boldsymbol{c} | \boldsymbol{c}', \boldsymbol{c}_1') = W(\boldsymbol{c}, \boldsymbol{c}_1 | \boldsymbol{c}_1', \boldsymbol{c}') = W(\boldsymbol{c}_1, \boldsymbol{c} | \boldsymbol{c}_1', \boldsymbol{c}'). \tag{30.3b}$$

[1] Die Bezeichnung $\boldsymbol{c}'$, $\boldsymbol{c}_1'$ für die Geschwindigkeiten nach dem Stoß ist, für sich betrachtet, unzureichend, weil sie nicht erkennen läßt, welche Massen in (22.17) einzusetzen sind. Man sollte besser z.B. $\boldsymbol{c}^{(ij)}$, $\boldsymbol{c}_1^{(ij)}$ schreiben. Gleichwohl ist die kurze Bezeichnung durch die Striche unmißverständlich. Welche Partner gemeint sind, ergibt sich stets eindeutig aus den Indices des Streuquerschnitts, mit dem der betreffende, $\boldsymbol{c}'$ und $\boldsymbol{c}_1'$ enthaltende Ausdruck multipliziert ist.

3. Die *Inversion*, d.h. die Umkehr der Raum- und Zeitkoordinaten, läßt die Geschwindigkeiten nach Betrag und Richtung ungeändert, vertauscht aber den Zustand vor und nach dem Stoß. Da sich an der Stoßwahrscheinlichkeit bei punktförmigen Teilchen durch Inversion nichts ändern kann, muß also ferner die Symmetrie

$$W(\mathbf{c}, \mathbf{c}_1 | \mathbf{c}', \mathbf{c}_1') = W(\mathbf{c}', \mathbf{c}_1' | \mathbf{c}, \mathbf{c}_1) \tag{30.4}$$

erfüllt sein, vgl. WATANABE [54].

Aus (30.1) erhält man nun durch Benutzung von (30.4) die Boltzmann-Gleichung in der Gestalt

$$\frac{\partial f}{\partial t} + \cdots = \iiint W(\mathbf{c}, \mathbf{c}_1 | \mathbf{c}', \mathbf{c}_1') (f' f_1' - f f_1)\, d\mathbf{c}_1\, d\mathbf{c}'\, d\mathbf{c}_1'. \tag{30.5}$$

Bei den Betrachtungen in Ziff. 36 werden, soweit sie ganz allgemein sind, keine spezielleren Kenntnisse über W benötigt. Will man aber dort den Hilbertschen Kern tatsächlich mittels des Streuquerschnitts σ aus Ziff. 23 ausdrücken, so ist es von Vorteil, zuerst W in Verbindung mit σ zu bringen. Das soll nun noch geschehen.

Dazu stellen wir folgende weiteren Eigenschaften von W fest:

4. *Galilei-Invarianz:* W kann sich nicht ändern beim Übergang zu einem anderen gleichförmig translatorisch bewegten Bezugssystem, d.h. W kann nur von den Differenzen der Geschwindigkeiten, deren es drei linear unabhängige gibt, abhängen. Wir können daher schreiben:

$$W(\mathbf{c}, \mathbf{c}_1 | \mathbf{c}', \mathbf{c}_1') = W(\mathbf{c} - \mathbf{c}_1, \mathbf{c}' - \mathbf{c}_1', \mathbf{c} + \mathbf{c}_1 - \mathbf{c}' - \mathbf{c}_1'). \tag{30.6}$$

Wenn diese Funktion W überdies invariant ist gegen Vorzeichenumkehr der drei als Argumente auftretenden Vektoren, so sind auch die Eigenschaften (30.3) und (30.4) erfüllt.

5. *Impulserhaltung.* W muß den Faktor $\delta_3\left(\frac{\mathbf{c} + \mathbf{c}_1 - \mathbf{c}' - \mathbf{c}_1'}{2}\right)$ enthalten. Dabei bedeutet δ_3 die Diracsche δ-Funktion des dreidimensionalen Raums ($\int \delta_3(\mathbf{c})\, d\mathbf{c} = 1$). Über (30.6) hinausgehend können wir also setzen

$$W(\mathbf{c}, \mathbf{c}_1 | \mathbf{c}', \mathbf{c}_1') = W(\mathbf{c} - \mathbf{c}_1, \mathbf{c}' - \mathbf{c}_1')\, \delta_3\left(\frac{\mathbf{c} + \mathbf{c}_1 - \mathbf{c}' - \mathbf{c}_1'}{2}\right), \tag{30.7}$$

wo nun nur noch eine Funktion von zwei Vektoren unbestimmt ist. Jetzt kann in (30.5) eine Integration eingeführt werden. Wir führen zu dem Zweck statt $\mathbf{c}'$, $\mathbf{c}_1'$ die neuen Variablen ein

$$\mathbf{c}_s' = \tfrac{1}{2}(\mathbf{c}' + \mathbf{c}_1'), \qquad \boldsymbol{g}' = \mathbf{c}' - \mathbf{c}_1'.$$

Bekanntlich ist

$$d\mathbf{c}'\, d\mathbf{c}_1' = d\mathbf{c}_s'\, d\boldsymbol{g}'.$$

Mit Benutzung der Abkürzungen

$$\mathbf{c}_s = \tfrac{1}{2}(\mathbf{c} + \mathbf{c}_1), \qquad \boldsymbol{g} = \mathbf{c} - \mathbf{c}_1$$

erhält man so aus (30.5) durch Integration über $\mathbf{c}_s'$:

$$\frac{\partial f}{\partial t} + \cdots = \iint W(\boldsymbol{g}, \boldsymbol{g}') (f' f_1' - f f_1)\, d\mathbf{c}_1\, d\boldsymbol{g}', \tag{30.8}$$

wobei jetzt $\mathbf{c}'$, $\mathbf{c}_1'$ als Abkürzungen für

$$\mathbf{c}' = \mathbf{c}_s + \tfrac{1}{2}\boldsymbol{g}', \qquad \mathbf{c}_1' = \mathbf{c}_s - \tfrac{1}{2}\boldsymbol{g}'$$

zu verstehen sind.

6. *Drehinvarianz.* W kann nur von dem Betrag der beiden Vektoren $\boldsymbol{g}$, $\boldsymbol{g}'$ und dem Winkel zwischen ihnen abhängen:

$$W(\boldsymbol{g}, \boldsymbol{g}') = W(g, g', \chi), \qquad \chi = \sphericalangle(\boldsymbol{g}\,\boldsymbol{g}'). \tag{30.9}$$

7. *Energieerhaltung.* W muß den Faktor $\delta_1\left(\frac{g^2 - g'^2}{2}\right)$ enthalten, wo δ_1 die Diracsche δ-Funktion des eindimensionalen Raums bedeutet ($\int \delta_1(g)\, dg = 1$). Wir schreiben

$$W(\boldsymbol{g}, \boldsymbol{g}') = \sigma(g, \chi)\, \delta_1\left(\frac{g^2 - g'^2}{2}\right). \tag{30.10}$$

Dann kann in (30.8) noch eine Integration, nämlich die über den Betrag g' allgemein ausgeführt werden:

$$\int_0^\infty \delta_1\left(\frac{g^2 - g'^2}{2}\right) g'^2\, dg' = g.$$

Indem wir das zu $d\boldsymbol{g}'$ gehörige Raumwinkelelement wie früher mit $d\boldsymbol{e}'$ bezeichnen, bekommen wir aus (30.8)

$$\frac{\partial f}{\partial t} + \cdots = \iint (f' f_1' - f f_1)\, \sigma\, g\, d\boldsymbol{e}'\, d\boldsymbol{c}_1,$$

d.i. die Boltzmann-Gleichung (28.5). Die in (30.10) eingeführte Größe σ bedeutet also in der Tat den differentiellen Streuquerschnitt. Von (30.10) zu (30.7) zurückgehend, erhalten wir den gewünschten Zusammenhang

$$\left.\begin{aligned} W(\boldsymbol{c}, \boldsymbol{c}_1 | \boldsymbol{c}', \boldsymbol{c}_1') &= \sigma\left(|\boldsymbol{c} - \boldsymbol{c}_1|, \arccos \frac{(\boldsymbol{c} - \boldsymbol{c}_1)(\boldsymbol{c}' - \boldsymbol{c}_1')}{|\boldsymbol{c} - \boldsymbol{c}_1| \cdot |\boldsymbol{c}' - \boldsymbol{c}_1'|}\right) \times \\ &\times \delta_3\left(\frac{\boldsymbol{c} + \boldsymbol{c}_1 - \boldsymbol{c}' - \boldsymbol{c}_1'}{2}\right) \delta_1\left(\frac{(\boldsymbol{c} - \boldsymbol{c}_1)^2 - (\boldsymbol{c}' - \boldsymbol{c}_1')^2}{2}\right). \end{aligned}\right\} \tag{30.11}$$

Die Übertragung dieser Betrachtungen auf Gemische macht keine Schwierigkeit. Anstelle von (30.1) tritt für die Verteilungsfunktion der i-ten Komponente die Beziehung

$$\frac{\partial f_i}{\partial t} + \cdots = \sum_j \iiint [W_{ij}(\boldsymbol{c}', \boldsymbol{c}_1' | \boldsymbol{c}, \boldsymbol{c}_1)\, f_i(\boldsymbol{c}')\, f_j(\boldsymbol{c}_1') - W_{ij}(\boldsymbol{c}, \boldsymbol{c}_1 | \boldsymbol{c}', \boldsymbol{c}_1')\, f_i(\boldsymbol{c})\, f_j(\boldsymbol{c}_1)]\, d\boldsymbol{c}_1\, d\boldsymbol{c}'\, d\boldsymbol{c}_1'. \tag{30.12}$$

Für die Stoßwahrscheinlichkeit W_{ij} gilt wieder nach Definition

$$W_{ij}(\boldsymbol{c}, \boldsymbol{c}_1 | \boldsymbol{c}', \boldsymbol{c}_1') = W_{ji}(\boldsymbol{c}_1, \boldsymbol{c} | \boldsymbol{c}_1', \boldsymbol{c}') \tag{30.13}$$

und ferner wegen der Inversibilität

$$W_{ij}(\boldsymbol{c}, \boldsymbol{c}_1 | \boldsymbol{c}', \boldsymbol{c}_1') = W_{ij}(\boldsymbol{c}', \boldsymbol{c}_1' | \boldsymbol{c}, \boldsymbol{c}_1). \tag{30.14}$$

Daher kann man (30.12) auch in der Form schreiben

$$\frac{\partial f_i}{\partial t} + \cdots = \sum_j \iiint W_{ij}(\boldsymbol{c}, \boldsymbol{c}_1 | \boldsymbol{c}', \boldsymbol{c}_1')\, [f_i(\boldsymbol{c}')\, f_j(\boldsymbol{c}_1') - f_i(\boldsymbol{c})\, f_j(\boldsymbol{c}_1)]\, d\boldsymbol{c}_1\, d\boldsymbol{c}'\, d\boldsymbol{c}_1'. \tag{30.15}$$

Dies ist die Erweiterung von (30.5) auf Gemische. Die übrigen, an (30.5) anschließenden Betrachtungen brauchen nicht wiederholt zu werden.

31. Transportgleichung und Erhaltungssätze für reine Gase. Aus den Erhaltungssätzen (22.1) und (22.2) beim Stoß ergeben sich mittels der Boltzmann-Gleichung die Erhaltungssätze für das Gas. Der Weg dazu führt über die Transportgleichung.

Es sei eine beliebige Funktion $\Psi(t, \boldsymbol{r}, \boldsymbol{c})$ betrachtet. Für sie entnimmt man aus der Boltzmann-Gleichung (28.5)

$$\int \Psi\, \mathscr{D}(f)\, d\boldsymbol{c} = \iiint \Psi (f' f_1' - f f_1)\, \sigma\, g\, d\boldsymbol{e}'\, d\boldsymbol{c}\, d\boldsymbol{c}_1.$$

$\mathscr{D}(f)$ ist die in (28.1) benutzte Abkürzung für die linke Seite von (28.5). Das dreifache Integral besteht, entsprechend den zwei Gliedern in der Klammer, aus zwei Teilen. Nun seien in dem ersten Teilintegral die Variablen in der Weise $(\boldsymbol{c}, \boldsymbol{c}_1; \boldsymbol{c}', \boldsymbol{c}_1') \to (\boldsymbol{c}', \boldsymbol{c}_1'; \boldsymbol{c}, \boldsymbol{c}_1)$ umbenannt. Da $\sigma' = \sigma$ und mit Benutzung des Liouvilleschen Satzes (22.22) erhält man so

$$\int \Psi\, \mathscr{D}(f)\, d\boldsymbol{c} = \iiint (\Psi' - \Psi)\, f f_1\, \sigma\, g\, d\boldsymbol{e}'\, d\boldsymbol{c}\, d\boldsymbol{c}_1.$$

Durch die weitere Umbenennung $(\boldsymbol{c}, \boldsymbol{c}_1; \boldsymbol{c}', \boldsymbol{c}_1') \to (\boldsymbol{c}_1, \boldsymbol{c}; \boldsymbol{c}_1', \boldsymbol{c}')$ entsteht daraus

$$\int \Psi\, \mathscr{D}(f)\, d\boldsymbol{c} = \iiint (\Psi_1' - \Psi_1)\, f f_1\, \sigma\, g\, d\boldsymbol{e}'\, d\boldsymbol{c}\, d\boldsymbol{c}_1,$$

so daß durch Addition sich ergibt

$$\int \Psi\, \mathscr{D}(f)\, d\boldsymbol{c} = n \left(\frac{\delta \overline{\Psi}}{\delta t}\right)_{\text{Stöße}} = \frac{1}{2} \iiint (\Psi' + \Psi_1' - \Psi - \Psi_1)\, f f_1\, \sigma\, g\, d\boldsymbol{e}'\, d\boldsymbol{c}\, d\boldsymbol{c}_1. \tag{31.1}$$

Falls $\Psi = \Psi(\boldsymbol{c})$ nur von der Geschwindigkeit abhängt, folgt daraus nach (28.1)

$$\frac{\partial n \overline{\Psi}}{\partial t} = - \frac{\partial n \overline{\Psi c_\mu}}{\partial x_\mu} + n \left(\overline{\frac{\partial \Psi b_\mu}{\partial c_\mu}} + \left(\frac{\delta \Psi}{\delta t}\right)_{\text{Stöße}}\right), \tag{31.1a}$$

wobei durch Überstreichen die Mittelwerte

$$\overline{\Psi} = \frac{1}{n} \int \Psi f \, d\boldsymbol{c}, \quad \text{usw.}$$

bezeichnet sind und im vorletzten Glied partiell integriert wurde.

Wegen $\frac{\partial(\boldsymbol{c}, \boldsymbol{c}_1)}{\partial(\boldsymbol{c}_S, \boldsymbol{g})} = -1$, $d\boldsymbol{e}' \, d\boldsymbol{c} \, d\boldsymbol{c}_1 = d\boldsymbol{e}' \, d\boldsymbol{c}_S \, d\boldsymbol{g} = d\boldsymbol{e}' \, d\boldsymbol{c}_S \, g^2 \, dg \, d\boldsymbol{e}$ kann man statt (31.1) auch schreiben

$$n \left(\frac{\delta \overline{\Psi}}{\delta t} \right)_{\text{Stöße}} = \frac{1}{2} \iiint (\Psi' + \Psi_1' - \Psi - \Psi_1) f f_1 \sigma g^3 \, dg \, d\boldsymbol{e} \, d\boldsymbol{e}' \, d\boldsymbol{c}_s. \tag{31.1b}$$

Dabei ist $\boldsymbol{c}_{(1)} = \boldsymbol{c}_{S} \,{}_{(}\!\pm\!{}_{)}\, \frac{1}{2} g \boldsymbol{e}$; $\boldsymbol{c}'_{(1)} = \boldsymbol{c}_S \,{}_{(}\!\pm\!{}_{)}\, \frac{1}{2} g \boldsymbol{e}'$ eingesetzt zu denken. Die Variablen $\boldsymbol{c}_S, g, \boldsymbol{e}, \boldsymbol{e}'$ sind für die Auswertung der Transportgleichung besonders geeignet (s. Ziff. 50).

Die *Transportgleichung* (31.1) bzw. (31.1a) hatte bereits MAXWELL 1866 ([*38*], Bd. 2, S. 26–78); die drei Terme rechts in (31.1a) entsprechen der Änderung pro Zeiteinheit der in der Volumeneinheit enthaltenen Größe $n\overline{\Psi}$ durch Zustrom, durch die äußere Kraft und durch die Molekularstöße.

Ist speziell $\Psi = \psi$ eine Stoßinvariante, was bedeuten soll, daß

$$\psi + \psi_1 = \psi' + \psi_1' \tag{31.2}$$

gilt, so hat man also nach (31.1)

$$\int \psi \mathscr{D}(f) \, d\boldsymbol{c} = 0 \tag{31.3}$$

oder ausführlich nach (31.1a), wenn speziell $\boldsymbol{b}$ nicht von $\boldsymbol{c}$ abhängt,

$$\frac{\partial}{\partial t} \int \psi f \, d\boldsymbol{c} + \frac{\partial}{\partial x_\mu} \int \psi c_\mu f \, d\boldsymbol{c} = b_\mu \int \frac{\partial \psi}{\partial c_\mu} f \, d\boldsymbol{c}. \tag{31.3a}$$

Das sind die *Erhaltungssätze.*

Es gibt die Stoßinvarianten [s. auch Gln. (22.1) und (22.2)]

$$\left.\begin{array}{lll} \text{Teilchenzahl} & \psi = 1\,, & \\ \text{Impuls} & \psi = m\, c_\mu & (\mu = 1, 2, 3)\,, \\ \text{Energie} & \psi = \frac{m}{2} c^2. & \end{array}\right\} \tag{31.4}$$

Aus (31.3a) erhält man daher mit

$$n = \int f \, d\boldsymbol{c} \qquad = \text{Teilchenkonzentration}, \ \varrho = n\, m = \text{Dichte}, \tag{31.5}$$

$$v_\mu = \frac{1}{n} \int c_\mu f \, d\boldsymbol{c} = \text{Gasgeschwindigkeit}, \tag{31.6}$$

$$\int \frac{m}{2} c^2 f \, d\boldsymbol{c} \qquad = \text{Energie pro Volumeneinheit im Laborsystem} \tag{31.7}$$

die Erhaltungsgleichung für die Teilchenanzahl (Kontinuitätsgleichung)

$$\frac{\partial n}{\partial t} + \frac{\partial n v_\mu}{\partial x_\mu} = 0 \quad \text{oder} \quad \frac{\partial \varrho}{\partial t} + \frac{\partial \varrho v_\mu}{\partial x_\mu} = 0, \tag{31.8}$$

die Bewegungsgleichung

$$\frac{\partial \varrho v_\mu}{\partial t} + \frac{\partial}{\partial x_\nu} \int m\, c_\mu c_\nu f \, d\boldsymbol{c} = \varrho\, b_\mu \tag{31.9}$$

und die Energiegleichung

$$\frac{\partial}{\partial t} \int \frac{m}{2} c^2 f \, d\boldsymbol{c} + \frac{\partial}{\partial x_\mu} \int \frac{m}{2} c^2 c_\mu f \, d\boldsymbol{c} = \varrho\, b_\mu v_\mu. \tag{31.10}$$

Die Gln. (31.8) bis (31.10) sollen noch umgeschrieben werden. Wir führen die „Eigengeschwindigkeit"

$$C_\mu = c_\mu - v_\mu \tag{31.11}$$

der Atome ein, d.i. ihre Geschwindigkeit im lokalen Ruhsystem des Gases, und finden hiermit für den Impuls- und den Energiestrom aus (31.9) und (31.10)

$$\int m\, c_\mu\, c_\nu\, f\, d\boldsymbol{c} = \varrho\, v_\mu\, v_\nu + p_{\mu\nu}, \tag{31.12}$$

$$\int \frac{m}{2} c^2\, c_\mu\, f\, d\boldsymbol{c} = q_\mu + p_{\mu\nu}\, v_\nu + \varrho\left(u + \frac{1}{2} v^2\right) v_\mu. \tag{31.13}$$

Dabei wurden eingeführt der Drucktensor

$$p_{\mu\nu} = \int m\, C_\mu\, C_\nu\, f\, d\boldsymbol{C}, \tag{31.14}$$

die spezifische Energie im lokalen, mitbewegten Bezugsystem, bzw. die Temperatur durch

$$u = \frac{3}{2}\frac{\mathrm{k}T}{m} = \frac{1}{\varrho}\int \frac{m}{2} C^2\, f\, d\boldsymbol{C} \tag{31.15}$$

und der Wärmestrom

$$q_\mu = \int \frac{m}{2} C^2\, C_\mu\, f\, d\boldsymbol{C}. \tag{31.16}$$

Ferner bedienen wir uns der Abkürzung

$$\frac{da}{dt} = \frac{\partial a}{\partial t} + v_\mu \frac{\partial a}{\partial x_\mu} \tag{31.17}$$

für den „substantiellen" Differentialquotienten irgendeiner in dem Gaskontinuum definierten Funktion $a(t, \boldsymbol{r})$ nach der Zeit.

Nach kurzer Umformung bekommt man hiermit die Beziehungen (31.8) bis (31.10) in der Gestalt

$$\frac{1}{n}\frac{dn}{dt} = -\frac{\partial v_\mu}{\partial x_\mu} \quad \text{oder} \quad \frac{1}{\varrho}\frac{d\varrho}{dt} = -\frac{\partial v_\mu}{\partial x_\mu} \tag{31.18}$$

$$\varrho\,\frac{dv_\mu}{dt} = -\frac{\partial p_{\mu\nu}}{\partial x_\nu} + \varrho\, b_\mu \tag{31.19}$$

$$\varrho\,\frac{d}{dt}\left(u + \tfrac{1}{2} v^2\right) = -\frac{\partial q_\mu}{\partial x_\mu} - \frac{\partial}{\partial x_\mu}(p_{\mu\nu}\, v_\nu) + \varrho\, b_\mu\, v_\mu. \tag{31.20}$$

Aus (31.20) erhält man durch Subtraktion der mit v_μ multiplizierten Gl. (31.19) auch

$$\varrho\,\frac{du}{dt} = -\frac{\partial q_\mu}{\partial x_\mu} - p_{\mu\nu}\frac{\partial v_\mu}{\partial x_\nu}. \tag{31.21}$$

Dies sind die Erhaltungssätze des Gases in der Form, wie sie in Ziff. 2 angegeben wurden. Sie sind physikalisch unmittelbar einleuchtend: die rechte Seite von (31.18) bedeutet bis aufs Vorzeichen die relative Änderung pro Zeiteinheit des mitbewegten Volumens pro Atom; die rechte Seite von (31.19) stellt die Kraft dar, welche die in der Volumeneinheit befindliche Masse ϱ beschleunigt; die rechte Seite von (31.21) gibt die der Masse ϱ in der Zeiteinheit zugeströmte Wärme und die vom Druck an ihr geleistete Arbeit.

Eine Bemerkung über den Drucktensor ist noch zu machen. Es werde durch

$$p = n\,\mathrm{k}T \tag{31.22}$$

rein rechnerisch der „statische“ Druck p eingeführt. Speziell im thermischen Gleichgewicht gilt also $p_{\mu\nu} = p\,\delta_{\mu\nu}$ zufolge der idealen Zustandsgleichung. Nach (31.14) und (31.15) ist allgemein und exakt

$$\tfrac{1}{3}\,p_{\mu\mu} = p. \tag{31.23}$$

Offenbar beruht dieser Zusammenhang wesentlich darauf, daß bei einatomigen Gasen zur spezifischen Energie gemäß (31.15) nur die translatorische kinetische Energie der Atome beiträgt. Aus (31.23) ergibt sich somit ohne weitere Rechnung nach (2.12) das Verschwinden der Volumviscosität einatomiger Gase. Bei mehratomigen Gasen entfällt (31.15) und daher auch (31.23); diese zeigen eine Volumviscosität, s. Ziff. 96 und 99.

Für die Diskussion der Wärmeleitung im ruhenden oder im Verhältnis zur Schallgeschwindigkeit langsam strömenden Gas ist es zweckmäßig die Energiegleichung (31.21) noch einmal umzuformen durch Einführung der spezifischen Enthalpie

$$i = u + \frac{p}{\varrho} = \frac{5}{2}\,\frac{\mathrm{k}T}{m}. \tag{31.24}$$

Nach (31.17) und (31.18) gilt

$$\varrho\,\frac{d}{dt}\,\frac{p}{\varrho} = \frac{\partial p}{\partial t} + \frac{\partial v_\mu p}{\partial x_\mu} = \frac{dp}{dt} + p\,\frac{\partial v_\mu}{\partial x_\mu}.$$

Addition dieser Gleichung zu (31.21) liefert

$$\varrho\,\frac{di}{dt} - \frac{dp}{dt} = -\frac{\partial q_\mu}{\partial x_\mu} - (p_{\mu\nu} - p\,\delta_{\mu\nu})\,\frac{\partial v_\mu}{\partial x_\nu}. \tag{31.25}$$

Bei langsamer Strömung kann man die Reibungsarbeit, d.i. das zweite Glied auf der rechten Seite von (31.25) vernachlässigen, und ferner ist dann nach (31.19)

$$\frac{\partial p}{\partial x_\mu} \approx \varrho\, b_\mu.$$

Somit hat man für das einatomige Gas

$$\frac{5}{2}\,n\,\mathrm{k}\,\frac{dT}{dt} - \frac{\partial p}{\partial t} \approx -\frac{\partial q_\mu}{\partial x_\mu} + \varrho\, b_\mu\, v_\mu. \tag{31.26}$$

Bei stationärer Strömung ist überdies $\partial p/\partial t = 0$, ebenso bei einem zwischen festen adiabatischen Wänden eingeschlossenen Gas. Gl. (31.26) zeigt daher, daß es bei der Wärmeleitung in Gasen vorwiegend auf die spezifische Wärme bei konstantem Druck, $\frac{5}{2}\mathrm{k}$ pro Atom, ankommt. Der letzte Term rechts in (31.26) gibt die bei der Strömung durch das äußere Kraftfeld bewirkte Temperaturänderung wieder; ein Beispiel ist die Abkühlung einer im Schwerefeld aufsteigenden Luftmasse.

32. Die Erhaltungssätze für Gasgemische. α) *Die allgemeine Transportgleichung.* Für Funktionen $\Psi_i(t, \boldsymbol{r}, \boldsymbol{c})$, welche für jede Komponente i $(i = 1, \ldots, K)$ definiert seien, entnimmt man aus den Boltzmann-Gleichungen (29.2) für Gemische

$$\int \Psi_i\,\mathscr{D}_i(f_i)\,d\boldsymbol{c} = \sum_j \iiint \Psi_i(\boldsymbol{c})\,[f_i(\boldsymbol{c}')\,f_j(\boldsymbol{c}_1') - f_i(\boldsymbol{c})\,f_j(\boldsymbol{c}_1)]\,\sigma_{ij}\,g\,d\boldsymbol{e}'\,d\boldsymbol{c}\,d\boldsymbol{c}_1.$$

$\mathscr{D}_i(f_i)$ ist die Abkürzung aus (29.1). Durch die Umbenennung $(\boldsymbol{c}, \boldsymbol{c}_1; \boldsymbol{c}', \boldsymbol{c}_1') \to (\boldsymbol{c}', \boldsymbol{c}_1'; \boldsymbol{c}, \boldsymbol{c}_1)$ im ersten Teilintegral auf der rechten Seite entsteht unter Benutzung von (22.22)

$$\int \Psi_i\,\mathscr{D}_i(f_i)\,d\boldsymbol{c} = \sum_j \iiint [\Psi_i(\boldsymbol{c}') - \Psi_i(\boldsymbol{c})]\,f_i(\boldsymbol{c})\,f_j(\boldsymbol{c}_1)\,\sigma_{ij}\,g\,d\boldsymbol{e}'\,d\boldsymbol{c}\,d\boldsymbol{c}_1.$$

Jetzt werde auf beiden Seiten dieser Gleichung auch über den Index i summiert und dann seien rechts die Umbenennungen $(\boldsymbol{c}, \boldsymbol{c}_1; \boldsymbol{c}', \boldsymbol{c}_1') \to (\boldsymbol{c}_1, \boldsymbol{c}; \boldsymbol{c}_1', \boldsymbol{c}')$ sowie $(ij) \to (ji)$ vorgenommen. Indem man die so erhaltene Gleichung zu der ursprünglichen addiert, erhält man als Verallgemeinerung von (31.1)

$$\left.\begin{aligned} &\sum_i \int \Psi_i \mathcal{D}_i(f_i)\, d\boldsymbol{c} \\ &= \tfrac{1}{2} \sum_{ij} \iiint [\Psi_i(\boldsymbol{c}') + \Psi_j(\boldsymbol{c}_1') - \Psi_i(\boldsymbol{c}) - \Psi_j(\boldsymbol{c}_1)]\, f_i(\boldsymbol{c})\, f_j(\boldsymbol{c}_1)\, \sigma_{ij}\, g\, d\boldsymbol{e}'\, d\boldsymbol{c}\, d\boldsymbol{c}_1. \end{aligned}\right\} \tag{32.1}$$

Das ist die Transportgleichung für das Gemisch.

β) Die Erhaltungssätze im Laboratoriumssystem. Bilden speziell die $\Psi_i = \psi_i$ einen Satz von Invarianten aller im Gemisch möglichen Zweierstöße, d.h. daß

$$\psi_i(\boldsymbol{c}) + \psi_j(\boldsymbol{c}_1) = \psi_i(\boldsymbol{c}') + \psi_j(\boldsymbol{c}_1') \qquad (i, j = 1, \ldots, K) \tag{32.2}$$

gilt, so hat man aus (32.1) als Analogon zu (31.3)

$$\sum_i \int \psi_i \mathcal{D}_i(f_i)\, d\boldsymbol{c} = 0 \tag{32.3}$$

oder ausführlich, wenn speziell $\boldsymbol{b}_i$ nicht von $\boldsymbol{c}$ abhängt,

$$\frac{\partial}{\partial t} \sum_i \int \psi_i f_i\, d\boldsymbol{c} + \frac{\partial}{\partial x_\mu} \sum_i \int \psi_i c_\mu f_i\, d\boldsymbol{c} = \sum_i b_{\mu i} \int \frac{\partial \psi_i}{\partial c_\mu} f_i\, d\boldsymbol{c}. \tag{32.3a}$$

Es gibt $(K+4)$ linear unabhängige Sätze ψ_i von Stoßinvarianten:

$$\left.\begin{aligned} &\text{a) Teilchenzahl jeder Komponente } k\text{:} \\ &\qquad \psi_i = \delta_i^{(k)} \quad (= 1 \text{ oder } 0 \text{ für } i = k \text{ oder } \neq k;\ k = 1, \ldots, K), \\ &\text{b) Impuls} \quad \psi_i = m_i c_\mu, \\ &\text{c) Energie} \quad \psi_i = \frac{m_i}{2} c^2. \end{aligned}\right\} \tag{32.4}$$

Diese Stoßinvarianten hat man der Reihe nach in (32.3a) einzusetzen, um die Erhaltungssätze des Gasgemisches zu bekommen.

Es sei zunächst die Teilchenzahl betrachtet. Wir setzen $\psi_i = \delta_i^{(k)}$ aus (32.4) in (32.3a) ein und schreiben nach Ausführung der Summation wieder i statt k. Mit den Definitionen der Teilchenkonzentration, der Dichte und der mittleren Geschwindigkeit für jede Komponente

$$n_i = \int f_i\, d\boldsymbol{c}, \quad \varrho_i = n_i m_i, \quad v_{\mu i} = \frac{1}{n_i} \int c_\mu f_i\, d\boldsymbol{c} \tag{32.5}$$

entstehen die Kontinuitätsgleichungen

$$\frac{\partial n_i}{\partial t} + \frac{\partial n_i v_{\mu i}}{\partial x_\mu} = 0 \quad \text{oder auch} \quad \frac{\partial \varrho_i}{\partial t} + \frac{\partial \varrho_i v_{\mu i}}{\partial x_\mu} = 0. \tag{32.6}$$

Für das Weitere ist es zweckmäßig hier auch gleich Teilchenkonzentration n und Dichte ϱ des Gemisches, sowie dessen mittlere Atommasse m einzuführen durch

$$n = \sum_i n_i, \quad \varrho = \sum_i \varrho_i = n\, m. \tag{32.7}$$

Als nächstes sei der Impuls betrachtet. Beim Einsetzen von $\psi_i = m_i c_\mu$ in (32.3a) tritt die Impulsdichte des Gases $\sum_i \varrho_i v_{\mu i}$ auf; sie führt zur Definition der lokalen, mittleren Massengeschwindigkeit des Gases oder der Gasgeschwindigkeit $\boldsymbol{v}(t, \boldsymbol{r})$ schlechthin:

$$v_\mu = \frac{1}{\varrho} \sum_i \varrho_i v_{\mu i} = \frac{1}{m n} \sum_i \int m_i c_\mu f_i\, d\boldsymbol{c}. \tag{32.8}$$

Mit dieser Definition erhält man aus (32.3a) die Bewegungsgleichung des Gases

$$\frac{\partial \varrho v_\mu}{\partial t} + \frac{\partial}{\partial x_\nu} \sum_i \int m_i c_\mu c_\nu f_i \, d\boldsymbol{c} = \sum_i \varrho_i b_{\mu i}. \tag{32.9}$$

Mittels der letzten Stoßinvarianten, der Energie, folgt aus (32.3a) die Energiegleichung in der Gestalt

$$\frac{\partial}{\partial t} \sum_i \int \frac{m_i}{2} c^2 f_i \, d\boldsymbol{c} + \frac{\partial}{\partial x_\mu} \sum_i \int \frac{m_i}{2} c^2 c_\mu f_i \, d\boldsymbol{c} = \sum_i \varrho_i b_{\mu i} v_{\mu i}. \tag{32.10}$$

γ) Die Erhaltungssätze im Bezugssystem der lokalen, mittleren Massengeschwindigkeit. Es sollen an Hand der „Eigengeschwindigkeit“ der Atome

$$C_\mu = c_\mu - v_\mu \tag{32.11}$$

ähnliche Umschreibungen der Gln. (32.6), (32.9) und (32.10) vorgenommen werden wie beim reinen Gas. Wir führen dazu entsprechend (32.11) noch die Diffusionsgeschwindigkeit

$$V_{\mu i} = v_{\mu i} - v_\mu = \int C_\mu \frac{f_i}{n_i} \, d\boldsymbol{C} \tag{32.12}$$

der Komponente i in dem mit der Geschwindigkeit $\boldsymbol{v}$ bewegten Bezugssystem ein. Nach (32.7) und (32.8) gilt

$$\sum_i \varrho_i V_{\mu i} = 0. \tag{32.13}$$

Ferner seien der Drucktensor

$$p_{\mu\nu} = \sum_i \int m_i C_\mu C_\nu f_i \, d\boldsymbol{C}, \tag{32.14}$$

die spezifische Energie im lokalen mitbewegten $\boldsymbol{v}$-System bzw. die Temperatur

$$u = \frac{3}{2} \frac{kT}{m} = \frac{1}{\varrho} \sum_i \int \frac{m_i}{2} C^2 f_i \, d\boldsymbol{C} \tag{32.15}$$

und der Wärmestrom im $\boldsymbol{v}$-System

$$q_\mu = \sum_i \int \frac{m_i}{2} C^2 C_\mu f_i \, d\boldsymbol{C} \tag{32.16}$$

definiert.

Anwendung dieser Definitionen ergibt zunächst anstatt (32.6) die Kontinuitätsgleichungen in der Form

$$\frac{d n_i}{d t} = - n_i \frac{\partial v_\mu}{\partial x_\mu} - \frac{\partial (n_i V_{\mu i})}{\partial x_\mu} \quad \text{oder} \quad \frac{d \varrho_i}{d t} = - \varrho_i \frac{\partial v_\mu}{\partial x_\mu} - \frac{\partial (\varrho_i V_{\mu i})}{\partial x_\mu}. \tag{32.17}$$

Wegen (32.13) folgt daraus übrigens durch Summation

$$\frac{d \varrho}{d t} = - \varrho \frac{\partial v_\mu}{\partial x_\mu}. \tag{32.18}$$

Der substantielle Differentialquotient d/dt ist weiterhin formal durch (31.17) gegeben.

Aus (32.9) folgt sodann mittels der Identität

$$\sum_i \int m_i c_\mu c_\nu f_i \, d\boldsymbol{c} = p_{\mu\nu} + \varrho v_\mu v_\nu$$

und durch Benutzung von (32.18) die Bewegungsgleichung in der Gestalt

$$\varrho \frac{d v_\mu}{d t} = - \frac{\partial p_{\mu\nu}}{\partial x_\nu} + \sum_i \varrho_i b_{\mu i}. \tag{32.19}$$

Um schließlich die Energiegleichung (32.10) umzuformen, hat man die Identitäten

$$\sum_i \int \frac{m_i}{2} c^2 f_i \, d\mathbf{c} = \varrho \left(u + \frac{1}{2} v^2\right),$$

$$\sum_i \int \frac{m_i}{2} c^2 c_\mu f_i \, d\mathbf{c} = q_\mu + p_{\mu\nu} v_\nu + \varrho \left(u + \frac{1}{2} v^2\right) v_\mu,$$

sowie (32.18) zu benutzen. Dies gibt die Energiegleichung zunächst in der Form

$$\varrho \frac{d}{dt}\left(u + \frac{1}{2} v^2\right) = -\frac{\partial}{\partial x_\mu}(q_\mu + p_{\mu\nu} v_\nu) + \sum_i \varrho_i b_{\mu i} v_{\mu i}, \tag{32.20}$$

woraus durch Verwendung von (32.19) folgt

$$\varrho \frac{du}{dt} = -\frac{\partial q_\mu}{\partial x_\mu} - p_{\mu\nu} \frac{\partial v_\nu}{\partial x_\mu} + \sum_i \varrho_i b_{\mu i} V_{\mu i}. \tag{32.21}$$

Die früheren Beziehungen (31.20) und (31.21) sind Spezialfälle der beiden letzten Gleichungen. Eine etwas andere Schreibweise der Energiegleichung (32.21) erhält man noch, indem man $\frac{3}{2} \frac{\mathrm{k} T}{m}$ anstelle von u schreibt und dm/dt ersetzt. Es ist nach (32.17)

$$\frac{dn}{dt} = -n \frac{\partial v_\mu}{\partial x_\mu} - \frac{\partial}{\partial x_\mu} \sum_i n_i V_{\mu i},$$

woraus nach (32.18) folgt

$$\frac{1}{m} \frac{dm}{dt} = \frac{1}{n} \frac{\partial}{\partial x_\mu} \sum_i n_i V_{\mu i}. \tag{32.22}$$

Setzt man dies in (32.21) ein, so entsteht

$$\frac{3}{2} n \mathrm{k} \frac{dT}{dt} = -\frac{\partial q_\mu}{\partial x_\mu} + \frac{3}{2} \mathrm{k} T \frac{\partial}{\partial x_\mu} \sum_i n_i V_{\mu i} - p_{\mu\nu} \frac{\partial v_\mu}{\partial x_\nu} + \sum_i \varrho_i b_{\mu i} V_{\mu i}. \tag{32.23}$$

Auch in dieser Form einer Temperaturgleichung findet sich der Energiesatz in der Literatur.

Wie vorher beim einfachen Gas, soll im Hinblick auf die Diskussion der Wärmeleitvorgänge in (32.21) die spezifische Enthalpie gemäß (31.24) eingeführt werden. Wegen

$$\varrho \frac{d}{dt} \frac{p}{\varrho} = \frac{\partial p}{\partial t} + \frac{\partial v_\mu p}{\partial x_\mu} = \frac{dp}{dt} + p \frac{\partial v_\mu}{\partial x_\mu}$$

ergibt sich aus (32.21) die Enthalpiegleichung

$$\varrho \frac{di}{dt} - \frac{dp}{dt} = -\frac{\partial q_\mu}{\partial x_\mu} - (p_{\mu\nu} - p\,\delta_{\mu\nu}) \frac{\partial v_\mu}{\partial x_\nu} + \sum_i \varrho_i b_{\mu i} V_{\mu i}. \tag{32.24}$$

Durch Anwendung von (32.22) kann diese auch als Temperaturgleichung geschrieben werden, nämlich

$$\frac{5}{2} n \mathrm{k} \frac{dT}{dt} - \frac{dp}{dt} = -\frac{\partial q_\mu}{\partial x_\mu} + \frac{5}{2} \mathrm{k} T \frac{\partial}{\partial x_\mu} \sum_i n_i V_{\mu i} - (p_{\mu\nu} - p\,\delta_{\mu\nu}) \frac{\partial v_\mu}{\partial x_\nu} + \sum_i \varrho_i b_{\mu i} V_{\mu i}. \tag{32.25}$$

Bei langsamer Strömung ist die Reibung zu vernachlässigen und nach (32.19) gilt dann

$$\frac{\partial p}{\partial x_\mu} = \sum_i \varrho_i b_{\mu i}.$$

Daher hat man in diesem Fall

$$\frac{5}{2} n \mathsf{k} \frac{dT}{dt} - \frac{\partial p}{\partial t} \approx -\frac{\partial q_\mu}{\partial x_\mu} + \frac{5}{2} \mathsf{k} T \frac{\partial}{\partial x_\mu} \sum_i n_i V_{\mu i} + \sum_i \varrho_i b_{\mu i} v_{\mu i} \tag{32.26}$$

als genäherte Temperaturgleichung.

δ) Diffusion und Wärmeleitung, vom Bezugssystem der lokalen mittleren Teilchengeschwindigkeit aus betrachtet. Neben dem in (γ) betrachteten Bezugssystem, in dem die mittlere Massengeschwindigkeit $\boldsymbol{v}$ verschwindet, hat man bei Gemischen als gleichberechtigt das lokale Koordinatensystem, in welchem die mittlere Teilchengeschwindigkeit $\boldsymbol{w}$ verschwindet. Das $\boldsymbol{v}$-System ist angemessen zur Beschreibung von Beschleunigungs- und Reibungsvorgängen. Das $\boldsymbol{w}$-System legt man zweckmäßig bei der Beschreibung von Diffusion und Wärmeleitung in langsam bewegten Gasen zugrunde, wo Beschleunigung und Reibung eine untergeordnete Rolle spielen.

Als Beispiel sei die Diffusion in einem Gasgemisch zwischen festen Wänden betrachtet. Wenn die Komponenten verschiedenes Molekulargewicht haben, ist bei Beginn und während der Diffusion die Dichte räumlich inhomogen und der Schwerpunkt der gesamten Gasmasse verschiebt sich langsam, so daß die Geschwindigkeit $\boldsymbol{v}$ während der Diffusion ständig etwas von Null verschieden ist. Die Schwerpunktsverschiebung ist von einem Druckgefälle begleitet. Dieses ist aber, wenn es sich nicht gerade um enge Capillaren handelt, sehr klein, derart, daß die mittlere Teilchenkonzentration n während des ganzen Diffusionsvorgangs praktisch konstant bleibt (vom Diffusionsthermoeffekt sei der Einfachheit halber abgesehen). Exakte Konstanz der mittleren Teilchenkonzentration würde heißen, daß die mittlere Teilchen-Geschwindigkeit $\boldsymbol{w}$ exakt verschwindet. Demnach ist es angebracht für die Beschreibung der Diffusionsprozesse n und $\boldsymbol{w}$ anstelle von ϱ und $\boldsymbol{v}$ zu verwenden, wie es schon in Ziff. 3 geschehen ist.

Es sollen im folgenden der Zusammenhang zwischen den Diffusionsgeschwindigkeiten in beiden Systemen sowie die Kontinuitätsgleichung im $\boldsymbol{w}$-System aufgestellt werden; anschließend geschieht das Entsprechende für den Wärmestrom.

Diffusionsgeschwindigkeiten $\boldsymbol{W}_i$; Kontinuitätsgleichungen. Die mittlere Teilchengeschwindigkeit $\boldsymbol{w}$ ist, in sinngemäßer Abänderung der Definition (32.8) der mittleren Massengeschwindigkeit $\boldsymbol{v}$, durch

$$w_\mu = \frac{1}{n} \sum_i n_i v_{\mu i} = v_\mu + \sum_i \gamma_i V_{\mu i} = v_\mu + \frac{1}{n} \int C_\mu \sum_i f_i \, d\boldsymbol{C} \tag{32.27}$$

gegeben [vgl. auch (32.12)]. Die Diffusionsgeschwindigkeiten im lokal mit $\boldsymbol{w}$ mitbewegten Bezugssystem sind

$$W_{\mu i} = v_{\mu i} - w_\mu = V_{\mu i} - \sum_l \gamma_l V_{\mu l} = \int C_\mu \left(\frac{f_i}{n_i} - \frac{1}{n} \sum_l f_l \right) d\boldsymbol{C}. \tag{32.28}$$

Durch eine mit der Geschwindigkeit $\boldsymbol{w}$ bewegte Fläche treten insgesamt keine Teilchen; es gilt

$$\sum_i n_i W_{\mu i} = 0. \tag{32.29}$$

Auch die Umkehrung von (32.27) und (32.28) wird benötigt. Nach (32.28) und wegen (32.13) ist

$$\sum_i \varrho_i W_{\mu i} = -\varrho \sum_l \gamma_l V_{\mu l}.$$

Also hat man nach (32.27)

$$v_\mu = w_\mu + \sum_i \frac{\varrho_i}{\varrho} W_{\mu i} \tag{32.30}$$

und nach (32.28)

$$V_{\mu i} = W_{\mu i} - \sum_l \frac{\varrho_l}{\varrho} W_{\mu l}. \tag{32.31}$$

Nun seien die Kontinuitätsgleichungen mittels der Diffusionsgeschwindigkeiten $\boldsymbol{W}_i$ formuliert. Wir schreiben uns (32.6) gemäß (32.28) auf:

$$\frac{\partial n_i}{\partial t} + \frac{\partial}{\partial x_\mu} n_i (w_\mu + W_{\mu i}) = 0.$$

Durch Summation folgt daraus nach (32.29)

$$\frac{\partial n}{\partial t} + \frac{\partial}{\partial x_\mu} n w_\mu = 0. \tag{32.32}$$

Für den Molenbruch $\gamma_i = n_i/n$ erhält man danach

$$\left(\frac{d\gamma_i}{dt}\right)_w = -\frac{1}{n} \frac{\partial}{\partial x_\mu} n_i W_{\mu i}. \tag{32.33}$$

Diese Gleichung zusammen mit (32.32), wofür auch

$$\left(\frac{dn}{dt}\right)_w = -n \frac{\partial w_\mu}{\partial x_\mu} \tag{32.34}$$

geschrieben werden kann, bildet das Äquivalent von (32.17). Anstelle des auf $\boldsymbol{v}$ bezogenen Differentialquotienten $\frac{d}{dt} = \frac{d}{dt} + v_\mu \frac{\partial}{\partial x_\mu}$ aus (31.17), steht jetzt der auf $\boldsymbol{w}$ bezogene Differentialquotient

$$\left(\frac{d}{dt}\right)_w = \frac{\partial}{\partial t} + w_\mu \frac{\partial}{\partial x_\mu} \tag{32.35}$$

Wärmestrom und Temperaturgleichung. Im $\boldsymbol{w}$-System findet insgesamt kein Teilchentransport statt [s. Gl. (32.29)]. Man wird also den Wärmestrom im $\boldsymbol{w}$-System erhalten, indem man vom Wärmestrom im $\boldsymbol{v}$-System den Enthalpiestrom durch Teilchendiffusion im $\boldsymbol{v}$-System abzieht:

$$q_\mu^{(w)} = q_\mu - \frac{5}{2} p (w_\mu - v_\mu) = q_\mu - \frac{5}{2} \mathrm{k} T \sum_i n_i V_{\mu i} = \sum_i \int \left(\frac{m_i}{2} C^2 - \frac{5}{2} \mathrm{k} T\right) C_\mu f_i \, d\boldsymbol{C}. \tag{32.36}$$

Die Zweckmäßigkeit oder Zwangsläufigkeit dieser Abspaltung wird sich später (Ziff. 57) deutlich zeigen. Damit nimmt die Temperaturgleichung (23.52) unter Verwendung der Differentiation (32.25) die Gestalt an:

$$\frac{5}{2} n \mathrm{k} \left(\frac{dT}{dt}\right)_w - \frac{dp}{dt} = -\frac{\partial q_\mu^{(w)}}{\partial x_\mu} - (p_{\mu\nu} - p\,\delta_{\mu\nu}) \frac{\partial v_\mu}{\partial x_\nu} + \sum_i \varrho_i b_{\mu i} V_{\mu i}. \tag{32.37}$$

In die Gln. (32.33) und (32.37) wird in Ziff. 56 und 57 die hydrodynamische Näherung für die Diffusionsgeschwindigkeiten und den Wärmestrom eingesetzt und so die ausgefüllte Form der Diffusions- und Temperaturgleichungen erhalten.

33. Entropiesatz (H-Theorem). Außer den Erhaltungssätzen läßt sich aus der Boltzmann-Gleichung eine weitere allgemeine Beziehung herleiten, welche die Form einer Ungleichung hat und den zweiten Hauptsatz für Gase darstellt.

Wir formulieren den gastheoretischen Entropiesatz, der von BOLTZMANN (1872) entdeckt und als H-Theorem bezeichnet wurde, gleich für Gasgemische.

Mit der spezifischen Entropie[1]

$$s = -\frac{\mathsf{k}}{\varrho}\sum_i \int f_i \log(f_i\,\alpha_i)\,d\boldsymbol{c}, \tag{33.1}$$

wo die α_i beliebige Konstanten sind von der Dimension f^{-1}, bekommt man wegen

$$[\log(f_i\,\alpha_i) + 1]\,d f_i = d\,[f_i \log(f_i\,\alpha_i)]$$

aus den Boltzmann-Gleichungen (29.2) durch Multiplikation mit $-\mathsf{k}(\log(f_i\,\alpha_i)+1)$, Integration über $\boldsymbol{c}$, Summation über i und ähnliche Umformungen, wie sie zu (32.1) führten[2], den Entropiesatz in der Form

$$\left.\begin{aligned}&\frac{\partial\varrho s}{\partial t} + \frac{\partial}{\partial x_\mu}\Big(-\mathsf{k}\sum_i\int c_\mu f_i \log(f_i\,\alpha_i)\,d\boldsymbol{c}\Big)\\ &= \frac{\mathsf{k}}{4}\sum_{ij}\iiint [f_i(\boldsymbol{c}')\,f_j(\boldsymbol{c}_1') - f_i(\boldsymbol{c})\,f_j(\boldsymbol{c}_1)][\log f_i(\boldsymbol{c}')\,f_j(\boldsymbol{c}_1') - \log f_i(\boldsymbol{c})\,f_j(\boldsymbol{c}_1)]\times\\ &\quad\times\sigma_{ij}\,g\,d\boldsymbol{e}'\,d\boldsymbol{c}\,d\boldsymbol{c}_1 \geqq 0.\end{aligned}\right\} \tag{33.2}$$

Dieser Integralausdruck, die lokale Entropieerzeugung, kann nicht negativ sein, da wegen

$$(x-y)(\log x - \log y) > 0 \quad \text{außer für } x = y$$

die Integranden nicht negativ sind.

Für die Entropie S des Gases, die als additive Größe durch

$$S = \int \varrho\, s\, d\boldsymbol{r}$$

definiert wird, gilt also bei räumlich homogenem Zustand

$$\frac{dS}{dt} \geqq 0. \tag{33.3}$$

---

[1] Dieser Entropieausdruck von BOLTZMANN ist der einzig mögliche, welcher die Anforderungen des zweiten Hauptsatzes erfüllt. Definiert man — für reines Gas — die Entropie zunächst unbestimmt durch

$$s = -\int G(f)\,d\boldsymbol{c},$$

so verlangt der zweite Hauptsatz gemäß der Boltzmann-Gleichung

$$[g(f) + g(f_1) - g(f') - g(f_1')]\,(f f_1 - f' f_1') \geqq 0.$$

Dabei ist $g = dG/df$ abgekürzt und f, f_1, f', f_1' können beliebige positive Werte annehmen. Nun sei $f' f_1'/f f_1 = x$ gesetzt. Dann muß gelten

$$g\left(\frac{f f_1}{f'}\,x\right) - [g(f) + g(f_1) - g(f')] \gtreqless 0 \quad \text{für} \quad x \gtreqless 1.$$

Wenn g stetig ist, muß für $x = 1$ gelten

$$g\left(\frac{f f_1}{f'}\right) = g(f) + g(f_1) - g(f').$$

Durch die Wahl $f' = 1$ entsteht daraus

$$g(f f_1) - g(1) = g(f) - g(1) + g(f_1) - g(1).$$

Das ist die Funktionalgleichung des Logarithmus.

[2] Auf einen wesentlichen Unterschied sei aber hingewiesen, der besonders deutlich wird, wenn man von der Boltzmann-Gleichung in der Form (30.1) bzw. (30.12) ausgeht. Für die Aufstellung der den Erhaltungssätzen zugrunde liegenden Beziehungen (32.1) benötigt man nur die triviale Symmetrie (30.3) bzw. (30.13) der Stoßwahrscheinlichkeiten W_{ij}, während für das H-Theorem in der Form (33.2) auch die Inversions-Symmetrie (30.4) bzw. (30.14) herangezogen werden muß. Deshalb läßt sich auch der obige Beweis des Theorems nicht unmittelbar auf mehratomige Gase übertragen (s. dazu Ziff. 93).

Dasselbe ergibt sich für die Entropie eines Gases, das zwischen spiegelnden, also adiabatischen Wänden[1] eingeschlossen ist, da in diesem Fall die über den Raum integrierte Divergenz des Entropiestroms, d.i. das zweite Glied links in (33.2), verschwindet. Wegen weiterer Diskussion der Entropieerzeugungsgleichung (33.2) und wegen des damit verknüpften Extremalprinzips sei auf ENSKOG (1917, S. 108), KOHLER (1947b) und SOMMERFELD ([*49*], S. 300) verwiesen. Vom rein-mathematischen Gesichtspunkt aus hat CARLEMAN (1933) das H-Theorem für ein Gas aus starren Kugeln untersucht; von MORGENSTERN (1955) wurde das verallgemeinert.

Aus (33.3) folgt, daß ein abgeschlossenes System einem Gleichgewichtszustand zustrebt. Wegen der Erhaltung von Teilchenzahl und Energie ist die Entropie beschränkt und strebt daher asymptotisch einem endlichen Maximalwert zu. In diesem Endzustand muß für die Verteilungsfunktion — wir schreiben F für das thermische Gleichgewicht — gelten

$$F_i(\mathbf{c})\, F_j(\mathbf{c}_1) = F_i(\mathbf{c}')\, F_j(\mathbf{c}_1'). \tag{33.4}$$

Daraus ergibt sich die Maxwell-Verteilung (s. nächste Ziffer). In diesem Zustand ist für einatomige Gase das detaillierte Gleichgewicht verwirklicht (detailed balance), bei dem jeder Reaktion — hier elastischer Stoß — eine ebenso häufige Gegenreaktion — hier der inverse Stoß — gegenübersteht. Bei mehratomigen Gasen mit rotierenden Molekülen dagegen gibt es einen inversen Stoß im allgemeinen nicht; es liegt ein cyclisches Gleichgewicht vor (BOLTZMANN [*3*], S. 235f.; TOLMAN [*51*], S. 140; WATANABE [*54*]); s.a. Teil H dieses Artikels.

In der einseitigen Änderung der Entropie und dem Erreichen eines Gleichgewichtszustands kommt am deutlichsten zum Ausdruck, daß die irreversible statistische Mechanik, vertreten durch die Boltzmann-Gleichung, nicht exakt dasselbe ist wie die reversible Punktmechanik in der klassischen oder quantentheoretischen Form. Darauf basierte der Umkehreinwand von LOSCHMIDT (1876). Der Unterschied kommt durch die Unordnungsannahme herein, welche bedeutet, daß in der Boltzmann-Gleichung ein (praktisch aber äußerst geringfügiger) Teil der punktmechanisch möglichen Bewegungen ignoriert wird. Bei Systemen mit vielen Freiheitsgraden ist dies angemessen und läßt das Wesentliche, die makroskopische Irreversibilität, klar hervortreten. Es sei auf die Diskussion in dem Buch von TOLMAN [*51*], S. 155 hingewiesen. An neuerer Literatur zu dem Problem der Irreversibilität seien neben den schon in Ziff. 28 erwähnten Arbeiten von KÜMMEL (1955, 1956) diejenigen von VAN KAMPEN (1954), VAN HOVE (1955) und TEMPERLEY (1956) zitiert. In der letztgenannten Arbeit wird an einem einfachen Modell gezeigt, daß die makroskopische Irreversibilität auf drei wesentlichen Voraussetzungen beruht: Mitteilung über die Phasen der Wellenfunktion, große Teilchenzahl, große Zellenzahl.

Um den Übergang von der reversiblen zur irreversiblen Beschreibung an einem Beispiel zu veranschaulichen, sei das Überströmen eines Gases aus einem Vorratsbehälter A in ein leeres Gefäß B beim plötzlichen Öffnen eines Lochs in der Trennwand betrachtet. Die im ersten Augenblick austretenden schnellen Moleküle werden keine Stöße machen, sondern geradlinig davonfliegen. Dabei ändert sich die Entropie nicht, denn auch für die noch in A verbliebene Hauptmenge verschwindet das Stoßintegral wegen des anfänglichen thermischen Gleichgewichts. Dasselbe würde gelten, wenn man alle Molekülgeschwindigkeiten

---

[1] Bei der adiabatischen Wand der Thermodynamik hat man es mit demselben idealisierten Gebilde zu tun; eine Wand, die in Strenge keine Wärme durchläßt, kann nur ein idealer Spiegel für die Moleküle sein.

zu Anfang exakt umkehrte. In einem etwas späteren Zeitpunkt aber finden in B, hauptsächlich in der Nähe des Lochs, wo die Konzentration am größten ist, Stöße statt; die Entropie nimmt zu. Ändert man in diesem späteren Zeitpunkt die Lagen der Moleküle etwas ab — größenordnungsmäßig um einen Moleküldurchmesser —, so wird dies ohne merklichen Einfluß auf den weiteren Vorgang sein, wie es die mit stetiger Verteilungsfunktion operierende Boltzmann-Gleichung annimmt. Nun mögen gedanklich in dem späteren Zeitpunkt alle Geschwindigkeiten an den Molekülen (auch diejenigen etwaiger innerer Freiheitsgrade) umgekehrt werden [quantenmechanisch: $\psi(t, \boldsymbol{r}_1, \ldots) \to \psi^*(-t, \boldsymbol{r}_1, \ldots)$]. Hält man bei dem Umkehren die Lagen exakt fest, so sollte sich B wieder völlig entleeren, die Entropie abnehmen. Dieser Zustand ist aber ganz exzeptionell insofern, als seine zukünftige Entwicklung äußerst empfindlich von Änderungen der Lagen der Moleküle abhängt, sein Wesentliches also gar nicht durch eine stetige Verteilungsfunktion $f(t, \boldsymbol{r}, \boldsymbol{c})$ beschrieben werden kann. Schon Verschiebungen um einen Moleküldurchmesser würden bewirken, daß Moleküle, welche beim Austritt aus dem Loch in dessen Nähe zusammengestoßen sind, sich nun beim Rückweg verfehlen, B also nicht entleert, sondern weiter gefüllt wird, und die Entropie zunimmt, wie es die Boltzmann-Gleichung besagt. Den genannten exzeptionellen Zustand, der sich durch keine praktikable Methode realisieren läßt, kann die Boltzmann-Gleichung, die im statistischen Mittel und für stetige Verteilungen gilt, auch nicht annähernd erfassen. Das ist aber kein Mangel, sondern ein Vorzug und entspricht durchaus den experimentellen Gegebenheiten.

Wenn es also auch zweifellos berechtigt ist, bei der Betrachtung von Systemen mit vielen Freiheitsgraden den Gesichtspunkt der Unordnung einzuführen, so ist die quantitative Fassung dieses Begriffs gleichwohl ein Problem, was besonders deutlich in der Theorie der komprimierten Gase zutage tritt (vgl. auch Ziff. 28, am Schluß).

34. Das thermische Gleichgewicht. Im thermischen Gleichgewicht verschwindet die lokale Entropieerzeugung aus (33.2). Wie schon erwähnt, ist notwendig und hinreichend dafür die Bedingung (33.4) des detaillierten Gleichgewichts, die auch so

$$\log F_i(\boldsymbol{c}) + \log F_j(\boldsymbol{c}_1) = \log F_i(\boldsymbol{c}') + \log F_j(\boldsymbol{c}_1') \tag{34.1}$$

geschrieben werden kann. Wenn die Verteilungsfunktion diese Eigenschaft hat, so verschwindet das Stoßintegral in der Boltzmann-Gleichung (29.2) und es muß daher auch gelten

$$\mathcal{D}_i(F_i) = 0. \tag{34.2}$$

Durch (34.1) und (34.2) ist die Verteilungsfunktion im wesentlichen eindeutig bestimmt.

Zunächst sei bemerkt, daß Gl. (34.1), die ja bei beliebigem Stoßvektor $\boldsymbol{e}'$ in (22.17) zu gelten hat, sicher dann erfüllt ist, wenn $\log F_i$ aus einer Linearkombination der Stoßinvarianten (32.4) besteht:

$$\log F_i = \xi_i + m_i c_\mu \eta_\mu + \frac{m_i}{2} c^2 \zeta. \tag{34.3}$$

Die Faktoren ξ_i, η_μ, ζ dürfen von $t, \boldsymbol{r}$, aber nicht von $\boldsymbol{c}$ abhängen. Indem ζ durch $-1/\mathrm{k}T$, η_μ durch $v_\mu/\mathrm{k}T$ ersetzt wird und ξ_i in n_i steckt, hat man auch

$$F_i = n_i \left(\frac{m_i}{2\pi \mathrm{k}T}\right)^{\frac{3}{2}} \mathrm{e}^{-\frac{m_i C^2}{2\mathrm{k}T}}, \qquad C_\mu = c_\mu - v_\mu, \tag{34.4}$$

d.i. die Maxwellsche Geschwindigkeitsverteilung eines einatomigen Gasgemisches. Darin sind n_i die lokalen Teilchenkonzentrationen, $\boldsymbol{v}$ die lokale Geschwindigkeit und T die lokale Temperatur. Das folgt aus den für F_i geltenden Identitäten [s. dazu auch Gl. (32.14) und (32.15)]

$$\left.\begin{aligned} &\int F_i\, d\boldsymbol{c} = n_i, && \sum_i \int c_\mu F_i\, d\boldsymbol{c} = n\, v_\mu \\ &p_{\mu\nu} = \sum_i n_i\, \mathrm{k} T \cdot \delta_{\mu\nu}, && \varrho\, u = \sum_i n_i \cdot \tfrac{3}{2}\, \mathrm{k} T, \end{aligned}\right\} \tag{34.5}$$

die man durch Ausführung der Integrationen bestätigt[1].

In (34.5) ist die ideale Zustandsgleichung, das Daltonsche Gesetz von der Additivität der Partialdrucke und der Gleichverteilungssatz der Translationsenergien enthalten. — Es scheint zunächst merkwürdig, daß die Boltzmann-Gleichung, welche die Zweierstöße berücksichtigt, zur idealen und nicht zu der durch den zweiten Virialkoeffizienten erweiterten Zustandsgleichung führt. Das liegt, worauf HIRSCHFELDER *et al.* ([*25*], S. 612) hinweisen, an einer in der Boltzmann-Gleichung enthaltenen Ungenauigkeit. Diese kommt zum Vorschein, wenn man auf dichtere Gase verallgemeinert, und besteht darin, daß die mit der Wegverkürzung beim Stoß verknüpfte Erhöhung der Stoßzahl außer Acht gelassen ist. Berücksichtigt man dies, so ergibt sich die richtige nichtideale Zustandsgleichung. Für die Theorie der Transportvorgänge bei mittleren Drucken ist dies aber belanglos.

Erwähnt sei auch noch, daß sich die Einzigkeit der Lösung (34.3) in bekannter Weise aus der Maximumsforderung für die lokale Entropiedichte ergibt, bei gleichzeitiger Festhaltung der lokalen Konzentrationen sowie der Impuls- und Energiedichte.

Jetzt zur Diskussion von (34.2): *Im kräftefreien* Fall ist diese Gleichung jedenfalls erfüllt, wenn n_i, $\boldsymbol{v}$ und T von t und $\boldsymbol{r}$ unabhängig sind. (34.4) stellt dann die Geschwindigkeitsverteilung in einem homogenen, strömenden Gasgemisch dar. Wie man leicht sieht, ist es mit (34.2) auch noch verträglich, daß das Gas in dem mit der Geschwindigkeit $\boldsymbol{v}$ bewegten Bezugssystem wie ein starrer Körper um irgendeine Achse rotiert (wobei dann auch die n_i ortsabhängig werden). In dem idealisierten Fall verschwindender Volumreibung, der bei einatomigen, nicht anregbaren Molekülen vorliegt, ist sogar die allseitige gleich-

---

[1] Es seien, zugleich zwecks späterer Rückverweisung, hier einige in der Gastheorie häufig vorkommende Integrale über die Maxwell-Verteilung

$$F = n\left(\frac{\beta}{\pi}\right)^{\frac{3}{2}} \mathrm{e}^{-\beta C^2}, \quad \beta = \frac{m}{2\mathrm{k}T}$$

notiert. Es ist

$$I(p) = \int \left(\sqrt{\beta}\, C\right)^p \frac{F}{n}\, d\boldsymbol{C} = \frac{2}{\sqrt{\pi}} \left(\frac{p+1}{2}\right)!$$

oder im einzelnen:

p gerade	0	2	4	…	p
$I(p)$	1	$\frac{3}{2}$	$\frac{15}{4}$	…	$\left(2 \cdot \frac{1}{2} \cdot \frac{3}{2} \cdots \frac{p+1}{2}\right)$
p ungerade	1	3	5	…	p
$I(p)$	$\frac{2}{\sqrt{\pi}}$	$\frac{4}{\sqrt{\pi}}$	$\frac{12}{\sqrt{\pi}}$	…	$\frac{2}{\sqrt{\pi}}\left(\frac{p+1}{2}\right)!$

Daraus folgt auch

$$\int \beta\, C_\mu C_\nu \frac{F}{n}\, d\boldsymbol{C} = \frac{1}{2}\, \delta_{\mu\nu},$$

$$\int \beta^2 C_\mu C_\nu C^2 \frac{F}{n}\, d\boldsymbol{C} = \frac{5}{4}\, \delta_{\mu\nu}.$$

mäßige Expansion [vgl. (2.10a)] ein nicht-dissipativer Zustand. Es ist nicht nötig, hier weiter darauf einzugehen (s. CHAPMAN und COWLING [4], S. 79; auch MEIXNER (1942)). — Im Fall eines *äußeren Kraftfelds* ist thermisches Gleichgewicht nur möglich, wenn eine zeitunabhängige Hamilton-Funktion existiert, wie beim statischen Schwerefeld oder elektromagnetischen Feld. Es sei durch

$$b_{\mu i} = \frac{e_i}{m_i}\left(E_\mu + \frac{1}{C_L} H_{\mu\nu} c_\nu\right) \tag{34.6}$$

die Beschleunigung gegeben, wo

$$E_\mu = -\frac{\partial \Psi}{\partial x_\mu}, \tag{34.7}$$

$$H_{\mu\nu} = -H_{\nu\mu} \tag{34.8}$$

das elektrische bzw. magnetische Feld, Ψ das skalare Potential, e_i die Ladung und C_L die Lichtgeschwindigkeit bedeuten. Als Lösung von (34.1) und (34.2) verifiziert man die Maxwell-Boltzmannsche Verteilungsfunktion

$$F_i = n_i(0)\left(\frac{m_i}{2\pi \mathrm{k} T}\right)^{\frac{3}{2}} \exp\left\{-\frac{1}{\mathrm{k}T}\left[\frac{m_i}{2} c^2 + e_i\left(\Psi(\boldsymbol{r}) - \Psi(0)\right)\right]\right\}. \tag{34.9}$$

Demnach hängt die Konzentration gemäß der barometrischen Höhenformel

$$n_i = n_i(0)\, \mathrm{e}^{-e_i[\Psi(r) - \Psi(0)]/\mathrm{k}T} \tag{34.10}$$

vom Ort ab. Im Gegensatz zu (34.4) kommt im allgemeinen jetzt, außer bei Vorliegen besonderer Symmetrie, nur die ruhende Maxwell-Verteilung ($\boldsymbol{v}=0$) in Frage. Daß (34.9) die Gl. (34.2) erfüllt, beruht wesentlich auf den Beziehungen (34.7) und (34.8), wie man beim Einsetzen erkennt.

Es seien noch einige, die Maxwell-Verteilung betreffenden Zahlenangaben notiert. Aus (34.4) entnimmt man für das mittlere Quadrat der Eigengeschwindigkeit einer Komponente

$$\overline{C_i^2} = \frac{1}{n_i}\int C^2 F_i\, d\boldsymbol{c} = \frac{3\mathrm{k}T}{m_i} = \frac{3\mathrm{R}T}{M_i} \tag{34.11}$$

und für den mittleren Betrag der Eigengeschwindigkeit

$$\overline{C}_i = \frac{1}{n_i}\int C F_i\, d\boldsymbol{c} = \sqrt{\frac{8\mathrm{k}T}{\pi m_i}} = \sqrt{\frac{8\mathrm{R}T}{\pi M_i}}. \tag{34.12}$$

Darin bedeutet

$$\mathrm{R} = 8{,}314 \cdot 10^7\ \mathrm{erg/grad\cdot mol} = 1{,}986\ \mathrm{cal_{15}/grad\cdot mol} \tag{34.13}$$

die molare Gaskonstante, M_i die Molmasse. Numerisch gilt

$$\left\{\begin{matrix}\sqrt{\overline{C_i^2}}\\ \overline{C}_i\end{matrix}\right\} = \left\{\begin{matrix}2{,}61\\ 2{,}41\end{matrix}\right\}\cdot 10^5\sqrt{T/273\,M_i} \quad (\mathrm{cm/sec}), \tag{34.14}$$

wo T bzw. M_i die Maßzahlen der in °K bzw. in Gramm gemessenen Größen sind. So ist beispielsweise bei 0° C für Sauerstoff $\overline{C} = 427$ m/sec, für Wasserstoff 1700 m/sec.

Experimentell geprüft wurde die Maxwell-Verteilung von STERN (1920). Neuerdings haben E. W. BECKER und HENKES (1956) mit einer Laufzeitmethode Geschwindigkeitsverteilungen studiert, auch solche nicht-Maxwellscher Art, die sich durch seitlichen Austritt der schnellen Teilchen aus einem Molekülstrahl ausbilden.

D. Die Eigenfunktionsmethode in der kinetischen Theorie.

In den folgenden Teilen D bis G werden Anwendungen der Boltzmann-Gleichung für einatomige Gase behandelt. Nach der in Ziff. 1 gegebenen Einteilung wären dabei mäßig verdünnte Gase (Gefäßabmessung $L \gg$ freie Weglänge l) und hochverdünnte Gase ($l \gg L$) als Grenzfälle zu unterscheiden.

Bei den hochverdünnten Gasen kann man das Stoßintegral rechts in der Boltzmann-Gleichung (28.5) in erster Näherung vernachlässigen und hat es dann mit einer homogenen partiellen Differentialgleichung erster Ordnung zu tun, die mit geeigneten Randbedingungen zu lösen ist. Setzt man die so erhaltene Lösung rechts in (28.5) ein, so liegt eine inhomogene Gleichung vor und man hat den ersten Schritt des Iterationsverfahrens von JAFFÉ (1930), das einer Reihenentwicklung nach steigenden Potenzen von L/l entspricht, und das neuerdings von KELLER (1948) sowie von WANG CHANG und UHLENBECK (1949, 1950) angewandt wurde. Auf die hochverdünnten Gase können wir in diesem Artikel nicht weiter eingehen.

Bei den mäßig verdünnten Gasen sind die Stöße ausschlaggebend. Falls sich der Gaszustand raumzeitlich nicht allzu stark ändert, muß das Stoßintegral fast verschwinden, die Verteilungsfunktion lokal nahezu maxwellisch sein. Hier stehen mehrere Näherungsverfahren zur Verfügung. Einmal das Verfahren von HILBERT-ENSKOG, das auf eine Entwicklung nach steigenden Potenzen von l/L hinausläuft und dessen erster Schritt in den Teilen E und F näher diskutiert wird. Ob die höheren Näherungen konvergieren, ist nicht bekannt. Zum zweiten sei das Momentenverfahren von GRAD (1949) genannt, welches im Artikel von GRAD in diesem Band behandelt wird. Eine Darstellung dieses Verfahrens durch F. BOPP findet sich auch in den Vorlesungen von SOMMERFELD [*49*], S. 302. Schließlich kann man, in Anlehnung an MAXWELLS Begriff der Relaxation, das Problem formal durch Entwicklung nach Eigenfunktionen zu lösen suchen, was in dem vorliegenden Teil D geschehen soll.

In speziellen Fällen können noch andere Methoden anwendbar sein. So hat PEKERIS (1955) gezeigt, daß die Berechnung des Selbstdiffusionskoeffizienten für starre Kugeln exakt auf die Lösung einer gewöhnlichen linearen Differentialgleichung zweiter Ordnung (für die Viscosität von vierter Ordnung) hinausläuft, und die Lösung auch numerisch ausgeführt. Schon BOLTZMANN war im Besitz dieser Differentialgleichungen, hatte sie aber nicht gelöst.

Weil die Boltzmann-Gleichung rechnerisch so kompliziert ist, wurden in neuerer Zeit vereinfachte mathematische Modelle dafür vorgeschlagen, z.B. von BHATNAGAR, GROSS und KROOK (1954). Näheres darüber findet sich in dem Artikel von GRAD in diesem Band.

35. Die lineare Näherung der Boltzmann-Gleichung. Das Stoßintegral der Boltzmann-Gleichung ist vom zweiten Grad in der Verteilungsfunktion. Diese Nichtlinearität ist wesentlich bei Gas-Zuständen, die sehr stark vom thermischen Gleichgewicht abweichen, z.B. bei einem wärmeleitenden Gas mit großen Temperaturunterschieden oder bei den in Überschallströmungen auftretenden Stoßwellen. Im ersten Fall hat man wenigstens noch „lokal" annähernd thermisches Gleichgewicht, im zweiten Fall, bei starken Stößen, ist in der Stoßfront die Verteilungsfunktion auch nicht annähernd maxwellisch (zur kinetischen Theorie der Stoßwellen s. MOTT-SMITH (1951), ZOLLER (1951), GRAD (1952) und das Buch von PATTERSON [*41*]).

In diesem Teil D sei nun aber angenommen, daß das Gas als Ganzes sich nahezu im thermischen Gleichgewicht befindet, die raumzeitlichen Änderungen des Zustands also relativ gering sind. Dann setzt man die Verteilungsfunktion zweckmäßig in der Form an

$$f(t, \boldsymbol{r}, \boldsymbol{c}) = f^{(0)}(c)\,[1 + \Phi(t, \boldsymbol{r}, \boldsymbol{c})]. \tag{35.1}$$

Dabei bedeutet

$$f^{(0)} = n_0 \left(\frac{\beta}{\pi}\right)^{\frac{3}{2}} e^{-\beta c^2} \quad \text{mit} \quad \beta = \frac{m}{2\,\mathrm{k}\,T_0} \tag{35.2}$$

eine ruhende Maxwell-Verteilung. Diese soll die exakte Verteilungsfunktion f immer und überall gut approximieren, so daß „im Wesentlichen“, das soll heißen für Molekülgeschwindigkeiten, die nicht groß gegen die charakteristische Geschwindigkeit $\sqrt{\mathrm{k}T_0/m}$ der Temperaturbewegung sind, der Korrekturfaktor wenig von 1 verschieden ist:

$$|\Phi| \ll 1 \quad \text{für} \quad c \lessapprox \sqrt{\mathrm{k}T_0/m}. \tag{35.3}$$

Ferner soll für $c \to \infty$ das Produkt $f^{(0)}\,\Phi$ hinreichend stark verschwinden, so daß alle physikalischen Größen, wie Teilchenzahl, Geschwindigkeit, Energie, Entropie endliche Werte haben. Letzteres bedeutet, wie man durch Einsetzen von (35.1) in die Definition der Entropie (33.1) sieht, daß

$$\int \Phi^2 f^{(0)}\, d\boldsymbol{c} \quad \textit{existiert.} \tag{35.4}$$

Geht man mit dem Ansatz (35.1) in die Boltzmann-Gleichung (28.5) für einfache Gase ein, so kann man die in Φ quadratischen Glieder im Stoßintegral unterdrücken. Das Stoßintegral wird zu einem linearen Funktional:

$$I(\Phi) = \frac{1}{n_0^2} \iint f^{(0)} f_1^{(0)}\, \delta(\Phi)\, \sigma g\, d\boldsymbol{e}'\, d\boldsymbol{c}_1, \tag{35.5}$$

wo

$$\delta(\Phi) = \Phi + \Phi_1 - \Phi' - \Phi_1'. \tag{35.6}$$

Die Boltzmann-Gleichung (28.5) mit $b_\mu = 0$[1] lautet dann näherungsweise

$$f^{(0)}\left(\frac{\partial \Phi}{\partial t} + c_\mu \frac{\partial \Phi}{\partial x_\mu}\right) = -\, n_0^2\, I(\Phi). \tag{35.7}$$

Diese lineare Integrodifferentialgleichung ist im Weiteren zu untersuchen. Wir wollen hier noch zwei allgemeine Eigenschaften des Operators I konstatieren, eine dritte wird in Ziff. 36 erörtert.

Es gilt für zwei beliebige Funktionen Φ und Ψ die Identität

$$\iiint f^{(0)} f_1^{(0)}\, \Psi\, \delta(\Phi)\, \sigma g\, d\boldsymbol{e}'\, d\boldsymbol{c}\, d\boldsymbol{c}_1 = \tfrac{1}{4} \iiint f^{(0)} f_1^{(0)}\, \delta(\Psi)\, \delta(\Phi)\, \sigma g\, d\boldsymbol{e}'\, d\boldsymbol{c}\, d\boldsymbol{c}_1. \tag{35.8}$$

Diese gewinnt man ähnlich wie bei (31.1) durch geeignete Umbenennung der Integrationsvariablen unter Benutzung der Energieerhaltung $g' = g$, aus welcher auch $f^{(0)\prime} f_1^{(0)\prime} = f^{(0)} f_1^{(0)}$ folgt, und des Liouvilleschen Satzes (22.22).

Somit hat das durch

$$[\Psi, \Phi] = \int \Psi\, I(\Phi)\, d\boldsymbol{c} = \frac{1}{4 n_0^2} \iiint f^{(0)} f_1^{(0)}\, \delta(\Psi)\, \delta(\Phi)\, \sigma g\, d\boldsymbol{e}'\, d\boldsymbol{c}\, d\boldsymbol{c}_1 \tag{35.9}$$

definierte, bestimmte Integral die Symmetrie- und Linearitätseigenschaft

$$[\Psi, \Phi] = [\Phi, \Psi]; \quad [\Psi, \Phi_{\mathrm{I}} + \Phi_{\mathrm{II}}] = [\Psi, \Phi_{\mathrm{I}}] + [\Psi, \Phi_{\mathrm{II}}]. \tag{35.10}$$

Als zweite Eigenschaft werde hervorgehoben, daß der Operator I positiv-semidefinit ist, d.h. daß für beliebiges Φ stets

$$[\Phi, \Phi] = \int \Phi\, I(\Phi)\, d\boldsymbol{c} \geqq 0 \tag{35.11}$$

gilt. Auch dies ergibt sich sofort aus der Identität (35.8) durch die Bemerkung, daß für $\Psi = \Phi$ der Integrand nicht negativ sein kann, daß aber Verschwinden für alle Werte von $\boldsymbol{c}$ und $\boldsymbol{c}_1$ tatsächlich vorkommt, nämlich dann und nur dann, wenn Φ gleich einer Linearkombination der Stoßinvarianten 1; $\boldsymbol{c}$; c^2 ist.

---

[1] Wir sehen in Teil D durchwegs von einem Kraftfeld ab und beschränken uns auf reine Gase. Den entgegengesetzten Fall eines räumlich homogenen Lorentzschen Gasgemisches im äußeren Kraftfeld behandelten Bayet, Delcroix und Dénisse (1954, 1955) nach der Eigenfunktionsmethode.

36. Der linearisierte Stoßoperator — ein linearer Integraloperator mit symmetrischem Kern. Wir geben der linearisierten Boltzmann-Gleichung (35.7) mit (35.5) an Hand von (30.5) die Gestalt

$$-f^{(0)}\left(\frac{\partial \Phi}{\partial t}+\cdots\right)=n_0^2\, I(\Phi)=\iiint w(\mathbf{c},\mathbf{c}_1|\mathbf{c}',\mathbf{c}_1')\,(\Phi+\Phi_1-\Phi'-\Phi_1')\,d\mathbf{c}_1\,d\mathbf{c}',\,d\mathbf{c}_1'. \tag{36.1}$$

Dabei ist zur Abkürzung

$$w(\mathbf{c},\mathbf{c}_1|\mathbf{c}',\mathbf{c}_1')=W(\mathbf{c},\mathbf{c}_1|\mathbf{c}',\mathbf{c}_1')\,[f^{(0)}f_1^{(0)}f^{(0)\prime}f_1^{(0)\prime}]^{\frac{1}{2}}\geq 0 \tag{36.2}$$

geschrieben. Diese Funktion einzuführen ist zweckmäßig, da auch w wieder die Symmetrien

$$w(\mathbf{c},\mathbf{c}_1|\mathbf{c}',\mathbf{c}_1')=w(\mathbf{c}_1,\mathbf{c}|\mathbf{c}_1',\mathbf{c}')\quad\text{und}\quad w(\mathbf{c},\mathbf{c}_1|\mathbf{c}',\mathbf{c}_1')=w(\mathbf{c}',\mathbf{c}_1'|\mathbf{c},\mathbf{c}_1) \tag{36.3}$$

besitzt, vgl. (30.3a) und (30.4).

Es ist rasch zu sehen, daß I eng verknüpft ist mit einem linearen Integraloperator mit symmetrischem Kern. Wir bemerken dazu, daß entsprechend den vier Summanden von $\Phi+\Phi_1-\Phi'-\Phi_1'$ auch I in vier Integrale zerfällt, und benennen die Integrationsvariablen $(\mathbf{c}_1,\mathbf{c}',\mathbf{c}_1')$ in dem Φ'-Integral um in $(\mathbf{c}',\mathbf{c}_1,\mathbf{c}_1')$, in dem Φ_1'-Integral um in $(\mathbf{c}',\mathbf{c}_1',\mathbf{c}_1)$. Dadurch wird aus (36.1)

$$\left.\begin{aligned}n_0^2\, I(\Phi)&=\Phi\cdot\iiint w(\mathbf{c},\mathbf{c}_1|\mathbf{c}',\mathbf{c}_1')\,d\mathbf{c}_1\,d\mathbf{c}'\,d\mathbf{c}_1+\\&+\iiint[w(\mathbf{c},\mathbf{c}_1|\mathbf{c}',\mathbf{c}_1')-w(\mathbf{c},\mathbf{c}'|\mathbf{c}_1,\mathbf{c}_1')-w(\mathbf{c},\mathbf{c}'|\mathbf{c}_1',\mathbf{c}_1)]\,\Phi_1\,d\mathbf{c}_1\,d\mathbf{c}'\,d\mathbf{c}_1'.\end{aligned}\right\} \tag{36.4}$$

Jetzt führen wir die Funktion ein

$$K(\mathbf{c},\mathbf{c}_1)=\iint[-w(\mathbf{c},\mathbf{c}_1|\mathbf{c}',\mathbf{c}_1')+w(\mathbf{c},\mathbf{c}'|\mathbf{c}_1,\mathbf{c}_1')+w(\mathbf{c},\mathbf{c}'|\mathbf{c}_1',\mathbf{c}_1)]\,d\mathbf{c}'\,d\mathbf{c}_1'. \tag{36.5}$$

Diese ist symmetrisch

$$K(\mathbf{c},\mathbf{c}_1)=K(\mathbf{c}_1,\mathbf{c}). \tag{36.6}$$

Es ist ja

$$K(\mathbf{c}_1,\mathbf{c})=\iint[-w(\mathbf{c}_1,\mathbf{c}|\mathbf{c}',\mathbf{c}_1')+w(\mathbf{c}_1,\mathbf{c}'|\mathbf{c},\mathbf{c}_1')+w(\mathbf{c}_1,\mathbf{c}'|\mathbf{c}_1',\mathbf{c})]\,d\mathbf{c}'\,d\mathbf{c}_1',$$

woraus durch Umbenennung $(\mathbf{c}',\mathbf{c}_1')$ in $(\mathbf{c}_1',\mathbf{c}')$ entsteht

$$K(\mathbf{c}_1,\mathbf{c})=\iint[-w(\mathbf{c}_1,\mathbf{c}|\mathbf{c}_1',\mathbf{c}')+w(\mathbf{c}_1,\mathbf{c}_1'|\mathbf{c},\mathbf{c}')+w(\mathbf{c}_1,\mathbf{c}_1'|\mathbf{c}',\mathbf{c})]\,d\mathbf{c}'\,d\mathbf{c}_1'.$$

Nach (36.3) ist dieser Ausdruck in der Tat mit dem Ausdruck (36.5) identisch. Ferner ist nach (36.5)

$$\int K(\mathbf{c},\mathbf{c}_1)\,d\mathbf{c}_1=\iiint w(\mathbf{c},\mathbf{c}_1|\mathbf{c}',\mathbf{c}_1')\,d\mathbf{c}_1\,d\mathbf{c}'\,d\mathbf{c}_1'. \tag{36.7}$$

Demnach hat der Stoßoperator (36.4) die Gestalt

$$n_0^2\, I(\Phi)=\int K(\mathbf{c},\mathbf{c}_1)\,(\Phi(\mathbf{c})-\Phi(\mathbf{c}_1))\,d\mathbf{c}_1 \tag{36.8}$$

oder auch

$$\left.\begin{aligned}n_0^2\, I(\Phi)&=\overline{K}(\mathbf{c})\,\Phi(\mathbf{c})-\int K(\mathbf{c},\mathbf{c}_1)\,\Phi(\mathbf{c}_1)\,d\mathbf{c}_1\\ \text{mit}\qquad \overline{K}(\mathbf{c})&=\int K(\mathbf{c},\mathbf{c}_1)\,d\mathbf{c}_1\geq 0.\end{aligned}\right\} \tag{36.9}$$

Wir haben so ohne spezielle Rechnung den von HILBERT 1912 [*23*], S. 272 zuerst für starre kugelige Atome ausgesprochenen Satz gewonnen, daß $I(\Phi)=0$ eine lineare homogene Integralgleichung zweiter Art mit symmetrischem Kern darstellt. Damit ist die Boltzmann-Gleichung in das Begriffssystem der Integralgleichungs-Theorie eingeordnet.

Gl. (36.5) gibt den Zusammenhang des Kerns $K(\mathbf{c},\mathbf{c}_1)$ mit $w(\mathbf{c},\mathbf{c}_1|\mathbf{c}',\mathbf{c}_1')$ oder nach (36.2) auch mit $W(\mathbf{c},\mathbf{c}_1|\mathbf{c}',\mathbf{c}_1')$ an. W geht seinerseits nach (30.11) auf den Streuquerschnitt zurück. Trägt man σ in (36.5) ein und führt die dort vorgeschriebenen Integrationen durch, soweit dies allgemein möglich ist, so findet man — die ganz zwangsläufigen Rechnungen seien hier übergangen — schließlich

$$\left.\begin{aligned}K(\mathbf{c},\mathbf{c}_1)=2\,\pi n_0^2\,\mathrm{e}^{-\beta(c^2+c_1^2)}\,g\int_0^\pi\Bigg\{\Bigg[\sigma\left(\frac{g}{\sin\frac{\chi}{2}},\chi\right)+\sigma\left(\frac{g}{\sin\frac{\chi}{2}},\pi-\chi\right)\Bigg]\frac{\mathrm{e}^{-\beta g^2\cot^2\frac{\chi}{2}}}{\sin^4\frac{\chi}{2}}\times\\ \times J_0\left(2i\beta\,|[\mathbf{c}\,\mathbf{c}_1]|\cot\frac{\chi}{2}\right)-\sigma(g,\chi)\Bigg\}\cdot\sin\chi\,d\chi.\end{aligned}\right\} \tag{36.10}$$

Dabei ist $g=|\mathbf{c}-\mathbf{c}_1|$, $\beta=m/2kT$ und J_0 die nullte Bessel-Funktion. Diese Formel, welche den von HILBERT berechneten Kern für starre Kugeln als Spezialfall enthält, ist von

ENSKOG (1917), S. 148 aufgestellt worden. Auf die Arbeiten von HILBERT und ENSKOG sowie auf HECKE (1922) sei auch verwiesen wegen der Diskussion der Singularitäten des Kerns, deren Beschaffenheit bekanntlich für die Anwendbarkeit der Theorie der Integralgleichungen wesentlich ist.

Die Übertragung der vorstehenden Überlegungen auf Gemische ist einfach. Durch den Ansatz

$$f_i = f_i^{(0)}(1 + \Phi_i) \tag{36.11}$$

erhält man aus (30.15) unter Vernachlässigung der in Φ quadratischen Glieder als Analogon zu (36.1)

$$-f_i^{(0)}\left(\frac{\partial \Phi_i}{\partial t} + \cdots\right) = \sum_j \iiint w_{ij}(\mathbf{c}, \mathbf{c}_1 | \mathbf{c}', \mathbf{c}_1')(\Phi_i + \Phi_{j1} - \Phi_i' - \Phi_{j1}')\, d\mathbf{c}_1\, d\mathbf{c}'\, d\mathbf{c}_1', \tag{36.12}$$

wobei

$$w_{ij} = W_{ij}\,[f_i^{(0)} f_{j1}^{(0)} f_i^{(0)\prime} f_{j1}^{(0)\prime}]^{\frac{1}{2}}$$

gesetzt ist. Nach (30.13) und (30.14) hat man die Symmetrien

$$w_{ij}(\mathbf{c}, \mathbf{c}_1 | \mathbf{c}', \mathbf{c}_1') = w_{ji}(\mathbf{c}_1, \mathbf{c} | \mathbf{c}_1', \mathbf{c}) = w_{ij}(\mathbf{c}', \mathbf{c}_1' | \mathbf{c}, \mathbf{c}_1).$$

Durch dieselbe Umbenennung der Integrationsvariablen wie vorher erhält man die linearisierten Boltzmann-Gleichungen in der Gestalt

$$-f_i^{(0)}\left(\frac{\partial \Phi_i}{\partial t} + \cdots\right) = \int K_i(\mathbf{c}, \mathbf{c}_1)(\Phi_i - \Phi_{i1})\, d\mathbf{c}_1 + \sum_{j \neq i}{}' \int K_{ij}(\mathbf{c}, \mathbf{c}_1)\, \Phi_{j1}\, d\mathbf{c}_1. \tag{36.13}$$

Dabei sind die Integralkerne K_i und K_{ij} gegeben durch

$$K_i(\mathbf{c}, \mathbf{c}_1) = K_i(\mathbf{c}_1, \mathbf{c}) = \iint [-w_{ii}(\mathbf{c}, \mathbf{c}_1 | \mathbf{c}', \mathbf{c}_1') + \sum_j w_{ij}(\mathbf{c}, \mathbf{c}' | \mathbf{c}_1, \mathbf{c}_1') + w_{ii}(\mathbf{c}, \mathbf{c}' | \mathbf{c}_1', \mathbf{c}_1)]\, d\mathbf{c}'\, d\mathbf{c}_1' \tag{36.14}$$

und

$$K_{ij}(\mathbf{c}, \mathbf{c}_1) = K_{ji}(\mathbf{c}_1, \mathbf{c}) = \iint [w_{ij}(\mathbf{c}, \mathbf{c}_1 | \mathbf{c}', \mathbf{c}_1') - w_{ij}(\mathbf{c}, \mathbf{c}' | \mathbf{c}_1', \mathbf{c}_1)]\, d\mathbf{c}'\, d\mathbf{c}_1'. \tag{36.15}$$

Das reine Gas ist als Spezialfall darin enthalten.

Mit dem Kern

$$\mathcal{K}_{ij} = (K_i + K_{ii})\,\delta_{ij} - K_{ij}$$

kann (36.13) auch in der Form

$$-f_i^{(0)}\left(\frac{\partial \Phi_i}{\partial t} + \cdots\right) = \sum_j \int \mathcal{K}_{ij}(\mathbf{c}, \mathbf{c}_1)(\Phi_j - \Phi_{j1})\, d\mathbf{c}_1 \tag{36.16}$$

geschrieben werden.

37. Ein mit dem linearen Stoßoperator verknüpftes Eigenwertproblem. Es sei die räumlich konstante, zeitlich exponentielle Lösung $\Phi = e^{-\omega t}\,\psi(\mathbf{c})$ der genäherten Boltzmann-Gleichung (35.7) betrachtet. Für die Funktion $\psi(\mathbf{c})$ gilt

$$n_0^2 I(\psi) = \omega f^{(0)} \psi. \tag{37.1}$$

ψ darf für $c \to \infty$ nicht zu stark anwachsen. Für physikalisch brauchbare Verteilungsfunktionen muß das mit der Entropie zusammenhängende Normierungsintegral existieren:

$$\int \psi^2 f^{(0)}\, d\mathbf{c} = \text{endlich}. \tag{37.2}$$

Es ist plausibel, daß (37.1) und (37.2) ein System von unendlich vielen Eigenfunktionen ψ_N mit diskreten Eigenwerten ω_N definieren, wobei $\omega_N \to \infty$ für $N \to \infty$, und daß dieses System vollständig ist insofern, als alle physikalisch sinnvollen Funktionen Φ, vgl. (35.4), nach den ψ_N entwickelt werden können. In Ziff. 38 bis 40 wird gezeigt, daß diese Eigenschaften jedenfalls einem speziellen Molekülmodell — den Maxwellschen Kraftzentren — zukommen.

Nimmt man die Existenz von Eigenfunktionen an, so ergibt sich sogleich, daß zwei zu verschiedenen Eigenwerten ω_N und $\omega_{N'}$ gehörige Funktionen ψ_N und

$\psi_{N'}$ orthogonal zueinander sind:

$$\int \psi_N \psi_{N'} f^{(0)} d\boldsymbol{c} = 0 \quad \text{für} \quad \omega_N \neq \omega_{N'}. \tag{37.3}$$

Dies erhält man in bekannter Weise, indem man (37.1), angeschrieben für ψ_N und $\psi_{N'}$, mit $\psi_{N'}$ bzw. ψ_N multipliziert, integriert und beide Gleichungen voneinander subtrahiert. Die Symmetriebeziehung (35.10) ergibt dann (37.3).

Die Eigenwerte ω_N stellen die reziproken Relaxationszeiten des Gases dar. Für Maxwellsche Kraftzentren können sie explizit angegeben werden [s. Gl. (39.10)]. Jedenfalls sind die ω_N reell und nicht negativ (die Abweichungen eines sich selbst überlassenen Gases vom thermischen Gleichgewicht nehmen nach der Boltzmann-Gleichung stets ab); denn aus (37.1) und (35.11) folgt

$$\omega_N \int \psi_N^2 f^{(0)} d\boldsymbol{c} = n_0^2 [\psi_N, \psi_N] \geq 0,$$

also ist

$$\omega_N \geq 0. \tag{37.4}$$

Über die Eigenfunktionen läßt sich auf Grund der Drehinvarianz des Operators I allgemein sagen, daß sie von der Form

$$\psi_N(\boldsymbol{c}) = \psi_{lr}(c)\, Y_{lm}(\vartheta, \varphi) \tag{37.5}$$

sind, wo c, ϑ, φ sphärische Polarkoordinaten im $\boldsymbol{c}$-Raum, $Y_{lm} \propto P_l^m(\cos\vartheta)\, e^{im\varphi}$ eine Kugelflächenfunktion und der Index r die radiale Eigenfunktionsnummer (etwa die radiale Knotenzahl) bedeuten. Nach einem bekannten Satz von Hecke stellt nämlich ein linearer drehinvarianter Operator Ω aus einer beliebigen Funktion mit der Winkelabhängigkeit Y_{lm} wieder eine gleichartige Funktion her (s. z.B. Schmeidler [47], S. 360):

$$\Omega[\psi(c)\, Y_{lm}] = Y_{lm} \Omega^{(l)}[\psi(c)], \quad \Omega \text{ linear, drehinvariant.} \tag{37.6}$$

Daraus ergibt sich die angegebene Gestalt der Eigenfunktionen und für die Eigenwerte die Indizierung

$$\omega_N = \omega_{lr}. \tag{37.5a}$$

Die Orthogonalitätsrelationen (37.3) sind für Eigenfunktionen mit verschiedenen Werten von l, m wegen der Orthogonalität der Kugelflächenfunktionen eo ipso erfüllt. Bei gleichen Werten $l=l'$, $m=m'$ aber ergibt sich eine Orthogonalitätsrelation für die „Radial"-Funktionen $\psi_{lr}(c)$, welche zu der Orthonormierungsrelation ergänzt werden soll:

$$\int \psi_{lr} \psi_{lr'} \frac{f^{(0)}}{n_0} d\boldsymbol{c} = \delta_{rr'}. \tag{37.7}$$

Für die Zwecke der Gastheorie ist es vorzuziehen, statt der Kugelflächenfunktionen die zugehörigen Tensoren zu benutzen. Bekanntlich ist der symmetrische irreduzible Tensor[1] $c_{\mu_1 \ldots \mu_l}$, welcher homogen vom Grad l in den c_μ ist, dem System der $2l+1$ linear unabhängigen Kugelflächenfunktionen Y_{lm} mit festem l äquivalent, d.h. seine Komponenten, von denen genau $2l+1$ linear unabhängig sind, lassen sich nach Abspalten eines Faktors c^l linear aus den Y_{lm} kombinieren. Die ersten dieser Tensoren, so normiert, daß das Glied $c_{\mu_1} c_{\mu_2} \ldots c_{\mu_l}$

[1] Irreduzibel bedeutet, daß jede Art von Spurbildung an dem Tensor nicht auf einen neuen Tensor führt, sondern Null ergibt:

$$c_{\mu_1 \mu_1 \mu_3 \ldots \mu_l} = 0, \text{ usw.}$$

den Koeffizienten 1 hat, lauten[1]

$$\left.\begin{array}{l} l=0:\ \text{Skalar } 1 \\ l=1:\ \text{Vektor } c_\mu \\ l=2:\ \ c_{\mu\nu}=\langle c_\mu c_\nu\rangle = c_\mu c_\nu - \tfrac{1}{3}c^2\,\delta_{\mu\nu} \\ l=3:\ c_{\mu\nu\lambda}=\langle\langle c_\mu c_\nu\rangle c_\lambda\rangle = c_\mu c_\nu c_\lambda - \tfrac{1}{5}c^2\,(c_\mu\,\delta_{\nu\lambda}+c_\nu\,\delta_{\lambda\mu}+c_\lambda\,\delta_{\mu\nu}) \\ \dots\dots\dots\dots\dots\dots \end{array}\right\} \qquad (37.8)$$

Die Relation (37.6), geschrieben mittels irreduzibler Tensoren, lautet

$$\Omega\,[\psi(c)\,c^{-l}\,c_{\mu_1\dots\mu_l}] = c^{-l}\,c_{\mu_1\dots\mu_l}\,\Omega^{(l)}[\psi(c)]\,, \qquad (37.9)$$

also beispielsweise

$$\Omega\left[\psi(c)\,\frac{c_\mu}{c}\right] = \frac{c_\mu}{c}\,\Omega^{(1)}\,[\psi(c)] \qquad (37.10)$$

$$\Omega\left[\psi(c)\,\frac{c_{\mu\nu}}{c^2}\right] = \frac{c_{\mu\nu}}{c^2}\,\Omega^{(2)}\,[\psi(c)]\,. \qquad (37.11)$$

Zum Beweis kann man auf die Kugelflächenfunktionen Bezug nehmen oder direkt folgendermaßen schließen: (37.10) ist selbstverständlich; wegen der Linearität und Drehinvarianz des Operators Ω muß durch Anwendung von Ω wieder ein Vektor entstehen; außer c_μ ist aber kein Vektor vorhanden. Um (37.11) einzusehen, bemerkt man, daß

$$\Omega\left[\psi(c)\,\frac{c_\mu c_\nu}{c^2}\right] = \frac{c_\mu c_\nu}{c^2}\,\Omega^{(2)}\,[\psi(c)] + \varphi(c)\,\delta_{\mu\nu}\,;$$

denn es muß ein symmetrischer Tensor zweiter Stufe entstehen und nur $c_\mu\,c_\nu$ sowie $\delta_{\mu\nu}$ stehen zur Verfügung. Nun werde beiderseits die Spur gebildet

$$\Omega\,[\psi(c)] = \Omega^{(2)}\,[\psi(c)] + 3\,\varphi(c)\,.$$

Diese Gleichung werde mit $\frac{1}{3}\,\delta_{\mu\nu}$ multipliziert und von der vorhergehenden subtrahiert. Dann fällt der $\delta_{\mu\nu}$-Tensor ganz weg und man hat (37.11). So erkennt man als allgemeines Charakteristikum der irreduziblen Tensoren $c_{\mu_1\dots\mu_l}$, daß sie sich bei einer linearen drehinvarianten Operation bis auf einen skalaren Faktor reproduzieren.

Statt (37.5) können also die Eigenfunktionen auch in der Tensorform, als Eigentensoren

$$\psi^{(r)}_{\mu_1\dots\mu_l} = \psi_{l\,r}(c)\cdot c^{-l}\,c_{\mu_1\dots\mu_l} \qquad (37.12)$$

geschrieben werden, wo r wieder die radiale Knotenzahl bedeutet. Nach (37.8) hat man für die ersten l-Werte

$$\left.\begin{array}{l} l=0:\ \ \psi^{(r)} = \psi_{0\,r}(c) \\ l=1:\ \ \psi^{(r)}_\mu = \psi_{1\,r}(c)\,\dfrac{c_\mu}{c} \\ l=2:\ \ \psi^{(r)}_{\mu\nu} = \psi_{2\,r}(c)\,\dfrac{\langle c_\mu c_\nu\rangle}{c^2} \\ \dots\dots\dots\dots \end{array}\right\} \qquad (37.13)$$

[1] Das Zeichen $\langle\ \rangle$ soll auf die Herstellung des symmetrisch-irreduziblen Tensors aus dem eingeklammerten Tensor hinweisen. Das allgemeine Glied von (37.8) ist übrigens durch

$$c_{\mu_1\dots\mu_l} = \frac{(-1)^l}{1\cdot3\cdot5\dots(2l-1)}\,c^{2l+1}\,\frac{\partial^l}{\partial c_{\mu_1}\dots\partial c_{\mu_l}}\,\frac{1}{c}$$

gegeben. Homogenität, Symmetrie und Irreduzibilität sind daraus ersichtlich.

Die Eigenwertgleichung (37.1) nimmt die Gestalt an

$$n_0^2\, I(\psi^{(r)}_{\mu_1\ldots\mu_l}) = \omega_{lr}\, f^{(0)}\, \psi^{(r)}_{\mu_1\ldots\mu_l}. \tag{37.14}$$

Auch bei Benutzung der tensoriellen Schreibweise sei die Orthonormierung (37.7) zugrunde gelegt. Dann hat man u. a. die folgenden Beziehungen

$$\left.\begin{aligned} l=0:&\quad \int \psi^{(r)}\,\psi^{(r')}\,\frac{f^{(0)}}{n_0}\,d\boldsymbol{c} = \delta_{rr'}\\ l=1:&\quad \int \psi^{(r)}_{\mu}\,\psi^{(r')}_{\mu}\,\frac{f^{(0)}}{n_0}\,d\boldsymbol{c} = \delta_{rr'}\\ l=2:&\quad \int \psi^{(r)}_{\mu\nu}\,\psi^{(r')}_{\mu\nu}\,\frac{f^{(0)}}{n_0}\,d\boldsymbol{c} = \frac{2}{3}\,\delta_{rr'}, \end{aligned}\right\} \tag{37.15}$$

letzteres wegen

$$\langle c_\mu\, c_\nu\rangle \langle c_\mu\, c_\nu\rangle = \tfrac{2}{3}\, c^4.$$

Einige Bemerkungen sind noch zu machen über den Eigenwert $\omega=0$. Er kommt bei beliebigem Molekülmodell vor, und zwar ist er stets fünffach entartet. Für die fünf Stoßinvarianten $\psi=1$; c^2; c_μ gilt ja $I(\psi)=0$. Weitere linear unabhängige Lösungen dieser Gleichung kann es aber nicht geben. Denn nach der Identität (35.8), genommen für $\Psi=\Phi=\psi$, folgt umgekehrt aus $I(\psi)=0$, daß $\psi+\psi_1-\psi'-\psi_1'=0$ sein muß, daß also jede Eigenfunktion zu $\omega=0$ auch Stoßinvariante ist. Es kann aber nicht mehr als fünf Stoßinvarianten geben; die Transformation (22.9) stellt sechs lineare Beziehungen dar und enthält zwei willkürliche Parameter, nämlich die beiden Polarwinkel des Einheitsvektors $\boldsymbol{e}'$. Also kann es nicht mehr als vier, von $\boldsymbol{e}'$ freie Beziehungen geben, eben Energie- und Impulserhaltung, wozu dann noch die triviale Invariante $\psi=1$ hinzukommt (Überlegung von Hecke, s. Hilbert [*23*], S. *277*). — Aus den fünf Stoßinvarianten können nun durch Linearkombination fünf Eigentensoren, nämlich zwei skalare und ein vektorieller, gebildet werden:

$$\left.\begin{aligned} \psi^{(0)} = \psi_{00} = 1;\quad \psi^{(1)} = \psi_{01} = \sqrt{\tfrac{2}{3}}\,(\tfrac{3}{2}-\beta c^2);\\ \psi^{(0)}_{\mu} = \psi_{10}\,\frac{c_\mu}{c} = \sqrt{\frac{2}{3}\,\beta}\; c_\mu. \end{aligned}\right\} \tag{37.16}$$

Die Funktionen ψ_{00}, ψ_{01}, ψ_{10} sind im Sinn von (37.7) orthonormiert.

38. Die kugelsymmetrischen Eigenfunktionen des Maxwell-Gases und die zugehörigen Relaxationszeiten. Der linearisierte Stoßoperator lautet nach (35.5)

$$I(\Phi) = \left(\frac{\beta}{\pi}\right)^3 \iint e^{-\beta(c^2+c_1^2)}\,\delta(\Phi)\,\sigma g\, d\boldsymbol{e}'\, d\boldsymbol{c}_1, \tag{38.1}$$

wobei zur Abkürzung

$$\delta(\Phi) = \Phi + \Phi_1 - \Phi' - \Phi_1' \tag{38.2}$$

eingeführt war und $f^{(0)}$ gemäß (35.2) ersetzt ist.

Unter Maxwell-Molekülen versteht man punktförmige Zentren, die sich mit der Kraft

$$K = \frac{\varkappa}{r^5}$$

abstoßen. Daß sich für dieses Kraftgesetz die Rechnungen sehr vereinfachen, wurde von Maxwell 1866 ([38], Bd. 2, S. 26ff.) entdeckt. Der Grund liegt darin, daß der Faktor σg im Stoßintegral hier ausnahmsweise nicht auch von g, sondern nur vom Ablenkwinkel χ oder, was dasselbe ist, von $\Theta=\frac{1}{2}(\pi-\chi)$ abhängt, vgl. (27.12) und (27.13). Danach ist

$$\int \ldots \sigma g \, d\boldsymbol{e}' = \int_0^{\pi/2} \int_0^{2\pi} \ldots \Gamma(\Theta) \sin\Theta \, d\Theta \, d\varepsilon. \tag{38.3}$$

Dank dieser Vereinfachung kann man das vollständige System der Eigenfunktionen und Eigenwerte explizit angeben. Schon MAXWELL hat ω_{11}, ω_{20}, ω_{30} berechnet ([*38*], Bd. 2, S. 694—696; s. auch BOLTZMANN [*3*], 1. Bd., S. 164—176). Die allgemeinen Formeln sind u. a. von WANG CHANG und UHLENBECK (1952) und, gestützt auf die Methode von BURNETT (1935a), von WALDMANN (1956) mitgeteilt worden. Da sie in der Literatur nicht leicht zugänglich sind, sollen sie im folgenden ausführlich hergeleitet werden, zuerst für den kugelsymmetrischen und in der nächsten Ziffer für den allgemeinen Fall. — Mit der Theorie des Maxwell-Gases befaßten sich ferner IKENBERRY und TRUESDELL (1956).

Gemäß der Methode von BURNETT werde zunächst $I(\mathrm{e}^{-\lambda c^2})$ berechnet, wo λ ein beliebiger Parameter ist. Zur Auswertung des Ausdrucks

$$\delta(\mathrm{e}^{-\lambda c^2}) = \mathrm{e}^{-\lambda c^2} + \mathrm{e}^{-\lambda c_1^2} - \mathrm{e}^{-\lambda c'^2} - \mathrm{e}^{-\lambda c_1'^2}$$

gehen wir von den Stoßgleichungen (22.9) aus und entnehmen diesen die Beziehungen

$$c_{(1)}'^2 = \tfrac{1}{2}(c^2 + c_1^2) \underset{(-)}{+} g\, \boldsymbol{c}_S \, \boldsymbol{e}'. \tag{38.4}$$

Nun seien für den Einheitsvektor $\boldsymbol{e}'$ die Polarkoordinaten χ, ε eingeführt in bezug auf $\boldsymbol{g}$ als Polarachse. γ sei der Winkel zwischen $\boldsymbol{c}_S$ und $\boldsymbol{g}$. Das Azimut ε werde von der Ebene $(\boldsymbol{c}_S \boldsymbol{g})$ aus gezählt, so daß für die üblichen zugehörigen cartesischen Koordinaten gilt:

$$\begin{aligned} \boldsymbol{g} &= (0, 0, g) \\ \boldsymbol{c}_S &= (c_S \sin\gamma, 0, c_S \cos\gamma) \\ \boldsymbol{e}' &= (\sin\chi\cos\varepsilon, \sin\chi\sin\varepsilon, \cos\chi). \end{aligned}$$

Demnach ist

$$g\,\boldsymbol{c}_S\,\boldsymbol{e}' = g\,c_S \sin\gamma \sin\chi \cos\varepsilon + g\,c_S \cos\gamma \cos\chi.$$

Durch Verwendung der Beziehungen

$$g\,c_S \sin\gamma = |[\boldsymbol{g}\,\boldsymbol{c}_S]| = \tfrac{1}{2}|[\boldsymbol{c} - \boldsymbol{c}_1, \boldsymbol{c} + \boldsymbol{c}_1]| = |[\boldsymbol{c}\,\boldsymbol{c}_1]| = \sqrt{c^2 c_1^2 - (\boldsymbol{c}\,\boldsymbol{c}_1)^2},$$

$$g\,c_S \cos\gamma = \boldsymbol{g}\,\boldsymbol{c}_S = \tfrac{1}{2}(\boldsymbol{c} - \boldsymbol{c}_1)(\boldsymbol{c} + \boldsymbol{c}_1) = \tfrac{1}{2}(c^2 - c_1^2)$$

und Benutzung des Winkels Θ aus (27.4) anstelle von $\chi = \pi - 2\Theta$ ergibt sich aus (38.4) rasch

$$\left.\begin{aligned} c'^2 &= c^2 \sin^2\Theta + c_1^2 \cos^2\Theta + \sqrt{c^2 c_1^2 - (\boldsymbol{c}\,\boldsymbol{c}_1)^2} \sin 2\Theta \cos\varepsilon \\ c_1'^2 &= c^2 \cos^2\Theta + c_1^2 \sin^2\Theta - \sqrt{c^2 c_1^2 - (\boldsymbol{c}\,\boldsymbol{c}_1)^2} \sin 2\Theta \cos\varepsilon. \end{aligned}\right\} \tag{38.5}$$

Damit sind die Quadrate der Geschwindigkeiten nach dem Stoß ausgedrückt durch die Geschwindigkeiten vor dem Stoß und durch die Winkel Θ, ε, welche den Stoß festlegen.

Nun wird (38.5) in den Ausdruck $\delta(e^{-\lambda c^2})$ und dieser in (38.1) eingesetzt, unter Benutzung der Abkürzung (38.3). So entsteht[1]

$$I_M(\mathrm{e}^{-\lambda c^2}) = \left(\frac{\beta}{\pi}\right)^{\frac{3}{2}} \mathrm{e}^{-\beta c^2} \int_0^{\pi/2} (A + A_1 - A' - A_1')\, \Gamma(\Theta) \sin\Theta \, d\Theta, \tag{38.6}$$

[1] Der Index M soll auf das Maxwell-Gas hinweisen.

wo die A folgende Bedeutung haben

$$A_{(1)} = \left(\frac{\beta}{\pi}\right)^{\frac{3}{2}} \int_0^{2\pi}\!\!\int e^{-\beta c_1^2 - \lambda c_{(1)}^2}\, d\varepsilon\, d\boldsymbol{c}_1 \tag{38.7}$$

$$A' = \left(\frac{\beta}{\pi}\right)^{\frac{3}{2}} \int_0^{2\pi}\!\!\int e^{-\beta c_1^2 - \lambda\left(\sin^2\Theta\, c^2 + \cos^2\Theta\, c_1^2 + \sin 2\Theta \sqrt{c^2 c_1^2 - (\boldsymbol{c}\boldsymbol{c}_1)^2}\cos\varepsilon\right)}\, d\varepsilon\, d\boldsymbol{c}_1 . \tag{38.8}$$

A_1' ergibt sich aus A' durch die Substitution $(\Theta, \varepsilon) \to \left(\frac{\pi}{2} - \Theta, \varepsilon + \pi\right)$.

Diese Integrale können elementar berechnet werden. Sofort sind anzugeben (s. S. 362, Anm. 1)

$$A = 2\pi\, e^{-\lambda c^2}, \quad A_1 = 2\pi\left(\frac{\beta}{\beta+\lambda}\right)^{\frac{3}{2}}. \tag{38.9}$$

Zur Berechnung von A' seien zunächst für $\boldsymbol{c}_1$ cartesische Koordinaten eingeführt, derart das $\boldsymbol{c} = (0, 0, c)$, $\boldsymbol{c}_1 = (c_{1x}, c_{1y}, c_{1z})$ gilt. Dann ist

$$\sqrt{c^2 c_1^2 - (\boldsymbol{c}\boldsymbol{c}_1)^2} = c\sqrt{c_{1x}^2 + c_{1y}^2} .$$

Wir schreiben zur Abkürzung

$$r = \sqrt{c_{1x}^2 + c_{1y}^2}$$

und haben also nach (38.8)

$$A' = \left(\frac{\beta}{\pi}\right)^{\frac{3}{2}} e^{-\lambda \sin^2\Theta\, c^2} \int_{-\infty}^{\infty} e^{-(\beta + \lambda\cos^2\Theta)\, c_{1z}^2}\, dc_{1z} \int_0^{2\pi}\!\!\iint_{-\infty}^{\infty} e^{-(\beta+\lambda\cos^2\Theta)\, r^2 - \lambda \sin 2\Theta\, c\, r \cos\varepsilon}\, d\varepsilon\, dc_{1x}\, dc_{1y} .$$

Das Integral über c_{1z} kann (s. S. 362, Anm. 1) sofort ausgerechnet werden. Für c_{1x}, c_{1y} schreiben wir vorübergehend

$$c_{1x} = r\cos\varphi, \quad c_{1y} = r\sin\varphi$$

und haben

$$A' = \frac{\beta}{\pi}\left(\frac{\beta}{\beta + \lambda\cos^2\Theta}\right)^{\frac{1}{2}} e^{-\lambda\sin^2\Theta\, c^2} \iint_0^{2\pi}\int_0^{\infty} e^{-(\beta+\lambda\cos^2\Theta)\, r^2 - \lambda\sin 2\Theta\, c\, r\cos\varepsilon}\, d\varepsilon\, d\varphi\, r\, dr . \tag{38.10}$$

Jetzt kehren wir wieder zu cartesischen Koordinaten zurück, indem wir setzen

$$r\cos\varepsilon = x, \quad r\sin\varepsilon = y .$$

Dadurch nimmt das dreifache Integral in (38.10) die Gestalt an

$$\iint_{-\infty}^{\infty}\int_0^{2\pi} e^{-(\beta+\lambda\cos^2\Theta)(x^2+y^2) - \lambda\sin 2\Theta\, c\, x}\, dx\, dy\, d\varphi .$$

Nun ist man fertig, denn die Integrationen über y und φ sind trivial und für diejenige über x braucht man auch nur noch die Umformung

$$\begin{aligned} &-(\beta + \lambda\cos^2\Theta)\, x^2 - \lambda\sin 2\Theta\, c\, x \\ &\quad = -(\beta + \lambda\cos^2\Theta)\left(x + \frac{\lambda\sin\Theta\cos\Theta}{\beta + \lambda\cos^2\Theta}\, c\right)^2 + \frac{(\lambda\sin\Theta\cos\Theta\, c)^2}{\beta + \lambda\cos^2\Theta} \end{aligned}$$

zu machen. Man erhält so (s. S. 362, Anm. 1)

$$A' = 2\pi\left(\frac{\beta}{\beta + \lambda\cos^2\Theta}\right)^{\frac{3}{2}} e^{-\frac{\beta\lambda\sin^2\Theta}{\beta+\lambda\cos^2\Theta} c^2} . \tag{38.11}$$

Das Integral A_1' schließlich gewinnt man ohne weitere Rechnung durch die Bemerkung, daß es [s. bei (38.8)] aus A' entsteht durch Vertauschen von $\sin\Theta$ und $\cos\Theta$, daß also nach (38.11)

$$A_1' = 2\pi\left(\frac{\beta}{\beta+\lambda\sin^2\Theta}\right)^{\frac{3}{2}} e^{-\frac{\beta\lambda\cos^2\Theta}{\beta+\lambda\sin^2\Theta}c^2} \tag{38.12}$$

sein muß.

Mit (38.9), (38.11) und (38.12) ist das Stoßintegral (38.6) berechnet; zusammengefaßt gilt

$$\left.\begin{aligned} I_M(e^{-\lambda c^2}) &= \left(\frac{\beta}{\pi}\right)^{\frac{3}{2}} e^{-\beta c^2}\int_0^{\pi/2}(\ldots)\,2\pi\Gamma\sin\Theta\,d\Theta,\\ \text{mit}\quad (\ldots) &= e^{-\lambda c^2} + \left(\frac{\beta}{\beta+\lambda}\right)^{\frac{3}{2}} - \left(\frac{\beta}{\beta+\lambda\cos^2\Theta}\right)^{\frac{3}{2}} e^{-\frac{\beta\lambda\sin^2\Theta}{\beta+\lambda\cos^2\Theta}c^2} -\\ &\quad - \left(\frac{\beta}{\beta+\lambda\sin^2\Theta}\right)^{\frac{3}{2}} e^{-\frac{\beta\lambda\sin^2\Theta}{\beta+\lambda\sin^2\Theta}c^2}. \end{aligned}\right\} \tag{38.13}$$

Um daraus die Eigenfunktionen und die Eigenwerte abzulesen, werde statt λ ein neuer, dimensionsloser Parameter s durch

$$\lambda = \frac{s}{1-s}\beta \tag{38.14}$$

eingeführt. Ferner seien die Abkürzungen

$$\Phi_l(\beta c^2, s) = (1-s)^{-\frac{3}{2}-l} e^{-\frac{s}{1-s}\beta c^2} \tag{38.15}$$

benutzt, also speziell

$$\Phi_0(\beta c^2, s) = (1-s)^{-\frac{3}{2}} e^{-\frac{s}{1-s}\beta c^2}, \tag{38.16}$$

und es werde geschrieben

$$s' = s\cdot\sin^2\Theta, \quad s_1' = s\cdot\cos^2\Theta. \tag{38.17}$$

Multiplizieren von (38.13) mit $(1-s)^{-\frac{3}{2}}$ ergibt

$$\left.\begin{aligned} I_M[\Phi_0(\beta c^2, s)] &= \left(\frac{\beta}{\pi}\right)^{\frac{3}{2}} e^{-\beta c^2}\times\\ &\times\int_0^{\pi/2}[\Phi_0(\beta c^2, s) + 1 - \Phi_0(\beta c^2, s') - \Phi_0(\beta c^2, s_1')]\,2\pi\Gamma\sin\Theta\,d\Theta. \end{aligned}\right\} \tag{38.18}$$

Die Funktion Φ_l ist die erzeugende Funktion der aus der Schrödingerschen Theorie des Wasserstoffatoms geläufigen Laguerreschen Polynome. In die Gastheorie wurden diese Polynome durch BURNETT (1935a) eingeführt. Die Indizierung ist bei BURNETT etwas anders als bei SCHRÖDINGER und man spricht in der englischen gastheoretischen Literatur gewöhnlich von den Sonine-Polynomen. Dem wollen wir folgen und das *Sonine-Polynom* $S_{l+\frac{1}{2}}^{(r)}$ definieren durch

$$S_{l+\frac{1}{2}}^{(r)}(x) = \frac{1}{r!}\left(\frac{\partial}{\partial s}\right)^r \Phi_l(x,s)\big|_{s\to 0}. \tag{38.19}$$

Die explizite Darstellung lautet[1] (r ganz, l beliebig)

$$\left.\begin{aligned} S_{l+\frac{1}{2}}^{(r)}(x) &= \left(l+\frac{1}{2}\right)\left(l+\frac{3}{2}\right)\cdots\left(l+\frac{1}{2}+r\right)\times\\ &\times\sum_{p=0}^{r}\frac{1}{p!\,(r-p)!\,(l+\frac{1}{2})(l+\frac{3}{2})\cdots(l+\frac{1}{2}+p)}(-x)^p, \end{aligned}\right\} \tag{38.20}$$

[1] Näheres über die Sonine-Polynome siehe bei CHAPMAN u. COWLING [4], S. 123 oder bei MAGNUS u. OBERHETTINGER [37], S. 85. Hier wird die Normierung nach CHAPMAN und COWLING benutzt.

also speziell

$$\left.\begin{aligned} S_{l+\frac{1}{2}}^{(0)} &= 1 \\ S_{l+\frac{1}{2}}^{(1)} &= l + \tfrac{3}{2} - x \\ S_{l+\frac{1}{2}}^{(2)} &= \tfrac{1}{2}\left(l + \tfrac{3}{2}\right)\left(l + \tfrac{5}{2}\right) - \left(l + \tfrac{5}{2}\right) x + \tfrac{1}{2} x^2. \end{aligned}\right\} \tag{38.21}$$

Schließlich sei noch die Orthonormierung für die Polynome mit gleichem l-Index notiert:

$$\int_0^\infty e^{-x} S_{l+\frac{1}{2}}^{(r)}(x)\, S_{l+\frac{1}{2}}^{(r')}(x)\, x^{l+\frac{1}{2}}\, dx = \frac{\left(l + \frac{1}{2} + r\right)!}{r!}\,\delta_{rr'}. \tag{38.22}$$

Die Definition (38.19) ist nun dem Ausdruck (38.18) ausgezeichnet angepaßt. Wenn man nämlich auf beiden Seiten in (38.18) die Differentiation $\frac{1}{r!}\left(\frac{\partial}{\partial s}\right)^r$ ausführt, die aus (38.17) folgenden Beziehungen

$$\frac{\partial}{\partial s} = \sin^2\Theta \frac{\partial}{\partial s'} = \cos^2\Theta \frac{\partial}{\partial s_1'} \tag{38.23}$$

beachtet und sodann $s \to 0$, d.h. auch $s', s_1' \to 0$ gehen läßt, so nimmt die rechte Seite von (38.18) eine einheitliche Abhängigkeit von c an, die von $e^{-\beta c^2} S_{\frac{1}{2}}^{(r)}(\beta c^2)$, und auf der linken Seite steht im Argument dasselbe Sonine-Polynom. Indem wir gemäß (35.2) wieder $f^{(0)}$ statt $e^{-\beta c^2}$ einführen, bekommen wir so aus (38.18)

$$n_0^2 I_M \left[S_{\frac{1}{2}}^{(r)}(\beta c^2)\right] = \omega_{0r} f^{(0)} S_{\frac{1}{2}}^{(r)}(\beta c^2). \tag{38.24}$$

Dabei ist

$$\left.\begin{aligned} \omega_{00} &= 0 \\ \omega_{0r} &= n_0 \cdot 2\pi \int_0^{\pi/2} \left(1 - \sin^{2r}\Theta - \cos^{2r}\Theta\right) \Gamma(\Theta) \sin\Theta\, d\Theta \\ &= n_0 \cdot 2\pi \int_0^\infty (\ldots)\, g\, b\, db \quad \text{für} \quad r > 0. \end{aligned}\right\} \tag{38.25}$$

Bei der Schreibweise mit dem Stoßparameter b sind (38.3) und (28.6) benutzt.

Gl. (38.24) hat die Form der Eigenwertgleichung (37.1) und wir haben also, vgl. Gl. (37.5) oder (37.13), in

$$\psi^{(r)} = \psi_{0r}(c) \propto S_{\frac{1}{2}}^{(r)}(\beta c^2) \tag{38.26}$$

den vollständigen Satz der *skalaren (kugelsymmetrischen) Eigenfunktionen des Maxwellschen Gases* vor uns. Die Orthogonalitätsrelation (38.22) der Sonine-Polynome erscheint als Spezialfall der allgemeinen Orthogonalitätsrelation (37.7), hier zunächst beide für $l = 0$. Durch (38.25) sind die zugehörigen Eigenwerte *(reziproke Relaxationszeiten)* gegeben; sie werden in Ziff. 40 weiter diskutiert.

39. Die allgemeinsten Eigenfunktionen und Relaxationszeiten des Maxwell-Gases. Um die Eigenfunktionen zu $l \neq 0$ zu finden, werde zunächst $I_M\left[e^{-\lambda(\boldsymbol{c}-\boldsymbol{v})^2}\right]$ berechnet. Dabei sei $\boldsymbol{v}$ eine beliebige, feste Geschwindigkeit. Wir setzen zur Abkürzung

$$\boldsymbol{C} = \boldsymbol{c} - \boldsymbol{v}, \qquad \boldsymbol{C}_1 = \boldsymbol{c}_1 - \boldsymbol{v}. \tag{39.1}$$

Die Gln. (38.6) bis (38.8) ändern sich dann nur insofern, als in den λ-haltigen Termen $\boldsymbol{C}$, $\boldsymbol{C}_1$ statt $\boldsymbol{c}$, $\boldsymbol{c}_1$ steht.

Die Integrale A, A_1 können wieder sofort ausgeführt werden; man erhält nach S. 362, Anm. 1

$$A = 2\pi\, e^{-\lambda(\boldsymbol{c}-\boldsymbol{v})^2}; \qquad A_1 = 2\pi \left(\frac{\beta}{\beta+\lambda}\right)^{\frac{3}{2}} e^{-\frac{\beta\lambda}{\beta+\lambda} v^2}. \tag{39.2}$$

Die Berechnung von A' ist etwas umständlicher als im Fall $l = 0$. Wir benutzen zunächst cartesische Koordinaten, derart, daß $\boldsymbol{C} = (0, 0, C)$, $\boldsymbol{v} = (v_x, 0, v_z)$,

$\boldsymbol{C}_1 = (C_{1x}, C_{1y}, C_{1z})$ wird, und schreiben zur Abkürzung

$$R = \sqrt{C_{1x}^2 + C_{1y}^2}\,; \qquad C_{1x} = R\cos\xi, \qquad C_{1y} = R\sin\xi\,.$$

Über C_{1z} kann ohne weiteres integriert werden und man erhält als Verallgemeinerung von (38.10)

$$\left.\begin{aligned} A' &= \frac{\beta}{\pi}\left(\frac{\beta}{\beta+\lambda\cos^2\Theta}\right)^{\frac12}\exp\left[-\lambda\sin^2\Theta\, C^2 - \frac{\beta}{\beta+\lambda\cos^2\Theta}(\lambda\cos^2\Theta\, v^2 + \beta v_x^2)\right]\times \\ &\times\iint_{0\,0}^{2\pi\,\infty}\!\!\int \exp[-(\beta+\lambda\cos^2\Theta)R^2 - 2\beta v_x R\cos\xi - \lambda\sin 2\Theta\, C R\cos\varepsilon]\, d\varepsilon\, d\xi\, R\, dR. \end{aligned}\right\}\quad(39.3)$$

Jetzt werde statt ξ eine neue Variable φ eingeführt durch

$$\xi = \varepsilon + \varphi, \quad \text{d.h.} \quad \cos\xi = \cos\varepsilon\cos\varphi - \sin\varepsilon\sin\varphi\,.$$

Da der Integrand periodisch in φ ist, können die Integrationsgrenzen beibehalten werden. Wir kehren wieder zu cartesischen Koordinaten zurück durch

$$R\cos\varepsilon = X, \qquad R\sin\varepsilon = Y\,.$$

Der Exponent des Integranden nimmt die Gestalt an

$$\begin{aligned} &-(\beta+\lambda\cos^2\Theta)\,[(X+\cdots)^2 + (Y+\cdots)^2] + \\ &+ \frac{1}{\beta+\lambda\cos^2\Theta}\,[(\lambda\sin\Theta\cos\Theta\cdot C)^2 + \beta^2 v_x^2 + \beta\,\lambda\sin 2\Theta\, C\, v_x\cos\varphi]\,. \end{aligned}$$

Über X und Y (von $-\infty$ bis $+\infty$) kann integriert werden mit dem Ergebnis

$$\left.\begin{aligned} A' &= 2\pi\left(\frac{\beta}{\beta+\lambda\cos^2\Theta}\right)^{\frac32}\exp\left[-\frac{\beta\lambda}{\beta+\lambda\cos^2\Theta}(\sin^2\Theta\, C^2 + \cos^2\Theta\, v^2)\right]\times \\ &\times J_0\left(i\,\frac{\beta\lambda\sin 2\Theta}{\beta+\lambda\cos^2\Theta}\,|[\boldsymbol{c}\,\boldsymbol{v}]|\right). \end{aligned}\right\}\quad(39.4)$$

Dabei wurde verwendet, daß

$$|C\, v_x| = |[\boldsymbol{C}\,\boldsymbol{v}]| = |[\boldsymbol{c}\,\boldsymbol{v}]|$$

und daß

$$\int_0^{2\pi}\exp\left[\frac{\beta\lambda\sin 2\Theta}{\beta+\lambda\cos^2\Theta}\,|[\boldsymbol{c}\,\boldsymbol{v}]|\cos\varphi\right]d\varphi = 2\pi J_0\left(\pm i\,\frac{\beta\lambda\sin 2\Theta}{\beta+\lambda\cos^2\Theta}\,|[\boldsymbol{c}\,\boldsymbol{v}]|\right). \quad (39.5)$$

J_0 bezeichnet die nullte Bessel-Funktion (vgl. MAGNUS-OBERHETTINGER [*37*], S. 26). Gl. (39.4) stellt die gesuchte Verallgemeinerung von (38.11) dar. Aus (39.4) erhält man den Ausdruck für A_1' durch Vertauschen von $\sin\Theta$ und $\cos\Theta$.

Vereinigt man (39.2) und (39.4) sowie den entsprechenden Ausdruck für A_1' in der auf $e^{-\lambda C^2}$ verallgemeinerten Gl. (38.6), so ist $I_M(e^{-\lambda C^2})$ soweit berechnet, wie es analytisch möglich ist. Nun sollen wieder die Parameter s, s', s_1' aus (38.14) und (38.17) benutzt werden und schließlich sei die verallgemeinerte Gl. (38.6) beiderseits mit $(1-s)^{-\frac32}e^{\lambda v^2}$ multipliziert. So entsteht als Verallgemeinerung von (38.18):

$$\left.\begin{aligned} I_M\left[\Phi_0(\beta c^2, s)\, e^{\frac{s}{1-s}2\beta\boldsymbol{c}\boldsymbol{v}}\right] &= \left(\frac{\beta}{\pi}\right)^{\frac32} e^{-\beta c^2}\int_0^{\pi/2}\left[\Phi_0(\beta c^2, s)\, e^{\frac{s}{1-s}2\beta\boldsymbol{c}\boldsymbol{v}} + e^{\frac{s^2}{1-s}\beta v^2} - \right. \\ &- \Phi_0(\beta c^2, s')\, e^{\frac{s'}{1-s'}2\beta\boldsymbol{c}\boldsymbol{v} + \frac{s s_1'}{(1-s)(1-s')}\beta v^2} J_0\left(i\,\frac{s\cdot\sin 2\Theta}{1-s'}\,\beta\,|[\boldsymbol{c}\,\boldsymbol{v}]|\right) - \\ &- \left.\cdots\right]\cdot 2\pi\Gamma\sin\Theta\, d\Theta\,. \end{aligned}\right\}\quad(39.6)$$

Das nicht angeschriebene Glied erhält man aus dem vorhergehenden durch Vertauschen von s' und s_1'; wegen Φ_0 siehe (38.16).

Um (39.6) auf die Form einer Eigenwertgleichung zu bringen, führen wir für $\boldsymbol{c}$ sphärische Polarkoordinaten c, ϑ, φ ein, wo ϑ der Winkel zwischen $\boldsymbol{c}$ und $\boldsymbol{v}$ ist. Es ist also $\boldsymbol{c}\boldsymbol{v} = c v \cos\vartheta$, $|[\boldsymbol{c}\,\boldsymbol{v}]| = c\,v \sin\vartheta$. Wir entwickeln in (39.6) links und im ersten Glied rechts den Faktor mit $\boldsymbol{c}\,\boldsymbol{v}$ im Exponenten gemäß der bekannten Formel (vgl. MAGNUS-OBERHETTINGER [37], S. 21]

$$e^{iz\cos\vartheta} = \sum_{l=0}^{\infty} i^l (2l+1) \left(\frac{\pi}{2z}\right)^{\frac{1}{2}} J_{l+\frac{1}{2}}(z)\, P_l(\cos\vartheta) \tag{39.7}$$

nach Legendre-Polynomen $P_l(\cos\vartheta)$. Im dritten Glied rechts in (39.6) sei statt der Funktion J_0 wieder das Integral aus (39.5) geschrieben, also

$$e^{\frac{s'}{1-s'} 2\beta \boldsymbol{c}\boldsymbol{v}} J_0(\ldots) = \frac{1}{2\pi} \int_0^{2\pi} \exp\left[\frac{s\cdot\sin\Theta\cdot 2\beta c v}{1-s'} (\sin\Theta\cos\vartheta + \cos\Theta\sin\vartheta\cos\varphi)\right] d\varphi\,.$$

Dieser Integrand werde ebenfalls gemäß (39.7) nach den Funktionen

$$P_l\left[\cos\left(\frac{\pi}{2} - \Theta\right)\cos\vartheta + \sin\left(\frac{\pi}{2} - \Theta\right)\sin\vartheta\cos\varphi\right]$$

entwickelt. Aus dem Additionstheorem der Kugelfunktionen folgt sodann, daß (vgl. MAGNUS-OBERHETTINGER [37], S. 55)

$$\int_0^{2\pi} P_l\left[\cos\left(\frac{\pi}{2} - \Theta\right)\cos\vartheta + \sin\left(\frac{\pi}{2} - \Theta\right)\sin\vartheta\cos\varphi\right] d\varphi = 2\pi\, P_l(\sin\Theta)\, P_l(\cos\vartheta)\,.$$

Damit ist (39.6) beiderseits nach $P_l(\cos\vartheta)$ entwickelt. Man kann die Summenzeichen über l nun weglassen, da ja der drehinvariante Operator I_M die Winkelabhängigkeit $P_l(\cos\vartheta)$ reproduziert. (39.6) zerfällt so in Gleichungen für $l = 0, 1$, usw. Indem wir die l-te dieser Gleichungen durch $(2l+1)\,(4s\beta v)^l$ dividieren, $v \to 0$ gehen lassen und bemerken, daß für $z \to 0$ gilt (vgl. MAGNUS-OBERHETTINGER [37], S. 18)

$$\left(\frac{\pi}{2z}\right)^{\frac{1}{2}} J_{l+\frac{1}{2}}(z) \approx \frac{l!}{(2l+1)!}\,(2z)^l,$$

erhalten wir aus (39.6) für $l > 0$ die Beziehungen

$$\left.\begin{aligned} &I_M\left[\Phi_l(\beta c^2, s)\, c^l\, P_l(\cos\vartheta)\right] = \left(\frac{\beta}{\pi}\right)^{\frac{3}{2}} e^{-\beta c^2} \int_0^{\pi/2} (\ldots)\, 2\pi\, \Gamma \sin\Theta\, d\Theta \cdot c^l\, P_l(\cos\vartheta) \\ &\text{mit} \\ &(\ldots) = \Phi_l(\beta c^2, s) - \Phi_l(\beta c^2, s') \sin^l\Theta\, P_l(\sin\Theta) - \Phi_l(\beta c^2, s_1') \cos^l\Theta\, P_l(\cos\Theta)\,. \end{aligned}\right\} \tag{39.8}$$

Von (39.8) gelangt man durch r-malige Differentiation nach dem Parameter s unter Beachtung von (38.19) und (38.23) schließlich zu der gesuchten *Eigenwertgleichung*

$$n_0^2\, I_M\left[S_{l+\frac{1}{2}}^{(r)}(\beta c^2)\, c^l\, P_l(\cos\vartheta)\right] = \omega_{l r}\, f^{(0)}\, S_{l+\frac{1}{2}}^{(r)}(\beta c^2)\, c^l\, P_l(\cos\vartheta)\,. \tag{39.9}$$

Dabei gilt für die *Eigenwerte*

$$\left.\begin{aligned} \omega_{l r} &= n_0 \cdot 2\pi \int_0^{\pi/2} \left[1 - \sin^{l+2r}\Theta\, P_l(\sin\Theta) - \cos^{l+2r}\Theta\, P_l(\cos\Theta)\right] \Gamma(\Theta) \sin\Theta\, d\Theta \\ &= n_0 \cdot 2\pi \int_0^{\infty} [\ldots]\, g\, b\, db\,. \end{aligned}\right\} \tag{39.10}$$

Darin ist (38.25) als der Spezialfall $l = 0$ enthalten.

Natürlich kann man in (39.9) statt $P_l(\cos\vartheta)$ auch allgemein die Kugelflächenfunktion $Y_{lm}(\vartheta,\varphi)$ einsetzen oder mit Benutzung des irreduziblen homogenen Tensors $c_{\mu_1\ldots\mu_l}$ vom Grad l aus (37.8) auch schreiben

$$n_0^2 I_M\left[S_{l+\frac{1}{2}}^{(r)}(\beta c^2)\, c_{\mu_1\ldots\mu_l}\right] = \omega_{lr} f^{(0)} S_{l+\frac{1}{2}}^{(r)}(\beta c^2)\, c_{\mu_1\ldots\mu_l}. \tag{39.11}$$

Für die durch (37.5) oder (37.12) definierten, nach (37.7) normierten Radialfunktionen $\psi_{lr}(c)$ gilt beim Maxwell-Gas

$$\psi_{lr}(c) = \left(\frac{\sqrt{\pi}\, r!}{2(l+\frac{1}{2}+r)!}\right)^{\frac{1}{2}} S_{l+\frac{1}{2}}^{(r)}(\beta c^2)\left(\sqrt{\beta}\, c\right)^l, \tag{39.12}$$

wie man aus (38.22) entnimmt.

Es sei noch bemerkt, daß diese Eigenfunktionen übereinstimmen mit den Schrödingerschen Eigenfunktionen des isotropen harmonischen Oszillators, in Polarkoordinaten separiert.

40. Diskussion der Ergebnisse von Ziff. 38 und 39. Zuerst sei das asymptotische Verhalten der Eigenwerte betrachtet. Da $\sin^2\Theta$, $\cos^2\Theta$, $|P_l| \leq 1$, folgt aus (39.10), daß

$$\omega_{lr} \to n_0 \cdot 2\pi \int_0^\infty g\, b\, db = \infty \quad \text{für} \quad l + 2r \to \infty. \tag{40.1}$$

Die Eigenwerte sind, wie in Ziff. 37 allgemein behauptet wurde, diskret und häufen sich nur bei $\omega = \infty$. Die zugehörigen Eigenfunktionen haben alle Polynomcharakter; die Sonine-„Funktionen" (Laguerre-Funktionen), welche die kontinuierlichen Zustände des Schrödingerschen Wasserstoffatoms beschreiben, treten in der Gastheorie nicht in Erscheinung. Das ist nicht anders zu erwarten, denn sie würden keine physikalisch sinnvollen Geschwindigkeits-Verteilungsfunktionen liefern (keinen genügend starken Abfall für $c \to \infty$).

Als nächstes bemerken wir, daß sich beim Maxwell-Gas nach (39.10) alle Eigenwerte zu $l > 0$ aus den Eigenwerten ω_{0r} linear kombinieren lassen. Man sieht sofort, indem man die Legendre-Polynome einsetzt, daß

$$\left.\begin{aligned} \omega_{1r} &= \omega_{0,r+1} \\ \omega_{2r} &= \tfrac{3}{2}\omega_{0,r+2} - \tfrac{1}{2}\omega_{0,r+1} \\ \omega_{3r} &= \tfrac{5}{2}\omega_{0,r+3} - \tfrac{3}{2}\omega_{0,r+2} \quad \text{usw.} \end{aligned}\right\} \tag{40.2}$$

Die Zahlenkoeffizienten sind die Koeffizienten des entsprechenden Legendre-Polynoms.

Die Eigenwerte des Maxwell-Gases sind temperaturunabhängig. Es ist ja nach (27.11) für reines Gas mit der Atommasse m

$$g\, b\, db = \sqrt{\frac{2\varkappa}{m}}\, s_0\, ds_0,$$

also

$$\omega_{0r} = n_0 \cdot 2\pi \sqrt{\frac{2\varkappa}{m}} \int_0^\infty (1 - \sin^{2r}\Theta - \cos^{2r}\Theta)\, s_0\, ds_0, \tag{40.3}$$

wo $\Theta = \Theta(s_0)$ nach (27.8). Die Temperatur kommt nicht vor.

Mit Benutzung des Winkels $\chi = \pi - 2\Theta$ hat man auch

$$\omega_{0r} = n_0 \cdot 2\sqrt{\frac{2\varkappa}{m}} \int_0^\infty \left[1 - 2^{-r}(1+\cos\chi)^r - 2^{-r}(1-\cos\chi)^r\right] s_0\, ds_0. \tag{40.4}$$

Also ist speziell

$$\omega_{02} = n_0 \cdot \sqrt{\frac{2\varkappa}{m}}\, A_2(5)\,, \tag{40.5}$$

wo der Zahlenkoeffizient $A_2(5)$ ein Spezialfall der für beliebiges Potenzgesetz definierten Integrale

$$A_l(\nu) = \int_0^\infty (1 - \cos^l \chi)\, s_0\, d s_0 \tag{40.6}$$

ist. Schon MAXWELL ([38], Bd. 2, S. 42) hat

$$A_1(5) = 0{,}422; \qquad A_2(5) = 0{,}436 \tag{40.7}$$

berechnet (dort etwas anders bezeichnet).

In Ziff. 42 wird gezeigt, daß für die Viscosität des Maxwellschen Gases gilt

$$\eta = \frac{p_0}{\omega_{20}} \qquad (p_0 = \text{Druck}).$$

Tabelle 3. *Einige Werte* $\omega_{lr} \cdot \frac{\eta}{p_0}$ *für das Maxwellsche Gas.*

l \ r	0	1	2	3
0	0	0	$\frac{2}{3}$	1
1	0	$\frac{2}{3}$	1	
2	1	$\frac{7}{6}$		
3	$\frac{3}{2}$			

Nach (40.2) ist $\omega_{20} = \frac{3}{2}\omega_{02}$, also hat man auch

$$\omega_{02} = \frac{2}{3}\frac{p_0}{\eta}\,. \tag{40.8}$$

Ferner findet man

$$\omega_{03} = \frac{p_0}{\eta}\,. \tag{40.9}$$

Die weiteren, nach (40.2) sich ergebenden Eigenwerte sind in Tabelle 3 angegeben. Die höheren Eigenwerte enthalten neue numerische Koeffizienten; so ist z.B.

$$\omega_{04} = \frac{p_0}{\eta}\left(1 + \frac{1}{6}\frac{A_4(5)}{A_2(5)}\right); \quad \omega_{05} = \frac{5}{6}\frac{p_0}{\eta}\left(1 + \frac{1}{2}\frac{A_4(5)}{A_2(5)}\right). \tag{40.10}$$

Ungefähr werden die Zahlenwerte aus Tabelle 3 auch bei allgemeinem Kraftgesetz gelten. Für Stickstoff bei 760 Torr ($= 1{,}01 \cdot 10^6$ dyn/cm²), 0° C folgt aus $\eta = 1{,}67 \cdot 10^{-4}$ g/cm sec

$$\frac{1}{\omega_{02}} = 2{,}5 \cdot 10^{-10}\ \text{sec}$$

als größte endliche Relaxationszeit.

Zum Abschluß seien noch die ersten Eigentensoren $\psi^{(r)}_{\mu_1 \ldots \mu_l}$ des Maxwell-Gases nach (39.12) notiert:

l \ r	0	1	2
0	1	$\sqrt{\frac{2}{3}}\left(\frac{3}{2} - \beta c^2\right)$	$\sqrt{\frac{8}{15}}\left(\frac{15}{8} - \frac{5}{2}\beta c^2 + \frac{1}{2}\beta^2 c^4\right)$
1	$\sqrt{\frac{2}{3}\beta}\, c_\mu$	$2\sqrt{\frac{\beta}{15}}\left(\frac{5}{2} - \beta c^2\right) c_\mu$	. . . .
2	$\frac{2}{\sqrt{15}}\beta c_{\mu\nu}$	. . . .	. . . .

(40.11)

41. Entwicklung der linearisierten Boltzmann-Gleichung nach Eigentensoren. Wir kehren zur allgemeinen Theorie zurück und entwickeln in zwei Schritten die Verteilungsfunktion nach den in Ziff. 37 definierten Eigenfunktionen im

c-Raum. Die dabei schließlich auftretenden Entwicklungskoeffizienten hängen noch von t und $\boldsymbol{r}$ ab, gruppieren sich in Skalare, Vektoren, irreduzible Tensoren zweiter Stufe usw. und gehorchen einem System von linearen, partiellen Differentialgleichungen erster Ordnung.

α) Entwicklung nach irreduziblen homogenen Tensoren. Zunächst werde die Verteilungsfunktion bezüglich ihrer Abhängigkeit von der Geschwindigkeitsrichtung einer Entwicklung unterworfen. Als Entwicklungsfunktionen dienen die in (37.8) eingeführten, nur richtungsabhängigen Größen $c^{-l} c_{\mu_1 \ldots \mu_l}$, welche den Kugelflächenfunktionen äquivalent sind. Es werde also für die Funktion Φ aus (35.1) angesetzt

$$\left.\begin{aligned} \boldsymbol{\Phi}(t, \boldsymbol{r}, \boldsymbol{c}) &= \sum_l \varphi_{\mu_1 \ldots \mu_l}(t, \boldsymbol{r}, c)\, c^{-l}\, c_{\mu_1 \ldots \mu_l} \\ &= \varphi + \varphi_\mu \frac{c_\mu}{c} + \varphi_{\mu\nu} \frac{c_{\mu\nu}}{c^2} + \cdots. \end{aligned}\right\} \tag{41.1}$$

Die Eigenschaft der $c_{\mu_1 \ldots \mu_l}$, symmetrisch und irreduzibel zu sein, überträgt sich auf die $\varphi_{\mu_1 \ldots \mu^l}$, wenn letztere eindeutig festgelegt sein sollen. Es ist also

$$\varphi_{\mu\nu} = \varphi_{\nu\mu}, \qquad \varphi_{\mu\mu} = 0, \qquad \text{usw.} \tag{41.1a}$$

Diese Feststellung ergibt sich aus dem

Satz: Das Verschwinden eines Skalars

$$b_{\mu_1 \ldots \mu_l}\, c_{\mu_1 \ldots \mu_l} = 0$$

hat, wenn der Tensor (c) symmetrisch-irreduzibel und $\neq 0$ ist, das identische Verschwinden des Tensors (b) (41.2)

$$b_{\mu_1 \ldots \mu_l} = 0$$

dann und nur dann zur Folge, wenn auch (b) symmetrisch-irreduzibel ist.

Geht man mit dem Ansatz (41.1) in die Boltzmann-Gleichung (35.7) ein, so tritt u. a. die Reihe auf

$$c_\mu \frac{\partial \Phi}{\partial x_\mu} = c\left(\frac{\partial \varphi}{\partial x_\mu} \frac{c_\mu}{c} + \frac{\partial \varphi_\mu}{\partial x_\nu} \frac{c_\mu c_\nu}{c^2} + \frac{\partial \varphi_{\mu\nu}}{\partial x_\lambda} \frac{c_{\mu\nu} c_\lambda}{c^3} + \cdots\right).$$

Diese Reihe muß, damit dann gliedweise verglichen werden kann, nach dem Satz (41.2) in eine Entwicklung nach symmetrisch-irreduziblen homogenen Tensoren zurückverwandelt werden. Das erste Glied rechts hat bereits die gewünschte Form. Für die beiden anderen Glieder notieren wir nach (37.8)

$$c_\mu c_\nu = \frac{c^2}{3} \delta_{\mu\nu} + c_{\mu\nu}; \qquad c_{\mu\nu} c_\lambda = \frac{c^2}{5}\left(c_\mu \delta_{\nu\lambda} + c_\nu \delta_{\mu\lambda} - \frac{2}{3} c_\lambda \delta_{\mu\nu}\right) + c_{\mu\nu\lambda}; \; \ldots$$

Also ist

$$\frac{\partial \varphi_\mu}{\partial x_\nu} \frac{c_\mu c_\nu}{c^2} = \frac{1}{3} \frac{\partial \varphi_\mu}{\partial x_\mu} + \frac{\partial \varphi_\mu}{\partial x_\nu} \frac{c_{\mu\nu}}{c^2} = \frac{1}{3} \frac{\partial \varphi_\mu}{\partial x_\mu} + \left\langle \frac{\partial \varphi_\mu}{\partial x_\nu} \right\rangle \frac{c_{\mu\nu}}{c^2}$$

und

$$\frac{\partial \varphi_{\mu\nu}}{\partial x_\lambda} \frac{c_{\mu\nu} c_\lambda}{c^3} = \frac{2}{5} \frac{\partial \varphi_{\mu\nu}}{\partial x_\nu} \frac{c_\mu}{c} + \frac{\partial \varphi_{\mu\nu}}{\partial x_\lambda} \frac{c_{\mu\nu\lambda}}{c^3} = \frac{2}{5} \frac{\partial \varphi_{\mu\nu}}{\partial x_\nu} \frac{c_\mu}{c} + \left\langle \frac{\partial \varphi_{\mu\nu}}{\partial x_\lambda} \right\rangle \frac{c_{\mu\nu\lambda}}{c^3}.$$

Dabei sind der symmetrisch-irreduzible Tensor zweiter Stufe

$$\left\langle \frac{\partial \varphi_\mu}{\partial x_\nu} \right\rangle = \frac{1}{2}\left(\frac{\partial \varphi_\mu}{\partial x_\nu} + \frac{\partial \varphi_\nu}{\partial x_\mu}\right) - \frac{1}{3} \frac{\partial \varphi_\lambda}{\partial x_\lambda} \delta_{\mu\nu} \tag{41.3}$$

und der analoge dritter Stufe eingeführt worden.

Nun kann man in der Boltzmann-Gleichung gliedweise vergleichen und erhält das System

$$\left.\begin{aligned}
&f^{(0)}\left(\frac{\partial\varphi}{\partial t} + \frac{1}{3}\,c\,\frac{\partial\varphi_\mu}{\partial x_\mu}\right) = -n_0^2\, I(\varphi)\\
&f^{(0)}\left(\frac{\partial\varphi_\mu}{\partial t} + c\,\frac{\partial\varphi}{\partial x_\mu} + \frac{2}{5}\,c\,\frac{\partial\varphi_{\mu\nu}}{\partial x_\nu}\right) = -n_0^2\, I^{(1)}(\varphi_\mu)\\
&f^{(0)}\left(\frac{\partial\varphi_{\mu\nu}}{\partial t} + c\left\langle\frac{\partial\varphi_\mu}{\partial x_\nu}\right\rangle + \cdots\cdots\right) = -n_0^2\, I^{(2)}(\varphi_{\mu\nu})\\
&\cdots\cdots\cdots\cdots\cdots\cdots
\end{aligned}\right\} \tag{41.4}$$

Wegen der Operatoren $I^{(l)}$ sei auf (37.9) verwiesen[1].

β) Entwicklung nach radialen Eigenfunktionen. Die in (41.1) benutzten Koeffizienten $\varphi_{\mu_1\ldots\mu_l}$ sind Funktionen des Betrages c der Geschwindigkeit. Wir entwickeln sie nach den radialen Eigenfunktionen $\psi_{l r}$ aus (37.12):

$$\varphi_{\mu_1\ldots\mu_l}(t, \boldsymbol{r}, c) = \sum_r a^{(r)}_{\mu_1\ldots\mu_l}(t, \boldsymbol{r}) \cdot \psi_{l r}(c)\,, \tag{41.5}$$

also im einzelnen

$$\varphi = \sum a^{(r)}\,\psi_{0r}\,,\qquad \varphi_\mu = \sum a^{(r)}_\mu\,\psi_{1r}\,,\qquad \varphi_{\mu\nu} = \sum a^{(r)}_{\mu\nu}\,\psi_{2r}\,,\ \ldots.$$

Nach (41.1) lautet die Entwicklung für Φ dann

$$\Phi = \sum_r \left(a^{(r)}\,\psi_{0r} + a^{(r)}_\mu\,\psi_{1r}\,\frac{c_\mu}{c} + a^{(r)}_{\mu\nu}\,\psi_{2r}\,\frac{c_{\mu\nu}}{c^2} + \cdots\right). \tag{41.6}$$

Dies sei in das System (41.4) eingesetzt. Dabei schreiben wir

$$c\,\psi_{l r} = \sum_{r'} c(l\,r\,|\,l'\,r')\,\psi_{l' r'}.$$

Nach der Orthonomierung (37.7) gilt für die Koeffizienten $c(lr|l'r')$

$$c(l\,r\,|\,l'\,r') = \int c\,\psi_{l r}\,\psi_{l' r'}\,\frac{f^{(0)}}{n_0}\,d\boldsymbol{c} = c(l'\,r'\,|\,l\,r)\,. \tag{41.7}$$

So gelangt man unter Benutzung von (37.14) zu folgendem System von linearen, partiellen Differentialgleichungen erster Ordnung:

$$\left.\begin{aligned}
&\left(\frac{\partial}{\partial t} + \omega_{0r}\right) a^{(r)} + \frac{1}{3}\sum_{r'} c(0\,r\,|\,1\,r')\,\frac{\partial a^{(r')}_\mu}{\partial x_\mu} = 0\\
&\left(\frac{\partial}{\partial t} + \omega_{1r}\right) a^{(r)}_\mu + \sum_{r'} c(1\,r\,|\,0\,r')\,\frac{\partial a^{(r')}}{\partial x_\mu} + \frac{2}{5}\sum_{r'} c(1\,r\,|\,2\,r')\,\frac{\partial a^{(r')}_{\mu\nu}}{\partial x_\nu} = 0\\
&\left(\frac{\partial}{\partial t} + \omega_{2r}\right) a^{(r)}_{\mu\nu} + \sum_{r'} c(2\,r\,|\,1\,r')\left\langle\frac{\partial a^{(r')}_\mu}{\partial x_\nu}\right\rangle + \cdots = 0\\
&\cdots\cdots\cdots\cdots\cdots\cdots
\end{aligned}\right\} \tag{41.8}$$

Das allgemeine Glied kann man sich gemäß Anm. 1 anschreiben. Auf ein (41.8) in mathematischer Beziehung analoges System wird man in der Quantenmechanik des Pol-Dipol-Teilchens geführt[2].

---

[1] Die allgemeine Zeile von (41.4) lautet, wie ohne Beweis angegeben sei, so:

$$f^{(0)}\left(\frac{\partial\varphi_{\mu_1\ldots\mu_l}}{\partial t} + c\left\langle\frac{\partial\varphi_{\mu_1\ldots\mu_{l-1}}}{\partial x_{\mu_l}}\right\rangle + \frac{l+1}{2l+3}\,c\,\frac{\partial\varphi_{\mu_1\ldots\mu_{l+1}}}{\partial x_{\mu_{l+1}}}\right) = -\,n_0^2\, I^{(l)}(\varphi_{\mu_1\ldots\mu_l})\,.$$

[2] L. WALDMANN: Z. Naturforsch. **8**a, 583 (1953).

Das System (41.8) enthält nur noch die von Raum und Zeit abhängigen Funktionen a. Die Eigenschaften des Stoßoperators stecken in den Eigenwerten ω und den Matrixelementen (41.7). Einige der letzteren können universell angegeben werden an Hand von (37.16), nämlich

$$\left.\begin{aligned} c(00\,|\,1r') &= \sqrt{\frac{3}{2\beta}}\,\delta_{0r'} \\ c(01\,|\,10) &= -\frac{1}{\sqrt{\beta}} \\ c(10\,|\,0r') &= \frac{1}{\sqrt{\beta}}\left(\sqrt{\frac{3}{2}}\,\delta_{0r'} - \delta_{1r'}\right). \end{aligned}\right\} \tag{41.9}$$

Beim Maxwell-Gas sind nach (39.12) alle radialen Eigenfunktionen bekannt. Mittels der Darstellung (38.19) der Sonine-Polynome durch ihre erzeugende Funktion findet man nach kurzer Rechnung

$$c(l, r\,|\,l-1, r') = \frac{1}{\sqrt{\beta}}\left(\sqrt{l+\frac{1}{2}+r}\,\delta_{rr'} - \sqrt{r+1}\,\delta_{r+1,r'}\right) \tag{41.10}$$

und daraus durch Umbenennungen auch

$$c(l, r\,|\,l+1, r') = \frac{1}{\sqrt{\beta}}\left(\sqrt{l+\frac{3}{2}+r}\,\delta_{rr'} - \sqrt{r}\,\delta_{r-1,r'}\right). \tag{41.11}$$

Diese Matrixelemente stimmen überein mit den entsprechenden wellenmechanischen Matrixelementen des isotropen harmonischen Oszillators, was nach dem am Schluß von Ziff. 39 Gesagten klar ist.

γ) *Zeitliches Verhalten von Mittelwerten.* Hier soll noch eine Bemerkung über das zeitliche Verhalten von Mittelwerten Platz finden, die ebensogut schon in Ziff. 37 hätte gemacht werden können. Es sei durch

$$f = f^{(0)}\left(1 + \sum_{l,r} a^{(r)}_{\mu_1\ldots\mu_l}(t)\,\psi^{(r)}_{\mu_1\ldots\mu_l}\right), \tag{41.12}$$

vgl. (41.6), ein räumlich homogener Zustand gegeben. Dann ist nach (41.8)

$$a^{(r)}_{\mu_1\ldots\mu_l}(t) = a^{(r)}_{\mu_1\ldots\mu_l}(0)\,\mathrm{e}^{-\omega_{lr}t}. \tag{41.13}$$

Dieses exponentielle Abklingen beschreibt die Einstellung des thermischen Gleichgewichts im räumlich homogenen Gas. Daraus folgt, daß auch der Mittelwert jedes Eigentensors

$$\overline{\psi^{(r)}_{\mu_1\ldots\mu_l}} = \int \psi_{\mu_1\ldots\mu_l}\,\frac{f}{n_0}\,d\mathbf{c}, \tag{41.14}$$

welcher wegen der Orthogonalitätsrelationen zu $a^{(r)}_{\mu_1\ldots\mu_l}$ proportional ist, exponentiell abfällt:

$$\overline{\psi^{(r)}_{\mu_1\ldots\mu_l}(t)} = \overline{\psi^{(r)}_{\mu_1\ldots\mu_l}(0)}\,\mathrm{e}^{-\omega_{lr}t}. \tag{41.15}$$

An diesem Verhalten der Mittelwerte hat MAXWELL zuerst die Relaxation studiert.

42. Zusammenhang von Relaxations- und Transporterscheinungen. Das Gleichungssystem (41.8), welches der Boltzmann-Gleichung äquivalent ist, soll näherungsweise gelöst werden. Zuvor sind die Ausdrücke für die makroskopischen Zustandsgrößen zu finden.

α) *Die Ausdrücke für Konzentration, Temperatur und Geschwindigkeit.* Die drei Erhaltungssätze werden durch die zu $(l, r) = (0, 0)$, $(0, 1)$, $(1, 0)$ gehörigen Gleichungen von (41.8) repräsentiert; denn die Eigenwerte ω_{lr} verschwinden allgemein für diese drei Fälle, die zugehörigen Gln. (41.8) sind also vom Typ der Erhaltungssätze. Die entsprechenden Funktionen $a^{(0)}, a^{(1)}, a^{(0)}_\mu$ müssen mit den Größen n, T, v_μ gleichwertig sein. Um den genauen Zusammenhang zu finden, berechnen wir die letzteren an Hand der Entwicklung (41.6) der Verteilungsfunktion. Man findet mittels (37.16) und der Orthonormierung (37.7) der Eigenfunktionen Folgendes:

$$n = \int f\, d\mathbf{c} = n_0 + \int \psi_{00}\, \Phi\, f^{(0)}\, d\mathbf{c} = n_0 (1 + a^{(0)}), \tag{42.1}$$

ferner

$$\frac{3}{2} n \mathrm{k} T = \int \frac{m}{2} c^2 f\, d\mathbf{c} = \frac{3}{2} n_0 \mathrm{k} T_0 + \mathrm{k} T_0 \int \left(\frac{3}{2}\psi_{00} - \sqrt{\frac{3}{2}}\,\psi_{01}\right) \Phi\, f^{(0)}\, d\mathbf{c}$$

$$= \frac{3}{2} n_0 \mathrm{k} T_0 \left(1 + a^{(0)} - \sqrt{\frac{2}{3}}\, a^{(1)}\right)$$

oder nach (42.1) auch

$$T = T_0 \left(1 - \sqrt{\frac{2}{3}}\, a^{(1)}\right) \tag{42.2}$$

und schließlich

$$v_\mu = \frac{1}{n_0} \int c_\mu f\, d\mathbf{c} = \frac{1}{n_0} \int \sqrt{\frac{3}{2\beta}}\, \psi_{10} \frac{c_\mu}{c}\, \Phi\, f^{(0)} d\,\mathbf{c} = \frac{1}{\sqrt{6\beta}}\, a^{(0)}_\mu . \tag{42.3}$$

Bei der letzten Integration ist von der Relation

$$\int F(c)\, \frac{c_\mu c_\nu}{c^2}\, d\mathbf{c} = \frac{1}{3} \int F(c)\, d\mathbf{c} \cdot \delta_{\mu\nu}$$

Gebrauch gemacht.

β) Die Erhaltungssätze. Die Gleichung $l = 0$, $r = 0$ von (41.8) lautet nach (41.9)

$$\frac{\partial a^{(0)}}{\partial t} + \frac{1}{\sqrt{6\beta}} \frac{\partial a^{(0)}_\mu}{\partial x_\mu} = 0$$

oder nach (42.1) und (42.3)

$$\frac{\partial n}{\partial t} + \frac{\partial n_0 v_\mu}{\partial x_\mu} = 0 .$$

Das ist die der linearisierten Boltzmann-Gleichung entsprechende Kontinuitätsgleichung.

Die Gleichung $l = 0$, $r = 1$ von (41.8) lautet nach (41.9)

$$\frac{\partial a^{(1)}}{\partial t} - \frac{1}{3\sqrt{\beta}} \frac{\partial a^{(0)}_\mu}{\partial x_\mu} + \frac{1}{3} \sum_{r'=1}^{\infty} c(01\,|\,1r')\, \frac{\partial a^{(r')}_\mu}{\partial x_\mu} = 0 .$$

Nach (42.2) und (42.3) heißt dies

$$\frac{3}{2} n_0 \mathrm{k} \frac{\partial T}{\partial t} + \frac{\partial}{\partial x_\mu} (p_0 v_\mu + q_\mu) = 0 ,$$

wobei

$$q_\mu = -\frac{p_0}{\sqrt{6}} \sum_{r'=1}^{\infty} c(01\,|\,1r')\, a^{(r')}_\mu , \qquad p_0 = n_0 \mathrm{k} T_0 . \tag{42.4}$$

Dies ist der Energiesatz; $\boldsymbol{q}$ bedeutet den Wärmestrom.

Die Gleichung $l = 1$, $r = 0$ von (41.8) schließlich lautet nach (41.9)

$$\frac{\partial a^{(0)}_\mu}{\partial t} + \sqrt{\frac{3}{2\beta}} \frac{\partial a^{(0)}}{\partial x_\mu} - \frac{1}{\sqrt{\beta}} \frac{\partial a^{(1)}}{\partial x_\mu} + \frac{2}{5} \sum_{r'=0}^{\infty} c(10\,|\,2r')\, \frac{\partial a^{(r')}_{\mu\nu}}{\partial x_\nu} = 0 .$$

Mittels (42.1) bis (42.3) folgt daraus

$$n_0 m \frac{\partial v_\mu}{\partial t} + \frac{\partial p_{\mu\nu}}{\partial x_\nu} = 0,$$

wobei

$$p_{\mu\nu} = p\,\delta_{\mu\nu} + p_0 \cdot \tfrac{2}{5} \sqrt{\tfrac{2}{3}\beta} \sum_{r'=0}^{\infty} c(1\,0\,|\,2r')\, a_{\mu\nu}^{(r')}. \tag{42.5}$$

Dies ist der Impulssatz; $p_{\mu\nu}$ bedeutet den Drucktensor.

γ) Eine erste Näherung für Wärmestrom und Drucktensor. Es sei ein Gas mit zeitlich langsam veränderlicher Verteilungsfunktion betrachtet. Dann kann man in (41.8) die Glieder $\partial a/\partial t$ gegen ωa vernachlässigen, falls $\omega \neq 0$; löst man nach den a auf, so erhält man folgende Größenordnungen

$a^{(0)},\ a^{(1)};\quad a_\mu^{(0)}$	proportional ω^0,	
$a_\mu^{(1)},\ a_\mu^{(2)}, \ldots;\ a_{\mu\nu}^{(0)},\ a_{\mu\nu}^{(1)}, \ldots$	proportional ω^{-1},	
$a^{(2)},\ a^{(3)}, \ldots;\ a_{\mu\nu\lambda}^{(0)},\ a_{\mu\nu\lambda}^{(1)}, \ldots$	proportional ω^{-2},	usw.

Bis zur Näherung ω^{-1} gilt also nach (41.8)

$$a_\mu^{(r)} \approx -\frac{1}{\omega_{1r}}\, c(1r\,|\,0\,1)\, \frac{\partial a^{(1)}}{\partial x_\mu} \qquad \text{für} \quad r = 1, \ldots, \tag{42.6}$$

$$a_{\mu\nu}^{(r)} \approx -\frac{1}{\omega_{2r}}\, c(2r\,|\,1\,0) \left\langle \frac{\partial a_\mu^{(0)}}{\partial x_\nu} \right\rangle \qquad \text{für} \quad r = 0, \ldots. \tag{42.7}$$

Setzt man (42.6) in (42.4) ein, so erhält man unter Benutzung von (42.2) für den Wärmestrom die Beziehung

$$q_\mu \approx -\lambda \frac{\partial T}{\partial x_\mu} \quad \text{mit} \quad \lambda = \frac{1}{2} n_0\, \mathrm{k} \sum_{r'=1}^{\infty} \frac{[c(0\,1\,|\,1r')]^2}{\omega_{1r'}}. \tag{42.8}$$

Und (42.7) ergibt, in (42.5) eingesetzt unter Benutzung von (42.3), für den Drucktensor die Beziehung

$$p_{\mu\nu} \approx p\,\delta_{\mu\nu} - \eta \cdot 2 \left\langle \frac{\partial v_\mu}{\partial x_\nu} \right\rangle \quad \text{mit} \quad \eta = \frac{1}{5} n_0 m \sum_{r'=0}^{\infty} \frac{[c(1\,0\,|\,2r')]^2}{\omega_{2r'}}. \tag{42.9}$$

Dabei ist nach (41.3)

$$2 \left\langle \frac{\partial v_\mu}{\partial x_\nu} \right\rangle = \frac{\partial v_\mu}{\partial x_\nu} + \frac{\partial v_\nu}{\partial x_\mu} - \frac{2}{3} \frac{\partial v_\lambda}{\partial x_\lambda}\, \delta_{\mu\nu}.$$

(42.8) und (42.9) stellen also die „hydrodynamische" Näherung für die Transportgrößen dar; an den Ausdrücken für den Wärmeleitkoeffizienten λ und die Viscosität η erkennt man, daß diese positiv sind, wie es der zweite Hauptsatz verlangt.

Für das Maxwell-Gas sind die Matrixelemente c bekannt; nach (41.11) gilt

$$c(0\,1\,|\,1r') = \frac{1}{\sqrt{\beta}} \left(\sqrt{\frac{5}{2}}\, \delta_{1r'} - \delta_{0r'} \right); \qquad c(1\,0\,|\,2r') = \frac{1}{\sqrt{\beta}} \sqrt{\frac{5}{2}}\, \delta_{0r'}.$$

Somit ist hier nach (42.8)

$$\lambda = \frac{5}{2} \frac{\mathrm{k}\, p_0}{m\, \omega_{11}} \quad \text{mit} \quad p_0 = n_0\, \mathrm{k} T_0 \tag{42.10}$$

und nach (42.9)

$$\eta = \frac{p_0}{\omega_{20}}. \tag{42.11}$$

Wegen $\omega_{11} = \frac{2}{3}\omega_{20}$ (vgl. Ziff. 40, Tabelle 3) hat man auch

$$\lambda = \frac{5}{2} c_v \eta, \quad \text{wo} \quad c_v = \frac{3}{2} \frac{\mathrm{k}}{m}. \tag{42.12}$$

Diese Beziehungen erinnern an die Gln. (4.13) und präzisieren sie. Bei beliebigem Kraftgesetz gelten die Relationen (42.10) und (42.11) natürlich nur näherungsweise.

Der Zusammenhang zwischen den Relaxationszeiten und den Transportkoeffizienten wurde von Maxwell 1866 gefunden, [*38*] Bd. 2, S. 26—78. Er wurde neuerdings von Kohler (1948) wieder studiert.

Die Fortsetzung des mit (42.6) und (42.7) begonnenen Verfahrens führt zu erweiterten Wärmeleitungs- und Reibungsgleichungen, die höhere Ableitungen von T und $\boldsymbol{v}$ nach den Ortskoordinaten enthalten. Man bekommt dann Näherungen, die verwandt sein dürften mit den höheren Näherungen der Enskogschen Theorie (Burnett 1935b). Ähnliche Überlegungen hat Herzfeld (1935) angestellt, wobei die Konvergenz des Verfahrens aber zweifelhaft blieb. Die höheren Näherungen interessieren auch für die Deutung der von Greenspan (1950, 1956) gemessenen Schallabsorption in einatomigen Gasen bei Wellenlängen, die vergleichbar mit der freien Weglänge sind. Eine Theorie dazu auf Grund der Burnettschen Näherung war schon vor den genauen Messungen von Wang Chang (1948) entwickelt worden. Kritisch betrachtet werden diese höheren Näherungen von Ikenberry und Truesdell (1956).

Die Formeln (42.8) und (42.9) erwecken den Eindruck, als müßte man, um den Wärmeleitkoeffizienten und die Viscosität anzugeben, das Eigenwertproblem des Stoßoperators lösen. Dies ist aber nur scheinbar. Tatsächlich lassen sich die beiden Reihen durch Integrale, die den inversen Stoßoperator enthalten (ω im Nenner!), ersetzen. Das ist leicht zu zeigen, kann aber übergangen werden, da im Teil E, bei der Besprechung der Chapman-Enskogschen Methode, sich diese andere Darstellung von selbst ergibt [s. (48.3) mit (47.2) bzw. (48.11) mit (47.3)]. Die Näherung (42.6) und (42.7) ist im wesentlichen der ersten Chapman-Enskogschen Näherung äquivalent.

E. Wärmeleitung und Reibung einfacher Gase.

In Teil D wurden die Transportvorgänge in reinen Gasen auf Grund einer Entwicklung der Verteilungsfunktion nach Eigenfunktionen des Stoßoperators behandelt. Hier soll zunächst die allgemeine Theorie noch einmal nach einer anderen Methode, der von Chapman und Enskog, aufgestellt werden, unter Beschränkung auf die erste Näherung, die man auch die hydrodynamische nennt (Abschnitt I). Wegen einer vollständigen Systematik der Enskogschen Approximationsmethode, deren Konvergenz übrigens nicht bewiesen ist, sei auf Chapman und Cowling [*4*], Kap. 7, verwiesen. — Anschließend wird die Theorie mit dem Experiment verglichen (Abschnitt II).

I. Allgemeine kinetische Theorie nach Enskog und Chapman.

43. Der Grundgedanke der hydrodynamischen Näherung. Der Grundgedanke des Verfahrens ist der folgende: Die Verteilungsfunktion wird in zwei additive Anteile zerlegt. Der erste Anteil, eine raum-zeitlich veränderliche Maxwell-Verteilung, ist so eingerichtet, daß er die richtigen Werte der lokalen Konzentration, Gasgeschwindigkeit und Energiedichte wiedergibt. Der zweite Anteil liefert dann die Transportgrößen, d.s. Wärme- und Impulsstrom. Beide Anteile sind näherungsweise durch den linearisierten Stoßoperator miteinander verknüpft derart, daß das Problem der Wärmeleitung und Reibung auf die Auflösung einer linearen inhomogenen Integralgleichung zweiter Art hinausläuft. Eine zentrale Rolle spielen dabei die mit den Erhaltungssätzen eng verknüpften Lösbarkeitsbedingungen dieser Integralgleichung. Beim Maxwell-Gas ist die exakte Lösung bekannt; im allgemeinen Fall geschieht die Lösung durch eine Reihenentwicklung nach den Eigentensoren des Maxwell-Gases.

44. Zerlegung der Verteilungsfunktion in zwei Anteile. Für die Verteilungsfunktion sei angesetzt

$$f(t,\boldsymbol{r},\boldsymbol{c}) = F(t,\boldsymbol{r},\boldsymbol{c})\,[1+\Phi^{(1)}(t,\boldsymbol{r},\boldsymbol{c})]. \tag{44.1}$$

Darin ist

$$F = n\left(\frac{\beta}{\pi}\right)^{\frac{3}{2}} \mathrm{e}^{-\beta C^2} \tag{44.2}$$

mit

$$\beta = \frac{m}{2\mathrm{k}T}, \qquad \boldsymbol{C} = \boldsymbol{c} - \boldsymbol{v} \tag{44.3}$$

eine Maxwell-Funktion zu den von t und $\boldsymbol{r}$ abhängigen Werten n, $\boldsymbol{v}$, T der lokalen Konzentration, Gasgeschwindigkeit und Temperatur. Der Ansatz (35.1) mit (35.2) unterscheidet sich von dem jetzigen insofern, als dort eine raumzeitlich konstante Maxwell-Funktion $f^{(0)}$ mit $\boldsymbol{v}=0$ als Faktor gewählt war.

Es wird also verlangt, daß

$$\left.\begin{aligned} n &= \int F\,d\boldsymbol{c} = \int f\,d\boldsymbol{c},\\ v_\mu &= \frac{1}{n}\int c_\mu F\,d\boldsymbol{c} = \frac{1}{n}\int c_\mu f\,d\boldsymbol{c},\\ \frac{3}{2}\mathrm{k}T &= \frac{1}{n}\int \frac{m}{2} C^2 F\,d\boldsymbol{c} = \frac{1}{n}\int \frac{m}{2} C^2 f\,d\boldsymbol{c}. \end{aligned}\right\} \tag{44.4}$$

Somit ist der zweite Anteil $F\Phi^{(1)}$ von f den fünf Beziehungen unterworfen

$$\int \psi F\,\Phi^{(1)}\,d\boldsymbol{c} = 0, \tag{44.5}$$

worin

$$\psi = 1; \quad m c_\mu \ \text{ oder } \ m C_\mu; \quad \frac{m}{2}c^2 \ \text{ oder } \ \frac{m}{2}C^2 \tag{44.6}$$

sein kann. Dieser Anteil trägt nichts zu den lokalen Werten von n, $\boldsymbol{v}$, T bei.

Umgekehrt trägt der erste Anteil, da er einem „lokalen thermischen Gleichgewicht" entspricht, nichts zum Wärmestrom und zum Reibungsdruck bei; es gilt nach (44.1) sowie (31.14) und (31.16)

$$p_{\mu\nu} = p\,\delta_{\mu\nu} + \int m\,C_\mu C_\nu F\,\Phi^{(1)}\,d\boldsymbol{c}, \tag{44.7}$$

$$q_\mu = \int \frac{m}{2} C^2 C_\mu F\,\Phi^{(1)}\,d\boldsymbol{c}. \tag{44.8}$$

Es soll noch die jetzige Aufteilung (44.1) mit dem früheren Ansatz (41.6) verglichen werden. Macht man eine Taylor-Entwicklung von F nach $n-n_0$, $T-T_0$ und $\boldsymbol{v}$ und bricht mit den linearen Gliedern ab, so erkennt man, daß die Funktion $\Phi^{(1)}$ aus (44.1) mit demjenigen Teil der Reihe (41.6) identisch ist, welcher die nichttrivialen, zu $\omega \neq 0$ gehörigen Eigenfunktionen enthält. Die restlichen fünf Glieder dieser Reihe aber stellen die Funktion $(F-f^{(0)})/f^{(0)}$ dar — linear genähert —, also die relative Abweichung der „lokalen" Maxwell-Verteilung F von der raumzeitlich konstanten ruhenden Maxwell-Verteilung $f^{(0)}$. Im Unterschied zu den Betrachtungen von Teil D wird bei der Enskogschen Methode nicht vorausgesetzt, daß die letztgenannte Abweichung immer und überall klein sei.

45. Die genäherte Boltzmann-Gleichung. Der Ansatz (44.1) ist nun in die Boltzmann-Gleichung (28.5) einzutragen. Ähnlich wie in Ziff. 35 sei dabei angenommen, daß „im Wesentlichen" $|\Phi^{(1)}| \ll 1$ ist. [Wegen eines Zahlenbeispiels

s. das nach (49.9) Gesagte.] Auf jeder Seite der Boltzmann-Gleichung werde nur das dementsprechend größte Glied beibehalten:

$$\mathscr{D}(F) \approx -n^2 I(\Phi^{(1)}), \tag{45.1}$$

wobei abgekürzt ist

$$\left.\begin{aligned} \mathscr{D} &= \frac{\partial}{\partial t} + c_\mu \frac{\partial}{\partial x_\mu} + b_\mu \frac{\partial}{\partial c_\mu} \\ &= \frac{d}{dt} + C_\mu \frac{\partial}{\partial x_\mu} + b_\mu \frac{\partial}{\partial c_\mu} \quad \text{mit} \quad \frac{d}{dt} = \frac{\partial}{\partial t} + v_\mu \frac{\partial}{\partial x_{x_\mu}} \end{aligned}\right\} \tag{45.2}$$

und

$$n^2 I(\Phi) = \iint F F_1 \cdot (\Phi + \Phi_1 - \Phi' - \Phi_1')\, \sigma g\, d\boldsymbol{e}'\, d\boldsymbol{c}_1. \tag{45.3}$$

Der Operator I hat wie derjenige aus (35.5), die Dimension sec², unterscheidet sich aber etwas von dem dortigen, da jetzt das von t und $\boldsymbol{r}$ abhängige F anstelle von $f^{(0)}$ steht.

Die genäherte Boltzmann-Gleichung (45.1) ist nicht mehr eine Integrodifferentialgleichung, sondern eine inhomogene Integralgleichung zweiter Art — s. Gl. (36.9) —, denn ihre linke Seite besteht aus einer bekannten Funktion von $\boldsymbol{c}$. Die Auflösung nach der Funktion $\Phi^{(1)}$ ist das Grundproblem der Kinetik der Transporterscheinungen.

46. Die Lösbarkeitsbedingungen. Damit (45.1) eine vernünftige Lösung $\Phi^{(1)}$ besitzt, hat die linke Seite fünf Bedingungen zu erfüllen. Für die fünf Funktionen ψ aus (44.6) ist ja $\psi + \psi_1 - \psi' - \psi_1' = 0$, also $I(\psi) = 0$. Nach (35.10) ist für beliebiges Φ

$$\int \psi I(\Phi)\, d\boldsymbol{c} = \int \Phi I(\psi)\, d\boldsymbol{c} = 0$$

und nach (45.1) muß somit gelten

$$\int \psi \mathscr{D}(F)\, d\boldsymbol{c} = 0. \tag{46.1}$$

Diese fünf Bedingungen sind notwendig und, wie in der Theorie der Integralgleichungen gezeigt wird, auch hinreichend für die Lösbarkeit von (45.1).

Die Bedingung (46.1) hat dieselbe Gestalt wie (31.3), nur ist das dortige f, die exakte Verteilungsfunktion, jetzt durch F, eine Maxwell-Funktion, ersetzt. Wir erhalten also die explizite Gestalt der Bedingung (46.1), indem wir in den exakten Erhaltungssätzen (31.18), (31.19), (31.21) den Drucktensor durch

$$p_{\mu\nu}^{(0)} = \int m C_\mu C_\nu F\, d\boldsymbol{C} = p\,\delta_{\mu\nu} \quad \text{mit} \quad p = n\mathrm{k}T$$

und den Wärmestrom durch

$$q_\mu^{(0)} = \int \frac{m}{2} C^2 C_\mu F\, d\boldsymbol{C} = 0$$

ersetzen. Das gibt

$$\left.\begin{aligned} \frac{1}{n}\frac{dn}{dt} &\approx -\frac{\partial v_\mu}{\partial x_\mu}, \\ \varrho \frac{dv}{dt} &\approx -\frac{\partial p}{\partial x_\mu} + \varrho b_\mu, \\ \frac{3}{2} n\mathrm{k} \frac{dT}{dt} &\approx -p \frac{\partial v_\mu}{\partial x_\mu} \quad \text{oder} \quad \frac{dT}{T\,dt} \approx -\frac{2}{3}\frac{\partial v_\mu}{\partial x_\mu}. \end{aligned}\right\} \tag{46.2}$$

Mittels dieser fünf Beziehungen, welche eine erste Annäherung an die exakten Erhaltungsgleichungen darstellen, kann man die Zeitableitungen von n, $\boldsymbol{v}$ und T aus $\mathscr{D}(F)$ eliminieren. Aus

$$\log F = \log n - \frac{3}{2}\log T - \frac{m(\boldsymbol{c} - \boldsymbol{v})^2}{2\mathrm{k}T}$$

entnimmt man

$$\frac{dF}{F} = \frac{dn}{n} + \left(\frac{m C^2}{2kT} - \frac{3}{2}\right)\frac{dT}{T} - \frac{m}{kT} C_\mu\, d(c_\mu - v_\mu).$$

Geht man damit in

$$\mathscr{D}(F) = \frac{dF}{dt} + C_\mu \frac{\partial F}{\partial x_\mu} + b_\mu \frac{\partial F}{\partial c_\mu}$$

ein und ersetzt die Zeitableitungen gemäß (46.2), so kommt zunächst

$$\mathscr{D}(F) \approx F \cdot \left[\left(\frac{m C^2}{2kT} - \frac{5}{2}\right) C_\mu \frac{\partial T}{T\,\partial x_\mu} + \frac{m}{kT}\left(C_\mu C_\nu \frac{\partial v_\mu}{\partial x_\nu} - C^2 \cdot \frac{1}{3}\frac{\partial v_\mu}{\partial x_\mu}\right)\right]. \tag{46.3}$$

Die äußere Kraft tritt hier nicht mehr in Erscheinung.

Nun werde die folgende Umformung, deren Notwendigkeit in der nächsten Ziffer erläutert ist, vorgenommen:

$$C_\mu C_\nu \frac{\partial v_\mu}{\partial x_\nu} - C^2 \cdot \frac{1}{3}\frac{\partial v_\mu}{\partial x_\mu} = C_\mu C_\nu \left\langle \frac{\partial v_\mu}{\partial x_\nu}\right\rangle = \langle C_\mu C_\nu\rangle \left\langle \frac{\partial v_\mu}{\partial x_\nu}\right\rangle.$$

Wie in (37.8) und (41.3) weist das $\langle\,\rangle$-Zeichen auf die Bildung des symmetrischen, irreduziblen Tensors hin. Es ist

$$\langle C_\mu C_\nu\rangle = C_\mu C_\nu - \tfrac{1}{3} C^2 \delta_{\mu\nu}, \qquad \langle C_\mu C_\mu\rangle = 0, \tag{46.4}$$

$$\left\langle \frac{\partial v_\mu}{\partial x_\nu}\right\rangle = \frac{1}{2}\left(\frac{\partial v_\mu}{\partial x_\nu} + \frac{\partial v_\nu}{\partial x_\mu}\right) - \frac{1}{3}\frac{\partial v_\lambda}{\partial x_\lambda}\delta_{\mu\nu}, \qquad \left\langle \frac{\partial v_\mu}{\partial x_\mu}\right\rangle = 0. \tag{46.5}$$

Trägt man dies unter Benutzung der Abkürzung β aus (44.3) in (46.3) ein, so nimmt die Boltzmann-Gleichung (45.1) die Form an

$$F \cdot \left[\left(\beta c^2 - \frac{5}{2}\right) C_\mu \frac{\partial T}{T\,\partial x_\mu} + 2\beta \langle C_\mu C_\nu\rangle \left\langle \frac{\partial v_\mu}{\partial x_\nu}\right\rangle\right] = -n^2 I(\Phi^{(1)}). \tag{46.6}$$

47. Die Gestalt der Funktion $\Phi^{(1)}$. Durch (46.6) allein ist $\Phi^{(1)}$ nicht eindeutig festgelegt, da zu irgendeiner Lösung eine beliebige Linearkombination der fünf Funktionen (44.6) addiert werden kann. Es waren aber an $\Phi^{(1)}$ außerdem von Anfang an die Forderungen (44.5) gestellt. Dadurch wird die Eindeutigkeit bewirkt. Wir schreiben für diese eindeutige Auflösung symbolisch

$$\Phi^{(1)} = -\frac{1}{n^2} I^{-1}[\mathscr{D}(F)].$$

Nun ist der reziproke Operator I^{-1} sicher drehinvariant, da ja weder durch I noch durch die Bedingungen (44.5) eine Vorzugsrichtung gegeben ist. Nach dem Satz (37.6) macht aber ein linearer drehinvarianter Operator aus Funktionen der Winkelabhängigkeit $\frac{C_\mu}{C}$, $\frac{\langle C_\mu C_\nu\rangle}{C^2}$, usw. wieder Funktionen derselben Winkelabhängigkeit. Folglich muß $\Phi^{(1)}$ die Gestalt haben:

$$\Phi^{(1)} = -\frac{1}{n}\left(A(C)\, C_\mu \frac{\partial T}{T\,\partial x_\mu} + B(C)\,\beta \langle C_\mu C_\nu\rangle \left\langle\frac{\partial v_\mu}{\partial x_\nu}\right\rangle\right). \tag{47.1}$$

Für diesen Schluß war die Einführung symmetrisch-irreduzibler Tensoren in (46.6) wesentlich.

Das Problem läuft also auf die Bestimmung zweier Funktionen $A(C)$, $B(C)$ einer Variablen hinaus. A und B sind von der Konzentration unabhängig. A/n und B/n stellen eine Art geschwindigkeitsabhängiger Stoß- (oder Relaxations-) Zeiten dar. Man kann (47.1) als zweites Glied einer Entwicklung der Verteilungsfunktion nach fallenden Potenzen von n auffassen.

Die Größen $\partial T/\partial x_\mu$ und $\langle \partial v_\mu / \partial x_\nu \rangle$ können beliebig gegeben sein. Geht man mit dem Ansatz (47.1) in die Boltzmann-Gleichung (46.6) ein, so darf man daher nach dem Satz (41.2) gliedweise vergleichen. Dadurch erhält man als Äquivalent von (46.6) die beiden einfacheren Integralgleichungen

$$I(A\,C_\mu) = \frac{F}{n}\left(\beta C^2 - \frac{5}{2}\right) C_\mu, \tag{47.2}$$

$$I(B\,\langle C_\mu C_\nu\rangle) = \frac{2F}{n} \langle C_\mu C_\nu \rangle. \tag{47.3}$$

Es muß noch festgestellt werden, welchen Bedingungen A und B nach (44.5) unterworfen sind. Man darf wieder gliedweise vergleichen und erhält

$$\int \psi F A\, C_\mu\, d\boldsymbol{C} = 0; \qquad \int \psi F B \langle C_\mu C_\nu \rangle\, d\boldsymbol{C} = 0.$$

Die Bedingung für A ist für $\psi = 1,\ \frac{m}{2} C^2$ identisch erfüllt, da hier der Integrand ungerade ist. $\psi = m C_\nu$ aber gibt

$$\int C_\nu F A\, C_\mu\, d\boldsymbol{C} = 0.$$

Nun gilt allgemein für eine beliebige Funktion $G(C)$

$$\int G(C)\, C_\mu C_\nu\, d\boldsymbol{C} = \tfrac{1}{3} \int G\, C^2\, d\boldsymbol{C} \cdot \delta_{\mu\nu}. \tag{47.4}$$

Denn dieses Integral muß ein Tensor zweiter Stufe sein; aus Isotropiegründen kommt nur ein Vielfaches des Einheitstensors $\delta_{\mu\nu}$ in Frage und durch Spurbildung $\mu = \nu$ auf beiden Seiten sieht man, daß der Zahlfaktor rechts stimmt. Somit muß A der Bedingung genügen

$$\int C^2 A F\, d\boldsymbol{C} = 0. \tag{47.5}$$

Die Bedingung für B hingegen ist für alle fünf Funktionen ψ aus (44.6) identisch erfüllt. Mit $\psi = 1,\ \frac{m}{2} C^2$ muß nämlich das zugehörige Integral wieder ein Vielfaches von $\delta_{\mu\nu}$ sein, es verschwindet aber identisch wegen $\langle C_\mu C_\mu \rangle = 0$. Mit $\psi = m C_\lambda$ ist der Integrand ungerade, so daß das Integral wieder gleich Null ist.

48. Allgemeine Darstellung des Wärmeleit- und des Reibungskoeffizienten. Nachdem durch (47.1) die Gestalt der Verteilungsfunktion festliegt, kann man nach (44.7) und (44.8) die des Drucktensors und des Wärmestroms angeben. Wir beschäftigen uns zunächst mit dem Letzteren.

α) *Der Wärmeleitkoeffizient.* Der B-Term von (47.1) liefert in (44.8) keinen Beitrag, da der zugehörige Integrand ungerade ist. Man bekommt einfach

$$q_\mu = -\int \frac{m}{2} C^2\, C_\mu F A\, C_\nu\, d\boldsymbol{C} \cdot \frac{\partial T}{T\, \partial x_\nu}.$$

Nach (47.4) folgt daraus das Wärmeleitungsgesetz

$$q_\mu = -\lambda \frac{\partial T}{\partial x_\mu}, \tag{48.1}$$

wo

$$\lambda = \frac{1}{3} \mathsf{k} \int \beta C^4 \frac{F}{n} A\, d\boldsymbol{C} \tag{48.2}$$

den Wärmeleitkoeffizienten bedeutet. Nach (47.5) kann man auch schreiben

$$\lambda = \frac{1}{3} \mathsf{k} \int \frac{F}{n}\left(\beta C^2 - \frac{5}{2}\right) C_\mu A\, C_\mu\, d\boldsymbol{C}, \tag{48.3}$$

woraus sich nach (47.2) die äquivalente Form ergibt

$$\lambda = \frac{1}{3}\,\mathsf{k}\int I(A\,C_\mu)\cdot A C_\mu\, d\boldsymbol{C} = \frac{1}{3}\,\mathsf{k}\,[A C_\mu, A C_\mu]. \tag{48.4}$$

Darin bedeutet [...] ein ähnliches bilineares, bestimmtes Integral, wie es schon früher [s. (35.9) bis (35.11)] verwendet wurde, nämlich

$$\left.\begin{aligned}[\Psi,\Phi] &= [\Phi,\Psi] = \int \Psi\cdot I(\Phi)\,d\boldsymbol{c}\\ &= \frac{1}{4n^2}\iiint F F_1\,\delta(\Psi)\,\delta(\Phi)\,\sigma g\,d\boldsymbol{e}'\,d\boldsymbol{c}\,d\boldsymbol{c}_1\end{aligned}\right\} \tag{48.5}$$

mit

$$\delta(\Psi) = \Psi + \Psi_1 - \Psi' - \Psi_1'; \qquad [\Psi,\Psi] \geq 0.$$

Es steht jetzt lediglich F statt des früheren $f^{(0)}$.

Man sieht aus (48.4), daß $\lambda > 0$ ist, wie es der zweite Hauptsatz verlangt.

β) Der Reibungskoeffizient. Zum Drucktensor (44.7) trägt der A-Term von $\Phi^{(1)}$ aus (47.1) nichts bei, da der zugehörige Integrand ungerade ist. Der B-Term dagegen liefert

$$p_{\mu\nu} = p\,\delta_{\mu\nu} - \int m C_\mu C_\nu \frac{F}{n} B\beta \langle C_{\mu'} C_{\nu'}\rangle\, d\boldsymbol{C} \left\langle\frac{\partial v_{\mu'}}{\partial x_{\nu'}}\right\rangle. \tag{48.6}$$

Nun gilt bei beliebiger Funktion $G(C)$

$$\int G(C)\, C_\mu C_\nu \langle C_{\mu'} C_{\nu'}\rangle\, d\boldsymbol{C} = \tfrac{1}{15}\int G C^4\, d\boldsymbol{C}\,(\delta_{\mu\mu'}\delta_{\nu\nu'} + \delta_{\mu\nu'}\delta_{\nu\mu'} - \tfrac{2}{3}\delta_{\mu\nu}\delta_{\mu'\nu'}). \tag{48.7}$$

Es handelt sich ja um einen Tensor vierter Stufe, welcher aus Isotropiegründen eine Linearkombination von $\delta_{\mu\nu}\delta_{\mu'\nu'}$, $\delta_{\mu\mu'}\delta_{\nu\nu'}$ und $\delta_{\mu\nu'}\delta_{\mu'\nu}$ sein muß. Die angeschriebene Linearkombination besitzt aber gerade die richtige Symmetrie — sie ist symmetrisch in den Indexpaaren $(\mu\nu)$ und $(\mu'\nu')$ — und hat die Eigenschaft, daß die Spur $\mu' = \nu'$ verschwindet, was wegen $\langle C_{\mu'} C_{\mu'}\rangle = 0$ sein muß. Um den Faktor rechts vor der Klammer zu finden, nimmt man beiderseits die Doppelspur $\mu = \mu'$, $\nu = \nu'$.

Somit vereinfacht sich (48.6) zu dem Reibungsgesetz

$$\left.\begin{aligned}p_{\mu\nu} &= p\,\delta_{\mu\nu} - \eta\cdot 2\left\langle\frac{\partial v_\mu}{\partial x_\nu}\right\rangle\\ &= p\,\delta_{\mu\nu} - \eta\left(\frac{\partial v_\mu}{\partial x_\nu} + \frac{\partial v_\nu}{\partial x_\mu} - \frac{2}{3}\frac{\partial v_\lambda}{\partial x_\lambda}\delta_{\mu\nu}\right),\end{aligned}\right\} \tag{48.8}$$

wo

$$\eta = \frac{1}{15}\, m\int \beta C^4 \frac{F}{n} B\, d\boldsymbol{C} \tag{48.9}$$

die Viscosität bedeutet. Wegen

$$\langle C_\mu C_\nu\rangle\langle C_\mu C_\nu\rangle = C_\mu C_\nu \langle C_\mu C_\nu\rangle = \tfrac{2}{3} C^4 \tag{48.10}$$

kann man statt (48.9) auch schreiben

$$\eta = \frac{1}{10}\,\mathsf{k}T\int 2\frac{F}{n}\beta\langle C_\mu C_\nu\rangle\, B\beta\langle C_\mu C_\nu\rangle\, d\boldsymbol{C}. \tag{48.11}$$

Nach (47.3) und (48.5) ergibt sich daraus die äquivalente Form

$$\left.\begin{aligned}\eta &= \tfrac{1}{10}\,\mathsf{k}T\int \beta^2 I(B\langle C_\mu C_\nu\rangle)\, B\langle C_\mu C_\nu\rangle\, d\boldsymbol{C}\\ &= \tfrac{1}{10}\,\mathsf{k}T\beta^2\,[B\langle C_\mu C_\nu\rangle, B\langle C_\mu C_\nu\rangle].\end{aligned}\right\} \tag{48.12}$$

Es ist also $\eta > 0$, wie es sein muß.

49. Entwicklung der Verteilungsfunktion nach Eigentensoren des Maxwellschen Gases. In Ziff. 37 wurden die Eigentensoren $\psi^{(r)}_{\mu_1\ldots\mu_l}$ für den linearisierten Stoßoperator bei beliebigem Molekülmodell eingeführt. Entwickelt man die gesuchte Verteilungsfunktion $\Phi^{(1)}$ aus (47.1) formal nach Eigentensoren desselben Modells, so gehen (47.2) und (47.3) in lineare Gleichungen über, die man sofort auflösen kann; man erhält die Ausdrücke (42.8) und (42.9) für den Wärmeleit- und den Reibungskoeffizienten.

Aber die Eigentensoren sind im allgemeinen nicht bekannt. Eine Ausnahme bildet das Maxwellsche Gas; nach (37.13) und (39.12) gilt hier für $l=1, 2$

$$\psi^{(r)}_\mu = \left(\frac{\sqrt{\pi}\, r!\, \beta}{2\,(\frac{3}{2}+r)!}\right)^{\frac{1}{2}} S^{(r)}_{\frac{3}{2}}(\beta C^2)\, C_\mu; \quad \psi^{(r)}_{\mu\nu} = \left(\frac{\sqrt{\pi}\, r!\, \beta^2}{2\,(\frac{5}{2}+r)!}\right)^{\frac{1}{2}} S^{(r)}_{\frac{5}{2}}(\beta C^2)\, \langle C_\mu C_\nu\rangle, \tag{49.1}$$

wo die Soninschen Polynome $S^{(r)}_{l+\frac{1}{2}}$ durch (38.19) gegeben sind. Es ist zweckmäßig $\Phi^{(1)}$ für beliebiges Molekülmodell nach den Tensoren (49.1) zu entwickeln (BURNETT 1935a). Die Funktionen A bzw. B aus (47.1) erscheinen dabei als Reihen von Sonine-Polynomen zu $l=1$ bzw. 2.

Zwei Vorbemerkungen sind nötig. Erstens kann man wegen

$$S^{(1)}_{\frac{3}{2}} = \tfrac{5}{2} - \beta C^2, \quad S^{(0)}_{\frac{5}{2}} = 1, \tag{49.2}$$

vgl. (38.21), die Integralgleichungen (47.2) und (47.3) so schreiben

$$I(A C_\mu) = -\frac{F}{n} S^{(1)}_{\frac{3}{2}} C_\mu, \tag{49.3}$$

$$I(B\langle C_\mu C_\nu\rangle) = 2\frac{F}{n} S^{(0)}_{\frac{5}{2}} \langle C_\mu C_\nu\rangle. \tag{49.4}$$

Zweitens sei an die Normierung (37.15) erinnert. Für $l=1, 2$ besagt sie nach (49.1), daß

$$\beta \int S^{(r)}_{\frac{3}{2}} C_\mu S^{(r')}_{\frac{3}{2}} C_\mu \frac{F}{n}\, d\boldsymbol{C} = \frac{2\,(\frac{3}{2}+r)!}{\sqrt{\pi}\, r!}\, \delta_{rr'}, \tag{49.5}$$

$$\beta^2 \int S^{(r)}_{\frac{5}{2}} \langle C_\mu C_\nu\rangle S^{(r')}_{\frac{5}{2}} \langle C_\mu C_\nu\rangle \frac{F}{n}\, d\boldsymbol{C} = \frac{4\,(\frac{5}{2}+r)!}{3\sqrt{\pi}\, r!}\, \delta_{rr'}. \tag{49.6}$$

Spezialfälle davon sind

$$\left.\begin{aligned} &\beta \int S^{(r)}_{\frac{3}{2}} C_\mu S^{(0)}_{\frac{3}{2}} C_\mu \frac{F}{n}\, d\boldsymbol{C} = \frac{3}{2}\,\delta_{r0}; \quad \beta \int S^{(r)}_{\frac{3}{2}} C_\mu S^{(1)}_{\frac{3}{2}} C_\mu \frac{F}{n}\, d\boldsymbol{C} = \frac{15}{4}\,\delta_{r1};\\ &\beta^2 \int S^{(r)}_{\frac{5}{2}} \langle C_\mu C_\nu\rangle S^{(0)}_{\frac{5}{2}} \langle C_\mu C_\nu\rangle \frac{F}{n}\, d\boldsymbol{C} = \frac{5}{2}\,\delta_{r0}. \end{aligned}\right\} \tag{49.7}$$

Wir betrachten nun zunächst das Maxwell-Gas und dann den allgemeinen Fall.

α) Das Maxwell-Gas. Die Eigenwertgleichungen für den Vektor und den Tensor aus (49.1) lauten nach (39.11), mit F statt $f^{(0)}$,

$$\frac{n^2}{\omega_{1r}} I_M(S^{(r)}_{\frac{3}{2}} C_\mu) = F\, S^{(r)}_{\frac{3}{2}} C_\mu,$$

$$\frac{n^2}{\omega_{2r}} I_M(S^{(r)}_{\frac{5}{2}} \langle C_\mu C_\nu\rangle) = F\, S^{(r)}_{\frac{5}{2}} \langle C_\mu C_\nu\rangle,$$

wo I_M speziell den Operator des Maxwell-Gases bezeichnet.

Vergleich mit (49.3) und (49.4) zeigt, daß für dieses Modell exakt gilt

$$A = -\frac{n}{\omega_{11}} S^{(1)}_{\frac{3}{2}} = \frac{n}{\omega_{11}}\left(\beta C^2 - \frac{5}{2}\right); \quad B = \frac{2n}{\omega_{20}} S^{(0)}_{\frac{5}{2}} = \frac{2n}{\omega_{20}}. \tag{49.8}$$

Die Nebenbedingung (47.5) ist nach (49.7) erfüllt.

Aus (48.3) und (48.11) folgen unter Benutzung von (49.2) und (49.7) wieder die Beziehungen (42.10) und (42.11) für den Wärmeleitkoeffizienten und die Viscosität des Maxwell-Gases. Drückt man mittels dieser Beziehung die Relaxationskonstanten ω_{11}, ω_{20} durch λ, η aus, setzt (49.8) in (47.1) und dies in (44.1) ein, so erhält man in hydrodynamischer Näherung als Verteilungsfunktion des inhomogenen Maxwell-Gases exakt

$$f = F \cdot \left\{1 - \frac{m}{n(\mathrm{k}T)^2}\left[\frac{2}{5}\lambda\left(\frac{mC^2}{2\mathrm{k}T} - \frac{5}{2}\right)C_\mu \frac{\partial T}{\partial x_\mu} + \eta \langle C_\mu C_\nu\rangle \left\langle\frac{\partial v_\mu}{\partial x_\nu}\right\rangle\right]\right\}. \tag{49.9}$$

Wir schätzen an Hand dieser Formel die Größenordnung von $\Phi^{(1)}$ ab. Für das wärmeleitende Gas erhält man mittels der Darstellung von λ durch l, die freie Weglänge, s. Gl. (4.13), aus (49.9)

$$\Phi^{(1)} \approx \left(\beta C^2 - \frac{5}{2}\right)\sqrt{\beta}\, C_\mu \cdot \frac{l}{T}\frac{\partial T}{\partial x_\mu}.$$

Es seien $l = 10^{-5}$ cm, etwa entsprechend Wasserstoff von 1 atm, 0° C und ein Temperaturgefälle von 30°/cm angenommen. Dann ist $l\,\partial T/T\,\partial x = 10^{-6}$, also $\Phi^{(1)} \approx 1$ für $\sqrt{\beta}C \approx 100$. Der zugehörige Exponentialfaktor in F beträgt e^{-10000}. Daraus ist zu ersehen, daß es auf die $\Phi^{(1)}$-Werte für so hohe Geschwindigkeiten praktisch nicht ankommt. Wesentlich sind nur die $\Phi^{(1)}$-Werte für $\sqrt{\beta}C \lesssim 1$ und sie sind in unserem Beispiel $\lesssim 10^{-6}$, so daß $(\Phi^{(1)})^2$ und sonstige höhere Glieder vernachlässigbar sind. Selbst bei extrem großen Temperaturgradienten, z. B. dem in der Brennzone einer Wasserstoffflamme — Dicke 0,02 cm bei 2000° Flammentemperatur, vgl. Clusius, Kölsch und Waldmann (1941) —, hat man bei Atmosphärendruck noch $l\,\partial T/T\,\partial x \approx \frac{1}{800}$ und ist somit berechtigt, die hydrodynamische Näherung anzuwenden. Aber offensichtlich versagt die Enskogsche Methode bei niedrigen Drucken. Im Knudsen-Gebiet kann der Fall $l\,\partial T/T\,\partial x \gg 1$ realisiert sein. Die erste Enskogsche Näherung stellt hier keine gute Approximation dar und die höheren Näherungen — abgesehen von deren Kompliziertheit — dürften das kaum ändern.

β) Allgemeines Molekülmodell. Nach dem eingangs Gesagten setzen wir im allgemeinen Fall an

$$A(C) = -\beta\sum_{r=1}^{\infty} A_r S_{\frac{3}{2}}^{(r)}(\beta C^2); \qquad B(C) = \beta\sum_{r=0}^{\infty} B_r S_{\frac{5}{2}}^{(r)}(\beta C^2). \tag{49.10}$$

Die Koeffizienten A_r, B_r haben wie A, B die Dimensionen cm^{-1} sec^{-1}. Es ist $A_0 = 0$ wegen der Nebenbedingung (47.5). Aus (49.3) und (49.4) entsteht

$$\beta\sum_s A_s\, I(S_{\frac{3}{2}}^{(s)} C_\mu) = \frac{F}{n} S_{\frac{3}{2}}^{(1)} C_\mu, \tag{49.11}$$

$$\beta\sum_s B_s\, I(S_{\frac{5}{2}}^{(s)}\langle C_\mu C_\nu\rangle) = 2\,\frac{F}{n} S_{\frac{5}{2}}^{(0)}\langle C_\mu C_\nu\rangle. \tag{49.12}$$

Nun werden (49.11) bzw. (49.12) mit $\beta S_{\frac{3}{2}}^{(r)} C_\mu$ bzw. $\beta^2 S_{\frac{5}{2}}^{(r)}\langle C_\mu C_\nu\rangle$ multipliziert und über $\boldsymbol{C}$ integriert. Nach (49.7) kommen rechts einfach $\frac{15}{4}\,\delta_{r1}$ bzw. $5\,\delta_{r0}$. Links treten die Matrizen auf

$$a_{rs} = \beta^2 [S_{\frac{3}{2}}^{(r)} C_\mu,\ S_{\frac{3}{2}}^{(s)} C_\mu] = a_{sr}, \tag{49.13}$$

$$b_{rs} = \beta^3 [S_{\frac{5}{2}}^{(r)}\langle C_\mu C_\nu\rangle,\ S_{\frac{5}{2}}^{(s)}\langle C_\mu C_\nu\rangle] = b_{sr}, \tag{49.14}$$

vgl. (48.5). Wegen $\delta(C_\mu) = 0$ (Impulssatz) ist $a_{0s} = 0$. Die Gln. (49.11) und (49.12) gehen dadurch in die beiden linearen Systeme über

$$\sum_{s=1}^{\infty} a_{rs} A_s = \tfrac{15}{4}\,\delta_{r1} \qquad (r = 1, \ldots), \tag{49.15}$$

$$\sum_{s=0}^{\infty} b_{rs} B_s = 5\,\delta_{r0} \qquad (r = 0, \ldots). \tag{49.16}$$

In der Einfachheit der rechten Seiten liegt der Vorteil der Burnettschen Entwicklung.

Aus (48.3) und (48.11) entnimmt man durch Einsetzen der Reihen (49.10) und Benutzung von (49.2) und (49.7) exakt

$$\lambda = \tfrac{5}{4}\,\mathrm{k}\,A_1; \qquad \eta = \tfrac{1}{4}\,m\,B_0. \tag{49.17}$$

Auch diese Beziehungen sind typisch für die Entwicklung nach Sonine-Polynomen.

Vernachlässigt man in (49.15) und (49.16) alle nichtdiagonalen Matrixelemente $s \neq r$, so kommt

$$[\lambda]_1 = \frac{75}{16}\,\frac{\mathrm{k}}{a_{11}}; \qquad [\eta]_1 = \frac{5}{4}\,\frac{m}{b_{00}}. \tag{49.18}$$

Für diese erste Näherung brauchen also nur die Diagonalelemente a_{11} und b_{00} (sec^{-1}) berechnet zu werden (s. Ziff. 50); die zugehörige Verteilungsfunktion ist durch (49.9) mit $[\lambda]_1$, $[\eta]_1$ statt λ, η gegeben, weil dann $A_{r\neq 1} = B_{r\neq 0} = 0$. Für das Maxwell-Gas ist diese erste Näherung die exakte Lösung.

Als zweite Näherung wären nach ENSKOG und CHAPMAN a_{12} und b_{01} zu berücksichtigen, alle übrigen nichtdiagonalen Matrixelemente wieder zu vernachlässigen. Ein anderes Verfahren für die höheren Näherungen, das auf der Entwicklung nach Potenzen der nichtdiagonalen Matrixelemente a_{rs}, b_{rs} beruht und rechnerisch einfacher zu behandeln ist, wurde neuerdings von KIHARA [32], aufgestellt. Wir kommen darauf in Ziff. 52γ kurz zurück.

50. Die erste Näherung für den Wärmeleit- und den Reibungskoeffizienten. Darstellung durch einen temperaturabhängigen, effektiven Streuquerschnitt (Ω-Integral). Es sollen jetzt die Näherungsformeln (49.18) für λ und η weiter ausgewertet werden. Dazu sind die durch (49.13), (49.14) und (49.2) definierten Ausdrücke

$$a_{11} = \beta^2\,[(\tfrac{5}{2} - \beta C^2)\,C_\mu,\,(\tfrac{5}{2} - \beta C^2)\,C_\mu], \tag{50.1}$$

$$b_{00} = \beta^3\,[\langle C_\mu C_\nu\rangle,\,\langle C_\mu C_\nu\rangle] \tag{50.2}$$

zu berechnen.

Nach (48.5) ist wegen $\delta(C_\mu) = 0$ ausführlich

$$a_{11} = \frac{\beta^4}{4}\iiint \left(\frac{\beta}{\pi}\right)^3 \mathrm{e}^{-\beta(C^2+C_1^2)}\,\delta(C^2 C_\mu)\,\delta(C^2 C_\mu)\,\sigma g\,d\boldsymbol{e}'\,d\boldsymbol{C}\,d\boldsymbol{C}_1. \tag{50.3}$$

Dieses Integral soll soweit wie möglich allgemein ausgeführt werden. Der Streuquerschnitt σ hängt vom Betrag g der Relativgeschwindigkeit ab. Über g kann demnach nicht allgemein integriert werden und man wird deshalb $\boldsymbol{g}$ und die Schwerpunktsgeschwindigkeit $\boldsymbol{C}_S$ als Integrationsvariable einführen:

$$\boldsymbol{C} = \boldsymbol{C}_S + \tfrac{1}{2}\boldsymbol{g}, \qquad \boldsymbol{C}' = \boldsymbol{C}_S + \tfrac{1}{2}\boldsymbol{g}',$$
$$\boldsymbol{C}_1 = \boldsymbol{C}_S - \tfrac{1}{2}\boldsymbol{g}, \qquad \boldsymbol{C}_1' = \boldsymbol{C}_S - \tfrac{1}{2}\boldsymbol{g}'.$$

Es ist

$$C^2 + C_1^2 = 2C_S^2 + \tfrac{1}{2}g^2; \qquad C^2 - C_1^2 = 2\boldsymbol{C}_S\boldsymbol{g}; \qquad C'^2 - C_1'^2 = 2\boldsymbol{C}_S\boldsymbol{g}'.$$

Somit gilt

$$\delta(C^2 C_\mu) = \tfrac{1}{2}\,\delta(C^2 g_\mu) = \tfrac{1}{2}\,(C^2 - C_1^2)\,g_\mu - \tfrac{1}{2}\,(C'^2 - C_1'^2)\,g_\mu' = C_{S\nu}(g_\nu g_\mu - g_\nu' g_\mu')$$

und demnach

$$\delta(C^2 C_\mu)\,\delta(C^2 C_\mu) = C_{S\nu} C_{S\lambda}\,(g_\nu g_\mu - g_\nu' g_\mu')\,(g_\lambda g_\mu - g_\lambda' g_\mu').$$

Durch Einsetzen in (50.3) und Benutzung der Formeln

$$d\boldsymbol{C}\,d\boldsymbol{C}_1 = d\boldsymbol{C}_S\,d\boldsymbol{g}; \qquad d\boldsymbol{g} = g^2\,dg\,d\boldsymbol{e}; \int \frac{(2\beta)^{\frac{5}{2}}}{\pi^{\frac{3}{2}}}\,\mathrm{e}^{-2\beta C_S^2}\,C_{S\nu} C_{S\lambda}\,d\boldsymbol{C}_S = \frac{1}{2}\,\delta_{\nu\lambda}$$

bekommt man

$$a_{11} = \frac{\beta^3}{8} \iint \left(\frac{\beta}{2\pi}\right)^{\frac{3}{2}} e^{-\frac{\beta}{2} g^2} (g^4 - (\boldsymbol{g}\boldsymbol{g}')^2)\, \sigma g^3\, dg\, d\boldsymbol{e}\, d\boldsymbol{e}'.$$

Dies kann weiter ausgerechnet werden. Wir beschreiben den Einheitsvektor $\boldsymbol{e}'$ durch seine Polarwinkel χ, ε in bezug auf $\boldsymbol{e}$. Die Integration über ε und sodann auch diejenige über $\boldsymbol{e}$ kann ausgeführt werden; sie liefert einfach den Faktor 2π bzw. 4π. Wegen $\boldsymbol{g}\boldsymbol{g}' = g^2 \cos\chi$ erhält man somit

$$a_{11} = 4\left(\frac{\beta}{2\pi}\right)^{\frac{1}{2}} 2\pi \int_0^{\pi}\int_0^{\infty} \left(\frac{\beta}{2}\right)^4 e^{-\frac{\beta}{2} g^2} g^7 (1 - \cos^2\chi)\, \sigma(\chi, g) \sin\chi\, d\chi\, dg. \tag{50.4}$$

Eine weitere Reduktion ist allgemein nicht möglich.

Das in (50.4) vorkommende zweifache Integral ist typisch für die Gaskinetik. Man führt, s. Chapman und Cowling [*4*], S. 157, verallgemeinernd die folgenden Integrale ein:

$$\Omega_{ij}^{(l,r)} = \sqrt{\frac{\mathrm{k}T}{2\pi m_{ij}}} \int_0^{\infty} e^{-\gamma^2} \gamma^{2r+3} Q_{ij}^{(l)}\, d\gamma. \tag{50.5}$$

Dabei ist γ der Betrag der dimensionslos gemachten Relativgeschwindigkeit

$$\gamma^2 = \frac{m_{ij}}{2\mathrm{k}T} g^2, \qquad m_{ij} = \frac{m_i m_j}{m_i + m_j}. \tag{50.6}$$

Ferner bedeutet

$$Q_{ij}^{(l)}(g) = 2\pi \int_0^{\pi} (1 - \cos^l\chi)\, \sigma_{ij}(\chi, g) \sin\chi\, d\chi \tag{50.7}$$

einen über alle Ablenkwinkel χ gemittelten, aber noch von der Relativgeschwindigkeit abhängigen Streuquerschnitt; speziell bei klassischer Mechanik kann man [s. (28.6)] auch schreiben

$$Q_{ij}^{(l)} = 2\pi \int_0^{\infty} (1 - \cos^l\chi_{ij})\, b\, db. \tag{50.8}$$

Die Definition der *Ω-Integrale* ist, im Hinblick auf die spätere Behandlung der Gemische, für den Stoß zweier verschiedener Moleküle gegeben. Beim reinen Gas können die unteren Indices wegbleiben. Die Größen $\sqrt{2\pi m_{ij}/\mathrm{k}T}\,\Omega_{ij}$ stellen einen temperaturabhängigen effektiven Streuquerschnitt dar; aus dem über den Ablenkwinkel gemittelten molekularen Streuquerschnitt Q_{ij} ergeben sie sich durch nochmalige Mittelung über alle Relativgeschwindigkeiten, entsprechend deren Häufigkeit bei der betreffenden Gastemperatur. — Bei klassischer Mechanik kann man die Massenabhängigkeit der Ω_{ij} allgemein angeben. Nach (27.4) und (27.5) hängt der Ablenkwinkel außer von b von dem Produkt $\frac{1}{2} m_{ij} g^2 = \mathrm{k}T\gamma^2$ ab. Das Integral in (50.5) ist also eine Funktion nur der Temperatur — von der Masse hängt es nicht ab — und man hat allgemein

$$\Omega_{ij}^{(l,r)} \propto \frac{1}{\sqrt{m_{ij}}}. \tag{50.9}$$

Schließlich sei noch die Rekursionsformel notiert:

$$\Omega_{ij}^{(l,r+1)} = \frac{d\Omega_{ij}^{(l,r)}}{d\log T} + \left(r + \frac{3}{2}\right)\Omega_{ij}^{(l,r)}. \tag{50.10}$$

Man beweist sie an Hand der Definition (50.5), indem man γ gemäß (50.6) durch g ersetzt.

Das Matrixelement a_{11} aus (50.4) wird nun, wie man ohne weiteres sieht, gemäß

$$a_{11} = \frac{2m}{kT} \Omega^{(2,2)} \tag{50.11}$$

durch ein Ω-Integral dargestellt. Somit ist nach (49.18)

$$[\lambda]_1 = \frac{25}{16} \frac{c_v kT}{\Omega^{(2,2)}} \quad \text{mit} \quad c_v = \frac{3}{2} \frac{k}{m}. \tag{50.12}$$

Die Berechnung von

$$b_{00} = \frac{\beta^3}{4} \iiint \left(\frac{\beta}{\pi}\right)^3 e^{-\beta(C^2+C_1^2)} \delta(C_\mu C_\nu)\, \delta(C_\mu C_\nu)\, \sigma g\, d\boldsymbol{e}'\, d\boldsymbol{C}\, d\boldsymbol{C}_1 \tag{50.13}$$

verläuft ähnlich wie die von a_{11}. Es ist

$$\delta(C_\mu C_\nu) = \delta(C_{S\mu} C_{S\nu} + \tfrac{1}{2} g_\mu C_{S\nu} + \tfrac{1}{2} C_{S\mu} g_\nu + \tfrac{1}{4} g_\mu g_\nu) = \tfrac{1}{2}(g_\mu g_\nu - g'_\mu g'_\nu),$$

also

$$\delta(C_\mu C_\nu)\, \delta(C_\mu C_\nu) = \tfrac{1}{2}(g^4 - (\boldsymbol{g}\boldsymbol{g}')^2).$$

Die Integration über die Schwerpunktsgeschwindigkeit kann ausgeführt werden und man sieht, daß der verbleibende Integrand derselbe ist wie in (50.4); es ist genau

$$b_{00} = a_{11}. \tag{50.14}$$

Nach (49.18) hat man also

$$[\eta]_1 = \frac{5}{8} \frac{kT}{\Omega^{(2,2)}}. \tag{50.15}$$

Vergleich mit (50.12) zeigt, daß in der ersten Näherung

$$\left[\frac{\lambda}{\eta}\right]_1 = \frac{5}{2} c_v \tag{50.16}$$

gilt, unabhängig vom molekularen Kraftgesetz und unabhängig davon, ob die klassische Mechanik oder die Quantenmechanik zur Berechnung des Streuquerschnitts benutzt wird. Beim Maxwell-Gas [vgl. (42.12)] ist diese Beziehung exakt erfüllt. Die Prandtl-Zahl (2.26) hat demnach bei einatomigen Gasen den universellen Wert $\frac{2}{3}$.

Die Enskog-Chapmanschen Formeln (50.12) und (50.15), zusammen mit (50.5) bis (50.8), stellen eines der wichtigsten Ergebnisse der kinetischen Theorie einatomiger Gase dar. Wie weit (50.16) mit dem Experiment übereinstimmt, wird in Ziff. 52 besprochen. Bei mehratomigen Gasen ist (50.15) noch näherungsweise anwendbar, dagegen treten infolge des Beitrags der innermolekularen Energie zum Wärmetransport unter Umständen erhebliche Änderungen an (50.12) und damit an (50.16) ein (s. Ziff. 96 und 98).

II. Auswertung der Theorie und Vergleich mit dem Experiment.

51. Die Auswertung des Viscositätsintegrals $\Omega^{(2,2)}$ und einiger anderer für spezielle Molekülmodelle. *α) Starre Kugeln.* Für starre Kugeln sind die Ω-Integrale elementar berechenbar. Sei d_{ij} der für den Stoß eines Moleküls i mit einem Molekül j gültige Durchmesser. Dann ist nach (26.6) $\sigma_{ij} = \frac{1}{4} d_{ij}^2$, und man findet mittels (50.8) und (50.5)

$$\Omega_{ij,\,\text{starr}}^{(l,r)} = \sqrt{\frac{\pi kT}{2m_{ij}}}\, d_{ij}^2\, f^{(l,r)} \tag{51.1}$$

mit

$$f^{(l,r)} = \frac{1}{4}\left[2 - \frac{1}{l+1}\left(1 + (-1)^l\right)\right](r+1)!\,.$$

Später werden speziell benötigt

$$f^{(1,1)} = 1; \quad f^{(1,2)} = 3; \quad f^{(1,3)} = 12; \quad f^{(2;2)} = 2. \tag{51.2}$$

Die Viscosität ist nach (50.15) in erster Näherung durch das Integral $\Omega^{(2,2)}$ gegeben. Für starre Kugeln vom Durchmesser d hat man

$$[\eta]_1 = \frac{5}{16\sqrt{\pi}} \frac{m\sqrt{\mathrm{k}T/m}}{d^2} = 0{,}176 \frac{\sqrt{m\,\mathrm{k}T}}{d^2}. \tag{51.3}$$

Tabelle 4. *Gemessene Viscosität η_0 einiger Gase bei 0° C; nach* (51.3) *daraus berechnete Moleküldurchmesser d_0; nach* (51.5) *berechnete freie Weglängen l_0 bei 760 Torr; Sutherland-Temperatur S_η, nebst erprobtem Gültigkeitsbereich (in °C) der zugehörigen Näherung* (51.10); *Exponent ν_0, aus S_η berechnet gemäß $\nu = 3 + 2T/S_\eta$ für $T = 273°$ K. Gemessene Wärmeleitkoeffizienten λ_0 bei 0°C.*

	He	Ne	A	Kr	X	H_2	N_2	O_2	CO_2	H_2O [1]
(μ-Poise) [2]	188,7	298	211	233	211	85,7	167,4	192,6	138,8	127,0
(Å)	2,18	2,58	3,66	4,14	4,88	2,72	3,78	3,62	4,62	4,14
$)^5 l_0$ (cm)	1,72	1,25	0,63	0,49	0,35	1,13	0,58	0,64	0,39	0,67
$_\eta$ (°K) [2]	80 (0—200)	60 (0—100)	140 (0—800)	190 (0—100)	250 (0—100)	75 (0—100)	105 (0—800)	125 (0—800)	250 (0—280)	650 (100—300)
	9,9	12,1	6,9	5,9	5,2	10,3	8,2	7,2	5,2	4,2
μ-cal / grad cm sec	336	109	38,9	21,2	12,4	417	58	59	33,7	55,1

[1] Alle Daten von Wasserdampf beziehen sich auf 100° C.

[2] Nach LANDOLT-BÖRNSTEIN [*34*]; die meisten der angeführten η-Werte stammen von TRAUTZ *et al.*

Aus (50.16) oder (50.12) folgt für den Wärmeleitkoeffizienten

$$[\lambda]_1 = \frac{75}{64\sqrt{\pi}} \frac{\mathrm{k}\sqrt{\mathrm{k}T/m}}{d^2} = 0{,}662 \frac{\mathrm{k}\sqrt{\mathrm{k}T/m}}{d^2}. \tag{51.4}$$

Früher wurde in der kinetischen Gastheorie viel benutzt die mittlere freie Weglänge

$$l = \frac{1}{\sqrt{2}\, n\, \pi\, d^2}. \tag{51.5}$$

Mit dieser, mit der mittleren Molekulargeschwindigkeit $\overline{C}$ aus (34.12) und der Dichte $\varrho = n m$ schreibt sich (51.3)

$$\eta = \frac{5\pi}{32} \varrho \overline{C}\, l = 0{,}491\, \varrho \overline{C}\, l. \tag{51.6}$$

In der älteren elementaren Theorie stand gewöhnlich $\frac{1}{3}$ anstelle des Faktors 0,491.

In Tabelle 4 sind einige Zahlenbeispiele aufgeführt. Die der willkürlichen Temperatur 0° C zugeordneten Durchmesser und freien Weglängen haben freilich nur qualitative Bedeutung; eine objektive Definition von Moleküldurchmessern und Werte dafür werden alsbald (s. u., Abschnitt δ) gegeben. In der Tabelle sind auch die Sutherland-Temperaturen S_η sowie die Exponenten ν enthalten, welche gemäß (51.10) und (51.8) die gemessene Temperaturabhängigkeit der Viscosität approximieren. Schließlich sind die Wärmeleitkoeffizienten angegeben.

β) Potenzkraft-Zentren. Legt man das Potenz-Kraftgesetz (25.2) zugrunde, so hat man zunächst die transzendente Funktion $\chi_{ij}(s_{0,ij}) = \pi - 2\Theta_{ij}$ gemäß (27.8) zu berechnen. Beim Einsetzen in (50.8) tritt die reine Zahl

$$A_l(\nu_{ij}) = \left[\int_0^\infty (1-\cos^l\chi)\, s_0\, d s_0\right]_{ij} \tag{51.7a}$$

auf, welche ebenfalls numerisch berechnet werden muß; wegen der Originalliteratur dazu s. CHAPMAN und COWLING [4], S. 172. Spezielle Werte sind in Tabelle 5 wiedergegeben. Im übrigen können dann die Integrale $\Omega^{(l,r)}$ auf die Γ-Funktion zurückgeführt werden und man erhält

$$\Omega_{ij}^{(l,r)} = \sqrt{\frac{\pi\,\mathrm{k}T}{2m_{ij}}}\left(\frac{\varkappa_{ij}}{2\mathrm{k}T}\right)^{\frac{2}{\nu_{ij}-1}}\left(r+1-\frac{2}{\nu_{ij}-1}\right)!\,A_l(\nu_{ij}). \tag{51.7b}$$

Tabelle 5. *Werte von $A_2(\nu)$ aus* (51.7a) *nach Chapman und Cowling* [4], *S. 172, und Werte von $\eta^*(\nu)$ aus* (51.8).

ν	5	7	9	11	15	25	∞
$A_2(\nu)$	0,436	0,357	0,332	0,319	0,309	0,306	0,333
$\eta^*(\nu)$	0,342	0,308	0,286	0,270	0,250	0,224	0,176

Es ist also

$$[\eta]_1 = \eta^*(\nu)\,\varkappa^{-\frac{2}{\nu-1}}\, m^{\frac{1}{2}}\,(\mathrm{k}T)^{\frac{1}{2}+\frac{2}{\nu-1}},$$

wo

$$\eta^* = \frac{5}{8\sqrt{\pi}}\,\frac{2^{\frac{2}{\nu-1}}}{\left(3-\frac{2}{\nu-1}\right)!\,A_2(\nu)}. \tag{51.8}$$

Spezielle Werte der reinen Zahl $\eta^*(\nu)$ sind ebenfalls in Tabelle 5 aufgeführt.

γ) Anziehungskraft, näherungsweise berücksichtigt. Für genügend hohe Temperatur kann die schwache Anziehung, die bei großem Molekülabstand z.B. nach dem Lennard-Jonesschen Ansatz (25.3) statthat, näherungsweise durch eine Störungsrechnung berücksichtigt werden, indem man von der gemäß β) exakt behandelten Bewegung im reinen Abstoßungsfeld als nullter Näherung ausgeht. Das ergibt als zu (25.3) gehörige Temperaturabhängigkeit der Viscosität

$$\eta(T,\varkappa,\varkappa') = \frac{\eta(T,\varkappa,0)}{1+(S_\eta/T)^{\frac{\nu-\nu'}{\nu-1}}}, \tag{51.9}$$

wo S_η eine Konstante von der Dimension einer Temperatur ist. (Näheres bei CHAPMAN und COWLING [4], S. 180—187.) Für $\nu=\infty$ hat man speziell die Sutherland-Formel

$$\eta(T) = \frac{\eta_{\mathrm{starr}}(T)}{1+S_\eta/T}. \tag{51.10}$$

Darin ist η_{starr} durch (51.3) gegeben. Wegen der weiteren Diskussion sei auf Ziff. 71, γ verwiesen.

δ) Exakte Rechnung für das Lennard-Jonessche (6; 12)-Potential. Die Näherung (51.9) ist nur bei genügend hoher Temperatur brauchbar. Wegen des großen theoretischen Interesses des Lennard-Jones-Kraftgesetztes (25.3) und der Anziehungskräfte haben daher HASSÉ und COOK (1929) die Viscosität ohne diese Einschränkung numerisch berechnet, für den Spezialfall $\nu=9$, $\nu'=5$. Seit der

Arbeit von F. LONDON (1930) über die Molekülkräfte (s. Ziff. 24) gibt man dem Lennard-Jonesschen (6; 12)-Potential[1]

$$\Phi = 4\varepsilon\left[\left(\frac{d}{r}\right)^{12} - \left(\frac{d}{r}\right)^{6}\right], \tag{51.11}$$

[vgl. (25.5)], den Vorzug. Dieses wurde in neuerer Zeit, unabhängig voneinander, durch KIHARA und KOTANI (1942), DE BOER und VAN KRANENDONK (1948), HIRSCHFELDER, BIRD und SPOTZ (1948, 1949) sowie ROWLINSON (1949) numerisch vollständig ausgewertet. Diese wichtigen Ergebnisse sind in Tabellenform auch in dem Buch von HIRSCHFELDER, CURTISS und BIRD [25] niedergelegt.

Zunächst ist $\chi(b^*, g^*)$ [s. (27.4)] zu berechnen, wo $b^* = b/d$, $g^* = g\sqrt{m_{\text{red}}/2\varepsilon}$ reduzierte Größen (reine Zahlen) sind. An Hand dieser Funktion sind neun verschiedene Ω-Integrale berechnet worden und in dem genannten Buch tabuliert in Abhängigkeit von der reduzierten Temperatur (reine Zahl)

$$T^* = \frac{T}{\varepsilon/\mathrm{k}}. \tag{51.12}$$

Es ist beispielsweise für Argon $\varepsilon/\mathrm{k} = 123°$ K, so daß hier zu 0° C der Wert $T^* = 2{,}22$ gehört. HIRSCHFELDER *et al.* geben für $\Omega^{(l,r)}$ reduzierte, dimensionslose Werte $\Omega^\star$ an, welche sie durch

$$\Omega^{(l,r)\star} = \frac{\Omega^{(l,r)}}{\Omega^{(l,r)}_{\text{starr}}} \tag{51.13}$$

definieren. Dabei ist $\Omega^{(l,r)}_{\text{starr}}$ der in (51.1) gegebene Wert für harte Kugeln vom Durchmesser d, wo nun aber mit d die charakteristische Länge des Lennard-Jones-Potentials (51.11) gemeint ist. Nach (50.15) sowie (51.3) und (51.13) gilt somit

$$[\eta]_1 = \frac{5}{16\sqrt{\pi}} \cdot \frac{\sqrt{m\,\mathrm{k}T}}{d^2\,\Omega^{(2,2)\star}(T^*)}. \tag{51.14}$$

Als Beispiel sind in Fig. 31 die vier wichtigsten $\Omega^\star(T^*)$ gezeichnet. $\Omega^{(1,1)}$ ist für den Diffusionskoeffizienten maßgebend (s. Ziff. 68); $\Omega^{(1,2)}$ und $\Omega^{(1,3)}$ werden für den Thermodiffusionsfaktor benötigt (s. Ziff. 69).

Nach (51.14) mit (51.12) und Fig. 31 kann die Viscosität für jedwedes Lennard-Jonessche (6; 12)-Gas von bekanntem m, d und ε in Abhängigkeit von der Temperatur angegeben werden.

ε) Exakte Rechnung für das (6; Exp)-Potential. Dieses Potential [s. (25.7)] hat ebenso wie das Lennard-Jones-Potential halbempirischen Charakter, ist aber gleichwohl realistischer und — wegen des weiteren (Steilheits-) Parameters s — anpassungsfähiger. Die Berechnung der Ω-Integrale für $s = 12$, 13, 14 und 15 ist MASON (1954) zu verdanken und durch graphische Extrapolation noch ergänzt worden von MASON und RICE (1954b) für $s = 16$ und 17. Wegen der Zahlenwerte sei auf die Originalarbeiten oder das Buch von HIRSCHFELDER *et al.* [25] verwiesen.

ζ) Prinzip der korrespondierenden Zustände für die Transportkoeffizienten, speziell für die Viscosität. Die Formel (51.14) verkörpert das Prinzip der korrespondierenden Zustände für die Viscosität. Diese läßt sich danach für jedes beliebige Lennard-Jones-Gas auf die einzige universelle Funktion $\Omega^{(2,2)\star}$ zurückführen, wenn nur die Parameter ε und d bekannt sind. Diese Reduktionsmöglichkeit beschränkt sich nicht auf das spezielle Potential (51.11); sie besteht bei beliebiger Wechselwirkung

$$\Phi = \varepsilon\,\varphi\left(\frac{r}{d}\right). \tag{51.15}$$

[1] Wir folgen der Bezeichnung von HIRSCHFELDER *et al.*, welche die Exponenten des Potentials (nicht die der Kraft) zur Charakterisierung benutzen.

Darin sollen ε und d wieder einen Energie- und Längenparameter, φ eine für alle zu vergleichenden Gase gemeinsame Funktion bedeuten. Offenbar ist aus Dimensionsgründen auch in diesem allgemeinen Fall

$$\frac{\eta\, d^2}{\sqrt{m\,\mathrm{k}T}} = F\left(\frac{\mathrm{k}T}{\varepsilon}\right), \tag{51.16}$$

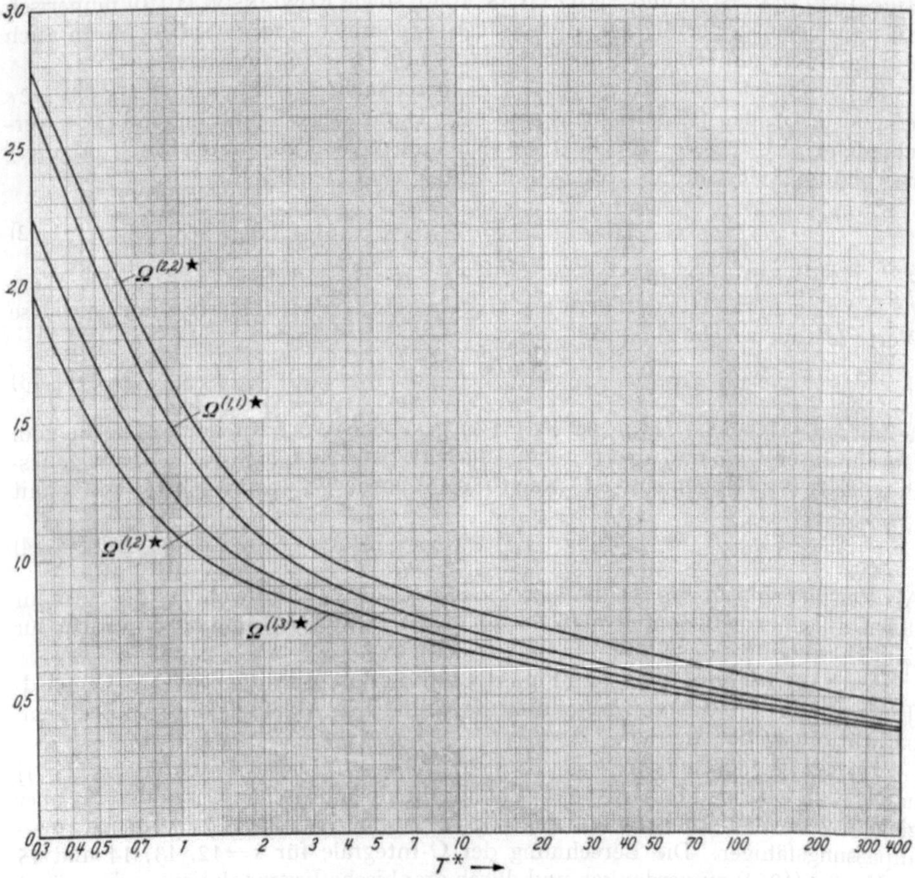

Fig. 31. $\Omega^{(l,r)\star}$ in Abhängigkeit von T^* für das (6; 12)-Potential, s. (51.11) bis (51.13); gezeichnet nach der Tabelle von HIRSCHFELDER, CURTISS und BIRD [25], S. 1126.

wo wieder F eine den Gasen gemeinsame Funktion bedeutet. Man kann die Parameter ε und d mit der kritischen Temperatur und dem kritischen Druck in Verbindung bringen:

$$T_{\mathrm{krit}} \propto \frac{\varepsilon}{\mathrm{k}}\,; \qquad p_{\mathrm{krit}} \propto \frac{\varepsilon}{d^3}\,, \tag{51.17}$$

letzteres, weil der Druck die Dimension Energie pro Volumeinheit hat. Die Proportionalitätsfaktoren werden von der Gestalt der Potentialfunktion φ abhängen. Demnach ist $T_{\mathrm{krit}}/p_{\mathrm{krit}} \propto d^3$ und man hat nach (51.16) auch

$$\frac{\eta}{\sqrt{m\,\mathrm{k}T}}\left(\frac{T_{\mathrm{krit}}}{p_{\mathrm{krit}}}\right)^{\frac{2}{3}} = G\left(\frac{T}{T_{\mathrm{krit}}}\right), \tag{51.18}$$

wo G eine eng mit F zusammenhängende, universelle Funktion ist. In dieser Form hat WEBER (1939) das Prinzip aufgestellt und gezeigt, daß die reduzierten Viskositäten, d.i. die linke Seite von (51.18), für die Gase Ne, A, Kr, X, N_2, O_2 und CO sich durch dieselbe Funktion der reduzierten Temperatur T/T_{krit} darstellen lassen. Da die Prüfung des Prinzips implizit in der Prüfung der spezielleren Beziehung (51.14) enthalten ist (s. die folgende Ziffer), brauchen wir hier nicht weiter darauf einzugehen.

Die Übertragung dieser Betrachtung auf die anderen Transportkoeffizienten bereitet keine Schwierigkeit. Es sei aber noch daran erinnert, daß dem Prinzip der korrespondierenden Zustände in der obigen Form die klassische Mechanik zugrunde liegt. Bei leichten Gasen und tiefen Temperaturen tritt in den universellen Funktionen F bzw. G eine weitere Variable auf, der quantenmechanische Parameter Λ (vgl. Ziff. 81).

52. Quantitativer Vergleich mit Messungen des Reibungs- und Wärmeleitkoeffizienten. Die Massenabhängigkeit von η, die in (51.3) und (51.14) zum Ausdruck kommt, wurde schon in Ziff. 4 besprochen. Hier soll die Temperaturabhängigkeit von η und λ diskutiert werden. Einige ergänzende Betrachtungen über höhere Näherungen und die eindeutige Bestimmbarkeit der Molekularkräfte schließen sich an.

Tabelle 6. *Gemessene Viscosität von Argon und Stickstoff, verglichen mit der nach dem Lennard-Jones(6; 12)-Potential berechneten.*

Auszug nach HIRSCHFELDER, CURTISS und BIRD [25], S. 562; dort auch Quellenangaben.

T (°K)		100	160	220	280	800	1500
η_A (μ-Poise)	exp.	83,9	129,8	173,9	214,5	462,1	677,8
	theor.	81,4	130,0	174,4	214,3	464,1	698,3
η_{N_2} (μ-Poise)	exp.	69,8	106,8	140,3	169,6	349,3	505,0
	theor.	68,7	107,0	140,2	169,3	352,8	526,8

α) Temperaturabhängigkeit von η. Viele $\eta(T)$-Werte sind der Trautzschen Schule zu verdanken [s. z.B. TRAUTZ und ZINK (1930)]. Eine umfangreiche neuere Arbeit stammt von JOHNSTON und GRILLY (1942).

In Tabelle 6 ist die berechnete Viscosität von Argon und Stickstoff mit der gemessenen verglichen; die für die Rechnung benutzten Parameter ε und d sind in Tabelle 7 enthalten. Wie man sieht, gibt das Lennard-Jones-Modell die Viscosität selbst bei dem zweiatomigen Stickstoff in weitem Temperaturbereich befriedigend wieder. Bei mehratomigen, nicht-kugelsymmetrischen Molekülen ist mit dem Potential (51.11) natürlich ein gewisser Mittelwert des wahren Potentials über die gegenseitige Orientierung gemeint.

In Tabelle 7 sind außerdem für einige andere Gase die Lennard-Jones-Parameter angegeben, welche $\eta(T)$ am besten darstellen. Der Parameter $r_m = \sqrt[6]{2}\, d = 1{,}122\, d$ gibt die Lage des Minimums des Lennard-Jones-Potentials an (s. Ziff. 25, Fig. 28d). Die Lennard-Jones-Parameter für He und H_2 wurden erstmals von DE BOER und MICHELS (1938) aus der Zustandsgleichung ermittelt und von DE BOER und KRANENDONK (1948) mit der Viscosität verglichen. SRIVASTAVA und SAXENA (1956) haben sich die Aufgabe gestellt, einen systematischen Zusammenhang der Parameter ε und d mit der Molekülmasse zu finden. — Zur Illustration des Prinzips der korrespondierenden Zustände ist in der Tabelle auch $\mathsf{k}T_{krit}/\varepsilon$ aufgeführt, was bei exakt ähnlichen Potentialkurven exakt konstant sein müßte. Besser wäre es, ε mit $\mathsf{k}T_{Boyle}$ zu vergleichen, doch standen diese

Tabelle 7. *Parameter ε und d für das Lennard-Jones (6; 12)-Potential aus (25.5), bestimmt aus der Viscosität der Gase. Wegen r_m s. Text.* Auszug nach HIRSCHFELDER, CURTISS und BIRD [25], S. 1110. T_{krit} nach D'ANS und LAX [8]. *Parameter s, ε und r_m für das (6; Exp)-Potential aus* (25.7). Nach MASON und RICE (1954a, b).

		He	Ne	A	Kr	X	H_2	N_2	O_2	CO	CO_2	CH_4	C_2H_4
(6; 12)	ε/k (°K) . . .	10,22	35,7	124	190	229	37,0	91,5	113	110	190	137	205
	d (Å)	2,556	2,789	3,418	3,61	4,055	2,928	3,681	3,433	3,590	3,996	3,882	4,232
	r_m (Å)	2,869	3,13	3,84	4,05	4,55	3,287	4,13	3,84	4,03	4,48	4,35	4,74
	T_{krit} (°K) . . .	5,20	44,46	150,76	209,36	289,76	33,26	126,06	154,4	132,96	304,16	190,66	282,7
	$kT_{\text{krit}}/\varepsilon = T^*_{\text{krit}}$	0,51	1,24	1,21	1,10	1,27	0,90	1,38	1,36	1,21	1,60	1,39	1,38
(6; Exp)	s	12,4	14,5	14,0	12,3	13	14,0	17,0		17,0		14,0	
	ε/k (°K) . . .	9,16	38,0	123	158	231	37,3	101		119		153	
	r_m (Å)	3,135	3,147	3,866	4,056	4,450	3,337	4,011		3,937		4,206	

Temperaturen nicht vollständig zur Verfügung[1]. Das starke Herausfallen von He und das schwächere von H_2 hat quantentheoretische Gründe; bei deren kritischer Temperatur macht sich die Wellennatur der Partikeln deutlich bemerkbar. Die Abweichung bei CO_2 ist nicht verwunderlich, da es sich hier um ein recht langgestrecktes Molekül handelt.

Die Parameter des (6; Exp)-Potentials sind für eine Reihe unpolarer Moleküle von MASON und RICE (1954a, b) an Hand von Kristalldaten, zweiten Virialkoeffizienten und Viscositäten ermittelt worden (wegen des experimentellen Materials s. die Originalarbeiten). Sie sind ebenfalls in Tabelle 7 aufgeführt und man sieht, daß die Werte von r_m und ε im großen ganzen in gutem Einklang sind mit den auf Grund des (6; 12)-Potentials gefundenen. Die Wiedergabe der Temperaturabhängigkeit der Viscosität ist bei beiden Potentialen gleich gut, außer bei He und H_2, wo das (6; Exp)-Potential sich als überlegen erweist und offenbar das Abstoßungspotential r^{-12} zu steil ist. Bemerkt sei noch, daß die Quantenkorrekturen an der Viscosität und übrigens auch an dem Wärmeleit- und Selbstdiffusionskoeffizienten dieser beiden Gase für $T > 200°$ K bzw. 250° K kleiner als 0,6% sind. Bei der Ermittlung der Parameter aus den Transportkoeffizienten für diese genügend hohen Temperaturen können also die klassisch berechneten Ω-Integrale verwendet werden.

β) Temperaturabhängigkeit von λ. Diese wird am besten diskutiert an Hand des Verhältnisses $\lambda/\eta c_v$, das nach (2.26) umgekehrt proportional der Prandtl-Zahl ist und nach (50.16) in der ersten Näherung für jedes Kraftgesetz den Wert 2,5 haben sollte, natürlich nur für einatomige Gase. Die vollständigsten Experimente darüber haben KANNULUIK und CARMAN (1952) gemacht (s. Tabelle 8).

Eine ähnliche Tabelle findet sich auch bei E. U. FRANCK (1951); dort weitere Zitate. Das dort und auch das bei KEESOM [33], S. 111, über Helium bei tiefen Temperaturen Gesagte ist nicht richtig, weil dabei der Wert $\lambda_{\text{He}}(20{,}9°\text{ K}) = 51{,}8\ \mu\text{cal}/°\text{ cm sec}$ von EUCKEN (1913) verwendet ist. Dieser mittels der Methode des koaxialen Drahts gewonnene λ-Wert ist offenbar zu klein. UBBINK und DE HAAS (1943b) finden mit der Plattenmethode $\lambda_{\text{He}}(20{,}1°\text{ K}) = 62{,}4\ \mu\text{cal}/°\text{ cm sec}$. Dieser Wert wurde für Tabelle 8 kombiniert mit $\eta_{\text{He}}(20{,}17°\text{ K}) = 34{,}98\ \mu$ Poise nach KAMERLINGH ONNES und WEBER (1913). Dadurch entfällt beim Helium für 20° K praktisch die Abweichung

[1] T_{Boyle} gibt die Temperatur an, bei welcher der 2. Virialkoeffizient verschwindet. Es ist fast genau $kT_{\text{Boyle}}/\varepsilon = 3{,}4$ für Ne, A, H_2, N_2, dagegen $= 2{,}5$ für He.

des Verhältnisses $\lambda/\eta c_v$ vom theoretischen Wert 2,5. Die in der früheren Literatur festgestellte Abweichung wurde zu Unrecht auf den Einfluß der Quantenmechanik geschoben; der Wert 2,5 muß ja — in erster Näherung — für jedweden Streuquerschnitt, klassisch oder quantenmechanisch, gelten.

Wie man aus Tabelle 8 sieht, gibt es aber doch einige systematische Abweichungen. Bei 100° C sind zwischen He und X fast 7% Unterschied. Das sollte eigentlich nicht sein. So groß ist die Ungenauigkeit der ersten Näherung nicht, wie anschließend gezeigt wird. Man möchte einen leichten systematischen Fehler in den Meßwerten für λ vermuten. Wenn die Diskrepanz reell wäre, würde sie eine ernste Schwierigkeit für die Theorie darstellen.

Etwas grob gesehen, ist aber die wichtige Beziehung (50.16) durch das Experiment bestätigt.

Tabelle 8. *$\lambda\eta/c_v$ für Edelgase bei etwa 1 atm nach Kannuluik und Carmann (1952).* Wegen He (20° K) s. Text.

T (°K)	He	Ne	A	X
20,1	2,40	—	—	—
90,19	2,43	2,49	2,50	—
194,66	2,42	2,48	2,52	—
273,16	2,44	2,48	2,51	2,58
373,16	2,44	2,48	2,51	2,61
491,16	2,43	2,44	2,50	2,58
579,06	2,42	2,47	2,48	2,54

γ) Die höheren Näherungen für η und λ. In Ziff. 50 wurden nur die ersten Näherungen $[\eta]_1$, $[\lambda]_1$ betrachtet, welche für das Maxwell-Gas exakt sind. Die höheren Näherungen für starre Kugeln sind bei CHAPMAN und COWLING ([*4*], S. 169) angegeben; es ist

$$\left(\frac{\lambda}{\eta\, c_v}\right)_{\text{starr, exakt}} = 2{,}522,$$

womit es bereits wahrscheinlich erscheint, daß auch für ein realistisches Potential die Korrektur gegenüber der ersten Näherung weniger als 1% beträgt. Das wird bestätigt durch die Rechnungen von HIRSCHFELDER *et al.* ([*25*], S. 1130) am (6; 12)-Potential, aus denen man etwa für $\lambda/\eta c_v$ bei $T^* = 2$ eine Korrektur von nur 0,7% entnimmt. Dasselbe liefert die ebenfalls dort (S. 609) erwähnte Korrektur nach KIHARA [*32*], welche besagt, daß

$$\frac{\lambda}{\eta} = \left[\frac{\lambda}{\eta}\right]_1 \left[1 + 0{,}034\left(\frac{4\Omega^{(2,3)\star}}{\Omega^{(2,2)\star}} - \frac{7}{2}\right)^2\right].$$

Noch weiter hat neuerdings MASON (1957) die Näherung getrieben.

δ) Lassen sich aus $\eta(T)$ die Molekularkräfte eindeutig bestimmen? Die Frage ist wohl zu verneinen. Das Lennard-Jones-Potential hat einen hohen Grad von Plausibilität, jedoch stecken darin quantentheoretische Überlegungen sowie andere Erfahrungen, z. B. die der Kondensation, welche auf Anziehung bei großem, Abstoßung bei kleinem Abstand schließen läßt. Nimmt man die Funktion (25.5) einschließlich der Vorzeichen als richtig an, so kann man nach HIRSCHFELDER *et al.* aus $\eta(T)$ recht genau die Parameter ε und d ermitteln. Aber man kann denselben Verlauf $\eta(T)$ unter Umständen auch mit reinen Abstoßungskräften bekommen, worauf CHAPMAN und COWLING ([*4*], S. 228) hinwiesen; der Zusammenhang von Viscosität und Potential dürfte nicht eindeutig umkehrbar sein.

Das kann man sich an dem einfachen Modell der elastischen Kugeln mit umgebendem Potentialtopf oder -berg klarmachen. Beim Topf hat man ungefähr das Sutherland-Modell und daher nimmt $\eta/\sqrt{T}$ mit T zu, s. (51.10). Beim Berg hat man bei niedriger Temperatur nach (51.3) $\eta/\sqrt{T} \propto d_a^{-2}$, bei hoher Temperatur $\eta/\sqrt{T} \propto d_i^{-2}$, wo $d_{a,i}$ den äußeren und inneren Durchmesser des Topfes bedeutet. Also ergibt auch diese Annahme ein Anwachsen von $\eta/\sqrt{T}$ mit der Temperatur.

Es ist daher wohl nicht möglich aus $\eta(T)$ die molekulare Wechselwirkung qualitativ zu erschließen. Setzt man aber ein Potential von der Art des Lennard-Jonesschen von vornherein voraus, so kann man dessen Parameter an Hand der Viscositätsmessungen eindeutig bestimmen, wie es zu Beginn dieser Ziffer dargelegt worden ist.

Es ist daher um so wichtiger, daß außer der Viscosität andere Daten, Diffusions- und Thermodiffusionskoeffizienten, zur Verfügung stehen, die das Bild vervollständigen und es erlauben, gewisse Kraftgesetze, z. B. Abstoßung nach reinem Potenzgesetz, mit Sicherheit auszuschließen (vgl. Ziff. 75, β).

F. Diffusion, Wärmeleitung und Reibung in Gasgemischen.

Die Theorie der Gemische, spezialisiert auf den binären Fall zweier Komponenten, ist in den grundlegenden Arbeiten von ENSKOG (1917) und CHAPMAN (1917) enthalten. Die Übertragung auf den polynären Fall vieler Komponenten haben HELLUND und UEHLING (1939) sowie HELLUND (1940) unter gleichzeitiger Berücksichtigung der Quantenstatistik vorgenommen. Neuerdings betrachteten CURTISS und HIRSCHFELDER (1949) nochmals die polynären Gemische und erweiterten die Anwendungen der Theorie.

Die Vorgänge in Gasgemischen sind infolge des Auftretens der Diffusion vielfältiger als diejenigen in einfachen Gasen. Entsprechend weitläufiger ist die theoretische Behandlung. Bei einem Gemisch von K Komponenten gibt es $\frac{1}{2}K(K-1)$ linear unabhängige Diffusions- und $K-1$ linear unabhängige Thermodiffusionskoeffizienten neben dem Wärmeleit- und dem Reibungskoeffizienten.

Die allgemeinen Grundlagen der Chapman-Enskogschen hydrodynamischen Näherung für polynäre einatomige Gemische werden im folgenden ungekürzt dargestellt. Dabei ergibt sich der allgemeine Zusammenhang der Thermodiffusion mit ihrem Gegenstück, der Diffusionswärme. Die Diskussion der Thermodiffusion und der Wärmeleitung wird durch die Einführung der Thermodiffusionsverhältnisse k_{Ti}, die für polynäre Gemische anscheinend bisher in der Literatur nicht benutzt wurden, besonders übersichtlich. Mit ihrer Hilfe läßt sich der (experimentelle) Wärmeleitkoeffizient formal exakt darstellen (Abschnitt I).

Im weiteren muß auf die lückenlose Berechnung der Transportkoeffizienten, wie sie im Teil E für die einfachen Gase vorgeführt wurde, verzichtet werden. Als Beispiel wird nur die erste Näherung für den gewöhnlichen (Konzentrations-) Diffusionskoeffizienten vollständig behandelt, während wir uns bei der Darstellung der Koeffizienten der Wärmeleitung, Thermodiffusion und Reibung durch Ω-Integrale auf die Andeutung des Gangs der Rechnung und die Mitteilung von Ergebnissen beschränken müssen (Abschnitt II).

Schließlich werden spezielle Anwendungen betrachtet und das Erfahrungsmaterial diskutiert (Abschnitte III bis V).

I. Formale Grundlagen der hydrodynamischen Näherung.

Für den Grundgedanken des Verfahrens gilt wieder das in Ziff. 43 Gesagte. Es wird vorausgesetzt, daß sich das Gas lokal stets nahezu im thermischen Gleichgewicht befindet. Für die Verteilungsfunktionen der Komponenten wird ein entsprechender Ansatz gemacht. Anschließend stellen wir die Lösbarkeitsbedingungen des zugehörigen genäherten Systems von linearen Integralgleichungen auf. Aus ihnen ergibt sich die genauere Gestalt der Verteilungsfunktionen, worauf allgemeine Formeln für die Transportkoeffizienten gewonnen werden. In Ziff. 56 und 57 betrachten wir Diffusion und Wärmeleitung zunächst vom Bezugssystem der mittleren Teilchengeschwindigkeit aus, was für die Anwendungen nützlich ist. Ziff. 58 ergänzt dies durch die Betrachtung des Bezugssystems der mittleren Massengeschwindigkeit. Den Abschluß des Abschnitts I bilden Ausführungen über die lokale Entropieerzeugung.

53. Ansatz für die Verteilungsfunktionen. Genäherte Boltzmann-Gleichungen. Für die Verteilungsfunktion der i-ten Komponente wird geschrieben

$$f_i(t, \boldsymbol{r}, \boldsymbol{c}) = F_i(t, \boldsymbol{r}, \boldsymbol{C})\left(1 + \Phi_i^{(1)}(t, \boldsymbol{r}, \boldsymbol{C})\right); \qquad i = 1, \ldots, K. \tag{53.1}$$

Dabei sind die

$$F_i = n_i\left(\frac{\beta_i}{\pi}\right)^{\frac{3}{2}} \mathrm{e}^{-\beta_i C^2} \quad \text{mit} \quad \beta_i = \frac{m_i}{\mathrm{k}T}, \qquad \boldsymbol{C} = \boldsymbol{c} - \boldsymbol{v} \tag{53.2}$$

Maxwell-Verteilungen mit gleicher Temperatur T und gleicher mittlerer Massengeschwindigkeit $\boldsymbol{v}$. Die Parameter n_i, $\boldsymbol{v}$, T seien so gewählt, daß

$$\left.\begin{aligned} n_k &= \sum_i \int \delta_i^{(k)} F_i\, d\boldsymbol{c} = \sum_i \int \delta_i^{(k)} f_i\, d\boldsymbol{c} \\ v_\mu &= \frac{1}{m\,n} \sum_i \int m_i c_\mu F_i\, d\boldsymbol{c} = \frac{1}{m\,n} \sum_i \int m_i c_\mu f_i\, d\boldsymbol{c} \\ \frac{3}{2} \mathsf{k} T &= \frac{1}{n} \sum_i \int \frac{m_i}{2} C^2 F_i\, d\boldsymbol{c} = \frac{1}{n} \sum_i \int \frac{m_i}{2} C^2 f_i\, d\boldsymbol{c}, \end{aligned}\right\} \tag{53.3}$$

s. dazu (32.5), (32.8) und (32.15). Die Funktionen $\Phi_i^{(1)}$ sind also den $K+4$ Bedingungen

$$\sum_i \int \psi_i F_i \Phi_i^{(1)}\, d\boldsymbol{c} = 0 \tag{53.4}$$

unterworfen, wo die ψ_i die $K+4$ Sätze von Stoßinvarianten aus (32.4) bedeuten:

$$\psi_i = \delta_i^{(k)}; \quad m_i c_\mu; \quad \frac{m_i}{2} c^2. \tag{53.5}$$

Für die Transportgrößen im $\boldsymbol{v}$-System, nämlich die Diffusionsgeschwindigkeiten, den Drucktensor und den Wärmestrom [s. (32.12), (32.14) und (32.16)] gilt offenbar

$$V_{\mu i} = v_{\mu i} - v_\mu = \int C_\mu \frac{F_i}{n_i} \Phi_i^{(1)}\, d\boldsymbol{c} \tag{53.6}$$

$$p_{\mu\nu} = \sum_i \int m_i C_\mu C_\nu F_i \Phi_i^{(1)}\, d\boldsymbol{c} \tag{53.7}$$

$$q_\mu = \sum_i \int \frac{m_i}{2} C^2 C_\mu F_i \Phi_i^{(1)}\, d\boldsymbol{c}. \tag{53.8}$$

Gl. (53.4) besagt also, daß die Anteile $F_i \Phi_i^{(1)}$ der Verteilungsfunktionen nichts beitragen zu den Werten der lokalen Konzentration jeder Komponente, der Gasgeschwindigkeit und der Temperatur, während (53.6) bis (53.8) bedeuten, daß die Maxwellschen Anteile F_i der Verteilungsfunktionen nichts zu den Transportgrößen beitragen.

Für später sind auch die Diffusionsgeschwindigkeiten und der Wärmestrom im $\boldsymbol{w}$-System zu notieren. Nach (32.28) und (32.36) hat man

$$W_{\mu i} = \int C_\mu \left(\frac{F_i}{n_i} \Phi_i^{(1)} - \frac{1}{n} \sum_l F_l \Phi_l^{(1)} \right) d\boldsymbol{c} \tag{53.9}$$

$$q_\mu^{(w)} = \sum_i \int \left(\frac{m_i}{2} C^2 - \frac{5}{2} \mathsf{k} T \right) C_\mu F_i \Phi_i^{(1)}\, d\boldsymbol{c}. \tag{53.10}$$

Das genäherte Boltzmann-Gleichungssystem für die $\Phi_i^{(1)}$ erhält man aus (29.2), indem man dort links die exakten Verteilungsfunktionen f_i durch die Maxwell-Funktionen F_i ersetzt und rechts Glieder in $(\Phi^{(1)})^2$ wegläßt:

$$\mathscr{D}_i(F_i) \approx - \sum_j n_i n_j I_{ij}(\Phi^{(1)}). \tag{53.11}$$

Die Abkürzungen haben folgende Bedeutung:

$$\left.\begin{aligned} \mathscr{D}_i &= \frac{\partial}{\partial t} + c_\mu \frac{\partial}{\partial x_\mu} + b_{\mu i} \frac{\partial}{\partial c_\mu} = \frac{d}{dt} + C_\mu \frac{\partial}{\partial x_\mu} + b_{\mu i} \frac{\partial}{\partial c_\mu} \\ \text{mit} \qquad & \frac{d}{dt} = \frac{\partial}{\partial t} + v_\mu \frac{\partial}{\partial x_\mu} \end{aligned}\right\} \tag{53.12}$$

und

$$\left.\begin{aligned} n_i n_j I_{ij}(\Phi) &= \iint F_i(\boldsymbol{C}) F_j(\boldsymbol{C}_1)\, \delta_{ij}(\Phi)\, \sigma_{ij}\, g\, d\boldsymbol{e}'\, d\boldsymbol{c}_1 \\ \text{mit} \qquad \delta_{ij}(\Phi) &= \Phi_i(\boldsymbol{c}) + \Phi_j(\boldsymbol{c}_1) - \Phi_i(\boldsymbol{c}') - \Phi_j(\boldsymbol{c}_1'). \end{aligned}\right\} \tag{53.13}$$

54. Die Lösbarkeitsbedingungen. Die genäherten Boltzmann-Gleichungen (53.11) sind ein System gekoppelter Integralgleichungen für die Unbekannten $\Phi_i^{(1)}$ und nur lösbar, wenn die linke Seite $K+4$ Orthogonalitätsbedingungen erfüllt.

Um diese zu formulieren, sei die dem Gemisch als Ganzem zugeordnete nur noch von Zeit und Ort abhängige Größe eingeführt

$$[\Psi, \Phi] = \sum_{i,j} \int \Psi_i \gamma_i \gamma_j I_{ij}(\Phi)\, d\boldsymbol{c} \quad \text{mit} \quad \gamma_i = \frac{n_i}{n}. \tag{54.1}$$

Sie ist bilinear

$$[\Psi, \Phi_{\mathrm{I}} + \Phi_{\mathrm{II}}] = [\Psi, \Phi_{\mathrm{I}}] + [\Psi, \Phi_{\mathrm{II}}], \text{ usw.} \tag{54.2}$$

und symmetrisch, denn es gilt nach (53.13)

$$\left.\begin{aligned} [\Psi, \Phi] &= \frac{1}{n^2} \sum_{i,j} \iiint F_i(\boldsymbol{C})\, F_j(\boldsymbol{C}_1)\, \Psi_i\, \delta_{ij}(\Phi)\, \sigma_{ij}\, g\, d\boldsymbol{e}'\, d\boldsymbol{c}\, d\boldsymbol{c}_1 \\ &= \frac{1}{4n^2} \sum_{i,j} \iiint F_i(\boldsymbol{C})\, F_j(\boldsymbol{C}_1)\, \delta_{ij}(\Psi)\, \delta_{ij}(\Phi)\, \sigma_{ij}\, g\, d\boldsymbol{e}'\, d\boldsymbol{c}\, d\boldsymbol{c}_1 = [\Phi, \Psi]. \end{aligned}\right\} \tag{54.3}$$

Dies ergibt sich durch dieselben Umbenennungen der Integrationsvariablen $\boldsymbol{c}$, $\boldsymbol{c}_1$ oder $\boldsymbol{C}$, $\boldsymbol{C}_1$ sowie der Indices i, j, wie sie zu (32.1) führten. Ferner sieht man, daß $[\Psi, \Psi]$ positiv-semidefinit ist. Verschwinden kann $[\Psi, \Psi]$ nur, wenn $\delta_{ij}(\Psi) = 0$, d.h. wenn die Ψ_i identisch verschwinden oder eine Linearkombination ψ_i der Stoßinvarianten (53.5) sind. Somit hat man

$$\left.\begin{aligned} [\Psi, \Psi] &> 0 \quad \text{für} \quad \Psi_i \neq 0, \ \neq \psi_i \\ \text{und} \qquad [\psi, \psi] &= 0. \end{aligned}\right\} \tag{54.4}$$

Für eine beliebige Linearkombination ψ_i der Stoßinvarianten gilt ferner nach (54.3) bei beliebigen Funktionen Φ_i

$$[\psi, \Phi] = 0. \tag{54.5}$$

Die Klammer $[\Psi, \Phi]$ ist von der Gesamtkonzentration n unabhängig, wenn dies für Ψ und Φ gilt.

Aus der Beziehung (54.5) ergeben sich für das System (53.11) als notwendige Lösbarkeitsbedingungen

$$\sum_i \int \psi_i \mathscr{D}_i(F_i)\, d\boldsymbol{c} = 0. \tag{54.6}$$

In der Theorie der linearen Integralgleichungen wird gezeigt, daß diese Bedingungen auch hinreichend für die Lösbarkeit sind. Es sind dieselben Gleichungen wie in (32.3), nur stehen jetzt die Maxwell-Funktionen F_i anstelle der exakten Verteilungsfunktionen f_i bei den früheren, zu den Erhaltungsgleichungen (32.17), (32.19) und (32.23) führenden Betrachtungen. Die Lösbarkeitsbedingungen müssen daher mit den zitierten Erhaltungsgleichungen identisch werden, wenn man darin die Transportgrößen $V_{\mu i}$, $p_{\mu\nu}$, q_μ durch ihre Näherungen nullter Ordnung, wie sie sich aus den Maxwell-Funktionen F_i ergeben, ersetzt:

$$V_{\mu i}^{(0)} = 0, \qquad p_{\mu\nu}^{(0)} = n\,\mathrm{k}T\,\delta_{\mu\nu} = p\,\delta_{\mu\nu}, \qquad q_\mu^{(0)} = 0.$$

Ausführlich lauten die Bedingungen (54.6) also

$$\left.\begin{aligned} \frac{dn_i}{dt} &= -\, n_i \frac{\partial v_\mu}{\partial x_\mu}, \\ \varrho \frac{dv_\mu}{dt} &= -\frac{\partial p}{\partial x_\mu} + \sum_i \varrho_i b_{\mu i}, \\ \frac{3}{2} n\,\mathrm{k} \frac{dT}{dt} &= -\, p \frac{\partial v_\mu}{\partial x_\mu}. \end{aligned}\right\} \tag{54.7}$$

Die Lösbarkeitsbedingungen (54.7) kann man benutzen um die Zeitableitungen der in F_i enthaltenen Parameter n_i, v_μ, T, welche links in (53.11) in den Ausdrücken $\mathcal{D}_i(F_i)$ vorkommen, zu eliminieren. Dann wird die linke Seite von (53.11) nur noch Ableitungen dieser Parameter nach den Ortskoordinaten enthalten und die Lösbarkeitsbedingungen bei beliebiger Vorgabe dieser Ortsableitungen identisch erfüllen. Um die erwähnte Elimination zu bewirken, schreibt man sich die logarithmische Ableitung der Maxwell-Funktion F_i auf:

$$\frac{dF_i}{F_i} = \frac{dn_i}{n_i} + \left(\beta_i C^2 - \frac{3}{2}\right)\frac{dT}{T} - 2\beta_i C_\mu d(c_\mu - v_\mu).$$

Ausführung der Differentiationen in $\mathcal{D}_i(F_i)$ mit Benutzung von (54.7) liefert sodann zunächst

$$\mathcal{D}_i(F_i) = F_i\left\{C_\mu\left[\frac{1}{n_i}\frac{\partial n_i}{\partial x_\mu} + \frac{2\beta_i}{\varrho}\left(-\frac{\partial p}{\partial x_\mu} + \sum_k \varrho_k b_{\mu k}\right) - 2\beta_i b_{\mu i} + \right.\right.$$
$$\left.\left. + \left(\beta_i C^2 - \frac{3}{2}\right)\frac{\partial T}{T\,\partial x_\mu}\right] + 2\beta_i\left[C_\mu C_\nu \frac{\partial v_\mu}{\partial x_\nu} - \frac{1}{3}C^2\frac{\partial v_\mu}{\partial x_\mu}\right]\right\}.$$

Definitionsgemäß ist $p = n\mathrm{k}T$, wo $n = \sum_i n_i$, also ließe sich der Gradient von p durch den von n und T ausdrücken. Da bei vielen Anwendungen aber der Gradient von p verschwindet, ist es zweckmäßig ihn beizubehalten und statt dessen auf die Variable n zu verzichten. Wir führen deshalb die Relativkonzentrationen oder Molenbrüche γ_i ein durch

$$\gamma_i = \frac{n_i}{n} \quad \text{mit} \quad \sum_i \gamma_i = 1, \tag{54.8}$$

haben dann

$$\frac{dn_i}{n_i} = \frac{d\gamma_i}{\gamma_i} + \frac{dp}{p} - \frac{dT}{T}$$

und erhalten damit anstelle von (53.11) das Integralgleichungssystem

$$F_i \cdot\left[C_\mu \frac{1}{\gamma_i} d_{\mu i} + \left(\beta_i C^2 - \frac{5}{2}\right) C_\mu \frac{\partial T}{T\,\partial x_\mu} + 2\beta_i \langle C_\mu C_\nu\rangle\left\langle\frac{\partial v_\mu}{\partial x_\nu}\right\rangle\right] = -\sum_j n_i n_j I_{ij}(\Phi^{(1)}). \tag{54.9}$$

Darin bedeutet

$$d_{\mu i} = \frac{\partial \gamma_i}{\partial x_\mu} + \left(\gamma_i - \frac{\varrho_i}{\varrho}\right)\frac{\partial p}{p\,\partial x_\mu} - \frac{\varrho_i}{p}\left(b_{\mu i} - \sum_k \frac{\varrho_k}{\varrho} b_{\mu k}\right). \tag{54.10}$$

Wegen $\sum \gamma_i = \sum \varrho_i/\varrho = 1$ ist

$$\sum_i d_{\mu i} = 0. \tag{54.11}$$

Die symmetrischen, irreduziblen Tensoren $\langle C_\mu C_\nu\rangle$, $\left\langle\frac{\partial v_\mu}{\partial x_\nu}\right\rangle$ wurden schon früher, in (46.4) und (46.5), erläutert.

Das Integralgleichungssystem (54.9) erfüllt, bei willkürlicher Vorgabe der Größen $d_{\mu i}$ [allerdings gemäß (54.11) eingeschränkt], $\frac{\partial T}{\partial x_\mu}$ und $\left\langle\frac{\partial v_\mu}{\partial x_\nu}\right\rangle$ die Lösbarkeitsbedingungen identisch und bestimmt die Funktionen $\Phi_i^{(1)}$ bis auf eine beliebige additive Linearkombination der Stoßinvarianten (53.5). Da aber die $\Phi_i^{(1)}$ außerdem den $K+4$ Bedingungen (53.4) genügen sollen, fällt auch diese Unbestimmtheit weg: (54.9) legt zusammen mit (53.4) die $\Phi_i^{(1)}$ eindeutig fest.

55. Die Gestalt der Funktionen $\Phi_i^{(1)}$. Integralgleichungen für ihre verschiedenen Anteile. Entsprechend der Linearität der Integraloperatoren I_{ij} in (54.9) hat man, analog wie in Ziff. 47, für $\Phi_i^{(1)}$ einen in den Tensoren $d_{\mu i}$, $\frac{\partial T}{\partial x_\mu}$, $\left\langle\frac{\partial v_\mu}{\partial x_\nu}\right\rangle$ linearen, skalaren Ausdruck anzusetzen

$$\Phi_i^{(1)} = -\frac{1}{n}\Big(\sum_k A_i^{(k)}(C)\, C_\mu\, d_{\mu k} + A_i(C)\, C_\mu \frac{\partial T}{T\,\partial x_\mu} + B_i(C)\,\beta_i \langle C_\mu C_\nu\rangle \left\langle\frac{\partial v_\mu}{\partial x_\nu}\right\rangle\Big). \tag{55.1}$$

Die Funktionssätze $A^{(k)}$, A, B haben die Dimension sec/cm³ und hängen nur vom Betrag der Eigengeschwindigkeit der Atome ab; sie sind von der Gesamtkonzentration n unabhängig.

Nun soll der Ausdruck (55.1) in das Gleichungssystem (54.9) eingesetzt und dann gliedweise verglichen werden. Dabei ist noch hinderlich, daß die $d_{\mu k}$ durch die Relation (54.11) linear abhängig sind. Man kann die $d_{\mu k}$ durch linear unabhängige Größen $X_{\mu k}$ darstellen mittels

$$d_{\mu k} = X_{\mu k} - \Lambda_k \sum_l X_{\mu l}. \tag{55.2}$$

Dabei ist lediglich zu verlangen, daß für die ansonsten willkürlichen Faktoren Λ_k

$$\sum_k \Lambda_k = 1 \tag{55.3}$$

gilt. Im übrigen sei die Wahl der Λ_k zunächst offen gelassen; sie wird in (56.4) getroffen werden. So erhält man aus (54.9) die getrennten Gleichungssysteme

$$\sum_j \gamma_j I_{ij}\Big[\Big(A^{(k)} - \sum_l A^{(l)} \Lambda_l\Big) C_\mu\Big] = \frac{F_i}{n_i} C_\mu \Big(\frac{\delta_{ik}}{\gamma_i} - \frac{\Lambda_i}{\gamma_i}\Big) \tag{55.4}$$

$$\sum_j \gamma_j I_{ij}(A\, C_\mu) = \frac{F_i}{n_i}\Big(\beta_i C^2 - \frac{5}{2}\Big) C_\mu \tag{55.5}$$

$$\sum_j \gamma_j I_{ij}(B\beta\langle C_\mu C_\nu\rangle) = 2\frac{F_i}{n_i}\beta_i \langle C_\mu C_\nu\rangle. \tag{55.6}$$

Durch (55.3) ist gerade die Lösbarkeit von (55.4) garantiert, wie man sich leicht überzeugt.

Dazu kommen noch die Nebenbedingungen (53.4). Die meisten davon sind durch den Ansatz (55.1) eo ipso erfüllt, weil die betreffenden Integranden ungerade in $\boldsymbol{C}$ sind; es bleibt nur zu fordern, daß erstens

$$\sum_i \int m_i C_\mu F_i \sum_k A_i^{(k)} C_\nu d_{\nu k}\, d\boldsymbol{C} = 0$$

oder auch nach (55.2)

$$\sum_i \int m_i C^2 F_i \Big(A_i^{(k)} - \sum_l A_i^{(l)} \Lambda_l\Big)\, d\boldsymbol{C} = 0 \quad \text{für alle } k \tag{55.7}$$

und analog dazu, daß zweitens

$$\sum_i \int m_i C^2 F_i A_i\, d\boldsymbol{C} = 0. \tag{55.8}$$

Durch (55.4) und (55.7) sind die Funktionssätze $A^{(k)} - \sum A^{(l)} \Lambda_l$ eindeutig bestimmt und Entsprechendes gilt für die A. Die B sind bereits durch (55.6) eindeutig festgelegt. Die Sätze $A^{(k)}$ selbst bleiben auch nach Wahl der Parameter Λ_l etwas unbestimmt; über die Linearkombination $\sum_l A^{(l)} \Lambda_l$ kann nämlich beliebig verfügt werden. Es ist keine Beschränkung der Allgemeinheit, jedoch zweckmäßig zu verlangen, daß

$$\sum_l A_i^{(l)} \Lambda_l = 0. \tag{55.9}$$

Dies ist künftig angenommen. Dadurch sind die $A^{(k)}$ nach getroffener Wahl der Parameter Λ_l eindeutig bestimmt.

56. Die Diffusionsgeschwindigkeiten und -gleichungen im Bezugssystem der mittleren Teilchengeschwindigkeit $\boldsymbol{w}$. Trägt man (55.1) in (53.9) ein, so entsteht

$$W_{\mu i} = -\sum_k D_{ik} d_{\mu k} - D_{Ti} \frac{\partial T}{T \partial x_\mu}, \tag{56.1}$$

wobei

$$D_{ik} = \frac{1}{3n} \int C^2 \left(\frac{F_i}{n_i} A_i^{(k)} - \frac{1}{n} \sum_l F_l A_l^{(k)} \right) d\boldsymbol{C}, \tag{56.2}$$

$$D_{Ti} = \frac{1}{3n} \int C^2 \left(\frac{F_i}{n_i} A_i - \frac{1}{n} \sum_l F_l A_l \right) d\boldsymbol{C}. \tag{56.3}$$

Die Diffusionskoeffizienten D_{ik} und die Thermodiffusionskoeffizienten D_{Ti} haben die Dimension cm²/sec.

Es ist möglich diese Koeffizienten mit Hilfe der in (54.1) definierten Klammer darzustellen. Zu dem Zweck sei (55.4) nochmals aufgeschrieben unter Verwendung von (55.9) und mit dem Index l statt i, sowie i statt k:

$$\sum_j \gamma_l \gamma_j I_{lj}(A^{(i)} C_\mu) = \frac{F_l}{n_l} C_\mu (\delta_{li} - \Lambda_l).$$

Diese Gleichung werde mit $\frac{1}{3} A_l^{(k)} C_\mu \, d\boldsymbol{C}$ durchmultipliziert, integriert und über l summiert. Dadurch erhält man

$$\frac{1}{3} [A^{(i)} C_\mu, A^{(k)} C_\mu] = \frac{1}{3} \int C^2 \left(\frac{F_i}{n_i} A_i^{(k)} - \frac{1}{n} \sum \frac{\Lambda_l}{\gamma_l} F_l A_l^{(k)} \right) d\boldsymbol{C}.$$

Jetzt werde eine spezielle Wahl der Faktoren Λ_l getroffen, und zwar so, daß die rechte Seite dieser Gleichung durch den Diffusionskoeffizienten (56.2) dargestellt werden kann. Dazu ist zu setzen

$$\Lambda_l = \gamma_l. \tag{56.4}$$

Der Integrabilitätsbedingung (55.3) ist dadurch ebenfalls genügt. Mit dieser Wahl der Λ_l hat man

$$D_{ik} = \frac{1}{3n} [A^{(k)} C_\mu, A^{(i)} C_\mu], \tag{56.5}$$

und in analoger Weise ergibt sich statt (56.3)

$$D_{Ti} = \frac{1}{3n} [A C_\mu, A^{(i)} C_\mu]. \tag{56.6}$$

Aus (56.5) und (54.3) folgt, daß die Matrix der Diffusionskoeffizienten symmetrisch ist[1]:

$$D_{ik} = D_{ki}. \tag{56.7}$$

Nach (54.4) ist übrigens $D_{ii} > 0$; die allgemeinen Vorzeicheneigenschaften der D_{ik} werden in Ziff. 61 näher besprochen. Die Diffusionskoeffizienten sind der Gesamtkonzentration umgekehrt proportional, wie es sein muß.

Da die $A_i^{(l)}$ nach (55.9) linear abhängig sind, gilt dies auch für die D_{ik} und die D_{Ti}. Mit der speziellen Wahl (56.4) lautet (55.9)

$$\sum_k A_i^{(k)} \gamma_k = 0. \tag{56.8}$$

---

[1] Hierin findet für den speziellen Fall des einatomigen Gasgemischs, ein Teil der Onsagerschen Reziprozitätsrelationen der irreversiblen Thermodynamik seinen Ausdruck, siehe auch S. 410, Fußnote 1.

Daher ist nach (56.5) und (56.6)

$$\left.\begin{aligned} \sum_k D_{ik}\gamma_k = \sum_i \gamma_i D_{ik} = 0\,, \\ \sum_i \gamma_i D_{Ti} = 0\,. \end{aligned}\right\} \tag{56.9}$$

Es gibt also $\frac{1}{2}K(K-1)$ linear unabhängige Diffusionskoeffizienten und $K-1$ linear unabhängige Thermodiffusionskoeffizienten.

Die Thermodiffusionskoeffizienten D_{Ti} wurden von MEIXNER (1943a) eingeführt bei der Untersuchung des Zusammenhangs von Thermodiffusion und Diffusionswärme in polynären Gemischen nach der Methode der irreversiblen Thermodynamik und unabhängig von WALDMANN (1947a) bei den entsprechenden gaskinetischen Betrachtungen. Es ist aber für das Folgende zweckmäßig, hier noch einen anderen Satz von $K-1$ linear unabhängigen Größen, die Thermodiffusionsverhältnisse k_{Ti} zu definieren[1] durch die Festsetzung

$$D_{Ti} = \sum_k D_{ik} k_{Tk}\,, \tag{56.10}$$

wobei

$$\sum_i k_{Ti} = 0 \tag{56.11}$$

gelten soll[2]. Mit Verwendung der k_T schreibt sich (56.1) so

$$W_{\mu i} = -\sum_k D_{ik}\left(d_{\mu k} + k_{Tk}\frac{\partial T}{T\,\partial x_\mu}\right). \tag{56.12}$$

Die Einführung der k_{Ti} erweist sich als nützlich bei der Definition des Wärmeleitkoeffizienten und des Diffusionswärmestroms (Ziff. 57).

Man kann gemäß (55.9) und (56.4) jetzt (55.4) die endgültige Gestalt geben

$$\sum_j \gamma_j I_{ij}(A^{(k)} C_\mu) = \frac{F_i}{n_i} C_\mu\left(\frac{\delta_{ik}}{\gamma_i} - 1\right). \tag{56.13}$$

Die Nebenbedingungen (55.9) und (55.7) lauten

$$\sum_k A_i^{(k)}\gamma_k = 0\,; \quad \sum_i \int m_i C^2 F_i A_i^{(k)}\, d\boldsymbol{C} = 0\,. \tag{56.14}$$

Dadurch sind die $A_i^{(k)}$ eindeutig festgelegt.

Zum Schluß seien die Diffusionsgleichungen in der Form, welche sie in der hydrodynamischen Näherung erhalten, aufgeschrieben. Dazu ist (56.1) oder (56.12) in (32.33) einzutragen. Das gibt

$$\left.\begin{aligned} \left(\frac{d\gamma_i}{dt}\right)_w &= \frac{1}{n}\frac{\partial}{\partial x_\mu} n_i\left(\sum_k D_{ik} d_{\mu k} + D_{Ti}\frac{\partial T}{T\,\partial x_\mu}\right) \\ &= \frac{1}{n}\frac{\partial}{\partial x_\mu} n_i \sum_k D_{ik}\left(d_{\mu k} + k_{Tk}\frac{\partial T}{T\,\partial x_\mu}\right). \end{aligned}\right\} \tag{56.15}$$

---

[1] Beim binären Gemisch reduzieren sich diese Größen auf das bekannte Chapmansche k_T, s. Ziff. 59.

[2] Die k_{Ti} sind durch (56.10) und (56.11) eindeutig mit den D_{Ti} verknüpft. Es verschwindet zwar wegen des Bestehens der linearen Beziehungen (56.9) die Determinante der D_{ik} (56.10) hat aber trotzdem Lösungen k_{Ti}, weil — ebenfalls nach (56.9) — die D_{Tk} orthogonal sind zu den Lösungen γ_k des zu (56.10) gehörigen, homogenen Gleichungssystems. Nach (56.10) bleiben die k_{Ti} unbestimmt um ein zu γ_i proportionales additives Glied. Durch (56.11) wird dieses festgelegt und die Eindeutigkeit erzielt.

Nach der Bedeutung (54.10) von $d_{\mu k}$ und wegen (56.9) ist ausführlich

$$\sum_k D_{ik} d_{\mu k} = \sum_k D_{ik} \left\{ \frac{\partial \gamma_k}{\partial x_\mu} - \frac{\varrho_k}{p} \left[b_{\mu k} + \frac{1}{\varrho} \left(\frac{\partial p}{\partial x_\mu} - \sum_l \varrho_l b_{\mu l} \right) \right] \right\}. \tag{56.16}$$

Der Diffusionsstrom — ohne Thermodiffusion — besteht also aus drei Anteilen, die vom Gradienten der Zusammensetzung, dem äußeren Feld und dem Druckgradienten herrühren. Die Diffusion infolge äußeren Felds beschreibt man häufig auch mittels der Beweglichkeiten; in (56.16) ist der bekannte, aus der Thermodynamik folgende Einsteinsche Zusammenhang von Diffusionskoeffizienten und Beweglichkeiten enthalten. Druckdiffusion tritt in strömenden Gasen auf, auch bei Abwesenheit äußerer Felder. Sie gibt beispielweise Anlaß zu der Entmischung in der Zentrifuge. Sie dürfte auch einigermaßen die Trennerscheinungen beschreiben, welche an Überschall-Gasgemisch-Strahlen beim Ausströmen durch Laval-Düsen quer zur Strahlrichtung beobachtet werden — Verfahren von E. W. BECKER, BIER und BURGHOFF (1955). Das gilt natürlich nur, wenn der Druck wenigstens in der Strahlachse hoch genug ist für die Anwendbarkeit der hydrodynamischen Näherung[1]. Andernfalls befindet sich das Gas durchweg im Knudsen-Gebiet und man hat die Vorstellungen der Molekularströmung zu benutzen.

Bei langsamer Strömung kann man Beschleunigung und Reibung vernachlässigen und hat nach der Bewegungsgleichung (32.19)

$$\frac{\partial p}{\partial x_\mu} \approx \sum_l \varrho_l b_{\mu l}. \tag{56.17}$$

Im thermischen Gleichgewicht bei ruhenden Wänden gilt das exakt. Damit kommt als genährte Diffusionsgleichung

$$\left(\frac{d\gamma_i}{dt}\right)_w \approx \frac{1}{n} \frac{\partial}{\partial x_\mu} n_i \sum_k D_{ik} \left(\frac{\partial \gamma_k}{\partial x_\mu} - \frac{\varrho_k}{p} b_{\mu k} + k_{Tk} \frac{\partial T}{T \partial x_\mu} \right). \tag{56.18}$$

57. Der Wärmestrom und die Temperaturgleichung im Bezugssystem der mittleren Teilchengeschwindigkeit $\boldsymbol{w}$. Der Wärmestrom aus (53.10)

$$q_\mu^{(w)} = \mathrm{k} T \sum_i \int (\beta_i C^2 - \tfrac{5}{2})\, C_\mu F_i \Phi_i^{(1)}\, d\boldsymbol{C}$$

hat im Integranden gerade die rechte Seite der Integralgleichung (55.5). Das zeigt die Zweckmäßigkeit der in (32.36) vorgenommenen Abspaltung des Diffusions-Enthalpiestroms im $\boldsymbol{v}$-System. Es treten also wieder Bildungen vom Typ der Klammer (54.1) auf, wenn der Ausdruck (55.1) für $\Phi_i^{(1)}$ in das Wärmestrom-Integral eingesetzt wird. Man erhält

$$q_\mu^{(w)} = -p \sum_k D_{Tk} d_{\mu k} - \lambda' \frac{\partial T}{\partial x_\mu}. \tag{57.1}$$

Darin bedeutet

$$\left.\begin{aligned} D_{Tk} &= \frac{1}{3n} \sum_i \int \left(\beta_i C^2 - \frac{5}{2}\right) C^2 \frac{F_i}{n} A_i^{(k)}\, d\boldsymbol{C} \\ &= \frac{1}{3n} [A^{(k)} C_\mu, A C_\mu] = \frac{1}{3n} [A C_\mu, A^{(k)} C_\mu] \end{aligned}\right\} \tag{57.2}$$

---

[1] Bei neuen Versuchen von E. W. BECKER et al. (1957) war tatsächlich die freie Weglänge um mindestens eine Größenordnung kleiner als die Strahldicke und die Bedingung für Druckdiffusion erfüllt.

wieder den Thermodiffusionskoeffizienten aus (56.3) und (56.6) — für den somit eine zweite äquivalente Darstellung existiert — dank der Symmetrierelation $[\Psi, \Phi] = [\Phi, \Psi]$ aus (54.3)[1]. Der „theoretische" Wärmeleitkoeffizient λ' ist durch

$$\lambda' = \frac{1}{3}\,\mathrm{k} \sum_i \int \left(\beta_i C^2 - \frac{5}{2}\right) C^2 \frac{F_i}{n} A_i \, d\boldsymbol{C} = \frac{1}{3}\,\mathrm{k}\,[A\,C_\mu, A\,C_\mu] \tag{57.3}$$

gegeben. Nach (54.4) ist $\lambda' > 0$.

Der Koeffizient λ' ist der Messung nicht unmittelbar zugänglich, da bei stationärer Wärmeleitung sich im allgemeinen infolge Thermodiffusion ein Gefälle der Relativkonzentrationen einstellt, also die $d_{\mu i} \neq 0$ sind. Im eingestellten Zustand verschwinden aber die Diffusionsgeschwindigkeiten $W_{\mu i}$ und deshalb ist es zweckmäßig (57.1) so umzuformen, daß anstelle der $d_{\mu k}$ die $W_{\mu k}$ stehen. Dazu braucht man nur in dem ersten Glied von (57.1) mittels (56.10), (56.7) und (56.1) zu schreiben

$$\sum_k D_{Tk} d_{\mu k} = \sum_i k_{Ti} \sum_k D_{ik} d_{\mu k} = -\sum_i k_{Ti} \left(W_{\mu i} + D_{Ti} \frac{\partial T}{T\,\partial x_\mu} \right).$$

Das liefert für den Wärmestrom

$$q_\mu^{(w)} = -\lambda \frac{\partial T}{\partial x_\mu} + p \sum_i k_{Ti} W_{\mu i}, \tag{57.4}$$

wobei der „experimentelle" oder eigentliche Wärmeleitkoeffizient λ durch

$$\lambda = \lambda' - n\,\mathrm{k} \sum_i k_{Ti} D_{Ti} \tag{57.5}$$

gegeben ist. Dieser ist im stationären Wärmeleitversuch am ruhenden Gas direkt meßbar, da das zweite Glied rechts in (57.4) hierbei exakt verschwindet. Der Unterschied von λ und λ' ist übrigens nicht groß; er ist ja nach (57.5) proportional dem Quadrat der Thermodiffusionskoeffizienten, die ihrerseits relativ klein sind.

Das zweite Glied rechts in (57.4) wird als Diffusionswärmestrom bezeichnet. Dieser ist die Ursache für die Temperaturdifferenzen, welche bei der Diffusion anfänglich gleich temperierter Gase ineinander auftreten (Diffusionsthermoeffekt, s. Ziff. 3). In der Arbeit von WALDMANN (1947a) über den Diffusionsthermoeffekt in polynären Gemischen war die Schreibweise (57.1) benutzt und das erste Glied rechts als Diffusionswärmestrom bezeichnet worden. Aus den genannten Gründen ist (57.4) vorzuziehen. HIRSCHFELDER *et al.* ([*25*], S. 490) benutzen einen Wärmeleitkoeffizienten, der dem obigen λ nahekommt, aber nicht exakt damit übereinstimmt. Da diese Autoren die Koeffizienten k_{Ti} nicht einführen, benutzen sie eine Näherung für die D_{ik}, lösen (56.10) auf und verwenden diese genäherten k_{Ti} bei der Berechnung von λ. Für die allgemeinen Überlegungen ist die Einführung der exakten k_{Ti} vorzuziehen.

Es soll noch der eigentliche Wärmeleitkoeffizient λ durch eine Klammer vom Typ (54.1) dargestellt werden. Dazu seien zunächst die Ausdrücke (56.5) und (56.6) für D_{ik} und D_{Ti} in die Definitionsgleichung (56.10) der k_{Ti} eingesetzt. Das gibt

$$[A^{(i)} C_\mu, A\,C_\mu] = \sum_k [A^{(i)} C_\mu, A^{(k)} C_\mu]\, k_{Tk}$$

oder wegen der Bilinearität (54.2) der Klammer auch

$$[A^{(i)} C_\mu, (A - \sum_k A^{(k)} k_{Tk})\, C_\mu] = 0. \tag{57.6}$$

[1] Hierin finden, für das einatomige Gasgemisch, die restlichen Onsagerschen Reziprozitätsrelationen der irreversiblen Thermodynamik ihren Ausdruck, s. auch S. 407, Fußnote 1.

Nun ist nach (57.5), (57.3) und (57.2)

$$\lambda = \tfrac{1}{3}\mathsf{k}\,([A\,C_\mu, A\,C_\mu] - \sum_i [A^{(i)} C_\mu, A\,C_\mu]\, k_{Ti}) = \tfrac{1}{3}\mathsf{k}\,[(A - \sum_i A^{(i)} k_{Ti})\, C_\mu, A\,C_\mu].$$

Daraus folgt wegen (57.6) die gewünschte Darstellung

$$\lambda = \tfrac{1}{3}\mathsf{k}\,[\tilde{A}\,C_\mu, \tilde{A}\,C_\mu] \quad \text{mit} \quad \tilde{A}_i = A_i - \sum_k A_i^{(k)} k_{Tk}. \tag{57.7}$$

Nach (54.4) ist $\lambda > 0$, was im Hinblick auf den zweiten Hauptsatz gefordert werden muß.

Zum Schluß sei die Temperaturgleichung in der hydrodynamischen Näherung notiert. Dazu ist (57.4) in (32.37) einzutragen. Indem wir gleichzeitig das Reibungsgesetz (60.1) vorwegnehmen, erhalten wir

$$\left.\begin{aligned} \frac{5}{2} n\,\mathsf{k}\left(\frac{dT}{dt}\right)_w - \frac{dp}{dt} &= \frac{\partial}{\partial x_\mu}\left(\lambda \frac{\partial T}{\partial x_\mu} - p \sum_i k_{Ti} W_{\mu i}\right) + \\ &\quad + 2\eta \left\langle \frac{\partial v_\mu}{\partial x_\nu} \right\rangle \left\langle \frac{\partial v_\mu}{\partial x_\nu} \right\rangle + \sum_i \varrho_i b_{\mu i} V_{\mu i}. \end{aligned}\right\} \tag{57.8}$$

Der vorletzte Term rechts, die Reibungswärme, ist stets positiv. Vernachlässigt man, entsprechend der Annahme langsamer Strömung, die Reibung und setzt

$$v_\mu \frac{\partial p}{\partial x_\mu} \approx \sum_i \varrho_i b_{\mu i} v_\mu,$$

dann lautet (32.37)

$$\frac{5}{2} n\,\mathsf{k}\left(\frac{dT}{dt}\right)_w - \frac{\partial p}{\partial t} \approx \frac{\partial}{\partial x_\mu}\left(\lambda \frac{\partial T}{\partial x_\mu} - p \sum_i k_{Ti} W_{\mu i}\right) + \sum_i \varrho_i b_{\mu i} v_{\mu i}. \tag{57.9}$$

Dies ist eine für die Diskussion des Diffusionsthermoeffekts geeignete Form der Temperaturgleichung.

58. Diffusion und Wärmeleitung, vom Bezugssystem der mittleren Massengeschwindigkeit aus betrachtet. Die Diffusionsgeschwindigkeiten $\boldsymbol{W}_i$ aus (56.1) und der Wärmestrom $\boldsymbol{q}^{(w)}$ aus (57.4) beziehen sich auf das System der mittleren Teilchengeschwindigkeit $\boldsymbol{w}$. Sie sind, wie in Ziff. 32, δ ausgeführt wurde, angepaßt der Beschreibung von Diffusions- und Wärmeleitvorgängen in Gasen mit Geschwindigkeitsunterschieden, die klein gegen die Schallgeschwindigkeit sind. Bei der Auswertung von Experimenten über Gas-Diffusion werden gewöhnlich die auf das $\boldsymbol{w}$-System bezüglichen Koeffizienten (56.5) und (56.6) ermittelt.

Der Vollständigkeit halber seien die entsprechenden Formeln angegeben für die Diffusionsgeschwindigkeiten $\boldsymbol{V}_i$ und den Wärmestrom $\boldsymbol{q}$ aus (53.6) und (53.8), welche sich auf das $\boldsymbol{v}$-System beziehen.

Nach dem Zusammenhang (32.31) und nach (56.1) ist

$$V_{\mu i} = -\sum_k \left(D_{ik} - \sum_l \frac{\varrho_l}{\varrho} D_{lk}\right) d_{\mu k} - \left(D_{Ti} - \sum_l \frac{\varrho_l}{\varrho} D_{Tl}\right) \frac{\partial T}{T\,\partial x_\mu}. \tag{58.1}$$

Es liegt nahe die neuen Koeffizienten einzuführen

$$D_{ik}^{(v)} = D_{ik} - \sum_l \frac{\varrho_l}{\varrho}(D_{li} + D_{lk}) + \sum_{lj} D_{lj} \frac{\varrho_l \varrho_j}{\varrho^2}, \tag{58.2}$$

$$D_{Ti}^{(v)} = D_{Ti} - \sum_l \frac{\varrho_l}{\varrho} D_{Tl}. \tag{58.3}$$

Diese haben die Eigenschaften

$$D_{ik}^{(v)} = D_{ki}^{(v)}; \quad \sum_i \varrho_i D_{ik}^{(v)} = 0; \quad \sum_i \varrho_i D_{Ti}^{(v)} = 0. \tag{58.4}$$

Ferner gilt

$$D_{Ti}^{(v)} = \sum_k D_{ik}^{(v)} k_{Tk}. \tag{58.5}$$

Das ergibt sich aus (56.10), indem man von $\sum_i k_{Ti} = 0$ Gebrauch macht. Die Thermodiffusionsverhältnisse k_T bleiben also bei der Transformation in das $\boldsymbol{v}$-System ungeändert. Wir notieren uns noch die (56.5) und (56.6) entsprechenden Beziehungen. Indem man diese in (58.2) und (58.3) einsetzt, erhält man

$$\left.\begin{aligned} D_{ik}^{(v)} &= \frac{1}{3n}\,[A^{(i)\prime}\,C_\mu,\, A^{(k)\prime}\,C_\mu] \\ D_{Ti}^{(v)} &= \frac{1}{3n}\,[A^{(i)\prime}\,C_\mu,\, A\,C_\mu] \\ \text{mit}\qquad A_i^{(k)\prime} &= A_i^{(k)} - \sum_l \frac{\varrho_l}{\varrho} A_i^{(l)}. \end{aligned}\right\} \tag{58.6}$$

[Direkt gelangt man zu diesen Funktionen, wenn man statt (56.4) die Wahl $\Lambda_l = \varrho_l/\varrho$ trifft.] Für die Diffusionsgeschwindigkeiten (58.1) erhält man mit den neuen Diffusionskoeffizienten

$$V_{\mu i} = -\sum_k D_{ik}^{(v)}\, d_{\mu k} - D_{Ti}^{(v)} \frac{\partial T}{T\,\partial x_\mu} = -\sum_k D_{ik}^{(v)} \left(d_{\mu k} + k_{Tk} \frac{\partial T}{T\,\partial x_\mu}\right). \tag{58.7}$$

Nach der Bedeutung (54.10) der $d_{\mu i}$ und nach (58.4) ist ausführlich

$$\sum_k D_{ik}^{(v)}\, d_{\mu k} = \sum_k D_{ik}^{(v)} \left[\frac{\partial \gamma_k}{\partial x_\mu} - \frac{1}{p}\left(\varrho_k\, b_{\mu k} - \gamma_k \frac{\partial p}{\partial x_\mu}\right)\right]. \tag{58.8}$$

Zum Schluß sei der Wärmestrom $\boldsymbol{q}$ im $\boldsymbol{v}$-System notiert. Nach dem Zusammenhang (32.36) der Wärmeströme $\boldsymbol{q}$ und $\boldsymbol{q}^{(w)}$, nach (57.4) und wegen $\sum_i k_{Ti} = 0$ ist

$$q_\mu = -\lambda \frac{\partial T}{\partial x_\mu} + p \sum_i \left(k_{Ti} + \frac{5}{2}\gamma_i\right) V_{\mu i}. \tag{58.9}$$

So kann man, wenn es angebracht erscheint, in einfacher Weise von dem einen Bezugssystem zum anderen übergehen. Wir werden von den Koeffizienten $D^{(v)}$ weiter kaum Gebrauch machen.

59. Diffusion und Wärmeleitung im binären Gemisch. Da es bei einem Gemisch mit K Komponenten $\frac{1}{2}K(K-1)$ linear unabhängige Diffusionskoeffizienten und $K-1$ linear unabhängige Thermodiffusionskoeffizienten gibt, hat man es beim binären Gemisch nur mit je einem solchen Koeffizienten zu tun.

Aus (56.9) folgt mit

$$-D_{12} = -D_{21} = D > 0, \tag{59.1}$$

daß

$$D_{11} - D_{12} = \frac{1}{\gamma_1} D; \quad D_{22} - D_{21} = \frac{1}{\gamma_2} D. \tag{59.2}$$

Ferner hat man nach (56.11)

$$k_{T1} = -k_{T2} = k_T = \alpha\gamma_1\gamma_2. \tag{59.3}$$

Der Thermodiffusionsfaktor α, der hiermit zugleich eingeführt ist, hat die angenehme Eigenschaft für $\gamma_{1,2} \to 0$ von Null verschieden und im Fall des Isotopengemischs vom Molenbruch unabhängig zu sein.

Setzt man dies in (56.12) ein, so kommt für die Teilchen-Diffusionsströme

$$n_1 W_{\mu 1} = -n_2 W_{\mu 2} = -n D\left(d_{\mu 1} + k_T \frac{\partial T}{T\,\partial x_\mu}\right) \tag{59.4}$$

oder auch

$$W_{\mu 1} - W_{\mu 2} = \frac{1}{\gamma_2} W_{\mu 1} = -\frac{1}{\gamma_1} W_{\mu 2} = -\frac{1}{\gamma_1\gamma_2} D\left(d_{\mu 1} + k_T \frac{\partial T}{T\,\partial x_\mu}\right). \tag{59.5}$$

Ausführlich ist nach der Bedeutung (54.10) von $\boldsymbol{d}_i$

$$n_1 W_{\mu 1} = -n D\left[\frac{\partial \gamma_1}{\partial x_\mu} + \frac{m_2 - m_1}{m}\gamma_1\gamma_2 \frac{\partial p}{p\,\partial x_\mu} + \frac{\varrho_1\varrho_2}{\varrho p}(b_{\mu 2} - b_{\mu 1}) + \alpha\gamma_1\gamma_2\frac{\partial T}{T\,\partial x_\mu}\right]. \quad (59.6)$$

Man sieht daraus, daß in der Zentrifuge die schwere Komponente sich am Rand anreichert und daß die Sedimentation im Schwerefeld, wo ja die Beschleunigung $\boldsymbol{b}$ für alle Komponenten die gleiche ist, ebenfalls lediglich durch Druckdiffusion zustandekommt. Im übrigen sei auf die Bemerkungen zu Gl. (56.16) verwiesen. Die Diffusionsgleichung (56.15) schließlich lautet

$$\left(\frac{d\gamma_1}{dt}\right)_w = -\left(\frac{d\gamma_2}{dt}\right)_w = \frac{1}{n}\frac{\partial}{\partial x_\mu} n D\left(d_{\mu 1} + k_T\frac{\partial T}{T\,\partial x_\mu}\right). \quad (59.7)$$

Für den Wärmestrom im $\boldsymbol{w}$-System entnimmt man aus (57.4)

$$q_\mu^{(w)} = -\lambda\frac{\partial T}{\partial x_\mu} + p\,k_T(W_{\mu 1} - W_{\mu 2}) \quad (59.8)$$

oder nach (59.5) und mit dem Thermodiffusionsfaktor aus (59.3) auch

$$q_\mu^{(w)} = -\lambda\frac{\partial T}{\partial x_\mu} + \alpha\,\mathrm{k}T\,n_1 W_{\mu 1}. \quad (59.9)$$

An Hand dieser Schreibweise ist deutlich zu sehen, wie sehr bei den diffundierenden Atomen die im thermischen Gleichgewicht geltende Äquipartition der Energie gestört ist. Bei Vertauschen zweier Atome 1, 2 durch Diffusion wird einseitig der Energiebetrag $\alpha k T$ transportiert. Er kann von der Größenordnung der Gleichgewichtsenergie $\frac{3}{2}\mathrm{k}T$ sein; bei N_2/H_2-Gemisch z.B. ist $\alpha\approx 0{,}3$. — Die genäherte Temperaturgleichung (57.9) lautet für das binäre Gemisch

$$\frac{5}{2}n\,\mathrm{k}\left(\frac{dT}{dt}\right)_w - \frac{\partial p}{\partial t} \approx \frac{\partial}{\partial x_\mu}\left(\lambda\frac{\partial T}{\partial x_\mu} - \alpha\,\mathrm{k}T\,n_1 W_{\mu 1}\right). \quad (59.10)$$

Von einem äußeren Kraftfeld ist dabei abgesehen.

Es sei noch kurz das $\boldsymbol{v}$-System betrachtet. Wir schreiben wieder

$$-D_{12}^{(v)} = -D_{21}^{(v)} = D^{(v)}. \quad (59.11)$$

Nach (58.2) mit (59.1) und (59.2) ergibt sich der Zusammenhang

$$D^{(v)} = \frac{m_1 m_2}{m^2} D. \quad (59.12)$$

Der Diffusionskoeffizient $D^{(v)}$ ist maßgebend für die Massen-Diffusionsströme im $\boldsymbol{v}$-System. Nach (58.7) gilt

$$\varrho_1 V_{\mu 1} = -\varrho_2 V_{\mu 2} = -\varrho D^{(v)}\left(d_{\mu 1} + k_T\frac{\partial T}{T\,\partial x_\mu}\right). \quad (59.13)$$

Die Diffusionsgleichungen (32.17) lauten, etwas umgeformt,

$$\frac{d}{dt}\frac{\varrho_1}{\varrho} = -\frac{d}{dt}\frac{\varrho_2}{\varrho} = \frac{1}{\varrho}\frac{\partial}{\partial x_\mu}\varrho D^{(v)}\left(d_{\mu 1} + k_T\frac{\partial T}{T\,\partial x_\mu}\right). \quad (59.14)$$

Dies entspricht der Diffusionsgleichung (59.7) im $\boldsymbol{w}$-System; doch ist jene Gleichung praktischer, weil die Konzentration n, im Gegensatz zur Dichte ϱ, bei Gas-Diffusionsversuchen meist konstant ist. Auch der Koeffizient D hat diesen Vorzug gegenüber $D^{(v)}$. Wie später, s. Gl. (68.3), gezeigt wird, hängt der binäre Diffusionskoeffizient D in erster Näherung nicht vom Mischungsverhältnis ab, was nach (59.12) für $D^{(v)}$ nicht gilt. — Der Wärmestrom im $\boldsymbol{v}$-System beträgt nach (58.9) und (59.13)

$$q_\mu = -\lambda\frac{\partial T}{\partial x_\mu} + p\left(k_T + \frac{5}{2}\frac{m_2 - m_1}{m}\gamma_1\gamma_2\right)(V_{\mu 1} - V_{\mu 2}). \quad (59.15)$$

60. Der Drucktensor und die Bewegungsgleichung. Wir kehren zu den polynären Gemischen zurück. Hinsichtlich des Drucktensors ergibt sich gegenüber den einfachen Gasen (s. Ziff. 48, β), formal nichts wesentlich Neues. Man findet durch Eintragen von (55.1) in (53.7) wieder das Reibungsgesetz (48.8)

$$p_{\mu\nu} = p\,\delta_{\mu\nu} - \eta \cdot 2 \left\langle \frac{\partial v_\mu}{\partial x_\nu} \right\rangle \tag{60.1}$$

mit der Viscosität

$$\left.\begin{aligned} \eta &= \frac{1}{15} \sum_i m_i \int \beta_i C^4 \frac{F_i}{n} B_i \, d\boldsymbol{C} \\ &= \frac{1}{10} \mathrm{k}T \sum_i \gamma_i \int 2 \frac{F_i}{n_i} \beta_i \langle C_\mu C_\nu \rangle B_i \beta_i \langle C_\mu C_\nu \rangle \, d\boldsymbol{C} \\ &= \frac{1}{10} \mathrm{k}T \, [B\beta \langle C_\mu C_\nu \rangle, B\beta \langle C_\mu C_\nu \rangle]\,. \end{aligned}\right\} \tag{60.2}$$

Ersichtlich ist $\eta > 0$. Eine Volumviscosität tritt bei einatomigen Gemischen ebensowenig auf wie beim reinen, einatomigen Gas.

Die Bewegungsgleichung (32.19) lautet in der hydrodynamischen Näherung

$$\varrho \frac{d v_\mu}{dt} = - \frac{\partial p}{\partial x_\mu} + 2 \frac{\partial}{\partial x_\nu} \eta \left\langle \frac{\partial v_\mu}{\partial x_\nu} \right\rangle + \sum_i \varrho_i b_{\mu i}. \tag{60.3}$$

61. Die lokale Entropieerzeugung in der hydrodynamischen Näherung. Vorzeichenaussagen über die Transportkoeffizienten. Die lokale Entropieerzeugung ist exakt durch den Integralausdruck rechts in (33.2) dargestellt. Sie sei mit $(\delta \varrho s/\delta t)_{\text{Stöße}}$ bezeichnet. Geht man dort mit der Zerlegung (53.1) der Verteilungsfunktion ein, entwickelt die Logarithmen nach Potenzen von $\Phi^{(1)}$ und vernachlässigt im Endresultat Glieder von höherer Ordnung als $(\Phi^{(1)})^2$, so erhält man

$$\left(\frac{\delta \varrho s}{\delta t}\right)_{\text{Stöße}} = \mathrm{k}\, n^2 \, [\Phi^{(1)}, \Phi^{(1)}] > 0\,. \tag{61.1}$$

Dabei ist [ ] die in (54.1) definierte, dem Gemisch als Ganzem zugeordnete Summe von bestimmten Integralen.

Diese physikalische Bedeutung des Klammersymbols wurde von ENSKOG (1917, 1929) gefunden und diskutiert. Die Betrachtung der Entropieerzeugung bildet den Ausgangspunkt der Thermodynamik der irreversiblen Prozesse, die von MEIXNER (1941, 1943a, b) im Anschluß an ONSAGER (1931) ausgebaut wurde. PRIGOGINE (1949) hat, auch für chemisch-reagierende Gasgemische, dargetan, daß der Gültigkeitsbereich der Thermodynamik irreversibler Prozesse zusammenfällt mit demjenigen der hydrodynamischen Näherung nach der kinetischen Gastheorie. Einer eingehenderen Diskussion bedarf bei diesen Betrachtungen vor allem die Definition der lokalen Entropiedichte im Nicht-Gleichgewichtszustand.

Um den der hydrodynamischen Näherung angemessenen Ausdruck für die Entropieerzeugung zu erhalten, gehen wir mit (54.9) in (54.1) ein. Das gibt

$$\begin{aligned} n^2 [\Phi^{(1)}, \Phi^{(1)}] = &- \sum_i \int \Phi_i F_i \times \\ &\times \left[C_\mu \frac{1}{\gamma_i} d_{\mu i} + \left(\beta_i C^2 - \frac{5}{2}\right) C_\mu \frac{\partial T}{T \partial x_\mu} + 2\beta_i \langle C_\mu C_\nu \rangle \left\langle \frac{\partial v_\mu}{\partial x_\nu} \right\rangle \right] d\boldsymbol{C} \end{aligned}$$

oder nach (53.6), (53.10) und (53.7)

$$n^2 [\Phi^{(1)}, \Phi^{(1)}] = - n \sum_i V_{\mu i} d_{\mu i} - \frac{1}{\mathrm{k}T} q_\mu^{(w)} \frac{\partial T}{T \partial x_\mu} - \frac{1}{\mathrm{k}T} (p_{\mu\nu} - p\,\delta_{\mu\nu}) \left\langle \frac{\partial v_\mu}{\partial x_\nu} \right\rangle. \tag{61.2}$$

Nach dem Zusammenhang (32.31) der beiden Sätze von Diffusionsgeschwindigkeiten und wegen $\sum_i d_{\mu i} = 0$ ist

$$\sum_i V_{\mu i} d_{\mu i} = \sum_i W_{\mu i} d_{\mu i}.$$

Führt man ferner den Wärmestrom nach (57.4) und den Drucktensor nach (60.1) auf die Gradienten von Temperatur und Geschwindigkeit zurück, so entsteht

$$n^2 [\Phi^{(1)}, \Phi^{(1)}] = -n \sum_i W_{\mu i} \left(d_{\mu i} + k_{Ti} \frac{\partial T}{T \partial x_\mu} \right) + \frac{\lambda}{kT^2} \frac{\partial T}{\partial x_\mu} \frac{\partial T}{\partial x_\mu} + \frac{2\eta}{kT} \left\langle \frac{\partial v_\mu}{\partial x_\nu} \right\rangle \left\langle \frac{\partial v_\mu}{\partial x_\nu} \right\rangle.$$

Nun seien noch die Diffusionsgeschwindigkeiten gemäß (56.12) ersetzt. So folgt als Endresultat

$$\left.\begin{aligned} n^2 [\Phi^{(1)}, \Phi^{(1)}] &= n \sum_{i,k} D_{ik} \tilde{d}_{\mu i} \tilde{d}_{\mu k} + \frac{\lambda}{kT^2} \frac{\partial T}{\partial x_\mu} \frac{\partial T}{\partial x_\mu} + \frac{2\eta}{kT} \left\langle \frac{\partial v_\mu}{\partial x_\nu} \right\rangle \left\langle \frac{\partial v_\mu}{\partial x_\nu} \right\rangle \\ \text{mit} \qquad \tilde{d}_{\mu i} &= d_{\mu i} + k_{Ti} \frac{\partial T}{T \partial x_\mu}; \qquad \sum_i \tilde{d}_{\mu i} = 0. \end{aligned}\right\} \tag{61.3}$$

Benutzung von (58.7) anstatt (56.12) liefert dasselbe mit $D_{ik}^{(v)}$ anstelle von D_{ik}.

Damit die Entropieerzeugung positiv ist, müssen die Transportkoeffizienten also notwendig die folgenden Eigenschaften haben:

a) Es seien weder Temperatur- noch Geschwindigkeitsgradienten vorhanden; es muß gelten

$$\sum_{i,k} D_{ik} x_i x_k = \sum_{i,k} D_{ik}^{(v)} x_i x_k \geqq 0 \quad \text{wenn} \quad \sum_i x_i = 0, \tag{61.4}$$

d.h. die Matrizen D_{ik} und $D_{ik}^{(v)}$ haben positiv-definit zu sein für alle „Vektoren" x_i, welche die Nebenbedingung $\sum x_i = 0$ erfüllen. Daraus folgt nicht notwendig, daß $D_{ii}, D_{ii}^{(v)} > 0$, jedoch trifft dies zu für die D_{ik} aus (56.5) bzw. die $D_{ik}^{(v)}$ aus (58.6).

b) Es seien $\frac{\partial v_\mu}{\partial x_\nu} = 0$ und $\tilde{d}_{\mu i} = 0$ gewählt; daraus folgt

$$\lambda > 0. \tag{61.5}$$

c) Es sei nur $\frac{\partial v_\mu}{\partial x_\nu} \neq 0$; daraus ergibt sich

$$\eta > 0. \tag{61.6}$$

Diese Eigenschaften der Koeffizienten sind auch hinreichend und garantieren, daß bei beliebiger Wahl der Gradienten die Entropieerzeugung positiv ausfällt. Es ergeben sich also insbesondere keine Vorzeicheneigenschaften und Einschränkungen für die k_{Ti}.

II. Methoden zur Berechnung der Verteilungsfunktionen und der Transportkoeffizienten.

In dem speziellen Fall des Lorentzschen Gemischs können bei beliebiger Wechselwirkung die Verteilungsfunktionen der Komponenten exakt berechnet werden. Für die Diffusions- und Thermodiffusionskoeffizienten gibt es einfache Integraldarstellungen. Beim Maxwellschen Gasgemisch von beliebiger Zusammensetzung sind ebenfalls die exakten Verteilungsfunktionen einfach gebaut.

Im allgemeinen Fall beliebiger Massen- und Mischungsverhältnisse sowie beliebiger Wechselwirkung muß man die Integralgleichungen der hydrodynamischen Näherung durch geeignete Reihenentwicklungen lösen. Das Ziel ist dabei, die Berechnung der Transportkoeffizienten auf Quadraturen des molekularen Streuquerschnitts (Ω-Integrale) zurückzuführen. Dies ist bei den Gemischen erheblich mühsamer als bei den reinen Gasen, so daß der Gang der Rechnung nur angedeutet werden kann und — als einfachstes Beispiel — lediglich der binäre Diffusionskoeffizient in erster Näherung vollständig behandelt wird.

62. Das Lorentz-Gasgemisch. Unter einem Lorentzschen Gemisch versteht man nach CHAPMAN ein Gas, welches eine schwere häufige Komponente hat — als 1 bezeichnet — und im übrigen nur leichte seltene Komponenten besitzt — als $i=2\ldots, K$ bezeichnet. Anläßlich der Untersuchung des Lorentz-Gemischs fand ENSKOG (1911, 1912) die Thermodiffusion[1].

Es soll also vorausgesetzt sein, daß

$$m_1 \gg m_{2,\ldots,K}, \qquad \gamma_1 \approx 1 \gg \gamma_{2,\ldots,K}. \tag{62.1}$$

Die Atome 1 sind im lokalen Schwerpunktssystem wegen ihrer großen Masse nahezu in Ruhe. Da die leichten Atome selten sind, stoßen sie fast nur mit schweren zusammen. In dem Stoßoperator rechts in (53.11) braucht man also für $i\neq 1$ fast nur das Glied $j=1$ der Summe zu berücksichtigen und es kann darin $\boldsymbol{C}_1' \approx \boldsymbol{C}_1$ gesetzt werden. Nach (53.13) gilt somit

$$\sum_j \gamma_j I_{ij}(\Phi^{(1)}) \approx \frac{F_i}{n_i} \int [\Phi_i^{(1)}(\boldsymbol{C}) - \Phi_i^{(1)}(\boldsymbol{C}')]\, \sigma_{i1}\, g\, d\boldsymbol{e}' \quad \text{für} \quad i \neq 1. \tag{62.2}$$

Ferner hat man $\boldsymbol{C}=\boldsymbol{g}$, der Relativgeschwindigkeit vor dem Stoß, und die Geschwindigkeit eines leichten Atoms nach dem Stoß ist durch

$$\boldsymbol{C}' = C\,\boldsymbol{e}', \qquad C'^2 = C^2 \tag{62.3}$$

gegeben [s. (22.17)].

Die Näherung (62.2), in die Boltzmann-Gleichungen (53.11) eingesetzt, läßt allerdings die $\Phi_i^{(1)}$ zunächst unbestimmt um eine beliebige additive Funktion von C. Tatsächlich würden im Grenzfall $m_1\to\infty$, $\gamma_1\to 1$ die leichten Atome bei den Stößen nur abgelenkt werden, ihre beliebige anfängliche Energieverteilung im Schwerpunktssystem aber für alle Zeiten behalten. Dieser Grenzfall ist zu sehr idealisiert. In Wirklichkeit wird beim Stoß eines leichten Atoms gegen ein schweres etwas Energie ausgetauscht. Auch Stöße der leichten untereinander wirken in demselben Sinn. Die in (62.2) ausgelassenen Glieder sorgen dafür, daß auch beim Lorentz-Gemisch aus beliebiger, anfänglicher Energieverteilung heraus in makroskopisch kurzer Zeit sich lokal eine nahezu Maxwellsche Verteilung einstellt. So kann man sicher sein, daß in der Verteilungsfunktion der hydrodynamischen Näherung die erwähnte beliebige additive Funktion von C verschwindet und auch beim Lorentz-Gemisch der Ansatz (55.1) für $\Phi_i^{(1)}$ gilt.

Nun werde das Integral (62.2) vorbereitend berechnet für eine Funktion der Form

$$\Phi_i^{(1)} = G_i(C)\, C_\mu.$$

Wir benutzen für $\boldsymbol{e}'$ die Polarwinkel χ, ε in bezug auf die Geschwindigkeitsrichtung $\boldsymbol{e}$ vor dem Stoß. Die Integration über das Azimut ε kann man ausführen:

$$\int_0^{2\pi} (\boldsymbol{e}-\boldsymbol{e}')\, d\varepsilon = 2\pi(1-\cos\chi)\,\boldsymbol{e}.$$

Es verbleibt ein Integral über den Ablenkwinkel χ, welches einen über alle möglichen Ablenkungen gemittelten Stoßquerschnitt darstellt. Wir kürzen ab

$$2\pi\int_0^{\pi} (1-\cos\chi)\,\sigma_{i1}\sin\chi\, d\chi = \frac{1}{n\, l_{i1}(C)}. \tag{62.4}$$

---

[1] Das Quasi-Lorentz-Gas (wenig schwere, viel leichte Teilchen) hat MASON (1957a) untersucht (Anm. bei der Korrektur).

l_{i1} ist eine geschwindigkeitsabhängige freie Weglänge für den Stoß i gegen 1. Damit hat man

$$\sum_j \gamma_j I_{ij}(G C_\mu) \approx \frac{F_i}{n_i} \frac{C}{n l_{i1}} G_i C_\mu \quad \text{für} \quad i \neq 1, \tag{62.5}$$

so daß also der Stoßoperator beim Lorentz-Gemisch die Funktion $G_i C_\mu$ lediglich mit einem vom Betrag der Geschwindigkeit abhängigen Faktor multipliziert.

Jetzt kann man aus (56.13) und (55.5) die Verteilungsfunktionen, welche die Diffusion beschreiben, direkt entnehmen. Für die leichten Komponenten ist

$$A_i^{(k)} = n \frac{l_{i1}}{C}\left(\frac{\delta_{ik}}{\gamma_i} - 1\right), \qquad A_i = n \frac{l_{i1}}{C}\left(\beta_i C^2 - \frac{5}{2}\right). \tag{62.6}$$

Für die schwere Komponente dagegen ist ganz grob größenordnungsmäßig

$$A_1^{(k)} \propto A_1 \propto \frac{1}{C Q_{11}}. \tag{62.7}$$

[wegen Q s. (50.7)].

Einsetzen in (56.2) für $i \neq 1$ unter Beibehaltung nur der Glieder $\propto 1/\gamma_i$ liefert die Matrix der *Diffusionskoeffizienten*

$$D_{ik} = D_{ii} \delta_{ik} \quad \text{für} \quad i, k \neq 1 \tag{62.8}$$

mit den Diagonalgliedern

$$D_{ii} = \frac{1}{\gamma_i} \cdot \frac{1}{3} \int C l_{i1} \frac{F_i}{n_i} d\boldsymbol{C} = \frac{1}{\gamma_i} \cdot \frac{1}{3} \overline{C l_{i1}}. \tag{62.9}$$

Dazu kommen, entweder wieder aus (56.2) oder aus (56.9), die Elemente der ersten Zeile und Spalte

$$-D_{i1} = \gamma_i D_{ii} = \tfrac{1}{3} \overline{C l_{i1}} \quad \text{für} \quad i \neq 1; \qquad D_{11} = 0. \tag{62.10}$$

Dabei sind Glieder, die im Grenzfall $\gamma_{i \neq 1} \to 0$ verschwinden, überall fortgelassen.

Für die *Thermodiffusionskoeffizienten* bekommt man aus (56.3), (62.6) und (62.7)

$$D_{T1} = 0; \qquad D_{Ti} = \tfrac{1}{3} \overline{C l_{i1} (\beta_i C^2 - \tfrac{5}{2})}.$$

Eine kurze Umformung zeigt, daß der Zusammenhang gilt

$$D_{Ti} = -T^3 \left(\frac{\partial}{\partial T} \frac{D_{i1}}{T^2}\right)_p = \gamma_i T^3 \left(\frac{\partial}{\partial T} \frac{D_{ii}}{T^2}\right)_{p,\gamma}. \tag{62.11}$$

Schließlich entnimmt man aus (56.10) für $i \neq 1$

$$D_{Ti} = D_{ii} k_{Ti} + D_{i1} k_{T1} = D_{ii} (k_{Ti} - \gamma_i k_{T1}).$$

Da $\sum_i k_{Ti} = 0$, sind die k_T als von gleicher Größenordnung zu behandeln. Das Glied $\gamma_i k_{T1}$ ist gegen k_{Ti} somit zu vernachlässigen und es gilt

$$k_{Ti} = \frac{D_{Ti}}{D_{ii}} = -\gamma_i \frac{D_{Ti}}{D_{i1}} = \gamma_i \left[\left(\frac{\partial \log D_{ii}}{\partial \log T}\right)_{p,\gamma} - 2\right] \quad \text{für} \quad i \neq 1. \tag{62.12}$$

Dazu kommt

$$k_{T1} = -\sum_{i=2}^{K} k_{Ti}. \tag{62.13}$$

Wir spezialisieren auf das binäre Gemisch. Der *binäre Diffusionskoeffizient* D aus (59.2) beträgt nach (62.10)

$$D = \tfrac{1}{3} \overline{C l_{21}}. \tag{62.14}$$

Für das Modell der starren Kugeln mit Durchmessern $d_{1,2}$ kann man die freie Weglänge nach (62.4) mit (26.6) sofort berechnen:

$$\frac{1}{l_{21}} = n\,\pi\,d_{12}^2, \quad d_{12} = \frac{d_1 + d_2}{2}.$$

Damit kommt nach (62.14) mit (34.12) für den binären Diffusionskoeffizienten des Lorentz-Gemischs

$$D_{\text{starr}} = \frac{4}{3\pi^{\frac{3}{2}}} \frac{\sqrt{kT/2m_2}}{n\,d_{12}^2}. \tag{62.15}$$

Der *Thermodiffusionsfaktor* α beträgt nach (62.12) und (59.3)

$$\alpha = 2 - \left(\frac{\partial \log D}{\partial \log T}\right)_p. \tag{62.16}$$

Für das Potenz-Kraftgesetz $K_{12} \propto r^{-\nu_{12}}$ hat man nach den Dimensionsformeln (4.14)

$$D \propto \frac{1}{p}\, T^{\frac{3}{2} + \frac{2}{\nu_{12}-1}}.$$

Somit gilt hier

$$\alpha = \frac{\nu_{12} - 5}{2(\nu_{12} - 1)}. \tag{62.17}$$

Daraus lassen sich einige wichtige Charakteristika der Thermodiffusion ablesen. Für Maxwell-Moleküle, $\nu_{12} = 5$, verschwindet sie. Das gilt nicht nur beim Lorentz-Gemisch, sondern bei beliebigem Massen- und Mischungsverhältnis, wie in der folgenden Ziffer deutlich wird. Im Fall „harter" Stöße, $\nu_{12} > 5$, ist $\alpha > 0$; das bedeutet nach (59.6), daß die schwere Komponente „1" im stationären Zustand ($\boldsymbol{W}_1 = 0$) das Gebiet der tieferen Temperatur bevorzugt. Wenn $\nu_{12} > 5$, erreicht α den größten Wert für $\nu_{12} = \infty$, d.h. bei starren Molekülen:

$$\alpha_{\text{starr}} = \tfrac{1}{2}. \tag{62.18}$$

Im Fall „weicher" Stöße, $\nu_{12} < 5$, dagegen reichert sich die schwere Komponente auf der warmen Seite an. Tatsächlich kommen beide Fälle vor. Das Potenz-Kraftgesetz ist aber zu stark idealisiert. Das zeigt sich darin, daß nach (62.17) der Thermodiffusionsfaktor temperaturunabhängig sein sollte. In Wirklichkeit ist das nicht der Fall. Der Betrag von α nimmt im allgemeinen mit der Temperatur zu; in vielen Fällen findet sogar ein Vorzeichenwechsel statt (s. Ziff. 75 und 77).

Auch die *Wärmeleitfähigkeit* des Lorentz-Gemischs, soweit sie von den leichten Komponenten stammt, ist mittels der Verteilungsfunktionen (62.6) zu berechnen. Dazu kommt — und zwar einfach additiv [s. z.B. WALDMANN (1950a)[1]] — die Wärmeleitfähigkeit λ_1 der reinen schweren Komponente. Der Wärmeleitkoeffizient des binären Lorentz-Gemischs läßt sich so schreiben:

$$\lambda = \lambda_1 + \left(\frac{5}{2} - \frac{d\,\alpha\,T}{dT}\right) n_2\, k\, D. \tag{62.19}$$

Der eingeklammerte Faktor beträgt $\frac{5}{2}$ für Maxwell-Moleküle und 2 für starre Kugeln.

63. Das Maxwell-Gasgemisch. Man kann das in Ziff. 37 betrachtete Eigenwertproblem in analoger Weise auch für Gemische mit K Komponenten formulieren. Anstelle jedes Eigentensors aus (37.12) tritt ein Satz von je K solcher Tensoren, die durch den Komponentenindex i unterschieden sind.

[1] Dort war $m_2 \gg m_1$ vorausgesetzt.

Beim Maxwell-Gasgemisch findet man nach der Methode von Ziff. 38 und 39 ähnlich wie beim reinen Maxwell-Gas, s. (39.12) oder (40.11), die Eigentensoren des linearisierten Stoßoperators. Die für $l=1,2$ lauten

$$\psi_{\mu,i}^{(r)} \propto S_{\frac{3}{2}}^{(r)}(\beta_i C^2)\, C_\mu, \quad \psi_{\mu\nu,i}^{(r)} \propto S_{\frac{5}{2}}^{(r)}(\beta_i C^2)\,\langle C_\mu C_\nu\rangle, \quad r=0,1,\ldots. \tag{63.1}$$

Wegen der Sonine-Polynome $S_{l+\frac{1}{2}}^{(r)}$ s. (38.21) oder speziell (49.2). Die Eigenwerte und die nichtangeschriebenen Proportionalitätsfaktoren in (63.1) ergeben sich durch Lösung einer Säkulargleichung vom Grad K, in welche die Konzentrationen n_i eingehen sowie numerisch zu berechnende Größen ähnlich den $A_l(5)$ aus (40.6). Es ist nicht nötig, das hier weiter auszuführen.

Die Integralgleichungen für die Funktionen $A_i^{(k)}$, A_i und B_i, welche die Verteilungsfunktion $\Phi_i^{(1)}$ gemäß (55.1) aufbauen, lauten nach (56.13), (55.5) und (55.6) mit Benutzung der Sonine-Polynome im Fall *beliebiger Wechselwirkung*

$$\sum_j \gamma_j I_{ij}(A^{(k)} C_\mu) = \frac{F_i}{n_i} S_{\frac{3}{2}}^{(0)}(\beta_i C^2)\, C_\mu \left(\frac{\delta_{ik}}{\gamma_i} - 1\right) \tag{63.2}$$

$$\sum_j \gamma_j I_{ij}(A\, C_\mu) = -\frac{F_i}{n_i} S_{\frac{3}{2}}^{(1)}(\beta_i C^2)\, C_\mu \tag{63.3}$$

$$\sum_j \gamma_j I_{ij}(B\,\beta\langle C_\mu C_\nu\rangle) = 2\frac{F_i}{n_i} S_{\frac{5}{2}}^{(0)}(\beta_i C^2)\,\beta_i\langle C_\mu C_\nu\rangle. \tag{63.4}$$

Beim *Maxwell-Gemisch* hat man nach dem über dessen Eigenwertgleichungen Gesagten zu setzen

$$A_i^{(k)} \propto S_{\frac{3}{2}}^{(0)} = 1, \qquad A_i \propto S_{\frac{3}{2}}^{(1)}, \qquad B_i \propto S_{\frac{5}{2}}^{(0)} = 1. \tag{63.5}$$

Damit ist man, bis auf Zahlenfaktoren im Besitz der exakten Verteilungsfunktionen der hydrodynamischen Näherung.

Insbesondere sieht man aus (63.5) und (56.3), daß wegen der Orthogonalität (49.5) der Sonine-Polynome die Thermodiffusionskoeffizienten D_{Ti} für das Maxwell-Gemisch exakt verschwinden.

64. Reihenentwicklung nach Sonine-Polynomen im allgemeinen Fall. Physikalische Bedeutung einiger Entwicklungskoeffizienten. Im Fall beliebiger Wechselwirkung wird man die Bestandteile $A_i^{(k)} C_\mu$, $A_i C_\mu$, $B_i\langle C_\mu C_\nu\rangle$ der Verteilungsfunktionen nach den Eigentensoren (63.1) entwickeln und demgemäß ansetzen:

$$A_i^{(k)}(C) = \beta_i \sum_{s=0}^{\infty} A_{i,s}^{(k)} S_{\frac{3}{2}}^{(s)}(\beta_i C^2) \tag{64.1}$$

$$A_i(C) = -\beta_i \sum_{s=0}^{\infty} A_{i,s} S_{\frac{3}{2}}^{(s)}(\beta_i C^2) \tag{64.2}$$

$$B_i(C) = \beta_i \sum_{s=0}^{\infty} B_{i,s} S_{\frac{5}{2}}^{(s)}(\beta_i C^2). \tag{64.3}$$

Die Entwicklungskoeffizienten $A_{i,s}^{(k)}$, $A_{i,s}$, $B_{i,s}$ — Dimension $\mathrm{cm}^{-1}\,\mathrm{sec}^{-1}$ — sind von der Gesamtkonzentration n unabhängig. Die Einfügung der Faktoren β_i ist im folgenden bequem.

Einigen der Entwicklungskoeffizienten kommt eine unmittelbare physikalische Bedeutung zu. Durch Einsetzen der Entwicklungen (64.1) bis (64.3) in die Gln. (56.2), (56.3), (57.2), (57.3) und (60.2) erhält man nämlich unter

Beachtung der Orthonormierungsrelationen (49.7) die Zusammenhänge

$$D_{ik} = \frac{1}{2n}\Big(A_{i,0}^{(k)} - \sum_l \gamma_l A_{l,0}^{(k)}\Big) \tag{64.4}$$

$$D_{Ti} = -\frac{1}{2n}\Big(A_{i,0} - \sum_l \gamma_l A_{l,0}\Big) = -\frac{5}{4n}\sum_k \gamma_k A_{k,1}^{(i)} \tag{64.5}$$

$$\lambda' = \tfrac{5}{4}\,\mathrm{k}\sum_i \gamma_i A_{i,1} \tag{64.6}$$

$$\eta = \tfrac{1}{4}\sum_i m_i \gamma_i B_{i,0}. \tag{64.7}$$

Die Hauptaufgabe besteht also in der Berechnung der vier Sätze von Koeffizienten $A_{i,0}^{(k)}$, $A_{i,0}$, $A_{i,1}$ und $B_{i,0}$. (Die zweite Darstellung der Thermodiffusionskoeffizienten wird nicht benutzt werden.) Zur Lösung dieser Aufgabe sollen die Integralgleichungen (63.2) bis (63.4) in lineare Gleichungssysteme verwandelt werden. Für diesen Zweck benötigte Klammerausdrücke werden in der folgenden Ziffer definiert.

65. Definition der Klammern $[\,]'_{ij}$, $[\,]''_{ij}$. Der linearisierte Stoßoperator I_{ij} aus (53.13) werde in zwei Anteile zerlegt

$$I_{ij}(\Phi) = I'_{ij}(\Phi_i) + I''_{ij}(\Phi_j), \tag{65.1}$$

welche jeweils nur von der Funktion Φ_i bzw. Φ_j abhängen:

$$n_i n_j I'_{ij}(\Phi_i) = \iint F_i(\boldsymbol{C})\, F_j(\boldsymbol{C}_1)\,[\Phi_i(\boldsymbol{c}) - \Phi_i(\boldsymbol{c}')]\,\sigma_{ij}\, g\, d\boldsymbol{e}'\, d\boldsymbol{c}_1 \tag{65.2}$$

$$n_i n_j I''_{ij}(\Phi_j) = \iint F_i(\boldsymbol{C})\, F_j(\boldsymbol{C}_1)\,[\Phi_j(\boldsymbol{c}_1) - \Phi_j(\boldsymbol{c}'_1)]\,\sigma_{ij}\, g\, d\boldsymbol{e}'\, d\boldsymbol{c}_1. \tag{65.3}$$

An Hand der Operatoren I' und I'' seien die folgenden bilinearen Klammerausdrücke definiert

$$[\Psi, \Phi]'_{ij} = \int \Psi_i I'_{ij}(\Phi_i)\, d\boldsymbol{c} = [\Phi, \Psi]'_{ij} \tag{65.4}$$

$$[\Psi, \Phi]''_{ij} = \int \Psi_i I''_{ij}(\Phi_j)\, d\boldsymbol{c} = [\Phi, \Psi]''_{ji}. \tag{65.5}$$

Die Symmetriebeziehungen ergeben sich, indem man ähnlich wie bei den zu (32.1) führenden Betrachtungen die Geschwindigkeiten vor und nach dem Stoß umbenennt.

Die auf das Gemisch als Ganzes bezügliche Klammer (54.1) stellt sich durch die auf Komponenten-Paare bezüglichen Klammern (65.4), (65.5) so dar:

$$[\Psi, \Phi] = \sum_{i,j} \gamma_i \gamma_j \big([\Psi, \Phi]'_{ij} + [\Psi, \Phi]''_{ij}\big) = [\Phi, \Psi].$$

66. Umwandlung der Integralgleichungen für die Verteilungsfunktionen in lineare Gleichungssysteme. Um die Integralgleichungen (63.2) bis (63.4) in unendliche lineare Gleichungssysteme zu überführen, seien (63.2) und (63.3) mit $\beta_i S_{\frac{3}{2}}^{(r)} C_\mu$, (63.4) mit $\beta_i S_{\frac{5}{2}}^{(r)} \langle C_\mu C_\nu \rangle$ durchmultipliziert und über alle Geschwindigkeiten $\boldsymbol{C}$ integriert.

Wegen der Orthonormierungsrelationen (49.7) der Sonine-Polynome bekommen die rechten Seiten den Faktor δ_{r0} bzw. δ_{r1}, verschwinden also zumeist.

Links treten beim Einsetzen der Entwicklungen (64.1) bis (64.3) die Ausdrücke auf (wegen der Klammern s. die vorige Ziffer)

$$a'_{ij,rs} = \beta_i^2\, [S_{\frac{3}{2}}^{(r)} C_\mu,\ S_{\frac{3}{2}}^{(s)} C_\mu]'_{ij} = a'_{ij,sr} \tag{66.1}$$

$$a''_{ij,rs} = \beta_i \beta_j\, [\ldots]''_{ij} = a''_{ji,sr} \tag{66.2}$$

sowie

$$b'_{ij,rs} = \beta_i^3 \, [S_{\frac{3}{2}}^{(r)}\langle C_\mu C_\nu\rangle, S_{\frac{3}{2}}^{(s)}\langle C_\mu C_\nu\rangle]'_{ij} = b'_{ij,sr} \tag{66.3}$$

$$b''_{ij,rs} = (\beta_i \beta_j)^{\frac{3}{2}} \, [\ldots]''_{ij} = b''_{ji,sr}. \tag{66.4}$$

Alle diese Matrixelemente haben die Dimension cm · sec. Nach dem Impulssatz gilt

$$a'_{ij,r0} + a''_{ij,r0} = 0 \tag{66.5}$$

oder wegen der Symmetrierelationen auch

$$a'_{ij,0r} + a''_{ji,0r} = 0. \tag{66.6}$$

Die Integralgleichungen, mit den zugehörigen Nebenbedingungen (56.14) und (55.8), nehmen die folgende Form an:

Diffusion

$$\left.\begin{aligned} &\sum_{j,s} \gamma_j (a'_{ij,rs} A_{i,s}^{(k)} + a''_{ij,rs} A_{j,s}^{(k)}) = \frac{3}{2}\left(\frac{\delta_{ik}}{\gamma_i} - 1\right)\delta_{r0} \\ \text{mit} \quad &\sum_j \varrho_j A_{j,0}^{(k)} = 0. \end{aligned}\right\} \tag{66.7}$$

Wärmeleitung und Thermodiffusion

$$\left.\begin{aligned} &\sum_{j,s} \gamma_j (a'_{ij,rs} A_{i,s} + a''_{ij,rs} A_{j,s}) = \tfrac{15}{4}\, \delta_{r1} \\ \text{mit} \quad &\sum_j \varrho_j A_{j,0} = 0. \end{aligned}\right\} \tag{66.8}$$

Reibung

$$\sum_{j,s} \gamma_j \left(b'_{ij,rs} \sqrt{m_i}\, B_{i,s} + b''_{ij,rs} \sqrt{m_j}\, B_{j,s}\right) = 5 \sqrt{m_i}\, \delta_{r0}. \tag{66.9}$$

In den Systemen (66.8) und (66.9) sind die früheren Gln. (49.15) und (49.16) für reine Gase als Spezialfälle enthalten.

Eine strenge Lösung dieser unendlichen Gleichungssysteme ist natürlich nicht möglich. Man muß die Summationen abbrechen, um praktisch verwendbare Beziehungen zu erhalten. Zuvor sollen in der nächsten Ziffer einige der Matrixelemente a', a'' und b', b'' explizit durch Ω-Integrale ausgedrückt werden.

67. Einige der Matrixelemente aus (66.1) bis (66.4). Ganz einfach zu berechnen ist das Matrixelement a' für $r = s = 0$. Nach (66.1), (65.2) und (65.4) ist

$$\begin{aligned} a'_{ij,00} &= \beta_i^2 \, [C_\mu, C_\mu]'_{ij} \\ &= \frac{1}{2} \beta_i^2 \iiint \left(\frac{\beta_i \beta_j}{\pi^2}\right)^{\frac{3}{2}} e^{-(\beta_i C^2 + \beta_j C_1^2)} (\boldsymbol{C} - \boldsymbol{C}')^2 \sigma_{ij}\, g \, d\boldsymbol{e}' \, d\boldsymbol{C} \, d\boldsymbol{C}_1. \end{aligned}$$

Wie bei den Integrationen in Ziff. 50 führt man die Schwerpunkts- und die Relativgeschwindigkeit gemäß (22.14) und (22.15) als Integrationsvariable ein. Es ist damit

$$\beta_i C^2 + \beta_j C_1^2 = (\beta_i + \beta_j)\, C_s^2 + \beta_{ij}\, g^2 \quad \text{wo} \quad \beta_{ij} = \frac{\beta_i \beta_j}{\beta_i + \beta_j}$$

und

$$(\boldsymbol{C} - \boldsymbol{C}')^2 = 2\left(\frac{m_j}{m_i + m_j}\right)^2 g^2 (1 - \cos\chi); \quad d\boldsymbol{C}\, d\boldsymbol{C}_1 = d\boldsymbol{C}_s\, d\boldsymbol{g}.$$

Nun kann die Integration ausgeführt werden und man bekommt nach kurzer Rechnung [s. auch (66.5)]

$$a'_{ij,00} = -a''_{ij,00} = \frac{4}{kT} \frac{m_i m_j}{m_i + m_j} \Omega_{ij}^{(1,1)}. \tag{67.1}$$

Dabei bedeutet Ω das Chapman-Integral aus (50.5). Diese Matrixelemente bestimmen die erste Näherung der Diffusionskoeffizienten, wie in der folgenden Ziffer dargelegt wird.

Alle anderen Matrixelemente enthalten im Integranden neben dem Exponentialfaktor Ausdrücke, die von höherem als dem zweiten Grad in den Geschwindigkeiten sind und man kann selbst bei den nächst-einfachen nicht mehr wie bei der Berechnung von (50.3) einfach die Erhaltungssätze ausnutzen, da es sich jetzt um die verkürzten (gestrichenen) Klammern (65.4) und (65.5) handelt. Deshalb ist für die Ausführung dieser Integrationen von BURNETT (1935a) ein eigenes Verfahren entwickelt worden, das auf den besonderen Eigenschaften der Sonine-Polynome beruht. Man geht aus von entsprechenden Klammerausdrücken für die erzeugenden Exponentialfunktionen (38.15) der Sonine-Polynome und stellt die gewünschten Klammern, welche die Polynome selbst enthalten, durch sukzessive Differentiationen gemäß (38.19) daraus her. Das Verfahren ist in erweiterter Form bei CHAPMAN und COWLING [4], Kap. 9, dargestellt. Wir geben deren Ergebnisse für diejenigen Matrixelemente an, welche im weiteren benötigt werden:

$$\left.\begin{aligned} a'_{ij,01} &= -a''_{ji,01} = -a''_{ij,10} = a'_{ij,10} \\ &= \frac{4}{kT}\,\frac{m_i m_j^2}{(m_i+m_j)^2}\left(\frac{5}{2}\,\Omega_{ij}^{(1,1)} - \Omega_{ij}^{(1,2)}\right) \end{aligned}\right\} \tag{67.2}$$

$$\left.\begin{aligned} a'_{ij,11} = \frac{4}{kT}\,\frac{m_i m_j}{(m_i+m_j)^3}\Big(&\frac{5}{4}\,(6m_i^2+5m_j^2)\,\Omega_{ij}^{(1,1)} - 5m_j^2\,\Omega_{ij}^{(1,2)} + \\ &+ m_j^2\,\Omega_{ij}^{(1,3)} + 2m_i m_j\,\Omega_{ij}^{(2,2)}\Big) \end{aligned}\right\} \tag{67.3}$$

$$a''_{ij,11} = \frac{4}{kT}\,\frac{(m_i m_j)^2}{(m_i+m_j)^3}\left(-\frac{55}{4}\,\Omega_{ij}^{(1,1)} + 5\Omega_{ij}^{(1,2)} - \Omega_{ij}^{(1,3)} + 2\Omega_{ij}^{(2,2)}\right) \tag{67.4}$$

$$b'_{ij,00} = \frac{8}{3kT}\,\frac{m_i m_j}{(m_i+m_j)^2}\left(5m_i\,\Omega_{ij}^{(1,1)} + \frac{3}{2}\,m_j\,\Omega_{ij}^{(2,2)}\right) \tag{67.5}$$

$$b''_{ij,00} = \frac{8}{3kT}\,\frac{(m_i m_j)^{\frac{3}{2}}}{(m_i+m_j)^2}\left(-5\Omega_{ij}^{(1,1)} + \frac{3}{2}\,\Omega_{ij}^{(2,2)}\right). \tag{67.6}$$

Die Matrixelemente (67.3) und (67.4) sind für den Wärmeleitkoeffizienten maßgebend, (67.5) und (67.6) für die Viscosität. In die Thermodiffusionskoeffizienten gehen die Elemente (67.1) bis (67.4) ein.

68. Die erste Näherung für die Konzentrations-Diffusionskoeffizienten. Darstellung durch Ω-Integral. Die Diffusionskoeffizienten erhält man durch Auflösen des linearen Gleichungssystems (66.7). In erster Näherung seien die Nichtdiagonalglieder $s \neq r$ der Matrizen a', a'' vernachlässigt. Das ist beim Maxwell-Gas exakt richtig. Für starre Kugeln gleicher Masse $m_i = m_j$ beträgt nach (67.1) und (67.2) sowie nach (51.1) und (51.2) das Verhältnis $a'_{ij,01}/a'_{ij,00} = -\frac{1}{4}$. Das ist nicht gerade sehr klein gegen 1; doch stellen die starren Kugeln einen extremen Grenzfall dar.

Diese Vernachlässigung bewirkt, daß für $r \neq 0$ alle $A_{i,r}^{(k)} = 0$ werden. Für $r = 0$ aber lautet (66.7) mit Benutzung von (64.4) und (66.5) so:

$$\sum_j \gamma_j\, a'_{ij,00}\,(D_{ik} - D_{jk}) \approx \frac{3}{4n}\left(\frac{\delta_{ik}}{\gamma_i} - 1\right). \tag{68.1}$$

Zwischen diesen Gleichungen besteht — bei festem k — eine lineare Beziehung. Zur eindeutigen Festlegung der D_{ik} muß man

$$\sum_j \gamma_j D_{jk} = 0 \tag{68.2}$$

hinzunehmen. Damit ist es dann möglich, die Diffusionskoeffizienten gemäß (67.1) durch Integrale $\Omega^{(1,1)}$ darzustellen. Für den Diffusions-Koeffizienten $D = D(1,2)$ des *binären Gemischs* gilt nach (59.2)

$$D = D(1,2) = \gamma_1 (D_{11} - D_{21}).$$

Die Gl. (68.1) für $k=1$, $i=2$ lautet hier

$$\gamma_1 a'_{21,00}(D_{21} - D_{11}) = -\frac{3}{4n}.$$

Somit erhält man die wichtige Formel

$$D(1,2) \approx [D(1,2)]_1 = \frac{3}{4n\, a'_{12,00}} = \frac{3\,\mathrm{k}T}{16 n\, m_{12}\,\Omega_{12}^{(1,1)}} \quad \text{mit} \quad m_{12} = \frac{m_1 m_2}{m_1 + m_2}. \tag{68.3}$$

Besonders bemerkenswert ist, daß der binäre Diffusionskoeffizient in erster Näherung nicht vom Mischungsverhältnis abhängt. Dies ist experimentell auf etwa 10% bestätigt (s. Ziff. 73). Für den Selbstdiffusionskoeffizienten gilt

$$D = \frac{3\,\mathrm{k}T}{8 n\, m\, \Omega^{(1,1)}}. \tag{68.4}$$

Jetzt kann man in (68.1) die Koeffizienten a durch die binären Diffusionskoeffizienten ersetzen und so die polynären D_{ik} auf die binären $D(i,k)$ zurückführen. Wir deuten das für das *ternäre Gemisch* an. Für die Diffusionsgeschwindigkeit werde geschrieben

$$W_{\mu 1} = \sum_k D_{1k} \frac{\partial \gamma_k}{\partial x_\mu} = -(D_{11} - D_{12}) \frac{\partial \gamma_2}{\partial x_\mu} - (D_{11} - D_{13}) \frac{\partial \gamma_3}{\partial x_\mu}, \quad \text{usw.} \tag{68.5}$$

Die Differenzen $D_{11} - D_{12}$, $D_{11} - D_{13}$, usw. sind durch (68.1) eindeutig bestimmt. Kurze Rechnung ergibt

$$\gamma_1 (D_{11} - D_{12}) = \frac{\gamma_1 D(1,2)\, D(2,3) + (\gamma_2 + \gamma_3)\, D(1,2)\, D(1,3)}{\gamma_1 D(2,3) + \gamma_2 D(3,1) + \gamma_3 D(1,2)}. \tag{68.6}$$

Alle übrigen Koeffizienten erhält man durch geeignete Vertauschung der Zahlen 1 bis 3.

Im allgemeinen Fall sind die polynären Diffusionskoeffizienten nach der Cramerschen Regel durch das Verhältnis von zwei Determinanten gegeben, welche die binären Koeffizienten und die Molenbrüche enthalten. Das findet man in der Arbeit von HELLUND (1940) oder auch in dem Buch von HIRSCHFELDER *et al.* [*25*] näher ausgeführt.

Wegen der weiteren Diskussion sei auf Ziff. 71 bis 73 verwiesen.

69. Wärmeleitkoeffizient und Thermodiffusionsverhältnisse. Für die Wärmeleitfähigkeit und die Thermodiffusionskoeffizienten werden die $A_{i,1}$ und die $A_{i,0}$ benötigt [s. (64.6) und (64.5)]. Man bekommt eine erste Näherung für die $A_{i,r}$, indem man in (66.8) die nicht-diagonalen Matrixelemente $s \neq r$ vernachlässigt. Das ergibt $A_{i,r} = 0$ für alle $r \neq 1$, was insbesondere bedeutet, daß in dieser Näherung die Thermodiffusionskoeffizienten verschwinden. Die Gleichung für $r=1$ lautet

$$\sum_j \gamma_j (a'_{ij,11} A_{i,1} + a''_{ij,11} A_{j,1}) \approx \tfrac{15}{4}. \tag{69.1}$$

Durch Auflösen dieser K linearen Gleichungen erhält man die $A_{i,1}$ und damit den *Wärmeleitkoeffizienten* λ', der in dieser Näherung verschwindender Thermodiffusion nicht von λ zu unterscheiden ist [vgl. (57.5)]. Nach der Cramerschen

Regel und dem Entwicklungssatz für Determinanten kann man auch schreiben

$$\lambda \approx -\tfrac{75}{16}\,\mathrm{k}\begin{vmatrix} a_{11,11} \cdots a_{1K,11} & 1 \\ \vdots \quad\quad \vdots & \vdots \\ a_{K1,11} \cdots a_{KK,11} & 1 \\ \gamma_1 \quad \cdots \gamma_K & 0 \end{vmatrix} : \begin{vmatrix} a_{11,11} \cdots a_{1K,11} \\ \vdots \quad\quad \vdots \\ a_{K1,11} \cdots a_{KK,11} \end{vmatrix}. \tag{69.2}$$

Dabei ist abgekürzt

$$a_{ij,rs} = \delta_{ij}\sum_l \gamma_l\, a'_{il,rs} + \gamma_j\, a''_{ij,rs} = \frac{\gamma_j}{\gamma_i}\, a_{ji,sr}. \tag{69.3}$$

Zusammen mit (67.3) und (67.4) bildet dies die Darstellung des Wärmeleitkoeffizienten durch Ω-Integrale. Die ausführliche Formel ist selbst im binären Fall ziemlich weitläufig (s. CHAPMAN und COWLING [4], S. 166). Wegen einer kurzen Besprechung experimenteller Ergebnisse sei auf Ziff. 80 verwiesen.

Da die Thermodiffusion in der eben betrachteten ersten Näherung von (66.8) verschwindet, ist es zu ihrer Behandlung nötig, mindestens einen Schritt weiterzugehen. Dieser soll darin bestehen, daß wir die nicht-diagonalen Matrixelemente $(r, s) = (0, 1)$ oder $(1, 0)$ von Null verschieden annehmen und dabei für die $A_{i,1}$ die durch (69.1) schon bekannte erste Näherung einsetzen. Für $r=0$ erhält man so, unter Benutzung von (64.5) und (66.6), das K-fache Gleichungssystem

$$\sum_j \gamma_j\, a'_{ij,00}\,(D_{Ti} - D_{Tj}) \approx \frac{1}{2n}\sum_j \gamma_j\,(a'_{ij,01}\, A_{i,1} - a'_{ji,01}\, A_{j,1}). \tag{69.4}$$

Zwischen diesen Gleichungen besteht eine lineare Beziehung. Zur Eindeutigkeit muß man

$$\sum_j \gamma_i\, D_{Tj} = 0 \tag{69.5}$$

hinzunehmen. Die rechten Seiten sind bekannt, so daß daraus die Thermodiffusionskoeffizienten D_{Ti} — in der ersten, nichtverschwindenden Näherung — berechnet werden können und, über (67.1) und (67.2), im Prinzip auf Ω-Integrale zurückgeführt sind.

Die Thermodiffusionsverhältnisse k_{Ti} lassen sich, in der vorliegenden Näherung, sogar direkt, ohne Auflösung des Systems (69.4), angeben. Dazu schreiben wir statt (56.10)

$$\sum_k (D_{ik} - D_{jk})\, k_{Tk} = D_{Ti} - D_{Tj}.$$

Diese Gleichung multiplizieren wir beidseitig mit $\gamma_j a'_{ij,00}$ und summieren über j. Dadurch bekommt ihre linke Seite nach (68.1) den Wert $\frac{3}{4n\gamma_i} k_{Ti}$ — unter Benutzung von $\sum_k k_{Tk} = 0$ — und rechts erscheint die linke Seite von (69.4). Mit den Abkürzungen

$$\alpha_{ij} \approx \tfrac{2}{3}\,(a'_{ij,01}\, A_{i,1} - a'_{ji,01}\, A_{j,1}) = -\,\alpha_{ji}, \tag{69.6}$$

die wir die *polynären Thermodiffusionsfaktoren* — in der ersten, nicht-verschwindenden Näherung[1] — nennen können, hat man somit für die *Thermodiffusionsverhältnisse* die einfache Darstellung

$$k_{Ti} = \gamma_i \sum_j \gamma_j\, \alpha_{ij}. \tag{69.7}$$

Wegen der Schiefsymmetrie der α_{ij} ist die Nebenbedingung $\sum_i k_{Ti} = 0$ identisch erfüllt.

---

[1] Wir deuten diese Näherung in dieser Ziffer durch das Zeichen $\approx$ an.

Es sei betont, daß die polynären α_{ij}, deren es beim K-fachen Gemisch $\frac{1}{2}K(K-1)$ linear unabhängige gibt, abgesehen vom binären Fall rein theoretische Größen sind und nicht eindeutig experimentell bestimmt werden können. Meßbar sind nur die $K-1$ Linearkombinationen k_{Ti}.

Gleichwohl sind die α_{ij} für die Diskussion der Thermodiffusion recht geeignet. Für eine erste Orientierung sei der Wert (67.2) für $a'_{ij,01}$ in (69.6) eingetragen. Das gibt mit der reduzierten Masse $m_{ij}=m_i m_j/(m_i+m_j)$

$$\alpha_{ij} \approx \frac{8m_{ij}^2}{3\,\mathrm{k}T}\left(\frac{5}{2}\Omega_{ij}^{(1,1)}-\Omega_{ij}^{(1,2)}\right)\left(\frac{A_{i,1}}{m_i}-\frac{A_{j,1}}{m_j}\right). \tag{69.8}$$

Darin kann man $\Omega^{(1,2)}$ gemäß (50.10) durch $\Omega^{(1,1)}$ und dessen Temperaturabhängigkeit ausdrücken und sodann $\Omega_{ij}^{(1,1)}$ nach (68.3) durch den binären Diffusionskoeffizienten $D(i,j)$ ersetzen. So ergibt sich

$$\alpha_{ij} \approx \left[\left(\frac{\partial \log D(i,j)}{\partial \log T}\right)_p - 2\right]\cdot\frac{m_{ij}}{2n\,D(i,j)}\left(\frac{A_{i,1}}{m_i}-\frac{A_{j,1}}{m_j}\right). \tag{69.9}$$

Der erste Faktor [...] in dieser Formel ist uns schon beim Lorentz-Gemisch in (62.16) begegnet. Er bewirkt, daß $\alpha_{ij}=0$ ist für das Maxwell-Gas und wechselt als Funktion der Temperatur unter Umständen das Vorzeichen (s. Ziff. 75 und 77). Der zweite Faktor in (69.9) verschwindet, wenn sich die Atome i und j weder in der Masse noch in der Wechselwirkung mit sämtlichen Komponenten des Gemischs unterscheiden: die Thermodiffusion ist ein Differenzeffekt. Der zweite Faktor ist ebenfalls eine reine Zahl. Er ist nach (64.6) dimensionell gleich $\lambda/n\mathrm{k}D$, d.h. dem Verhältnis von Temperaturleitfähigkeit [s. (2.24)] zu Diffusionskoeffizient. Er hängt durch die $A_{i,1}$ von den Mischungsverhältnissen ab und kann unter Umständen als Funktion von γ das Vorzeichen wechseln (s. Ziff. 76).

Zum Schluß sei die explizite Formel für den *binären Thermodiffusionsfaktor* notiert (s. Chapman und Cowling [*4*], S. 167). α hängt, als reine Zahl, nur von Verhältnissen von Ω-Integralen ab. Im Hinblick auf die Tabellen von Hirschfelder *et al.* [*25*] benutzen wir die Abkürzungen, welche diese Autoren eingeführt haben [s. auch Gl. (51.13)]:

$$\left.\begin{aligned}
A^\star &= \frac{1}{2}\,\frac{\Omega_{12}^{(2,2)}}{\Omega_{12}^{(1,1)}} = \frac{\Omega_{12}^{(2,2)\star}}{\Omega_{12}^{(1,1)\star}}\\
B^\star &= \frac{1}{3}\,\frac{5\Omega_{12}^{(1,2)}-\Omega_{12}^{(1,3)}}{\Omega_{12}^{(1,1)}} = \frac{5\Omega_{12}^{(1,2)\star}-4\Omega_{12}^{(1,3)\star}}{\Omega_{12}^{(1,1)\star}}\\
C^\star &= \frac{1}{3}\,\frac{\Omega_{12}^{(1,2)}}{\Omega_{12}^{(1,1)}} = \frac{\Omega_{12}^{(1,2)\star}}{\Omega_{12}^{(1,1)\star}}.
\end{aligned}\right\} \tag{69.10}$$

Diese Größen, die sich von den A, B, C, von Chapman und Cowling lediglich um Zahlenfaktoren unterscheiden, enthalten nur die Wechselwirkung verschiedenartiger Partner. Sie sind nach Definition exakt $=1$ für starre Kugeln. $A^\star$ und $B^\star$ sind im allgemeinen nicht sehr von 1 verschieden (s. Ziff. 75, Fig. 33). Ferner werden die Verhältnisse benötigt

$$E_1 = \frac{1}{5M_2}\,\frac{\Omega_{11}^{(2,2)}}{\Omega_{12}^{(1,1)}} = \frac{2}{5}\sqrt{\frac{2}{M_2}}\left(\frac{d_1}{d_{12}}\right)^2\frac{\Omega_{11}^{(2,2)\star}}{\Omega_{12}^{(1,1)\star}} = \frac{2}{3}\,\frac{\varrho_1 D(1,2)}{\eta_1}; \quad E_2 = \cdots. \tag{69.11}$$

[Die zuletzt notierte Identität für E_1 folgt aus (50.15) und (68.3).] E_2 entsteht aus E_1 durch Vertauschen der unteren Indices. E_1 und E_2 enthalten jeweils auch eine Wechselwirkung gleichartiger Partner. Die Abkürzungen $M_{1,2}$ bedeuten die folgenden Massenverhältnisse

$$M_1 = \frac{m_1}{m_1 + m_2}, \qquad M_2 = \frac{m_2}{m_1 + m_2}. \tag{69.12}$$

Der Ausdruck für die erste nicht-verschwindende Näherung von α lautet nun

$$\alpha \approx (6C^\star - 5) \frac{S_1\gamma_1 - S_2\gamma_2}{Q_1\gamma_1^2 + Q_2\gamma_2^2 + Q_{12}\gamma_1\gamma_2}. \tag{69.13}$$

Dabei ist

$$\left.\begin{aligned} S_1 &= M_1 E_1 - \tfrac{8}{5} M_1 M_2 A^\star + 3(M_1 - M_2) M_2 \\ Q_1 &= E_1 [6 M_2^2 + M_1^2 (5 - \tfrac{12}{5} B^\star) + \tfrac{16}{5} M_1 M_2 A^\star] \\ Q_{12} &= 3(M_1 - M_2)^2 (5 - \tfrac{12}{5} B^\star) + \tfrac{8}{5} M_1 M_2 A^\star (11 - \tfrac{12}{5} B^\star) + 2 E_1 E_2. \end{aligned}\right\} \tag{69.14}$$

Die Zahlen S_2, Q_2 ergeben sich aus S_1, Q_1 durch Vertauschen der Indices 1 und 2.

Dieser Ausdruck ist ziemlich kompliziert, was aber bei der Mannigfaltigkeit der Erscheinungen nicht anders zu erwarten ist. In Ziff. 74 und 75 wird gezeigt, wie man sich an Hand spezieller Fälle rasch Einsicht in das Wesentliche verschaffen kann.

70. Der Reibungskoeffizient. Die erste Näherung für η ergibt sich ganz analog wie diejenige für λ. Man läßt in (66.9) alle nicht-diagonalen Elementes $s \neq r$ fort und hat dann

$$\sum_j \gamma_j (b'_{ij,00} \sqrt{m_i}\, B_{i\,0} + b''_{ij,00} \sqrt{m_j}\, B_{j,0}) \approx 5 \sqrt{m_i}\,. \tag{70.1}$$

Durch Auflösen dieser K linearen Gleichungen erhält man die $B_{i,0}$ und damit nach (64.7) den Reibungskoeffizienten. Nach der Cramerschen Regel und dem Entwicklungssatz für Determinanten kann man auch schreiben

$$\eta \approx -\frac{5}{4} \begin{vmatrix} b_{11,00} & \cdots & b_{1K,00} & \sqrt{m_1} \\ \vdots & & \vdots & \vdots \\ b_{K1,00} & \cdots & b_{KK,00} & \sqrt{m_K} \\ \sqrt{m_1}\,\gamma_1 & \cdots & \sqrt{m_K}\,\gamma_K & 0 \end{vmatrix} : \begin{vmatrix} b_{11,00} & \cdots & b_{1K,00} \\ \vdots & & \vdots \\ b_{K1,00} & \cdots & b_{KK,00} \end{vmatrix} \tag{70.2}$$

Dabei ist zur Abkürzung gesetzt

$$b_{ij,rs} = \delta_{ij} \sum_l \gamma_l\, b'_{il,rs} + \gamma_j\, b''_{ij,rs} = \frac{\gamma_j}{\gamma_i}\, b_{ji,sr}. \tag{70.3}$$

Damit hat man, über (67.5) und (67.6), im Prinzip die Viscosität durch Ω-Integrale ausgedrückt.

Die Beziehung $\lambda/\eta \approx \frac{5}{2} c_v$, welche für reine Gase gilt [s. Gl. (50.16)], überträgt sich nicht auf Gemische, wie der Vergleich von (69.2) und (70.2) zeigt.

Wegen der weiteren Diskussion sei auf Ziff. 78 und 79 verwiesen.

III. Theoretische und experimentelle Ergebnisse für den Diffusionskoeffizienten.

In diesem Abschnitt wird für einige spezielle Molekülmodelle der Diffusionskoeffizient angegeben und anschließend mit Meßwerten verglichen. Besonders aufschlußreich ist die Betrachtung der Selbstdiffusion. Diese eröffnet, zusätzlich zur Viscosität, eine Möglichkeit zur Bestimmung der Wechselwirkung (fast) gleichartiger Moleküle. Der Vergleich von beiden zeigt die Überlegenheit des Lennard-Jonesschen Modells über das Potenzkraft-Modell. Zum Schluß wird die Abhängigkeit des Diffusionskoeffizienten binärer Gemische vom Mischungsverhältnis diskutiert. Hier gibt es noch Unstimmigkeiten.

71. Der Diffusionskoeffizient für spezielle Molekülmodelle. Erster Vergleich mit dem Experiment. Nach (68.3) ist die erste Näherung für den Diffusionskoeffizienten im wesentlichen durch das Integral $\Omega_{12}^{(1,1)}$ bestimmt. Wir betrachten, wie bei der Viscosität in Ziff. 51, wieder die Modelle der starren Kugeln, der Potenzkraft-Zentren und Anziehungskräfte nach LENNARD-JONES.

α) Starre Kugeln. Es sei $d_{12} = (d_1 + d_2)/2$ der für den Stoß der beiden Partner maßgebende Moleküldurchmesser. Dann ist nach (51.1) und (51.2) sowie (68.3) der binäre Diffusionskoeffizient in erster Näherung gegeben durch

$$[D(1,2)]_1 = \frac{3\sqrt{kT/2m_{12}}}{8\sqrt{\pi}\, n\, d_{12}^2} \quad \text{mit} \quad m_{12} = \frac{m_1 m_2}{m_1 + m_2}. \tag{71.1}$$

Tabelle 9. *Gemessene Selbstdiffusionskoeffizienten D_0 einiger Gase*[1] *bei 0° C, 760 Torr; nach (71.2) daraus berechnete Moleküldurchmesser d_0.*

	Ne	A	Kr	X	H_2	N_2	O_2	CO_2
D_0 (cm²/sec) . . .	0,452	0,156	0,079	0,048	1,28	0,185	0,187	0,0974
d_0 (Å)	2,42	3,48	4,05	4,65	2,56	3,50	3,35	4,30

Im Fall der Selbstdiffusion[2] hat man $2m_{12} = m$, $d_{12} = d$ und daher

$$D \approx 0{,}212 \frac{\sqrt{kT/m}}{n\, d^2}. \tag{71.2}$$

Vergleich mit (51.3) ergibt für die Zahl $\varrho D/\eta$, wieder in erster Näherung, den Wert

$$\left(\frac{\varrho D}{\eta}\right)_{\text{starr}} = \frac{6}{5}. \tag{71.3}$$

Mit der freien Weglänge l aus (51.5) und der mittleren Molekulargeschwindigkeit $\overline{C}$ aus (34.12) kann man für den Selbstdiffusionskoeffizienten starrer Kugeln auch schreiben

$$D = \frac{3\pi}{16} \overline{C}\, l = 0{,}590\, \overline{C}\, l. \tag{71.4}$$

Die Tabelle 9 gibt einige Zahlenbeispiele. Die nach der Formel für starre Kugeln berechneten Durchmesser haben natürlich nur qualitative Bedeutung. Sie sind alle um 5 bis 10% kleiner als die Durchmesser, welche in analoger Weise aus der Viscosität entnommen und in Tabelle 4 (Ziff. 51) aufgeführt sind.

β) Potenzkraft-Zentren nach (25.2). Hier tritt bei der Bildung von $\Omega^{(1,1)}$ das Integral $A_1(\nu)$ [vgl. (51.7a, b)] auf, das numerisch berechnet werden muß. Es ergibt sich

$$D = D^*(\nu)\, \varkappa^{-\frac{2}{\nu-1}}\, p^{-1}\, m^{-\frac{1}{2}}\, (kT)^{\frac{3}{2}+\frac{2}{\nu-1}} \tag{71.5}$$

---

[1] H_2 nach HARTECK und SCHMIDT (1933); Kr nach GROTH und HARTECK (1941); X nach VISNER (1951), mit $S_D = 190$ bzw. 250° K auf 0° C umgerechnet, s. (71.8); die übrigen nach WINN (1950).

[2] Von Selbstdiffusion spricht man dann, wenn sich die beiden Molekülsorten nur sehr wenig unterscheiden, z.B. isotop oder gar isomer sind. Für eine exakt einheitliche Substanz aus nicht-unterscheidbaren Teilchen kann man keine Selbstdiffusion definieren, s. auch Ziff. 83.

mit

$$D^* = \frac{3}{8\sqrt{\pi}} \frac{2^{\frac{2}{\nu-1}}}{\left(2 - \frac{2}{\nu-1}\right)!\,A_1(\nu)}. \tag{71.6}$$

Dies ist der Ausdruck für den Selbstdiffusionskoeffizienten. Bei Fremddiffusion wäre $\varkappa_{12}$ und ν_{12} zu schreiben und $2m_{12}$ anstelle von m zu setzen. Numerische Ergebnisse für einige ν-Werte sind in Tabelle 10 zusammengestellt.

Wie der Vergleich mit (51.8) zeigt, gilt ferner

$$\frac{\varrho D}{\eta} = \frac{D^*}{\eta^*} = \frac{3}{5}\left(3 - \frac{2}{\nu-1}\right)\frac{A_2(\nu)}{A_1(\nu)}. \tag{71.7}$$

Dieses Verhältnis kann an Hand der Werte aus Tabelle 5 (Ziff. 51) und Tabelle 10 angegeben werden und ist in letzterer mitaufgeführt.

Tabelle 10. *Werte für $A_1(\nu)$ aus (51.7a) nach Chapman und Cowling [4], S. 172. Werte von $D^*(\nu)$ nach (71.6). Werte von $\varrho D/\eta$ nach (71.7).*

ν	5	7	9	11	15	∞
$A_1(\nu)$	0,422	0,385	0,382	0,383	0,393	0,500
$D^*(\nu)$	0,532	0,460	0,410	0,380	0,338	0,212
$\frac{\varrho D}{\eta}$	1,55	1,49	1,43	1,41	1,35	1,20

γ) Anziehungskraft, näherungsweise berücksichtigt. Hier gelten analoge Bemerkungen, wie sie in Ziff. 51, γ gemacht wurden bei der Behandlung der Viscosität. Insbesondere erhält man wieder eine Sutherland-Formel der Gestalt

$$D(T) = \frac{D_{\text{starr}}(T)}{1 + S_D/T}. \tag{71.8}$$

Es ist zu betonen, daß die Sutherland-Konstante S_η aus (51.10) und S_D im allgemeinen nicht übereinstimmen. Man kann im Gegenteil, wenn man beide kennt, auf den Exponenten der Anziehungskraft schließen und die maximale Anziehungsenergie bei Berührung ermitteln. Eine solche Diskussion zeigt dann aber deutlich, daß die Sutherland-Formeln sich zwar zur empirischen Interpolation eignen, jedoch nicht modellmäßig aufzufassen sind. HUTCHINSON (1949) findet durch Gegenüberstellung der Viscosität und der Selbstdiffusion von Argon, daß der Exponent $\nu' = 16$ und die Anziehungsenergie $\mathrm{k} \times 500°$ K sein sollte, was sicher beides zu verwerfen ist. Ähnliche Kritik haben auch schon CHAPMAN und COWLING geübt.

δ) Das Lennard-Jonessche (6; 12)-Potential. Mit den reduzierten Ω-Integralen von HIRSCHFELDER *et al.* [s. Gl. (51.13) sowie nach (68.3) und (71.1)] schreibt sich der binäre Diffusionskoeffizient wie folgt:

$$[D(1,2)]_1 = \frac{3\sqrt{\mathrm{k}T/2m_{12}}}{8\sqrt{\pi}\, n\, d_{12}^2\, \Omega_{12}^{(1,1)\star}}. \tag{71.9}$$

Mit d_{12} ist die charakteristische Länge in dem Lennard-Jones-Potential (51.11) für zwei verschiedenartige Moleküle gemeint. Das reduzierte Integral $\Omega^\star$ ist eine Funktion nur der reduzierten Temperatur [s. Gl. (51.12)]

$$T_{12}^* = \frac{T}{\varepsilon_{12}/\mathrm{k}}. \tag{71.10}$$

[ε_{12} bedeutet die charakteristische Energie aus dem Potential (51.11), genommen für zwei verschiedenartige Moleküle] und kann aus den Tabellen bei HIRSCHFELDER *et al.* [25] entnommen werden. In Fig. 31 (Ziff. 51) ist es graphisch dargestellt. Somit kann man durch Analyse von binären Diffusionskoeffizienten die Parameter der Wechselwirkung von verschiedenartigen Molekülen ermitteln.

Oft benutzt man als Faustregel

$$d_{12} = \tfrac{1}{2}(d_1 + d_2), \qquad \varepsilon_{12} = \sqrt{\varepsilon_1 \varepsilon_2}. \tag{71.11}$$

Theoretische Gründe dafür hat ZENER (1931) angegeben. Akzeptiert man diese Regel, so kann man mittels der Wechselwirkungsparameter gleichartiger Moleküle, deren einige in Tabelle 7 (Ziff. 52) aufgeführt sind, den Diffusionskoeffizienten verschiedenartiger Moleküle sofort angeben. Empirische Modifikationen von (71.11) haben SRIVASTAVA und SAXENA (1956) angegeben.

Als Beispiel sei die Diffusion von N_2/H_2 betrachtet. Nach (71.1) und (71.9) bis (71.11), nach Tabelle 7 (Ziff. 52) und Fig. 31 (Ziff. 51) findet man für $T = 293{,}2°$ K, $p = 760$ Torr

d_{12}	ε_{12}/k	D_{starr}	T_{12}^*	$\Omega_{12}^{(1,1)\star}$	D_{theor}
3,305 Å	58,1° K	0,625 cm²/sec	5,05	0,841	0,745 cm²/sec

(71.12)

Das paßt gut zu dem Wert $D_{\text{exp}} = 0{,}76$ cm²/sec, der übereinstimmend von BOARDMAN und WILD (1937) sowie von WALDMANN (1944) gemessen wurde beim selben T und p.

Das bedeutet aber nicht, daß die für d_{12} und ε_{12} angenommenen Werte nun schon die richtigen seien. Der Diffusionskoeffizient bei mindestens einer weiteren Temperatur müßte gemessen sein und mit den berechneten verglichen werden. Mangels experimenteller Daten wird man die Thermodiffusion heranziehen. Das geschieht in Ziff. 76 und führt zu den merklich anderen Werten (76.1). Da letztere aus dem Experiment stammen, sind sie vorzuziehen und es ist befriedigend, daß sie einen noch etwas besseren Wert für den theoretischen Diffusionskoeffizienten geben, nämlich $D = 0{,}750$ cm²/sec. Der Fall zeigt, daß die Regel (71.11) nicht immer genaue Werte der Wechselwirkungskonstanten liefert. Erwähnt sei noch, daß auch die Viscosität keine Entscheidung zwischen den Konstanten-Sätzen (71.12) und (76.1) ermöglicht, s. Ziff. 79; der auf die Thermodiffusion sich stützende Vergleich ist überlegen.

Andere Fälle sind in dem Buch von HIRSCHFELDER *et al.* [25], S. 578, berechnet. Wir werden uns beim weiteren Vergleich mit dem Experiment in der nächsten Ziffer auf die Selbstdiffusion beschränken. Wegen der einfacheren Verhältnisse — man hat es nur mit den Parametern d und ε der Wechselwirkung fast gleichartiger Moleküle zu tun — und weil für den Selbstdiffusionskoeffizienten die vollständigsten Meßreihen vorliegen, kann hier der Vergleich der Theorie mit dem Experiment am schärfsten angestellt werden. Man macht diesen Vergleich zweckmäßig an Hand der dimensionslosen Größe $\varrho D/\eta$. Diese ist im wesentlichen durch das Verhältnis $A^\star$ [s. Gl. (69.10)] gegeben; nach (51.14) und (71.9) ist nämlich

$$\frac{D\varrho}{\eta} = \frac{6}{5} A^\star. \tag{71.13}$$

$A^\star$ hat für starre Kugeln nach Definition den Wert 1 und hängt wie die $\Omega^\star$ nur von der Temperatur, nicht von der Masse ab. Die Temperaturabhängigkeit ist aber in weitem Bereich gering; $A^\star$ hat für die Lennard-Jonessche (6; 12)-Wechselwirkung praktisch den konstanten Wert 1,1 [s. Fig. 33 (Ziff. 75)]. Wie weit das experimentell bestätigt ist, wird in der nächsten Ziffer behandelt.

72. Experimentelle Ergebnisse für die Temperaturabhängigkeit der Selbstdiffusion und des Verhältnisses $\varrho D/\eta$. Die erste Messung der Selbstdiffusion eines Gases haben HARTECK und SCHMIDT (1933) gemacht. Sie untersuchten die

Diffusion von Para- in Normalwasserstoff bei Zimmertemperatur und darunter nach den in Ziff. 13 und 14 geschilderten Methoden; Tabelle 11 zeigt ihre Ergebnisse. Das dimensionslose Verhältnis $\varrho D/\eta$ hat bei den beiden höheren Temperaturen den theoretischen Wert von etwa 1,3; bei 20° K liegt der Wert erheblich darunter. Das kann man nicht auf den Einfluß der Quanteneffekte zurückführen. Dieser ist geringer, wie die letzte Zeile der Tabelle zeigt. HARTECK und SCHMIDT betonen, daß die stationäre Methode nicht sehr präzis ist. Der Meßdruck bei 20° K war übrigens nur 1,5 Torr, das Gas also ideal.

Tabelle 11. ϱ Dichte des idealen Gases, D Selbstdiffusionskoeffizient von Wasserstoff, beide bei 760 Torr; η Viscosität. Nach HARTECK und SCHMIDT (1933). Vergleich von $\varrho D/\eta$ mit dem theoretischen Wert $\frac{6}{5}A^\star$ für (6; 12)-Potential.

T (°K)	20,4	85	273
$10^3\,\varrho$ (g/cm³)	1,20[1]	0,289	0,0899
D (cm²/sec) .	0,00816	0,172	1,27
η (μ-Poise) .	8,6 [10,80][2]	37,7 [38,9][2]	84,4
$\varrho D/\eta$	1,14 [0,91][3]	1,32 [1,28][3]	1,35
$\frac{6}{5}A^\star$	1,27[4]	1,31[5]	1,33[5]

[1] Gegenüber der Originalarbeit geändert.
[2] [ ]-Werte nach E. W. BECKER und MISENTA (1955).
[3] [ ]-Werte mittels der eingeklammerten η-Werte berechnet.
[4] Quantentheoretisch nach COHEN et al. (1956).
[5] Klassisch.

Eine neuere Arbeit über die Selbstdiffusion von Gasen, die besonders reich an experimentellen Ergebnissen ist, stammt von WINN (1950). WINN benutzte stabile Isotope — die Methode ist in Ziff. 13 beschrieben — und hat bei verschiedenen Temperaturen gemessen. Diese Versuche ermöglichen eine genaue Nachprüfung der Beziehung (71.13). WINN hat auch, was schon HUTCHINSON (1949) tat, die kleine Korrektur wegen der Verschiedenheit der Isotopenmassen angebracht und auf „wahre“ Selbstdiffusion umgerechnet.

Tabelle 12. ϱ Dichte des idealen Gases, D Selbstdiffusionskoeffizient von Argon, beides bei 760 Torr; η Viscosität. Vergleich von $\varrho D/\eta$ mit dem theoretischen Wert $\frac{6}{5}A^\star$, s. (71.13), für (6; 12)-Potential; $T^\star$ = reduzierte Temperatur aus (51.12) mit $\varepsilon/k = 124°$ K.

T (°K)	77,7	90,2	194,7	273,2	295,2	326,7	353,2
$10^3\,\varrho$ (g/cm³)	6,30	5,42	2,51	1,79	1,66	1,50	1,39
D (cm²/sec)	0,0134	0,0180	0,0830	0,156	0,178	0,212[1]	0,249
η (μ-Poise)[2]	67,51	76,4	155	211	223	242	257
$\varrho D/\eta$	1,25	1,28	1,34	1,32	1,32	1,31	1,35
$T^\star$	0,63	0,73	1,57	2,20	2,38	2,64	2,85
$\frac{6}{5}A^\star$	1,320	1,320	1,315	1,313	1,313	1,313	1,313

[1] Nach HUTCHINSON (1949); alle übrigen D-Werte nach WINN (1950).
[2] Nach JOHNSTON und GRILLY (1942).

Es sei als Beispiel das Argon herausgegriffen. WINN untersuchte die Diffusion von ^{40}A/^{36}A, HUTCHINSON die des β^--aktiven ^{41}A gegen ^{40}A, welch letzteres den Hauptbestandteil des natürlichen Argons bildet. Die Ergebnisse sind in Tabelle 12 zusammengefaßt. Daraus ist zu sehen, daß tatsächlich $\varrho D/\eta$ in weitem Temperaturbereich fast konstant ist und dem theoretischen Wert $\frac{6}{5}A^\star$ für das Lennard-Jonessche (6; 12)-Modell recht nahekommt[1]. Dieses schöne Ergebnis

[1] In der Arbeit von WINN ist der Meßdruck nicht angegeben. Argon ist bei 80 bis 90° K und 1 atm bereits merklich nicht-ideal. Für Tabelle 12 wurde die Annahme gemacht, daß der Druck stets niedrig genug war für die Anwendbarkeit der idealen Gasgleichung (d.h. daß die ϱ- und D-Werte zu 77,7 und 90,2° K lediglich auf 1 atm umgerechnete Werte darstellen, nicht die wirklichen). Wenn hingegen der Meßdruck etwa 1 atm betrug, so würden sich die ersten beiden Werte für $\varrho D/\eta$ um schätzungsweise 10 bzw. 6% erhöhen.

zeigt, wie außerordentlich wertvoll die Ausarbeitung der kinetischen Theorie des Lennard-Jones-Modells war. Man muß sich dabei vergegenwärtigen, daß die quantitative Deutung der Versuchsergebnisse etwa mittels des Potenz-Modells ganz unmöglich ist. Der gemessene Wert von $\varrho D/\eta \approx 1{,}35$ würde nach Tabelle 10 (Ziff. 71) auf die Potenz $\nu \approx 15$ schließen lassen, in deutlichem Widerspruch zu dem Wert $\nu_0 \approx 7$, s. Tabelle 4 (Ziff. 51), den man zur Erklärung der Temperaturabhängigkeit der Viscosität von Argon braucht. Durch das Lennard-Jonessche Modell dagegen sind, wie gezeigt, Viscosität und Selbstdiffusion sowie deren Temperaturabhängigkeit unter einem einheitlichen Gesichtspunkt quantitativ darstellbar[1].

Tabelle 13. *Vergleich von Dichte ϱ, Selbstdiffusion D — beides bei 760 Torr — und Viscosität η. M = Molekulargewicht.*

	Ne [1]	A [1]	Kr [2]	X [3]	H_2 [4]	N_2 [1,7]	O_2 [1,7]	HCl [5]	HBr [5]	CO_2 [1,7]	CH_4 [1]	UF_6 [6]
T (°K)	273	273	294	300	273	273	273	295	295	273	273	303
M	20	40	83,7	131	2	28	32	36,5	80,9	44	16	352
$10^3\,\varrho$ (g/cm³)	0,893	1,79	3,46	5,33	0,090	1,25	1,43	1,51	3,34	1,96	0,715	14,2
D (cm²/sec)	0,452	0,156	0,091	0,058	1,28	0,185 (0,172)	0,187 (0,175)	0,125	0,079	0,0974 (0,096)	0,206	0,0165
η (μ-Poise)	298	211	249	232	85,7	167	193	144	186	139	103	179
$\varrho D/\eta$	1,35	1,32	1,27	1,33	1,34	1,39 (1,29)	1,39 (1,30)	1,32	1,42	1,38 (1,36)	1,42	1,31

[1] WINN (1950) [Ne und A sind auch von GROTH und SUSSNER (1944) gemessen].
[2] GROTH und HARTECK (1941). Der in dieser Arbeit angegebene D-Wert für X ist offenbar zu klein; der dortige ϱ-Wert wäre zu berichtigen.
[3] VISNER (1951).
[4] HARTECK und SCHMIDT (1933).
[5] BRAUNE und ZEHLE (1941).
[6] NEY und ARMISTEAD (1947).
[7] WINTER (1951); dessen Werte sind eingeklammert.

Dieser Erfolg bestätigt sich bei anderen Gasen. In Tabelle 13 sind eine Reihe weiterer Selbstdiffusionskoeffizienten mit den zugehörigen Viscositäten verglichen. Irgendwelche auffällige Diskrepanzen sind nicht festzustellen. Bei den mehratomigen, nichtkugelsymmetrischen Molekülen kann man ohnedies keine strikte Übereinstimmung erwarten. Das haben besonders MASON und RICE (1954a und b) betont, welche die Selbstdiffusion an Hand des (6; Exp)-Potentials diskutierten.

Erwähnt sei schließlich noch, daß beim Selbstdiffusionskoeffizienten die Ungenauigkeit der ersten Näherung gering ist; der Unterschied gegenüber der zweiten Näherung beträgt nur etwa 1%.

73. Die Abhängigkeit des binären Diffusionskoeffizienten vom Molenbruch. Diese Frage hat eine lange Geschichte. Die älteren, auf dem Begriff der freien Weglänge basierenden Theorien ergaben eine viel zu starke Abhängigkeit vom Mischungsverhältnis. Eine Ausnahme bildete von Anfang an MAXWELLs Theorie des Modells $\nu = 5$, welche einen konstanten Diffusionskoeffizienten lieferte. Daß die Abhängigkeit tatsächlich gering ist, wurde für eine Reihe von Gaspaaren von stark verschiedenem Molekulargewicht, z.B. N_2/H_2, CO_2/H_2, A/He experimentell bestätigt, vgl. LONIUS (1909). Nach der Chapman-Enskogschen Theorie

[1] Bei Ausdehnung des Temperaturbereichs machen sich jedoch auch hier gewisse Unzulänglichkeiten bemerkbar. Nach AMDUR und SCHATZKI (1957) dürfte für die Edelgase die Mulde des (6; 12)-Potentials zu breit sein (Anm. bei der Korrektur).

war dies verständlich, denn auch diese ergibt, wie die Maxwellsche, wenigstens in erster Näherung, s. (68.3), Unabhängigkeit vom Mischungsverhältnis. Die beobachtete geringe Abhängigkeit war in dem Sinn, daß wenig leichte Moleküle in viel schweren etwa 10% schneller diffundieren als wenig schwere in viel leichten.

Es ist zu erwarten, daß sich dies auf Grund der zweiten Näherung für den Diffusionskoeffizienten erklären läßt (Chapman 1917). Die zweite Näherung wird für den Grenzfall $\gamma_1 = 0$ (oder 1) sehr einfach. Man entnimmt aus (66.7), angeschrieben für $r = 0$ und 1, nach wenigen Zeilen

$$[D]_2 = [D]_1 \, [1 + \Delta(\gamma_1)] \tag{73.1}$$

mit

$$\Delta(0) = \frac{(a'_{12,01})^2}{a'_{12,00}\, a'_{12,11}}; \quad \Delta(1) = \frac{(a'_{21,01})^2}{a'_{21,00}\, a'_{21,11}}. \tag{73.2}$$

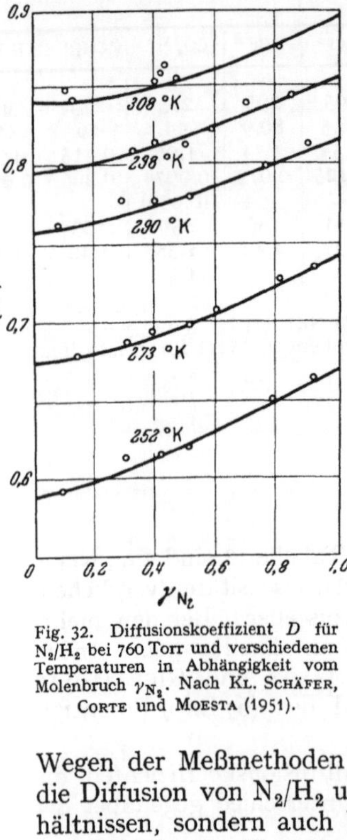

Fig. 32. Diffusionskoeffizient D für N_2/H_2 bei 760 Torr und verschiedenen Temperaturen in Abhängigkeit vom Molenbruch γ_{N_2}. Nach Kl. Schäfer, Corte und Moesta (1951).

Trägt man darin die Ausdrücke aus Ziff. 67 ein und benutzt die in (69.10) definierten Abkürzungen $A^\star$, $B^\star$, $C^\star$ für die dortigen massenunabhängigen Verhältnisse von reduzierten Ω-Integralen, so kommt

$$\Delta(0) = \frac{(6C^\star - 5)^2}{30\left(\frac{m_1}{m_2}\right)^2 + 25 - 12B^\star + 16\,\frac{m_1}{m_2}\,A^\star}. \tag{73.3}$$

Den anderen Wert $\Delta(1)$ erhält man daraus durch Vertauschen der beiden Massen. Wenn $m_1 \gg m_2$, so wird also

$$\Delta(0) = 0; \quad \Delta(1) = \frac{(6C^\star - 5)^2}{25 - 12B^\star}. \tag{73.4}$$

Die Korrektur für $\gamma = 1$ sollte, wenn sie bereits exakt wäre, zu dem Diffusionskoeffizienten (62.14) des Lorentz-Gemisches führen. Daran läßt sich die Güte der zweiten Näherung prüfen (s. Chapman und Cowling [4], S. 196). Für starre Kugeln ist $B^\star = C^\star = 1$, daher $\Delta(1) = 0{,}077$. Nach (62.15) ist hier die exakte Korrektur $\Delta = 0{,}13$, also fast doppelt so groß wie diejenige, welche gemäß (73.1) zur zweiten Näherung führt.

Neuere Versuche stammen von Kl. Schäfer, Corte und Moesta (1951), sowie von Nettley (1954). Wegen der Meßmethoden s. Ziff. 13 und 17. Die erstgenannten Autoren haben die Diffusion von N_2/H_2 und CO_2/H_2 nicht nur bei verschiedenen Mischungsverhältnissen, sondern auch bei verschiedenen Temperaturen untersucht (s. Fig. 32 und Tabelle 14). Nettleys Ergebnisse für N_2/H_2 bei 288° K stimmen mit denen von Schäfer *et al.* ziemlich gut überein. In den älteren Versuchen, s. Lonius (1909), war in diesem Fall die entgegengesetzte Abhängigkeit vom Mischungsverhältnis gefunden worden. Waldmann (1947b) (s. Ziff. 15) konnte keine Abhängigkeit feststellen. Die Methode von Nettley ist dem Problem besonders angepaßt. Leider liegen Messungen nur bei einer einzigen Temperatur vor.

Der Vergleich der Ergebnisse von Schäfer *et al.* mit der Theorie ist interessant. Wir ziehen dazu das Lennard-Jonessche (6; 12)-Potential heran, für welches $B^\star$ und $C^\star$ aus Fig. 33 (Ziff. 75) zu entnehmen sind. Mit dem Wert $\varepsilon_{12}/k = 58°$ K, s. (71.12), findet man die in Tabelle 14 angegebenen reduzierten Temperaturen $T^\star$. Die zugehörigen Werte von $6C^\star - 5$ sind in der Tabelle mit aufgeführt. $B^\star$ ist konstant $= 1{,}09$. Damit ergibt sich nach (73.3), oder praktisch ebensogut nach

(73.4), die letzte Zeile der Tabelle, die nach (73.1) mit dem experimentellen Ergebnis für den relativen Unterschied der Diffusionskoeffizienten übereinstimmen sollte.

Wie man sieht, kann von einer quantitativen Übereinstimmung keine Rede sein und der Temperaturgang ist bei beiden Zahlenreihen sogar entgegengesetzt. Die Gründe hierfür sind unklar. Wertvoll wäre es für den Grenzfall $m_1 \gg m_2$,

Tabelle 14. $D(\gamma)$ = Diffusionskoeffizient von N_2/H_2 bei 760 Torr; γ = Molenbruch von N_2; $\Delta(1) - \Delta(0)$ = relativer Unterschied der Diffusionskoeffizienten $D(1)$ und $D(0)$ nach der Theorie des Lennard-Jones-(6; 12)-Potentials. Messungen von Kl. SCHÄFER, CORTE und MOESTA (1951).

T (°K)		252	273	290	298	308
experimentell	$D(0)\ \frac{\text{cm}^2}{\text{sec}}$	$0{,}58_5$	$0{,}67_5$	0,76	$0{,}79_5$	0,84
	$D(1)\ \frac{\text{cm}^2}{\text{sec}}$	0,67	0,74	0,82	0,85	$0{,}88_5$
	$\frac{D(1) - D(0)}{D(0)}$	$0{,}14_5$	0,095	0,080	0,070	0,050
theoretisch	T^*	4,35	4,70	5,00	5,13	5,30
	$6C^* - 5$	0,563	0,580	0,590	0,595	0,600
	$\Delta(1) - \Delta(0)$	0,024	0,025	0,026	0,027	0,027

$\gamma_1 \approx 0$, der dem Lorentz-Gemisch gerade entgegengesetzt ist, eine genaue Theorie zu haben, um allen Zweifeln an der Güte der zweiten Näherung enthoben zu sein[1]. Ferner wären Versuche mit Edelgasgemischen erwünscht.

IV. Theoretische und experimentelle Ergebnisse für den Thermodiffusionsfaktor.

Der Thermodiffusionsfaktor hängt besonders empfindlich vom molekularen Kraftgesetz ab, empfindlicher als die Koeffizienten der Reibung und der Diffusion. Das geht schon daraus hervor, daß er für Maxwellsche Moleküle verschwindet und in anderen Fällen beiderlei Vorzeichen haben kann. Wie ein Blick auf die Formeln (62.16) oder (69.9) zeigt, kann man sagen, daß die Natur im Thermodiffusionsfaktor die Ableitung des Diffusionskoeffizienten nach der Temperatur zur Verfügung stellt. Diese Ableitung durch direkte Messung der Diffusion bei verschiedenen Temperaturen mit gleicher Genauigkeit zu ermitteln, ist ganz aussichtslos. Die Analyse der Thermodiffusion stellt daher die schärfste Probe für jeden Wechselwirkungsansatz dar. Die Potenzkraft-Zentren z.B. können bestimmt ausgeschlossen werden. Sie führen zu einem bei Isotopen streng temperaturunabhängigen Thermodiffusionsfaktor, was der Wirklichkeit nicht entspricht. Das Lennard-Jonessche (6; 12)-Potential stellt die Beobachtungen qualitativ einigermaßen gut dar. Jedoch werden bei der Diskussion der Temperaturabhängigkeit der Thermodiffusion die Grenzen auch dieses Modelles sichtbar und das (6; Exp)-Modell erweist sich als wertvoll wegen seiner größeren Anpassungsfähigkeit.

Aus dem genannten Grund ist die Thermodiffusion in neuerer Zeit, über die grundlegenden Arbeiten von ENSKOG und CHAPMAN hinaus, vielfach theoretisch und experimentell bearbeitet worden. Eine Gesamtdarstellung des Gebiets vermittelt die Monographie von GREW und IBBS [*18*]. Dort findet man auch ein umfassendes Verzeichnis der gemessenen α-Werte und der Literatur.

---

[1] Siehe Anm. 1, S. 416.

Weil die Thermodiffusion ein subtiler Differenzeffekt ist, gibt es keine einwandfreie elementare Theorie dafür. Es ist kein Zufall, daß erst die exakte Kinetik von ENSKOG und CHAPMAN zu ihrer Entdeckung führte. Die vollständigste elementare Theorie stammt von FÜRTH (1942) und ist in dem eben erwähnten Buch von GREW und IBBS ([*18*], S. 118), dargestellt. Eine einwandfreie Dimensionsbetrachtung über die Thermodiffusion hat FRANKEL (1940) angestellt.

74. Der Thermodiffusionsfaktor für spezielle binäre Gemische. Um rasch einen Überblick über die Erscheinungen zu gewinnen, wollen wir folgende speziellen Fälle betrachten:

I. $m_1 \gg m_2$.

Die Moleküle der einen Sorte sind viel schwerer als die der anderen. Wir werden das Augenmerk besonders richten auf die Endwerte[1] des Thermodiffusionsfaktors $\alpha\,(\gamma_1 = 0)$ und $\alpha\,(\gamma_2 = 0)$.

II. $m_1 = m_2, \quad \Phi_{11} \approx \Phi_{22} \approx \Phi_{12}$.

Die Moleküle sind isobar; die Wechselwirkungen zwischen gleich- und verschiedenartigen Partnern sind wenig verschieden.

III. $m_1 \approx m_2, \quad \Phi_{11} = \Phi_{22} = \Phi_{12}$.

Der relative Massenunterschied ist klein. Die Moleküle sind isotop, alle Wechselwirkungen sind gleich. Dieser Fall ist für die Erschließung der Wechselwirkung der wichtigste, weil nur eine einzige vorkommt.

Aus der umfassenden Formel (69.13) sollen nun diese drei Fälle herausgeschält werden. Zuvor noch eine allgemeine Bemerkung. Nach (69.13) besteht α aus zwei Faktoren. Diese Eigenschaft trat uns von Anfang an, bereits in (69.8) und (69.9), entgegen. Der erste Faktor

$$6C^\star - 5 = 2\left[2 - \left(\frac{\partial \log D(1,2)}{\partial \log T}\right)_p\right] \tag{74.1}$$

hängt nur von der Wechselwirkung verschiedenartiger Partner und der Temperatur ab, nicht von den Massen und dem Mischungsverhältnis. Er verschwindet für Maxwellsche Moleküle und bestimmt in der Hauptsache die Temperaturabhängigkeit von α. Der zweite Faktor, der Bruch in (69.13), hängt im allgemeinen auch von der Wechselwirkung gleichartiger Moleküle ab sowie vom Massen- und Mischungsverhältnis. — Stets handelt es sich hier und in den folgenden Ziffern bei den mit α bezeichneten theoretischen Werten stillschweigend um die erste, nicht verschwindende Näherung des Thermodiffusionsfaktors.

Jetzt zu den Spezialfällen.

Fall I. Hier ist $M_1 \approx 1$, $M_2 \approx 0$ — wegen $M_{1,2}$ siehe (69.12) — und man entnimmt aus (69.13), (69.14) und (69.11) sogleich die Endwerte

$$\left.\begin{aligned} \alpha\,(\gamma_1 = 0) &= (6C^\star - 1)\,\frac{5}{4\sqrt{2}}\left(\frac{d_{12}}{d_{22}}\right)^2 \frac{\Omega_{12}^{(1,1)\star}}{\Omega_{22}^{(2,2)\star}} \\ \alpha\,(\gamma_2 = 0) &= (6C^\star - 1)\,\frac{5}{25 - 12B^\star}\,. \end{aligned}\right\} \tag{74.2}$$

An Hand des zweiten Endwerts kann man die Güte der ersten Näherung von α, verdeutlicht $[\alpha]_1$, nachprüfen. Vom Lorentz-Gemisch her, s. (62.16), ist

$$\alpha_{\text{exakt}} = 2 - \left(\frac{\log D}{\log T}\right)_p = \frac{1}{2}\,(6C^\star - 1)$$

[1] Bezeichnung nach CHAPMAN (1940).

bekannt. Es ist also in dem vorliegenden doppelten Spezialfall

$$\alpha_{\text{exakt}} = [\alpha]_1 f_\alpha \quad \text{mit} \quad f_\alpha = \tfrac{1}{10}(25 - 12B^\star). \tag{74.3}$$

Der Faktor f_α bildet einen Anhaltspunkt für die Güte der Näherung.

Die weitere Diskussion stellen wir bis zur Behandlung spezieller Modelle (Ziff. 75) zurück. Ein ungefähres Beispiel für Fall I bildet das Gaspaar N_2/H_2, das in Ziff. 76 näher besprochen wird.

Fall II. Hier ist $M_1 = M_2 = \frac{1}{2}$. Mit den Abkürzungen aus der großen Arbeit von CHAPMAN (1940) über die Charakteristika der Thermodiffusion

$$s_1^2 = d_1^2 \Omega_{11}^{(2,2)\star}, \quad s_2^2 = d_2^2 \Omega_{22}^{(2,2)\star}, \quad s_{12}^2 = d_{12}^2 \Omega_{12}^{(2,2)\star} \tag{74.4}$$

wird nach (69.10) und (69.11)

$$\frac{E_1}{A^\star} = \frac{4}{5}\frac{s_1^2}{s_{12}^2}, \quad \frac{E_2}{A^\star} = \frac{4}{5}\frac{s_2^2}{s_{12}^2}. \tag{74.5}$$

Wenn die Wechselwirkungen wenig verschieden sind, kann man in dem Nenner in (69.13) deren Unterschied überhaupt vernachlässigen und erhält sogleich

$$\alpha = \frac{10(6C^\star - 5)}{55 + 16A^\star - 12B^\star}\left[\frac{s_1^2 - s_2^2}{s_1^2 + s_2^2} + \frac{s_1^2 + s_2^2 - 2s_{12}^2}{s_1^2 + s_2^2}(\gamma_1 - \gamma_2)\right]. \tag{74.6}$$

Man hat also bei Isobaren, entsprechend den beiden Summanden in der eckigen Klammer, zweierlei Querschnitt-Effekte zu unterscheiden. Bei gleicher Häufigkeit beider Molekülsorten, $\gamma_1 = \gamma_2$, kommt es nur auf die effektiven Stoßquerschnitte gleichartiger Moleküle an; die größeren Moleküle wandern, falls $6C^\star - 5 > 0$ ist, zur kalten Seite. Bei verschiedener Häufigkeit kommt es auch auf den effektiven Stoßquerschnitt verschiedenartiger Moleküle an.

Ein Beispiel, bei dem der erste Effekt überwiegt, ist das Gaspaar D_2/He: das größere Deuterium — man kann denselben Durchmesser wie für H_2 annehmen, vgl. Tabelle 7 (Ziff. 52) — geht tatsächlich zur kalten Seite mit $\alpha = 0{,}02$ um 340° K nach MURPHEY (1947). Schon früher hatten IBBS und UNDERWOOD (1927) den Isobarenfall, unter anderem N_2 und CO, experimentell auf indirekte Weise untersucht. Da nach SMITH (1922) oder nach TRAUTZ und MELSTER (1930) diese Gase jedenfalls von 0° bis 100° C dieselbe Viscosität und denselben Wärmeleitkoeffizienten haben, können die Gemische nicht mit der Wärmeleitzelle analysiert werden. IBBS und UNDERWOOD beobachteten daher die Thermodiffusion jedes dieser Gase gegen ein drittes und fanden mit N_2 als Partner jeweils dasselbe α wie mit CO als Partner. Man entnimmt aus ihrer Arbeit für etwa 50° C und $\gamma_1 = \gamma_2$

$$\alpha(N_2/H_2) = \alpha(CO/H_2) = 0{,}37; \quad \alpha(N_2/CO_2) = \alpha(CO/CO_2) = 0{,}060$$
$$\alpha(N_2/N_2O) = \alpha(CO/N_2O) = 0{,}057.$$

Daraus schlossen die Verfasser, daß das N_2- und das CO-Molekül in ihren Wechselwirkungseigenschaften übereinstimmen und daher $\alpha(N_2/CO) = 0$ sein muß. Dies wurde von WALL und HOLLEY (1940) durch Trennrohrversuche unter Benutzung einer chemischen Analysenmethode direkt bewiesen[1]. In dieser aufschlußreichen Arbeit werden noch andere Isobaren-Gemische untersucht. Stets ergaben sich recht kleine Thermodiffusionsfaktoren, z.B. $\alpha(CO/C_2H_4) = -0{,}014$ für 100° C und $\gamma_1 = \gamma_2$[2]. Das stimmt überein mit dem Ergebnis für das Gemisch aus N_2 und C_2H_4 (Äthylen), s. Fig. 40 (Ziff. 77), welches mittels des Diffusionsthermoeffekts gewonnen, siehe (3.12), wurde.

Wir fahren in der Diskussion von (74.6) fort. Der zweite Effekt, beruhend auf einer Verschiedenheit des gemittelten, effektiven Querschnitts gleichartiger und des effektiven Querschnitts verschiedenartiger Moleküle, äußert sich in einer Abhängigkeit des Thermodiffusionsfaktors vom Mischungsverhältnis. Dieser Effekt kann, wenn er im Vergleich zu dem ersterwähnten stark genug ist, einen

[1] Die von WINTER (1950) angegebene Diskrepanz besteht nicht.

[2] Das Vorzeichen von α ist bei WALL und HOLLEY (1940) und ebenso bei WALDMANN (1949) umgekehrt wie in diesem Artikel, (s. 3.12), gewählt.

Vorzeichenwechsel von α in Abhängigkeit von γ bewirken. Bedingung für den Vorzeichenwechsel bei Isobaren ist, wie man an den Endwerten sieht, daß entweder

$$s_{12}^2 > s_1^2, s_2^2 \quad \text{oder} \quad s_{12}^2 < s_1^2, s_2^2. \tag{74.7}$$

Ein Beispiel für den Vorzeichenwechsel der Isothermen $\alpha(\gamma)$ hat Grew (1944) in dem Gaspaar $^{20}Ne/NH_3$ gefunden. Jedoch kommt der Vorzeichenwechsel nicht gemäß (74.7) zustande — dieses Gemisch ist gar nicht isobar — sondern durch Zusammenwirken des Massen- und der Durchmessereffekte: das schwerere Neon ist kleiner als Ammoniak, so daß sich Masseneffekt und erster Durchmessereffekt kompensieren; nur der zweite Durchmessereffekt, der $\propto \gamma_1 - \gamma_2$ ist, bleibt übrig. Siehe dazu auch Ziff. 76. Ein Beispiel, das die Bedingung (74.7) erfüllt, scheint bisher nicht bekannt zu sein. Das Gemisch NH_3/CH_4 kommt vielleicht in Frage, s. Ziff. 78.

Fall III. Hier ist $M_1 \approx M_2 \approx \frac{1}{2}$, $E_1 = \frac{2}{5}\sqrt{\frac{2}{M_2}}A^\star$, E_2 analog und man entnimmt aus (69.14) ohne Mühe

$$S_1 - S_2 = \tfrac{3}{5}(2A^\star + 5)(M_1 - M_2).$$

Der Thermodiffusionsfaktor ist vom Mischungsverhältnis fast unabhängig und man bekommt für $\gamma_1 = \gamma_2 = \frac{1}{2}$ aus (69.13) sogleich

$$\alpha = \frac{15(6C^\star - 5)(2A^\star + 5)}{2A^\star(55 + 16A^\star - 12B^\star)} \cdot \frac{m_1 - m_2}{m_1 + m_2}. \tag{74.8}$$

Die erste ausgedehnte Diskussion dieser wichtigen Formel für den Isotopen-Thermodiffusionsfaktor hat Clark Jones (1940, 1941) vorgenommen.

Die Formel (74.8) überträgt sich nach Hellund (1940) und Clark Jones (1941) sogar auf polynäre Isotopengemische. Man hat hierzu nur α_{ij} statt des obigen $\alpha = \alpha_{12}$ zu schreiben und die Indices der Massen sinngemäß zu ändern. Die so erhaltenen polynären Thermodiffusionsfaktoren sind dann in (69.7) einzusetzen. Die Anwendung davon auf die Diffusionswärme von vielkomponentigen Isotopengemischen hat Waldmann (1946a) gemacht. Diese Formeln waren ferner von Nutzen bei der Deutung der Vorgänge, welche sich im Thermodiffusions-Trennrohr abspielen, wenn in dem vielkomponentigen Gemisch ein lokales Isotopen-Austauschgleichgewicht sich einstellt. Das hat Clusius (1950) bei Versuchen zur Reindarstellung von Stickstoff studiert, einen theoretischen Beitrag dazu lieferte Waldmann (1950c); vgl. auch Ziff. 19.

Noch einfacher als (74.8) ist die erste Näherung von Kihara [*32*] für den Isotopen-Thermodiffusionsfaktor. Sie lautet

$$\alpha = \frac{15}{16}\,\frac{6C^\star - 5}{A^\star} \cdot \frac{m_1 - m_2}{m_1 + m_2} \tag{74.9}$$

und ist leicht aus (69.8) oder (69.9) zu erraten. Dazu macht man (ohne Begründung!) die Annahme, daß die dortigen Entwicklungskoeffizienten $A_{i,1}$ bzw. $A_{j,1}$ mit denen für die reinen Komponenten übereinstimmen. Dann kann man sie nach (64.6) auf die Wärmeleitkoeffizienten der reinen Gase zurückführen gemäß

$$\lambda_i = \tfrac{5}{4}\,\mathrm{k}\,A_{i,1}$$

und kennt damit die Massenabhängigkeit (für $m_i \approx m_j$):

$$\frac{A_{i,1}}{m_i} - \frac{A_{j,1}}{m_j} = -\frac{12}{5}\,\frac{\lambda}{m\,\mathrm{k}}\,\frac{m_i - m_j}{m_i + m_j}.$$

Somit erhält man aus (69.8) oder (69.9) mit der Abkürzung $C^\star$ aus (69.10)

$$\alpha_{ij} = \frac{3}{10}(6C^\star - 5)\frac{\lambda}{n\,\mathrm{k}\,D} \cdot \frac{m_i - m_j}{m_i + m_j} = -\frac{3}{5}\left(\frac{\partial \log D/T^2}{\partial \log T}\right)_p \frac{\lambda}{n\,\mathrm{k}\,D} \cdot \frac{m_i - m_j}{m_i + m_j}. \tag{74.10}$$

Drückt man gemäß (50.12) und (68.4) λ und D durch ihre Ω-Integrale aus und benutzt die Abkürzung $A^\star$ aus (69.10), so entsteht — im binären oder polynären Fall — genau die Kiharasche Näherung (74.9)[1]. Die Schreibweise (74.10) hat vielleicht eine gewisse Bedeutung im Hinblick auf mehratomige Isotopen-Gemische.

75. Der Thermodiffusionsfaktor für spezielle Molekülmodelle in speziellen Fällen. Wir betrachten α) die starren Kugeln, β) die Potenzkraftzentren und γ) das Lennard-Jones-Modell.

α) *Starre Kugeln.* Hier sind nach Definition alle $\Omega^\star = 1$ und daher nach (69.10) und (69.11)

$$A^\star = B^\star = C^\star = 1\,; \qquad E_1 = \frac{2}{5}\sqrt{\frac{2}{M_2}}\left(\frac{d_1}{d_{12}}\right)^2, \qquad E_2 \text{ analog.} \tag{75.1}$$

Für das spezielle Gemisch (I) aus Ziff. 74 hat man nach (74.2) als Endwerte

$$\alpha(\gamma_1 = 0) = 0{,}885\left(\frac{d_{12}}{d_{22}}\right)^2; \qquad \alpha(\gamma_2 = 0) = 0{,}384. \tag{75.2}$$

Wenn also die Moleküldurchmesser ungefähr gleich sind, so ist der Thermodiffusionsfaktor für wenig schwere Moleküle in viel leichten etwa doppelt so groß wie im umgekehrten (Lorentzschen) Fall. Für N_2/H_2 sind die beiden Endwerte experimentell $\alpha = 0{,}5$ bzw. 0,20 bei 293° K (s. Ziff. 76). Also ist hier tatsächlich das genannte Verhältnis ungefähr verwirklicht. Zugleich sieht man aber an dem Beispiel, daß das Modell der starren Kugeln zu große Absolutwerte für den Thermodiffusionsfaktor liefert. Das gilt ganz allgemein.

Für das spezielle Gemisch (II) aus Ziff. 74 — Isobare mit relativ wenig verschiedenen Moleküldurchmessern — hat man nach (74.6), mit $d_{12} = \frac{1}{2}(d_1 + d_2)$ wie es starren Kugeln entspricht,

$$\alpha = \frac{20}{59}\,\frac{d_1 - d_2}{d_1 + d_2}\left[1 + \frac{1}{2}\,\frac{d_1 - d_2}{d_1 + d_2}\,(\gamma_1 - \gamma_2)\right]. \tag{75.3}$$

Das zweite Glied in der Klammer, das auf den zweiten „Querschnitt-Effekt" von Ziff. 74 zurückgeht, stellt bei starren ungefähr gleich großen Kugeln lediglich eine kleine Korrektur dar. Die schweren Moleküle gehen hier stets zur kalten Seite. Da die relativen Durchmesserdifferenzen von Isobaren gewöhnlich von der gleichen Größenordnung sind wie die relativen Massendifferenzen von (nicht zu leichten) Isotopen, ist vorauszusehen, daß auch die α für Isobaren höchstens von derselben Größenordnung wie die für Isotopen sein werden.

Für das spezielle Gemisch (III) aus Ziff. 74 — Isotope von relativ wenig verschiedenen Massen — gilt nach (74.8)

$$\alpha = \frac{105}{118}\cdot\frac{m_1 - m_2}{m_1 + m_2}. \tag{75.4}$$

Diese Formel wurde von Furry, Jones und Onsager (1939) in ihrer bekannten Arbeit über das Trennrohr angegeben.

Eine Bemerkung über beliebige Gemische sei noch angefügt. Bei starren Kugeln hängt der Thermodiffusionsfaktor vom Molenbruch, sowie vom Verhältnis der Durchmesser und der Massen ab, nicht aber von der Temperatur. Im Fall beliebiger Gemische und beliebiger Wechselwirkung, kommt auch noch die Abhängigkeit von der Temperatur hinzu, aber wie sich herausstellt, s. Fig. 36 (Ziff. 76), derart, daß in

$$\alpha = \alpha_{\text{starr}}(\gamma)\,R_T \tag{75.5}$$

[1] Die obige Überlegung erspart einem die Berechnung der Matrixelemente $a'_{ij,11}$ und $a''_{ij,11}$ aus Ziff. 67. Leider fehlt aber die Begründung für die Annahme über die $A_{i,1}$.

der Faktor R_T sich fast nicht mit dem Molenbruch, sondern lediglich mit der Temperatur ändert. Deshalb ist diese Produktzerlegung, welche CLARK JONES (1940) eingeführt hat, sehr zweckmäßig für die Diskussion der theoretischen und experimentellen Ergebnisse über die Thermodiffusion. Zur Berechnung von α_{starr} hat man für die Wechselwirkung typische Längen zu nehmen, z.B. die Parameter d_1, d_2 und d_{12} des Lennard-Jones-Potentials. Der Faktor R_T, welcher stets < 1 ist, bringt die Abweichung der Moleküle von den starren Kugeln zum Ausdruck („Härtefunktion"). Er wird im folgenden öfters benutzt.

β) Potenzkraft-Zentren. Die Ω-Integrale für dieses Modell sind durch (51.7a, b) gegeben. Nach (69.10) können sogleich gebildet werden:

$$\left.\begin{aligned} A^\star &= \frac{1}{3}\left(3 - \frac{2}{\nu_{12}-1}\right)\frac{A_2(\nu_{12})}{A_1(\nu_{12})}, \qquad B^\star = \frac{1}{3}\,\frac{(3\nu_{12}-5)(\nu_{12}+1)}{(\nu_{12}-1)^2}, \\ 6C^\star - 5 &= \frac{\nu_{12}-5}{\nu_{12}-1}. \end{aligned}\right\} \tag{75.6}$$

Tabelle 15. *Korrekturfaktor f_α, der nach (74.3) beim Lorentz-Gemisch von der ersten Näherung für α zum exakten Wert führt. Potenzkraft-Modell.*

ν_{12}	5	7	9	11	15	∞
f_α	1	1,075	1,13	1,16	1,19	1,30

Diese Größen sind von der Temperatur exakt unabhängig. Dagegen sind E_1 und E_2, in welche nach (69.11) auch ν_{11} und ν_{22} eingehen, nur im Isotopenfall temperaturunabhängig.

An Hand von (74.3) diskutieren wir zunächst die Güte der ersten Näherung für den Thermodiffusionsfaktor. In Tabelle 15 sind einige Korrekturfaktoren aufgeführt, die sich sofort an Hand des $B^\star$-Werts aus (75.6) ergeben. Am größten — 30% — ist die Abweichung der ersten Näherung vom exakten Wert bei starren Kugeln. Auf die Ungenauigkeit der ersten Näherung ist also beim Vergleich mit dem Experiment sehr wohl zu achten. Sie ist erheblich größer als die Ungenauigkeit der ersten Näherung für die Viscosität oder den Diffusionskoeffizienten.

Tabelle 16. *Werte der Funktion $g(\nu)$ und R_T aus (75.7) und (75.5) für das Potenzkraft-Modell.*

ν	5	7	9	11	15	∞
$g(\nu)$	0,816	0,843	0,865	0,882	0,906	1,000
R_T	0	0,281	0,432	0,529	0,647	1,000

Im übrigen beschränken wir die Diskussion auf den Isotopenfall. Nach CLARK JONES (1940) schreiben wir (74.8) in der Form (75.5), wo nun

$$\alpha_{\text{starr}} = \frac{105}{118}\cdot\frac{m_1-m_2}{m_1+m_2}\,; \qquad R_T = \frac{\nu-5}{\nu-1}\,g(\nu). \tag{75.7}$$

Zahlenwerte für die Isotopenfunktion $g(\nu)$ kann man an Hand von (74.8), (75.6) und den in Tabelle 5 (Ziff. 51) sowie Tabelle 10 (Ziff. 71) aufgeführten Werten von A_2 und A_1 berechnen. Einige sind in Tabelle 16 angegeben. Befriedigend an diesem Ergebnis ist, daß der α-Wert auch bei recht harten Molekülen, z.B. $\nu = 15$, merklich kleiner ausfällt als bei starren Kugeln. Trotzdem ist das Potenzkraft-Modell auch qualitativ ganz unzureichend: R_T ist in Wirklichkeit stark von der Temperatur abhängig.

γ) Das Lennard-Jonessche (4; 8)- und (6; 12)-Potential. Wegen des eben erwähnten Mangels des Potenzkraft-Modells hat CLARK JONES (1941) die Thermodiffusion für die Lennard-Jones (4; 8)-Wechselwirkungsenergie

$$\Phi = 4\varepsilon\left[\left(\frac{d}{r}\right)^8 - \left(\frac{d}{r}\right)^4\right] \tag{75.8}$$

ausgearbeitet. Selbstverständlich sind Φ, ε und d sinngemäß mit Indices zu versehen, je nachdem um welche Partner es sich handelt. CLARK JONES wählte diesen Ansatz, weil er dabei die Vorarbeit benutzen konnte, welche HASSÉ und

Cook (1929) geleistet hatten. Das Ergebnis für die Isotopen-Thermodiffusion ist, in Gestalt des Faktors R_T aus (75.5), in Fig. 34 wiedergegeben.

Die $\Omega^\star$-Integrale für das (6; 12)-Modell [s. (51.11)] sind bereits in Fig. 31 (Ziff. 51) gezeichnet. Die für die Thermodiffusion wichtigen Verhältnisse $A^\star$, $B^\star$, $C^\star$ sind — wieder nach den Tabellen von Hirschfelder *et al.* [25] — in Fig. 33 eigens dargestellt.

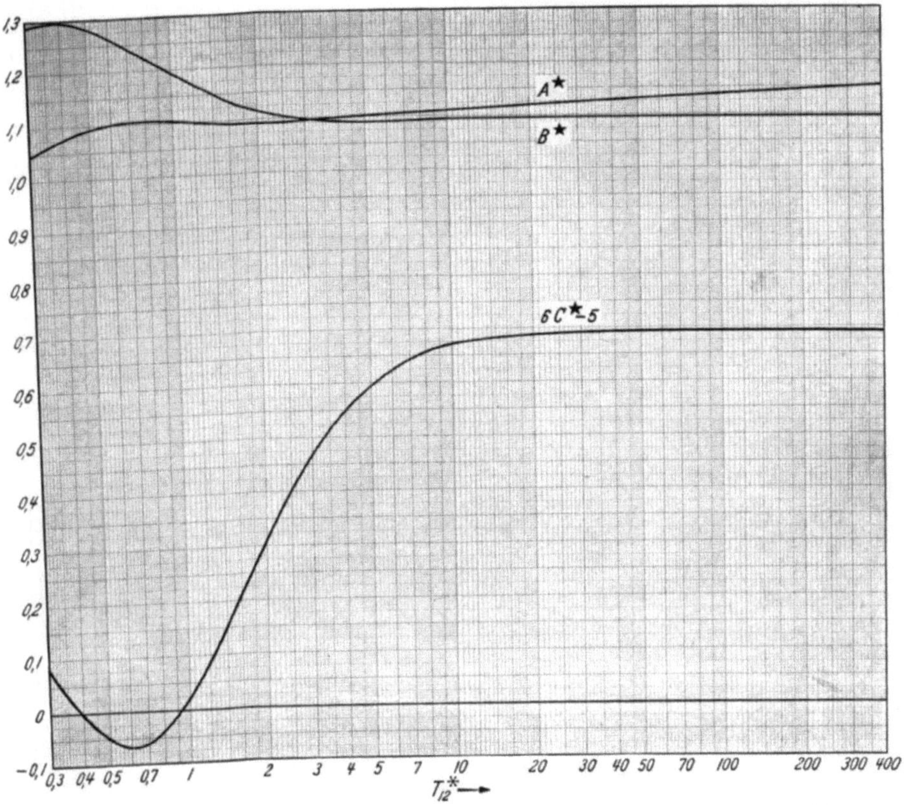

Fig. 33. Die Funktion $A^\star$, $B^\star$, $6C^\star - 5$ aus (69.10) für das Lennard-Jones-(6; 12)-Modell in Abhängigkeit von der reduzierten Temperatur $T_{12}^* = \mathrm{k}\,T/\varepsilon_{12}$. Autoren s. Text.

Der daraus nach (74.8) und (75.5) sich ergebende Faktor R_T ist aus Fig. 34 zu ersehen. Das auffälligste Merkmal, in dem der große Fortschritt gegenüber dem Potenzkraft-Modell zum Ausdruck kommt, ist die starke Temperaturabhängigkeit von α und der bei $T^\star = 1{,}88$ bzw. $0{,}93$ stattfindende Vorzeichenwechsel; er wird durch den Faktor $6C^\star - 5$, s. (74.8), bewirkt. Die Temperaturabhängigkeit von $\alpha/(6C^\star - 5)$ beträgt bei Isotopen nur wenige Prozent.

Die Korrektur f_α, welche die Güte der ersten Näherung im Lorentzschen Grenzfall bestimmt, ist in Tabelle 17 angegeben. Der Vergleich mit Tabelle 15 zeigt, daß sich gegenüber dem Potenzkraft-Modell, wenn man ν geeignet als temperaturabhängig auffaßt, keine größenordnungsmäßige Änderung ergibt. Doch ist zu bemerken, daß beim Lorentz-Gemisch der Fehler der ersten Näherung von α besonders groß ist. Bei der Isotopen-Thermodiffusion ist er glücklicherweise erheblich kleiner, nur etwa 2% für $T^\star = 1$ bis 4 und etwa 5% bei hohen $T^\star$-Werten, s. Fig. 35, nach dem nun zu besprechenden (6; Exp)-Potential. Das gleiche ergaben die Rechnungen von Saxena (1956) am (6; 12)-Potential.

δ) Das (6; Exp)-Potential. Der Lennard-Jonessche Potenzansatz für das Abstoßungspotential hat keine reale physikalische Bedeutung. Realistischer ist der aus quantenmechanischen Näherungen hervorgehende Exponentialansatz

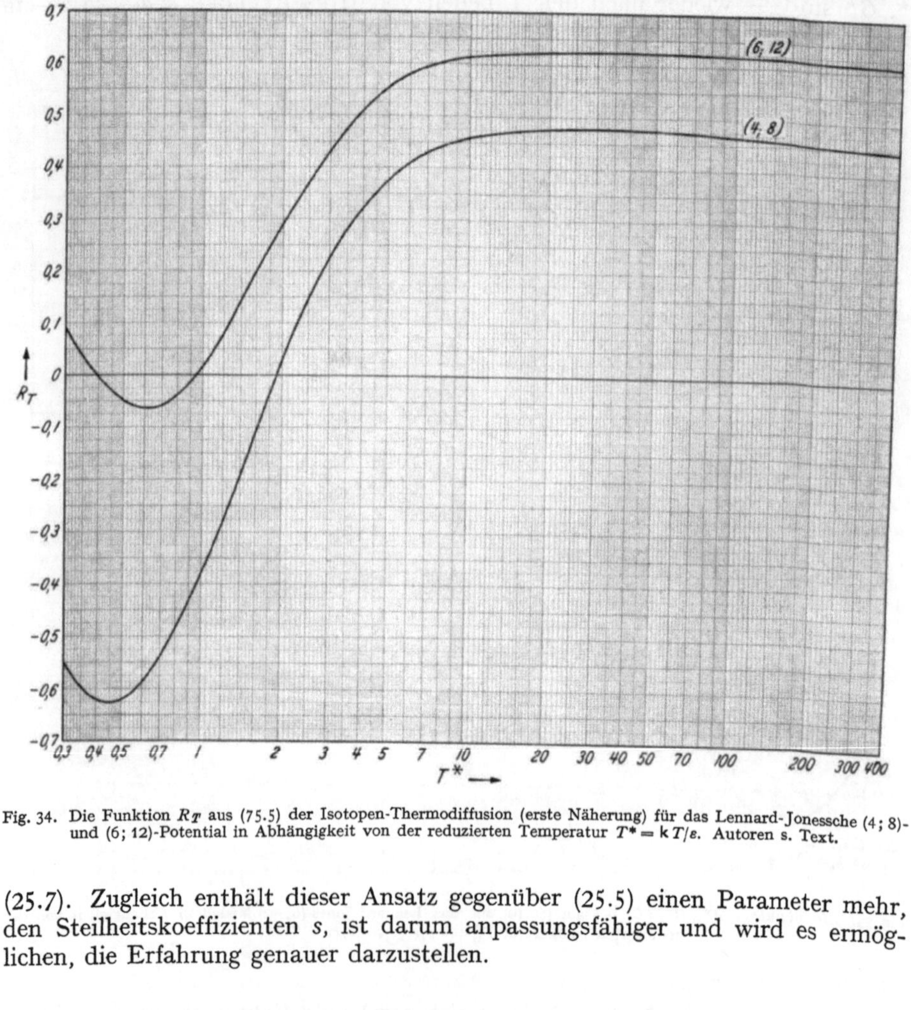

Fig. 34. Die Funktion R_T aus (75.5) der Isotopen-Thermodiffusion (erste Näherung) für das Lennard-Jonessche (4; 8)- und (6; 12)-Potential in Abhängigkeit von der reduzierten Temperatur $T^* = kT/\varepsilon$. Autoren s. Text.

(25.7). Zugleich enthält dieser Ansatz gegenüber (25.5) einen Parameter mehr, den Steilheitskoeffizienten *s*, ist darum anpassungsfähiger und wird es ermöglichen, die Erfahrung genauer darzustellen.

Tabelle 17. *Korrekturfaktor f_α, der nach* (74.3) *beim Lorentz-Gemisch von der ersten Näherung für α zum exakten Wert führt. Lennard-Jones-*(6; 12)*-Modell.*

T^*_{12}	0,30	1	3	30	300
f_α	0,953	1,070	1,179	1,186	1,186

Die von MASON (1954) sowie von MASON und RICE (1954b) an Hand des Ansatzes (25.7) mittels eines Rechenautomaten erhaltenen R_T-Werte für die Isotopen-Thermodiffusion sind in Fig. 35 dargestellt. MASON hat auch die zweite Näherung für R_T berechnet, so daß man zugleich einen Begriff von der Güte der ersten hat.

Das auffälligste Ergebnis dieser Rechnungen ist, daß ein Vorzeichenwechsel nur bei „flacher“ Abstoßung, $s=12$, stattfindet, s. dazu auch (25.8), während er bei „steiler“ Abstoßung, $s \gtrapprox 15$, unterbleibt.

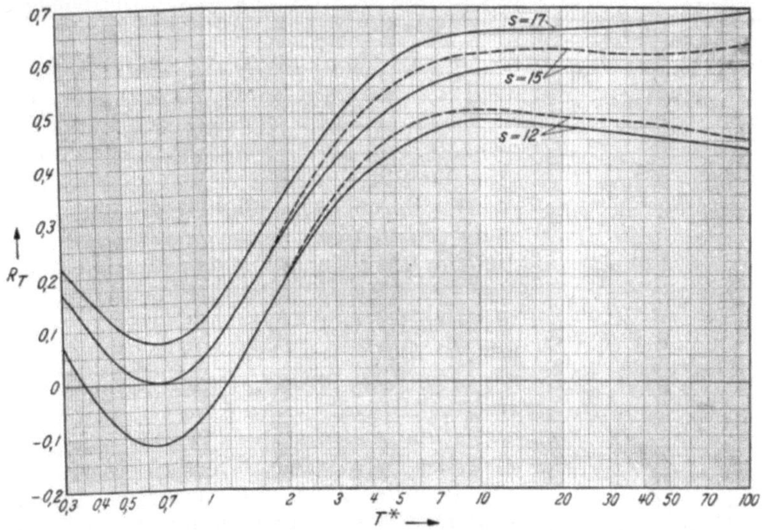

Fig. 35. R_T aus (75.5) für Isotopen-Thermodiffusion und (6; Exp)-Potential (25.7) in Abhängigkeit von der reduzierten Temperatur $T^* = kT/\varepsilon$. s Steilheitsparameter. Ausgezogene Kurven: 1. Näherung nach CHAPMAN und COWLING (74.8). Gestrichelte Kurven: 2. Näherung. Nach MASON (1954) sowie MASON und RICE (1954b).

76. Die Abhängigkeit des Thermodiffusionsfaktors vom Molenbruch. Vergleich mit dem Experiment. *α) Massen stark verschieden.* Die Abhängigkeit $\alpha(\gamma)$ ist dann besonders ausgeprägt, wenn die beiden Molekülsorten des Gemisches stark verschiedene Massen haben. Als Beispiel sei das Gaspaar N_2/H_2 betrachtet. Fig. 36 zeigt die Meßwerte. Zum Vergleich mit der Theorie ist auch die erste Näherung für α eingezeichnet.

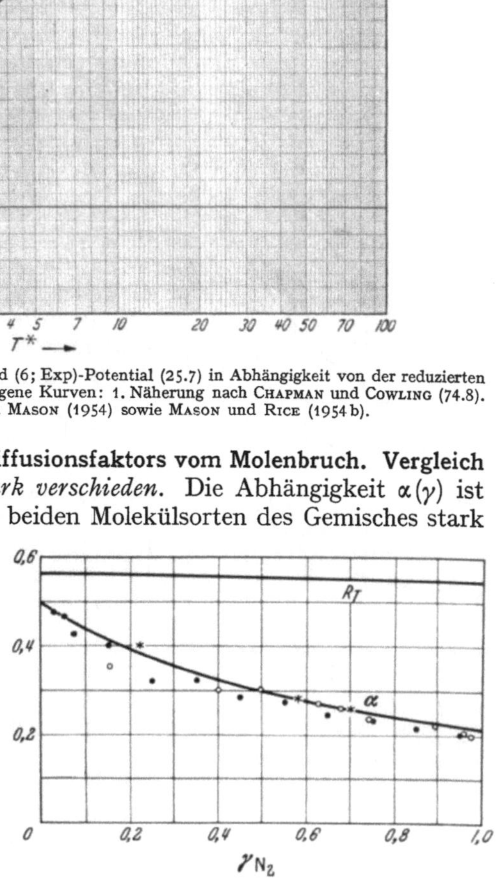

Fig. 36. Meßwerte für den Thermodiffusionsfaktor α von N_2/H_2 in Abhängigkeit vom Molenbruch γ_{N_2} bei 293° K. ○ IBBS (1925); ∗ IBBS, GREW und HIRST (1929); ● WALDMANN (1947b; Diffusionsthermoeffekt). Ausgezogene Kurven: theoretisch mit Lennard-Jonesschem (6; 12)-Potential.

Sie wurde aus (69.13) mittels des (6; 12)-Potentials nach folgendem Verfahren berechnet. Zunächst sind aus den experimentellen Werten $\eta_1 = 177$ μ-Poise, $\eta_2 = 90{,}5$ μ-Poise, $D(1,2) = 0{,}76$ cm²/sec bei 293° K, 760 Torr gemäß (69.11) zu entnehmen: $E_1 = 3{,}34$, $E_2 = 0{,}470$. Fast dieselben E-Werte findet man theoretisch mittels der d und ε für N_2, H_2 aus Tabelle 7, (Ziff. 52) sowie von d_{12} und ε_{12} aus (71.12). Insofern ergibt sich also noch kein Widerspruch zu der Regel (71.11). Man kann nun aber auch α selbst berechnen und findet etwa 20% zu große Werte für jedes Mischungsverhältnis. Das liegt in überwiegendem Maß an dem mit T^* stark veränderlichen Faktor $6C^* - 5$. Der in (71.12) angenommene Wert von ε_{12} ist entschieden zu groß. Eine andere Wahl hierfür beeinflußt aber auch den Durchmesser d_{12}. Wir wählen nach Probieren

$$d_{12} = 3{,}16 \text{ Å}; \quad \frac{\varepsilon_{12}}{k} = 83{,}7° \text{ K}. \tag{76.1}$$

Damit erhält man theoretisch $E_1 = 3{,}30$, $E_2 = 0{,}477$ in guter Übereinstimmung mit den obigen experimentellen Werten. Weiter ergeben sich nun theoretisch mit $T_1^* = 3{,}20$, $T_2^* = 8{,}80$,

$T_{12}^* = 3{,}50$ die Werte

$$S_1 = 3{,}18, \quad S_2 = -2{,}50, \quad Q_1 = 7{,}72, \quad Q_2 = 2{,}55, \quad Q_{12} = 9{,}41$$

und die nach (69.13) dazu gehörigen α zeigen nach Fig. 36 gute Übereinstimmung mit den Messungen.

Ebenfalls in Fig. 36 eingezeichnet ist die Funktion R_T aus (75.5); α_{starr} wurde mit denselben d-Werten berechnet wie α. Man sieht, daß R_T im ganzen Intervall nur um etwa 2% abnimmt; bei weniger verschiedenen Massen ist die Veränderlichkeit noch geringer. Deshalb ist es zweckmäßig, Thermodiffusionsfaktoren mit α_{starr} zu reduzieren.

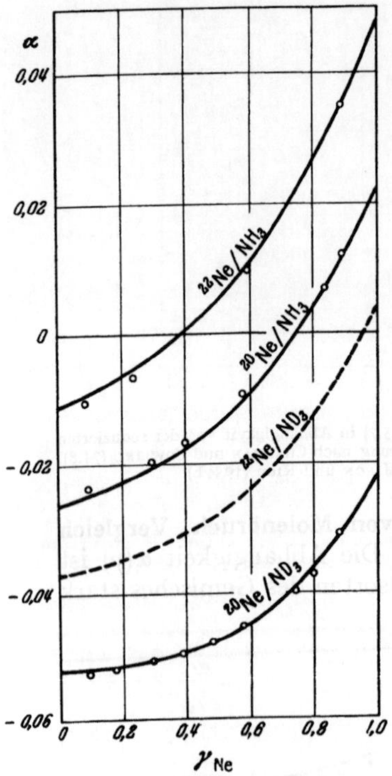

Fig. 37. Meßwerte für den Thermodiffusionsfaktor Neon/Ammoniak in Abhängigkeit vom Molenbruch γ_{Ne} bei 383° K. Ausgezogene und gestrichelte Kurven: theoretisch mit Lennard-Jonesschem (6; 12)-Potential. Nach Clusius und Huber (1955b).

Das Beispiel lehrt, daß die Thermodiffusion eine recht genaue Ermittlung von ε_{12} (und d_{12}) gestattet. Bei der Diskussion des Diffusionskoeffizienten in Ziff. 71 fiel dagegen die erhebliche Abweichung von der Regel (71.11) nicht auf. Das ist der großen Empfindlichkeit des Faktors $6C^\star - 5$ zu verdanken. Ein Wert für ε_{12}, der um 10% kleiner als der in (76.1) ist, gäbe einen 5% größeren Wert für α, was schon außerhalb der Meßfehler liegt. Da wir aber die erste Näherung von α benutzten, und es sich überdies nicht um einatomige Gase handelt, wird man dies doch als die ungefähre Fehlergrenze der Bestimmung von ε_{12} anzusehen haben. Jedenfalls zeigt die Analyse der Thermodiffusion, in der Gegenüberstellung von (71.12) und (76.1), daß die Regel (71.11) in unserem Beispiel zu hohe Werte für d_{12} und zu niedrige Werte für ε_{12} liefert. Zu einem ähnlichen Ergebnis gelangten Clusius und Huber (1955a) für das Gaspaar H_2/CO_2; ihre Diskussion ist noch genauer, da sie die Kiharaschen Formeln für α verwenden.

β) Massen wenig verschieden. Wie schon in Ziff. 74 erwähnt, kann in diesem Fall α in Abhängigkeit von γ unter Umständen das Vorzeichen wechseln. Das hat Grew (1944) bei Trennrohrversuchen am Gemisch Ne/NH_3 tatsächlich beobachtet. Clusius und Huber (1955b) haben diese Versuche bestätigt und durch Benutzung deuterierten Ammoniaks erweitert, zum Teil unter Benutzung der Trennschaukel (s. Ziff. 19). Fig. 37 zeigt ihre Ergebnisse. Man sieht daran besonders schön die gleichzeitige Wirksamkeit des Massen- und der Durchmesser-Effekte. Das Neonatom ist kleiner als das Ammoniakmolekül, aber in keinem Fall leichter. ^{22}Ne geht gegenüber dem leichtesten Ammoniak NH_3 daher überwiegend, nämlich für $\gamma_{\text{Ne}} > 0{,}4$, nach der kalten Seite (α positiv). Hier dominiert der Masseneffekt. ^{20}Ne dagegen geht gegenüber dem schwersten, nämlich isobaren Ammoniak ND_3 für jedes Mischungsverhältnis nach der warmen Seite (α negativ). Hier sind allein die Durchmesser-Effekte von (74.6) vorhanden und zwar muß s_{12} zwischen s_1 und s_2 liegen, da α das Vorzeichen nicht wechselt. Die starke Krümmung der Kurven in Fig. 37 zeigt, daß die Näherung (74.6) nicht mehr gut ist wegen zu großen Unterschieds der Durchmesser. Die von Clusius und Huber für die theoretische Auswertung benutzten Durchmesser sind für

Neon (1) und Ammoniak (2)

$$d_1 = 2{,}79\ \text{Å}, \quad d_{12} = 2{,}87\ \text{Å}, \quad d_2 = 3{,}70\ \text{Å}.$$

Wegen der weiteren Einzelheiten sei auf die Originalarbeit verwiesen.

77. Die Abhängigkeit des Thermodiffusionsfaktors von der Temperatur. Vergleich mit Messungen. Es sollen betrachtet werden α) die Gemische der Edelgase, β) Gemische, die wenigstens eine mehratomige Komponente enthalten, γ) Isotopengemische.

α) Gemische der Edelgase. Die Temperaturabhängigkeit des Thermodiffusionsfaktors haben erstmals IBBS und GREW (1931) experimentell untersucht, und zwar an den drei binären Gemischen von He, Ne und A im Bereich von −180 bis 0° C. Es ergab sich, daß α oder R_T [s. (75.5)] in allen Fällen mit wachsender Temperatur zunimmt. Die Versuche wurden von ATKINS, BASTICK und IBBS (1939) auf alle binären Edelgasgemische von 15 bis 100° C ausgedehnt.

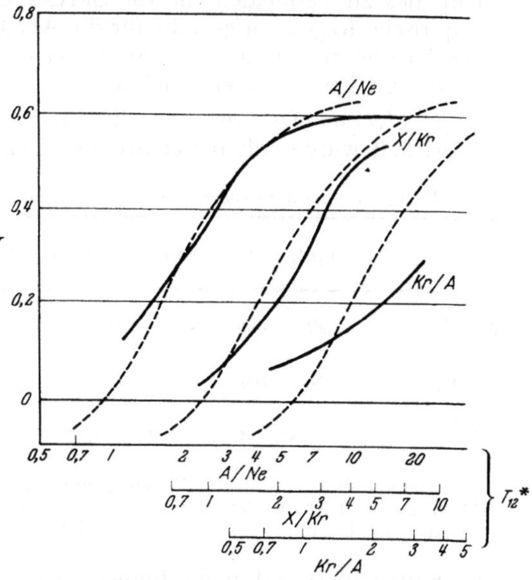

Fig. 38. R_T für Edelgasgemische in Abhängigkeit von $T_{12}^* = \mathrm{k}\,T/\varepsilon_{12}$. Ausgezogene Kurven: Messungen. Gestrichelte Kurven: theoretisch nach Lennard-Jones-(6; 12)-Potential. Nach Tabelle 2 und Fig. 6 der Arbeit von GREW, JOHNSTON und NEAL (1954).

Sodann hat GREW (1947) den Temperaturbereich auf −180 bis 400° C erweitert und alle binären Gemische der Edelgase von He bis X mit Ausnahme von X/Kr gemessen. Einen Vergleich mit der Theorie des Lennard-Jones-(6; 12)-Modells, die HIRSCHFELDER, BIRD und SPOTZ (1948) entwickelten, haben erstmals GREW (1949) und WINTER (1950) angestellt. In der Arbeit von GREW, JOHNSON und NEAL (1954) geht die Temperatur von 18,5 bis 1250° K und ist unter anderem auch X/Kr untersucht. Die Meßergebnisse für einige Edelgasgemische sind in Fig. 38 dargestellt. Zur Reduktion der Meßwerte α auf R_T wird α_{starr} benötigt; dieses berechnen die Autoren mit den von HIRSCHFELDER *et al.* (1949) [s. auch Tabelle 7 (Ziff. 52)] angegebenen, aus der Viscosität entnommenen Durchmessern der reinen Gase unter Benutzung der Regel (71.11). Auch der für den Abszissenmaßstab benötigte Wert von ε_{12} ist so gewonnen. Folgende Werte liegen der Fig. 38 zugrunde:

	A/Ne	Kr/A	X/Kr
α_{starr} ($\gamma = 0{,}5$)	0,332	0,313	0,218
$\varepsilon_{12}/\mathrm{k}$ (°K)	66,5	153	209

Man sieht, daß bei den Edelgasen R_T mit wachsender Temperatur einem konstanten Wert von etwa 0,6 zustrebt, daß aber im ganzen die Temperaturabhängigkeit durch das Lennard-Jones-(6; 12)-Potential nicht exakt wiedergegeben wird. Insbesondere ist bei keinem einzigen binären Gemisch einatomiger Gase eine Vorzeichenumkehr von α beobachtet worden. Durch das von MASON ausgewertete

(6; Exp)-Potential, s. Fig. 35 (Ziff. 75), ließe sich durch Wahl eines genügend hohen Werts für den Parameter $s(\gtrsim 15)$ wohl eine bessere Darstellung erzielen und ein ins einzelne gehender Vergleich auf dieser Grundlage wäre erwünscht.

Hier kann einstweilen nur konstatiert werden, daß die Thermodiffusion, besonders was ihre Temperaturabhängigkeit anbelangt, die schärfste Probe für einen Wechselwirkungsansatz darstellt und daß sie speziell die Ungenauigkeit auch des Lennard-Jonesschen (6; 12)-Potentials erkennen läßt. Allerdings darf nicht übersehen werden, daß dabei auch die Ungenauigkeit der Regel (71.11) eingeht. Um dies zu vermeiden, haben SRIVASTAVA und SAXENA (1956) allgemeinere, empirische Regeln aufgestellt für die Abhängigkeit der d_{12} und ε_{12} von der Masse der Partner und damit die Funktion $\alpha(m)$ bei fester Temperatur und $\gamma=0,5$ für die Edelgas-Gemische diskutiert.

β) Mehratomige Gase. Die kinetische Theorie, soweit sie in diesem Kapitel entwickelt wurde, gilt für einatomige Gemische. Es ist eigentlich nicht erlaubt,

Tabelle 18. *Vorzeichenwechsel von α. Temperaturen $T_{12}^{(0)}$, für welche $\alpha=0$. Autoren s. Text.*

	D_2/H_2	O_2/N_2	A/N_2	A/O_2	A/C_2H_4	O_2/C_2H_4	CO_2/N_2	$^{15}NH_3/^{14}NH_3$	$^{13}C^{16}O_2/^{12}C^{16}O_2$	$^{12}C^{16}O^{17}O$ oder $^{12}C^{16}O^{18}O/^{12}C^{16}O_2$
$T_{12}^{(0)}$ (°K)	31	90	115	145	210	370	205	290	300	378
$\frac{\varepsilon_{12}}{k}$ (°K)[1]	33,3	102	107	118	159	152	132	203	190	190
$\frac{k\,T_{12}^{(0)}}{\varepsilon_{12}}$	0,93	0,88	1,07	1,23	1,32	2,44	1,55	1,43	1,58	1,99

[1] Die ε_{12} sind aus Ziff. 52, Tabelle 7 mittels der Regel $\varepsilon_{12}=\sqrt{\varepsilon_1\varepsilon_2}$ berechnet. Der ε-Wert für Wasserstoff ist der Arbeit von GREW, JOHNSON und NEAL (1954) entnommen; derjenige für Ammoniak stammt von CLUSIUS und HUBER (1955b), s. S. 443.

sie unverändert auf mehratomige Gase anzuwenden, indem man diesen eine kugelsymmetrische molekulare Wechselwirkung zuordnet und nur elastische Stöße annimmt. Gleichwohl sei aber diese Inkonsequenz, die schon einige Male begangen wurde, hier wieder hingenommen, weil sich ein großer Teil des Erfahrungsmaterials auf mehratomige Gase bezieht und ein Vorzeichenwechsel von α mit abnehmender Temperatur nur bei Gemischen beobachtet worden ist, die wenigstens eine mehratomige, z.B. eine zweiatomige Komponente enthalten.

Zunächst sei dieser Vorzeichenwechsel besprochen. Ihn fanden als erste WATSON und WOERNLEY (1943) bei Trennrohrversuchen an $^{15}NH_3/^{14}NH_3$, siehe Tabelle 18. Dann wurde er von WALDMANN (1947d, 1949) mittels des Diffusionsthermoeffekts in mehreren Fällen beobachtet. Die Feststellung des Vorzeichenwechsels nach der Diffusionswärmemethode ist sehr einfach und sicher. Ebenfalls nach ihr haben WALDMANN und E. W. BECKER (1948) qualitativ die Vorzeichenumkehr bei He/H_2 zwischen 20 und 90° K (allerdings bei etwa 1 atm) gefunden, was von DE TROYER, VAN ITTERBEEK und RIETVELD (1951) durch Thermodiffusionsversuche bestätigt zu sein scheint. Auch D_2/H_2 kehrt um nach Aussage der letztgenannten Autoren, deren Resultate an D_2/H_2 von GREW *et al.* (1954) qualitativ bestätigt und quantitativ verbessert wurden.

Interessant ist die Isotopenthermodiffusion des Kohlendioxyds, welche E. W. BECKER und DÖRNENBURG (1950), sowie E. W. BECKER und BEYRICH (1952) untersucht haben. Die Umkehrtemperatur von α hängt stark davon ab, ob ein symmetrisches Molekül ($^{13}C^{16}O_2$) oder ein unsymmetrisches Molekül ($^{12}C^{16}O^{17}O$ oder $^{12}C^{16}O^{18}O$) gegen das symmetrische $^{12}C^{16}O_2$ diffundiert. Daraus, daß für die beiden unsymmetrischen Moleküle der Vorzeichenwechsel bei genau derselben Temperatur liegt, könnte man versucht sein zu folgern, daß der Unterschied zwischen den Umkehrtemperaturen der symmetrischen und der unsymmetrischen

Moleküle durch das Ausfallen der ungeraden Rotationsniveaus bei den ersteren, also durch einen Quanteneffekt bedingt sei. Dieser Schluß trifft aber nicht zu. Sowohl die Differenz der Massen als auch die der Trägheitsmomente ist für das Paar $^{12}C^{16}O^{18}O/^{12}C^{16}O_2$ doppelt so groß wie für das Paar $^{12}C^{16}O^{17}O/^{12}C^{16}O_2$. Wenn also der Thermodiffusionsfaktor bei mehratomigen Isotopen-Molekülen nach der klassischen Theorie eine lineare Funktion dieser Differenzen ist, so unterscheiden sich die α's in beiden Fällen lediglich um den Faktor 2 und die Umkehrtemperatur muß für die beiden unsymmetrischen Kohlendioxyde in der Tat genau dieselbe sein. Bei dem Paar $^{13}C^{16}O_2/^{12}C^{16}O_2$ dagegen verschwindet die Differenz der Trägheitsmomente, der entsprechende Beitrag zu α entfällt und das Vorzeichen wird bei einer anderen Temperatur umkehren. Daß die Verschiebung der Umkehrtemperatur so groß ist, s. Tabelle 18, überrascht allerdings zunächst. Es ist aber von der Theorie der einatomigen Gase her bekannt, daß kleine Änderungen des molekularen Kraftgesetzes sich in der Umkehrtemperatur sehr deutlich bemerkbar machen. Wenn auch detaillierte Rechnungen

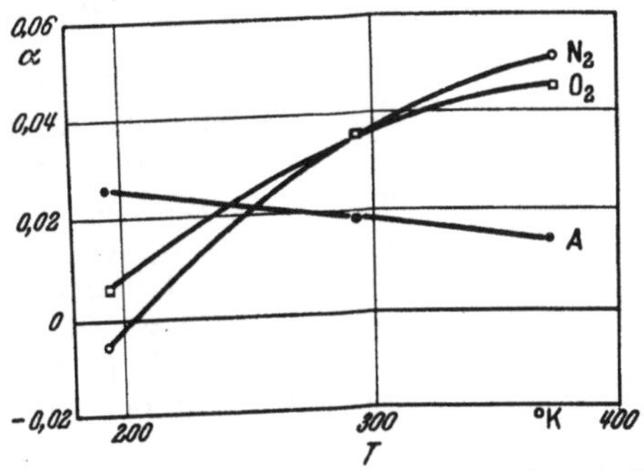

Fig. 39. Thermodiffusionsfaktor von CO_2/N_2, O_2, A nach WALDMANN (1949). Mischungsverhältnis etwa 1:1.

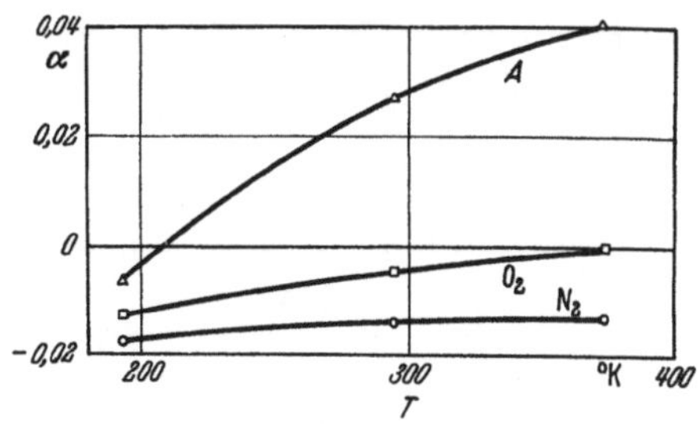

Fig. 40. Thermodiffusionsfaktor von N_2, O_2, A/C_2H_4 nach WALDMANN (1949). Mischungsverhältnis etwa 1:1.

nicht möglich sind, so scheint es demnach doch, daß man zur Erklärung der Beobachtungen von BECKER und Mitarbeiter keine Quanteneffekte heranzuziehen hat, was bei der hohen Rotationsanregung des Kohlendioxyds unmotiviert wäre.

Ähnliche Experimente an Kohlenmonoxyd haben DE VRIES, HARING und SLOTS (1956) angestellt.

Theoretisch hat WINTER (1950) einige Beispiele der Vorzeichenumkehr diskutiert. Diese Diskussion ist in Tabelle 18 vervollständigt. Es ergibt sich keine Konstanz des Verhältnisses $kT_{12}^{(0)}/\varepsilon_{12}$. Das ist nicht sehr verwunderlich, da dieses Verhältnis stark vom Kraftgesetz abhängt [$=1,88$ bzw. 0,93 für das Lennard-Jones-(4; 8)- bzw. -(6; 12)-Potential]. Überdies geht die Ungenauigkeit der Regel $\varepsilon_{12}=\sqrt{\varepsilon_1\,\varepsilon_2}$ ein.

Nun besprechen wir die Temperaturabhängigkeit von α noch genauer. In den Fig. 39 bis 41 sind einige der experimentellen Ergebnisse im einzelnen dargestellt. Wie immer in diesem Artikel bedeutet $\alpha>0$, daß das erstgenannte Gas sich im Gebiet der niedrigen Temperatur anreichert bzw. sich bei der Diffusion abkühlt.

An Fig. 39 ist auffällig, daß α für CO_2/A mit wachsender Temperatur merklich abnimmt, was keines der für einatomige Gase gültigen theoretischen Modelle erklären würde. Wegen der Besonderheit dieser Erscheinung wäre die Nachprüfung der Messungen durch Thermodiffusionsversuche erwünscht. (Für die Interpretation der stationären Diffusionswärmeversuche ist die Kenntnis der Wärmeleitfähigkeit und des Diffusionskoeffizienten erforderlich, was zu Unsicherheiten führt, wenn diese Daten nicht vollständig bekannt sind.) Eine ähnliche Abnahme von α hatte MURPHEY (1947) an D_2/He und He/H_2 beobachtet. Jedoch ist diese im letztgenannten Fall durch neue, noch unveröffentlichte Versuche von GREW nicht bestätigt; GREW u. Mitarb. finden zwischen 90 und 800° K konstantes α für He/H_2[1].

[1] Freundliche briefliche Mitteilung von Herrn Dr. K. E. GREW.

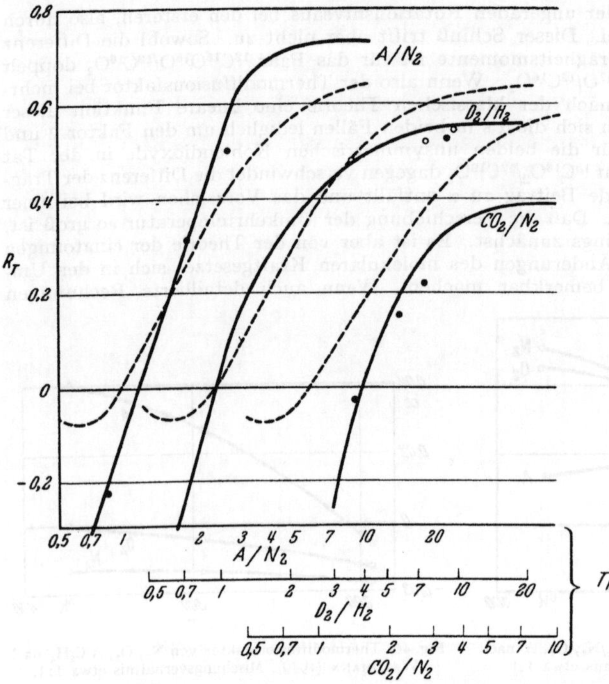

Fig. 41. R_T als Funktion von $T_{12}^* = kT/\varepsilon_{12}$. Ausgezogene Kurven nach Messungen von GREW et al. (1954), ○ nach MURPHEY (1947), ● nach WALDMANN (1949). Gestrichelte Kurven: theoretisch für das Lennard-Jones-(6; 12)-Potential. Nach GREW, JOHNSON und NEAL (1954).

Fig. 40 zeigt besonders deutlich auch den Einfluß des Massenunterschieds auf den Thermodiffusionsfaktor. Das Wechselwirkungspotential der Moleküle N_2, O_2 und A mit Äthylen kann man ganz grob als gleich annehmen; dasselbe sei für die Wechselwirkung von N_2, O_2, A untereinander vorausgesetzt. Bei dem Isobaren-Gemisch N_2/C_2H_4 geht das kleinere N_2-Molekül zur warmen Seite (es erwärmt sich im Diffusionsthermoeffekt; $\alpha < 0$). In den beiden anderen Fällen aber geht der zwar geometrisch kleinere, jedoch schwerere Partner O_2 und A bei genügend hoher Temperatur zur kalten Seite. Man ist geneigt den Vorzeichenwechsel in den beiden Fällen dem zweiten Faktor in der Produktdarstellung (69.13) von α zuzuschreiben. Denn wenn der erste Faktor $6C^* - 5$ das Vorzeichen umkehren würde, so hätte man das bei der mutmaßlichen Ähnlichkeit der Kraftgesetze auch für das Gemisch mit N_2 zu erwarten im Widerspruch zum experimentellen Befund.

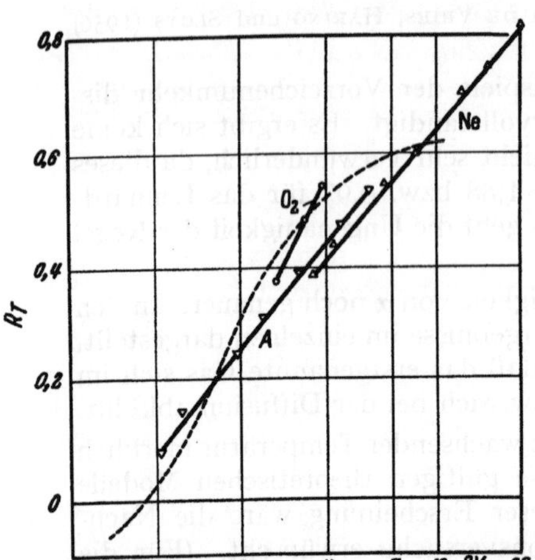

Fig. 42. R_T für Isotopen-Thermodiffusion in Abhängigkeit von $T^* = kT/\varepsilon$. Ausgezogene Kurven: Messungen von STIER (1942) △ an $^{22}Ne/^{20}Ne$, ▽ an $^{40}A/^{36}A$, und von WHALLEY, WINTER und BRISCOE (1949) ○ an $^{16}O^{18}O/^{16}O_2$. Gestrichelte Kurve: theoretisches R_T für Lennard-Jones-(6; 12)-Potential.

Fig. 41 zeigt nochmals die teilweise Unzulänglichkeit des Lennard-Jones-(6; 12)-Potentials, die schon aus Fig. 38 hervorging. Der Fig. 41 liegen folgende Werte zugrunde [nach (71.11); der ε-Wert für D_2 bzw. H_2 ist etwas anders als der aus Tabelle 7 (Ziff. 52)]:

	A/N_2	D_2/H_2	CO_2/N_2
α_{starr} ($\gamma = 0{,}5$)	0,141	0,276	0,236
ε_{12}/k (°K)	106	33,3	132

Bei niedrigen Werten von kT/ε_{12} wäre offenbar nach Fig. 34 (Ziff. 75) das (4; 8)-Potential besser geeignet. Aber im ganzen muß man wohl sagen, daß man von der für einatomige Gase gültigen Theorie und den zugehörigen einfachen Methoden nicht zu viel verlangen darf hinsichtlich des komplizierteren Falls der mehratomigen Gasgemische.

γ) Isotopengemische. Isotopengemische sind bei Trennrohrversuchen besonders häufig benutzt worden, aber exakte α-Werte, zumal in Abhängigkeit von der Temperatur, sind wegen der experimentellen Schwierigkeit doch nur für wenige Fälle bekannt. Neben dem schon in Fig. 41 aufgeführten Beispiel D_2/H_2 sind besonders die Messungen von STIER (1942) an Neon und Argon — nach dem Vorbild von NIER (1940) — sowie die von WHALLEY, WINTER und BRISCOE (1949) an Sauerstoff zu nennen. Die Ergebnisse zeigt Fig. 42. Die ihr zugrunde liegenden ε-Werte sind aus Tabelle 7 (Ziff. 52) zu entnehmen. Die Zuordnung von α-Wert und Temperatur wurde von den Autoren in allen drei Fällen nach der Formel (18.4) von BROWN vorgenommen. Die Stierschen Ergebnisse für Neon und Argon bei 329° K hat übrigens MANN (1948) bestätigt und gezeigt, daß R_T für $^{14}N^{15}N/^{14}N_2$ praktisch denselben Wert hat wie für $^{40}A/^{36}A$. Auffällig an Fig. 42 ist, daß keine Konstanz von R_T für Neon bei hoher Temperatur beobachtet wurde. Vielleicht sind zu hohe α-Werte gemessen. Diese Vermutung ist schon von GREW und IBBS in ihrer Monographie geäußert worden, auf welche wegen weiterer Einzelheiten verwiesen sei. So hat man leider für die Isotopenthermodiffusion, die theoretisch am besten zu übersehen ist, bisher nur ziemlich spärliche Vergleichsmöglichkeiten mit geeignetem experimentellen Material. Dessen Vermehrung wäre erwünscht.

V. Theoretische und experimentelle Ergebnisse für den Reibungs- und den Wärmeleitkoeffizienten binärer Gemische.

Diese Koeffizienten sind, abgesehen von technischen Anwendungen, für die Gasanalyse von Interesse (vgl. Ziff. 12). Zur genauen Ermittlung der Wechselwirkung ungleichartiger Moleküle sind sie weniger geeignet als die Koeffizienten der Diffusion und Thermodiffusion, weil zur Viscosität und Wärmeleitfähigkeit in erster Linie die Wechselwirkung gleichartiger Teilchen beiträgt. Wir werden uns im folgenden darauf beschränken, die Abhängigkeit dieser Koeffizienten vom Mischungsverhältnis näher zu betrachten. Bezüglich der Temperaturabhängigkeit ergeben sich keine wesentlich neuen Gesichtspunkte gegenüber den in Ziff. 51 und 52 dargelegten Verhältnissen bei einfachen Gasen.

78. Der theoretische Ausdruck für die Viscosität des binären Gemisches. Bedingung für ein Maximum von $\eta(\gamma)$. Die Viscosität des binären Gemischs ist in erster Näherung nach (70.2) als Quotient zweier rein quadratischer Formen der Molenbrüche γ_1 und γ_2 gegeben. Die ausführliche Darstellung durch Ω-Integrale, die man durch Einsetzen der Elemente b' und b'' aus (67.5) und (67.6) in (70.3) erhält, ist bei CHAPMAN und COWLING [4], S. 167 angegeben.

Wir schreiben das Ergebnis in etwas anderer Form auf, welche die Abweichung von der Linearität direkt erkennen läßt und zweckmäßig ist für die Entscheidung der Frage, wann die Viscosität in Abhängigkeit vom Molenbruch γ_1 (oder γ_2) ein Maximum annehmen kann. Es gilt

$$\eta = \eta_1\gamma_1 + \eta_2\gamma_2 + \frac{R_1\gamma_1 + R_2\gamma_2}{\frac{1}{\eta_1}P_{11}\gamma_1^2 + \frac{2}{\eta_{12}}P_{12}\gamma_1\gamma_2 + \frac{1}{\eta_2}P_{22}\gamma_2^2}\cdot\gamma_1\gamma_2, \tag{78.1}$$

wobei

$$\left.\begin{aligned} P_{11} &= 1 + \tfrac{3}{5}A^\star\frac{M_1}{M_2}, \qquad P_{12} = 1 + \frac{3}{5}A^\star\frac{\eta_{12}^2}{4M_1M_2\eta_1\eta_2}, \\ R_1 &= \left(1 - \frac{\eta_2}{\eta_1}\right)P_{11} + 2\left[1 - \frac{\eta_1}{\eta_{12}} + \frac{3}{5}A^\star\left(\frac{\eta_{12}}{4M_1M_2\eta_1} - 1\right)\right]. \end{aligned}\right\} \tag{78.2}$$

Die Zahlen P_{22} und R_2 erhält man durch Vertauschen der Indices 1 und 2. $A^\star$ und $M_{1,2}$ sind in (69.10) und (69.12) erklärt. η_1 und η_2 bedeuten die Viscositäten der reinen Gase [vgl. (50.15)], η_{12} ist eine der Wechselwirkung ungleichartiger Moleküle in entsprechender Weise zugeordnete fiktive Viscosität:

$$\eta_1 = \frac{5}{8}\,\frac{kT}{\Omega_{11}^{(2,2)}}, \qquad \eta_{12} = \frac{5}{8}\,\frac{kT}{\Omega_{12}^{(2,2)}} = \frac{5}{3A^\star}\,n\,m_{12}\,D(1,2). \tag{78.3}$$

Die zuletzt genannte Identität folgt aus (68.3) und (69.10).

Aus (78.1) ist ersichtlich, daß die Viscosität an den Intervall-Enden $\gamma_1 = 0$ oder 1 die Werte η_2 bzw. η_1 annimmt, wie es sein muß. Das letzte Glied rechts in (78.1) kann im übrigen positiv oder negativ sein. Die Neigung der $\eta(\gamma_1)$-Kurve an den Intervall-Enden läßt sich unmittelbar ablesen:

$$\left.\begin{aligned}\left(\frac{d\eta}{d\gamma_1}\right)_{\gamma_1=0} &= \eta_1 - \eta_2 + \frac{\eta_2 R_2}{P_{22}}\\ &= \frac{2\eta_2}{1+\frac{3}{5}A^\star\frac{M_2}{M_1}}\left[1-\frac{\eta_2}{\eta_{12}}+\frac{3}{5}A^\star\left(\frac{\eta_{12}}{4M_1M_2\eta_2}-1\right)\right].\end{aligned}\right\} \tag{78.4}$$

Das Entsprechende für $\gamma_1 = 1$ ergibt sich daraus durch Vertauschen der Indices 1 und 2 sowie Umkehr des Vorzeichens.

Ein Maximum tritt dann auf, wenn die eckige Klammer in (78.4) und der durch Vertauschen der Indices 1, 2 daraus entstehende Ausdruck positiv sind. Wenn beide negativ sind, so hat $\eta(\gamma_1)$ ein Minimum. Auch das scheint in einem Fall, bei NO_2/N_2O_4, beobachtet zu sein [vgl. SCHRÖER (1936)].

Man erhält rasch einen Einblick in das Wesentliche durch Betrachtung zweier Grenzfälle:

α) Isobare, $m_1 = m_2$. Hierfür gilt nach (78.4) wegen $M_1 = M_2 = \frac{1}{2}$

$$\left(\frac{d\eta}{d\gamma_1}\right)_{\gamma_1=0} = 2\,\frac{\eta_2 + \frac{3}{5}A^\star\eta_{12}}{\eta_{12}(1+\frac{3}{5}A^\star)}\,(\eta_{12}-\eta_2) \tag{78.5}$$

und Entsprechendes für $\gamma_1 = 1$. Mit den in (74.4) definierten effektiven Viscositäts-Durchmessern s_1, s_2 und s_{12} läßt sich dies auch so schreiben:

$$\left(\frac{d\eta}{d\gamma_1}\right)_{\gamma_1=0} = 2\,\frac{s_{12}^2 + \frac{3}{5}A^\star s_2^2}{s_2^2(1+\frac{3}{5}A^\star)}\cdot\frac{s_2^2 - s_{12}^2}{s_{12}^2}\,\eta_2. \tag{78.6}$$

Das Entsprechende für $\gamma_1 = 1$ erhält man wieder durch Vertauschen der Indices und Vorzeichenumkehr.

Ein Maximum liegt also vor, wenn $\eta_{12} > \eta_1, \eta_2$ oder, was dasselbe ist, $s_{12}^2 < s_1^2, s_2^2$. Die gleiche Bedingung gilt für den Vorzeichenwechsel des isothermen Thermodiffusionsfaktors bei Isobaren-Gemischen [s. (74.7)].

Im allgemeinen liegt s_{12} zwischen s_1 und s_2 [vgl. die Regel (71.11)]. Dem entspricht es, daß bei keinem der von TRAUTZ und MELSTER (1930) untersuchten Isobaren-Gemische, nämlich N_2/CO, N_2/C_2H_4 und CO/C_2H_4, ein Maximum beobachtet worden ist. In allen drei Fällen verlaufen vielmehr die Isothermen $\eta(\gamma)$ praktisch linear.

Hingegen fanden JUNG und SCHMICK (1930) bei dem nahezu isobaren Gemisch CH_4/NH_3 ein schwaches Maximum, s. Fig. 43 (Ziff. 79). Nach (74.7) und (78.6) möchte man also schließen, daß bei diesem Gaspaar, ähnlich wie bei Ne/NH_3, der isotherme Thermodiffusionsfaktor $\alpha(\gamma)$ das Vorzeichen wechselt. Eine Unsicherheit für diesen Schluß liegt aber, abgesehen von dem vorhandenen kleinen Massenunterschied, in dem Umstand, daß man die für einatomige Gase gültigen Formeln verwenden muß.

β) Stark verschiedene Massen, $m_1 \ll m_2$. In diesem Fall ist $M_1 \ll M_2 \approx 1$ und man hat, wenn die Moleküldurchmesser und sonstigen Wechselwirkungs-Konstanten von gleicher Größenordnung sind, wegen $\eta \propto \sqrt{m}$ außerdem die Beziehung $\eta_1, \eta_{12} \ll \eta_2$. Damit ergibt sich nach (78.4)

$$\left(\frac{d\eta}{d\gamma_1}\right)_{\gamma_1=0} = \frac{1}{2}\left[1 - \frac{20}{3A^\star}\left(\frac{\eta_2}{\eta_{12}}\right)^2 \frac{m_1}{m_2}\right]\eta_{12}; \quad \left(\frac{d\eta}{d\gamma_1}\right)_{\gamma_1=1} = -\frac{3}{10} A^\star \frac{m_2}{m_1} \eta_{12}. \quad (78.7)$$

Die Steigung für $\gamma_1 = 1$ ist demnach negativ, ein Minimum kann sicher nicht vorkommen. Für $\gamma_1 = 0$ ist die Steigung positiv, wenn $\eta_2^2 m_1 / \eta_{12}^2 m_2$ nicht zu groß ist. Dann tritt ein Maximum auf. So ist es bei He/A-Gemisch. Im anderen Fall, der bei H_2/N_2-Gemisch verwirklicht ist, fällt die Kurve $\eta(\gamma_1)$ monoton ab. Näheres darüber findet sich in der folgenden Ziffer.

Schließlich sei noch auf die Arbeiten von Buddenberg und Wilke (1949) sowie von Wilke (1950) hingewiesen (s. auch Hirschfelder *et al.* [*25*], S. 533). Diese Autoren stellen vereinfachte Formeln auf für die Viscosität binärer oder auch polynärer Gemische, in welche außer den Molenbrüchen nur die Viscositäten der reinen Gase und die Massenverhältnisse eingehen. Trotz dieser Vereinfachung fällt der Vergleich mit dem Experiment sehr günstig aus und liefert bei binären Gemischen auch in extremen Fällen eine Darstellung, die fast ebensogut ist wie diejenige durch die konsequente Chapmansche Formel.

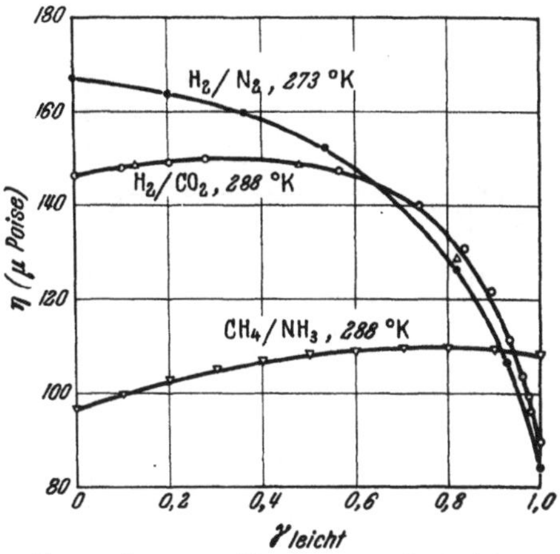

Fig. 43. Gemessene Viscosität von Gasgemischen: H_2/CO_2 (○ nach Puluj und △ nach Breitenbach), von H_2/N_2 (● nach Kleint) und von CH_4/NH_3 (▽ nach Jung und Schmick, 1930) in Abhängigkeit vom Molenbruch γ_{H_2} bzw. γ_{CH_4}. Entnommen aus den Landolt-Börnsteinschen Tabellen.

79. Vergleich von gemessenen und berechneten Viscositäten binärer Gemische.

Es können nur einige, hauptsächlich neuere Messungen besprochen werden; das weitere experimentelle Material, das zu einem beträchtlichen Teil von Trautz und Mitarbeitern stammt, ist in den Landolt-Börnsteinschen Tabellen zusammengefaßt. Wegen weiterer Literatur s. auch den Artikel von Herzfeld ([*25*], S. 196).

Wir betrachten zunächst Gasgemische mit großem relativen Massenunterschied. Wie Fig. 43 zeigt, tritt bei H_2/CO_2 ein Maximum auf (was von Trautz und Kurz 1931 nachgemessen wurde), bei H_2/N_2 dagegen nicht. Das ist auf Grund von (78.7) verständlich, weil sowohl η_2 als auch m_1/m_2 beim erstgenannten Gaspaar kleiner ist als beim zweiten.

Für einen quantitativen Vergleich mit der Theorie des Lennard-Jonesschen (6; 12)-Modells sei für H_2/N_2 bei 273° K angenommen

$$\eta_1 = 84{,}1; \quad \eta_2 = 167{,}1; \quad \eta_{12} = 83{,}5\ \mu\text{Poise}; \quad A^\star = 1{,}10.$$

Die beiden ersten sind die experimentellen Viscositäten der reinen Gase aus Fig. 43, der Wert von η_{12} ist nach (78.3) mit den Konstanten aus (71.12) berechnet. η_{12} kann — mit dem gleichen Ergebnis — auch aus dem Diffusionskoeffizienten entnommen werden. Sodann berechnet man nach (78.4) $d\eta/d\gamma_1 = -10{,}7$ bzw. $-315\ \mu$Poise für $\gamma_1 = 0$ bzw. 1, während hierfür aus der Fig. 43 als ungefähre experimentelle Werte -16 bzw. $-350\ \mu$Poise zu entnehmen sind. Die Übereinstimmung ist ausreichend, wenn man bedenkt, daß kleine Ungenauigkeiten der Meßwerte die Differentialquotienten stark beeinflussen. Jedenfalls ist der monotone Abfall von $\eta(\gamma)$ im Fall H_2/N_2 damit theoretisch begründet. Erwähnt sei noch, daß sich an den Ergebnissen der Rechnung nahezu nichts ändert, wenn man statt der Konstanten aus

(71.12) die besseren aus (76.1) wählt. Dies zeigt, daß mittels der Viscosität keine so genaue Festlegung der Wechselwirkungsparameter möglich ist wie mit der Thermodiffusion. Dasselbe war in Ziff. 71 für die Diffusion festgestellt worden.

Bei He/A-Gemisch tritt ein Maximum auf, das erstmals von TÄNZLER (Dissertation Halle 1906) beobachtet wurde. TRAUTZ und KIPPHAN bzw. BINKELE (1929, 1930) haben dies auch bei höheren Temperaturen gemessen und neuerdings wurde das Gemisch bei tiefen Temperaturen untersucht durch RIETVELD, VAN ITTERBEEK und VAN DEN BERG (1953) (s. Fig. 44).

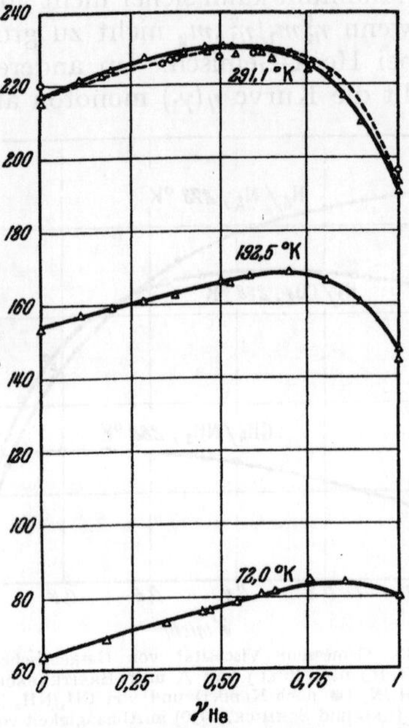

Fig. 44.

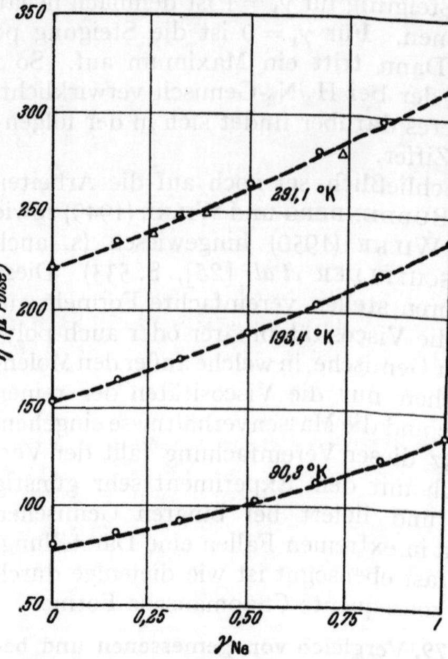

Fig. 45.

Fig. 44. Ausgezogene Kurven: gemessene Viscosität von He/A-Gemisch, △ nach RIETVELD, VAN ITTERBEEK und VAN DEN BERG (1953), ○ nach TRAUTZ und KIPPHAN (1929) sowie TRAUTZ und BINKELE (1930). Gestrichelte Kurve berechnet von RIETVELD et al. nach dem Lennard-Jones-(6; 12)-Modell und der Regel (71.11).

Fig. 45. Viscosität von Ne/A-Gemisch, ○ nach RIETVELD und VAN ITTERBEEK (1956), △ nach TRAUTZ und BINKELE (1930), in Abhängigkeit vom Molenbruch γ_{Ne}. Gestrichelte Kurven: berechnet von RIETVELD et al. nach dem Lennard-Jones-(6; 12)-Modell.

Es seien wieder die zur Berechnung der Kurvenneigung an den Intervall-Enden nötigen Daten notiert. Mit $d_{12} = 2{,}987$ Å, $\varepsilon_{12}/k = 36°$ K [aus der Regel (71.11) und aus Tabelle 7, Ziff. 52] und den experimentellen Viscositäten der reinen Gase hat man für He/A bei 291° K

$$\eta_1 = 191{,}4; \quad \eta_2 = 217{,}2; \quad \eta_{12} = 161 \text{ (μPoise)}; \quad A^\star = 1{,}107.$$

Daraus berechnet man nach (78.4) $d\eta/d\gamma_1 = 27$ bzw. -297 μPoise für $\gamma_1 = 0$ bzw. 1, während aus der experimentellen Kurve die Werte 37 bzw. -200 μPoise bei 291° K zu entnehmen sind. Wieder ist die Übereinstimmung ausreichend. Die berechneten und gemessenen Werte der Viscosität selbst fallen im übrigen für mittlere Molenbrüche recht gut zusammen, wie Fig. 44 zeigt.

In der vorigen Ziffer wurde dargelegt, daß bei Gemischen, deren Partner Molekulargewichte gleicher Größenordnung haben, gewöhnlich die Viscositäts-Isothermen kein Maximum haben. Das Gaspaar CH_4/NH_3 bildet eine Ausnahme (s. Fig. 43). Dagegen repräsentiert unter anderem das Gemisch Ne/A den Normalfall ohne Maximum, wie Fig. 45 zeigt.

Die dortigen berechneten Kurven sind teilweise den Messungen angeglichen nach dem folgenden Verfahren. Aus der Regel (71.11) und Tabelle 7 (Ziff. 52) entnimmt man $d_{12} = 3{,}103$ Å, $\varepsilon_{12}/k = 66{,}5°$ K. Der letztere Wert wurde beibehalten, die Durchmesser d_1 und d_2

aber, abweichend von Tabelle 7 so gewählt, daß für die reinen Gase die berechneten und gemessenen Viscositäten übereinstimmen; dann wurde nach (71.11) das jeweilige d_{12} gewonnen.

Die geringe Temperaturabhängigkeit der d-Werte (s. Tabelle 19) bestätigt nochmals die Brauchbarkeit des Lennard-Jones-Potentials.

Tabelle 19. *Lennard-Jones-(6; 12)-Parameter* d_1 *für* Ne, d_2 *für* A, *welche den von Rietveld und van Itterbeek berechneten Kurven aus Fig.* 45 *zugrunde liegen.* $d_{12} = \frac{1}{2}(d_1 + d_2)$.

T (°K)	291	193	90
d_1 (Å) . .	2,778	2,794	2,786
d_2 (Å) . .	3,426	3,450	3,325
d_{12} (Å) . .	3,102	3,122	3,056

80. Die Wärmeleitfähigkeit von Gasgemischen. Ebenso wie die Viscosität wird auch der Wärmeleitkoeffizient eines binären einatomigen Gemischs durch den Quotienten zweier rein quadratischer Formen der Molenbrüche γ_1 und γ_2 dargestellt. Man kann diesen Ausdruck, der bei Chapman und Cowling [4], S. 166, explizit angegeben ist, wieder auf die zu (78.1) analoge Form bringen. Der dabei im Nenner des letzten Glieds auftretende quadratische Ausdruck stimmt überein mit dem Nenner des Thermodiffusionsfaktors [s. Gl. (69.13)], ist also ein anderer als bei der Viscosität. Auch die Koeffizienten des Zählers sind keineswegs verwandt mit den entsprechenden Größen in dem Ausdruck für die Viscosität.

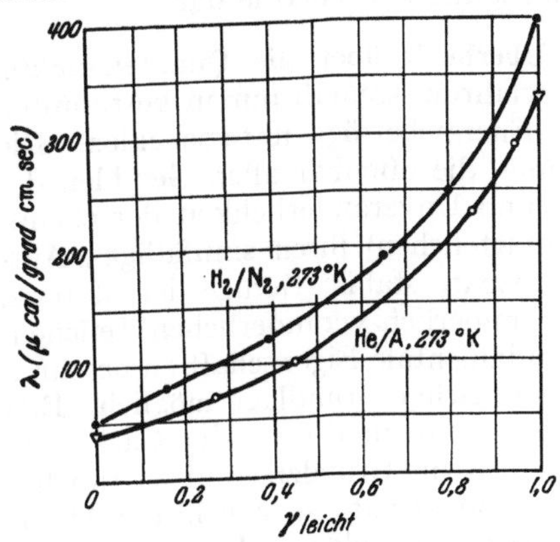

Fig. 46. Gemessener Wärmeleitkoeffizient von Gasgemischen: He/A ○ nach Wachsmuth (1908), ▽ nach Curie und Lepape (1931); H_2/N_2 nach Ibbs und Hirst (1929) in Abhängigkeit von γ_{He} bzw. γ_{H_2}.

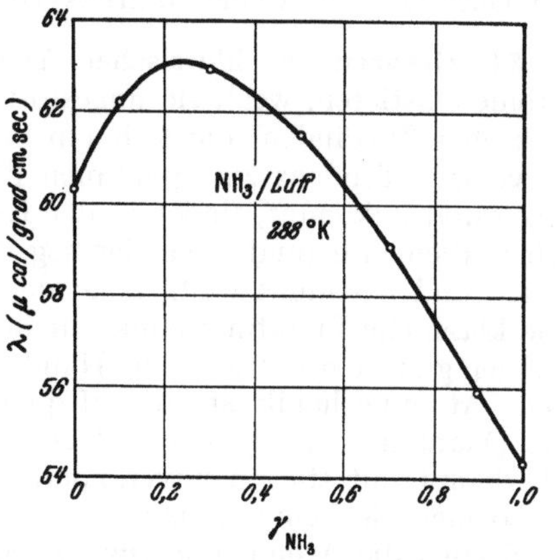

Fig. 47. Gemessener Wärmeleitkoeffizient von NH_3/Luft-Gemisch in Abhängigkeit vom Molenbruch γ_{NH_3}. Nach Grüss und Schmick (1928) [der Zusammenstellung von Bock (1951) entnommen].

Somit gibt es keine einfache und genaue Beziehung zwischen der Viscosität und dem Wärmeleitkoeffizienten eines Gemischs. Die für reine, einatomige Gase gültige Beziehung (50.16) überträgt sich nicht auf die Gemische.

Eine genaue Auswertung der Formeln für einatomige Gemische haben Cohen, Offerhaus und de Boer (1954) anläßlich der Diskussion der Transporteigenschaften der Helium-Isotope bei tiefen Temperaturen vorgenommen (vgl. Ziff. 85).

In Fig. 46 und 47 sind einige experimentelle Ergebnisse dargestellt. Im Fall des He/A-Gemischs ist die Theorie der einatomigen Gase in Strenge anwendbar. Bei der Wärmeleitfähigkeit mehratomiger Gemische hingegen spielen, anders als bei deren Viscosität, die inneren Freiheitsgrade der Moleküle eine wichtige Rolle. Die Euckensche Beziehung (s. Ziff. 98) berücksichtigt diese näherungsweise bei reinen Gasen. Zur Verallgemeinerung haben Hirschfelder *et al.* [25], S. 536 die Euckensche Beziehung in halbempirischer Weise mit der Formel für das einatomige Gemisch kombiniert; wegen der Einzelheiten sei auf ihre

Darstellung verwiesen. — Ebenso wie die Isothermen der Viscosität (vgl. Fig. 43 und 44) können nach Fig. 47 auch diejenigen der Wärmeleitfähigkeit ein Maximum besitzen. Der Schluß von der Wärmeleitfähigkeit auf die Zusammensetzung ist dann nicht mehr eindeutig. GRÜSS und SCHMICK (1928) haben auch an dem Gemisch NH_3/CO ein Maximum festgestellt. Bei der weitgehenden Ähnlichkeit des CO- und N_2-Moleküls ist verständlich, daß man in beiden Fällen qualitativ dasselbe beobachtet.

G. Quantenmechanische Effekte bei den Transportvorgängen in leichten Gasen.

Die Quanteneffekte bei den Transporterscheinungen in einatomigen Gasen gehören dem Tieftemperaturgebiet an. Sie waren erst einer fortgeschrittenen experimentellen Technik zugänglich; ihre theoretische Deutung erfordert einen recht großen Aufwand an subtilen quantenmechanischen Berechnungen. Aus beiden Gründen spielten diese Effekte — anders als die Probleme, welche die spezifische Wärme der mehratomigen Gase aufwarf — historisch für die Entwicklung der Quantentheorie keine Rolle. Aber ihre in den letzten Jahren erreichte weitgehende Klärung stellt eine wichtige Ergänzung des Gesamtbilds der Gaskinetik und eine eindrucksvolle Bestätigung der Theorie dar.

81. Grenzen der klassischen Theorie. Überblick über die Quanteneffekte. Atome existieren, wenn sie ungestört sind, erfahrungsgemäß nur in bestimmten diskreten Zuständen, die sich um endliche Energiebeträge unterscheiden. Die universelle Tendenz zur spontanen Einstellung, die offenbar allen Gebilden der Mikrowelt zukommt, findet in der Existenz und Unveränderlichkeit der chemischen Elemente (und auch der sog. Elementarteilchen) ihren sinnfälligen Ausdruck und bestimmt weitgehend das beobachtbare stabile Gefüge der Materie. Die klassische Punktmechanik mit ihren kontinuierlich veränderlichen Teilchenbahnen gibt kein Mittel an die Hand diese fundamentale Eigenschaft zu beschreiben. Atommechanik ist nur auf quantentheoretischer Grundlage möglich. Erst die Quantentheorie gab eine Begründung dafür, daß man es in der Gaskinetik überhaupt mit starren unveränderlichen Atomen im Grundzustand zu tun hat — solange die Temperatur zur Anregung nicht ausreicht —, obwohl die Atome (und auch die Atomkerne usw.) zusammengesetzte Gebilde sind.

Die Quantenmechanik ordnet den Elementarteilchen, hier den Elektronen und Atomkernen, Wellenfunktionen zu, die in bekannter Weise statistisch interpretiert werden. Im Grundzustand eines Atoms oder Moleküls oder in einem nicht zu hoch angeregten Zustand ist die zugeordnete Wellenlänge der Elektronen von der gleichen Größenordnung wie die Ausdehnung des Atoms selbst, so daß die Wellennatur der Elektronen (Beugungserscheinungen) ausschlaggebend die atomare Struktur bestimmt.

Auch die Wechselwirkung zweier Atome oder Moleküle wird durch die Elektronen vermittelt und ist, wie in Ziff. 24 dargelegt wird, auf quantenmechanischer Grundlage verständlich. Ist aber einmal das Wechselwirkungspotential bekannt, so kann man die Dynamik der Relativbewegung der zwei Atome in den meisten Fällen klassisch behandeln, wie es in Ziff. 26 und 27 und bei den Anwendungen auf die Gaskinetik in den Teilen E und F dieses Artikels geschieht. Voraussetzung hierfür ist, daß die dieser Relativbewegung bei der betreffenden Temperatur im Mittel zugeordnete de Broglie-Wellenlänge klein ist gegen die Länge d (Atomdurchmesser), welche das Wechselwirkungspotential charakterisiert. Seien m_i, m_j die Massen der beiden Atome, $m_{ij} = m_i\, m_j/(m_i + m_j)$ die reduzierte Masse,

$\bar{g}$ ihre mittlere Relativgeschwindigkeit, h das Wirkungsquantum, so ist die mittlere Wellenlänge durch

$$\overline{\lambda_{ij}} = \frac{h}{m_{ij}\,\bar{g}}$$

gegeben oder, indem wir etwas willkürlich $\frac{1}{2} m_{ij}\, \bar{g}^2 = \mathrm{k} T$ setzen, durch

$$\overline{\lambda_{ij}} = \frac{h}{\sqrt{2 m_{ij}\, \mathrm{k} T}}\,. \tag{81.1}$$

Numerisch, mit dem reduzierten Molekulargewicht M_{ij} und der Temperatur in °K, gilt

$$\overline{\lambda_{ij}} = \frac{43{,}8}{\sqrt{2 M_{ij}\, T}}\ \text{Å}. \tag{81.2}$$

Je schwerer das Gas und je höher die Temperatur, desto besser ist also die Annäherung durch die klassische Mechanik.

Tabelle 20. *Mittlere de Broglie-Wellenlänge $\bar{\lambda}$ in Å nach* (81.2) *für einige reine Gase bei verschiedenen Temperaturen.*

T (°K)	4	20	80	273
^4He; D_2	10,95	4,90	2,46	1,33
H_2 . .		6,94	3,47	1,88
$^{14}N_2$.			0,930	0,502

Tabelle 21. *Der quantenmechanische Parameter Λ^* aus* (81.3) *für einige reine Gase*[1].

	^3He	^4He	H_2	D_2	$^{14}N_2$
ε/k (°K)	10,22		37,0		79,8
d (Å) . .	2,556		2,928		3,749
Λ^*. . .	3,10	2,67	1,73	1,23	0,247

[1] Die Werte ε, d für He stammen von DE BOER und MICHELS (1938), die für H_2 von DE BOER (1940). Sie wurden aus dem zweiten Virialkoeffizienten bei hohen Temperaturen bestimmt.

Umgekehrt werden sich Quanteneffekte beim molekularen Stoß um so deutlicher zeigen, je leichter das Gas und je tiefer die Temperatur ist. Die Quanteneffekte in der Gaskinetik wurden zuerst von UEHLING und UHLENBECK (1933) und von UEHLING (1934) allgemein theoretisch behandelt und gleichzeitig von MASSEY und MOHR (1933) speziell am Modell der starren Kugeln studiert. DE BOER und MICHELS (1939 und spätere Arbeiten) haben die Rechnungen für das realistischere Lennard-Jones-Modell durchgeführt. Da die verschiedenen Gase bis zu verschiedenen Minimaltemperaturen herab experimentell untersucht werden können, ist es zweckmäßig, zur Charakterisierung eine für das betreffende Gas typische Temperatur ins Auge zu fassen, etwa die kritische Temperatur oder die das Lennard-Jonessche (6; 12)-Potential, s. (25.5), kennzeichnende Temperatur ε/k. Vergleicht man die nach (81.1) zu der letzteren Temperatur gehörige Wellenlänge mit der Längenkonstanten d aus dem Lennard-Jones-Potential, so erhält man, auf Gemische verallgemeinernd, die dimensionslosen Parameter

$$\Lambda_{ij}^* = \frac{h}{d_{ij}\sqrt{2 m_{ij}\, \varepsilon_{ij}}}\,. \tag{81.3}$$

Sie charakterisieren das Gasgemisch in bezug auf die Quantenmechanik: falls eines der $\Lambda^* \gtrsim 1$ ist, so machen sich an der unteren Temperaturgrenze Quanteneffekte bemerkbar; falls alle $\Lambda^* \ll 1$, kann für alle in Betracht kommenden Temperaturen die klassische Näherung benutzt werden. In den Tabellen 20 und 21 sind zur Erläuterung einige Zahlenwerte für reine Gase notiert. Genau kann man den Gültigkeitsbereich der klassischen Theorie natürlich erst auf Grund der vollständigen quantenmechanischen Rechnung abstecken, deren Prinzip in den folgenden Ziffern geschildert wird. Dies vorwegnehmend, sei bemerkt, daß bei He für $T > 200°$ K und bei H_2 für $T > 250°$ K die Quanteneffekte an der Viscosität, der Wärmeleitung und der Diffusion weniger als 0,6% ausmachen (Zahlenangaben nach MASON und RICE 1954a).

Worin bestehen die Quanteneffekte? Die Form (28.5) oder (29.2) der Boltzmann-Gleichung ist, wie schon früher betont wurde, bei hinreichend verdünnten Gasen allgemein auch bei Benutzung der Quantenmechanik zuständig. Die Quanteneffekte kommen in den differentiellen Streuquerschnitten $\sigma(g, \chi)$ bzw. $\sigma_{ij}(g, \chi)$ zum Ausdruck. Qualitativ kann man, mit einigem Vorbehalt, über das Verhältnis des quantenmechanischen zum klassischen Streuquerschnitt etwa folgendes sagen. Bei starren Kugeln ist ersterer größer und die quantenmechanischen Reibungs-, Wärmeleitungs- und Diffusionskoeffizienten sind kleiner als die klassischen (MASSEY und MOHR 1933). Bei Vorhandensein von Anziehungskräften (z.B. beim Lennard-Jones-Modell) ergibt sich das Umgekehrte: die Atome wirken unter Umständen transparenter (Ramsauer-Effekt) und die Transportkoeffizienten sind gegenüber den klassischen vergrößert. Die erste Aufgabe der Theorie ist es daher, das quantenmechanische Beugungsproblem hinreichend genau zu lösen. Die Lösung ist eindeutig, wenn es sich um den Stoß zweier verschiedenartiger (unterscheidbarer) Teilchen handelt (Boltzmann-Statistik). Bei gleichartigen (nicht-unterscheidbaren) Teilchen gibt es zu denselben Versuchsbedingungen rein rechnerisch zunächst zwei linear unabhängige Wellenfunktionen, eine in den Schwerpunktskoordinaten der beiden Teilchen symmetrische und eine antisymmetrische. Das hat zuerst MOTT (1930) am Fall des Elektronenstoßes studiert, die ersten detaillierten, aber noch nicht ganz konsequenten Anwendungen davon auf die Gaskinetik haben HALPERN und GWATHMEY (1937) gemacht. Je nach Art der Teilchen ist nur die eine oder die andere Lösung physikalisch realisiert (Einstein-Bose- oder Fermi-Dirac-Statistik). Die Auswahl hängt von den weiteren Freiheitsgraden der Teilchen ab, bei Elementarteilchen vom Spin, bei zusammengesetzten Teilchen vom gesamten inneren Aufbau[1]. Kennt man den Streuquerschnitt, so kann man die gaskinetischen Koeffizienten nach den Methoden von Teil E und F berechnen. Besonders eklatant äußern sich die Quanteneffekte beim Vergleich von ^{3}He und ^{4}He. Diese beiden Gase haben dasselbe molekulare Kraftgesetz, aber verschiedene Masse und Statistik. Klassisch sollte unabhängig von der Temperatur $\eta(^3\text{He}) = \sqrt{\frac{3}{4}}\,\eta(^4\text{He}) = 0{,}865\,\eta(^4\text{He})$ sein. Tatsächlich ist aber experimentell nach E.W. BECKER, MISENTA und SCHMEISSNER (1954) und theoretisch nach DE BOER und COHEN (1951) sowie BUCKINGHAM und SCRIVEN (1952) bei tiefen Temperaturen die Viscosität des leichten Gases größer als die des schweren; beispielsweise gilt $\eta(^3\text{He}) = 1{,}21\,\eta(^4\text{He})$ bei 4,2° K.

Eine Bemerkung sei noch angefügt über Quanteneffekte, die in Gasen von hohem Druck zu erwarten sind (Gasentartung). Eine merkliche Besetzung der Quantenzustände der Translation — man denke z.B. an das Elektronengas im Metall — prägt sich in der Streuwahrscheinlichkeit aus; als Endzustände beim Stoß kommen, etwa im Fall der Fermi-Statistik, nur unbesetzte Zustände in Frage. Dementsprechend haben UEHLING und UHLENBECK (1933) im Stoßintegral der Boltzmann-Gleichung den Ausdruck $f' f'_1 - f f_1$ durch

$$f' f'_1 (1 + \Theta f)(1 + \Theta f_1) - f f_1 (1 + \Theta f')(1 + \Theta f'_1) \qquad (81.4)$$

ersetzt, wobei

$$\Theta = \begin{cases} \pm (h/m)^3 & \text{für nicht-unterscheidbare Teilchen } (+\text{ Bose-Statistik; } -\text{ Fermi-Statistik}) \\ 0 & \text{für unterscheidbare Teilchen (Boltzmann-Statistik).} \end{cases}$$

[1] Natürlich hat man dann von der auf mehratomige Gase erweiterten Boltzmann-Gleichung auszugehen, siehe Teil H.

Haben die Teilchen innere Freiheitsgrade, z.B. Spin, so ist das nicht a priori durch Hinzufügen eines statistischen Gewichtsfaktors zu berücksichtigen, sondern man hat das Gas — wenigstens zunächst — als Gemisch zu behandeln (siehe Teil H dieses Artikels). Setzt man den Ausdruck (81.4) gleich null, so erhält man die bekannten Verteilungsfunktionen des thermischen Gleichgewichts für das wechselwirkungsfreie Gas nach den verschiedenen Statistiken. Gleichwohl ist der Ansatz (81.4) zu eng für eine konsequente Theorie. Der Integrand des Stoßintegrals ist jetzt vom vierten Grad in der Funktion f. Dann müssen aber im allgemeinen auch die Terme dritten und vierten Grades, herrührend von der mechanischen Wechselwirkung, also den Dreier- und Viererstößen, berücksichtigt werden. Daß diese Terme keineswegs zu vernachlässigen sind, wird bestätigt durch eine Abschätzung, die DE BOER (1943) auf Grund der Enskogschen Formeln für dichte Gase gemacht hat. Die Modifikation (81.4) soll daher außerhalb der weiteren Erörterungen in diesem Artikel bleiben. Tatsächlich macht sich die Entartung in Gasen von Atmosphärendruck erst bei sehr tiefen Temperaturen ($<1^\circ$ K für Helium) bemerkbar und ist daher praktisch nicht von Belang. Bei flüssigem Helium hingegen ist die Gasentartung ausschlaggebend, was neuerdings BUCKINGHAM und TEMPERLEY (1950) betonten; doch geht man bei der Behandlung dieses Problems zweckmäßig nicht von der Gaskinetik, sondern von anderen theoretischen Bildern aus[1].

Über die Quanteneffekte der Gastheorie sind in den letzten Jahren zwei zusammenfassende Berichte erschienen. Der eine, von DE BOER und BIRD, findet sich als Kap. 10 in dem Buch von HIRSCHFELDER *et al.* [25], der andere stammt von DE BOER [2] und behandelt etwas spezieller vor allem den Fall des Heliums.

82. Die Streuung nach der Quantenmechanik[2]. *α) Die Schrödinger-Gleichung.* Die Schrödinger-Gleichung für die Relativbewegung zweier Teilchen mit den Massen m_i, m_j und dem vom Abstand abhängigen Wechselwirkungspotential $\Phi_{ij}(r)$ lautet

$$\left(-\frac{\hbar^2}{2m_{ij}}\Delta+\Phi_{ij}\right)\psi=E_{\text{rel}}\,\psi. \tag{82.1}$$

Dabei ist $m_{ij}=m_i m_j/(m_i+m_j)$ die reduzierte Masse und

$$E_{\text{rel}}=\tfrac{1}{2}\,m_{ij}\,g^2 \tag{82.2}$$

die kinetische Energie der Relativbewegung [vgl. (22.16)]. g bezeichnet die Relativgeschwindigkeit bei großem Abstand. Für das Potential ist $\Phi(\infty)=0$ vorausgesetzt.

Wegen der Drehinvarianz des Hamilton-Operators in (82.1) sind die Lösungen von (82.1) entartet. Man kann sie nach Drehimpulsen l ordnen; die Wellenfunktion ψ läßt sich in Polarkoordinaten r, χ, ε separieren:

$$\psi=R_l(r)\,Y_{lm}\quad\text{mit}\quad Y_{lm}\propto P_l^m(\cos\chi)\,e^{im\varepsilon}. \tag{82.3}$$

Darin bedeutet Y_{lm} die Kugelflächenfunktion. Mit der Wellenzahl

$$k=\frac{m_{ij}\,g}{\hbar}\quad\text{oder}\quad E_{\text{rel}}=\frac{\hbar^2}{2m_{ij}}k^2 \tag{82.4}$$

lautet die Gleichung für die Radialfunktion R_l

$$\frac{d^2 r R_l}{dr^2}+\left[k^2-\frac{l(l+1)}{r^2}-\frac{2m_{ij}\,\Phi_{ij}}{\hbar^2}\right]r\,R_l=0. \tag{82.5}$$

[1] Vgl. R. P. FEYNMAN: Application of Quantum Mechanics to Liquid Helium (in C. J. GORTER, Progress in Low Temperature Physics, I, Amsterdam 1955).

[2] Wegen einer ausführlichen Darstellung sei auf das Buch von MOTT und MASSEY [39] und die Beiträge von MASSEY zu Bd. XXXVI dieses Handbuches verwiesen.

Um diese Gleichung dimensionslos zu machen, ist es zweckmäßig, den im Potential vorkommenden „Atomdurchmesser“ d_{ij} als Längeneinheit zu wählen. Für die spezielle Lennard-Jones-Wechselwirkung (25.5) erhält man mit $r^* = r/d_{ij}$, $k^* = k d_{ij}$ die Gleichung

$$\frac{d^2 r^* R_l}{d r^{*2}} + \left[k^{*2} - \frac{l(l+1)}{r^{*2}} - \frac{16\pi^2}{\Lambda_{ij}^{*2}}\left(r^{*-12} - r^{*-6}\right)\right] r^* R_l = 0. \tag{82.6}$$

Hierin ist Λ_{ij}^* der in (81.3) eingeführte quantenmechanische Parameter.

β) Die Streuphasen δ_l. Die Gl. (82.5) ist — eindeutig bis auf einen konstanten Faktor — zu lösen, ausgehend von den Randbedingungen an der Stelle $r = 0$. Man braucht, um den Streuquerschnitt zu erhalten, nur das asymptotische Verhalten der Wellenfunktion für $r \to \infty$ zu kennen. Für $r \to \infty$ ist jede Lösung rein harmonisch wie diejenige im kräftefreien Fall. Für diesen gilt asymptotisch

$$\left.\begin{aligned}(r R_l)_{\text{kräftefrei}} &\propto \sin\left(k r - \frac{\pi}{2} l\right) \propto \\ &\propto e^{-i\left(k r - \frac{\pi}{2} l\right)} - e^{i\left(k r - \frac{\pi}{2} l\right)}.\end{aligned}\right\} \tag{82.7}$$

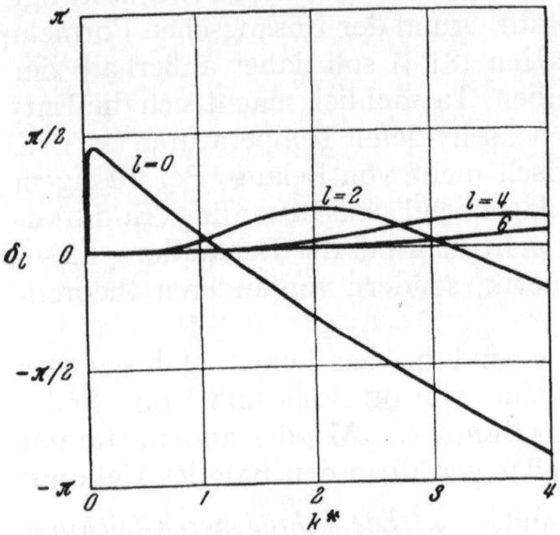

Fig. 48. Streuphasen δ_l für ^{4}He in Abhängigkeit von $k^* = k d$ für das Lennard-Jones-(6; 12)-Potential, siehe (82.6) mit $\Lambda^* = 2{,}67$. Nach DE BOER und MICHELS (1939).

Man kann diese Wellenfunktion auffassen als Überlagerung einer einlaufenden und einer auslaufenden Welle, wobei sich letztere aus der ersteren effektiv durch Reflexion an der Stelle $r_l \approx l/k$ ergibt. Das ist für große l genau der kürzeste Abstand, bis zu dem sich klassisch zwei Teilchen vom Drehimpuls $\hbar\, l(l+1)$ nähern. Durch die Wechselwirkung wird diese Stelle verschoben, durch Abstoßung nach größeren, durch Anziehung nach kleineren r-Werten. Das wird in dem asymptotischen Verhalten

$$r R_l \propto e^{-i\left(k r - \frac{\pi}{2} l\right)} - e^{2 i \delta_l} e^{i\left(k r - \frac{\pi}{2}\right)} \propto \sin\left(k r - \frac{\pi}{2} l + \delta_l\right) \tag{82.8}$$

durch die Phasenverschiebung $\delta_l(k)$ zum Ausdruck gebracht. Auch (82.8) bedeutet wieder eine Reflexion mit gleichbleibendem Betrag der Amplitude, wie es die Erhaltung der Teilchenzahl verlangt. Zur eindeutigen Festlegung der Phasen ist es zweckmäßig zu verlangen, daß

$$\delta_l = 0 \quad \text{für} \quad k \to \infty. \tag{82.9}$$

Diese Festsetzung ist jedenfalls möglich, wenn das Potential Φ überall endlich ist.

Die Schrödinger-Gleichung (82.1) kann, von wenigen Spezialfällen abgesehen, nur numerisch integriert werden. Unter Benutzung des Lennard-Jonesschen (6; 12)-Potentials, s. Gl. (82.6), haben diese Aufgabe DE BOER und MICHELS (1939) behandelt für ^{4}He, DE BOER, VAN KRANENDONK und COMPAAN (1950) für ^{3}He und COHEN *et al.* (1956) für H_2. Fig. 48 zeigt die Ergebnisse für ^{4}He. Für $k^* = 0$ sind nach der Figur alle $\delta_l = 0$. Das gilt stets dann, wenn keine gebundenen Zustände, also keine Eigenlösungen von (82.1) mit $E_{\text{rel}} < 0$ existieren und wenn die Festsetzung (82.9) getroffen wird. Wenn aber s_l gebundene Zustände mit dem Drehimpuls l vorhanden sind, so gilt

$$\delta_l = s_l \pi \quad \text{für} \quad k \to 0. \tag{82.10}$$

(s. MOTT und MASSEY [39], Kap. II). Eine Ausnahme hiervon hat man, wenn es zu $l=0$ den Eigenwert $E_{\text{rel}}=0$ gibt. Dann ist

$$\delta_0 = \frac{\pi}{2} + s_0 \pi, \tag{82.10a}$$

wo s_0 die Anzahl der gebundenen Zustände zu $l=0$ mit $E_{\text{rel}} < 0$ bezeichnet. Daraus, daß dieser Ausnahmefall beim ^{4}He nahezu verwirklicht ist, erklärt sich der steile Anstieg von δ_0 bei kleinen k^*-Werten. Bei ^{3}He, welches das gleiche Potential wie ^{4}He, aber kleinere Masse hat, gibt es keinen gebundenen Zustand; hier sind alle $\delta_l(0)=0$. Bei H_2 gibt es zu $l=0$ und 1 je einen gebundenen Zustand; hier ist $\delta_0(0)=\delta_1(0)=\pi$, s. dazu die Arbeit von COHEN *et al.* (1956).

γ) Der Streuquerschnitt. Um den differentiellen Streuquerschnitt (vgl. Ziff. 23) zu erhalten, hat man eine Lösung der Schrödinger-Gleichung aufzustellen, die asymptotisch aus einer ebenen Welle und einer auslaufenden Kugelwelle besteht. Nach dem bekannten Verfahren von FAXÉN und HOLTSMARK (1927) kann man eine solche Lösung durch Superposition der Partialwellen (82.3) gewinnen. Man geht aus von der Rayleighschen Darstellung der ebenen Welle durch Kugelwellen der kräftefreien Teilchen. Für $r\to\infty$ lautet diese Darstellung

$$e^{ikz} \approx \frac{i}{2kr} \sum_{l=0,1,\ldots} (2l+1)\, i^l \left[e^{-i\left(kr-\frac{\pi}{2}l\right)} - e^{i\left(kr-\frac{\pi}{2}l\right)} \right] P_l(\cos\chi). \tag{82.11}$$

Dabei bezeichnet $z=r\cos\chi$ die Koordinate in der Fortschreitungsrichtung der Welle und P_l die Legendre-Polynome in der Normierung $P_l(1)=1$. Der [ ]-Ausdruck stimmt überein mit dem asymptotischen Ausdruck (82.7). Nun sollen mit denselben Koeffizienten, die (82.11) zugrundeliegen, die Kugelwellen der wechselwirkenden Teilchen superponiert werden. Nach (82.8) gibt dies, wieder für $r\to\infty$, die folgende Wellenfunktion

$$\psi \approx \frac{i}{2kr} \sum_{l=0,1,\ldots} (2l+1)\, i^l \left[e^{-i\left(kr-\frac{\pi}{2}l\right)} - e^{2i\delta_l} e^{i\left(kr-\frac{\pi}{2}l\right)} \right] P_l(\cos\chi). \tag{82.12}$$

Indem man $e^{2i\delta_l} = 1 + e^{2i\delta_l} - 1$ schreibt und (82.11) benutzt, bekommt man

$$\psi \approx e^{ikz} + \frac{e^{ikr}}{kr}\, a(\chi, k) \tag{82.13}$$

mit

$$a(\chi, k) = \frac{1}{2i} \sum_{l=0,1,\ldots} (2l+1)\left(e^{2i\delta_l(k)} - 1\right) P_l(\cos\chi). \tag{82.14}$$

Die Funktion ψ aus (82.13) stellt die Lösung mit den gesuchten Eigenschaften dar. $a(\chi)$ ist die winkelabhängige Amplitude der Streuwelle, die nun unmittelbar den Streuquerschnitt liefert:

$$\sigma_{ij}(\chi, k) = \left|\frac{a_{ij}}{k}\right|^2 = \frac{1}{4k^2} \left| \sum_{l=0,1,\ldots} (2l+1)\left(e^{2i\delta_{l,ij}(k)} - 1\right) P_l(\cos\chi) \right|^2 \tag{82.15}$$

Dabei haben wir zum Schluß durch die Indices (ij) an σ, a und δ_l wieder zum Ausdruck gebracht, daß es sich um zwei verschiedene Teilchen handelt[1]. Die Formel (82.15) ist anzuwenden bei unterscheidbaren Teilchen, z. B. beim Stoß ^{3}He/^{4}He oder p-H_2/o-H_2 (Boltzmann-Statistik).

[1] Eine Verwechslung des Index i und der imaginären Einheit i ist nicht zu befürchten.

δ) Symmetrisierung bei gleichartigen Teilchen. Es seien zwei Teilchen betrachtet von gleicher Zusammensetzung und gleichem inneren Zustand. Auch die räumliche Orientierung[1] soll zu den Merkmalen des inneren Zustands gehören. Beispiele sind zwei ^{4}He-Atome oder zwei ^{3}He-Atome gleicher Kernspinstellung oder zwei Ortho-H_2-Moleküle von gleicher Stellung der Kernspins und der inneren Bahndrehimpulse. Zwei solche Teilchen lassen sich durch kein physikalisches Mittel voneinander unterscheiden. Die Wellenfunktion, die das aus den beiden Teilchen bestehende System beschreibt, ist daher in den Orts-(Schwerpunkts-)Koordinaten der Teilchen entweder symmetrisch oder antisymmetrisch. Der erste Fall (Bose-Statistik) liegt vor, wenn die Teilchen aus einer geraden Anzahl von Fermionen bestehen (z.B. ^{4}He, H_2). Den zweiten Fall (Fermi-Statistik) hat man bei einer ungeraden Anzahl von Fermionen als Bausteine (z.B. ^{3}He). Wenn aber die gleich zusammengesetzten Teilchen verschiedenen inneren Zustand haben, dann gibt es immer genau eine Wellenfunktion, die symmetrisch, und eine, die antisymmetrisch ist in den Ortskoordinaten der beiden Teilchen. Beispiele sind zwei ^{3}He-Atome von verschiedener Kernspinstellung oder ein Para-H_2- und ein Ortho-H_2-Molekül oder zwei Ortho-H_2-Moleküle verschiedener Orientierung. Angewandt auf Stoßprozesse, bei denen sich der innere Zustand nicht ändert[2], bildet das Gesagte den Inhalt des Satzes von Ehrenfest und Oppenheimer (1931), vgl. auch Falk (1955).

Die Wellenfunktion (82.13) besitzt keine Symmetrie, doch läßt sich diese sogleich herstellen. Es ist ja

$$z = z_1 - z_2 = r\cos\chi,$$

wo die Indices 1, 2 die beiden nicht-unterscheidbaren Teilchen bezeichnen. Deren Vertauschung führt also z, χ in $-z, \pi-\chi$ über. Damit erhält man als symmetrische (bzw. antisymmetrische) Wellenfunktion

$$\psi_{s(a)} = \frac{1}{\sqrt{2}}\left[e^{ikz} \underset{(-)}{+} e^{-ikz} + \frac{e^{ikr}}{kr}\left(a(\chi) \underset{(-)}{+} a(\pi-\chi)\right)\right].$$

Der Streuquerschnitt für gleichartige Teilchen mit gleichem inneren Zustand (nicht-unterscheidbare Teilchen) ist also durch

$$\sigma_{s(a)} = \frac{1}{2k^2}\,|a(\chi) \underset{(-)}{+} a(\pi-\chi)|^2 = \frac{1}{2k^2}\left|\sum_{\substack{l=0,2,\ldots\\(l=1,3,\ldots)}} (2l+1)\,(e^{2i\delta_l}-1)\,P_l(\cos\chi)\right|^2 \tag{82.16}$$

gegeben. σ_s ist bei Bose-, σ_a bei Fermi-Statistik anzuwenden.

Zwei gleichartige Teilchen mit verschiedenem inneren Zustand (unterscheidbare Teilchen) treffen sich nach der Unordnungsannahme der Gastheorie im Mittel ebenso oft mit symmetrischer wie mit antisymmetrischer Ortsfunktion. Ihr mittlerer differentieller Streuquerschnitt sei mit σ_u bezeichnet; er beträgt also

$$\sigma_u = \tfrac{1}{2}(\sigma_s + \sigma_a). \tag{82.17}$$

Bemerkt sei, daß dieser Querschnitt mit dem Querschnitt (82.15) für verschiedenartige Teilchen von derselben Masse und Wechselwirkung nicht übereinstimmt.

83. Darstellung der Wirkungsquerschnitte für Viscosität und Diffusion durch die Streuphasen. Nach (50.15) und (68.3) sind die Viscosität und der binäre Diffusionskoeffizient durch

$$\eta = \frac{5}{8}\,\frac{kT}{\Omega^{(2,2)}}, \qquad D(1,2) = \frac{3kT}{16n\,m_{12}\,\Omega_{12}^{(1,1)}} \tag{83.1}$$

---

[1] Wir nennen die $2J+1$ entarteten Zustände zur inneren Drehimpulsquantenzahl J Zustände verschiedener räumlicher Orientierung.

[2] Dazu muß die Wechselwirkung jedenfalls kugelsymmetrisch sein.

gegeben. Die Ω-Größen erhält man nach (50.5) durch Integration der beiden Wirkungsquerschnitte

$$\left.\begin{aligned} &\text{Diffusion:} && Q_{12}^{(1)} = 2\pi \int_0^\pi (1-\cos\chi)\,\sigma_{12}\sin\chi\,d\chi \\ &\text{Viscosität:} && Q_{s(a)}^{(2)} = 2\pi \int_0^\pi (1-\cos^2\chi)\,\sigma_{s(a)}\sin\chi\,d\chi\,. \end{aligned}\right\} \tag{83.2}$$

Dabei ist für die Diffusion jedenfalls der für unterscheidbare Teilchen gültige Streuquerschnitt (82.15) zu benutzen. Nur bei irgendwie unterscheidbaren Teilchen hat der Begriff der Diffusion einen physikalischen Sinn; eine Selbstdiffusion im wörtlichen Sinn gibt es nicht, und es ist nicht angebracht, einen Selbst-Diffusionskoeffizienten mittels eines der symmetrischen Querschnitte aus (82.16) zu definieren. Das betonten schon HALPERN und GWATHMEY (1937). Bei der Viscosität reiner Gase dagegen ist im allgemeinen eine Linearkombination der symmetrisierten Querschnitte zu verwenden; das wird in der folgenden Ziffer näher erläutert.

Um die Querschnitte Q durch die Streuphasen darzustellen, hat man die σ aus (82.15) bzw. (82.16) über den Ablenkwinkel zu integrieren. Das Ergebnis lautet in der Schreibweise von H. A. KRAMERS[1] [vgl. DE BOER und COHEN (1951)]

$$Q_{12}^{(1)} = \frac{4\pi}{k^2} \sum_{l=0,1,\ldots} (l+1)\sin^2(\delta_{l+1}-\delta_l)_{12} \tag{83.3}$$

$$Q_{s(a)}^{(2)} = \frac{8\pi}{k^2} \sum_{\substack{l=0,2,\ldots \\ (l=1,3,\ldots)}} \frac{(l+1)(l+2)}{2l+3}\sin^2(\delta_{l+2}-\delta_l)\,. \tag{83.4}$$

Zur Berechnung der Viscosität von Gemischen und des Thermodiffusionsfaktors wird auch der Querschnitt

$$Q_{12}^{(2)} = \frac{4\pi}{k^2} \sum_{l=0,1,\ldots} \frac{(l+1)(l+2)}{2l+3}\sin^2(\delta_{l+2}-\delta_l)_{12} \tag{83.5}$$

benötigt. Man sieht, daß speziell gilt

$$Q_{12}^{(2)} = \tfrac{1}{2}\left(Q_{12,s}^{(2)} + Q_{12,a}^{(2)}\right). \tag{83.6}$$

Die beiden Querschnitte rechts in (83.6) sind dabei lediglich als Abkürzungen für entsprechende Summen gemäß (83.5), genommen über die geraden bzw. die ungeraden l, zu verstehen; eine eigentliche physikalische Bedeutung haben sie

[1] Die ursprünglichen komplizierteren Formeln von MASSEY und MOHR (1933) lauteten

$$Q_{12}^{(1)} = \frac{4\pi}{k^2}\sum_l \left[(2l+1)\sin^2\delta_l - 2(l+1)\cos(\delta_l-\delta_{l+1})\sin\delta_l\sin\delta_{l+1}\right]_{12}$$

$$Q_{s(a)}^{(2)} = \frac{16\pi}{k^2}\sum_l \left[\frac{2l^3+3l^2-l-1}{(2l-1)(2l+3)}\sin^2\delta_l - \frac{(l+1)(l+2)}{2l+3}\cos(\delta_l-\delta_{l+2})\sin\delta_l\sin\delta_{l+2}\right].$$

Man erhält daraus die Schreibweise von KRAMERS durch Addition der Identitäten

$$\frac{4\pi}{k^2}\sum_l \left[-l\sin^2\delta_l + (l+1)\sin^2\delta_{l+1}\right]_{12} = 0,$$

bzw.

$$\frac{8\pi}{k^2}\sum_l \left[-\frac{l(l-1)}{2l-1}\sin^2\delta_l + \frac{(l+2)(l+1)}{2l+3}\sin^2\delta_{l+2}\right] = 0.$$

Die Summationen sind dabei wie in (83.3) und (83.4) gemeint.

im Fall unterscheidbarer Teilchen nicht. Die Beziehung (83.6) ist nicht selbstverständlich, weil für die differentiellen Streuquerschnitte σ die analoge Relation nicht besteht.

Es seien noch einige Bemerkungen über den klassischen Grenzfall angefügt. Für genügend schwere oder genügend energiereiche Teilchen gehen die quantenmechanischen Querschnitte (83.3) und (83.4) oder (83.5) in die entsprechenden klassischen Querschnitte (50.8) über. Um dies zu sehen, ändern wir (82.11) und (82.12) ab, indem wir statt der vollständigen Reihen zunächst die Summen über $l_0 - l_1 \leq l \leq l_0 + l_1$ nehmen. Zudem sei $l_0 \gg l_1 \gg 1$. (82.11) stellt dann eine Welle dar, die nur innerhalb eines Hohlzylinders um die z-Achse eine merkliche Amplitude hat. Den Radius b des Hohlzylinders erhält man aus der Gegenüberstellung des quantenmechanischen und des klassischen Ausdrucks für den Drehimpuls:

$$l_0 \hbar = b\, m_{ij}\, g, \quad \text{d.h.} \quad b = \frac{l_0}{k}. \tag{83.7}$$

In der Streuamplitude (82.14) werde nun

$$\delta_l = \delta_{l_0} + \delta'_{l_0}(l - l_0) \quad \text{mit} \quad \delta'_{l_0} = \delta_{l_0+1} - \delta_{l_0}$$

gesetzt. Ferner benutzen wir die für große l geltende asymptotische Darstellung der Legendreschen Polynome[1]

$$P_l(\cos\chi) \approx \frac{e^{i\chi l} + e^{-i\chi l}}{\sqrt{l \sin\chi}}.$$

Die Summe (82.14) zerfällt damit in vier Teilsummen, welche über die Exponentialglieder

$$e^{i(2\delta'_{l_0} \pm \chi)l} \quad \text{und} \quad e^{\pm i\chi l}$$

erstreckt werden. Die nicht angeschriebenen Koeffizienten sind langsam mit l veränderlich. Bei der Summation über l zerstören sich daher die Glieder gegenseitig, außer wenn

$$\chi = \mp 2\delta'_{l_0} \quad \text{und} \quad \chi = 0.$$

Das sind die Richtungen, in denen eine merkliche Streuamplitude $a(\chi)$ zu beobachten ist. Die klassische Streurichtung ist also

$$\chi = 2|\delta'_{l_0}| = 2|\delta_{l_0+1} - \delta_{l_0}| = |\delta_{l_0+2} - \delta_{l_0}|. \tag{83.8}$$

Die große Amplitude für $\chi = 0$ geht im Primärstrahl unter. Ersetzt man nun in (83.3) und (83.4) l und δ durch b und χ (bzw. χ_{12}) gemäß (83.7) und (83.8) und geht zum Integral über, so kommt

$$Q_{12}^{(1)} = 4\pi \int_0^\infty \sin^2 \tfrac{1}{2}\chi_{12}\, b\, db,$$

$$Q^{(2)} = 2\pi \int_0^\infty \sin^2 \chi\, b\, db.$$

Das sind die klassischen Formeln, vgl. (50.8). — Eine eingehendere Untersuchung dieses Grenzübergangs haben DE BOER und BIRD (1954) angestellt (s. auch Kap. 10 des Buches von HIRSCHFELDER *et al.* [*25*]). Diese Autoren verwenden die WKB-Methode und erhalten eine bei hohen Temperaturen konvergente Entwicklung der Transportkoeffizienten nach geraden Potenzen der Planckschen Konstanten. Der Einfluß der Statistik ist nach dieser Methode nicht

[1] Vgl. MAGNUS und OBERHETTINGER [*37*], S. 71.

untersucht worden. Man hätte dazu die Auswertung mittels der Eulerschen Summenformel so zu verfeinern, daß sich ein Unterschied ergibt, je nachdem ob man die Summenformel auf $Q_s^{(2)}$ oder $Q_a^{(2)}$ aus (83.4) anwendet.

84. Über Gase, die aus gleichartig zusammengesetzten Partikeln mit verschiedenen inneren Zuständen bestehen. Wir betrachten ein Gas, dessen Moleküle aus den gleichen Elementarteilchen bestehen. Die Temperatur sei so niedrig, daß eine Anregung von höheren Rotations- oder Oszillationszuständen der Moleküle praktisch nicht stattfindet. Trotzdem können die Moleküle verschiedene innere Zustände[1] haben, nämlich dann, wenn sie nicht-kugelsymmetrisch sind und sich daher räumlich verschieden orientieren können oder wenn eine Entartung durch den Kernspin vorliegt.

Die Wechselwirkung ist in solchen Fällen nicht in Strenge kugelsymmetrisch. In der Näherung kugelsymmetrischer Wechselwirkung aber bleibt die Orientierung der inneren Bahndrehimpulse und der Kernspins jedes Moleküls beim Stoß erhalten. In dieser Näherung läßt sich das Gas also auffassen als ein Gemisch einatomiger Komponenten, deren Teilchenzahl je für sich konstant ist. Die Anzahl K der Komponenten des Gemisches ist gleich der Anzahl der möglichen inneren Zustände eines Moleküls. Es gilt also für dieses Gas das Boltzmann-Gleichungssystem (29.2), wobei f_i die Verteilungsfunktion der Moleküle im i-ten inneren Zustand bezeichnet. Für die Stöße zweier Moleküle in gleichen Zuständen ist der Streuquerschnitt σ_s oder σ_a aus (82.16) maßgebend, für zwei Moleküle in verschiedenen Zuständen der Streuquerschnitt σ_u aus (82.17). Es ist demnach in (29.2) zu verwenden

$$\left.\begin{aligned} &\sigma_{ii} = \begin{cases} \sigma_s & \text{bei Bose-Teilchen} \\ \sigma_a & \text{bei Fermi-Teilchen} \end{cases} \\ \text{und}\quad &\sigma_{ij} = \sigma_u = \tfrac{1}{2}(\sigma_s + \sigma_a) \quad \text{wenn} \quad i \neq j. \end{aligned}\right\} \tag{84.1}$$

Nun sei angenommen, daß zur Zeit $t=0$

$$f_i = \frac{1}{K} f \quad \text{wobei} \quad i = 1, \ldots, K \tag{84.2}$$

gilt. Es sollen also anfänglich alle Orientierungen der Kernspins und der inneren Bahndrehimpulse überall gleich häufig sein (isotrope Verteilung). Nach dem Boltzmann-Gleichungssystem gilt dann aber (84.2) auch zu jeder anderen Zeit, die Gln. (29.2) werden alle identisch, die Summation läßt sich ausführen und man erhält eine Boltzmann-Gleichung der Art (28.5) für ein reines Gas. Die darin vorkommende Verteilungsfunktion f beschreibt die Translationsbewegung der Moleküle unter Ignorierung der verschiedenen Orientierungen und zufolge (84.1) tritt darin als resultierender Streuquerschnitt auf

$$\left.\begin{aligned} \bar{\sigma} &= \frac{1}{K}\sigma_{s(a)} + \left(1 - \frac{1}{K}\right)\sigma_u, \\ &= \frac{K+1}{2K}\sigma_{s(a)} + \frac{K-1}{2K}\sigma_{a(s)} \quad \text{für Bose- (Fermi-) Teilchen}. \end{aligned}\right\} \tag{84.3}$$

Damit haben wir in allgemeiner Form den von COHEN *et al.* (1956) am Beispiel des Ortho-Wasserstoffs ausgesprochenen Satz, der angibt, daß und in welcher Weise ein Gemisch der geschilderten Art durch ein fiktives reines Gas ersetzt werden kann.

[1] Über den inneren Zustand s. Ziff. 82, δ.

Es sei betont, daß gleiche Häufigkeit der inneren Zustände wesentlich ist. Ist das nicht der Fall, dann ist die Reduktion auf reines Gas nur teilweise möglich. Gibt es z.B. zwei Klassen von inneren Zuständen (Ortho- und Para-Modifikation) und ist die Anzahl der Moleküle in zwei Zuständen verschiedener Klasse verschieden, so kann das Gas als Gemisch von zwei fiktiven reinen Gasen aufgefaßt werden. Dieses Gemisch kann sich auf ein fiktives reines Gas reduzieren, nämlich dann, wenn alle Zustände unabhängig von ihrer Klassenzugehörigkeit gleich häufig vorkommen. Dies sei als *Iso-Gemisch* bezeichnet. Für dieses Gemisch verschwindet die Thermodiffusion, da alle Zustände gleichwertig sind.

Nun seien die experimentell untersuchten Fälle im einzelnen betrachtet.

1. ^{4}He. Dies ist ein Bose-Teilchen ohne Drehimpuls ($K=1$). Man hat ein in Strenge einatomiges Gas und es ist

$$\bar{\sigma}_{^4\mathrm{He}} = \sigma_s. \tag{84.4}$$

2. ^{3}He. Dies ist ein Fermi-Teilchen mit Kernspin $\frac{1}{2}$ ($K=2$). Nach (84.3) ist daher

$$\bar{\sigma}_{^3\mathrm{He}} = \tfrac{3}{4}\,\sigma_a + \tfrac{1}{4}\,\sigma_s. \tag{84.5}$$

3. H_2. Dies ist ein Bose-Teilchen mit zwei Kernspins $\frac{1}{2}$. Es gibt für das Einzelmolekül drei symmetrische Kernspinfunktionen (Ortho-Modifikation[1]) und eine antisymmetrische (Para-Modifikation). Der Para-Grundzustand hat den inneren Bahndrehimpuls $L=0$ und $K=1$, daher ist

$$\bar{\sigma}_{\mathrm{p-H_2}} = \sigma_s \quad \text{(nicht-rotierend)}. \tag{84.6}$$

Der Ortho-Grundzustand hat $L=1$. Für reinen o-H_2 im Grundzustand ist daher $K=9$ und nach (84.3) gilt

$$\bar{\sigma}_{\mathrm{o-H_2}} = \tfrac{5}{9}\,\sigma_s + \tfrac{4}{9}\,\sigma_a. \tag{84.7}$$

Ein Gemisch von Ortho- und Para-H_2 bei tiefer Temperatur kann als binäres Gemisch aufgefaßt werden, wobei für den Stoß zweier Moleküle verschiedener Klasse der Streuquerschnitt

$$\sigma_{\mathrm{op}} = \tfrac{1}{2}\,(\sigma_s + \sigma_a) \tag{84.8}$$

maßgebend ist. Das Gemisch läßt sich aber wieder auf ein reines Gas reduzieren, wenn für die Molenbrüche speziell gilt

$$\gamma_{\mathrm{o-H_2}} = \tfrac{9}{10}, \quad \gamma_{\mathrm{p-H_2}} = \tfrac{1}{10} \quad \text{(Iso-Gemisch)}. \tag{84.9}$$

Dieses Gemisch ist verschieden vom Normal-Wasserstoff, für welchen $\gamma_{\mathrm{o-H_2}} = \frac{3}{4}$ gilt. Der Streuquerschnitt des Iso-Wasserstoffs ergibt sich aus (84.3) mit $K=10$ zu

$$\bar{\sigma}_{\mathrm{i-H_2}} = \tfrac{11}{20}\,\sigma_s + \tfrac{9}{20}\,\sigma_a. \tag{84.10}$$

4. D_2. Dies ist ein Bose-Teilchen mit zwei Kernspins 1. Es gibt also für das Einzelmolekül sechs symmetrische Kernspinfunktionen (Ortho) und drei antisymmetrische (Para). Der Ortho-Grundzustand hat $L=0$, also $K=6$; der Para-Grundzustand hat $L=1$, also $K=9$. Somit ist nach (84.3)

$$\left.\begin{aligned} \bar{\sigma}_{\mathrm{o-D_2}} &= \tfrac{7}{12}\,\sigma_s + \tfrac{5}{12}\,\sigma_a \quad \text{(nicht-rotierend)},\\ \bar{\sigma}_{\mathrm{p-D_2}} &= \tfrac{5}{9}\,\sigma_s + \tfrac{4}{9}\,\sigma_a. \end{aligned}\right\} \tag{84.11}$$

Für das Gemisch beider Modifikationen wird außerdem (84.8) benötigt. Das Gemisch der Zusammensetzung

$$\gamma_{\mathrm{p-D_2}} = \tfrac{9}{15}, \quad \gamma_{\mathrm{o-D_2}} = \tfrac{6}{15} \quad \text{(Iso-Gemisch)} \tag{84.12}$$

[1] Vgl. A. EUCKEN [*13*], S. 254.

kann aber wieder als reines Gas mit dem Streuquerschnitt

$$\bar{\sigma}_{\mathrm{i-D_2}} = \tfrac{8}{15}\,\sigma_s + \tfrac{7}{15}\,\sigma_a \qquad (84.13)$$

aufgefaßt werden. Für Normal-Deuterium ist $\gamma_{\mathrm{p-D_2}} = \frac{1}{3}$.

Damit sind die allgemeinen Unterlagen für die Anwendungen bereitgestellt, die in den folgenden Ziffern besprochen werden. Abschließend sei noch einmal der Näherungscharakter der vorstehenden Ansätze betont. Für eine umfassende Begründung und Verschärfung hätte man von der Gastheorie für Teilchen mit inneren Freiheitsgraden auszugehen, vgl. Kapitel H, besonders Ziff. 92 und 94.

85. Viscosität und Wärmeleitfähigkeit der beiden Heliumisotope bei tiefen Temperaturen. Das Helium als leichtes einatomiges Gas mit zwei Isotopen von verschiedener Statistik gibt die beste Möglichkeit zu einem Vergleich von Theorie und Experiment, worauf HALPERN (1951) hinwies.

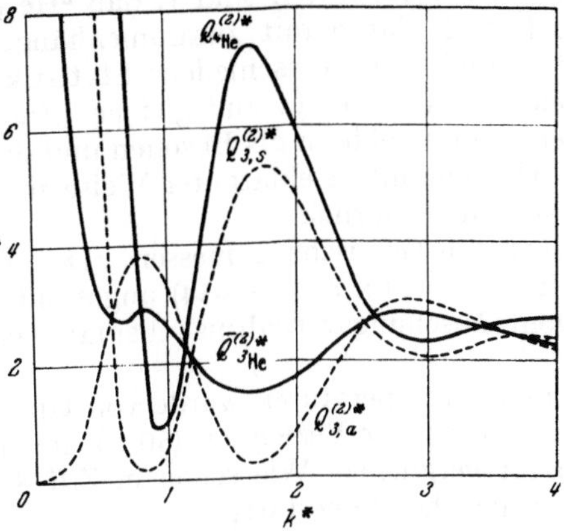

Fig. 49.

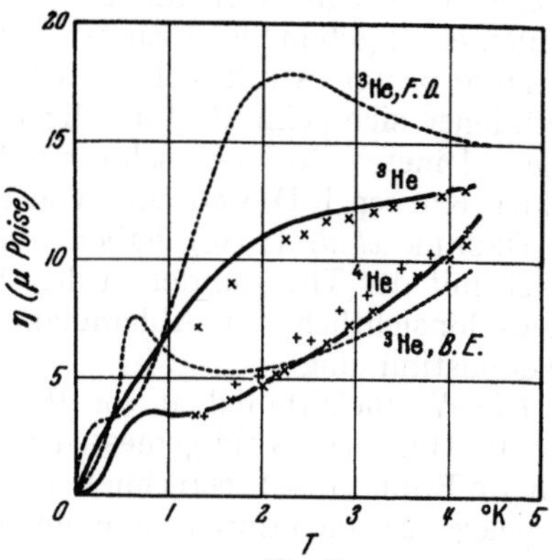

Fig. 50.

Fig. 49. Querschnitte $Q^{(2)*} = Q^{(2)}/d^2$, vgl. (85.1), für ^{4}He und ^{3}He in Abhängigkeit von $k^* = kd$, berechnet für das Lennard-Jones-(6; 12)-Potential. (Parameter s. Tabelle 21, Ziff. 81.) Nach DE BOER und COHEN (1951).

Fig. 50. Ausgezogene Kurven: von DE BOER und COHEN (1951) berechnete Viscosität der Helium-Isotope in Abhängigkeit von der Temperatur. Punktierte Kurven: Viscosität eines fiktiven ^{3}He ohne Kernspin, mit $\Omega^{(2)}_{3,s}$ (B. E.) bzw. $\Omega^{(2)}_{3,a}$ (F. D.) berechnet. × Meßpunkte von BECKER et al. (1954). + Meßpunkte von v. ITTERBEEK et al. (1953). Nach COHEN, OFFERHAUS und DE BOER (1954).

Von der Berechnung der Streuphasen δ_l war in Ziff. 82 die Rede. Mittels der δ_l können nach (83.4) zunächst die Viscositätsquerschnitte $Q^{(2)}_s$ und $Q^{(2)}_a$ berechnet werden. Gemäß (84.4) und (84.5) ist sodann

$$Q^{(2)}_{^4\mathrm{He}} = Q^{(2)}_{4,s}; \qquad \bar{Q}^{(2)}_{^3\mathrm{He}} = \tfrac{3}{4}\,Q^{(2)}_{3,a} + \tfrac{1}{4}\,Q^{(2)}_{3,s}. \qquad (85.1)$$

Zur Verdeutlichung ist die Massenzahl als Index beigefügt.

Für das Isotopengemisch braucht man außerdem noch, s. Gl. (83.3) und (83.5),

$$Q^{(1)}_{34} \quad \text{und} \quad Q^{(2)}_{34} = \tfrac{1}{2}\left(Q^{(2)}_{34,s} + Q^{(2)}_{34,a}\right). \qquad (85.2)$$

Aus diesen Größen findet man gemäß (50.5) die zugehörigen Ω-Integrale und hat dann damit gemäß (83.1) die Viscosität der reinen Gase und den Diffusionskoeffizienten des Isotopengemischs, sowie nach (50.16) auch die Wärmeleitfähigkeit.

Die Viscosität von ^{4}He, ^{3}He und von deren Gemisch ist unter Zugrundelegung des Lennard-Jones-(6; 12)-Potentials von DE BOER (1943), von DE BOER und COHEN (1951) und von COHEN, OFFERHAUS und DE BOER (1954) berechnet worden. BUCKINGHAM und SCRIVEN (1952) haben unter Benutzung des Exponentialpotentials, vgl. (25.6), unabhängig quantitativ fast dieselben Ergebnisse erhalten. Fig. 49 zeigt das charakteristische Oszillieren der quantenmechanischen Querschnitte bei Veränderung der Wellenzahl. Die Arbeit von DE BOER und BIRD (1954) bringt außerdem für ^{4}He einen Vergleich mit dem entsprechenden klassischen Querschnitt; dieser ist bei kleinen k-Werten größer als der quantenmechanische Querschnitt, bei großen k-Werten etwas kleiner.

Fig. 50 zeigt die berechnete Viscosität der Helium-Isotope und vergleicht sie mit den Messungen von E. W. BECKER, MISENTA und SCHMEISSNER (1954)[1] an den getrennten Isotopen; auch die Messungen von V. ITTERBEEK, SCHAPINK, VAN DEN BERG und V. BEEK (1953) an ^{4}He sind mit eingezeichnet. Wie man sieht ist — im Gegensatz zu dem was nach der klassischen Theorie zu erwarten gewesen wäre — die Viscosität für das leichte Isotop unterhalb von 4° K größer, nämlich etwa doppelt so groß wie die des schweren[2].

Die Chapman-Enskogsche Theorie einschließlich der Annahmen über die Statistik der Helium-Atome ist damit auch bei tiefen Temperaturen experimentell bestätigt. Die Gründe für das Verhältnis der Viscositäten sind 1. daß ^{3}He eine größere de Broglie-Wellenlänge als ^{4}He hat, 2. das damit zusammenhängende Fehlen eines gebundenen Zustands bei ^{3}He und 3. die verschiedene Statistik. — Letztere bringt es mit sich, daß bei gleichem Kraftgesetz und gleicher Energie Fermionen sich beim Stoß im allgemeinen weniger nähern als Bosonen und daher einen kleineren Streuquerschnitt haben. Das kommt in eklatanter Weise in den beiden Kurven F.D. und B.E. von Fig. 50 zum Ausdruck.

BECKER *et al.* (1954) haben auch Isotopen-Gemische gemessen. Der Vergleich mit der Theorie, der in der Arbeit von COHEN *et al.* (1954) angestellt ist, ergab hinsichtlich der Abhängigkeit vom Mischungsverhältnis ebenfalls volle Übereinstimmung.

Die Wärmeleitfähigkeit von ^{4}He bei tiefen Temperaturen wurde von UBBINK und DE HAAS (1943b)[3] gemessen und auf Grund der Beziehung (50.16) mit der von DE BOER (1943) berechneten Viscosität verglichen. Wie schon in Ziff. 52, β dargelegt, ist auch durch diese Messungen die Theorie bestätigt.

86. Viscosität und Wärmeleitfähigkeit von Wasserstoff und Deuterium im Ortho- und Parazustand bei tiefen Temperaturen. Auch für Wasserstoff und Deuterium ist, wie Tabelle 20 und 21 aus Ziff. 81 zeigen, die de Broglie-Wellenlänge bei tiefen Temperaturen erheblich größer als der Moleküldurchmesser. Dies äußert sich besonders im Verhältnis der Viscositäten beider Isotope, wovon schon in Ziff. 4, γ die Rede war. Hier soll als weiterer interessanter Punkt für beide Gase der Viscositätsunterschied der Ortho- und Paramodifikation behandelt werden. Dabei bestehen, tiefe Temperaturen vorausgesetzt, keine Bedenken gegen die Anwendung der Theorie einatomiger binärer Gemische mit kugelsymmetrischen Potentialen für die Wechselwirkung (oo), (pp) und (op). Es sind ja nur Moleküle mit den inneren Rotationsquantenzahlen $L=0$ und 1 vorhanden, zwischen denen kein Übergang möglich ist, und auch die räumliche Orientierung der Moleküle mit $L=1$ wird im Mittel isotrop sein. Es ist aber zunächst anzunehmen, daß alle drei genannten Potentiale irgendwie verschieden sind[4].

Im folgenden sei aber der Unterschied der Potentiale vernachlässigt. Dann sind nach Ziff. 84 die grundlegenden Streuquerschnitte jeweils auf zwei Querschnitte σ_s und σ_a zurückzuführen. Natürlich sind σ_s wie auch σ_a für Wasserstoff

[1] Wegen der Meßmethode s. Ziff. 8.

[2] E. W. BECKER und MISENTA (1955) haben die Viscosität von ^{3}He auch bei 14 bis 20° K gemessen. In diesem Bereich ist bereits $\eta_{^4\mathrm{He}} > \eta_{^3\mathrm{He}}$.

[3] Wegen der Meßmethode s. Ziff. 10, β.

[4] Die Verschiedenheit wird durch die Nicht-Kugelsymmetrie der Elektronenwolke der orientierten Moleküle hervorgerufen. Die relative Molekülaufweitung durch die Rotation beträgt nach FALK und MANN (1955) nur 1 ‰ pro Rotationsquantenzahl und ist zu vernachlässigen.

und Deuterium wegen der Verschiedenheit der de Broglie-Wellenlängen verschieden. Die benötigten Streuquerschnitte $\bar{\sigma}_o, \bar{\sigma}_p$ und σ_{op} wurden schon in (84.6) bis (84.8) und (84.11) notiert. Ein etwaiger Unterschied der Viscosität des Ortho- und Para-Gases geht nach dieser Vorstellung, da ja die de Broglie-Wellenlängen für beide Modifikationen jeweils exakt gleich sind, allein auf die Verschiedenheit der Symmetrisierungsvorschrift für die beiden Modifikationen zurück.

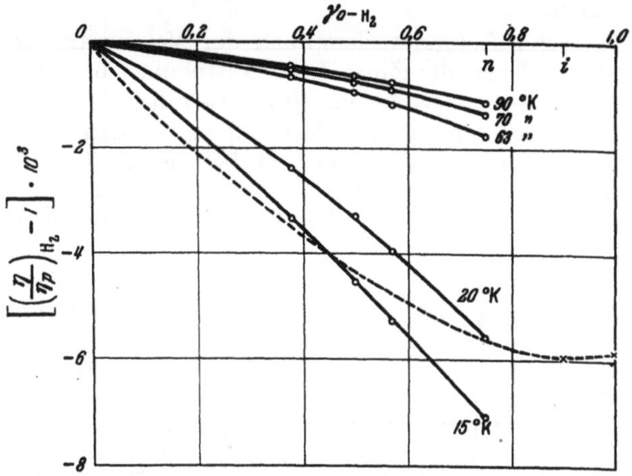

Fig. 51. Relative Differenz der Viscosität von H_2 gegen die von p-H_2 in Abhängigkeit vom Molenbruch $\gamma_{o\text{-}H_2}$ bei verschiedenen Temperaturen. Ausgezogen: Meßkurven von BECKER und STEHL (1952a). Gestrichelt: halbempirische Isotherme für 20° K, s. Text zu Gl. (86.3) *n*: normales, *i*: Iso-Gemisch, s. (84.9).

Die Experimente[1] von E. W. BECKER und STEHL (1952a) an H_2 und von BECKER, MISENTA und STEHL (1953) an D_2 ergaben in der Tat bei tiefen Temperaturen einen Unterschied der Viscositäten, und zwar hat in beiden Fällen die nichtrotierende Modifikation ($L = 0$) die größere Viscosität. Bei 20° K ist der relative Viscositätsunterschied des Normal-Gemisches gegen die Modifikation mit $L = 0$ beim H_2 etwa $-6\,^0/_{00}$, beim D_2 etwa $-0{,}6\,^0/_{00}$ (siehe Fig. 51 und 52)[2].

Es soll zunächst festgestellt werden, was dieses experimentelle Ergebnis über die Größen $\Omega_s^{(2,2)}$ und $\Omega_a^{(2,2)}$ aussagt, mittels deren sich die Viscosität der reinen Ortho- und Para-Gase darstellen läßt. Indem man die aus (84.6), (84.7) und (84.11) folgenden analogen Beziehungen für die Ω-Integrale in die Viscositätsformel (83.1) einsetzt, erhält man

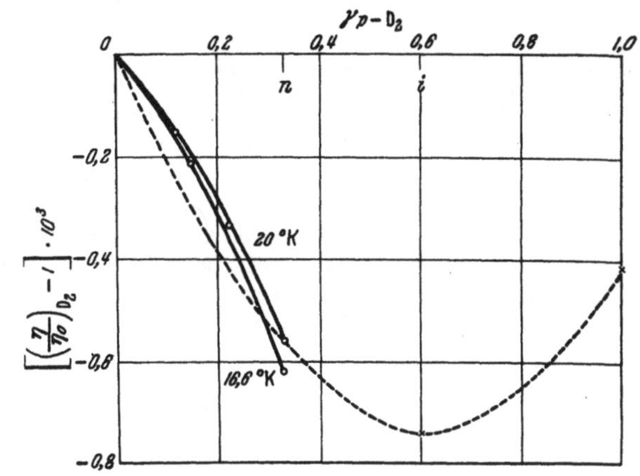

Fig. 52. Relative Differenz der Viscosität von D_2 gegen die von o-D_2 in Abhängigkeit vom Molenbruch $\gamma_{p\text{-}D_2}$ bei verschiedenen Temperaturen. Ausgezogen: Meßkurven von BECKER, MISENTA und STEHL (1953). Gestrichelt: wie bei Fig. 51. *n*: normales, *i*: Iso-Gemisch, s. (84.12).

$$\left(\frac{\eta_o}{\eta_p}\right)_{H_2} = \frac{\Omega_s^{(2,2)}}{\frac{5}{9}\Omega_s^{(2,2)} + \frac{4}{9}\Omega_a^{(2,2)}};$$

$$\left(\frac{\eta_p}{\eta_o}\right)_{D_2} = \frac{\frac{7}{12}\Omega_s^{(2,2)} + \frac{5}{12}\Omega_a^{(2,2)}}{\frac{5}{9}\Omega_s^{(2,2)} + \frac{4}{9}\Omega_a^{(2,2)}}.$$

Da sich Ω_s und Ω_a wenig unterscheiden, folgt daraus

$$\left(\frac{\eta_o}{\eta_p}\right)_{H_2} - 1 = \frac{4}{9}\,\frac{\Omega_s^{(2,2)} - \Omega_a^{(2,2)}}{\Omega_a^{(2,2)}}; \qquad \left(\frac{\eta_p}{\eta_o}\right)_{D_2} - 1 = \frac{1}{36}\,\frac{\Omega_s^{(2,2)} - \Omega_a^{(2,2)}}{\Omega_a^{(2,2)}}. \tag{86.1}$$

[1] Wegen der Meßmethode s. Ziff. 6.

[2] Kürzlich wurde auch die Viskosität von HD bis zu 14° K herab gemessen von RIETVELD und VAN ITTERBEEK (1957).

Der Vergleich mit dem Experiment verlangt also, daß für beide Isotopen

$$\Omega_a^{(2,2)} > \Omega_s^{(2,2)} \tag{86.2}$$

sein muß.

Dieser Befund widersprach der Voraussage von HALPERN und GWATHMEY (1937), denen die ganze Problemstellung zu verdanken ist, die aber mit harten Kugeln gerechnet hatten. Für diese gilt $\Omega_a^{(2,2)} < \Omega_s^{(2,2)}$ und so ergab sich zunächst theoretisch ein falsches Vorzeichen für den Viscositätsunterschied. Aber das Modell der harten Kugeln ist bei tiefen Temperaturen schlecht. Tatsächlich finden COHEN, OFFERHAUS, VAN LEEUWEN, ROOS und DE BOER (1955, 1956) für ein Lennard-Jonessches (6; 12)-Potential mit den Konstanten von DE BOER und MICHELS (1938) (s. Tabelle 21 in Ziff. 81) das richtige Vorzeichen[1] und die richtige Größenordnung des Effekts (s. Fig. 53). Damit dürfte der Effekt in seinen wesentlichen Zügen verstanden sein.

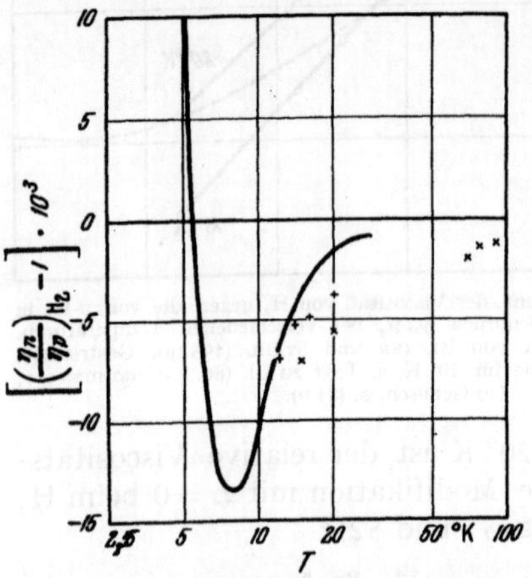

Fig. 53. Relative Viscositätsdifferenz für Normal- und Parawasserstoff in Abhängigkeit von der Temperatur. Ausgezogene Kurve: theoretisch nach COHEN et al. (1956). × Meßpunkte von BECKER und STEHL (1952a); + unveröffentlichte Messungen von v. ITTERBEEK und COREMANS. Aus COHEN, OFFERHAUS, VAN LEUWEN, ROOS und DE BROER (1956).

Schon früher hatte MIYAKO (1942) das Problem unter Benutzung desselben Lennard Jones-Potentials behandelt, aber für 20° K einen 20mal zu großen Effekt von entgegengesetztem Vorzeichen gefunden. Das lag wohl daran, daß die Miyakoschen Streuphasen doch noch nicht genau oder nicht zahlreich genug waren. Die kleine relative Differenz der Integrale Ω_s und Ω_a erfordert sehr hohe Präzision der Ω-Werte selbst[2]. MIYAKO hatte auch einen etwa 20%igen Unterschied der zweiten Virialkoeffizienten von n- und p-H_2 gefunden, im Gegensatz zu den Experimenten von KL. SCHÄFER (1937) und den Berechnungen von COHEN *et al.* (1956), die beide keinen Unterschied ergaben innerhalb 1‰.

In der Arbeit von COHEN *et al.* (1956) werden ferner die Gemische der beiden Wasserstoffmodifikationen ausführlich behandelt. Dabei bestätigte sich eine Erkenntnis, die schon FALK und MANN (1955) hatten. Diese gehen aus von der geschilderten einfachen Chapman-Enskogschen Theorie mit einem einzigen Wechselwirkungspotential. Sie finden, daß die Isothermen der Viscositätsdifferenz theoretisch die entgegengesetzte Krümmung haben wie nach den Messungen von E. W. BECKER und STEHL, wenn vorausgesetzt wird, daß für die Viscositätsdifferenz selbst theoretisch und experimentell das Vorzeichen übereinstimmt.

Die Feststellung von FALK und MANN ist rasch folgendermaßen einzusehen. Man kann ohne Benutzung der etwas komplizierten Viscositätsformel für binäre Gemische die Viscosität für das in Ziff. 84 definierte Iso-Gemisch angeben. Denn dieses Gemisch läßt sich gemäß (84.10) bzw. (84.13) als fiktives reines Gas betrachten. So kann man seine Viscosität, ähnlich wie diejenige des reinen Ortho- und Para-Gases, in einfacher Weise durch $\Omega_s^{(2,2)}$ und $\Omega_a^{(2,2)}$ darstellen. Man findet

[1] Das ist besonders bemerkenswert im Hinblick auf die Ergebnisse für ^{3}He. Für dieses ist, wie man aus Ziff. 85, Fig. 50 entnehmen kann, nämlich von 2 bis 4° K $\Omega_a^{(2,2)}$ etwa 50% kleiner als $\Omega_s^{(2,2)}$.

[2] Die Rechnungen der Amsterdamer Gruppe lieferten diese Differenz mit einer Genauigkeit von ungefähr 5% (freundliche briefliche Mitteilung von Prof. J. DE BOER). — Daß HALPERN und GWATHMEY wie auch MIYAKO für die Gemische nicht die konsequenten Formeln der Chapman-Enskogschen Theorie benutzten, ist für die Vorzeichenfrage ohne Belang.

sofort mit Benutzung auch von (86.1)

$$\left.\begin{aligned} \left(\frac{\eta_i}{\eta_p}\right)_{H_2} - 1 &= \frac{9}{20}\,\frac{\Omega_s^{(2,2)} - \Omega_a^{(2,2)}}{\Omega_a^{(2,2)}} = \frac{81}{80}\left[\left(\frac{\eta_o}{\eta_p}\right)_{H_2} - 1\right] \\ \left(\frac{\eta_i}{\eta_o}\right)_{D_2} - 1 &= \frac{1}{20}\,\frac{\Omega_s^{(2,2)} - \Omega_a^{(2,2)}}{\Omega_a^{(2,2)}} = \frac{9}{5}\left[\left(\frac{\eta_p}{\eta_o}\right)_{D_2} - 1\right]. \end{aligned}\right\} \tag{86.3}$$

Falls (86.2) erfüllt ist, gilt demnach

$$\eta_{i-H_2} < \eta_{o-H_2} < \eta_{p-H_2}; \qquad \eta_{i-D_2} < \eta_{p-D_2} < \eta_{o-D_2}. \tag{86.4}$$

Das bedeutet, daß die theoretischen Viscositäts-Isothermen in Fig. 51 und 52 ein Minimum, also positive Krümmung haben[1], während die Krümmung der experimentellen Isothermen negativ ist. In den Figuren ist als gestrichelte Kurve freihändig je eine halbempirische Isotherme eingezeichnet, die dem experimentellen η-Wert bei 20° K für das Normal-Gemisch angepaßt ist und die das Verhältnis der Viscosität des indifferenten Gemischs zu der von reinem o—H_2 bzw. reinem p—D_2 gemäß (86.3) richtig wiedergibt. Wegen der genauen theoretischen Kurven für H_2 sei auf die Arbeit von COHEN *et al.* (1956) verwiesen. Für D_2 ist die vollständige Theorie nicht durchgeführt.

Die hinsichtlich der Isothermen-Krümmung bestehende Diskrepanz könnte von seiten der Theorie nur dadurch weiter analysiert werden, daß man die Nicht-Kugelsymmetrie der Wasserstoff- bzw. Deuterium-Moleküle berücksichtigt, auf die eingangs hingewiesen wurde. Danach hat man es mit mehreren Potentialen zu tun. Ansätze in dieser Richtung sind schon von FALK und MANN (1955) gemacht worden.

Zum Schluß sei noch kurz die Wärmeleitfähigkeit betrachtet. Da bei tiefer Temperatur beide Modifikationen die spezifische Wärme des einatomigen Gases haben, ist für die (fiktiven) reinen Gase, also für das Ortho- und Para-Gas sowie deren indifferentes Gemisch, die Wärmeleitfähigkeit einfach proportional der Viscosität. Somit lassen sich die voranstehenden theoretischen Betrachtungen insoweit ohne weiteres auf die Wärmeleitfähigkeit übertragen. Man hat insbesondere dieselben relativen Differenzen für die reine Ortho- und Para-Modifikation zu erwarten. Die bisher vorliegenden Experimente von UBBINK (1948) und von WALDMANN und BECKER (1948) ergaben allerdings keinen meßbaren Unterschied. Den letztgenannten Autoren zufolge sollte der relative Unterschied in λ für Para- und Normal-Wasserstoff kleiner als $1^0/_{00}$ sein. Doch war möglicherweise die Meßgenauigkeit überschätzt worden. Weitere Versuche wären erwünscht.

87. Thermodiffusion der leichten Gase bei tiefen Temperaturen. Wir besprechen die Thermodiffusion der beiden Heliumisotope und die von Ortho- gegen Parawasserstoff.

α) *Helium-Isotope.* Der Thermodiffusionsfaktor für Isotope von kleinem relativen Massenunterschied ist durch (74.8) oder (74.9) gegeben. Die dortigen Größen $A^\star$, $B^\star$, $C^\star$, welche Abkürzungen für die Verhältnisse von Ω-Integralen darstellen, s. (69.10), sind für den Stoß $^4He/^3He$ von BUCKINGHAM und SCRIVEN (1952) sowie von COHEN, OFFERHAUS und DE BOER (1954) unter Zugrundelegung der Quantenmechanik berechnet worden[2]. Es ergab sich $A^\star \approx B^\star \approx 1{,}1$ bei 4° K, ähnlich wie nach der klassischen Theorie (vgl. Fig. 33 in Ziff. 75). Die Größe $6C^\star - 5$ aber ist gegen den klassischen Wert stark verändert, wie Fig. 54 zeigt. Ihr Betrag ist bedeutend größer und der Vorzeichenwechsel, klassisch bei $T^\star =$

[1] Das Minimum liegt vermutlich beim Iso-Gemisch.
[2] COHEN *et al.* geben $A = \frac{2}{5} A^\star$, $B = \frac{3}{5} B^\star$ und $C = \frac{6}{5} C^\star$ an.

0,93, liegt bei $T^* = 0{,}08$. Mit diesem C^*-Wert erhält man sodann aus (74.8) (s. COHEN *et al.*)

$$\alpha_{\text{theor.}}\,({}^4\text{He}/{}^3\text{He}) = \begin{cases} 0{,}098 \\ 0{,}044 \end{cases} \text{bei} \begin{matrix} 4^\circ\,\text{K} \\ 1^\circ\,\text{K}. \end{matrix} \tag{87.1}$$

Dies sind ziemlich große, gut meßbare Werte, die eine Anreicherung von ^{4}He an der kalten Seite bedeuten. Bisher wurden am Helium bei tiefen Temperaturen keine Thermodiffusionsversuche gemacht.

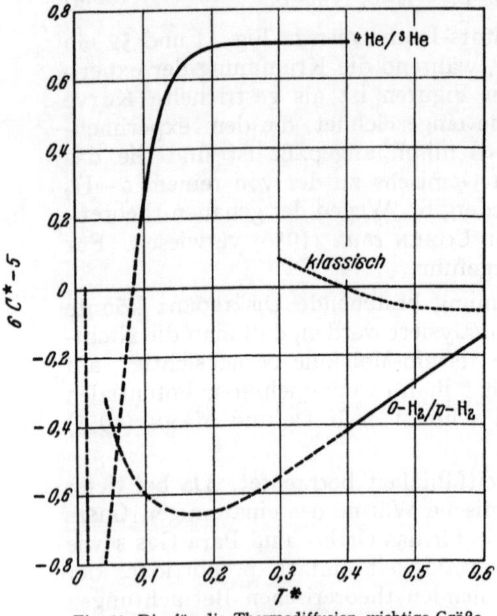

Fig. 54. Die für die Thermodiffusion wichtige Größe $6\,C^* - 5$, s. (74.8) und (74.6), in Abhängigkeit von $T^* = \mathrm{k}\,T/\varepsilon$ für Helium und Wasserstoff mit Lennard-Jones-(6; 12)-Potential. Benutzte Parameter siehe in Tabelle 21 (Ziff. 81). Ausgezogen: experimentell zugänglicher Bereich ($T > 1^\circ$ K für He, $> 15^\circ$ K für H_2). Nach COHEN, OFFERHAUS und DE BOER (1954) und nach COHEN, OFFERHAUS, VAN LEEUWEN, ROOS und DE BOER (1956).

β) Ortho- und Parawasserstoff[1]. Nach den Betrachtungen von Ziff. 84 kann Wasserstoff bei genügend tiefer Temperatur ($T < 40^\circ$ K) als binäres Gemisch aufgefaßt werden. Dessen Komponenten sind isobar. Daher ist Formel (74.6) für den Thermodiffusionsfaktor zuständig.

Es sei $o-H_2$ als Gas 1, $p-H_2$ als Gas 2 bezeichnet. Für die effektiven Moleküldurchmesser s aus (74.4) findet man mittels (84.6) bis (84.8)

$$\frac{s_1^2}{s_{12}^2} - 1 = \frac{1}{18}\,\frac{\Omega_s^{(2,2)} - \Omega_a^{(2,2)}}{\Omega_a^{(2,2)}},$$

$$\frac{s_2^2}{s_{12}^2} - 1 = \frac{1}{2}\,\frac{\Omega_s^{(2,2)} - \Omega_a^{(2,2)}}{\Omega_a^{(2,2)}}.$$

Durch Einsetzen in (74.6) ergibt sich mit Benutzung von (86.1)

$$\left.\begin{aligned} \alpha = \frac{25(6C^* - 5)}{2(55 + 16A^* - 12B^*)} \times \\ \times\left[1 - \left(\frac{\eta_o}{\eta_p}\right)_{H_2}\right]\left(\frac{9}{10} - \gamma_{o-H_2}\right), \end{aligned}\right\} \tag{87.2}$$

vgl. WALDMANN (1946b). Wie zu erwarten, ist also $\alpha = 0$ für das Iso-Gemisch $\gamma_{o-H_2} = \frac{9}{10}$, bei dem die neun Ortho- und die eine Parakomponete gleichhäufig sind, siehe Ziff. 84. Ein Gemisch mit noch höherem $o-H_2$-Gehalt sollte durch Thermodiffusion bei tiefer Temperatur nicht hergestellt werden können.

Aus Fig. 51 (Ziff. 86) und aus Fig. 54 entnimmt man $1 - \left(\frac{\eta_o}{\eta_p}\right)_{H_2} \approx 7 \cdot 10^{-3}$, $6C^* - 5 \approx -0{,}45$ für $T = 15^\circ$ K. Nach COHEN *et al.* (1956) ist ferner $A^* \approx B^* \approx 1$. Somit hat man

$$\alpha_{\text{theoret}}\,(o-H_2/p-H_2,\ 15^\circ\,\text{K}) \approx -0{,}7 \cdot 10^{-3}\left(\tfrac{9}{10} - \gamma_{o-H_2}\right). \tag{87.3}$$

Dieser Wert ist etwa hundertmal kleiner als der für Helium. Das ist verständlich, weil die Thermodiffusion im Fall des Wasserstoffs ein reiner Quanteneffekt ist — Massen und Kräfte sind gleich — während beim Helium ein beträchtlicher Massenunterschied besteht. — Messungen der Diffusionswärme durch WALDMANN und E. W. BECKER (1948) ergaben, daß jedenfalls $|\alpha(H_2, 20^\circ\,\text{K})| < 0{,}01$ ist.

Bei höheren Temperaturen haben BECKER, HENKES und SEIFERT (1955) die Thermodiffusion von Ortho- gegen Parawasserstoff mit dem Trennrohr

[1] Wegen der Diffusion von $p-H_2$ gegen $n-H_2$ siehe Ziff. 72.

untersucht. Sie finden $\alpha = 1{,}1 \cdot 10^{-3}$ für 90° K und $\gamma_{\mathrm{O-H_2}} = 0{,}75$. Der Orthowasserstoff geht an die kalte Seite. Die Formel (87.2) mit dem experimentellen Wert für den Zähigkeitsunterschied aus Fig. 51 (Ziff. 86) und mit dem zu $T^* = 2{,}43$ gehörigen C^*-Wert aus Fig. 33 (Ziff. 75) würde $\alpha = 10^{-5}$ ergeben, also zwar das richtige Vorzeichen, aber viel zu kleinen Betrag. Doch ist bei 90° K die Anwendung der Formel (87.2) nicht statthaft, da das Para-Niveau $L = 2$ schon merklich angeregt ist. Das äußert sich in dem großen relativen Unterschied der Wärmeleitfähigkeiten von $1 - \left(\frac{\lambda_{\mathrm{O}}}{\lambda_{\mathrm{P}}}\right)_{\mathrm{H_2}} \approx 0{,}14$, welcher überdies mit der Temperatur rasch veränderlich ist. Einsetzen der λ in (87.2) anstelle der η ergibt $\alpha \approx 5 \cdot 10^{-3}$, also einen zu großen Wert. Doch wäre auch dieses Vorgehen unbegründet. Die Theorie von α in diesem Gebiet, welche die Übergangsmöglichkeiten zwischen den Niveaus $L = 0$ und 2 berücksichtigen müßte, ist nicht entwickelt. Auch die in (87.2) behauptete spezielle Abhängigkeit der Thermodiffusion vom Molenbruch dürfte bei höherer Temperatur entfallen und statt dessen die Möglichkeit der Reindarstellung von Orthowasserstoff bestehen. Auf Grund der Versuche von BECKER *et al.* läßt sich darüber aber nichts aussagen, da die Autoren nur Normal-Wasserstoff $\gamma_{\mathrm{O-H_2}} = 0{,}75$ untersuchten.

H. Mehratomige Gase.

Im Abschnitt I dieses Kapitels wird die allgemeine Theorie auf quantenmechanischer Grundlage behandelt. Die hauptsächlichen Ergebnisse sind die Boltzmann-Gleichung für ein Gas aus rotierenden Molekülen und das zugehörige H-Theorem (Ziff. 92 und 93). Im Abschnitt II werden spezielle Anwendungen besprochen.

I. Zur allgemeinen Theorie.

88. Diskrete Zustände. Die Gleichungen des unelastischen Stoßes. Es sei ein mehratomiges Molekül im Ruhsystem seines Schwerpunkts betrachtet. Der (innere) Drehimpuls des Moleküls werde durch die ganz- oder halbzahlige Quantenzahl J charakterisiert; seine Einstellung (Orientierung) in bezug auf eine beliebig wählbare, raumfeste z-Richtung sei durch die Quantenzahl M gekennzeichnet. Bei gegebenen J kann M die $2J+1$ Werte

$$M = -J, -J+1, \ldots, J-1, J \tag{88.1}$$

annehmen. Wenn kein äußeres Feld vorhanden ist — was im folgenden stets angenommen wird — haben diese $2J+1$ Zustände die gleiche Energie (Richtungsentartung). Zur Festlegung des Molekülzustands werden im allgemeinen weitere innere Quantenzahlen nötig sein, die wir durch V symbolisieren (Vibrationsquantenzahl; jedoch ist darin unter Umständen auch eine weitere „drehinvariante" Kreisel-Drehimpulsquantenzahl enthalten). Die zugehörigen inneren Orts- und die Spinkoordinaten sollen durch X zusammengefaßt werden. Für die inneren Eigenfunktionen des ruhenden Moleküls schreiben wir demgemäß

$$\Psi = \Psi_{V,J,M}(X) = \Psi_I(X), \quad \text{orthonormiert.} \tag{88.2}$$

Die Zusammenfassung der sämtlichen inneren Molekülquantenzahlen zu dem einen Index I dient zur Kürze. Die innere Energie des Moleküls

$$E_I = E_{V,J} \tag{88.3}$$

hängt, entsprechend der Richtungsentartung, nicht vom Index M ab. Bei Wahl einer anderen z-Richtung transformieren sich die $2J+1$ Funktionen (88.2), welche zu einem bestimmten Satz der „drehinvarianten" Quantenzahlen V, J gehören und durch den Index M unterschieden sind, in bekannter Weise linear in sich, s. z. B. Wigner [55], S. 180.

Nun sollen zwei chemisch gleichartige, aber in verschiedenen inneren Zuständen I, I_1 befindliche Moleküle mit (Schwerpunkts-) Geschwindigkeiten $\boldsymbol{c}$, $\boldsymbol{c}_1$ zusammenstoßen. Die Quantenzahlen und Geschwindigkeiten nach dem im allgemeinen unelastischen Stoß seien durch einen Strich gekennzeichnet. Mit den Relativgeschwindigkeiten $\boldsymbol{g} = \boldsymbol{c} - \boldsymbol{c}_1$, $\boldsymbol{g}' = \boldsymbol{c}' - \boldsymbol{c}_1'$ lautet dann der Energiesatz im gemeinsamen Schwerpunktssystem der beiden Moleküle (reduzierte Masse!)

$$E = \frac{m_{\text{red}}}{2} g^2 + E_I + E_{I_1} = \frac{m_{\text{red}}}{2} g'^2 + E_{I'} + E_{I_1'}. \tag{88.4}$$

Als Verallgemeinerung der Beziehungen (22.9) und (22.10) für den elastischen Stoß hat man demnach jetzt die *Stoßgleichungen*

$$\boldsymbol{c}'_{(1)} = \boldsymbol{c}_S \,{}_{(\pm)}\, \tfrac{1}{2} g' \boldsymbol{e}' \tag{88.5}$$

und umgekehrt

$$\boldsymbol{c}_{(1)} = \boldsymbol{c}_S \,{}_{(\pm)}\, \tfrac{1}{2} g \boldsymbol{e}. \tag{88.5'}$$

Dabei bedeuten $\boldsymbol{e}$, $\boldsymbol{e}'$ wieder Einheitsvektoren in den Richtungen von $\boldsymbol{g}$, $\boldsymbol{g}'$; $\boldsymbol{c}_S$ bezeichnet die Schwerpunktsgeschwindigkeit [s. Gl. (22.3)]. Für die Beträge der Relativgeschwindigkeiten ist nach dem Energiesatz eingesetzt zu denken

$$\left.\begin{aligned} &g' = g_{I' I_1'}(E) = \sqrt{\frac{2}{m_{\text{red}}}(E - E_{I'} - E_{I_1'})} \quad \text{bzw.} \quad g = g_{I I_1}(E) \\ &\text{oder auch} \\ &g' = \sqrt{g^2 + \delta} \quad \text{bzw.} \quad g = \sqrt{g'^2 - \delta} \quad \text{mit} \quad \delta = \frac{2}{m_{\text{red}}}(E_I + E_{I_1} - E_{I'} - E_{I_1'}). \end{aligned}\right\} \tag{88.6}$$

Welche dieser Beziehungen man zu benutzen hat, hängt von der Wahl der unabhängigen Variablen, E oder g, ab. Statt g und g' werden wir häufig die Wellenzahlen

$$k = \frac{m_{\text{red}}}{\hbar} g = \frac{m_{\text{red}}}{\hbar} g_{I I_1}(E) = k_{I I_1}(E) \quad \text{und} \quad k' = \frac{m_{\text{red}}}{\hbar} g' = k_{I' I_1'}(E) \tag{88.7}$$

verwenden.

In Ziff. 91 werden wir das Lorentzsche Gas, s. auch Ziff. 62, behandeln. Beim Stoß gegen einen unendlich schweren, ruhenden Streuer, welcher keine Energie aufnehmen kann, lautet der Energiesatz

$$E = \frac{m}{2} c^2 + E_I = \frac{m}{2} c'^2 + E_{I'}. \tag{88.8}$$

Die Spezialisierung von (88.5) lautet

$$\boldsymbol{c}' = c' \boldsymbol{e}' \quad \text{mit} \quad c' = c_{I'}(E) = \sqrt{\frac{2}{m}(E - E_{I'})}. \tag{88.9}$$

Analoges gilt für die Geschwindigkeit $\boldsymbol{c}$. Statt c, c' werden auch die Wellenzahlen

$$k = \frac{m}{\hbar} c = \frac{m}{\hbar} c_I(E) = k_I(E), \quad k' = \frac{m}{\hbar} c' = k_{I'}(E) \tag{88.10}$$

benutzt.

Schließlich haben wir noch den *Liouvilleschen Satz* zu notieren. Nach der Energiegleichung (88.4) ist für gegebene Quantenzahlen I, I_1, I', I_1'

$$d g^2 = d g'^2.$$

Somit gilt nach (22.21)

$$d\boldsymbol{e}'\, d\boldsymbol{c}\, d\boldsymbol{c}_1 / g = d\boldsymbol{e}\, d\boldsymbol{c}'\, d\boldsymbol{c}_1' / g'. \tag{88.11}$$

Dieser Liouvillesche Satz für den allgemeinen unelastischen Stoß wird häufig bei Integralumformungen benötigt. Man hat

$$d\boldsymbol{e}'\,d\boldsymbol{c}/c = d\boldsymbol{e}\,d\boldsymbol{c}'/c' \tag{88.12}$$

als Spezialisierung für das Lorentz-Gas.

89. Die quantenmechanische Streuamplitude. Das Schattentheorem. Bewegungsumkehr. *α) Streuung von Kugelwellen. Die S-Matrix*[1]. Es seien zwei Moleküle mit inneren Zuständen I, I_1, vgl. (88.2), betrachtet, welche sich im gemeinsamen Schwerpunktssystem als einlaufende Kugelwelle mit den Bahndrehimpuls-Quantenzahlen l, m und der Gesamtenergie E nähern. In der auslaufenden Welle werden dann auch beliebige andere Bahndrehimpulse l', m' und beliebige andere innere Zustände I', I_1' vertreten sein.

Wir führen die asymptotischen aus- bzw. einlaufenden Kugelwellen ein

$$\varphi_\beta^{(\pm)} = \Psi_{I I_1}(X, X_1)\,\frac{e^{\pm i k_{I I_1} r'}}{\sqrt{k_{I I_1}}\, r'}\, Y_{lm}(\pm \boldsymbol{e}'), \tag{89.1}$$

wobei abkürzend

$$\Psi_{I I_1}(X, X_1) = \Psi_I(X)\,\Psi_{I_1}(X_1) \tag{89.2}$$

geschrieben ist. Hierin bezeichnet $\boldsymbol{r}' = r'\boldsymbol{e}'$ den Ortsvektor vom einen Molekülschwerpunkt zum anderen. Wegen $k_{I I_1}(E)$ sei auf (88.7) verwiesen. Y_{lm} bezeichnet die orthonormierten Kugelflächenfunktionen:

$$\int Y_{lm}^*(\boldsymbol{e}')\,Y_{l'm'}(\boldsymbol{e}')\,d\boldsymbol{e}' = \delta_{lm,\,l'm'} \tag{89.3}$$

mit konventioneller Phasenwahl, derart daß die Relation

$$(-1)^{l+m}\,Y_{l-m}^*(\boldsymbol{e}') = (-1)^l\,Y_{lm}(\boldsymbol{e}') = Y_{lm}(-\boldsymbol{e}') \tag{89.4}$$

statthat. Die Zusammenfassung der Indices gemäß

$$\beta = (I I_1 l m) \tag{89.5}$$

dient der Kürze.

Der Index β der einlaufenden Welle $\varphi_\beta^{(-)}$ kann nun also beliebig vorgegeben werden. Die Gesamtwelle stellt sich dann asymptotisch, für $r' \to \infty$, so dar:

$$\varphi_\beta = \varphi_\beta^{(-)} - \sum_{\beta'} \varphi_{\beta'}^{(+)}\,S_{\beta'\beta}(E) = \varphi_\beta^{(-)} - \varphi_\beta^{(+)} - \sum_{\beta'} \varphi_{\beta'}^{(+)}\,(S_{\beta'\beta} - \delta_{\beta'\beta}). \tag{89.6}$$

Die hierin vorkommende Streumatrix S ist durch die Wechselwirkung der beiden Moleküle bestimmt. Ist sie bekannt, so lassen sich alle den Stoß betreffenden physikalischen Fragen beantworten. Ohne Wechselwirkung gilt $S_{\beta'\beta} = \delta_{\beta'\beta}$ oder in Matrixschreibweise: $\mathsf{S} = 1$. In der Quantenmechanik wird allgemein gezeigt, daß bei orthonormierten Ψ_I die S-Matrix unitär ist:

$$\sum_\beta S_{\beta'\beta}\,S_{\beta''\beta}^* = \sum_\beta S_{\beta\beta'}^*\,S_{\beta\beta''} = \delta_{\beta'\beta''} \tag{89.7}$$

oder mit der Adjungierten

$$S_{\beta\beta'}^\dagger = S_{\beta'\beta}^* \tag{89.8}$$

auch kurz in Matrixschreibweise

$$\mathsf{S}\,\mathsf{S}^\dagger = \mathsf{S}^\dagger\,\mathsf{S} = 1\,. \tag{89.9}$$

Die Unitarität von S gewährleistet die Teilchenerhaltung beim Stoß. Die Drehimpulserhaltung zieht weitere allgemeine Eigenschaften von S nach sich. Da wir

[1] Quantenmechanische Matrizen (Operatoren) werden in Teil H dieses Artikels, wenn sie ohne Indices geschrieben sind, zwecks Betonung ihres Charakters durch Blockschrift bezeichnet. Beispiele: S, a, $\mathsf{f}(\boldsymbol{c})$, O.

von diesen im folgenden keinen Gebrauch machen, können sie unerörtert bleiben. Näheres darüber findet sich in einer Arbeit von TAKAYANAGI (1954).

Ausführungen über die Berechnung der Streumatrix mehratomiger Moleküle können hier nicht gemacht werden. Wegen der Kompliziertheit des Problems gibt es bisher nur wenige Arbeiten darüber. Auch sind bisher — außer den „beladenen Kugeln" (loaded spheres) von JEANS [27], S. 92 und 192, und den klassischen rauhen Kugeln von PIDDUCK (1922), (s. CHAPMAN und COWLING [4], Kap. 11) keine weiteren, halbwegs einfachen, klassischen Modellvorstellungen für unelastische Stöße entwickelt worden. Die klassische Mechanik des Stoßes starrer, konvexer, nichtkugelförmiger Körper und die zugehörige Gaskinetik behandelten CURTISS (1956) sowie CURTISS und MUCKENFUSS (1957). Mit der quantenmechanischen Beschreibung mehratomiger Moleküle, insbesondere der Abtrennung der Rotationsbewegung, befaßten sich CURTISS, HIRSCHFELDER und ADLER (1950) sowie CURTISS (1953a). Die Streuung eines Atoms an einem zweiatomigen Molekül behandelten CURTISS und ADLER (1952) und entwickelten ein Störungsverfahren, welches gut ist bei Stößen niedriger Energie und wenn die elastischen Stöße viel häufiger sind als die unelastischen. CURTISS (1953b) hat diese Betrachtungen auf den Stoß von zwei zweiatomigen Molekülen übertragen. TAKAYANAGI und KANEKO (1954) haben den Stoß eines Sauerstoffmoleküls mit einem Heliumatom theoretisch untersucht; TAKAYANAGI und OHNO (1955) haben mit der quantenmechanischen Behandlung des Stoßes von Wasserstoff-Molekülen gegeneinander begonnen. HORNIG und HIRSCHFELDER (1956) haben diskutiert, inwieweit der Begriff der zwischenmolekularen Kräfte bei unelastischen Stößen noch anwendbar ist.

β) Streuung von ebenen Wellen. Die Streuamplituden-Matrix $\mathfrak{a}(\boldsymbol{e}', \boldsymbol{e})$. Die Rayleighsche Zerlegung der ebenen Welle mit dem Wellenvektor $\boldsymbol{k} = k\boldsymbol{e}$ in Kugelwellen mit der Phasenwahl (89.4) lautet asymptotisch bekanntlich

$$e^{i\boldsymbol{k}\boldsymbol{r}'} \approx \frac{2\pi i}{k r'} \sum_{lm} \left[e^{-ikr'} Y_{lm}(-\boldsymbol{e}') - e^{ikr'} Y_{lm}(\boldsymbol{e}')\right] Y^*_{lm}(\boldsymbol{e}).$$

Für zwei Moleküle, welche die inneren Zustände I, I_1 haben und sich im gemeinsamen Schwerpunktssystem ohne Wechselwirkung als ebene Welle mit der Einfallsrichtung $\boldsymbol{e}$ und mit der Gesamtenergie E bewegen, lautet die Rayleighsche Zerlegung demnach asymptotisch

$$\psi^{(0)}_{II_1} = \Psi_{II_1} e^{ik_{II_1}\boldsymbol{e}\boldsymbol{r}'} \approx \frac{2\pi i}{\sqrt{k_{II_1}}} \sum_{lm} \left(\varphi^{(-)} - \varphi^{(+)}\right)_{II_1 lm} Y^*_{lm}(\boldsymbol{e}). \tag{89.10}$$

Die zugehörige Welle mit Wechselwirkung entsteht dadurch, daß man die Kugelwellen mit Wechselwirkung (89.6) mittels der aus (89.10) ersichtlichen Koeffizienten $(2\pi i/\sqrt{k_{II_1}})\, Y^*_{lm}(\boldsymbol{e})$ linear kombiniert. So bekommt man

$$\psi_{II_1} = \Psi_{II_1} e^{ik_{II_1}\boldsymbol{e}\boldsymbol{r}'} + \sum_{I'I'_1} \Psi_{I'I'_1} a_{I'I'_1, II_1}(\boldsymbol{e}', \boldsymbol{e}, E) \frac{e^{ik_{I'I'_1}r'}}{\sqrt{k_{II_1} k_{I'I'_1}}\, r'}, \tag{89.11}$$

wobei

$$a_{I'I'_1, II_1}(\boldsymbol{e}', \boldsymbol{e}, E) = \frac{2\pi}{i} \sum_{\substack{lm \\ l'm'}} Y_{l'm'}(\boldsymbol{e}') \left(S_{\beta'\beta}(E) - \delta_{\beta'\beta}\right) Y^*_{lm}(\boldsymbol{e}). \tag{89.12}$$

Diese Lösung des Streuproblems ist dadurch gekennzeichnet, daß sie sich für $r' \to \infty$ wie eine in der Richtung $\boldsymbol{e}$ fortschreitende ebene Welle verhält, welche von einer auslaufenden Kugelwelle mit der winkelabhängigen, dimensionslosen Streuamplituden-Matrix $\mathfrak{a}$ begleitet ist[1].

Für die Berechnung der Erwartungswerte irgendwelcher beobachtbaren Größen benötigt man auch die zu (89.11) gehörige komplex-konjugierte Welle:

$$\psi^*_{II_1} = e^{-ik_{II_1}\boldsymbol{e}\boldsymbol{r}'} \Psi^*_{II_1} + \sum_{I'I'_1} \frac{e^{-ik_{I'I'_1}r'}}{\sqrt{k_{II_1} k_{I'I'_1}}\, r'} a^\dagger_{II_1, I'I'_1}(\boldsymbol{e}, \boldsymbol{e}', E)\, \Psi^*_{I'I'_1}. \tag{89.13}$$

[1] Der Ansatz (89.11) ist nicht ganz allgemein. Er ignoriert die Nichtunterscheidbarkeit der Bausteine (Elektronen, Nukleonen) der beiden Moleküle und die Möglichkeit von deren Austausch beim Stoß. Auf die entsprechende Erweiterung des Ansatzes muß hier verzichtet werden.

Dabei ist die adjungierte Streuamplitude $\mathfrak{a}^\dagger$ gemäß

$$\left.\begin{aligned} a^\dagger_{I I_1, I' I'_1}(\boldsymbol{e}, \boldsymbol{e}', E) &= a_{I' I'_1, I I_1}(\boldsymbol{e}', \boldsymbol{e}, E)^* \\ &= -\frac{2\pi}{i} \sum_{\substack{l\,m \\ l'\,m'}} Y_{lm}(\boldsymbol{e}) \left(S^\dagger_{\beta\beta'}(E) - \delta_{\beta\beta'}\right) Y^*_{l'm'}(\boldsymbol{e}') \end{aligned}\right\} \quad (89.14)$$

definiert.

γ) Das Schattentheorem. Als Schattentheorem — auch optisches Theorem genannt — werde der bekannte Zusammenhang zwischen dem über alle Streuwinkel gemittelten Absolutquadrat der Streuamplitude und dem Imaginärteil der Streuamplitude in der Vorwärtsrichtung bezeichnet. Dieses Theorem läßt sich sogleich an Hand von (89.12) und (89.14) ableiten.

Durch Einsetzen dieser Darstellungen von $\mathfrak{a}$ und $\mathfrak{a}^\dagger$ und Benutzung der Orthonormierungsrelation (89.3) der Kugelflächenfunktionen sowie der Unitarität (89.9) der **S**-Matrix erhält man

$$\left.\begin{aligned} &\int \sum_{I' I'_1} a^\dagger_{I I_1, I' I'_1}(\boldsymbol{e}, \boldsymbol{e}', E)\, a_{I' I'_1, I'' I''_1}(\boldsymbol{e}', \boldsymbol{e}'', E)\, d\boldsymbol{e}' \\ &\quad = \int \sum_{I' I'_1} a_{I I_1, I' I'_1}(\boldsymbol{e}, \boldsymbol{e}', E)\, a^\dagger_{I' I'_1, I'' I''_1}(\boldsymbol{e}', \boldsymbol{e}'', E)\, d\boldsymbol{e}' \\ &\quad = \frac{2\pi}{i} \left[a(\boldsymbol{e}, \boldsymbol{e}'', E) - a^\dagger(\boldsymbol{e}, \boldsymbol{e}'', E)\right]_{I I_1, I'' I''_1}. \end{aligned}\right\} \quad (89.15)$$

In Matrix-Kurzschrift lauten diese Beziehungen

$$\left.\begin{aligned} &\int \mathfrak{a}^\dagger(\boldsymbol{e}, \boldsymbol{e}', E)\, \mathfrak{a}(\boldsymbol{e}', \boldsymbol{e}'', E)\, d\boldsymbol{e}' = \int \mathfrak{a}(\boldsymbol{e}, \boldsymbol{e}', E)\, \mathfrak{a}^\dagger(\boldsymbol{e}', \boldsymbol{e}'', E)\, d\boldsymbol{e}' \\ &\quad = \frac{2\pi}{i} \left[\mathfrak{a}(\boldsymbol{e}, \boldsymbol{e}'', E) - \mathfrak{a}^\dagger(\boldsymbol{e}, \boldsymbol{e}'', E)\right]. \end{aligned}\right\} \quad (89.15')$$

Dies gilt im allgemeinen Fall des unelastischen Stoßes rotierender Moleküle.

Speziell für die beiderseitigen Diagonalelemente $I'', I''_1 = I, I_1, \boldsymbol{e}'' = \boldsymbol{e}$ von (89.15) hat man

$$\left.\begin{aligned} &\frac{1}{4\pi} \int \sum_{I' I'_1} |a_{I' I'_1, I I_1}(\boldsymbol{e}', \boldsymbol{e}, E)|^2\, d\boldsymbol{e}' = \frac{1}{4\pi} \int \sum_{I' I'_1} |a_{I I_1, I' I'_1}(\boldsymbol{e}, \boldsymbol{e}', E)|^2\, d\boldsymbol{e}' \\ &\quad = \frac{1}{2i} \left[a(\boldsymbol{e}, \boldsymbol{e}, E) - a^*(\boldsymbol{e}, \boldsymbol{e}, E)\right]_{I I_1, I I_1}. \end{aligned}\right\} \quad (89.16)$$

Dies heißt mit Worten, daß der über alle Ablenkwinkel gemittelte Gesamt-Streuquerschnitt gleich ist dem Imaginärteil der Streuamplitude in der Vorwärtsrichtung.

Dieses Theorem gilt unter sehr allgemeinen Voraussetzungen, vgl. WICK (1949) und LAX (1950). Statt die Unitarität der **S**-Matrix heranzuziehen, kann man auch direkt von der Kontinuitätsgleichung ausgehen, vgl. FEENBERG (1932) oder GLAUBER und SCHOMAKER (1953).

*δ) Invarianz des Schattentheorems gegen eine **c**-abhängige unitäre Transformation der Drehimpulsfunktionen Ψ_I.* Weil es für den Beweis des H-Theorems (Ziff. 93) benötigt wird, soll hier noch untersucht werden, wie sich die Streuamplitude und das Schattentheorem verhalten beim Übergang zu neuen Basis-Funktionen $\breve{\Psi}_I$. Diese sollen mit den ursprünglichen Funktionen (88.2) durch eine Transformation $\mathsf{U}(\boldsymbol{c})$ verknüpft sein gemäß

$$\breve{\Psi}_I = \sum_{I'} \Psi_{I'}\, U_{I' I}(\boldsymbol{c}), \qquad \breve{\Psi}_I^* = \sum_{I'} U^\dagger_{I I'}(\boldsymbol{c})\, \Psi^*_{I'} \quad (89.17)$$

oder kurz in Matrix-Schreibweise

$$\breve{\Psi} = \Psi\, \mathsf{U}(\boldsymbol{c}), \qquad \breve{\Psi}^* = \mathsf{U}^\dagger(\boldsymbol{c})\, \Psi^*. \quad 89.18)$$

Die Transformation soll unitär sein, die Umkehrung also lauten

$$\Psi = \check{\Psi}\,\mathsf{U}^\dagger(\boldsymbol{c}), \quad \Psi^* = \mathsf{U}(\boldsymbol{c})\,\check{\Psi}^*. \tag{89.19}$$

Die Abhängigkeit von $\boldsymbol{c}$ bedeutet, daß die Transformation verschieden gewählt werden kann je nach der Linear-Geschwindigkeit des Moleküls. Jedoch soll die Transformation stets nur Zustände gleicher innerer Energie [s. Gl. (88.3)] miteinander verknüpfen, also nur auf die magnetische Quantenzahl M wirken:

$$U_{II'} = \delta_{VJ,V'J'}\,U^{(V,J)}_{MM'}. \tag{89.20}$$

Nun sei ein Molekül-Paar betrachtet mit den Lineargeschwindigkeiten $\boldsymbol{c}, \boldsymbol{c}_1$. Hier gilt entsprechend zu (89.17), unter Benutzung der Abkürzung (89.2),

$$\check{\Psi}_{II_1} = \sum_{I'I_1'} \Psi_{I'I_1'}\,U_{I'I}(\boldsymbol{c})\,U_{I_1'I_1}(\boldsymbol{c}_1). \tag{89.21}$$

Wir schreiben kürzer

$$U_{II'}(\boldsymbol{c})\,U_{I_1I_1'}(\boldsymbol{c}_1) = U_{II_1,I'I_1'}(\boldsymbol{c},\boldsymbol{c}_1) \tag{89.22}$$

oder in Matrix-Schreibweise [direktes Produkt, s. z.B. Wigner [55], S. 19]

$$\mathsf{U}(\boldsymbol{c}) \times \mathsf{U}(\boldsymbol{c}_1) = \mathsf{U}(\boldsymbol{c},\boldsymbol{c}_1). \tag{89.22'}$$

Gleiches gilt dann für die adjungierten Matrizen $\mathsf{U}^\dagger$. Damit lauten (89.21) und die weiteren aus (89.18) und (89.19) hervorgehenden Beziehungen in Matrix-Schreibweise einfach

$$\check{\Psi} = \Psi\,\mathsf{U}(\boldsymbol{c},\boldsymbol{c}_1), \qquad \check{\Psi}^* = \mathsf{U}^\dagger(\boldsymbol{c},\boldsymbol{c}_1)\,\Psi^*, \tag{89.23}$$

$$\Psi = \check{\Psi}\,\mathsf{U}^\dagger(\boldsymbol{c},\boldsymbol{c}_1), \qquad \Psi^* = \mathsf{U}(\boldsymbol{c},\boldsymbol{c}_1)\,\check{\Psi}^*. \tag{89.24}$$

Um die transformierte Streuamplitude $\check{\mathfrak{a}}$ zu finden, hat man den Stoß eines mit dem inneren Zustand $\check{\Psi}$ einfallenden Molekülpaares zu betrachten. Man erhält nach (89.23) die zugehörige Wellenfunktion, indem man (89.11) von rechts mit $\mathsf{U}(\boldsymbol{c},\boldsymbol{c}_1)$ multipliziert und gleichzeitig in der Streuwelle — dem zweiten Term rechts in (89.11) — gemäß (89.24) Ψ durch $\check{\Psi}$ substituiert. Somit behält (89.11) seine Form bei, nämlich

$$\check{\psi}_{II_1} = \check{\Psi}_{II_1}\,e^{ik_{II_1}\boldsymbol{e}\boldsymbol{r}} + \sum_{I'I_1'} \check{\Psi}_{I'I_1'}\,\check{a}_{I'I_1',II_1}(\boldsymbol{c}',\boldsymbol{c}_1'\,|\,\boldsymbol{c},\boldsymbol{c}_1)\,\frac{e^{ik_{I'I_1'}r}}{\sqrt{k_{II_1}k_{I'I_1'}}\,r}, \tag{89.25}$$

wenn man als transformierte Amplitude

$$\check{\mathfrak{a}}(\boldsymbol{c}',\boldsymbol{c}_1'\,|\,\boldsymbol{c},\boldsymbol{c}_1) = \mathsf{U}^\dagger(\boldsymbol{c}',\boldsymbol{c}_1')\,\mathfrak{a}(\boldsymbol{e}',\boldsymbol{e},E)\,\mathsf{U}(\boldsymbol{c},\boldsymbol{c}_1) \tag{89.26}$$

einführt. Analog ist (89.13) forminvariant, wenn die adjungierte Matrix $\mathfrak{a}^\dagger$ gemäß

$$\check{\mathfrak{a}}^\dagger(\boldsymbol{c},\boldsymbol{c}_1\,|\,\boldsymbol{c}',\boldsymbol{c}_1') = \mathsf{U}^\dagger(\boldsymbol{c},\boldsymbol{c}_1)\,\mathfrak{a}^\dagger(\boldsymbol{e},\boldsymbol{e}',E)\,\mathsf{U}(\boldsymbol{c}',\boldsymbol{c}_1'), \tag{89.27}$$

also in derselben Weise wie $\mathfrak{a}$ transformiert wird. Die Geschwindigkeiten $\boldsymbol{c},\ldots,\boldsymbol{c}_1'$ sollen dabei jeweils zu den Richtungen $\boldsymbol{e}, \boldsymbol{e}'$ und der Energie E passend gewählt werden nach Maßgabe der Indices V, J bzw. V', J' an $\mathfrak{a}$ oder $\mathfrak{a}^\dagger$, vgl. (89.20).

Die Forminvarianz des Schattentheorems (89.15') ergibt sich nun ohne weiteres. Man hat lediglich diese Beziehung von links mit $\mathsf{U}^\dagger(\boldsymbol{c},\boldsymbol{c}_1)$, von rechts mit $\mathsf{U}(\boldsymbol{c}'',\boldsymbol{c}_1'')$ zu multiplizieren und in den Integralen zwischen den Faktoren $\mathfrak{a}$ und $\mathfrak{a}^\dagger$ die Einheit gemäß

$$1 = \mathsf{U}(\boldsymbol{c}',\boldsymbol{c}_1')\,\mathsf{U}^\dagger(\boldsymbol{c}',\boldsymbol{c}_1')$$

zu ersetzen. So bekommt man aus (89.15′)

$$\left.\begin{aligned}\int \breve{a}^\dagger(\boldsymbol{c}, \boldsymbol{c}_1 | \boldsymbol{c}', \boldsymbol{c}_1')\, \breve{a}(\boldsymbol{c}', \boldsymbol{c}_1' | \boldsymbol{c}'', \boldsymbol{c}_1'')\, d\boldsymbol{e}' \\ = \int \breve{a}(\boldsymbol{c}, \boldsymbol{c}_1 | \boldsymbol{c}', \boldsymbol{c}_1')\, \breve{a}^\dagger(\boldsymbol{c}', \boldsymbol{c}_1' | \boldsymbol{c}'', \boldsymbol{c}_1'')\, d\boldsymbol{e}' \\ = \frac{2\pi}{i}\left[\breve{a}(\boldsymbol{c}, \boldsymbol{c}_1 | \boldsymbol{c}'', \boldsymbol{c}_1'') - \breve{a}^\dagger(\boldsymbol{c}, \boldsymbol{c}_1 | \boldsymbol{c}'', \boldsymbol{c}_1'')\right].\end{aligned}\right\} \tag{89.28}$$

Damit ist die Invarianz des Schattentheorems gegenüber der geschwindigkeitsabhängigen unitären Transformation (89.17) gezeigt.

ε) Die Symmetrierelation der Bewegungsumkehrbarkeit. Bei punktförmigen Teilchen ohne innere Freiheitsgrade und ohne äußeres Magnetfeld ergibt sich aus der Umkehrbarkeit der Bewegung[1] (Invarianz der Bewegungsgleichungen gegen Zeitumkehr) die folgende Symmetrie der Streuamplitude

$$a(\boldsymbol{e}, \boldsymbol{e}', E) = a(-\boldsymbol{e}', -\boldsymbol{e}, E). \tag{89.29}$$

Um dies abzuleiten, bemerkt man zunächst, daß im vorliegenden Fall der Schrödingersche Hamilton-Operator reell ist, $H^* = H$, und daß daher, falls ψ einen stationären Zustand der Energie E bedeutet, auch ψ^* einen solchen darstellt. Dies wendet man auf die Welle (89.6) an. Durch geeignete Linearkombination und Benutzung der Unitarität der S-Matrix erhält man daraus zunächst die Welle

$$\hat{\varphi}_\beta = -(\varphi\, \mathsf{S}^\dagger)_\beta = \varphi_\beta^{(+)} - \sum_{\beta'} S^*_{\beta\beta'}\, \varphi_{\beta'}^{(-)}.$$

Nach dem Gesagten ist dann auch

$$\hat{\varphi}^*_\beta = \varphi_\beta^{(+)*} - \sum_{\beta'} S_{\beta\beta'}\, \varphi_{\beta'}^{(-)*}$$

eine mögliche Welle. Nun gilt aber nach der Relation (89.4) für die aus- bzw. einlaufenden Kugelwellen (89.1)

$$(-1)^{l+m}\, \varphi_{-\beta}^{(\pm)*} = \varphi_\beta^{(\mp)}, \quad \text{wo } -\beta = (l, -m). \tag{89.30}$$

Demnach ist

$$(-1)^{l+m}\, \hat{\varphi}^*_{-\beta} = \varphi_\beta^{(-)} - \sum_{\beta'} (-1)^{l+m-l'-m'}\, S_{-\beta-\beta'}\, \varphi_{\beta'}^{(+)}.$$

Dies stimmt hinsichtlich der einlaufenden Welle mit der ursprünglichen Lösung (89.6) überein, muß also mit ihr identisch sein. Daher gilt für die Streumatrix

$$(-1)^{l+m-l'-m'}\, S_{-\beta-\beta'} = S_{\beta'\beta}. \tag{89.31}$$

Verwendung dieser Beziehung in der Darstellung (89.12) der Streuamplitude a liefert sodann die eingangs erwähnte Symmetrie (89.29).

Diese Betrachtungen sind auf Teilchen mit innerem Drehimpuls, den wir mit $\hbar\, \vec{\mathsf{s}}$ bezeichnen, zu übertragen. Wir benutzen $s_z = -J, \ldots, J$ als Spinkoordinate des Teilchens, womit die Variable X aus (88.2) spezifiziert ist. Aus der zugehörigen Darstellung der Spin-Matrizen[2] sieht man, daß für diese gilt

$$\vec{\mathsf{s}}^* = -\,\mathsf{u}^{-1}\, \vec{\mathsf{s}}\, \mathsf{u},$$

wo der unitäre Operator u durch

$$\mathsf{u}\, \Psi(s_z) = \mathrm{e}^{-i\pi s_z}\, \Psi(-s_z) \tag{89.32}$$

gegeben ist [s. WIGNER (1932)]. Für den Bahndrehimpuls $\vec{\mathsf{l}}$ der Relativbewegung der Teilchen, der mit u vertauschbar ist, gilt ebenfalls

$$\vec{\mathsf{l}}^* = -\,\vec{\mathsf{l}} = -\,\mathsf{u}^{-1}\, \vec{\mathsf{l}}\, \mathsf{u}$$

und somit für den Hamilton-Operator[3] bei Spin-Bahnkopplung proportional $\vec{\mathsf{l}}\,\vec{\mathsf{s}}$

$$\mathsf{H}^* = \mathsf{u}^{-1}\, \mathsf{H}\, \mathsf{u}.$$

---

[1] Diesen Ausdruck hat LÜDERS (1952) geprägt.

[2] Zum Beispiel M. BORN und P. JORDAN, Elementare Quantenmechanik, Berlin 1930, S. 142.

[3] Der konjugierte Operator H*, welcher sich auf eine ganz bestimmte Darstellung bezieht, ist nicht zu verwechseln mit dem adjungierten. H ist natürlich selbst-adjungiert (hermitisch). — Von einem äußeren Magnetfeld haben wir wieder abgesehen.

Die Zeitumkehrbarkeit besagt somit, daß bei Spinteilchen zugleich mit ψ auch $\mathsf{u}\,\psi^*$ eine stationäre Lösung der Schrödinger-Gleichung ist.

Dies werde nun zunächst auf die Eigenfunktionen des ruhenden Moleküls angewandt und liefert bei geeigneter Wahl der Phasen den Zusammenhang

$$e^{i\pi M}\,\mathsf{u}\,\Psi^*_{-I} = \Psi_I, \quad \text{wo} \quad -I = (V, J, -M)\,. \tag{89.33}$$

Die erste Relation (89.4) für die Kugelflächenfunktion kann als Spezialfall davon angesehen werden. Für die Kugelwellen aus (89.1) ergibt sich so als Verallgemeinerung der Beziehungen (89.30)

$$(-1)^{l+m}\,e^{i\pi(M+M_1)}\,\mathsf{u}\,\mathsf{u}_1\,\varphi^{(\pm)*}_{-\beta} = \varphi^{(\mp)}_{\beta}, \quad \text{wo} \quad -\beta = (-I, -I_1, l, -m)\,. \tag{89.34}$$

Der auf die Spinkoordinate s_{z1} des Partner-Teilchens wirkende unitäre Operator ist sinngemäß mit u_1 bezeichnet.

Die weiteren Überlegungen verlaufen jetzt formal ganz ähnlich wie bei spinlosen Teilchen. Anstelle von (89.31) ergibt sich

$$e^{i\pi(l+m+M+M_1-l'-m'-M'-M'_1)}\,S_{-\beta-\beta'} = S_{\beta'\beta}\,. \tag{89.35}$$

Für die Streumatrix $\mathfrak{a}$ aus (89.12) erhält man anstelle von (89.29) in dem jetzigen allgemeineren Fall sogleich die Symmetrie-Relation

$$a_{II_1,I'I'_1}(\boldsymbol{e}, \boldsymbol{e}', E) = e^{i\pi(M+M_1-M'-M'_1)}\,a_{-I'-I'_1,-I-I_1}(-\boldsymbol{e}', -\boldsymbol{e}, E)\,. \tag{89.36}$$

Wir werden sie in Ziff. 95 benutzen. — Diese Zusammenhänge gelten sehr allgemein, auch wenn neue Teilchen beim Stoß entstehen [s. COESTER (1951) und LÜDERS (1952)].

90. Beschreibung des Gases durch die Verteilungsmatrix $\mathsf{f}(t, \boldsymbol{r}, \boldsymbol{c})$. Zur statistischen Beschreibung eines verdünnten einatomigen Gases dient die in Ziff. 28 eingeführte Verteilungsfunktion. Bei mehratomigen Gasen tritt an deren Stelle eine Verteilungsmatrix $\mathsf{f}(t, \boldsymbol{r}, \boldsymbol{c})$. Eng hängen mit dieser der statistische Operator und die Dichtematrix zusammen, vgl. v. NEUMANN [*40*], S. 157 oder TOLMAN [*51*], § 78.

Um das Zustandekommen und die Bedeutung der f-Matrix zu erläutern, sei der Einfachheit alber ein Gas betrachtet, dessen sämtliche Moleküle stets die gleichen inneren Quantenzahlen V und J besitzen, also nach (88.3) einheitliche innere Energie haben und sich nur durch die Orientierung ihres Eigendrehimpulses — zugehörige Quantenzahl M, vgl. (88.1) — unterscheiden. Ein beliebig herausgegriffenes Molekül werde durch ein räumlich begrenztes, auf eins normiertes Wellenpaket beschrieben gemäß

$$\psi = \sum_M \Psi_M\,\alpha_M \cdot e^{ik\boldsymbol{e}\boldsymbol{r}}\,A(\boldsymbol{r}, t) \qquad \text{mit} \qquad \iint \psi^*\psi\,d\boldsymbol{r}\,dX = 1\,. \tag{90.1}$$

Der Amplitudenfaktor A, welcher die Gestalt des mit der Gruppengeschwindigkeit $\boldsymbol{c}$ fortschreitenden Wellenpakets angeben soll, sei gemäß

$$\int A^* A\,d\boldsymbol{r} = 1 \tag{90.2}$$

normiert. Die Amplituden α_M, mit welchen die orthonormierten Zustände Ψ_M, vgl. (88.2), am tatsächlichen inneren Zustand des betrachteten Moleküls — aufgefaßt als reiner Fall — beteiligt sind, müssen dann ihrerseits die Normierungsbedingung

$$\sum_M \alpha^*_M\,\alpha_M = 1 \tag{90.3}$$

erfüllen.

Nun fragen wir nach dem Erwartungswert irgendeines auf den Index M wirkenden Operators $\mathsf{O}(\boldsymbol{r}, \boldsymbol{c})$[1], z.B. einer Drehimpulskomponente, zunächst für das Einzelmolekül, sodann für das Gas. Wir stellen O gemäß

$$\mathsf{O}\,\Psi_M = \sum_{M'} \Psi_{M'}\,O_{M'M} \tag{90.4}$$

[1] Die Variablen $\boldsymbol{r}$ und $\boldsymbol{c}$ sollen — als klassische Parameter (s. die folgende Anmerkung) — eventuell in O eingehen (z.B. beim Bahndrehimpuls).

durch eine Matrix dar. Für ein Molekül im Zustand (90.1), dessen Ort $\boldsymbol{r}$ und dessen Geschwindigkeit $\boldsymbol{c}$ gut definiert sind[1], hat man daher für den Erwartungswert von O

$$\langle \mathsf{O} \rangle = \iint \psi^* \mathsf{O} \psi \, d\boldsymbol{r} \, dX = \sum_{M,N} \alpha_N^* O_{NM} \alpha_M . \tag{90.5}$$

Zwecks Erzielung kurzer und invarianter Schreibweise ist es angebracht, diese Formel in die Matrizenschreibweise zu setzen. Dazu führt man gemäß

$$\alpha_M \alpha_N^* = P_{MN} \tag{90.6}$$

den (Projektions-) Operator P ein. Für P gilt nach (90.3)

$$\sum_M P_{MM} = \mathrm{Sp}\,[\mathsf{P}] = 1 . \tag{90.7}$$

Das im folgenden oft verwendete Symbol Sp bezeichnet die Spurbildung. Damit lautet (90.5) kurz so:

$$\langle \mathsf{O} \rangle = \sum_{M,N} O_{NM} P_{MN} = \mathrm{Sp}\,[\mathsf{O}\,\mathsf{P}] = \mathrm{Sp}\,[\mathsf{P}\,\mathsf{O}] . \tag{90.8}$$

Um den Übergang zum Gas zu vollziehen, greifen wir zunächst die in einer Volumeinheit enthaltenen Moleküle mit Geschwindigkeiten $\boldsymbol{c}$ in $d\boldsymbol{c}$ heraus, unterscheiden sie durch den oberen Index (i) und definieren die Verteilungsmatrix f gemäß

$$f_{MN}(t, \boldsymbol{r}, \boldsymbol{c}) \, d\boldsymbol{c} = \sum_i (\alpha_M \alpha_N^*)^{(i)} \tag{90.9}$$

oder kürzer

$$\mathsf{f}(t, \boldsymbol{r}, \boldsymbol{c}) \, d\boldsymbol{c} = \sum_i \mathsf{P}^{(i)} . \tag{90.10}$$

f hängt auch vom Ort und von der Zeit ab, da die herausgegriffenen Moleküle ständig wechseln. Wir bezeichnen den lokalen, gemittelten Erwartungswert von O pro Molekül mit $\overline{O}$ und haben nach (90.5) und (90.9) in summa über alle n Moleküle der Volumeinheit

$$n\overline{O}\,(t, \boldsymbol{r}) = \sum_{d\boldsymbol{c}} \sum_i \langle \mathsf{O} \rangle^{(i)} = \sum_{M,N} \int O_{NM}(\boldsymbol{c}) \, f_{MN}(t, \boldsymbol{r}, \boldsymbol{c}) \, d\boldsymbol{c} . \tag{90.11}$$

In Matrix-Kurzschrift lautet diese Beziehung

$$\overline{O} = \frac{1}{n} \int \mathrm{Sp}\,[\mathsf{O}\,\mathsf{f}] \, d\boldsymbol{c} . \tag{90.12}$$

Wenn f bekannt ist, so können mittels dieser Beziehung alle interessierenden physikalischen Größen, z.B. Dichten und Ströme von Impuls, Drehimpuls usw. berechnet werden. Speziell entnimmt man mit $\mathsf{O} = 1$ für die Teilchenkonzentration

$$n = \sum_M \int f_{MM} \, d\boldsymbol{c} = \int \mathrm{Sp}\,[\mathsf{f}] \, d\boldsymbol{c} . \tag{90.13}$$

Die f-Matrix ist hermitisch [s. (90.9)] und ihrem Wesen nach positiv-semidefinit. Das heißt, daß ihre $2J+1$ Eigenwerte immer und überall reell und nichtnegativ sind. Gleiches gilt, bei beliebiger Wahl der Basisfunktionen Ψ_M, für die Diagonalelemente f_{MM}. Die weiteren Betrachtungen in den beiden nächsten Ziffern zielen nun darauf ab, etwas über die zeitliche Änderung der f-Matrix zu erfahren.

---

[1] Die Unschärferelation zwischen $\boldsymbol{r}$ und $\boldsymbol{c}$ wird also ignoriert. An dieser Stelle behalten die Betrachtungen gewisse Züge der klassischen Mechanik. Bei hinreichend verdünnten Gasen ist die Darstellung der einzelnen Moleküle durch hinreichend getrennte und hinreichend langsam zerfließende Wellenpakete möglich. Hier wird man für die Translationsbewegung gewiß die klassische Näherung anwenden dürfen [halbquantisierte Systeme, vgl. KOPPE (1955)].

91. Die Boltzmann-Gleichung für das Lorentz-Gas aus rotierenden Molekülen. In dieser Ziffer nehmen wir an, daß die Gasmoleküle, welche alle gleichen Vibrationszustand und gleichen Betrag, jedoch verschiedene Orientierung des Eigendrehimpulses haben mögen, gegen ruhende, unendlich schwere Streuzentren, aber nicht gegeneinander stoßen (Lorentz-Gas). Die genannten Voraussetzungen über den inneren Zustand der Teilchen treffen zu für Elementarteilchen mit Spin oder z.B. für Ortho-Wasserstoff bei tiefen Temperaturen. Wir vollziehen die Aufstellung der Boltzmann-Gleichung in den folgenden drei Schritten: Zunächst

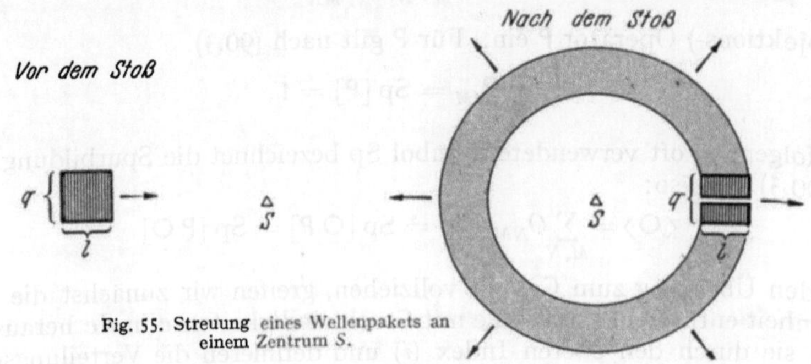

Fig. 55. Streuung eines Wellenpakets an einem Zentrum S.

untersuchen wir die Änderung des Erwartungswerts $\langle \mathsf{O} \rangle$ eines Operators für das Einzelmolekül, wenn dieses — im Ortsraum dargestellt durch ein Wellenpaket — an einem Streuzentrum vorbeistreicht. Sodann übertragen wir diese Betrachtung auf das Gas und gelangen zur Maxwellschen Transportgleichung. Aus dieser gewinnen wir schließlich durch eine einfache Umformung die gewünschte Boltzmann-Gleichung.

α) Änderung von $\langle \mathsf{O} \rangle$ beim Einzelstoß. Das Wellenpaket, welches das betrachtete Molekül vor dem Stoß darstellt, ist durch (90.1) gegeben. Es sei im Innern gut homogen und habe die Gestalt eines Zylinders vom Querschnitt q und der Länge l (s. Fig. 55). Nach dem Stoß kommt zu dem Paket, welches geradlinig weiterläuft und in Fig. 55 durch starke Schraffur angedeutet ist, eine Hohlkugelwelle hinzu, deren Amplitude aus (89.11) bzw. (89.13) zu entnehmen ist. Der gesamte Streuvorgang des Wellenpakets wird also durch Ausdrücke der Form

$$\left.\begin{aligned} \psi &= \sum_M \Psi_M \alpha_M \, e^{i k \boldsymbol{e} \boldsymbol{r}'} A + \sum_{M'} \Psi_{M'} \alpha'_{M'} \frac{e^{i k r'}}{k r'} A' \\ \psi^* &= A^* e^{-i k \boldsymbol{e} \boldsymbol{r}'} \sum_N \alpha_N^* \Psi_N^* + A'^* \frac{e^{-i k r'}}{k r'} \sum_{N'} \alpha'^*_{N'} \Psi^*_{N'} \end{aligned}\right\} \tag{91.1}$$

dargestellt. Dabei ist der Aufpunkt mit $\boldsymbol{r}'$ bezeichnet und vorübergehend abgekürzt

$$\alpha'_{M'} = \sum_M a_{M'M}(\boldsymbol{e}', \boldsymbol{e})\, \alpha_M, \quad \alpha'^*_{N'} = \sum_N \alpha_N^* a^\dagger_{NN'}(\boldsymbol{e}, \boldsymbol{e}'). \tag{91.2}$$

Für die orts- und zeitabhängigen Amplituden $A, \ldots, A'^*$, welche langsam veränderlich sein sollen, gilt offenbar:

$$\left.\begin{aligned} &\text{vor dem Stoß:} \quad |A| = \frac{1}{\sqrt{q l}} \text{ innerhalb des Zylinders } q l, \text{ sonst } = 0, \\ &\qquad\qquad\quad\;\; |A'| = 0, \end{aligned}\right\} \tag{91.3}$$

$$\left.\begin{aligned} &\text{nach dem Stoß:} \quad |A| = \frac{1}{\sqrt{q l}} \text{ innerhalb des Zylinders } q l, \text{ sonst } = 0, \\ &\qquad\qquad\qquad |A'| = \frac{1}{\sqrt{q l}} \text{ innerhalb der Hohlkugel, sonst } = 0. \end{aligned}\right\} \tag{91.4}$$

Die Phasenfaktoren der A brauchen nicht aufgeführt zu werden. Sie sind für A und A' gemeinsam. Auf gemeinsame Phasenfaktoren kommt es aber nicht an; sie heben sich bei der Bildung von Erwartungswerten heraus. Man könnte im folgenden, ohne einen rechnerischen Fehler zu begehen, statt der Amplituden $A, \ldots, A'^*$ gleich deren Beträge schreiben.

Für den Erwartungswert einer beliebigen Observablen $\mathsf{O}(\boldsymbol{c})$ des Einzelmoleküls erhält man mittels (90.5), (90.4) und (91.1) unter Benutzung der Orthonormierung der Funktionen Ψ_M zunächst die folgende Darstellung durch ein Raumintegral:

$$\left.\begin{aligned}\langle \mathsf{O}\rangle = \int \Big(A^* \, \mathrm{e}^{-ik\boldsymbol{e}\boldsymbol{r}'} \alpha_N^* \, \delta_{NM''} + A'^* \frac{\mathrm{e}^{-ikr'}}{k r'} \alpha_{N'}'^* \, \delta_{N'M''}\Big) \times \\ \times \Big(O_{M''M}(\boldsymbol{c}) \, \alpha_M \, \mathrm{e}^{ik\boldsymbol{e}\boldsymbol{r}'} A + O_{M''M'}(\boldsymbol{c}') \, \alpha_{M'}' \frac{\mathrm{e}^{ikr'}}{k r'} A'\Big) d\boldsymbol{r}'.\end{aligned}\right\} \tag{91.5}$$

Hierin ist über alle (doppelten) Indices nach dem Ausmultiplizieren der Klammern zu summieren. Nun seien α', α'^* gemäß (91.2) durch α, α^* ersetzt und die entstehenden Produkte gemäß (90.6) durch den Projektionsoperator P ausgedrückt. Man erhält so anstelle von (91.5) in Matrix-Schreibweise

$$\left.\begin{aligned}\langle \mathsf{O}\rangle = \mathrm{Sp} \int \Big[A^* A \, \mathsf{O}(\boldsymbol{c}) \, \mathsf{P} + \frac{A'^* A'}{k^2 r'^2} \mathsf{O}(\boldsymbol{c}') \, \mathfrak{a}(\boldsymbol{e}', \boldsymbol{e}) \, \mathsf{P} \, \mathfrak{a}^\dagger(\boldsymbol{e}, \boldsymbol{e}') + \\ + A^* A' \frac{\mathrm{e}^{ik(r' - \boldsymbol{e}\boldsymbol{r}')}}{k r'} \mathsf{O}(\boldsymbol{c}') \, \mathfrak{a}(\boldsymbol{e}', \boldsymbol{e}) \, \mathsf{P} + \\ + A'^* A \frac{\mathrm{e}^{-ik(r' - \boldsymbol{e}\boldsymbol{r}')}}{k r'} \mathsf{O}(\boldsymbol{c}) \, \mathsf{P} \, \mathfrak{a}^\dagger(\boldsymbol{e}, \boldsymbol{e}')\Big] d\boldsymbol{r}'.\end{aligned}\right\} \tag{91.6}$$

Dieser Ausdruck soll jetzt weiter ausgewertet werden, vor und nach dem Stoß.

Vor dem Stoß ist keine Streuwelle vorhanden — $A' = 0$ — und es ergibt sich mittels (91.3) aus (91.6) sogleich

$$\langle \mathsf{O}\rangle = \mathrm{Sp}\,[\mathsf{O}(\boldsymbol{c}) \, \mathsf{P}]. \tag{91.7}$$

Nach dem Stoß ist $A' \neq 0$ innerhalb der Hohlkugel (s. Fig. 55) und es sind alle vier Summanden rechts in (91.6) zu berücksichtigen. Der erste Summand gibt den Beitrag des unverändert weitergelaufenen Wellenpakets, d.i. wieder (91.7). Der zweite Summand gibt den Beitrag der Hohlkugelwelle; die Integration über r' kann mittels (91.4) sogleich ausgeführt werden und es entsteht so der erste Term rechts in (91.8). Die beiden letzten Summanden enthalten die Interferenzfaktoren $\mathrm{e}^{\pm ikr'(1 - \boldsymbol{e}\boldsymbol{e}')}$, welche für die Auslöschung eines Teils der Primärintensität sorgen. Diese Faktoren haben in der Vorwärtsrichtung $\boldsymbol{e}' = \boldsymbol{e}$ den Wert 1 und oszillieren im übrigen rasch, so daß bei der Integration über $\boldsymbol{r}'$ nur das in Fig. 55 unschraffiert gelassene Schattengebiet in der Nähe der Zylinderachse merklich beiträgt und in O wie in $\mathfrak{a}$ und $\mathfrak{a}^\dagger$ statt $\boldsymbol{e}'$ die Einfallsrichtung $\boldsymbol{e}$ genommen werden darf. Man findet mit Benutzung von $\boldsymbol{e}\,\boldsymbol{e}' = \cos\chi$, $d\boldsymbol{r}' = 2\pi r'^2 \, dr' \sin\chi \, d\chi$ und von (91.4) durch partielle Integration nach χ

$$\begin{aligned}\int A^* A' \frac{\mathrm{e}^{ik(r' - \boldsymbol{e}\boldsymbol{r}')}}{k r'} d\boldsymbol{r}' &= \frac{2\pi}{i k^2} \int_0^\infty \int_0^\pi A^* A' \frac{\partial}{\partial \chi} \mathrm{e}^{ikr'(1 - \cos\chi)} \, dr' \, d\chi \\ &= \frac{2\pi}{i k^2} \int_0^\infty [A^* A' \, \mathrm{e}^{ikr'(1 - \cos\chi)}]_0^\pi \, dr' - \cdots \approx \frac{2\pi i}{k^2 q}.\end{aligned}$$

Das durch ... angedeutete Integral über $d\boldsymbol{r}'$, dessen Integrand den für ein genügend ausgedehntes Wellenpaket beliebig kleinen Faktor $\partial A^* A'/\partial\chi$ hat und proportional $1/r'^2$ ist, kann weggelassen werden. Somit erhält man

$$\left.\begin{aligned} \delta\langle \mathsf{O}\rangle &= \langle \mathsf{O}\rangle_{\text{nach}} - \langle \mathsf{O}\rangle_{\text{vor}} \\ &= \frac{1}{k^2 q}\,\mathrm{Sp}\Big[\int \mathsf{O}(\boldsymbol{c}')\,\mathfrak{a}(\boldsymbol{e}',\boldsymbol{e})\,\mathsf{P}\,\mathfrak{a}^\dagger(\boldsymbol{e},\boldsymbol{e}')\,d\boldsymbol{e}' - \\ &\qquad - \mathsf{O}(\boldsymbol{c})\,\frac{2\pi}{i}\big(\mathfrak{a}(\boldsymbol{e},\boldsymbol{e})\,\mathsf{P} - \mathsf{P}\,\mathfrak{a}^\dagger(\boldsymbol{e},\boldsymbol{e})\big)\Big]. \end{aligned}\right\} \tag{91.8}$$

Dieser Ausdruck für die Änderung des Erwartungswerts $\langle \mathsf{O}\rangle$ beim Vorbeistreichen des Wellenpakets am Streuer bildet die Grundlage für die weiteren Überlegungen.

β) Die Maxwellsche Transportgleichung. Wir interessieren uns jetzt für die Änderung pro Zeiteinheit, welche der lokale statistische Mittelwert $\overline{O}$ für das Gas [s. Gln. (90.11) oder (90.12)] erfährt. Ohne Stöße gilt offenbar, analog wie beim einatomigen Gas [s. Gl. (31.1 a)][1],

$$\frac{\partial n\overline{O}}{\partial t} + \frac{\partial n\overline{c_\mu O}}{\partial x_\mu} \equiv \int \mathrm{Sp}\Big[\mathsf{O}(\boldsymbol{c})\Big(\frac{\partial \mathsf{f}}{\partial t} + c_\mu\frac{\partial \mathsf{f}}{\partial x_\mu}\Big)\Big]\,d\boldsymbol{c} = 0.$$

Die Gesamtänderung pro Volumeinheit infolge der Stöße erhält man durch Addition all der Änderungen (91.8), welche bei den einzelnen Molekülen der Volumeinheit eintritt. Bei dieser Addition entsteht zunächst gemäß (90.10) aus P die Verteilungsmatrix $\mathsf{f}\,d\boldsymbol{c}$. Über $\boldsymbol{c}$ ist sodann zu integrieren. Ferner ist zu berücksichtigen, daß (91.8) die Änderung von $\langle \mathsf{O}\rangle$ beim Überstreichen eines einzigen Streuers gibt. Tatsächlich überstreicht jedes Wellenpaket pro Zeiteinheit qcn_1 Streuzentren, wo n_1 deren Konzentration beze chnet. Somit hat man bei Berücksichtigung der Stöße

$$\left.\begin{aligned} &\int \mathrm{Sp}\Big[\mathsf{O}(\boldsymbol{c})\Big(\frac{\partial \mathsf{f}(\boldsymbol{c})}{\partial t} + c_\mu\frac{\partial \mathsf{f}(\boldsymbol{c})}{\partial x_\mu}\Big)\Big]\,d\boldsymbol{c} \\ &\qquad = n_1\int \mathrm{Sp}\Big[\int \mathsf{O}(\boldsymbol{c}')\,\mathfrak{a}(\boldsymbol{e}',\boldsymbol{e},E)\,\mathsf{f}(\boldsymbol{c})\,\mathfrak{a}^\dagger(\boldsymbol{e},\boldsymbol{e}',E)\,d\boldsymbol{e}' - \\ &\qquad\quad - \mathsf{O}(\boldsymbol{c})\cdot\frac{2\pi}{i}\big(\mathfrak{a}(\boldsymbol{e},\boldsymbol{e},E)\,\mathsf{f}(\boldsymbol{c}) - \mathsf{f}(\boldsymbol{c})\,\mathfrak{a}^\dagger(\boldsymbol{e},\boldsymbol{e},E)\big)\Big]\,\frac{c\,d\boldsymbol{c}}{k^2}. \end{aligned}\right\} \tag{91.9}$$

Dies ist in invarianter Matrix-Schreibweise die *Maxwellsche Transportgleichung für das Lorentz-Gas* aus rotierenden Molekülen.

Man entnimmt aus ihr sogleich die Erhaltungssätze für Teilchenzahl ($O=1$) und Energie ($O=\frac{1}{2}mc^2$). In diesen Fällen ist O keine Matrix, sondern eine klassische Zahl und Sp kann hinter O gesetzt werden. Wegen der Vertauschbarkeit zweier Faktoren in der Spur

$$\mathrm{Sp}\,(\mathfrak{a}\,\mathfrak{b}) = \mathrm{Sp}\,(\mathfrak{b}\,\mathfrak{a}) \tag{91.10}$$

kann in dem letzten Term in (91.9) der Faktor f ausgeklammert werden. Dann läßt sich das Schattentheorem (89.15′), mit $\boldsymbol{e}''=\boldsymbol{e}$, anwenden und man erhält, unter nochmaliger Benutzung der Regel (91.10), aus (91.9)

$$\frac{\partial n\overline{O}}{\partial t} + \frac{\partial n\overline{c_\mu O}}{\partial x_\mu} = n_1\int\!\!\int \big(O(\boldsymbol{c}') - O(\boldsymbol{c})\big)\,\mathrm{Sp}\,[\mathfrak{a}(\boldsymbol{e}',\boldsymbol{e})\,\mathsf{f}(\boldsymbol{c})\,\mathfrak{a}^\dagger(\boldsymbol{e},\boldsymbol{e}')]\,d\boldsymbol{e}'\,\frac{c\,d\boldsymbol{c}}{k^2}. \tag{91.11}$$

Dies macht, wegen $O(\boldsymbol{c}') = O(\boldsymbol{c})$, die genannten Erhaltungssätze evident.

γ) Die Boltzmann-Gleichung. Die Transportgleichung (91.9) gestattet es, auch die zeitliche Änderung der f-Matrix selbst zu ermitteln. Dazu benennen

[1] Statt des dortigen Ψ heißt es jetzt O; von einem äußeren Kraftfeld sehen wir jetzt ab.

wir in dem Doppelintegral auf der rechten Seite, welches bilinear in $\mathfrak{a}$, $\mathfrak{a}^\dagger$ ist, die Variablen $\boldsymbol{c}$, $\boldsymbol{c}'$ bzw. $\boldsymbol{e}$, $\boldsymbol{e}'$ um in $\boldsymbol{c}'$, $\boldsymbol{c}$ bzw. $\boldsymbol{e}'$, $\boldsymbol{e}$ und benutzen den Liouvilleschen Satz (88.12). Die Transportgleichung lautet dann

$$\left.\begin{aligned}&\int \mathrm{Sp}\left[\mathsf{O}(\boldsymbol{c})\left(\frac{\partial \mathsf{f}(\boldsymbol{c})}{\partial t}+\cdots\right)\right] d\boldsymbol{c}\\&\quad= n_1\int \mathrm{Sp}\left[\mathsf{O}(\boldsymbol{c})\left\{\int \mathfrak{a}(\boldsymbol{e},\boldsymbol{e}')\,\mathsf{f}(\boldsymbol{c}')\,\mathfrak{a}^\dagger(\boldsymbol{e}',\boldsymbol{e})\,d\boldsymbol{e}'-\frac{2\pi}{i}\big(\mathfrak{a}(\boldsymbol{e},\boldsymbol{e})\,\mathsf{f}(\boldsymbol{c})-\mathsf{f}(\boldsymbol{c})\,\mathfrak{a}^\dagger(\boldsymbol{e},\boldsymbol{e})\big)\right\}\right]\frac{c\,d\boldsymbol{c}}{k^2}\end{aligned}\right\}\tag{91.9'}$$

und zwar muß dies gelten für beliebige Matrizen $\mathsf{O}(\boldsymbol{c})$. Dies ist nur möglich, wenn

$$\left.\begin{aligned}&\frac{\partial \mathsf{f}(\boldsymbol{c})}{\partial t}+c_\mu\frac{\partial \mathsf{f}(\boldsymbol{c})}{\partial x_\mu}\\&\quad=\frac{n_1 c}{k^2}\left\{\int \mathfrak{a}(\boldsymbol{e},\boldsymbol{e}',E)\,\mathsf{f}(\boldsymbol{c}')\,\mathfrak{a}^\dagger(\boldsymbol{e}',\boldsymbol{e},E)\,d\boldsymbol{e}'-\frac{2\pi}{i}\big(\mathfrak{a}(\boldsymbol{e},\boldsymbol{e},E)\,\mathsf{f}(\boldsymbol{c})-\mathsf{f}(\boldsymbol{c})\,\mathfrak{a}^\dagger(\boldsymbol{e},\boldsymbol{e},E)\big)\right\}.\end{aligned}\right\}\tag{91.12}$$

Damit hat man die quantentheoretische *Boltzmann-Gleichung für das Lorentz-Gas* mit rotierenden Molekülen (WALDMANN 1957). Sie ist als Matrixgleichung ebenso wie die Transportgleichung (91.9) invariant beim Übergang zu anderen orthogonalen Drehimpulsfunktionen Ψ_M. Die Reihenfolge der Faktoren in (91.12) ist wesentlich, da f, $\mathfrak{a}$ und $\mathfrak{a}^\dagger$ im allgemeinen nicht vertauschbar sind. Das erste Glied rechts in (91.12) stellt den Zuwachs von $\mathsf{f}(\boldsymbol{c})$ durch die Streuprozesse dar, bei denen vor dem Stoß die Geschwindigkeit irgendeine Richtung $\boldsymbol{e}'$ und danach die Richtung $\boldsymbol{e}$ hat; dieses Glied ist in den $\mathfrak{a}$ quadratisch und besitzt Integralform. Das zweite Glied rechts in (91.12) stellt die Abnahme von $\mathsf{f}(\boldsymbol{c})$ durch die Streuprozesse dar, bei denen vor dem Stoß die Geschwindigkeit die Richtung $\boldsymbol{e}$ und danach irgendeine Richtung hat. Es enthält die Streuamplituden $\mathfrak{a}$, $\mathfrak{a}^\dagger$ in der Vorwärtsrichtung $\boldsymbol{e}'=\boldsymbol{e}$ und ist in diesen linear. Es kann im allgemeinen Fall nicht, wie man es vom einatomigen Gas her gewohnt wäre, als ein dem ersten Glied ähnlicher Integralausdruck geschrieben werden. MÜHLSCHLEGEL und KOPPE[1] haben auf Grund der Gleichung (91.12) die Theorie der Vielfachstreuung polarisierter Elektronen ausgearbeitet.

Wegen der weiteren Diskussion sei auf die folgende Ziffer und auf Ziff. 94 verwiesen.

92. Die Boltzmann-Gleichung für ein beliebiges Gas aus rotierenden Molekülen. In der vorigen Ziffer wurde die Boltzmann-Gleichung für den einfachen Fall des Lorentz-Gases aufgestellt. Der allgemeine Fall läßt sich, mit einigen Erweiterungen, weitgehend analog behandeln. Wieder sei aber angenommen, daß die Teilchen gleichen Vibrationszustand und gleichen Betrag des Drehimpulses, Quantenzahl J, haben und daß kein äußeres Kraftfeld vorhanden ist, also $(2J+1)$-fache Entartung der Drehimpulszustände vorliegt.

α) Änderung von $\langle \mathsf{O}\rangle$ beim Einzelstoß. Wir betrachten zunächst ein wechselwirkungsfreies Molekülpaar, dessen Partner die Geschwindigkeiten $\boldsymbol{c}$, $\boldsymbol{c}_1$ und die Drehimpulsfunktionen $\sum\limits_M \Psi_M(X)\,\alpha_M(\boldsymbol{c})$, $\sum\limits_{M_1} \Psi_{M_1}(X_1)\,\alpha_{M_1}(\boldsymbol{c}_1)$ haben[2]. Der Ortsvektor vom einen Molekülschwerpunkt zum anderen sei $\boldsymbol{r}'=r'\,\boldsymbol{e}'$, die Relativgeschwindigkeit $\boldsymbol{g}=g\boldsymbol{e}$; wegen der zugehörigen Wellenzahl k siehe (88.7). Wir gehen in das Schwerpunktsystem und schreiben für das zugehörige Wellenpaket [s. auch (89.2)],

$$\psi^{(0)}=\sum_{MM_1}\Psi_{MM_1}\alpha_{MM_1}(\boldsymbol{c},\boldsymbol{c}_1)\,e^{ik\boldsymbol{e}\boldsymbol{r}'}A(\boldsymbol{r}',t)\quad\text{mit}\quad\iiint\psi^{(0)*}\psi^{(0)}\,d\boldsymbol{r}'\,dX\,dX_1=1.\tag{92.1}$$

[1] B. MÜHLSCHLEGEL u. H. KOPPE: Z. Physik 1958 (im Erscheinen).

[2] Die $\boldsymbol{c}$ im Argument dienen lediglich zur Unterscheidung zweier α-Sätze. Es ist natürlich nicht gemeint, daß alle Moleküle mit der Geschwindigkeit $\boldsymbol{c}$ den gleichen α-Satz haben.

Dabei ist der Kürze halber das direkte Produkt

$$\alpha_{MM_1}(\boldsymbol{c}, \boldsymbol{c}_1) = \alpha_M(\boldsymbol{c})\, \alpha_{M_1}(\boldsymbol{c}_1) \tag{92.2}$$

eingeführt. Die Normierung der Amplituden A, α sei entsprechend wie in (90.2) und (90.3) gewählt.

Bei Vorhandensein von Wechselwirkung kommt eine Streuwelle hinzu. Mittels Gl. (89.11) bzw. (89.13) erhält man

$$\left.\begin{aligned}\psi &= \sum_{MM_1} \Psi_{MM_1} \alpha_{MM_1} e^{ik\boldsymbol{e}\boldsymbol{r}'} A + \sum_{M'M'_1} \Psi_{M'M'_1} \alpha'_{M'M'_1} \frac{e^{ikr'}}{k r'}\\ \psi^* &= A^* e^{-ik\boldsymbol{e}\boldsymbol{r}'} \sum_{NN_1} \alpha^*_{NN_1} \Psi^*_{NN_1} + A'^* \frac{e^{-ikr'}}{k r'} \sum_{N'N'_1} \alpha'^*_{N'N'_1} \Psi^*_{N'N'_1}.\end{aligned}\right\} \tag{92.3}$$

Für A, A' gelten dabei wieder (91.3) und (91.4), und es wurde abgekürzt

$$\left.\begin{aligned}\alpha'_{M'M'_1} &= \sum_{MM_1} a_{M'M'_1, MM_1}(\boldsymbol{e}', \boldsymbol{e}, E)\, \alpha_{MM_1}\\ \alpha'^*_{N'N'_1} &= \sum_{NN_1} \alpha^*_{NN_1}\, a^\dagger_{NN_1, N'N'_1}(\boldsymbol{e}, \boldsymbol{e}', E).\end{aligned}\right\} \tag{92.4}$$

Die Formeln (92.3 und (92.4) unterscheiden sich von (91.1) und (91.2) lediglich durch die Verdopplung der Indices. Deshalb können die Überlegungen von Ziff. 91 nun weitgehend im folgenden benutzt werden.

Es soll für ein Wellenpaket der Art (92.3) zunächst der Erwartungswert eines Operators $\mathsf{O}(\boldsymbol{c}, \boldsymbol{c}_1)$ berechnet werden, welcher auf die magnetischen Quantenzahlen beider Moleküle wirkt. Wir führen, in Analogie zu (90.6), entsprechend der jetzigen Indexverdopplung, den Projektionsoperator eines Molekülpaars

$$\alpha_{MM_1}(\boldsymbol{c})\, \alpha^*_{NN_1}(\boldsymbol{c}_1) = P_{MM_1, NN_1}(\boldsymbol{c}, \boldsymbol{c}_1) \tag{92.5}$$

ein. Nach (92.2) hängt er mit den Projektionsoperatoren der Einzelmoleküle nach Art des direkten Produkts zusammen:

$$P_{MM_1, NN_1}(\boldsymbol{c}, \boldsymbol{c}_1) = P_{MN}(\boldsymbol{c})\, P_{M_1N_1}(\boldsymbol{c}_1). \tag{92.6}$$

Ferner führen wir für eine beliebige, auf ein Molekülpaar bezügliche Matrix $\mathfrak{a}$, welche also zwei Indexpaare haben soll, die folgenden Spurbildungen ein:

$$\left.\begin{aligned}&(\mathrm{Sp}_1\, a)_{MN} = \sum_{M_1} a_{MM_1, NM_1}; \qquad (\mathrm{Sp}\, a)_{M_1N_1} = \sum_{M} a_{MM_1, MN_1}\\ \text{sowie}\quad &\mathrm{Sp}\,\mathrm{Sp}_1\, \mathfrak{a} = \sum_{MM_1} a_{MM_1, MM_1} = \mathrm{Sp}_1\,\mathrm{Sp}\, \mathfrak{a}.\end{aligned}\right\} \tag{92.7}$$

$\mathrm{Sp}_1\, \mathfrak{a}$ stellt also eine Matrix mit zwei einfachen Indices dar, während die Doppelspur eine gewöhnliche Zahl ist. Damit erhalten wir, durch unmittelbare Übertragung von (91.8) auf den jetzigen Fall, als Änderung des Erwartungswerts $\langle \mathsf{O}(\boldsymbol{c}, \boldsymbol{c}_1)\rangle$ beim Stoß

$$\left.\begin{aligned}\delta\langle \mathsf{O}(\boldsymbol{c}, \boldsymbol{c}_1)\rangle = \frac{1}{k^2 q}\, \mathrm{Sp}\,\mathrm{Sp}_1 \Big[\int \mathsf{O}(\boldsymbol{c}', \boldsymbol{c}'_1)\, \mathfrak{a}(\boldsymbol{e}', \boldsymbol{e})\, \mathsf{P}(\boldsymbol{c}, \boldsymbol{c}_1)\, \mathfrak{a}^\dagger(\boldsymbol{e}, \boldsymbol{e}')\, d\boldsymbol{e}' -\\ - \mathsf{O}(\boldsymbol{c}, \boldsymbol{c}_1) \cdot \frac{2\pi}{i} \big(\mathfrak{a}(\boldsymbol{e}, \boldsymbol{e})\, \mathsf{P}(\boldsymbol{c}, \boldsymbol{c}_1) - \mathsf{P}(\boldsymbol{c}, \boldsymbol{c}_1)\, \mathfrak{a}^\dagger(\boldsymbol{e}, \boldsymbol{e})\big)\Big].\end{aligned}\right\} \tag{92.8}$$

Im weiteren interessiert uns nur ein Operator $\mathsf{O}(\boldsymbol{c})$, welcher sich auf ein einziges Molekül, nämlich dasjenige mit der Geschwindigkeit $\boldsymbol{c}$ bzw. $\boldsymbol{c}'$ bezieht, und welcher demgemäß auch nur auf die diesem zugeordneten Indices wirkt. Wir nehmen also speziell

$$O_{MM_1, NN_1}(\boldsymbol{c}, \boldsymbol{c}_1) = O_{MN}(\boldsymbol{c})\, \delta_{M_1N_1}$$

an. Für einen solchen Operator und eine beliebige Matrix a mit zwei Indexpaaren gilt nach (92.7)

$$\operatorname{Sp}\operatorname{Sp}_1[\mathsf{O}(\boldsymbol{c},\boldsymbol{c}_1)\,\mathsf{a}] = \sum_{M M_1 N} O_{MN}(\boldsymbol{c})\, a_{NM_1,MM_1} = \operatorname{Sp}[\mathsf{O}(\boldsymbol{c})\operatorname{Sp}_1\mathsf{a}]. \tag{92.9}$$

Angewandt auf (92.8) ergibt dies

$$\left.\begin{aligned}\delta\langle\mathsf{O}(\boldsymbol{c})\rangle = \frac{1}{k^2 q}\operatorname{Sp}\Big[\int \mathsf{O}(\boldsymbol{c}')\operatorname{Sp}_1\big(\mathsf{a}(\boldsymbol{e}',\boldsymbol{e})\,\mathsf{P}(\boldsymbol{c},\boldsymbol{c}_1)\,\mathsf{a}^\dagger(\boldsymbol{e},\boldsymbol{e}')\big)\,d\boldsymbol{e}' - \\ - \mathsf{O}(\boldsymbol{c})\cdot\frac{2\pi}{i}\operatorname{Sp}_1\big(\mathsf{a}(\boldsymbol{e},\boldsymbol{e})\,\mathsf{P}(\boldsymbol{c},\boldsymbol{c}_1) - \mathsf{P}(\boldsymbol{c},\boldsymbol{c}_1)\,\mathsf{a}^\dagger(\boldsymbol{e},\boldsymbol{e})\big)\Big].\end{aligned}\right\} \tag{92.10}$$

β) Die Maxwellsche Transportgleichung. Für die zeitliche Änderung des statistischen Mittelwerts $n\overline{O}$ für das Gas ohne Stöße gilt dasselbe wie im Lorentzschen Fall (s. Ziff. 91, β). Um den Beitrag der Stöße zu erhalten, haben wir (92.10) zu summieren über alle Molekülpaare mit Geschwindigkeiten $\boldsymbol{c}$, $\boldsymbol{c}_1$ in $d\boldsymbol{c}\,d\boldsymbol{c}_1$, welche in der Volumeinheit enthalten sind. Dabei entstehen nach (92.6) und (90.10) aus den P die Verteilungsmatrizen f:

$$\sum_{i,i_1} P^{(i,i_1)}_{MM_1,NN_1}(\boldsymbol{c},\boldsymbol{c}_1) = \sum_{i,i_1} P^{(i)}_{MN}(\boldsymbol{c})\,P^{(i_1)}_{M_1N_1}(\boldsymbol{c}_1) = f_{MN}(\boldsymbol{c})\,f_{M_1N_1}(\boldsymbol{c}_1)\,d\boldsymbol{c}\,d\boldsymbol{c}_1.$$

In Analogie zu (92.6) führen wir zwecks Kürze die Paar-Verteilungsmatrix ein gemäß

$$f_{MM_1,NN_1}(\boldsymbol{c},\boldsymbol{c}_1) = f_{MN}(\boldsymbol{c})\,f_{M_1N_1}(\boldsymbol{c}_1) \tag{92.11}$$

oder in Matrix-Schreibweise

$$\mathsf{f}(\boldsymbol{c},\boldsymbol{c}_1) = \mathsf{f}(\boldsymbol{c})\times\mathsf{f}(\boldsymbol{c}_1). \tag{92.11'}$$

Indem wir noch beachten, daß ein Molekül $\boldsymbol{c}$ in der Zeiteinheit mit allen Molekülen $\boldsymbol{c}_1$ zusammenstößt, welche im Volumen qg enthalten sind ($g = |\boldsymbol{c}-\boldsymbol{c}_1|$), und schließlich über alle Geschwindigkeiten integrieren, erhalten wir aus (92.10) die *Maxwellsche Transportgleichung*

$$\left.\begin{aligned}&\int \operatorname{Sp}\Big[\mathsf{O}(\boldsymbol{c})\Big(\frac{\partial\mathsf{f}(\boldsymbol{c})}{\partial t} + c_\mu\frac{\partial\mathsf{f}(\boldsymbol{c})}{\partial x_\mu}\Big)\Big]\,d\boldsymbol{c} \\ &= \iint \operatorname{Sp}\Big[\int \mathsf{O}(\boldsymbol{c}')\operatorname{Sp}_1\big(\mathsf{a}(\boldsymbol{e}',\boldsymbol{e},E)\,\mathsf{f}(\boldsymbol{c},\boldsymbol{c}_1)\,\mathsf{a}^\dagger(\boldsymbol{e},\boldsymbol{e}',E)\big)\,d\boldsymbol{e}' - \\ &\quad - \mathsf{O}(c)\cdot\frac{2\pi}{i}\operatorname{Sp}_1\big(\mathsf{a}(\boldsymbol{e},\boldsymbol{e},E)\,\mathsf{f}(\boldsymbol{c},\boldsymbol{c}_1) - \mathsf{f}(\boldsymbol{c},\boldsymbol{c}_1)\,\mathsf{a}^\dagger(\boldsymbol{e},\boldsymbol{e},E)\big)\Big]\frac{g\,d\boldsymbol{c}\,d\boldsymbol{c}_1}{k^2}.\end{aligned}\right\} \tag{92.12}$$

Die Erhaltungssätze für Teilchenzahl, Energie und Impuls lassen sich daraus ähnlich wie in Ziff. 91 mittels des Schattentheorems (89.15′) ablesen. Man braucht nur noch zu bemerken, daß bei der Umbenennung $\boldsymbol{c}, \boldsymbol{c}_1, \boldsymbol{c}', \boldsymbol{c}_1' \to \boldsymbol{c}_1, \boldsymbol{c}, \boldsymbol{c}_1', \boldsymbol{c}'$ die Einheitsvektoren $\boldsymbol{e}, \boldsymbol{e}'$ ihre Vorzeichen umkehren und daß natürlich gilt

$$a_{M_1'M',M_1M}(-\boldsymbol{e}',-\boldsymbol{e}) = a_{M'M_1',MM_1}(\boldsymbol{e}',\boldsymbol{e}), \qquad \text{für } \mathsf{a}^\dagger \text{ analog.}$$

Damit erhält man für ein klassisches O

$$\left.\begin{aligned}\frac{\partial n\overline{O}}{\partial t} + \cdots = \iiint \big(O(\boldsymbol{c}') + O(\boldsymbol{c}_1') - O(\boldsymbol{c}) - O(\boldsymbol{c}_1)\big)\times \\ \times\operatorname{Sp}\operatorname{Sp}_1[\mathsf{a}(\boldsymbol{e}',\boldsymbol{e})\,\mathsf{f}(\boldsymbol{c},\boldsymbol{c}_1)\,\mathsf{a}^\dagger(\boldsymbol{e},\boldsymbol{e}')]\frac{g\,d\boldsymbol{e}'\,d\boldsymbol{c}\,d\boldsymbol{c}_1}{k^2}\end{aligned}\right\} \tag{92.13}$$

und hat die genannten Erhaltungssätze.

γ) Die Boltzmann-Gleichung. Diese läßt sich nach dem bereits in Ziff. 91 benutzten Gedanken sofort aus der Transportgleichung entnehmen. Wir vertauschen in dem dreifachen Integral rechts in (92.12) die gestrichenen und die ungestrichenen Variablen und benutzen den Liouvilleschen Satz (88.11). Dann kann man rechts $\mathfrak{O}(\boldsymbol{c})$ ausklammern und somit beiderseits auf die Integration über $\boldsymbol{c}$ und die Spurbildung Sp verzichten. So erhält man die *Boltzmann-Gleichung*

$$\left.\begin{aligned} \frac{\partial \mathfrak{f}(\boldsymbol{c})}{\partial t} + c_\mu \frac{\partial \mathfrak{f}(\boldsymbol{c})}{\partial x_\mu} = \int \mathrm{Sp}_1\Big[\int \mathfrak{a}(\boldsymbol{e}, \boldsymbol{e}', E)\, \mathfrak{f}(\boldsymbol{c}', \boldsymbol{c}_1')\, \mathfrak{a}^\dagger(\boldsymbol{e}', \boldsymbol{e}, E)\, d\boldsymbol{e}' - \\ - \frac{2\pi}{i}\big(\mathfrak{a}(\boldsymbol{e}, \boldsymbol{e}, E)\, \mathfrak{f}(\boldsymbol{c}, \boldsymbol{c}_1) - \mathfrak{f}(\boldsymbol{c}, \boldsymbol{c}_1)\, \mathfrak{a}^\dagger(\boldsymbol{e}, \boldsymbol{e}, E)\big)\Big] \frac{g\, d\boldsymbol{c}_1}{k^2}\,. \end{aligned}\right\} \quad (92.14)$$

Dabei ist die Paar-Verteilungsmatrix $\mathfrak{f}(\boldsymbol{c}, \boldsymbol{c}_1)$ nach (92.11) bzw. (92.11′) eine Abkürzung für das direkte Produkt $\mathfrak{f}(\boldsymbol{c}) \times \mathfrak{f}(\boldsymbol{c}_1)$ von zwei Einzel-Verteilungsmatrizen. Auch an die Bedeutung (92.7) des Zeichens Sp_1 sei nochmals erinnert.

Die Gl. (92.14) kann als stochastische Matrixgleichung bezeichnet werden. Der stochastische Charakter wird an der Hauptachsenform (93.15) deutlich werden.

Zur Diskussion von (92.14) verweisen wir zunächst auf das, was im Anschluß an die Boltzmann-Gleichung (91.12) für das Lorentz-Gas gesagt wurde. Einige weitere Punkte seien hier noch gestreift. Oben unter *β)* war nicht vom Erhaltungssatz des Drehimpulses die Rede. Wir haben den Drehimpulssatz unter phänomenologischem Gesichtspunkt in Ziff. 2δ besprochen. Er kann im Rahmen der Kinetik verdünnter Gase nur näherungsweise formuliert werden, vgl. CHAPMAN und COWLING [*4*], S. 202. Spin und Bahndrehimpuls treten in dieser Näherung als entkoppelt auf. Unter Benutzung der klassischen Mechanik haben CURTISS (1956) sowie CURTISS und MUCKENFUSS (1957) die kinetische Theorie für Gase aus nicht-kugelsymmetrischen starren Molekülen entwickelt. CURTISS erhielt eine Gleichung für die Änderung der Drehimpuls-Dichte, die mit derjenigen übereinstimmt, welche CHAPMAN und COWLING [*4*] an Hand des Modells der rauhen Kugeln aufstellten. — Die zweite Bemerkung bezieht sich auf äußere Felder, von denen wir in den obigen Betrachtungen abgesehen haben. Äußere magnetische oder elektrische Felder können eine Ausrichtung nichtkugelsymmetrischer Moleküle bewirken. Ein Beispiel bietet der paramagnetische Sauerstoff O_2. Diese Ausrichtung zeigt sich in einer Feldabhängigkeit der Transportkoeffizienten, die von SENFTLEBEN (1930) entdeckt wurde. Das experimentelle Material über den Fremdfeldeinfluß auf die Transportvorgänge in Gasen ist in dem Artikel von HERZFELD [*22*], S. 223 zusammengefaßt; Beiträge zur Theorie haben von LAUE (1935) und vor allem ZERNIKE und VAN LIER (1939) geliefert. Im Rahmen der oben dargestellten quantenmechanischen Gaskinetik entsteht die interessante Frage, wie man die durch das Feld bewirkte graduelle Aufhebung der Entartung durch eine Erweiterung der Fundamentalgleichung (92.14) zum Ausdruck bringen kann. Wir können auf diese Orientierungseffekte hier nicht näher eingehen. — Eine letzte Bemerkung bezieht sich auf die Nichtunterscheidbarkeit. Bei gleichartigen Teilchen, die wir ja voraussetzen, wären in (92.14) strenggenommen symmetrisierte Streuamplituden einzusetzen. Sodann könnte man eine genauere Untersuchung anstellen, inwieweit die Ansätze aus Ziff. 84 von dem jetzigen allgemeinen Standpunkt aus gerechtfertigt bzw. ungenau sind. Eine eingehendere Diskussion auch dieses Punktes muß hier unterbleiben.

93. Das H-Theorem. Um den Entropiesatz zu beweisen und das thermische Gleichgewicht zu finden, werden wir eine Hauptachsen-Transformation der Verteilungsmatrix $\mathfrak{f}(t, \boldsymbol{r}, \boldsymbol{c})$ vornehmen. Die Boltzmann-Gleichung (92.14) ist, als Matrixgleichung, form-invariant gegen eine feste Transformation der orthogonalen Basisfunktionen Ψ_M, also z.B. unempfindlich gegen eine feste Drehung der Raumkoordinaten. Darüber hinaus ändert sie ihr Aussehen aber auch nur geringfügig, nämlich nur auf der linken Seite, vgl. (93.10), wenn eine variable unitäre

Transformation $\mathsf{U}(t, \boldsymbol{r}, \boldsymbol{c})$ der Ψ_M vorgenommen wird, wie sie schon in Ziff. 89, δ betrachtet wurde. Dies muß im folgenden zunächst dargelegt werden.

α) Die Transformation der Boltzmann-Gleichung bei unitärer Transformation $\mathsf{U}(t, \boldsymbol{r}, \boldsymbol{c})$ *der Drehimpulsfunktionen* Ψ_M. Die Transformationsformeln der Ψ und Ψ^* sind schon in (89.17) oder (89.18) notiert. Wir wiederholen sie:

$$\check{\Psi} = \Psi\,\mathsf{U}(\boldsymbol{c}), \qquad \check{\Psi}^* = \mathsf{U}^\dagger(\boldsymbol{c})\,\Psi^*. \tag{93.1}$$

Die unitäre Matrix $\mathsf{U}(\boldsymbol{c})$ kann, ohne daß dies in der Schreibweise angedeutet sei, auch von t and $\boldsymbol{r}$ abhängen. Die Transformationsformeln der Amplituden α, α^* aus (90.1) ergeben sich aus denen für Ψ, Ψ^* durch die Bemerkung, daß ψ, ψ^* invariant sind:

$$\alpha = \mathsf{U}(\boldsymbol{c})\,\check{\alpha}, \qquad \alpha^* = \check{\alpha}^*\,\mathsf{U}^\dagger(\boldsymbol{c}).$$

Daraus erhält man auf Grund der Definition (90.9) der Verteilungsmatrix deren Transformationsgesetz:

$$\mathsf{f}(\boldsymbol{c}) = \mathsf{U}(\boldsymbol{c})\,\check{\mathsf{f}}(\boldsymbol{c})\,\mathsf{U}^\dagger(\boldsymbol{c}). \tag{93.2}$$

Hiermit läßt sich — zweckmäßig zunächst durch Anschreiben mit Indices — auch die Transformation der Paar-Verteilungsmatrix (92.11) angeben. Sie lautet

$$\mathsf{f}(\boldsymbol{c}, \boldsymbol{c}_1) = \mathsf{U}(\boldsymbol{c}, \boldsymbol{c}_1)\,\check{\mathsf{f}}(\boldsymbol{c}, \boldsymbol{c}_1)\,\mathsf{U}^\dagger(\boldsymbol{c}, \boldsymbol{c}_1), \tag{93.3}$$

wo $\mathsf{U}(\boldsymbol{c}, \boldsymbol{c}_1)$ die in (89.22) eingeführte Matrix mit zwei Indexpaaren bedeutet. Die Transformation der Streuamplituden ist bereits in (89.26) und (89.27) angegeben. Wir notieren uns hier die zugehörige Umkehrung

$$\left.\begin{aligned} \mathfrak{a}(\boldsymbol{e}, \boldsymbol{e}') &= \mathsf{U}(\boldsymbol{c}, \boldsymbol{c}_1)\,\check{\mathfrak{a}}(\boldsymbol{c}, \boldsymbol{c}_1 | \boldsymbol{c}', \boldsymbol{c}_1')\,\mathsf{U}^\dagger(\boldsymbol{c}', \boldsymbol{c}_1'), \\ \mathfrak{a}^\dagger(\boldsymbol{e}', \boldsymbol{e}) &= \mathsf{U}(\boldsymbol{c}', \boldsymbol{c}_1')\,\check{\mathfrak{a}}^\dagger(\boldsymbol{c}', \boldsymbol{c}_1' | \boldsymbol{c}, \boldsymbol{c}_1)\,\mathsf{U}^\dagger(\boldsymbol{c}, \boldsymbol{c}_1). \end{aligned}\right\} \tag{93.4}$$

Damit ist man nun in der Lage in der Boltzmann-Gleichung (92.14) sämtliche Matrizen durch deren Transformierte zu ersetzen.

Man sieht sogleich — unter Benutzung der Unitarität von $\mathsf{U}(\boldsymbol{c}, \boldsymbol{c}_1)$ — daß sich der in $\mathfrak{a}$ bilineare Ausdruck $\mathfrak{a}\,\mathsf{f}\,\mathfrak{a}^\dagger$ im Integranden von (92.14) wie f selbst transformiert:

$$\left.\begin{aligned} &\mathfrak{a}(\boldsymbol{e}, \boldsymbol{e}')\,\mathsf{f}(\boldsymbol{c}', \boldsymbol{c}_1')\,\mathfrak{a}^\dagger(\boldsymbol{e}', \boldsymbol{e}) \\ &\quad = \mathsf{U}(\boldsymbol{c}, \boldsymbol{c}_1)\,\check{\mathfrak{a}}(\boldsymbol{c}, \boldsymbol{c}_1 | \boldsymbol{c}', \boldsymbol{c}_1')\,\check{\mathsf{f}}(\boldsymbol{c}', \boldsymbol{c}_1')\,\check{\mathfrak{a}}(\boldsymbol{c}', \boldsymbol{c}_1' | \boldsymbol{c}, \boldsymbol{c}_1)\,\mathsf{U}^\dagger(\boldsymbol{c}, \boldsymbol{c}_1). \end{aligned}\right\} \tag{93.5}$$

Natürlich ist auch

$$\mathfrak{a}(\boldsymbol{e}, \boldsymbol{e})\,\mathsf{f}(\boldsymbol{c}, \boldsymbol{c}_1) = \mathsf{U}(\boldsymbol{c}, \boldsymbol{c}_1)\,\check{\mathfrak{a}}(\boldsymbol{c}, \boldsymbol{c}_1 | \boldsymbol{c}, \boldsymbol{c}_1)\,\check{\mathsf{f}}(\boldsymbol{c}, \boldsymbol{c}_1)\,\mathsf{U}^\dagger(\boldsymbol{c}, \boldsymbol{c}_1) \tag{93.6}$$

und Analoges für $\mathsf{f}\,\mathfrak{a}^\dagger$. — Sei ferner A eine beliebige Matrix mit zwei Indexpaaren, welche sich wie $\mathsf{f}(\boldsymbol{c}, \boldsymbol{c}_1)$ transformiert:

$$\mathsf{A} = \mathsf{U}(\boldsymbol{c}, \boldsymbol{c}_1)\,\check{\mathsf{A}}\,\mathsf{U}^\dagger(\boldsymbol{c}, \boldsymbol{c}_1). \tag{93.7}$$

Dann gilt die Rechenregel:

$$\mathsf{U}^\dagger(\boldsymbol{c})\,(\mathrm{Sp}_1\,\mathsf{A})\,\mathsf{U}(\boldsymbol{c}) = \mathrm{Sp}_1\,\check{\mathsf{A}}. \tag{93.8}$$

Man beweist diese Regel leicht durch Anschreiben in Komponenten — wegen Sp_1 siehe (92.7). Tatsächlich sind nun nach (93.5) und (93.6) die Matrizenprodukte rechts in der Boltzmann-Gleichung (92.14) sämtlich vom Transformations-Typus der Matrix A aus (93.7). Multipliziert man also die rechte Seite von (92.14) von links mit $\mathsf{U}^\dagger(\boldsymbol{c})$, von rechts mit $\mathsf{U}(\boldsymbol{c})$ und zieht diese Faktoren unter die Integrale, so behält nach (93.8) die rechte Seite der Boltzmann-Gleichung ihre ursprüngliche Gestalt bei, mit den Transformierten $\check{\mathsf{f}}$, $\check{\mathsf{a}}$, $\check{\mathsf{a}}^\dagger$ anstelle von f, a, $\mathsf{a}^\dagger$.

Die linke Seite von (92.14) ist, um die Gleichheit zu wahren, ebenfalls von links mit $\mathsf{U}^\dagger(\boldsymbol{c})$ und von rechts mit $\mathsf{U}(\boldsymbol{c})$ zu multiplizieren. Wir kürzen ab

$$\mathscr{D} = \frac{\partial}{\partial t} + c_\mu \frac{\partial}{\partial x_\mu} \tag{93.9}$$

und haben dann als linke Seite von (92.14)

$$\begin{aligned}\mathsf{U}^\dagger \mathscr{D}(\mathsf{U}\check{\mathsf{f}}\mathsf{U}^\dagger)\,\mathsf{U} &= \mathsf{U}^\dagger\mathscr{D}(\mathsf{U})\,\check{\mathsf{f}} + \mathscr{D}(\check{\mathsf{f}}) + \check{\mathsf{f}}\,\mathscr{D}(\mathsf{U}^\dagger)\,\mathsf{U}\\ &= \mathscr{D}(\check{\mathsf{f}}) + \mathsf{U}^\dagger\mathscr{D}(\mathsf{U})\,\check{\mathsf{f}} - \check{\mathsf{f}}\,\mathsf{U}^\dagger\mathscr{D}(\mathsf{U}).\end{aligned}$$

Dabei wurde die Unitarität $\mathsf{U}^\dagger\mathsf{U} = 1$ benutzt. Die beiden letzten Terme haben die Form eines Kommutators und geben die Änderung von f infolge Änderung der Basisfunktionen an.

Somit lautet die transformierte Boltzmann-Gleichung

$$\left.\begin{aligned}&\mathscr{D}(\check{\mathsf{f}}(\boldsymbol{c})) + \mathsf{U}^\dagger(\boldsymbol{c})\,\mathscr{D}(\mathsf{U}(\boldsymbol{c}))\,\check{\mathsf{f}}(\boldsymbol{c}) - \check{\mathsf{f}}(\boldsymbol{c})\,\mathsf{U}^\dagger(\boldsymbol{c})\,\mathscr{D}(\mathsf{U}(\boldsymbol{c}))\\ &\quad= \int \mathrm{Sp}_1\Big[\int \check{\mathsf{a}}(\boldsymbol{c},\boldsymbol{c}_1|\boldsymbol{c}',\boldsymbol{c}_1')\,\check{\mathsf{f}}(\boldsymbol{c}',\boldsymbol{c}_1')\,\check{\mathsf{a}}^\dagger(\boldsymbol{c}',\boldsymbol{c}_1'|\boldsymbol{c},\boldsymbol{c}_1)\,d\boldsymbol{e}' -\\ &\quad - \frac{2\pi}{i}\big(\check{\mathsf{a}}(\boldsymbol{c},\boldsymbol{c}_1|\boldsymbol{c},\boldsymbol{c}_1)\,\check{\mathsf{f}}(\boldsymbol{c},\boldsymbol{c}_1) - \check{\mathsf{f}}(\boldsymbol{c},\boldsymbol{c}_1)\,\check{\mathsf{a}}^\dagger(\boldsymbol{c},\boldsymbol{c}_1|\boldsymbol{c},\boldsymbol{c}_1)\big)\Big]\frac{g\,d\boldsymbol{c}_1}{k^2}.\end{aligned}\right\} \tag{93.10}$$

Damit hat man den Ausgangspunkt für die weitere Diskussion.

β) Hauptachsentransformation der Verteilungsmatrix. Die Boltzmann-Gleichung gestattet es die Verteilungsmatrix $\mathsf{f}(t, \boldsymbol{r}, \boldsymbol{c})$ zu berechnen, wenn $\mathsf{f}(t=0, \boldsymbol{r}, \boldsymbol{c})$ bekannt ist. f ist hermitisch (und definit). Es ist daher im Prinzip möglich, eine Transformation (93.2) mit speziellem $\mathsf{U}(t, \boldsymbol{r}, \boldsymbol{c})$ anzugeben, daß stets, überall und für alle Geschwindigkeiten $\check{\mathsf{f}}$ die Hauptachsenform hat:

$$\left.\begin{aligned}\check{f}_{MN}(\boldsymbol{c}) &= F_M(\boldsymbol{c})\,\delta_{MN} \geqq 0;\\ \check{f}_{MM_1,NN_1}(\boldsymbol{c},\boldsymbol{c}_1) &= F_M(\boldsymbol{c})\,F_{M_1}(\boldsymbol{c}_1)\,\delta_{MN}\,\delta_{M_1N_1} \geqq 0.\end{aligned}\right\} \tag{93.11}$$

Für diese spezielle Transformation verschwinden die Diagonalelemente des Kommutators auf der linken Seite von (93.10). Es ist ja

$$[\mathsf{U}^\dagger\mathscr{D}(\mathsf{U})\,\check{\mathsf{f}} - \check{\mathsf{f}}\,\mathsf{U}^\dagger\mathscr{D}(\mathsf{U})]_{MN} = \sum_{M'} U^\dagger_{MM'}\,\mathscr{D}(U_{M'N})\,(F_M - F_N),$$

woraus man die Behauptung ersieht[1]. Jetzt setzen wir auch in allen übrigen Gliedern von (93.10) die speziellen Matrizen (93.11) ein und betrachten die Diagonalelemente $M = N$. So entsteht die Gleichung

$$\left.\begin{aligned}\mathscr{D}(F_M(\boldsymbol{c})) = \int\sum_{M_1}\Big[&\int\sum_{M',M_1'} F_{M'}(\boldsymbol{c}')\,F_{M_1'}(\boldsymbol{c}_1')\,|A_{MM_1,M'M_1'}(\boldsymbol{c},\boldsymbol{c}_1|\boldsymbol{c}',\boldsymbol{c}_1')|^2\,d\boldsymbol{e}'\\ &- F_M(\boldsymbol{c})F_{M_1}(\boldsymbol{c}_1)\cdot\frac{2\pi}{i}\big(A(\boldsymbol{c},\boldsymbol{c}_1|\boldsymbol{c},\boldsymbol{c}_1) - A^\dagger(\boldsymbol{c},\boldsymbol{c}_1|\boldsymbol{c},\boldsymbol{c}_1)\big)_{MM_1,MM_1}\Big]\frac{g\,d\boldsymbol{c}_1}{k^2}.\end{aligned}\right\} \tag{93.12}$$

Dabei wurde jetzt A, $A^\dagger$ statt $\check{a}$, $\check{a}^\dagger$ geschrieben um deutlich zu machen, daß es sich um die Streuamplituden für das jeweilige Hauptachsensystem von f handelt.

[1] Darin findet die Extremaleigenschaft der Eigenwerte F_M ihren Ausdruck.

Dank des Übergangs zu diagonalem $\check{\mathsf{f}}$ konnten im zweiten Term auf der rechten Seite von (93.12), anders als in (93.10), die F ausgeklammert werden. Dadurch wird nun die Anwendung des Schattentheorems möglich. In Ziff. 89, δ haben wir dessen Invarianz bei beliebiger Transformation $\mathsf{U}(\boldsymbol{c})$ gezeigt. Schreibt man (89.28) für die Diagonalglieder hin, so tritt in dem dortigen Integranden das Absolutquadrat von A auf:

$$\sigma_{M'M_1', MM_1}(\boldsymbol{c}', \boldsymbol{c}_1' | \boldsymbol{c}, \boldsymbol{c}_1) = \frac{1}{k^2} | A_{MM_1, M'M_1'}(\boldsymbol{c}, \boldsymbol{c}_1 | \boldsymbol{c}', \boldsymbol{c}_1')|^2 \geq 0, \qquad (93.13)$$

bzw. Analoges mit Vertauschung von Gestrichenem und Ungestrichenem. Der Faktor $1/k^2$ wurde in die Definition von σ aufgenommen, damit dieses die Dimension einer Fläche bekommt; A ist ja ebenso wie a dimensionslos, vgl. (89.11). Tatsächlich hat σ, wie sich sogleich zeigen wird, die Bedeutung eines Streuquerschnitts, und zwar bezeichnen, wie die Gegenüberstellung mit (89.11) lehrt, die gestrichenen Größen in (93.13) den Zustand vor dem Stoß, die ungestrichenen den Zustand danach. Mit (93.13) schreiben sich die Diagonalglieder des Schattentheorems (89.28) so:

$$\left.\begin{aligned}
&\sum_{M'M_1'} \int \sigma_{MM_1, M'M_1'}(\boldsymbol{c}, \boldsymbol{c}_1 | \boldsymbol{c}', \boldsymbol{c}_1')\, d\boldsymbol{c}' \\
&\quad = \sum_{M', M'_1} \int \sigma_{M'M_1', MM_1}(\boldsymbol{c}', \boldsymbol{c}_1' | \boldsymbol{c}, \boldsymbol{c}_1)\, d\boldsymbol{c}' \\
&\quad = \frac{2\pi}{i k^2} \big(A(\boldsymbol{c}, \boldsymbol{c}_1 | \boldsymbol{c}, \boldsymbol{c}_1) - A^{\dagger}(\boldsymbol{c}, \boldsymbol{c}_1 | \boldsymbol{c}, \boldsymbol{c}_1)\big)_{MM_1, MM_1}.
\end{aligned}\right\} \qquad (93.14)$$

Dies verwenden wir nun in (93.12) und erhalten für die diagonalen Hauptachsen-Komponenten der Boltzmann-Gleichung die Beziehung

$$\left.\begin{aligned}
\frac{\partial F_M(\boldsymbol{c})}{\partial t} + c_\mu \frac{\partial F_M(\boldsymbol{c})}{\partial x_\mu} = \sum_{M_1 M' M_1'} \iint [F_{M'}(\boldsymbol{c}') F_{M_1'}(\boldsymbol{c}_1') - F_M(\boldsymbol{c}) F_{M_1}(\boldsymbol{c}_1)] \times \\
\times\, \sigma_{M'M_1', MM_1}(\boldsymbol{c}', \boldsymbol{c}_1' | \boldsymbol{c}, \boldsymbol{c}_1)\, g\, d\boldsymbol{c}'\, d\boldsymbol{c}_1.
\end{aligned}\right\} \qquad (93.15)$$

Hier zeigt sich deutlich der stochastische Charakter der Boltzmann-Gleichung für rotierende Moleküle. In der Tat ist (93.15) genau von dem Typ der klassischen Gl. (29.2) für einatomige Gemische, mit dem einzigen Unterschied, daß jetzt beim Stoß gewissermaßen eine Umwandlung der Gemischkomponenten ineinander stattfinden kann. Aber man muß sich darüber klar sein, daß (93.15) zur Lösung des gaskinetischen Problems nicht ausreicht. Der Streuquerschnitt σ ist ja nicht ein für allemal bekannt, sondern nach (93.13) durch die Streuamplituden A im Hauptachsensystem von f gegeben. σ weiß man also erst, wenn das kinetische Problem mittels der Matrix-Gleichung (92.14) bereits gelöst ist.

Gleichwohl reicht (93.15) zum Beweis des H-Theorems aus.

γ) Entropie und H-Theorem. Nach v. NEUMANN [*40*], S. 202 definieren wir, in Weiterbildung des Boltzmannschen Ansatzes (33.1), die Entropiedichte für ein Gas mit rotierenden Molekülen durch folgenden Ansatz, über dessen Brauchbarkeit das H-Theorem entscheiden wird:

$$\varrho s = -\mathsf{k} \int \mathrm{Sp}\,[\mathsf{f}(\boldsymbol{c}) \ln \mathsf{f}(\boldsymbol{c})]\, d\boldsymbol{c} = -\mathsf{k} \sum_M \int F_M(\boldsymbol{c}) \ln F_M(\boldsymbol{c})\, d\boldsymbol{c}. \qquad (93.16)$$

Darin bezeichnet ϱ die Massendichte, s die lokale spezifische Entropie und k die Boltzmann-Konstante. Das erste Integral — mit der Spur — hat, weil der Logarithmus keine Potenzreihendarstellung zuläßt, lediglich formale Bedeutung und bringt die Unitärinvarianz von s zum Ausdruck. Die zweite Schreibweise mit der Summe von Integralen, welche die Eigenwerte F_M enthalten, gibt die eigentliche konkrete Definition der Spur von f ln f und damit der Entropie.

Um den Beweis des H-Theorems zu führen, notieren wir noch die Identität

$$\sum_{MM_1M'M'_1} \iiint (F_{M'} F_{M'_1} - F_M F_{M_1})\, \sigma_{M'M'_1, MM_1}\, g\, d\boldsymbol{e}'\, d\boldsymbol{c}\, d\boldsymbol{c}_1 = 0. \qquad (93.17)$$

Die Variablen $\boldsymbol{c}, \ldots, \boldsymbol{c}'_1$ sind hier — wie auch in den weiteren Formeln dieser Ziffer — der Kürze halber weggelassen; man hat sie sich in den F und in σ jeweils nach Maßgabe von deren Indices $M, \ldots, M'_1$ eingesetzt zu denken. Die Identität (93.17) ergibt sich sofort, indem man im zweiten Teilintegral die gestrichenen und ungestrichenen Variablen $\boldsymbol{c}, \ldots$ und Indices $M, \ldots$ vertauscht (Umbenennung) und sodann den Liouvilleschen Satz (88.11) sowie die Integralrelation (93.14) benutzt. Wie der Anblick von (93.15) zeigt, sorgt die Identität (93.17) für die Teilchenerhaltung. Das ist nicht verwunderlich, denn (93.14) geht ja auf die Unitarität der Streumatrix zurück, vgl. Ziff. 89γ.

Wir gehen über zur Berechnung der Entropieänderung infolge der Stöße. Unter Benutzung der Teilchenerhaltung und einer Umbenennung erhält man mittels (93.15)

$$\begin{aligned}
\left(\frac{\delta \varrho s}{\delta t}\right)_{\text{Stöße}} &= -k \sum_M \int \ln F_M \left(\frac{\delta F_M}{\delta t}\right)_{\text{Stöße}} d\boldsymbol{c} \\
&= -k \sum_{M, \ldots, M'_1} \iiint \ln F_M \,[F_{M'} F_{M'_1} - F_M F_{M_1}]\, \sigma_{M'M'_1, MM_1}\, g\, d\boldsymbol{e}'\, d\boldsymbol{c}\, d\boldsymbol{c}_1 \\
&= -\frac{k}{2} \sum_{M, \ldots, M'_1} \iiint \ln (F_M F_{M_1})\, [F_{M'} F_{M'_1} - F_M F_{M_1}]\, \sigma_{M'M'_1, MM_1}\, g\, d\boldsymbol{e}'\, d\boldsymbol{c}\, d\boldsymbol{c}_1 .
\end{aligned}$$

Zur weiteren Umformung werde in der letzten Form der logarithmische Faktor in die eckige Klammer hineinmultipliziert. Dadurch zerfällt das dreifache Integral in zwei Teilintegrale. In dem zweiten Teilintegral vertauschen wir wieder Gestrichenes und Ungestrichenes (Umbenennung) und benutzen den Liouvilleschen Satz (88.11) mit $g' = g$ sowie die Relation (93.14). Dadurch entsteht

$$\left(\frac{\delta \varrho s}{\delta t}\right)_{\text{Stöße}} = -\frac{k}{2} \sum_{M, \ldots, M'_1} \iiint \ln \left(\frac{F_M F_{M_1}}{F_{M'} F_{M'_1}}\right) F_{M'} F_{M'_1}\, \sigma_{M'M'_1, MM_1}\, g\, d\boldsymbol{e}'\, d\boldsymbol{c}\, d\boldsymbol{c}_1 . \qquad (93.18)$$

Wäre nun eine „detailed balance"-Symmetrie der Art $\sigma_{M'M'_1, MM_1} = \sigma_{MM_1, M'M'_1}$ erfüllt, so könnte man den Ausdruck (93.18) nach dem Boltzmannschen Verfahren nochmals nach geeigneter Umbenennung derart umformen, daß ein nichtnegativer Integrand entstünde, genau so wie in der Beziehung (33.2) für das einatomige Gemisch. Damit wäre das H-Theorem bewiesen. Bei den rotierenden Molekülen gilt aber allgemein statt der „detailed balance" nur die schwächere integrale Symmetrierelation (93.14), welche nichts weiter als die Teilchenerhaltung besagt[1]. Doch reicht, und das ist entscheidend, auch diese bereits aus, um (93.18) derart umzuschreiben, daß ein nichtnegativer Integrand auftritt. Dazu folgen wir einem Gedanken von W. PAULI, vgl. STÜCKELBERG (1952). Wir

[1] Auch die Heranziehung der Symmetrierelation (89.36) der Zeitumkehr würde es nicht ermöglichen, bei mehratomigen Gasen dieselbe Beweismethode anzuwenden wie bei den einatomigen. Es ist bemerkenswert, daß von der Symmetrie (89.36) bei der zu (93.20) führenden Überlegung kein Gebrauch gemacht wird. — In der klassischen Theorie sind die beiden Fälle, einatomige und mehratomige Gase, ebenfalls verschieden zu behandeln. Zum Beweis des H-Theorems dienen im letzteren Fall die Boltzmannschen Cyclen (s. BOLTZMANN [3] II, S. 235 und auch TOLMAN [51], S. 140).

multiplizieren die Identität (93.17) mit dem Faktor $k/2$ und subtrahieren sie von (93.18) mit dem Ergebnis:

$$\left.\begin{aligned}\left(\frac{\delta \varrho s}{\delta t}\right)_{\text{Stöße}} = \frac{k}{2} \sum_{M,\ldots,M_1'} \iiint \left(\frac{F_M F_{M_1}}{F_{M'} F_{M_1'}} - 1 - \ln \frac{F_M F_{M_1}}{F_{M'} F_{M_1'}}\right) \times \\ \times F_{M'} F_{M_1'} \sigma_{M'M_1', MM_1}\, g\, d\boldsymbol{e}'\, d\boldsymbol{c}\, d\boldsymbol{c}_1 .\end{aligned}\right\} \tag{93.19}$$

Weil $F, \sigma \geqq 0$ sind und wegen

$$x - 1 - \ln x > 0 \quad (\text{bzw.} = 0) \quad \text{für positive } x \neq 1 \quad (\text{bzw.} = 1),$$

hat man jetzt einen nichtnegativen Integranden. Damit ist das H-Theorem

$$\left(\frac{\delta \varrho s}{\delta t}\right)_{\text{Stöße}} \geqq 0 \tag{93.20}$$

bewiesen.

δ) Das thermische Gleichgewicht. Nach (93.19) nimmt durch die Stöße die Entropiedichte solange zu, bis

$$F_M(\boldsymbol{c})\, F_{M_1}(\boldsymbol{c}_1) = F_{M'}(\boldsymbol{c}')\, F_{M_1'}(\boldsymbol{c}_1') \tag{93.21}$$

gilt, und zwar für alle Werte $\boldsymbol{c}', \boldsymbol{c}_1', M', M_1'$, welche mit beliebigen Werten von $\boldsymbol{c}, \boldsymbol{c}_1, M, M_1$ nach dem Impuls- und Energiesatz verträglich sind. Wenn (93.21) gilt, dann ist das lokale thermische Gleichgewicht erreicht. Nun hängt voraussetzungsgemäß [s. Gl. (88.3)] die Energie eines Moleküls nicht von der magnetischen Quantenzahl M ab. Daher hat (93.21) Gültigkeit für beliebiges M bei festem M_1, M', M_1'. Das bedeutet, daß im thermischen Gleichgewicht

$$F_M(\boldsymbol{c}) = F(\boldsymbol{c}), \quad \text{d.h.} \quad f_{MN}(\boldsymbol{c}) = F(\boldsymbol{c})\, \delta_{MN} \tag{93.22}$$

ist, daß sich also die Moleküldrehimpulse isotrop im Raum verteilen. Jetzt schließt man wie bei den einatomigen Gasen — Ziff. 34 — weiter und stellt fest, daß im thermischen Gleichgewicht die Maxwell-Verteilung

$$[f_{MN}(\boldsymbol{c})]_{\text{th. Gl.}} = \frac{n}{2J+1} \left(\frac{m}{2\pi k T}\right)^{\frac{3}{2}} e^{-\frac{m C^2}{2kT}}\, \delta_{MN} \quad \text{mit} \quad \boldsymbol{C} = \boldsymbol{c} - \boldsymbol{v} \tag{93.23}$$

realisiert ist[1].

94. Näherung der Boltzmann-Gleichung durch Annahme isotroper Orientierung der molekularen Drehimpulse. Durch die quantentheoretische Boltzmann-Gleichung (92.14) dürften die kinetischen Vorgänge in einem Gas mit rotierenden Molekülen von festem Drehimpulsbetrag exakt erfaßt werden. Dank der Zustandsbeschreibung durch eine Verteilungsmatrix $f_{MN}(\boldsymbol{c})$ ist es im Prinzip möglich, anisotrope Richtungsverteilungen der Molekülachsen zu studieren. Solche Anisotropien werden — auch ohne äußeres richtendes Kraftfeld — beispielsweise in einem inhomogen strömenden Gas auftreten. Es sei an die Strömungsdoppelbrechung erinnert, welche an Flüssigkeiten aus langgestreckten Makromolekülen leicht zu beobachten ist. Ähnliche Erscheinungen gibt es vielleicht beim Vorhandensein von Temperaturgradienten. Auch der Barnett-Effekt, vgl. Ziff. 2δ, gehört zu dieser Gruppe von Erscheinungen.

---

[1] Die obige Schlußweise und das Ergebnis (93.23) sind allerdings nicht ganz genau, da in der dargelegten Theorie der verdünnten Gase die Erhaltung des Gesamtdrehimpulses beim Einzelstoß nicht exakt berücksichtigt werden kann. Strenggenommen findet bei einem als Ganzes einschließlich der Wände rotierenden Gas im thermischen Gleichgewicht eine teilweise Ausrichtung der Eigendrehimpulse der Moleküle statt, vgl. Ziff. 2δ.

Doch sind das bei Gasen gewiß äußerst subtile Effekte. Normalerweise sind, wenn nicht gerade ein starkes äußeres Richtfeld da ist (Senftleben-Effekt, s. am Schluß von Ziff. 92), die Molekülachsen nahezu isotrop orientiert und man darf die vereinfachende Annahme machen, daß die Verteilungsmatrix der Einheit proportional ist:

$$f_{MN}(\boldsymbol{c}) \approx \frac{1}{2J+1} f(\boldsymbol{c})\, \delta_{MN}. \tag{94.1}$$

Dann ist lediglich eine gewöhnliche Funktion $f(\boldsymbol{c})$ zu bestimmen und die Matrix-Gleichung (92.14) muß in eine klassische Boltzmann-Gleichung übergehen. Um dies zu sehen, bemerkt man, daß jetzt im zweiten Glied rechts von (92.14) der Faktor $f(\boldsymbol{c}, \boldsymbol{c}_1)$ ausgeklammert und somit das Schattentheorem (89.15) angewandt werden kann. Danach nehmen wir in (92.14) beiderseits die Spur Sp über die noch freien Indices. Aus (94.1) folgt

$$\sum_{M=-J}^{J} f_{MM}(\boldsymbol{c}) \approx f(\boldsymbol{c}).$$

Unter Beachtung von (92.11) erhält man so für die gewöhnliche Funktion f aus (94.1) die folgende klassische Boltzmann-Gleichung

$$\frac{\partial f(\boldsymbol{c})}{\partial t} + c_\mu \frac{\partial f(\boldsymbol{c})}{\partial x_\mu} \approx \iint [f(\boldsymbol{c}')\, f(\boldsymbol{c}_1') - f(\boldsymbol{c})\, f(\boldsymbol{c}_1)]\, \overline{\sigma}(\boldsymbol{e}', \boldsymbol{e}, E)\, g\, d\boldsymbol{e}'\, d\boldsymbol{c}_1. \tag{94.2}$$

Der effektive Streuquerschnitt $\overline{\sigma}$ ist durch

$$\left.\begin{aligned} \overline{\sigma}(\boldsymbol{e}', \boldsymbol{e}, E) &= \frac{1}{(2J+1)^2 k^2} \operatorname{Sp} \operatorname{Sp}_1 [\mathfrak{a}(\boldsymbol{e}, \boldsymbol{e}', E)\, \mathfrak{a}^\dagger(\boldsymbol{e}', \boldsymbol{e}, E)] \\ &= \frac{1}{(2J+1)^2 k^2} \sum_{M, \ldots, M_1'} |a_{MM_1, M'M_1'}(\boldsymbol{e}, \boldsymbol{e}', E)|^2 \end{aligned}\right\} \tag{94.3}$$

gegeben. Er hat keinen Index mehr. Deshalb kann er, da keine Raumrichtung ausgezeichnet ist, nur vom Winkel χ zwischen $\boldsymbol{e}'$ und $\boldsymbol{e}$ abhängen[1]:

$$\overline{\sigma}(\boldsymbol{e}', \boldsymbol{e}, E) = \overline{\sigma}(\chi, g) = \overline{\sigma}(\boldsymbol{e}, \boldsymbol{e}', E). \tag{94.4}$$

Somit stimmt (94.2) vollkommen überein mit der Boltzmann-Gleichung (28.5) für das einatomige Gas.

Die klassische Formel (94.2) tritt also nur in Verbindung mit der Relation (94.4) der „detailed balance" auf. Für die einzelnen Rotationszustände von Molekülen (oder Elementarteilchen) hingegen gibt es keine allgemeine „detailed balance"-Relation. Es ist bezeichnend, daß bei den rotierenden Molekülen die strenge Boltzmann-Gleichung (92.14) eine wesentlich andere Gestalt als (94.2) hat.

95. Die Boltzmann-Gleichung für Gase aus anregbaren Molekülen. *α) Allgemeines.* Bei den Betrachtungen von Ziff. 90 ab postulierten wir einheitlichen inneren Zustand der Moleküle, von der Orientierung abgesehen. Die Moleküle sollten also nicht-anregbar und die Stöße elastisch sein, wenngleich mit Orientierungsänderungen verknüpft. Diese Voraussetzung sei jetzt fallengelassen.

Wir erwähnen im folgenden statt der beiden „drehinvarianten" inneren Quantenzahlen V, J, vgl. Gl. (88.2), zwecks Kürze nur die eine, J, und schreiben die Streuamplitude aus (89.11) in der Weise an

$$a_{I'I_1', II_1}(\boldsymbol{e}', \boldsymbol{e}, E) = a^{(J'J_1', JJ_1)}_{M'M_1', MM_1}(\boldsymbol{e}', \boldsymbol{e}, E). \tag{95.1}$$

[1] Der Wechsel von der Variablen E zur Variablen g geschieht im Hinblick auf die frühere Schreibweise (23.3) für den Streuquerschnitt des einatomigen Gases.

Ähnlich wie in (92.3) werde wieder ein Wellenpaket nebst Streuwellen betrachtet. In dem einfallenden „ebenen" Paket sollen nur zwei ganz bestimmte Zustände J, J_1, jedoch mit allen möglichen verschiedenen Orientierungen M, M_1, vertreten sein. Die elastische Streuwelle, eine Hohlkugelwelle wie in Fig. 55, enthält dann ebenfalls nur diese beiden Zustände, wieder mit allen möglichen Orientierungen. Von den außerdem auftretenden unelastischen Hohlkugelwellen, d.s. die Streuwellen mit $J', J_1' \neq J, J_1$, sei angenommen, daß sie nicht zu lange nach dem Stoß, nämlich bevor der nächste Stoß geschieht, voneinander und von der elastischen Streuwelle räumlich vollständig getrennt sind. Voraussetzung dafür ist, daß die verschiedenen Energieniveaus E_J hinreichend weit auseinander liegen. Dann hat man auch nach dem Stoß wieder nur Wellenpakete mit einheitlichem Zustand J', J_1'. Einen Schatten gibt es nur durch Interferenz des ebenen Pakets mit der elastischen Streuwelle. Die Folge ist, daß in der Verallgemeinerung von (92.10) nur diejenigen Matrixelemente des Projektionsoperators P vorkommen, welche bezüglich der J-Indices diagonal sind. Dies überträgt sich auf die Maxwellsche Transportgleichung und auf die Boltzmann-Gleichung. In diese gehen nur die in J diagonalen Elemente der Verteilungsmatrix $f_{JM,KN}$ ein, die wir mit

$$f_{JM,JN} = f_{MN}^{(J)} \tag{95.2}$$

bezeichnen. Damit entsteht aus (92.14) die verallgemeinerte Boltzmann-Gleichung

$$\left.\begin{aligned} \frac{\partial \mathfrak{f}^{(J)}(\boldsymbol{c})}{\partial t} + \cdots = \sum_{J_1 J' J_1'} \int \mathrm{Sp}_1 \Big[\int \mathfrak{a}^{(JJ_1, J'J_1')}(\boldsymbol{e}, \boldsymbol{e}', E)\, \mathfrak{f}^{(J'J_1')}(\boldsymbol{c}', \boldsymbol{c}_1')\, \mathfrak{a}^{\dagger\,(J'J_1', JJ_1)}(\boldsymbol{e}', \boldsymbol{e}, E)\, d\boldsymbol{e}' - \\ - \frac{2\pi}{i} \big(\mathfrak{a}^{(JJ_1, JJ_1)}(\boldsymbol{e}, \boldsymbol{e}, E)\, \mathfrak{f}^{(JJ_1)}(\boldsymbol{c}, \boldsymbol{c}_1) - \mathfrak{f}^{(JJ_1)}(\boldsymbol{c}, \boldsymbol{c}_1)\, \mathfrak{a}^{\dagger\,(JJ_1, JJ_1)}(\boldsymbol{e}, \boldsymbol{e}, E) \Big] \frac{g\, d\boldsymbol{c}_1}{k^2}. \end{aligned}\right\} \tag{95.3}$$

Die Spurbildung bezieht sich auf die M-Indices der Matrizen $\mathfrak{f}^{(J)}$ und $\mathfrak{a}^{(JJ_1, J'J_1')}$ bzw. ihrer Produkte.

β) Kugelsymmetrische Näherung. Wie in der vorigen Ziffer betrachten wir jetzt die Näherung der Isotropie:

$$f_{MN}^{(J)}(\boldsymbol{c}) \approx \frac{1}{2J+1} f_J(\boldsymbol{c})\, \delta_{MN}. \tag{95.4}$$

Für die Konzentration der Teilchen mit der Quantenzahl J — ohne Rücksicht auf die Orientierung — gilt dann

$$n_J = \mathrm{Sp} \int \mathfrak{f}^{(J)}\, d\boldsymbol{c} = \int f_J(\boldsymbol{c})\, d\boldsymbol{c}. \tag{95.5}$$

Wir gehen mit dieser Näherung in (95.3) ein und nehmen beiderseits die Spur Sp. So entsteht als kugelsymmetrische Näherung der Boltzmann-Gleichung[1]

$$\frac{\partial f_J}{\partial t} + \cdots = \sum_{J_1 J' J_1'} \iint \left[f_{J'} f_{J_1'} - f_J f_{J_1} \frac{(2J'+1)(2J_1'+1)}{(2J+1)(2J_1+1)} \right] g'^2\, \bar{\sigma}_{J'J_1', JJ_1}(\boldsymbol{e}', \boldsymbol{e}, E)\, \frac{d\boldsymbol{e}'\, d\boldsymbol{c}_1}{g}. \tag{95.6}$$

Dabei wurde das Schattentheorem (89.15) benutzt und der effektive Streuquerschnitt

$$\bar{\sigma}_{J'J_1', JJ_1}(\boldsymbol{e}', \boldsymbol{e}, E) = \frac{1}{(2J'+1)(2J_1'+1)\, k'^2} \sum_{M, \ldots, M_1'} \left| a_{MM_1, M'M_1'}^{(JJ_1, J'J_1')}(\boldsymbol{e}, \boldsymbol{e}', E) \right|^2 \tag{95.7}$$

eingeführt. Dieser Streuquerschnitt gibt die Übergangswahrscheinlichkeit an aus einem Anfangszustand $I', I_1', \boldsymbol{e}'$ in irgendeinen der $(2J+1)(2J_1+1)$ Endzustände mit gegebenen $J, J_1, \boldsymbol{e}$, und zwar gemittelt über alle $(2J'+1)(2J_1'+1)$

---

[1] Die Variablen $\boldsymbol{c}, \ldots, \boldsymbol{c}_1'$ sind entsprechend den Indices $J, \ldots, J_1'$ eingesetzt zu denken.

verschiedenen anfänglichen Orientierungen M', M_1'. Nach dem Schattentheorem ist übrigens

$$\sum_{J'J_1'} \int (2J'+1)(2J_1'+1)\, k'^2 \bar{\sigma}_{J'J_1', JJ_1}(\boldsymbol{e}', \boldsymbol{e})\, d\boldsymbol{e}' = \sum_{J'J_1'} \int (2J+1)(2J_1+1)\, k^2 \bar{\sigma}_{JJ_1, J'J_1'}(\boldsymbol{e}, \boldsymbol{e}')\, d\boldsymbol{e}',$$

so daß man anstelle von (95.6) auch schreiben kann

$$\frac{\partial f_J}{\partial t} + \cdots = \sum_{J_1 J' J_1'} \iint \left[f_{J'} f_{J_1'} g'^2 \bar{\sigma}_{J'J_1', JJ_1}(\boldsymbol{e}', \boldsymbol{e}) - f_J f_{J_1} g^2 \bar{\sigma}_{JJ_1, J'J_1'}(\boldsymbol{e}, \boldsymbol{e}')\right] \frac{d\boldsymbol{e}'\, d\boldsymbol{c}_1}{g}. \quad (95.8)$$

Mittels des Liouvilleschen Satzes (88.11) liest man daraus sofort die Erhaltungssätze ab.

Aus der Relation der Zeitumkehr

$$|a_{II_1, I'I_1'}(\boldsymbol{e}, \boldsymbol{e}', E)|^2 = |a_{-I'-I_1', -I-I_1}(-\boldsymbol{e}', -\boldsymbol{e}, E)|^2$$

vgl. (89.36), ergibt sich für den Streuquerschnitt zunächst die Beziehung

$$(2J'+1)(2J_1'+1)\, g'^2 \bar{\sigma}_{J'J_1', JJ_1}(\boldsymbol{e}', \boldsymbol{e}, E) = (2J+1)(2J_1+1)\, g^2 \bar{\sigma}_{JJ_1, J'J_1'}(-\boldsymbol{e}, -\boldsymbol{e}', E).$$

Da in dem Streuquerschnitt keine Raumrichtung mehr ausgezeichnet ist — über sämtliche M ist ja summiert —, muß überdies gelten[1]

$$\bar{\sigma}_{J'J_1', JJ_1}(\boldsymbol{e}', \boldsymbol{e}, E) = \bar{\sigma}_{J'J_1', JJ_1}(\chi, g') \quad \text{mit} \quad \chi = \sphericalangle(\boldsymbol{e}', \boldsymbol{e}).$$

Damit hat man die erweiterte „detailed balance"-Relation

$$(2J'+1)(2J_1'+1)\, g'^2 \bar{\sigma}_{J'J_1', JJ_1}(\chi, g') = (2J+1)(2J_1+1)\, g^2 \bar{\sigma}_{JJ_1, J'J_1'}(\chi, g). \quad (95.9)$$

Es sei betont, daß nur die Zeitumkehrbarkeit und die Drehinvarianz, nicht aber die Spiegelung benutzt wurde. Die Betrachtungen gelten also z.B. auch für Moleküle mit einem asymmetrischen Kohlenstoffatom; insofern unterscheiden sie sich von denjenigen WATANABES [*54*][2].

Durch Verwendung dieser Relation nimmt die Boltzmann-Gleichung für das Gas mit anregbaren Molekülen in kugelsymmetrischer Näherung schließlich die Gestalt[3] an

$$\frac{\partial f_J}{\partial t} + \cdots = \sum_{J_1 J' J_1'} \iint \left[f_{J'} f_{J_1'} \frac{(2J+1)(2J_1+1)}{(2J'+1)(2J_1'+1)} - f_J f_{J_1}\right] \bar{\sigma}_{JJ_1, J'J_1'}(\chi, g)\, g\, d\boldsymbol{e}'\, d\boldsymbol{c}_1. \quad (95.10)$$

Eine eingehende Diskussion des physikalischen Inhalts dieser Gleichung findet sich in der folgenden Ziffer. Auch die zugehörige Gleichgewichtsverteilung ist dort angegeben [s. Gl. (96.9) mit (96.5)].

Durch diese Betrachtung ist die Theorie der allgemeinen mehratomigen Moleküle näherungsweise auf die wesentlich einfachere Theorie der kugelsymmetrischen mehratomigen Moleküle zurückgeführt. Unmöglich ist selbstverständlich die Zurückführung auf kugelsymmetrische, nicht-anregbare Moleküle, d.h. auf einatomige Gase. Ausdrücklich stellen das MASON und RICE (1954b)

---

[1] Siehe auch Anm. 1 von S. 490.

[2] In älteren Arbeiten findet sich häufig eine Begründung der „detailed balance" mittels der ersten Näherung der Störungsrechnung. Aus der Hermitezität der Wechselwirkung schloß man auf die Symmetrie der Übergangswahrscheinlichkeit bezüglich des kompletten Anfangs- und Endzustands (einschließlich Orientierung), erhielt also eine Relation, die über die Aussage (95.9) noch hinausgeht. Diese Begründung war aber nicht stichhaltig, weil bei Streuproblemen die erste Näherung der Störungsrechnung im allgemeinen keineswegs verbindlich ist.

[3] Siehe auch Anm. 1 von S. 491.

fest, die belegen, daß man zur genäherten Wiedergabe der Eigenschaften mehratomiger Gase mittels der für einatomige Gase aufgestellten Formeln bei jedem Transportphänomen andere Lennard-Jones-Parameter braucht.

96. Die formale Theorie der hydrodynamischen Näherung für Gase, die aus kugelsymmetrischen anregbaren Molekülen bestehen. Die formale Enskog-Chapmansche Theorie einatomiger Gase haben WANG CHANG und UHLENBECK (1951) in hydrodynamischer Näherung auf Gase übertragen, die aus kugelsymmetrischen anregbaren Molekülen bestehen. Nach den Betrachtungen aus Ziff. 95 lassen sich diese Ergebnisse in guter Näherung auch für Gase aus beliebigen, nicht-kugelsymmetrischen Molekülen übernehmen. Wir werden den folgenden Betrachtungen die Boltzmann-Gleichung (95.10) zugrundelegen.

Der Unterschied zwischen einem reinen Gas aus mehratomigen kugelsymmetrischen Molekülen und einem Gemisch einatomiger Gase, wie es in Ziff. 32 und 53—61 behandelt wurde, besteht u.a. darin, daß im ersten Fall nur ein einziger Teilchen-Erhaltungssatz existiert und daß die Energie des Moleküls aus zwei Anteilen, der Energie der Translation und derjenigen der inneren Bewegung, besteht. Anstelle der *Stoßinvarianten* (32.4) hat man jetzt

$$\psi_J = 1; \quad \psi_J = m c_\mu; \quad \psi_J = \frac{m}{2} c^2 + E_J. \tag{96.1}$$

Ebenso wie in (95.10) sind dabei durch J die Quantenzahl des innermolekularen Drehimpulsbetrags und etwaige weitere, drehinvariante innere Quantenzahlen des Moleküls (Vibration) zusammengefaßt.

Die *Definitionen* der Konzentration, Gasgeschwindigkeit und des Drucks mittels der Verteilungsfunktionen f_J bedürfen nach dem in Ziff. 32 Gesagten weiter keiner Erläuterung:

$$\left.\begin{aligned} n &= \sum_J \int f_J \, d\mathbf{c}, \quad v_\mu = \frac{1}{n} \sum_J \int c_\mu f_J \, d\mathbf{c}, \\ p_{\mu\nu} &= \sum_J \int m C_\mu C_\nu f_J \, d\mathbf{c} \quad \text{mit} \quad C_\mu = c_\mu - v_\mu. \end{aligned}\right\} \tag{96.2}$$

Dagegen gibt es bei den energetischen Größen einige Änderungen gegenüber dem früheren. Die spezifische Energie (pro Masseneinheit) ist gegeben durch, bzw. die Temperatur im Nichtgleichgewicht sei definiert durch [s. auch (32.15)],

$$u = \frac{1}{m}\left[\frac{3}{2}\,\mathrm{k}T + \overline{E}(T)\right] = \frac{1}{\varrho} \sum_J \int \left(\frac{m}{2} C^2 + E_J\right) f_J \, d\mathbf{c}. \tag{96.3}$$

Dabei ist unter $\overline{E}(T)$ die mittlere innermolekulare Energie im thermischen Gleichgewicht bei der Temperatur T zu verstehen[1]. Es ist also

$$\overline{E}(T) = \frac{1}{n} \sum_J n_J(T)\, E_J \tag{96.4}$$

mit der Abkürzung

$$n_J(T) = n\,(2J+1)\,\mathrm{e}^{-E_J/\mathrm{k}T} \Big/ \sum_{J_0} (2J_0+1)\,\mathrm{e}^{-E_{J_0}/\mathrm{k}T}. \tag{96.5}$$

[1] CHAPMAN und COWLING [4], S. 206, definieren die Temperatur im Nichtgleichgewicht durch die Forderung

$$\tfrac{1}{3} p_{\mu\mu} = n\,\mathrm{k}T.$$

Die physikalischen Ergebnisse sind natürlich unabhängig von der Temperaturdefinition.

Nebenbei sei die aus den beiden letzten Gleichungen folgende Beziehung für die spezifische Wärme notiert:

$$\frac{d\overline{E}}{dT} = \frac{1}{kT^2}\left[\frac{1}{n}\sum_J n_J E_J^2 - (\overline{E})^2\right] = \frac{1}{kT^2}\left[\overline{E^2} - (\overline{E})^2\right]. \tag{96.6}$$

Der Wärmestrom schließlich ist gegeben durch

$$q_\mu = \sum_J \int \left(\frac{m}{2} C^2 + E_J\right) C_\mu f_J \, d\boldsymbol{c}. \tag{96.7}$$

Die Erhaltungsgleichungen sind formal dieselben wie beim einatomigen Gas. Wir brauchen lediglich die an Stelle der Temperaturgleichung (32.23) tretende Beziehung

$$n\left(\frac{3}{2}k + \frac{d\overline{E}}{dT}\right)\frac{dT}{dt} = -\frac{\partial q_\mu}{\partial x_\mu} - p_{\mu\nu}\frac{\partial v_\mu}{\partial x_\nu} \tag{96.8}$$

zu notieren.

Was nun die *hydrodynamische Näherung* anbelangt, so verlaufen die allgemeinen Überlegungen weitgehend parallel zu denen in Teil F, I dieses Artikels. Man setzt

$$f_J = F_J(1 + \Phi_J^{(1)}) \quad \text{mit} \quad F_J = n_J\left(\frac{\beta}{\pi}\right)^{\frac{3}{2}} e^{-\beta C^2}, \quad \beta = \frac{m}{kT} \tag{96.9}$$

und hat nach (95.10) näherungsweise

$$\frac{\partial F_J}{\partial t} + \cdots = \mathcal{D}(F_J) = -\sum_{J_1 J' J_1'} n_J n_{J_1} I_{JJ_1, J'J_1'}(\Phi^{(1)}) \tag{96.10}$$

mit

$$\left.\begin{aligned} n_J n_{J_1} I_{JJ_1, J'J_1'}(\Phi) = \iint F_J(\boldsymbol{C}) F_{J_1}(\boldsymbol{C}_1)\,[\Phi_J(\boldsymbol{c}) + \Phi_{J_1}(\boldsymbol{c}_1) - {}\\ - \Phi_{J'}(\boldsymbol{c}') - \Phi_{J_1'}(\boldsymbol{c}_1')]\,\overline{\sigma}_{JJ_1, J'J_1'}(\chi, g)\, g\, d\boldsymbol{e}'\, d\boldsymbol{c}_1. \end{aligned}\right\} \tag{96.11}$$

Damit $\Phi^{(1)}$ eindeutig festgelegt ist, wird verfügt, daß $F_J\Phi_J^{(1)}$ nichts zu den lokalen Werten von Konzentration, Impuls- und Energiedichte beitragen soll:

$$\left.\begin{aligned} &\sum_J \int F_J \Phi_J^{(1)}\, d\boldsymbol{c} = 0;\\ &\sum_J \int m\, C_\mu F_J \Phi_J^{(1)}\, d\boldsymbol{c} = 0;\\ &\sum_J \int \left(\frac{m}{2}C^2 + E_J\right) f_J\, \Phi_J^{(1)} d\boldsymbol{c} = 0. \end{aligned}\right\} \tag{96.12}$$

Aus (96.9) mit (96.5)entnimmt man nun

$$\frac{dF_J}{F_J} = \frac{dn}{n} + \left[\beta C^2 - \frac{3}{2} + \frac{1}{kT}(E_J - \overline{E})\right]\frac{dT}{T} - 2\beta C_\mu d(c_\mu - v_\mu).$$

Damit kann man die linke Seite von (96.10) im einzelnen aufschreiben [man benutzt für $\mathcal{D}$ die zweite Schreibweise aus (53.12) mit $b_{\mu J} = b_\mu$]. Die Ableitungen von n, $\boldsymbol{v}$ und T nach der Zeit eliminiert man mittels der Lösbarkeitsbedingungen von (96.10), die wieder mit den Erhaltungssätzen in nullter Näherung identisch sind:

$$\frac{dn}{dt} = -n\frac{\partial v_\mu}{\partial x_\mu}; \quad \varrho\frac{dv_\mu}{dt} = -\frac{\partial p}{\partial x_\mu} + \varrho\, b_\mu; \quad n\left(\frac{3}{2}k + \frac{d\overline{E}}{dT}\right)\frac{dT}{dt} = -p\frac{\partial v_\mu}{\partial x_\mu}. \tag{96.13}$$

So findet man

$$\left.\begin{aligned}\mathscr{D}(F_J) = F_J\Big\{&\Big(\beta C^2 - \frac{5}{2} + \frac{E_J - \overline{E}}{\mathsf{k}T}\Big) C_\mu \frac{\partial T}{T\,\partial x_\mu} + 2\beta \langle C_\mu C_\nu\rangle \Big\langle\frac{\partial v_\mu}{\partial x_\nu}\Big\rangle + \\ &+ \frac{1}{m\,c_v}\Big[\Big(\frac{2}{3}\beta C^2 - 1\Big)\frac{d\overline{E}}{dT} - \frac{E_J - \overline{E}}{T}\Big]\frac{\partial v_\mu}{\partial x_\mu}\Big\},\end{aligned}\right\} \tag{96.14}$$

wobei

$$c_v = \frac{1}{m}\Big(\frac{3}{2}\mathsf{k} + \frac{d\overline{E}}{dT}\Big) \tag{96.15}$$

die spezifische Wärme bedeutet. Die symmetrischen, irreduziblen Tensoren $\langle C_\mu C_\nu\rangle$, $\langle \partial v_\mu/\partial x_\nu\rangle$ sind in (46.4) und (46.5) erklärt. Neu gegenüber dem einatomigen Gas ist vor allem das Glied proportional der Divergenz $\partial v_\mu/\partial x_\mu$. Wie alsbald deutlich wird, hat es zur Folge, daß die Spur des Reibungsdruck-Tensors nicht mehr verschwindet, womit das Auftreten einer Volumviscosität zusammenhängt.

Entsprechend der Form (96.14), welche die linke Seite der Integralgleichungen (96.10) annimmt, setzt man nun an

$$\Phi_J^{(1)} = -\frac{1}{n}\Big[A_J(C)\, C_\mu \frac{\partial T}{T\,\partial x_\mu} + B_J(C)\,\beta\langle C_\mu C_\nu\rangle\Big\langle\frac{\partial v_\mu}{\partial x_\nu}\Big\rangle + D_J(C)\,\frac{\partial v_\mu}{\partial x_\mu}\Big] \tag{96.16}$$

und erhält für die Funktionssätze A, B, D die Gleichungen

$$\sum_{J_1 J' J_1'} \frac{n_{J_1}}{n} I_{J J_1, J' J_1'}\left\{\begin{array}{l} A\, C_\mu \\ B\,\beta\langle C_\mu C_\nu\rangle \\ D \end{array}\right\} = \frac{F_J}{n_J}\cdot\left\{\begin{array}{l}\Big(\beta C^2 - \frac{5}{2} + \frac{E_J - \overline{E}}{\mathsf{k}T}\Big) C_\mu \\ 2\beta\langle C_\mu C_\nu\rangle \\ \frac{1}{m\,c_v}\Big[\Big(\frac{2}{3}\beta C^2 - 1\Big)\frac{d\overline{E}}{dT} - \frac{E_J - \overline{E}}{T}\Big].\end{array}\right\} \tag{96.17}$$

Die meisten der Nebenbedingungen (96.12) sind dabei identisch erfüllt. Lediglich die Bedingungen

$$\sum_J \int m\,C^2 F_J A_J\, d\boldsymbol{C} = 0;\quad \sum_J \int F_J D_J\, d\boldsymbol{C} = 0;\quad \sum_J \int\Big(\frac{m}{2}C^2 + E_J\Big) F_J D_J\, d\boldsymbol{C} = 0 \tag{96.18}$$

müssen ausdrücklich gestellt werden. Man überzeugt sich, daß die rechten Seiten von (96.17) die Lösbarkeitsbedingungen erfüllen, d.h. orthogonal sind zu den Lösungen (96.1) der homogenen Integralgleichung; dabei hat man von (96.6) Gebrauch zu machen.

Das Problem ist nun auf die Bestimmung der Funktionen A_J, B_J, D_J reduziert, die durch (96.17) mit (96.18) eindeutig festgelegt sind. Im Besitze dieser Lösungen kann man die *Transportkoeffizienten* angeben. Zunächst erhält man durch Einsetzen von (96.16) in (96.7) für den Wärmestrom

$$q_\mu = -\lambda\frac{\partial T}{\partial x_\mu} \quad\text{mit}\quad \lambda = \frac{\mathsf{k}}{3}\sum_J\int\Big(\beta C^2 - \frac{5}{2} + \frac{E_J - \overline{E}}{\mathsf{k}T}\Big) C^2 \frac{F_J}{n} A_J\, d\boldsymbol{C}. \tag{96.19}$$

Mit dem Klammersymbol

$$[\Psi, \Phi] = \sum_{J J_1 J' J_1'} \int \Psi_J \frac{n_J n_{J_1}}{n^2} I_{J J_1, J' J_1'}(\Phi)\, d\boldsymbol{c} = [\Phi, \Psi];\quad [\Phi, \Phi] \geqq 0 \tag{96.20}$$

hat man somit nach (96.17) für den Wärmeleitkoeffizienten auch

$$\lambda = \frac{\mathsf{k}}{3}[A\, C_\mu, A\, C_\mu] > 0. \tag{96.21}$$

Für den Drucktensor aus (96.2) bekommt man

$$p_{\mu\nu} = p\,\delta_{\mu\nu} - \eta\cdot 2\left\langle\frac{\partial v_\mu}{\partial x_\nu}\right\rangle - \eta_V \frac{\partial v_\lambda}{\partial x_\lambda}\,\delta_{\mu\nu} \tag{96.22}$$

mit $p = n\,\mathrm{k}T$, mit der Scher-Viscosität

$$\eta = \frac{m}{15}\sum_J \int \beta C^4 \frac{F_J}{n} B_J\,d\boldsymbol{C} = \frac{\mathrm{k}T}{10}\,[B\beta\,\langle C_\mu C_\nu\rangle\,,\; B\beta\,\langle C_\mu C_\nu\rangle] > 0 \tag{96.23}$$

und der Volumviscosität

$$\eta_V = \frac{m}{3}\sum_J \int C^2 \frac{F_J}{n} D_J\,d\boldsymbol{C} = \mathrm{k}T\,[D, D] > 0\,. \tag{96.24}$$

Bei der letzten Umformung hat man die Nebenbedingungen (96.18) zu benutzen.

Für die *lokale Entropieerzeugung* gilt nach wie vor (61.1) mit der Bedeutung (96.20) des Klammersymbols. Durch Einsetzen von (96.16) erhält man unter Benutzung von (96.17), (96.12) und (96.15) als Verallgemeinerung von (61.2)

$$n^2\,[\Phi^{(1)}, \Phi^{(1)}] = -\frac{1}{\mathrm{k}T}\left(q_\mu \frac{\partial T}{T\,\partial x_\mu} + (p_{\mu\nu} - p\,\delta_{\mu\nu})\frac{\partial v_\mu}{\partial x_\nu}\right). \tag{96.25}$$

Nach (96.19) und (96.22) kann man dafür auch

$$n^2\,[\Phi^{(1)}, \Phi^{(1)}] = \frac{1}{\mathrm{k}T}\left(\frac{\lambda}{T}\frac{\partial T}{\partial x_\mu}\frac{\partial T}{\partial x_\mu} + 2\eta\left\langle\frac{\partial v_\mu}{\partial x_\nu}\right\rangle\left\langle\frac{\partial v_\mu}{\partial x_\nu}\right\rangle + \eta_V\left(\frac{\partial v_\mu}{\partial x_\mu}\right)^2\right) \tag{96.26}$$

schreiben. Dies stellt die Übertragung von (61.3) auf ein reines mehratomiges Gas mit kugelsymmetrischem Streuquerschnitt dar.

Das Auftreten der *Volumviscosität* bedeutet, daß — anders als im einatomigen Fall — bei allseitiger Kompression oder Expansion des Gases eine Energiedissipation stattfindet. Diese Energiedissipation macht sich in einer zusätzlichen Schallabsorption bemerkbar, welche zu der „klassischen", durch λ und η bedingten Schallabsorption hinzukommt [TISZA (1942), s. auch Ziff. 99]. Das Wesen dieser Energiedissipation tritt durch die folgende Überlegung von WANG CHANG und UHLENBECK (1951) schön hervor. Man betrachtet die Dichte der Translationsenergie

$$\varrho\,u_{\text{trans}} = \sum_J \int \frac{m}{2} C^2 f_J\,d\boldsymbol{c} = \frac{1}{2}\,p_{\mu\mu}\,,$$

also nach (96.22)

$$\varrho\,u_{\text{trans}} = \frac{3}{2}\left(n\,\mathrm{k}T - \eta_V\frac{\partial v_\mu}{\partial x_\mu}\right), \tag{96.27}$$

und definiert eine Translationstemperatur durch

$$\varrho\,u_{\text{trans}} = \frac{3}{2}\,n\,\mathrm{k}\,T_{\text{trans}}\,. \tag{96.28}$$

Somit ist

$$T_{\text{trans}} - T = -\frac{\eta_V}{n\,\mathrm{k}}\frac{\partial v_\mu}{\partial x_\mu}\,. \tag{96.29}$$

Die Volumviscosität ist also maßgebend für den Unterschied der Translationstemperatur und der tatsächlichen Temperatur des Gases. Im Auftreten eines solchen Unterschieds liegt die Irreversibilität des Kompressions- und Dilatations-Vorgangs begründet. Man kann weiter auch die Dichte der innermolekularen Energie betrachten [s. auch (96.3)]

$$\varrho\,u_{\text{intra}} = \varrho(u - u_{\text{trans}}) = n\,\overline{E}\,(T) + \frac{3}{2}\,\eta_V\frac{\partial v_\mu}{\partial x_\mu} \tag{96.30}$$

und eine entsprechende innere Temperatur durch

$$\varrho\, u_{\text{intra}} = n \bar{E}\,(T_{\text{intra}}) \tag{96.31}$$

definieren. Dann gilt, wenn $T_{\text{intra}} \approx T$, nach den beiden letzten Gleichungen

$$T_{\text{intra}} - T \approx \frac{3}{2}\,\frac{\eta_V}{n\,(d\bar{E}/dT)}\,\frac{\partial v_\mu}{\partial x_\mu} \tag{96.32}$$

und nach (96.29) mit (96.15)

$$T_{\text{trans}} - T_{\text{intra}} \approx -\frac{m\,c_v}{d\bar{E}/dT}\cdot T\,\frac{\eta_V}{p}\,\frac{\partial v_\mu}{\partial x_\mu}. \tag{96.33}$$

Somit kann man η_V auch als Maß für den Unterschied der Translations- und der inneren Temperatur ansehen, wenn er klein ist. Die Größe

$$\frac{\eta_V}{p} = \tau_{\text{intra}} = \frac{1}{n}\,[D, D] \tag{96.34}$$

hat die Dimension einer Zeit. Sie stellt die Relaxationszeit dar, mit der sich in einem sich selbst überlassenen Gas die Temperaturen der Translation und der innermolekularen Bewegung einander angleichen. —

Die Beschreibung von Schallwellen durch die hydrodynamische Näherung (einschließlich Volumviscosität) ist nur dann möglich, wenn die Schallfrequenz klein ist zu allen reziproken Relaxationszeiten τ^{-1} des Gases. Die Schall*dispersion*, welche bei Frequenzen von der Größe τ^{-1} merklich ist, kann nicht hydrodynamisch beschrieben werden. Deshalb betrachten WANG CHANG und UHLENBECK (1951) noch eine andere Art der Näherung, bei welcher sie voraussetzen, daß der Austausch von translatorischer und innerer Energie, gemessen an dem Austausch der Translationsfreiheitsgrade unter sich, langsam stattfindet[1] und daher zunächst ganz vernachlässigt werden kann (Unterdrückung der entsprechenden Streuquerschnitte). Durch eine Störungsrechnung wird dann im nächsten Schritt der Austausch (unelastische Stöße) berücksichtigt. So erhalten die Autoren eine kinetische Begründung des Ansatzes von HERZFELD und RICE (1928) für das Relaxationsgesetz der innermolekularen Temperatur. Dieser Ansatz ermöglicht es bekanntlich, die Schalldispersion, nicht nur die Absorption, zu beschreiben (KNESER (1931), RUTGERS (1933), s. auch RICHARDS [*44*], RICHARDSON [*45*]). Wegen der Einzelheiten sei auf die Originalarbeit von WANG CHANG und UHLENBECK (1951) verwiesen.

Schließlich sei bemerkt, daß die Übertragung der obigen Betrachtungen auf Gasgemische keine Schwierigkeit hat. Da dabei die Vorstellung kugelsymmetrischer Moleküle mit den entsprechenden "detailed balance"-Relationen zugrunde liegt, erhält man im wesentlichen wieder denselben Zusammenhang von Thermodiffusion und Diffusionswärme wie beim einatomigen Gemisch (vgl. Ziff. 57).

II. Spezielle Anwendungen.

97. Das Modell der rauhen Kugeln. Dieses Modell für rotierende Moleküle wurde von BRYAN (1894) aufgestellt. PIDDUCK (1922) hat die Methoden der exakten Gastheorie darauf angewandt. Unter idealer Rauhigkeit wird verstanden, daß die Relativgeschwindigkeit der zwei Kugeloberflächen-Punkte, welche bei der Berührung zusammenfallen, sich im Augenblick der Berührung genau umkehrt (was bei ideal glatten Kugeln nur für die Komponente der Relativgeschwindigkeit senkrecht zur gemeinsamen Tangentialebene im Berührungspunkt gilt). Das Modell hat den Vorteil, daß man keine Winkelvariablen benötigt;

[1] Das trifft für die Oszillation generell zu, für die Rotation zumeist nicht.

der Vektor der Winkelgeschwindigkeit kommt als einzige zusätzliche Variable in der klassischen Verteilungsfunktion gegenüber dem einatomigen Fall hinzu.

Die Berechnungen von PIDDUCK sind ausführlich in dem Buch von CHAPMAN und COWLING [4], Kap. 11, wiedergegeben. Wir geben die Resultate an. Es sei d der Durchmesser der Kugeln und I das Trägheitsmoment der radialsymmetrisch verteilten Masse in bezug auf eine Achse der Kugel. Das Trägheitsmoment geht in Form der dimensionslosen Konstanten

$$K = \frac{4I}{m\,d^2}, \quad 0 \leq K \leq \frac{2}{3} \tag{97.1}$$

in die Transportkoeffizienten ein. Der für K angegebene Spielraum entspricht den beiden Grenzfällen, daß die ganze Masse im Kugelmittelpunkt konzentriert bzw. gleichmäßig auf der Kugeloberfläche verteilt ist. Die Koeffizienten der Wärmeleitung, Reibung und Selbstdiffusion haben, dargestellt durch die entsprechenden Koeffizienten (51.4), (51.3) und (71.2) der einatomigen starren Kugeln von gleichem Durchmesser und gleicher Masse, folgende Werte

$$\left.\begin{aligned} \lambda &= \lambda_{\text{einatomig}} \cdot \frac{12\,(1+K)^2\,(37+151\,K+50\,K^2)}{25\,(12+75\,K+101\,K^2+102\,K^3)} \\ \eta &= \eta_{\text{einatomig}} \cdot \frac{(1+K)^2}{1+\frac{13}{6}K} \\ D &= D_{\text{einatomig}} \cdot \frac{1+K}{1+2K}\,. \end{aligned}\right\} \tag{97.2}$$

Nach PIDDUCK's Methode hat ferner KOHLER (1947a) die Volumviscosität η_V für die rauhen Kugeln berechnet und als Zusammenhang mit deren Scher-Viscosität η gefunden

$$\eta_V = \frac{1+\frac{13}{6}K}{10\,K}\,\eta\,. \tag{97.3}$$

Aus (97.2) entnimmt man, mit $c_v = 3\,\mathrm{R}/M$, M = Molgewicht, für die rauhen Kugeln auch

$$f = \frac{\lambda}{\eta\,c_v} = f_{\text{einatomig}} \cdot \frac{(6+13\,K)\,(37+151\,K+50\,K^2)}{25\,(12+75\,K+101\,K^2+102\,K^3)} \tag{97.4}$$

$$\frac{D\varrho}{\eta} = \left(\frac{D\varrho}{\eta}\right)_{\text{einatomig}} \cdot \frac{1+\frac{13}{6}K}{(1+K)\,(1+2K)}\,, \tag{97.5}$$

wobei nach (50.16) und (71.3)

$$f_{\text{einatomig}} = 2{,}5\,; \quad \left(\frac{D\varrho}{\eta}\right)_{\text{einatomig}} = \frac{6}{5} \tag{97.6}$$

gilt. In Tabelle 22 sind die Transportkoeffizienten für die beiden extremalen Massenverteilungen eigens aufgeführt.

Tabelle 22. *Die Transportkoeffizienten für ein Gas aus rauhen, starren Kugeln, verglichen mit denen des einatomigen Gases aus starren Kugeln.* Wegen K siehe (97.1).

	$\lambda_{\text{rauh}}/\lambda_{\text{einat.}}$	$\eta_{\text{rauh}}/\eta_{\text{einat.}}$	$D_{\text{rauh}}/D_{\text{einat.}}$	$(\eta_V/\eta)_{\text{rauh}}$	f_{rauh}	$(D\varrho/\eta)_{\text{rauh}}$
$K=0$	1,48	1	1	∞	1,85	1,20
$K=\frac{2}{3}$	1,56	1,13	0,71	0,367	1,71	0,755

An der Tabelle fällt zunächst auf, daß der Wärmeleitkoeffizient auch im Fall $K\to 0$ gegenüber dem des einatomigen Gases stark erhöht ist. Das ist verständlich auf Grund des klassischen Gleichverteilungssatzes der Energie, nach welchem die Rotationsenergie im thermischen Gleichgewicht unabhängig vom Trägheitsmoment ist. Die Kugeln mit $K \to 0$ rotieren eben sehr schnell und transportieren dadurch fast dieselbe Wärme wie die langsam rotierenden Kugeln, deren Masse auf der Oberfläche verteilt ist. Dagegen wirkt sich in Reibung und Diffusion die sehr schnelle Rotation der Kugeln mit $K=0$ bemerkenswerterweise gar nicht aus. Das Verhältnis f und die Volumreibung wird in den beiden folgenden Ziffern diskutiert.

98. Der Zusammenhang des Wärmeleitkoeffizienten mit der Viscosität (Euckensche Beziehung). Die Experimente zeigen, daß bei den mehratomigen Gasen das Verhältnis $f = \lambda/\eta\, c_v$ kleiner ist als bei den einatomigen, wo es nach MAXWELL den Wert 2,5 hat. Die Theorie des Modells der rauhen Kugeln (s. vorige Ziffer) gibt eine Begründung für diese Beobachtung.

EUCKEN (1913) hatte schon früher eine einfache Formel aufgestellt, welche die Versuchsergebnisse in großen Zügen befriedigend wiedergibt und die auch heute noch das praktisch wichtigste Resultat bei den mehratomigen Gasen sein dürfte. Die Formel lautet

$$\frac{\lambda}{\eta} = 2{,}5\, c_{v,\,\mathrm{trans}} + c_{\mathrm{intra}}, \tag{98.1}$$

wo $c_{v,\,\mathrm{trans}} = \frac{3}{2}\frac{\mathsf{R}}{M}$ (R = molare Gaskonstante, M = Molmasse) den durch die Translation der Moleküle bedingten Anteil der spezifischen Wärme bei konstantem Volumen bedeutet und c_{intra} den durch die innermolekulare Bewegung, Rotation und Oszillation, bedingten Anteil. Die Begründung für diesen Ansatz lautet etwa so: (98.1) ist richtig für einatomige Gase ($c_{\mathrm{intra}} = 0$). Der Faktor 2,5 bringt hierbei zum Ausdruck, daß die schnellen Moleküle, die wegen kürzerer Flugzeit stärker zur Reibung beitragen als die langsamen, erst recht stärker zur Wärmeleitung beitragen. Denn im Fall der Wärmeleitung ist das, was transportiert wird, nämlich die kinetische Energie, an sich schon bei den schnellen Molekülen größer als bei den langsamen. Soviel über die einatomigen Gase. Was nun aber die mehratomigen anbelangt, so besteht hier bei der innermolekularen Energie keine oder jedenfalls keine so ausgesprochene Korrelation mit der Molekülgeschwindigkeit. Deshalb ist bei c_{intra} kein Faktor 2,5 zu erwarten; EUCKEN wählt 1 als Faktor. Damit hat man die Beziehung (98.1). Man kann sie auch in der Form

$$f = \frac{\lambda}{\eta\, c_v} = 1 + \frac{9}{4}\,\frac{R}{C_{v,\,\mathrm{mol}}} \tag{98.2}$$

schreiben, wo $C_{v,\,\mathrm{mol}} = M\, c_v$ die Molwärme des Gases bedeutet.

Eine exakte Begründung einer etwas modifizierten Euckenschen Formel in einem Grenzfall geben CHAPMAN und COWLING [4], Kap. 13. Sie betrachten den Fall recht langsamen Energieaustausches zwischen der Translation und der inneren Bewegung. Dann kann man von einer Diffusion angeregter Moleküle sprechen. Mit dieser Diffusion ist ein Transport von innerer Energie verknüpft. Anstelle des Faktors 1 bei c_{intra} in der ursprünglichen Eucken-Formel (98.1) tritt so der etwas größere Faktor $D\varrho/\eta$. Dadurch erhält man etwas größere f-Werte als nach (98.1). Die Annahme langsamen Austauschs trifft aber nur bei der Oszillationsbewegung zu. Die Rotation tauscht im allgemeinen schon nach wenigen Stößen die Energie mit der Translation aus und man kann, da die „Individualität" der Moleküle so rasch verloren geht, nicht gut von einem Diffusionsvorgang sprechen. Der Vergleich mit den Versuchen und auch der Vergleich mit der Theorie der rauhen Kugeln zeigt, daß bei rotierenden Molekülen die ursprüngliche Eucken-Formel besser ist als ihre Modifikation im Sinne der Diffusionstheorie.

Tabelle 23. *Meßwerte des Wärmeleitkoeffizienten bzw. der Viscosität von Wasserstoff nach Ubbink (1948) bzw. nach v. Itterbeek und v. Paemel (1940) sowie Becker und Misenta (1955). Molwärme für* n-H_2 *nach Farkas [14], S. 19.*

T (°K)	15	20	60	70	80	90	273
λ (μ-cal/grad cm sec) .	29	38,2	96,7	110,2	124	137,5	417
η (μ-Poise)	8,1	11,0	28	33,3	37	40,5	83,6
$C_{v,\,\text{mol}}$ (cal/grad) . .	2,98	2,98	3,02	3,08	3,19	3,28	4,85

Es sei nun kurz das experimentelle Material betrachtet. Besonders instruktiv ist das Beispiel des Wasserstoffs. Er hat bei Temperaturen bis 40° K die Molwärme des einatomigen Gases; dagegen sind bei 273° K die zwei Rotationsfreiheitsgrade nahezu voll angeregt. Dieser Anstieg der Molwärme ist mit einem entsprechenden Abfall des Verhältnisses f gemäß (98.2) verknüpft. Tabelle 23 gibt die aus Meßwerten interpolierten Daten; Fig. 56 zeigt die Temperaturabhängigkeit von f.

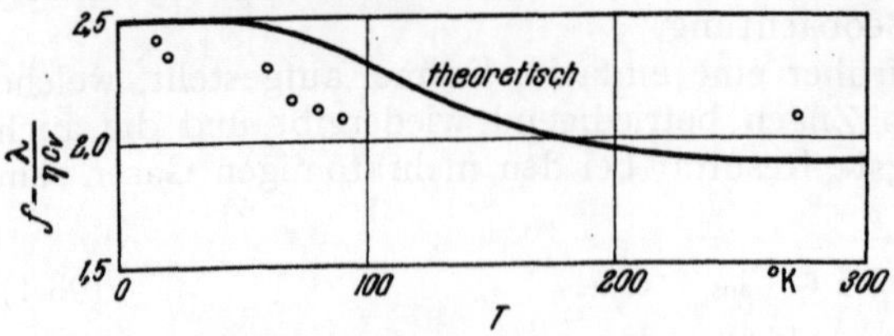

Fig. 56. Vergleich der aus den Daten von Tabelle 23 entnommenen f-Werte für Wasserstoff [o] mit den aus (98.2) folgenden theoretischen f-Werten [ausgezogene Kurve].

In Tabelle 24 sind einige weitere experimentelle Daten für f angegeben und den nach (98.2) berechneten Werten gegenübergestellt. Wegen einatomiger Gase s. Tabelle 8 (Ziff. 52).

Tabelle 24. *Wärmeleitkoeffizient* λ *einiger mehratomiger Gase bei 273° K.* $f=\lambda/\eta\, c_v$; f_{theor} *nach* (98.2). (Auszug nach CHAPMAN und COWLING [4], Kap. 13; dort Literaturangaben.)

	H_2*	D_2*	N_2	O_2	CO	CO_2	Cl_2	CH_4	NH_3	SO_2	C_2H_4
λ (μ-cal/grad cm sec)	417	302	58	58,5	55,9	35,2	18,3	72,1	52,2	20,6	40,7
f	2,09	2,10	1,97	1,91	1,91	1,67	1,79	1,73	1,41	1,49	1,44
f_{theor}	1,92	1,90	1,91	1,89	1,91	1,68	1,80	1,70	1,70	1,64	1,56

* Nach Ubbink (1948).

Eine umfassende Durchsicht des experimentellen Materials hat EUCKEN (1940) gegeben. Eine Tabelle zur Temperaturabhängigkeit von f findet sich auch in der Arbeit von E. U. FRANCK (1951). Die Übereinstimmung bei dipollosen Gasen ist recht gut. Bei Dipol-Gasen, d.h. bei den letzten drei Fällen in Tabelle 24, dagegen liefert die einfache Theorie entschieden zu hohe f-Werte. Es hat den Anschein, als ob bei den Dipolgasen infolge der starken, beim Stoß wirkenden Drehmomente anstelle des von den kugelsymmetrischen, einatomigen Molekülen stammenden Faktors 2,5 in (98.1) ein kleinerer Faktor zu treten hätte. Eine einwandfreie Theorie gibt es bisher nicht[1].

Eine weitere Besonderheit der Dipolgase liegt in der verhältnismäßig starken Zunahme ihres Wärmeleitkoeffizienten mit dem Druck, schon bei 1 atm. Dies ist in Ziff. 5 besprochen.

[1] Der Erklärungsversuch von SCHÄFER (1943) ging von unzutreffenden Voraussetzungen über die für die Zähigkeit, Wärmeleitung und Diffusion einatomiger Gase maßgebenden Wirkungsquerschnitte (Ω-Integrale) aus. — Erst recht existieren keine konkreten Überlegungen auf den in Ziff. 95 geschilderten exakten Grundlagen.

Interessant ist ferner das Verhalten der mehratomigen Gase beim Energietransport, wenn der Druck soweit erniedrigt wird, daß sich die Erscheinungen der freien Weglänge bemerkbar machen. Da der Energieaustausch zwischen Translation und Oszillation langsam ist, hat man letzterer eine größere freie Weglänge zuzuordnen als ersterer. Die hydrodynamische Näherung versagt hinsichtlich der Oszillation bereits bei höherem Druck als hinsichtlich der Translation. Das hat zuerst UBBELOHDE (1935) theoretisch untersucht. Näheres findet man in dem Artikel von HERZFELD [*22*], S. 210; neuere Literatur ist von WALDMANN [*53*], S. 203 und besonders in dem Bericht von KL. SCHÄFER [*46*] besprochen.

99. Die Volumviscosität. Die Volumviscosität η_V ist in (2.11) phänomenologisch eingeführt worden. In (96.22) bis (96.24) ergab sie sich zwangsläufig aus der exakten kinetischen Theorie. In (97.3) ist der von KOHLER (1947a) für rauhe Kugeln berechnete Wert angegeben.

Für einatomige Gase gilt $\eta_V = 0$ (Stokessche Beziehung). Aber schon STOKES[1] selbst betonte, daß die gewöhnlichen Reibungsexperimente an Flüssigkeiten und Gasen, bei denen die Dichte nahezu konstant oder langsam veränderlich ist, nicht beweisen, daß allgemein $\eta_V = 0$ wäre. Um die Existenz der Volumviscosität deutlich zu machen, muß man rasche Kompressions- und Dilatationsvorgänge betrachten, also die Schallabsorption. Wegen ihrer starken Kompressibilität zeigen die Gase die Wirkung der Volumreibung schon im Hör-Schall-Gebiet, während sie bei Flüssigkeiten bisher nur für Ultraschall nachgewiesen ist. Schon DUFF (1898) beobachtete in Luft bei 6800 Hertz den Absorptionskoeffizienten $\beta = 4{,}2 \cdot 10^{-5}\,\mathrm{cm}^{-1}$ [Definition von β siehe bei Gl. (99.2)], während die auf Grund nur der gewöhnlichen Scher-Viscosität η berechnete Absorption den Wert $\beta = 0{,}62 \cdot 10^{-5}\,\mathrm{cm}^{-1}$ hat.

Darauf, daß die Volumreibung bei nicht zu hohen Frequenzen (so daß die Hydrodynamik noch anwendbar ist) die Schallabsorption zu beschreiben gestattet, hat TISZA (1942) hingewiesen und gezeigt, daß im allgemeinen bei mehratomigen Gasen und auch bei Flüssigkeiten $\eta_V \gg \eta$ gilt. Bei niedrigen Frequenzen ergibt die durch die Volumreibung ergänzte Hydrodynamik dieselbe Frequenzabhängigkeit der Schallabsorption wie die Relaxationstheorie, welche auf der Idee der verzögerten Einstellung der inneren Freiheitsgrade beruht (s. dazu die am Schluß von Ziff. 96 zitierten Arbeiten). Der Zusammenhang von beidem ist nach TISZA durch die Formel

$$\eta_V = \varrho\, a^2\, c_{\mathrm{intra}} \left(\frac{1}{c_v} - \frac{1}{c_p}\right) \tau_{\mathrm{intra}} \tag{99.1}$$

gegeben. Dabei bezeichnen ϱ die Dichte, a die Schallgeschwindigkeit, c_v, c_p die spezifischen Wärmen, c_{intra} den Anteil der innermolekularen Bewegung daran und τ_{intra} die Relaxationszeit. Je größer letztere ist, desto größer ist die Volumviscosität. Daß der Austausch von Oszillationsenergie bei einem Stoß sehr gering ist, haben LANDAU und TELLER (1936) theoretisch gezeigt. Durch spektroskopische Experimente wies zuerst DWYER (1939) nach, daß der Vibrationszustand in der Tat über Tausende von Stößen hinweg erhalten bleibt.

Der Absorptionskoeffizient für die Schallamplitude beträgt nach der hydrodynamischen Theorie von STOKES und KIRCHHOFF [vgl. KOHLER (1948)]

$$\beta = \frac{\omega^2}{2a^3}\left[\frac{4}{3}\,\frac{\eta}{\varrho} + \frac{\eta_V}{\varrho} + \left(\frac{1}{c_v} - \frac{1}{c_p}\right)\frac{\lambda}{\varrho}\right]. \tag{99.2}$$

---

[1] Math. Phys. Papers, Bd. 1, S. 88.

Die Schallwelle ist dabei durch $\varrho - \varrho_0 \propto e^{-\beta x} \cos \omega \left(t - \frac{x}{a}\right)$ gegeben. Der Beitrag der Wärmeleitfähigkeit λ zur Absorption β ist verhältnismäßig klein. Setzt man nun die bei kleinen ω gemessene Schallabsorption in (99.2) ein, so erhält man η_V. Einige solche Werte gibt Tabelle 25. In den aufgeführten Fällen handelt es sich um Gase mit relativ rascher Einstellung der inneren Freiheitsgrade (Rotation). In Fällen dagegen, wo die Einstellung langsam geschieht, kann η_V um mehrere Zehnerpotenzen größer als η sein. Genau das besagt auch die Formel (97.3) für die rauhen Kugeln; wenn $K \gg 1$ ist, hat man rasche Rotation, langsame Einstellung und große Volumviscosität.

Tabelle 25. *Scher-Viscosität η und Volumviscosität η_V bei etwa 290° K, letztere von Kohler (1948) aus Schallabsorptionsmessungen von Keller (1940) entnommen.*

	η (μ-Poise)	η_V (μ-Poise)	η_V/η
N_2	174,5	110	0,6
NH_3	98,2	130	1,3

Die Volumreibung trägt zur lokalen Entropieerzeugung bei. Pro Raum- und Zeiteinheit ist nach (61.1) mit (96.26) und (96.13) dieser Beitrag

$$\left(\frac{\delta \varrho s}{\delta t}\right)_{\text{Stöße}} = \frac{\eta_V}{T} \left(\frac{\partial v_\mu}{\partial x_\mu}\right)^2 = \frac{\eta_V}{T} \left(\frac{d\varrho}{\varrho\, dt}\right)^2. \tag{99.3}$$

In langsam strömenden Gasen ist die zugehörige Erwärmung sehr klein im Vergleich zu der Temperaturänderung, welche gemäß der Adiabatengleichung bei Dichteänderung eintritt. Deshalb ist sie nicht direkt experimentell nachzuweisen. Erst recht gilt das für die fast inkompressiblen Flüssigkeiten. Bei periodischen Dichteschwankungen $\varrho = \varrho_0 + \varrho_1 \cos \omega t$ hat man aber im Zeitmittel

$$\overline{\left(\frac{\delta \varrho s}{\delta t}\right)}_{\text{Stöße}} = \frac{1}{2} \frac{\eta_V}{T} \omega^2 \left(\frac{\varrho_1}{\varrho_0}\right)^2. \tag{99.4}$$

Nun ist zwar beim Schall $\varrho_1/\varrho_0 \ll 1$, aber für genügend hohe Frequenz wird trotzdem die Entropieerzeugung durch Volumreibung merklich.

Es gibt zur Bestimmung der Volumviscosität noch eine andere Methode, welche nicht den Weg über die Schallabsorption geht. Diese Methode, die ECKART (1948) ausgedacht hat, benutzt den Schallwind, d.h. die Strömungen, welche in einem von Schallwellen durchsetzten Medium infolge des akustischen Strahlungsdrucks sich ausbilden. LIEBERMANN (1949) hat diesbezügliche Versuche an einigen Flüssigkeiten angestellt und in groben Zügen die mittels Absorption gefundenen η_V-Werte bestätigt.

Einen zusammenfassenden Bericht über die Volumsviscosität haben KARIM und ROSENHEAD [*30*] gegeben. Dort findet man weitere Literatur.

Verzeichnis der durchgängig benutzten Symbole.

Für *Tensoren* (einschließlich Vektoren) wird im allgemeinen die Komponenten-Schreibweise benutzt, z.B. $(x_\mu;\, p_{\mu\nu})$. Es gilt die Summationsvorschrift, vgl. S. 298, Fußnote 1. Vektoren sind häufig auch durch Fettdruck bezeichnet ($\boldsymbol{r}$). Quantenmechanische Matrizen (Operatoren) werden in Teil H durch Blockschrift bezeichnet (z.B. a, f, O).

Folgende *Klammer-Symbole* finden Verwendung:

$\langle\ \rangle$ Bildung des symmetrisch-irreduziblen Tensors, z.B. $\langle c_\mu c_\nu \rangle$, s. (37.8), oder $\left\langle \frac{\partial v_\mu}{\partial x_\nu} \right\rangle$, s. (46.5). In Teil H auch Erwartungswert eines quantenmechanischen Operators im reinen Fall, s. (90.5).

$[\Psi, \Phi]$ Bestimmtes, bilineares, positiv-semidefinites Integral bzw. Summe von solchen. Siehe (35.9) oder (48.5) für einatomige reine Gase, (54.1) für einatomige Gemische, (96.20) für mehratomige reine Gase.

$[\Psi, \Phi]'_{ij}$, $[\Psi, \Phi]''_{ij}$ Bestimmte bilineare Integrale für einatomige Gemische, s. (65.4) und (65.5).

Es folgen die *Buchstabensymbole* mit Angabe einschlägiger Gleichungen des Textes.

Lateinisch:

$a^{(r)}_{\mu_1\ldots\mu_l}(t, \boldsymbol{r})$ Entwicklungskoeffizienten aus (41.5).

a_{rs} Matrix (49.13).

$a'_{ij,rs}$, $a''_{ij,rs}$, $a_{ij,rs}$ Matrizen (66.1), (66.2) und (69.3).

$a(\chi, k)$ Streuamplitude für einatomige Moleküle, (82.13).

$\mathfrak{a}(\boldsymbol{e}', \boldsymbol{e}, E)$, $\mathfrak{a}^{+}(\boldsymbol{e}, \boldsymbol{e}', E)$ Streuamplituden-Matrix für mehratomige Moleküle, (89.11), und ihre Adjungierte, (89.13).

$\breve{\mathfrak{a}}(\boldsymbol{c}', \boldsymbol{c}_1' | \boldsymbol{c}, \boldsymbol{c}_1)$, $\breve{\mathfrak{a}}^{+}(\boldsymbol{c}, \boldsymbol{c}_1 | \boldsymbol{c}', \boldsymbol{c}_1')$ dieselben transformiert, (89.26) und (89.27).

$A(C)$; $A_i^{(k)}(C)$, $A_i(C)$ und $A_J(C)$ Bestandteile der hydrodynamisch genäherten Verteilungsfunktionen (47.1); (55.1) und (96.16).

A_r Entwicklungskoeffizienten aus (49.10).

$A^{(k)}_{i,s}$, $A_{i,s}$ Entwicklungskoeffizienten aus (64.1) und (64.2).

$A_l(\nu)$ Zahlenkoeffizienten für das Potenz-Kraftgesetz, (40.6) und (51.7a).

$A^\star$ Abkürzung aus (69.10).

A Amplitudenfaktor eines Wellenpakets, (90.1), (92.1).

$A_{MM_1, M'M_1'}(\boldsymbol{c}, \boldsymbol{c}_1 | \boldsymbol{c}', \boldsymbol{c}_1')$ Streuamplitude für rotierende Moleküle im Hauptachsensystem von $\mathfrak{f}$, (93.12).

$\boldsymbol{b} = (b_1, b_2, b_3) = (b_\mu)$ Äußere Kraft pro Masseneinheit (Beschleunigung), (2.6); $\boldsymbol{b}_i$ entsprechend für Gemische, s. a. (34.6).

b Stoßparameter, siehe bei (26.1).

b_{rs} Matrix (49.14).

$b'_{ij,rs}$, $b''_{ij,rs}$, $b_{ij,rs}$ Matrizen (66.3), (66.4) und (70.3).

$B(C)$; $B_i(C)$ und $B_J\,C)$ Bestandteile der hydrodynamisch genäherten Verteilungsfunktionen (47.1); (55.1) und (96.16).

B_r Entwicklungskoeffizienten aus (49.10).

$B_{i,s}$ Entwicklungskoeffizienten aus (64.3).

$B^\star$ Abkürzung aus (69.10).

$\boldsymbol{c} = (c_1, c_2, c_3) = (c_\mu)$ Geschwindigkeit eines Moleküls im Laboratoriumssystem.

$\boldsymbol{c}_1$ Geschwindigkeit eines anderen (Partner-) Moleküls (vor dem Stoß).

$\boldsymbol{c}'$, $\boldsymbol{c}_1'$ Geschwindigkeit der zwei Moleküle nach dem Stoß, (22.9), (22.18), (88.5).

$\boldsymbol{c}_S$ Schwerpunktsgeschwindigkeit eines Molekülpaars, (22.3).

$c_{\mu\nu}$; allgemein $c_{\mu_1\ldots\mu_l}$ Homogener irreduzibler Tensor zweiter bzw. l-ter Stufe, (37.8).

$c(lr | l'r')$ Matrixelement nach (41.7).

$c_v\,(c_p)$ Spezifische Wärme bei konstantem Volumen (Druck), (2.18) und (2.20).

$\boldsymbol{C}$ Molekulargeschwindigkeit im lokalen Ruhsystem des Gases, (31.11) und (32.11).

$\overline{C}$ Mittlere Molekulargeschwindigkeit im thermischen Gleichgewicht, (34.12).

$C^\star$ Abkürzung aus (69.10).

d oder d_{ij} Moleküldurchmesser (25.1); Parameter des Lennard-Jones-Potentials (25.5).

$\boldsymbol{d}_i$ Abkürzung aus (54.10).

$d\boldsymbol{r} = dx_1\,dx_2\,dx_3$, $d\boldsymbol{c} = dc_1\,dc_2\,dc_3$, etc. Volumelemente des Orts- und Geschwindigkeitsraums.

$d\boldsymbol{e}$, $d\boldsymbol{e}'$ Zu den Relativgeschwindigkeiten $\boldsymbol{g}$, $\boldsymbol{g}'$ vor und nach dem Stoß gehörige Raumwinkelelemente, (22.20) und (23.4).

$\frac{d}{dt} = \frac{\partial}{\partial t} + v_\mu \frac{\partial}{\partial x_\mu}$ Zeitliche Differentiation im $\boldsymbol{v}$-System, (2.3), (31.17).

$\left(\frac{d}{dt}\right)_w = \frac{\partial}{\partial t} + w_\mu \frac{\partial}{\partial x_\mu}$ Zeitliche Differentiation im $\boldsymbol{w}$-System, (3.9), (32.35).

$D = D(1,2)$ Binärer Diffusionskoeffizient im $\boldsymbol{w}$-System, (3.10) und (59.1).

D_{ik} Polynäre Diffusionskoeffizienten im $\boldsymbol{w}$-System, (56.1) und (56.2).

D_{Ti} Polynäre Thermodiffusionskoeffizienten im $\boldsymbol{w}$-System, (56.1) und (56.3).

$D^{(v)}_{ik}$ Diffusionskoeffizienten im $\boldsymbol{v}$-System, (58.7) und (59.12).

$D_{Ti}^{(v)}$ Thermodiffusionskoeffizienten im $\boldsymbol{v}$-System, (58.7).
$D_J(C)$ Bestandteil der hydrodynamisch genäherten Verteilungsfunktionen (96.16).
$\mathcal{D}, \mathcal{D}_i$ Differentialoperatoren (28.1) und (29.1).
$\boldsymbol{e}, \boldsymbol{e}'$ Einheitsvektoren in Richtung der Relativgeschwindigkeit vor und nach dem Stoß, (22.8).
E_I Innere Energie eines mehratomigen Moleküls im stationären Quantenzustand I, (88.3).
E Gesamtenergie beim Stoß im Schwerpunktssystem, (88.4).
$\bar{E}(T)$ Mittlere innermolekulare Energie, (96.6).
E_1, E_2 Abkürzungen aus (69.11).
$f(t, \boldsymbol{r}, \boldsymbol{c})$ Verteilungsfunktion des einatomigen reinen Gases, Ziff. 28. Zuweilen wird abgekürzt $f(t, \boldsymbol{r}, \boldsymbol{c}') = f', f(\ldots, \boldsymbol{c}_1) = f_1, f(\ldots, \boldsymbol{c}_1') = f_1'$.
$f_i(t, \boldsymbol{r}, \boldsymbol{c})$ Dasselbe für einatomige Gemische, Ziff. 29.
$\mathsf{f}(t, \boldsymbol{r}, \boldsymbol{c})$; $\mathsf{f}(\ldots, \boldsymbol{c}, \boldsymbol{c}_1)$; $\mathsf{f}^{(J)}(\ldots, \boldsymbol{c})$ Verteilungsmatrizen für Gase mit rotierenden Molekülen, (90.9), (92.11), (95.2).
$f_J(t, \boldsymbol{r}, \boldsymbol{c})$ Genäherte Verteilungsfunktion für mehratomige reine Gase, Ziff. 95.
$f^{(0)}$ Verteilungsfunktion für *ruhendes* einatomiges Gas im thermischen Gleichgewicht, (35.2).
F, F_i sowie F_J Verteilungsfunktionen im thermischen Gleichgewicht, (44.2), (34.4) und (96.9), zugehörige lokale Gasgeschwindigkeit $\boldsymbol{v}$.
$F_M(t, \boldsymbol{r}, \boldsymbol{c})$ Eigenwerte der Verteilungsmatrix f, (93.11).
$f = \dfrac{\lambda}{\eta c_v}$ Abkürzung aus (97.4).
$\boldsymbol{g}, \boldsymbol{g}'$ Relativgeschwindigkeit zweier Moleküle vor und nach dem Stoß, (22.4).
h Plancksches Wirkungsquantum; $\hbar = h/2\pi$.
i Spezifische Enthalpie, (2.20), (31.24).
i Bezeichnet als Index die Komponente eines einatomigen Gasgemischs.
I (auch I_1, I', I_1') Bezeichnet als Index die Quantenzustände der Moleküle eines mehratomigen reinen Gases, (88.2).
$I(\Phi)$ Linearisierter Stoßoperator aus (35.5) oder (45.3).
$I_{ij}(\Phi)$ Dasselbe für einatomige Gemische, (53.13).
$I_{JJ_1, J'J_1'}(\Phi)$ Dasselbe für mehratomige reine Gase, kugelsymmetrische Moleküle, (96.11).
J Drehimpulsquantenzahl, (88.2).
$J_0(z)$, $J_{l+\frac{1}{2}}(z)$, Bessel-Funktionen.
k Boltzmann-Konstante.
k, k' Wellenzahl, (82.4), (88.7); $k^* = kd$.
k_T Thermodiffusionsverhältnis (3.10), (56.10) und (56.12) sowie (59.3)
K Temperaturleitfähigkeit, (2.24).
K Anzahl der Komponenten eines Gasgemischs, (53.1).
K Trägheitsmoment einer rauhen Kugel, gemessen in Einheiten $m(d/2)^2$, (97.1).
$K(r)$ Kraft zwischen zwei Atomen, Ziff. 25.
l Freie Weglänge, (1.2) und (51.5).
l Index der Eigentensoren (37.12) und der Sonine-Polynome (38.19).
l und l' Bahndrehimpuls-Quantenzahlen, (82.3), (89.1).
l Länge des einfallenden Wellenpakets aus (91.1).
L Charakteristische Gefäßabmessung, (1.1).
m Masse eines Moleküls bei reinem Gas; bei Gemischen mittlere Masse gemäß (32.7).
m_i Masse eines Moleküls der Komponente i aus einem Gasgemisch.
m_{ij} Reduzierte Masse eines Molekülpaars (ij), (22.16).
m (auch m') magnetische Bahn-Quantenzahl für die Relativbewegung zweier Atome oder Moleküle, (82.3) und (89.1).
M Molmasse oder Molekulargewicht.

M magnetische Quantenzahl des Eigendrehimpulses eines Moleküls, (88.1).
M_1, M_2 Massenbrüche, (69.12).
n Konzentration (Gesamt-Teilchenzahl pro Volumeinheit), (31.5) und (96.2).
n_0 Ein raumzeitlich konstanter Vergleichswert für n, (35.2).
n_i Konzentration der i-ten Komponente eines Gasgemischs, (32.5).
n_J bzw. $n_J(T)$ Abkürzungen aus (95.5) bzw. (96.5).
N Gesamtzahl der Moleküle einer Gasmasse.
O quantenmechanischer Operator, (90.4).
p Statischer Druck (im Nichtgleichgewichtszustand ist p lediglich Abkürzung für $n \mathrm{k} T$), (31.22).
p_0 Ein raumzeitlich konstanter Vergleichswert für p, (42.4).
$p_{\mu\nu}$ Drucktensor (2.5), (2.33), (31.14), (32.14) und (96.2).
$P_l(x)$ Legendre-Polynom; $P_l(1) = 1$.
P Projektionsoperator, (90.6), (92.6).
$\boldsymbol{q} = (q_1, q_2, q_3) = (q_\mu)$ Wärmestrom für einatomige reine Gase (31.16), für einatomige Gemische (32.16), gemeint im Bezugssystem der mittleren Massengeschwindigkeit $\boldsymbol{v}$; für mehratomige reine Gase (96.7).
$\boldsymbol{q}^{(w)}$ Wärmestrom für Gemische im Bezugssystem der mittleren Teilchengeschwindigkeit $\boldsymbol{w}$, (32.36) und (57.4).
q Querschnitt des einfallenden Wellenpakets aus (91.1)
$Q_{ij}^{(l)}$ Gemittelter Streuquerschnitt, (50.7).
Q_1, Q_2, Q_{12} Abkürzungen aus (69.14).
$\boldsymbol{r} = (x_1, x_2, x_3) = (x_\mu)$ Ortsvektor.
$\boldsymbol{r}'$ Relativkoordinaten eines Molekülpaares, siehe bei (89.1).
r Abstand zweier Moleküle, Ziff. 25.
r_m Abstand zweier Moleküle, für welchen die Wechselwirkungsenergie ihr Minimum annimmt, (25.7).
r Index der Eigentensoren (37.12) bzw. der Sonine-Polynome (38.19) (radiale Knotenzahl).
R Universelle molare Gaskonstante, (34.13).
R_T Temperaturabhängiger Bestandteil des Thermodiffusionsfaktors, (75.5).
s Spezifische Entropie, (33.1) und (93.16).
$s_{(ij)}^*$ Abkürzung aus (27.5)
$s_{0(ij)}$ Abkürzung aus (27.7).
s Steilheitsparameter des (6; Exp)-Potentials (25.7).
s, s', s_1' Parameter aus (38.14) und (38.17).
s_1, s_2, s_{12} Abkürzungen aus (74.4).
S Entropie des Gases, (33.3).
$S_{l+\frac{1}{2}}^{(r)}(x)$ Sonine-Polynom, (38.19) bis (38.22).
S_η, S_D Sutherland-Konstanten, (51.9) und (71.8).
S_1, S_2 Abkürzungen aus (69.14).
$\mathsf{S}, \mathsf{S}^\dagger$ S-Matrix für mehratomige Moleküle (89.6), und ihre Adjungierte.
Sp; Sp_1 Spurbildung, (90.7), (92.7).
t Zeit.
T Absolute (kinetische) Temperatur, (31.15), (32.15) und (96.3).
T_0 Ein raumzeitlich konstanter Vergleichswert für T, (35.2).
$T_{(12)}^*$ Reduzierte Temperatur nach (51.12) bzw. (71.10).
u Spezifische Energie (31.15), (32.15) und (96.3).
$\mathsf{U}(\boldsymbol{c})$ bzw. $\mathsf{U}(\boldsymbol{c}, \boldsymbol{c}_1)$ unitäre Transformation, (89.18) bzw. (89.22).
$\boldsymbol{v} = (v_1, v_2, v_3) = (v_\mu)$ Gasgeschwindigkeit, (31.6) und (96.2), bzw. mittlere Massengeschwindigkeit bei Gemischen, (32.8).
$\boldsymbol{V}_i$ Diffusionsgeschwindigkeit im $\boldsymbol{v}$-System, (3.5) oder (32.12).

V Streuvolumen, Ziff. 23.

V Vibrationsquantenzahl, (88.2).

$\boldsymbol{w} = (w_1, w_2, w_3) = (w_\mu)$ Mittlere Teilchengeschwindigkeit bei Gemischen, (3.4) oder (32.27).

$\boldsymbol{W}_i$ Diffusionsgeschwindigkeit im $\boldsymbol{w}$-System, (3.6) oder (32.28).

$W_{(ij)}(\boldsymbol{c}, \boldsymbol{c}_1 | \boldsymbol{c}', \boldsymbol{c}_1')$ Wahrscheinlichkeitsfunktion, (30.1) und (30.12).

$x_1, x_2, x_3 = (x_\mu)$ Ortskoordinaten.

X oder X_1 innere Koordinaten eines mehratomigen Moleküls, (88.2)

Y_{lm} Normierte Kugelflächenfunktion, (89.3) und (89.4).

Griechisch:

α bzw. α_{ij} Thermodiffusionsfaktor, (3.11), (59.3) und (69.7).

α_M; α_{MM_1} Amplitudenfaktoren, (90.1), (92.1).

$\beta = \frac{m}{2kT_0}$ bzw. $\frac{m}{2kT}$ Abkürzung aus (35.2) bzw. (44.3) und (96.9).

$\beta_i = \frac{m_i}{2kT}$ Abkürzung aus (53.2).

β Abkürzung für einen Satz von Quantenzahlen, (89.5).

$\gamma_i = \frac{n_i}{n}$ Molenbruch der Komponente i eines Gemischs, (3.7) oder (54.8).

γ Dimensionslos gemachte Relativgeschwindigkeit, (50.6).

$\Gamma_{(ij)}$ Abkürzung aus (27.12).

$\delta_{\mu\nu}, \delta_i^{(k)}$ Kroneckersches Symbol ($= 1$ für $\mu = \nu$ bzw. $i = k$; sonst $= 0$).

$\delta(\Phi)$ bzw. $\delta_{ij}(\Phi)$ Abkürzungen nach (35.6) bzw. (53.13).

δ_l Streuphasen, (82.8).

Δ Laplace-Operator.

ε Azimutwinkel von $\boldsymbol{g}'$, (23.4).

ε oder ε_{ij} Energie-Parameter der molekularen Wechselwirkung, (25.5) und (25.7).

η Viscosität, vgl. (2.11), (42.9), (48.9), (60.2) und (96.23).

η_V Volumviscosität, vgl. (2.11), (96.24).

Θ Winkel zwischen $\boldsymbol{g}$ und $\boldsymbol{e}$, (27.4).

$\varkappa$ Konstante des Potenz-Kraftgesetzes (25.2).

λ Wärmeleitkoeffizient (2.22), (42.8), (48.2), (57.5) und (96.19).

λ' „Theoretischer" Wärmeleitkoeffizient für Gemische, (57.3).

λ Willkürlicher Parameter, siehe vor Gl. (38.4).

Λ_i In (55.2) eingeführte, in (56.4) spezifizierte Faktoren.

$\Lambda, \Lambda^*_{(ij)}$ Quantenmechanische Parameter aus (4.3b) und (81.3).

$\mu, \nu, \ldots (= 1, 2, 3)$ Beziehen sich als Indices von Tensorkomponenten auf die drei Raumrichtungen eines cartesischen Koordinatensystems; für griechische Indices gilt die Summationsvorschrift von S. 298, Fußnote 1.

$\mu = \eta/\varrho$ Kinematische Zähigkeit, (2.14).

ν Als Index siehe bei μ.

ν Exponent des Potenz-Kraftgesetzes (25.2).

ϱ bzw. ϱ_i Dichte des Gases bzw. einer Komponente, (31.5) und (32.7) bzw. (32.5).

σ bzw. σ_{ij} Differentieller Streuquerschnitt für den Stoß zweier gleichartiger bzw. verschiedenartiger einatomiger Moleküle, (23.3).

$\sigma_{M'M_1'}, {}_{MM_1}$ Dasselbe für zwei gleichartige mehratomige Moleküle, (93.13).

$\sigma_s, \sigma_a, \sigma_u$ Auf symmetrische (anti-; un-) Weise gebildeter quantenmechanischer Streuquerschnitt, (82.16) und (82.17).

$\bar{\sigma}\ldots$ Resultierender Streuquerschnitt, (84.3), (94.3), (95.7).

$\tau, \tau_{\text{intra}}$ Relaxationszeiten (Stoßzeiten) (4.12), (96.34).

$\varphi_{\mu_1 \ldots \mu_l}(t, \boldsymbol{r}, \boldsymbol{c})$ Entwicklungskoeffizienten aus (41.1).

φ zusammengesetzte Kugelwelle, (89.6).

$\varphi^{(+)}, \varphi^{(-)}$ aus- bzw. einlaufende Kugelwelle, (89.1).

$\Phi(t, \boldsymbol{r}, \boldsymbol{c})$ Relative Abweichung der Verteilungsfunktion f von $f^{(0)}$, (35.1). Wegen Φ' usw. siehe bei $f(r, \boldsymbol{r}, \boldsymbol{c})$.

$\Phi^{(1)}(t, \boldsymbol{r}, \boldsymbol{c})$ Relative Abweichung der Verteilungsfunktion f von der lokalen Gleichgewichtsverteilung F, (44.1); wegen $\Phi^{(1)\prime}$ usw. siehe bei $f(t, \boldsymbol{r}, \boldsymbol{c})$.

$\Phi_i^{(1)}(t, \boldsymbol{r}, \boldsymbol{c})$ Dasselbe für Gemische, (53.1).

$\Phi_J^{(1)}(t, \boldsymbol{r}, \boldsymbol{c})$ Dasselbe für mehratomige reine Gase, (96.9).

$\Phi(r)$ oder $\Phi_{ij}(r)$ Potentielle Energie der molekularen Wechselwirkung für einatomige Gase, Ziff. 25.

$\Phi_l(\beta c^2, s)$ Erzeugende Funktion der Sonine-Polynome, (38.15).

χ Winkel zwischen $\boldsymbol{g}$ und $\boldsymbol{g}'$, (23.3).

ψ, ψ_i, ψ_J Stoßinvarianten (31.2), (32.2), (96.1).

$\psi_N(\boldsymbol{c}), \psi_{lr}(c)$ Eigenfunktionen aus (37.5).

$\psi^{(\cdot)}_{\mu_1 \ldots \mu_l}$ Eigentensoren aus (37.12).

ψ Schrödingersche Wellenfunktion in Teil G und H.

Ψ bzw. Ψ_i Eine beliebige Funktion von $t, \boldsymbol{r}, \boldsymbol{c}$, (31.1) bzw. (32.1). Wegen Ψ' usw. siehe bei $f(t, \boldsymbol{r}, \boldsymbol{c})$.

Ψ_I bzw. Ψ_{II_1} innere Eigenfunktionen eines Moleküls bzw. Molekülpaares, (88.2) und (89.2).

ω bzw. ω_{lr} Abklingkonstante (reziproke Relaxationszeit), (37.1) bzw. (37.5a) allgemein und (38.25) sowie (39.10) für das Maxwell-Gas.

$\Omega, \Omega^{(l)}$ Lineare drehinvariante Operatoren, (37.6).

$\Omega_{ij}^{(l,r)}$ Abkürzung nach (50.5), (Ω-Integral).

$\Omega_{ij}^{(l,r)\star}$ Reduzierte Ω-Integrale nach (51.13).

$\Omega_s^{(l,r)}, \Omega_a^{(l,r)}$ Nach (50.5) und (83.2) mittels symmetrisierter quantenmechanischer Streuquerschnitte gebildete Ω-Integrale.

Literaturverzeichnis.

Originalarbeiten.

Hinter jedem Zitat ist die Ziffer dieses Artikels (in einigen Fällen die Seite) angegeben, in welcher die betreffende Arbeit zitiert wird.

Amdur, I., u. A. L. Harkness: J. Chem. Phys. **22**, 664 (1954) (Ziff. 23).
—, and E. A. Mason: J. Chem. Phys. **22**, 670 (1954) (Ziff. 23).
—, and H. Pearlman: J. Chem. Phys. **9**, 503 (1941) (Ziff. 23, 25).
—, and T. F. Schatzki: J. Chem. Phys. **27**, 1049 (1957) (Ziff. 72).
Archer, C. T.: Nature, Lond. **138**, 286 (1936) (Ziff. 4).
Atkins, B. E., R. E. Bastick and T. L. Ibbs: Proc. Roy. Soc. Lond., Ser. A **172**, 142 (1939) (Ziff. 77).
Bayet, M., J.-L. Delcroix et J.-F. Dénisse: J. Phys. Radium **15**, 795 (1954); **16**, 274 (1955) (Ziff. 35).
Becker, E. W.: Z. Naturforsch. **5**a, 457 (1950) (Ziff. 5, 21).
—, and W. Beyrich: J. Phys. Chem. **56**, 911 (1952) (Ziff. 18, 77).
— — K. Bier, H. Burghoff u. F. Zigan: Z. Naturforsch. **12**a, 609 (1957) (Ziff. 56).
—, u. K. Bier: Z. Naturforsch. **9**a, 975 (1954) (Ziff. 23).
— — u. H. Burghoff: Z. Naturforsch. **10**a, 565 (1955) (Ziff. 56).
—, u. E. Dörnenburg: Naturwiss. **37**, 165 (1950) (Ziff. 77).
—, u. W. Henkes: Z. Physik **146**, 320 (1956) (Ziff. 34).
— — u. U. Seifert: Z. physik. Chem. **5**, 90 (1955) (Ziff. 19, 87).
—, u. R. Misenta: Z. Physik **140**, 535 (1955) (Ziff. 4, 72, 85, 98).
— — u. F. Schmeissner: Z. Physik **137**, 126 (1954) (Ziff. 8, 81, 85).
— — u. O. Stehl: Z. Physik **136**, 457 (1953) (Ziff. 86).
—, u. A. Schulzeff: Naturwiss. **35**, 218 (1948) (Ziff. 21).
—, u. O. Stehl: Z. Physik **133**, 615 (1952a) (Ziff. 6, 86).
—, u. O. Stehl: Z. angew. Phys. **4**, 20 (1952b) (Ziff. 6).
—, W. Vogell u. F. Zigan: Z. Naturforsch. **8**a, 686 (1953) (Ziff. 5).
Becker, R.: Z. Physik **8**, 321 (1922) (Ziff. 1).
Bhatnagar, P. L., E. P. Gross and M. Krook: Phys. Rev. **94**, 511 (1954) (S. 364).
Bleick, W. E., and J. E. Mayer: J. Chem. Phys. **2**, 252 (1934) (Ziff. 25).
Blüh, G. u. O.: Z. Physik **90**, 12 (1934) (Ziff. 18).

Boardman, L. E., and N. E. Wild: Proc. Roy. Soc. Lond., Ser. A **162**, 511 (1937) (Ziff. 13, 71).
Bock, H.: Ann. Physik **8**, 134 (1951) (Ziff. 80).
Boer, J. de: Diss. Amsterdam 1940 (Ziff. 81).
— Physica, Haag **10**, 348 (1943) (Ziff. 81, 85).
—, u. R. B. Bird: Physica, Haag **20**, 185 (1954) (Ziff. 83, 85).
—, u. E. G. D. Cohen: Physica, Haag **17**, 993 (1951) (Ziff. 81, 83, 85).
—, u. J. van Kranendonk: Physica, Haag **14**, 442 (1948) (Ziff. 51, 52).
— — u. K. Compaan: Physica, Haag **16**, 545 (1950) (Ziff. 82).
—, u. A. Michels: Physica, Haag **5**, 945 (1938) (Ziff. 52, 81, 86).
— — Physica, Haag **6**, 409 (1939) (Ziff. 81, 82).
Bogoljuboff, N.: J. Phys. USSR. **10**, 265 (1946) (S. 345).
Boltzmann, L.: Ber. Wien. Akad. **66**, 275 (1872) (Ziff. 1, S. 345, Ziff. 33).
Bond, W. N.: Proc. Phys. Soc. Lond. **49**, 205 (1937) (Ziff. 6, 7).
Born, M., and H. S. Green: Proc. Roy. Soc. Lond., Ser. A **188**, 10 (1946) (Ziff. 1, S. 345).
— — Proc. Roy. Soc. Lond., Ser. A **190**, 455 (1947) (Ziff. 1, S. 345).
—, u. J. E. Mayer: Z. Physik **75**, 1 (1932) (Ziff. 25).
Braune, H., u. F. Zehle: Z. phys. Chem., Abt. B **49**, 247 (1941) (Ziff. 13, 72).
Brown, H.: Phys. Rev. **58**, 661 (1940) (Ziff. 18).
Brueckner, K. A., and K. M. Watson: Phys. Rev. **102**, 19 (1956) (Ziff. 28).
Bryan: Brit. Assoc. Reports p. 83, 1894 (Ziff. 97).
Buckingham, R. A.: Proc. Roy. Soc. Lond., Ser. A **168**, 264 (1938) (Ziff. 25).
—, and J. Corner: Proc. Roy. Soc. Lond., Ser. A **189**, 118 (1947) (Ziff. 25).
—, and R. A. Scriven: Proc. Phys. Soc. Lond. A **65**, 376 (1952) (Ziff. 81, 85, 87).
—, and H. N. Temperley: Phys. Rev. **78**, 482 (1950) (Ziff. 81).
Buddenberg, J. W., and C. R. Wilke: Industr. Engng. Chem. **41**, 1345 (1949) (Ziff. 78).
Burnett, D.: Proc. Lond. Math. Soc. **39**, 385 (1935a) (Ziff. 38, 49, 67).
— Proc. Lond. Math. Soc. **40**, 382 (1935b) (Ziff. 42).
Carleman, T.: Acta math., Stockh. **60**, 91 (1933) (Ziff. 33).
Chapman, S.: Phil. Trans. Roy. Soc. Lond. **216**, 279 (1916) (Ziff. 1).
— Phil. Trans. Roy. Soc. Lond. **217**, 115 (1917) (Ziff. 1, 3, S. 402, Ziff. 73).
— Phil. Mag. **38**, 182 (1919) (Ziff. 19).
— Proc. Roy. Soc. Lond., Ser. A **177**, 38 (1940) (Ziff. 74).
—, and F. W. Dootson: Phil. Mag. **33**, 248 (1917) (Ziff. 3).
Clark Jones, R.: Phys. Rev. **58**, 111 (1940) (Ziff. 74, 75).
— Phys. Rev. **59**, 1019 (1941) (Ziff. 25, 74, 75).
Clausius, R.: Ann. Physik **105**, 239 (1858) (Ziff. 1, 25).
Cleave, A. B. van, and O. Maass: Canad. J. Res. **12**, 57 (1935) (Ziff. 4).
Clusius, K.: Helv. chim. Acta **33**, 2134 (1950) (Ziff. 74).
—, u. G. Dickel: Naturwiss. **26**, 546 (1938) (Ziff. 19).
— — Z. phys. Chem., Abt. B **44**, 397, 451 (1939) (Ziff. 19).
— — Z. phys. Chem., Abt. B **52**, 348 (1942) (Ziff. 19).
—, u. M. Huber: Z. Naturforsch. **10**a, 230 (1955a) (Ziff. 19, 76).
— — Z. Naturforsch. **10**a, 556 (1955b) (Ziff. 76, 77).
— — H. Hürzeler u. E. Schumacher: Z. Naturforsch. **11**a, 702 (1956) (Ziff. 19).
— K. Kölsch u. L. Waldmann: Z. phys. Chem., Abt. A **189**, 131 (1941) (Ziff. 49).
—, u. L. Waldmann: Naturwiss. **30**, 711 (1942) (Ziff. 3).
Coester, F.: Phys. Rev. **84**, 1259 (1951) (Ziff. 89).
Cohen, E. G. D., M. J. Offerhaus u. J. de Boer: Physica, Haag **20**, 501 (1954) (Ziff. 80, 85, 87).
— — J. M. J. van Leeuwen, B. W. Ross u. J. de Boer: Physica, Haag **21**, 737 (1955); **22**, 791 (1956) (Ziff. 82, 84, 86, 87).
Corbett, J. W., and W. W. Watson: Phys. Rev. **101**, 519 (1956) (Ziff. 19).
Couette, M.: Ann. chim. phys. **21**, 433 (1890) (Ziff. 7).
Curie, M., and A. Lepape: J. Phys. Radium **2**, 392 (1931) (Ziff. 11, 80).
Curtiss, C. F.: J. Chem. Phys. **21**, 1199 (1953a) (Ziff. 89).
— J. Chem. Phys. **21**, 2045 (1953b) (Ziff. 89).
— J. Chem. Phys. **24**, 225 (1956) (Ziff. 89, 92).
—, and F. T. Adler: J. Chem. Phys. **20**, 249 (1952) (Ziff. 89).
—, and J. O. Hirschfelder: J. Chem. Phys. **17**, 550 (1949) (S. 402).
— — u. F. T. Adler: J. Chem. Phys. **18**, 1638 (1950) (Ziff. 89).
—, and C. Muckenfuss: J. Chem. Phys. **26**, 1619 (1957) (Ziff. 89, 92).
Daynes, H. A.: Proc. Roy. Soc. Lond., Ser. A **97**, 273 (1920) (Ziff. 12).
Debye, P.: Ann. Phys. **36**, 284 (1939) (Ziff. 19).
Duff, A. W.: Phys. Rev. **6**, 129 (1898) (Ziff. 99).

DUFOUR, L.: Ann. Phys. **148**, 490 (1873) (Ziff. 3).
DWYER, R. J.: J. Chem. Phys. **7**, 40 (1939) (Ziff. 99).
ECKART, C.: Phys. Rev. **73**, 68 (1948) (Ziff. 99).
EHRENFEST, P., and J. R. OPPENHEIMER: Phys. Rev. **37**, 333 (1931) (Ziff. 82).
EMMETT, P. H., and J. F. SHULTZ: J. Amer. Chem. Soc. **54**, 3780 (1932) (S. 327).
ENGELHARDT, H., u. H. SACK: Phys. Z. **33**, 724 (1932) (Ziff. 6).
ENSKOG, D.: Phys. Z. **12**, 533 (1911) (Ziff. 3, 62).
— Ann. Phys. **38**, 731 (1912) (Ziff. 62).
— Diss. Uppsala 1917 (Ziff. 1, 3, 33, 36, S. 402, Ziff. 61).
— Ark. Mat., Astronom. Fys. **16**, 1 (1921) (Ziff. 1).
— Kungl. Svenska Vetenskapsakad. Handl. **63**, Nr. 4 (1922) (Ziff. 1, 5).
— Z. Physik **54**, 498 (1929) (Ziff. 61).
EUCKEN, A.: Phys. Z. **12**, 1101 (1911a) (Ziff. 10).
— Ann. Physik **34**, 185 (1911b) (Ziff. 10).
— Phys. Z. **14**, 324 (1913) (Ziff. 52, 98).
— Forsch. Arb. Ing.-Wes. **11**, 6 (1940) (Ziff. 98).
EVETT, A. A., and H. MARGENAU: Phys. Rev. **90**, 1021 (1953) (Ziff. 25).
FALK, G.: Z. Physik **142**, 297 (1955) (Ziff. 82).
—, u. A. MANN: Z. Physik **142**, 277 (1955) (Ziff. 86).
FAXÉN, H., u. J. HOLTSMARK: Z. Physik **45**, 307 (1927) (Ziff. 82).
FEENBERG, E.: Phys. Rev. **40**, 40 (1932) (Ziff. 89).
FRANCK, E. U.: Z. Elektrochem. **55**, 636 (1951) (Ziff. 10, 52, 98).
FRANKEL, S. P.: Phys. Rev. **57**, 661 (1940) (S. 434).
FÜRTH, R.: Proc. Roy. Soc. Lond., Ser. A **179**, 461 (1942) (S. 434).
FUES, E.: Z. Physik **118**, 409 (1941) (Ziff. 2).
FURRY, W. H., R. CLARK JONES and L. ONSAGER: Phys. Rev. **55**, 1083 (1939) (Ziff. 19, 75).
GARDINER, W. C., u. KL. SCHÄFER: Z. Elektrochem. **60**, 588 (1956) (Ziff. 10).
GILCHRIST, L.: Phys. Z. **14**, 160 (1913) (Ziff. 7).
GLAUBER, R., and V. SCHOMAKER: Phys. Rev. **89**, 667 (1953) (Ziff. 89).
GOLDSCHMIDT, R.: Phys. Z. **12**, 417 (1911) (Ziff. 10).
GRAD, H.: Comm. Pure a. Appl. Math. **2**, 331 (1949) (S. 364).
— Comm. Pure a. Appl. Math. **5**, 257 (1952) (Ziff. 35).
GREEN, M. S.: J. Chem. Phys. **25**, 836 (1956) (S. 345).
GREENSPAN, M.: J. Acoust. Soc. Amer. **22**, 568 (1950) (Ziff. 42).
— J. Acoust. Soc. Amer. **28**, 644 (1956) (Ziff. 42).
GREGORY, H., and C. T. ARCHER: Phil. Mag. **1**, 593 (1926) (Ziff. 5).
GREW, K. E.: Phil. Mag. **35**, 30 (1944) (Ziff. 74, 76).
— Nature, Lond. **156**, 267 (1945) (Ziff. 5, 18).
— Proc. Roy. Soc. Lond., Ser. A **189**, 402 (1947) (Ziff. 12, 77).
— Proc. Phys. Soc. Lond. A **62**, 655 (1949) (Ziff. 18, 77).
— F. A. JOHNSON and W. E. J. NEAL: Proc. Roy. Soc. Lond., Ser. A **224**, 513 (1954) (Ziff. 25, 77).
GRINTEN, W. VAN DER: Naturwiss. **27**, 317 (1939) (Ziff. 19).
GROTH, W., u. P. HARTECK: Z. Elektrochem. **47**, 167 (1941) (Ziff. 71, 72).
—, u. E. SUSSNER: Z. phys. Chem. **193**, 296 (1944) (Ziff. 12, 72).
GRÜSS, H., u. H. SCHMICK: Wiss. Veröff. Siemens-Konzern **7**, 202 (1928) (Ziff. 80).
HAASE, R.: Z. Physik **127**, 1 (1950) (Ziff. 21).
HALPERN, O.: Phys. Rev. **82**, 561 (1951) (Ziff. 85).
—, and E. GWATHMEY: Phys. Rev. **52**, 944 (1937) (Ziff. 81, 83, 86).
HARRISON, G. E.: Proc. Roy. Soc. Lond., Ser. A **161**, 80 (1937) (Ziff. 18).
— Proc. Roy. Soc. Lond., Ser. A **181**, 93 (1942) (Ziff. 18).
HARTECK, P., u. H. W. SCHMIDT: Z. phys. Chem., Abt. B **21**, 447 (1933) (Ziff. 4, 14, 71, 72).
HASSÉ, H. R., and W. R. COOK: Proc. Roy. Soc. Lond., Ser. A **125**, 196 (1929) (Ziff. 25, 51, 75).
HEATH, H. R., T. L. IBBS and N. E. WILD: Proc. Roy. Soc. Lond., Ser. A **178**, 380 (1941) (Ziff. 4).
HECKE, E.: Math. Z. **12**, 274 (1922) (Ziff. 36).
HEITLER, W., u. F. LONDON: Z. Physik **44**, 455 (1927) (Ziff. 24).
HELLUND, E. J.: Phys. Rev. **57**, 319, 328 (1940) (S. 402, Ziff. 68, 74).
—, and E. A. UEHLING: Phys. Rev. **56**, 818 (1939) (S. 402).
HERCUS, E. O., and T. H. LABY: Proc. Roy. Soc. Lond., Ser. A **95**, 190 (1919) (Ziff. 10).
—, and D. M. SUTHERLAND: Proc. Roy. Soc. Lond., Ser. A **145**, 599 (1934) (Ziff. 10).
HERTZ, G.: Z. Physik **19**, 35 (1923) (Ziff. 14).
HERZFELD, K.: Ann. Physik **23**, 465, 476 (1935) (Ziff. 42).
—, and F. O. RICE: Phys. Rev. **31**, 691 (1928) (Ziff. 96).
HIBY, J. W., u. M. PAHL: Z. Physik **129**, 517; **130**, 348 (1951) (Ziff. 23).

HIRSCHFELDER, J. O., R. B. BIRD and E. L. SPOTZ: J. Chem. Phys. **16**, 968 (1948) (Ziff. 25, 51, 77).
— — — Chem. Rev. **44**, 205 (1949) (Ziff. 25, 51).
HOEPPLER, F.: Z. techn. Phys. **14**, 165 (1933) (Ziff. 9).
HOLLERAN, E. M.: J. Chem. Phys. **21**, 1901 (1953) (Ziff. 18).
HORNIG, J. F., and J. O. HIRSCHFELDER: Phys. Rev. **103**, 908 (1956) (Ziff. 89).
HOVE, L. VAN: Physica, Haag **21**, 517 (1955) (Ziff. 33).
HUTCHINSON, F.: J. Chem. Phys. **17**, 1081 (1949) (Ziff. 71, 72).
IBBS, T. L.: Proc. Roy. Soc. Lond., Ser. A **99**, 385 (1921) (Ziff. 12).
— Proc. Roy. Soc. Lond., Ser. A **107**, 470 (1925) (Ziff. 12, 18).
—, and K. E. GREW: Proc. Phys. Soc. Lond. **43**, 142 (1931) (Ziff. 77).
— —, and A. HIRST: Proc. Phys. Soc. Lond. **41**, 456 (1929) (Ziff. 76).
—, and A. HIRST: Proc. Roy. Soc. Lond., Ser. A **123**, 134 (1929) (Ziff. 80).
—, and L. UNDERWOOD: Proc. Phys. Soc. Lond. **39**, 227 (1927) (Ziff. 74).
IKENBERRY, E., and C. TRUESDELL: J. Rat. Mech. a. Analysis **5**, 1 (1956) (Ziff. 38, 42).
ITTERBEEK, A. VAN, u. A. CLAES: Physica, Haag **5**, 938 (1938) (Ziff. 4).
—, u. O. VAN PAEMEL: Physica, Haag **5**, 1004 (1938) (Ziff. 8).
— — Physica, Haag **7**, 265 (1940) (Ziff. 4, 8, 98).
— —, u. J. VAN LIERDE: Physica, Haag **13**, 231 (1947) (Ziff. 8, 18).
— F. W. SCHAPINK, G. J. VAN DEN BERG u. H. J. M. VAN BEEK: Physica, Haag **19**, 1158 (1953) (Ziff. 85).
IRVING, J. H., and J. G. KIRKWOOD: J. Chem. Phys. **18**, 817 (1950) (Ziff. 1, S. 345).
JAFFÉ, G.: Ann. Physik **6**, 195 (1930) (S. 364).
JOHNSTON, H. L., and E. R. GRILLY: J. Phys. Chem. **46**, 948 (1942) (Ziff. 52, 72).
JUNG, G., u. H. SCHMICK: Z. phys. Chem., Abt. B **7**, 130 (1930) (Ziff. 78, 79).
KAMERLINGH ONNES, H., u. S. WEBER: Proc. Amsterdam **15**, 1396 (1913) (Ziff. 8, 52).
KAMPEN, N. G. VAN: Physica, Haag **20**, 603 (1954) (Ziff. 33).
KANNULUIK, W. G., and E. H. CARMAN: Proc. Phys. Soc. Lond. B **65**, 701 (1952) (Ziff. 10, 52).
—, and L. H. MARTIN: Proc. Roy. Soc. Lond., Ser. A **141**, 144 (1933) (Ziff. 10).
— — Proc. Roy. Soc. Lond., Ser. A **144**, 496 (1934) (Ziff. 10).
KELLER, H.: Phys. Z. **41**, 386 (1940) (Ziff. 99).
KELLER, J. B.: Comm. Pure a. Appl. Math. **1**, 275 (1948) (S. 364).
KELLSTRÖM, G.: Phil. Mag. **23**, 313 (1937) (Ziff. 7).
KIHARA, T., and M. KOTANI: Proc. Phys. Math. Soc. Japan **24**, 76 (1942) (Ziff. 51).
KIRKWOOD, J. G.: J. Chem. Phys. **14**, 180 (1946) (Ziff. 1, S. 345).
KNESER, H. O.: Ann. Physik **11**, 761 (1931) (Ziff. 96).
KNUDSEN, M.: Ann. Phys. **28**, 75 (1909) (Ziff. 1).
— Ann. Phys. **34**, 593 (1911) (Ziff. 1).
KOGA, T.: J. Chem. Phys. **23**, 2275 (1955) (Ziff. 28).
KOHLER, M.: Z. Physik **124**, 757 (1947a) (Ziff. 97, 99).
— Z. Physik **124**, 772 (1947b) (Ziff. 33).
— Z. Physik **125**, 715 (1948) (Ziff. 42, 99).
KOPPE, H.: Z. Physik **141**, 103 (1955) (Ziff. 90).
KRAMERS, H. A., u. J. KISTEMAKER: Physica, Haag **10**, 699 (1943) (Ziff. 1).
KÜMMEL, H.: Z. Physik **143**, 219 (1955) (Ziff. 28, 33).
— Z. Naturforsch. **11**a, 15 (1956) (Ziff. 28, 33).
KUNDT, A., u. E. WARBURG: Ann. Physik **155**, 337, 525 (1875a) (Ziff. 1).
— — Ann. Physik **156**, 177 (1875b) (Ziff. 5).
LABY, T. H.: Proc. Roy. Soc. Lond., Ser. A **144**, 494 (1934) (Ziff. 10).
LAMBERT, J. D., E. N. STAINES and S. D. WOODS: Proc. Roy. Soc. Lond. Ser. A **200**, 262 (1950) (Ziff. 5).
LANDAU, L., u. E. TELLER: Phys. Z. Sowjet. **10**, 34 (1936) (Ziff. 99).
LAUE, M. v.: Ann. Physik **23**, 1 (1935) (Ziff. 92).
LAX, M.: Phys. Rev. **78**, 306 (1950) (Ziff. 89).
LENNARD-JONES, J. E.: Proc. Roy. Soc. Lond., Ser. A **106**, 441, 463 (1924) (Ziff. 25).
LIEBERMANN, L. N.: Phys. Rev. **75**, 1415 (1949) (Ziff. 99).
LONDON, F.: Z. phys. Chem., Abt. B **11**, 222 (1930) (Ziff. 24, 51).
LONIUS, A.: Ann. Physik **29**, 664 (1909) (Ziff. 13, 73).
LORENTZ, H. A.: Proc. Amsterdam Acad. **7**, 438 (1905) (Ziff. 1).
LOSCHMIDT, J.: Ber. Wien. Akad. **61**, 367 (1870) (Ziff. 13).
— Ber. Wien. Akad. **73**, 139 (1876) (Ziff. 33).
LÜDERS, G.: Z. Physik **133**, 325 (1952) (Ziff. 89).
MANN, A. K.: Phys. Rev. **73**, 412 (1948) (Ziff. 77).
MADELUNG, E.: Ann. Physik **43**, 417 (1943) (Ziff. 2).
MALLOCK, A.: Proc. Roy. Soc. Lond., Ser. A **45**, 126 (1888) (Ziff. 7).

MARROT, R.: J. Math. pures appl. **25**, 93 (1946) (S. 345).
MASON, E. A.: J. Chem. Phys. **22**, 169 (1954) (Ziff. 25, 51, 75).
— J. Chem. Phys. **27**, 75 (1957) (Ziff. 52).
— J. Chem. Phys. **27**, 782 (1957a) (Ziff. 62).
—, and W. E. RICE: J. Chem. Phys. **22**, 522 (1954a) (Ziff. 52, 72, 81).
— — J. Chem. Phys. **22**, 843 (1954b) (Ziff. 51, 52, 72, 75, 95).
MASSEY, H. S. W., and C. B. O. MOHR: Proc. Roy. Soc. Lond., Ser. A **141**, 434 (1933) (Ziff. 81, 83).
MAWOOD, G. E.: Physica, Haag **5**, 374 (1938) (Ziff. 8).
MEISSNER, W. u. G.: Ann. Physik **36**, 303 (1939) (Ziff. 2).
MEIXNER, J.: Ann. Physik **39**, 333 (1941) (Ziff. 61).
— Ann. Physik **41**, 409 (1942) (Ziff. 34).
— Ann. Physik **43**, 244 (1943a) (Ziff. 56, 61).
— Z. phys. Chem., Abt. B **53**, 235 (1943b) (Ziff. 61).
MELVILLE, H. W., and J. L. BOLLAND: Proc. Roy. Soc. Lond., Ser. A **160**, 384 (1937) (Ziff. 12).
MEYER, O. E.: Ann. Physik **125**, 564 (1865) (Ziff. 5).
MICHELS, A., u. A. BOTZEN: Physica, Haag **18**, 605 (1952) (Ziff. 10).
— — Physica, Haag **19**, 585 (1953) (Ziff. 5).
— — u. W. SCHUURMAN: Physica, Haag **20**, 1141 (1954) (Ziff. 5).
— — A. S. FRIEDMAN u. J. V. SENGERS: Physica, Haag **22**, 121 (1956) (Ziff. 5, 10).
—, and A. O. GIBSON: Proc. Roy. Soc. Lond., Ser. A **134**, 288 (1931) (Ziff. 5).
— S. J. JACOBS u. Z. I. SLAWSKI: Physica, Haag **20**, 223, 1157 (1954) (Ziff. 2).
— A. C. J. SCHIPPER u. W. H. RINTOUL: Physica, Haag **19**, 1011 (1953) (Ziff. 4).
MILVERTON, S. W.: Phil. Mag. **17**, 397 (1934) (Ziff. 5).
MIYAKO, R.: Proc. Phys.-Math. Soc. Japan **24**, 852 (1942) (Ziff. 86).
MORGENSTERN, D.: J. Rat. Mech. a. Analysis **4**, 533 (1955) (Ziff. 33).
MOTT, N. F.: Proc. Roy. Soc. Lond., Ser. A **126**, 259 (1930) (Ziff. 81).
MOTT-SMITH, H. M.: Phys. Rev. **82**, 885 (1951) (Ziff. 35).
MURPHEY, B. F.: Phys. Rev. **72**, 834 (1947) (Ziff. 74, 77).
NERNST, W.: Boltzmann-Festschrift, S. 904. Leipzig 1904 (Ziff. 5).
NETTLEY, P.T.: Proc. Phys. Soc. Lond. B **67**, 753 (1954) (Ziff. 12, 17, 73).
NEY, E.P., and F.C. ARMISTEAD: Phys. Rev. **71**, 14 (1947) (Ziff. 13, 72).
NIER, A.O.: Phys. Rev. **56**, 1009 (1939) (Ziff. 18, 19).
— Phys. Rev. **57**, 338 (1940) (Ziff. 18, 19, 77).
NÖLLER, H.G., G.W. OETJEN u. R. JAECKEL: Z. Naturforsch. **4**a, 101 (1949) (Ziff. 23).
NOTHDURFT, W.: Ann. Physik **28**, 157 (1937) (Ziff. 4).
ONSAGER, L.: Phys. Rev. **37**, 405; **38**, 2265 (1931) (Ziff. 61).
ORNSTEIN, L. S., u. J.M.W. MILATZ: Physica, Haag **6**, 1139 (1939) (Ziff. 28).
PEKERIS, C.L.: Proc. Nat. Acad. Sci. **41**, 661 (1955) (S. 364).
PIDDUCK, F.B.: Proc. Roy. Soc. Lond., Ser. A **101**, 101 (1922) (Ziff. 89, 97).
PRIGOGINE, I.: Physica, Haag **15**, 272 (1949) (Ziff. 61).
RAPP, I.M.: Phys. Rev. **2**, 363 (1913) (Ziff. 6).
RAYLEIGH, Lord: Proc. Roy. Soc. Lond., Ser. A **66**, 68 (1900) (Ziff. 4).
RIETVELD, A.O., u. A. VAN ITTERBEEK: Physica, Haag **22**, 785 (1956) (Ziff. 79).
— — Physica, Haag **23**, 838 (1957) (Ziff. 4, 86).
— — u. G. J. VAN DEN BERG: Physica, Haag **19**, 517 (1953) (Ziff. 79).
ROBB, W.L., and H.G. DRICKAMER: J. Chem. Phys. **19**, 1504 (1951) (Ziff. 5).
ROWLINSON, J.S.: J. Chem. Phys. **17**, 101 (1949) (Ziff. 51).
RUTGERS, A.I.: Ann. Physik **16**, 350 (1933) (Ziff. 96).
SACHSSE, H., u. K. BRATZLER: Z. phys. Chem., Abt. A **171**, 331 (1935) (Ziff. 12).
SAXENA, S.C.: J. Chem. Phys. **24**, 1209 (1956) (Ziff. 75).
SCHÄFER, KL.: Z. phys. Chem., Abt. B **36**, 85 (1937) (Ziff. 86).
— Z. phys. Chem., Abt. B **53**, 149 (1943) (Ziff. 98).
— H. CORTE u. H. MOESTA: Z. Elektrochem. **55**, 662 (1951) (Ziff. 13, 73).
—, u. O.R. FOZ GAZULLA: Z. phys. Chem., Abt. B **52**, 299 (1942) (Ziff. 5).
SCHLEIERMACHER, A.: Ann. Physik **34**, 623 (1888) (Ziff. 10).
SCHMAHL, N.G., u. J. SCHEWE: Z. Elektrochem. **46**, 203 (1940) (Ziff. 18).
SCHOTTKY, W.F.: Z. Elektrochem. **56**, 889 (1952) (Ziff. 10).
SCHRÖER, E.: Z. phys. Chem., Abt. B **34**, 161 (1936) (Ziff. 78).
SCHUMACHER, E., u. K. CLUSIUS: Helv. chim. Acta **36**, 961 (1953) (Ziff. 19).
SENFTLEBEN, H.: Phys. Z. **31**, 822, 961 (1930) (Ziff. 6, 92).
SLATER, J. C.: Phys. Rev. **32**, 349 (1928) (Ziff. 25).
—, and J. G. KIRKWOOD: Phys. Rev. **37**, 682 (1931) (Ziff. 25).
SLAWSKI, Z. I., A. MICHELS u. S. J. JACOBS: Physica, Haag **20**, 210 (1954) (Ziff. 2).
SMITH, L. J.: Z. Proc. Phys. Soc. Lond. **34**, 155 (1922) (Ziff. 74).

SMOLUCHOWSKI, M. v.: Ann. Physik **64**, 101 (1898) (Ziff. 1, 10).
SRIVASTAVA, B. N., and M. P. MADAN: J. Chem. Phys. **21**, 807 (1953) (Ziff. 18).
—, u. S. C. SAXENA: Physica, Haag **22**, 253 (1956) (Ziff. 52, 71, 77).
STERN, O.: Z. Physik **2**, 49; **3**, 417 (1920) (Ziff. 34).
STIER, L. G.: Phys. Rev. **62**, 548 (1942) (Ziff. 19, 77).
STÜCKELBERG, E. C. G.: Helv. phys. Acta **25**, 577 (1952) (Ziff. 93).
SUTHERLAND, W.: Phil. Mag. **36**, 507 (1893) (Ziff. 25).
TAKAYANAGI, K.: Progr. Theor. Phys. **11**, 557 (1954) (Ziff. 89).
—, and S. KANEKO: Sci. Rep. Saitama Univ. A **1**, 111 (1954) (Ziff. 89).
—, and K. OHNO: Progr. Theor. Phys. **13**, 243 (1955) (Ziff. 89).
TAYLOR, T. I., and R. B. BERNSTEIN: J. Amer. Chem. Soc. **69**, 2076 (1947) (Ziff. 19).
—, and W. SPINDEL: J. Chem. Phys. **16**, 635 (1948) (Ziff. 19).
TEMPERLEY, H. N. V.: Proc. Cambridge Phil. Soc. **52**, 712 (1956) (Ziff. 33).
TIMMERHAUS, K. D., and H. G. DRICKAMER: J. Chem. Phys. **19**, 1242 (1951) (Ziff. 5).
TIMIRIAZEFF, A.: Ann. Physik **40**, 971 (1913) (Ziff. 7).
TISZA, L.: Phys. Rev. **61**, 531 (1942) (Ziff. 96, 99).
TORREY, H. C.: Phys. Rev. **47**, 644 (1935) (Ziff. 4).
TRAUTZ, M., u. H. E. BINKELE: Ann. Physik **5**, 561 (1930) (Ziff. 79).
—, u. K. F. KIPPHAN: Ann. Physik **2**, 743 (1929) (Ziff. 79).
—, u. F. KURZ: Ann. Physik **9**, 981 (1931) (Ziff. 79).
—, u. A. MELSTER: Ann. Physik **7**, 409 (1930) (Ziff. 74, 78).
—, u. W. WEIZEL: Ann. Phys. **78**, 305 (1925) (Ziff. 6).
—, u. R. ZINK: Ann. Physik **7**, 427 (1930) (Ziff. 52).
TROYER, A. DE, A. VAN ITTERBEEK u. A. O. RIETVELD: Physica, Haag **17**, 938 (1951) (Ziff. 77).
UBBELOHDE, A. R.: J. Chem. Phys. **3**, 219, 362 (1935) (Ziff. 98).
UBBINK, J. B.: Physica, Haag **14**, 165 (1948) (Ziff. 4, 86, 98).
—, u. W. J. DE HAAS: Physica, Haag **10**, 451 (1943a) (Ziff. 10).
— — Physica, Haag **10**, 465 (1943b) (Ziff. 10, 52, 85).
UEHLING, E. A.: Phys. Rev. **46**, 917 (1934) (Ziff. 81).
—, and G. E. UHLENBECK: Phys. Rev. **43**, 552 (1933) (Ziff. 1, S. 345, Ziff. 81).
ULSAMER, J.: Z. VDI **80**, 537 (1936) (Ziff. 10).
VISNER, S.: Phys. Rev. **82**, 297 (1951) (Ziff. 71, 72).
VRIES, A. E. DE, A. HARING u. W. SLOTS: Physica, Haag **22**, 247 (1956) (Ziff. 77).
WACHSMUTH, J.: Phys. Z. **9**, 235 (1908) (Ziff. 80).
WAERDEN, B. L. v. D.: Z. Naturforsch. **12**a, 583 (1957) (Ziff. 19).
WALDMANN, L.: Z. Physik **114**, 53 (1939) (Ziff. 19).
— Z. Physik **121**, 501 (1943) (Ziff. 3).
— Naturwiss. **32**, 222 (1944) (Ziff. 4, 71).
— Z. Naturforsch. **1**, 12 (1946a) (Ziff. 74).
— Z. Naturforsch. **1**, 483 (1946b) (Ziff. 87).
— Z. Physik **124**, 175 (1947a) (Ziff. 20, 56, 57).
— Z. Physik **124**, 2 (1947b) (Ziff. 15, 20, 73, 76).
— Z. Physik **124**, 30 (1947c) (Ziff. 21).
— Z. Naturforsch. **2**a, 358 (1947d) (Ziff. 77).
— Z. Naturforsch. **4**a 105 (1949) (Ziff. 21, 74, 77).
— Z. Naturforsch. **5**a, 322 (1950a) (Ziff. 62).
— Z. Naturforsch. **5**a, 327 (1950b) (Ziff. 16).
— Z. Naturforsch. **5**a, 399 (1950c) (Ziff. 74).
— Z. Naturforsch. **11**a, 523 (1956) (Ziff. 38).
— Z. Naturforsch. **12**a, 660 (1957) (Ziff. 91).
—, u. E. W. BECKER : Z. Naturforsch. **3**a, 180 (1948) (Ziff. 77, 86, 87).
WALL, F. T., and C.E. HOLLEY: J. Chem. Phys. **8**, 949 (1940) (Ziff. 74).
WANG CHANG, C. S.: Appl. Phys. Lab. Johns Hopkins Univ. Report CM 467, 1948 (Ziff. 42).
—, and G. E. UHLENBECK: Eng. Res. Inst. Univ. Michigan, Report CM 579, Nov. 1949 (S. 364).
— — Eng. Res. Inst. Univ. Michigan, Report CM 654, Dez. 1950 (S. 364).
— — Eng. Res. Inst. Univ. Michigan Report CM 681, Juli 1951 (Ziff. 96).
— — Eng. Res. Inst. Univ. Michigan, Project M 999, Okt. 1952 (Ziff. 38).
WATSON, K. M.: Phys. Rev. **102**, 12 (1956) (Ziff. 28).
WATSON, W. W., and D. WOERNLEY: Phys. Rev. **63**, 181 (1943) (Ziff. 77).
WEBER, S.: Ann. Physik **54**, 325 (1917) (Ziff. 10).
— Ann. Physik **82**, 479 (1927) (Ziff. 10).
— Physica, Haag **6**, 562 (1939) (Ziff. 51).
WESTENBERG, A. A., and R. E. WALKER: J. Chem. Phys. **26**, 1753 (1957) (Ziff. 14).
WHALLEY, E.: Trans. Faraday Soc. **47**, 815 (1951) (S. 327).

WHALLEY, E., E. R. S. WINTER and H. V. A. BRISCOE: Trans. Faraday Soc. **45**, 1085 (1949) (Ziff. 77).
WICK, G. C.: Phys. Rev. **75**, 1459 (1949) (Ziff. 89).
WIGNER, E.: Nachr. Ges. Wiss. Göttingen, math.-physik. Kl. **1932**, 546 (Ziff. 89).
WILKE, C. R.: J. Chem. Phys. **18**, 517 (1950) (Ziff. 78).
WINN, E. B.: Phys. Rev. **80**, 1024 (1950) (Ziff. 13, 71, 72).
WINTER, E. R. S.: Trans. Faraday Soc. **46**, 81 (1950) (Ziff. 74, 77).
— Trans. Faraday Soc. **47**, 342 (1951) (Ziff. 72).
WOBSER, R., u. FR. MÜLLER: Kolloid-Beih. **52**, 165 (1941) (Ziff. 9).
WOGAU, M. v.: Ann. Phys. **23**, 345 (1907) (Ziff. 13).
ZOLLER, K.: Z. Physik **130**, 1 (1951) (Ziff. 35).
ZENER, C.: Phys. Rev. **37**, 556 (1931) (Ziff. 71).
ZERNIKE, F., u. C. VAN LIER: Physica, Haag **6**, 961 (1939) (Ziff. 92).

Zusammenfassende Darstellungen.

Lehrbücher, Monographien, Review-Artikel und Tabellenwerke.

[1] BOER, J. DE: Molecular Distribution and Equation of State of Gases. Reports on Progress in Physics, **12**, 305. 1948.
[2] BOER, J. DE: Transport Properties of Gaseous Helium at Low Temperatures. In C. J. GORTER, Progress in Low Temperature Physics I. Amsterdam 1955.
[3] BOLTZMANN, L.: Vorlesungen über Gastheorie I. und II. Teil. Leipzig 1895.
[4] CHAPMAN, S., and T. G. COWLING: Mathematical Theory of Non-Uniform Gases. Cambridge 1939 (Ergänzung dazu 1951).
[5] CLARK JONES, R., and W. H. FURRY: The Separation of Isotopes by Thermal Diffusion. Rev. Mod. Phys. **18**, 151 (1946).
[6] COWLING, T. G.: Magnetohydrodynamics. New York 1957.
[7] CRANK, J.: The Mathematics of Diffusion. Oxford 1956.
[8] D'ANS, J., u. E. LAX: Taschenbuch für Chemiker und Physiker. Berlin 1943.
[9] DAYNES, H. A.: Gas Analysis by Measurement of Thermal Conductivity. Cambridge 1933.
[10] DUSHMAN, S.: Scientific Foundations of Vacuum Technique. New York 1949.
[10a] EBBINGHAUS, E.: Gasanalyse durch Messung des Wärmeleitvermögens. In J. HENGSTENBERG et al., Messen und Regeln in der chemischen Technik. Berlin-Göttingen-Heidelberg 1957.
[11] ENSKOG, D.: Kinetische Theorie der Vorgänge in mäßig verdünnten Gasen. I. Allgemeiner Teil. Diss. Uppsala 1917.
[12] ESTERMANN, I.: Molecular Beam Technique. Rev. Mod. Phys. **18**, 300 (1946).
[13] EUCKEN, A.: Lehrbuch der chemischen Physik, Bd. II/1. Leipzig 1943. Bd. I. Leipzig 1949.
[14] FARKAS, A.: Orthohydrogen, Parahydrogen and Heavy Hydrogen. Cambridge 1935.
[15] FLEISCHMANN, R., u. H. JENSEN: Das Trennrohr. Ergebn. exakt. Naturw. **20**, 121 (1942).
[16] FRANK, E. U.: Wärmeleitung in hochverdichteten Gasen. Chem.-Ing.-Techn. **25**, 238 (1953).
[17] FRANK, P., u. R. v. MISES: Die Differential- und Integralgleichungen der Mechanik und Physik, 2. Aufl., Bd. II. Braunschweig 1935.
[18] GREW, K. E., and T. L. IBBS: Thermal Diffusion in Gases. Cambridge 1952.
[19] HARTMANN, H.: Theorie der chemischen Bindung. Berlin 1954.
[20] HELLMANN, H.: Einführung in die Quantenchemie. Leipzig u. Wien 1937.
[21] HERZFELD, K. F.: Kinetische Theorie der Wärme. In MÜLLER-POUILLETs Lehrbuch der Physik, Bd. III/2. Braunschweig 1925.
[22] HERZFELD, K. F.: Freie Weglänge und Transporterscheinungen in Gasen. In EUCKEN-WOLFs Hand- und Jahrbuch der chemischen Physik, Bd. 3/2. Leipzig 1939.
[23] HILBERT, D.: Grundzüge einer allgemeinen Theorie der linearen Integralgleichungen. Leipzig 1912.
[24] HILSENRATH, J., et. al.: Tables of Thermal Properties of Gases. National Bureau of Standards Circular 564, 1955.
[25] HIRSCHFELDER, J. O., C. F. CURTISS and R. B. BIRD: Molecular Theory of Gases and Liquids. New York 1954.
[26] JÄGER, G.: Die kinetische Theorie der Gase und Flüssigkeiten. In GEIGER-SCHEELs Handbuch der Physik, Bd. 9. Berlin 1926.
[27] JEANS, J. H.: The Dynamical Theory of Gases. Cambridge 1904.

[28] JEANS, J. H.: Dynamische Theorie der Gase. Nach der 4. engl. Aufl. übersetzt von R. FÜRTH. Braunschweig 1926.
[29] JOST, W.: Diffusion in Solids, Liquids, Gases. New York 1952.
[30] KARIM, S. M., and L. ROSENHEAD: The Second Coefficient of Viscosity of Liquids and Gases. Rev. Mod. Phys. **24**, 108 (1952).
[31] KENNARD, E. H.: Kinetic Theory of Gases. New York 1938.
[32] KIHARA, T.: Nichtideale Gase [Japanisch]. Tokyo 1949.
[33] KEESOM, W. H.: Helium. Amsterdam 1942.
[34] LANDOLT-BÖRNSTEIN: Physikalisch-Chemische Tabellen, 5. Aufl. Berlin 1923 — 1936.
[35] LIENEWEG, F.: Temperaturmessung. In Krönerts Handb. d. technischen Betriebskontrolle, Bd. III. Leipzig 1944.
[36] LORENTZ, H. A.: The Theory of Electrons. Leipzig 1909.
[37] MAGNUS, W., u. F. OBERHETTINGER: Formeln und Sätze für die speziellen Funktionen der mathematischen Physik. Berlin 1943.
[38] MAXWELL, J. C.: Scientific Papers, 2 Bde., hrsgg. von W. D. NIVEN. Cambridge 1890.
[38a] MISES, R. v.: Wahrscheinlichkeit, Statistik und Wahrheit, 3. Aufl. Wien 1951.
[39] MOTT, N. F., and H. S. W. MASSEY: Theory of Atomic Collisions, 2. Aufl. Oxford 1949.
[40] NEUMANN, J. v.: Mathematische Grundlagen der Quantenmechanik. Berlin 1932.
[41] PATTERSON, G. N.: Molecular Flow of Gases. New York 1956.
[42] PAULING, L.: The Nature of the Chemical Bond, 2. Aufl. New York 1940.
[43] RAMSEY, N. F.: Molecular Beams. Oxford 1956.
[44] RICHARDS, W. T.: Supersonic Phenomena. Rev. Mod. Phys. **11**, 36 (1939).
[45] RICHARDSON, E. G.: Absorption and Velocity of Sound in Vapors. Rev. Mod. Phys. **27**, 15 (1955).
[46] SCHÄFER, KL.: Probleme der Wärmeleitung in Gasen bei niedrigem Druck und der Energieübertragung an festen Oberflächen. Fortschrt. chem. Forsch. **1**, 61 (1949).
[47] SCHMEIDLER, W.: Integralgleichungen mit Anwendungen in Physik und Technik I. Leipzig 1955.
[48] SCHMIDT, E.: Einführung in die technische Thermodynamik, 4. Aufl. Berlin-Göttingen-Heidelberg 1950.
[49] SOMMERFELD, A.: Vorlesungen über theoretische Physik, Bd. V: Thermodynamik und Statistik. Wiesbaden 1952.
[50] SPITZER jr., L.: Physics of Fully Ionized Gases. New York 1956.
[51] TOLMAN, R. C.: Principles of Statistical Mechanics. Oxford 1938.
[52] WALDMANN, L.: Die Temperaturerscheinungen bei der Diffusion. Z. Naturforsch. **1**, 59 (1946).
[53] WALDMANN, L.: Transporterscheinungen in Gasen. In Fiat Rev. of German Science, Bd. 10. Physik der Flüssigkeiten und Gase, hrsg. von E. KAPPLER. Wiesbaden 1948.
[54] WATANABE, S.: Symmetry of Physical Laws I. Rev. Mod. Phys. **27**, 26 (1955).
[55] WIGNER, E.: Gruppentheorie und ihre Anwendung auf die Quantenmechanik der Atomspektren. Braunschweig 1931.

Berichtigungen:

Zu Gl. (92.3): In der ersten Zeile ist im zweiten Glied rechts der Faktor A' anzufügen.

Zu Gl. (92.5): Diese soll lauten

$$\alpha_{MM_1}(\mathbf{c}, \mathbf{c}_1)\, \alpha^*_{NN_1}(\mathbf{c}, \mathbf{c}_1) = P_{MM_1NN_1}(\mathbf{c}, \mathbf{c}_1).$$

Zu Gl. (92.13): Die rechte Seite ist mit dem Faktor $\frac{1}{2}$ zu versehen.

In der auf (93.11) folgenden Gleichung: $F_N - F_M$ statt $F_M - F_N$.

Allgemeine Vakuumphysik.

Von

R. Jaeckel.

Mit 120 Figuren.

1. Übersicht. Im internationalen Schrifttum über Vakuumphysik erfolgen die Druckangaben in den verschiedensten Maßeinheiten. Tabelle 2 bringt eine Zusammenstellung der gebräuchlichsten und der Umrechnungsfaktoren zum Übergang von einer Einheit auf die andere. Im folgenden sollen Drucke stets in Torr

Tabelle 1.

	Grobvakuum	Feinvakuum	Hochvakuum	Ultrahochvakuum
Druckbereich (Torr)	$760 \ldots 1$	$1 \ldots 10^{-3}$	$10^{-3} \ldots 10^{-6}$	$10^{-6} \ldots 10^{-11}$
Anzahl der Teilchen (pro cm^3)	$10^{19} \ldots 10^{16}$	$10^{16} \ldots 10^{13}$	$10^{13} \ldots 10^{10}$	$10^{10} \ldots 10^{5}$
Art der Strömung	Strömungskontinuum	Übergang zur Molekularströmung	Molekularströmung	Molekularströmung
Mittlere freie Weglänge	kleiner als die Gefäßdimensionen	kleiner als die oder gleich den Gefäßdimensionen	normalerweise größer als die Gefäßdimensionen	größer als die Gefäßdimensionen
Wiederbedeckungszeit für monoatomare Schicht (sec)			$10^{-3} \ldots 10^{0}$	$10^{0} \ldots 10^{5}$
Wärmeleitung und Reibung	unabhängig vom Druck	vom Druck abhängig[1]	proportional zum Druck	praktisch Null
Maßgebend für die Dimensionierung der Pumpe (zum Evakuieren eines Behälters)	Gefäßvolumen	Gefäßvolumen und Gefäßform	Größe und Beschaffenheit der Oberfläche	Größe und Beschaffenheit der Oberfläche
Anhalt für Dimensionierung der Pumpe	Auspumpzeit-Kurven		zum schnellen Auspumpen: 1000 l/sec/m^2 [2]	
Geeignete Pumpenart		Dampfstrahl-, Roots-, Treibdampfpumpe	Diffusionspumpe	Verdampfer-Ionenpumpe

1 Maßgebend ist das Verhältnis der Gefäßdimensionen zur mittleren freien Weglänge.
2 Zum Aufrechterhalten: 100 l/sec/m^2.

angegeben werden (falls keine anderweitige Angabe erfolgt). In der Vakuumphysik erstrecken sich die vorkommenden Druckwerte über viele Zehnerpotenzen von 10^3 bis herunter zu 10^{-11} bzw. 10^{-14} Torr. Für manche Zwecke hat sich daher eine grobe Übersicht nach Vakuumbereichen mit einer Einteilung gemäß Tabelle 1 in Grobvakuum, Feinvakuum, Hochvakuum und Ultrahochvakuum als zweckmäßig erwiesen. Diese grobe Einteilung nach runden Zehnerpotenzen in Torreinheiten hat vor allem den Vorteil der Einfachheit und genügt den meisten

Tabelle 2. *Druckeinheiten*

$$\% \text{ Vakuum} = \frac{760 - p_{\text{Torr}}}{760} \cdot 100;$$

	Torr	dyn/cm²	Millibar	Bar
1 Torr = 1 mm Quecksilbersäule	1	$1{,}33322 \cdot 10^3$	1,33322	$1{,}33322 \cdot 10^{-3}$
1 dyn/cm² = 1 Mikrobar (μb)	$0{,}75006 \cdot 10^{-3}$	1	10^{-3}	10^{-6}
1 Millibar (mb) = 10^3 dyn/cm²	0,75006	10^3	1	10^{-3}
1 Bar (b) = 10^6 dyn/cm²	750,06	10^6	10^3	1
1 kg/m² ≈ 1 mm Wassersäule	$0{,}73556 \cdot 10^{-1}$	$0{,}980665 \cdot 10^2$	$0{,}980665 \cdot 10^{-1}$	$0{,}980665 \cdot 10^{-4}$
1 kg/cm² = 1 at	735,56	$0{,}980665 \cdot 10^6$	$0{,}980665 \cdot 10^3$	0,980665
1 Atm. = 760 Torr	760	$1{,}01325 \cdot 10^6$	$1{,}01325 \cdot 10^3$	1,01325
1 lb per sq. inch (Engl. Pfund/Quadrat-Zoll)	$5{,}1715 \cdot 10^1$	$0{,}68948 \cdot 10^5$	$0{,}68948 \cdot 10^2$	$0{,}68948 \cdot 10^{-1}$
1 Mikron (μ) = $1 \cdot 10^{-3}$ Torr	10^{-3}	1,33322	$1{,}33322 \cdot 10^{-3}$	$1{,}33322 \cdot 10^{-6}$
1 inch of mercury (Engl. Zoll Quecksilbersäule)	25,400	$0{,}33864 \cdot 10^5$	$0{,}33864 \cdot 10^2$	$0{,}33864 \cdot 10^{-1}$

Umrechungstabellen für je zwei Druckeinheiten

praktischen Erfordernissen. Demgegenüber erscheinen Bemühungen zur Verfeinerung dieser Einteilung z. B. in Bereiche wie Mikrotorr, Millitorr- usw.[1] oder nach speziellen Fixpunkten (z. B. dem Tripelpunkt des Wassers)[2] weniger empfehlenswert, da bei ihnen gerade der Vorteil der Einfachheit der obengenannten Einteilung wieder verlorengeht.

Fig. 1 zeigt ein allgemeines Schema für eine Vakuumapparatur. Die wesentlichen Bauteile, Pumpen, Verbindungsleitungen, Vakuummeßinstrumente, sind

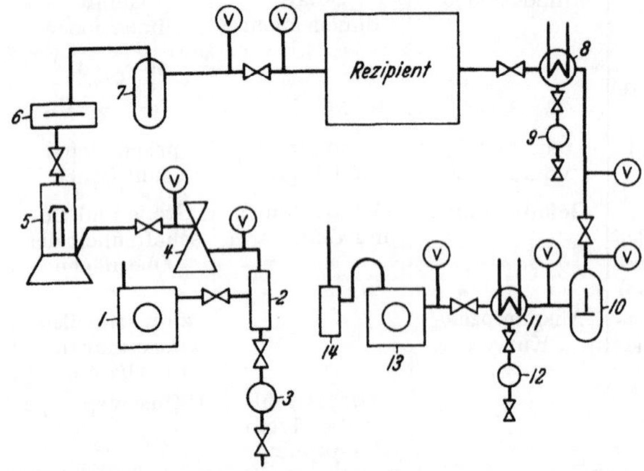

Fig. 1. Allgemeines Schema für eine Vakuumapparatur. *1* Rotierende Ölluftpumpe (Gasballastpumpe); *2* saugseitiger Abscheider; *3* Wechselvorlage; *4* Dampfstrahlpumpe; *5* Diffusionspumpe; *6* Baffle; *7* Kühlfalle; *8* gekühlter Kondensator; *9* Wechselvorlage; *10* Roots-Pumpe; *11* mit Kühlmittel gekühlter Zwischenkondensator; *12* Wechselvorlage, *13* rotierende Ölluftpumpe (Gasballastpumpe); *14* druckseitiger Abscheider. (V) Vakuummeßinstrumente.

allen Vakuumapparaturen gemeinsam. Die Art des hier als Rezipient angegebenen Vakuumbehälters richtet sich nach den Erfordernissen des speziellen Falles. Der Rezipient kann eine kleine Glasröhre für elektronische Versuche oder auch ein großer Metallbehälter für technische Verfahren sein. Auch spezielle Appara-

[1] K. SCHAEFER: Vak. Techn. **4**, 24 (1955); **7**, 21 (1958).

[2] E. L. HOLLAND-MERTEN: Techn. **12**, 10, 693 (1957). — Vak. Techn. **7**, 1, 16 (1958).

(*s. auch DIN 1314*).

$$p_{\text{Torr}} = 760\left(1 - \frac{\%\,\text{Vakuum}}{100}\right).$$

kg/m² (mm WS)	kg/cm² (at)	Atm (760 Torr)	lb per sq. inch	Mikron (μ)	inch of mercury
13,5951	$1{,}35951 \cdot 10^{-3}$	$1{,}31579 \cdot 10^{-3}$	$1{,}9337 \cdot 10^{-2}$	10^{3}	$3{,}9370 \cdot 10^{2}$
$1{,}01972 \cdot 10^{-2}$	$1{,}01972 \cdot 10^{-6}$	$0{,}98692 \cdot 10^{-6}$	$1{,}4503 \cdot 10^{-5}$	0,75006	$2{,}9530 \cdot 10^{-5}$
10,1972	$1{,}01972 \cdot 10^{-3}$	$0{,}98692 \cdot 10^{-3}$	$1{,}4503 \cdot 10^{-2}$	$0{,}75006 \cdot 10^{3}$	$2{,}9530 \cdot 10^{-2}$
$1{,}01972 \cdot 10^{4}$	1,01972	0,98692	$1{,}4503 \cdot 10^{1}$	$0{,}75006 \cdot 10^{6}$	$2{,}9530 \cdot 10^{1}$
1	10^{-4}	$0{,}96784 \cdot 10^{-4}$	$1{,}4223 \cdot 10^{-3}$	$0{,}73556 \cdot 10^{2}$	$2{,}8959 \cdot 10^{-3}$
10^{4}	1	0,96784	$1{,}4223 \cdot 10^{1}$	$0{,}73556 \cdot 10^{6}$	$2{,}8959 \cdot 10^{1}$
$1{,}03323 \cdot 10^{4}$	1,03323	1	$1{,}4695 \cdot 10^{1}$	$0{,}760 \cdot 10^{6}$	$2{,}9921 \cdot 10^{1}$
$0{,}70307 \cdot 10^{3}$	$0{,}70307 \cdot 10^{-1}$	$0{,}68046 \cdot 10^{-1}$	1	$5{,}1715 \cdot 10^{4}$	2,0360
$1{,}35951 \cdot 10^{-2}$	$1{,}35951 \cdot 10^{-6}$	$1{,}31579 \cdot 10^{-6}$	$1{,}9337 \cdot 10^{-5}$	1	$3{,}9370 \cdot 10^{-5}$
$1{,}34532 \cdot 10^{3}$	$0{,}34532 \cdot 10^{-1}$	$0{,}33421 \cdot 10^{-1}$	0,49115	$2{,}5400 \cdot 10^{4}$	1

ineinander; siehe HÜTTE Bd. 1, Anhang.

turen wie Elektronenmikroskope, Teilchenbeschleuniger und dergleichen sollen hierunter verstanden werden. Auf den Aufbau des Rezipienten soll im folgenden nicht näher eingegangen werden. Die übrigen Teile einer Vakuumapparatur entsprechen etwa den Angaben des Schemas der Fig. 1, wobei in einfachen Fällen die dort angegebenen Bauteile teilweise weggelassen werden können, in komplizierteren dagegen noch zusätzliche Bauteile hinzukommen. Im folgenden sollen nun die wesentlichen Elemente, Vakuummeßinstrumente, Pumpen und die Verbindungsleitungen näher behandelt werden. Einzelheiten, die übrigen Bauteile betreffend, entnehme man den Spezialwerken über Vakuumphysik (s. Literaturverzeichnis am Ende des Beitrages) bzw. den Katalogen der Lieferfirmen.

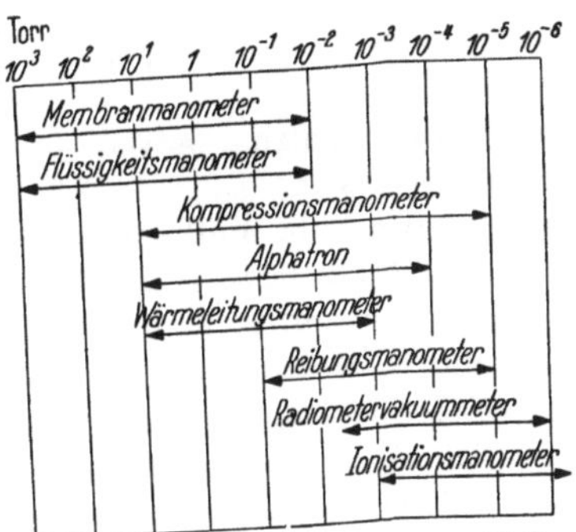

Anzeige beruht auf	ist abhängig von Gasart	des Totaldruckes von Gasen und Dämpfen
Druck	nein	ja
Druck	nein	ja
Druck nach vorheriger Kompression	nein	nein
Ionisation durch α-Strahlen	ja	ja
Wärmeleitung im Gasraum	ja	ja
Innere Reibung der Gase	ja	ja
Thermischer Molekulardruck	nein	ja
Ionisation durch Elektronen	ja	ja

Fig. 2. Meßbereiche der verschiedenen Vakuummeterarten.

I. Vakuummessung.

In technischen Vakuumanlagen erreicht man heute Vakua von 10^{-6} bis 10^{-7} Torr. Im Laboratorium (Röhren) kommt man dagegen bis in das Gebiet des Ultrahochvakuums, d.h. bis zu Drucken von 10^{-11} Torr und darunter (s. den Artikel von ALPERT in diesem Bande).

Eine unmittelbare Ausnutzung der Druckkräfte auf die Wandung eines Vakuumgefäßes zur Messung ist nur bei hohen Drucken, d.h. im Gebiet des Grobvakuums möglich. Bei niedrigeren Drucken sind die auf die Wandung wirkenden Druckkräfte zu klein. Man muß daher auf eine indirekte Druckmessung zurückgreifen. Dies geschieht in der Weise, daß physikalische Vorgänge, die vom Druck abhängig sind, gemessen werden, und daß aus diesen Messungen auf den jeweils herrschenden Druck zurückgeschlossen wird. Eine Übersicht über die üblicherweise zu Druckmessungen im Vakuum herangezogenen physikalischen Vorgänge gibt die Fig. 2.

2. Genauigkeit von Vakuummessungen. Da der Gesamtvakuumbereich mehr als 14 Zehnerpotenzen umfaßt, ist es notwendig, daß der Meßbereich eines einzelnen Instrumentes möglichst groß ist, um eine übermäßige Vielzahl von erforderlichen Geräten herabzusetzen. Dies bedingt aber gleichzeitig einen Verzicht auf prozentuale Meßgenauigkeit für jedes Einzelgerät. Eine Abhilfe ist bis zu einem gewissen Grade durch die Verwendung von Meßbereichschaltungen möglich. Darüber hinaus ist jede Vakuummessung mit einer Vielzahl von Fehlerquellen behaftet. So tritt bei jedem Pumpvorgang ein Druckabfall innerhalb der Leitungen des Vakuumsystems auf, so daß bei jeder Druckangabe wesentlich darauf geachtet werden muß, an welcher Stelle sie gemessen wurde. Die Gasartabhängigkeit der Anzeige von indirekt messenden Vakuummetern bringt dadurch eine Unsicherheit mit sich, daß die Gaszusammensetzung im Vakuumraum in den meisten Fällen nicht genau bekannt ist. Die Kompressionsmanometer messen nur den Druck der vorhandenen Permanentgase richtig. Das Vorhandensein von kondensierbaren Dämpfen kann hier zu erheblichen Fehlern Anlaß geben. Andere Vakuummeter, wie z.B. das Vakuummeter nach PENNING (s. S. 533) zeigen den Effekt der Gasaufzehrung und setzen damit den gemessenen Druck herab im Gegensatz zu anderen Fällen, bei denen durch Gasabgabe im Vakuummeter selbst ein zu hoher Druck gemessen wird. Berücksichtigt man alle diese Fehlerquellen, so ist es einleuchtend, daß an Vakuummessungen keineswegs dieselben Genauigkeitsansprüche gestellt werden können, wie es sonst in der Meßtechnik üblich ist. Eine bis auf den Faktor 2 genaue Vakuummessung kann im allgemeinen schon als gut bezeichnet werden und ist auch völlig ausreichend, da in sehr vielen Fällen schon die Kenntnis der Zehnerpotenz des vorhandenen Druckes genügt. Genaue Druckmessungen werden vor allem für die Bestimmung der Sauggeschwindigkeit von Pumpen benötigt.

Auch bei Vakuummessungen kann eine unmittelbare Skalenablesung oder sogar laufende Registrierung der Messungen wünschenswert sein. Dies geschieht jedoch in vielen Fällen nur unter Verzicht auf Genauigkeit, da Meßmethoden, die eine Hilfsmessung erfordern, häufig genauer sind. Um kurzzeitige Änderungen des Druckes in Vakuumapparaturen verfolgen zu können, benötigt man Geräte mit möglichst kurzer Einstelldauer. Nach dem oben über die Kleinheit der zur Verfügung stehenden Kräfte Gesagten, bedingt dies, daß hier trägheitsfreie, d.h. leichte Systeme erforderlich sind. Hochempfindliche Geräte mit kurzer Einstelldauer sind also mit technisch robusten, z.B. erschütterungsunempfindlichen Systemen prinzipiell nicht erzielbar.

3. Membranmanometer[1]. Bei diesen Vakuummetern wird der Gasdruck anhand der elastischen Verformung eines Teiles der Gefäßwand, die von ihm hervorgerufen wird, gemessen. Diese Meßmethode stellt also eine unmittelbare Gasdruckmessung dar. Fig. 3 zeigt ein solches Membranmanometer. Für höhere Drucke können an Stelle der Membran Bourdon-Röhren und für niedrigere Drucke

[1] W. WUEST: Archiv techn. Messen V 1343—3 (1943).

bis herunter zu etwa 1 Torr Membrandosen Anwendung finden. Bei all diesen Messungen über die elastische Verformung der Gefäßwand kommt es sehr darauf an, die Fehlerquellen wie Kriechen, elastische Nachwirkung, Hysterese, Temperaturfehler möglichst klein zu halten. Dies ist im wesentlichen eine Frage des Membranmaterials, wozu Kupferberyllium besonders geeignet ist.

Bei den obengenannten Membranmanometern wird die Verformung der Gefäßwand über ein Hebelsystem auf einen Zeigerausschlag zur unmittelbaren Ablesung übertragen. Eine empfindlichere Anordnung, die es auch gestattet, noch niedrigere Drucke zu messen, erhält man, wenn man die Membrandurchbiegung nicht auf mechanischem, sondern elektrischem Wege[1] mißt bzw. kompensiert. Bis zu welch niedrigen Drucken man mit solchen Anordnungen gelangt, ist eine Frage der Membrandicke. Diese muß jedenfalls so fest sein, daß sie einer einseitigen Belastung mit dem Druck einer Atmosphäre standhält. Die durch die Membrandicke gegebene Grenze für die Druckmessung dürfte etwa bei 10^{-2} Torr liegen.

Fig. 3. Membranmanometer mit Zeigerablesung durch Hebelübertragung, Meßbereich 10 bis 760 Torr.

Will man nicht absolute Drucke, sondern nur kleine Druckdifferenzen messen, so kann man mit der Membrandicke sehr weit heruntergehen und erhält in elektrischen Brückenschaltungen hochempfindliche Meßinstrumente, die bis zu Druckdifferenzen von etwa 10^{-5} Torr reichen[2]. Eine sehr wesentliche Fehlerquelle ist bei diesen hochempfindlichen Membranmanometern die Notwendigkeit, die Temperatur sehr genau konstant zu halten.

4. Flüssigkeitsmanometer, insbesondere U-Rohrmanometer. *α) Höhere Drucke.* Zur Messung höherer Drucke werden vielfach bis herab zu Werten von 1 Torr die bekannten U-Rohrmanometer in offener oder mit einem Schenkel geschlossener Bauart verwendet. Eine spezielle Abart, die es in einfacher Weise gestattet, die Höhendifferenz des Flüssigkeitsstandes in beiden Schenkeln auf eine Zeigerdrehung zu übertragen, ist der Ringwaagedruckmesser. Als Manometerflüssigkeit kommt vor allem Quecksilber in Frage. Für die Messung kleiner Drucke hat dieses den Nachteil eines hohen spezifischen Gewichtes. Ersatzweise wurde daher schon die Verwendung von borwolframsaurem Cadmium (spezifisches Gewicht 3,28) vorgeschlagen[3]. Auch hochmolekulare Flüssigkeiten mit niedrigem Dampfdruck (s. unter Treibmittel der Diffusionspumpen) wie Apiezonöl, Octoil, Butylphthalat und andere sind schon von einigen Autoren[4] benutzt worden. Diese Manometerflüssigkeiten haben gegenüber dem Quecksilber den Vorteil des niedrigen Dampfdruckes; dem steht der Nachteil gegenüber, daß sich Gase in diesen Flüssigkeiten sehr stark lösen, so daß sie bei jedem Evakuierungsvorgang erneut unter starkem Schäumen erst wieder entgast werden müssen.

β) Kleine Drucke. Hier werden zur Messung gelegentlich U-Rohrmanometer angewandt, deren abgeschlossener Schenkel nicht senkrecht, sondern schräg steht.

---

[1] C. Brinkmann: Arch. Elektrotechn. **32**, 59 (1938). — A. Haase: Z. VDI **80**, 563 (1936).

[2] E. W. Becker: Z. angew. Phys. **4**, 1, 20 (1952). — N. Warmoltz: Appl. Sci. Res., Sect. B **4**, 329 (1955).

[3] A. Kolb: Z. VDI **85**, 625 (1941).

[4] K. C. D. Hickmann: Rev. Sci. Instrum. **5**, 161 (1934).

Der Meßbereich dieser Vakuummeterart reicht von 0,3 bis 100 Torr. Die Meßgenauigkeit beträgt etwa ±0,3 Torr. Eine andere Konstruktion verwendet die Überschichtung einer schweren Manometerflüssigkeit (Quecksilber) mit einer leichten (Apiezonöl)[1]. (Meßbereich 0,2 bis 10 Torr, Meßgenauigkeit ±0,1 Torr.) Zur genauen Messung der Verschiebung der Quecksilbermenisken in U-Rohrmanometern werden auch gelegentlich in beiden Manometerschenkeln elektrische Widerstandsdrähte ausgespannt, so daß die Lage der Quecksilberspiegel auf eine elektrische Widerstandsmessung zurückgeführt werden kann[2]. Eine weitere Methode zur genauen Messung der Lage der Quecksilberspiegel in U-Rohrmanometern bedient sich eines kleinen Schwimmers auf der Quecksilberoberfläche, dessen Lage durch Spiegelablesung genau gemessen werden kann[3], und schließlich gehört zu dieser Gruppe von Geräten auch noch das Tauchglockenmanometer, bei dem eine Tauchglocke teilweise in eine Manometerflüssigkeit eintaucht und bei dem die Tauchtiefe der Glocke von dem Druckunterschied zwischen der Unterseite und der Oberseite der Tauchglocke abhängt.

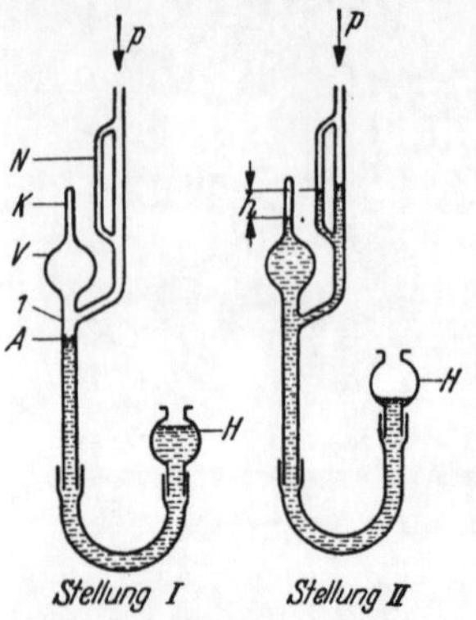

Fig. 4. Vakuummeter nach McLeod. Stellung I vor der Messung, Stellung II während der Messung.

5. Kompressionsmanometer. α) *Vakuummeter nach McLeod*[4]. Bei dieser Vakuummeterart werden kleine Drucke, die an der Niveaudifferenz zweier Quecksilbermenisken nicht mehr abgelesen werden können, dadurch der Messung zugänglich gemacht, daß eine gewisse Menge Gas, die sich in einem großen Volumen unter dem zu messenden Druck p befindet, in ein kleineres Volumen zusammengedrängt wird. Dadurch wird der Druck entsprechend dem Verhältnis der Volumina vergrößert und nunmehr meßbar. Aus dem gemessenen Druck und dem Verhältnis der Volumina kann dann rückwärts der ursprüngliche Gasdruck berechnet werden.

Die Durchführung der Messungen erfolgt gemäß Fig. 4. Zunächst befindet sich das Gerät in Stellung *I*. Durch Heben der Quecksilberkugel H steigt der Quecksilbermeniskus A bis zu der Verzweigungsstelle *1*. In diesem Augenblick wird das in dem Volumen der großen Kugel V und der Kapillaren K befindliche Gas abgesperrt und bei weiterem Heben der Quecksilberkugel aus dem Volumen V in die Kapillare K komprimiert, bis die Stellung *II* erreicht ist. Der Gasdruck in dem Restvolumen der Kapillare K kann jetzt an der Höhendifferenz h zwischen den Quecksilberständen der Kapillare K und der Vergleichskapillare N abgelesen werden. Zur Vermeidung von Fehlern durch Kapillardepression müssen die beiden Kapillaren K und N denselben Querschnitt haben.

Die Durchführung der Messung erfolgt nach zwei verschiedenen Methoden.

a) Lineare Teilung der Meßskala, Fig. 5. Hier sind an der Kompressionskapillare K Marken angebracht, die angeben, den wievielten Teil des Gesamtvolumens V von Kugel und Kapillare der Raum über der Meßmarke V_c in der Kapillare K ausmacht. Hebt man den Quecksilberspiegel bis zu einer solchen Marke V_c und liest außerdem die Höhendifferenz h der Quecksilberspiegel in Kompressionskapillare K und Vergleichskapillare N ab, so gilt für den Gasdruck p

[1] O. Beeck: Rev. Sci. Instrum. **6**, 399 (1935).

[2] H. Klumb u. Th. Haase: Z. techn. Phys. **13**, 372 (1932).

[3] J. E. Strader u. H. M. Ryder: Phys. Rev. **13**, 321 (1919).

[4] McLeod: Phil. Mag. **48**, 110 (1874). — H. Ebert: Heft 11 d. Verfahrens- u. Meßkde. Braunschweig: Vieweg 1951.

vor der Kompression die Beziehung

$$p\,V = (p + h)\,V_c, \qquad p = h\,\frac{V_c}{V - V_c},$$

also für $p \ll h$; $V_c \ll V$:

$$p = h\,\frac{V_c}{V}.$$

Fig. 5 zeigt den Meßbereich eines solchen Vakuummeters in Abhängigkeit vom Gesamtvolumen V und den übrigen im einzelnen angegebenen geometrischen Abmessungen.

b) Quadratische Teilung der Meßskala (Fig. 6). Hier wird der Quecksilberspiegel in der Vergleichskapillare N bis zur selben Höhe wie das abgeschmolzene Ende der Meßkapillare K gehoben. Für den Gasdruck p gilt jetzt die Beziehung (d = Kapillardurchmesser):

$$p = h\,\frac{V_c}{V - V_c}; \qquad V_c = h\,\frac{\pi}{4}\,d^2,$$

also für $p \ll h$; $V_c \ll V$:

$$p = h^2\,\frac{\pi}{4}\,\frac{d^2}{V}.$$

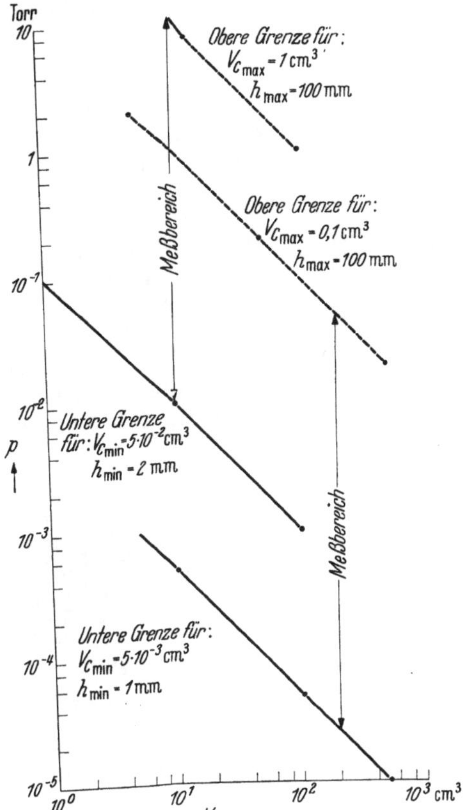

a

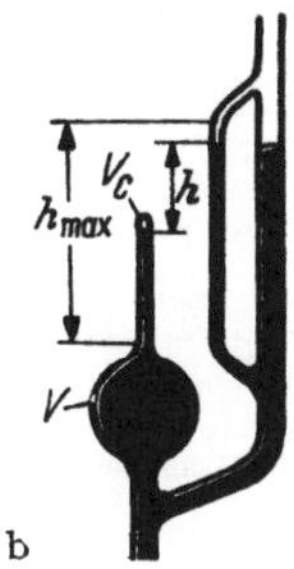

b

Fig. 5 a u. b. Vakuummeter nach McLeod mit linearer Teilung der Meßskala. Obere und untere Grenze des Meßbereiches in Abhängigkeit vom Gesamtvolumen V.

Fig. 6 zeigt für diese Meßmethode die Abhängigkeit des Meßbereiches vom Gesamtvolumen V und den übrigen geometrischen Abmessungen.

Zur Erweiterung des Meßbereiches eines McLeods wurde auch vorgeschlagen, die Abmessungen für die Kapillaren abzusetzen, d.h. an eine Kapillare $K1$ mit großem Querschnitt noch eine Kapillare $K2$ mit engerem Querschnitt anzuschließen.

Die Vakuummeter nach McLeod messen nur den Partialdruck der Permanentgase richtig, während Dämpfe bei dem Kompressionsvorgang kondensiert werden und damit die Messung verfälschen können. Sie sollten daher vorzugsweise als Standardinstrumente zur Eichung anderer Vakuummeterarten benutzt werden, bei denen die Eichkurve nicht unmittelbar aus den geometrischen Abmessungen abgelesen werden kann.

Wegen ihrer Verwendung als Standardinstrumente kommt es bei den McLeods auf äußerste Sauberkeit an. Dies gilt sowohl für das Quecksilber wie die Glasoberflächen. Insbesondere in den Kapillaren sollten die Oberflächen im Hinblick auf das Vorhandensein sehr starker Kapillarkräfte genau definiert sein. Diesem Ziel diente z. B. künstliche Aufrauhung[1] oder Behandlung mit verdünnter Flußsäure[2] sowie Beheizung der Kapillaren von außen.

Es ist verschiedentlich versucht worden, den Meßbereich der McLeods nach niedrigen Drucken hin zu erweitern. Ein Irrweg in dieser Hinsicht ist das sog. Klebevakuum, das aber in Wirklichkeit keinen Aufschluß über die Erreichung niedriger Drucke gibt, sondern vielmehr ein Maß dafür ist, in welchem Grade die oben aufgeführten Bedingungen der absoluten Sauberkeit eingehalten wurden

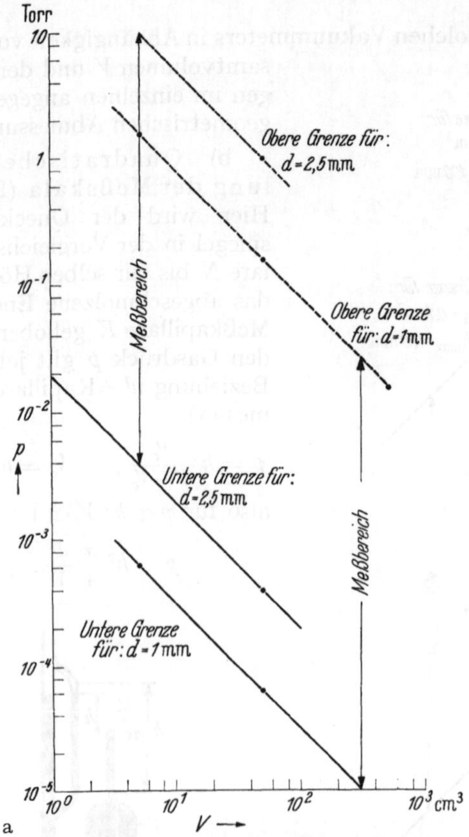

a

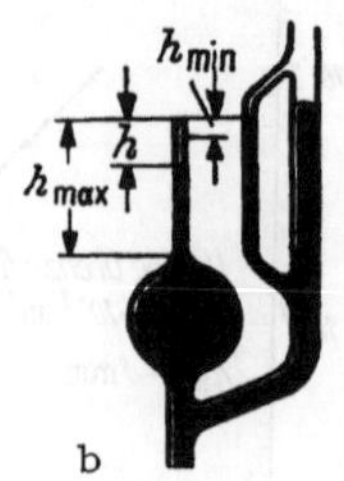

b

Fig. 6 a u. b. Vakuummeter nach McLEOD mit quadratischer Teilung der Meßskala. Obere und untere Grenze des Meßbereiches in Abhängigkeit vom Gesamtvolumen V.

bzw. nicht eingehalten wurden. Günstiger in dieser Hinsicht liegen zwei Anordnungen, bei denen einmal der Druck in der Meßkapillare nicht durch visuelle Ablesung, sondern durch ein dort eingebautes Wärmeleitungsmanometer[3] bestimmt wird. Statt des Wärmeleitungsmanometers kann auch ein Membranvakuummeter[4] in der Meßkapillare K untergebracht werden.

β) Drehbare Kompressionsmanometer[5]. Bei diesen erfolgt die Kompression eines abgesperrten Gasvolumens zwecks Messung des dadurch erhöhten Druckes nicht durch Heben einer Quecksilberkugel, sondern durch Drehen des gesamten Meß-

[1] P. ROSENBERG: Rev. Sci. Instrum. **10**, 131 (1939).
[2] G. HAASE: Z. techn. Phys. **24**, 27, 53 (1943).
[3] H. TANNER: J. Phys. Chem. **34**, 1113 (1930).
[4] F. JAMASUKI u. Z. JOSIDA: Proc. Phys.-Math. Soc. Japan **15**, 400 (1933).
[5] A. AXELBANK: Rev. Sci. Instrum. **21**, 511 (1950).

systems. Verschiedene Ausführungsformen wurden angegeben von VON REDEN[1], REIFF[2], MOSER[3] und GAEDE[4]. Die Fig. 7 zeigt das drehbare Kompressionsmanometer in der Ausführungsform des Vakuskops nach GAEDE. Der Meßbereich dieses Vakuskops reicht nach niedrigen Drucken nicht weiter als zu $1 \cdot 10^{-2}$ Torr. Dies rührt daher, daß die Vakuummeter nur mit verhältnismäßig kleinen Quecksilbermengen gefüllt werden können, und daß daher zur Messung kleiner Drucke sehr enge Kapillaren erforderlich wären. Diese können aber nicht benutzt werden, da in solchen die Kapillarkräfte unverhältnismäßig groß würden und dadurch zu Verfälschungen der Messung Anlaß geben könnten.

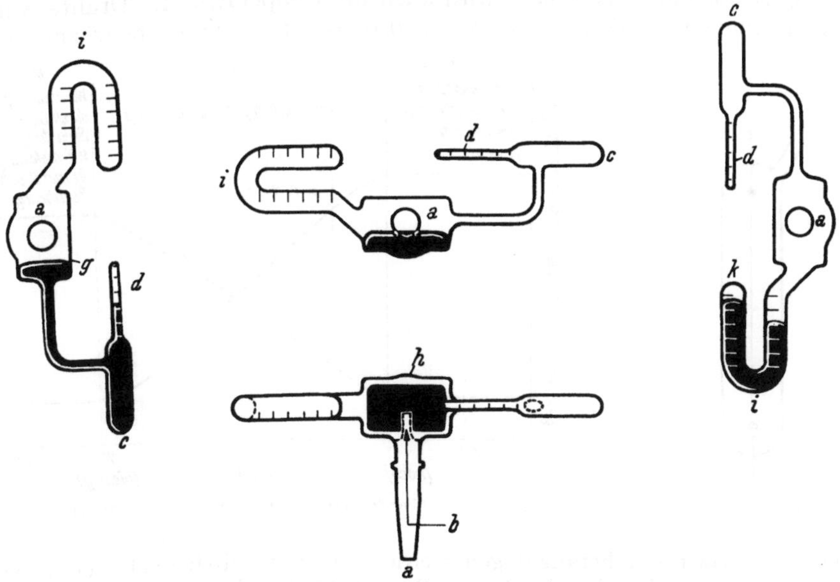

Fig. 7. Vakuskop nach GAEDE, Mitte Ruhestellung, rechts Meßstellung als U-Rohr-Manometer, links Meßstellung als Kompressionsmanometer. *a* Konusschliff zum drehbaren Anschluß an die Vakuumapparatur; *b* Quecksilbereinfüllstutzen; *d* Kompressionskapillare; *c* zu komprimierendes Volumen; *g* Quecksilbermeniskus; *i* U-Rohr-Manometer.

6. Wärmeleitungsmanometer. Die Wärmeleitung durch Gase, die bei hohen Drucken druckunabhängig ist, wird bei niedrigen Drucken dann druckabhängig, wenn die mittlere freie Weglänge der Gasmoleküle vergleichbar wird mit den Dimensionen der wärmeabführenden Teile. In diesem Gebiet kann also die Wärmeleitung zur Druckmessung benutzt werden.

α) Wärmeströmung durch den Gasraum in verschiedenen Druckbereichen. Fig. 8 zeigt die grundsätzliche Anordnung eines Wärmeleitungsmanometers. Sie besteht aus einem dünnen Draht mit dem Radius r_1, der in einem weiten Zylinder mit dem Radius r_2 ($r_1 \ll r_2$) ausgespannt ist. Der Draht kann elektrisch geheizt werden. Bei konstanter Energiezufuhr stellt sich die Temperatur des Drahtes so ein, daß Gleichgewicht herrscht zwischen der zugeführten elektrischen Energie (Q_{elektr}) einerseits und der durch Wärmeableitung durch das Gas hindurch (Q_{Gas}), durch Wärmestrahlung (Q_{Strahl}) und durch Wärmeableitung durch die metallischen Stromzuführungen (Q_{Metall}) abgeführten Energie.

[1] v. REDEN: Z. Instrumentenkde **24**, 52 (1911).
[2] H. J. REIFF: Z. Instrumentenkde **34**, 97 (1914).
[3] MOSER: Phys. Z. **36**, 1 (1935).
[4] W. v. MEYEREN: Z. phys. Chem., Abt. A **160**, 272 (1932).

Solange die mittlere freie Weglänge Λ der Gasmoleküle klein ist, $\Lambda \ll r_1$, ist die Wärmeabfuhr durch das Gas und damit auch die gesamte Wärmeabfuhr unabhängig vom Druck. Das gleiche gilt für die sich einstellende Temperatur des Drahtes (s. Fig. 9, Bereich I). Wird dann aber bei weiterhin abnehmendem Druck Λ mit r_1 vergleichbar, so nimmt die Wärmeabfuhr durch das Gas hindurch mit fallendem Druck ab. Die Temperatur des Drahtes steigt entsprechend (Fig. 9, Bereich *II*). Bei ganz niedrigen Drucken wird die Wärmeableitung durch das Gas hindurch (Q_{Gas}) klein gegen die Wärmeabfuhr durch Strahlung (Q_{Strahl}) und gegen die Wärmeableitung in den metallischen Stromzuführungen (Q_{Metall}) (s. Fig. 9, Bereich *III*). Nunmehr wird auch die Temperatur des Drahtes wiederum unabhängig vom Druck. Sie kann also weiterhin nicht mehr zur Druck-

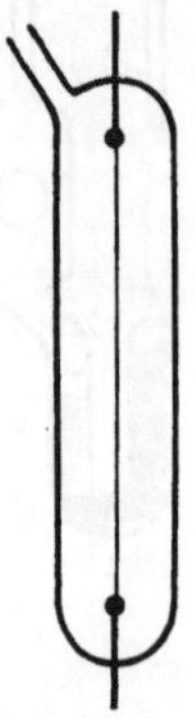

Fig. 8. Wärmeleitungsmanometer.

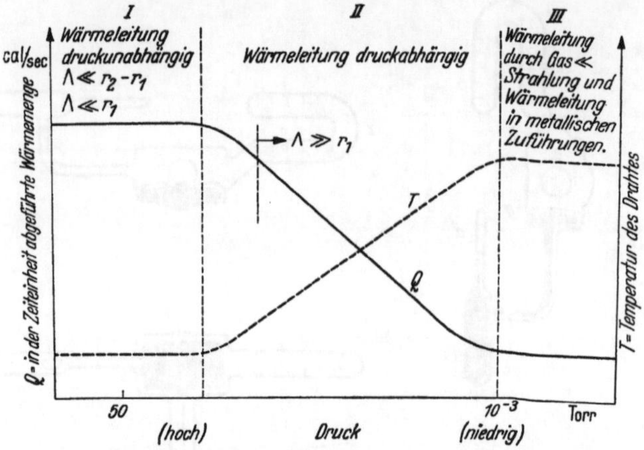

Fig. 9. Übersicht über die Wärmeleitung in verschiedenen Druckbereichen.

messung im Gasraum herangezogen werden. Der Arbeitsbereich der Wärmeleitungsmanometer ist also durch den Bereich *II* der Fig. 9 fixiert.

Einige Zahlenwerte sollen dies noch näher erläutern. Der Übergang von Bereich *I* zu Bereich *II* erfolgt dann, wenn die mittlere freie Weglänge Λ vergleichbar wird mit dem Drahtradius r_1. Der Draht habe beispielsweise einen Radius $r_1 = 0{,}1$ mm, dem entspricht in Luft von 20° C eine freie Weglänge $\Lambda = 0{,}1$ mm, wenn der Druck p einen Wert von $p = 0{,}5$ Torr hat. Nach hohen Drucken zu wird also der Arbeitsbereich eines solchen Wärmeleitungsmanometers etwa bei 0,5 Torr enden. Zur Berechnung der unteren Grenze des Meßbereiches betrachten wir die Wärmebilanz des Drahtes. Danach muß die zugeführte elektrische Energie (Q_{elektr}) gleich sein der Summe der drei anderen Energieanteile:

$$Q_{\text{elektr}} = Q_{\text{Gas}} + Q_{\text{Strahl}} + Q_{\text{Metall}}.$$

Für einatomige Gase gilt für Q_{Gas} folgende Beziehung:

$$Q_{\text{Gas}} = \gamma p \sqrt{\frac{2\mathsf{R}}{\pi M T_1}}\,(T_2 - T_1)\, 2\pi r_1 l,$$

T_1 = Außentemperatur, T_2 = Drahttemperatur, M = Molekulargewicht, r_1 = Drahtradius, l = Drahtlänge.

Darin sollen als Beispiel folgende Werte eingesetzt werden:

$\gamma = 1$; $T_1 = 273°\,\text{K}$; $T_2 = 423°\,\text{K}$; $M = 29$; $p = 10^{-3}$ Torr, $2\pi r_1 l = 1\,\text{cm}^2$.

Unter der Annahme, daß der Draht überall die Temperatur T_2 besitzt, ergibt sich daraus

$$Q_{\text{Gas}} = 3{,}9 \cdot 10^{-4} \left[\frac{\text{cal}}{\text{sec}}\right].$$

Zum Vergleich hiermit berechnen wir die Wärmeabgabe durch Strahlung Q_{Strahl}. Unter der Voraussetzung, daß der Draht wie ein schwarzer Körper strahlt, hat diese den Wert:

$$Q_{\text{Strahl}} = \sigma (T_2^4 - T_1^4)\, 2\pi r_1 l; \quad \sigma = 1{,}355 \cdot 10^{-12}\,[\text{cal}\,\text{cm}^{-2}\,\text{sec}^{-1}\,\text{grad}^{-4}],$$

also in unserem Zahlenbeispiel:

$$Q_{\text{Strahl}} = 3{,}6 \cdot 10^{-2}\,[\text{cal}\,\text{sec}^{-1}].$$

Für $p = 10^{-3}$ Torr wird also das Verhältnis

$$\frac{Q_{\text{Gas}}}{Q_{\text{Strahl}}} = 1{,}08 \cdot 10^{-2}.$$

Die Wärmeleitung durch das Gas (Q_{Gas}) beträgt also bei diesem Druck nur noch einige Prozent der Wärmeabfuhr durch Strahlung (Q_{Strahl})! Wir befinden uns daher an der Grenze zwischen Bereich *II* und Bereich *III* der Fig. 9. In diesem Gebiet werden also durch geringfügige Änderungen der Oberflächeneigenschaften des Drahtes und der dadurch bedingten Änderung der Wärmeabstrahlung erhebliche Veränderungen der Eichkurve eines Wärmeleitungsmanometers hervorgerufen (s. weiter unten bei der Diskussion der technischen Geräte, S. 526/527).

Obgleich bei einem Druck von 10^{-3} Torr die Wärmeabfuhr durch das Gas nur noch einige Prozent der Wärmeabfuhr durch Strahlung ausmacht und hierdurch allein bereits das Ende des Arbeitsbereiches der Wärmeleitungsmanometer gegeben ist, wollen wir anschließend der Vollständigkeit halber auch noch in Betracht ziehen, welchen Einfluß die Wärmeableitung durch die metallischen Stromzuführungen, Q_{Metall}, hat. Machen wir hierzu die vereinfachende Annahme, daß die Drahttemperatur T_2 von der Mitte aus linear nach beiden Seiten auf T_1 abnimmt, so hat Q_{Metall} den Wert

$$Q_{\text{Metall}} = 2\lambda r_1^2 \pi \frac{T_2 - T_1}{l/2},$$

λ = Wärmeleitzahl des Drahtmaterials, angenommen $\lambda = 0{,}1$.

Damit ergibt sich

$$\frac{Q_{\text{Gas}}}{Q_{\text{Metall}}} = \gamma p \sqrt{\frac{\mathsf{R}}{2\pi M T_1}}\, \frac{1}{\lambda}\, \frac{l^2}{r_1}; \qquad p \text{ in } \frac{\text{dyn}}{\text{cm}^2}.$$

Um in den Zahlenwerten des oben angegebenen Beispiels zu bleiben, müssen wir hier einsetzen $r_1 = 0{,}1$ cm, $l = 15$ cm, $p = 10^{-3}$ Torr:

$$\frac{Q_{\text{Gas}}}{Q_{\text{Metall}}} = 2{,}92 \cdot 10^{-1}.$$

Die Wärmeableitung durch das Gas hindurch ist also bei einem Druck von 10^{-3} Torr nur noch 29% der Wärmeabführung durch die metallischen Stromzuführungen. Durch einen längeren Draht ließe sich letztere zwar noch herabsetzen, da aber bei einem Druck von 10^{-3} Torr die Wärmeabstrahlung allein schon die Wärmeströmung durch das Gas hindurch bei weitem überwiegt, so würde dies zu keiner wesentlichen Erweiterung des Meßbereiches führen. Demgegenüber kann es von Vorteil sein, einen dünneren Draht zu verwenden, weil dadurch die obere Grenze des Meßbereiches weiter nach höheren Drucken zu verschoben wird.

β) Technische Geräte. Bei den technischen Geräten der Wärmeleitungsmanometer kommen die beiden folgenden Prinzipien zur Anwendung. Die Widerstandsmanometer nach Pirani nutzen die Temperaturabhängigkeit des Drahtwiderstandes selbst zur Temperaturmessung aus. Bei den thermoelektrischen Vakuummetern wird die Temperatur des Drahtes mittels eines Thermoelementes gemessen.

Die grundsätzliche Schaltung eines *Widerstandsmanometers nach* Pirani[1] zeigt Fig. 10. Die Meßröhre *M* liegt in einer Wheatstoneschen Brückenschaltung. Der Brückenstrom wird dabei zunächst so abgeglichen, daß das Galvanometer nicht ausschlägt. Ändert sich infolge von Druckänderungen in der Meßröhre *M* der Widerstand, so zeigt das Galvanometer einen Ausschlag. Als Eichkurve des Gerätes wird der Ausschlag des Galvanometers in Abhängigkeit vom Druck

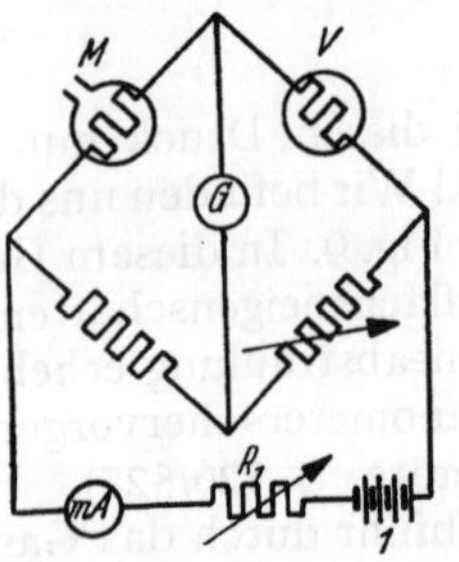

Fig. 10. Schaltung eines Widerstandsmanometers nach Pirani.

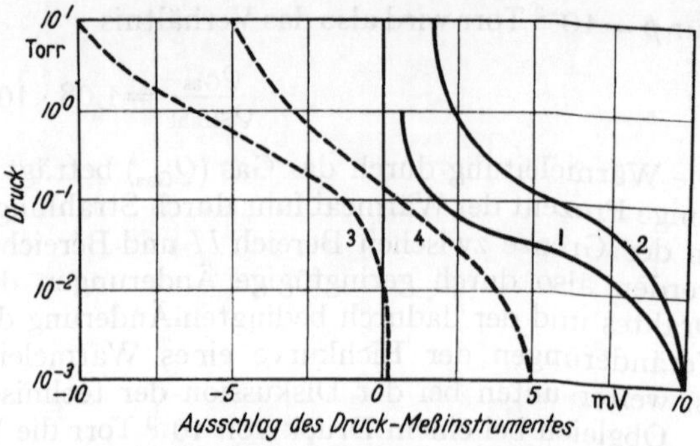

Fig. 11. Eichkurven für verschiedene Arten von thermoelektrischen Vakuummetern.

aufgetragen, wobei nach den obigen Darlegungen der Arbeitsbereich von etwa 1 bis 10^{-3} Torr reicht (Fig. 9, Bereich *II*). Da die Temperatur des Drahtes in der Meßröhre nicht nur von dem Wärmegleichgewicht in dieser, sondern auch von der Außentemperatur abhängt, benutzt man eine bei konstantem Druck abgeschmolzene Vergleichsröhre *V*, die aber sonst völlig gleich gebaut ist wie die Meßröhre *M*, um diese Temperaturschwankungen infolge Änderung der Außentemperatur möglichst weitgehend zu kompensieren.

Die Empfindlichkeit solcher Geräte hängt natürlich von der Temperaturabhängigkeit des Widerstandsdrahtes ab. Man ist daher bestrebt, hier Materialien mit möglichst hohen Temperaturkoeffizienten des Widerstandes zu benutzen, wie sie beispielsweise in *Halbleitern (Thermistorvakuummeter)*[2] vorliegen.

Bei den *thermoelektrischen Vakuummetern*[3] wird vielfach das Thermoelement selber als wärmeabführender Draht benutzt, dessen Wärmegleichgewicht zur Druckmessung dient. Dies geschieht in der Weise, daß die Energiezufuhr zum Draht mittels Wechselstroms erfolgt, so daß davon unabhängig die Thermogleichspannung des Thermoelementes und damit seine Temperatur gemessen werden kann. Ein solches thermoelektrisches Vakuummeter hat einen Meßbereich, der etwa mit dem des oben behandelten Pirani-Manometers zusammenfällt (Fig. 11, Kurve 1 und 2). Trifft man dagegen die Anordnung so, daß in der Nähe des kälteren Thermoelementes eine wärmere Fläche innerhalb des Gasraumes

---

[1] M. v. Pirani: Verh. dtsch. phys. Ges. **4**, 686 (1906). — J. H. Leck: J. Sci. Instrum. **33**, 181 (1956).

[2] H. Gruber: Glas u. Hochv.technn. **1**, 15, 302 (1951).

[3] W. Voege: Phys. Z. **7**, 498 (1906). — W. H. W. I. Moll u. H. C. Burger: Z. techn. Phys. **21**, 199 (1940). — H. Schlitt: Z. angew. Phys. **8**, 216 (1956).

angebracht ist, so daß die Wärmezufuhr zum Thermoelement vom Gasdruck in der Röhre abhängig ist, so erhält man einen Meßbereich, der gegenüber den oben beschriebenen beiden Anordnungen etwas nach höheren Drucken zu verschoben liegt (s. Fig. 11, Kurve 3). Fig. 12 zeigt eine Kombination dieser beiden Meßprinzipien: ein unmittelbar durch Stromdurchgang beheiztes Thermoelement *2* zusammengeschaltet mit einem zweiten Thermoelement *1*, dem von der beheizten Fläche *3* durch den Gasraum hindurch Wärme zugeführt wird. Das Meßinstrument *5* mißt die Summe beider Thermospannungen. Der Meßbereich dieses Gerätes erstreckt sich also über den gesamten Bereich der beiden Thermoelemente *1* und *2* und ist daher gegenüber dem Einzelelement erweitert (s. Fig. 11, Kurve 4).

Alle Wärmeleitungsmanometer haben gemeinsam, daß sich der Meßbereich nach hohen Drucken zu nicht wesentlich über einige Torr hinaus ausdehnen läßt, und daß die Eichkurve an der unteren Grenze sehr stark abhängig ist

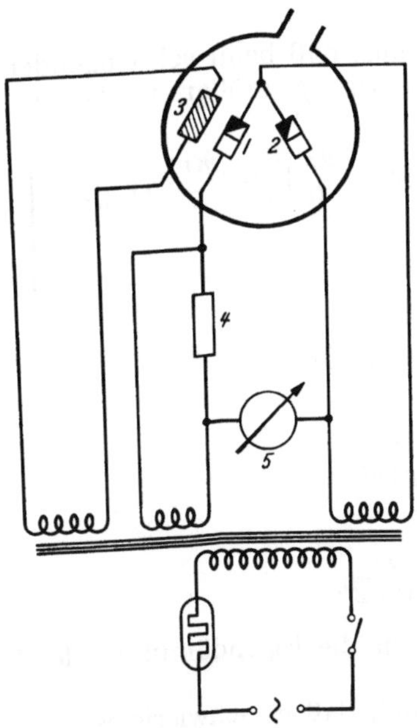

Fig. 12. Schaltung eines thermoelektrischen Vakuummeters.

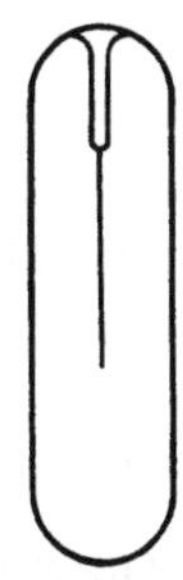

Fig. 13. Quarzfadenpendel als Reibungsmanometer.

von den Wärmestrahlungseigenschaften der wärmeabführenden Drähte. Sie ist daher sehr empfindlich gegen Verschmutzungen der Oberfläche. Dies hat in unangenehmer Weise zur Folge, daß sich die Eichkurven der Wärmeleitungsmanometer am unteren Ende ihres Bereiches während der Benutzung leicht ändern.

7. Reibungsmanometer[1]. Die innere Reibung der Gase, die ebenso wie die Wärmeleitung bei höheren Drucken druckunabhängig ist, läßt sich bei niedrigen Drucken, wenn die mittlere freie Weglänge vergleichbar mit den Gefäßdimensionen wird, zur Messung des Gasdruckes ausnutzen.

Die einfachste Form eines solchen Reibungsmanometers, das *Quarzfadenpendel*[2], zeigt Fig. 13. Läßt man den frei aufgehängten Quarzfaden in einem Gasraum, in dem der Druck gemessen werden soll, schwingen, so werden die Schwingungen durch die Gasreibung gedämpft. Die Halbwertszeit t_H ist um so größer, je kleiner die Dämpfung, d.h. die Gasreibung r und damit der Druck p ist.

[1] J. L. Hoog: Proc. Amer. Acad. **42**, 115 (1906). — O. Heiligbrunner: Vak. Techn. 5, 133 (1956).
[2] J. Langmuir: Phys. Rev. **1**, 337 (1913). — G. Wetterer: Z. techn. Phys. **20**, 281 (1939).

Für eine solche Schwingung gelten die folgenden Beziehungen:

$$m_1 \frac{d^2 x}{dt^2} + r \frac{dx}{dt} + Kx = 0, \tag{7.1}$$

worin m_1 die Masse des Quarzpendels ist. Die Lösung dieser Gleichung lautet:

$$x = x_0 e^{-\beta t} \sin \sqrt{\frac{K}{m_1}}\, t \quad \text{mit} \quad \beta = \frac{r}{2m_1},$$

also

$$t_H = 2 \frac{m_1}{r} \ln 2. \tag{7.2}$$

Die Reibungskonstante r ergibt sich daraus, daß beim schwingenden Pendel der Druck der Gasmoleküle auf seine Vorderseite p_v größer als auf die Hinterseite p_h ist:

$$\left.\begin{aligned} r \frac{dx}{dt} &= p_v - p_h = \frac{mn}{3}\left[\left(\bar{c} + \frac{dx}{dt}\right)^2 - \left(\bar{c} - \frac{dx}{dt}\right)^2\right] = \frac{mn}{3}\, 4\bar{c}\, \frac{dx}{dt}; \\ \bar{c} &= \sqrt{\frac{8\mathrm{R}T}{\pi M}}; \quad n = \frac{pN}{\mathrm{R}T}. \end{aligned}\right\} \tag{7.3}$$

Daraus folgt

$$r = \frac{4}{3}\sqrt{\frac{8M}{\pi \mathrm{R} T}}\, p = B p \sqrt{M}, \tag{7.4}$$

und nach (7.2):

$$t_H = \text{const}\, \frac{m_1}{p\sqrt{M}} \tag{7.5}$$

oder genauer nach Haber-Kerschbaum[1]:

$$t_H = \text{const}\, \frac{D}{p\sqrt{M + Z'}}, \tag{7.6}$$

mit D = Quarzfadendurchmesser, Z' = Maß für die Eigendämpfung des Fadens, const = 122.

Die Quarzfadenpendel haben den Nachteil, daß es schwierig ist, eine reine Pendelschwingung anzustoßen. Im allgemeinen werden gleichzeitig eine Anzahl Oberschwingungen angeregt. Außerdem schwingt der Faden nicht in einer Ebene, sondern beschreibt Lissajoussche Figuren. Zur Abhilfe benutzt Coolidge[2] ein Doppelfadenpendel (s. Fig. 14). Die Eichkurve solcher Doppelfadenpendel geht bei niedrigen Drucken (s. Fig. 15) dann in eine Horizontale über, wenn die Eigendämpfung z' des Pendels größer wird, als die Gasreibung r. Brüche[3] sucht diesen Übelstand durch Hinzufügen eines Reibungsblättchens am unteren Ende des Pendels (Fig. 14, Mitte und rechts) zu vermeiden. Dadurch erhält man Eichkurven (s. Fig. 15), die sich weiter zu niedrigen Drucken hin erstrecken. Man muß dagegen aber den Nachteil in Kauf nehmen, daß die Eichkurven jetzt auch erst bei wesentlich niedrigeren Drucken beginnen als die des einzelnen Quarz-

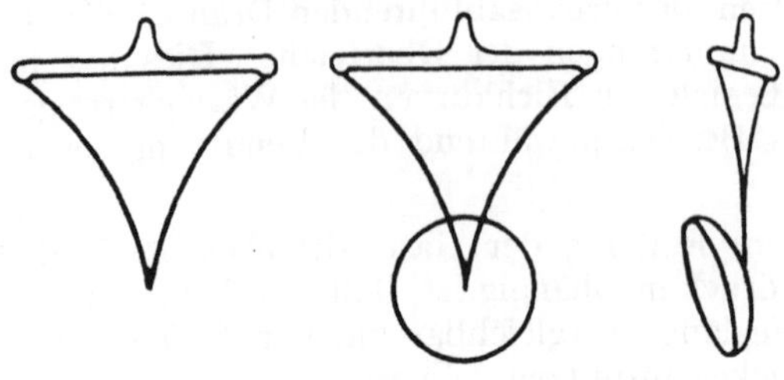

Fig. 14. Doppelfadenpendel (links), dasselbe mit Reibungsblättchen (Mitte und rechts).

[1] F. Haber u. F. Kerschbaum: Z. Elektrochem. **20**, 296 (1914).
[2] A. S. Coolidge: J. Amer. chem. Soc. **45**, 1637 (1923).
[3] E. Brüche: Ann. Phys. **79**, 695 (1926).

fadens. Liegt doch die obere Grenze des Arbeitsbereiches für den einzelnen Quarzfaden bei solchen Drucken, für die die mittlere freie Weglänge der Gasmoleküle mit dem Fadendurchmesser vergleichbar wird, während bei den Blättchenmanometern die mittlere freie Weglänge der Gasmoleküle mit dem wesentlich größeren Blättchendurchmesser vergleichbar sein muß, also bei niedrigeren Drucken liegt (s. Fig. 15).

Zusammenfassend kann gesagt werden: Alle Quarzfadenmanometer weisen einige Nachteile auf:

1. Die Messung des Druckes ist abhängig von der Gasart;

2. Die Geräte sind anfällig gegen äußere Erschütterungen;

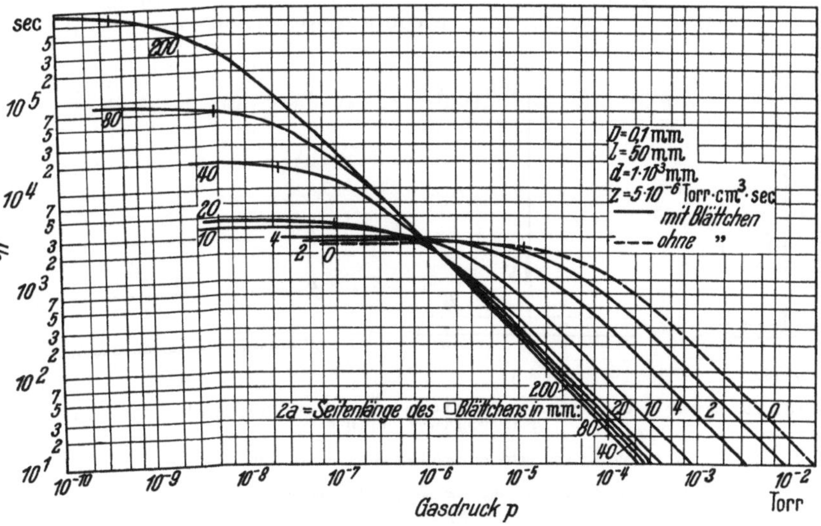

Fig. 15. Eichkurven für Doppelfadenpendel mit Blättchen verschiedener Seitenlänge. Halbwertszeit (t_H) in Abhängigkeit vom Gasdruck (p) für verschiedene Seitenlängen der Blättchen ($2a$).

3. Die Druckmessung kann nicht unmittelbar durch eine Zeigerablesung ausgeführt werden, sondern erfordert eine Hilfsmessung (Zeitmessung mit Stoppuhr).

Demgegenüber stehen folgende Vorteile:

1. Das Manometer besteht aus Quarz bzw. aus Glas. Es besitzt infolgedessen eine außerordentliche chemische Widerstandsfähigkeit.

2. Das Fehlen von Metallteilen im Vakuum hat eine geringe Gasabgabe und hohe Korrosionsbeständigkeit zur Folge.

3. Das Fehlen von heißen Oberflächen im Vakuum (im Gegensatz beispielsweise zum Wärmeleitungs- und Ionisationsmanometer) vermeidet eine Rückwirkung auf das zu messende Gas (z.B. durch thermische Zersetzung).

Von den verschiedensten Autoren ist immer wieder versucht worden, den Reibungsmanometern eine technisch brauchbare Form zu geben. Fig. 16 zeigt eins der verwendeten Grundprinzipien. Eine an einem Faden aufgehängte Platte kann zwischen zwei feststehenden Platten Drehschwingungen ausführen, deren Amplitude mittels des Spiegels S bestimmt wird. Fig. 17 zeigt eine zweite Form nach LANGMUIR und DUSHMAN[1]. Hier ist eine Platte M an einem Faden drehbar

[1] J. LANGMUIR: Phys. Rev. **1**, 337 (1913). — A. TINNRIAZEFF: Ann. Phys. **40**, 971 (1913). — S. DUSHMAN: Phys. Rev. **5**, 212 (1915). — W. E. BRIGGS: Vac. Symposium Trans. 1954, 1.

aufgehängt. Unter ihr ist eine drehbar gelagerte Platte A angebracht. Wird die Platte A in schnelle Drehungen versetzt, so wird die Platte M infolge Gasreibung mitgenommen. Der Drehwinkel, um den die Platte M gedreht wird, kann mittels des Spiegels S abgelesen werden. Er ist ein Maß für die auftretenden Reibungskräfte und damit für den Gasdruck.

Trotz der verschiedenartigsten Bemühungen hat sich das Reibungsmanometer bis heute nicht in die Technik eingeführt. KLUMB und SCHWARZ[1] haben neuerlich noch eine Form eines Reibungsmanometers angegeben, das nach Art eines Schleifenoszillographen arbeitet. Bei ihm ist eine drehbar ausgespannte Drahtschleife, an der ein Spiegel befestigt ist, zwischen den Polen eines Magneten gelagert. Die Schwingungen der Schleife werden mittels des Spiegels auf eine Photozelle übertragen, deren Strom wiederum als Rückkopplung durch die

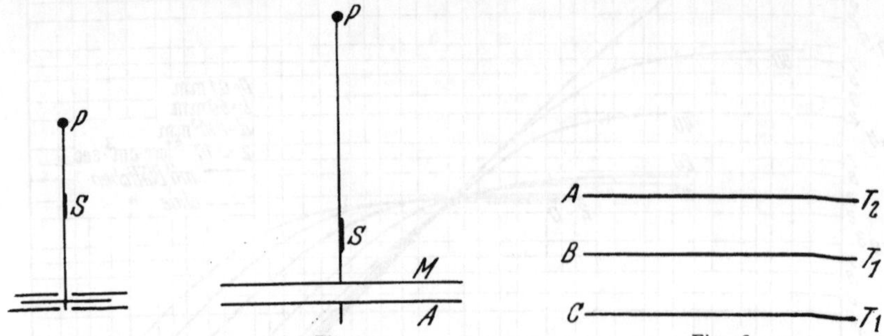

Fig. 16. Fig. 17. Fig. 18.

Fig. 16. Reibungsmanometer. Die obere und untere Platte stehen fest. Zwischen ihnen kann die mittlere Platte gedämpfte Drehschwingungen ausführen.

Fig. 17. Molekularmanometer nach LANGMUIR und DUSHMAN.

Fig. 18. Prinzip eines Radiometervakuummeters.

Schleife geschickt wird, wodurch die Schleifenschwingungen (entgegen der Dämpfung durch die Gasreibung) entdämpft werden können. Die auftretenden Amplituden sind also ein Maß für die Dämpfung und damit für den Gasdruck; sie können an der Spiegelablesung beobachtet werden.

8. Vakuummessung durch thermischen Molekulardruck. Radiometervakuummeter[2]. Zum Verständnis der Wirkungsweise dieser Vakuummeterart betrachten wir Fig. 18. Eine bewegliche Fläche B der Temperatur T_1 sei zwischen zwei festen Flächen A mit der Temperatur T_2 und C mit der Temperatur T_1 angebracht. Dabei sei $T_2 > T_1$. Bei niedrigen Drucken, wenn die freie Weglänge Λ groß st gegen den Plattenabstand d, haben die von oben auf die Fläche B auftreffenden Gasmoleküle ihren letzten Stoß auf der Fläche A erlitten und demgemäß eine Geschwindigkeit c_2, die der Temperatur T_2 entspricht. Die von unten auf B auftreffenden Moleküle haben ihren letzten Stoß auf der Fläche C erlitten und infolgedessen eine Geschwindigkeit c_1, die der Temperatur T_1 entspricht. Da also sekundlich auf die Platte B von oben ein größerer Impuls übertragen wird als von unten, wird die Platte B das Bestreben haben, sich in Richtung auf C hin in Bewegung zu setzen.

Der Druck, den die Gasmoleküle auf die obere Seite von B ausüben, errechnet sich folgendermaßen. Zwischen A und B möge ein Molekülstrom mit der Dichte

[1] H. KLUMB: Phys. Bl. **5**, 380 (1949). — H. SCHWARZ: A.T.M.—V 1341—4 (1952).

[2] M. KNUDSEN: Ann. Phys. **31**, 633 (1910). — H. KLUMB u. E. WEISSMANN: Glas u. Hochv.techn. **2**, 266 (1953). — E. WEISSMANN: Vak. Techn. **4**, 152 (1955).

$n_1 c_1/3$ von unten nach oben und ein Molekülstrom mit der Dichte $n_2 c_2/3$ von oben nach unten fliegen. Da der Zustand stationär sein soll, gilt außerdem

$$\uparrow \frac{n_1 c_1}{3} = \downarrow \frac{n_2 c_2}{3} = \frac{n c_1}{6}; \tag{8.1}$$

$n =$ Zahl der Gasmoleküle pro cm³, und weiterhin

$$p = \frac{m n c_1^2}{3}.$$

Daraus errechnet sich der Druck, den die Gasmoleküle auf die obere Seite von B ausüben zu

$$\left.\begin{aligned} \frac{m n_1 c_1^2}{3} + \frac{m n_2 c_2^2}{3} &= \frac{m n_1 c_1}{3}(c_1 + c_2) = \frac{m n c_1}{6}(c_1 + c_2) \\ &= \frac{m n c_1^2}{6}\left(1 + \frac{c_2}{c_1}\right) = \frac{p}{2}\left(\sqrt{\frac{T_2}{T_1}} + 1\right). \end{aligned}\right\} \tag{8.2}$$

Der Drucküberschuß zwischen der Ober- und Unterseite von B ergibt sich daher nach KNUDSEN zu

$$K = \frac{p}{2}\left(\sqrt{\frac{T_2}{T_1}} + 1\right) - p = \frac{p}{2}\left(\sqrt{\frac{T_2}{T_1}} - 1\right). \tag{8.3}$$

Die praktische Ausführung eines Radiometervakuummeters zeigt Fig. 19. Ein leichter Rahmen B ist an einem Torsionsfaden F aufgehängt. Seine Stellung kann mittels des Spiegels S abgelesen werden. Dem Rahmen B stehen auf entgegengesetzten Seiten die beheizten Bänder A gegenüber. Die von diesen Bändern A kommenden Gasmoleküle üben auf den Rahmen B eine Kraft in Pfeilrichtung aus. Für den Druck in dieser Anordnung gilt die Beziehung

$$p = \frac{4\pi^2 I D}{r F t^2 d} \frac{T_2}{T_2 - T_1} \,[\text{dyn cm}^{-2}], \tag{8.4}$$

$I =$ Trägheitsmoment des Rahmens B in g cm²;
$r =$ mittlerer Radius des beweglichen Rahmens B in cm;
$F =$ Oberflächen von B, die A gegenüber stehen, in cm²;
$t =$ Eigenschwingungszeit des Rahmens B in sec;
$D =$ Skalenausschlag in cm;
$d =$ Skalenentfernung in cm.

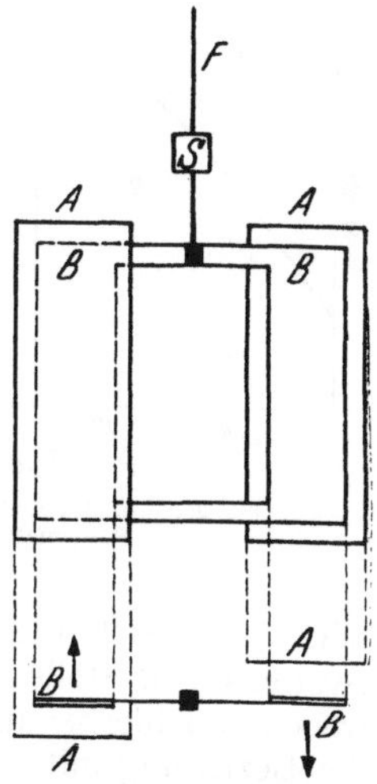

Fig. 19. Schema eines Radiometervakuummeters.

Bei den bisherigen Ableitungen wurde vorausgesetzt, daß sich die Gasmoleküle bei jedem Stoß an einer festen Wand ins Temperaturgleichgewicht setzen, d.h., daß der Akkommodationskoeffizient für den Temperaturausgleich γ den Wert 1 hat. Ist dies nicht der Fall, so gilt nach SMOLUCHOWSKI[1] die folgende Formel:

$$K = \left(1 + \frac{\gamma_2 - \gamma_1}{\gamma_1 + \gamma_2 - \gamma_1\gamma_2}\right)\frac{p}{2}\left(\sqrt{\frac{T_2}{T_1}} - 1\right). \tag{8.5}$$

Darin ist γ_2 der Akkommodationskoeffizient an den heißen und γ_1 der Akkommodationskoeffizient an den kalten Flächen.

Wir haben im Vorhergehenden außerdem stillschweigend vorausgesetzt, daß der Abstand der Platten d klein gegen ihre Ausdehnung b, und die freie Weglänge

[1] M. v. SMOLUCHOWSKI: Ann. Phys. 35, 983 (1911).

der Gasmoleküle Λ groß gegen die Manometerdimensionen sei. Diese Voraussetzung ist nach FREDLUND[1] solange berechtigt wie:

$$\Lambda \geqq 10\,d.$$

Für den Fall, daß der Plattenabstand d nicht klein ist gegen die Plattenabmessungen, gibt SPIWAK[2] die folgende Beziehung an

$$\left.\begin{aligned} K &= C\,\frac{p}{2}\left(\sqrt{\frac{T_2}{T_1}}-1\right),\\ C &= \frac{1}{\pi}\left[\frac{2\left(1+\frac{d^2}{b^2}\right)b^2-d^2}{b^2\sqrt{1+\frac{d^2}{b^2}}}-\frac{d}{b}\right]\operatorname{arc\,tan}\frac{b}{d}\,. \end{aligned}\right\} \tag{8.6}$$

Den grundsätzlichen Verlauf der Eichkurve von Radiometervakuummetern zeigt Fig. 20. Bei kleinen Drucken, solange die freie Weglänge groß ist gegen die Vakummeterdimensionen, gilt für die Kraft auf das bewegliche System Gl. (8.3). Diese und damit der Ausschlag ist also proportional zum Druck (Bereich I). Ist bei hohen Drucken die freie Weglänge nicht mehr groß gegen die Vakuummeterdimensionen, so wird die Druckkraft auf das bewegliche System nicht mehr durch den thermischen Molekulardruck bestimmt, sondern es tritt die Erscheinung der thermischen Gleitung der Gasmoleküle ein. Der Vorgang der thermischen Gleitung (thermische Effusion) bewirkt, daß sich die Gasmoleküle längs der festen Oberfläche des beweglichen Systems von Stellen niedrigerer Temperatur zu Stellen höherer Temperatur bewegen und hierbei auf das bewegliche System eine Druckkraft K_2 ausüben, für die die Beziehung gilt:

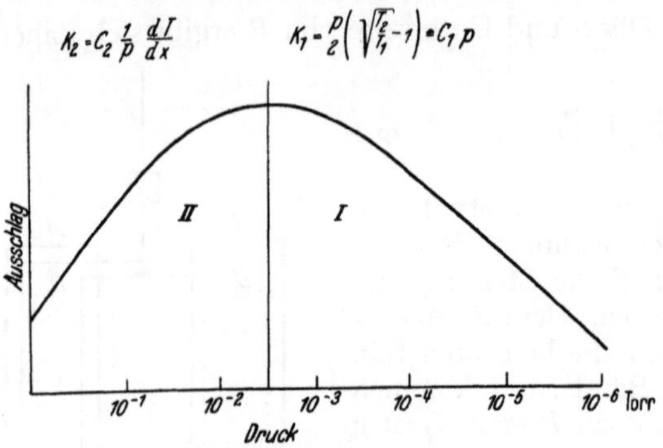

Fig. 20. Grundsätzliche Form der Eichkurve eines Radiometervakuummeters.

$$K_2 = C_2\,\frac{1}{p}\,\frac{dT}{dx}\,. \tag{8.7}$$

Im Bereich höherer Drucke (Bereich II) ist also die Druckkraft K_2 auf das bewegliche System und damit der Ausschlag umgekehrt proportional zum Druck. Zwischen den Bereichen I und II liegt ein Maximum des Ausschlages.

Die Radiometervakuummeter haben gegenüber den übrigen Fein- und Hochvakuummetern (abgesehen von den Membranmanometern) den entscheidenden Vorteil, daß sie nicht nur den Totaldruck von Gasen und Dämpfen messen, sondern daß innerhalb des Bereiches I diese Messung auch unabhängig von der Gasart ist (falls nicht durch den Akkommodationskoeffizienten eine kleine Korrektur erforderlich wird). Dieser bedeutende Vorteil der Radiometervakuummeter rührt daher, daß die Messung nicht auf dem Umweg über eine von der Gasart abhängige Größe (wie beispielsweise die Wärmeleitfähigkeit) erfolgt, sondern daß unmittelbar der Druck p gemessen wird.

[1] E. FREDLUND: Ann. Phys. **14**, 617 (1932).
[2] G. SPIWAK: Z. Physik **77**, 123 (1932).

Trotz dieses großen prinzipiellen Vorteils dieser Vakuummeterart haben sie sich als technische Geräte bisher nicht einführen können. An Bemühungen der verschiedensten Autoren hierzu hat es zwar nicht gefehlt. Genannt seien besonders: KNUDSEN (s. S. 530), der zuerst diese Vakuummeterart angegeben hat, eine besondere Ausführungsform von RIEGGER[1], das Molvakuummeter von GAEDE[2], das die Radiometerkräfte nicht an einem Zeigerausschlag, sondern über eine Schwingungsdauer mißt und dessen System gleichzeitig auch als Reibungsmanometer Verwendung finden kann. Schließlich haben sich in neuerer Zeit insbesondere KLUMB (s. S. 530) und seine Schüler um diese Vakuummeterart bemüht. Daß alle diese Anstrengungen trotzdem keine technischen Formen haben einbringen können, liegt daran, daß die Geräte 1. in fester Verbindung mit der Vakuumapparatur eine sehr genaue Justierung der Lage verlangen, und daß sie 2. außerdem außerordentlich anfällig sind gegen Störungen durch äußere Erschütterungen. Hier genügt z.B. schon die Erschütterung durch die rotierende Vorpumpe oder gar das Sieden des Treibmittels in den Diffusionspumpen.

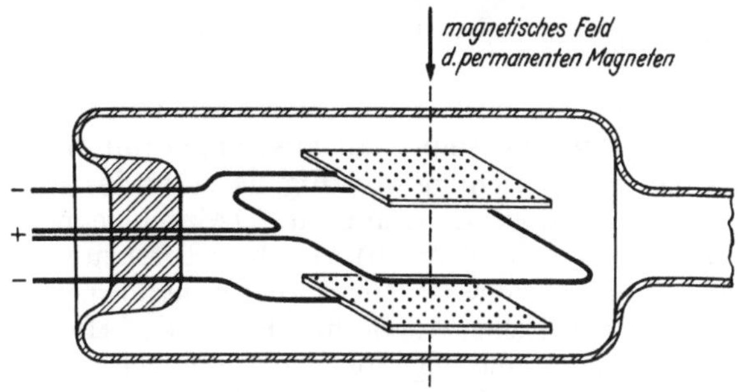

Fig. 21. Meßröhre eines Philips-Vakuummeters nach PENNING.

9. Vakuummeter nach PENNING[3] (Philips-Vakuummeter). Elektrische Entladungen in Röhren mit kalter Kathode erlöschen im allgemeinen beim Unterschreiten von Drucken unter 10^{-2} bis 10^{-3} Torr. Bringt man eine solche Entladungsröhre in geeigneter Weise in ein Magnetfeld, so kann man dadurch die Bahn der Elektronen verlängern und damit ihre Ionisationswahrscheinlichkeit so stark erhöhen, daß auch bei niedrigen Drucken noch stabile Entladungen möglich sind, die außerdem die Eigenschaft haben, daß die Stromstärke der Entladung von dem herrschenden Gasdruck abhängig ist. Ein auf diesem Prinzip beruhendes Vakuummeter wurde von PENNING[3] angegeben. Das Meßsystem (Fig. 21) besteht aus zwei einander gegenüberstehenden plattenförmigen Kathoden und einer ringförmigen Anode, die zwischen beiden angebracht ist. Das ganze Meßsystem wird in einem Magnetfeld untergebracht, dessen Feldlinien etwa senkrecht zu den plattenförmigen Kathoden verlaufen und das eine Stärke von etwa 350 Ø haben soll. (eDise Anordnung steht damit in Analogie zu einer vorher schon von GAEDE[4] und VON MEYEREN benutzten, die allerdings mit heißer Kathode arbeitete.) Legt man zwischen Anode und Kathode eine Gleichspannung von etwa 2000 V an, so genügt schon die natürliche Raumionisation, um eine Entladung zu zünden und aufrechtzuerhalten. Die Elektronen werden durch das Magnetfeld auf spiralförmige Bahnen gezwungen und pendeln vielfach zwischen den Kathoden hin und her, bis sie schließlich durch Stöße aus der Bahn herausgeworfen werden und dann die ringförmige Anode erreichen können. Unter den angegebenen Bedingungen erhält man bei Drucken zwischen 10^{-3} und 10^{-5} Torr Ströme zwischen 10 μA und 1 mA, deren Stärke ein Maß für den

[1] H. RIEGGER: Z. techn. Phys. **1**, 16 (1920).
[2] W. GAEDE: Z. techn. Phys. **15**, 644 (1934).
[3] F. M. PENNING: Physica, Haag **4**, 71 (1937). — M. VARIAC: J. Sci. Instrum. **32**, 346 (1955). — R. HAEFER: Acta phys. Austriaca **9**, 200 (1955).
[4] W. GAEDE: DRP 716721.

herrschenden Gasdruck ist. Das Gerät kann auch mit Wechselspannung betrieben werden, da es infolge seiner stark unsymmetrischen Elektroden selbst als Gleichrichter wirkt. Im Hinblick darauf, daß die Entladung bei sehr niedrigen Drucken nur schwer zündet, benutzt McHWRAITH[1] eine Hilfskathode mit Wolframglühdraht.

Das Meßsystem kann entweder in einen Glaskolben oder in ein Metallgefäß eingebaut werden. Glassysteme zeigen besonders stark den Effekt der Gasaufzehrung[2]. Es ist in diesen Fällen besonders wichtig, Meßröhre und Rezipient durch eine möglichst kurze und weite Leitung miteinander zu verbinden, wenn man Fehlmessungen des Druckes vermeiden will. In dieser Hinsicht günstiger sind offene Meßsysteme ohne Mantel, die zusammen mit dem Permanentmagneten in den Vakuumrezipienten, in dem der Druck gemessen werden soll, eingebaut werden können. Systeme, die in ein Metallgefäß eingebaut sind, zeigen häufig infolge Erwärmung durch die Gasentladung eine starke Gasabgabe, wodurch die Druckmessung gegenüber der Gasaufzehrung im umgekehrten Sinne verfälscht wird.

Es sind eine große Anzahl von Varianten für die Elektrodenformen angegeben worden. PENNING selbst hat zusammen mit NIENHUIS[3] eine solche beschrieben, bei der in Abweichung von Fig. 21 statt des Anodenringes ein ganzer Anodenzylinder zur Anwendung kommt. Dadurch gelingt es, den Meßbereich noch weiter nach niedrigen Drucken (10^{-4} bis 10^{-6} Torr) zu verschieben. Allen in der Literatur beschriebenen verschiedenen Elektrodenformen gemeinsam ist die starke Unsymmetrie zwischen Anode und Kathode. Hiervon sei nur noch eine Form erwähnt, bei der eine kugelförmige Anode sich innerhalb eines geschlossenen Metallmantels, der als Kathode dient, befindet. Bei genügender Asymmetrie kann sogar das Magnetfeld in Fortfall geraten.

Bei allen Vakuummetern der eben beschriebenen Art ist darauf zu achten, daß sie sehr starke Neigung zur Erzeugung von hochfrequenten elektrischen Schwingungen zeigen.

Obgleich die Philips-Vakuummeter keine sehr genauen Vakuummessungen ermöglichen, haben sie sich doch als technische Geräte außerordentlich stark eingebürgert. Dies hat verschiedene Gründe. Sie haben vor allen Dingen eine außerordentlich kleine Störanfälligkeit (kein Durchbrennen eines Glühfadens, keine empfindliche Strommessung, Erschütterungsunempfindlichkeit, keine Gefahr des Durchbrennens von Schaltungselementen bei höheren Drucken und Lufteinbrüchen und so fort). Auch in der Bedienung sind sie außerordentlich einfach.

10. Ionisationsmanometer. Beim Durchgang von geladenen Teilchen, z.B. Elektronen, durch ein verdünntes Gas ist die spezifische Ionisation s, d.h. die pro Zentimeter Weg erzeugte Zahl der Ionenpaare unter sonst gleichen Bedingungen proportional zum Gasdruck. Die Messung der Ionisation kann also zur Druckbestimmung ausgenutzt werden. Hierauf beruhen die Ionisationsmanometer.

α) Triode[4]. Im einfachsten Falle verwendet man zur Messung eine normale Dreielektrodenröhre (Triode). Fig. 22 zeigt die Meßanordnung. Die von der Glühkathode F ausgehenden Elektronen werden in Richtung auf den positiv

[1] CH. G. MCHWRAITH: Rev. Sci. Instrum. **18**, 683 (1947).

[2] W. v. MEYEREN: Z. Physik **91**, 727 (1934). — H. SCHWARZ: Rev. Sci. Instrum. **25**, 924 (1954). — J. R. YOUNG: J. Appl. Phys. **27**, 926 (1956).

[3] F. M. PENNING u. K. NIENHUIS: Philips techn. Rdsch. **11**, 116 (1949).

[4] H. RUKOP u. J. HAUSER: Vortragsreihe Elektrotechn. Ver., Berlin 1919. — H. SIMON: Z. techn. Phys. **5**, 221 (1924). — S. DUSHMAN u. C. H. FOUND: Phys. Rev. **17**, 7 (1921). — J. H. LECK u. A. RIDDOCH: Brit. J. Appl. Phys. **7**, 153 (1956).

aufgeladenen Auffangzylinder A beschleunigt und erzeugen bei Zusammenstößen mit Gasmolekülen positive Ionen, die von dem schwach negativ aufgeladenen Gitter G aufgefangen werden. Der Strom i_- zu dem positiv geladenen Metallzylinder A ist also ein Maß für die Zahl der Elektronen, die in der Zeiteinheit den Gasraum passieren und der Strom i_+ zu dem negativ geladenen Gitter G ein Maß für die durch diese Elektronen erzeugten positiven Ionen.

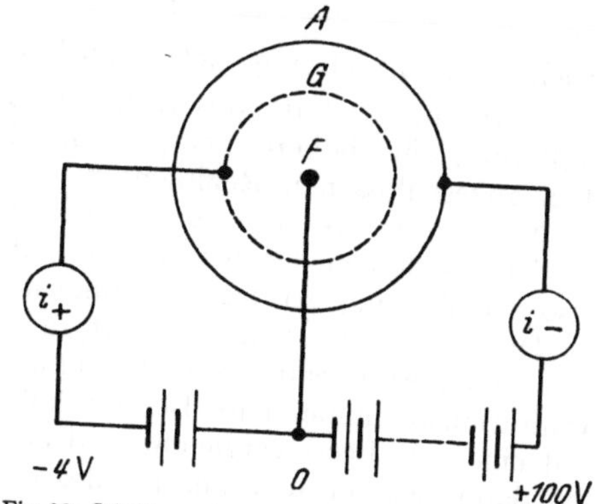

Fig. 22. Schaltung einer Triode als Ionisationsmanometer.

Für das Verhältnis der beiden Ströme gilt dann

$$\frac{i_+}{i_-} = \frac{d}{\Lambda_e}, \qquad (10.1)$$

d = Bahnlänge der Elektronen in der Röhre; Λ_e = freie Weglänge der Elektronen.

Darin ist Λ_e umgekehrt proportional zum herrschenden Gasdruck p, der also auf diese Weise gemessen werden kann.

Vergleicht man auf diese Weise ermittelte Werte für den Vakuumfaktor i_+/i_- mit Messungen, die unmittelbar mit druckanzeigenden Geräten, beispielsweise einem McLeod unter Verwendung eines definierten Füllgases (z.B. N_2) bestimmt wurden, so ergibt sich, daß die oben durchgeführte grobe Berechnung des Vakuumfaktors zu niedrige Werte liefert. Dies kommt vor allem daher, daß die

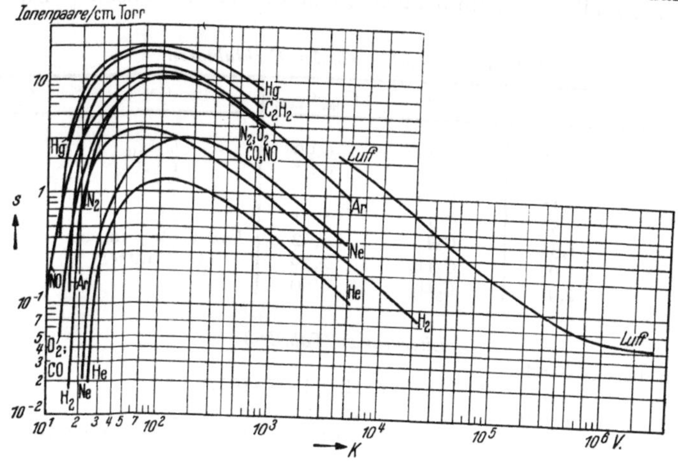

Fig. 23. Spezifische Ionisation s für verschiedene Gase in Abhängigkeit von der Energie der Elektronen (in eV).

verwandte mittlere freie Weglänge kein richtiges Maß für die wirkliche Zahl der ionisierenden Stöße darstellt. Zur Berechnung der tatsächlichen Ionisation muß die spezifische Ionisation s, die selber wiederum von der Energie K der Elektronen abhängt, eingesetzt werden. Fig. 23 zeigt die Abhängigkeit der spezifischen Ionisation s von der Energie der Elektronen für verschiedene Füllgase.

Für eine genauere Berechnung müssen wir daher in der Gleichung

$$\frac{i_+}{i_-} = C \cdot p, \qquad (10.2)$$

zur Berechnung von C die spezifische Ionisation $s(v)$ über den gesamten Elektronenweg zwischen Kathode und Anode integrieren:

$$\frac{i^+}{i^-} = C \cdot p = p \int_F^A s(v)\, dx. \tag{10.3}$$

Auch diese Berechnung liefert noch keine exakten Werte, und zwar

1. weil in der Nähe des Glühfadens, d.h. aber mehr oder weniger im ganzen Meßraum, eine höhere Temperatur, d.h. eine kleinere Dichte, herrscht als in dem angeschlossenen Rezipienten, in dem das Vakuum gemessen werden soll,

2. weil die positiven Ionen nicht nur von dem negativ geladenen Gitter G, sondern auch von der Kathode F aufgefangen werden. Die letzteren entgehen also der Messung, so daß der Ionenstrom i_+ zu klein gemessen wird.

Die geschilderte Schwierigkeit, den Eichfaktor C eines Ionisationsmanometers aus seinen Abmessungen zu berechnen, wird dadurch umgangen, daß man das Ionisationsmanometer im Bereich höherer Drucke (zwischen 10^{-3} und 10^{-4} Torr) mit einem McLeod vergleicht und auf diese Weise den Eichfaktor C bestimmt, der dann nachher auch bei der Messung niedrigerer Drucke mit dem Ionisationsmanometer zur Druckberechnung benutzt werden kann, wozu vielfach die handlichere Form

$$p = C_1 \frac{i_+}{i_-} \tag{10.4}$$

benutzt wird. Da C bzw. C_1 natürlich von der Gasart abhängen, sind hierüber von verschiedenen Autoren Vergleichsmessungen durchgeführt worden, deren Ergebnisse in Tabelle 3 zusammengestellt sind. Die technischen Daten für die

Tabelle 3. *Verhältnis der Eichfaktoren.*

Gas	C_{Gas}/C_{Luft}	Gas	C_{Gas}/C_{Luft}	Gas	C_{Gas}/C_{Luft}
H_2	0,43	H_2O	0,90	C_4H_{10}	4,85
D_2	0,40	O_2	0,95	CF_2Cl_2	2,9
He	0,18	Luft	1	SF_6	2,5
Ne	0,30	CO	1,11	Hg	3,4
Ar	1,32	CH_4	1,42	Diffusionspumpen-treibmittel	9 bis 13
Kr	1,9	C_2H_6	2,79		
Xe	2,9	C_3H_8	4,60		

Literatur zu Tabelle 3.

ALPERT in diesem Bande.
H. MOESTA u. R. RENN: Vak. Techn. **6**, H. 2/3, 35 (1957).
G. REICH: Z. angew. Phys. **9**, H. 1, 23—29 (1957).
Private Mitteilung SOMMERMEYER (Freiburg) und ZIOCK u. BÄCHLER (Leybold, Köln).

gebräuchlichsten Röhren für Ionisationsmanometermessungen zeigt Tabelle 4, einschließlich des auf Luft bezogenen Eichfaktors C. Die Angabe Schaltung A bzw. Schaltung B bedeutet nach SIMON folgendes. Schaltung A entspricht Fig. 22, sie ist die unempfindlichere. In Schaltung B wird der Metallzylinder als Auffangelektrode benutzt und schwach negativ aufgeladen, während das Gitter G als Anode dient und positiv vorgespannt wird. In Schaltung B gelangen die Elektronen von der Glühkathode nicht unmittelbar auf die Anode, sondern führen verschiedene Pendelungen um das als Anode geschaltete Gitter durch, wodurch ihr Weg verlängert und damit die Ionisationswahrscheinlichkeit erhöht wird.

Tabelle 4. *Betriebsdaten einiger Ionisationsmanometerröhren.*

Röhre	Hersteller	Schaltung	i_- mA	C für Luft	Heizstrom A	V_G Volt	V_A Volt
RE 11	Telefunken	A	1	0,74	0,5	−2	+100
RE 11	Telefunken	B	1	3,7	0,5	+100	−10
JM 1	Telefunken	A	1	2,3	1,5	−10	+200
JM 1	Telefunken	B	1	10	1,5	+200	−10
VG-1	Distillation Products	B	5	40	2,75	+150	−25
D 79510	Western Electric	B	5	15	—	+200	−10
45	Sylvania	B	5	6,7	—	+200	−10
NRC 507	Nat. Res.	B	5	20	5	+150	−20
VG 1 A	Distillation Products	B	5	24	3,5−5	+150	−25
VG 2	Distillation Products	B	10	3,2	3−4	+150	−25

Zur praktischen Anwendung werden die Triodensysteme der Ionisationsmanometer sowohl in Glasröhren eingeschmolzen als auch in Metallkolben untergebracht und schließlich sind auch offene Systeme, die unmittelbar innerhalb der Vakuumapparatur angebracht werden können, gebräuchlich. Zur Vermeidung des Übelstandes, daß der Glühfaden der Kathoden im Gebrauch z.B. bei Lufteinbrüchen und dergleichen leicht durchbrennt, sind Systeme entwickelt worden, die entweder das Neueinziehen eines Glühfadens in leichter Weise ermöglichen, oder die einen ganzen Vorrat von Wolframdraht derart enthalten, daß nach Durchbrennen eines Glühfadens das Wolfram aus dem Vorrat in einfacher Weise wieder nachgezogen werden kann. Da bei der Messung sehr niedriger Drucke die Ionenströme natürlich sehr klein sind, ist darauf zu achten, daß keine Isolationsfehler und die dadurch bedingten Kriechströme die Messungen verfälschen.

Der Meßbereich der Ionisationsmanometer reicht von 10^{-3} bis etwa $5 \cdot 10^{-8}$ Torr. Bei größeren Drucken als 10^{-3} Torr brennt entweder der Glühfaden sehr schnell durch oder seine Lebensdauer ist doch zumindestens sehr stark begrenzt. Nach niedrigen Drucken zu liegt die Grenze bei etwa $5 \cdot 10^{-8}$ Torr. Dies rührt daher, daß die auf die Anode auftreffenden Elektronen dort weiche Röntgenstrahlen auslösen, die wiederum an dem Ionenauffänger Elektronen frei machen und so einen zu großen Ionenstrom vortäuschen. Das hat zur Folge, daß Ionenströme, die kleineren Drucken als etwa $5 \cdot 10^{-8}$ Torr entsprechen, nicht mehr gemessen werden können, es sei denn, man treffe besondere Vorkehrungen, um auch das Gebiet des Ultrahochvakuums zu erfassen, wie sie in dem Artikel von Alpert in diesem Bande behandelt sind.

Das Ionisationsmanometer weist eine Reihe unverkennbarer Mängel auf. So kann der heiße Glühfaden durch chemische Reaktion mit dem Füllgas zerstört oder dieses selbst durch Zersetzungserscheinungen an dem heißen Glühfaden in seiner Zusammensetzung gegenüber dem Raum, in dem das Vakuum gemessen werden soll, verändert werden. Außerdem neigen alle Ionisationsmanometer mehr oder weniger stark zu Gasaufzehrung, wodurch der Druck in der Meßröhre gegenüber der angeschlossenen Apparatur (wenn auch in wesentlich geringerem Maß wie bei dem Philips-Vakuummeter) erniedrigt wird. Trotz dieser Mängel ist der Gebrauch des Ionisationsmanometers als genau anzeigendem Hochvakuummeßinstrument außerordentlich weit verbreitet. Es hat gegenüber dem vielfach benutzten Philips-Vakuummeter den Vorzug größerer Genauigkeit. Im übrigen kann das Meßsystem mit kleinen räumlichen Abmessungen an jeder beliebigen Stelle der Vakuumapparatur angebracht werden. Es ist selbst weitgehend erschütterungsunempfindlich, und da die Messung auf eine elektrische Messung zurückgeführt wird, kann entfernt von dem eigentlichen Meßsystem an jeder beliebigen

Stelle des Versuchsraumes abgelesen oder sogar registriert werden, wofür bewährte Standardinstrumente der elektrischen Meßtechnik in genügender Auswahl zur Verfügung stehen. Es hat darüber hinaus noch den Vorteil, daß es allen Druckschwankungen in der Vakuumapparatur unverzüglich folgt, wobei seine Ansprechzeit durch diejenige der elektrischen Meßinstrumente gegeben ist.

Für technische Messungen ist es allerdings erforderlich, nicht den Vakuumfaktor i_+/i_- zu bestimmen, sondern den Elektronenstrom i_- konstant zu halten und nur den Ionenstrom i_+ zu messen, der dann ein unmittelbares Maß für den Druck liefert. Hierfür sind die verschiedensten elektrischen Schaltungen angegeben worden. Fig. 24 zeigt das Prinzip einer solchen[1]. Die von der heißen Glühkathode *2* ausgehenden Elektronen werden von dem positiv vorgespannten Gitter *3* aufgefangen, dieser Elektronenstrom wird mit dem Verstärker *7* verstärkt

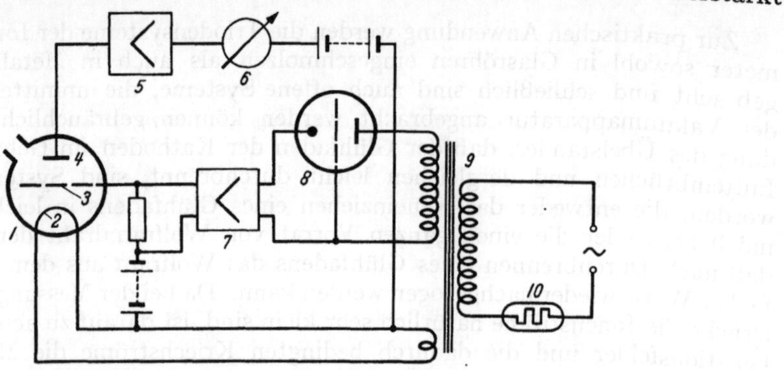

Fig. 24. Schaltung eines Ionisationsmanometers zur Stabilisierung des Elektronenstromes.

und belastet über die Endverstärkerröhre *8* die Sekundärseite des Transformators *9*. Auf der Sekundärseite dieses Transformators wird außerdem der Heizstrom für die Glühkathode *2* abgegriffen. Schwankungen des Elektronenstromes i_- führen also zu einer veränderten Belastung der Sekundärseite des Transformators, so daß für die auch sekundärseitig aufgebrachte Heizwicklung nur noch größere oder kleinere Energiebeträge zur Verfügung stehen. Der Heizstrom des Glühfadens *2* kann also auf diese Weise geregelt und damit der Elektronenstrom i_- zur Elektrode *3* stabilisiert werden. Bei stabilisiertem Elektronenstrom ist aber der Ionenstrom i_+, der zur Auffangelektrode *4* gelangt, ein unmittelbares Maß für den Druck. Er wird mittels des Verstärkers *5* verstärkt und kann an dem Instrument *6* abgelesen oder mit demselben registriert werden.

β) Alphatron[2]. Bei diesem Gerät erzeugen die von einem Radiumpräparat (zeitlich praktisch konstanter α-Strahler) ausgehenden Alphastrahlen in einer Ionisationskammer eine Ionenmenge, die vom Druck in der Kammer abhängig und diesem weitgehend proportional ist. Die Druckmessung kann also auf die Messung der in der Zeiteinheit von den Alphastrahlen in der Kammer erzeugten Ionenmenge zurückgeführt werden.

Fig. 25 zeigt im Schema die praktische Ausführung eines solchen Meßgerätes. In der Meßkammer befindet sich ein Radiumpräparat *1* mit etwa $2 \cdot 10^{-4}$ g Radium, das gut gasdicht abgeschlossen sein muß, damit die gebildete Radiumemanation

[1] K. Ziock: Phys. Bl. **5**, 523 (1949).

[2] Shephard, Roberts: Rev. Sci. Instrum. **10**, 181 (1939). — G. L. Mellen u. J. R. Downing: Rev. Sci. Instrum. **17**, 218 (1946). — W. Steckelmacher: J. Sci. Instrum. **21**, 195 (1950). — R. H. Vacca: Vac. Symposium Trans. 1956, 93. New York: Pergamon Press 1957.

nicht entweichen kann. Die von dem Präparat ausgehenden Alphastrahlen bilden in der Ionisationskammer Ionen, die von der vierarmigen Sammelelektrode *2* aufgefangen werden. Deren Ströme sind sehr klein. Sie haben bei einem Druck von 10^{-3} Torr einen Wert von etwa $2 \cdot 10^{-13}$ A. Im Hinblick auf die Kleinheit dieser Ströme muß die Durchführung der Sammelelektrode durch das Gehäuse der Kammer mit einem auf demselben Potential befindlichen Schutzring zur Vermeidung von Kriechströmen umgeben sein. Die Ionenströme werden dann weiterhin mit dem Verstärker *3* soweit verstärkt, bis sie mittels eines Zeiger-

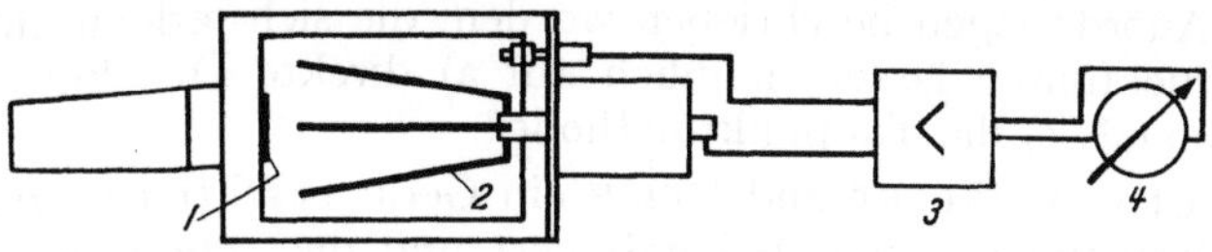

Fig. 25. Alphatron. Meßröhre mit radioaktivem Präparat *1* und Ionenfänger *2*, Verstärker für den Ionenstrom *3*, Anzeigegerät *4*.

instrumentes *4* abgelesen oder registriert werden können. Im Hinblick auf die Kleinheit der primären Ionenströme wird zweckmäßigerweise die erste Röhre des Verstärkers mit ihrem Gitter unmittelbar mit dem Ionenauffänger verbunden; sie bildet also abweichend von dem übrigen Teil des Verstärkers *3* eine Einheit mit der Meßkammer. Um zu vermeiden, daß die von den Alphateilchen an den Kammerwänden ausgelösten Sekundärelektronen mitgemessen werden und damit die Messung verfälschen, muß die Sammelelektrode ein solches Potential gegenüber der Kammerwand haben, daß sie wirklich die positiven Ionen und nicht etwa die gleichzeitig mitgebildeten Elektronen sammelt.

Fig. 26 zeigt die von der Gasart abhängigen Eichkurven eines solchen Alphatrons. Die Grenzen seines Meßbereiches liegen zwischen etwa 10 Torr und 10^{-3} bis 10^{-4} Torr. Die obere Grenze ist dadurch gegeben, daß bei höheren Drucken teilweise Rekombinationen der gebildeten Ionen stattfindet, so daß sich diese also dadurch der Messung entziehen. Die untere Grenze des Meßbereiches ist bedingt

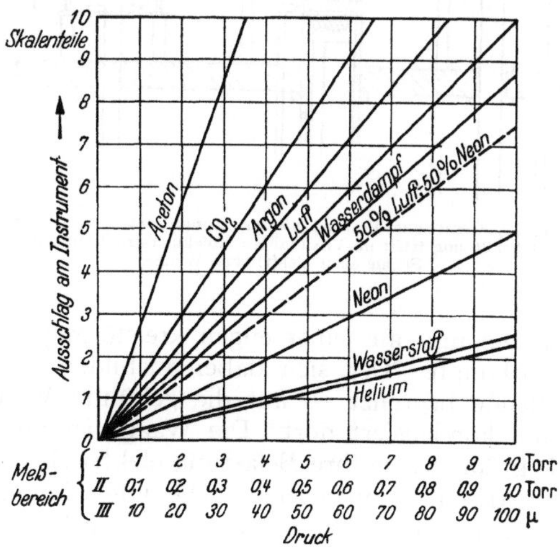

Fig. 26. Eichkurven eines Alphatrons für verschiedene Füllgase.

1. durch die Größe der Meßkammer und die Stärke des Radiumpräparates;

2. durch einen Nullstrom von etwa 10^{-15} A, der von den Sekundärelektronen herrührt, die die Alphateilchen in der Sammelelektrode auslösen, und

3. durch den Isolationsstrom des Gitters der Eingangsröhre des Verstärkers. (Hier wird also zweckmäßigerweise eine Elektrometerröhre benutzt.)

Das Alphatron ergänzt mit seinem Meßbereich die Hochvakuummeter (Philips-Vakuummeter und Ionisationsmanometer mit Glühkathode) nach hohen Drucken zu. Es hat gegenüber den Wärmeleitungsmanometern den Vorteil, daß sein Meßbereich bis zu höheren Drucken reicht und daß auch die Eichkurve am unteren Ende des Meßbereiches wesentlich unveränderlicher ist als die der Wärmeleitungsmanometer.

11. Dampfdruckmessungen. Für die gesamte Vakuumphysik ist es wesentlich, nicht nur den Gasdruck in den Vakuumbehältern messen zu können, sondern auch Kenntnis zu besitzen von den Dampfdrucken der in Hochvakuumapparaturen benutzten Substanzen (s. z.B. weiter unten die Dampfdruckkurven der Treibmittel für Diffusions- und Dampfstrahlpumpen). Zur Messung solcher kleiner Dampfdrucke, die vor allem im Gebiet unter 10^{-2} Torr liegen, sind eine Vielzahl von Anordnungen beschrieben worden, die sich jedoch auf drei Grundprinzipien zurückführen lassen, nämlich auf a) direkte Druckmessung[1], b) die Effusionsmethode[2], c) die Taupunktmethode[3].

Neuerlich ist von HERLET und REICH ein Gerät beschrieben worden, das die beiden Grundprinzipien a und b kombiniert und das schon sehr weittragende Ergebnisse[4] geliefert hat. Dieses Gerät hat den Vorteil, daß man nicht nur eine Aussage über den Dampfdruck, sondern auch über das Molekulargewicht der Substanz erhält und damit eine Möglichkeit zur Substanzkontrolle gewinnt. Außerdem hat es einen über mehrere Zehnerpotenzen reichenden Meßbereich. Es gestattet mit relativ stabilen und wenig störanfälligen Instrumenten bis herunter zu Drucken von 10^{-6} Torr zu messen.

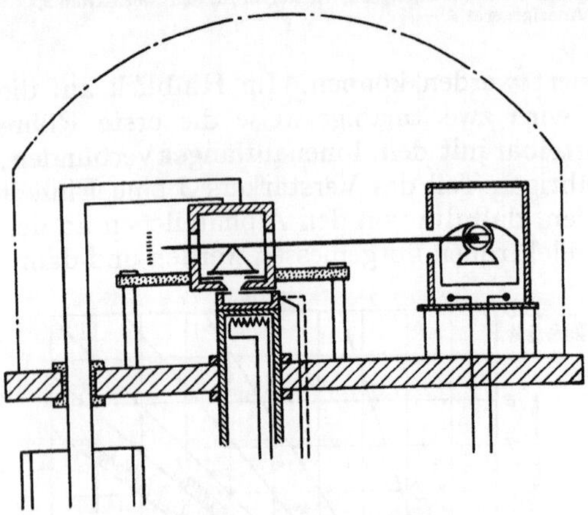

Fig. 27. Gerät zur Messung kleiner Dampfdrucke. Aus einer Ofenöffnung trifft im Vakuum ein Molekularstrahl auf die Schale einer elektrischen Waage.

Sein Meßprinzip zeigt Fig. 27. Es wird von den Verfassern etwa folgendermaßen beschrieben: Die Substanz wird in einem Ofen auf konstanter Temperatur gehalten und aus einer engen kreisförmigen Öffnung des Ofens ins Hochvakuum verdampft. Der sich dabei ausbildende Molekularstrahl trifft — durch eine Blende begrenzt — auf die gekühlte Waagschale einer empfindlichen Waage und kondensiert dort. Die Waage ist im Prinzip eine elektrische Drehwaage. Man kann also ihre Belastung elektrisch messen und außerdem ihren Ausschlag elektrisch kompensieren. Die Ofenöffnung kann freigegeben und abgedeckt werden. Bei Öffnung des Ofens wird die Waagschale durch den Impuls des Molekularstrahles um einen Betrag angehoben, der dem Dampfdruck p der Substanz im Ofen proportional ist. Bei Schließung sinkt die Waagschale um den

[1] H. MAYER: Z. Physik **67**, 264 (1931). — G. SEYDEL: Z. techn. Phys. **16**, 107 (1935). — K. NEUMANN u. E. VÖLKER: Z. phys. Chem., Abt. A **161**, 33 (1932). — F. H. VERHOCK u. A. L. MARSHALL: J. Amer. Chem. Soc. **61**, 2737 (1939).

[2] M. KNUDSEN: Ann. d. Phys. **28**, 75 (1909). — M. VOLMER: Z. phys. Chem., Bodenstein-Festband **1931**, 863. — E. MIESCHER: Helv. phys. Acta **14**, 507 (1941). — F. METZGER: Helv. phys. Acta **16**, 323 (1943). — G. WESSEL u. W. PAUL: Z. Physik **124**, 691 (1948). — A. HERLET: Phys. Bl. **5**, 523 (1949). — R. JAECKEL u. G. REICH: Forsch.ber. 332 Wirtsch. u. Verk. Min. NRW, Köln u. Opladen 1956.

[3] S. F. KAPPF u. R. B. JACOBS: Rev. Sci. Instrum. **18**, 581 (1947).

[4] A. HERLET u. G. REICH: Z. angew. Phys. **9**, 14 (1957). — G. REICH: Z. angew. Phys. **9**, 23 (1957). — K. G. GÜNTHER: Glastechn. Ber. **31**, 9 (1958). — H. HOYER u. W. PEPERLE: Z. Elektrochem. **62**, 61 (1958).

gleichen Betrag. Während der Öffnungszeit wird die Waagschale durch Kondensation des Dampfes laufend schwerer. Diese Gewichtszunahme ist proportional der Verdampfungsgeschwindigkeit G. Aus der Aufzeichnung der Stellung der Waage als Funktion der Zeit erhält man also unabhängig voneinander zwei Größen, die dem Dampfdruck p und der Verdampfungsgeschwindigkeit G proportional sind. Da ferner der Dampfdruck p und die Verdampfungsgeschwindigkeit G durch die Hertzsche Beziehung

$$G = 5{,}8 \cdot 10^{-2} p \sqrt{\frac{M}{T}} \; [\mathrm{g\,sec^{-1}\,cm^{-2}}] \tag{11.1}$$

p = Dampfdruck [Torr], M = Molekulargewicht, T = Temperatur [°K]

gekoppelt sind, ist damit gleichzeitig eine Bestimmung des Molekulargewichtes M verbunden.

Die Methode hat als wesentliche Voraussetzung die, daß die auf die Waagschale auftreffenden Dampfmoleküle dort vollständig kondensieren. Eingehende Messungen haben gezeigt, daß diese Bedingung jedenfalls für eine große Anzahl von höhermolekularen organischen Substanzen erfüllt ist.

Zur Bestimmung der Verdampfungsgeschwindigkeit G haben neuerlich TREVOY und TORPEY[1] ein Gerät beschrieben, das im Prinzip aus einem Flügelrad sehr kleiner Abmessungen besteht, das auf harten, weitgehend reibungsfreien Spitzen gelagert ist, und mit dem ein kleiner Permanentmagnet verbunden ist. Bei der Drehung des Flügelrades bewegt sich der Permanentmagnet vor einer feststehenden Spule und erzeugt in dieser durch Induktion elektrische Spannungen, die der Drehgeschwindigkeit proportional sind. Mit diesem Gerät kann also der Molekülstrom, der von einer im Vakuum verdampfenden Oberfläche ausgeht und auf das Windrad trifft, in seiner Intensität gemessen und daraus die Verdampfungsgeschwindigkeit ermittelt werden.

12. Lecksuche und Leckmessung. Bekanntlich ist jeder Vakuumphysiker genötigt, für die Auffindung von undichten Stellen der Vakuumapparatur und ihre Beseitigung einen erheblichen Zeitaufwand zu treiben. Da jede Vakuumapparatur größere oder kleinere Undichtigkeiten aufweist, ist es unzweckmäßig, von dichten oder undichten Vakuumapparaturen zu sprechen. Richtig ist es dagegen, die Undichtigkeit einer Vakuumapparatur zahlenmäßig anzugeben. Sie wird gemessen als die pro Zeiteinheit durch Undichtigkeiten aus der äußeren Atmosphäre eindringende Gasmenge U gegeben durch Druck mal Volumen pro Zeit; d.h. also Torr · l/sec. Welche Undichtigkeit U für eine Apparatur noch zugelassen werden kann, hängt ganz vom Verwendungszweck ab. Wir geben weiter unten noch einige Zahlenwerte für tragbare Undichtigkeiten an.

Zur Messung der Undichtigkeiten einer Vakuumapparatur (Leckmessung) wurde früher ausschließlich die Methode angewendet, die Apparatur zunächst soweit wie möglich auszupumpen und dann den Druckanstieg in Abhängigkeit von der Zeit zu messen. Die Messung kann aber durch Gasabgabe (z.B. von den Wänden der Vakuumapparatur) oder durch die Verdampfung von Restfeuchtigkeiten wesentlich verfälscht werden. Ein Kriterium, ob solche Fehlmessungen vorliegen, ist darin zu finden, daß man nach einer Druckanstiegsmessung die Apparatur erneut evakuiert und diese Druckanstiegsmessungen mehrmals wiederholt. Zeigen diese Wiederholungen denselben Verlauf, so ist anzunehmen, daß keine wesentliche Verfälschung der Messung vorliegt. Außerdem zeigt der Druckanstieg im Falle der Verfälschung durch Dämpfe einen Sättigungswert.

[1] D. J. TREVOY u. W. A. TORPEY: Rev. Sci. Instrum. **24**, 676 (1953).

Für die Größe der für eine Vakuumapparatur noch zulässigen Undichtigkeit U gelten etwa folgende Richtwerte, die aber ihrerseits noch von der Größe der angeschlossenen Pumpe abhängen: Für Arbeiten im Feinvakuum sind im allgemeinen Undichtigkeiten von 10^{-3} Torr l/sec, im Hochvakuum von 10^{-6} Torr l/sec, und im Ultrahochvakuum von 10^{-13} Torr l/sec tragbar. Die Undichtigkeiten von abgeschmolzenen Elektronenröhren (nachziehenden Röhren) müssen kleiner als 10^{-11} Torr l/sec sein.

Zur Lokalisierung der gröberen Lecks, die zunächst beseitigt werden müssen, damit die Undichtigkeit den zulässigen Wert unterschreitet, gibt es verschiedene Verfahren. So kann man versuchen, verdächtige Stellen mit Ölen, Wachsen oder Dichtungslacken abzudichten und anschließend zu prüfen, ob bei laufender Pumpe niedrigere Druckwerte erzielt werden als vorher, d.h. ob die Undichtigkeit geringer geworden ist. Auch die Tatsache, daß die meisten Vakuummeterarten in ihrer Anzeige von der Gasart abhängig sind, kann man in der Weise ausnutzen, daß man die verdächtigen Stellen mit verdampfbaren Flüssigkeiten wie z.B. Alkohol, Tetrachlorkohlenstoff, Trichloräthylen oder dergleichen abpinselt. Trifft man hierbei die undichte Stelle, so dringt die Flüssigkeit in die Vakuumapparatur ein, verdampft dort und kann durch die plötzliche Änderung der Vakuummeteranzeige festgestellt werden. Auch die Tatsache, daß Sauerstoffbeladung die Elektronenemission von glühenden Wolframdrähten herabsetzt, kann durch Absprühen verdächtiger Stellen mit Sauerstoff unter Verwendung eines Ionisationsmanometers benutzt werden. Eine viel angewandte Technik besteht darin, daß man die Vakuumapparatur oder Teile derselben vor der Inbetriebnahme unter Wasser hält, innen unter Druckluft setzt und die aus undichten Stellen heraustretenden Luftbläschen beobachtet. Analog kann man auch die innen unter Druck gesetzte Apparatur außen mit Schaummitteln, wie z.B. Nekal, abpinseln. Eine sehr empfindliche von DÄLLENBACH[1] angegebene Methode besteht darin, daß man die Apparatur innen unter einen geringen Überdruck von Ammoniakgas setzt und außen auf die verdächtigen Stellen in Mercuronitrat getränktes Papier bringt. An undichten Stellen zeigt das Papier nach einiger Zeit Schwärzung. Dieses Verfahren ist deswegen sehr empfindlich, weil man durch längeres Warten die austretenden Mengen über große Zeiten integrieren kann. Zur Ausübung dieses Verfahrens kann man auch die gesamte Apparatur mit angefeuchtetem Ozalidpapier (für Lichtpausen bekannt) umwickeln.

Die modernen Verfahren gestatten es, die beiden Aufgaben der Lecksuche und der Leckmessung gleichzeitig zu lösen. Ihr einheitliches Merkmal besteht darin, daß man die zu prüfende Vakuumapparatur (s. z.B. Fig. 28), und zwar insbesondere deren für Undichtigkeiten verdächtige Stellen, außen mit einem Testgas absprüht, das normalerweise nicht in der atmosphärischen Luft vorhanden ist. In der Vakuumapparatur ist dann ein Meßgerät untergebracht, das als spezifischer Indicator für das Testgas nur auf dieses anspricht. Trifft man beim Absprühen der Vakuumapparatur eine undichte Stelle, so dringt durch diese Undichtigkeit anstelle von atmosphärischer Luft Testgas in die Vakuumapparatur ein und der spezifische Indikator zeigt dann und nur dann einen Ausschlag, dessen Größe gleichzeitig ein Maß für die Größe der Undichtigkeit darstellt. Zur Messung der gesamten Undichtigkeit einer Vakuumapparatur kann man diese in einem zweiten Behälter unterbringen und den Zwischenraum mit dem Testgas fluten. Außerdem sollte der Zwischenraum ein Leck von einstellbarer oder bekannter Undichtigkeit (U_n) enthalten. Die Undichtigkeit (U_x) des Prüflings ergibt sich dann aus dem Indikatorausschlag bei Anschluß von Testleck und Prüf-

[1] W. DÄLLENBACH: DRP 503073; DRP 554038.

ling (A_1) und von Prüfling allein (A_x) zu:

$$U_x = \frac{A_x}{A_1 - A_x} U_n \text{ [Torr l/sec]}.$$

Darin ist

$$U = p_0 \cdot L,$$

p_0 = Atmosphärendruck, L = Gasdurchlässigkeit des Lecks.

Ein wesentlicher Vorteil des Verfahrens besteht darin, daß man kleine Lecks gleichzeitig neben großen Lecks nachweisen und die Größe der einzelnen Lecks

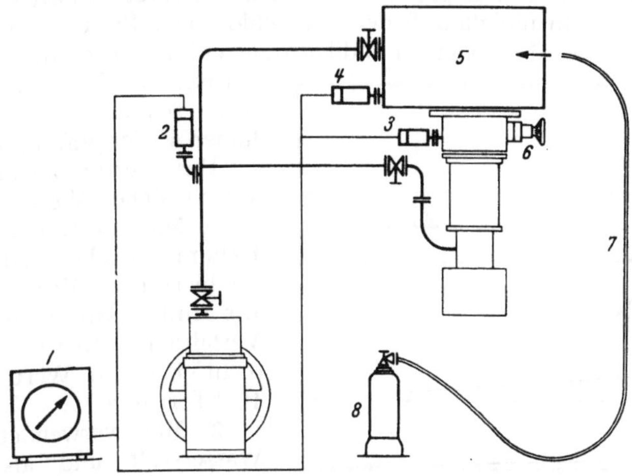

Fig. 28. Leck-Such- und Meßverfahren. Absprühen der Vakuumapparatur mit einem Testgas während an die Vakuumapparatur selbst eine Meßröhre angebracht ist, die spezifisch nur auf das benutzte Testgas anspricht. *1* Anzeigegerät; *2, 3, 4* Meßröhren, die speziell auf das Testgas ansprechen; *5* Vakuumrezipient; *6* Hochvakuumventil zwischen Diffusionspumpe und Rezipient *5*; *7* bewegliche Schlauchleitung zum Absprühen der Vakuumapparatur mit dem Testgas; *8* Stahlflasche mit Testgasvorrat.

zahlenmäßig bestimmen kann. Die Größe und der zeitliche Verlauf des Ausschlages am Indikator A hängt dabei außer von der Undichtigkeit U noch von dem Verhältnis:

$$\frac{\text{Sauggeschwindigkeit der Pumpe}}{\text{Volumen des Prüflings}} = \frac{S}{V}$$

ab. Es ist unmittelbar einleuchtend, daß die Zeit zwischen dem Besprühen eines Lecks, bis zur Bereitschaft für einen neuen Test auch von dem Verhältnis S/V abhängt, da bei dem Überstreichen eines Lecks mit der Sprühdüse der Prüfling zunächst mit dem Testgas überflutet wird. Die Abhängigkeit des Partialdruckes p des Testgases in der Vakuumapparatur und damit des Ausschlages A des Indikators von der Zeit t ergibt sich folgendermaßen. Der Zustrom an Testgas in den Prüfling mit dem Volumen V ist gegeben durch

$$d(p\,V) = (p_0 L - p\,S)\,dt.$$

$$\text{für} \quad t < T: \quad p_t = \frac{L}{S} p_0 \left(1 - e^{-\frac{S}{V} t}\right),$$

$$\text{für} \quad t > T: \quad p_t = \frac{L}{S} p_0 \left(1 - e^{-\frac{S}{V} T}\right) e^{-\frac{S}{V}(t-T)},$$

$$\text{für} \quad \begin{matrix} T \to \infty \\ t \to T \end{matrix} \quad p_\infty = \frac{L}{S} p_0.$$

Dabei ist T die Zeit, während der das Leck besprüht wird. Praktisch sollte die Bewegung der Sprühdüse über die Oberfläche des Prüflings mit einer Geschwindigkeit von etwa 1 cm/sec erfolgen.

Fig. 29 zeigt den Indikatorausschlag A in der Abhängigkeit von der Zeit t für $T=1$ sec und verschiedene Werte von S/V in Prozenten des Gleichgewichtsausschlages $p_\infty = \frac{S}{L} p_0$. Der Gleichgewichtsausschlag ist natürlich um so größer, je kleinere Werte die Sauggeschwindigkeit der angeschlossenen Pumpe hat. Andererseits benötigt man für eine schnelle Testbereitschaft möglichst große Werte von S/V. Ausschlagsgröße und schnelle Testbereitschaft sind also bezüglich der Sauggeschwindigkeit S der angeschlossenen Pumpe miteinander konkurrierende Größen. Je größer der Indikatorausschlag, um so länger ist die Zeit bis zur nächsten Testbereitschaft. Diese ist um so günstiger, je kleiner das Volumen V der Vakuumapparatur ist.

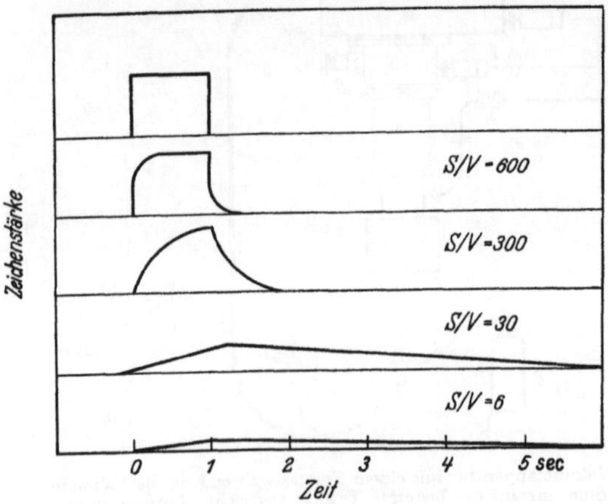

Fig. 29. Signalform im Lecksucher in Abhängigkeit von dem Verhältnis Sauggeschwindigkeit der Diffusionspumpe S zum Volumen V des Prüflings (Besprühzeit $T=1$ sec).

Das Testgasverfahren wird in verschiedener Weise ausgeübt:

1. Man verwendet als Testgas Helium und als spezifischen Indikator hierfür ein Massenspektrometer. Die untere Grenze der mit diesem Verfahren noch nachweisbaren Undichtigkeit hat Werte von 10^{-8} bis 10^{-9} Torr l/sec [1].

2. Man verwendet als Testgas Wasserstoff und als spezifischen Indikator entweder ein Penning-Vakuummeter (Philips-Vakuummeter) oder ein normales Ionisationsmanometer, die beide von der übrigen Vakuumapparatur durch eine Wand von hocherhitztem Palladium getrennt sein müssen. Durch die Verwendung der glühenden Palladiumwand sprechen in beiden Fällen die Vakuummeter speziell nur auf das Testgas Wasserstoff an. Die kleinste mit diesem Verfahren noch erfaßbare Undichtigkeit hat einen Wert von 10^{-7} Torr l/sec [2].

3. Man kann als Testgas ein Halogen und als spezifischen Indikator die Emission von Alkaliionen von einer glühenden Platinanode benutzen. Die untere Grenze für die noch erfaßbaren Undichtigkeiten liegt hier bei 10^{-6} Torr l/sec. Sie reicht für alle technischen Vakuumanlagen im Fein- und Hochvakuumgebiet aus [3].

II. Die Pumpen.

a) Allgemeine Fragen.

13. Charakteristische Größen. Zur Charakterisierung einer Pumpe dienen folgende Größen:

[1] H. A. THOMAS, H. A. WILLIAMS u. I. A. HIPPLE: Rev. Sci. Instrum. **17**, 368 (1946). — O. A. NIER, C. M. STEVENS, A. HASTRULID u. T. A. ABBOT: J. Appl. Phys. **18**, 30 (1947). — R. B. JACOBS u. H. F. ZUHR: J. Appl. Phys. **18**, 34 (1947). — D. E. CHARPENTIER: Vac. Symposium Trans. 1956, 114. New York: Pergamon Press 1957. — R. E. MOODY: Vac. Symposium Trans. 1956, 119. New York: Pergamon Press 1957. — J. PETERS u. F. RAIBLE: Vac. Symposium Trans. 1956, 107. New York: Pergamon Press 1957.

[2] H. NELSON: Rev. Sci. Instrum. **16**, 273 (1945). — F. M. PENNING u. K. NIENHUIS: Philips techn. Rdsch. **11**, 116 (1949).

[3] Brit. Thomson Houston Co.: Activities 1952, 81. — Dest. Prod. Inc.: J. Appl. Phys. **23**, 12 (1952).

1. Die Förderleistung oder Saugleistung Q. Dies ist die von der Pumpe in der Zeiteinheit geförderte Gasmenge gemessen in g/sec oder Torr l/sec.

2. Die Sauggeschwindigkeit, das ist das von der Pumpe in der Zeiteinheit geförderte Gasvolumen S [in l/sec oder m^3/h] gemessen beim Druck p an der Saugseite der Pumpe.

Danach ist die Förderleistung

$$Q = p\,S.$$

3. Endvakuum oder Enddruck einer Pumpe heißt derjenige Druck, der sich nach längerem Pumpen an der Saugseite stationär einstellt, wenn die Pumpe vom Rezipienten abgetrennt ist und kein zusätzliches Gas von außen eingelassen

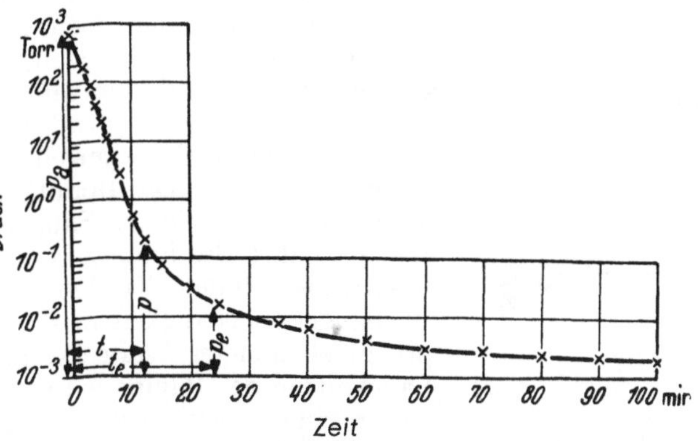

Fig. 30. An einem Behälter von bekanntem Rauminhalt gemessene Kurve für die Auspumpzeit zur Bestimmung der Sauggeschwindigkeit der angeschlossenen Pumpe.

wird. Dabei ist der Enddruck der Totaldruck, der sich aus dem Restgasdruck und dem Restdampfdruck zusammensetzt.

4. Der Vorvakuumdruck. Bei Pumpen, die (wie Dampfstrahl- und Diffusionspumpen) nicht gegen den Atmosphärendruck zu fördern vermögen, bezeichnet der Vorvakuumdruck denjenigen an dem Vorvakuumanschluß der Pumpe maximal zulässigen Druck, bei dessen Überschreitung die charakteristische Dampfströmung, die diesen Pumpvorgang bewirkt, zusammenbricht.

Die Messung der Sauggeschwindigkeit erfolgt bei mechanischen Pumpen in der Weise, daß man mit der Pumpe einen Behälter vom Rauminhalt V evakuiert und dabei die erreichten Druckwerte p in Abhängigkeit von der Zeit t aufnimmt (s. Fig. 30). Dabei ist darauf zu achten, daß die Pumpe durch eine möglichst kurze und weite Leitung mit dem Behälter verbunden ist, da sonst längs der Leitung Druckgefälle auftreten (s. unten), die die Messung verfälschen würden. Nehmen wir an, daß das von der Pumpe pro Zeiteinheit geförderte Volumen S klein ist gegen den Rauminhalt des Behälters V, so ist nicht dieses geförderte Volumen S, sondern nur die Druckabnahme dp unmittelbar erfaßbar. Es gilt die Beziehung

$$p\,V = (p + dp)\,(V + S\,dt) \tag{13.1}$$

oder

$$\frac{dp}{dt}\,V = -\,S\,p \tag{13.2}$$

oder

$$S = V\left(-\,\frac{d\ln p}{dt}\right). \tag{13.3}$$

Die Kurve für die Abhängigkeit der Sauggeschwindigkeit vom Druck ergibt sich also durch Differentiation der Druck-Zeit-Kurve. Die Sauggeschwindigkeit ist in solchen Bereichen konstant, in denen die Druck-Zeit-Kurve (Fig. 30) geradlinig verläuft. Bei Annäherung an das Endvakuum geht die Sauggeschwindigkeit $S \to 0$; die Druck-Zeit-Kurve geht asymptotisch gegen die Horizontale.

Andererseits kann man beim Vorliegen der Kurve für die Sauggeschwindigkeit in Abhängigkeit vom Druck die Evakuierungszeit für einen Behälter vorgegebener Größe V durch Integration berechnen:

$$t_e = V \int_{p_e}^{p_a} \frac{d \ln p}{S}. \tag{13.4}$$

Die vorstehend angegebene Methode zur Bestimmung der Sauggeschwindigkeit gilt exakt nur im Grobvakuumgebiet. Im Gebiet des Fein- und erst recht des Hochvakuums ist aber die Zahl der Moleküle, die sich frei im Gasraum bewegen, gleich oder sogar kleiner als die Zahl derjenigen Moleküle, die von den Wänden des Behälters adsorbiert sind, selbst wenn man nur das Vorliegen einer monoatomaren Schicht annimmt. Die Benutzung der eben besprochenen Methode zur Sauggeschwindigkeitsmessung führt also in diesem Gebiet infolge der Gasabgabe von den Wänden zu Fehlmessungen. Die Sauggeschwindigkeitsmessung erfolgt hier in prinzipiell anderer Weise (s. Fig. 31). An den Saugstutzen der Pumpe wird ein Vakuummeter zur Messung des Ansaugdruckes p und ein Nadelventil *1* angeschlossen. Die pro Zeiteinheit durch das Nadelventil eingelassene kleine Luftmenge v kann nach Schließen des Hahnes *2* am Fortschreiten der Sperrflüssigkeit *4* in der kalibrierten Kapillare *3* gemessen werden. Nach Umrechnung der eingelassenen Gasmenge v vom Atmosphärendruck (760 Torr) auf den Ansaugdruck p ergibt sich also für die Sauggeschwindigkeit S die Beziehung

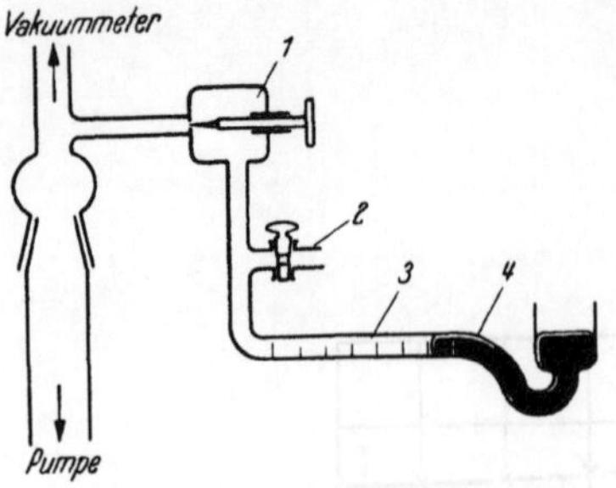

Fig. 31. Anordnung zur Sauggeschwindigkeitsmessung im Fein- und Hoch-Vakuum mittels einer kalibrierten Kapillare. *1* Nadelventil zum Gaseinlaß; *2* Absperrhahn; *3* kalibrierte Kapillare; *4* Sperrflüssigkeit.

$$S = \frac{760}{p} v. \tag{13.5}$$

Aber auch diese Methode zur Bestimmung der Sauggeschwindigkeit im Fein- und Hochvakuumgebiet ist mit einer erheblichen Problematik verbunden. Hierauf haben insbesondere DAYTON[1] und neuerlich BLEARS[2] und VENEMA[3] hingewiesen[4]. Von einem entscheidenden Einfluß ist vor allem die Lage und Form des Gaseinlasses relativ zur Lage des Vakuummeßinstrumentes. Theoretisch korrekt müßte zwischen Gaseinlaß und Sauganschluß der Pumpe ein so großes Ausgleichsgefäß angeschlossen werden, daß keinerlei Gasströmung auftritt und der Druck der Gase unabhängig von einer durch die Strömung bedingten Vorzugsgeschwindigkeit der Gasmoleküle gemessen würde. Die Einschaltung eines so großen Behälters ist aber infolge der oben genannten Gasabgabe durch die Wände wiederum nicht möglich. OATLEY[5] versucht die auftretenden Schwierig-

[1] B. B. DAYTON: Industr. Engng. Chem. **40**, 795 (1948). — Rev. Sci. Instrum. **19**, 533 (1948).

[2] J. BLEARS u. R. W. HILL: Rev. Sci. Instrum. **19**, 847 (1948); s. auch C. E. RUFER: C.V.T. Vac. Symposium Trans. 1956.

[3] A. VENEMA: Vacuum **4**, 272 (1954).

[4] Siehe auch den Sitzungsbericht der Arbeitsgemeinschaft Vakuum. Phys. Verh. **6**, 28 (1955).

[5] C. W. OATLEY: Brit. J. Appl. Phys. **5**, 358 (1954).

keiten dadurch zu umgehen, daß er nicht den Ansaugdruck und die eingelassene Gasmenge, sondern den durch die Gasströmung an einer Blendenöffnung auftretenden Druckabfall mißt.

Allen Messungen der Sauggeschwindigkeit gemeinsam ist die Schwierigkeit, daß in den gemessenen Wert unmittelbar die Größe des gemessenen Ansaugdruckes eingeht. Dadurch ist eine erhebliche Unsicherheit der absoluten Größe bedingt, da Absolutmessungen von Drucken im Fein- und Hochvakuumgebiet nach dem oben unter Vakuummeßinstrumenten Gesagten mit erheblichen Fehlern behaftet sind.

b) Mechanische Pumpen.

Bei diesen beruht der Pumpvorgang auf einer periodischen Vergrößerung und Verkleinerung des Schöpfraumes mittels mechanischer Vorgänge (mechanische Pumpen). Sie vermögen durchwegs gegen den vollen Atmosphärendruck zu komprimieren. Ihr Endvakuum setzt sich zusammen aus dem Restgasdruck und dem Restdampfdruck. Der Restgasdruck ist teilweise bedingt durch den Gasrücktritt von der Druck- auf die Saugseite infolge des schädlichen Raumes. Er rührt außerdem von den Undichtigkeiten (z. B. zwischen den einzelnen gegeneinander beweglichen Teilen der Pumpen) her. Der Restdampfdruck stammt im wesentlichen vom Dichtungsmittel. Fig. 32 gibt eine Übersicht über die Arbeitsbereiche der verschiedenen Pumpentypen.

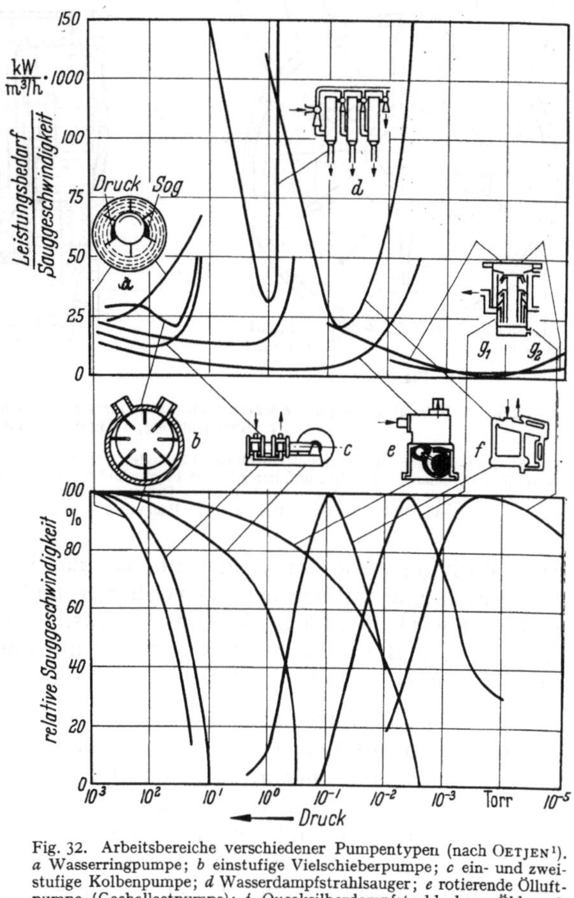

Fig. 32. Arbeitsbereiche verschiedener Pumpentypen (nach OETJEN[1]). *a* Wasserringpumpe; *b* einstufige Vielschieberpumpe; *c* ein- und zweistufige Kolbenpumpe; *d* Wasserdampfstrahlsauger; *e* rotierende Ölluftpumpe (Gasballastpumpe); *f* Quecksilberdampfstrahl- bzw. Öldampfstrahlpumpe; g_1 Treibdampfpumpe (Booster); g_2 Diffusionspumpe.

14. Flüssigkeitsringpumpen.

Ihre bekanntesten Vertreter sind die Wasserringpumpen. Die Pumpwirkung kommt dadurch zustande, daß ein aus verschiedenen Kammern bestehender Rotor sich, exzentrisch gelagert, in einem zylindrischen Gehäuse dreht und daß ein durch die Fliehkraft bedingter Flüssigkeitsring die verschiedenen Kammern des Rotors nach außen absperrt. Wie Fig. 33 zeigt, werden dadurch die einzelnen Kammern des Rotors bei seiner Drehung periodisch vergrößert und wieder verkleinert, was bei

[1] G. W. OETJEN: Ullmanns Encyklopädie der technischen Chemie, Bd. 1, S. 126d. 3. Aufl. München u. Berlin: Urban & Schwarzenberg 1951.

entsprechendem Anschluß an Saug- und Druckstutzen der Pumpe zur Pumpwirkung ausgenutzt werden kann.

15. Kolben- oder Stiefelpumpen. Diese Pumpenart wird ein- oder zweistufig gebaut. Der Pumpvorgang kommt dadurch zustande, daß ein hin und her gehender Kolben einen Schöpfraum periodisch vergrößert und verkleinert. Dieser Schöpfraum wird durch gesteuerte oder selbsttätige Unter- und Überdruckventile abwechselnd an die Saug- oder Drucköffnung angeschlossen (Fig. 34).

16. Die Töpler-Pumpe (s. Fig. 35). Dies war lange Zeit hindurch die einzige Pumpe, mit der es gelang, in das Fein- und Hochvakuumgebiet vorzudringen.

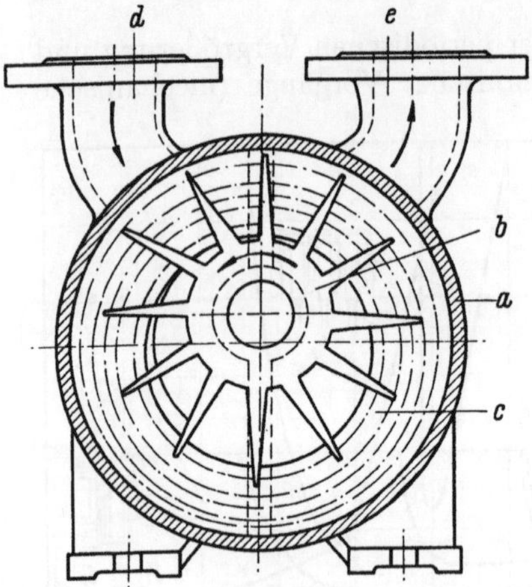

Fig. 33.

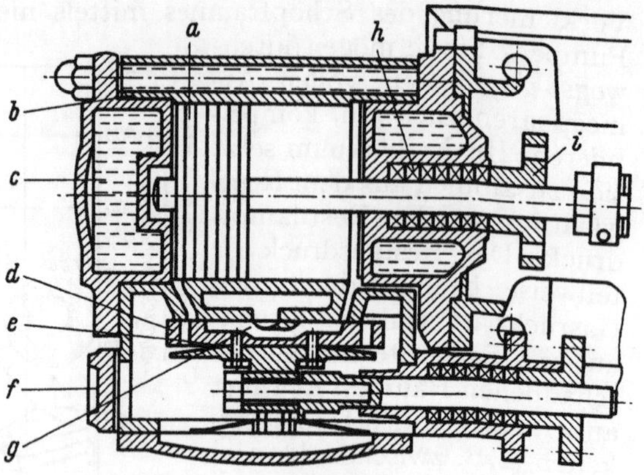

Fig. 34.

Fig. 33. Flüssigkeitsringpumpe schematisch (nach OETJEN[1]). *a* Gehäuse; *b* exzentrisch gelagertes Zellenrad; *c* Flüssigkeitsring; *d* Ansaugöffnung; *e* Auspuff.

Fig. 34. Kolben- oder Stiefelpumpe. *a* Kolben; *b* Zylinder. Die Gaszuführung erfolgt über den T-förmigen Raum in der Wand des Zylinders *b*. Der hin und her gehende Flachschieber. *d* stellt abwechselnd die Verbindung des Schöpfraumes mit der Ansaugöffnung oder dem Auspuff *f* her. *g* Druckausgleichskanal zur Herabsetzung der Wirkung des schädlichen Raumes. Dichtung des Kolbens *a* durch Packungen *h* und die Stopfbuchse *i* (Schema nach OETJEN).

Ihre Wirkungsweise beruht darauf, daß das in einem Behälter *A* vorhandene Gas durch Quecksilber hieraus verdrängt und auf die Druckseite der Pumpe gefördert wird. Durch Senken des Quecksilberspiegels kann dann anschließend dieser Raum *A* wiederum evakuiert und mit der Saugöffnung der Pumpe verbunden werden, worauf das Spiel aufs neue beginnt.

17. Vielschieberpumpen (s. Fig. 36). Bei diesen Pumpen dreht sich ein Rotor, exzentrisch gelagert, in einem zylindrischen Gehäuse. In seinen Schlitzen sind beweglich Schieber angebracht, die durch die Fliehkraft gegen die zylindrische Gehäusewandung gedrückt werden und auf diese Weise zusammen mit dem Rotor eine große Anzahl von Kammern bilden, die bei der Drehung periodisch vergrößert und verkleinert werden. Durch Anschluß an Saug- und Drucköffnung der Pumpe in den geeigneten Zeitmomenten kommt damit die Pumpwirkung zustande. Diese Pumpenart hat trotz des relativ hohen Enddruckes vor allem Bedeutung als Schnelläuferpumpe, d.h. also, mit diesem Pumpentyp kann bei verhältnismäßig geringen Abmessungen trotzdem eine große Sauggeschwindigkeit erzielt werden. Eine Kurve der Sauggeschwindigkeit in Abhängigkeit vom Ansaugdruck ist in Fig. 37 wiedergegeben.

[1] Siehe Fußnote 1, S. 547.

18. Rotierende Ölluftpumpen[1]. Dieser Pumpentyp ist von besonderer Bedeutung für das Fein- und Hochvakuumgebiet. Er überbrückt einen sehr großen Druckbereich. Die rotierenden Ölluftpumpen vermögen einerseits gegen Atmosphärendruck zu arbeiten und erreichen andererseits *einstufig* Restgasdrucke von $2 \cdot 10^{-3}$ Torr, Restdampfdrucke je nach der Art des Öles von etwa 10^{-2} Torr, *zweistufig* Restgasdrucke von 10^{-5} Torr, Restdampfdrucke je nach der Art des Öles von etwa 10^{-3} Torr. Die hohen Endvakuumwerte, die mit ihnen zu erreichen sind, beruhen vor allem auf folgendem:

1. Durch die Ölüberlagerung des Ventils wird verhindert, daß atmosphärische Luft durch die Dichtungsflächen des Ventils in den Schöpfraum der Pumpe zurücktreten kann.

2. Aus dem Ölvorrat über dem Ventil wird durch einen Kanal laufend Öl in den Schöpfraum der Pumpe eingelassen und dient dort zur Ölabdichtung der gegeneinander bewegten Flächen und damit zur Trennung der Räume mit verschiedenen Drucken.

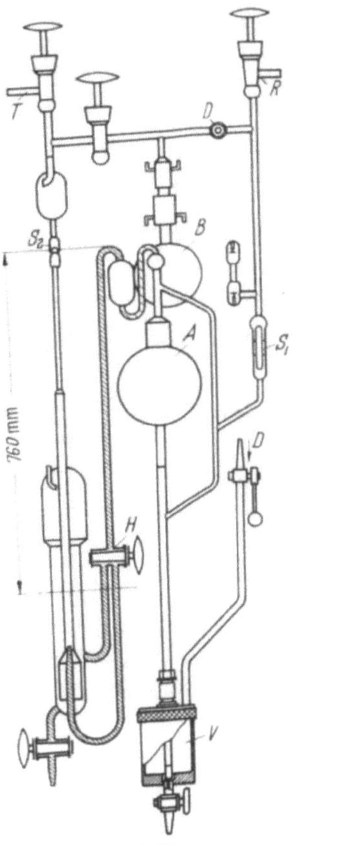

Fig. 35.

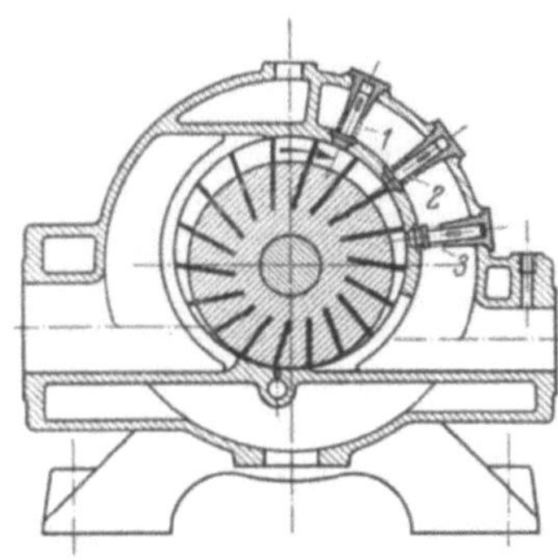

Fig. 36.

Fig. 35. Töpler-Pumpe (nach KOHLRAUSCH, Praktische Physik, 20. Aufl., Bd. 1, S. 261. Stuttgart 1955). Das angesaugte Gas tritt durch den Hahn *R* und das Rückschlagventil S_1 in die Kugel *A*. Durch Lufteinlaß bei *D* kann aus dem Raum *V* Quecksilber in die Kugel *A* gedrückt werden. Hierdurch wird das in *A* angesammelte Gas über den Zweiweg-Hahn *H* und das Kugelventil S_2 sowie den Hahn *T* nach außen gedrückt.

Fig. 36. Vielschieberpumpe (KLEIN, SCHANZLIN und BECKER).

3. Durch die Ölauffüllung des schädlichen Raumes wird dessen Einfluß und die dadurch bedingte Verschlechterung des Endvakuums weitestgehend ausgeschaltet (vgl. Fig. 38). Dabei tritt folgender Vorgang ein: Die Schieber *3* schieben nicht nur Gas, sondern auch das durch den Kanal *8* in den Schöpfraum tretende Öl vor sich her. Beim Endvakuum sind die geförderten Gasmengen nur noch so gering, daß ihr Druck nicht mehr zur Öffnung des Ventils *5* ausreicht. Durch den Kanal *8* dringt aber schließlich soviel Öl in den Schöpfraum, daß in der obersten Stellung der Ventilkanal *7*, d.h. der schädliche Raum zwischen

[1] W. GAEDE: Z. Naturforsch. **2a**, 233 (1947). — W. H. KOHL: Iron Age **164**, 59 (1949). — R. JAECKEL u. H. G. NOELLER: Z. VDI **94**, 797 (1952). — W. C. FRYE: C.V.T. Vac. Symposium Trans. 1954.

dem Ventil *5* und dem vorderen der Schieber *3* ganz mit Öl gefüllt ist. Im schädlichen Raum sind also jetzt die Gasreste durch Öl verdrängt. Außerdem dient

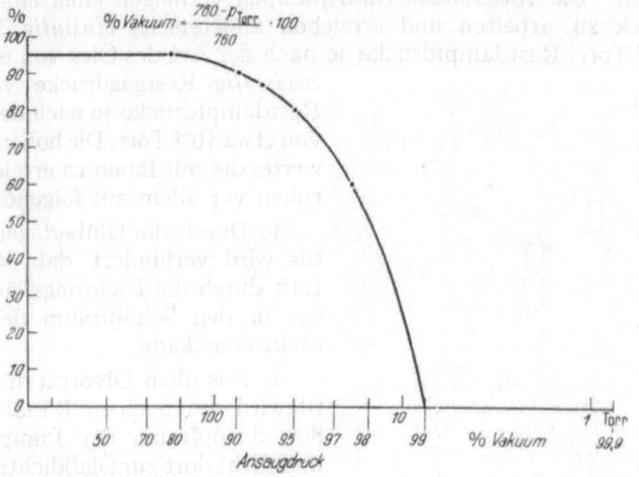

Fig. 37. Abhängigkeit der Sauggeschwindigkeit vom Ansaugdruck bei einer Vielschieberpumpe. Ordinaten: Sauggeschwindigkeit in Prozenten der theoretisch aus Schöpfraum mal Tourenzahl berechneten.

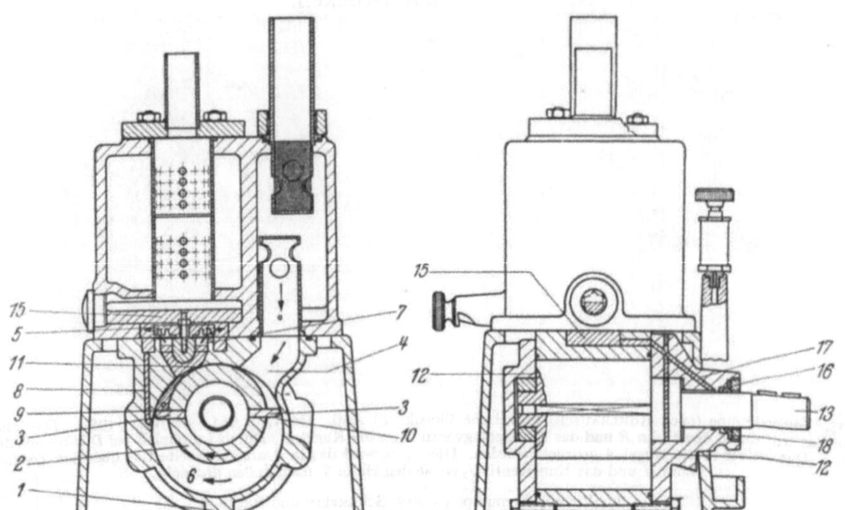

Fig. 38. Einstufige rotierende Ölluftpumpe (Gasballastpumpe, Leybold). Endvakuum: $2 \cdot 10^{-3}$ ($5 \cdot 10^{-2}$ Torr) ohne Gasballast, besser als 1 Torr mit Gasballast. Sauggeschwindigkeit: 2 bis 10 m³/h. Wirkungsweise: In dem zylindrischen Gehäuse *1* dreht sich ein exzentrisch gelagerter, geschlitzter Rotor *2* in Richtung des Pfeiles. In dem Schlitz des Rotors befinden sich durch Federn auseinandergedrückte Schieber *3*, die bei der Drehung an der Gehäusewand entlanggleiten und dabei die an der Saugöffnung *4* eingedrungene Luft vor sich herschieben, um sie schließlich durch das Druckventil *5* wieder auszustoßen. Im Raum *4* herrscht also ein niedrigerer Druck als im Raum *6* und im Raum *6* ein niedrigerer Druck als im Raum *7*. Ölrückführung durch den Kanal *8* in den Raum *6*. Die Schieber *3* fördern also nicht nur Gas sondern auch Öl. Dadurch erfolgt Ölabdichtung an den Flächen *9*, *10*, *11* und *12* (Ziffer *12* nur in der rechten Zeichnung angegeben).

dies nicht komprimierbare Öl zur zwangsläufigen Öffnung des Ventils. Die Wirkung des schädlichen Raumes wird also weitgehend herabgesetzt. Der noch übrigbleibende Einfluß rührt von den im Öl gelösten Gasresten her. Die Einstellung der richtigen Ölumlaufmenge ist kritisch und dabei temperaturabhängig. Zuviel Öl verschlechtert infolge der gelösten Gasreste das Endvakuum. Zu wenig Öl füllt den Ventilkanal nicht vollständig aus. Der Ölumlauf hat zur Folge, daß beim Endvakuum, wenn die Pumpe praktisch kein Gas, sondern nur noch Öl fördert, ein außerordentlich hartes Öffnen des Ventils, der sog. *Ölschlag,* auftritt.

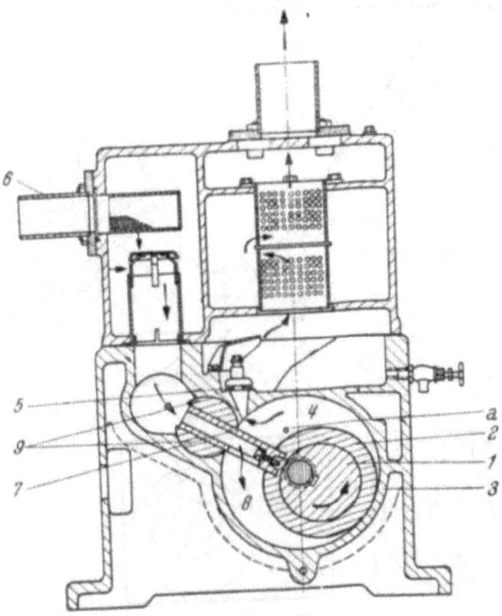

Fig. 39.

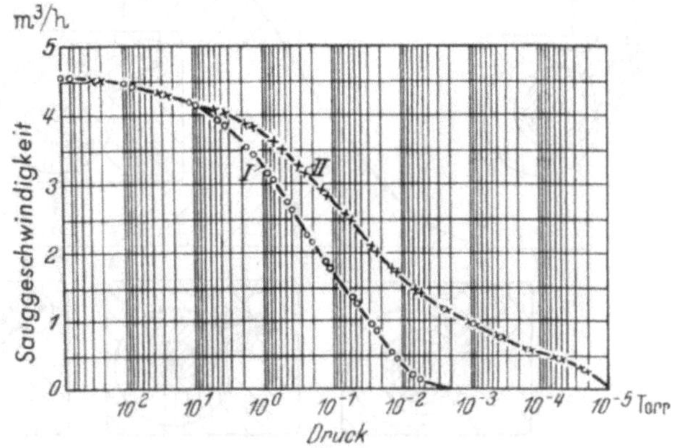

Fig. 40.

Fig. 39. Schnitt durch eine Pumpe mit exzentrisch rotierendem Kolben (Drehkolbenpumpe, Leybold, Modell XIV). Wirkungsweise: Ein Kolben *2*, der von einem Exzenter *1* mitgenommen wird, gleitet längs der Gehäusewand *3* und schiebt dabei die in dem Raum *4* befindliche Gasmenge vor sich her, bis sie durch das Ventil *5* ausgestoßen wird. Das abzusaugende Gas tritt bei *6* ein und gelangt durch die Bohrung des Schiebers *7* in den Raum *8*. Der Schieber *7* ist mit dem Kolben *2* fest verbunden und gleitet zwischen den Lamellen *9*, die ihrerseits in dem Gehäuse *3* drehbar sind, hin und her.

Fig. 40. Sauggeschwindigkeit einer einstufigen rotierenden Ölluftpumpe (Kurve I) und einer zweistufigen rotierenden Ölluftpumpe (Kurve II) in Abhängigkeit vom Ansaugdruck.

Diese unangenehme Erscheinung, die als hartes klapperndes Geräusch zu beobachten ist, kann dadurch weitgehend vermieden werden, daß unter der Ventilplatte des Ventils *5* ein kleines Luftpolster gehalten wird[1]. Die rotierenden Ölluftpumpen werden für kleine Sauggeschwindigkeiten als Drehschieberpumpen (s. Fig. 38) und für große Sauggeschwindigkeiten (Fig. 39) als Drehkolbenpumpen gebaut.

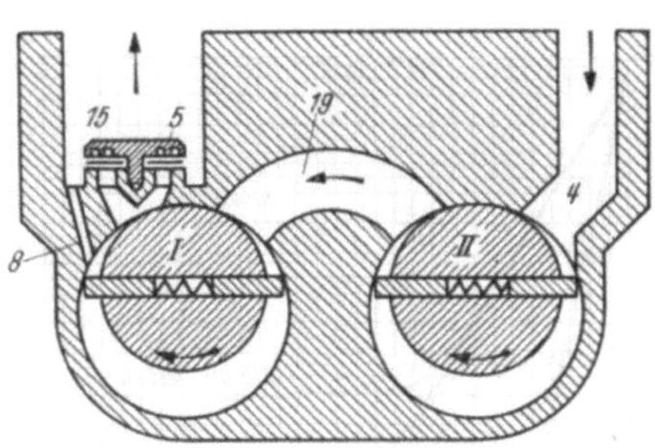

Fig. 41. Schema einer zweistufigen rotierenden Ölluftpumpe.

Den Arbeitsbereich der einstufigen Drehschieberpumpen zeigt Kurve I der Fig. 40. Dort ist die Sauggeschwindigkeit der einstufigen Drehschieberpumpen in Abhängigkeit vom Druck bei Betrieb ohne Gasballast angegeben (über Gasballast s. weiter unten). Bei den zweistufigen Drehschieberpumpen, deren Prinzip das Schema der Fig. 41 zeigt, befinden sich zwei gleichartige Pumpsysteme auf derselben Achse und sind vakuummäßig hintereinandergeschaltet. Das Pumpsystem II arbeitet dabei ohne Auspuffventil. Den Arbeitsbereich dieser zweistufigen Drehschieberpumpen zeigt Kurve II der Fig. 40. Ihr Vorzug besteht

[1] Siehe Vortrag auf der Physikertagung Heidelberg 1957, K. Brill: Phys. Verh. **8**, 7, 201 (1957).

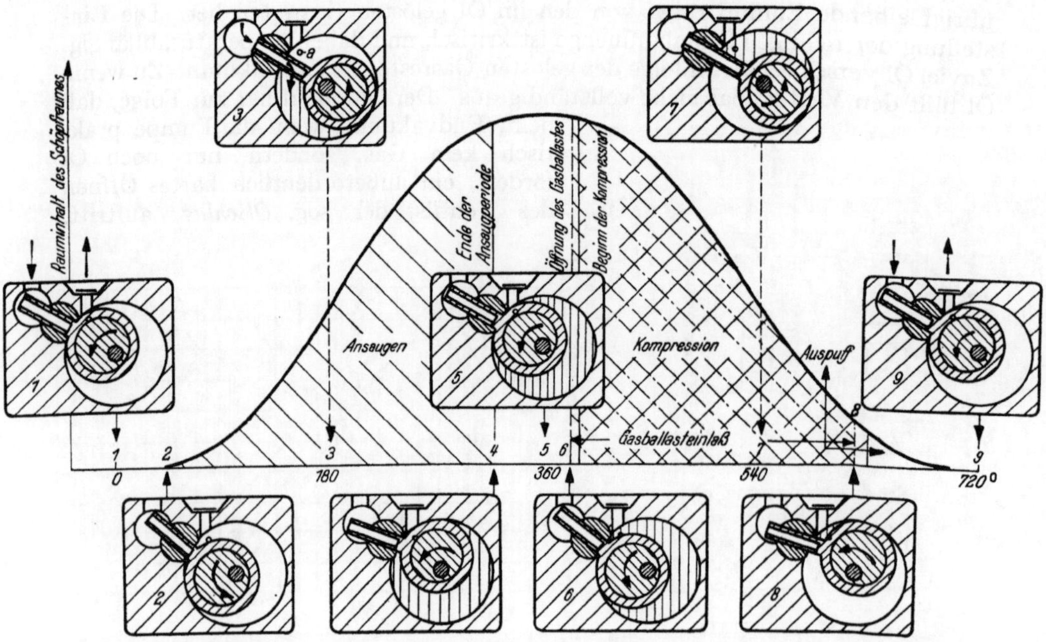

Fig. 42. Wirkungsweise der Drehkolbenpumpen. Als Abszisse ist der Drehwinkel des Exzenters und als Ordinate der zu jeder Exzenterstellung gehörige Rauminhalt des Schöpfraumes aufgetragen. Die Darstellungen zeigen schematisch das Unterteil der Pumpe (Fig. 39) in einigen charakteristischen Exzenterstellungen. In diesen Darstellungen ist auch die Einlaßöffnung für den Gasballast (Schema 3, mit *a* bezeichnet) angegeben, dessen Öffnung und Schließung durch die Stirnfläche des Pumpenkolbens gesteuert wird. Stellung 1: Oberer Totpunkt. Stellung 2: Der Schlitz im Saugkanal des Schiebers wird gerade freigegeben. Beginn der Ansaugperiode. Stellung 3: Unterer Totpunkt. Der Schlitz am Saugkanal ist ganz frei. Das abzusaugende Gas (Pfeile) kann frei in den Schöpfraum (schraffiert gezeichnet) gelangen. Stellung 4: Der Schlitz am Saugkanal wird gerade wieder durch die Lamellen verschlossen. Ende der Ansaugperiode. Stellung 5: Oberer Totpunkt. Maximaler Rauminhalt des Schöpfraumes. Stellung 6: Noch vor dem Beginn der Kompressionsperiode beginnt die Freigabe der Gasballastöffnung durch die Stirnfläche des Pumpenkolbens. Beginn des Gasballasteinlasses. Unmittelbar danach beginnt die Kompressionsperiode. Stellung 7: Gasballastöffnung ist ganz frei. Stellung 8: Ende des Gasballasteinlasses. Unmittelbar danach ist das kritische Kompressionsverhältnis (von 9) (s. Fig. 50) erreicht. Der Auspuff beginnt. Stellung 9: Ende einer Pumpperiode.

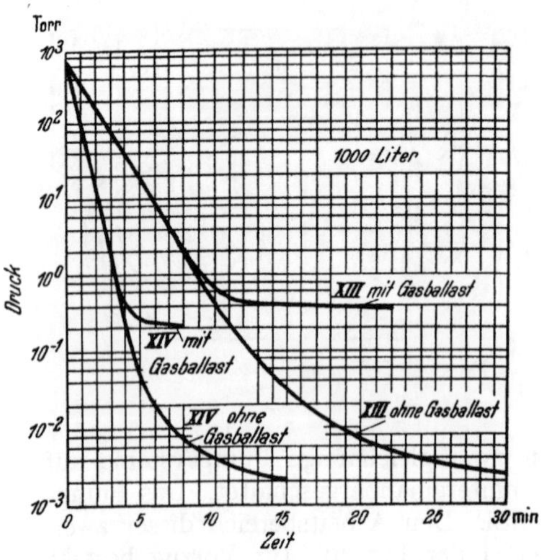

Fig. 43. Auspumpzeiten eines 1000 Liter-Behälters mit einstufigen Drehkolbenpumpen bei Betrieb mit und ohne Gasballast.

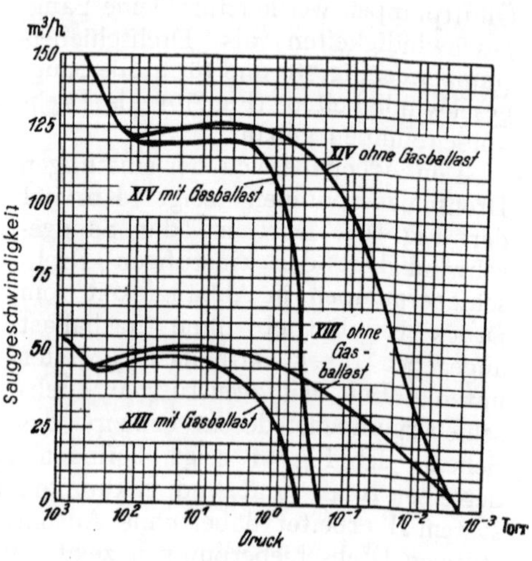

Fig. 44. Sauggeschwindigkeiten von einstufigen Drehkolbenpumpen mit und ohne Gasballast in Abhängigkeit vom Ansaugdruck.

außer der Erzielung niedriger Restgasdrucke darin, daß sie im Feinvakuumgebiet größere Sauggeschwindigkeiten haben als die einstufigen Pumpen, und zwar sowohl bei Betrieb mit als auch ohne Gasballast.

Die Wirkungsweise der Drehkolbenpumpen (Fig. 39), d.h. das Zustandekommen eines periodisch vergrößerten und verkleinerten Schöpfraumes und damit ihre Pumpwirkung geht aus Fig. 42 hervor. Die Fig. 43 zeigt die Evakuierungszeitkurven für solche Pumpen und Fig. 44 die Sauggeschwindigkeit in Abhängigkeit vom Ansaugdruck. Bei diesen großen Pumpen werden zweistufige Anordnungen nicht wie bei den Drehschieberpumpen aus zwei gleich großen hintereinandergeschalteten Einzelpumpen kombiniert, sondern durch Hintereinanderschaltung einer Pumpe großer Sauggeschwindigkeit (auf der Saugseite *S* 150) (s. Fig. 45) mit einer kleineren Type auf der Druckseite (*S* 5). Eine solche Anordnung ist möglich und zweckmäßig infolge der Vorkompression der Gase in der saugseitigen Pumpe *S* 150.

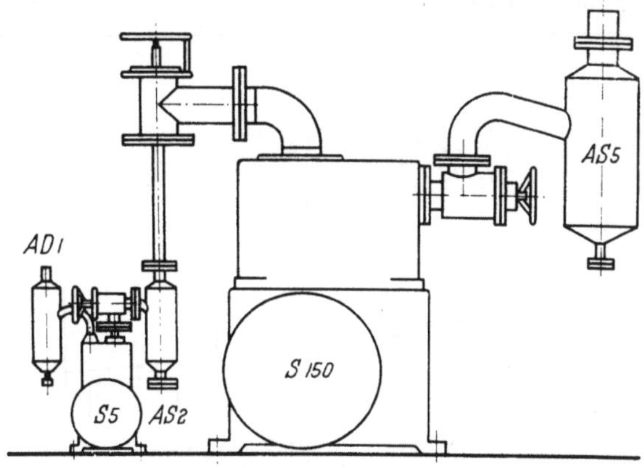

Fig. 45. Pumpenkombination. Bestehend aus einer Pumpe großer Sauggeschwindigkeit auf der Saugseite (*S 150*) und einer vorgeschalteten Gasballastpumpe mit kleinerer Sauggeschwindigkeit (*S 5*) als Vorpumpe. *AS 5* Flüssigkeitsabscheider auf der Saugseite, *AS 2* Flüssigkeitsabscheider im Zwischenvakuum, *AD 1* Flüssigkeitsabscheider auf der Druckseite.

Infolgedessen haben die Gase auf der Saugseite der druckseitigen Pumpe nur noch ein kleineres Volumen. Zur Vermeidung von gefährlichen Überdrucken beim Anlaufen einer solchen Pumpenkombination (Fig. 45) muß zwischen den beiden Pumpen ein Überdruckventil angebracht werden.

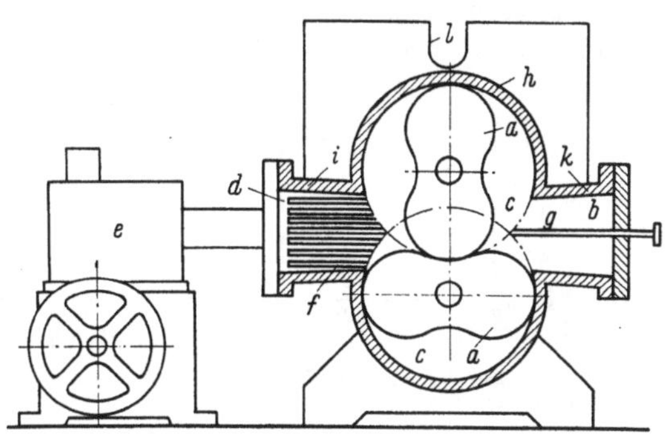

Fig. 46. Pumpenanordnung bestehend aus einer Roots-Pumpe mit Luftkühler im Auslaß und einer Gasballastvorpumpe. *a* achtförmiger Kolben; *b* Saugseite; *c* Schöpfraum; *d* Vorvakuumseite; *e* Gasballastvorpumpe; *f* Luftkühlung; *g* Thermoelement zur Temperaturmessung am Kolben (nach dem Stillsetzen der Pumpe); *h* Thermoelement zur Messung der Temperatur der Gehäusewand; *i*, *k* Druckmeßstellen; *l* Manometer zum Messen des Druckunterschiedes zwischen *i* und *k* (nach NOELLER und THEES).

19. Roots-Pumpen[1]. Die Roots-Pumpen, die schon lange als Verdichter bei höheren Drucken Verwendung fanden, haben neuerlich erheblich an Bedeutung gewonnen als Feinvakuumpumpen hoher Sauggeschwindigkeit, insbesondere mit einer kleinen rotierenden Ölluftpumpe auf der Druckseite als Vorpumpe (s. Fig. 46). Die Wirkungsweise der Roots-Pumpen geht aus Fig. 47 hervor. Bei ihnen drehen sich zwei achtförmige Kolben *a* gegenläufig in einem Gehäuse und wälzen sich dabei aufeinander und auf den Gehäusewandungen ab. Die Drehung

[1] A. LORENZ: Phys. Verh. **5**, 8, 182 (1954). — H. G. NOELLER: C.V.T. Vac. Symposium Trans. 1956. — C. M. VAN ATTA: C.V.T. Vac. Symposium Trans. 1956. — H. G. NOELLER u. R. THEES: Z. VDI **99**, 613 (1957).

der beiden achtförmigen Kolben *a* erfolgt dabei mit gleicher Umdrehungsgeschwindigkeit und mit konstanter Phasenverschiebung durch zwangsläufigen Antrieb über die beiden Achsen. Das auf der Saugseite *b* eindringende Gas wird bei der Drehung der achtförmigen Kolben in einem Schöpfraum *c*, der zwischen Kolben und Gehäusewandung gebildet wird, eingeschlossen. Bei weiterer Drehung kommt dann der Schöpfraum *c* mit dem Auspuffraum *d* in Verbindung. Das dort unter höherem Druck stehende Gas schlägt in den Schöpfraum *c* zurück und wird anschließend wieder zusammen mit dem von der Saugöffnung her geförderten Gas an der Auspuffseite ausgestoßen. Fig. 48 zeigt die periodische Verkleinerung des mit der Auspuffseite verbundenen Schöpfraumes. Da die beiden achtförmigen Kolben bei der weiteren Drehung ineinandergreifen, ist das freie Volumen des Schöpfraumes bei dem Übergang von der Druck- auf die Saugseite wesentlich kleiner als umgekehrt, so daß eine dauernde Förderung von der Saug- auf die Druckseite resultiert. In Fig. 49 ist die Sauggeschwindigkeit einer Roots-Pumpe in Abhängigkeit vom Ansaugdruck zusammen mit der Sauggeschwindigkeit der

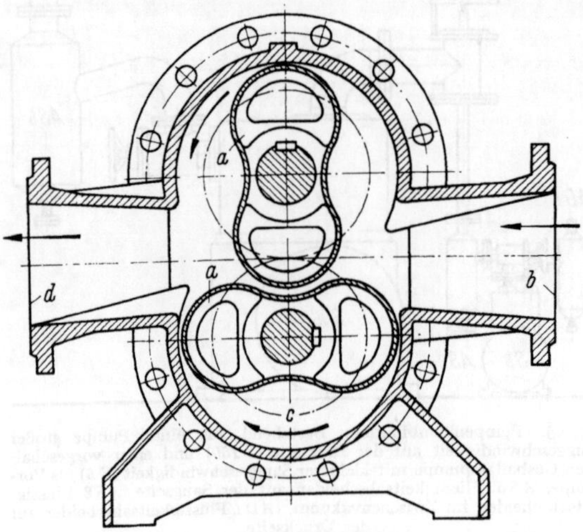

Fig. 47. Schnitt durch eine Roots-Pumpe (nach NOELLER und THEES). *a* achtförmiger Kolben; *b* Eintrittsquerschnitt; *c* Schöpfraum; *d* Austrittsquerschnitt.

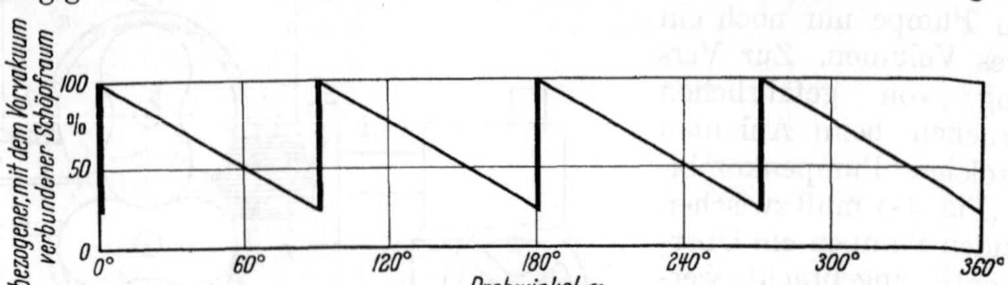

Fig. 48. Abhängigkeit des mit dem Vorvakuum verbundenen Schöpfraumes einer Roots-Pumpe vom Drehwinkel φ (nach NOELLER und THEES). Schöpfraum bezogen auf den maximal mit dem Vorvakuum verbundenen Schöpfraum.

Vorpumpe aufgetragen. Der Arbeitsbereich dieser Pumpentype reicht also etwa von 10^{-3} bis 10 Torr. Da bei dieser Pumpenart keine Flüssigkeiten zur Abdichtung der gegeneinander bewegten Teile eingesetzt werden, kann sie im Hinblick auf die geringe Reibungsarbeit als Schnelläuferpumpe gebaut werden. Sie ist daher besonders geeignet für hohe Sauggeschwindigkeiten bei kleinen Abmessungen. Die Abdichtung der druckseitigen von den saugseitigen Räumen erfolgt nur durch die Strömungswiderstände, die durch die relativ engen Spalte zwischen den abtrennenden Teilen gegeben sind. Die Strömungswiderstände sind natürlich besonders groß bei niedrigen Drucken, daher rührt die besondere Eignung der Pumpen als Feinvakuumpumpen. Andererseits bedingt dies aber auch, daß die Kompression in der Pumpe nicht allzu hoch getrieben werden kann. Die maximal erreichbaren Werte sind mit abhängig von der zweckmäßigerweise im Auspuffstutzen (s. Fig. 46) unterzubringenden Kühlung. Bei schwacher Kühlung erreicht man Kompressionsverhältnisse (Druck auf der Auspuffseite zu Druck auf der Saugseite) von 10 zu 1, bei guter Kühlung ein Mehrfaches dieses Wertes. Dabei ist ferner zu beachten, daß beim Anlauf einer Pumpenkombination nach Fig. 46 (solange also auf der Saugseite der Roots-Pumpe noch hohe Drucke vorliegen) zwischen den beiden Pumpen erhebliche Überdrucke gegenüber einer Atmosphäre auftreten können. Um diesen Übelstand und die dadurch bedingten Gefahren zu vermeiden, dient die Umwegleitung (l in Fig. 46) zwischen Druck und Saugseite der Roots-Pumpen, in der ein Überdruckventil angebracht ist, das sich beim Auftreten von unzulässig hohen Kompressionsdrucken öffnet und über die Umwegleitung l zu einem Druckausgleich führt.

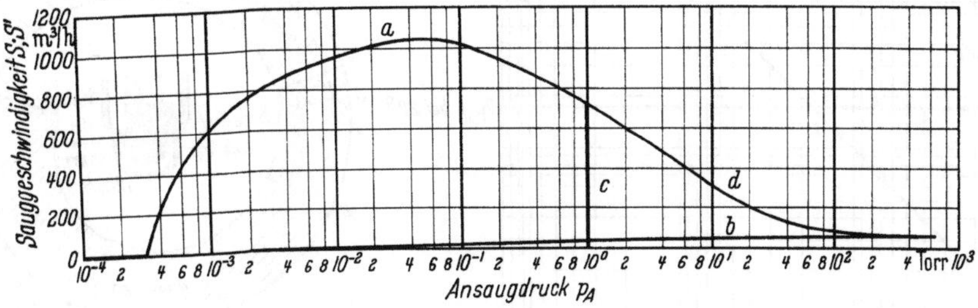

Fig. 49. Abhängigkeit der Sauggeschwindigkeit einer Roots-Pumpe mit Gasballastvorpumpe vom Ansaugdruck (nach NOELLER und THEES). a Verlauf der Sauggeschwindigkeit S der Roots-Pumpe; b Verlauf der Sauggeschwindigkeit S' der Gasballastpumpe; c Übergang von a auf b durch Abschalten der Roots-Pumpe bei Erreichen des maximal zulässigen Ansaugdruckes; d Verlauf von S mit Steuerung der theoretischen Sauggeschwindigkeit oder des volumetrischen Wirkungsgrades.

20. Absaugen von Dämpfen, der Gasballast[1]. Alle bis auf Atmosphärendruck komprimierenden mechanischen Pumpen, insbesondere auch die bewährten rotierenden Ölluftpumpen, haben den Nachteil, daß man zwar mit ihnen sehr gut Permanentgase absaugen kann, daß aber beim Absaugen von Dämpfen, z.B. von gesättigtem Wasserdampf, meistens eine Störung des Pumpvorganges eintritt. Da aber Wasserdampf bei Vakuumarbeiten in sehr vielen Fällen vorhanden ist, bedeutet dies eine ernste Komplikation. Sie rührt daher, daß die Arbeitstemperatur der Pumpen in den meisten Fällen unter 100° C liegt, daß also Wasserdampf bei der Kompression in der Pumpe schließlich gesättigt wird und, bevor er einen Wert von 760 Torr erreicht und damit in der Lage ist, das Auspuffventil zu öffnen, kondensiert. Näheres hierüber zeigt Fig. 50. Zur Verhinderung dieser störenden Komplikation bzw. zur Beseitigung der dadurch bedingten Mängel sind verschiedene Verfahren angewandt worden, z.B. Zentrifugierung des mit Wasser emulgierten Pumpenöls, Heizen der Pumpen auf Temperaturen über 100°, ständige Ölerneuerung und anderes. Am besten bewährt hat sich die von GAEDE angegebene Methode des Gasballastes. Diese unterscheidet sich von den anderen Methoden dadurch, daß sie nicht nachträglich den schädlichen Einfluß des kondensierten Wasserdampfes herabzusetzen versucht, sondern daß sie vielmehr eine Kondensation von Wasserdampf in den Pumpen vollständig vermeidet. Sie besteht darin, daß in den Schöpfraum der Pumpe dauernd eine dosierte Menge Frischluft (der sog. Gasballast) eingelassen wird und dadurch ein Druck von 760 Torr im Pumpraum erreicht wird, noch bevor der Wasserdampf auf den der Pumpentemperatur entsprechenden Sättigungsdruck komprimiert ist und damit zur Kondensation kommen kann. Dieser Frischlufteinlaß beginnt unmittelbar, nachdem der Schöpfraum vom Ansaugstutzen abgetrennt ist, wodurch eine zu weit gehende Verschlechterung des Endvakuums vermieden wird.

[1] W. GAEDE: l. c. — R. THEES: Vak. Techn. 6, 160 (1957).

Die Wirkungsweise der Gasballastmethode erläutert Fig. 50. Fig. 51 zeigt vergleichsweise den praktischen Kompressionsvorgang in einer Gasballastpumpe.

Die zur Verhinderung der Kondensation erforderliche Gasballastmenge B (l/sec) bei Atmosphärendruck p_{at} (Torr) hat GAEDE folgendermaßen berechnet:

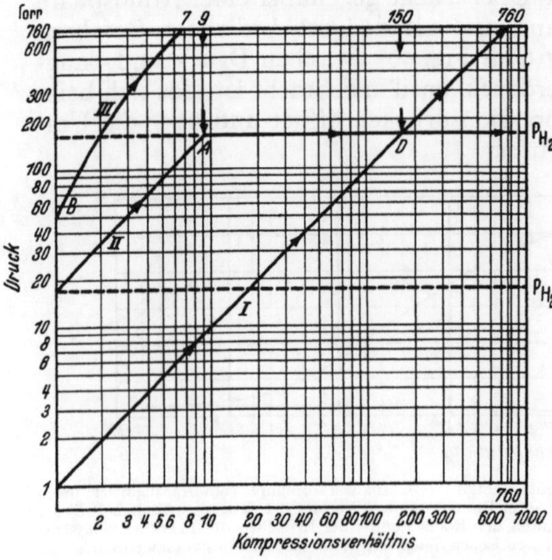

Fig. 50.

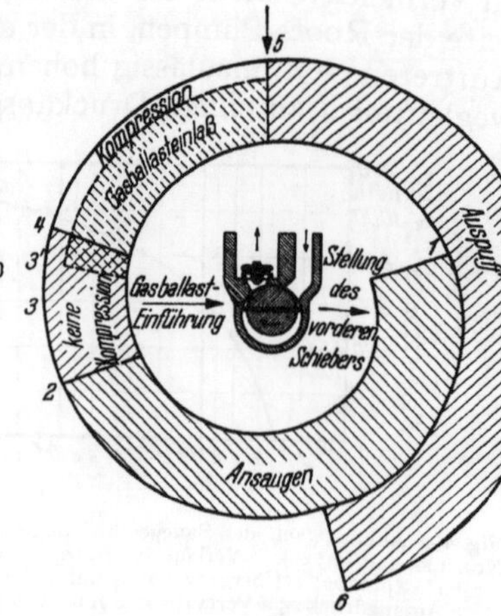

Fig. 51.

Fig. 50. Druckanstieg bei der Kompression in rotierenden Pumpen. Kompressionsverhältnis gleich Volumen des Schöpfraumes im Augenblick der Absperrung von der Saugseite zum Restvolumen in einem bestimmten Augenblick des Pumpvorganges. Kurve I: Beim Absaugen von permanenten Gasen steigt der Druck (Anfangsdruck im Beispiel 1 Torr) proportional zum Kompressionsverhältnis an, bis bei einem Druck von 760 Torr das Auspuffventil geöffnet wird. Kurve II und III: Absaugen von Dämpfen: Im Beispiel bei 20° C gesättigter Wasserdampf (Sättigungsdruck 17,5 Torr, Betriebstemperatur der Pumpe 60° C). Kurve II: Absaugen von Dämpfen mit normalen rotierenden Ölluftpumpen. Druck steigt zunächst proportional zum Kompressionsverhältnis an, bis (Punkt A) der Sättigungsdruck entsprechend der Betriebstemperatur der Pumpe erreicht ist (im Beispiel 150 Torr bei Kompressionsverhältnis 9). Bei weiterer Kompression Kondensation bei konstantem Druck. Druck von 760 Torr wird nicht erreicht. Auspuffventil bleibt geschlossen. Kurve III: Absaugen von Dämpfen mit Gasballastpumpen. Schon vor dem Beginn der Kompression wird Frischluft (Gasballast) in den Schöpfraum der Pumpe eingelassen. Beim Einsetzen der Kompression (Punkt B) ist also im Schöpfraum ein Dampf-Luft-Gemisch unter erhöhtem Druck (im Beispiel 50 Torr) vorhanden. Bei zunehmendem Kompressionsverhältnis steigt der Druck einerseits infolge der Kompression des Dampf-Luft-Gemisches und andererseits durch die dauernde weitere Gasballastzufuhr, bis schließlich (im Beispiel bei einem Kompressionsverhältnis 7) ein Druck von 760 Torr erreicht und damit das Auspuffventil geöffnet wird. Bei genügender Gasballastzufuhr geschieht dies, bevor der Dampf bis zur Sättigung komprimiert ist (im Beispiel bei einem Kompressionsverhältnis 9). Die Dämpfe werden also ohne Kondensation wieder aus der Pumpe herausgespült.

Fig. 51. Pumpvorgang in einer Gasballastpumpe. Die Spirale um die schematisch gezeichnete Pumpe gibt die Vorgänge in der Pumpe in Abhängigkeit von der Stellung des vorderen der beiden Schieber an. In Stellung I wird der Raum unter diesem Schieber vom Saugstutzen abgetrennt und der Raum über diesem Schieber bei weiterem Drehen der Pumpe allmählich vergrößert. Dargestellt ist das, was mit dem Raum über dem Schieber in Abhängigkeit von seiner Stellung geschieht. Stellung 1 bis Stellung 2: Ansaugen. Stellung 2: Abtrennung vom Saugstutzen durch den hinteren Schieber. Stellung 2 bis Stellung 3: Weitere Vergrößerung des Schöpfraumes. Stellung 3 bis Stellung 4: Schöpfraum wird wieder auf den Wert von Stellung 2 verkleinert. Stellung 3': Beginn des Gasballasteinlasses. Stellung 4 bis Stellung 5: Kompression unter dauerndem weiteren Gasballasteinlaß. Stellung 5: Das Auspuffventil öffnet sich. Stellung 5 bis Stellung 6: Auspuff der geförderten Dämpfe und des Gasballastes.

Für einen gasfreien Dampf ($p_{g,1} = 0$) gilt

$$S_1 p_{d,1} = S_5 p_{d,5}; \qquad S = S_1 = V_1 n \,[\text{l/sec}] \quad S_5 = V_5 n \tag{20.1}$$

$$p_{d,5} \leqq p_s \tag{20.2}$$

$$B p_{at} = S_5 p_{g,5}; \qquad p_{g,5} = p_{at} - p_{d,5} \tag{20.3}$$

$$B p_{at} = S_5 (p_{at} - p_{d,5}). \tag{20.4}$$

Hierbei bedeuten:

p_{at} = Atmosphärendruck (Torr),
p_g = Druck des Gases (Torr),
V = Schöpfraumvolumen (l),
n = Tourenzahl (sec^{-1}),
p_d = Druck des Dampfes (Torr),
p_s = Sättigungsdruck des Dampfes bei der Betriebstemperatur der Pumpe (Torr).

Die Zahlen beziehen sich auf die Schieberstellung in Fig. 51, also Stellung 1 Beginn der Ansaugperiode, Stellung 5 Beginn des Auspuffs nach Erreichen des Atmosphärendrucks im Schöpfraum.

Mit diesen Ansätzen ergibt sich weiterhin

$$B = S\,p_{d,1}\left(\frac{1}{p_{d,5}} - \frac{1}{p_{\text{at}}}\right), \tag{20.5}$$

$$B \geq S\,p_{d,1}\left(\frac{1}{p_s} - \frac{1}{p_{\text{at}}}\right). \tag{20.6}$$

Die erforderliche Gasballastmenge B ist also um so größer, je größer die Sauggeschwindigkeit der Pumpe S und je höher der Wasserdampfdruck am Ansaugstutzen $p_{d,1}$ ist. Sie kann um so kleiner bleiben, je höhere Werte der Sättigungsdruck des Wasserdampfes bei der Temperatur der Pumpe (p_s) hat. Die Gasballastmenge B wächst jedoch wieder mit dem Barometerstand p_{at}; sie wird Null ($B=0$) für den Fall $p_s = p_{\text{at}}$ (geheizte Pumpe).

Wird nicht gesättigter Wasserdampf, sondern ein Gas-Dampf-Gemisch abgesaugt, so gilt in Analogie zu oben

$$p_{\text{at}}\,B = S_5\left[p_{\text{at}} - (p_{g,1} + p_{d,1})\,\frac{S}{S_5}\right], \tag{20.7}$$

$$B = S_5\left[1 - \frac{p_{g,1} + p_{d,1}}{p_{\text{at}}}\,\frac{S}{S_5}\right]; \qquad \frac{S}{S_5} = \frac{p_{d,5}}{p_{d,1}}, \tag{20.8}$$

$$B \geq S\,p_{d,1}\left[\frac{1}{p_s} - \frac{1 + \frac{p_{g,1}}{p_{d,1}}}{p_{\text{at}}}\right]. \tag{20.9}$$

Die erforderliche Gasballastmenge B hängt in derselben Weise von der Sauggeschwindigkeit der Pumpe S, dem Wasserdampfdruck am Ansaugstutzen $p_{d,1}$ und dem Sättigungsdruck des Wasserdampfes bei der Temperatur der Pumpe p_s ab wie oben. Als neu kommt hinzu die Abhängigkeit von dem Verhältnis $p_{g,1}/p_{d,1}$, d.h. von der dem Dampf auf der Saugseite beigemischten Gasmenge. Die erforderliche Gasballastmenge ist jetzt um so kleiner, je größer die zugesetzte Gasmenge ist. $B=0$ wird erreicht für

$$\frac{p_{g,1}}{p_{d,1}} \geq \frac{p_{\text{at}}}{p_s} - 1. \tag{20.10}$$

Bei genügendem Gaszusatz ist also kein Gasballast mehr erforderlich (vgl. Fig. 52, Bereich A). Es ist üblich, die Gasballastmenge B so einzustellen, daß ein Druck für den gesättigten Wasserdampf $p_{d,1} \geq 30$ Torr zulässig ist. In den Fig. 52 und 53 sind für einstufige Drehschieberpumpen bzw. für Pumpenkombinationen beim Ansaugen von verschiedenen Wasserdampf-Luft-Gemischen (p_d Partialdruck des Wasserdampfes, p_g Partialdruck der Luft) die Arbeitsbereiche angegeben. Auf den Koordinatenachsen sind der Gaspartialdruck bzw. der Dampfpartialdruck aufgetragen. Die eingezeichneten Symbole geben an, unter welchen Bedingungen die Pumpen laufen müssen, damit bei dem vorliegenden Dampf-Gas-Gemisch der abgesaugte Dampf nicht in der Pumpe kondensiert wird.

Fig. 52 gilt für einstufige Drehschieberpumpen. Im Gebiet A ist der Anteil der nicht kondensierbaren Gase so groß, daß die Pumpe ohne Gasballast laufen kann. Im Gebiet B ist der Wasserdampfanteil so groß, daß die Pumpe selbst bei Betrieb mit Gasballast einen auf der Saugseite vorgeschalteten Kondensator zur Niederschlagung des Wasserdampfes erfordert. Dagegen ist im Gebiet C der

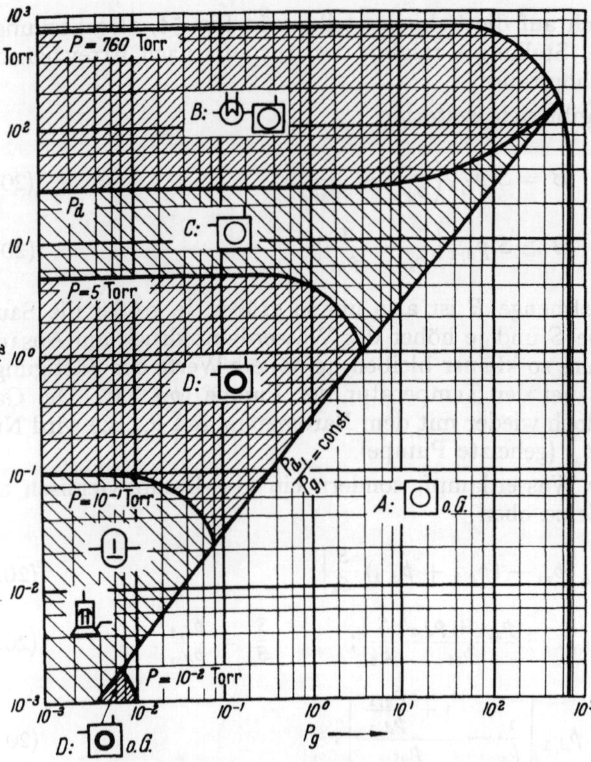

Fig. 52. Einsatzgebiete einstufiger und zweistufiger Gasballastpumpen nach dem Drehschieberprinzip (nach THEES). p_g Permanentgasdruck; p_d Wasserdampfdruck.

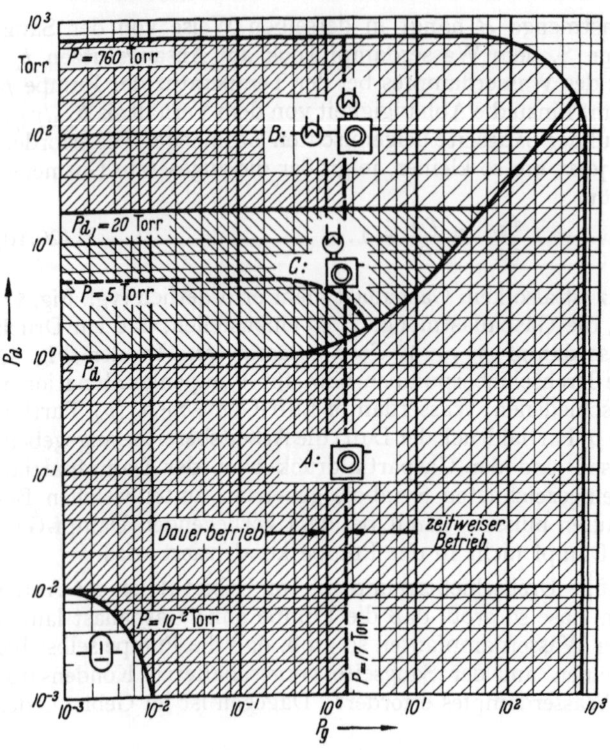

Fig. 53. Einsatzgebiete für eine Pumpenkombination bestehend aus einer Drehkolbenpumpe großer Sauggeschwindigkeit mit einer Gasballastpumpe kleiner Sauggeschwindigkeit als Vorpumpe nach THEES. Die Trennungslinie zwischen Dauerbetrieb und zeitweisem Betrieb ist irrtümlich bei $p = 1{,}7$ Torr statt wie richtig $p = 17$ Torr gezeichnet.

Wasserdampfanteil geringer; die Pumpe kann ohne Kondensator, muß jedoch mit Gasballast betrieben werden. Einstufige Pumpen erzielen unter diesen Bedingungen Endvakuumdrucke von etwa 5 Torr. Will man noch niedrigere Drucke erreichen, so muß man zweistufige Drehschieberpumpen im Gebiet *D* oder schließlich bei den ganz niedrigen Drucken Roots-Pumpen einsetzen.

In Analogie hierzu zeigt Fig. 53 die Arbeitsbereiche für Pumpenkombinationen. Der wesentliche Unterschied gegenüber einstufigen Gasballastpumpen besteht darin, daß bei diesen Pumpenkombinationen auch bei Betrieb der Vorpumpe mit Gasballast saugseitig nur Drucke unter 1 Torr zulässig sind. Dieser Wert ist zu vergleichen mit dem oben für einstufige Gasballastpumpen angegebenen Wert von 20 (30) Torr. Nun erfolgt aber schon in der vakuumseitigen Pumpe eine Vorkompression, so daß beim Ansaugen von Drucken über 1 Torr der Wasserdampfdruck im Zwischenvakuum, d.h. auf der Ansaugseite der Vorpumpe größere Werte als 20 Torr annimmt und infolgedessen dann schließlich in der Vorpumpe kondensiert. Die Höhe dieser Vorkompression errechnet sich aus dem Verhältnis der Sauggeschwindigkeiten der beiden hintereinandergeschalteten Pumpen und ergibt den bereits genannten Wert für die Wasserdampfverträglichkeit von 1 Torr. Will man Wasserdampfdrucke zwischen 1 und 20 Torr absaugen, so muß man zwischen den beiden Pumpen der Kombination einen Zwischenkondensator einschalten (Fig. 53, Gebiet C). Bei hohen Wasserdampfdrucken (Gebiet B) ist außerdem ein Wasserdampfkondensator auf der Saugseite der großen Pumpe zusätzlich erforderlich. Dagegen ist im Gebiet A der Permanentgasanteil so groß bzw. der Wasserdampfdruck so klein, daß Betrieb ohne Kondensatoren möglich ist. Mechanische Pumpen, insbesondere rotierende Ölluftpumpen sind speziell für das Fein- und Hochvakuumgebiet geeignet. Sie sollten im Grobvakuumgebiet nur für zeitweisen Betrieb zur Anwendung kommen (s. Fig. 53, Trennungsstrich bei 17 Torr).

c) Strömungspumpen.

Hierzu gehören: Molekularluftpumpen, Zentrifugalverdichter, Flüssigkeitsstrahlpumpen, Dampfstrahlpumpen und Diffusionspumpen.

Fig. 54 gibt eine Übersicht über die Arbeitsbereiche verschiedener Pumpenarten. Allen Strömungspumpen gemeinsam ist, daß eine einstufige Pumpe nicht unmittelbar aus dem Feinvakuumgebiet bis auf Atmosphärendruck komprimieren kann. Hierzu muß man also entweder mehrere Stufen derselben Pumpart hintereinanderschalten oder vor die eigentliche Pumpe noch eine Vorpumpe setzen. Eine Ausnahme machen nur die Flüssigkeitsstrahlpumpen, die z.B. als Wasserstrahlpumpen auch gegen Atmosphärendruck arbeiten können. Mehrstufig zwecks Kompression gegen Atmosphärendruck werden im allgemeinen die Zentrifugalverdichter und die Wasserdampfstrahlpumpen gebaut, dagegen

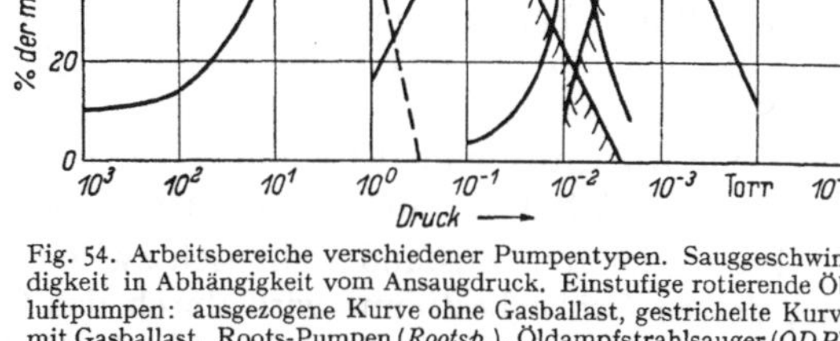

Fig. 54. Arbeitsbereiche verschiedener Pumpentypen. Sauggeschwindigkeit in Abhängigkeit vom Ansaugdruck. Einstufige rotierende Ölluftpumpen: ausgezogene Kurve ohne Gasballast, gestrichelte Kurve mit Gasballast. Roots-Pumpen (*Rootsp.*), Öldampfstrahlsauger (*ODP*), Öldiffusionspumpen (*DO*), Öldiffusionspumpen mit erhöhter Heizleistung und nach hohen Drucken zu verschobenem Arbeitsbereich (*OT*) (in der angelsächsischen Literatur als Booster bezeichnet).

arbeiten Molekularluftpumpen, Öldampfstrahlpumpen, Quecksilberdampfstrahlpumpen, Öldiffusionspumpen und Quecksilberdiffusionspumpen besser gegen eine Vorpumpe.

Die Pumpwirkung kommt bei allen Strömungspumpen auf grundsätzlich andere Weise zustande wie bei den mechanischen Pumpen, und zwar derart, daß dem abzupumpenden Gas eine zusätzliche Geschwindigkeit in einer Vorzugsrichtung erteilt wird, die es von der Saugseite zur Druckseite treibt. Diese Zusatzgeschwindigkeit erhält es entweder von bewegten festen Körpern (Molekularluftpumpe und Zentrifugalverdichter) oder von einer Flüssigkeits- oder Dampfströmung [Flüssigkeitsstrahlpumpen, Wasser(dampf)strahlpumpen, Öldampfstrahlpumpen, Quecksilberdampfstrahlpumpen, Diffusionspumpen].

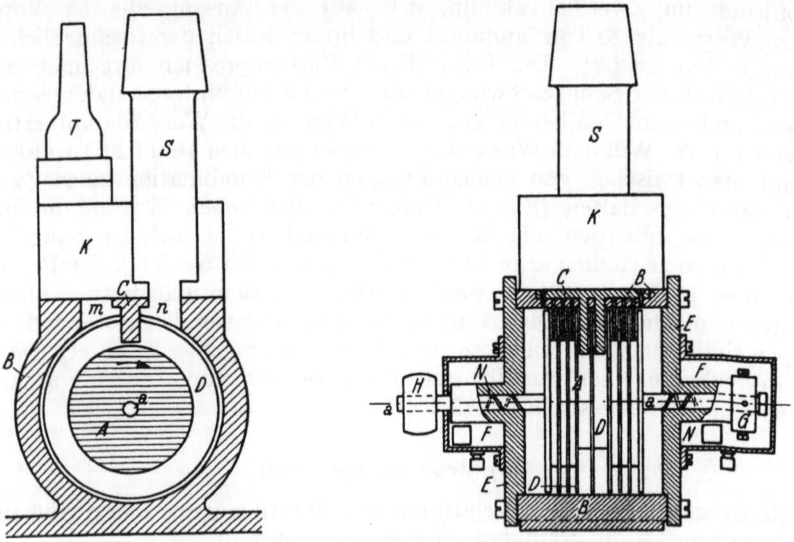

Fig. 55. Molekularluftpumpe nach GAEDE. *A* schnell rotierende Trommel; *B* Gehäuse; *D* in die Trommel eingestochene Nuten; *n* Ansaugöffnung; *m* Auspufföffnung; *C* Nasen, die die Nuten zwischen *m* und *n* absperren; *S* Saugstutzen; *K*, *T* Auspuff; *a* Achse; *H* Antriebscheibe.

21. Molekularluftpumpen[1]. Fig. 55 zeigt eine Molekularluftpumpe, in der ursprünglich von GAEDE angegebenen Form. Ihre Arbeitsweise beruht darauf, daß sich eine Trommel *A* in einem Gehäuse *B* dreht. In diese Trommel *A* sind Nuten *D* eingestochen. Durch die Drehung der Trommel *A* wird Gas von der Ansaugöffnung bei *n* mitgenommen und zur Drucköffnung *m* gefördert. Zwischen den Öffnungen *m* und *n* ragt eine Nase *C* in die Nut *D*, wodurch diese verengt und der Strömungswiderstand zwischen *m* und *n* vergrößert wird. Auf diese Weise wird eine Rückströmung der geförderten Gase von der Drucköffnung *m* unmittelbar zur Saugöffnung *n* weitgehend herabgesetzt.

Im stationären Zustand stellt sich zwischen *m* und *n* ein Druckgefälle p_1, p_2 ein, dessen Größe bestimmt wird durch das Gleichgewicht zwischen dem durch Reibung an der rotierenden Trommel in der Zeiteinheit geförderten Gasvolumen v und der durch das Druckgefälle p_1, p_2 bedingten Gasrückströmung entgegen

[1] W. GAEDE: Ann. d. Phys. **41**, 337 (1913). — F. HOLWECK: C. R. Acad. Sci., Paris **177**, 43 (1923). — M. SIEGBAHN: Ark. Mat. Astronom. Fys. B **30**, Nr. 2 (1944). — R.R. JACOBS: J. Appl. Phys. **22**, 217 (1951). — W. BECKER: Vortr. 1. int. Kongress, f. Vak. Namur 1958.

der Pumprichtung (η = Viskosität, M = Molekulargewicht):

$$v = \frac{p_1 - p_2}{p} \frac{r^3}{l} \left(\frac{\pi r}{8\eta} \frac{p_1 + p_2}{2} + \frac{8}{3} \sqrt{\frac{\pi R T}{2M}} \right). \tag{21.1}$$

Diese Formel gilt nur für kreisförmige Nuten; für die praktisch allein üblichen rechteckigen Nuten gibt RISCH[1] eine entsprechend modifizierte Formel.

Bei hohen Drucken überwiegt die innere Reibung im Gasraum, die äußere Reibung an den Wänden. Das heißt, in Gl. (21.1) ist das erste Glied der Klammer groß gegen das zweite. Da außerdem gilt

$$p = \frac{p_1 + p_2}{2}, \tag{21.2}$$

läßt sich diese Größe aus der Gleichung eliminieren, und es folgt, daß bei konstanter Sauggeschwindigkeit v im stationären Zustand die Druckdifferenz $p_1 - p_2$ konstant ist.

Anders bei niedrigen Drucken, dem eigentlichen Arbeitsbereich der Molekularpumpen. Hier überwiegt die äußere Reibung an den Wänden die Reibung im Gas. In Gl. (21.1) ist also auch das zweite Glied der Klammer groß gegen das erste, und außerdem gilt die folgende Beziehung:

$$v = F \frac{u}{2} \qquad \begin{array}{l}(F = \text{Nutquerschnitt}, \\ u = \text{Umfangsgeschwindigkeit der Trommel}).\end{array} \tag{21.3}$$

Für den stationären Endzustand gelten also hier die Beziehungen

$$v = F \frac{u}{2} = \frac{(p_1 - p_2)}{p} \frac{8 r^3}{3 l} \sqrt{\frac{\pi R T}{2M}} \quad \text{mit} \quad p_1 \gg p_2 \quad \text{und} \quad p \approx p_1, \tag{21.4}$$

$$\frac{p_2}{p_1} = 1 - F \frac{u}{2} \cdot \frac{3 l}{8 r^3} \sqrt{\frac{2M}{\pi R T}}. \tag{21.5}$$

Bei niedrigen Drucken ist also im stationären Zustand nicht die Druckdifferenz, sondern das Druckverhältnis p_1/p_2 von Vorvakuumdruck zu Endvakuumdruck konstant, und zwar um so größer, je größer die Umfangsgeschwindigkeit u der Trommel und je größer die Wurzel aus dem Molekulargewicht, $\sqrt{M}$, ist. Hierin unterscheiden sich also die Molekularluftpumpen prinzipiell von den weiter unten behandelten Diffusionspumpen, die gerade die gegenläufige Abhängigkeit vom Molekulargewicht M aufweisen.

Für die Berechnung der Sauggeschwindigkeit S einer Molekularluftpumpe mit der Umfangsgeschwindigkeit u der Trommel und dem Nutquerschnitt F:

$$S = F \frac{u}{2} - \left(1 - \frac{p_2}{p_1}\right) \frac{8}{3} \sqrt{\frac{\pi R T}{2M}} \frac{r^3}{l} \tag{21.6}$$

berücksichtigt RISCH außer der Rückströmung durch die Fördernut infolge des Druckgefälles $p_1 - p_2$ auch noch die Rückströmung durch die an sich versperrte Nut zwischen m und n, und zwar sowohl infolge des Druckgefälles $p_1 - p_2$ als auch infolge Mitnahme durch den Rotor. Er gelangt dadurch zu den folgenden

[1] R. RISCH: Schweiz. Arch. **14**, 279 (1948).

Beziehungen

$$\left.\begin{aligned}
S &= 3{,}04\cdot 10^{-20}\,\frac{Z}{p}\,[\mathrm{l/sec}] = \frac{3{,}04\cdot 10^{-20}}{p}\,[Z_1-(Z_2+Z_3+Z_4)] \\
Z_1 &= 2{,}52\cdot 10^{17}\,\frac{a^2 b^2 D^2 n\, p \cos\alpha}{(b+a)(D+a)\,T} && \text{Mitnahme in Fördernut durch Rotor} \\
Z_2 &= 8{,}06\cdot 10^{22}\,\frac{a^2 b^2\, p\cos\alpha}{\sqrt{MT}\,(b+a)\,(D+a)\,Q} && \text{Rückströmung in Fördernut durch Druckgefälle } p_1-p_2 \text{ bei Molekularströmung} \\
Z_3 &= 8\cdot 10^{23}\,\frac{h^2 D\, p\cos^3\alpha}{f\,Q\sqrt{MT}} && \text{dasselbe in kleiner Nut} \\
Z_4 &= 7{,}9\cdot 10^{17}\,\frac{h\,D^2 n\, p\sin\alpha\cos\alpha}{T} && \text{Mitnahme durch Rotor in kleiner Nut}
\end{aligned}\right\}\qquad(21.7)$$

a Tiefe der Nut [cm]
b Breite der Nut [cm]
α Steigungswinkel der Nut
Q Anzahl der Windungen der Nut zwischen zwei Punkten, deren Drucke sich wie 1:10 verhalten
f Breite des Kammes zwischen zwei Nuten [cm]
D Durchmesser des Rotors [cm]
n Drehzahl des Rotors [min^{-1}]
h Spaltweite zwischen Rotor und Stator [cm]

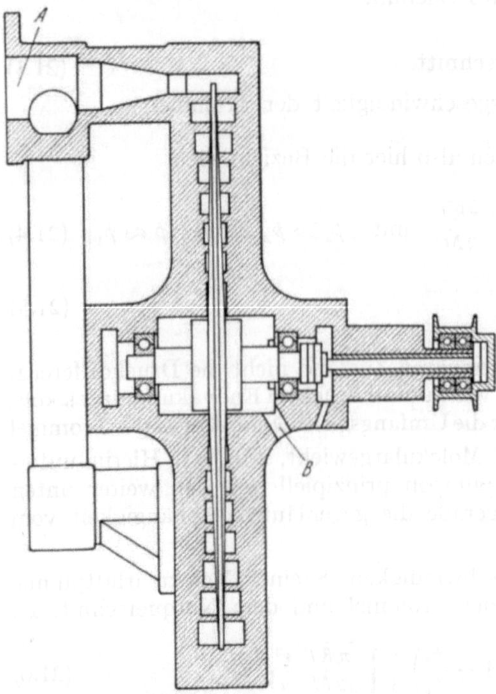

Fig. 56. Molekularluftpumpe vom Scheibentyp nach v. FRIESEN[1]. *A* Ansaugöffnung; *B* Vorvakuumanschluß.

Obgleich die Molekularluftpumpen die ältesten technischen Fein- bzw. Hochvakuumpumpen sind, haben sie doch bis heute nur verhältnismäßig selten Anwendung gefunden. Die Anzahl der vorliegenden Konstruktionen ist sehr klein. Dies rührt daher, daß an die Pumpen sehr hohe Forderungen im Hinblick auf mit ganz geringem Spiel trotz hoher Tourenzahlen ineinanderlaufende Teile gestellt werden müssen. Infolgedessen sind die Pumpen gegen mechanische Beschädigungen sehr anfällig. Zur technischen Ausführung nach Fig. 55 ist weiterhin zu bemerken, daß eine Anzahl Nuten, die in dieselbe Trommel eingestochen sind, druckmäßig hintereinandergeschaltet werden, und zwar derart, daß die Nutquerschnitte von der Saugseite zur Druckseite hin immer kleiner werden. Die Lager der Pumpe sind dabei gleichzeitig mit den druckseitigen Anschlüssen (Vorvakuum) verbunden. Dieses Kleinerwerden der Nutquerschnitte ist erforderlich, da die Nutabmessungen möglichst nicht größer sein sollen als die freie Weglänge des geförderten Gases und diese Größe mit zunehmendem Druck zur Auspuffseite der Pumpe hin immer kleiner wird.

[1] ST. V. FRIESEN: Rev. Sci. Instrum. **11**, 362 (1940).

Man hat versucht, die mechanische Empfindlichkeit der ursprünglichen Gaedeschen Konstruktion dadurch zu verringern, daß man die Nuten nicht in die umlaufende Trommel, sondern in das Gehäuse eingestochen hat. Eine solche nach GAEDE und HOHLWECK bezeichnete Konstruktion hat sich aber auch nicht durchsetzen können.

Von SIEGBAHN wurde ein Pumpentyp beschrieben (s. Fig. 56), bei dem statt einer umlaufenden Trommel eine umlaufende Scheibe in Anwendung kommt. Das hat gewisse Vorteile, insbesondere den, daß man bei großem Scheibendurchmesser am äußeren Umfang der Scheibe eine große Umfangsgeschwindigkeit erzielt. Dort erfolgt auch die Gaszufuhr in die in das Gehäuse eingeschnittenen Nuten, die spiralisch zur Achse hin verlaufen und dabei allmählich einen immer kleineren Querschnitt bekommen. Auch hier ist die Druckseite mit der Lagerseite verbunden. Fig. 57 zeigt die Sauggeschwindigkeit in Abhängigkeit vom Ansaugdruck für eine Molekularluftpumpe.

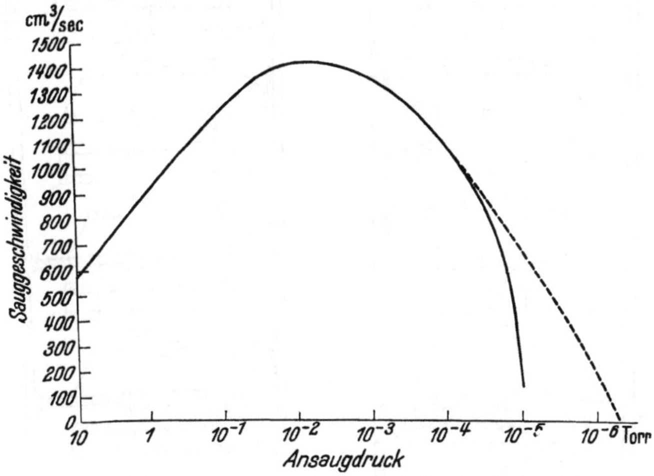

Fig. 57. Sauggeschwindigkeit in Abhängigkeit vom Ansaugdruck für eine Molekularluftpumpe nach GAEDE.

22. Zentrifugalverdichter. Bei dieser Pumpenart erfolgt die Verdichtung des angesaugten Gases (s. Fig. 58) dadurch, daß es zunächst durch Fliehkräfte beschleunigt und daß dann anschließend die so gewonnene kinetische Energie in Düsen in Druck umgewandelt wird. Die Pumpen werden fast ausschließlich als mehrstufige Pumpen mit einer größeren Anzahl von hintereinandergeschalteten Stufen gebaut, da das Verdichtungsverhältnis pro Stufe nicht sehr groß ist. Sie kommen vor allem für das Grob- und Feinvakuumgebiet als Pumpen sehr hoher Sauggeschwindigkeit mit Werten von 10000 m^3/h und einem mehrfachen hiervon in Frage. Näheres s. in der einschlägigen Literatur[1].

23. Flüssigkeitsstrahlpumpen. Die bekanntesten Vertreter dieser Pumpenart sind die Wasserstrahlpumpen (Fig. 59). Die Pumpwirkung kommt hier dadurch zustande, daß ein aus einer Düse austretender Wasserstrahl gleichzeitig in Turbulenz versetzt wird. Infolge der Turbulenz wird er dann im Mischraum mit dem angesaugten Gas durchmischt und das Wasser-Gas-Gemisch anschließend in einer erweiterten Staudüse unter Druckzunahme bis auf Atmosphärendruck verlangsamt. Der mit dieser Pumpenart erreichbare Endvakuum-Totaldruck ist durch den Dampfdruck der Treibflüssigkeit bei der Temperatur der Pumpe

[1] Zum Beispiel PFLEIDERER, Strömungsmaschinen, Berlin: Springer 1952.

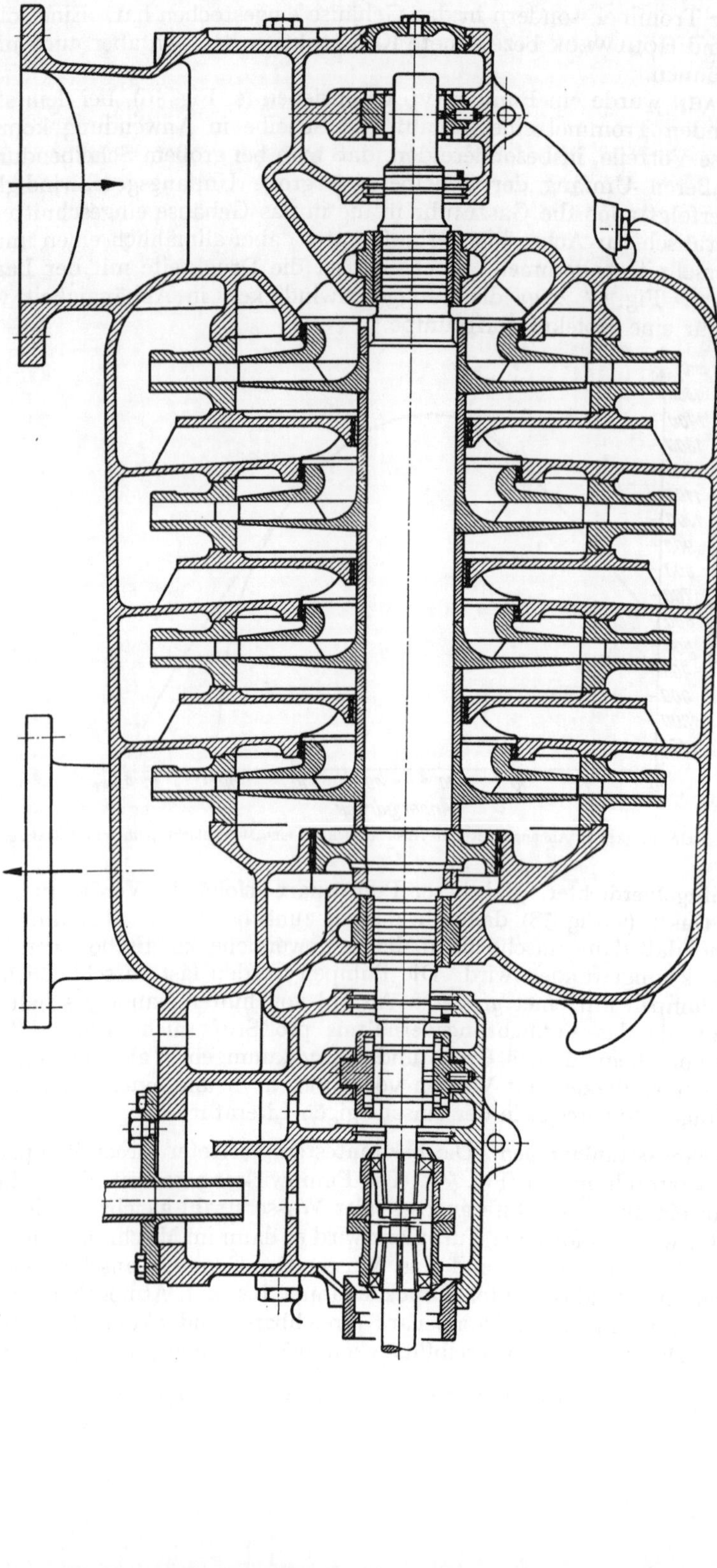

Fig. 58. Mehrstufiger Zentrifugalverdichter (Brown Boveri). Das angesaugte Gas wird in der Nähe der Achse zugeführt, durch Fliehkräfte verdichtet und in einer mitumlaufenden Treibdüse durch Expansion weiter beschleunigt. Anschließend tritt es in die ruhende Staudüse, in der es unter Geschwindigkeitsabnahme verdichtet wird. Das so komprimierte Gas wird dann einer zweiten analog gebauten Stufe wiederum in der Nähe der Achse zugeführt. Die Querschnitte der Düsen nehmen von Stufe zu Stufe von der Saugseite in Richtung auf die Druckseite in ihrem Querschnitt ab, da die Volumina der geförderten Gase infolge der Kompression in den einzelnen Stufen immer kleiner werden.

gegeben. Er liegt also bei Wasserstrahlpumpen etwa bei 12 Torr. Die Sauggeschwindigkeiten liegen für kleine Pumpen aus Glas bei etwa 100 l/h und reichen für große Metallpumpen bis zu 1500 l/h. Der Wasserverbrauch beträgt etwa 1 l Wasser mit einem Mindestwasserdruck von 1 Atü pro 0,6 l angesaugte Luft.

Eine sehr unangenehme Eigenschaft dieser Pumpenart ist die Rückschlaggefahr, sobald der Wasserdruck einen Mindestwert unterschreitet. Bei kleinen Pumpen kann diese Rückschlaggefahr durch enge Querschnitte im Mischraum herabgesetzt werden, was aber natürlich den erreichbaren Wert der Sauggeschwindigkeit verkleinert.

24. Wasserdampfstrahlpumpen (Luftstrahlpumpen[1]). Bei dieser Pumpenart erfolgt die Mitnahme des angesaugten Gases durch einen Wasserdampfstrom, der in einer Expansionsdüse zunächst beschleunigt wird. In einem anschließenden Mischraum wird dann Wasserdampf und angesaugtes Gas miteinander vermischt und in einer weiterhin folgenden Staudüse dieses Gemisch unter Druckzunahme wieder verlangsamt. Die Pumpen kommen meistens mehrstufig zur Anwendung (s. Fig. 60), wobei jeweils zwischen zwei Stufen ein Kondensor zur Niederschlagung des Wasserdampfes eingeschaltet ist, so daß jede folgende Stufe also nur das geförderte Gas, nicht aber auch noch den Wasserdampf der Vorstufe abpumpen muß. Mit solchen mehrstufigen Pumpen lassen sich durchaus Drucke von 0,1 Torr und sogar noch darunter erzielen, wenn die letzten Stufen ohne Kondensor arbeiten. Sie werden bis zu sehr großen Sauggeschwindigkeiten gebaut.

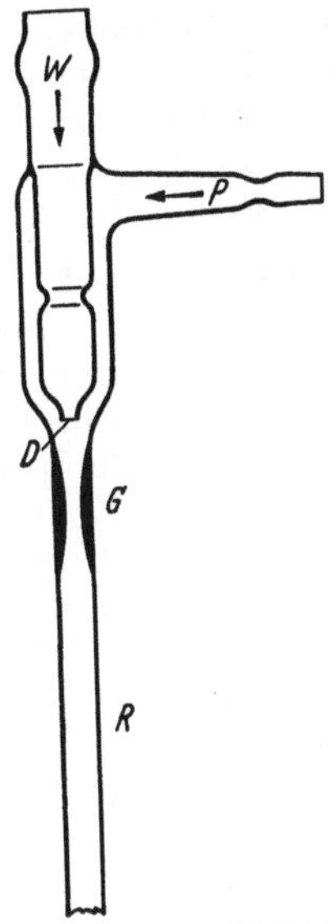

Fig. 59. Wasserstrahlpumpe aus Glas. *W* Wasserzufuhr; *D* Mündung der Treibdüse; *G* Staudüse; *R* Wasserfallrohr; *P* Anschlußstutzen für das abzusaugende Medium.

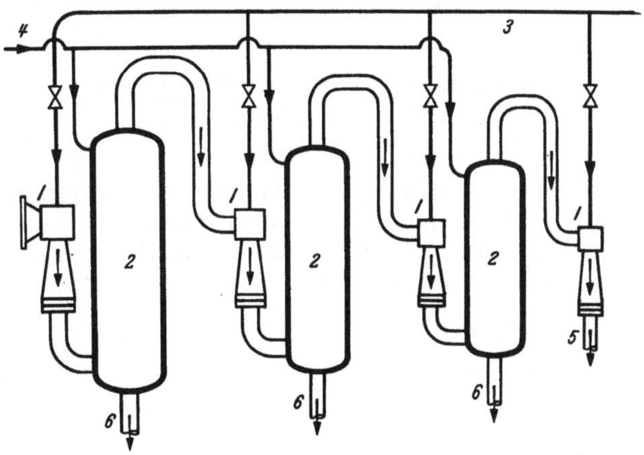

Fig. 60. Vierstufiger Wasserdampfstrahlsauger. *1* Strahldüsen; *2* wassergekühlte Kondensoren zur Niederschlagung des Treibdampfes zwischen den einzelnen Stufen; *3* Dampfzufuhrleitung; *4* Kühlwasserzuleitung; *5* Auspuff in die freie Atmosphäre; *6* barometrische Fallrohre für das Kondenswasser.

Anstelle der Verwendung von Wasserdampf als Treibmittel kommt auch die Verwendung von Luft, die vorher mittels eines Kompressors verdichtet wurde, in Frage. Was von beiden vorteilhafter ist, ist eine Frage der Kosten, die sich je nach Lage der örtlichen Verhältnisse erheblich ändern können.

[1] F. D. Berkeley: Chem. Engng. **61**, 11 (1954). — W. P. Schildlowsky: Nachr. Akad. Wiss. UdSSR., Abt. techn. Wiss. **1954**, 10, 119. — Y. Takashima u. M. Hasegawa: Chem. Engng. (Japan) **18**, 427 (1954).

Vom Verfasser wurde eine Abart des Luftstrahlsaugers[1] beschrieben. Danach erzeugt man mit einer Vorpumpe (Gasballastpumpe) zunächst ein Zwischenvakuum. Dann kann man einen Luftstrahlsauger mit atmosphärischer Luft betreiben, die durch die Düsen des Saugers ins Zwischenvakuum strömt und dabei weitere Gase bei einem niedrigeren Druck als dem des Zwischenvakuums ansaugt. Die Betriebsluft des Saugers und die zusätzlich angesaugten Gase müssen von der Vorpumpe fortgeschafft werden. Die Firma Siemens-Schuckert benutzt einen Luftstrahlsauger mit Wasserringvorpumpe.

Eine spezielle Anordnung insbesondere für Laboratorien besteht in der Verwendung einer Wasserstrahlvorpumpe mit einer nachgeschalteten Wasserdampfstrahlstufe. Mit solchen Anordnungen erreicht man leicht Drucke von 1 bis 3 Torr bei Kompression gegen Atmosphärendruck.

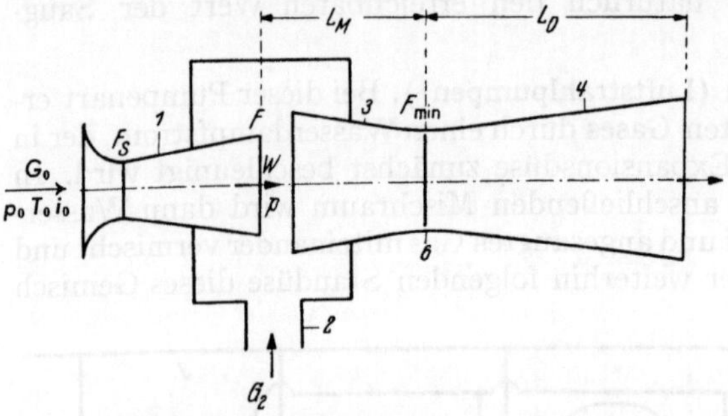

Fig. 61. Schema eines Dampfstrahlsaugers. *1* Treibdüse, Laval-Düse mit engstem Querschnitt F_s; *2* Ansaugstutzen; *6* Staudüse mit verengtem Teil *3* und erweiterter Teil *4*, engster Querschnitt $F_{\min}$.

25. Thermodynamik der Düsenvorgänge und Überschallströmungen. Bei den Dampfstrahl- und Diffusionspumpen kommt die Pumpwirkung dadurch zustande, daß Dampf mit dem Anfangszustand (Druck p_0, Temperatur T_0, Enthalpie i_0), aus dem Siedegefäß durch eine Treibdüse *1* (Fig. 61) ausströmt und dabei auf die Geschwindigkeit w beschleunigt wird, wobei gleichzeitig der Druck auf den Wert p herabgesetzt wird. In einem anschließenden Mischraum mischt sich das angesaugte Gas (G_2) mit dem strömenden Treibdampf, und dieses Gemisch tritt dann in eine Staudüse *4* ein, in der es unter Druckerhöhung auf den Vorvakuumdruck wieder an Geschwindigkeit bis zum Werte $w = 0$ verliert. Zur Behandlung der Vorgänge in den Diffusions- und Dampfstrahlpumpen seien daher zunächst einige Dinge aus der Physik der Düsenströmungen und der Überschallströmungen kurz zusammengestellt.

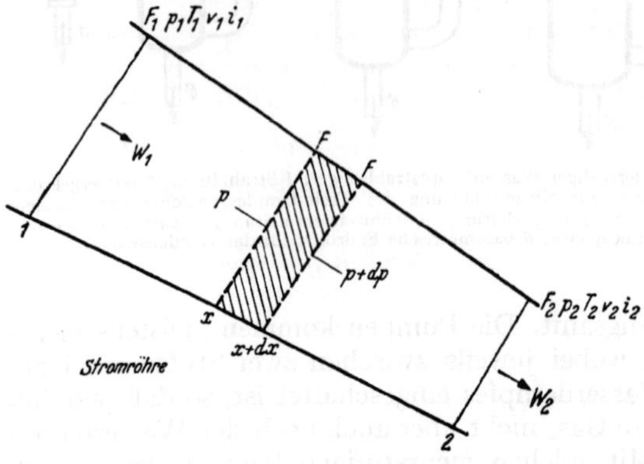

Fig. 62. Stromröhre.

α) Ausströmung aus Düsen (insbesondere Treibdüsen). Für die Durchflußmenge G durch jeden Querschnitt F einer Stromröhre (Düse) gilt die Kontinuitätsgleichung (vgl. Fig. 62), d.h. die Relationen

$$G \equiv F_1 w_1 \frac{1}{v_1} = F_2 w_2 \frac{1}{v_2} = \frac{F w}{v}\,; \quad f = \frac{F}{G}\,; \quad v = \frac{1}{\varrho}\,; \tag{25.1}$$

$$\frac{w f}{v} = 1\,; \quad \frac{dv}{v} = \frac{df}{f} + \frac{dw}{w}\,, \tag{25.2}$$

[1] D.B.P. 800086 angem. 2. 10. 48.

und außerdem der Energiesatz in der Form

$$q_{1,2} = i_2 - i_1 - \int_1^2 v\,dp = i_2 - i_1 + \frac{w_2^2 - w_1^2}{2}. \tag{25.3}$$

v = spezifisches Volumen; f = spezifischer Querschnitt;

q = zugeführte Wärmemenge; ϱ = spezifisches Gewicht.

Alle kleinen Buchstaben bezeichnen Größen pro Gramm Substanz.

Für die adiabatische Ausströmung aus einer Düse (Fig. 63) ergibt sich daraus

$$\frac{w_1^2 - w_0^2}{2} = i_0 - i_1. \tag{25.4}$$

Für $w_1 \gg w_0$ folgt:

$$\frac{w_1^2}{2} = c_p(T_0 - T_1) = \frac{\varkappa}{\varkappa - 1} p_0 v_0 \left[1 - \left(\frac{p_1}{p_0}\right)^{\frac{\varkappa-1}{\varkappa}}\right] \tag{25.5}$$

für $p_1 = 0$:

$$w_{\max} = \sqrt{2\frac{\varkappa}{\varkappa - 1} p_0 v_0} = \sqrt{2\frac{\varkappa}{\varkappa - 1}\frac{\mathsf{R}}{M} T_0}. \tag{25.6}$$

c_p = spezifische Wärme bei konstantem Druck;

$\varkappa$ = Adiabatenexponent;

R = Gaskonstante;

M = Molekulargewicht, ab β) M = Machsche Zahl.

Dieser Wert von $w_{\max}$ kann nicht mit einfachen konisch verengten Düsen, sondern nur mit Lavaldüsen erreicht werden (s. hierzu Fig. 64). Im Anfangszustand hat der Dampf ein endliches spezifisches Volumen v_0, die Geschwindigkeit $w_0 = 0$,

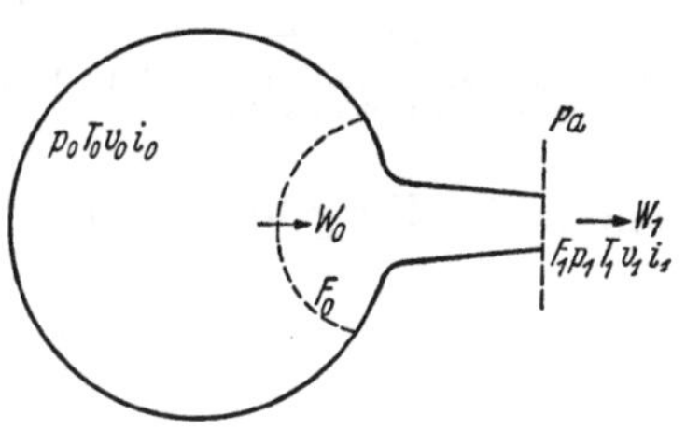

Fig. 63. Ausströmung durch verengte Düse.

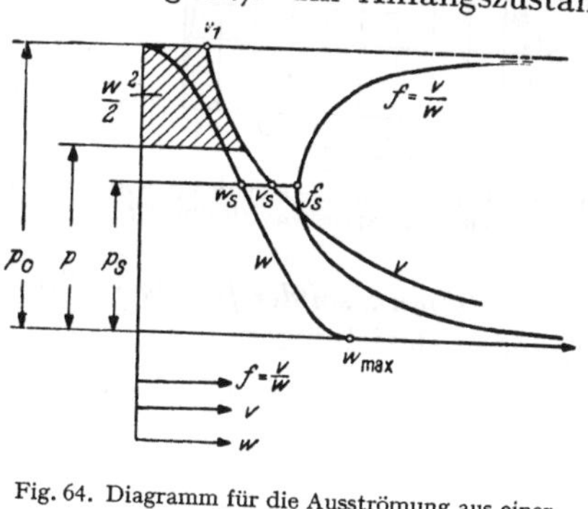

Fig. 64. Diagramm für die Ausströmung aus einer Laval-Düse.

und es ist daher zufolge Gl. (25.2) der spezifische Querschnitt $f \to \infty$ erforderlich. Mit längs einer Adiabate wachsendem v und abnehmendem p wächst w und nimmt daher endliche Werte ebenso wie f an. w hat als obere Grenze $w_{\max}$, während v stetig weiter wächst. Schließlich geht also wieder $f \to \infty$. f hat also im Endlichen ein Minimum. Um möglichst große Geschwindigkeitswerte mit einer Düse zu erreichen, muß diese also so gestaltet sein, daß ihr Querschnitt f erst ab- und dann wieder zunimmt *(Lavaldüse)*. Den Zustand im engsten Querschnitt f_s bezeichnet man als kritischen Zustand. Alle Zustandsgrößen p_s, v_s w_s

sollen durch den Index s gekennzeichnet sein. Hier ist die Strömungsgeschwindigkeit w gleich der örtlichen Schallgeschwindigkeit[1] c^*:

$$w_s = c^*, \tag{25.7}$$

$$c^* = \sqrt{\varkappa p_s v_s} = w_s = \sqrt{\frac{2\varkappa}{\varkappa+1} \frac{p_0}{\varrho_0}}. \tag{25.8}$$

Für Überschallgeschwindigkeiten $w > c^* = w_s$ sind Lavaldüsen erforderlich.

Die Durchflußmenge G für Düsen berechnet sich folgendermaßen:

$$G = F\psi \sqrt{2\frac{p_0}{v_0}}; \qquad F\psi = \text{const}, \tag{25.9}$$

$$\psi = \left(\frac{p}{p_0}\right)^{1/\varkappa} \sqrt{\frac{\varkappa}{\varkappa-1}\left[1-\left(\frac{p}{p_0}\right)^{\frac{\varkappa-1}{\varkappa}}\right]} \quad \text{(vgl. Fig. 65)}, \tag{25.10}$$

$$\frac{p_s}{p_0} = \left(\frac{2}{\varkappa+1}\right)^{\frac{\varkappa}{\varkappa-1}}, \tag{25.11}$$

$$\Theta = \frac{\varrho w}{\varrho^* w_s}. \tag{25.12}$$

Die zur Erzielung eines bestimmten Expansionsverhältnisses p/p_0 erforderliche Erweiterung einer Lavaldüse F^*/F zeigt Fig. 66.

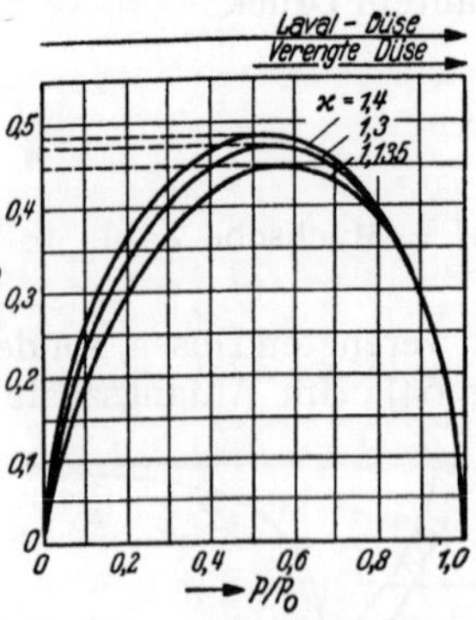

Fig. 65. Diagramm für die Funktion ψ in Abhängigkeit vom Expansionsverhältnis p/p_0.

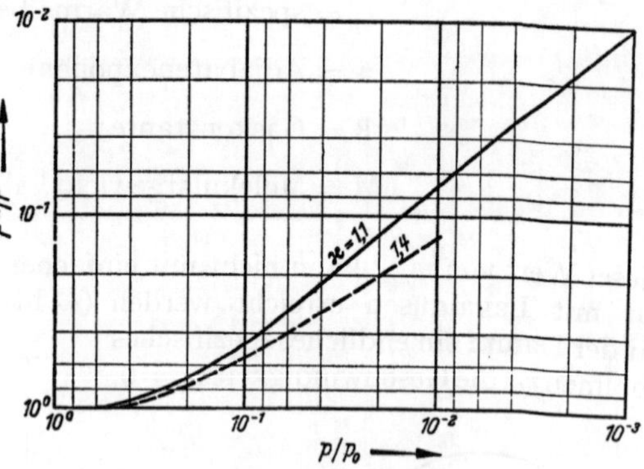

Fig. 66. Reziprokwert der Düsenerweiterung F^*/F als Funktion des Expansionsverhältnisses p/p_0.

β) Strömungsbilder für ebene Überschallströmungen[2]. Hier gelten zunächst die folgenden Beziehungen:

$$\text{Machscher Winkel } \alpha: \quad \sin\alpha = \frac{c}{w} = \frac{1}{M}, \tag{25.13}$$

$$\frac{c^*}{w} = \frac{1}{M^*}, \tag{25.14}$$

$$M^2 = \frac{2}{\varkappa-1}\left[\left(\frac{p_0}{p}\right)^{\frac{\varkappa-1}{\varkappa}} - 1\right]; \qquad M^{*2} = \frac{\varkappa+1}{\varkappa-1}\left[1-\left(\frac{p_0}{p}\right)^{-\frac{\varkappa-1}{\varkappa}}\right], \tag{25.15}$$

$$M^{*2} = \frac{(\varkappa+1)\,M^2}{(\varkappa-1)\,M^2+2}. \tag{25.16}$$

[1] Der Index s gilt für alle Größen die sich auf den engsten Querschnitt einer Lavaldüse beziehen, ein Stern * für alle Größen, die sich auf den kritischen Zustand (Strömungsgeschwindigkeit = Schallgeschwindigkeit) beziehen.

[2] Vgl. auch R. SAUER „Einführung in die theoretische Gasdynamik", Springer 1951.

Bei Reibungsfreiheit ($\eta = 0$) kann man ansetzen:

$$\mathfrak{w} = \operatorname{grad} \varphi, \tag{25.17}$$

und außerdem muß w der Bernoullischen Gleichung und der Kontinuitätsgleichung gehorchen.

Zur graphischen Ermittlung der Strömungsbilder wird vielfach das Charakteristiken-Diagramm nach PRANDTL und BUSEMANN angewandt. Zu seiner Erläuterung diene etwa die folgende Herleitung. Bei kleinen Störungen läßt sich die Ermittlung von w durch Linearisierung vereinfachen. Für die ungestörte Strömung gelte: $\bar{\varphi}$, $\bar{\mathfrak{w}}$. Hinter der Störung lasse sich die Strömung $\mathfrak{w}$ darstellen als

$$\mathfrak{w} = \bar{\mathfrak{w}} + \mathfrak{w}'; \qquad \varphi = \bar{\varphi} + \varphi'. \tag{25.18}$$

Für den Fall einer ebenen Überschallströmung $\bar{M} > 1$ gilt dann:

$$\frac{\partial^2 \varphi'}{\partial x^2} (\bar{M}^2 - 1) - \frac{\partial^2 \varphi'}{\partial y^2} = 0. \tag{25.19}$$

Die Lösung dieser Differentialgleichung hat die Form:

$$\varphi' = F_1(y - x \tan\bar{\alpha}) + F_2(y + x \tan\bar{\alpha}) \tag{25.20}$$

mit

$$\tan\bar{\alpha} = \frac{1}{\sqrt{\bar{M}^2 - 1}}. \tag{25.21}$$

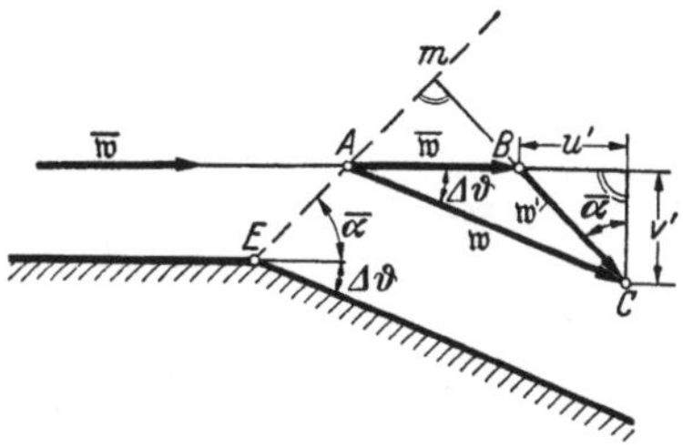

Fig. 67. Strömung um eine flache Ecke.

Machsches Netz aus Linien:

$$y - x \tan\bar{\alpha} = \text{const}$$

und

$$y + x \tan\bar{\alpha} = \text{const}.$$

Durch Einführung neuer rechtwinkliger Koordinaten $\bar{u}, \bar{v}$

$$\bar{u} = \bar{w} \cos\alpha, \qquad \bar{v} = \bar{w} \sin\alpha \tag{25.22}$$

ergibt sich:

$$\frac{\bar{u}^2}{w_{\max}^2} + \frac{\bar{v}^2}{c^{*2}} = 1. \tag{25.23}$$

Dies ist die Adiabatenellipse (Fig. 69).

Strömung um eine flache konvexe Ecke. Hier gelten die folgenden Beziehungen (Fig. 67):

$$u' = + \Delta\vartheta \bar{u} \tan\alpha, \tag{25.24}$$

$$v' = - \Delta\vartheta \bar{u}; \tag{25.25}$$

daher

$$\frac{u'}{v'} = - \tan\alpha, \tag{25.26}$$

d.h. $\mathfrak{w}'$ steht senkrecht auf der Machschen Linie (m).

$$\frac{v'}{\bar{u} + u'} \approx \frac{v'}{\bar{u}} = - \Delta\vartheta, \tag{25.27}$$

$$\Delta\mathfrak{w} \approx u' = \bar{u} \tan\alpha \, \Delta\vartheta, \tag{25.28}$$

$$\Delta p = p - \bar{p} = -\int_{\bar{w}}^{w} \varrho w \, dw = - \bar{\varrho} \bar{w} (w - \bar{w}) = - \varrho' \bar{w} u' = - \varrho^2 \bar{u}^2 \tan\bar{\alpha} \Delta\vartheta. \tag{25.29}$$

Hierbei ist α der Machsche Winkel zwischen Strömungsrichtung und Mach-Linie und ϑ der Ablenkwinkel der Stromlinien.

Jetzt besteht zwischen dem Stromlinienbild (Machsches Netz $\mathfrak{m}$) und dem Geschwindigkeitsbild ($\mathfrak{w}$) der in Fig. 68 gezeigte Zusammenhang. Zur schritt-

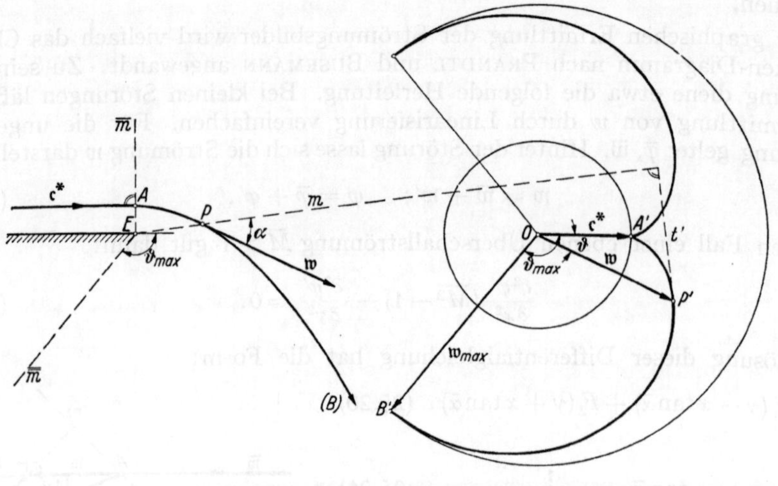

Fig. 68. Links Stromlinienbild (Machsches Netz $\mathfrak{m}$); rechts Geschwindigkeitsbild $\mathfrak{w}$ (hier sind die Geschwindigkeiten nach Größe und Richtung in Polarkoordinaten aufgetragen).

weisen Konstruktion des Stromlinienbildes auf Grund der Adiabatenellipse (Fig. 69) bzw. des Geschwindigkeitsbildes (vgl. Fig. 68 rechte Seite) dient das Charakteristikendiagramm nach Prandtl-Busemann (vgl. Fig. 70 und 70a, bzw.

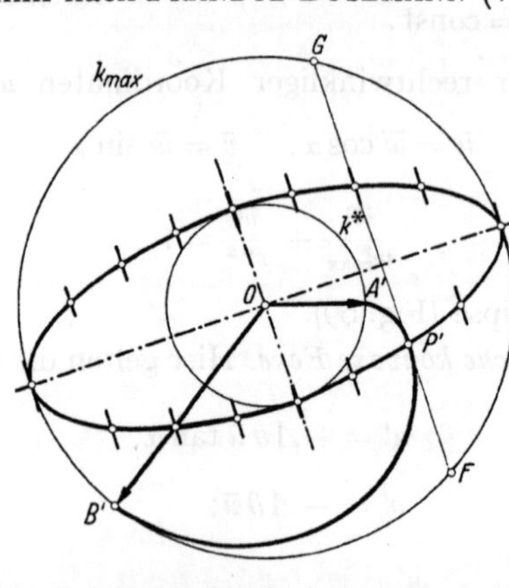

Fig. 69. Adiabatenellipse. Der kleine Kreis k^* entspricht der kritischen Geschwindigkeit $w = c^*$, der große Kreis k_{max} der Maximalgeschwindigkeit w_{max}.

71 und Tabelle 6). Mit Hilfe dieser Charakteristikendiagramme läßt sich beispielsweise der Expansionskegel bei der Ausströmung aus der Mündung einer Lavaldüse zeichnerisch ermitteln und mit photographisch aufgenommenen Strömungsbildern vergleichen (s. Fig. 72).

Verdichtungsstöße. Hat die Störung nicht die Form einer konvexen Ecke, sondern diejenige einer flachen konkaven Ecke, so tritt an Stelle der Expansion

Fig. 70. Charakteristikendiagramm nach PRANDTL-BUSEMANN für hochmolekulare Substanzen (wie organische Dämpfe als Treibmittel für Diffusions- und Dampfstrahlpumpen ($\varkappa = 1{,}1$) nach KUTSCHER. Bezüglich der Bedeutung der in diesem Diagramm verwandten Mach-Koordinaten s. Fig. 70a.

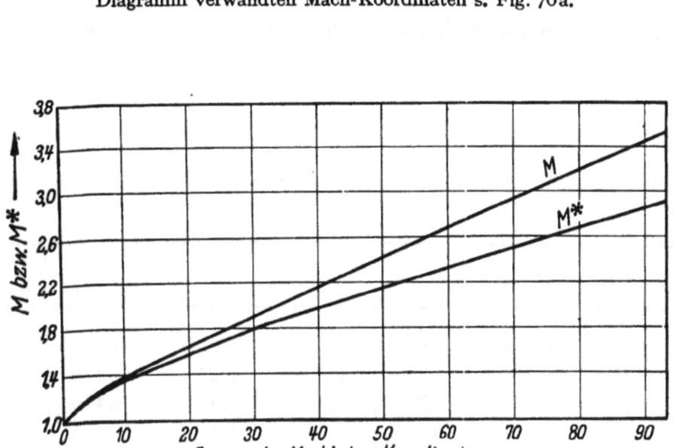

Fig. 70a. Werte für die in Fig. 70 benutzten Mach-Koordinaten. (Für jeden Schnittpunkt zweier Kurven der Fig. 70 sind die zugehörigen beiden Mach'schen Koordinaten zu addieren.)

längs einer Machschen Linie (m) eine Verdichtung (Verdichtungsstoß längs einer Stoßlinie s) auf. Vor der Stoßlinie s mögen die Bezeichnungen p, T, w gelten, hinter der Stoßlinie s: $\hat{p}$, $\hat{T}$, $\hat{w}$. Zwischen ihnen bestehen die folgenden Beziehungen (Fig. 73):

$$\varrho w_n = \hat{\varrho} \hat{w}_n, \quad w_t = \hat{w}_t, \tag{25.30}$$

$$i = \hat{i}_0; \quad T_0 = \hat{T}_0, \quad \frac{p_0}{\varrho_0} = \frac{\hat{p}_0}{\hat{\varrho}_0}, \tag{25.31}$$

$$c^* = \bar{c}^*, \quad w_{\max} = w^*_{\max}, \tag{25.32}$$

$$\frac{\Delta p}{\Delta \varrho} = \frac{\hat{p} - p}{\hat{\varrho} - \varrho} = \varkappa \frac{p + \frac{\Delta p}{2}}{\varrho + \frac{\Delta \varrho}{2}}, \tag{25.33}$$

$$\Delta p = \varrho w^2 \sin^2 \sigma \frac{\Delta \varrho}{\Delta \varrho + \varrho}, \tag{25.34}$$

$$\frac{\varrho + \Delta \varrho}{\varrho} = \frac{\tan \sigma}{\tan(\sigma - \vartheta)}. \tag{25.35}$$

Hierdurch ist der Verdichtungsstoß bestimmt (Stoßpolare):

$$\frac{\Delta p}{\Delta \varrho} = w_n \hat{w}_n = c^{*2} - \frac{\varkappa - 1}{\varkappa + 1} w_t^2. \tag{25.36}$$

Senkrechter Verdichtungsstoß.

$$\vartheta = 0; \quad \sigma = \frac{\pi}{2} \tag{25.37}$$

$w > c^*$ Überschallströmung; $\hat{w} < c^*$ Unterschallströmung.

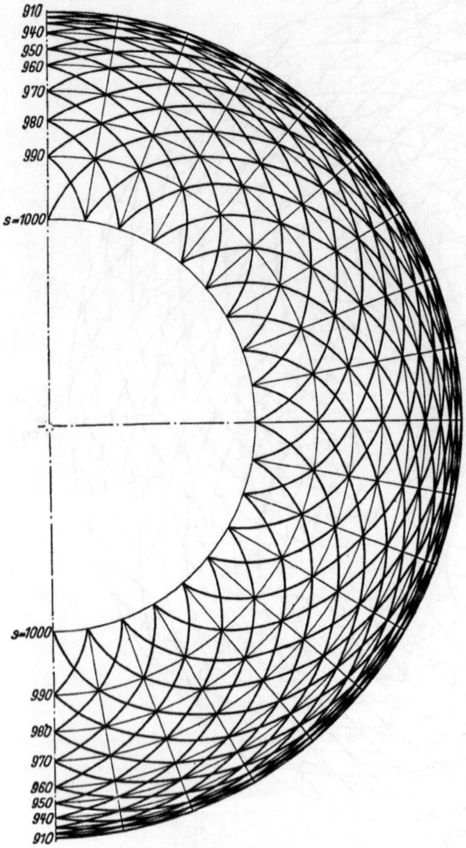

Fig. 71. Charakteristikendiagramm nach PRANDTL-BUSEMANN für Quecksilberdampf oder einatomige Gase ($\varkappa$ = 1,67) nach WASSERRAB. Bezüglich der Koordinaten s s. Tabelle 6.

Stoßpolarendiagramm nach Busemann (Fig. 75). Zu gegebenem w und σ liefert die Stoßpolare (Fig. 74) $\hat{w}$ und $\hat{p}_0/p_0$; zu gegebenem w und $\hat{w}$ liefert die Stoßpolare σ

Tabelle 6. *Hg-Dampf und einatomige Gase. Kenngrößen einer ebenen Überschallströmung* (nach WASSERRAB) ($\varkappa = 1,67$).

s	ϑ	ω	α	p/p_0	$M = w/c$	$M^* = w/c^*$	$\frac{1}{2} \frac{\varrho w^2}{p_0}$	$\frac{p}{p_0} + \frac{2}{1} \frac{\varrho w^2}{p_0}$	Θ
1000	0°	0,00°	90,00°	0,488	1,000	1,000	0,405	0,893	1,000
990	10°	57,5°	42,5°	0,285	1,395	1,255	0,470	0,755	0,911
980	20°	78,0°	32,0°	0,178	1,72	1,41	0,470	0,648	0,774
970	30°	95,5°	24,5°	0,102	2,14	1,55	0,383	0,485	0,596
960	40°	111,5°	18,5°	0,051	2,60	1,66	0,287	0,339	0,436
950	50°	126°	14°	0,0244	3,19	1,76	0,207	0,232	0,292
940	60°	140°	10°	0,0093	4,05	1,85	0,125	0,135	0,172
910	90°	180°	0°	0	∞	2,00	0	0	0,00

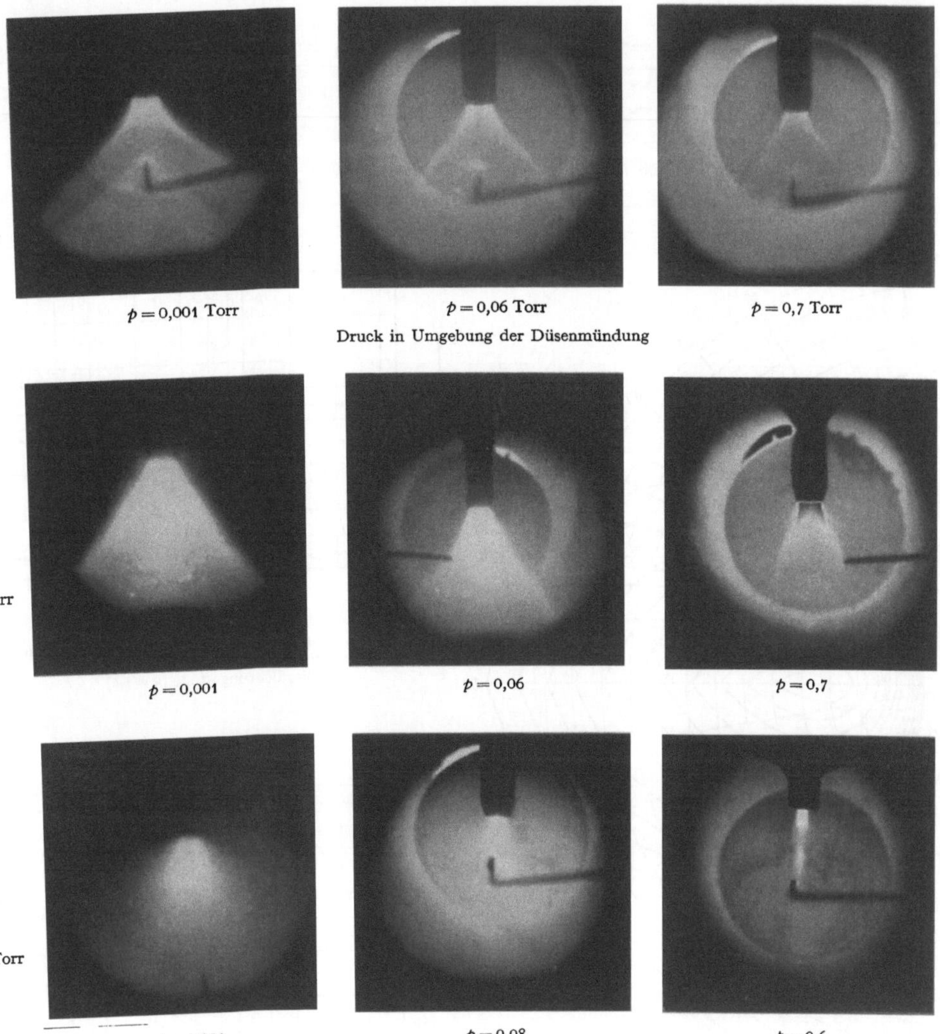

Fig. 72. Photographische Aufnahmen von Quecksilber-Überschallströmungen beim Austritt aus einer Laval-Düse nach H. G. NOELLER: Z. angew. Phys. 7, 218 (1955).

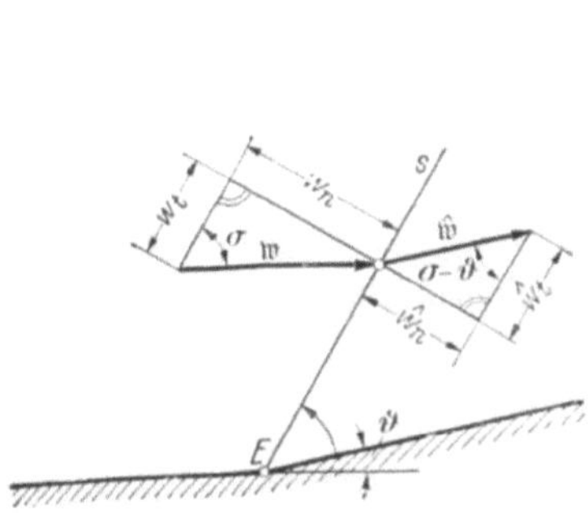

Fig. 73. Verdichtungsstoß bei der Strömung um eine konkave Ecke.

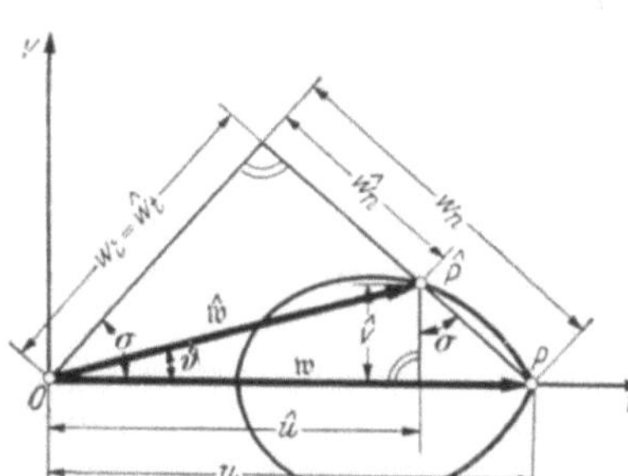

Fig. 74. Stoßpolare für Verdichtungsstöße.

und $\hat{p}_0/p_0$.

$$w_t = \hat{w}_t = u\cos\sigma; \qquad w_n = u\sin\sigma, \tag{25.38}$$

$$\hat{w}_n = u\sin\sigma - \frac{\hat{v}}{\cos\sigma}; \qquad \tan\sigma = \frac{u-\hat{u}}{\hat{v}}, \tag{25.39}$$

$$\hat{v}^2\left[\frac{c^{*2}}{u} + \frac{2}{\varkappa+1}u - u^*\right] = (u-\hat{u})^2\left(\hat{u} - \frac{c^{*2}}{u}\right). \tag{25.40}$$

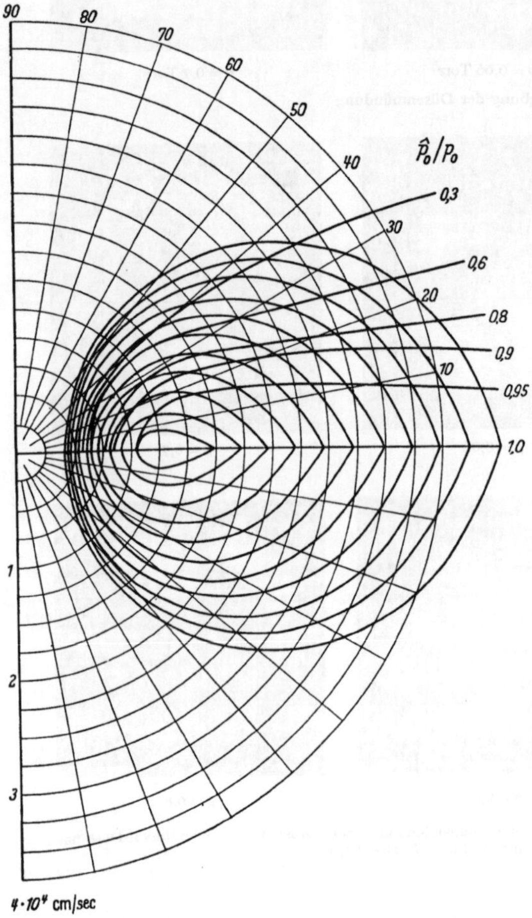

Fig. 75. Stoßpolarendiagramm nach Busemann für hochmolekulare Dämpfe ($\varkappa = 1{,}1$).

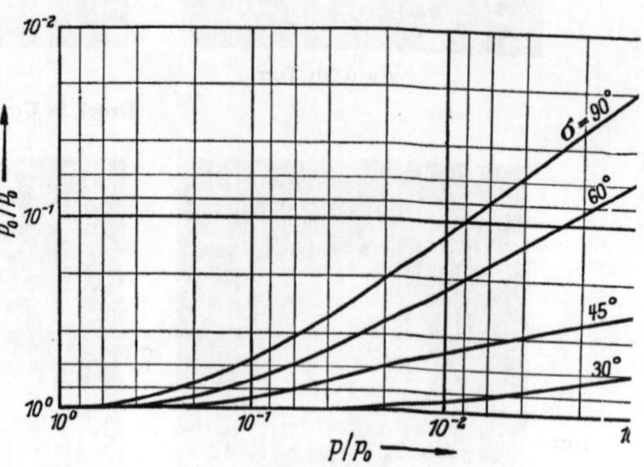

Fig. 76. Ruhedruckverluste $\hat{p}_0/p_0$ in Abhängigkeit vom Expansionsverhältnis p/p_0 für schräge Verdichtungsstöße ($\varkappa = 1{,}1$).

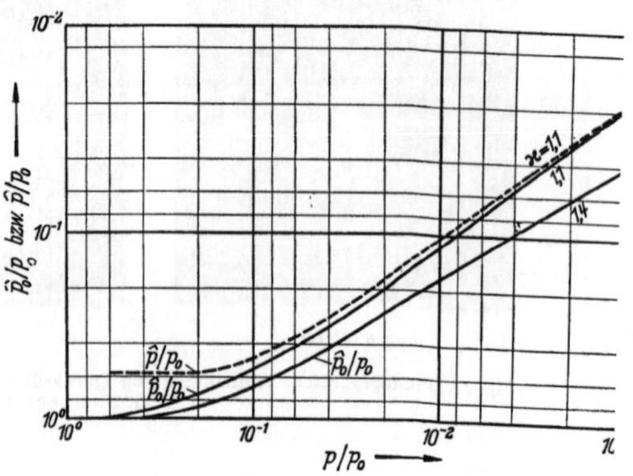

Fig. 77. Ruhedruckverluste $\hat{p}_0/p_0$ und Druckverluste $\hat{p}/p_0$ in Abhängigkeit vom Expansionsverhältnis p/p_0 für senkrechte Verdichtungsstöße.

Die Stoßpolare ist gültig für $w > c^*$, für $w = w_{\max}$ geht sie in einen Kreis über. Für die Druckwerte vor und hinter den Verdichtungsstößen gelten die folgenden Bezeichnungen:

p = Druck vor einem Verdichtungsstoß,

$\hat{p}$ = Druck hinter einem Verdichtungsstoß,

p_0 = Kesseldruck, von dem aus die Expansion auf den Druck p vor dem Verdichtungsstoß erfolgt ist,

$\hat{p}_0$ = zu $\hat{p}$ gehöriger (rechnerischer) Wert des Kesseldruckes.

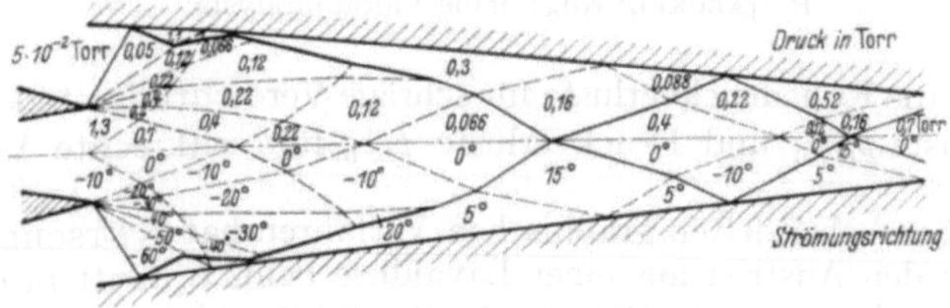

Fig. 78. Austritt der Strömung aus der Laval-Düse bei niedrigem Außendruck. Hochvakuumdruck: $5 \cdot 10^{-2}$ Torr; Vorvakuumdruck: 0,7 Torr. Der niedrige äußere Gasdruck bedingt starke Ablenkung der Strömung beim Austritt aus der Laval-Düse und damit auch einen großen Winkel beim Auftreffen auf die Staudüsenwand. Dort wird ein gerader Stoß ausgelöst, hinter dem sich ein Druck von 1,1 Torr einstellt. Dieser Druck ist höher als der Vorvakuumdruck. Daher folgt die Expansion sowohl in den Hochvakuumraum als auch in Richtung Vorvakuum. In Strömungsrichtung bildet sich eine neue freie Strahlgrenze aus. Infolge des hohen Außendruckes von 1,1 Torr ist der Winkel zwischen Strahlgrenze und Staudüsenwand klein genug, so daß beim erneuten Auftreffen des Strahles auf die Staudüsenwand eine Ablenkung in Staudüsenrichtung über einen schrägen Stoß möglich ist, hinter dem sich ein Druck von 0,5 Torr einstellt. Der schräge Stoß läuft bis in den Strahlkern hinein und schließlich weiter, bis er wieder die Staudüsenwand trifft

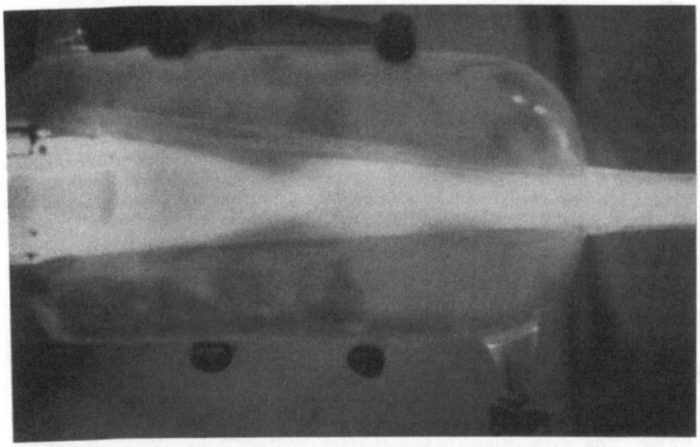

Fig. 79. Öldampfstrahl mit schrägen Verdichtungsstößen. Hochvakuumdruck = 10^{-2} Torr, Vorvakuumdruck = 0,1 Torr

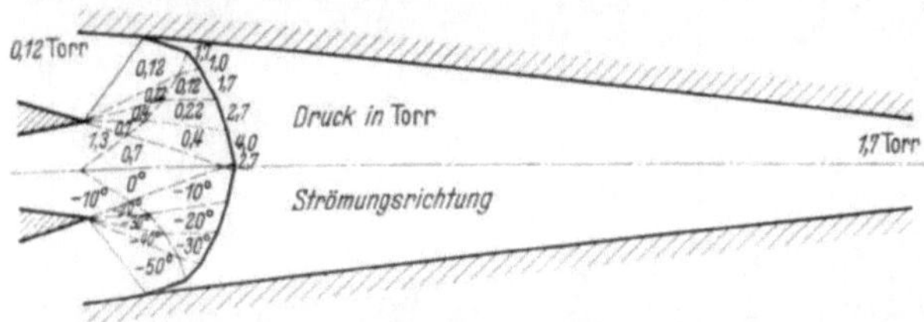

Fig. 80. Austritt der Strömung aus der Laval-Düse bei starkem äußerem Unterdruck (0,1 Torr) im Hochvakuumraum und hohem Vorvakuumdruck (1,7 Torr). Der Strahl trifft unter einem großen Winkel auf die Staudüsenwand. Es wird ein gerader Verdichtungsstoß ausgelöst, der sich bis in den Kern des Strahles fortsetzt, was auf Grund des hohen Vorvakuumdruckes möglich ist. Die Überschallströmung bricht somit bereits am Staudüseneingang ab. Es folgt Unterschallströmung bis zum Vorvakuumraum.

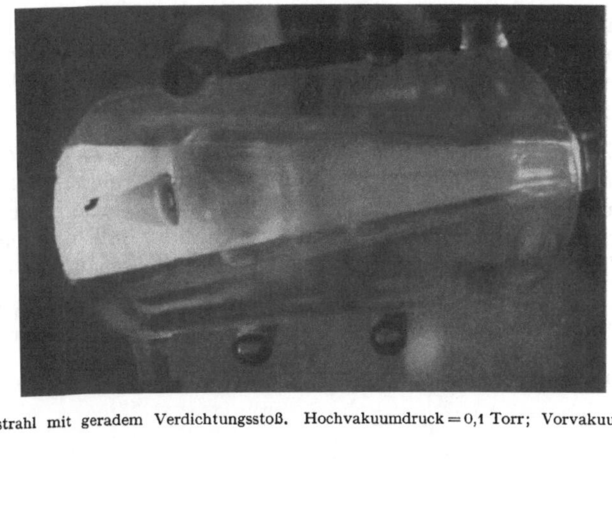

Fig. 81. Öldampfstrahl mit geradem Verdichtungsstoß. Hochvakuumdruck = 0,1 Torr; Vorvakuumdruck = 1,1 Torr

Fig. 76 zeigt die Ruhedruckverluste für schräge Verdichtungsstöße und Fig. 77 Ruhedruckverluste $\hat{p}_0/p_0$ und Druckverluste $\hat{p}/p_0$ für senkrechte Verdichtungsstöße.

Nach den oben behandelten graphischen Verfahren hat Kutscher[1] die Strömungsbilder für den Austritt aus einer Lavaldüse beim Eintritt in eine konisch verengte Staudüse konstruiert. Die Fig. 78, 80 und 82 zeigen die von Kutscher konstruierten ebenen Strömungsbilder für verschiedene Drucke in der Umgebung

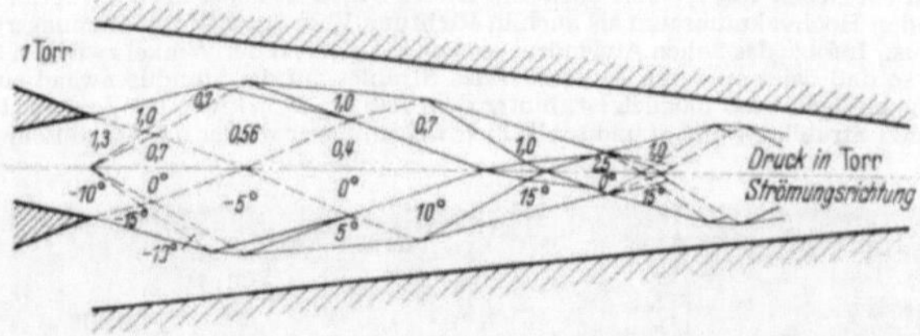

Fig. 82. Austritt der Strömung aus der Laval-Düse gegen Unterdruck (1 Torr). Randbedingung: freie Strahlgrenze (Außendruck). Bei Austritt des Strahles nur geringe Ablenkung der Stromlinien nach außen. Die Verdünnungswellen, die in den Strahl hineinlaufen, werden zunächst nicht durch Verdichtungswellen oder -stöße abgeschwächt. Das bedingt Einstellung von Drucken im Strahlinneren, die wesentlich geringer sind als der Außendruck p_g. Daher wird der Strahl durch den Außendruck stark komprimiert. Es bildet sich eine Einschnürung des Strahles. Der Druck im Strahlinneren wird höher als der Außendruck, was Expansion zur Folge hat. Es kommt zu erneuter Überexpansion, so daß wieder Kompression folgen muß.

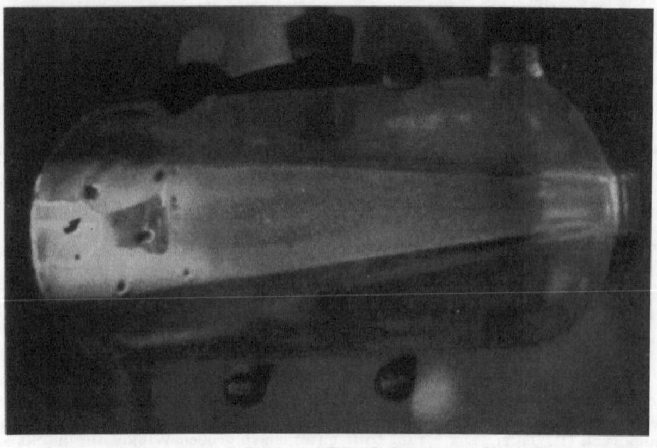

Fig. 83. Ausbildung eines freien Strahles, der aus der Laval-Düse bei geringem Unterdruck im Hochvakuumraum gegenüber dem Dampfdruck an der Düsenmündung austritt, ohne die Staudüsenwand zu erreichen. Hochvakuumdruck 1,0 Torr, Vorvakuumdruck 3,9 Torr.

der Düsenmündung (Ansaugdrucke) und verschiedene Drucke am Austritt der konisch verengten Staudüse (Vorvakuumdrucke). Diese sind zu vergleichen mit den photographisch von Kutscher aufgenommenen Strömungsbildern (Fig. 79, 81 und 83) für Öldampfströmungen ($\varkappa = 1{,}1$).

26. Theorie der Diffusions- und Dampfstrahlpumpen[2]. *α) Phänomenologischer Teil.* Die Diffusions- und Dampfstrahlpumpen haben sowohl in ihrem technischen Aufbau als auch in ihrer physikalischen Wirkungsweise viele gemeinsame Züge. Die Pumpwirkung in ihnen kommt dadurch zustande, daß in einem Siedegefäß ein flüssiges Treibmittel verdampft wird. Der Dampf expandiert dann in einer

[1] H. Kutscher: Z. f. angew. Phys. **2**, 229—234 (1955). Siehe hierzu auch ältere Untersuchungen von B. B. Dayton: Rev. Sci. Instrum. **19**, 793 (1948) und Industr. Engng. Chem. **40**, 795 (1948).

[2] R. Jaeckel, H. G. Noeller u. H. Kutscher: Vak. Techn. **3**, 1 (1954).

Treibdüse auf hohe Strömungsgeschwindigkeiten und niedrige Druckwerte. In dem anschließenden Mischraum wird das angesaugte Gas mit dem treibenden Dampf vermischt und strömt dann weiterhin durch die gekühlte Staudüse, in der das Treibdampf-Gas-Gemisch auf den Vorvakuumdruck komprimiert wird. Der Treibdampf wird schließlich kondensiert und läuft durch das Steigrohr in das Siedegefäß zurück. Charakteristische Unterschiede zwischen den beiden

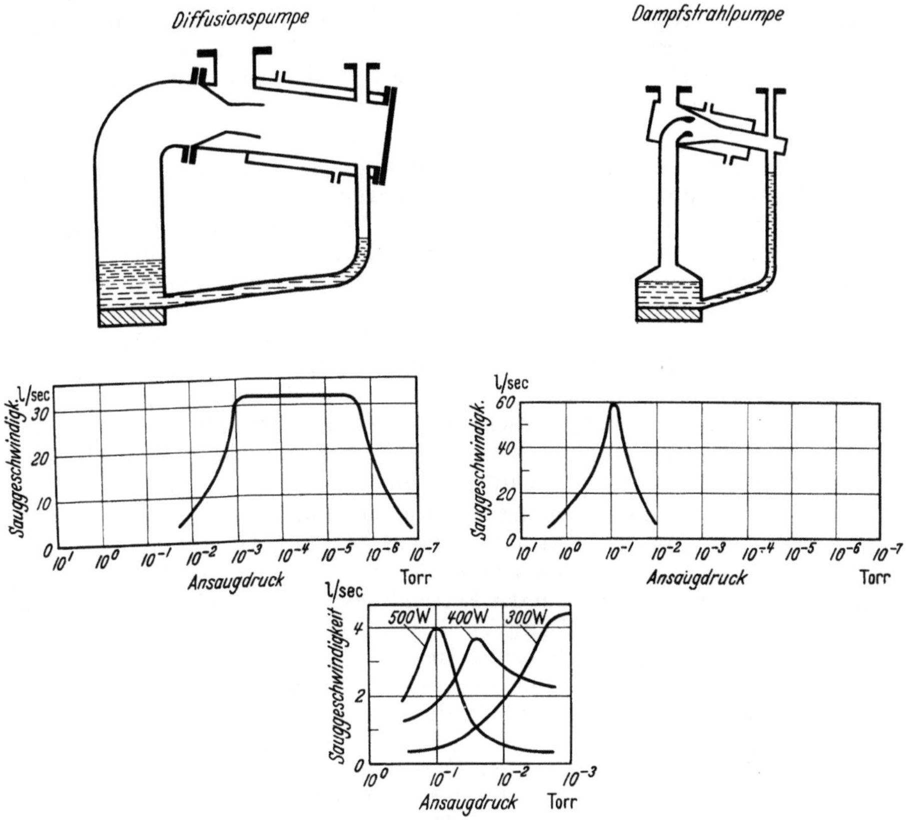

Fig. 84. Vergleich zwischen Diffusions- und Dampfstrahlpumpe. Schema des Aufbaus, Verlauf der Sauggeschwindigkeitskurven.

Pumpenarten bestehen darin, daß der Dampfdruck im Siedegefäß bei den Dampfstrahlpumpen wesentlich höhere Werte hat als bei den Diffusionspumpen. Infolgedessen muß auch das gleichzeitig als Steigrohr dienende Rücklaufrohr für den kondensierten Treibdampf länger sein als bei den Diffusionspumpen. Außerdem ist infolge des höheren Treibdampfdruckes der zulässige Vorvakuumdruck bei den Dampfstrahlpumpen höher und der Arbeitsbereich der Pumpen ist nach höheren Drucken hin verschoben (s. Fig. 84, welche die Sauggeschwindigkeit in Abhängigkeit vom Ansaugdruck darstellt). In ihrem technischen Aufbau unterscheiden sich beide Pumpenarten auch noch dadurch, daß bei den Dampfstrahlpumpen eine echte Staudüse zur Anwendung kommt, mit der also ein Wiederanstieg des Druckes im Dampf-Gas-Gemisch auf den Vorvakuumdruck erzielt wird, während bei den Diffusionspumpen der Vorvakuumdruck im wesentlichen durch den Dampfdruck des Treibdampfes beim Austritt aus der Treibdüse gegeben ist.

Betrachtet man bei beiden Pumpenarten die Kurven, welche die Sauggeschwindigkeit in Abhängigkeit vom Ansaugdruck darstellen, so sieht man bei den Diffusionspumpen, daß diese Kurven vom Endvakuum mit der Sauggeschwindigkeit Null rasch auf einen festen Wert ansteigen, der über mehrere Zehnerpotenzen konstant bleibt und erst wieder an der oberen Grenze des Arbeitsbereiches der Pumpen zu Null abfällt. Dagegen sind die Sauggeschwindigkeitskurven bei den Dampfstrahlpumpen nach höheren Drucken zu verschoben. Außerdem steigt bei ihnen der Wert für die Sauggeschwindigkeit von niedrigen Drucken kommend rasch auf einen Maximalwert an, fällt dann aber anschließend gleich wieder steil ab, so daß die Kurve einen spitzen, im Gegensatz zu dem plateauförmigen Verlauf bei den Diffusionspumpen zeigt.

Die beiden Pumpenarten zeigen aber nicht nur viele gemeinsame Züge, sie lassen sich auch durch Variation der Betriebsbedingungen ineinander überführen. So zeigt Fig. 84 unten die Kurve für die Sauggeschwindigkeit in Abhängigkeit vom Ansaugdruck für eine Dampfstrahlpumpe für verschiedene Heizleistungen. Man sieht, daß die Kurve für die niedrigste Heizleistung (300 Watt) durchaus den Charakter der Kurven bei Diffusionspumpen trägt, während sie mit zunehmender Heizleistung von 300 Watt über 400 Watt auf 500 Watt allmählich in einen spitzeren Verlauf entsprechend dem Charakter der Kurven bei den Dampfstrahlpumpen übergeht. Die beiden Pumpenformen sind also Grenzwerte ein und desselben Pumpentypus für verschiedene Betriebsbedingungen, die Diffusionspumpe für niedrige Treibdampfdrucke und niedrige Ansaugdrucke, die Dampfstrahlpumpe für hohe Dampfdrucke und hohe Ansaugdrucke. Dazwischen liegt der angelsächsisch als Booster benannte Pumpentyp.

Wie schon GAEDE[1] in seinen ersten Veröffentlichungen erkannt hatte, erfolgt bei den Diffusionspumpen die Mitnahme des angesaugten Gases durch den Treibdampf in der Weise, daß das angesaugte Gas durch Diffusion in den Treibdampf eindringt. Es kommt also nur auf das Partialdruckgefälle des Permanentgases an, während der Partialdruck des Dampfes viele Zehnerpotenzen höher sein kann als der des angesaugten Gases. Immerhin ist bei den Diffusionspumpen der Treibdampfdruck so niedrig und die mittlere freie Weglänge im Treibdampf infolgedessen so groß, daß das Gas durch Diffusion von beiden Seiten bis zum Kern des Treibdampfstrahles diffundieren kann (s. Fig. 85, linke Hälfte). Wie NOELLER[2] gezeigt hat, erfolgt aber auch bei den Dampfstrahlpumpen der Eintritt des angesaugten Gases in den Treibdampf durch Diffusion. Nur kann hier infolge des höheren Treibdampfdruckes und der für die Durchmischung zur Verfügung stehenden Zeiten, die durch die Geschwindigkeit der Treibdampfströmung bedingt sind, das Gas durch Diffusion nicht bis zum Kern des Treibdampfstromes, sondern nur in dessen Randschichten (Fig. 85, rechte Hälfte) vordringen.

Durch diese unterschiedliche Vermischung von angesaugtem Gas und Treibdampf erklärt sich auch der verschiedenartige Verlauf der Kurven für die Sauggeschwindigkeit in Abhängigkeit vom Ansaugdruck bei den beiden Pumpenarten. Betrachten wir hierzu Fig. 86. In der linken Hälfte sind die Vorgänge bei den *Diffusionspumpen*, in der rechten Hälfte die Vorgänge bei den Dampfstrahlpumpen schematisch dargestellt. Bei den Diffusionspumpen ist der Ansaugdruck für den größten Teil ihres Arbeitsbereiches klein gegen den Dampfdruck des Treibdampfes bei dem Austritt aus der Treibdüse. Infolgedessen ist auch das Strömungsbild für den Treibdampf unabhängig vom Ansaugdruck. Die Sauggeschwindigkeit ist, wie wir weiter unten sehen werden, nur durch die thermische

[1] W. GAEDE: Ann. Phys. **41**, 289 (1913); **46**, 357 (1915). — Z. techn. Phys. **4**, 337 (1923).
[2] H. G. NOELLER: Z. angew. Phys. **7**, 226 (1955).

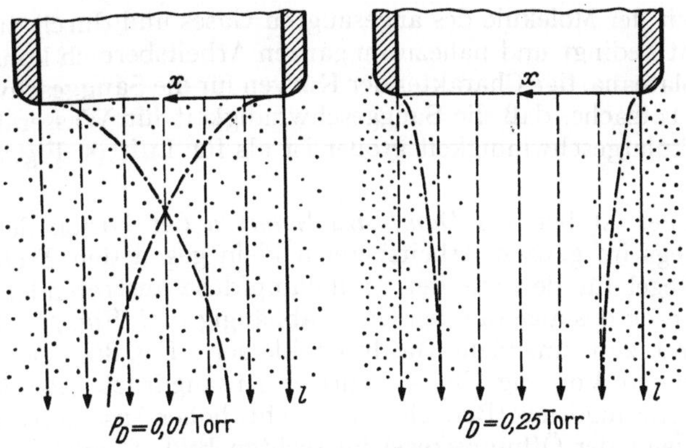

Fig. 85. Zur Frage der Durchmischung von angesaugtem Gas und Dampfstrom bei Diffusions- und Dampfstrahlpumpen (nach NOELLER). Links: niedriger Druck, Diffusionspumpe: das angesaugte Gas dringt durch Diffusion bis zum Kern des Dampfstromes vor. Rechts: hoher Druck, Dampfstrahlpumpe: das angesaugte Gas kann durch Diffusion nur in die Randpartien des Dampfstromes eindringen.

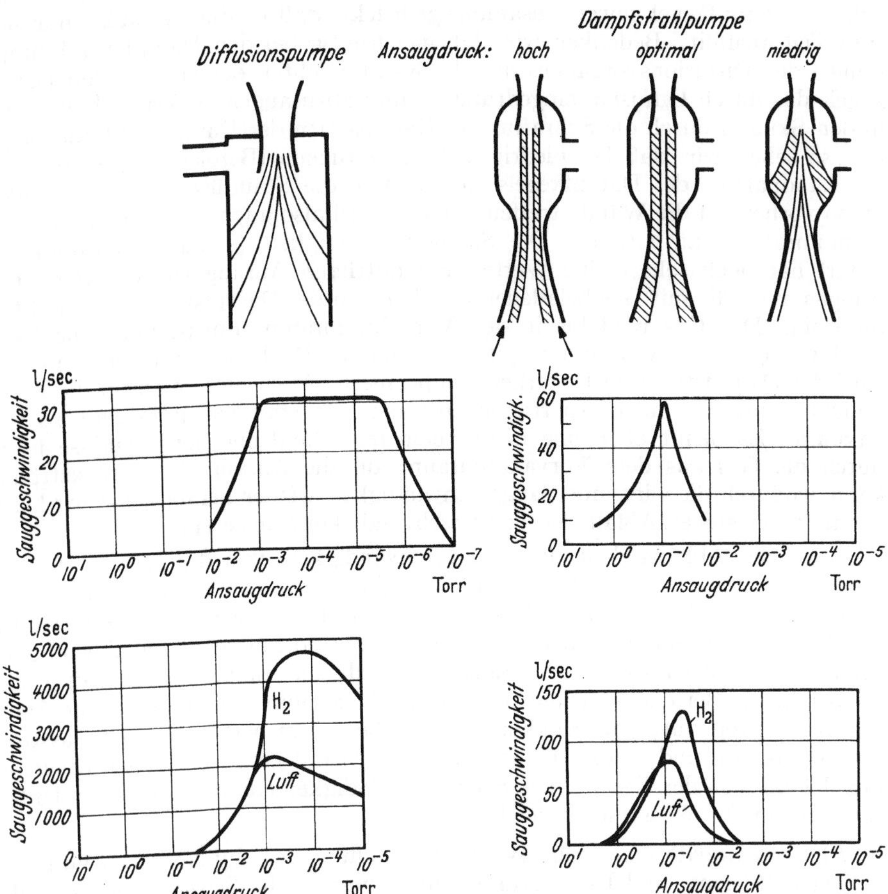

Fig. 86. Vergleich zwischen Diffusions- und Dampfstrahlpumpe. Sauggeschwindigkeit bei verschiedenen Ansaugdrucken und für verschiedene Gase.

Geschwindigkeit der Moleküle des angesaugten Gases und durch die Treibdampfgeschwindigkeit bedingt und nahezu im ganzen Arbeitsbereich konstant. Hieraus resultiert der plateauartige Charakter der Kurven für die Sauggeschwindigkeit und außerdem die Tatsache, daß die Sauggeschwindigkeit für Wasserstoff mit seiner größeren Molekulargeschwindigkeit größer ist als für Luft (s. Fig. 86, Diagramm links unten).

Demgegenüber ist bei den *Dampfstrahlpumpen* der Druck des angesaugten Gases keineswegs im ganzen Arbeitsbereich klein gegen den Treibdampfdruck. Der Öffnungskegel für den aus der Treibdampfdüse austretenden Dampfstrom ist also vom Druck des angesaugten Gases abhängig (vgl. Fig. 72). Der Öffnungswinkel wird mit zunehmendem Gasdruck kleiner. Fig. 86, oben rechts, zeigt schematisch diesen Vorgang [links niedriger Ansaugdruck (Bereich *a*), in der Mitte mittlerer Ansaugdruck (Bereich *b*) und rechts hoher Ansaugdruck (Bereich *c*)]. Infolgedessen zeigt der Öffnungskegel im rechten Bild (Bereich *a*) einen großen Winkel. Er trifft schon außerhalb der Staudüse auf die Wandung der Pumpe. Bei mittlerem Ansaugdruck (mittlere Skizze) (Bereich *b*) wird der Öffnungskegel zusammengedrückt, füllt aber die Staudüse noch in ihrem ganzen Querschnitt vollständig aus. Bei hohem Ansaugdruck (linke Skizze) (Bereich *c*) wird schließlich der Dampfkegel soweit zusammengedrückt, daß er die Staudüse nur noch zum Teil ausfüllt. Bedenken wir, daß gleichzeitig bei den Dampfstrahlpumpen infolge des Diffusionsvorganges nur die Randschichten des strömenden Dampfkegels das durch Diffusion eingedrungene und abzusaugende Gas enthalten (was in den Skizzen durch die Schraffur der Randpartien der Dampfkegel angedeutet ist), so sehen wir, daß bei niedrigen Ansaugdrucken (Bereich *a*) (rechtes Bild) die Randpartien des Dampfkegels gar nicht in die Staudüse gelangen, sondern schon vorher auf die Wände treffen. Damit wird also auch kein Gas in den Vorvakuumraum transportiert. Die Sauggeschwindigkeit (s. das Diagramm darunter) hat noch sehr kleine Werte. Bei mittleren Ansaugdrucken (Bereich *b*) dringen auch die mit Gas beladenen Randpartien des Dampfstromes in die Staudüse ein. Das Gas wird bis in den Vorvakuumraum transportiert; die Sauggeschwindigkeitskurve zeigt ein spitzes Maximum. Bei hohen Drucken (Bereich *c*) wird der Dampfstrom noch stärker komprimiert. Auch hier gelangen die Randschichten mit dem zu transportierenden Gas in die Staudüse und damit ins Vorvakuum. Es bleibt aber eine nicht versperrte Randzone der Staudüse übrig, durch die Gas aus dem Vorvakuumraum auf die Ansaugseite zurückströmen kann, wodurch die wirksame Sauggeschwindigkeit wieder herabgesetzt wird. Daher tritt ein steiler Abfall der Sauggeschwindigkeitskurve ein.

Auch die Abhängigkeit der Sauggeschwindigkeit von der Gasart (Fig. 86, unteres rechtes Diagramm) bei den Dampfstrahlpumpen bestätigt diese Vorstellung. Bei niedrigen Gasdrucken kommt es für die Größe der Sauggeschwindigkeit (in Analogie zu den Diffusionspumpen) nur auf die Molekulargeschwindigkeit des angesaugten Gases an. Die Sauggeschwindigkeit ist bei Wasserstoff größer als bei Luft. Bei den Drucken jenseits des Maximums ist die wirksame Sauggeschwindigkeit durch die Konkurrenz zwischen Transport durch den Gasstrom ins Vorvakuum und Rückströmung längs der Wand gegeben. Dieser Effekt ist unabhängig von der Gasart. Die Sauggeschwindigkeitskurve zeigt keinen Unterschied mehr für Wasserstoff und Luft.

β) Vorvakuumbeständigkeit. Bei der Behandlung der physikalischen Vorgänge in den Diffusions- und Dampfstrahlpumpen besteht die Schwierigkeit, daß der Hauptvorgang häufig von einer Reihe von Nebenerscheinungen begleitet wird, so daß es unter Umständen schwierig ist, die Haupt- und Nebenvorgänge in ihrer gegenseitigen Bedeutung richtig einzugruppieren. Viele Diskussionen im Schrifttum rühren daher, daß Nebenvorgänge als die wesentlichen angesehen und dann die gesamten physikalischen Vorgänge in den Pumpen unter diesem Aspekt behandelt wurden.

Fig. 87 möge schematisch den grundsätzlichen Aufbau einer Diffusionspumpe wiedergeben. Dabei tritt ein von A kommender Dampfstrom aus dem Siedegefäß durch die Lavaldüse, die ihre engste Stelle bei 1 hat. Er wird in dieser unter Druckabnahme bis zur Düsenmündung a beschleunigt und expandiert. Er trifft schließlich auf die gekühlte Wand D, wird kondensiert und fließt von dort aus in das Siedegefäß der Pumpe zurück. Die übrig bleibenden abgepumpten

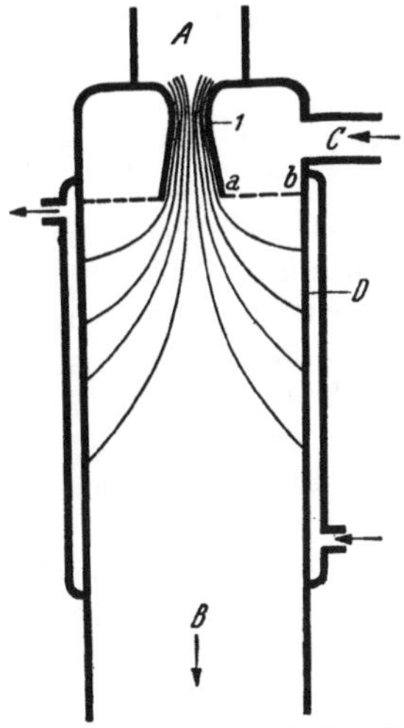

Fig. 87. Grundsätzlicher Aufbau einer Diffusionspumpe.

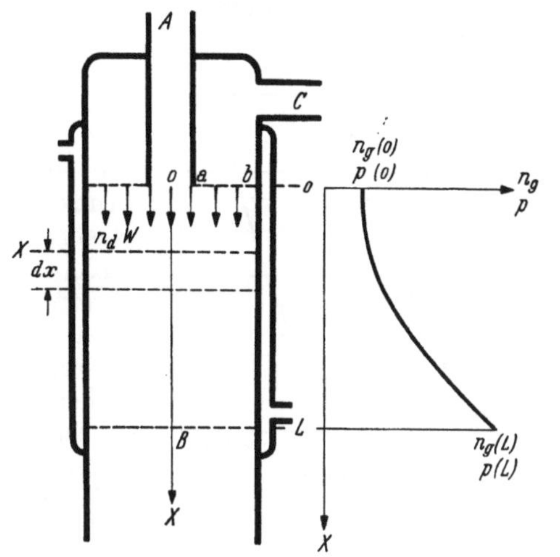

Fig. 88. Schema (erste Näherung) der Vorgänge in einer Diffusionspumpe; Diagramm rechts: Verlauf des Partialdruckes der permanenten Gase (Gasdichte) im Dampfstrom.

Gase sammeln sich im Vorvakuumraum bei B und werden zur Vorpumpe abgeführt. Das angesaugte Gas tritt bei C ein und gelangt dann (durch die Fläche ab) durch Diffusion (s. Ziff. 26α) in den Dampfstrom. Es wird von diesem mitgenommen und dabei schließlich auf den Vorvakuumdruck komprimiert.

Das Schema einer Diffusionspumpe (Fig. 87) wollen wir nach dem Vorgang von Matricon[1] noch weiter vereinfachen (s. Fig. 88). Danach sollen sich von der Grenzfläche ab des Dampfstromes weg in Richtung auf das bei B angeschlossene Vorvakuum $n_d w$ Dampfmoleküle mit einer räumlichen Dichte n_d und einer Molekulargeschwindigkeit w bewegen. Das anzusaugende Gas dringt bei C in den Raum oberhalb der Fläche ab ein und habe dort eine Dichte $n_g(0)$ entsprechend einem Restgasdruck $p_g(0)$. Das Restgas tritt infolge Diffusion durch die Fläche ab in den Dampfstrom ein und wird von demselben in Richtung auf das Vorvakuum mitgenommen und gleichzeitig komprimiert. Für irgendeine Schnittfläche durch den Dampfstrom an der Stelle x besteht Gleichgewicht zwischen der Anzahl der Permanentgasmoleküle, die einerseits durch den Dampfstrom von der Ansaugseite C zur Vorvakuumseite B mitgenommen werden und denjenigen Gasmolekülen, die aus dem Vorvakuumraum B infolge der dort herrschenden höheren Restgasdichte $n_g(L)$ entgegen der Dampfströmung in Richtung auf die Ansaugseite diffundieren. Dabei soll die Nebenbedingung gelten

$$n_d \gg n_g. \tag{26.1}$$

[1] M. Matricon: J. Phys. Radium (7) **2**, 86 (1931); **3**, 127 (1932).

Dann ergibt sich die Gleichgewichtsbedingung

$$n_g w = D \frac{d n_g}{d x}. \tag{26.2}$$

Darin ist D der Diffusionskoeffizient der Gasmoleküle im Dampf, für den wir setzen können

$$D_0 = D n_d. \tag{26.3}$$

Einsetzen in Gl. (26.2) ergibt

$$\frac{d n_g}{n_g} = \frac{d p}{p} = \frac{n_d w}{D_0} d x. \tag{26.4}$$

Für den stationären Fall des Endvakuums, d.h. also für den Fall, daß bei C kein Gaseinlaß erfolgt, hat diese Differentialgleichung die Lösung

$$\frac{p(L)}{p(0)} = \exp\left(\frac{n_d \cdot w}{D_0} L\right). \tag{26.5}$$

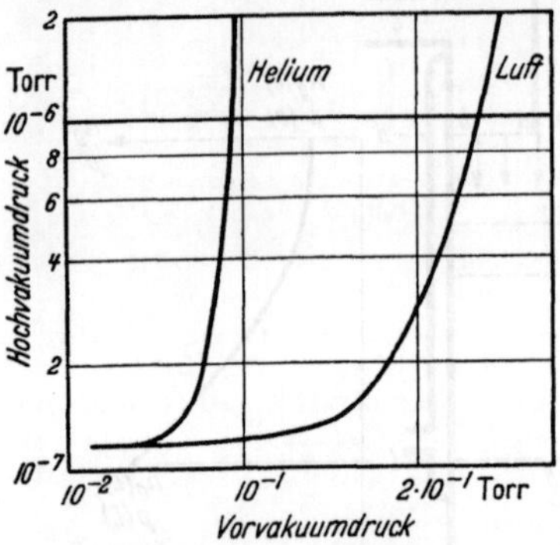

Fig. 89. Vorvakuumbeständigkeit für verschiedene Gase nach DAYTON.

Sie gibt das Verhältnis von Permanentgasdruck auf der Vorvakuumseite, $p(L)$, und Restgasdruck auf der Ansaugseite, $p(0)$, und außerdem für jede Stelle x innerhalb des Dampfstromes den örtlichen Wert für die Restgasdichte n_g und den entsprechenden Restgasdruck p an (s. in Fig. 88 das rechts stehende Diagramm).

Die Länge des Diffusionsweges L im Dampfstrom ist keineswegs eine konstante Größe, die bei einer bestimmten Diffusionspumpe von den Betriebsbedingungen unabhängig ist. L ist vielmehr, wie schon DAYTON[1] gezeigt hat, abhängig sowohl vom Dampfdruck im Siedegefäß, als auch von den beiden Größen Vorvakuumdruck $p(L)$ und Druck an der Saugseite $p(0)$.

Das Verhältnis des Vorvakuumdruckes $p(L)$ zum Restgasdruck $p(0)$ ist nach Gl. (26.5) um so größer, je größer $n_d w$ ist, es wächst also mit zunehmender Dichte n_d und Geschwindigkeit w der Dampfmoleküle im Dampfstrom. Beide Größen sind nach dem Verhalten von Dampfströmungen in Düsen um so größer, je höher der Dampfdruck beim Eintritt in die Treibdampfdüse, d.h. je höher die Temperatur im Siedegefäß der Pumpe ist. Die Vorvakuumbeständigkeit einer Diffusionspumpe wächst also mit der aufgewandten Heizleistung. Außerdem nimmt der Wert für den Restgasdruck $p(0)$ mit zunehmendem Vorvakuumdruck $p(L)$ höhere Werte an. Das Verhältnis von Hochvakuumdruck zu Vorvakuumdruck in seiner Abhängigkeit von der Gasart ist durch den Diffusionskoeffizienten D_0 gegeben. Bei gleichem Vorvakuumdruck ergeben sich also beispielsweise für Helium mit seinem großen Diffusionskoeffizienten höhere Werte für den Hochvakuumdruck als für Luft (s. Fig. 89).

Dem oben geschilderten Mechanismus für die Einstellung eines festen Verhältnisses zwischen Vorvakuumdruck $p(L)$ und Restgasdruck $p(0)$ können verschiedene Nebeneffekte überlagert sein, die die klare Kenntnis des Haupteffektes erschweren.

Als solche Nebeneffekte sind zu nennen:

1. Die Aufnahme von Gasmolekülen in das flüssige Treibmittel innerhalb des Vorvakuumraumes. Das so im Treibmittel gelöste Gas gelangt dann mit dem

[1] B. B. DAYTON: Rev. Sci. Instrum. **11**, 793 (1948).

Treibmittel in das Siedegefäß und wird nach der Verdampfung des Treibmittels im Dampfstrom mit- und in die Treibdüsen abgeführt. Es führt somit zu einer Erhöhung der Gasmenge im Treibdampfstrom beim Austritt aus der Mündung der Treibdüse und damit weiterhin auch zu einer Erhöhung der Restgasdichte auf der Ansaugseite.

2. Dasselbe kann bewirkt werden durch irgendwelche Gasquellen auf der Ansaugseite, wie z. B. an den Wänden adsorbierte Gase, die bei der Evakuierung allmählich abgegeben werden, oder Lecks in den Gefäßwänden auf der Saugseite. Auch die Entstehung von nicht kondensierbaren Zersetzungsprodukten des Treibmittels kann sich in gleicher Weise auswirken.

3. Errechnet man aus Gl. (26.5) für den normalen technischen Betrieb einer Diffusionspumpe für Luft als Restgas das Verhältnis $p(L)/p(0)$, so ergibt sich ein sehr großer Wert, d. h. die Rückdiffusion von Luft aus dem Vorvakuum durch den entgegenströmenden Treibdampf auf die Saugseite ist nur sehr schwach. Dies hat neuerlich HIESINGER[1] geprüft und bestätigt, indem er auf der Vorvakuumseite radioaktive Gasisotope eingelassen hat. Er konnte dabei auf der Hochvakuumseite keine Zunahme von radioaktiven Gasen messen, trotz der sehr hohen Empfindlichkeit dieser Methode.

4. Mißt man auf der Hochvakuumseite nicht den Restgaspartialdruck sondern den Totaldruck, so kann dieser wesentlich höhere Werte haben, und zwar einerseits wegen des Dampfdruckes des Treibmittels und andererseits infolge von Zersetzungsprodukten im Treibmittel. In diesem Falle ergeben sich für die Abhängigkeit des Hochvakuumdruckes vom Vorvakuumdruck Kurven, die über weite Bereiche des Vorvakuumdruckes von diesem unabhängig sind, d. h. also bei konstanten Werten des Hochvakuumdruckes verlaufen. Vgl. hierzu auch neuere Messungen von FLORESCU[2].

5. Schließlich ist noch bei der Anwendung der Gl. (26.5) zu beachten, daß die Dampfdichte n_d nicht an jeder Stelle x des Dampfstromes gleich ist, und daß sie selbst längs eines Querschnittes durch den Dampfstrom verschiedene Werte annehmen kann. Insbesondere wird der Wert für die Dampfdichte n_d in unmittelbarer Nähe der gekühlten Wand am kleinsten sein. Die Gefahr eines Gasdurchbruches von der Vor- auf die Hochvakuumseite der Pumpe ist also auch hier am größten (vgl. dazu die Messungen von ALEXANDER[3]). Dieser Sachverhalt kann auch so ausgedrückt werden, daß die Länge der Diffusionsstrecke L [Gl. (26.5)] hier die kleinsten Werte hat[4].

[1] L. HIESINGER: Vak. Techn. **6**, 65 (1957).

[2] N. A. FLORESCU [Invest. Physicae **1952**, Nr. 1 u. 2; **1953**, Nr. 1; **1955**, Nr. 2. — Vacuum **4**, 30 (1954)] glaubt darüber hinaus den oben geschilderten Hauptmechanismus über das Gleichgewicht zwischen Mitnahme von Gasmolekülen durch den Treibdampf und Rückdiffusion der Gasmoleküle vom Vorvakuum zum Hochvakuum bestreiten zu müssen. Er setzt an Stelle dessen eine Modellvorstellung über den Stoß zwischen den Treibdampfmolekülen und den Molekülen des angesaugten Gases, wodurch die ersteren die letzteren in Richtung auf das Vorvakuum vor sich hertreiben. Vergleicht man das Ergebnis dieser Modellvorstellung bei FLORESCU mit der oben gewonnenen Gl. (26. 5), so sieht man daß der Ansatz von FLORESCU zu prinzipiell dem gleichen Ergebnis führt wie Gl. (26.5).

[3] P. ALEXANDER: J. Sci. Instrum. **21**, 216 (1944); **23**, 11 (1946).

[4] Eine Anzahl von Autoren (ALEXANDER, FLORESCU usw.) haben sich bemüht zu zeigen, daß der hier durch Gl. (26.2) beschriebene Mechanismus eines Diffusionsvorganges für die Einstellung des Verhältnisses vom Vorvakuumdruck zum Hochvakuumdruck nicht zutreffe. Sie ziehen daraus aber weiterhin noch den Schluß, daß demnach also die von GAEDE beschriebenen Pumpen keine „Diffusions"-Pumpen seien. Tatsächlich hat aber GAEDE die von ihm zuerst beschriebenen Pumpen als Diffusionspumpen bezeichnet im Hinblick auf den oben beschriebenen Durchmischungsvorgang zwischen angesaugtem Gas und treibendem Dampf. Dies geht schon daraus hervor, daß er gleichzeitig für die Breite der Ringfläche *a b* (s. Fig. 88) eine Maximalbedingung angegeben hat, von der später noch die Rede sein wird.

Außer dem Begriff der Vorvakuumbeständigkeit als Verhältnis von Vorvakuumdruck zu Hochvakuumdruck, der oben benutzt wurde, ist viellfach auch noch der Begriff des maximal zulässigen Vorvakuumdruckes gebräuchlich. Hierunter ist derjenige Vorvakuumdruck zu verstehen, der auf der Vorvakuumseite maximal zulässig ist, ohne daß der Druck auf der Hochvakuumseite einen fest vorgegebenen Wert überschreitet.

Bei den *Dampfstrahlpumpen* ist eine Abhängigkeit des Druckes auf der Vakuumseite vom Vorvakuumdruck, wie sie hier für die Diffusionspumpen beschrieben wurde, im allgemeinen nicht festzustellen. Dies rührt daher, daß bei den Dampfstrahlpumpen die Dampfdichte im Treibdampf im allgemeinen so groß ist, daß eine Rückdiffusion vom Vorvakuum auf die Hochvakuumseite dadurch praktisch unterbunden wird, und daß andererseits infolge des Vorliegens von Überschallströmungen eine Ausbreitung des Vorvakuumdruckes entgegen dieser Überschallströmung bis auf die Saugseite nicht eintreten kann. Das hat zur Folge, daß bei den Dampfstrahlpumpen bei kleinen und mittleren Vorvakuumdrucken der Hochvakuumdruck vom Vorvakuumdruck unabhängig ist und daß ein Einfluß des Vorvakuumdruckes erst dann eintritt, wenn bei hohen Vorvakuumdrucken die Treibdampfströmung so stark eingeschnürt wird, daß zwischen Strömung und Wand der Düse eine direkte Rückströmung vom Vorvakuum zum Hochvakuum möglich ist.

γ) Sauggeschwindigkeit. Wir betrachten hierzu Fig. 88. Auf die Diffusionsfläche $a \ldots b$ vom Querschnitt F treffen von C aus $\frac{1}{4}\bar{c} F \nu$ Moleküle pro Sekunde auf, wenn mit ν die Anzahl der Gasmoleküle/cm³ bei C und mit $\bar{c}$ die mittlere Molekulargeschwindigkeit des Gases bezeichnet wird. Durch Rückdiffusion aus dem Dampfstrom treten in umgekehrter Richtung $\frac{1}{4}\bar{c} F n_g^s(0)$ Moleküle/sec durch $a \ldots b$ nach oben. Dabei ist $n_g^s(0)$ die Anzahl der Gasmoleküle/cm³ im Dampfstrom unmittelbar unterhalb $a \ldots b$ bei Sauggeschwindigkeit[1].

Für die Gesamtzahl N der Moleküle, die pro Sekunde mehr durch $a \ldots b$ von oben nach unten gehen als umgekehrt, ergibt sich also:

$$N = \tfrac{1}{4} F \bar{c}\,[\nu - n_g^s(0)]. \tag{26.6}$$

Die Sauggeschwindigkeit S ist das pro Sekunde abgesaugte Volumen. Sie ergibt sich aus der Beziehung

$$N = \nu S, \quad S = \frac{1}{4} F \bar{c} \left[1 - \frac{n_g^s(0)}{\nu}\right]. \tag{26.7}$$

Die spezifische Sauggeschwindigkeit s (= Sauggeschwindigkeit pro cm² der Diffusionsfläche) hat also den Wert

$$s = \frac{S}{F} = \frac{1}{4} \bar{c} \left[1 - \frac{n_g^s(0)}{\nu}\right]. \tag{26.8}$$

Zur Elimination von $n_g^s(0)$ aus Gl. (26.8) benutzen wir die für $0 \leqq x \leqq L$ gültige Differentialgleichung

$$w \cdot n_g^s(x) - D \frac{d n_g^s(x)}{dx} = \nu s. \tag{26.9}$$

Ihre Lösung lautet:

$$n_g^s(x) = \frac{\nu s}{w} + \left\{n_g^s(0) - \frac{\nu s}{w}\right\} e^{\frac{w}{D} x} \tag{26.10}$$

[1] Hier und im folgenden wird grundsätzlich zwischen zwei Zuständen der Diffusionspumpe unterschieden: „bei Endvakuum", d.h. bei geschlossenem Ansaugstutzen C, „bei Sauggeschwindigkeit", d.h. wenn durch den geöffneten Ansaugstutzen C Gase angesaugt werden, die Pumpe also eine meßbare Sauggeschwindigkeit hat. Der obere Index s bedeutet, daß sich die betreffende Größe auf den Zustand bei Sauggeschwindigkeit bezieht. Beim Endvakuum hat die Sauggeschwindigkeit nach Gl. (26.12) den Wert Null.

Hierin setzen wir $x = L$ und lösen nach $n_g^s(0)$ auf:

$$n_g^s(0) = \frac{\nu s}{w} + \left\{n_g^s(L) - \frac{\nu s}{w}\right\} e^{-\frac{w}{D}L}. \tag{26.11}$$

Einsetzen dieses Wertes von $n_g^s(0)$ in Gl. (26.8) und Auflösung nach s ergibt:

$$s = \frac{\bar{c}}{4} \frac{1 - \frac{n_g^s(L)}{\nu} e^{-\frac{w}{D}L}}{1 + \frac{1}{4}\frac{\bar{c}}{w}\left(1 - e^{-\frac{w}{D}L}\right)} = \frac{\bar{c}}{4} \frac{1 - \frac{n_g^s(L)}{\nu}\frac{n_g(0)}{n_g(L)}}{1 + \frac{1}{4}\frac{\bar{c}}{w}\left(1 - e^{-\frac{w}{D}L}\right)}. \tag{26.12}$$

Die Größe $\frac{\nu}{n_g^s(L)}$ in Gl. (26.12) kann dabei folgende Werte annehmen:

$$1 > \frac{\nu}{n_g^s(L)} \geqq \frac{n_g(0)}{n_g(L)} \ll 1. \tag{26.13}$$

Beim Endvakuum ist[1]

$$\frac{\nu}{n_g^s(L)} = \frac{n_g(0)}{n_g(L)}, \quad \text{d.h.} \quad s = 0. \tag{26.14}$$

Mit wachsendem Ansaugdruck wächst auch $\frac{\nu}{n_g^s(L)}$. Es wird sehr schnell

$$\frac{n_g^s(L)}{\nu} \frac{n_g(0)}{n_g(L)} \ll 1, \tag{26.15}$$

d.h. die spezifische Sauggeschwindigkeit s erreicht den Maximalwert

$$s = \frac{\bar{c}}{4} \frac{1}{1 + \frac{1}{4}\frac{\bar{c}}{w}}. \tag{26.16}$$

Nähert sich schließlich der Ansaugdruck der Diffusionspumpe dem Vorvakuumdruck, so nimmt der Wert von s wieder ab, da jetzt die Bedingungen (26.1), d.h. daß

$$n_d \gg n_g,$$

nicht mehr erfüllt ist. Diesen charakteristischen Verlauf eines breiten Maximums zeigen alle Sauggeschwindigkeitskurven von Diffusionspumpen.

Die spezifische Sauggeschwindigkeit s hängt also nach Gl. (26.16) vor allem von dem Verhältnis $w/\bar{c}$ der Dampfgeschwindigkeit w zur Molekulargeschwindigkeit $\bar{c}$ des Gases ab. Und zwar ist die spezifische Sauggeschwindigkeit s um so größer, je größer die Dampfgeschwindigkeit w im Vergleich zur Molekulargeschwindigkeit $\bar{c}$ des Gases ist. Für den Maximalwert der Sauggeschwindigkeitskurve ergeben sich folgende Werte:

$$\text{für} \quad w = \frac{\bar{c}}{4}: \qquad s = \frac{1}{8}\bar{c}, \tag{26.17}$$

$$\text{für} \quad w \gg \frac{\bar{c}}{4}: \qquad s_{\max} = \frac{1}{4}\bar{c}. \tag{26.18}$$

Mit

$$\bar{c} = \sqrt{\frac{8RT}{M}}$$

[1] „Beim Endvakuum" $\nu = n_g(0)$ muß s natürlich den Wert Null haben. Vgl. diesbezüglich die hier angegebenen Formeln mit den Ableitungen von MATRICON.

folgt als obere praktisch nicht erreichbare Grenze für die spezifische Sauggeschwindigkeit

$$s_{max} = \frac{1}{4}\sqrt{\frac{8\,R\,T}{M}}\,. \tag{26.19}$$

Für Luft ($M = 28{,}8$) von 20° C ergibt sich daraus

$$s_{max} = 11{,}6 \cdot 10^3 \left[\frac{cm^3}{cm^2\,sec}\right]. \tag{26.20}$$

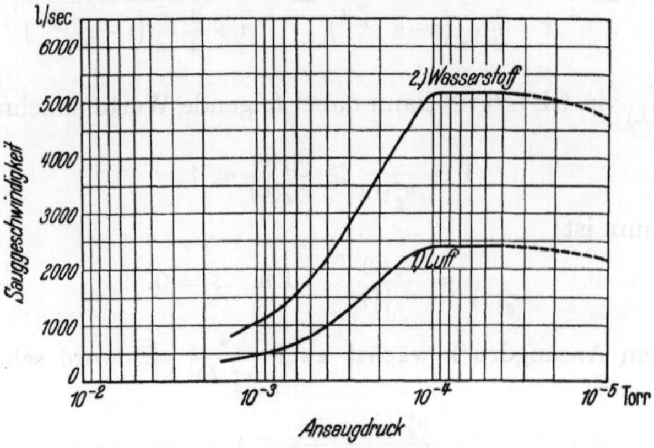

Fig. 90. Abhängigkeit der Sauggeschwindigkeit einer Öldiffusionspumpe vom Ansaugdruck bei verschiedenen Gasen.

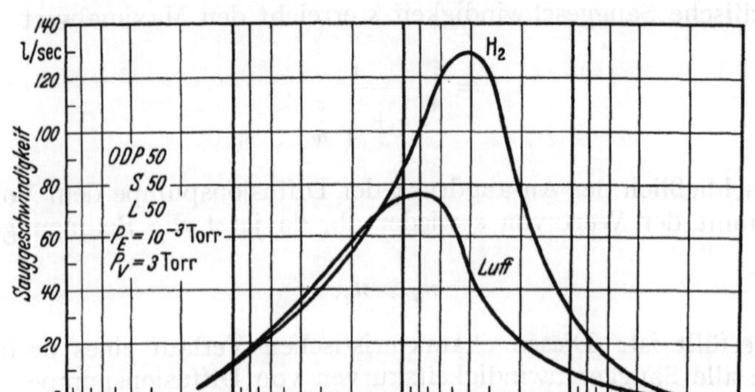

Fig. 91. Abhängigkeit der Sauggeschwindigkeit einer Dampfstrahlpumpe vom Ansaugdruck bei verschiedenen Gasen.

Dieser Wert von s_{max} läßt sich aber nicht realisieren, da er durch die höchstmöglichen Werte für die Dampfgeschwindigkeit w begrenzt wird. Nach den Ausführungen in Ziff. 25 liegt für eine Lavaldüse der Wert der Dampfgeschwindigkeit w zwischen der Schallgeschwindigkeit und der Maximalgeschwindigkeit w_{max}. Er steht in einem festen Verhältnis zu w_{max}, das durch die Düsenerweiterung gegeben ist, wobei w_{max} mit der Temperatur des Siedegefäßes (proportional T) zunimmt. In der Praxis wird man kaum wesentlich größere Werte als $w = 20000$ cm/sec erzielen können. Daraus ergibt sich nach Gl. (26.16) für Luft von 20° C:

$$s_{max} = \frac{462}{4}\,\frac{1}{1+\frac{1}{4}\,\frac{462}{200}} \cdot 10^2 = 7{,}3 \cdot 10^3 \left[\frac{cm^3}{cm^2\,sec}\right]. \tag{26.21}$$

Dieser Wert für die praktische obere Grenze der Sauggeschwindigkeit einer Diffusionspumpe für Luft steht in Einklang mit allen bisherigen Erfahrungen.

Setzt man in Gl. (26.16) für $\bar{c}$ die Werte für die mittlere Molekulargeschwindigkeit von Luft bzw. Wasserstoff ein und außerdem für die Dampfgeschwindigkeit einen Wert von $w \approx 20000$ cm/sec, so ergibt sich

$$\frac{s_{\mathrm{H}_2}}{s_{\mathrm{Luft}}} = 1{,}8.$$

Dies ist zu vergleichen mit den Werten der Fig. 90. Da bei niedrigen Ansaugdrucken für die Wirksamkeit der *Dampfstrahlpumpen* ein analoger Mechanismus wie bei den Diffusionspumpen wirksam ist, so ist es einleuchtend, daß sich bei den Dampfstrahlpumpen die Sauggeschwindigkeiten für verschiedene Gasarten

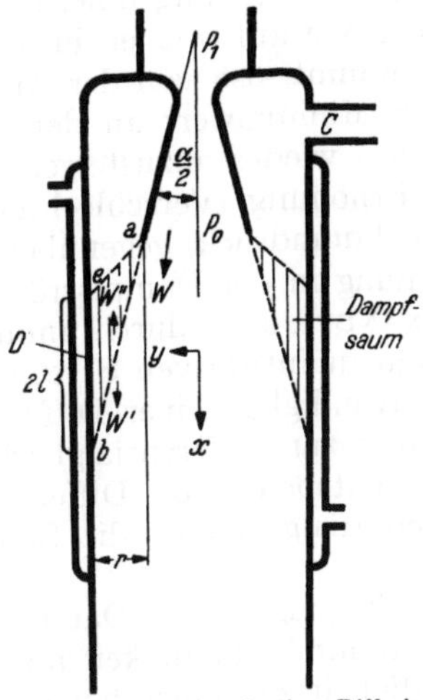

Fig. 92. Dampfverteilung in einer Diffusionspumpe (zweite Näherung).

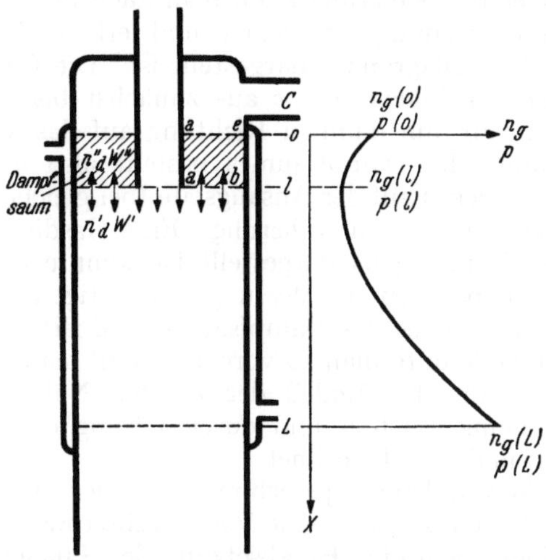

Fig. 93. Schema (zweite Näherung) der Vorgänge in einer Diffusionspumpe. Diagramm rechts: Verlauf des Partialdrucks der permanenten Gase (Gasdichte) im Dampfstrom.

bei niedrigen Ansaugdrucken unterhalb des Wertes für das Maximum der Sauggeschwindigkeit analog zueinander verhalten wie bei den Diffusionspumpen (vgl. hierzu die Meßergebnisse in Fig. 91).

δ) Zweite Näherung, Dampfsaum. Nach den bisherigen Darlegungen ist die spezifische Sauggeschwindigkeit s unabhängig von der Dampfdichte n_d und die Sauggeschwindigkeit S proportional zur Größe der Diffusionsfläche F. Wie die Messungen an technischen Diffusionspumpen zeigen, gilt beides nur in einer ersten Näherung. Für ein tieferes Verständnis des wirklichen Verhaltens der Diffusionspumpen müssen wir daher das vereinfachte Modell der ersten Näherung durch ein verbessertes Modell (zweite Näherung) ersetzen. Hierzu betrachten wir Fig. 92, die den schematischen Aufbau einer Diffusionspumpe wiedergibt. Danach ist ein zentraler Dampfstrom, der aus der Lavaldüse austritt, von einer ringförmigen Diffusionsfläche $a \ldots b$ umgeben. Nehmen wir einmal an, daß alle aus der Düsenmündung austretenden Dampfmoleküle nur eine Geschwindigkeitskomponente w in der Fortschreitungsrichtung des Dampfstromes hätten, so würde dieser etwa durch die Linie $a - b$ begrenzt. Nun enthält aber der mit der Geschwindigkeit w nach abwärts strömende Dampf auch Moleküle mit einer Geschwindigkeitskomponente der thermischen Molekularbewegung, die größer ist als w und die außerdem in entgegengesetzter Richtung weist. Diese Dampfmoleküle können daher die Fläche $a \ldots b$ nach rückwärts überschreiten und damit in den Raum zwischen $a \ldots b$ und $a \ldots e$ gelangen. Sie bilden dort einen in entgegengesetzter Richtung zum Hauptstrom strömenden Dampfsaum. Unter Berücksichtigung dieser Tatsache ergibt sich für die Wirkungsweise der Diffusionspumpen etwa das Schema der Fig. 93. Von der Fläche $a \ldots b$ nach abwärts bewegt sich ein Dampfstrom mit der Dampfdichte n'_d und der Geschwindigkeit w'. Zwischen der Fläche $a \ldots b$ und der Fläche $a \ldots e$ im Dampfsaum bewegt sich ein Dampfstrom mit der Dichte n''_d und der Geschwindigkeit w'' nach aufwärts. Die angesaugten Gasmoleküle treten also durch die Fläche $a \ldots e$ in einen entgegenströmenden Dampf (Dampfsaum) ein, durch den sie hindurch diffundieren müssen, bis sie unterhalb der Fläche $a \ldots b$ in einen Dampfstrom gelangen, der sie mitnimmt und in das Vorvakuum abführt. Durch analoge Ansätze, wie bei der ersten Näherung kann man auch hier die Gasdichteverteilung innerhalb der Dampfströmung errechnen und erhält dann einen Verlauf wie er in Fig. 93 (rechtes Diagramm) dargestellt ist. Die Gasdichte nimmt also von der Ansaugseite der Fläche $a \ldots e$ aus zunächst bis auf einen Minimalwert an der Stelle $a \ldots b$ ab, um dann in Richtung auf das Vorvakuum wieder anzusteigen. Der Einfluß des Dampfsaumes besteht also in einer Erhöhung (Verschlechterung) des Druckes auf der Ansaugseite (bei gleichem Vorvakuumdruck) gegenüber dem Modell der ersten Näherung. Eine Gasdichteverteilung in der Dampfströmung, wie sie in Fig. 93 dargestellt ist, konnte von ALEXANDER auch durch unmittelbare experimentelle Messungen bestätigt werden. Zur quantitativen Berechnung des Einflusses des Dampfsaumes auf die Sauggeschwindigkeit einer Diffusionspumpe könnte man so vorgehen, daß man in Analogie zur Differentialgleichung (26.9) für das Modell der zweiten Näherung die entsprechende Differentialgleichung aufstellt und durch Lösung dieser Differentialgleichung die Sauggeschwindigkeit berechnet.

Wir wollen uns jedoch mit folgender Abschätzung begnügen. Der Dampfsaum wird ebenso, wie er eine Verschlechterung der Vorvakuumbeständigkeit mit sich bringt, auch eine Herabsetzung der Sauggeschwindigkeit zur Folge haben. Als Verhaltensvorschrift ergibt sich daraus, daß Diffusionspumpen nach Möglichkeit so zu bauen sind, daß der Einfluß des Dampfsaumes nicht in Erscheinung tritt. Dies wird immer dann der Fall sein, wenn die mittlere freie Weglänge der angesaugten Gasmoleküle im Dampfsaum so groß ist, daß sie denselben durchfliegen können, ohne in ihm Stöße zu erleiden. Sie gelangen dann unmittelbar in die nach abwärts gerichtete Dampfströmung unterhalb der Fläche $a \ldots b$, wo also dieselben Verhältnisse vorliegen wie bei dem Modell der ersten Näherung.

Die mittlere freie Weglänge im Dampfsaum hängt ab von der Dampfdichte an der Düsenmündung und von der nach abwärts gerichteten Dampfströmungsgeschwindigkeit. Die mittlere freie Weglänge ist um so größer, je kleiner die Dampfdichte in der Düsenmündung und je größer die Dampfströmungsgeschwindigkeit ist. Die Dampfdichte und Dampfgeschwindigkeit sind aber beide abhängig von der Temperatur im Siedegefäß (s. die obigen Ausführungen über die Thermodynamik der Düsenvorgänge), und zwar wachsen beide Größen mit zunehmender Temperatur. Sie sind also bezüglich der mittleren freien Weglänge konkurrierende Größen. Erhöht man daher bei einer Diffusionspumpe die Heizleistung und damit die Temperatur im Siedegefäß, so wird zunächst die Dampfströmungsgeschwindigkeit wachsen und damit der Einfluß des Dampfsaumes auf die Sauggeschwindigkeit zurückgehen. Die Werte für die Sauggeschwindigkeit steigen (vgl. Fig. 94). Dies gilt solange, wie die Dampfdichte in der Düsenmündung absolut genommen noch kleine Werte hat, so daß der Einfluß der Dampfgeschwindigkeit überwiegt. Wächst aber mit weiterhin zunehmender Heizleistung der Absolutwert der Dampfdichte in der Düsenmündung, so überwiegt schließlich ihr Einfluß denjenigen der Strömungsgeschwindigkeit auf den Dampfsaum, so daß also die Sauggeschwindigkeit wieder herabgesetzt wird.

Wir wollen schließlich noch den Einfluß der Diffusionsfläche F, d.h. der Ringfläche $a \ldots e$ zwischen Düsenmündung und Kühlwassermantel (Fig. 92 und 93) betrachten.

Nach dem bisher Gesagten sollte man annehmen, daß die Sauggeschwindigkeit S proportional zur Größe der Diffusionsfläche F, d.h. also zur Größe des Kegelmantels zwischen Treibdüse und Kühlmantel anwüchse. Die Erfahrung zeigt aber, daß dies nicht der Fall ist (s. Fig. 95).

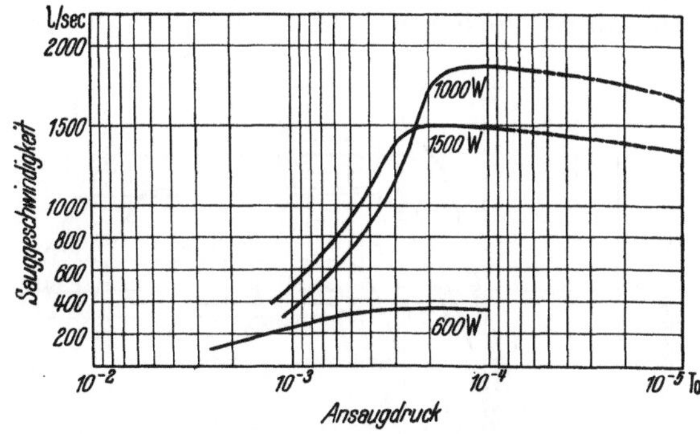

Fig. 94. Abhängigkeit der Sauggeschwindigkeit einer Öldiffusionspumpe von der Heizleistung.

Zur Erklärung hierfür betrachten wir noch einmal Fig. 92. Danach hat der Dampfsaum nicht überall dieselbe Breite. Er ist in unmittelbarer Nähe der Treibdüsenmündung schmaler als in der Nähe des Kühlwassermantels, wo er seinen größten Wert $2l$ hat. Versucht man nun durch Vergrößerung des Durchmessers des Kühlwassermantels die Diffusionsfläche F zu vergrößern, so vergrößert man damit auch die mittlere Breite des Dampfsaumes und setzt damit die Sauggeschwindigkeit herab. Besonders deutlich wird dies, wenn man nicht die Sauggeschwindigkeit S selber, sondern die spezifische Sauggeschwindigkeit s

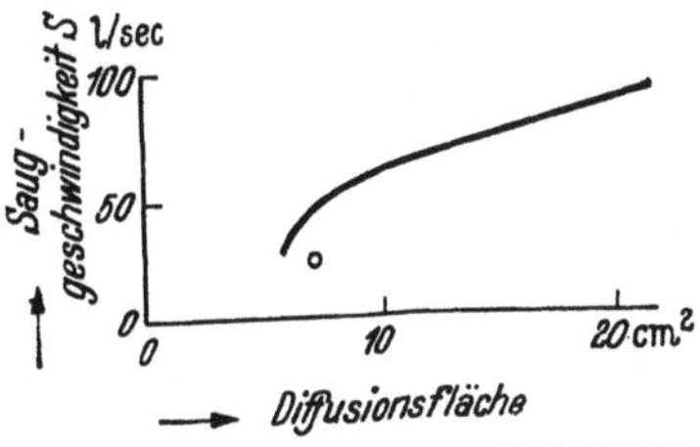

Fig. 95. Abhängigkeit der Sauggeschwindigkeit S von der Diffusionsfläche F (nach COPLEY u. Mitarb.).

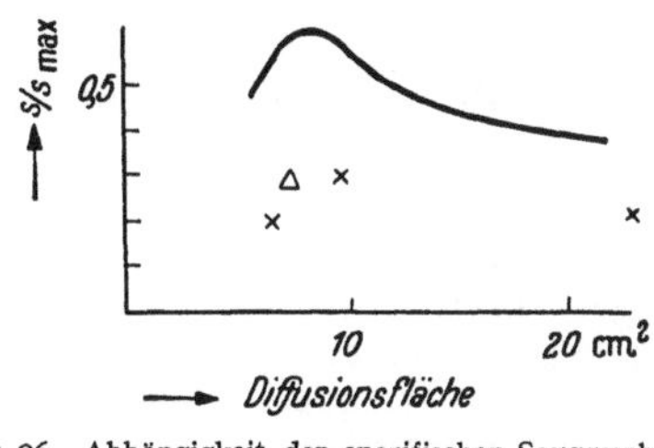

Fig. 96. Abhängigkeit der spezifischen Sauggeschwindigkeit s von der Diffusionsfläche F (nach COPLEY u. Mitarb.).

betrachtet (s. Fig. 96). Für kleine Werte der Diffusionsfläche hat s den Wert, der sich aus dem Modell der ersten Näherung ergibt; d.h. also der Dampfsaum ist zunächst ohne Einfluß. Mit zunehmender Größe der Diffusionsfläche wächst allmählich der Einfluß des Dampfsaumes. Der Wert für die spezifische Sauggeschwindigkeit fällt.

Zur Illustration des oben Gesagten betrachten wir schließlich noch die Fig. 97, in der für eine Diffusionspumpe die Sauggeschwindigkeit in Abhängigkeit vom Ansaugdruck dargestellt ist, und zwar für verschiedene Heizleistungen als Parameter. Man sieht daraus, wie mit zunehmender Heizleistung der Maximalwert der Sauggeschwindigkeit (breites Plateau) erst steigt und dann wieder abnimmt, in Analogie zu unseren Überlegungen bei Fig. 94. Außerdem nimmt aber auch der Wert des erreichbaren Hochvakuumdruckes zunächst mit wachsender

Heizleistung zu. Nach hohen Drucken zu reicht die Sauggeschwindigkeit mit wachsender Heizleistung zu immer größeren Drucken. Diese beiden Erscheinungen sind verständlich durch unsere Ausführungen über die Vorvakuumbeständigkeit von Diffusionspumpen. Schließlich steigt für sehr hohe Heizleistung auch der Druck auf der Hochvakuumseite wieder an. Dies ist aus den bisherigen Ausführungen noch nicht zu erklären. Es rührt daher, daß es sich bei der untersuchten Pumpe um eine Öldiffusionspumpe handelt, bei der der Hochvakuumdruck weitgehend durch den Dampfdruck des Treibmittels bedingt ist. Wir werden weiter unten sehen, daß diese Treibmittel bei zu starker Heizung sich thermisch zersetzen und damit Zerfallsprodukte bilden, die einen höheren Dampfdruck haben als das ursprüngliche Treibmittel.

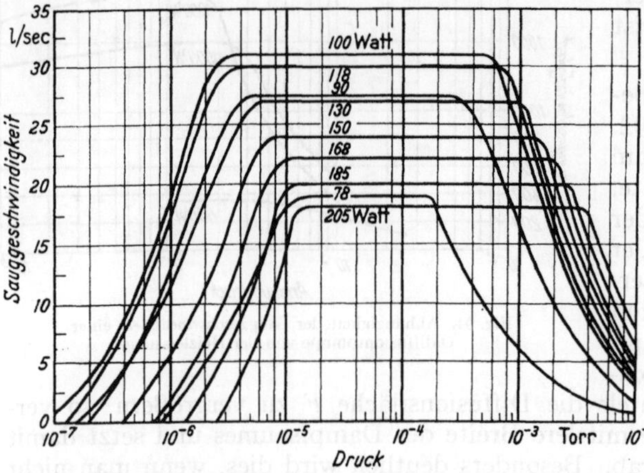

Fig. 97. Sauggeschwindigkeitskurven für eine dreistufige fraktionierende Öldiffusionspumpe in Abhängigkeit von der Heizleistung (DPI).

27. Technik der Dampfstrahl- und Diffussionspumpen. Wie in Ziff. 26 ausgeführt wurde, sind für Diffusionspumpen gute Vorvakuumbeständigkeit und hohe spezifische Sauggeschwindigkeit einander widersprechende Forderungen. Technische Ausführungen von Diffusionspumpen werden daher meistens mehrstufig gebaut (vgl. z.B. Fig. 98), und zwar in der Weise, daß mehrere hintereinander geschaltete Einzelpumpen in einem Gehäuse zu einer Einheit vereinigt werden. Dadurch kann man die erste Stufe, die der Ansaugöffnung am nächsten gelegen ist, so auslegen, daß sie bei zwar geringer Vorvakuumbeständigkeit eine möglichst gute spezifische Sauggeschwindigkeit besitzt. Die folgenden Stufen benötigen dann wegen der inzwischen erfolgten Vorkompression nur noch eine kleinere Sauggeschwindigkeit. Sie können dafür für günstigere Vorvakuumwerte bemessen werden. Durch Verteilung der einander widersprechenden Aufgaben auf verschiedene Stufen erhält man auf diese Weise Pumpen, die trotz guter Vorvakuumbeständigkeit auch hohe Werte der Sauggeschwindigkeit zeigen. Allgemein gilt, daß eine Erhöhung der Heizleistung bei der Diffusionspumpe auch den zulässigen Vorvakuumdruck nach höheren Werten verschiebt. Für die Erzielung günstiger Sauggeschwindigkeiten ist eine wirksame Kühlung entscheidend. Die Pumpen werden daher in der Mehrzahl der Fälle mit Wasserkühlung betrieben, wobei die Kühlwassertemperatur möglichst unter 20° C liegen soll. Der Kühlwasserdurchfluß ist (am besten automatisch) so zu regeln, daß die Temperatursteigerung des Kühlwassers beim Durchfluß nicht wesentlich mehr als 1° C beträgt. An Stelle der Wasserkühlung wird für Pumpen kleinerer Leistung gelegentlich auch Luft- oder Ölkühlung angewandt. Wegen des schlechteren Wärmeüberganges sind bei diesen Kühlmedien dann aber baulich besondere Maßnahmen erforderlich.

Als *Werkstoffe* für den Bau der Dampfstrahl- und Diffusionspumpen kommen Glas, Quarz und Metalle in Frage. Quarz hat gegenüber Glas den Vorteil der größeren Temperaturwechselfestigkeit. Glas und Quarz haben den Vorzug geringerer Korrosionsanfälligkeit (Verwendung in chemischen Laboratorien). Wegen der Bruchgefahr beschränkt sich ihre Verwendung aber ausschließlich auf das Laboratorium. Alle technischen Pumpen, insbesondere Pumpen großer Sauggeschwindigkeit, werden aus Metall hergestellt.

Als Treibmittel für die Diffusions- und Dampfstrahlpumpen verwendet man Quecksilberdampf und die Dämpfe von hochmolekularen organischen Substanzen (Öle). Man unterscheidet dementsprechend

1. *Quecksilber*-Diffusions- und Dampfstrahl-Pumpen,

2. *Öl*-Diffusions- und Dampfstrahl-Pumpen.

α) Quecksilber-Diffusions- und Dampfstrahl-Pumpen. Fig. 98 zeigt eine mehrstufige Quecksilberdiffusionspumpe aus Metall. Diese Pumpen werden vielfach so gebaut, daß die hintereinander geschalteten Pumpstufen zusammen mit einem gemeinsamen Steigrohr für den Quecksilberdampf eine aus dem Pumpenkörper herausziehbare Einheit bilden, wobei der Pumpenkörper selbst dann den Kühlwassermantel und das Verdampfergefäß für das Quecksilber zusammen mit den Anschlüssen an der Saugseite und der Vorvakuumseite trägt. Die Pumpstufen werden zweckmäßig so konstruiert, daß die saugseitige Stufe als einfache Umkehrdüse ausgebildet ist und als Diffusionspumpe arbeitet, während die anschließend folgenden vorvakuumseitigen Stufen als echte Dampfstrahlpumpen arbeiten und aus einer Treibdüse, einem Mischraum und einer Staudüse bestehen. Fig. 99 zeigt eine analoge Pumpe aus Glas. Zum Unterschied von der Metallpumpe sind hier die vorvakuumseitigen Dampfstrahlstufen mit Ringdüsen ausgerüstet. Kurve E in Fig. 100 zeigt die Sauggeschwindigkeit in Abhängigkeit vom Ansaugdruck für die Pumpe der Fig. 98. Nach hohen Drucken zu beginnt der Arbeitsbereich der Quecksilberdiffusionspumpen zwischen 10^{-2} und 10^{-3} Torr und reicht in technischen Anordnungen bis etwa 10^{-7} Torr. Darüber hinaus lassen sich aber auch bei ganz besonders sorgfältigen Vorkehrungen im Laboratorium mit Quecksilberdiffusionspumpen Restgasdrucke von etwa 10^{-9} Torr erreichen. Vorausgesetzt ist natürlich in allen Fällen, daß der Quecksilberdampf, dessen Druck bei Zimmertemperatur etwa einen Wert von 10^{-3} Torr hat, durch geeignete (mit flüssiger Luft gekühlte) Kühlfallen ausgefroren wird.

Da die Kurven für die Sauggeschwindigkeit in Abhängigkeit vom Ansaugdruck bei den Quecksilberdampfstrahlstufen bei etwa 10^{-1} Torr ein spitzes Maximum zeigen, kann man bei Quecksilberdiffusionspumpen mit vorgeschalteten Dampfstrahlstufen durch Steigerung der Leistung der Dampfstrahlstufen (z.B. durch Vermehrung der Düsenzahl je Stufe) die Kurve für die Sauggeschwindigkeit in Abhängigkeit vom Ansaugdruck so verändern, daß sie sich weiter in das Gebiet höherer Ansaugdrucke erstreckt (s. Fig. 100, Kurve M). Die Fig. 101 zeigt im Schema die zu dieser Kurve gehörige Pumpe im Schnitt. Insbesondere sieht man, daß die Zahl der Düsen in den Dampfstrahlstufen vermehrt ist. Die technische Pumpe enthält in der mittleren Pumpstufe sechs parallel und in der vorvakuumseitigen Pumpstufe zwei parallel geschaltete Düsen.

Der maximal zulässige Vorvakuumdruck hat bei einstufigen Quecksilberdiffusionspumpen (die aber nur selten verwendet werden) einen Wert von 0,1 Torr. Bei Vorschaltung einer Dampfstrahlstufe erhöht sich dieser Wert auf einige Torr bis 10 Torr und bei Vorschaltung von zwei Dampfstrahlpumpen auf 10 bis 20 Torr.

β) Öldampfstrahlpumpen. Wie bereits erwähnt wurde, benutzen die Öldampfstrahlpumpen als Treibmittel die Dämpfe von hochmolekularen organischen Substanzen (Öle). Von diesen Treibmitteln und den an sie zu stellenden Forderungen wird weiter unten bei den Öldiffusionspumpen noch ausführlich die Rede sein. Die Treibmittel für die Öldampfstrahlpumpen müssen eine möglichst hohe thermische Beständigkeit besitzen, und außerdem soll ihr Dampfdruck bei der Temperatur des Siedegefäßes möglichst hohe Werte haben. Demgegenüber kann die Forderung nach einem extrem niedrigen Dampfdruck bei der Temperatur der Kondensationsfläche zurücktreten, da hier geringere Ansprüche als bei den Öldiffusionspumpen ausreichen. Bewährt haben sich hierfür vor allem Clophene (Pentachlordiphenyl-ähnliche Substanzen) und außerdem höher siedende Mineralölfraktionen.

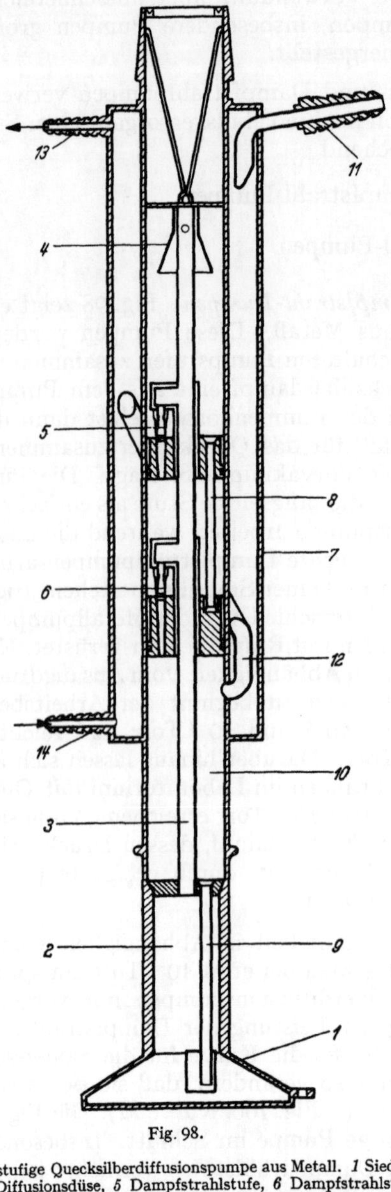

Fig. 98.

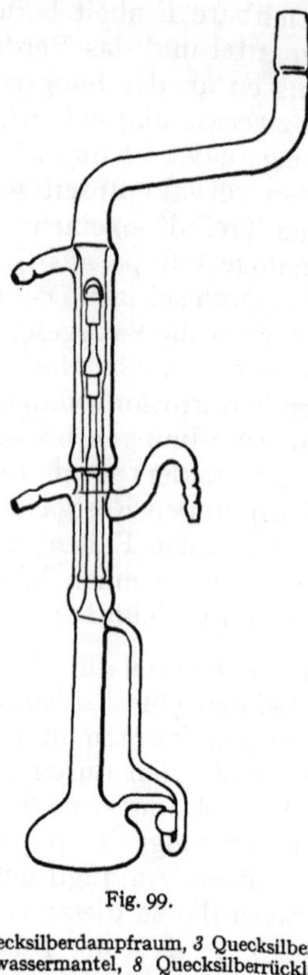

Fig. 99.

Fig. 98. Dreistufige Quecksilberdiffusionspumpe aus Metall. *1* Siedegefäß, *2* Quecksilberdampfraum, *3* Quecksilberdampfsteigrohr, *4* Diffusionsdüse, *5* Dampfstrahlstufe, *6* Dampfstrahlstufe, *7* Kühlwassermantel, *8* Quecksilberrücklaufrohr, *9* Quecksilberüberlauf, *11* Vorvakuumanschluß, *13* Kühlwasserabfluß, *14* Kühlwasserzufluß.

Fig. 99. Dreistufige Quecksilberdiffusionspumpe aus Glas.

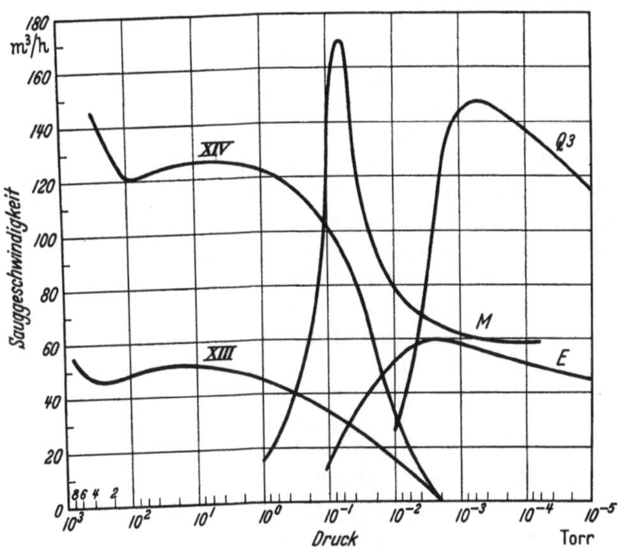

Fig. 100.

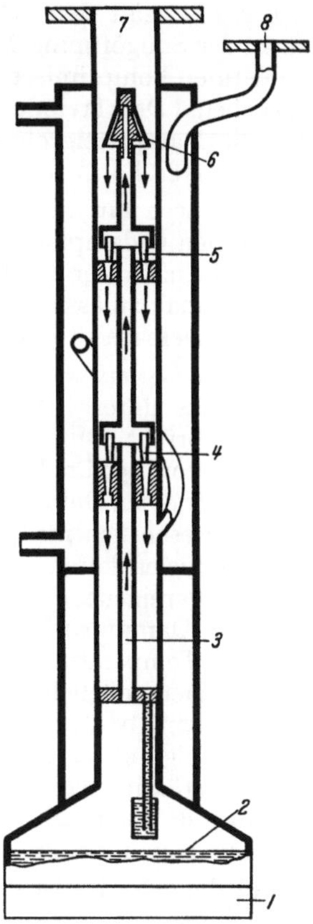

Fig. 101.

Fig 100. Sauggeschwindigkeit in Abhängigkeit vom Ansaugdruck für verschiedene Pumpenarten: XIII und XIV einstufige rotierende Ölluftpumpen, *M* Quecksilberdampfstrahlpumpe, *E* Quecksilberdiffusionspumpe, *Q 3* dreistufige fraktionierende Öldiffusionspumpe.

Fig. 101. Quecksilberdiffusions- und -dampfstrahlpumpe aus Metall. *1* Heizkörper, *2* Quecksilbersiedegefäß, *3* Quecksilberdampfsteigrohr, *4* zwei parallel geschaltete Quecksilberdampfstrahldüsen, *5* sechs parallel geschaltete Quecksilberdampfstrahldüsen, *6* Diffusionsdüse, *7* Hochvakuumanschluß, *8* Vorvakuumanschluß.

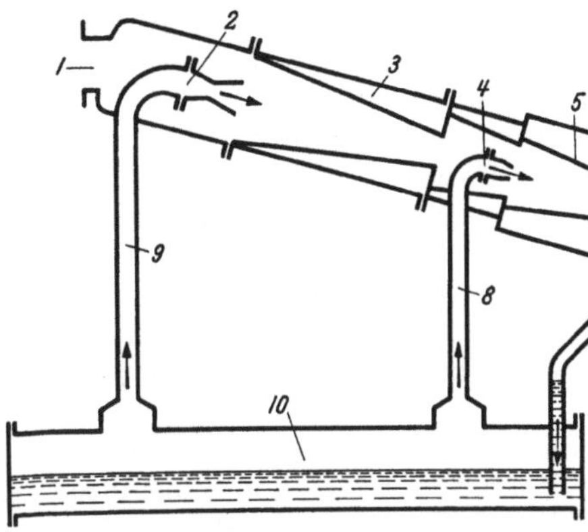

Fig. 102. Öldampfstrahlsauger. *2* Treibdampfdüse der ersten Stufe, *3* Staudüse und Kühlmantel der ersten Stufe, *4* Treibdüse der zweiten Stufe, *5* Staudüse der zweiten Stufe.

Fig. 102 zeigt den schematischen Schnitt durch einen Öldampfstrahlsauger. Die an der Saugöffnung *1* eintretenden Gase werden in ein oder mehreren Dampfstrahlstufen komprimiert und durch den Vorvakuumanschluß *6* an die Vorpumpe abgegeben. Der Treibdampf für die Treibdüsen gelangt aus dem Siedegefäß *10* über die Dampfsteigrohre *9* und *8* zu den Treibdüsen und läuft schließlich in kondensierter Form durch das Rücklaufrohr *7* in das Siedegefäß *10* zurück.

Die Kurve für die Sauggeschwindigkeit in Abhängigkeit vom Ansaugdruck hat bei den Öldampfstrahlsaugern ein spitzes Maximum bei etwa 10^{-1} Torr mit Werten je nach der Größe des Saugers zwischen 50 l/sec und einigen 100 l/sec. Die maximal zulässigen Vorvakuumdrucke liegen zwischen 1 und 3 Torr, die hierbei erzielbaren totalen Enddrucke bei etwa 10^{-3} Torr.

γ) Öldiffusionspumpen. Mit Quecksilberdiffusionspumpen kann man nur Totaldrucke unter 10^{-3} Torr erreichen, wenn man tiefgekühlte Fallen zur Anwendung bringt. Burch[1] hat als erster darauf hingewiesen, daß man durch Verwendung von hochmolekularen organischen Substanzen (Ölen), die bei Zimmertemperatur einen Dampfdruck unter 10^{-5} Torr haben, als Treibmittel in Diffusionspumpen wesentlich niedrigere Totaldrucke (ohne Verwendung von tiefgekühlten Fallen) erzielen kann. Diese *Treibmittel (Öle)* sollen möglichst folgende Eigenschaften vereinen: Sie sollen bei der Kühlwassertemperatur einen möglichst niedrigen Dampfdruck (10^{-5} Torr und darunter) haben. Außerdem sollen sie bei dieser Temperatur flüssig sein und bei der Temperatur des Siedegefäßes (etwa 200° C) einen möglichst hohen Dampfdruck (im Hinblick auf die Vorvakuumbeständigkeit) besitzen. Bei dieser Temperatur des Siedegefäßes sollen sie jedoch geringe Neigung zu thermischem Zerfall zeigen und außerdem möglichst oxydationsbeständig sein. In Tabelle 7 sind die wichtigsten und bekanntesten Substanzen, die bisher als Treibmittel Verwendung gefunden haben, zusammen mit den Angaben über ihre Dampfdruckwerte zusammengestellt.

Für die geforderten Eigenschaften kommen vor allen Dingen Substanzen mit Molekulargewichten zwischen etwa $M = 300$ und $M = 700$ in Frage. Die Kurven für die Dampfdrucke in Abhängigkeit von der Temperatur finden sich für einige der aufgeführten Substanzen in den Fig. 103 bis 106. Vergleichsweise zeigt Fig. 107 die Dampfdrucktemperaturkurven für einige der in der Hochvakuumtechnik für Schliffe und Hähne gebräuchlichen Dichtungsfette.

Es ist nun aber keineswegs so, daß die aus der Tabelle 7 und den Figuren 103 bis 107 für Zimmertemperatur zu entnehmenden Dampfdruckwerte auch gleichzeitig den Enddruck ergeben, der in Diffusionspumpen bei Verwendung des betreffenden Treibmittels erzielt werden kann. Der Enddruck setzt sich vielmehr zusammen aus dem Dampfdruck des Treibmittels bei Zimmertemperatur und dem Partialdruck der Restgase. Hinzu kommen noch Zersetzungsprodukte des Treibmittels, die sich durch thermische Zersetzung und Oxydation bilden, und die im allgemeinen höhere Dampfdruckwerte haben, als das Treibmittel selbst. Sie sind in der Regel bei Zimmertemperatur flüchtig, können aber in tiefgekühlten Fallen niedergeschlagen werden. Der Enddruck, der mit Öldiffusionspumpen erzielt werden kann, hängt also sehr stark von dem verwendeten Treibmittel, den Betriebsbedingungen und außerdem der Vorbehandlung ab. Meistens ist der Dampfdruck des Treibmittels bzw. seiner Zersetzungsprodukte größer als der Druck der Restgase.

Die verschiedenen Treibmittel lassen sich einteilen in chemische einheitliche Substanzen und solche, die aus Gemischen von Substanzen verschiedener chemischer Zusammensetzung bestehen. Zu den ersteren, die vor allen Dingen von

[1] C. R. Burch: Proc. Roy. Soc. Lond., Ser. A **123**, 271 (1929).

Tabelle 7. *Diffusionspumpentreibmittel.*

Treibmittel	Chemische Zusammensetzung	A	B	T^{-5} °C	T^{-2} °C	p bei 25° C[1]	Literatur
Butylphthalat	Di-n-butylphthalat	14,215	4680	18	81	$3,3 \cdot 10^{-5}$	1
Octylphthalat	Di-n-octylphthalat	15,94	6035	65	128	$4 \cdot 10^{-8}$	5
Nonylphthalat	Di-n-nonylphthalat	16,41	5690	73	146	$1 \cdot 10^{-8}$	5
Octoil	Di-2-octylhexylphthalat	15,116	5590	54	128	$2,3 \cdot 10^{-7}$	1
		15,90	6157	67	134	$3,3 \cdot 10^{-8}$	5
Narcoil 40	Di(3,5,5-trimethylhexylphthalat)	15,88	5936	57	124	$6 \cdot 10^{-8}$	5
Amoil	i-Di-amylphthalat	13,60	4610	22	93	$1,3 \cdot 10^{-5}$	1
Amoil S	i-D-amylsebacat	14,40	5190	25	114	$1 \cdot 10^{-6}$	4
Octoil S	Di-2-äthylhexylsebacat	14,26	5514	50	142	$2 \cdot 10^{-8}$	3
m-Cr	Tri-m-cresylphosphat	13,982	5373	50	141	$9 \cdot 10^{-8}$	2
p-Cr	Tri-p-cresylphosphat	15,223	5926	52	144	$2 \cdot 10^{-8}$	2
b-S	Di-Benzylsebacat	15,775	6320	64	155	$4 \cdot 10^{-9}$	2
Apiezon A	Gemisch aus Kohlenwasserstoffen			37	110	$2 \cdot 10^{-6}$	3
Apiezon B				50	127	$4 \cdot 10^{-7}$	3
Apiezon C		14,67	5925	77	160	$1 \cdot 10^{-8}$	6
Arochlor	ähnlich Pentachlordiphenyl			27	93	$8 \cdot 10^{-6}$	3
Littonoil	gerade Kohlenwasserstoffketten C_nH_{2n}			57	132	$1,4 \cdot 10^{-7}$	3
Chlophen A 40	chloriertes Benzol	13,15	4135	0	67	$2 \cdot 10^{-4}$	6
Silikon DC 703	halborganische Verbindung des Siliciums	14,32	6165	83	153	$5 \cdot 10^{-9}$	6
Diffelen						$5 \cdot 10^{-8}$	
Diffelen V						$1,5 \cdot 10^{-8}$	

Angegeben sind die Konstanten der Dampfdruckformel Log $P = A - B/T$, wobei P den Dampfdruck in Mikron (10^{-3} Torr) und T die absolute Temperatur bedeuten. T^{-5} und T^{-2} geben die Temperatur in °C an, bei denen ein Dampfdruck von 10^{-5} bzw. 10^{-2} Torr herrscht. Außerdem werden noch die Dampfdrucke bei Zimmertemperatur (25° C) (p in Torr) aufgeführt.

Literatur zur Tabelle 7.

1 K. C. D. Hickman, J. C. Hacker u. N. D. Embree: Industr. Engn. Chem. **9**, 264 (1937).
2 F. H. Verhoek u. A. L. Marshall: J. Amer. Chem. Soc. **61**, 2737 (1939).
3 Metropolitan Vickers (1946).
4 Values of constants deduced by S. F. Kapff of Distillation Products, Inc.
5 H. Hoyer u. W. Peperle: Z. Elektrochem. **62**, 61 (1958).
6 A. Herlet u. G. Reich: Z. angew. Phys. **9**, 1 (1957), insbesondere S. 14—23.

Hickman[1] zuerst angegeben wurden, gehören die Ester organischer Säuren mit höheren Alkoholen, zu den letzteren z. B. die Apiezonöle, die aus Mineralölen durch Abtrennung der leicht flüchtigen Fraktionen mittels Destillation im Feinvakuum gewonnen werden. Diese Treibmittel wurden zuerst von Burch[2] angewandt. In bezug auf die Beständigkeit im Pumpenbetrieb gegen Bildung von Zerfallsprodukten sind die letzteren den ersteren im allgemeinen überlegen, schon allein deswegen, weil die ersteren bei der Anwesenheit von geringen Mengen von Wasserdampf (was unter den meisten Betriebsbedingungen der Fall ist), durch Verseifung, d. h. unter Rückbildung von höherem Alkohol einerseits und organischen Säuren andererseits zerfallen.

Um den *Einfluß des Treibmittels* allein *auf* den mit Diffusionspumpen *erzielbaren Enddruck* zu untersuchen, ist es zweckmäßig, einstufige Öldiffusionspumpen (s. z. B. Fig. 108) zu verwenden, da in mehrstufigen, insbesondere fraktionierenden

1 K. C. D. Hickman: J. Phys. Chem. **34**, 637 (1930). — Hickman u. C. R. Sanford: Rev. Sci. Instrum. **1**, 140 (1930).
2 C. R. Burch: Proc. Roy. Soc. Lond., Ser. A **123**, 271 (1929).

Pumpen der Einfluß des Treibmittels allein zum Teil durch die Eigenschaften der Pumpe überdeckt werden kann. Fig. 109 zeigt den Einfluß der Temperatur

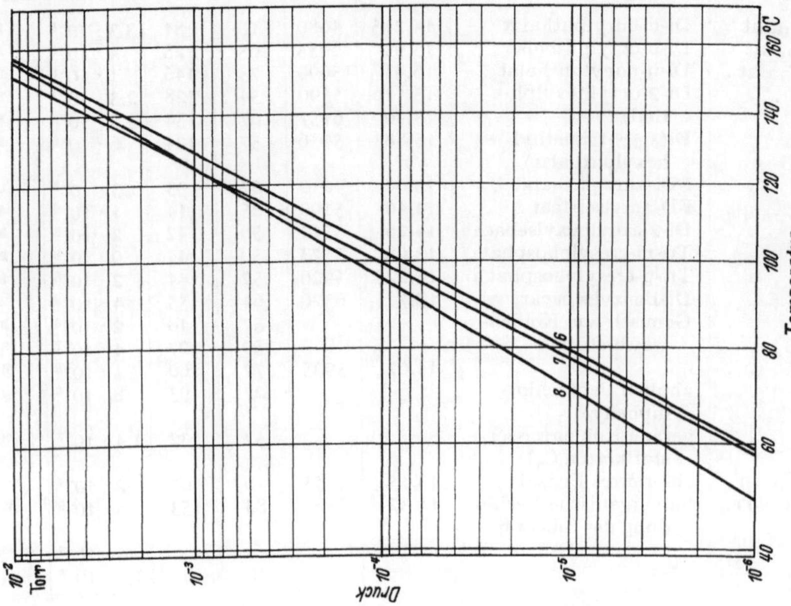

Fig. 104. Dampfdruckkurven von: *6* Leyboldöl F, *7* Leyboldöl H, *8* Apiezonöl E.

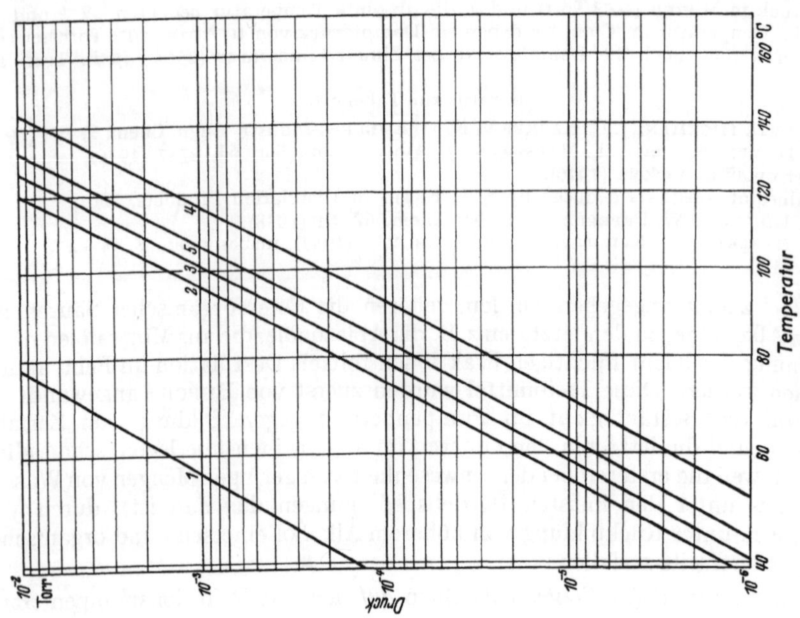

Fig. 103. Dampfdruckkurven von: *1* Butylphthalat, *2* Octoil, *3* Narcoil 40, *4* Octoil S, *5* Di-n-octylphthalat.

im Siedegefäß (bzw. der elektrischen Leistung des Heizkörpers) auf den mit Öldiffusionspumpen erzielbaren Enddruck. Bei niedriger Temperatur, d.h. kleiner Heizleistung von etwa 200 Watt, sinkt der Enddruck infolge des Auspumpvorganges nach etwa 210 min auf einen stationären Endwert von etwa $7 \cdot 10^{-6}$ Torr.

Bei Erhöhung der Heizleistung auf 250 Watt sinkt der Enddruck noch etwas weiter infolge der durch die höhere Temperatur und den daraus folgenden höheren

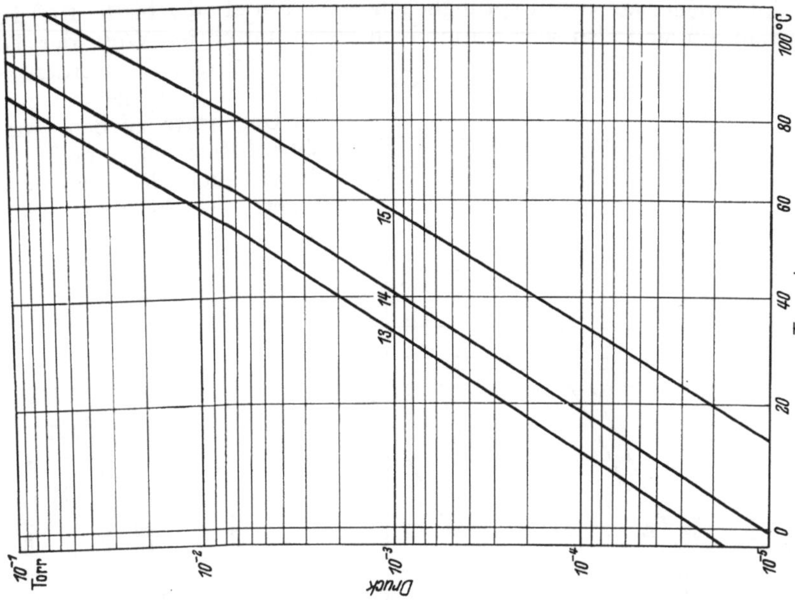

Fig. 106. Dampfdruckkurven von: *13* Clophen A 30, *14* Clophen A 40, *15* Clophen A 50

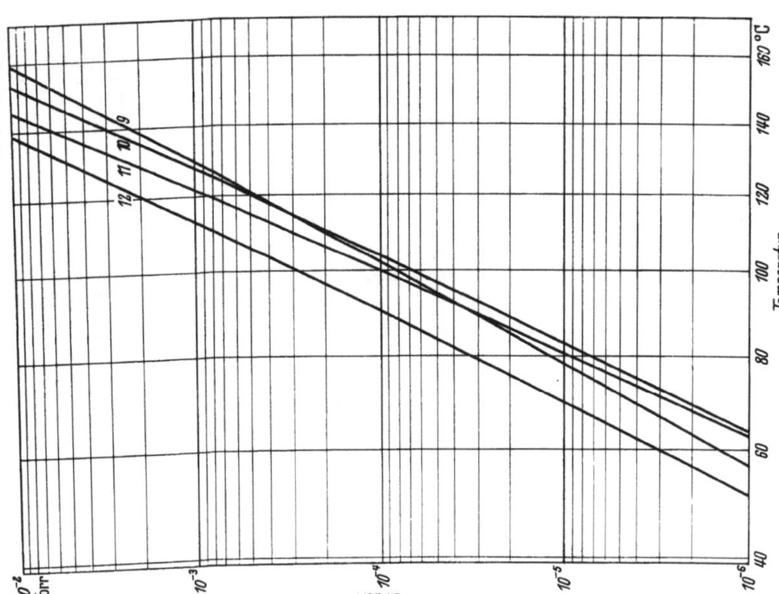

Fig. 105. Dampfdruckkurven von: *9* Apiezonöl C, *10* Silicon DC 703, *11* Siliconöl, *12* Tricresylphosphat.

Dampfdruck verbesserten Vorvakuumbeständigkeit der Pumpe. Bei noch weiterer Erhöhung der Temperatur (bzw. der Heizleistung auf 300 Watt) steigt aber der Enddruck infolge des jetzt einsetzenden thermischen Zerfalls wieder an. Bei weiterer Erhöhung der Heizleistung auf 350 bzw. 400 Watt erreicht der

Enddruck sogar nur Werte von etwa $2 \cdot 10^{-5}$ Torr. Setzt man dann anschließend die Heizleistung wieder auf 200 Watt herunter, so hört der thermische Zerfall auf. Die gebildeten Zerfallsprodukte werden ins Vorvakuum abgeführt und der Enddruck sinkt wieder auf den ursprünglichen Wert von etwa 7 bis $8 \cdot 10^{-6}$ Torr. Öldiffusionspumpen sind also sehr empfindlich gegen unrichtige Wahl der Temperatur im Siedegefäß. Diese muß genau und stationär eingestellt werden. Öldiffusionspumpen können daher nur mit elektrischer Heizung betrieben werden, im Gegensatz zu den Quecksilberdiffusionspumpen, bei denen auch die weniger definierte Gasheizung möglich ist.

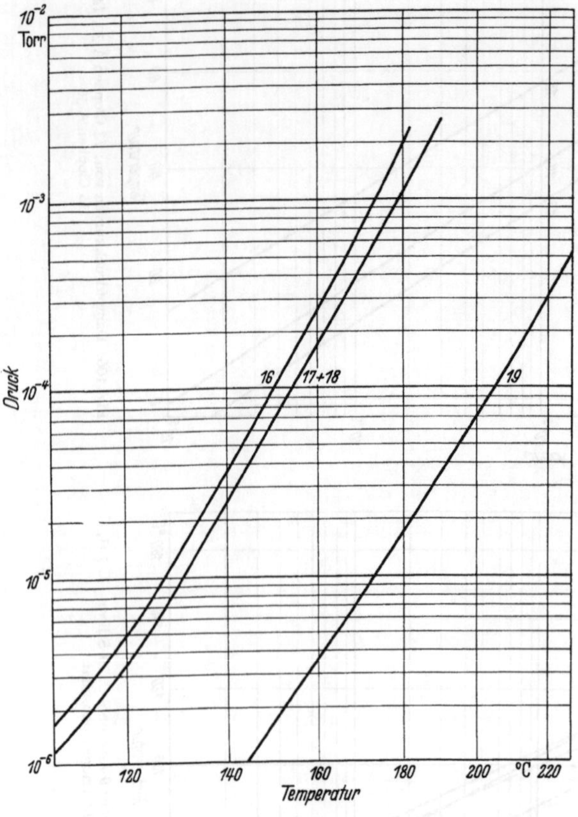

Fig. 107. Dampfdruckkurven von: *16* Apiezonfett P, *17* Apiezonfett R, *18* Apiezonfett S, *19* Siliconfett.

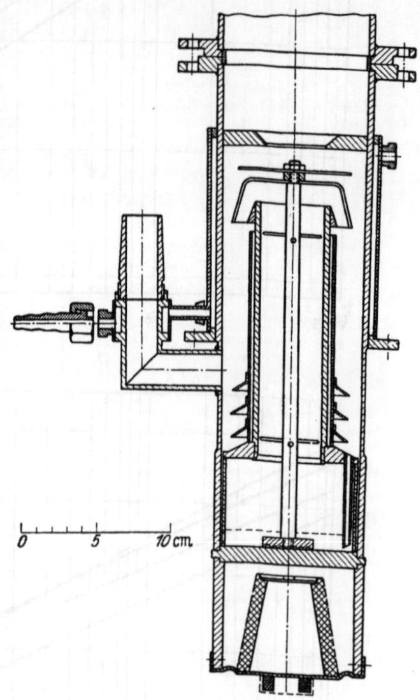

Fig. 108. Einstufige Öldiffusionspumpe aus Metall.

Für den Enddruck von Öldiffusionspumpen bedeuten mehr oder weniger starke *Lufteinbrüche* in die Vakuumapparatur und damit in die Pumpe eine erhebliche Gefahr. Durch die Anwesenheit des Luftsauerstoffes können die heißen Treibdämpfe oxydiert werden, wodurch Produkte entstehen, die im allgemeinen einen höheren Dampfdruck haben als das Treibmittel selbst. Dadurch wird also der erzielbare Enddruck der Pumpe verschlechtert. Es dauert dann, je nach der Stärke des Lufteinbruches eine längere oder kürzere Zeit, bis diese Produkte mit hohem Dampfdruck in das Vorvakuum abdestillieren, wodurch sich der Enddruck der Öldiffusionspumpe allmählich wieder verbessert. Fig. 110 zeigt den Enddruck einer einstufigen Öldiffusionspumpe in Abhängigkeit von der Zeit bei verschieden starken Lufteinbrüchen. Man sieht, daß je nach der Stärke des Lufteinbruches sich das ursprüngliche Endvakuum nur sehr langsam wieder einstellt.

GAEDE[1] hat zuerst den Gedanken ausgesprochen, daß man durch bewußte Züchtung von Destillationsprozessen eine Selbstreinigung der Öle in den Diffusionspumpen von solchen störenden Zersetzungsprodukten herbeiführen könne, um dadurch die Anfälligkeit der Pumpen gegen solche Lufteinbrüche und damit das Betriebsverhalten der Pumpen zu verbessern. Etwas später wurde von PILON[2] eine mehrstufige Pumpe mit einer entsprechenden Dampf- und Flüssigkeitsführung der Treibmittel zur Herbeiführung einer Destillationsreinigung beschrieben. Offenbar unabhängig von den beiden erstgenannten hat dann HICKMAN[3] eingehend *mehrstufige Öldiffusionspumpen mit Fraktioniereinrichtung* zur Selbstreinigung der Öle durch Destillationsprozesse in den Pumpen beschrieben und ihr Verhalten eingehend untersucht. Fig. 111 zeigt im Prinzip eine solche mehrstufige Öldiffusionspumpe mit Fraktioniereinrichtung in einer Form, wie sie heute vielfach

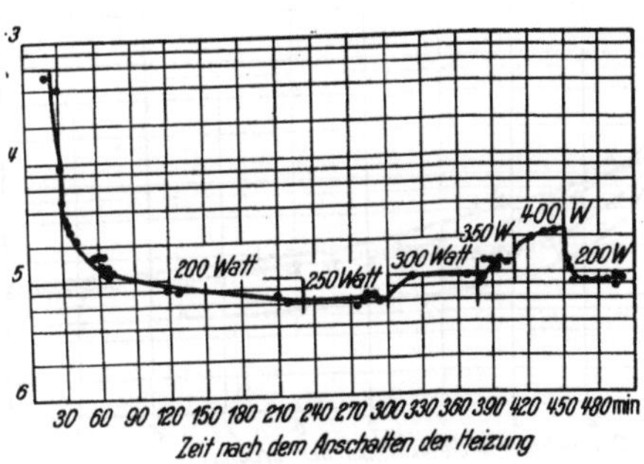

Fig. 109. Endvakuumdrucke einer mit Octoil betriebenen zweistufigen Diffusionspumpe aus Metall ohne Fraktionierung (Leybold) bei verschiedenen Heizleistungen.

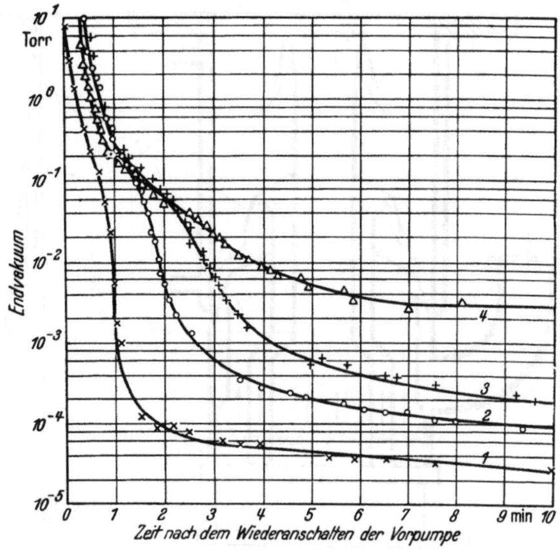

Fig. 110. Anfangsanstieg des Endvakuums nach Lufteinbrüchen in einer einstufigen Diffusionspumpe aus Metall (Leybold). Lufteinbrüche: *1* 10 Torr, 10 min; *2* 420 Torr, 2 min; *3* 760 Torr, 2 min; *4* 760 Torr, 10 min.

in technischem Gebrauch ist. Der aus den drei übereinander angeordneten Düsen austretende Treibdampf schlägt sich auf der wassergekühlten Kondensationsfläche *1* nieder und gelangt in flüssiger Form in das Siedegefäß *5*. Dieses ist durch drei zylindrische Rohre, die gleichzeitig als Dampfsteigrohre dienen, in drei konzentrische Räume unterteilt, die nur unterhalb des Flüssigkeitsspiegels miteinander in Verbindung stehen. Das flüssige Treibmittel gelangt zunächst in den äußeren der drei konzentrischen Räume. Dort verdampfen die leichtflüchtigsten Bestandteile mit dem höchsten Dampfdruck. Ihr Dampf dient dann zum Betrieb der untersten Düse, die dem Vorvakuum am nächsten gelegen ist. Das so von den flüchtigsten Bestandteilen befreite Treibmittel, gelangt in flüssiger Form in den mittleren der drei Teile des Siedegefäßes. Dort verdampfen die mittelschweren Fraktionen und dienen zum Betrieb der mittleren Dampfdüse. In den innersten Teil des Siedegefäßes gelangen also schließlich nur noch die schwerstsiedenden Bestandteile mit dem niedrigsten Dampfdruck, die dann zu dem Betrieb der obersten, dem Hochvakuum am nächsten gelegenen Düse dienen. Nur diese Anteile mit dem niedrigsten Dampfdruck beeinflussen also den mit der

[1] W. GAEDE: Z. techn. Phys. **13**, 210 (1932).
[2] Comp. Générale de Radiologie, Erf.: H. Pilon, Franz. Pat. 765978 angem. 16. 3. 33.
[3] K. C. D. HICKMAN: J. Franklin Inst. **221**, 215, 383 (1936).

Pumpe erzielbaren Enddruck. Wenn nun zeitweise durch Lufteinbrüche Bestandteile mit hohem Dampfdruck erzeugt werden oder wenn diese von vornherein als Verunreinigung im Treibmittel enthalten sind, so werden sie durch den eben beschriebenen Reinigungsvorgang durch fraktionierte Destillation aus der innersten, dem Hochvakuum am nächsten gelegenen Stufe entfernt und in die Vorvakuumstufen verschoben, so daß sie ohne schädlichen Einfluß auf den Enddruck der Öldiffusionspumpen bleiben.

Erst durch die Einführung dieser mehrstufigen fraktionierenden Diffusionspumpen sind die Öldiffusionspumpen genügend betriebssicher geworden, so daß

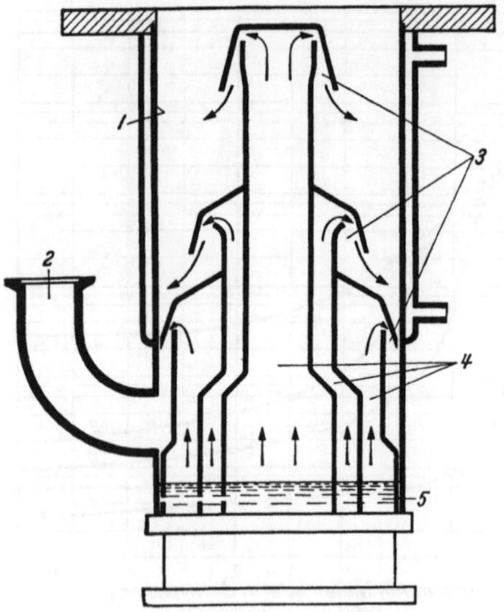

Fig. 111. Dreistufige Fraktionsöldiffusionspumpe aus Metall. *1* Kühlwassermantel, *2* Vorvakuumanschluß, *3* Diffusionsdüsen, *4* getrennte Dampfsteigrohre für die einzelnen Diffusionsstufen, *5* Siedegefäß.

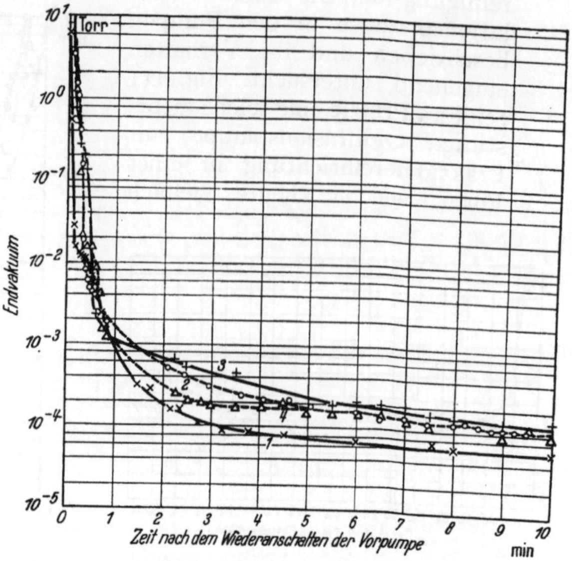

Fig. 112. Anfangsanstieg des Endvakuums nach Lufteinbrüchen in einer dreistufigen Öldiffusionspumpe aus Metall mit Fraktioniereinrichtung. Lufteinbrüche: *1* 10 Torr, 10 min; *2* 420 Torr, 2 min; *3* 760 Torr, 2 min; *4* 760 Torr, 4 min.

sich ihre Anwendung allgemein durchgesetzt hat. Fig. 112 zeigt, wie schnell sich das Endvakuum nach verschieden *starken Lufteinbrüchen* bei einer fraktionierenden Öldiffusionspumpe wieder einstellt zum Vergleich mit der Fig. 110, die das analoge Verhalten einer Pumpe ohne Fraktioniereinrichtung zeigt. Man sieht ohne weiteres die starke Verbesserung durch die Fraktioniereinrichtung. Abgesehen hiervon zeigen ganz allgemein die Öle in Metallpumpen eine stärkere Neigung zur Bildung von leichtflüchtigen Bestandteilen als in Glaspumpen. Die besten Endvakuumwerte erhält man daher mit mehrstufigen fraktionierenden Öldiffusionspumpen aus Glas. Eine solche Pumpe zeigt Fig. 113.

Die mit Öldiffusionspumpen erzielbaren Enddrucke haben etwa folgende Werte. Unter technischen Bedingungen ohne besondere Vorkehrungen erreicht man mit Metallpumpen ohne weiteres Enddrucke von 10^{-5} bis 10^{-6} Torr. Bei besonders sorgfältigem Arbeiten und mit hochwertigen Treibmitteln, lassen sich aber auch in Metallpumpen Vakua von einigen 10^{-7} Torr erzielen. Die tiefsten mit mehrstufigen fraktionierenden Öldiffusionspumpen aus Glas mit Octoil S erzielten Enddrucke liegen bei $5 \cdot 10^{-8}$ Torr. Die höchsten zulässigen Vorvakuumdrucke haben bei einstufigen Öldiffusionspumpen einen Wert von $5 \cdot 10^{-2}$ Torr, bei mehrstufigen Öldiffusionspumpen von einigen zehntel Torr. Es gilt aber ganz allgemein, daß zur Erzielung eines möglichst niedrigen Enddruckes auch möglichst niedrige Vorvakuumdrucke empfehlenswert sind.

Da das *Betriebsverhalten* der Öldiffusionspumpen und das mit ihnen erzielbare Endvakuum sehr stark von der Art der benutzten Öle sowie der Art und den Betriebsbedingungen der verwandten Pumpen abhängen, ist es Gegenstand zahlreicher und gründlicher Untersuchungen insbesondere von HICKMAN, JAECKEL, LATHAM u.a.[1] gewesen. Eine besondere Rolle hierbei spielt die Bildung von Zersetzungsprodukten, insbesondere Crackprodukten auf das Verhalten der Pumpen bzw. deren Anteile am erzielbaren Enddruck (im Vergleich zum Dampfdruck

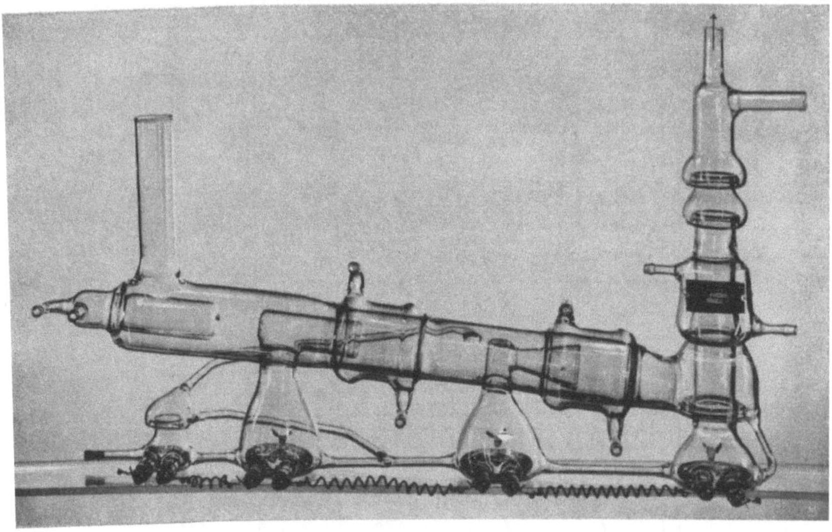

Fig. 113. Dreistufige Fraktionsöldiffusionspumpe aus Glas (DPI).

des Treibmittels selbst). REICH[2] hat diese Frage in folgender Weise einer Klärung näher zu bringen versucht. Er benutzte eine Öldiffusionspumpe aus Metall mit einer aufgesetzten Meßkammer, in der sich ein Ionisationsmanometer befand. Pumpe und Meßkammer konnten mittels einer wärmeübertragenden Flüssigkeit durch einen Thermostaten auf die gleiche Temperatur gebracht werden. Er hat dann mit dem Ionisationsmanometer den Ionisationsstrom in Abhängigkeit von der Temperatur gemessen und die gemessene Ionisation in Abhängigkeit von der Temperatur mit dem vorher auf direktem Wege bestimmten Dampfdruck in Abhängigkeit von der Temperatur verglichen (s. Fig. 114). Die untere Gerade zeigt den Dampfdruck, die obere Kurve die Ionisation in Abhängigkeit von der Temperatur. Wie man sieht, verlaufen beide Kurven bei höheren Temperaturen parallel, d.h. Ionisation und Dampfdruck sind proportional zueinander. Ihr Verhältnis (Tabelle 8, Spalte 2) ergibt die relative Empfindlichkeit des (mit Luft geeichten) Ionisationsmanometers für den Treibmitteldampf. Die Zusammensetzung des Treibdampfes ist zunächst bei allen Temperaturen dieselbe. Bei niedrigen Temperaturen jedoch wird der Einfluß des Treibdampfes auf den Gesamtdruck bzw. die Ionisation so klein, daß die Ionisation von (mit Wasserkühlung) nicht kondensierbaren Stoffen überwiegt. Die Kurve biegt

[1] R. JAECKEL: Z. techn. Phys. **23**, 177 (1942). — B.B. DAYTON: Industr. Engng. Chem. **40**, 795 (1948). — K.C.D. HICKMAN: Rev. Sci. Instrum. **22**, 141 (1951). — D. LATHAM, B. POWER u. N.T.M. DENNIS: Vacuum **1**, 97 (1951); **2**, 33 (1952).

[2] G. REICH: Z. angew. Phys. **9**, 23 (1957).

gegenüber der Geraden für die Ionisation des Treibdampfes nach oben ab, und zwar um einen von der Temperatur abhängigen Betrag Δp. Diesen über-

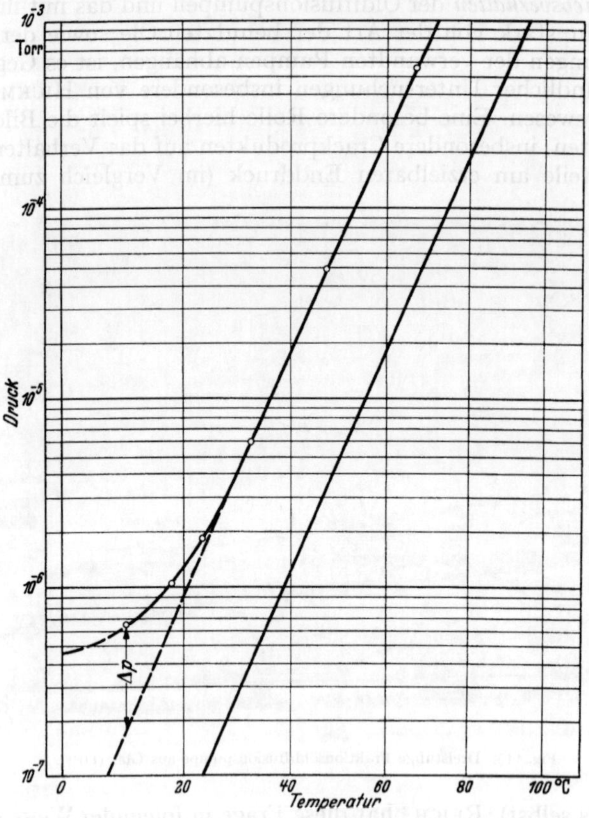

Fig. 114. Temperaturdruckkurve von Octoil (nach REICH). Untere Kurve: Dampfdruckmessung; obere Kurve: Messung mit dem Ionisationsmanometer.

schüssigen Druckanteil kann man leichtflüchtigen Zersetzungsprodukten zuschreiben. Die Tabelle 8 zeigt analoge Messungen für verschiedene Treibmittel,

Tabelle 8.

1 Treibmittel	2 rel. Empf.	3 p_e (10° C) (Torr)	4 Δp (Torr)	5 p_d (10° C) (Torr)	6 M	7 M rel. Empf.
Dibutylphthalat .	9,15	$2,2 \cdot 10^{-4}$	—	$2,4 \cdot 10^{-6}$	278	30,4
Narcoil 40	13,1	$3,5 \cdot 10^{-7}$	$2,4 \cdot 10^{-7}$	$8,5 \cdot 10^{-9}$	418	31,9
Octoil	13	$6 \cdot 10^{-7}$	$4,2 \cdot 10^{-7}$	$1,4 \cdot 10^{-8}$	390	30
Clophen A 40 . . .	—	$2 \cdot 10^{-7}$	—	—	~260	—
Leybold-Öl F . . .	—	$9 \cdot 10^{-7}$	$4,5 \cdot 10^{-7}$	—	~400	—
Apiezon-Öl C . . .	—	$7 \cdot 10^{-7}$	$5,8 \cdot 10^{-7}$	—	~450	—
Silicon DC 703 . .	—	$7 \cdot 10^{-7}$	$6 \cdot 10^{-7}$	—	~570	—

Spalte 3 den durch Ionisation gemessenen Druck und Spalte 5 den unmittelbar gemessenen Dampfdruck. Die Differenz zwischen beiden zeigt Spalte 4. Dies ist also der Anteil Δp, der leichtflüchtigen Produkte, deren Anteil bei niedrigen Drucken den Treibdampfanteil überwiegt. Die Messungen von REICH stehen also im Einklang damit, daß bei Öldiffusionspumpen der gemessene Enddruck im allgemeinen höher liegt, als der aus dem Dampfdruck des Treibmittels ermittelte.

In anderer Weise versuchen REICH und NÖLLER[1] dasselbe Problem zu klären. Sie haben auf der Saugseite einer Öldiffusionspumpe ein kleines Massenspektrometer in Form eines Omegatrons angebracht (s. den folgenden Artikel von ALPERT in diesem Bande). Mit Hilfe dieses Omegatrons wird das Massenspektrum des Dampfes auf der Saugseite einer Öldiffusionspumpe untersucht (s. Fig. 115). Es zeigen sich durchaus Anteile mit Massen, die kleiner sind als das Molekulargewicht der als Treibmittel benutzten Substanz. Sie dürften also von Zersetzungsprodukten des Treibmittels herrühren, da sie andererseits für Gasreste wesentlich zu hohe Massenzahlen haben. Ihre Massen drängen sich in einem Bereich zusammen, der etwa zwischen $\frac{1}{2}$ und $\frac{1}{5}$ der Treibmittelmasse liegt.

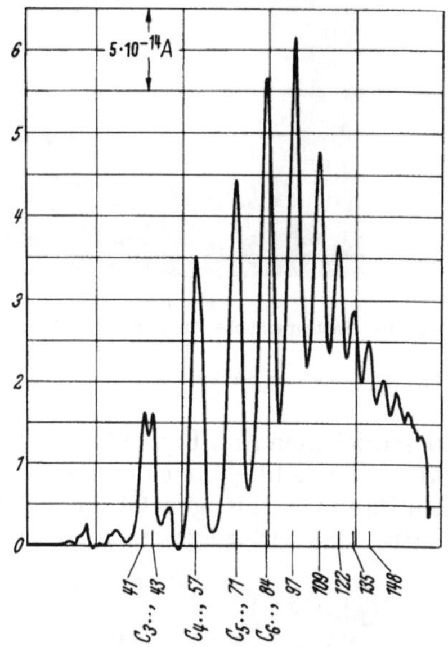

Fig. 115. Messung des Massenspektrums von Diffelen mit dem Omegatron (nach NÖLLER und REICH). Druck $2 \cdot 10^{-7}$ Torr, Baffletemperatur 13° C.

Aber auch ohne das Vorhandensein von Crack-Produkten kann sich im Vakuumraum eine höhere Dichte der Öldampfmoleküle einstellen, als dem Treibdampfdruck bei der Temperatur der Kondensationsfläche entspricht. Dies rührt daher, daß die Treibdampfmoleküle im Dampfstrom eine wesentlich höhere Temperatur haben als auf der Kondensationsfläche. Fliegen nun aus dem Treibdampfstrom Öldampfmoleküle unmittelbar in den Vakuumraum zurück, ohne inzwischen auf gekühlte Kondensationsflächen aufzutreffen, so müssen sie zu einer Erhöhung der Dampfdichte im Vakuumraum beitragen. Dieser Effekt kann nur durch Anbringung von geeigneten Kondensationsflächen (Baffles) verhindert werden. Einen wichtigen Hinweis für die Gestaltung dieser Baffles gewinnt man aus der Lage der Quelle für diese *Öldampfrückströmung*. RUF und WINKLER[2] haben eine interessante Methode zur Bestimmung der örtlichen Quellen für diese Öldampfrückströmung angegeben. Sie benutzen dabei die schon vorher von anderer Seite beschriebene Lochkameramethode zur Abbildung der Quellen von Molekularstrahlen durch diese Molekularstrahlen selbst und ihre Niederschlagung auf einer hierfür geeigneten Fläche. Besonders interessant ist bei der Methode von RUF und WINKLER die Art und Weise, in der sie die durch die Lochkameramethode abgebildeten Molekularstrahlen auf der Auffangfläche sichtbar machen. Sie benutzen hierzu ein halbfertiges Interferenzfilter, das aus zwei teildurchlässigen Silberschichten mit einer Zwischenschicht aus Thoriumfluorid besteht, und zwar in der Weise, daß sie zunächst eine teildurchlässige Silberschicht und einen Teil der Thoriumfluoridschicht aufdampfen. Die so vorbehandelte Platte wird zur Auffangung der aus Öldampfmolekülen bestehenden Molekularstrahlen benutzt. Die exponierte Platte wird schließlich erneut mit Thoriumfluorid bedampft und schließlich eine zweite halbdurchlässige Silberschicht

[1] G. REICH u. H. G. NÖLLER: Z. angew. Phys. 9, 617 (1957).

[2] J. RUF u. O. WINKLER: In M. AUWÄRTER, Ergebnisse der Hochvakuumtechnik usw., S. 207. Stuttgart: Wissenschaftliche Verlagsgesellschaft 1957.

aufgebracht. Das so hergestellte Interferenzfilter zeigt an den Stellen, an denen die Dampfmoleküle niedergeschlagen wurden, gut begrenzte farbige Konturen. Die Ergebnisse dieses Verfahrens sind in Fig. 116 bis 119 zusammengestellt. Fig. 116 bringt eine photographische Aufnahme der obersten Diffusionsumkehrdüse, die Fig. 117 bis 119 Lochkameraaufnahmen mit Molekularstrahlen der oben

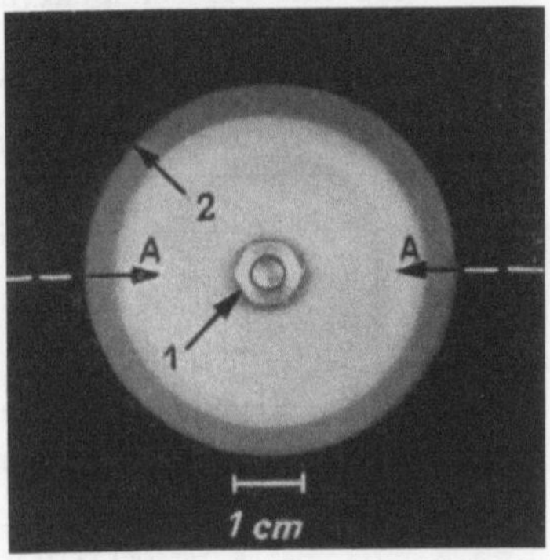

Fig. 116.

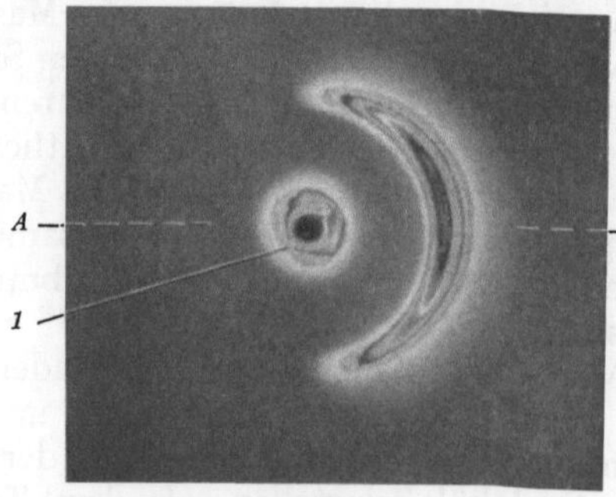

Fig. 117.

Fig. 116. Photographische Aufnahme der Diffusionsumkehrdüse einer Öldiffusionspumpe (nach RUF und WINKLER in Draufsicht. *1* Befestigungsschraube, *2* äußerster Düsenrand.

Fig. 117. Aufnahme der Düse der Fig. 116 mit der Molekularstrahl-Lochkamera-Methode (nach RUF und WINKLER).

beschriebenen Art. Fig. 117 zeigt als Quelle für die Öldampfmolekularstrahlen eine kleine Undichtigkeit einer Befestigungsschraube (*1*) und außerdem einen schmalen Rand von flüssigem Öl, das sich an dem Rand des Düsenhutes niedergeschlagen hatte. Fig. 118 zeigt einen wesentlich verbreiterten Kondensationsrand (*3*) und

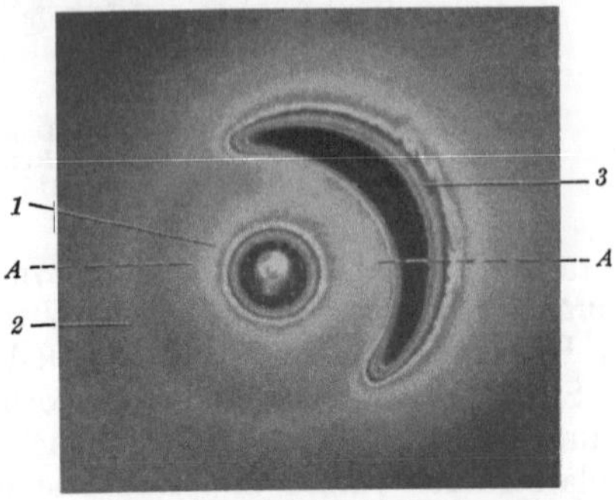

Fig. 118.

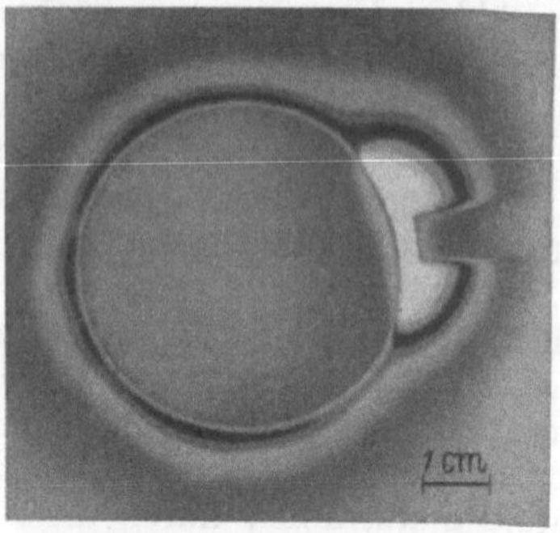

Fig. 119.

Fig. 118. Dasselbe wie Fig. 117 jedoch mit stärkerer Ölkondensation (*3*) am Düsenrand (*2*).

Fig. 119. Dasselbe wie Fig. 118, jedoch mit einem kräftigen Öltropfen am Düsenrand. Der Ausschnitt im Öltropfen kommt durch eine Blende zwischen Tropfen und Lochkamera zustande. Die scharfe Begrenzung zeigt den einwandfrei geradlinigen Verlauf der Molekularstrahlen in der Aufnahmeordnung.

Fig. 119 einen Kondensationsrand rings um die Umkehrdüse und außerdem einen dicken Tropfen an der rechten Seite. Es ergibt sich also daraus als besondere Quelle für die Ölrückströmung an der obersten Düse in flüssiger Form kondensiertes Treibmittel. Von POWER und CRAWLEY[1] und anderen Autoren wurde die gesamte Öldampfrückströmung ins Vakuum mit entsprechend konstruierten Mikrowaagen gemessen. Ein Mittel, um auch diese Öldampfrückströmung noch vom Vakuum fernzuhalten, besteht in der Anbringung von (gekühlten) Prallflächen (Baffles),

[1] B.D. POWER u. D. J. CRAWLEY: Vacuum 4, 415 (1954).

die den geraden Flug von Öldampfmolekülen ins Vakuum verhindern, die aber so in der Saugleitung der Pumpe angebracht sind, daß sie einen möglichst geringen Strömungswiderstand darstellen (s. z.B. Fig. 120).

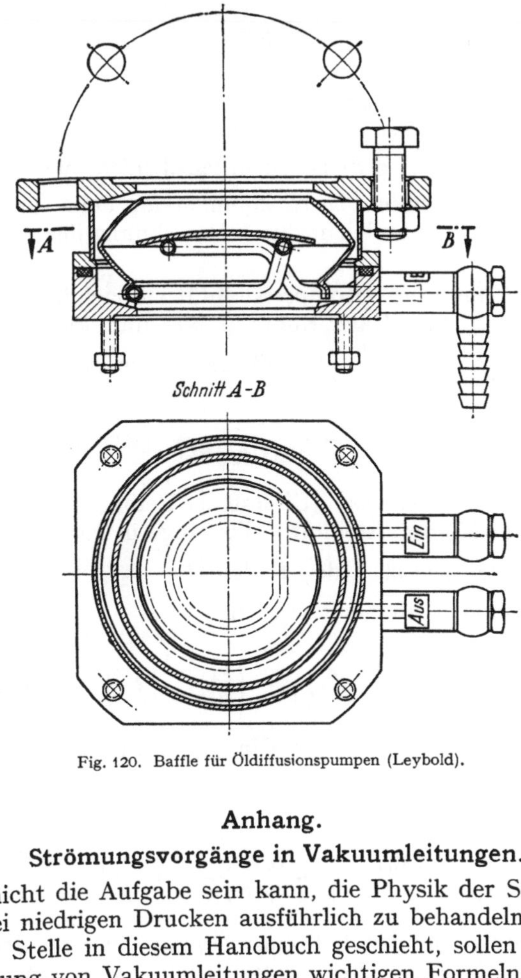

Fig. 120. Baffle für Öldiffusionspumpen (Leybold).

Anhang.

Strömungsvorgänge in Vakuumleitungen.

Da es hier nicht die Aufgabe sein kann, die Physik der Strömungsvorgänge in Leitungen bei niedrigen Drucken ausführlich zu behandeln, um so mehr, als dies an anderer Stelle in diesem Handbuch geschieht, sollen hier nur kurz die für die Berechnung von Vakuumleitungen wichtigen Formeln zusammengestellt werden. Wir benutzen dabei die folgenden Formelzeichen:

Zeichen	Größe	Maßeinheit
W	Leitungswiderstand	sec/l
S	Durchflußvolumen beim Druck p	l/sec
p_1	Druck am Anfang der Leitung	Torr
p_2	Druck am Ende der Leitung	Torr
G	Massendurchsatz	g/sec
r	Leitungs-Radius	cm
d	Leitungs-Durchmesser	cm
l	Leitungs-Länge	cm
F	Leitungsquerschnitt	cm^2
U	Querschnittsumfang	cm
v	spezifisches Volumen	cm^3/g
Λ	freie Weglänge	cm

Strömungswiderstand W einer Leitung:

$$W = \frac{p_1 - p_2}{p\, S}, \quad p = \frac{p_1 + p_2}{2}. \tag{1}$$

$$S \cdot p = \frac{G}{M} \mathsf{R} T. \tag{2}$$

$$W = \frac{1}{G} \frac{(p_1 - p_2)\, M}{\mathsf{R} T}. \tag{3}$$

1. Bei hohen Drucken ($\Lambda \ll d$) ist zu unterscheiden zwischen

α) Turbulenten Strömungen. Diese kommen im Fein- und Hoch-Vakuumgebiet nicht in Frage, wie man leicht durch Berechnung der Reynoldsschen Zahl zeigen kann.

β) Laminaren Strömungen. (Kompressibilität vernachlässigbar, maßgebender Einfluß der Reibung.) Für *lange* Rohre ($l \gg r$) und kleine Druckdifferenzen zwischen den Enden $\left(p_1 - p_2 \ll \frac{p_1 + p_2}{2}\right)$ gilt die Hagen-Poisseuillesche Gleichung

$$W = \frac{10^3\, l}{r^4} \frac{8\eta}{\pi} \frac{2}{1333\,(p_1 + p_2)} \quad [\mathrm{sec/l}]. \tag{4}$$

γ) Gasdynamischen Strömungen. (Einfluß der Reibung vernachlässigbar, maßgebender Einfluß der Kompressibilität.) Vorgänge in ***kurzen*** Rohren ($l \sim r$) bei hohen Druckdifferenzen zwischen den Enden $\left(p_1 - p_2 \geq \frac{p_1}{2}\right)$ können in Analogie zu den Düsenströmungen behandelt werden

$$G = F_2 \left(\frac{p_2}{p_1}\right)^{\frac{1}{\varkappa}} \sqrt{\frac{\varkappa}{\varkappa - 1}\left[1 - \left(\frac{p_2}{p_1}\right)^{\frac{\varkappa - 1}{\varkappa}}\right] 2\, \frac{p_1}{v_1}} \quad [\mathrm{g/sec}]. \tag{5}$$

δ) Das Zwischengebiet zwischen den Gln. (4) und (5) wird beschrieben durch die Gleichung:

$$G = F_2 \frac{M r^2 \frac{\varkappa}{\varkappa + 1}}{\mathsf{R} T_1\, 8\eta\, l} \frac{\left[p_1^2 - p_2^2 \left(\frac{p_1}{p_2}\right)^{\frac{\varkappa - 1}{\varkappa}}\right]}{\left[1 + \frac{G}{F_2} \frac{1}{6\eta\varkappa} \frac{r^2}{l} \ln \frac{p_1}{p_2}\right]} \quad [\mathrm{g/sec}]. \tag{6}$$

Der Massendurchsatz G für den Fall einer *Öffnung in* einer *dünnen Wand* bei großem Druckunterschied zwischen beiden Seiten ergibt sich aus Gl. (6) mit $p_2 \ll p_1$ und $l \to 0$ zu:

$$G = F p_1 \sqrt{\frac{3}{4} \varkappa \frac{M}{\mathsf{R} T_1}} \quad [\mathrm{g/sec}]. \tag{7}$$

2. Niedrige Drucke ($\Lambda \gg d$). Hier sind folgende Fälle zu unterscheiden:

α) Lange Rohre

$$W = \frac{3\, l\, U \cdot 10^3}{8 F^2} \sqrt{\frac{\pi M}{2 \mathsf{R} T}} \quad [\mathrm{sec/l}] \tag{8}$$

beliebige Gase, beliebige Querschnittsform.

$$W = \frac{l}{100\, r^3} \quad [\mathrm{sec/l}] \tag{9}$$

für Luft bei 20° C, kreisförmiger Querschnitt.

β) Kurze Rohre

$$W=\left(\frac{3lU}{16F}+1\right)\sqrt{\frac{2\pi M}{\mathrm{R}T}}\,\frac{10^3}{F}\quad[\mathrm{sec/l}] \tag{10}$$

beliebige Gase, beliebige Querschnittsform.

$$W=\frac{\frac{3l}{8r}+1}{36{,}3\,r^2}\quad[\mathrm{sec/l}] \tag{11}$$

für Luft bei 20° C, kreisförmiger Querschnitt.

γ) Öffnung in dünner Wand

$$W=\frac{10^3}{F}\sqrt{\frac{2\pi M}{\mathrm{R}T}}\quad[\mathrm{sec/l}] \tag{12}$$

beliebige Gase, beliebige Querschnittsform.

$$W=\frac{1}{36{,}3\,r^2}\quad[\mathrm{sec/l}] \tag{13}$$

für Luft bei 20° C, kreisförmiger Querschnitt.

3. Gesamtes Druckgebiet. Für lange kreisrunde Rohre und beliebige Gase gilt:

$$W=\frac{l}{r^3}\,\frac{10^3}{1333\,\frac{\pi r}{8\eta}\,\frac{p_1+p_2}{2}+\frac{8}{3}\sqrt{\frac{\pi\mathrm{R}T}{2M}}}\quad[\mathrm{sec/l}]. \tag{14}$$

4. Zusammengesetzte Leitungen. Reihenschaltung von Leitungen mit den Einzelwiderständen $W_1, W_2, \ldots, W_n$:

$$W_{\mathrm{gesamt}}=W_1+W_2+\cdots W_n. \tag{15}$$

Parallelschaltung von Leitungsstücken mit den Widerständen $W_1, W_2, \ldots, W_n$:

$$L_1=\frac{1}{W_1};\qquad L_2=\frac{1}{W_2};\qquad L_n=\frac{1}{W_n}; \tag{16}$$

$$L=L_1+L_2+\cdots L_n; \tag{17}$$

$$\left.\begin{aligned}W_{\mathrm{gesamt}}&=\frac{1}{L}=\frac{1}{L_1+L_2+\cdots L_n}\\&=\frac{1}{\frac{1}{W_1}+\frac{1}{W_2}+\cdots\frac{1}{W_n}}.\end{aligned}\right\} \tag{18}$$

Zusammenfassende Literatur.

α) Bücher.

ARDENNE, M. v.: Tabellen der Elektronenphysik, Ionenphysik und Übermikroskopie. Berlin: Deutscher Verlag der Wissenschaften 1956. Bd. II, Abschn. C, Kinetische Gastheorie und Hochvakuumtechnik.

Committee on Vac. Techn. Inc. Boston 9 Mass.: Vac. Symposium Transact. 1953, 1954, 1955, 1956, 1957. New York u. London: Pergamon Press.

DAVY, J. R.: Ind. High Vac., London: Bitman & Sons 1951.

DIELS, K., u. R. JAECKEL: Leybold Vak. Taschenbuch. Berlin: Springer 1958.

— — Abschn. Vakuum in ANGERER-EBERT, Technische Kunstgriffe bei physikalischen Untersuchungen, 12. Aufl. Braunschweig: Vieweg 1958.

DUNOYER, L.: Le vide et ses applications. Paris: Press. Univ. de France 1951.

Dushman, S.: Scient. Foundations of Vac. techn. New York: J. Wiley & Sons 1949.
Guthrie, A., and R. K. Wakerling: Vacuum Equipment and Technique New York-London-Toronto: McGraw-Hill 1949.
Heinze, W.: Einführung in die Vakuumtechnik. Berlin: VEB Verl. Techn. 1955.
Holland-Merten, E. L.: Handbuch der Vakuumtechnik. Halle 1950.
Jaeckel, R.: Kleinste Drucke, ihre Messung und Erzeugung. Berlin-Göttingen-Heidelberg: Springer; München: Bergmann 1950.
Laporte: Hochvakuum, seine Erzeugung, Messung und Anwendung. Halle: Knapp 1950.
Leblanc, M.: La technique du vide. Paris: Librairie Armand Colin 1951.
Loevinger, R., and A. Guthrie: Vacuum Equipment and Technique. New York-London-Toronto: McGraw Hill Book Co. 1949.
Martin, L. H., and R. D. Hill: A Manual of Vacuum Practice. Melbourne: Univ. Press 1949.
Mönch, G. Ch.: Hochvakuumtechnik. Pößneck: Rudolf Lang 1950.
Oetjen, G. W.: Ullmanns Enzyklopädie der technischen Chemie, 3. Aufl., Bd. 1, Kapitel Vakuumtechnik.
Reiman, A. L.: Vacuum Technique. London: Chapman & Hall 1952.
Wagner, G.: Erzeugung und Messung von Hochvakuum, Aufbau und Betrieb von Hochvakuumapparaten. Wien: Franz Deuticke 1950.
Yarwood, J.: High Vacuum Technique. London: Chapman & Hall 1948.

β) Vakuummessung.

Leck, J. H.: Pressure Measurement in Vacuum Systems. London: Chapman & Hall 1957.
Schwarz, H.: Arch. techn. Messen V, 1341 (2) (1951); **5**, 1341 (3), 1341 (4), 1341 (5) (1952).

γ) Werkstoffe der Vakuumtechnik.

Espe, W.: Werkstoffkunde der Elektrotechnik. Berlin: Akademische Verlagsgesellschaft 1954.
Espe, W., u. M. Knoll: Werkstoffkunde der Hochvakuumtechnik. Berlin: Springer 1936.
Smithels, C. J.: Metals Reference Book. London 1949.
Steyskal, H.: Arbeitsverfahren und Stoffkunde der Hochvakuumtechnik. Mosbach (Baden): Physik-Verlag 1955.

Production and Measurement of Ultrahigh Vacuum.

By

D. ALPERT.

With 27 Figures.

I. Introduction.

1. Ultrahigh vacuum. In recent years, developments in experimental techniques have made it possible to achieve and to measure pressures which are many orders of magnitude lower than were previously considered possible. Thus, while pressures in the range between 10^{-6} mm Hg and 10^{-8} mm Hg have for many years been called "high vacuum", it is now possible to evacuate an enclosure at room temperature to a total pressure of less than 10^{-11} mm Hg. Although this value of pressure corresponds to a particle density of only 10^5 per cm^3, it still does not appear to be an ultimate limit defined by the physical properties of the materials or the techniques used. Since the lower limit is not yet determined, it seems reasonable to define ultrahigh vacuum as the region of pressures below 10^{-8} mm Hg.

In the period between 1920 and 1950, the principal advances in vacuum technology were in the size of the vacuum systems used and in the speed with which they could be evacuated. Over that period of time, the ultimate pressures which were attainable did not greatly change, and the manometers for the measurement of pressure were essentially the same in sensitivity. In the early '30s, a number of workers[1] carried out experiments involving the physics of surfaces which required the lowest possible pressures. They found indications that under certain circumstances the "vacuum conditions" were considerably better than usual, even though the manometers did not show a corresponding change. There is now clear-cut evidence that several experimentalists achieved total pressures in the ultrahigh vacuum range using the techniques then available. However, considerable confusion remained in the general literature about the ultimate pressures which could be achieved. When means for reliably measuring pressures below 10^{-8} mm Hg became available in the late '40s and early '50s, it was easily demonstrated that some of the existing techniques were capable of achieving considerably lower pressures than were previously thought attainable. At the same time, it became possible and desirable to develop a number of new vacuum techniques and components to broaden and extend the range of useful ultrahigh vacuum technology. This close relationship between the measurement and the production of ultrahigh vacuum will be apparent in much of the presentation in this paper. After a brief discussion of some of the areas of application of ultrahigh vacuum techniques, it is proposed to review (1) the present state of the measurement of very low pressures, (2) the factors affecting the attainment of ultrahigh vacuum, (3) some means for evacuation to very low pressures, and (4) the design of ultrahigh vacuum systems and components.

2. Areas of application. The extension of the range of high vacuum was motivated by requirements in a number of fields of scientific investigation. Among

[1] These experiments will be discussed in the text of this paper.

the most important of these have been the fields of surface physics and surface chemistry. It has been known for many years[1] that the work function of a metal surface may be appreciably affected if the surface is altered by the deposition of a fraction of a monolayer of other atoms or molecules. It follows that the adsorption or deposition of even a fraction of a molecular layer would greatly affect the measurement of any surface property which depends on the work function of the material. Thus, careful investigations of thermionic emission[2,3] photoelectric emission[4], and field emission[5] all demonstrated the very large effects which adsorbed layers of atoms or molecules had on the particular phenomenon being studied. At a pressure of 10^{-6} mm Hg, an adsorbable gas may form a monolayer on a clean surface in about one second; hence, an experiment designed to study the properties of a newly cleaned surface would either have to be carried out in a very short time or in improved vacuum conditions. Considerable progress in the direction of better vacua was made prior to 1940. (For a more complete discussion, see [7], Chap. 10.) In his studies of the contact difference of potential between various metals, P. A. ANDERSON[6] developed methods for preparing a clean surface and evacuating the system such that changes in the work function were not observed for at least an hour after evaporating a clean surface. In 1937, NOTTINGHAM[2] used the rate of deactivation of a thoriated tungsten filament as a measurement of the magnitude of the adsorbable gas pressure. He showed that "this deactivation time served as a better indication of the condition of the tube than any measurements" obtained with an ionization gauge. Since 1950, it has been widely recognized that in various studies in these fields it is necessary and desirable to utilize ultrahigh vacuum techniques. At 10^{-11} mm Hg, the time interval required for the adsorption of a monolayer of gas is extended to many hours; and experiments on clean metals are made quite feasible. In some vacuum systems, the time to form a monolayer has been extended to weeks or months.

Another field of investigation which is greatly dependent on the condition of the surfaces involves the interactions of excited atoms and ions with the surfaces in question ([*1*], Chap. 9). In a very interesting series of papers, HAGSTRUM[7] has investigated the ejection of electrons from a number of metals by ions and metastable atoms and has measured the effect of adsorption at surfaces. In this case, the change in work function can account for only a small fraction of the observed change in the probability for secondary electron ejection.

The investigations which developed from the invention of the field emission microscope[8] by E. W. MÜLLER in 1937 involve a broad area of surface physics and surface chemistry which can, for the most part, be carried out only under conditions of uncontaminated surfaces. This unique tool has been utilized in a number of experiments including electron field emission, surface migration,

[1] See, for example, the work of LANGMUIR and his associates: I. LANGMUIR, J. Amer. Chem. Soc. **35**, 105 (1913), found O_2 to affect the thermionic emission from tungsten at temperatures below 2200° K. I. LANGMUIR and K. H. KINGDON: Science, Lancaster, Pa. **57**, 58 (1923). — Phys. Rev. **34**, 129 (1929). — J. B. TAYLOR and I. LANGMUIR: Phys. Rev. **44**, 423 (1933).

[2] W. B. NOTTINGHAM: J. Appl. Phys. **8**, 762 (1937).

[3] S. T. MARTIN: Phys. Rev. **56**, 947 (1939).

[4] A. L. HUGHES and L. A. DUBRIDGE: Photoelectric Phenomena. New York: McGraw Hill-Book Company, Inc. 1932.

[5] E. W. MÜLLER: Z. Physik **102**, 734 (1936).

[6] P. A. ANDERSON: Phys. Rev. **47**, 958 (1935).

[7] H. D. HAGSTRUM: Phys. Rev. **89**, 244 (1953); **104**, 1516 (1956). These two papers contain a number of the relevant references.

[8] For a discussion, see the review article by R. H. GOOD jr. and E. W. MÜLLER, this Encyclopedia, Vol. XXI.

catalysis, and the adsorption and desorption of gases on metal surfaces[1] [2], [3]. Most of the applications of this beautiful technique make use of the variation of field emission from the various crystalline faces of a small metallic point. Its sensitivity to contamination obviously calls for and has made necessary the use of ultrahigh vacuum technology.

Another broad area of surface physics which has attained considerable interest recently, and which is possible to investigate only under ultrahigh vacuum conditions, is the study of semiconductor surfaces. In the case of semiconductors, the accumulation of charge at various sites on the surface may have very significant effects on such physical properties of the surface as the hole-electron recombination rate, contact potential, and surface conductance. There is good evidence that adsorbed materials constitute one of the major sources for the accumulation of surface charge. In this case, one of the chief difficulties is the establishment of reasonable criteria for determining whether a surface is, in the rigorous sense, truly clean. FARNSWORTH[2] and his colleagues have utilized ion bombardment to remove adsorbed gas from the semiconductor surfaces and have used interesting electron diffraction techniques to obtain information about the condition of the surface.

The major motivation for the work by the author and his colleagues at the Westinghouse Research Laboratories [4] to [6], grew out of requirements in the field of gaseous electronics. In this field, the evacuation of the vacuum system is just the preliminary to the usual experiment in which the gas is then introduced into the system, the degree of evacuation setting one obvious limitation on gas purity. It has for many years been recognized that small amounts of impurity may greatly affect the electrical properties of a gas. For example, KRUITHOF and PENNING[3] showed that the breakdown voltage for neon gas is seriously affected by less than one part in 10^6 of a specified impurity. Other experiments are affected by even smaller impurities[4]. Utilizing ultrahigh vacuum techniques, M. A. BIONDI has shown that it is possible to carry out experiments in helium with an impurity of less than one part in 10^9 [5]. There is increasing evidence that experiments in many other fields either have been or could be greatly aided by the availability of ultrahigh vacuum. In some cases, it is necessary to reduce the density of particles in the volume to the ultrahigh vacuum range. In other cases, particularly those requiring stable, clean surfaces, the measured pressure is only an index to the condition of the surfaces during the course of the experiment.

II. Measurement of very low pressures.

3. X-ray effect in conventional ionization gauges. From the time of its initial development by BUCKLEY[6] and DUSHMAN[7], the ionization gauge has been the most widely used instrument for the measurement of low pressures. While it has certain limitations which will be discussed presently, this gauge has been characterized by its simplicity of construction and operation, and its ability to read pressures lower than were possible with other means. The conventional

[1] W. P. DYKE and W. W. DOLAN: Electronics and Electron Physics, Vol. 8. New York: Academic Press 1956.

[2] H. E. FARNSWORTH: Low-energy electron diffraction studies of cleaned and gas-covered germanium (100) surfaces. Semiconductor Surface Physics, p. 23. Philadelphia: University of Pennsylvania Press 1956.

[3] A. A. KRUITHOF and F. M. PENNING: Physica, Haag **4**, 430 (1937).

[4] For a discussion of some of these problems, see L. B. LOEB, this Encyclopedia, Vol. XXI.

[5] M. A. BIONDI: Rev. Sci. Instrum. **22**, 535 (1951).

[6] O. E. BUCKLEY: Proc. Nat. Acad. Sci. U.S.A. **2**, 683 (1916).

[7] S. DUSHMAN and C. G. FOUND: Phys. Rev. **17**, 7 (1921).

ionization gauge [7] structurally resembles a triode vacuum tube. A filament serves as a source of electrons. However, the grid serves as the electron collector and operates at a positive potential (100 to 200 volts) with respect to the cathode, while the cylindrical plate is negative by 10 to 50 ev. Electrons from the cathode, before being collected at the grid, are capable of ionizing gas molecules both inside the grid structure and between the grid and the negatively charged outer cylinder. The number of ions produced per electron is assumed proportional to the gas density, and the current to the ion collector is used as an indication of the pressure. That is, for a constant value of accelerating voltage in excess of the ionization potential, the number of positive ions formed should vary linearly with the pressure and the electron current,

$$i_c = K\, i_g\, p \tag{3.1}$$

where i_c is the ion current, i_g is the electron current to the grid, and p is the pressure in mm Hg. The sensitivity, K, of a given ion gauge is the proportionality constant in the basic ionization gauge equation above:

$$K = \frac{(i_c/i_g)}{p}. \tag{3.2}$$

A typical value of K for nitrogen is 10 (mm Hg)$^{-1}$, while a normal operating electron current is 10 ma; under these conditions the ion current at a pressure of 10^{-8} mm Hg is 10^{-9} amps. Hence, if the ability to read the ion current were the only limitation on the ionization gauge pressure range, modern electrometers would make it possible to read pressures orders of magnitude lower than those which have thus far been obtained.

The standard ionization gauge as used prior to 1948 was not capable of indicating pressures below 10^{-8} mm Hg. There is good reason, however, to believe that pressures considerably below this value were attained much earlier. As mentioned previously (Sect. 2), ANDERSON and NOTTINGHAM found that although the ionization gauge had reached its lowest ion current corresponding to 10^{-8} mm Hg, significant changes in the state of vacuum were still indicated by the variation in thermionic emission from a thoriated tungsten filament. These observations were among the first indications that vacuum better than 10^{-8} mm were attainable and, at the same time, served to heighten the mistrust of the ionization gauge as an instrument for the measurement of low pressures. It was not until considerably later that the low-pressure limit of the ionization gauge was clearly recognized as such in the literature.

It was NOTTINGHAM who first recognized a lower limit in the range of the ionization gauge and, in 1947[1], put forth a hypothesis to account for it. He suggested that there exists a residual current to the ion collector which is completely independent of the pressure. This current is caused by soft x-rays created when the electrons from the filament strike the grid. At the ion collector, the x-rays in turn release photoelectrons which are accelerated to the grid, thus constituting a current of the same sign as that for ions arriving at the collector. At the usual operating voltage for an ionization gauge, the ratio of "x-ray" current to grid current is 1 or 2×10^{-7}, and the corresponding pressure reading 1 or 2×10^{-8} mm Hg. Thus the pressure in a conventional ionization gauge must exceed 10^{-8} mm Hg in order to produce an ion current appreciably in excess of the residual current.

4. Bayard-Alpert ionization gauge. Subsequent to 1947, a number of investigators were independently studying the causes for the low pressure limitation

[1] W. B. NOTTINGHAM: MIT Conference on Physical Electronics, 1947.

on the ordinary ionization gauge. In 1950, BAYARD and ALPERT [4] substantiated NOTTINGHAM's x-ray hypothesis and described an ionization gauge with a low pressure limit at least one hundred times lower than previously available. Within a few months after the publication of this paper, independent investigations were reported by J. J. LANDER[1] and by G. H. METSON[2] who demonstrated the x-ray effect and suggested methods for substantially reducing this limitation.

The original evidence for a residual current independent of the pressure had come from the observation that the conventional gauge did not indicate pressures below 10^{-8} mm Hg even though the "vacuum conditions" seemed to improve with careful processing. Independent evidence for the existence of a residual current was obtained by several experimenters from other measured characteristics of the ionization gauge. If the ion collector current, i_c, is measured for varying grid potentials, v_g, at a given pressure and electron current, the curve showing i_c *vs.* v_g has a characteristic shape for pressures above 10^{-7} mm Hg. This curve, Fig. 1a upper curve, is similar to those for the gas ionization efficiency by electron impact[3] and is typical of data taken on an RCA type 1949 gauge. Note that i_c rises rapidly with v_g up to 200 volts, and varies slowly with grid voltages above this value. When the pressure is lowered considerably (ion gauge reading 10^{-8} mm Hg), the curve of i_c *vs.* v_g is radically different as shown by the lower curve in Fig. 1a. This residual curve continues to rise rapidly with grid potential, the slope on a log-log plot falling between 1.5 and 2. In the intermediate range of pressures (10^{-7} to 10^{-8} mm Hg), the measured curve corresponds to a superposition of the "gas ionization" curve and the "residual" curve. These results are in agreement with unpublished results of NOTTINGHAM and are interesting to compare with the early data obtained by JAYCOX and WEINHART[4].

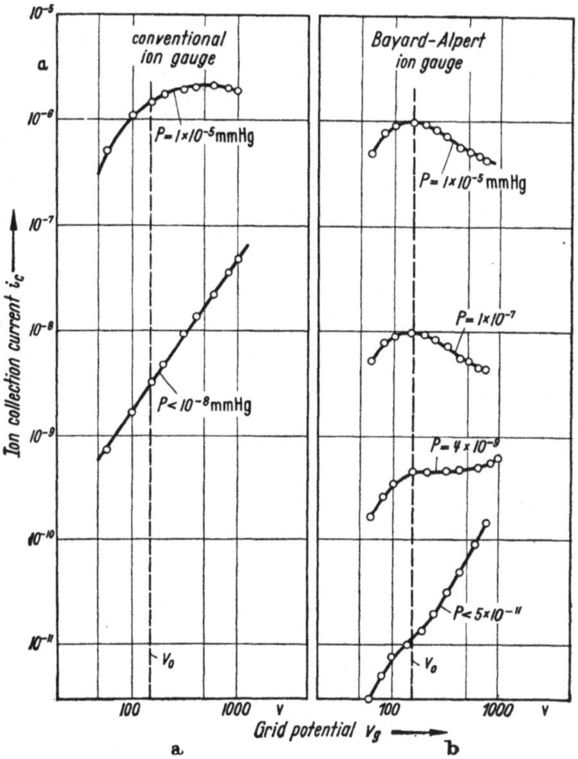

Fig. 1a and b. (a) Typical data for conventional gauge (RCA type 1949). (b) Typical data for Bayard-Alpert gauge. V_0 is the normal operating voltage (150 volt).

[1] J. J. LANDER: Rev. Sci. Instrum. **21**, 672 (1950).

[2] G. H. METSON: Brit. J. Appl. Phys. **2**, 46 (1951).

[3] See, for example, K. T. COMPTON and C. C. VAN VOORHIS: Phys. Rev. **27**, 724 (1946).

[4] E. K. JAYCOX and H. W. WEINHART: Rev. Sci. Instrum. **2**, 401 (1931). JAYCOX and WEINHART indicated their pressure to be 10^{-8} mm. However, when their data on the variation of ion collector current with grid accelerating voltage are plotted in the manner suggested above, it results in a curve almost identical to the "residual curve" of Fig. 1a. This constitutes evidence that the actual pressure in their tube was considerably below 10^{-8} mm Hg.

The Bayard-Alpert ionization gauge was the product of an experiment which was proposed to examine the Nottingham hypothesis that the residual current is due to photoelectrons leaving the ion collector. A design for an ion gauge was sought which would operate in the usual manner but for which the ion collector would intercept only a small fraction of the x-rays produced at the grid. The modified gauge, Fig. 2, has the usual three elements, but is inverted from the conventional arrangement; the filament is outside the cylindrical grid while the ion collector, now a very fine wire, is suspended within the grid. The volume enclosed by the grid is made comparable to that between the grid and ion collector of an ordinary ionization gauge. The positive potential on the grid forms a potential well for the ions created within the grid-enclosed volume so that many of them are collected at the center wire; the ion collection efficiency is thus comparable to that of the conventional ionization gauge. On the other hand, the solid angle which the ion collector presents to x-rays from the grid is at least one hundred times smaller than that for the conventional cylindrical collector.

The i_c *vs.* v_g characteristics of a Bayard-Alpert gauge are shown in Fig. 1b. At higher pressures, the curve for this gauge is similar to the RCA gauge, but it is evident that the modified gauge continues to have typical "gas ionization" characteristics at much lower pressures. At 4×10^{-9} mm Hg, the collector current is still predominantly due to gas ionization. Even at the lowest pressures which were attained, less than 5×10^{-11} mm Hg, it is clear (lowest curve) that the gauge still had not reached the point where the gas effect is not observable at normal operating voltages. At higher grid voltages, the "residual effect" is clearly the predominant one. The ratio of residual currents for the two types of ion gauges is found to be equal to the calculated ratio of the solid angles which they present to x-radiation. These results clearly substantiated the x-ray hypothesis and provided ionization gauges of simple design to cover a very large range of pressure.

A unique feature of the Bayard-Alpert gauge is the distribution of potential in the ionizing region. Since the collector is coaxial with the cylindrical grid, one has a logarithmic potential distribution between grid and ion collector with the greatest change in potential immediately adjacent to the collector wire. It follows that the ionizing electrons retain most of their energy throughout their trajectory within the grid volume. For the particular geometry used at present, the equipotentials are such that an incoming electron with 150 ev energy falls below an efficient ionizing energy (50 ev) only when it penetrates to a distance of about 0.010 in. from the collector wire. This represents a fractional ion collection volume of only 1% in which the ionizing cross section is very small. In the conventional ionization gauge, quite the opposite is true; the electrons are all decelerated so that they are inefficient for ionization over a sizeable portion of their trajectory. This means that in the Bayard-Alpert gauge the ionizing electrons do not vary too much in energy. Nottingham has used this property to identify the residual gas by measuring the appearance potential for ionization.

Two designs of the ionization gauge are shown in Fig. 2. The electrode structure is very similar in both, but in the so-called single-ended ionization gauge all of the electrodes are inserted through one press. The second tungsten filament is preferably electrically separated from the first and has at least two desirable uses aside from its value as a spare. It can be used in connection with the flash filament method; in addition, the electron emission characteristics of this filament can be used as a separate indication of vacuum conditions. The glass skirt around the ion collector lead-in serves several purposes: It prevents x-rays from striking the large diameter tungsten lead-in wire and this maintains a lower solid angle for their interception; it lengthens the surface leakage path and prevents current from the charged walls of the gauge from reaching the ion collector. If for any reason a metallic film is evaporated or otherwise deposited on the glass walls of the gauge, the glass skirt prevents the formation of a direct path to the ion collector. The ion collector (a 0.006-inch diameter tungsten wire) can be gradually tapered by an electrolytic etching process so as to give the minimum x-ray solid angle[1], and at the same time to retain enough stiffness

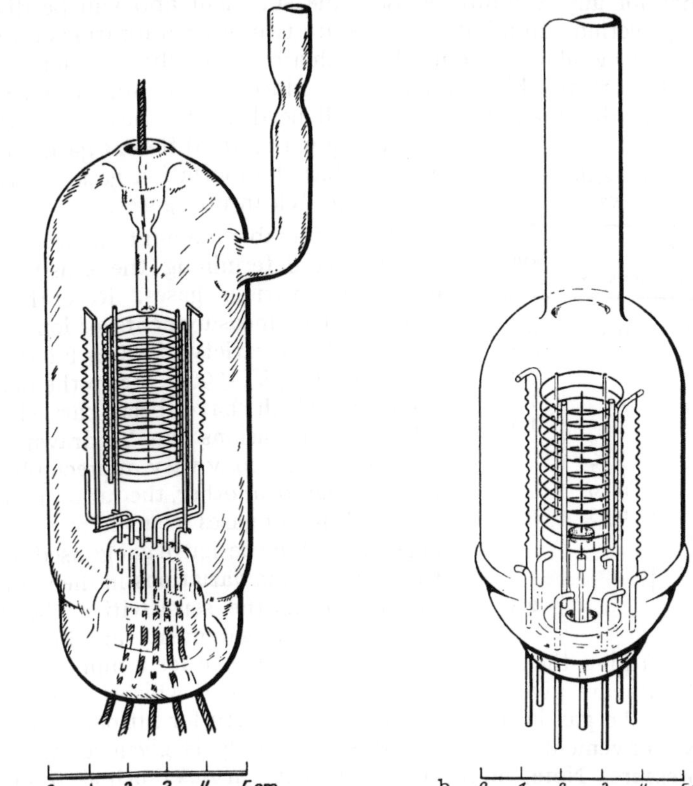

Fig. 2 a and b. (a) Standard form of Bayard-Alpert gauge. (b) Single-press design with central collector lead. This design, developed by R. S. Buritz, has a somewhat lower sensitivity than that of (a). However the tubulation may be made considerably larger and the design can be fabricated with automatic machinery.

to prevent large mechanical deformation or oscillations. The grid structure is helically wound 0.005-inch molybdenum wire spot welded to molybdenum supporting members. The filaments are made from 0.006-inch diameter tungsten wire.

The Bayard-Alpert gauge has a minimum of metallic surface so that it is easily outgassed by electron bombardment of grid and collector to temperatures above 1200° C with considerably less power than would be required for a conventional gauge. It may be operated on a standard power supply, and its over-all reliability and ease of outgassing are advantageous not only in the special field of ultrahigh vacuum, but over its entire pressure range.

[1] The data presented in this section were taken with such a collector. The commercial version of the gauge (WL-5966) has a straight 0.005-inch collector wire.

5. Ionization gauge sensitivity. The x-ray effect in the conventional ionization gauge constitutes one restriction which serves to limit its usefulness and previously has thrown doubt upon its dependability. This departure from linearity at very low pressures is considerably reduced in the Bayard-Alpert gauge, but it, like other ionization gauges, becomes non-linear and finally saturates at higher pressures (above 10^{-3} mm Hg). While the high pressure limit is not of specific concern in the ultrahigh vacuum range, the overall calibration of an ionization gauge is of importance for any accurate pressure measurement and will be discussed in a forthcoming section. In addition, there are two other limitations on ionization manometers in general which must be evaluated carefully in order to define or to minimize the errors which they may introduce into the measurement of pressure. These two disadvantages, which will be discussed next, are (1) that the sensitivity varies for different gases and vapors, and (2) that the ionization gauge may serve as a pump to "clean up" gases.

Table 1. *Values of sensitivity ratios for various gases.*

Gas	DUSHMAN and YOUNG [2]	WAGENER and JOHNSON [3]	SCHULZ
H_2	0.47	0.53	0.42
He	0.16		0.21
Ne	0.24		0.33
N_2	1	1	1
A	1.19		1.5
CO		1.07	
CO_2		1.37	
H_2O		0.89	
O_2		0.85	
Kr	1.9		
Xe	2.7		
Hg	3.4		
SF_6			2.5

A considerable amount of work has been done[1-3] on determining the sensitivity of ion gauges for various gases. Recently, SCHULZ[4] has also made measurements of ion gauge sensitivities for a variety of gases for the Bayard-Alpert gauge (WL-5966). From the many measurements which have been carried out, the following conclusions can be drawn: (a) While the sensitivities may vary appreciably from one gauge design to another, the ratio of sensitivities for the different gases seems to be essentially independent of the design of gauge used; (b) though there are certain variations in the methods used for measuring the sensitivities, the results of the various measurements are in good agreement. Tables 1 give the sensitivities of a given gauge for a number of common gases. The data are given in terms of a quantity R which is the ratio of the sensitivity for the gas in question to the sensitivity for nitrogen. This ratio is particularly convenient since the pressure usually is given as an equivalent nitrogen pressure. Note, however, that R may vary as much as 10 or 20% depending on the accelerating potential used in the calibration.

While Table 1 demonstrates graphically the inherent limitation of the ionization gauge resulting from the variation of sensitivity, it should be observed that the sensitivity for many of the gases falls within an approximate factor of two of the gauge sensitivity for nitrogen. The important exceptions are helium and neon, which have low sensitivities and for which the pressure may therefore be underestimated, and mercury, which has a high sensitivity and for which the pressure may therefore be overestimated.

HUTHSTEINER and DUSHMAN ([7], Chap. 6) have also made measurements on some of the vapors of organic liquids and solids such as iodine ($R=5.5$), cadmium ($R=2.4$), silicone oil ($R=2.7$), naphthalene ($R=0.85$), and Octoil-S

1 S. DUSHMAN and C. G. FOUND: Phys. Rev. **23**, 734 (1924).
2 S. DUSHMAN and A. H. YOUNG: Phys. Rev. **68**, 278 (1945).
3 S. WAGENER and C. B. JOHNSON: J. Sci. Instrum. **28**, 278 (1951).
4 G. J. SCHULZ: J. Appl. Phys. **1957** (in press). The sensitivity of the Bayard-Alpert gauge for nitrogen is 12 (mm Hg)$^{-1}$.

($R=5$). Thus it is seen that even including some vapors of high molecular weight, the sensitivity ratio does not vary by an order of magnitude.

In the case of gases or vapors which can react chemically or electrochemically (as ions) with the filament or other metallic elements, the measurement of the sensitivity of the ion gauge becomes difficult. It is known that oil vapors used in diffusion pumps are decomposed by the cathode into various molecules of lower molecular weight, so that the gas density in the gauge is greater than that in the rest of the system. The ions of these molecules obviously contribute toward the observed positive ion current, and the magnitude of the additional current must depend on the rate at which the decomposition products are formed and removed. For accurate studies of processes involving chemically active gases, the ionization gauge is not adequate and even the mass spectrometer introduces difficulties by causing cracking of the molecules in the course of making measurements. In most studies at low pressures, however, it is essential that the system be sealed off or otherwise isolated from major sources of contamination as a prerequisite for achieving pressures below 10^{-8} mm Hg. Any oil vapor remaining in the system will therefore be decomposed (ultimately to H_2, CO, CO_2, or to various hydrocarbons) and removed quickly by the gauge itself, the ion gauge reading being a reasonable measure of the removal rate. In general, accurate measurement of the pressures of chemically active gases is a difficult problem. The objective in most experiments is to eliminate such gases, and the ionization gauge gives a reasonably reliable index of the success in doing so.

From the foregoing considerations, it can be seen that while the ionization gauge does suffer from a variation of sensitivity with the gas in question, reasonably accurate pressure determinations can be obtained if the composition of the gas mixture is known. In addition, since the sensitivity to many of the common gases is very close to that for nitrogen, pressure measurements on mixtures of unknown composition are sufficiently accurate in many experiments. In view of the extreme simplicity of the gauge and its associated electrical circuits, its wide usage is easily understood.

6. Linearity and calibration of the ionization gauge. From previous discussion, it is clear that as the pressure approaches the x-ray limit, the gauge departs from linearity. That is, the ion collector current as measured is equal to the sum of the ion current and the x-ray induced photoelectric current. As the gauge reading decreases below 2 or 3×10^{-10} mm Hg, the electron current becomes an appreciable fraction of the total collector current and must be subtracted out to get a measure of the pressure.

It was important to demonstrate that the gauge is linear above 10^{-10} mm Hg. As described in Ref. [*6*], the linearity of the Bayard-Alpert ionization gauge was checked in a straightforward way.

The vacuum system used in the experiment is shown in Fig. 3 and is typical of those described in Ref. [*5*]. If, at a time $t=0$, we expose one side of a constriction of conductance, S_1, to a pressure, p_A, of a given gas, the rate of rise of pressure in the volume, V_1, is given by KNUDSEN's laws of gas flow:

$$\frac{dp_1}{dt} = \frac{S_1}{V_1} p_A \quad \text{if} \quad p_1 \ll p_A \tag{6.1}$$

so that

$$p_1 = at + p_{10}, \tag{6.2}$$

where $a=(S_1/V_1)\,p_A$ and p_{10} = the residual pressure in V_1 at time $t=0$. A linear increase of the ion gauge reading in volume V_1 with time constitutes a check on the linearity of the gauge. This method has, in fact, been used by DUSHMAN and FOUND[1] to indicate the linearity of a gauge above 10^{-6} mm Hg. If one were to try to extend this method down to 10^{-10} mm Hg, it would be necessary to adjust the conductance, S_1, so that the rate of rise in V_1 would be sufficiently small ($\approx 5\times 10^{-10}$ mm Hg per minute) to give enough time to take meaningful readings between 10^{-10} and 10^{-9} mm Hg. With such a small rate of rise, however,

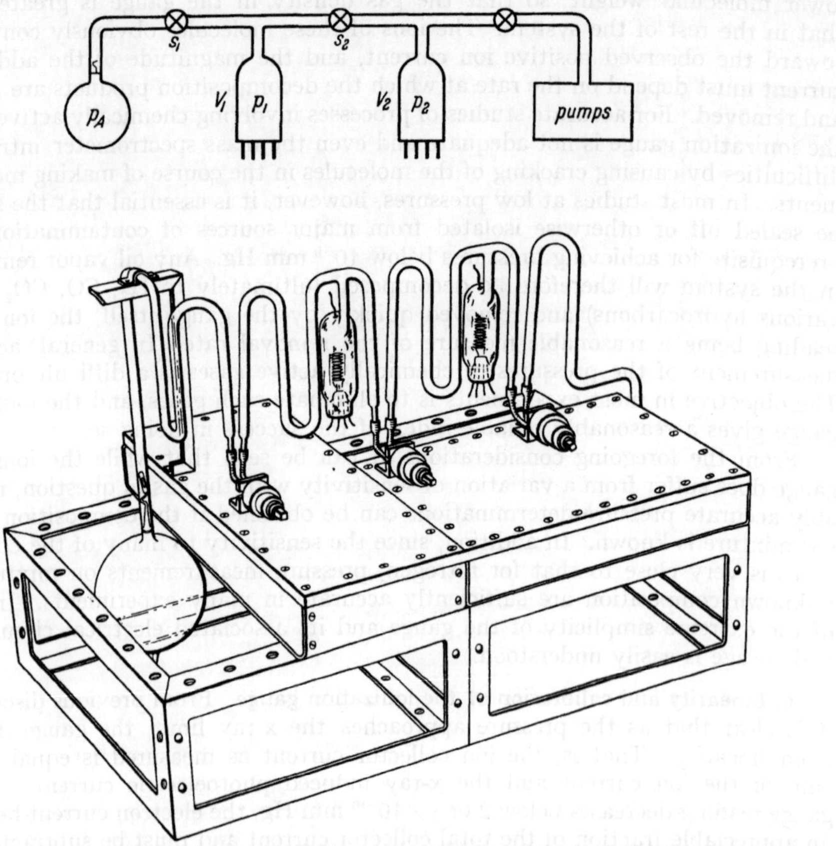

Fig. 3. Line drawing and schematic diagram of ultrahigh vacuum system for experiment to check ionization gauge linearity.

it would take about three years to cover the entire range of the gauge. Hence, a second valve[2] with conductance, S_2, and an ionization gauge were added as shown. The rate of rise of p_2, the pressure in volume, V_2, is now given by

$$\frac{dp_2}{dt} = \frac{S_2}{V_2}\, p_1 = \frac{S_2}{V_2}\,(a t + p_{10})\,. \tag{6.3}$$

For $t \geq 1$ second, $p_{10} \ll a t$. The solution of (6.3) is then of the form

$$p_2 = p_{20} + b t^2 \tag{6.4}$$

where p_{20} is the residual pressure in volume, V_2, and b is a constant of proportionality. Thus the pressure in V_2 is expected to rise quadratically in time, and

[1] S. DUSHMAN and C. G. FOUND: Phys. Rev. **15**, 7 (1921).

[2] The constrictions used were standard all-metal vacuum valves. D. ALPERT: Rev. Sci. Instrum. **22**, 536 (1951).

it is possible to take readings at very low pressures with a low rate of rise and still cover the entire range in a reasonable time. A plot of $p_2 - p_{20}$ vs. t^2 should thus give a straight line if the gauge is linear.

The experimental data are plotted in Fig. 4 for argon with an electron grid current of 10 milliamperes. Readings were taken by intermittently applying the operating voltages to the gauge for intervals of two or three seconds, and were automatically recorded. It is seen that the points follow a quadratic variation with time over almost seven decades of pressure, thus indicating linearity over a very wide range. At pressures above 10^{-3} mm Hg, the gauge showed departure from linearity, the shape of the curve depending strongly on the electron current to the grid. A calibration of the gauge for its linear region now requires a comparison with an absolute manometer at only a single point to determine its sensitivity.

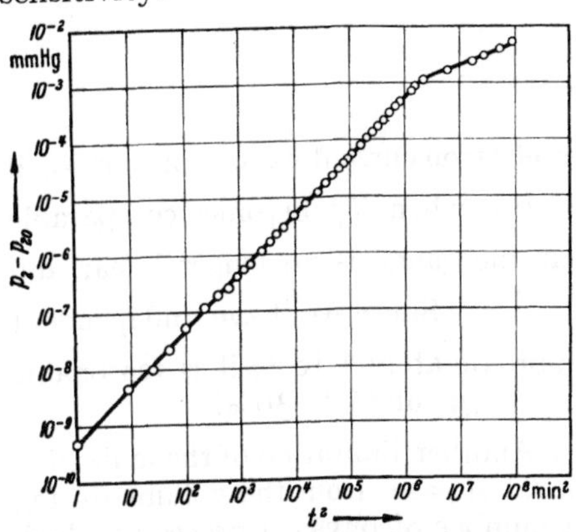

Fig. 4. $p_2 - p_{20}$ versus t^2 for argon.

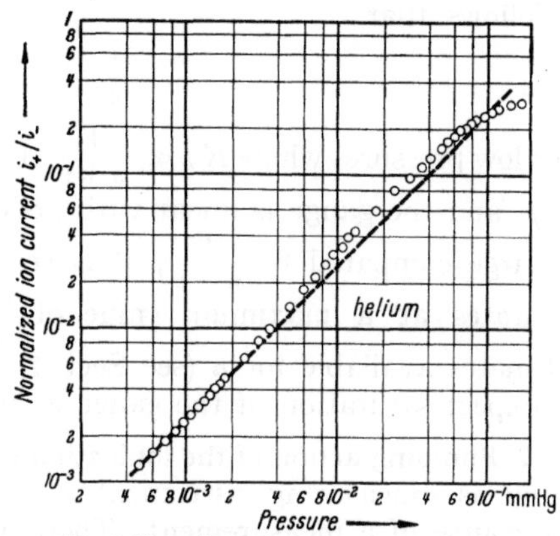

Fig. 5. Normalized ion current versus pressure (Bayard-Alpert ionization gauge).

Because calibration with a McLeod gauge involves a comparison with the Bayard-Alpert ionization gauge in its upper pressure range, and because some applications of the gauge require more accurate calibration, BURITZ[1] and SCHULZ[2] further investigated the linearity of the gauge at pressures above 10^{-5} mm Hg. BURITZ found that the linearity curves varied considerably with emission current. He also found that at an emission current of 10 ma the reproducibility of the observations was rather poor, probably due to the pumping action of the gauge and space charge effects. At electron currents of 0.1 ma or less, the performance of the gauge was much improved. Under these conditions (0.1 ma), BURITZ and SCHULZ observed a departure from linearity (which depended somewhat on the gas) for helium as shown in Fig. 5. Above 10^{-3} mm Hg, there was a gradual increase in sensitivity by as much as 20%, followed by a saturation of the i_c/i_g ratio. By enclosing the ends of the grid with a fine wire structure, SCHULZ eliminated the variation from a straight line and thus showed that the observed effect is due to a decrease in ion collection efficiency at lower pressures[3].

Saturation of the gauge reading at high pressure occurs with the gases studied when the ion current reaches about 20 to 30% of the electron current. A probable order-of-magnitude explanation of this saturation effect which occurs at higher pressures is as follows: Let us make the assumption that only the electrons which

[1] R. S. BURITZ: Rev. Sci. Instrum. **1957** (in press).

[2] G. J. SCHULZ: J. Appl. Phys. **28**, 1149 (1957).

[3] See W. B. NOTTINGHAM: 1954 Vacuum Symposium Transactions, p. 76. Boston: Committee on Vacuum Techniques, Inc. 1955.

leave the filament and are accelerated to grid potential are efficient for ionization. While this assumption is not strictly valid, the rather crude argument given here will not be greatly affected by a sizeably correction. Call this current due to electrons with full energy, i_0. The current recorded at the grid is then i_0 plus a current, which is equal to the ion current, i_+, due to the electrons released by ionization impact. In addition to the above currents, the grid collects essentially all of the electrons created by ionization in the volume outside the grid. If n ions are formed outside the grid for each ion within the grid, the total grid current is

$$i_g = i_0 + i_+ + n i_+ . \tag{6.5}$$

But since the ion current is given by [Eq. (3.1)]

$$i_+ = i_0 K p , \tag{6.6}$$

it follows that

$$\frac{i_+}{i_g} = \frac{K p}{1 + (n+1) K p} .$$

For low pressures where $Kp \ll \frac{1}{n+1}$, the ion-electron current ratio is proportional to p, and the gauge is linear with pressure. But when Kp becomes comparable or large compared to $\frac{1}{n+1}$, it is seen that the gauge is no longer linear and saturates at a maximum value of $\frac{i_+}{i_g} = \frac{1}{n+1}$. Since we have independent estimates available for n (see Sect. 7) of approximately 3 to 5, it is reasonable to expect saturation of the gauge at values of i_+/i_g of 15 to 30%.

7. Pumping action of the ionization gauge. Another limitation of the ionization gauge as a measuring instrument is its removal of gases from the volume during the course of a measurement. There are a number of physical processes which may cause a change in pressure in a system during the course of measurement, and it is necessary to assess these carefully to ascertain the specific limitations which they impose on the measurement of pressure. (For a general discussion, see [7], Chap. 10.) An understanding of the pumping action is also of obvious interest from another point of view. As will be discussed later, the ion gauge can be used as a very effective pump for the achievement of ultrahigh vacuum. We will briefly review the limitations imposed on the measurement of pressure and, in a later section, discuss in some detail recent studies of the mechanism of gas removal.

As usually operated for the measurement of ultrahigh vacuum, the ionization gauge is turned on for regular operation either during bake-out of the vacuum system or shortly thereafter. The electrodes are then heated to temperatures of 1200° C or higher by passing a current through them or (more easily in the case of the standard Bayard-Alpert gauge) by electron bombardment at several hundred volts. After "outgassing" the gauge in this fashion, during which an appreciable amount of metal may be vaporized or sputtered to the glass envelope, the ion gauge is turned on for "normal" measurement operation.

The ionization gauge may proceed to remove gases by any of several mechanisms. Leaving a detailed discussion of the phenomena to later sections, we will list some of the principal removal mechanisms and then make estimates of the maximum pumping speeds to be expected from each:

1. The surfaces of the electrodes or the envelope may remove certain gases by physical or chemical processes. For example, when the filament is heated to its normal operating temperature, interactions with chemically active gases

may take place which result in gas removal. Such a process accounts for the rapid removal of oxygen, which forms unstable oxides of tungsten at the hot filament and which, in turn, are deposited at the walls.

2. When the accelerating potentials are applied to the electrodes of the gauge, a second type of removal takes place which is usually referred to as "ion-pumping" or "electrical removal". In this case, the predominant removal process is the entrapment of the ions collected at negatively charged electrodes or walls. This mechanism is a predominant one for the noble gases and an important one for most of the molecular gases.

3. The electrons from the filament are also capable of causing excitation or dissociation of molecules within the gauge. In the Bayard-Alpert gauge (as in other gauges), the electrons are retarded after passing through the grid and have rather low energies during an appreciable portion of their trajectories exterior to the grid. The resulting "active" neutral gases may then be removed by attachment or combination at the walls of the gauge.

The rate of pressure reduction by any of these mechanisms is expected to be proportional to the pressure in the enclosure[1]. Hence, by analogy with the equation for the pumping speed of a conventional pump (see [7], p. 128),

$$\frac{dp}{dt} = -\frac{S}{V}(p - p_u), \tag{7.1}$$

where S is the pumping speed for a given mechanism,
V is the volume of the system,
p_u is the ultimate pressure attainable for a given mechanism.

If we assume $p_u \ll p$, Eq. (7.1) becomes

$$\frac{dp}{dt} = -\frac{S}{V}p. \tag{7.2}$$

The solution of (7.2) is

$$p = p_0 \,\mathrm{e}^{-\frac{S}{V}t} = p_0 \,\mathrm{e}^{-\frac{t}{\tau}}, \tag{7.3}$$

where p_0 is the pressure at time $t=0$ and $\tau = V/S$ is the characteristic time corresponding to a given mechanism. It is the time required to pump down from a given value to 1/e of that value.

If more than one mechanism takes place simultaneously, the total pumping speed, S, is equal to the sum of the individual speeds, and the total characteristic time constant is given by:

$$\frac{1}{\tau} = \frac{1}{\tau_i} + \frac{1}{\tau_c} + \cdots$$

where τ_i is the characteristic time for ion pumping, τ_c is the characteristic time for chemical reaction with the filament, etc.

A reasonable evaluation of the τ's is required in order to assess the importance of the pumping action of the ionization gauge:

(1) In the case of gases removed upon striking a surface, the maximum pumping speed, as shown in Sect. 12, is of the order of 10 liter/sec/cm² of surface, the exact value depending on the temperature and the molecular weight of the gas. For gases which are removed upon chemical interaction with the hot tungsten silament, the maximum pumping speed is determined by the area of the heated surface. In a typical gauge, this is approximately 0.2 cm². Thus the maximum

[1] These derivations were originally put forth in Ref. [5].

pumping speed, assuming immediate removal of each molecule upon collision with the filament, is 2 liter/sec (for oxygen). In actual cases, the "accomodation coefficient" or probability of an interaction upon collision with the filament varies exponentially with filament temperature, and may be considerably less than unity. Thus, for oxygen at the operating filament temperature of 2300° K, the maximum pumping speed to be expected is about 0.2 liter/sec.

(2) A first approximation to the ion-pumping speed of a gauge can be made by assuming that all the ions collected (in usual gauge operation) are permanently removed from the volume. The ion-pumping speed, S_i, can be defined by rewriting Eq. (7.2) in terms of the molecular density

$$\frac{dn}{dt} = -\frac{S_i}{V} n. \tag{7.4}$$

Since $V\frac{dn}{dt}$ is the total removal rate of molecules, $\frac{dN}{dt}$, the pumping speed is given by

$$S = -\frac{dN/dt}{n}. \tag{7.5}$$

But dN/dt, the number of ions removed per second is related to the sensitivity, K, of the ionization gauge by Eq. (3.1), since

$$\frac{dN}{dt} = a\,i_c, \tag{7.6}$$

where a is the number of electronic charges per unit current (for singly charged ions), and

$$i_c = K i_g p. \tag{3.1}$$

Rewriting (3.1) in terms of the molecular density

$$i_c = \frac{i_g K n}{b},$$

where b is a conversion factor, the number of molecules per unit pressure at a given temperature. Therefore

$$S_i = \frac{dN/dt}{n} = \frac{a}{b} i_g K. \tag{7.7}$$

If i_c and i_g are measured in amperes, $a = 6.3 \times 10^{18}$ sec^{-1}. If p is measured in mm Hg, K is given in (mm Hg)$^{-1}$, and $b = 3.2 \times 10^{19}$ molecules per liter at $p =$ 1 mm Hg at room temperature. The maximum pumping speed in liters per second is then given by

$$S_i = 0.2\, i_g K \text{ liter/sec}. \tag{7.8}$$

Thus the maximum ion pumping speed for a sensitivity $K = 10$ mm^{-1} and $i_g =$ 10 milliamperes is $S_i = 0.02$ liter/sec.

As will be indicated in Sect. 17, pumping speeds considerably higher than this have been observed for some gases in a Bayard-Alpert gauge. This has been attributed [5] to the ionization of gas outside the accelerating grid, with subsequent collection at the negatively charged walls of the gauge. If n ions are formed outside the grid for each one created within the grid, the speed would be given by

$$S_i = 0.2\,(n+1)\, i_g K. \tag{7.9}$$

Estimates of n have varied from 3 to 10, with corresponding pumping speeds of the order of 0.01 liter/sec/ma.

(3) In the case of removal of "active" neutral gases, it is very difficult to estimate a reasonable pumping speed, since the processes involved have not been studied in detail. In most cases, the cross sections for excitation of molecular metastables or for the dissociation of molecules by electron impact are less than those for ionization, usually considerably less (see [*1*], Chap. 4). Hence, one would conclude that the pumping speed by these mechanisms would be less than that for ion-pumping.

An evaluation of the relative importance of the above processes for various gases will be made in Sects. 15 and 16. For the present discussion, it is sufficient to indicate that processes (1) and (2) seem to be the dominant mechanisms for pump ing at very low pressures. The pumping speed for gases which react chemically with the hot filament may be of the order of 1 liter/sec, while that for ion pumping as high as 0.1 liter/sec for an emission current of 10 ma. For a system volume of one liter, the corresponding characteristic times are

$$\tau_c = 1 \text{ second for chemical removal, and}$$

$$\tau_i = 10 \text{ seconds for ion pumping.}$$

From these considerations, it is clear that the pumping action of an ionization gauge may make very appreciable changes in pressure during the course of a measurement. If the gauge is to be used to measure pressures without significantly altering them, one should take readings during intervals small compared to the pumping times. In the above cases, it is seen that sizeable effects are quite probable unless readings are taken for very short time intervals. In the case of chemical removal at the filament, a longer τ_c is possible only if one lowers the temperature of the filament (to reduce the pumping speed) or changes the ionization gauge design to limit the area of reaction. In the case of ion-pumping, however, it is seen from Eq. (7.6) that the pumping speed can be proportionately reduced by lowering the ionizing electron current. Hence, it is usually desirable to operate the gauge at emission currents of 0.1 ma or less. By taking the pumping speed into account, accurate measurements of pressure are quite feasible, and variations of pressure within the system caused by ion gauge pumping can readily be computed.

8. Extension of the lower limit of the Bayard-Alpert gauge. Nottingham modifications of gauge. On the basis of previous discussion, the lower limit of the ionization gauge can be extended by increasing the ratio of the sensitivity to the x-ray induced residual current. This would seem easy to accomplish at first glance. As a matter of fact, one of the gaps in the understanding of the operation of the gauge is its rather surprisingly low sensitivity using known ionization cross sections as a basis for comparison. At low pressures, it seems reasonable to assume that essentially all of the electrons collected at the grid originate at the ionization gauge filament. The sensitivity of a gauge can be defined as the product of the average path length of these ionizing electrons and the ionization efficiency, E_i (the number of ions produced per centimeter path per electron for a gas pressure of 1 mm). From measured values of the sensitivity, K, and of the ionization efficiency[1], one can estimate the average length of the trajectory of an ionizing electron in a gauge. For a standard Bayard-Alpert (WL-5966), the average electron collected at the grid of a Bayard-Alpert gauge travels only 1.5 cm before being collected. This value is surprisingly low, since the optical transparency of the grid is quite large and one might expect the average electron

[1] See, for example, K. T. COMPTON and C. C. VAN VOORHIS: Phys. Rev. **27**, 724 (1926).

to make several traversals through the grid before being collected. The reasons for this low sensitivity are therefore rather obscure. One factor, which according to SCHULZ can account for perhaps a 20% effect, is the loss of ions from the ends of the open grid structure. Another possibility is that some of the electrons may travel to the grid in an unspecified path outside the grid volume and, hence, without causing ionization which is recorded at the collector.

NOTTINGHAM[1] has proposed a design which is similar to the original Bayard-Alpert gauge, but which has two added features which increase the sensitivity of the gauge: (1) the addition of grid-like enclosed ends for the accelerating grid, and (2) the installation of a screen grid in the form of an open cylinder surrounding the other elements of the gauge. He found that the addition of these two features increased the sensitivity by a factor of 2 to 3. The enclosure of the accelerating grid, as has been previously discussed, serves to improve the efficiency of ion collection, especially at low pressures. The screen grid, which is maintained at a large negative potential, must account for most of the observed increase in sensitivity. Electrons which emerge from the grid volume are repelled by the strong radial field and presumably travel a longer effective path length (about 4 cm average path length). The screen grid also serves to shield the gauge from the influence of charges on the glass wall of the ionization gauge.

It is not clear in detail how the potential at the glass wall affects ionization gauge sensitivity. The wall potential is, however, a very important parameter in the operational and pumping characteristics of the gauge. It is well-known that the potential of the surface of adjacent insulators may vary widely with the potentials applied to the electrodes. This potential change is associated with the variation of the coefficient of secondary emission of electrons with incident electron energy[2]. Thus as the accelerating grid voltage in the ionization gauge is increased, the potential of the glass walls may vary from a value close to that of the hot filament to a rather high positive potential for electron energies above 200 volts. In normal operation, the walls of a Bayard-Alpert gauge float at a potential very close to that of the filament. Thus the ions formed outside the grid are normally collected at the wall. As will be discussed in a succeeding section, this greatly affects the pumping speed of the gauge. Further studies are required for a detailed understanding of the effect of wall charge on gauge sensitivity.

Concerning other methods of increasing the ratio of sensitivity to x-ray current in a Bayard-Alpert gauge, relatively little has been reported. Preliminary experiments[3] aimed at lowering the x-ray effect by reducing the diameter of the ion collector wire resulted in a sizeable reduction of the sensitivity, even with the grid volume enclosed at both ends. Thus there may be a critical size for the wire below which the probability of collecting ions goes down as rapidly as does the x-ray effect.

An eventual possible limitation on the extension of the gauge to lower pressures has been investigated by ALPERT and BURITZ [*6*]. This is due to the finite vapor pressure of the tungsten filament at an operating temperature of 2300° K. That is, the pressure reading due to the tungsten atoms sublimating from the hot filament and passing through the grid volume en route to the walls of the gauge is of the order of 10^{-12} mm Hg. This could be lowered by operating at

[1] W. B. NOTTINGHAM: 1954 Vacuum Symposium Transactions, p. 76. Boston: Committee on Vacuum Techniques, Inc. 1955.

[2] See, for example, K. G. MCKAY: Secondary Electron Emission. In: Advances in Electronics I. New York: Academic Press Inc. 1948.

[3] R. S. BURITZ and D. ALPERT: Unpublished.

lower filament temperatures, but ultimately would constitute a basic limitation on the ionization gauge.

CONN and DAGLISH[1] reported another modification of the Bayard-Alpert gauge, this one in the direction of simplifying the original structure. As shown in Fig. 6, the accelerating electrode structure in this case is an open plane grid structure. It is certainly easier to construct and can be enclosed in a smaller

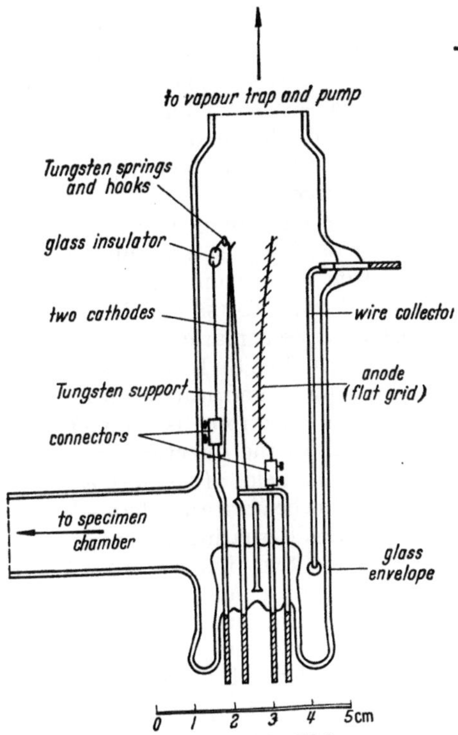

Fig. 6. Conn and Daglish ion gauge.

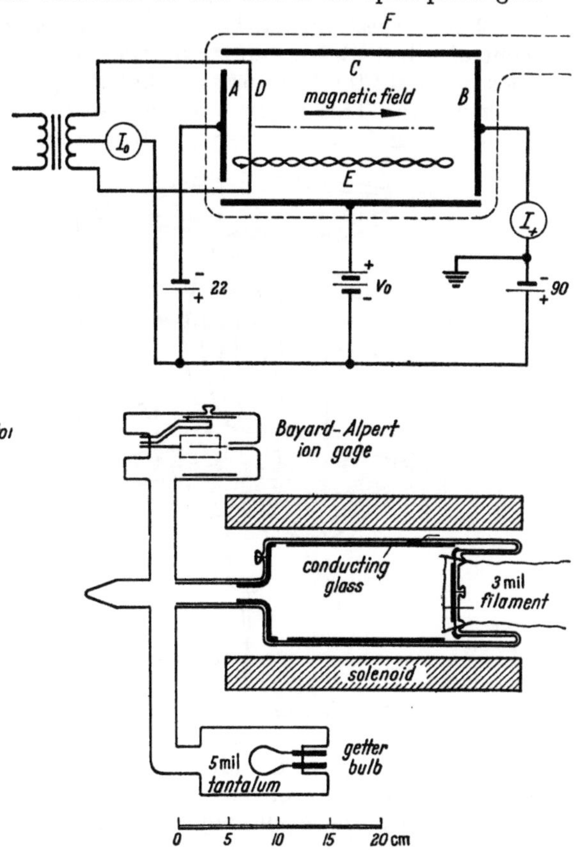

Fig. 7. Houston ion gauge.

envelope. On the other hand, it is clear from the preceding discussion that the sensitivity of the gauge in this case would seem to be very dependent on the wall potential of the envelope. It seems probable that the sensitivity would be reasonably high only if the walls floated at a positive potential with respect to the collector. This may occur at the recommended accelerating voltages. The authors claim a sensitivity only a "little less" than that of a BAYARD-ALPERT gauge.

9. Houston ionization gauge. J. M. HOUSTON has recently reported[2] a new ultrahigh vacuum gauge which is useful in the pressure range below 10^{-8} mm Hg and is probably linear to much lower pressures than the Bayard-Alpert gauge. By utilizing electric and magnetic fields to contain the ionizing electrons, average electron trajectories of 10^8 cm are obtained, thus allowing exceedingly small filament emission and providing high sensitivity. A sketch with a schematic diagram is shown in Fig. 7. While the upper limit of the usable pressure range

[1] G. K. T. CONN and H. N. DAGLISH: J. Sci. Instrum. **31**, 95 (1954).
[2] J. M. HOUSTON: Bull. Amer. Phys. Soc. (II) **1**, 301 (1956).

is quite low (it begins to depart from linearity in the 10^{-8} mm Hg range), it is unique in that it seems to have a lower limit several orders of magnitude below the Bayard-Alpert gauge. Houston reports a preliminary experiment in which pressures of about 2×10^{-12} mm Hg are attained in a gauge which utilizes its own high pumping speed to reach this pressure.

10. Mass spectrometers for the measurement of ultrahigh vacuum. The development of ultrahigh vacuum techniques has brought forth an increasing need to know not only the total pressure but the partial pressures of the various gases in an enclosure. This requires a mass spectrometer which has first a very high sensitivity, and second a very low background spectrum. This means, in general, an instrument capable of high temperature outgassing and with a reasonably high pumping conductance between the system to be studied and the ion source.

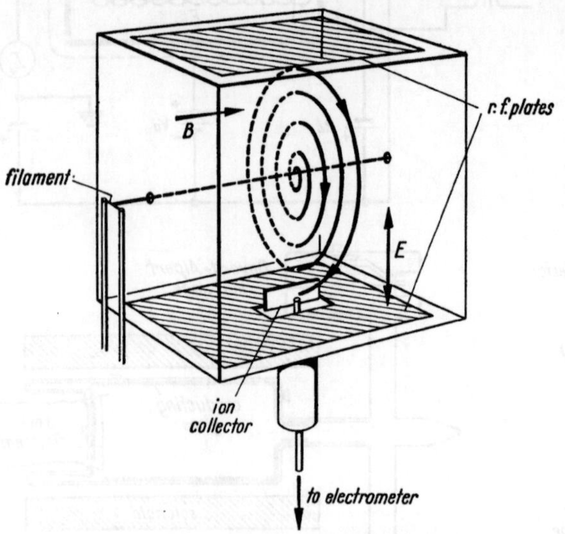

Fig. 8. Schematic diagram of simplified omegatron.

One instrument which has been used [*6*] is a simplified version of the omegatron of SOMMERS, THOMAS and HIPPLE[1]. A schematic view is shown in Fig. 8. In this case, a narrow cylindrical beam of ionizing electrons is formed parallel to the direction of a magnetic field. Ions formed along the beam are accelerated by the alternating potential between two flat *rf* plates which correspond to the dees of a cyclotron. When the value of the impressed radio frequency is equal to the cyclotron frequency $\omega = eH/M$, the ions of mass M, and charge e, are accelerated in orbits of increasing size and eventually strike the ion collector.

The sensitivity of the omegatron is inherently quite high, i.e., comparable to that of an ionization gauge. Its relatively simple mechanical construction makes it readily possible to attain pressures of approximately 10^{-9} mm Hg in a system with the omegatron operating. The principal limitation found in this work was the occurrence of "ghost" peaks in the instrument at multiples of the frequency for a given charge-to-mass ratio. Nevertheless, the gauge has proved valuable for studies of the permeation of helium into the vacuum enclosure and for measurement of the partial pressure of other gases for which the mass identification is unambiguous. The omegatron can easily be used as a helium leak detector for ultrahigh vacuum systems.

EDWARDS[2], WAGENER[3] and others have done further work on the omegatron for use as mass spectrometer and partial pressure ionization gauge. EDWARDS has made a variety of measurements of omegatron characteristics; he specifies a detection limit of approximately 10^{-11} mm Hg.

---

[1] H. SOMMERS, H. A. THOMAS and J. A. HIPPLE: Phys. Rev. **82**, 697 (1951).
[2] A. G. EDWARDS: Brit. J. Appl. Phys. **6**, 44 (1955).
[3] S. WAGENER: Private communication.

J. H. REYNOLDS[1] has described a high sensitivity mass spectrometer for noble gas analysis. Capable of bakeout, the instrument has been operated at total background pressures of 5×10^{-10} mm Hg or less as read on an ion gauge and is capable of analyzing very small background pressures. For some gases, it can read partial pressures well below 10^{-12} mm Hg. Further discussion of REYNOLDS' instrument and results is given in Sect. 14.

11. Flash-filament technique. In 1948, APKER[2] suggested a method for using an ionization gauge to indicate the state of vacuum in a system when the total pressure was less than 10^{-8} mm Hg. This technique, often called the "flash filament" method, has been further investigated by NOTTINGHAM[3] by MOLNAR[3] and by BECKER and HARTMAN [8].

The procedure consists of incorporating an auxiliary filament within the vacuum system, often within the ionization gauge. The filament is first cleaned by momentarily flashing to a temperature above 1500° K; for tungsten this is often at about 2300° K. After a time Δt, during which it is kept at a "cold" temperature, T_0, which may be room temperature or higher, the filament is again flashed. The second flash suddenly releases the molecules adsorbed on the filament during the time, Δt, and the resultant burst of gas registers as a momentary change in the reading of the ionization gauge, giving a so-called peak pressure indication[4]. This peak pressure reading at first increases linearly with Δt, corresponding to the accumulation of adsorbed molecules on the flash filament; for sufficiently large Δt, the peak pressure approaches a constant value and shows very small increases with further increase in Δt. This behavior is assumed to be due to the ready adsorption of molecules at early times when less than a monolayer is adsorbed on the filament, the rate depending on the pressure of adsorbable gas in the volume. The leveling-off point is taken to correspond to the time when the flash filament is covered with a monolayer of the molecules. BECKER and HARTMAN [8] have shown that for nitrogen on tungsten, a monolayer corresponds to a coverage by 1.25×10^{14} molecules/cm². One can calculate the rate at which molecules arrive at the filament surface at a given pressure (on the basis of kinetic theory) and estimate the number of molecules per cm² in a complete monolayer. The flash filament procedure gives $(\Delta t)_M$, the time required to cover the filament with a monolayer. From these numbers one can estimate the adsorbable gas pressure in the vacuum system even though it lies below the measurable range of the ionization gauge. A rough check on these assumptions, and hence a calibration of the flash filament technique, is possible in a straightforward way by raising the pressure to a point where it can be read directly with the ion gauge and then comparing this value with the flash filament measurement.

The advantage of the flash filament method lies in its great simplicity. It was the first technique which became widely used in the measurement of pressures below 10^{-8} mm Hg. A disadvantage of the method, as far as pressure measurement is concerned, is that it discriminates severely between different types of gases; in particular, it indicates only the partial pressures of certain adsorbable gases. Another difficulty is that it requires a considerable time interval for a single pressure determination. Nevertheless, the flash filament technique has been of great help in ultrahigh vacuum researches.

[1] J. H. REYNOLDS: Rev. Sci. Instrum. **27**, 928 (1956).

[2] L. APKER: Industr. Engng. Chem. **40**, 846 (1948).

[3] W. B. NOTTINGHAM and J. P. MOLNAR carried out a number of interesting, but unpublished, investigations in this field. They reported some of these at the MIT Conference on Physical Electronics in 1949 and 1950.

[4] Δt must be sufficiently large so that the adsorbed gas will give a peak pressure greater than the minimum value observable on the ionization gauge.

BECKER and HARTMAN [2] have used the flash filament technique to measure the sticking probabilities of nitrogen and oxygen on tungsten. They demonstrated the interesting result that the sticking probability, approximately 0.5 for nitrogen at room temperature, is essentially constant until a "monolayer" of gas is adsorbed. At constant gas pressure, this results in a linear increase with time in the quantity adsorbed until a monolayer is adsorbed.

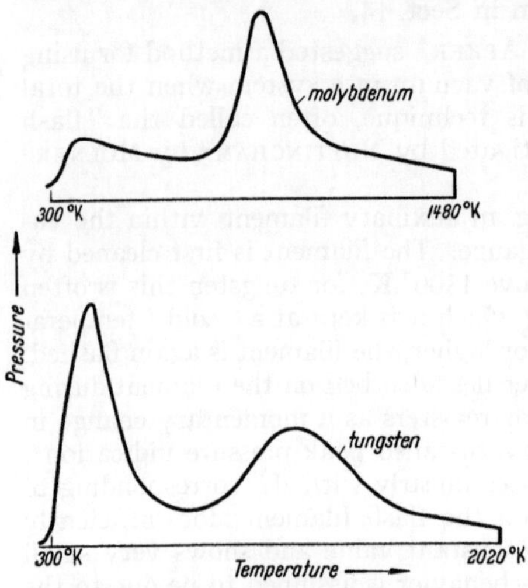

Fig. 9. HAGSTRUM'S measurements. Typical oscillo scope traces taken with the adsorption spectrometer for a molybdenum and a tungsten filament: $\Delta t_e =$ 25 min in each case. Note the two regions on the temperature scale at which gas is desorbed most readily.

HAGSTRUM has described [9] an interesting "adsorption spectrometer" devised by MOLNAR which indicates the desorption rate from a flash filament as a function of temperature. The information is displayed on an oscilloscope by heating the filament with a constant current source and indicating the pressure versus the filament temperature (calibrated as a function of filament resistance). A typical trace is shown in Fig. 9. This method can readily be applied to the study of adsorption activation energies and desorption rates. EHRLICH[1] has used this technique to study the states of adsorbed gases on tungsten. Thus, the flash filament technique has provided not only an effective method for measuring very low pressures but also a very useful tool in the study of physical and chemical adsorption.

III. Factors affecting attainment of ultrahigh vacuum.

12. Basic vacuum equations. The basic parameters affecting the attainment of ultrahigh vacuum are, generally speaking, the same as those which govern the production of high vacuum. Since derivations are available in a number of references, it is not proposed to develop in detail the basic equations of vacuum practice. For a thorough discussion, the reader is referred to GUTHRIE and WAKERLING [10], or S. DUSHMAN [7], Chap. 3). However, it may be useful to specify or derive some of the principal equations in order to better relate the discussion which follows.

For an ideal gas[2] at rest, the number of molecules striking a unit area per unit time is given by

$$\nu = \frac{1}{4}\, n\, \bar{v} \tag{12.1}$$

where n is the molecular density, and $\bar{v}$ is the average molecular velocity.

From kinetic theory [7], [10]

$$\bar{v} = \sqrt{\frac{8\,\mathrm{k}\,T}{\pi m}} \tag{12.2}$$

where m is the molecular mass, k is BOLTZMANN'S constant 1.38×10^{-16} dyne-cm/degree, and T is the absolute temperature.

[1] G. EHRLICH: J. Phys. Chem. **60**, 1388 (1956). — J. Phys. Chem. Solids **1**, 3 (1956).
[2] For sufficiently low pressures, i.e., in the entire high vacuum range, the ideal gas laws are obeyed with considerable accuracy, and the transport of gases follows KNUDSEN'S equations for "molecular flow".

Thus

$$\nu = n \sqrt{\frac{kT}{2\pi m}}. \tag{12.3}$$

The volume of gas striking unit area per unit time is ν/n and, hence, the quantity of gas, Q, striking area A, per unit time is given by

$$Q = pV = Ap \sqrt{\frac{kT}{2\pi m}}. \tag{12.4}$$

If one has an aperture of area A (small compared to the vessel and to the mean free path of the gas), the net flow of gas from one side at pressure p_1 to the other side at p_2 is just the difference between the flows in each direction

$$Q = \sqrt{\frac{kT}{2\pi m}} (p_1 - p_2) A. \tag{12.5}$$

The conductance of the aperture is defined as

$$C = \frac{Q}{p_1 - p_2} = \sqrt{\frac{kT}{2\pi m}} A. \tag{12.6}$$

For oxygen at room temperature (A in cm^2)

$$C = 11.1\, A \text{ liter per second.} \tag{12.7}$$

The pumping speed of the aperture may be defined as

$$S = \frac{Q}{p_1}. \tag{12.8}$$

If $p_2 = 0$, the pumping speed $S = C$. Thus, the maximum speed for any type of pump which involves the kinetic flow of gas through an orifice or striking a surface is approximately 11 liter/sec/cm² for oxygen and varies inversely with the square root of the molecular weight.

In an analogous way, the pumping speed, S, for any pump is defined by the expression

$$Q = Sp \tag{12.9}$$

where Q is the quantity of gas removed from a volume per second, and p is the pressure in the system. In general, if several pumping mechanisms take place simultaneously, the individual pumping speeds add and the effective S is given by $S_1 + S_2 + \cdots$. If the pump is separated from the system by a constriction or baffle of conductance C, the pumping speed of the overall system is given by

$$\frac{1}{S} = \frac{1}{S_p} + \frac{1}{C}, \tag{12.10}$$

where S_p is the pumping speed of the pump.

The rate of reduction of pressure in a system of volume, V, is given by

$$\frac{dp}{dt} = -\frac{S}{V} p. \tag{12.11}$$

If a quantity of gas or vapor is backstreaming from the pump at a constant rate, Q_p, we have

$$\frac{dp}{dt} = -\frac{S}{V} p + \frac{Q_p}{V}. \tag{12.12}$$

Another way ([7], p. 128) of defining the pumping speed is by the relation:

$$\frac{dp}{dt} = -\frac{S}{V}(p - p_u) \tag{12.13}$$

where p_u is the ultimate pressure attainable with the pump. It is seen that Eqs. (12.12) and (12.13) are equivalent if

$$p_u = \frac{Q_p}{S}. \tag{12.14}$$

That is, we can consider a pump with an ultimate attainable pressure, p_u, as one which has a constant backstreaming of gas, Q_p. This way of looking at the ultimate attainable pressure is particularly appropriate for diffusion pumps in which it has been shown that if backstreaming is strongly reduced, the pumps retain their speeds far below the published values. If the total influx of gas (which may vary with time t) is $Q(t)$, then Eq. (12.12) becomes

$$\frac{dp}{dt} = -\frac{S}{V}p + \frac{Q(t)}{V}. \tag{12.15}$$

But $Q(t)/V = r(t)$, the rate of rise of pressure in the absence of any pumping. Therefore,

$$\frac{dp}{dt} = -\frac{S}{V}p + r(t). \tag{12.15a}$$

If $r(t)$ is a constant, the solution of Eq. (12.15a) is

$$p = r\tau + (p_0 - r\tau)\,\mathrm{e}^{-\frac{t}{\tau}} \tag{12.16}$$

where $\tau = V/S$ is the characteristic pumping time, the time required to pump the system to $1/e$ of its initial value, and p_0 is the pressure at time $t = 0$. The equilibrium pressure, p_e, at sufficiently long times is then given by

$$p_e = r\tau, \tag{12.17}$$

$$= \frac{Q}{S}. \tag{12.18}$$

Q is then the total rate of influx of gas into the system. In general, it will be made up of at least three major constituents:

1. Q_p backstreaming or back diffusion from pumps,
2. Q_L influx from leaks, and
3. Q_W influx as a result of desorption, decomposition, or diffusion through the walls of the system.

The problem of achieving ultrahigh vacuum is summarized by the statement that the rate of gas influx divided by the pumping speed of the system must be small ($\ll 10^{-8}$ mm Hg).

ALPERT and BURITZ [6] showed that in the case of vacuum-tight bakeable systems which could be outgassed at temperatures above 400° C, the major influx of gas was due to atmospheric helium diffusing through the glass walls of the system at a rate corresponding to 3 to 4×10^{-11} mm Hg/min or approximately 5×10^{-13} mm Hg/sec. Since the volume was approximately one liter, this corresponded to a Q of 5×10^{-13} mm liter/sec. To reach a pressure of, say, 2×10^{-10} mm Hg, an effective pumping speed of only 0.002 liter/sec was required, a speed readily obtained by the ion pumping of a Bayard-Alpert ionization gauge. In the case of conventional high vacuum equipment, it is not uncommon to use a diffusion pump of, say, 10 liter/sec to evacuate a volume of one liter to achieve a pressure of 10^{-6} mm Hg. In this case, the hypothetical total flux into the system would be 10^{-5} mm liter/sec or 10^8 times that in the above case. This comparison illustrates the importance of evaluating both the rate of gas influx and the pumping speed in a discussion of the production of ultrahigh vacuum. It also emphasizes the value of a measurement of the rate of pressure rise in the absence of pumping as a very useful indication of the state of vacuum in a system.

A few general comments on scaling factors may be of interest. From Eq. (12.11) it is seen that the rate of pressure reduction is proportional to the pumping speed and inversely proportional to the volume of the system. For this reason, there is often a tendency to design pumping systems with the system volume as the most important scaling factor. From (12.18), however, it is seen that the ultimate equilibrium pressure in a system is dependent on the pumping speed, S, and on Q, the rate of influx into the system. Omitting Q_p, the back diffusion of gases from the pumps, which is a property of the pump itself, the principal sources of gas are leaks in the vacuum enclosure and the evolution of gas from the vacuum envelope. In a general way, both of these are proportional to the total surface area of the vacuum enclosure. (If the system contains many electrodes and other surfaces, a separate internal Q should to be evaluated.) For this reason, it is convenient to define a "specific influx"

$$q = \frac{Q_w}{A} \tag{12.19}$$

which gives the net influx of gas per unit area of envelope surface. This may be the most important parameter in ultrahigh vacuum system design.

To illustrate this point, a conventional vacuum system at a pressure of 10^{-7} mm Hg has approximately 3×10^9 molecules/cm^3 in the system volume. At this pressure, the walls usually have at least one monolayer (often as high as 10) or 10^{14} to 10^{15} molecules physically adsorbed per cm^2 of surface. Thus for a system in the shape of a 10 cm cube, there are about 10^6 molecules adsorbed on the surface for each one in the volume. This large reservoir of loosely bound molecules may supply gas to the volume for long periods of time, even without the usual replenishment from leaks, pumps, and gas diffusing in from the body of the solids.

From the data mentioned above, the specific influx for a tight glass ultrahigh vacuum system has been found to be as low as 2×10^{-15} mm liter/sec/cm^2. For a conventional unbaked system, it is not uncommon to observe q's of 2×10^{-7} mm liter/sec/cm^2.

13. Evolution or influx of gas into vacuum systems. On the basis of the previous discussion, it is important to evaluate the various mechanisms for the influx of gas into a system. Among the most important processes are the diffusion of gases through solids and the evolution of gas from surfaces by desorption or decomposition. While they are obviously essential to an understanding of ultrahigh vacuum problems, it is beyond the scope of this article to present a comprehensive discussion of these interrelated phenomena[1]. Despite the voluminous literature on the subject, very little data are available from which rates of gas evolution data can be calculated. Several gases are often given off simultaneously from the same material by different processes, and relatively few combinations

[1] The reader is again referred for a general discussion to references such as S. DUSHMAN [7], Chap. 8 and 9; R. M. BARRER: Diffusion in and through Solids (Cambridge: The University Press 1951); A. R. MILLER: Adsorption of Gases on Solids (Cambridge: The University Press 1949); R. JAECKEL [*11*], Chap. D.

of gases and metals have been quantitatively studied. For these reasons, the discussion will be limited to a brief description of some of the physical mechanisms involved followed by a resume of some studies of residual gases in vacuum systems. From these, it is possible to obtain rough estimates of the values for q, the specific influx of gas into the volume of a vacuum system.

The adsorption of gases on solids may, in general, be divided into two classifications—physical adsorption and chemical adsorption[1]. In physical adsorption the adsorbed molecule is held predominantly by relatively weak van der Waals forces. Interaction between solid and gas is sufficiently weak, so that no significant distortion of the adsorbed molecule takes place, nor is there significant electron-sharing or electron transfer between the adsorbed molecule and the surface. For the common diatomic molecules such as nitrogen, hydrogen, and carbon monoxide, the energy of attachment is of the order of 1–6 kcal/mol. It follows that desorption takes place rapidly at relatively low temperatures, and the large majority of such molecules are removed upon bakeout at a few hundred degrees.

In the case of chemisorption, the binding to the solid surface is of the same order as those in ordinary chemical bonds and the heat of adsorption is roughly ten times that for physical adsorption. Thus, certain clean metals, e.g., tungsten can chemisorb nitrogen and oxygen at temperatures above 1000°K. Although chemisorption has been under investigation for many years, only a relatively small number of systems, notably nitrogen and oxygen on tungsten[2,3,4] [3], have been studied under conditions which eliminated large extraneous effects due to impurities and contaminants. Generally speaking, the bakeout at a few hundred degrees leaves at least a monolayer of gas chemisorbed at the surface and to get atomically clean surfaces, outgassing at higher temperatures or other procedures are necessary.

Some of the gases to be expected in observable quantities in a high vacuum system are CO, CO_2, water vapor, nitrogen, hydrogen, and oxygen. In a study of the total amounts of gas given off by heating such metals as tungsten, molybdenum, and nickel, NORTON and MARSHALL[5] showed that CO, CO_2, N_2, and H_2 are given off in sizeable quantities for almost any of the common metals used in vacuum systems. Unfortunately, relatively little information is available with which to calculate the expected rates at which gases are evolved. In some cases (hydrogen on molybdenum), the gas comes off at low temperatures, presumably due to desorption of physically adsorbed gas. In other cases, the rate of evolution is determined by diffusion from within the solid (CO in nickel), and in some cases high temperatures are required to dissociate chemisorbed (nitrogen on tungsten) layers.

Oxygen may originate from the dissociation of oxide layers on various metal surfaces. A number of detailed studies have been made on the oxides of various metals[6]. WAGENER[7] used available information on the free energy of formation

[1] For a discussion, see, for example, A. WHEELER: Chemisorption on Solid Surfaces. In: Structure and Properties of Solid Surfaces, Chap. 13. Chicago: The University of Chicago Press 1953.

[2] Ref. [2], p. 136.

[3] G. EHRLICH: J. Phys. Chem. Solids **1**, 3 (1956).

[4] E. A. GULBRANSEN and W. S. WYSONG: Amer. Inst. Mining Met. Engrs., Inst. of Metals Div. **175**, 611 (1948).

[5] F. J. NORTON and A. L. MARSHALL: Trans. Amer. Inst. Mining Met. Engrs. **156**, 351 (1944).

[6] For a review and bibliography, see E. A. GULBRANSEN: Trans. Amer. Electrochem. Soc. **82**, 375 (1942).

[7] S. WAGENER: Vacuum **3**, 11 (1953).

of various oxides to estimate an upper limit for the dissociation pressure for the oxides of Ni, Mo, and W of 10^{-10} mm Hg at 1000°K. While this is not large, it still represents an appreciable amount of gas for some ultrahigh vacuum systems containing heated electrodes.

In some important cases, for example, water vapor from glass and carbon monoxide from nickel, the rate of evolution is determined by diffusion processes within the body of the solid material. LANGMUIR[1] was one of the first to show that large quantities of water vapor (as many as 30 monolayers) are released from the surface of glass, along with substantial quantities of CO_2 and nitrogen, upon bakeout at temperatures of 400 to 500° C. In a recent study, B. J. TODD[2] verified much of the earlier work and further showed that after the initial release of the water vapor from the surface (where it is probably stored in the form of hydrates), the subsequent release of water is governed by diffusion from the body of the glass. He has made measurements of the diffusion constants for various glasses over a wide range of temperatures. At room temperature, the influx of water due to diffusion is relatively small. TODD's data give values for borosilicate glass which correspond to a specific influx of H_2O of 3×10^{-18} mm liter/sec/cm^2, and, indeed, water vapor is not a large factor in most ultrahigh vacuum system. At 100° C, however, the rate of diffusion is higher by three orders of magnitude and the influx to be expected in a glass system is comparable to that of helium.

In the case of carbon monoxide, it is likely that in most vacuum systems the gas arises from a number of possible sources. In a widely quoted experiment, SMITHELLS and RANSLEY[3] showed that the large amount of CO which is found to be emitted from nickel at high temperature is not dissolved as such in the material. It is formed by the diffusion of carbon (dissolved in sizeable quantities in nickel) to the metal surfaces at which point the carbon reduces the oxides present to form CO. Smaller quantities of CO_2, N_2, and H_2 are also given off. On the basis of this process, the amount of CO given off by a clean nickel rod of substantial purity, 1 mm in diameter and 4 cm long, may be as high as 5×10^{-7} mm liter/sec at 925° C, and as high as 5×10^{-11} mm liter/sec at 475° C. Thus, a long-term release of CO is to be expected in tubes containing nickel or nickel-bearing alloys operating at elevated temperatures.

BLEARS[4] proposes an alternate mechanism for the formation of CO in a vacuum system which contains a tungsten filament. He states that if a tungsten filament is operated in hydrocarbon vapor, it may be converted to tungsten carbide and that in the presence of water vapor, a reaction at the filament results in the formation of carbon monixide. With either of the above mechanisms, a sizeable influx of CO may be expected in the vicinity of heated electrodes, and many authors agree that it is the most prevalent gas evolved in a clean vacuum system. This seems to be true in systems utilizing either mercury or oil vapor diffusion pumps; in both cases some organic impurities may arise from the mechanical backing pumps.

The permeation of atmospheric gases through the envelope of a vacuum system may also be a major source of influx at very low pressures. NORTON[5]

[1] I. LANGMUIR: J. Amer. Chem. Soc. **40**, 1361 (1918).
[2] B. J. TODD: J. Appl. Phys. **26**, 1238 (1955).
[3] C. J. SMITHELLS and C. E. RANSLEY: Proc. Roy. Soc. Lond. **155**, 195 (1936). More recent measurements of the diffusion constants have been made by J. J. LANDER, H. E. KERN and A. L. BEACH: J. Appl. Phys. **23**, 1305 (1952).
[4] J. BLEARS: J. Sci. Instrum. **28**, 36 (1951).
[5] F. J. NORTON: J. Amer. Ceram. Soc. **36**, 90 (1953).

and others[1] have made measurements of permeation rates of helium through various glasses. His results obtained with the use of a mass spectrometer, for the permeation rates of helium through soda lime glass, Pyrex no. 7740 glass, and fused silica glass are shown in Fig. 10. NORTON also made measurements of the permeation rates of a number of gases through fused silica glass at 700° C. The values for the permeation rate (in the units of Fig. 10) are: for helium, 2.1×10^{-8}; hydrogen, 2.1×10^{-9}, deuterium, 1.7×10^{-9}; neon, 4.2×10^{-10}; argon, oxygen, and nitrogen, less than 10^{-15}. The process for helium is pictured as the diffusion of helium atoms via vacancies in the irregular lattice structure in fused silica or glass. In the case of crystalline silica, RAYLEIGH found the permeation rate too small to measure, down by at least seven orders of magnitude. Because measured values for permeation rates among earlier workers varied by as much as a factor of 20, ROGERS, BURITZ, and ALPERT[2] also measured the permeation rate for helium in borosilicate glass (Corning 7740). Their results are in good agreement with the results of NORTON. They also obtained values for the diffusion coefficient, D, and solubility, S, for the case of helium in glass. These quantities are related[3] to the permeation rate, k, by the expression

$$k = D\,S. \tag{13.1}$$

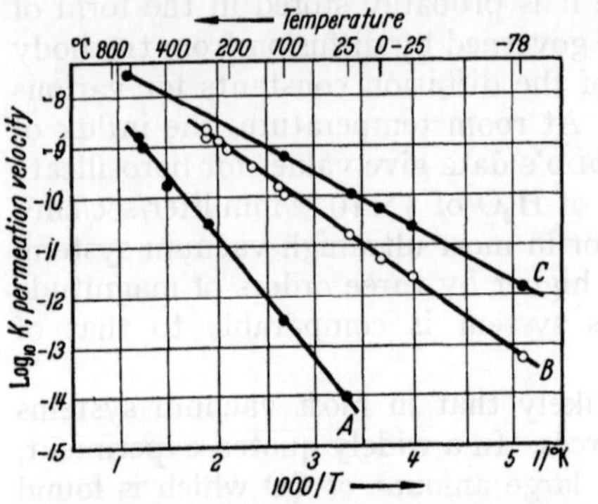

Fig. 10. NORTON's measurements of the permeation velocity K of helium diffusing through various glasses. The permeation velocity, $\log_{10} K$, is plotted against the reciprocal of the absolute temperature, $1000/T$. The permeation velocity K is in units of cm^3 gas (N.T.P.) sec/cm^2 area/mm thickness/cm Hg (gas pressure difference). Curve A, soda-lime glass No. 0080; B, chemical Pyrex brand glass No. 7740; C, fused silica.

They obtained the interesting results that (1) the solubility of helium in glass is approximately one percent, i.e., the gas density within the glass is one percent of that of the helium atmosphere with which it is in equilibrium, and (2) this solubility is quite insensitive to temperature between −78° C and 400° C. This latter property accounts for the fact that the rate of rise of helium in a vacuum system establishes equilibrium allmost immediately after a change in temperature.

Most of the information concerning the residual gases in high vacuum system have come from mass spectrometer studies. BLEARS [*12*] has applied the mass spectrometer to the measurement of the partial pressures of various gases in a demountable system using an oil diffusion pump and in a "very high vacuum" bakeable system using a mercury diffusion pump. He provided very interesting data of the characteristic spectrum of gases found in high vacuum systems under varying conditions. First of all, he showed that prior to bakeout the initial rate of pressure reduction of such gases as H_2O, CO, C_2H_5, C_3H_7, and many other hydrocarbons proceeds at a rate much slower (by a factor of 4000) than the speed (approximately 5 liter/sec) of the pumping system. That is, all of these gases seemed to have large reservoirs of supply; the rate of removal was almost identical for most of the above gas constituents. This could be explained by saying that the removal rate was determined by the binding forces of the molecules to the surfaces. Upon mild baking to only 100° C, the partial pressures of each of the

[1] W. D. URRY: J. Amer. Chem. Soc. **54**, 3887 (1932). — J. A. WILLIAMS and J. B. FERGUSON: J. Amer. Chem. Soc. **46**, 635 (1924).

[2] W. A. ROGERS, R. S. BURITZ and D. ALPERT: J. Appl. Phys. **25**, 868 (1954).

[3] See R. M. BARRER: Diffusion in and through Solids. Cambridge: The University Press 1951.

above gases and, hence, the rate of removal were greatly increased. Upon subsequent cooling, the equilibrium pressure was considerably reduced.

After repeated and prolonged bakeout (at least 48 hours), BLEARS measured the partial pressures of the principal constituents; the results are given in Table 2.

Table 2. *Units 10^{-9} mm Hg.*

Pump fluid	Apiezon C oil	Mercury			*
Baking temp °C	100° C	100° C	100° C	250° C	375° C
Trap temp °C	−78	−78	−170	−78	−170
Gas	partial pressure				
Hydrogen	4×10^{-9} mm Hg	—	—	—	—
Water	4	19	6	2.2	0.01
Carbon monoxide	3.1	9	1.0	0.1	0.15
Oxygen	0.5	1	1.8	0.5	0.1
Carbon dioxide	0.9	1.3	1.7	0.1	0.08
ΣCH	10	4	1.3	1.2	
Mercury	—	205	15	224	0.006
Total of principal constituents (excluding mercury)	2.3×10^{-8} mm Hg	3.4×10^{-8}	1.2×10^{-8}	4.1×10^{-9}	5×10^{-10}

* These are the data of J. H. REYNOLDS.

It is readily seen that the principal gases are removed by the bakeout to higher temperatures. A number of other mass spectrometer studies have been made with systems baked at high temperatures. J. H. REYNOLDS[1] measured the background peaks in a very sensitive mass spectrometer continuously pumped by mercury diffusion pumps and utilizing two traps, a dry ice trap between fore an diffusion pumps, and a liquid nitrogen trap between diffusion pump and bakeout valve. The entire system was baked out at 375° C for a 12-hour period and then pumped for a similar period. After two such bakeouts, pressures as low as 10^{-10} mm Hg were achieved, but with the spectrometer running continuously, the pressure rose to about 4 or 5×10^{-10} mm Hg as measured in an adjacent Bayard-Alpert gauge. Assuming that the total pressure in the mass spectrometer tube was approximately equal to that in the ionization gauge, the estimated partial pressures are shown in column 5 of Table 2 for comparison with BLEARS' results. REYNOLDS observed a rather surprisingly large carbon peak (comparable to that of oxygen) which is not listed in the table.

Since most of the residual gases in such continuously pumped systems are adsorbable on tungsten, the flash filament technique should give a reasonable estimate of the total pressure in the systems. HAGSTRUM [*9*] attained pressures as indicated by the flash filament technique as low as 2 to 3×10^{-11} mm Hg in a system continuously pumped by a mercury diffusion pump.

ALPERT and BURITZ [*6*] and CARMICHAEL[2] utilized an omegatron type of mass spectrometer in a sealed-off system at a total residual pressure of 10^{-9} mm Hg and found the principal constituent, aside from helium, to be mass 28. Water and oxygen were not detectable. Similar results were found by SHELTON[3].

A first approximation to the rate of evolution of gas in the above systems may be obtained by substituting values for the equilibrium pressure, p_e, and the pumping speed, S, into Eq. (12.18),

$$Q = p_e S.$$

[1] J. H. REYNOLDS: Rev. Sci. Instrum. **27**, 928 (1956).

[2] J. H. CARMICHAEL: Private communication.

[3] H. SHELTON: The Massitron, An Ionization Gauge that Provides Mass Analysis (MIT, Cambridge, Report on 15th Annual Conference on Physical Electronics, March 1955), p. 170.

Making the reasonable assumption (see Sect. 19) that the diffusion pump speeds are maintained unchanged at very low pressures, one can obtain values for the influx rate of the various constituents. In the case of CO, for example, the total influx rate is 5×10^{-8} mm liter/sec for BLEARS' system baked at 100° C, and 4×10^{-11} mm liter/sec in the well baked system of REYNOLDS.

In systems sealed off from diffusion pumps, much lower pressures of adsorbable gases have been observed than those reported above. A number of workers have reported flash filament measurements of the adsorbable gas in a sealed-off tube of less than 10^{-13} mm Hg. VARNERIN and WHITE[1] enclosed one small vacuum system completely within another to eliminate the influx of helium. They then obtained indications that the total pressure as measured on an ionization gauge was probably less than 10^{-11} mm Hg, including inert and adsorbable gas. In this system, which had been baked for several 12-hour intervals, there was approximately 30 cm² of metal surface (copper and Kovar). From this equilibrium pressure and an estimate of the pumping speed of the ionization gauge, one obtains a value for the upper limit of the degassing rate from the metal of about 3×10^{-15} mm liter/sec/cm² of surface. D. J. GROVE[2] has made the estimate based on experience with a number of systems utilizing stainless steel as the principal envelope material, that the degassing rate of a well-baked (400 to 475° C) metal system is approximately 5×10^{-15} mm liter/sec/cm², which he believes is largely carbon monixide.

Summary.

1. A large number of mechanisms are responsible for the influx or evolution of gases into the volume of a vacuum enclosure. Aside from evolution from pumping systems, the major sources of gas are leaks in the various envelope sections, diffusion through solids, and desorption or dissociation at the surfaces. Many of these processes are complex and interrelated, and are understood only on an integral or empirical basis.

2. The major fraction of gases adsorbed on the surfaces are rather loosely bound and are readily desorbed at temperatures of 250 to 500° C, with strong evidence that the higher temperatures are required for lowest attainable pressures. For a well-outgassed system, the specific influx rate is not greater than 5×10^{-15} mm liter/sec/cm².

3. Much of the past data on gas-surface reactions are confused or misleading because a technology for properly carrying out the experiments was not available. At the present time, many further studies are possible and desirable for a better understanding of the processes involved.

IV. The use of getters in the production of very high vacuum.

14. Removal of gases by adsorption on "getter" materials. One of the oldest and most widely used methods of evacuation to low pressures is the use of getters in which the gas molecules are removed from the volume by adhering to a gas absorbing material. Starting before the turn of the century, the use of getters has developed into a practice that makes possible the mass production of most electronic vacuum tubes. Getters that have been commonly used include barium, strontium, tantalum, thrium, zirconium, and uranium. Their principal utilization has occurred in the evacuation of sealed-off tubes, and their application in the attainment of ultrahigh vacuum in sealed-off experimental tubes has already

[1] L. J. VARNERIN and D. H. WHITE: J. Appl. Phys. **25**, 1007 (1954).

[2] D. J. GROVE: Private communication.

been mentioned[1]. In recent years, gettering by evaporated metals has also been used to lower partial pressure of adsorbable gases in continuously pumped systems. J. A. BECKER[2], for example, has used evaporated tantalum to attain ultrahigh vacua in his work on field emission. While the typical getter is selective in the gases it will pump, a number of workers have successfully applied gettering alone or in conjunction with electrical discharges to attain ultrahigh vacuum.

It is not proposed in this paper to give a detailed discussion of the literature in this broad field of activity. General surveys have been made by S. DUSHMAN ([7], Chap. 11, Sect. 2), by ESPE, KNOLL, and WILDER[3], and by JAECKEL [11]. Important work at low pressures has recently been carried out on zirconium-gas[4] and titanium-gas[5] reactions, and on a number of more volatile metals[6,7]. The rate of reduction of pressure by adsorption at a surface is given by an expression similar to that in Eq. (12.4), i.e.,

$$\frac{dp}{dt} = -\frac{SA}{V}p = -\frac{\alpha A}{V}\sqrt{\frac{kT}{2\pi m}}\,p \tag{14.1}$$

where α is the sticking probability, the fraction of incident molecules adhering to the getter. The maximum theoretical value of S is that for $\alpha = 1$. Typical values for gettering rates are given in Table 3 from S. WAGENER [13] for barium at room temperature. The barium in this case is evaporated on a thin mica disk in a carefully baked-out ultrahigh vacuum system. This is an important precaution since the gettering rate for a given gas is strongly affected by the adsorption of other gases.

Table 3.

Gas	Maximum theoretical rate liter/sec/cm²	Measured rate (material: barium) liter/sec/cm²	Measured rate / max. theor. rate
CO	11.9	5	0.4
CO_2	8.8	5	0.6
O_2	11.1	0.3	0.02
H_2	44.6	0.05	0.001
N_2	11.9	0.001	0.0003

The data in Table 3 were taken with a coverage of only 2 to 5% of a monolayer to eliminate saturation effects which are usually not severe for less than several monolayers with the above gas reactions. The ratios in the last column in Table 3 can be taken as approximate values of the sticking or adsorption probability α. The probabilities obtained above for carbon monoxide and carbon dioxide are of the same order as that found by BECKER and HARTMAN [8] for the adsorption of nitrogen on clean tungsten, and by J. A. BECKER and R. G. BRANDES[8] for oxygen on tungsten (approximately 0.2). Stout measured a value of α of 1.6×10^{-5} for hydrogen on a "clean" titanium metal sheet.

In most cases of contact gettering, the getter takes up more gas than would cover its apparent surface with a monomolecular layer before showing signs of saturation. The gas may be either physically adsorbed or chemically combined with the getter; in either case adsorption is often followed by diffusion into the interior of the getter material. In general, a rise in temperature increases the

[1] P. A. ANDERSON: Phys. Rev. **47**, 958 (1935).
[2] J. A. BECKER: Bell Syst. Techn. J. **30**, 907 (1951).
[3] W. ESPE, M. KNOLL and M. P. WILDER: Electronics **23**, 80 (1950).
[4] E. A. GULBRANSEN and K. F. ANDREWS: Trans. Amer. Inst. Mining Met. Engrs., J. Metals **185**, 515 (1949).
[5] E. A. GULBRANSEN and K. F. ANDREWS: Trans. Amer. Inst. Mining Met. Engrs., J. Metals **185**, 741 (1949).
[6] V. L. STOUT and M. D. GIBBONS: J. Appl. Phys. **26**, 1488 (1955).
[7] S. WAGENER: Vacuum **3**, 11 (1953). — J. Phys. Chem. **60**, 567 (1956). — [13].
[8] J. A. BECKER and R. G. BRANDES: J. Chem. Phys. **23**, 1323 (1953).

rate of diffusion of gas into the solid and also increases the rate of the desorption of gas from the surface. Thus for a given metal[1] some gases may be selectively sorbed at temperatures at which others are desorbed.

One of the important questions in the attainment of ultrahigh vacua is the rate of desorption at very low pressures. However, few quantitative experiments have been done in this field. WAGENER[2] shows that in the case of carbon monoxide on barium the rate of desorption of a saturated sample at 300° C is only one-hundredth the rate of sorption of an unsaturated getter at room temperature. He has also measured the gettering rates of O_2, H_2O, and CO_2 by barium and finds them unaffected by variation of pressure down to 10^{-7} mm Hg. From these results, he concludes that the rate of desorption for this type for molecular chemisorption is negligible at room temperature down to pressures below 10^{-7} mm Hg.

An important factor in the measurement of gettering rates is that the instrument used to measure the gettering rate may greatly affect the results. In most recent experiments, the gas pressure is measured with an ionization gauge. LUKIRSKY and PTIZYN[3] and REIMANN[4] demonstrated that the effect of the presence of an ionizing electron beam is so great that they suggested that different names be used for the gettering with or without an adjacent electron beam, i.e., "contact gettering" and "discharge gettering". It follows that the measurement of gettering rates with an ionization gauge often represents a mixture of the two processes.

There are significant differences in the results for gettering efficiencies recently reported by workers using other means for measuring pressures. Thus, KOBAYASKI and FURAYA[5], using Pirani gauges for measurement, give values for the gettering rate of oxygen on barium approximately an order of magnitude less than those of WAGENER, while HAASE[6], employing a Knudsen gauge, found the rates of gettering for all gases were down by several orders of magnitude at pressures below 10^{-6} mm Hg. These results would seem to indicate that the results of WAGENER are largely due to the so-called "discharge gettering". There is considerable reason[7] to question this assumption, however. Since the adsorption of impurities might greatly alter the gettering rate for a specified gas, it is necessary to assure not only that the measuring instruments do not influence the results, but also that the vacuum conditions are such as to eliminate the effects of impurities. Under clean vacuum conditions, the results of BECKER[8], and others would seem to indicate that gettering rates can be maintained in the ultrahigh vacuum range.

15. Discharge gettering. The cleanup of "active" neutral molecules. The so-called "discharge" gettering is very closely related to "ionization gauge pumping" which is discussed elsewhere in this article. As mentioned previously, it was found by LUKIRSKY and PTIZYN and by REIMANN that large increases were observed in the gettering rates of nitrogen, oxygen, hydrogen, and carbon monoxide on metals like barium, magnesium, and calcium when in the presence of a beam of ionizing electrons. It has already been shown that sizeable pumping speeds (of the order of 0.01 liter/sec/ma) are found for various molecular gases

[1] V. L. STOUT and M. D. GIBBONS: J. Appl. Phys. **26**, 1488 (1955).
[2] S. WAGENER: Vacuum **3**, 11 (1953).
[3] P. I. LUKIRSKY and S. W. PTIZYN: Z. Physik **71**, 339 (1931).
[4] A. L. REIMANN: Phil. Mag. **16**, 673 (1933); **18**, 1117 (1934).
[5] A. K. KOBAYASKI and S. F. FURAYA: J. Phys. Soc. Japan **6**, 238 (1951).
[6] G. H. HAASE: Z. angew. Phys. **2**, 188 (1950).
[7] See recent work of S. WAGENER: Nature, Lond. **173**, 684 (1954). — J. Phys. Chem. **60**, 567 (1956).
[8] J. A. BECKER: Bell Syst. Techn. J. **30**, 907 (1951).

in an ionization gauge when the electron beam is turned on. It is not unexpected, therefore, that appreciable changes in gettering rates are observed when a gauge is turned on in the vicinity of a getter surface. While few quantitative data are available, the rates as measured by these workers for "discharge gettering" are essentially the same as those observed for ion pumping. MORGULIS[1] reversed the above procedure of adding the effect of an ionization gauge to that of the gettering materials by applying a barium getter to the walls of a Bayard-Alpert gauge. He observed only small differences from the normal gauge in the pumping speed and ultimate pressure of the system, at most a factor of two. This is not surprising, since the efficiency of ion collection has been found to approach unity for molecular gases even in the absence of getters. Also, the difference between a gauge with and without added getter material is not great. VARNERIN and CARMICHAEL[2] have shown that unless extreme precautions are taken to prevent it, there is formed on the inner surface of an ionization gauge an evaporated film which has an important effect on ion pumping. Thus while one might expect that the getter might combine with active gases and affect the rate of reemission or the total amount of gas collected before saturation set in, one would also expect the initial removal process to be similar to that for ion-pumping.

One of the interesting results found in the studies of "discharge" gettering by LUKIRSKY and PTYZIN[3], and by REIMANN[4] is the enhanced cleanup in the presence of an electron beam which they attribute to the formation of metastable molecules in the cases of nitrogen and carbon monoxide on magnesium. The enhanced cleanup was found with electron energies below the ionization potential; onset occurred in the vicinity of the first excitation potential. The cleanup was accompanied by an electrical current from the getter surface which was attributed to the ejection of electrons from the surface by the metastable molecules ([*1*], p. 565). The effect was small compared with the cleanup when the electron energy exceeded the ionization potential.

S. WAGENER[5] also made measurements of gettering rates of barium with nitrogen, oxygen, and carbon monoxide, using a suppressor ion gauge of the Metson type[6]. He found that when an electron beam of about 10 milliamperes and 100 volts was turned on, there was an appreciable cleanup of the order of one liter/sec for oxygen and 0.05 liter/sec for nitrogen which he attributed to the removal of neutral "active" atoms or molecules. The evidence for this was (a) the gettering action did not vary significantly with the potential applied to the gettering surface, thus showing that the arrival of ions at the getter is not an appreciable factor; (b) in the case of nitrogen, the rate of removal depended on the physical location of the getter in the vacuum system. In the case of oxygen, the cleanup was more rapid but did not vary with the location of the getter material. He concluded from other evidence that the "active" constituents are atoms in the case of oxygen and metastable molecules in the case of nitrogen.

The dissociation of hydrogen ([*1*], Chap. 4) molecules by low energy electron bombardment was demonstrated by GLOCKLER, BAXTER, and DALTON[7] and by HUGHES and SKELLET[8]. In both cases, they demonstrated the dissociation into atomic hydrogen by a fall in hydrogen pressure due to attachment to the walls.

[1] N. D. MORGULIS: J. techn. Phys. USSR. **25**, 1667 (1955).
[2] L. J. VARNERIN and J. H. CARMICHAEL: J. Appl. Phys. **26**, 782 (1955).
[3] P. I. LUKIRSKY and S. W. PTIZYN: Z. Physik **71**, 339 (1931).
[4] A. L. REIMANN: Phil. Mag. **16**, 673 (1933); **18**, 1117 (1934).
[5] S. WAGENER: Br. J. Appl. Phys. **2**, 134 (1951).
[6] G. H. METSON: Br. J. Appl. Phys. **2**, 46 (1951).
[7] G. GLOCKLER, W. P. BAXTER and R. H. DALTON: J. Amer. Chem. Soc. **49**, 58 (1927).
[8] A. L. HUGHES and A. M. SKELLET: Phys. Rev. **30**, 11 (1927).

The onset potential occurred at about 11.5 ev and was attributed to excitation of the unstable triplet states. The dissociation of oxygen molecules by low energy electron impact has also been observed by GLOCKLER and WILSON[1] and others. These observations verify the formation of active "atomic" states of oxygen and hydrogen. It remains for further investigations to estimate the relative probability for their occurrence in an ionization gauge. It seems difficult to account for the magnitude of WAGENER'S results for oxygen on this basis.

In the case of nitrogen, dissociation into neutral atoms by electron impact has not been observed. Hence, the "active" particles are thought to be metastable molecules.

It is interesting to note that HAGSTRUM and TATE[2] found a cleanup effect which they attributed to the formation at a hot filament of "activated" oxygen molecules, perhaps in the $^1\Delta$ or $^1\Sigma$ metastable form ([*1*], p. 257). They then postulate that these molecules possess sufficient activation energy to react with substances of low vapor pressure adhering to the glass walls of the tube and form CO, CO_2, and H_2O.

V. Ion pumping and the use of ion pumps for the production of very low pressures.

16. Ion pumping and the use of ionization gauges as pumps. The removal of gases in the presence of an ionizing beam was first discovered almost a century ago. (For a general discussion, see [*7*], Chap. 10.) Much of the early work was done in gas discharge tubes at relatively high pressures in which it was found that sputtered deposits on glass played an important role. The work was complicated not only by the sputtering process, but by the discharge and the interpretation of the results was difficult. Working with sealed-off vacuum tubes containing hot cathodes, some of the earliest investigators[3,4] came to the the conclusion that the rate of cleanup of nitrogen and hydrogen in an enclosure was proportional to the rate of ionization and was independent of pressure.

W. VON MEYEREN[5] investigated the phenomenon for air and hydrogen at lower pressures in a hot cathode ionization tube enclosed by a glass envelope. He proposed that the ions formed were accelerated to and attached predominantly to the glass walls, although it is likely, in view of later investigations[6], that metal films may have been present and played an essential role. Von Meyeren found that as the tube continued to clean up the gas, the rate of cleanup decreased, a phenomenon which he ascribed to the reemission of gas from the surfaces. He also found a moderate increase in the pumping speed of the device with increasing ion energy between 400 and 800 ev. The total amount of hydrogen or nitrogen which could be taken up by ion pumping with the energies involved ($\sim$500 ev) was of the order of a few monolayers of gas on the surfaces. SCHWARZ[7] also made measurements of cleanup in an ionization gauge and found results similar to those of VON MEYEREN. He demonstrated an efficient removal of nitrogen and argon at the normal operating potentials of an ionization gauge (approximately 110 volt). The rate of cleanup was rather rapid, corresponding to the removal

[1] G. GLOCKLER and J. L. WILSON: J. Amer. Chem. Soc. **54**, 4544 (1932).
[2] H. D. HAGSTRUM and J. T. TATE: Phys. Rev. **57**, 1071 (1940).
[3] A. L. HUGHES: Phil. Mag. **41**, 778 (1921).
[4] N. R. CAMPBELL and E. G. NEW: Phil. Mag. **48**, 553 (1924).
[5] W. v. MEYEREN: Z. Physik **84**, 531 (1933).
[6] L. J. VARNERIN and J. H. CARMICHAEL: J. Appl. Phys. **26**, 782 (1955).
[7] H. SCHWARZ: Z. Physik **117**, 23 (1940); **122**, 437 (1944).

of a substantial fraction of the ions collected. SCHWARZ also observed the phenomenon of "interchange" of molecules at the surface. That is, if nitrogen were cleaned up by entrapment at the electrode surface and then argon was introduced and accelerated to the wall as ions, the nitrogen was ejected from the surface and replaced by the argon.

Recently, there have been a number of further papers on the removal of gases by ion-pumping. BLOOMER and HAINE[1] found that much of the residual gas cleaned up in a sealed-off ionization gauge was collected at the glass walls of the gauge, and that the efficiency of collection of a Bayard-Alpert gauge was of the order of 10 times that of conventional gauges.

VARNERIN and CARMICHAEL[2] demonstrated that in the case of helium, the Bayard-Alpert gauge does not pump effectively unless there is a thin film of metal which is evaporated or sputtered to the glass walls of the gauge in the course of ordinary use. In the absence of this film, the helium is apparently reemitted from the glass in times short compared to the characteristic time for ion-pumping.

J. R. YOUNG[3] verified the above results of VARNERIN and CARMICHAEL for helium, but found that the nitrogen was effectively collected on various surfaces. The efficiency of ion trapping for nitrogen was very close to unity; that is, almost every ion collected at either the walls or the ion collector was initially removed from the volume. For helium the efficiency was down by a factor of four. YOUNG also proposed a phenomenological theory to explain his results.

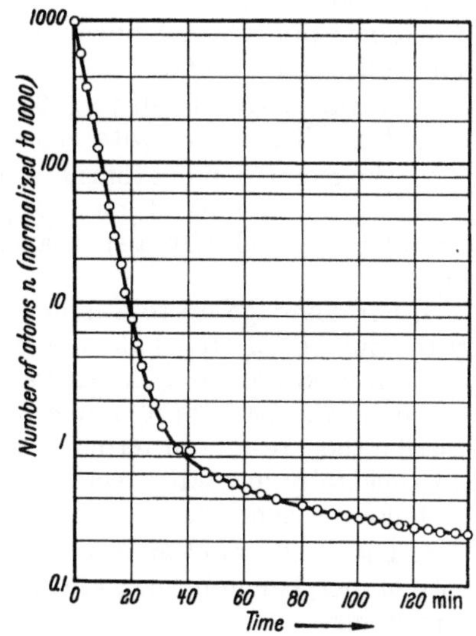

Fig. 11. Experimental pumping curve for 1100 ev helium ions on molybdenum. Number of atoms (n) in the enclosure as a function of time. Initial number $n_0 = 8.5 \times 10^{15}$.

VARNERIN and CARMICHAEL have recently reported[4] another study of the mechanism of removal of helium on molybdenum by ion bombardment. The data of various observers are rather similar. They all find that initially, when the potentials are applied, the pressure in a system decreases exponentially (see Sect. 7). After some removal has taken place, the curve departs from an exponential and indicates a saturation of the pumping effect. Typical data, in this case the results of VARNERIN and CARMICHAEL, are shown in Fig. 11. This is a plot of the reduction in the number of atoms in the volume due to the acceleration of 1100 ev singly charged helium ions into the molybdenum surface. It will be noted that initially the number of atoms decreases exponentially with time, as is to be expected if the rate of ionization is proportional to the gas density. In this part of the curve, η, the trapping efficiency, or number of atoms removed per ion collected, is less than one and varies with ion energy from about 0.3 at 150 ev to about 0.8 at 2100 ev. After some pumping has taken place, the curve departs from an exponential and indicates a saturation of the pumping effect.

[1] R. N. BLOOMER and M. E. HAINE: Vacuum **3**, 128 (1953).
[2] L. J. VARNERIN and J. H. CARMICHAEL: J. Appl. Phys. **26**, 782 (1955).
[3] J. R. YOUNG: J. Appl. Phys. **27**, 926 (1956).
[4] L. J. VARNERIN and J. H. CARMICHAEL: J. Appl. Phys. **1957** (in press).

This saturation, which has been generally observed in such experiments, has been explained in different ways. YOUNG and VON MEYEREN ascribe it to the depletion of available trapping sites, and, in addition, assume the possibility of a spontaneous reemission of gas at a rate proportional to the number of trapped gas atoms. VARNERIN and CARMICHAEL show that a reemission mechanism does describe the observed saturation, but indicate that in the case of helium a detailed study of the pumpdown and reemission process does not confirm the assumptions of Young and VON MEYEREN concerning depletion of trapping sites and spontaneous emission.

By studying the removal process during ion bombardment and the reemission of helium from the molybdenum in the absence of ion bombardment, VARNERIN and CARMICHAEL found:

1. In the range of 150 to 2600 ev, the trapping efficiency, η, varies between 0.3 and 0.8; it increases with increasing voltage.

2. The reemission of atoms is a function of the *time* when the ion was trapped, as well as the number of ions trapped; thus the reemission is *not* exponential as would be expected with a spontaneous emission rate proportional to the number trapped.

3. For coverage up to a monolayer of gas, the trapping of ions is unaffected by the number of atoms already trapped.

While VARNERIN and CAMRMICHAEL are not able to explain the reemission phenomenon on the basis of a simple physical trapping mechanism, they are able to describe their results in terms of a phenomenological theory which utilizes their experimentally measured reemission vs. time curves. On this basis they come to the following conclusions:

1. The electrical cleanup phenomenon for the case of helium ions in the presence of molybdenum surfaces can be described simply in terms of ionic pumping or trapping.

2. The reemission of trapped gas atoms plays an important role. The gas reevolved in a time scale short compared to experimental observation time results in an apparent trapping efficiency less than unity. The long-time reemission of gas gives rise to saturation effects. The short-time and long-time reemission of gas seem to be governed by the same physical mechanism.

3. A study of the reemission process by direct observation of rates of reemission points to a complicated physical mechanism. That is, it is not a capture within the solid followed by simple diffusion into the volume.

BROWN and LECK[1] made studies with a mass spectrometer of the desorption of gas which had previously been ionized and then accelerated to a cathode surface with energies of the order of 1000 ev. The gases studied were helium, neon, argon, nitrogen, oxygen, and krypton; the cathode materials were nickel, molybdenum, and aluminum. While the residual vacuum conditions were never very good, certain interesting results were obtained. They found that, first of all, the rates of sorption and desorption for most gases were independent of the chemical nature of the gas or the metal surface. Hence, they concluded that the molecules are probably bound to the surface by van der Waals forces. Secondly, they found that with the exception of helium and hydrogen there was virtually no desorption at room temperature. The gases, in general, were desorbed upon bakeout between 100 and 200° C, though in some cases, e.g., argon, some of the gas was retained up to 1000° C. They also verified earlier results that the maximum quantity of gas adsorbed corresponds approximately to one monolayer at the

[1] E. BROWN and J. H. LECK: Br. J. Appl. Phys. 6, 161 (1955).

cathode surface. As Brown and Leck point out, with the simple picture of a van der Waals attraction, it is difficult to understand why a gas adsorbed in the ionic state is held much more strongly than that adsorbed as neutral molecules; physical adsorption of the noble gases is found only at temperatures below $-80°$ C, whereas the surface layer formed by ion bombardment is stable up to approximately 100° C.

TRENDELENBURG and CARMICHAEL have also used the mass spectrometer to make studies of the ejection of previously entrapped noble gas ions by ionic impact. In a preliminary report[1], they found evidence for more than one state of adsorption for noble gas ions, and observed the interchange of noble gas ions at the surface. Thus far, however, little can be said about the physical mechanism for the ion entrapment of noble gases. It is not clear whether the ions are driven into interstitial positions in the metal lattice or whether the entire phenomenon takes place at grain boundaries or dislocations at the surface. The fact that helium ions have a smaller reemission probability at energies of 2600 volt than at 100 ev seems to imply a penetration of the ions beyond the lattice surface.

In the case of molecular ions, it is also very difficult to propose a detailed physical mechanism, and further work is necessary to establish a reasonable understanding of the observed results.

In addition to removal by ion-pumping, it has been indicated in Sect. 7 that the ionization gauge may remove gases by chemical reaction with the hot cathode. SCHWARZ[2] found that in the case of hydrogen, a rapid cleanup was initiated as soon as the filament in a gauge was heated to operating temperatures even without applying accelerating voltages to the electrodes[3]. His data were explained as due to the dissociation of hydrogen molecules at the cathode and subsequent removal by adsorption of the atomic hydrogen at the walls of the gauge.

In the case of oxygen, LANGMUIR[4] demonstrated a very rapid cleanup upon striking a hot tungsten filament. He explained it as due to the formation of tungsten oxide (WO_3) which distills to the walls as fast as it is formed. The probability for this reaction was found to vary with temperature with an activation energy of about 26500 calories per mole oxygen. At 2300° K, the "accommodation coefficient" is about 0.1 (see [7], p. 674). Hence, the pumping speed for a filament of 0.2 cm² area would be expected to be approximately 0.2 liter/sec.

Many processes may take place in an ionization gauge to remove hydrocarbons and organic vapors. Decomposition at the hot tungsten filament is combined with dissociation upon electron impact to make an understanding very difficult to achieve. BLEARS[5] and RIDDIFORD[6] have made quantitative measurements of the effect of the gauge on the measurement of organic vapors used in diffusion pumps. As Riddiford points out, a very rapid pumping speed of several liters per second is sometimes observed, but the explanation requires the assumption that many thousands of monolayers of oil decomposition products be absorbed on the glass surface without contributing substantially to the pressure reading.

---

1 E. A. TRENDELENBURG and J. H. CARMICHAEL: Bull. Amer. Phys. Soc. **2**, 35 (1957).

2 H. SCHWARZ: Z. Physik **122**, 437 (1944).

3 G. J. SCHULZ (private communication), working with hydrogen at higher pressures, did not observe this type of pumping; he measured a removal of gas only when accelerating potentials were applied, i.e., by ion pumping and dissociation by electron impact. This absence of an effect at the hot filament could be due to a rapid saturation of the surfaces at the higher pressures.

4 I. LANGMUIR: J. Amer. Chem. Soc. **35**, 105 (1913); **37**, 1139 (1915).

5 J. BLEARS: Proc. Roy. Soc. Lond. **188**, 62 (1946).

6 L. RIDDIFORD: J. Sci. Instrum. **28**, 375 (1951).

17. Measurement of ionization gauge pumping speeds. Relatively few measurements of the pumping speed of ionization gauges have been published. In the older literature, the measured speed varied considerably with the condition of the gauge, the amounts of unmeasured impurities, etc. Recently, however, it has been possible to use relatively pure gases and to establish reasonable criteria for the cleanliness of the gauge before making measurements.

RIDDIFORD and BILLS, ALLEN, and CARLETON[1] have reported measurements of the pumping speed of ionization gauges for oxygen. Both papers demonstrated the reaction with the hot tungsten filament to be the principal removal mechanism. As indicated previously, the mechanism involves an interaction with the hot tungsten filament to form an oxide of tungsten which is unstable and distills to the walls. At the normal operating temperature (2300° K) of a Bayard-Alpert gauge filament, this results in an expected pumping speed of the order of 0.2 liter/sec for oxygen (see Sect. 16). While the two investigations were carried out at different filament temperatures, their results are in reasonable agreement with the above estimate.

J. R. YOUNG[2] and BILLS, ALLEN and CARLETON[1] have reported measurements on the commercial version of the Bayard-Alpert gauge (WL-5966) for helium and nitrogen. They are compared in Table 4 with unpublished values

Table 4.

Gas	Sensitivity K	Computed maximum pumping speed liter/sec	Measured pumping speed liter/sec	Experimenter
He	2.5	0.005	0.004	J. R. YOUNG
			0.003	VARNERIN and CARMICHAEL
N_2	12	0.024	0.1	J. R. YOUNG
N_2	12	0.024	0.1 to 0.2	BILLS, ALLEN, CARLETON
N_2	12	0.024	0.16	C. G. MATLAND

obtained at the Westinghouse Research Laboratories by VARNERIN and CARMICHAEL and by C. G. MATLAND. The measured values of pumping speed are the maximum values determined at the initial portion of the pumpdown curve for an emission current of 10 ma. The major removal mechanism for both helium and nitrogen is assumed to be ion pumping, and Table 4 includes for comparison the maximum speed as computed from Eq. (7.8). It is seen that the pumping speed for helium is approximately equal to the computed value, while that for nitrogen is four to eight times the computed value. That is, four to eight nitrogen molecules are removed for each one collected at the ion collector. The large value for nitrogen can be understood on the basis (a) that three to seven ions are formed outside the grid for each one formed within it, or (b) that substantial amounts of "active" metastable nitrogen molecules are formed and permanently removed to the walls. YOUNG made measurements of the iron current collected at the outside wall and found values 5 to 10 times the currents to the ion collector. A fraction of his measured current (presumably a small fraction) could be due to the ejection of electrons from the walls by molecular excitation in various forms, i.e., by photoelectric effect, metastable molecules, etc. However, it is difficult to account for a major effect due to excitation, and YOUNG'S results constitute evidence for the formation and collection of nitrogen ions with a trapping efficiency, η, close to unity.

[1] D. G. BILLS, F. G. ALLEN and N. P. CARLETON: Phys. Rev. **99**, 1662 (1955).
[2] J. R. YOUNG: J. Appl. Phys. **27**, 926 (1956).

If we assume from the results in nitrogen that the number of ions formed outside the grid is four times that formed within it, the measured value of the pumping speed for helium gives a relatively low trapping efficiency of approximately $\eta = 0.2$ in the ionization gauge. VARNERIN and CARMICHAEL[1] report a value of $\eta = 0.34$ for helium on molybdenum. Thus the results are in general agreement with the ion pumping theory within a factor of 2. This is probably the limit of accuracy thus far available in ion-pumping measurements.

The question arises as to how much gas can be removed by ion pumping before saturation takes place. VON MEYEREN[2] found that the total amount of hydrogen or nitrogen which could be taken up by ion pumping was of the order of a few monolayers of gas. This would saturate a gauge in a relatively short time at high pressures, about one hour at 10^{-4} mm Hg. At pressures below 10^{-9} mm Hg, gauges have maintained ultrahigh vacuum for 75 days without signs of saturation.

The ultimate pressures attainable depend not only on the effective τ_i for ion pumping as measured at the onset of pumping, but upon the reemission of gas from the collector surfaces (see Eq. (12.15)]. The reemission depends on a number of parameters including temperature. Thus the very low values (approximately 4×10^{-11} mm Hg) of the ultimate pressures for helium reported by ALPERT and BURITZ [*6*] were obtained only with the ionization gauge immersed in liquid nitrogen[3].

The pumping speed of an ionization gauge is of the order of 0.01 liter/sec/ma for the most common gases found in a vacuum system. While such speeds are small compared to those of conventional pumps, they are quite adequate to evacuate an ultrahigh vacuum system of moderate volume ($\sim$1 liter). The ultimate vacuum attainable is of the order of 10^{-10} mm Hg for helium and less than 10^{-11} mm Hg for most adsorbable gases.

18. Other ion pumps. A number of pumps have been proposed[4,5,6] which involve the transport of gases by ionization and subsequent acceleration in electromagnetic fields to an exit port. In most of these, the return path for neutral atoms is large and the ultimate pressures thus far achieved are in the range 10^{-5} mm Hg to 5×10^{-6} mm Hg. In the Foster, Lawrence, and Lofgren pump, the ionization is carried out by electrons from a hot cathode which are contained as in the Philips gauge[7] by an axial magnitec field and an appropriate electric field. GUREWITSCH and WESTENDORP[8] have also described an ionic pump in which the electrons oscillate in a magnetic field in an arrangement similar to that of the Philips ionization gauge. In this case, the electrons are formed as a result of gas ionization in the discharge and the ions are driven into the negative electrodes made of carbon, titanium, or other metals. The lowest pressure reported is about 5×10^{-7} mm Hg.

---

[1] L. J. VARNERIN and J. H. CARMICHAEL: J. Appl. Phys. **1957** (in press).

[2] W. v. MEYEREN: Z. Physik **84**, 531 (1933).

[3] This not only reduced the reemission of helium; it also reduced the permeation of helium through the walls of the gauge. Since the major portion of the system was still at room temperature, however, the general conclusions of Ref. [*6*] are valid. With the gauge at room temperature, the lowest attainable pressures for a glass system of one liter volume are of the order of 2×10^{-10} mm Hg.

[4] R. CHAMPEIX: Le Vide **30**, 942 (1950).

[5] H. SCHWARZ: Rev. Sci. Instrum. **24**, 371 (1953).

[6] J. S. FOSTER, E. O. LAWRENCE and E. J. LOFGREN: Rev. Sci. Instrum. **24**, 388 (1953).

[7] F. M. PENNING: Physica, Haag **4**, 71 (1937).

[8] A. M. GUREWITSCH and N. F. WESTENDORP: Rev. Sci. Instrum. **25**, 389 (1954).

J. M. HOUSTON'S[1] new ionization gauge, which has previously been described in this paper, is also interesting as an ion pump. Since the electrons attain very long path lengths when contained in the magnetic field, the gauge maintains relatively high pumping speeds down to extremely low pressures. HOUSTON reports a pumping speed for helium of 0.1 liter/sec, or 25 times that of the Bayard-Alpert gauge. In preliminary experiments, he has achieved pressures of 2×10^{-12} mm Hg.

HERB, DAVIS, DIVATIA and SAXON[2] have combined the pumping action of a highly active getter with ion-pumping into the surface to achieve very high pumping speeds. This ingenious device evaporates titanium more or less continuously and is capable of speeds of 2000 to 3000 liter/sec for adsorbable gases. By utilizing an ionizing electron beam and driving the ions into the titanium deposited on the walls, speeds of hundreds of liters per second are attainable for noble gases. Ultimate pressures thus far attained with these "Evapor-ion" pumps are of the order of 10^{-7} mm Hg. It is not clear at this time whether considerably lower pressures can be achieved. The systems and pumps thus far used have not been baked at high temperatures, and the specific limitations have not been examined. J. C. SWARTZ[3] has made measurements of the residual gases in such a system, but further investigation with a baked out system seems necessary to determine the ultimate equilibrium pressures which can be achieved.

VI. Diffusion pumps for the production of ultrahigh vacuum.

19. Mercury vapor pumps. The pumps used in the early work of ANDERSON and NOTTINGHAM were mercury diffusion pumps. However, in both laboratories the best vacua were attained only after sealing the tube off from the pumps, and in some cases using barium getters and immersing the entire tube in liquid nitrogen. In work published in 1938, ANDERSON utilized oil diffusion pumps, but again sealed the tube off and utilized getters to achieve ultrahigh vacua.

More recently, NOTTINGHAM showed that very high vacua were attainable with mercury diffusion pumps continually pumping. A schematic diagram of a system based on the investigations of NOTTINGHAM[4] is shown in Fig. 12. The entire system, including both traps, are initially evacuated, then liquid nitrogen is put around trap Tr-1 and the remaining system is baked. Then Tr-2 is filled with liquid nitrogen. If the tube is to be isolated from the pumps, the construction is first heated to the highest possible temperature then liquid nitrogen is applied to Tr-2 and the tube is sealed off.

In these cases, it was often desirable to have two liquid nitrogen traps in series, one of which was baked at high temperatures along with the rest of the system, while the second was kept refrigerated during the latter part of the bakeout period. MOLNAR[5], BECKER and HARTMAN [*8*], and HAGSTRUM [*9*], at the Bell Telephone Laboratories, also used mercury diffusion pumps to achieve pressures in the ultrahigh vacuum range, and after continuous pumping for a period of time, achieved pressures at which the Bayard-Alpert gauge approached its x-ray limit. Their systems were isolated from the pumps by one or more liquid nitrogen traps and usually also by a ground glass cutoff valve. REYNOLDS[6] also used a mercury diffusion pump for his sensitive mass spectrometer (Sect. 13).

[1] J. M. HOUSTON: Bull. Amer. Phys. Soc. **2**, 1, 301 (1956).
[2] R. G. HERB, R. H. DAVIS, A. S. DIVATIA and D. SAXON: Phys. Rev. **89**, 897 A (1953).
[3] J. C. SWARTZ: Evapor-ion Pump Characteristics. 1955 Vacuum Symposium Transactions, p. 83. Boston: Committee on Vacuum Techniques, Inc. 1956.
[4] The system shown was designed and used by L. APKER [*7*].
[5] J. P. MOLNAR: Private communication.
[6] J. H. REYNOLDS: Rev. Sci. Instrum. **27**, 928 (1956).

Thus there is ample evidence that mercury pumps, together with adequate trapping at liquid nitrogen temperatures, are effective in reaching ultrahigh vacuum.

20. Oil diffusion pumps. Oil diffusion pumps have been widely used since 1928 when BURCH[1] showed that mercury could be replaced by certain high-boiling-point hydrocarbons as the pump fluid. He showed that with such oils as Apiezon A or B pressures as low as 10^{-7} mm Hg could be achieved without the use of liquid air traps. In 1936, HICKMAN[2] described some new fractionating pump fluids with which it was possible to attain equivalent nitrogen pressures below 10^{-7} mm Hg as read on an ionization gauge. While considerable doubt surrounded the interpretation of the ionization gauge readings[3], oil diffusion pumps were used for years to attain high vacuum.

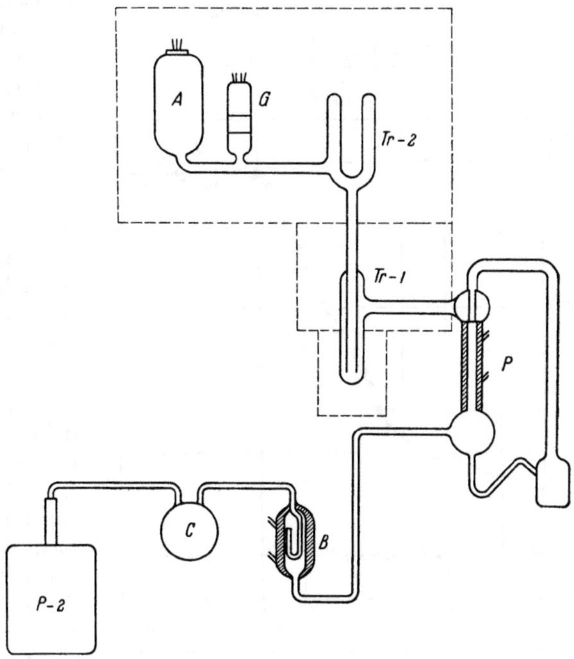

Fig. 12. Schematic diagram of exhaust system for production of extremely low pressure (NOTTINGHAM, APKER). A tube to be evacuated; G ionization gauge *Tr*-1 and *Tr*-2, liquid air traps *P*-1, mercury-vapor pump and *P*-2 rotary oil pump. Bulbs *B* and *C* are used to prevent diffusion of oil vapor into the mercury diffusion pump.

In the work on ultrahigh vacuum at Westinghouse Research Laboratories, oil diffusion pumps have been used almost exclusively. In an early experiment, it was demonstrated [5] that at pressures below 10^{-7} or 10^{-8} mm Hg, the oil diffusion pump becomes a source for contamination of a clean vacuum system. The pump used was a two-stage water-cooled glass oil diffusion pump, of the type designed by HICKMAN, manufactured by Consolidated Electrodynamics Corporation and utilizing their recommended pumping fluid, Octoil-S. A schematic diagram of the experiment, together with the resulting data, is shown in Fig. 13. A vacuum system was set up with two ionization gauges separated by a vacuum valve so that one gauge was connected to the portion of the system which could be isolated, while the other was in the vacuum line connected with the diffusion pump. After bakeout of the entire system at 420° C, the ion gauges were outgassed; then the valve was closed and both gauges run continuously at normal operating conditions (emission current, $i_e = 10$ milliamperes). Fig. 13 shows the pressures as a function of time after closing the valve for the isolated part of the system (dashed line *A*) and for the portion in contact with the pump (solid curve *B*). Initially both systems came down to pressures in the ultrahigh vacuum range. After several hours, however, the pressure adjacent to the diffusion

[1] C. R. BURCH: Nature, Lond. **122**, 729 (1928). — Proc. Roy. Soc. Lond., Ser. A **123**, 271 (1929).

[2] K. C. D. HICKMAN: J. Franklin Inst. **221**, 215, 383 (1936).

[3] See J. BLEARS: Proc. Roy. Soc. Lond., Ser. A **188**, 62 (1946).

pump began to rise and returned to the initial value of approximately 10^{-7} mm Hg, while the pressure in the isolated system remained at its lowest value for the duration of the experiment. The above results were qualitatively interpreted as follows. At the outset of the experiment, both ion gauges were pumping and reduced the pressure in each to the ultrahigh vacuum range. The baked glass lead from the pump to the system also acted as a partial baffle for returning oil vapor. After a short time, the back-diffusion of contaminants from the pumps (such as oil vapor and other hydrocarbons from catalytic cracking of the oil) caused a rise in pressure in ion gauge B. With the ion gauge still pumping, the pressure continued to rise, the rate depending on the quantity of oil products directly adsorbed on the clean glass tubulation and the rate of the ionization pumping action. After two or three days, an equilibrium pressure of 10^{-7} mm Hg to 10^{-8} mm Hg was reached. In the isolated portion of such systems, the ion pumping has maintained pressures as low as 2 or 3×10^{-10} mm Hg for months.

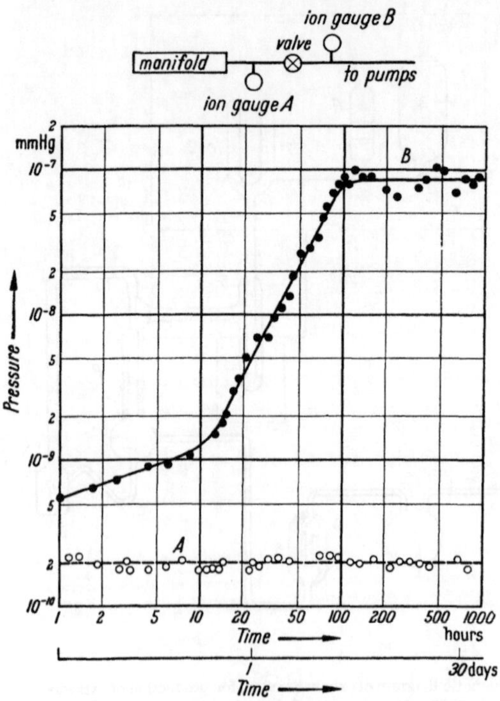

Fig. 13. Curve A, pressure versus time in sealed-off portion of system. Curve B, pressure versus time in ion gauge in contact with diffusion pump.

This experiment indicated that at pressures below 10^{-7} mm Hg, an oil diffusion pump constitutes a source of contaminating gases and vapors. On the other hand, it had been recognized by a number of workers[1,2,3] on the basis of mass spectrometer data, that specified gases may be removed by diffusion pumps to partial pressures far below 10^{-8} mm Hg. An experiment carried out by CARMICHAEL and ALPERT[4] demonstrated that oil diffusion pumps of the type described above could achieve pressures below 10^{-10} mm Hg if an adequate trap was used to prevent or reduce back-diffusion of contaminants from the pumps. The experiment consisted of adding a gas source and valve and inserting an isolating trap[5] between the vacuum valve and the manifold of Fig. 13, and measuring the rate of evacuation and the ultimate pressure attainable in the system. After system bakeout, helium was introduced to a pressure of the order of 10^{-7} mm Hg; then the valve to the diffusion pump was opened (at time $t=0$) and pressure measured as a function of time thereafter. The data are shown in Fig. 14 demonstrating that the oil diffusion pump continued to evacuate to a pressure below 10^{-10} mm Hg with a constant pumping speed. The characteristic pumping time was measured to be 3.3 sec, equivalent to the computed conductance of the vacuum system.

[1] R. E. FOX: Private communication.
[2] J. BLEARS: J. Sci. Instrum. **28**, 36 (1951).
[3] M. P. GARFUNKEL and A. WEXLER: Rev. Sci. Instrum. **25**, 170 (1954).
[4] J. H. CARMICHAEL and D. ALPERT: J. Appl. Phys. **1957** (in press).
[5] The trap used was the copper foil trap, D. ALPERT, Rev. Sci. Instrum. **24**, 1004 (1953), and described in Sect. 25.

This corresponds to a characteristic pumping time at least twenty times smaller than that for ionization gauge pumping. From Eq. (12.17), $p_e = r\tau$, and one would therefore expect a pressure considerably below 8×10^{-11} mm Hg, the ultimate pressure observed. Such a lower pressure was not reached in this experiment because the rate of rise, r, due to the leakage of helium from the gas source through the valve was 2×10^{-11} mm Hg/sec, or forty times the usual residual rate of rise due to diffusion through the glass. With smaller rates of rise, total pressures estimated to be 10^{-11} mm Hg or less have been obtained in this manner. Thus the experiment showed that the diffusion pump is capable of attaining pressures in the 10^{-11} mm Hg region and that there is no significant change in the pumping speed at these pressures.

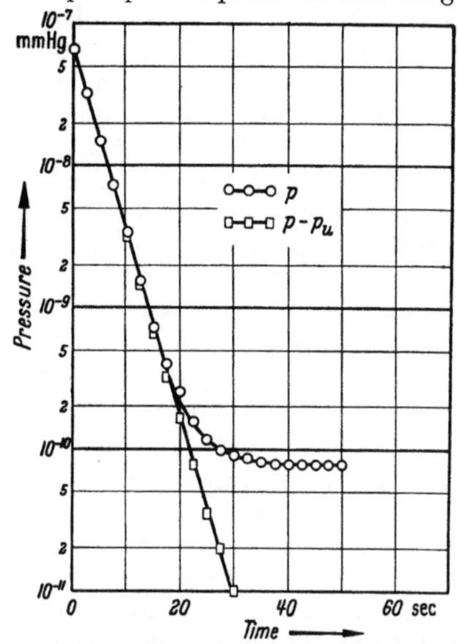

Fig. 14. Pump-down of helium by oil diffusion pumps. The characteristic pumping time τ = 3.3 sec. The rate of rise of pressure in system $\sim 2\times10^{-11}$ mm Hg/sec. This large rate of rise (approximately 40 times that due to diffusion of atmospheric helium through the glass envelope) was due to the large leakage through the vacuum valve used in this experiment.

This demonstration of the pumping effectiveness of oil diffusion pumps when adequately baffled emphasizes the conclusion that one of the principal reasons for higher ultimate pressures in conventional systems is due to the back-diffusion of oil vapor and other products from the pumps. In some cases, back-diffusion of some of the lighter gases through the pump has caused a limitation on the attainment of high vacuum. However, DAYTON[1] has shown that this limitation can be reduced by a proper selection of pump and pump fluid. A more serious source of back-diffusion comes from the diffusion pump itself either as oil vapor or in the form of various organic products produced by thermal dissociation of the oil. JAECKEL[2] found that the ultimate pressures attainable with metal pumps was significantly higher than those with glass pumps. He attributed this to a greater thermal dissociation at the metal surfaces. Other workers have observed similar results. It is not clear whether the entire effect is due to the catalytic cracking of the oil, or whether the jet design[3] of most metal pumps is poor from the standpoint of backstreaming. In any case, the amount of gas evolved from most metal pumps is very appreciable. It is not certain, at this time, whether scaled-up versions of the copper foil traps would be applicable to metal pumps. With traps refrigerated by liquid nitrogen, however, it is possible to attain pressures in the ultrahigh vacuum range even with metal oil diffusion pumps of relatively high speeds.

VII. Ultrahigh vacuum systems and components.

21. System requirement. On the basis of previous discussion, the ultimate equilibrium pressure in a vacuum system is given by $p_e = Q/S$ [Eq. (12.18)], where Q is the total rate of influx of gas, and S is the combined pumping speed due to all pumping mechanisms. From Sect. 12, Q is the sum of Q_L, the influx due to

[1] B. B. DAYTON: Rev. Sci. Instrum. **19**, 793 (1948).
[2] R. JAECKEL: Z. techn. Phys. **23**, 177 (1942).
[3] L. G. PARRATT and E. L. JOSSEM: Rev. Sci. Instrum. **23**, 188 (1952).

leaks, Q_p the back-diffusion from the pumps, and Q_W the influx due to desorption and dissociation and diffusion through the walls of the system. To attain the ultrahigh vacuum range, the ratio of each of the Q's to S must be made small compared to 10^{-8} mm Hg. To reduce Q_L, it is evident that techniques are required for leak detection which can assure very low influx rates. Similarly, to attain sufficiently low Q_p, it is necessary to develop new pumps or to utilize effective traps to prevent backstreaming of high vapor pressure components into the system. Since the desorption of loosely bound molecular layers from the walls is a major source of gas evolution in systems, it is clear from Sect. 13 that to lower Q_W to reasonable values, it is necessary to bake the system to temperatures between 250 and 500° C.

The requirement of high temperature bakeout rules out, for the most part, the use of rubber gaskets, greases, many of the conventional means for making demountable seals, conventional stopcocks, etc. The basic minimum for a system, and one that has been in use for a number of years, is a glass system capable of high temperature bakeout, a mercury-vapor pump, and adequate liquid nitrogen traps as described in Sect. 19.

In many experiments, however, it has proved necessary to build manipulative systems of greater flexibility [*10*]. In recent years, a number of components, such as stopcocks, demountable joints, flexible couplings, etc., have been developed which are applicable to ultrahigh vacuum systems, that is, (a) they are bakeable at high temperatures, (b) they do not evolve large amounts of gas in use. A number of such components will now be discussed.

22. Vacuum valves. In many systems, one requires not only the initial evacuation of the system to ultrahigh vacuum conditions, but also the ability to isolate sections of the system or to transfer gases from one portion of the system to another without introducing large impurities. The use of ordinary greased stopcocks or mercury cutoffs for this purpose is very questionable since they are not bakeable. Not only is the vapor pressure of the stopcock grease in some cases objectionable, but the amount of occluded non-condensible gas evolved with each rotation of a stopcock may be hundreds of times greater than the amount of residual gas in the system.

A number of variable leaks have been proposed which are useful for introducing gas into a system. One which has been widely used is the porcelain rod leak developed by HAGSTRUM and WEINHART[1]. Controlling the leak rate through a porous porcelain rod by variation of the degree of immersion in a column of mercury, this leak has a conductance which is variable between 3×10^{-7} and 3×10^{-9} liter/sec. After the leak and connecting tubing have been carefully torched, it is possible to attain good ultrahigh vacuum conditions[2] by isolating the leak from the system with a liquid nitrogen trap. R. T. BAYARD[3] utilized a simple leak which was capable of high temperature bakeout and had a variable conductance between less than 10^{-14} and about 10^{-7} liter/sec. It consisted of a small Kovar tubing of approximately 1/16-inch diameter, 0.010-inch wall thickness clamped flat between a C-clamp with hardened jaws. The Kovar retained enough elasticity to open slightly when the C-clamp was released.

In general, the problem encountered with leaks of very small maximum conductance is that the system volume on one side of the leak cannot, in general, be carefully evacuated. For this reason, it is desirable to have a valve which

[1] H. D. HAGSTRUM and H. W. WEINHART: Rev. Sci. Instrum. **21**, 394 (1950).

[2] See H. D. HAGSTRUM: Phys. Rev. **104**, 1516 (1956), for a discussion of impurities introduced with gases.

[3] R. T. BAYARD: Private communication.

can withstand high temperature bakeout and has both a reasonably large maximum conductance when open and a needle valve action with very low minimum conductance.

The ultrahigh vacuum valve developed by ALPERT[1] satisfies these conditions. As shown in the schematic diagram of Fig. 15, the valve is designed in two sections—a valve assembly which is sealed into the vacuum system, and a driver

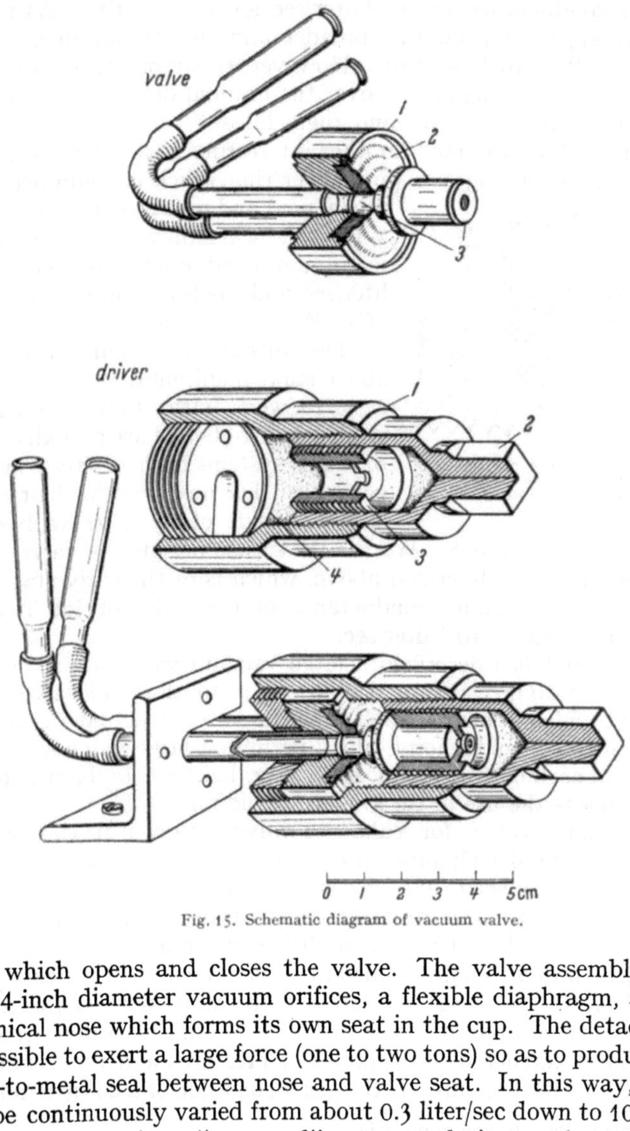

Fig. 15. Schematic diagram of vacuum valve.

mechanism which opens and closes the valve. The valve assembly contains a cup with 1/4-inch diameter vacuum orifices, a flexible diaphragm, and a highly polished conical nose which forms its own seat in the cup. The detachable driver makes it possible to exert a large force (one to two tons) so as to produce a vacuum tight metal-to-metal seal between nose and valve seat. In this way, the conductance can be continuously varied from about 0.3 liter/sec down to 10^{-9} or 10^{-10} liter/sec. The quantity of gas "squeezed" out upon closing a valve to its minimum conductance is approximately 5×10^{-10} mm liter or less.

A number of valves have recently been developed which are based on this design but have provided important improvements. BROWN and COYLE[2] enlarged

[1] D. ALPERT: Rev. Sci. Instrum. **22**, 536 (1951).

[2] S. C. BROWN and J. E. COYLE: Rev. Sci. Instrum. **23**, 570 (1952).

the valve and utilized an all-monel construction with a copper valve seat. BURITZ[1] showed that the copper seat was unnecessary and found similar values of minimum conductance with a monel nose and a monel valve seat formed by machining a 0.030 to 0.040-inch chamfer. CARMICHAEL[2] then showed that an all-monel valve was very effective if the chamfer of the seat was extremely small, 0.003 to 0.005 inch. The large pressures developed with a given force lowered the minimum conductance to 10^{-14} liter/sec for a new valve. As the valve was used, the seat apparently became broader and the minimum conductance deteriorated to 10^{-11} to 10^{-12} liter/sec. However, this represented an improvement of one to two orders of magnitude over the original design. It had a number of advantages, including simplicity and ruggedness.

BILLS and ALLEN[3] provided another interesting improvement in the form of a new valve seat consisting of a pure silver ring which is contained between the monel nose and monel seat as shown in Fig. 16. With this seat, they were able to provide a maximum conductance of the order of one liter/sec and to obtain minimum conductances of 10^{-14} liter/sec or less.

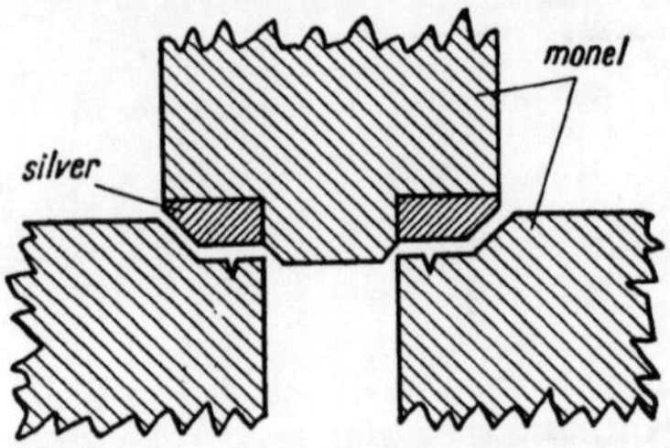

Fig. 16. Cross section of BILLS and ALLEN's metal valve seal. Dimensions have been distorted to show details of construction.

The ultrahigh vacuum valves described above make manipulative systems feasible and easy to work with; they can withstand repeated bakeouts and are an indispensable part of many systems both as valves and controlled gas leaks. The next step in their development is the design of a valve with considerably larger maximum conductance. W. J. LANGE[4] has designed a valve based on the same principles as those described above, which is in the early stages of testing, and which has a maximum conductance of the order of 100 liter/sec, and a minimum conductance of 10^{-9} liter/sec.

F. L. REYNOLDS[5] has described a large vacuum valve which may also have applications in ultrahigh vacuum systems. It is an all metal valve and contains an indium seat which is melted each time the valve is actuated. Basically, it is an internal controlled solder joint which permits a large valve aperture without huge forces for sealing the metal seat. For bakeout to higher temperatures REYNOLDS suggests the use of tin in place of indium.

The conductance values for leaks and valves as given in this section and in the discussion of small leak rates elsewhere assume "molecular flow" through the leak. KNUDSEN[6] showed that for gas pressures at which the mean free path is equal to or large compared to the diameter of a tube or orifice, the influx through the tube is proportional to the pressure difference across it, and the conductance is defined as the proportionality constant. For pressures in which the mean free path of the molecules is small compared to the dimensions of the tube, the gas flow follows the equations of "viscous flow"[7]. In this range, the gas influx varies with the difference of the squares of pressures and the "equivalent conductance" varies with pressure. If the gas on one side of a constriction or valve is

---

[1] R. S. BURITZ: Private communication.
[2] J. H. CARMICHAEL: Private communication.
[3] D. G. BILLS and F. G. ALLEN: Rev. Sci. Instrum. **26**, 654 (1955).
[4] W. J. LANGE: Private communication.
[5] F. L. REYNOLDS: All-Metal Vacuum Valve Using an Indium Seat. 1955 Vacuum Symposium Transactions, p. 74. Boston: Committee on Vacuum Techniques 1956.
[6] M. KNUDSEN: Ann. Phys. **28**, 75, 999 (1909).
[7] Ref. [7], Chap. 2. See also, R. E. HONIG: J. Appl. Phys. **16**, 646 (1945).

at a pressure close to atmospheric, the transport of gas, in general, represents a combination of the two types of flow. For this reason, the flow through capillaries is usually given in terms of the gas influx for a given pressure rather than in terms of the conductance. In the case of porous diaphragms with capillary diameters of the order of 10^{-5} cm or less, DUSHMAN [7], Chap. 2, Sect. 7) has shown that molecular flow predominates for pressures up to a sizeable fraction of an atmosphere. For the valves and leaks described in this paper with minimum conductances of 10^{-7} liter/sec or less, the dimensions of the opening are also very small, and it is highly likely that the equations of molecular flow apply. This has been empirically verified in the data of Fig. 4 for pressures up to the millimeter range. The conductance will, of course, vary with the molecular weight of the gas and an equivalent leakage conductance is usually given in terms of the conductance for nitrogen.

23. Demountable seals for ultrahigh vacuum systems. The literature on high vacuum technology has many references to demountable seals[1] [*10*], [*11*]. Since an ultrahigh vacuum system must be outgassed at high temperatures, the conventional gaskets of rubber[2] and lead or tin must be ruled out. There remain, however, a large number of possibilities utilizing various combinations of metalsand ceramics.

Metal gaskets of lead, tin, aluminum, copper, etc. have been in use for many years ([*14*], p. 320) but relatively few have been tested under ultrahigh vacuum conditions. It has been noted elsewhere that a seal which passes inspection with a conventional helium leak detector is not necessarily free of leaks. To attain a pressure of, say, 2×10^{-10} mm, with a pumping speed of 0.1 liter/sec, the rate of influx through a leak must be less than 2×10^{-11} mm liter/sec and, hence, the conductance must be less than $\sim3\times10^{-14}$ liter/sec. Thus the seal can, in general, be tested only in an ultrahigh vacuum system or with a sensitive mass spectrometer.

Two basic types of demountable seals have been successfully used for ultrahigh vacuum: (a) fine wire O-ring gaskets with flat flanges, and (b) variations of flat gaskets with knife edge flanges.

Fine wire O-ring gaskets of aluminum have been reported by BALDOCK and SITES[3] and of copper by SPROULL, DASH, TYLER, and MOORE[4]. However, VAN HEERDEN[5] has reported difficulties in attaining leak rates less than 10^{-9} mm liter/sec (or conductances less than 10^{-12} liter/sec). He reports requirements of an extreme degree of flange smoothness for the case of copper and failure in the case of aluminum.

A very successful seal which has subsequently been used widely in demountable ultrahigh vacuum systems was originally proposed by HICKAM[6] for use in an all-metal mass spectrometer. It consists of a gold wire gasket of 0.020 to 0.030-inch diameter wire welded into a ring and clamped between flat monel or stainless steel flanges. It makes a joint which can be repeatedly baked at temperatures as high as 475° C and with leak rates which are not measurable in an ultrahigh vacuum system. The gasket may be used either between flat flanges or clamped in the corner of a pair of mating stepped flanges.

[1] J. STRONG: Procedures in Experimental Physics, Chap. 3 and 13. New York: Prentice-Hall, Inc. 1944.

[2] F. J. NORTON has also measured relatively large permeation rates of H_2, He, and N_2 through rubber (Gen. Elec. Review, September 1952).

[3] R. BALDOCK and J. SITES: Phys. Rev. **81**, 294 (1951).

[4] R. L. SPROULL, W. C. DASH, W. W. TYLER and A. R. MOORE: Rev. Sci. Instrum. **22**, 410 (1951).

[5] P. J. VAN HEERDEN: Rev. Sci. Instrum. **26**, 1130 (1955).

[6] W. M. HICKAM: Rev. Sci. Instrum. **20**, 472 (1949).

An excellent and repetitive ultrahigh vacuum joint was proposed by PATTEE[1] which is particularly useful for inserting replaceable electrodes or for connecting two sections of tubing. As shown in Fig. 17, the joint is made by clamping a flat replaceable copper gasket between two knife-edge ridges in a steel tube or flange. The gasket is machined from OFHC copper rod, and the flat surfaces are polished with fine emery paper as are the knife edges. The latter are machined with a 45° wedge angle with a radius of 0.005-inch at the tip. Since adjustment of the electrodes is permitted by the deformation of the copper, the length of the gasket can be varied up to 1 cm in length to provide for such motion.

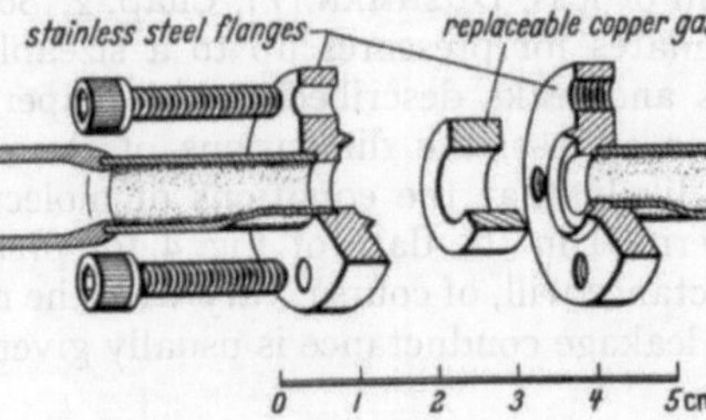

Fig. 17. Pattee demountable vacuum joint.

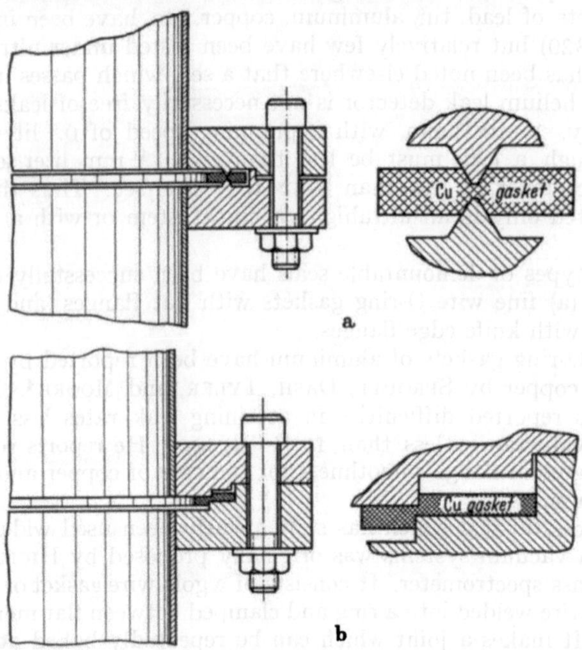

Fig. 18a and b. (a) Typical knife edge flange arrangement utilizing a flat copper gasket. (b) Step type demountable vacuum joint.

A number of workers have proposed other variations of the knife-edge vacuum seal[2-4]. Fig. 18a shows a typical knife-edge flange arrangement utilizing

[1] H. H. PATTEE jr.: Rev. Sci. Instrum. **25**, 1132 (1954).
[2] P. J. VAN HEERDEN: Rev. Sci. Instrum. **26**, 1130 (1955).
[3] G. HEES, W. EATON and J. LECK: The Knife Edge Vacuum Seal. 1955 Vacuum Symposium Transactions, p. 75. Boston: Committee on Vacuum Techniques, Inc. 1956.
[4] C. G. MATLAND and W. J. LANGE: Private communication.

a flat gasket. The gasket material is usually OFHC copper annealed in a hydrogen atmosphere. MATLAND and LANGE found that for a 0.045-inch thick copper gasket the knife edge could have any shape from a rounded edge 0.005-inch in diameter to a flat edge from 0.004 to 0.030-inch across. Such gaskets make excellent reproducible seals with copper, but rather questionable ones with aluminum.

While the knife-edge seal is quite dependable and reproducible, it has the drawback that it is difficult to machine, since the mating edges must be accurately concentric and of the same diameter. If the edge is even slightly damaged, the flange is rather difficult to repair. A much simpler design which provides the same type of joint has been described by LANGE and ALPERT[1]. The sharp edge in this case is provided by "steps" in the flange as shown in Fig. 18b. This flange design is self-aligning, easier to machine (the steps which form the seal may or may not overlap), and is much easier to repair in case of surface damage. In most copper gasket joints, it is found that after the first bakeout the clamping bolts become loose unless spring lock washers are used. It is therefore usually desirable to tighten the bolts after bakeout for mechanical rigidity.

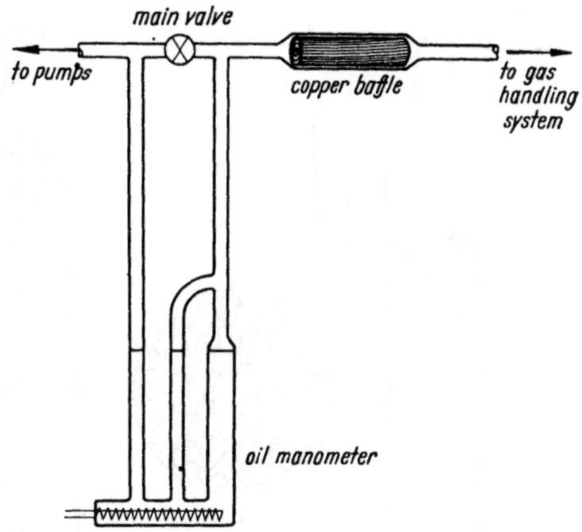

Fig. 19. Oil manometer connected to the gas handling vacuum system. The heater consists of a Nichrome wire, 10 ohms resistance.

The gold and copper gaskets have proved quite satisfactory in many applications. In some cases, seals between ceramics and metals[2] have been made in a similar way, and a large number of combinations of materials is undoubtedly possible. Utilizing these seals, ultrahigh vacuum joints have been made with flanges up to 12 inches diameter, and it seems feasible to develop systems with the same degree of flexibility as is practicable in conventional systems.

24. Manometers for the measurement of higher pressures. In the field of gaseous electronics, it is often necessary to measure pressures in the region above 10^{-2} mm Hg without introducing impurities into the system from the manometer itself. Two manometers have been described in the literature which are useful for this purpose. The first is that described by ALPERT, MATLAND, and MCCOUBREY[3] which utilizes a thin diaphragm as a pressure sensitive element and as a means for separating the baked portion of the vacuum system from the MCLEOD gauge or liquid manometer. With the pressure on both sides of the manometer equal, e.g., with both systems evacuated, the diaphragm assumes an equilibrium position as indicated by its capacity to an electrical probe. When a sample of pure gas is introduced into the clean system, air is simultaneously let into the manometer system until the diaphragm reassumes its null position. The liquid (or other absolute) manometer then directly reads the pure gas pressure.

BIONDI[4] has designed an oil manometer which has been used with several systems and is capable of measuring pressures between 0.1 mm Hg and 30 mm Hg

[1] W. J. LANGE and D. ALPERT: Rev. Sci. Instrum. **28**, 726 (1957).
[2] M. J. HIGATSBERGER and W. W. ERBE: Rev. Sci. Instrum. **27**, 110 (1956).
[3] D. ALPERT, C. G. MATLAND and A. O. MCCOUBREY: Rev. Sci. Instrum. **21**, 571 (1950).
[4] M. A. BIONDI: Rev. Sci. Instrum. **24**, 990 (1953).

with an accuracy of ± 0.01 mm Hg. The manometer is shown in Fig. 19. By utilizing metal valves, a heater to degas the oil, and a copper trap to isolate it from the system, the systems are capable of attaining pressures of 10^{-9} mm Hg with a rate of influx of approximately 2×10^{-12} mm-liter/sec.

With these manometers, it is possible to make absolute pressure measurements at high pressures. When used in conjunction with the ionization gauge, it is possible to cover the entire range of pressures with a minimum of contamination of a sample of pure gas.

25. Traps. To attain ultrahigh vacua with a diffusion pump, it is necessary to prevent the backstreaming of pump vapors into the system. In the case of mercury vapor pumps, this is usually accomplished by the use of one or more traps refrigerated to at least liquid nitrogen temperatures ($-196°$ C). It is desirable to bake at least one trap with the rest of the system to maintain low influx rates. After the trap is filled, it quickly condenses many monolayers of mercury which in turn adsorb sizeable quantities of gas. In the final stages of evacuation, the trap may be pumping most gases more effectively than the diffusion pump. Two considerations are of great importance:

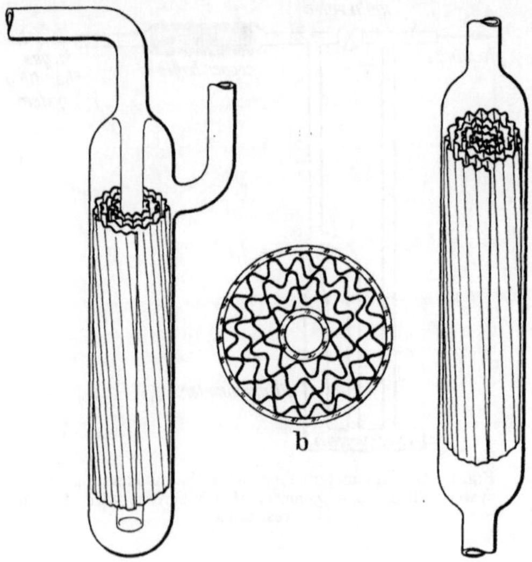

Fig. 20 a—c. Copper foil trap.

(1) A number of gases such as carbon monoxide, carbon dioxide, methane, and other hydrocarbons with appreciable vapor pressures at $-196°$ C, may be reevolved after several monolayers of gas have been adsorbed on the walls of the trap.

(2) If during the course of an experiment the liquid nitrogen level drops and the temperature rises in a portion of the trap, large quantities of gas may be reevolved into the system.

As a consequence of (1), the traps are often reheated during the course of a run to remove and pump away the "saturated" gases. It is helpful to isolate the system during this procedure if a vacuum valve is available. To minimize (2), various devices have been utilized either to maintain the liquid nitrogen at a constant level or to reduce the effect of a change in refrigerant level. The corrugated copper foil trap[1] was originally proposed to reduce the effect of changes in the liquid air level and to greatly increase the surface area and trapping probability. This trap consists of a standard reentrant glass trap into which corrugated copper has been inserted as shown in Fig. 20a. In its original form, it was intended to utilize the high thermal conductivity of the metallic insert so that the entire trapping surface would remain cold despite fluctuations in the liquid level. Using liquid nitrogen, the trap proved quite satisfactory, and pressure readings in the ultrahigh vacuum range were attained. It was subsequently found that the "copper foil" trap was a very efficient means for reducing backstreaming from oil diffusion pumps even at room temperature. That is, the clean

[1] D. ALPERT: Rev. Sci. Instrum. **24**, 1004 (1953).

copper is an efficient adsorbing surface for backstreaming oil vapor even at room temperature, and for many applications, the reentrant form has been dropped and a straight tube version used as shown in Fig. 20c. Such traps have been successfully used over a period of years even though little is known about the adsorption mechanism. Used in conjunction with glass oil diffusion pumps, the traps have continued to act as effective oil baffles at room temperatures for periods of one to two months. To "reactivate" the trap, it is only necessary to bake the trap at a few hundred degrees centigrade along with the rest of the vacuum system.

E. THOMAS[1] has described a variation of this trap for mercury pumps using stainless steel. His trap is shown in Fig. 21 which shows a cross section of the entire assembly (Fig. 21a) together with a sketch of the corrugated stainless steel (Fig. 21b).

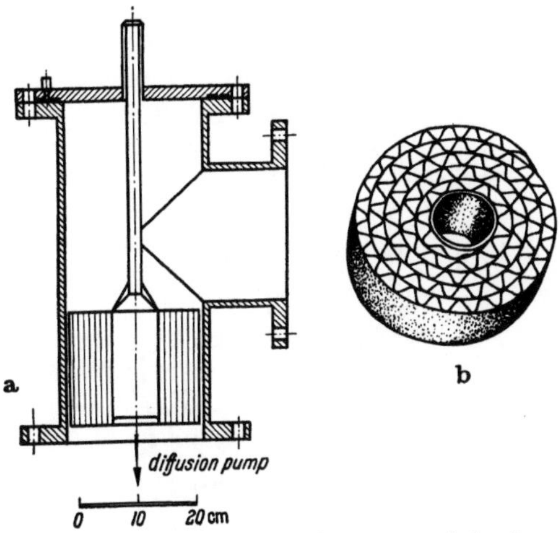

Fig. 21 a and b. Stainless steel liquid nitrogen trap designed by E. THOMAS.

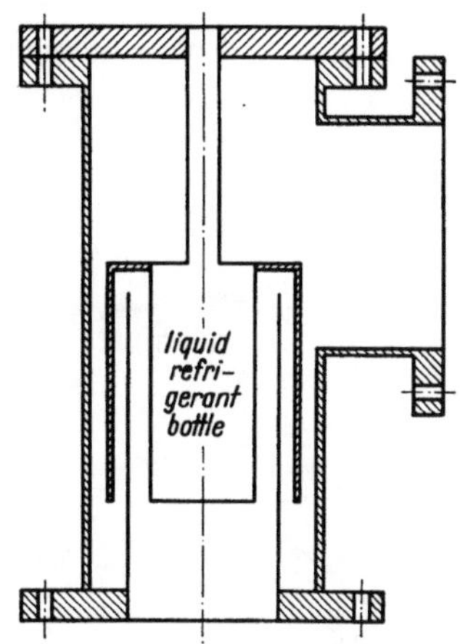

Fig. 22. Schematic diagram of trap designed by R. F. POST.

One of the most common traps utilized in large vacuum systems is the simple metal bottle suspended by a long narrow neck within a metal vacuum system. This leaves much of the envelope between pump and system at room temperature. Such traps are often inefficient not only because of the rather low probability for capturing condensible vapors, but also because of the possibility for oil to creep along the warm surface into the vacuum system. A trap suggested by R. F. POST[2] eliminates this difficulty without unduly restricting the conductance of the trap. As shown in the schematic diagram of Fig. 22, the heavy sleeve is cooled by conduction from the liquid nitrogen bottle, while the thin sleeve is cooled by radiation. Hence, cold surface barriers are present against both volume and surface flow.

It has been recognized for many years that in principle very high vacua are attainable through the use of liquid helium as a refrigerant since the vapor pressure of non-helium gas is very low at temperatures near the boiling point of liquid helium. To achieve ultrahigh vacuum at such temperatures, it is only necessary to pump out the helium[3] remaining in the vacuum enclosure.

[1] E. THOMAS: Vacuum **3**, 413 (1953).
[2] R. F. POST: Private communication.
[3] M. P. GARFUNKEL and A. WEXLER: Rev. Sci. Instrum. **25**, 170 (1954).

The use of liquid helium is, therefore, of considerable interest as a pump or trap, though its feasibility has not been evaluated for many experiments in this field. A most interesting application has been made by GOMER and HULM[1] in studies of field emission and surface mobility of oxygen on tungsten by the use of the field emission microscope. It is described in some detail in Ref. [3]. The experiments were carried out by immersing the entire "vacuum system", in this case a small Müller field emission microscope, in a liquid helium dewar at 2 to 4° K. Under these circumstances, the partial pressures of all possible contaminants are reduced to very low values without bakeout of the system.

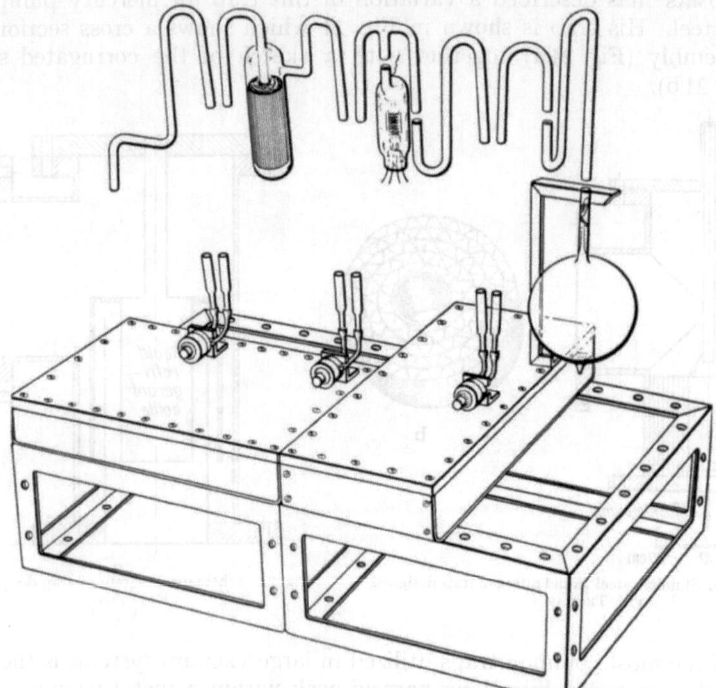

Fig. 23. Vacuum valves, ionization gauges, and gas source ready for assembly on module units.

26. Design and operation of vacuum systems. A considerable discussion of the typical operation of systems utilizing mercury diffusion pumps has been given in Ref. [7], Chap. 10, Sect. 12 and in Ref. [9]. The design of vacuum systems developed over a period of years at the Westinghouse Research Laboratories has been described in some detail in Ref. [5] and will be discussed very briefly in this paper.

The Westinghouse systems utilize oil diffusion pumps or ion gauge pumping to evacuate glass and metal systems of 1 to 10 liter volume; the system design and procedures have become widely used to achieve pressures between 10^{-9} and 10^{-10} mm Hg in a routine way without refrigerants or chemical getters. Fig. 23 shows a system of the type described in Ref. [5] in which the vacuum valves, copper trap, ionization gauge, and gas bottle are ready for assembly on a series of module units. These units, 16 inches square, form the bases for standard-sized furnaces with which the systems can be baked out at temperatures

[1] R. GOMER and J. K. HULM: J. Chem. Phys. **20**, 1500 (1952). — J. Amer. Chem. Soc. **75**, 4114 (1953).

as high as 500° C. To increase the size or complexity of the system, another module is bolted to the bench and additional vacuum components are attached,

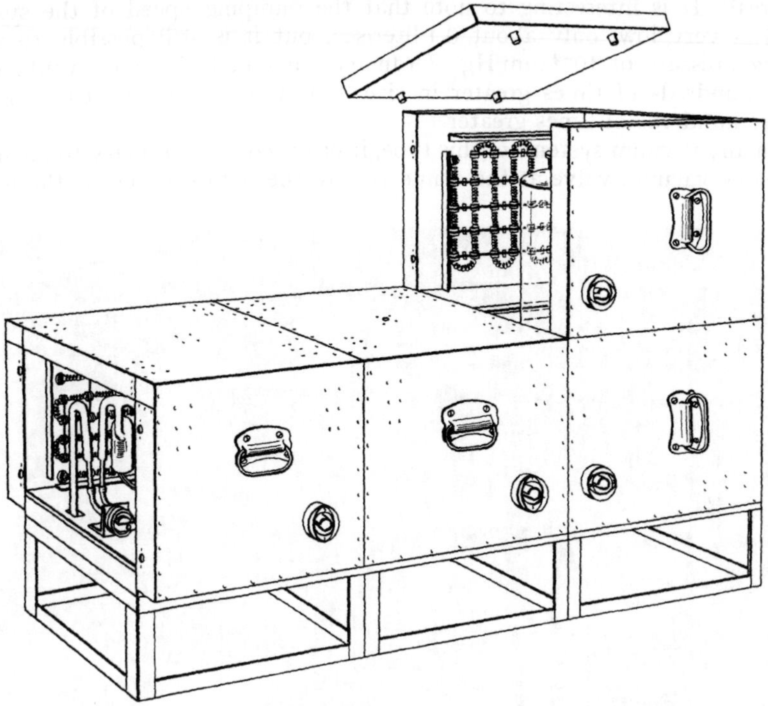

Fig. 24. Standard furnaces in place.

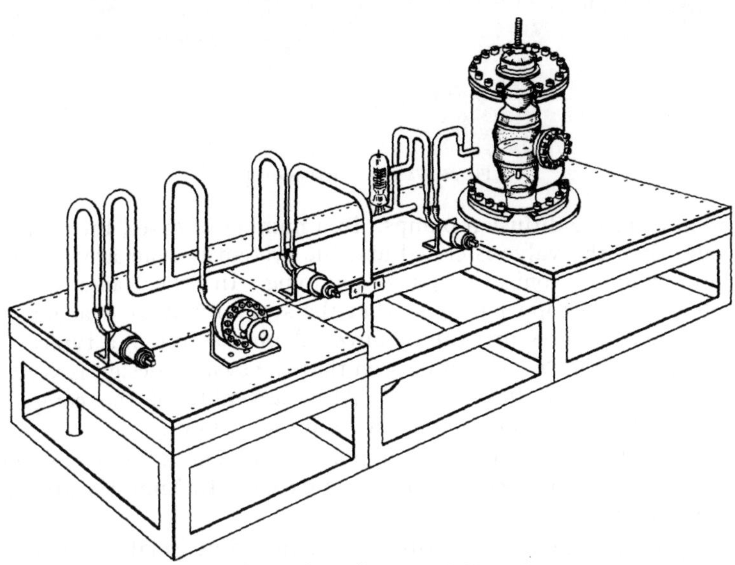

Fig. 25. Demountable chamber for studies of electrical breakdown in pure gases.

usually with a simple vertical glass seal. Fig. 24 shows the standard furnaces in place. Each unit covers a single base module; with this design the system can be extended in either the horizontal or vertical direction. Fig. 25 shows a stainless steel vacuum tank (10 liter-volume) assembled with four demountable gold gasket joints and used in studies of electrical breakdown in gases at atmospheric pressures[1]. It is interesting to note that the pumping speed of the system in Fig. 25 is very low, only about 0.1 liter/sec, but it is still possible to reach a working pressure of 10^{-9} mm Hg. Ordinary vacuum techniques would call for pumps hundreds of times greater in size, but would probably attain ultimate pressure hundreds of times greater.

In many vacuum systems of this type, it has become customary to incorporate a bakeable vacuum valve at the connection to the pump to permit the isolation

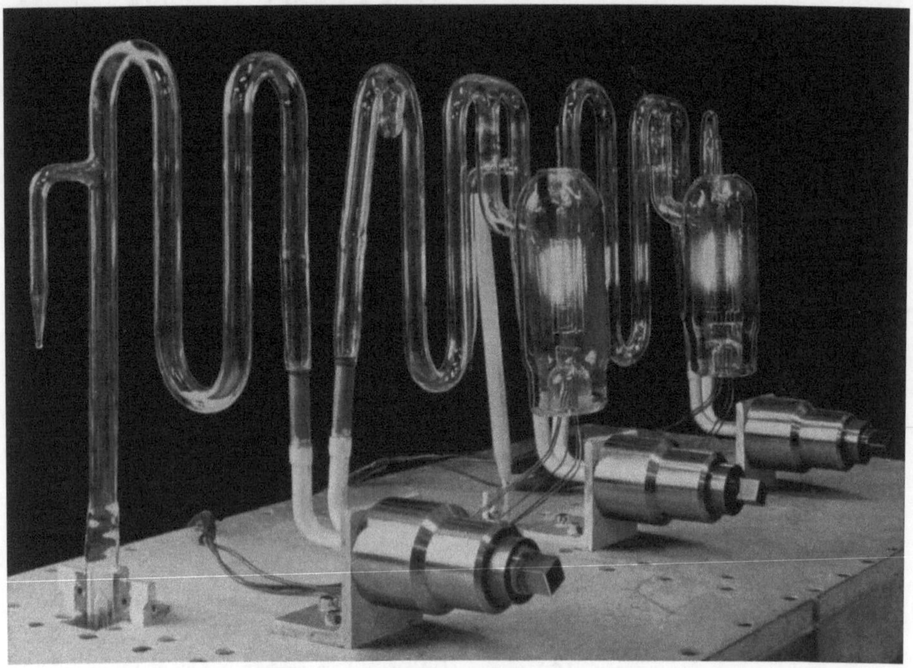

Fig. 26. Vacuum system assembly.

of the system from the diffusion pumps. In a number of cases, the entire system is baked out; then the valve is closed and the system is evacuated to the ultrahigh vacuum range by ionization gauge pumping. In other cases, the isolation valve is followed by a copper trap, so that the system may be evacuated by the pumps when very low pressures are required. A photograph of a typical vacuum system is shown in Fig. 26. It can be seen that the system design is very flexible and easily assembled or disassembled. After assembly of the components is completed, the evacuation procedure is roughly as follows:

(1) Start mechanical and diffusion pumps and check for large leaks with thermocouple or ionization gauge if pressure exceeds 10^{-5} mm Hg after an hour or two of pumping.

(2) Set up furnace and bake entire system, including isolating valve and trap at temperature between 400 and 500° C for several hours (usually overnight).

(3) Remove furnace; outgas elements of ionization gauge at elevated temperature (above 1200° C) for one hour or more.

[1] M. MENES: Phys. Rev. **98**, 561 A (1955).

(4) Close isolating valve and check for small leaks if ionization gauge running continuously does not reduce pressure to less than 10^{-9} mm Hg. Mesure rate of pressure rise in system.

(5) If trap is used, step (4) is necessary only if pressure does not reduce below 5×10^{-10} mm Hg.

Perhaps the most graphic demonstration of the relative ease with which ultrahigh vacua may be achieved with these methods is the rapidity with which a system can be brought into operation. In a number of instances, the above procedure has been followed and the system has attained a pressure in the 10^{-10} mm Hg range within 24 hours after initial assembly.

27. Detection of leaks. As indicated in Ref. [5], it might seem that the problem of eliminating leaks in a system which is to attain vacua one hundred or one thousand times better than 10^{-8} mm Hg might be hundreds or thousands of times more difficult. This has not proved to be the case for two reasons: (1) It has been found empirically that most of the influx in conventional systems is due to desorption or backdiffusion from the pumps and not due to a very large number of small leaks. (2) With the ability to achieve and to measure lower pressure has come also the ability to detect leaks which are proportionately smaller. Thus, while the maximum leak which can be tolerated in the systems described above is only about 10^{-14} liter/sec, the greater sensitivity of the ionization gauge makes it possible to locate such leaks by using standard methods of leak detection. For example, the application of acetone, ether, or carbon tetrachloride to a leak in a system operating at an equilibrium pressure of 5×10^{-9} mm Hg may either give a very rapid rise in pressure to a value above 10^{-8} mm Hg or temporarily close the leak and reduce the pressure to 10^{-9} mm Hg or less. A leak of this size would be undetectable in a conventional system due to the low pressure limitation of the ionization gauge and the high vapor pressure of various contaminants. In fact, it is readily possible to detect leaks in this manner which cannot be detected with a helium leak detector even after the location of the leak is known.

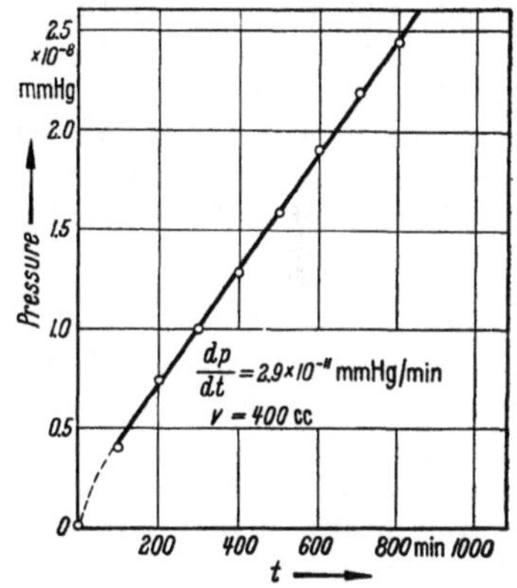

Fig. 27. Rate of rise of pressure in a typical vacuum system.

A standard method for ascertaining the presence of leaks is the "shut-off" test. In this procedure, the vacuum system is isolated from the diffusion pumps, and the measured rate of pressure rise is an index of the amount of leakage into the system. On the basis of our previous discussion, the pressure in an ultrahigh vacuum system is greatly affected by running the ionization gauge continuously. To get a meaningful rate of rise, it is therefore necessary not only to isolate the system from the diffusion pumps, but also to reduce the ion gauge pumping to a minimum. This can be done either by running the gauge at very low electron emission currents or by taking readings intermittently.

A typical rate of rise taken in this fashion is shown in Fig. 27. This is a typical "residual" rate of rise—that is, the lowest rate of pressure rise attainable however carefully the system has been baked and checked for leaks. After the first few minutes, the rate of pressure rise is approximately 3×10^{-11} mm/min or 5×10^{-13} mm/sec. This corresponds to a rate of rise of one micron per century.

For a system of one liter volume, the total influx is therefore about 5×10^{-13} mm liter/sec. The measured rate of rise in a system is therefore compared with a "residual" rate which is of the order of that quoted above. Its measurement is particularly useful in distinguishing a true leak from a virtual leak due to contamination with a high equivalent vapor pressure within the system. In the former case, the rate of rise is constant over a long period of time, while in the latter the pressure usually rises to a rather high asymptotic value which generally signifies the need for further bakeout. Thus the rate of rise of pressure in the system is an important index of its condition and often distinguishes between different sources of gas influx.

Using these straightforward methods, the criteria for leak-free operation are quite firmly established and the means for detecting leaks are considerably more sensitive than previously possible.

28. Concluding remarks. The ultrahigh vacuum systems described in this paper have, for the most part, been rather small, with volumes between 1 and 10 liters and pumping speeds between 0.05 liter/sec and 5 liter/sec. Most of them have been assembled with glass as the major vacuum envelope, though a sizeable number of components such as the valves, mass spectrometer tubes, etc. have been fabricated with metal envelopes.

An increasing interest is now being shown in the design of larger ultrahigh vacuum equipment. Some rather large systems have already been fabricated with metal envelopes using demountable copper and gold gasket seals. There seems to be good reason to believe that considerably larger systems can be built. Some of the components, such as demountable seals or traps, can be scaled upward in size in a straightforward way. In other cases, such as the various methods of pumping, further investigation is required. Most diffusion pumps present the problem that with increasing size and speed, the amount of backstreaming increases disproportionately, and it may be necessary to devise more adequate trapping or to redesign the pumps to achieve very low pressures.

Considerable further work should also be carried out to determine the ultimate limitations of ion pumps or "getter-ion" pumps; these may prove eventually to be of great value for high speed systems.

Concerning the attainment of very low pressures, it is already possible to achieve pressures which are lower than those measurable with the Bayard-Alpert gauge. New manometers, such as that of HOUSTON, are required for the measurement of these lower pressures. In addition, there is an increasing need to measure the partial pressures of the individual gas constituents in a vacuum system. This calls for bakeable mass spectrometers with very high sensitivity.

It is a pleasure to acknowledge the help of many of my colleagues in the Physics Department of the Westinghouse Research Laboratories. In particular, I wish to thank J. H. CARMICHAEL and W. J. LANGE for reading the manuscript and making a number of valuable suggestions.

References.

[1] MASSEY, H. S. W., and E. H. S. BURHOP: Electronic and Ionic Impact Phenomena. Oxford: The Clarendon Press 1952.

[2] BECKER, J. A.: Adsorption on Metal Surfaces and Its Bearing on Catalysis. In: Advances in Catalysis VII. New York: Academic Press, Inc. 1955.

[3] GOMER, R.: Field Emission Microscopy and Some Applications to Catalysis and Chemisorption. In: Advances in Catalysis VII. New York: Academic Press, Inc. 1955.

[4] BAYARD, R. T., and D. ALPERT: Extension of the Low Pressure Range of the Ionization Gauge. Rev. Sci. Instrum. **21**, 571 (1950).
[5] ALPERT, D.: New Developments in the Production and Measurement of Ultrahigh Vacuum. J. Appl. Phys. **24**, 860 (1953).
[6] ALPERT, D., and R. S. BURITZ: Ultrahigh Vacuum II. Limiting Factors on the Attainment of Very Low Pressures. J. Appl. Phys. **25**, 202 (1954).
[7] DUSHMAN, S.: Scientific Foundations of Vacuum Technique. New York: John Wiley & Sons, Inc. 1949.
[8] BECKER, J. A., and C. D. HARTMAN: Field Emission Microscope and Flash Filament Techniques. J. Phys. Chem. **57**, 157 (1953).
[9] HAGSTRUM, H. D.: Instrumentation and Experimental Procedure for Studies of Electron Ejection by Ions and Ionization by Electron Impact. Rev. Sci. Instrum. **24**, 1122 (1953).
[10] GUTHRIE, A., and R. K. WAKERLING: Vacuum Equipment and Techniques. New York: McGraw-Hill Book Company, Inc. 1949.
[11] JAECKEL, R.: Kleinste Drucke. Berlin: Springer 1950.
[12] BLEARS, J.: Application of the Mass Spectrometer to High Vacuum Problems. J. Sci. Instrum. **28**, 36 (1951).
[13] WAGENER, S.: The Production of Very High Vacua By the Use of Getters. Proc. Inst. Electr. Engrs. **99** (III), 135 (1952).
[14] ESPE, W., u. M. KNOLL: Werkstoffkunde der Hochvakuumtechnik. Berlin: Springer 1936.

Sachverzeichnis.

(Deutsch-Englisch.)

Bei gleicher Schreibweise in beiden Sprachen sind die Stichwörter nur einmal aufgeführt.

Subject Index.

(English-German.)

Where English and German spelling of a word is identical the German version is omitted.

MIX
Papier aus verantwortungsvollen Quellen
Paper from responsible sources
FSC® C105338

If you have any concerns about our products,
you can contact us on
ProductSafety@springernature.com

In case Publisher is established outside the EU,
the EU authorized representative is:
Springer Nature Customer Service Center GmbH
Europaplatz 3, 69115 Heidelberg, Germany

Printed by Libri Plureos GmbH
in Hamburg, Germany